Dear [illegible]

[illegible] [illegible] [illegible] [illegible] [illegible] 2001

Harold [illegible]

Principles and Practice of Proton Beam Therapy

Medical Physics Monograph No. 37

Principles and Practice of Proton Beam Therapy

Indra J. Das and Harald Paganetti

Editors

American Association of Physicists in Medicine
2015 Summer School Proceedings
Colorado College
Colorado Springs, Colorado
June 14–18, 2015

Published for the
American Association of Physicists in Medicine
by Medical Physics Publishing, Inc.

To order American Association of Physicists in Medicine (AAPM) publications, contact:

Medical Physics Publishing, Inc.
4555 Helgesen Dr.
Madison, WI 53718
Phone: (800) 442-5778 or (608) 224-4508
Fax: (608) 224-5016
E-mail: mpp@medicalphysics.org
Web: www.medicalphysics.org

Published by:
Medical Physics Publishing, Inc.
Madison, Wisconsin

Published for:
American Association of Physicists in Medicine (AAPM)
One Physics Ellipse
College Park, MD 20740-3846
Phone: (301) 209-3350
Fax: (301) 209-0862

Library of Congress Control Number: 2015941398

ISBN hardcover book: 978-1-936366-43-9
ISBN eBook: 978-1-936366-44-6

Printed in the United States of America.

Contents

DOSIMETRY

ACCPETANCE AND COMMISSIONING

OPERATION

TREATMENT PLANNING AND DELIVERY

Preface

Proton therapy has been used in radiation therapy for over 70 years, but the widespread interest in its clinical use has been realized only in the last decade, as reflected in its exponential growth. Since its inception, the AAPM has organized summer schools on every aspect of medical physics except proton therapy. Consequently, when we approached the AAPM in 2011 on this topic, there was overwhelming support from the committee. This summer school fills the proton therapy gap by focusing on the physics of proton therapy, including beam production, proton interactions, biology, dosimetry, treatment planning, quality assurance, commissioning, motion management, and uncertainties.

The 2015 summer school and this book are tailored mainly to the clinical physicists who will use this as a textbook in the fast-growing field of proton therapy. Furthermore, this book provides up-to-date references to the scientific literature on each aspect of proton therapy covered in the book's 28 chapters. There is also an appendix on nomenclature. Even though the focus of this book is on proton therapy, the content provided will also be valuable to those practicing heavy charged particle beam therapy.

Working on this summer school and book has been gratifying, but credit goes to many individuals behind the scenes. We would like to thank William Parker for his support for the summer school proposal when he was chairing the committee. We greatly acknowledge the members of the summer school subcommittee, but most importantly we thank Holly Lincoln, Diana Cody, Robin Miller, and Vrinda Narayan for their trust in us. Thanks also to the AAPM board for approving the proton therapy summer school in November of 2013.

This book would not be possible without support from our national and international faculties who volunteered their time writing chapters and participating in the summer school. Our sincere thanks go to Karen MacFarland of the AAPM for supporting us and coordinating the 2015 summer school for over two years now. We also thank Jacqueline Ogburn of the AAPM for organizing the SAM modules, which have become an important aspect of the AAPM. Our special thanks to Todd Hanson of Medical Physics Publishing who single-handedly edited and typeset this book. Finally, we acknowledge our families for their support while we were spending many weekends working on the summer school and book.

We hope readers will enjoy the fruits of our volunteer labor and appreciate the time and commitment from our authors, the AAPM, and the Medical Physics Publishing staff.

Indra J. Das
Harald Paganetti

June, 2015
Colorado Springs, Colorado

List of Contributors

Francesca Albertini, Ph.D.
Centre for Proton Therapy
Paul Scherrer Institut
Switzerland

Bijan Arjomandy, Ph.D.
McLaren Proton Therapy Center
Flint, MI
Arjomandy_2000@yahoo.com

Sam Beddar, Ph.D.
Professor, Department of Radiation Physics
UT MD Anderson Cancer Center
Houston, TX

Alessandra Bolsi, M.Sc.
Centre for Proton Therapy
Paul Scherrer Institut
Switzerland

Thomas Bortfeld, Ph.D.
Department of Radiation Oncology
Massachusetts General Hospital
Boston, MA
TBORTFELD@mgh.harvard.edu

Marc Bussiere, M.Sc.
Department of Radiation Oncology
Massachusetts General Hospital and Harvard Medical School
Boston, MA

David Craft, Ph.D.
Department of Radiation Oncology
Massachusetts General Hospital
Boston, MA

Juliane Daartz, Ph.D.
Department of Radiation Oncology
Francis H. Burr Proton Therapy Center
Massachusetts General Hospital
Boston, MA

Indra J. Das, Ph.D.
Professor and Director of Medical Physics
Department of Radiation Oncology
Indiana University School of Medicine
Indianapolis, IN
idas@iupui.edu

Ludovic De Marzi, M.Sc.
Medical Physicist, Centre de Protonthérapie d'Orsay
Institut Curie
Paris, France

Lei Dong, Ph.D.
Scripps Proton Therapy Center
San Diego, CA
Dong.Lei@scrippshealth.org

Martijn Engelsman, Ph.D.
HollandPTC and Delft University of Technology
Delft, The Netherlands
M.Engelsman@tudelft.nl

Jonathan B. Farr, Ph.D., D.Sc.
Chief of Radiation Physics and Associate Member
Department of Radiation Oncology
St. Jude Children's Research Hospital
Memphis, TN
jonathan.farr@stjude.org

Archana Singh Gautam, M.Sc.
Senior Medical Physicist, Department of Radiation Physics
University of Texas MD Anderson Cancer Center
Houston, TX

Annelise Giebeler, Ph.D.
Scripps Proton Therapy Center
San Diego, CA

Michael T. Gillin, Ph.D.
Department of Radiation Physics
University of Texas MD Anderson Cancer Center
Houston, TX

Bram L. Gorissen, Ph.D.
Department of Radiation Oncology
Massachusetts General Hospital
Boston, MA

Nisy Elizabeth Ipe, Ph.D.
Consultant
Shielding Design, Dosimetry, and Radiation Protection
San Carlos, CA
nisy@comcast.net

Antje-Christin Knopf, Ph.D.
The Institute of Cancer Research
and The Royal Marsden NHS Foundation Trust
London, UK
Antje.Knopf@icr.ac.uk

Jon J. Kruse, Ph.D.
Assistant Professor of Medical Physics
Department of Radiation Oncology
Mayo Clinic
Rochester, MN
kruse.jon@mayo.edu

Katja Langen, Ph.D.
Associate Professor
Department of Radiation Oncology
University of Maryland
Baltimore, MD
klangen@som.umaryland.edu

Heng Li, Ph.D.
Department of Radiation Physics
The University of Texas MD Anderson Cancer Center
Houston, TX
hengli@mdanderson.org

Haibo Lin, Ph.D.
Medical Physicist, Department of Radiation Oncology
University of Pennsylvania
Philadelphia, PA
Haibo.Lin2@uphs.upenn.edu

Tony Lomax, Ph.D.
Centre for Proton Therapy
Paul Scherrer Institut
Switzerland
tony.lomax@psi.ch

Alejandro Mazal, Ph.D
Head of Medical Physics
Institut Curie
Paris, France
alejandro_mazal@hotmail.com

James E. McDonough, Ph.D.
Department of Radiation Oncology
University of Pennsylvania
Philadelphia, PA

Radhe Mohan, Ph.D.
The University of Texas MD Anderson Cancer Center
Houston, TX
rmohan@mdanderson.org

Shinichiro Mori, Ph.D.
National Institute of Radiological Sciences
Research Center for Charged Particle Therapy
Chiba, Japan

Vadim P. Moskvin, Ph.D.
Proton Therapy Research Physicist, Division of Radiation Oncology
Department of Radiological Sciences
St. Jude Children's Research Hospital
Memphis, TN

Harald Paganetti, Ph.D.
Professor and Director of Physics Research
Department of Radiation Oncology
Massachusetts General Hospital and Harvard Medical School
Boston, MA
hpaganetti@partners.org

Hugo Palmans, Ph.D.
EBG MedAustron GmbH
Wiener Neustadt, Austria, and
National Physical Laboratory
Teddington, UK
hugo.palmans@npl.co.uk

Mark Pankuch, Ph.D.
Director of Medical Physics, Cadence Health Proton Center
Warrenville, IL

Annalisa Patriarca, M.Sc.
Medical Physicist
Centre de Protonthérapie d'Orsay, Institut Curie
Paris, France

Falk Poenisch, Ph.D.
Department of Radiation Physics
The University of Texas MD Anderson Cancer Center
Houston, TX

Jerimy Polf, Ph.D.
Assistent Professor
Department of Radiation Oncology
University of Maryland
Baltimore, MD

Narayan Sahoo, Ph.D.
Department of Radiation Physics
The University of Texas MD Anderson Cancer Center
Houston, TX
nsahoo@mdanderson.org

Marco Schippers, Ph.D.
Paul Scherrer Insitut
Villigen, Switzerland
marco.schippers@psi.ch

Jan Schuemann, Ph.D.
Massachusetts General Hospital
Boston, MA

Reinhard Schulte, M.D.
Professor, Division of Radiation Research,
Loma Linda University,
Loma Linda, CA

Jiajian Shen, Ph.D.
Assistant Professor, Department of Radiation Oncology
Mayo Clinic
Phoenix, AZ

C. Sunil, Ph.D.
Health Physics Division
Bhabha Atomic Research Centre
Mumbai, India

Jan Unkelbach, Ph.D.
Department of Radiation Oncology
Massachusetts General Hospital
Boston, MA

Stanislav Vatnitsky, Ph.D.
MedAustron GmbH,
Wiener Neustadt, Austria
stanislav.vatnitsky@medaustron.at

Damien Weber, M.D.
Centre for Proton Therapy
Paul Scherrer Institut
Switzerland

Claas Wessels, M.Sc.
Medical Physicist
Centre de Protonthérapie d'Orsay, Institut Curie
Paris, France

Brian Winey, Ph.D.
Assistant Professor, Department of Radiation Oncology
Massachusetts General Hospital Medical School
Boston, MA
Winey.Brian@MGH.Harvard.edu

Richard Wu, M.S.
Senior Medical Physicist, Department of Radiation Physics
UT MD Anderson Cancer Center
Houston, TX

Xiaodong Zhang, Ph.D.
Department of Radiation Physics
The University of Texas MD Anderson Cancer Center
Houston, TX

Timothy C. Zhu, Ph.D.
Professor, Department of Radiation Oncology
University of Pennsylvania
Philadelphia, PA
timzhu@uphs.upenn.edu

X. Ronald Zhu, Ph.D.
Department of Radiation Physics
The University of Texas MD Anderson Cancer Center
Houston, TX
xrzhu@mdanderson.org

Anthony Zietman, MD, FASTRO
Professor, Department of Radiation Oncology
Massachusetts General Hospital
Boston, MA
azietman@partners.org

Chapter 1

Introduction and History of Proton Therapy

Indra J. Das, Ph.D.[1] and Harald Paganetti, Ph.D.[2]

[1]Professor and Director of Medical Physics, Department of Radiation Oncology,
Indiana University School of Medicine
Indianapolis, IN

[2]Professor and Director of Physics Research, Department of Radiation Oncology,
Massachusetts General Hospital and Harvard Medical School
Boston, MA

1.1 Introduction

Soon after the discovery of x-rays by Roentgen in 1896, the radioactivity by Henri Becquerel in 1896, and the extraction of radium and polonium by Madame Curie in 1898, the new field of scientific investigation now known as radiation science took place. For their discoveries, Roentgen, Becquerel, and the Curies were awarded Nobel prizes in subsequent years. Figure 1–1 shows the scientists who made modern radiation science possible so it could benefit mankind.

Most of the modern fundamental radiation physics discoveries took place in the early 20th century. The pace for high-energy beams was growing rapidly during the

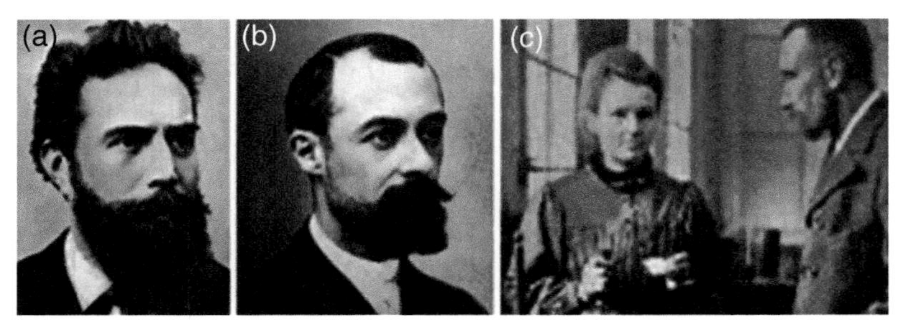

Figure 1–1 (a) Wilhelm C. Roentgen, (b) Henri Becquerel, and (c) Marie and Pierre Curie.

early part of the century, with a 10-fold increase in beam energy every six years between 1920 and 1960 [1]. The design of high-energy devices led to many discoveries and created a vast network of basic and fundamental research that spilled over to the medical sciences. Proton beam therapy is one such area that is a product of early innovation.

The medical use of radiation was immediately realized after the discoveries of x-rays and radioactivity. Today, the majority of cancer patients receive combined treatments, including surgery, chemotherapy, and radiation therapy. Nearly 40% of all patients receive radiation therapy at some point during the course of their cancer treatment. Over the last decades, treatment techniques have evolved, and radiation therapy has become more complex with the introduction of computerized treatment planning in the 1980s and the introduction of image guidance in the last decade, to name just two examples. Furthermore, different radiation modalities have been introduced over time. The dominant aim when introducing new modalities was to increase dose conformity (e.g., the introduction of protons or heavy ion therapy) or the increase in biological effect (e.g., the introduction of neutrons and heavy ion therapy).

As an introduction to this book, this chapter seeks to provide historical perspective. More details on some of the historical aspects can be found in several publications [2–4]. Chu [5] has provided detailed educational materials for the evolution of particle beams leading to the current status. A concise description is provided here as a segue for this book.

1.2 The Discovery of the Proton

Ernest Rutherford (Figure 1–2a), a British physicist working on alpha particle scattering, showed that there is a positive charge at the core of every atom, i.e., the nucleus. For this discovery, he received Nobel prize in chemistry in 1908. During alpha particle irradiation of nitrogen gas, he was amazed to see that in every experiment he was able to get a positive charge, which he later named a proton based on the Greek word *proto*, which means first. His intuitive views led to the following equation:

$$N^{17} + \alpha = O^{17} + H^{+} \tag{1.1}$$

Figure 1–2 (a) Ernest Rutherford, (b) William H. Bragg, and (c) Ernest Orlando Lawrence.

This equation is significant in terms of the building blocks of the periodic table, indicating the first and primary particle in every nucleus. In 1919, he concluded that the positive charge associated with his experiment was nothing but the nucleus of a hydrogen atom, and he coined the term "proton." He also postulated that the nucleus might contain a type of neutral particle, which was discovered later by Chadwick in 1932 and is known as the neutron. A detailed discussion on the modern understanding of the proton and its composition in terms of quarks can be found elsewhere [6].

1.3 The Stopping Power Concept by Bragg

William Bragg (Figure 1–2b), an Australian physicist, was trying to understand the ionizing property of alpha particles. He investigated the ionization produced in air and how far these particles traveled. He published his experimental findings on the stopping power of radiation in gases [7,8]. These elegant findings are still valid and define ionization, stopping power, and range with values very close to today's values. To honor his contribution to the field of ionization, showing the large increase in energy deposition at the end of a particle beam's range, the curve is known as the *Bragg peak* curve. Details on stopping power and range are presented in Chapter 3, and its clinical consequences in are covered in chapters 4 and 5.

1.4 The History of Particle Accelerators

1.4.1 Cyclotrons

During the atomic age of the early 1900s, there was a competition focused on gaining high-energy radiation beams. In the summer of 1928, a young faculty member named Earnest Orlando Lawrence (Figure 1–2c) left Yale University to join the University of California–Berkeley to work on a collaborative project with a chemist and a mechanical engineer. This association turned out to be very fruitful because Lawrence was able to make significant gains in particle acceleration. He was able to build a device 11 inches in diameter to slingshot a proton beam to very high energy. He called this device a "cyclotron," i.e., a device that accelerated particles in a circle [9]. To increase the energy, he started the design and construction of cyclotron with a larger diameter. In 1936, he was able to build a 37-inch cyclotron to accelerate deuterons and alpha particles to energies of 8 MeV and 16 MeV, respectively. This was a golden age for radiation experiments and an era of artificial radioactivity [10]. The desire to achieve higher-energy particle beams led to the development of even bigger cyclotrons. In 1939, a 60-inch cyclotron was built, for which Lawrence was awarded the Nobel prize in physics in 1939. Lawrence collaborated with many eminent scientists of the time, including medical doctors. He died at the early age of 57 in 1958, but he left a legacy of cyclotron physics that has created a new avenue to understand the nature and use of these beams for medical purposes.

The success of the cyclotron for accelerating high-energy charged particles led to the understanding of nuclear physics by breaking the nucleus and creating artificial isotopes, which were first discovered by Irène Joliot-Curie in 1934. This created a desire for most academic institutions to acquire such a machine to produce the isotopes that were finding applications in astrophysics, nuclear physics, and in medicine for diagnosis and therapy. Harvard University started a program in 1935 and other universities—like Princeton, Massachusetts Institute of Technology (MIT), Yale, and Cornell—also pursued acquiring cyclotrons.

1.4.2 The Use of Cyclotrons for Medical Use

The potential of using proton beams for cancer treatment was suggested in 1946 by Robert Wilson [11]. His suggestion to use protons (he also extended his thoughts to heavy ions) was based on the physics of protons—with their finite range in tissue resulting in a Bragg peak as they slow down during penetration in tissue. The physics of proton beams was well understood at that time. Furthermore, the tools to generate high-energy proton beams were in place, i.e., the cyclotron [12] and the synchrotron [13]. Wilson's paper triggered a series of radiobiological experiments using proton beams in the early 1950s. Tobias, Anger, and Lawrence published their work on biological studies on mice using protons in 1952 [14]. It didn't take long for the first patient treatments to happen, which ultimately led to the proton therapy uses we have today.

With the popularity of the Berkeley laboratory for the development of the cyclotron—plus the atomic and nuclear research during World War II via physics and chemistry experiments—medical use of radiation also become a necessity. Most influential universities in the world had joined the research efforts to design machines to accelerate particles. Harvard University built its first cyclotron in 1937 for nuclear physics research. One of Lawrence's graduate students, Robert Wilson, joined Harvard University in the midst of the war. He wrote the classic paper convincing all that the Bragg peak associated with a proton beam could be used for patient treatment [11]. Figure 1–3 shows the Bragg curve that was advocated by Wilson. Soon after the war, Berkley used a 184-inch cyclotron to treat the first patient with a proton beam in

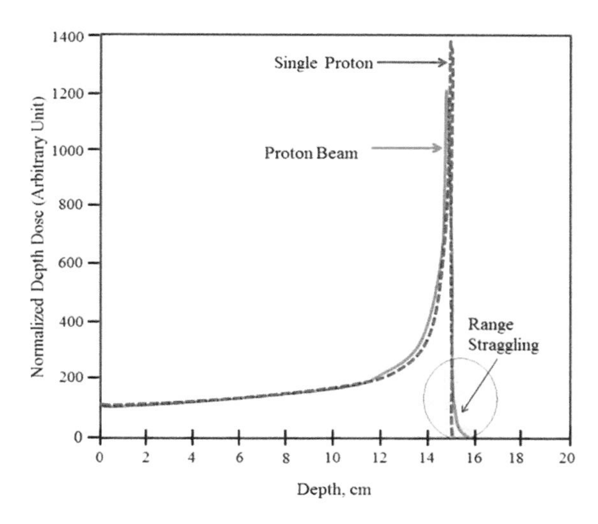

Figure 1–3 Depth–dose curve of a proton beam, showing a large peak and the end of range now known as Bragg peak. Also range straggling is shown when many protons form a beam. Adapted from reference [11].

1954, and then with He^{++} in 1957. Uppsala University in Sweden treated its first patient with a proton beam in 1957. After the war, Harvard decided to install a second cyclotron (the first one was sent to Los Alamos for war research) to be used for nuclear physics and, later, treating patients. In fact, the first patient was treated at Harvard on May 25, 1961, for neurosurgical irradiation. The Harvard Cyclotron Laboratory (HCL) in Cambridge, Massachusetts was originally built for nuclear physics experiments. A detail historical account of the HCL is provided in a book by Wilson [15]. It includes a long list of references and detailed seminal and historical landmarks in the evolution of proton beam therapy in Boston.

1.4.3 Synchrotrons

Based on the theory of relativity, particles gain mass as energy is increased. Consequently, the regular cyclotron fails to accelerate particles as they become out of sync in the cavity. A different approach was needed, and it was attempted early on in 1949, soon after World War II. Chapter 6 provides details of particle acceleration. For heavy charged particles, a synchrotron was developed. The detailed characteristics of a synchrotron are provided by Adruini et al. [16]. Currently several centers in Japan, the United States, and Germany are running particle beam therapy based on synchrotrons.

1.4.4 Clinically Based Accelerators

With the success of high-energy physics research, the University of Chicago and Femilab decided to investigate proton beams for neutron production, and these two facilities became the most important places for neutron physics research. However, later research on proton beams for medical use became more interesting compared to neutron research. Uppsala University in Sweden also started working on a cyclotron for nuclear physics and later moved to patient treatment. In 1957, Uppsala University built a synchrocyclotron capable of producing 185 MeV protons that was used for fractionated radiation treatment. The clinical results related to neurological treatments were reported in 1963 [17]. The medical use of proton beams also started in Dubna, Russia in 1967; Chiba, Japan in 1979; and Somerset West, South Africa in 1993. Most of these facilities were mainly used for physics research, but some beam time was given for clinical work. The first dedicated facility for particle therapy was built in South Africa. In the United States, the first dedicated hospital-housed facility was built in 1990 at Loma Linda Medical Center in California. Today, dedicated accelerators for proton therapy are commercially available from several vendors.

1.5 The Evolution of Proton Therapy

The seminal paper by Wilson did more than introduce the idea of using protons for cancer treatments. It also described how the beam could be shaped to conform to a target by utilizing a rotating wheel of variable thickness to generate a spread-out Bragg peak (SOBP), although this term was not used until much later [11,18].

The first patient was treated with protons at the Lawrence Berkeley Laboratory (LBL), Berkeley, California in 1954 [19]. However, proton beams were utilized very differently compared to modern-day proton treatments using Bragg peak. In fact, a 340 MeV proton beam was used, penetrating the patient and using the plateau region of the depth-dose curve with a cross-firing technique, i.e., similar to rotational treatments today. The Bragg peak was not utilized because of the inability to predict the range accurately. Targeting of radiation therapy beams was done based on bony landmarks alone. Due to these limitations, protons were applied to treat the pituitary gland for hormone suppression in patients with metastatic breast cancer. Between 1954 and 1957, 30 patients were treated using large, single-fraction doses [19]. In the late 1950s, fractionated delivery (three times a week) was introduced [20].

Not long after the first patient treatments at the LBL, patient treatment started in 1957 at the Gustav Werner Institute in Uppsala, Sweden on their 185 MeV cyclotron [21–23]. The fractionation regimen of administering high doses per fraction had to be chosen because of difficulties in securing beam time at the cyclotron. Other than at LBL, the Bragg peak was adopted using large fields from range-modulated beams [22,24,25]. A rotating wheel technique was applied to produce SOBPs [26–28]. Thus, this was the first use of proton therapy along the lines suggested by Wilson. At the Gustav Werner Institute, range modulation to produce a SOBP was pioneered by using a ridge filter [22,29,30]. Pre-clinical work toward the introduction of proton therapy at the Harvard Cyclotron Laboratory (HCL) started in 1959 [31]. The 160 MeV beam offered sufficient range to reach most sites in the body [32,33].

1.5.1 The 1960s

The number of patients treated with protons was still very low in the 1950s and early 1960s. Thanks to radiobiological experiments at LBL, there was awareness of the potential difference in radiobiological effect when comparing protons with conventional radiation. Several groups thus engaged in experiments to deduce the relative biological effectiveness (RBE) of proton beams using *in vitro* as well as *in vivo* endpoints (see Chapter 5). A significant number of mice experiments were done at LBL [34], and chromosome aberrations in bean roots were studied at Gustav Werner Institute [35]. A large radiobiology program was launched at the HCL starting with studies on mortality in mice [36] and skin reactions on primates [37], followed by a series of *in vitro* and *in vivo* experiments, building the basis for today's practice of using a clinical RBE of 1.1 [38–41].

Patient treatments were refined as well. The HCL began with the treatment of intracranial lesions using single fractions with small beams using a single scattering technique to broaden the beam. The first patient was treated in 1961 [31]. The Gustav Werner Institute (Sweden) was instrumental in the development of proton radiosurgery. By 1968, 69 patients had been treated for intracranial lesions [17,42]. In the same time period, a large clinical proton therapy program was started at the HCL in collaboration with the Massachusetts General Hospital (MGH) using the Bragg peak for radiosurgery. Due to a funding problem associated with physics and space radiation research at the HCL, the proton therapy program was in danger of being terminated in the late 1960s, but it eventually survived due to grants from the National Cancer Institute (NCI) in 1971 and the National Science Foundation in 1972.

Early adopters of proton therapy came from the Soviet Union. A facility in Dubna at the Joint Institute for Nuclear Research (JINR) started proton therapy treatments in 1967, followed by the Institute of Theoretical and Experimental Physics (ITEP) in Moscow in 1968 [43–47]. The program at ITEP was the largest of these programs and allowed treatments with up to 200 MeV protons, which was used mainly in combination with a ridge filter to create depth–dose distributions.

1.5.2 The 1970s

By the 1970s, various beam delivery techniques to produce an SOBP and a broad beam were in place. The beam broadening was mainly done with a single scattering foil, which limited the achievable flatness, field size, and beam efficiency. The introduction of the double-scattering technique was another milestone, as it allowed achieving parallel beam, producing a flat dose distribution with high efficiency [48]. The idea was based on existing devices for heavy ion and electron beams [49].

The program at HCL increased in size in the early 1970s. By 1975, 732 patients had undergone pituitary irradiation alone [50]. Thus, the program at HCL was the largest in existence and was formally established as MGH Radiation Oncology in 1973, starting with the treatment of a four-year-old male with a posterior pelvic sarcoma. The treatment options were expanded toward using protons for skull base sarcomas and head-and-neck region carcinomas using fractionated proton therapy [51]. Treatment of melanoma started in 1975 [52] after tests had been made using monkeys

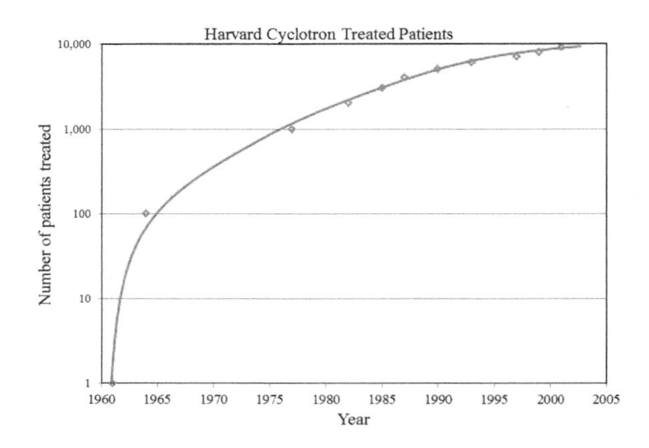

Figure 1-4 Patient treated at HCL over time indicating the growth and popularity of for patient treatments. Data adapted from reference [15].

[53,54]. The first prostate cancer treatments were done in the late 1970s at the HCL [55]. Instrumental for the further advancements of proton therapy was the award of a large research grant by the NCI in 1976 to MGH Radiation Oncology. This allowed extensive studies on medical, biological, and physical aspects of proton therapy. Figure 1–4 shows the growth of the patient numbers at MGH through HCL. It shows close to 10,000 patients were treated before the implementation of a commercial machine at the MGH.

The Russian proton program also expanded. A nuclear physics research facility near St. Petersburg in Gatchina started treating mainly intracranial diseases using Bragg curve plateau irradiation with a 1 GeV proton beam in 1975 [56]. Also at the ITEP facility, the majority of treatments irradiated the pituitary glands of breast and prostate cancer patients using the plateau of the Bragg curve [43,57], but by 1981, 575 patients with various indications had been treated with Bragg peak dose distributions [43].

At the end of the 1970s, Japan joined the proton therapy community. In 1979 the National Institute of Radiological Sciences (NIRS) at Chiba started treatment using a 70 MeV beam [58]. However, of the 29 patients treated between 1979 and 1984, only 11 received proton therapy alone. Most patients received a boost irradiation of protons following by either photon or neutron therapy.

The 1970s also saw plenty of research toward more precise treatment planning. Imaging for diagnosis and planning was done with x-rays and later with CT. Thus, the imaging modality used photons just as the therapeutic treatment beam. With protons it was realized clearly that additional information was desired because of the impact of density variations for each beam path [51,59,60]. The early targets for proton therapy were mainly pituitary adenomas and arteriovenous malformation, which could be visualized on x-rays using contrast material to visualize the vasculature [42,61]. The treatment of sites in heterogeneous areas, such as the head and neck region, would require additional information to obtain densities in the beam path [59]. When CT imaging became available in 1973, it was adopted in proton therapy planning before it eventually was used in conventional therapy [62–64].

1.5.3 The 1980s and 1990s

Major efforts on not only establishing proton therapy, but on improving its delivery and efficacy, were launched in the 1980s and early 1990s in several continents. Examples are the start of proton therapy at the Particle Radiation Medical Science Center in Tsukuba (Japan) in 1983, the Paul Scherrer Institute (PSI) (Switzerland) in 1984, the facility at Clatterbridge (UK) in 1989, in Orsay (France) in 1991, and at the iThemba Labs (South Africa) in 1993. By July 1993, 12,914 patients had been treated with proton therapy. Nearly half of these patients were treated in Boston at the HCL and 25% in the Soviet Union. During the same time, the radiobiological consequences of proton therapy were being explored [65].

The proton therapy community was very active in research, particularly for treatment planning. The reason was twofold. First, most proton centers were located at a research laboratory and second, proton therapy made it necessary to look into more precise planning and delivery in order to utilize its theoretical dosimetric advantage (see Chapter 2). The first computerized treatment planning program was developed for proton therapy [66–69]. Other developments included the beams-eye-view and the dose–volume histogram. New ways for patient positioning were developed because the finite beam range required a more precise patient setup [70].

Research also focused on new delivery methods. A method using rotating dipoles instead of a scattering system in order to produce a uniform dose distribution was considered [48]. Similarly, a technique called wobbling, using magnetic fields to broaden the beam without a double scattering system, was developed at Berkeley for heavy ion therapy to reduce the material in the beam path that led to secondary radiation in double scattering systems [71]. Already in the late 1970s and early 1980s there were studies on the clinical implications of pencil beam scanning [68,72]. The basic concept of using beam scanning in three dimensions for clinical proton beam delivery dates back to 1977 [73]. It was well understood that scanning not only increased the beam efficiency due to fewer beam shaping absorbers in the treatment head, but also the sparing of structures proximal to the SOBP due to variable modulation [68]. The value of beam scanning was recognized in the late 1970s and early 1980s. Spot-by-spot delivery using scanning and conforming the dose to a target volume was first introduced at NIRS using a 70 MeV beam. The main motivation for this technique was to improve the range of the beam by removing a scattering system. Initially two-dimensional scanning was applied in combination with a range-modulating wheel [58]. Later, three-dimensional scanning was developed using two scanning magnets and an automatic range degrader to change the spot energy [58, 74–77]. Various studies on scanning techniques, such as spot scanning and continuous scanning, were done in the early 1980s at LBL, and continuous scanning in three dimensions without collimator was introduced in the early 1990s [78].

While the early applications of proton therapy were driven by what could be treated safely, it later became clear that proton therapy had clear niches where it had advantageous outcomes compared to photon therapy [79]. Clinical efficacy of proton therapy was demonstrated in otherwise poorly manageable diseases, e.g., for chordoma and chondrosarcoma of the skull base and the spine [79,80]. In addition, choroidal melanomas became the most commonly treated tumor at the HCL [81]. Overall, by the mid 1980s the majority of proton treatments were intracranial radiosurgery treatments [82,83].

In the 1980s, all of the existing centers were based at research labs, which had several significant disadvantages. Nursing staff and clinicians had to travel from their hospital, and patient care other than treatment—such as diagnostic imaging and often also treatment planning—had to be done off-site with personnel not necessarily familiar with the treatment operation. Most importantly, treatments had to compete for beam time with research, and the patient numbers were thus very limited. A major milestone would be the building of the first hospital-based facility.

1.6 Evolution of Machines for Hospital-based Proton Therapy

The first hospital-based facility started treatments in 1990 at the Loma Linda University Medical Center (LLUM₵) in California [84]. Their synchrotron was developed in collaboration with Fermilab [85], and the gantries were designed by a group from the HCL [86]. The hospital-based facility at Loma Linda would soon not only treat the biggest share of proton therapy patients, it would also signal that proton therapy was ready for prime time and had made it from research labs into the health care environment. But still, all facilities up to this time had been developed and financed in part by research money. Furthermore, the facilities had all unique designs.

In the late 1990s, the first commercial proton therapy equipment from a vendor was installed at the MGH, financed in part by funds from the NCI. With its first treatment in 2001, MGH transferred its proton therapy program from the research environment at the HCL to the main hospital campus. Conversion of an existing physics cyclotron for medical use added a third facility in the United States at Indiana University in early 2000. This facility included a lot of indigenous advances, such as uniform scanning [87], which is now being used in commercial systems [88].

Subsequently, the number of patients treated with protons increased significantly, and so did the interest of the radiation oncology community. Today more than 100,000 patients have been treated with proton therapy. Figure 1–5 shows the number of patients and the number of facilities as a function of time. It shows exponential growth in both machines and the number of patients being treated worldwide.

The growth of proton therapy has caught the interest of many vendors seeking to provide therapy solutions. Some of the systems in the world have been built locally, like the ones at Loma Linda, PSI (Switzerland), and MPRI (Indiana). However, there are now more than 10 commercial vendors in the fray to provide particle beam therapy. New acceleration technologies based on dielectric wall acceleration [89,90], laser plasma acceleration [91–93], and linear accelerators are works in progress, and

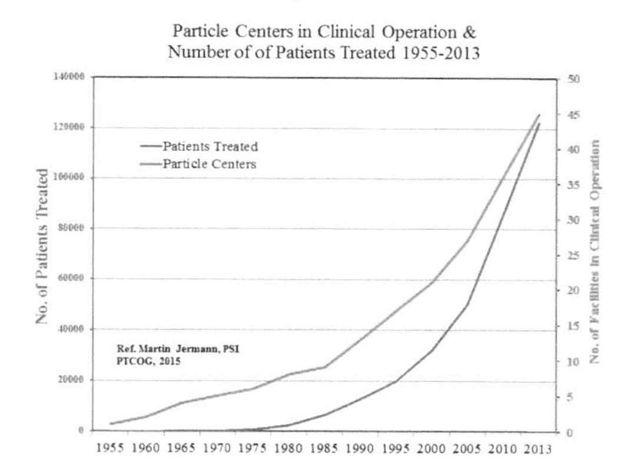

Figure 1–5 The number of patients treated (left axis) and the number of facilities in operation (right axis) from 1955 to 2014. (Courtesy of Martin Jermann, PSI, Switzerland.)

hopefully these techniques could provide badly needed compactness and cost reduction in particle beam therapy.

1.7 Historical Review of Beam-modifying Devices

1.7.1 Beam Broadening for Passively Scattered Delivery

As most proton beams from accelerators are relatively narrow, they are hard to use for large tumors and thus require broadening of the beam. Most centers initially used single scattering foils for broadening. This technique can only achieve a flat beam profile for treating relatively small lesions, and it lacks efficiency. Based on the idea of scattering foils used in electron therapy [49], the double-scattering technique was introduced in proton therapy by Koehler et al. in 1977 [48]. By choosing various materials for the first and second scatterer, it became possible to produce homogeneous lateral dose distributions in clinically acceptable sizes. The design requires a combination of materials with high and low atomic numbers to ensure broadening of the beam while maintaining uniform stopping power [94,95]. The double-scattering technique is being used today at all proton therapy centers treating large lesions without the use of beam scanning. In fact, a better term would be triple-scattering as the range modulator (see below) adds another component of the scattering system. For small lesions—such as in radiosurgery or for the treatment of ocular melanoma—single scattering systems are being used.

1.7.2 Beam Broadening by Scanning (Uniform Scanning)

The beam can also be broadened by a magnetic sweeping system. The idea of using rotating dipoles instead of a scattering system was already proposed in the 1970s [48].

The principle of magnetic beam scanning emerged already in the early 1960s when the idea to magnetically deflect proton beams for treatment was first published [22]. The system was not meant to scan the tumor with individual beamlets, as in beam scanning, but to replace the scattering system using a sweeping magnetic field. The principle was developed at Berkeley and led to uniform scanning or wobbling [71]. It was adopted and implemented at Indiana University in uniform scanning by magnetically sweeping the beam in the horizontal and vertical directions with specialized magnetic fields with a given frequency to provide uniform wide-field proton beam [88]. A description of uniform scanning is reported by Farr et al. [87]. An intercomparison of uniform scanning with commercial systems shows similar characteristics [88]. This technique provides unique treatment capabilities and reduced neutron dose compared to double scattering systems [96–98].

1.7.3 Depth Modulation for Passively Scattered Delivery

Modulator Wheel and Ridge Filter

Early uses of proton therapy had been done mainly without beam modulation. For the treatment of pituitary adenoma and hormonal disorders, the beam penetrated through the patient so that the Bragg peak itself was not used, as one was mainly interested in the favorable lateral penumbra of proton beams [19,20]. Another reason why the Bragg peak was not utilized lay in the limited imaging and, thus, planning capabilities to localize a tumor for treatment. The design of the rotating wheel—consisting of steps of variable thickness for creating a spread-out Bragg peak (SOBP)—was first published in 1975 by Koehler et al. [28,48]. It was later combined with the double scattering technique to provide uniform dose distributions for a certain treatment volume [48].

Uppsala University was first to describe the use of ridge filters to form an SOBP for depth modulation when treating relatively large tumors [22]. Ridge filters are comb-like devices with variable vertical thickness. Figure 1–6 shows several types of devices that are used to create an SOBP [99]. Modern treatments using passive scattering beams still use one of these forms of beam modification. Institutional variations have been adopted to provide uniform doses in depth [100].

1.7.4 Pencil Beam Scanning

Even though the concept of magnetically sweeping technology was known in particle beam, its usage for pencil-beam scanning did not get implemented soon. The pencil-beam scanning concept has a lot of merits because it reduces secondary dose (e.g., neutrons) and allows better dose conformity proximal to the target (see Chapters 23 Treatment Planning for Intensity-modulated Proton Therapy). The clinical implications of beam scanning were analyzed in the late 1970s and early 1980s [68,72]. In 1990 it was developed conceptually at PSI for spot scanning [101]. Pencil-beam scanning was implemented in a Moscow hospital in 1994 [102] and was refined further at PSI as shown by Pedroni et al. [103]. In parallel to intensity-modulated radiotherapy (IMRT), it was realized that intensity-modulated proton therapy (IMPT) can be imple-

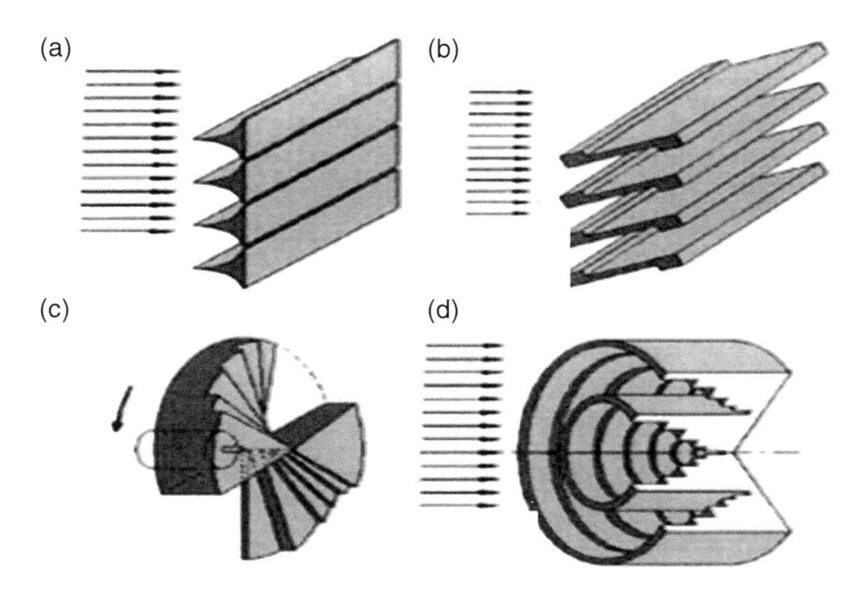

Figure 1–6 Various types of devices for creating an SOBP. (a) and (b) grating type, (c) propeller wheel, and (d) spiral ridge filter. (From reference [99]).

mented with pencil-beam scanning [104,105]. Proton beam scanning and its use in IMPT is currently being performed in Boston and Houston [104,106–109], and other centers are contemplating using it with help from vendors. The concept of beam scanning promises significant improvements in dose conformity (depending on the spot size), and many vendors are moving toward offering solutions for beam scanning only.

1.8 Current Technology

Proton therapy was introduced initially to improve target dose conformity. Today, mainly due to IMPT, the advantage in target dose conformity is not maintained for all sites. There is an advantage certainly for head and neck cancers. What still is and always will be advantageous when using protons is the advantage in integral dose, i.e., proton therapy reduces dose to most critical structures. Because each photon dose distribution can be duplicated by stacking individual pristine Bragg peaks, it is mathematically clear that the integral dose with protons is independent of the proton or photon treatment technique.

The debate about the place of proton therapy in radiation oncology is tied to health care costs [110–113]. If proton therapy would cost the same as photon therapy, it would be unwise to treat with the extra integral dose that photon treatments cause.

This assumes that the dose bath is not advantageous because of the uncertainties in defining the correct target location. A detailed discussion of clinical perspective by a clinician is provided in Chapter 2.

Furthermore, for many treatment sites, the dose reduction to organs at risk achievable with protons may not matter in terms of toxicities. It is here, where the debate about the pros and cons of clinical trials starts [114–117]. Prostate cancer is in the center of this controversy [117,118], while the clinical significance of the integral dose advantage is undisputed in the pediatric patient population [119]. Clinical trials to prove the efficacy and justify the cost are the subject of debates [113,120–123]. Randomized clinical trials are currently ongoing specifically for prostate, lung, and breast treatments. Hopefully this can be resolved in future by compiling all the clinical data that are being accumulated.

Proton therapy technology and delivery will change significantly in the years to come. More and more centers are moving toward beam scanning. The capability of intensity modulation will offer dose-sculpting capabilities that are impossible to achieve with photon beams. Furthermore, the field of proton therapy is expected to catch up with photon therapy when it comes to in-room imaging. Cone-beam CT is still not standard in proton therapy.

1.9 Historical View of the Particle Therapy Organization PTCOG

As particle beams became clinically relevant, the exchange of ideas and the dissemination of knowledge were needed. An ad hoc committee was formed and called the Proton Therapy Co-operative Group (PTCOG). The name was later changed to Particle Therapy Co-operative Group [124]. The first meeting of around 30 people was held in Boston on September 18, 1985. Later it was held nearly every six months with nearly 50 people. It was later rotated throughout various centers that had particle beams, including the United States, Canada, Germany, France, England, Japan, South Africa, Switzerland, and Sweden [5]. Starting in 2007, PTCOG has been held once a year, with the location based on voting by a steering committee. A growing population of attendee (close to 900 recently) attend separate educational and scientific meetings. This transition took place at PTCOG-47 which was held in Wanjie, China in 2007. The PTCOG provides a vast amount of educational information that can be accessed from its website [124]. In 2013, PTCOG started to publish the *International Journal of Particle Therapy* (IJPT) at http://www.theijpt.org/. PTCOG keeps track of the number of patients treated worldwide and of new centers that are operational and being planned. By the end of 2013, over 122,000 patients have been treated with particle therapy, with the majority of them treated with protons.

1.10 Summary/Conclusion

Proton beams have been utilized in research for over 100 years, and their clinical application is growing steadily since the first treatment in 1954. The physical characteristics are well suited to spare normal tissues and provide uniform and maximum dose to tumor. The popularity of proton therapy is growing at an exponential rate as shown in Figure 1–5 with a growth of the number of facilities and the number of patients treated worldwide. The increasing number of scientific publications indicates maturity and acceptability in most institutions, even though the financial burden is hampering widespread utilization. As the technology is rapidly improving, many vendors are aiming at providing more cost-effective treatments. There are still many challenges in the areas of infrastructures, understanding biology, treatment planning, motion management, post-treatment verification, and clinical outcome where the proton community has to muster its energy.

References

1. Swanson WP, Thomas RH. Dosimetry for radiological protection at high-energy particle accelerators. In: Kase KR, Bjärngard BE, Attix FH, editors. The Dosimetry of Ionizing Radition, Volume III. San Diego: Academic Press, 1990.
2. ICRU Report 78. Prescribing, recording, and reporting proton beam therapy. International Commission on Radiation Units and Measurements, 2007.
3. Paganetti H. Proton Therapy Physics. In: Proton therapy physics. Boca Raton, FL: CRC Press, 2012.
4. Brahme A. Recent advances in light ion radiation therapy. *Int J Radiation Oncol Biol Phys* 2004;58:603–616.
5. Chu WT. PTCOG, from 1985 to present and future. http://www.ptcog.ch/; PTCOG-50; 2011.
6. Suit H. Proton: The particle. *Int J Radiat Oncol Biol Phys* 2013;87:555–561.
7. Bragg WH, Kleeman R. On the a particles of radium and their loss of range in passing through various atoms and molecules. *Phil Mag J Sci* 1905;10:318–340.
8. Bragg WH, Kleeman R. On the ionization curves of radium. *Phil Mag J Sci* 1904;8:726–738.
9. Lawrence EO, Livingston MS. The production of high speed light ions without the use of high voltages. *Phys Rev* 1932;40:19–35.
10. Lawrence EO, Livingston MS, White MG. The disintegration of lithium by swiftly moving protons. *Phys Rev* 1932;42:150–151.
11. Wilson RR. Radiological use of fast protons. *Radiology* 1946;47:487–491.
12. Lawrence EO, Edlefson NE. On the production of high speed protons. *Science* 1930;72:376–377.
13. Oliphant MO. The acceleration of particles to very high energies. *Classified memo submitted to DSIR, United Kingdom; now in University of Birmingham Archive,* 1943.
14. Tobias CA, Anger HO, Lawrence JH. Radiological use of high energy deuterons and alpha particles. *Am J Roentgenol Radium Ther Nucl Med* 1952;67:1–27.
15. Wilson RR. A Brief History of Harvard University Cyclotrons, Cambridge: Harvard University Press, 2004.
16. Arduini G, Cambria R, Canzi C, Gerardi F, Gottschalk B, Leone R, Sangaletti L, Silari M. Physical specifications of clinical proton beams from a synchrotron. *Med Phys* 1996;23:939–951.
17. Larsson B, Leksell L, Rexed B. The use of high-energy protons for cerebral surgery in man. *Acta Chir Scandinavia* 1963;125:1–7.
18. Wilson RR. Range, straggling, and multiple scattering of fast protons. *Physical Review* 1947;74:385–386.
19. Lawrence JH. Proton irradiation of the pituitary. *Cancer* 1957;10:795–798.
20. Tobias CA, Lawrence JH, Born JL, McCombs R, Roberts JE, Anger HO, Low-Beer BVA, Huggins C. Pituitary irradiation with high energy proton beams: A preliminary report. *Cancer Research* 1958;18:121–134.
21. Larsson B, Kihlman BA. Chromosome aberrations following irradiation with high-energy protons and their secondary radiation: A study of dose distribution and biological efficiency using root-tips of vicia faba and allium cepa. *Int J Radiat Biol* 1960;2:8–19.
22. Larsson B. Pre-therapeutic physical experiments with high energy protons. *Br J Radiol* 1961;34:143–151.
23. Leksell L, Larsson B, Andersson B, Rexed B, Sourander P, Mair W. Lesions in the depth of the brain produced by a beam of high energy protons. *Acta radiol* 1960;54:251–264.

24. Falkmer S, Fors B, Larsson B, Lindell A, Naeslund J, Stenson S. Pilot study on proton irradiation of human carcinoma. *Acta radiol* 1962;58:33–51.
25. Fors B, Larsson B, Lindell A, Naeslund J, Stenson S. Effect of high energy protons on human genital carcinoma. *Acta Radiol Ther Phys Biol* 1964;2:384–398.
26. Koehler AM. Dosimetry of proton beams using small silicon detectors. *Radiat Res* 1967;7:s53–s63.
27. Koehler AM, Preston WM. Protons in radiation therapy. *Radiology* 1972;104:191–195.
28. Koehler AM, Schneider RJ, Sisterson JM. Range modulators for protons and heavy ions. *Nucl Instr Meth* 1975;131:437–440.
29. Blokhin SI, Gol'din LL, Kleinbok Ia L, Lomanov MF, Onosovskii KK, Pavlonskii LM, Khoroshkov VS. [dose field formation on proton beam accelerator itef]. *Med Radiol (Mosk)* 1970;15:64–68.
30. Karlsson BG. Methods for calculating and obtaining some favorable dosage distributions for deep therapy with high energy protons. *Strahlentherapie* 1964;124:481–492.
31. Kjellberg RN, Koehler AM, Preston WM, Sweet WH. Stereotaxic instrument for use with the bragg peak of a proton beam. *Confin Neurol* 1962;22:183–189.
32. Das IJ, Moskvin VP, Zhao Q, Cheng CW, Johnstone PA. Proton therapy facility planning from a clinical and operational model. *Technol Cancer Res Treat* 2014;10.7785/tcrt.2012.500444 (eprint).
33. Suzuki K, Gillin MT, Sahoo N, Zhu XR, Lee AK, Lippy D. Quantitative analysis of beam delivery parameters and treatment process time for proton beam therapy. *Med Phys* 2011;38:4329–4337.
34. Ashikawa JK, Sondhaus CA, Tobias CA, Kayfetz LL, Stephens SO, Donovan M. Acute effects of high-energy protons and alpha particles in mice. *Radiat Res Suppl* 1967;7:312–324.
35. Larsson B. Blood vessel changes following local irradiation of the brain with high-energy protons. *Acta Soc Med Ups* 1960;65:51–71.
36. Dalrymple GV, Lindsay IR, Hall JD, Mitchell JC, Ghidoni JJ, Kundel HL, Morgan IL. The relative biological effectiveness of 138-mev protons as compared to cobalt-60 gamma radiation. *Radiation Research* 1966;28:489–506.
37. Dalrymple GV, Lindsay IR, Ghidoni JJ, Hall JD, Mitchell JC, Kundel HL, Morgan IL. Some effects of 138-mev protons on primates. *Radiat Res* 1966;28:471–488.
38. Hall EJ, Kellerer AM, Rossi HH, Yuk-Ming PL. The relative biological effectiveness of 160 mev protons. Ii. Biological data and their interpretation in terms of microdosimetry. *Int J Radiat Onol Biol Phys* 1978;4:1009–1013.
39. Robertson JB, Williams JR, Schmidt RA, Little JB, Flynn DF, Suit HD. Radiobiological studies of a high-energy modulated proton beam utilizing cultured mammalian cells. *Cancer* 1975;35:1664–1677.
40. Tepper J, Verhey L, Goitein M, Suit HD. In vivo determinations of RBE in a high energy modulated proton beam using normal tissue reactions and fractionated dose schedules. *Int J Radiat Onol Biol Phys* 1977;2:1115–1122.
41. Todd P. Radiobiology with heavy charged particles directed at radiotherapy. *Eur J Cancer* 1974;10:207–210.
42. Larsson B, Leksell L, Rexed B, Sourander P, Mair W, Andersson B. The high-energy proton beam as a neurosurgical tool. *Nature* 1958;182:1222–1223.
43. Chuvilo IV, Goldin LL, Khoroshkov VS, Blokhin SE, Breyev VM, Vorontsov IA, Ermolayev VV, Kleinbock YL, Lomakin MI, Lomanov MF, et al. Itep synchrotron proton beam in radiotherapy. *Int J Radiat Oncol Biol Phys* 1984;10:185–195.
44. Khoroshkov VS, Goldin LL. Medical proton accelerator facility. *Int J Radiat Oncol Biol Phys* 1988;15:973–978.
45. Dzhelepov VP, Komarov VI, Savchenko OV. [development of a proton beam synchrocyclotron with energy from 100 to 200 mev for medico-biological research]. *Med Radiol (Mosk)* 1969;14:54–58.
46. Khoroshkov VS, Barabash LZ, Barkhudarian AV, Gol'din LL, Lomanov MF, Pliashkevich LN, Onosovskii KK. [a proton beam accelerator for radiation therapy]. *Med Radiol (Mosk)* 1969;14:58–62.
47. Dzhelepov VP, Savchenko OV, Komarov VI, Abasov VM, Goldin LL, Onossovsky KK, Khoroshkov VS, Lomanov MF, Blokhin NN, Ruderman AI, Astrakhan BV, Vajnberg MS, Minakova EI, Kisileva VN. Use of ussr proton accelerators for medical purposes. *IEEE Transact Nucl Science* 1973;20:268–270.

48. Koehler AM, Schneider RJ, Sisterson JM. Flattening of proton dose distributions for large-field radiotherapy. *Med Phys* 1977;4:297–301.
49. Sanberg G. Electron beam flattening with an annular scattering foil. *IEEE Transact Nucl Science* 1973;20:1025.
50. Kjellberg RN, Kliman B. Bragg peak proton treatment for pituitary-related conditions. *Proc R Soc Med* 1974;67:32–33.
51. Suit HD, Goitein M, Tepper J, Koehler AM, Schmidt RA, Schneider R. Exploratory study of proton radiation therapy using large field techniques and fractionated dose schedules. *Cancer* 1975;35:1646–1657.
52. Gragoudas ES, Goitein M, Koehler AM, Verhey L, Tepper J, Suit HD, Brockhurst R, Constable IJ. Proton irradiation of small choroidal malignant melanomas. *Am J Ophthalmol* 1977;83:665–673.
53. Constable IJ, Goitein M, Koehler AM, Schmidt RA. Small-field irradiation of monkey eyes with protons and photons. *Radiat Res* 1976;65:304–314.
54. Constable IJ, Roehler AM. Experimental ocular irradiation with accelerated protons. *Invest Ophthalmol* 1974;13:280–287.
55. Shipley WU, Tepper JE, Prout GR, Jr., Verhey LJ, Mendiondo OA, Goitein M, Koehler AM, Suit HD. Proton radiation as boost therapy for localized prostatic carcinoma. *JAMA* 1979;241:1912–1915.
56. Abrosimov NK, Gavrikov YA, Ivanov EM, Karlin DL, Khanzadeev AV, Yalynych NN, Riabov GA, Seliverstov DM, Vinogradov VM. 1000 mev proton beam therapy facility at petersburg nuclear physics institute synchrocyclotron. *Journal of Physics: Conference Series* 2006;41:424–432.
57. Savinskaia AP, Minakova EI. [proton hypophysectomy and the induction of mammary cancer]. *Med Radiol (Mosk)* 1979;24:53–57.
58. Kanai T, Kawachi K, Kumamoto Y, Ogawa H, Yamada T, Matsuzawa H, Inada T. Spot scanning system for proton radiotherapy. *Med Phys* 1980;7:365–369.
59. Goitein M. The measurement of tissue heterodensity to guide charged particle radiotherapy. *Int J Radiat Oncol Biol Phys* 1977;3:27–33.
60. Suit HD, Goitein M, Tepper JE, Verhey L, Koehler AM, Schneider R, Gragoudas E. Clinical experience and expectation with protons and heavy ions. *Int J Radiat Oncol Biol Phys* 1977;3:115–125.
61. Kjellberg RN, Nguyen NC, Kliman B. [the bragg peak proton beam in stereotaxic neurosurgery]. *Neurochirurgie* 1972;18:235–265.
62. Goitein M. Compensation for inhomogeneities in charged particle radiotherapy using computed tomography. *Int J Radiat Oncol Biol Phys* 1978;4:499–508.
63. Goitein M. Computed tomography in planning radiation therapy. *Int J Radiat Oncol Biol Phys* 1979;5:445–447.
64. Munzenrider JE, Pilepich M, Rene-Ferrero JB, Tchakarova I, Carter BL. Use of body scanner in radiotherapy treatment planning. *Cancer* 1977;40:170–179.
65. Raju MR. Proton radiobiology, radiosurgery and radiotherapy. *Int J Radiat Onol Biol Phys* 1995;67:237–259.
66. Goitein M, Abrams M, Gentry R, Urie M, Verhey L, Wagner M. Planning treatment with heavy charged particles. *Int J Radiat Onol Biol Phys* 1982;8:2065–2070.
67. Goitein M, Abrams M. Multi-dimensional treatment planning: I. Delineation of anatomy. *Int J Radiat Oncol Biol Phys* 1983;9:777–787.
68. Goitein M, Chen GTY. Beam scanning for heavy charged particle radiotherapy. *Med Phys* 1983;10:831–840.
69. Goitein M, Miller T. Planning proton therapy of the eye. *Med Phys* 1983;10:275–283.
70. Verhey LJ, Goitein M, McNulty P, Munzenrider JE, Suit HD. Precise positioning of patients for radiation therapy. *Int J Radiat Onol Biol Phys* 1982;8:289–294.
71. Chu WT, Curtis SB, LLacer J, Renner TR, Sorensen RW. Wobbler facility for biomedical experiments at the bevalac. *IEEE Transact Nucl Science* 1985;NS-32:3321–3323.
72. Grunder HA, Leemann CW. Present and future sources of protons and heavy ions. *Int J Radiat Oncol Biol Phys* 1977;3:71–80.
73. Leemann C, Alonso J, Grunder H, Hoyer E, Kalnins G, Rondeau D, Staples J, Voelker F. A 3-dimensional beam scanning system for particle radiation therapy. *IEEE Transact Nucl Science* 1977;NS-24:1052–1054.
74. Kanai T, Kawachi K, Matsuzawa H, Inada T. Three-dimensional beam scanning for proton therapy. *Nuc Instr Methods* 1983;214:491–496.

75. Kawachi K, Kanai T, Matsuzawa H, Inada T. Three dimensional spot beam scanning method for proton conformation radiation therapy. *Acta Radiol Suppl* 1983;364:81–88.
76. Kawachi K, Kanai T, Matsuzawa H, Kutsutani-Nakamura Y, Inada T. [proton radiotherapy facility using a spot scanning method]. *Nippon Igaku Hoshasen Gakkai Zasshi* 1982;42:467–475.
77. Hiraoka T, Kawashima K, Hoshino K, Kawachi K, Kanai T, Matsuzawa H. [dose distributions for proton spot scanning beams: Effect by range modulators]. *Nippon Igaku Hoshasen Gakkai Zasshi* 1983;43:1214-1223.
78. Chu WT, Ludewigt BA, Renner TR. Instrumentation for treatment of cancer using proton and light-ion beams. *Review of Scientific Instruments* 1993;64:2055–2122.
79. Suit H, Goitein M, Munzenrider J, Verhey L, Blitzer P, Gragoudas E, Koehler AM, Urie M, Gentry R, Shipley W, Urano M, Duttenhaver J, Wagner M. Evaluation of the clinical applicability of proton beams in definitive fractionated radiation therapy. *Int J Radiat Oncol Biol Phys* 1982;8:2199–2205.
80. Austin-Seymour M, Munzenrieder JE, Goitein M, Gentry R, Gragoudas E, Koehler AM, McNulty P, Osborne E, Ryugo DK, Seddon J, Urie M, Verhey L, Suit HD. Progress in low-let heavy particle therapy: Intracranial and paracranial tumors and uveal melanomas. *Radiat Res* 1985;104:S219–S226.
81. Gragoudas ES, Seddon JM, Egan K, Glynn R, Munzenrider J, Austin-Seymour M, Goitein M, Verhey L, Urie M, Koehler A. Long-term results of proton beam irradiated uveal melanomas. *Ophthalmol*1987;94:349–353.
82. Kjellberg RN, Davis KR, Lyons S, Butler W, Adams RD. Bragg peak proton beam therapy for arteriovenous malformation of the brain. *Clin Neurosurg* 1983;31:248–290.
83. Kjellberg RN, Hanamura T, Davis KR, Lyons SL, Adams RD. Bragg-peak proton-beam therapy for arteriovenous malformations of the brain. *N Engl J Med* 1983;309:269–274.
84. Slater JM, Archambeau JO, Miller DW, Notarus MI, Preston W, Slater JD. The proton treatment center at loma linda university medical center: Rationale for and description of its development. *Internati Int J Radiat Onol Biol Phys* 1992;22:383–389.
85. Cole F, Livdahl PV, Mills F, Teng L. Design and application of a proton therapy accelerator. *Proc 1987 IEEE Particle Accelerator Conference* 1987; Piscataway, NJ: IEEE Press. 1985–1987.
86. Koehler AM. Preliminary design study for a corkscrew gantry. *Harvard Cyclotron Laboratory report,* 1987.
87. Farr JB, Mascia AE, Hsi WC, Allgower CE, Jesseph F, Schreuder AN, Wolanski M, Nichiporov DF, Anferov V. Clinical characterization of a proton beam continuous uniform scanning system with dose layer stacking. *Med Phys* 2008;35:4945–4954.
88. Nichiporov D, Hsi W, Farr J. Beam characteristics in two different proton uniform scanning systems: A side-by-side comparison. *Med Phys* 2012;39:2559–2568.
89. Schippers JM, Lomax AJ. Emerging technologies in proton therapy. *Acta Oncol* 2011;50:838–850.
90. Zschornack G, Ritter E, Schmidt M, Schwan A. Electron beam ion sources for use in second generation synchrotrons for medical particle therapy. *The Review of scientific instruments* 2014;85:02B702.
91. Schwoerer H, Pfotenhauer S, Jackel O, Amthor KU, Liesfeld B, Ziegler W, Sauerbrey R, Ledingham KW, Esirkepov T. Laser-plasma acceleration of quasi-monoenergetic protons from microstructured targets. *Nature* 2006;439:445–448.
92. Bulanov SS, Brantov A, Bychenkov VY, Chvykov V, Kalinchenko G, Matsuoka T, Rousseau P, Reed S, Yanovsky V, Krushelnick K, Litzenberg DW, Maksimchuk A. Accelerating protons to therapeutic energies with ultraintense, ultraclean, and ultrashort laser pulses. *Med Phys* 2008;35:1770–1776.
93. Muramatsu M, Kitagawa A. A review of ion sources for medical accelerators (invited). *Rev Scient Instr* 2012;83:02B909.
94. Gottschalk B. On the scattering power of radiotherapy protons. *Med Phys* 2010;37:352–367.
95. Gottschalk B, Koehler AM, Schneider RJ, Sisterson JM, Wagner MS. Multiple Coulomb scattering of 160 mev protons. *Nucl Instr Meth Phys Res B* 1993;74:467–490.
96. Anferov VA. Scan pattern optimization for uniform proton beam scanning. *Med Phys* 2009;36:3560–3567.
97. Anferov V. Analytic estimates of secondary neutron dose in proton therapy. *Phys Med Biol* 2010;55:7509–7522.

98. Hecksel D, Anferov V, Fitzek M, Shahnazi K. Influence of beam efficiency through the patient-specific collimator on secondary neutron dose equivalent in double scattering and uniform scanning modes of proton therapy. *Med Phys* 2010;37:2910–2917.
99. Kostjuchenko V, Nichiporov D, Luckjashin V. A compact ridge filter for spread out bragg peak production in pulsed proton clinical beams. *Med Phys* 2001;28:1427–1430.
100. Akagi T, Higashi A, Tsugami H, Sakamoto H, Masuda Y, Hishikawa Y. Ridge filter design for proton therapy at hyogo ion beam medical center. *Phys Med Biol* 2003;48:N301–N312.
101. Blattmann H, Coray A, Pedroni E, Greiner R. Spot scanning for 250 MeV protons. *Strahlenther Onkol* 1990;166:45–48.
102. Khoroshkov VS, Onosovsky KK, Klenov GI, Zink S. Moscow hospital-based proton therapy facility design. *Am J Clin Oncol* 1994;17:109–114.
103. Pedroni E, Bacher R, Blattmann H, Bohringer T, Coray A, Lomax A, Lin S, Munkel G, Scheib S, Schneider U, et al. The 200-mev proton therapy project at the Paul Scherrer institute: Conceptual design and practical realization. *Med Phys* 1995;22:37–53.
104. Trofimov A, Bortfeld T. Beam delivery sequencing for intensity modulated proton therapy. *Phys Med Biol* 2003;48:1321–1331.
105. Lomax AJ, Bohringer T, Bolsi A, Coray D, Emert F, Goitein G, Jermann M, Lin S, Pedroni E, Rutz H, Stadelmann O, Timmermann B, Verwey J, Weber DC. Treatment planning and verification of proton therapy using spot scanning: Initial experiences. *Med Phys* 2004;31:3150–3157.
106. Kooy HM, Clasie BM, Lu HM, Madden TM, Bentefour H, Depauw N, Adams JA, Trofimov AV, Demaret D, Delaney TF, Flanz JB. A case study in proton pencil-beam scanning delivery. *Int J Radiat Oncol Biol Phys* 2010;76:624–630.
107. Zhu XR, Poenisch F, Lii M, Sawakuchi GO, Titt U, Bues M, Song X, Zhang X, Li Y, Ciangaru G, Li H, Taylor MB, Suzuki K, Mohan R, Gillin MT, Sahoo N. Commissioning dose computation models for spot scanning proton beams in water for a commercially available treatment planning system. *Med Phys* 2013;40:041723.
108. Gillin MT, Sahoo N, Bues M, Ciangaru G, Sawakuchi G, Poenisch F, Arjomandy B, Martin C, Titt U, Suzuki K, Smith AR, Zhu XR. Commissioning of the discrete spot scanning proton beam delivery system at the university of Texas M.D. Anderson cancer center, proton therapy center, houston. *Med Phys* 2010;37:154–163.
109. Smith A, Gillin M, Bues M, Zhu XR, Suzuki K, Mohan R, Woo S, Lee A, Komaki R, Cox J, Hiramoto K, Akiyama H, Ishida T, Sasaki T, Matsuda K. The M. D. Anderson proton therapy system. *Med Phys* 2009;36:4068–4083.
110. Goitein M, Jermann M. The relative costs of proton and x-ray radiation therapy. *Clinical Oncology* 2003;15:S37-50.
111. Lundkvist J, Ekman M, Ericsson SR, Jonsson B, Glimelius B. Proton therapy of cancer: Potential clinical advantages and cost-effectiveness. *Acta Oncol* 2005;44:850–861.
112. Peeters A, Grutters JP, Pijls-Johannesma M, Reimoser S, De Ruysscher D, Severens JL, Joore MA, Lambin P. How costly is particle therapy? Cost analysis of external beam radiotherapy with carbon-ions, protons and photons. *Radiother Oncol* 2010;95:45–53.
113. Lodge M, Pijls-Johannesma M, Stirk L, Munro AJ, De Ruysscher D, Jefferson T. A systematic literature review of the clinical and cost-effectiveness of hadron therapy in cancer. *Radiother Oncol* 2007;83:110–122.
114. Glimelius B, Montelius A. Proton beam therapy—do we need the randomised trials and can we do them? *Radiother Oncol* 2007;83:105–109.
115. Goitein M, Cox JD. Should randomized clinical trials be required for proton radiotherapy? *J Clin Oncol* 2008;26:175–176.
116. Goitein M. Trials and tribulations in charged particle radiotherapy. *Radiother Oncol* 2010;95:23–31.
117. Brada M, Pijls-Johannesma M, De Ruysscher D. Current clinical evidence for proton therapy. *Cancer J* 2009;15:319–324.
118. Konski A, Speier W, Hanlon A, Beck JR, Pollack A. Is proton beam therapy cost effective in the treatment of adenocarcinoma of the prostate? *J Clin Oncol* 2007;25:3603–3608.
119. Jagsi R, DeLaney TF, Donelan K, Tarbell NJ. Real-time rationing of scarce resources: The northeast proton therapy center experience. *J Clin Oncol* 2004;22:2246–2250.
120. Suit H, Goldberg S, Niemierko A, Trofimov A, Adams J, Paganetti H, Chen GTY, Bortfeld T, Rosenthal S, Loeffler J, Delaney T. Proton beams to replace photon beams in radical dose treatments. *Acta Oncologica* 2003;42:800–808.

121. Ju M, Berman AT, Vapiwala N. The evalotion of proton beam therapy: Insights from early trails and tribulations. *Int J Radiat Oncol Biol Phys* 2014;90:733–735.
122. Suit H, Kooy H, Trofimov A, Farr J, Munzenrider J, DeLaney T, Loeffler J, Clasie B, Safai S, Paganetti H. Should positive phase iii clinical trial data be required before proton beam therapy is more widely adopted? No. *Radiother Oncol* 2008;86:148–153.
123. Olsen D, Bruland O, Frykholm G, Norderhaug I. Proton therapy—a systematic review of clinical effectiveness. *Radiother Oncol* 2007;83:123–132.
124. Jarmann M. Http://www.Ptcog.Ch/; particle therapy cooperative oncology group.

Chapter 2

The Practicing Clinician's Perspective on Proton Beam Therapy

Anthony Zietman, MD, FASTRO

Professor, Department of Radiation Oncology,
Massachusetts General Hospital,
Boston, MA

2.1 Introduction

Proton beam therapy is being developed for clinicians to use to benefit their patients, and as spectacular as the technology is, it must be judged from this patient-centered perspective. What benefits does it bring to patients? Can these be quantified? What do they mean to patients? Are these benefits over and above the benefits derived from other kinds of radiation therapy (photon and electrons)? Are these benefits greater than those derived from other, newer non-radiation therapies?

There is a second context of importance, and that is societal. In an era of finite resources, can proton therapy's investment, cost, and the time be reasonably or legitimately justified? These issues shift in time and depend as much upon national economics and local ethics as technological efficiencies. These questions are also much more difficult for physicians and physicists to answer as they do not lend themselves to our usual tools of investigation. To some degree, we must forge ahead regardless, but these questions will always exist in the background and, for the foreseeable future, will be a measure by which proton beam therapy will be judged.

In this chapter we will look briefly at the history of proton therapy as a natural evolution of radiation therapy. We will then ask what are the proven benefits of this therapy to date? What new indications invite this therapy? How can the proton community assess this technology to the satisfaction of those who will pay for it, and how can they quantify its value to patients? We will also explore the ethical conundrum of the clinical trial, our standard measure of effectiveness, in the context of new technology and debate whether or not this cumbersome tool should be applied to proton therapy.

2.2 Proton Beam on the Radiation Therapy Continuum

Since the earliest years of the 20th century, radiation has been recognized to be one of the most "tumoricidal" agents known to man. The first 100 years of radiation therapy research were devoted to finding means of delivering this extraordinary agent to its target (i.e., the cancer cell) with accuracy. In the beginning, this meant either direct application of radioactive sources to superficial tumors (e.g., skin), or it meant the direct insertion of similar sources into tumors (e.g., prostate) or cavities containing tumors (e.g., uterus). With the ability to generate x-ray beams of ever-increasing energy, the therapy went through a succession of developmental eras (kilovoltage, supervoltage, megavoltage), which were each associated with new opportunities and applications for therapy. Over the last 20 years a revolution in imaging and computing power has allowed the goal of accurate radiation delivery to be achieved.

The drivers to technology development have largely been the clinicians themselves as they increasingly learned the consequences of misapplied radiation from their experience with patients. The acute mucosal reactions of radiation limit the tolerability and the achievable dose of the given treatment. They also limit the ability of radiation to be used in combination with other systemic agents, which may have equal or greater value in terms of cure by eradicating metastatic disease. The early pioneers soon recognized the late reactions to radiation, which we now know largely result from the effects on mesenchymal-derived tissues and blood vessels with occlusion and chronic tissue hypoxia. These are wide ranging in scope and severity, and they may include fibrosis and soft tissue necrosis. In some locations, these may be trivial, but in others, such as brain or spinal cord, they can be catastrophic. By the 1920s it was recognized that therapeutic maneuvers—such as delivery with a low dose rate or by fractionation—could reduce the likelihood of these late effects developing. However, it was not until the most recent era—when fractionation could be combined with high-energy treatment, good visualization of the target, and optimized planning—that we have been able to substantially reduce the incidence of these dread radiation effects.

Another late effect soon recognized was the induction of cancer by radiation. Therapy pioneers themselves suffered skin malignancies, leukemias, and sarcomas as a result of their efforts. The radium dial workers suffering osteosarcomas of the jaw brought wider public attention to these risks. Ultimately, the consequences of the atomic bombs dropped in a spirit of experimentation in the Marshall Islands and in anger on Japan greatly increased awareness of radiation-induced malignancy. As survival improves dramatically among cancer patients, so they are now living long enough to experience second malignancies induced by their therapy [1]. This has proven to be the case particularly for pediatric patients, and it is well illustrated by the experience of Ewing sarcoma survivors [2]. The interaction of radiation with chemotherapy to induce second malignancies is well recognized, as is the relationship to the volume of normal tissue irradiated. Malignancies rarely arise outside the radiation field. The relationship between tumor induction and radiation dose is more complex and whole-body radiation experiments suggest there may be a dose threshold beyond which the incidence goes down. It is of note, however, that those developing rectal and bladder cancers after prostate irradiation do so in the high-dose regions of irradiation (anterior rectal wall and trigone of the bladder) [3].

While increasing dose and increasing field size have been associated with an increased risk of complications, dose in particular has also been associated with an increased chance for cure. Suit and others recognized half a century ago that there is no "tumoricidal threshold dose," and that radiation kill is probabilistic [4]. Dose and tumor eradication in a population of patients or experimental animals have a relationship described by a sigmoidal curve [5]. The slope of this curve depends upon the number of clonogens in the tumor and upon local environmental issues affecting sensitivity, such as hypoxia. The more radiation dose that can be given, the higher up this dose–response curve the treating clinician or scientist can climb. This has been demonstrated for many tumors in man, such as prostate and head and neck cancers and, if the data could be developed, likely obtains for them all. This understanding of the relationship between dose and clonogen number underpins the concept of "shrinking fields" that is widely practiced in radiation oncology today. Treatment begins with a larger field that encompasses all known disease, both macroscopic and microscopic. Once sufficient dose to eradicate microscopic volumes of cancer is delivered, the field can then be shrunk and treatment continued to a higher dose to the gross tumor. Thus graded doses are delivered according to the likely clonogen number, and normal tissues are, to some degree, spared the higher doses.

The entire direction of radiotherapeutic thought and ingenuity for over 100 years has been to combine all of this knowledge into a philosophy that aims to deliver high doses of radiation to gross cancer and lower doses to microscopic cancer, thus eradicating all known disease, and to minimize the treatment of normal tissues, thus avoiding early and late complications. Proton therapy fits very clearly and comfortably onto this arrow of progress and, in the eyes of many, represents its sharp tip (Figure 1–1).

The series of events leading to the development of protons as a therapeutic tool are well known. The historical perspective of proton beam has been described in Chapter 1, but a short summary is worth repeating here. In 1929, Ernest Lawrence invented the cyclotron as a way to accelerate nuclear particles to very high speeds. In 1946, Lawrence's protégé, Robert R. Wilson, professor of physics at Harvard and designer of Harvard's cyclotron, first proposed using protons for the treatment of can-

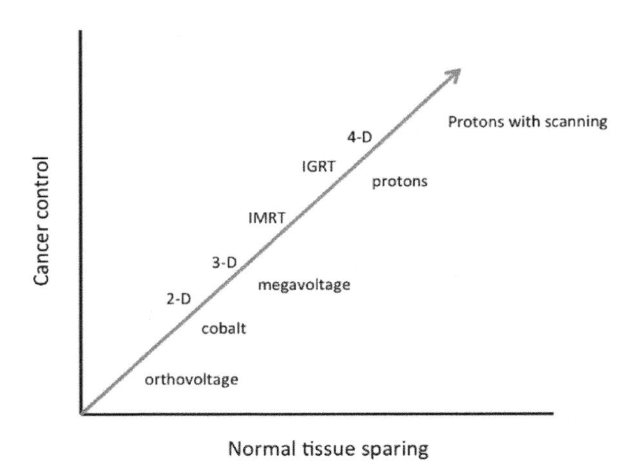

Figure 2–1 The arrow of progress in radiation therapy.

cer [6]. The first patient was treated with protons at Berkeley for pituitary tumor in 1954, and Europe followed with treatment starting in 1957 in Uppsala, Sweden.

The first patient was treated at the Harvard Cyclotron in a single-dose, stereotactic neurosurgical program of Kjellberg in 1961. In just over a decade, Suit, Goitein, and others had initiated a fractionated proton radiation program for many cancers, including those of the eye, skull base, and prostate. In 1988, the U.S. Food and Drug Administration (FDA) approved protons for selected cancers and, shortly after this, the patient-dedicated Loma Linda Medical Center synchrotron opened. The development of "off the shelf" equipment by a number of vendors, a sound conceptual basis for treatment, a healthcare system that now reimbursed treatment, and the initiation of a number of financing plans that allowed new centers to overcome the huge cost of initial installation have led, in less than a decade, to the current situation. There are now 42 centers operating worldwide (14 of which are in the United States), and many more are being planned.

In retrospect, this can be seen as reflecting previous quantum-leap innovations in radiation therapy, such as linear accelerators, CT simulators, or IMRT. All of these technologies promised to increase cure and reduce complications through improved targeting and delivery. The difference this time is two-fold. Firstly, there is the issue of cost, and establishing a proton center is a uniquely costly undertaking. No other medical device has ever cost this much; no other medical device even comes close. The second is an issue of timing. While timing is generally not of relevance to the clinician scientist seeking the truth, it is critical for the application of proton therapy. We are in an era of cost containment and evidence-based medicine. In an attempt to reduce costs, anything that cannot be justified by evidence will come under scrutiny. If research cannot demonstrate the value to patients by the criteria of outcomes, then that therapy will not be reimbursed. It will wither on the vine and die. That is the dilemma for clinicians wishing to employ proton beam therapy in their practice.

2.3 What Do Clinicians Require from Proton Therapy?

Clinicians require from protons many things that they sought from cobalt, megavoltage, intensity modulation, and image guidance when they were introduced:

1. delivery of higher radiation doses to tumor with a reduction in the high dose volume
2. delivery of lower doses to normal tissues and a reduction in the integral dose
3. the ability to safely integrate established and novel systemic agents
4. the ability to hypofractionate

Without reiterating in too much detail what will be spelled out in far greater detail in other chapters, it is clear that the potential clinical advantage for passively scattered protons when compared with photons appears to be physical rather than biological. The absence of exit dose beyond the Bragg peak yields a marked reduction in integral dose. When compared with 3D conformal photons, there are undoubtedly improved dose distributions compared to 3D conformal photons, and when compared to intensity-modulated photons, there is a lower integral dose and absence of exit dose. Although IMRT can spare selected normal tissues, this is at the cost of an increased dose to other normal tissues as the radiation is "displaced." There can be no advantage to the patient to irradiate normal tissue, even if it is to only low or moderate doses. These advantages are grounded in sound principles, and they are clearly demonstrated by a large literature of comparative planning experiments. The bigger question in this era of evidence-based medicine is, "Does this dosimetric advantage translate into detectable and meaningful patient gains?" If it does not, then it simply represents an intellectual exercise.

2.4 Current Clinical Applications for Proton Beam and Evidence

2.4.1 Pediatric Tumors

Two important factors distinguish children from adults in terms of the late effects of radiation therapy. Firstly, they have a higher risk of secondary malignancies [7], and secondly, they have a particular susceptibility to effects of radiation on organ growth and function [8]. The latter can cause significant medical morbidity and devastating cosmetic outcomes. Growth abnormalities and second cancers can both result from delivered doses at the *lower end* of the clinical range. The lower integral dose achieved by proton beam reduces the overall volume of irradiated tissue and, thus, has the potential to ameliorate both of these situations. A report by Chung et al. [9] compared patients treated with proton radiation between 1973 and 2001 at the Harvard Cyclotron with temporally matched patients treated with photons in the Surveillance, Epidemiology, and End Results (SEER) program cancer registry. Reporting at a median follow-up of seven years, second malignancies had occurred in 5.2% of

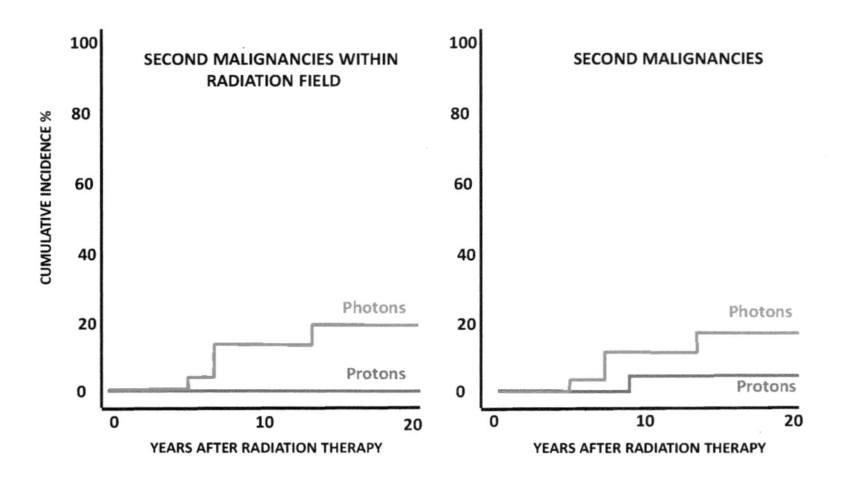

Figure 2–2 The cumulative incidence of second malignancies in patients receiving either photon or proton beam treatment for retinoblastoma [10].

patients treated with protons as compared with 7.5% of those receiving photons (adjusted hazard ratio 0.52, $p < 0.01$). These data have been criticized because the additional malignancies occurred in the first five years after treatment, considerably earlier than experience tells us we might ordinarily see radiation-induced malignancies. A second smaller retrospective study employing less rigorous methods compared retinoblastoma patients treated at Massachusetts General Hospital with protons to retinoblastoma patients treated at Boston Children's Hospital between 1986 and 2011 [10]. A reduction in in-field malignancies was seen at 10 years, a more plausible outcome (Figure 2–2).

2.4.1.1 Medulloblastoma-craniospinal Irradiation

Craniospinal irradiation (CSI) is often given for cancers with a propensity to disseminate throughout the entire neuroaxis, such as medulloblastoma. In many ways, this disease is the "poster boy" for proton therapy because the consequences of irradiating such a large volume of normal tissue and such a range of different organs in a child may be particularly severe. Dosimetric studies have clearly shown the substantial reduction in dose to normal tissues [11,12] with proton CSI when compared to photon CSI. Risk modeling studies suggest a powerful reduction in the risk of radiation-induced malignancies in patients undergoing proton CSI, as compared to conventional or IMRT CSI. A recent dosimetric study focused on the CSI effect on breast tissue and found close to a 100-fold reduction when proton CSI was compared to photon [13]. It has been calculated that by reducing or eliminating the late toxicities of CSI, protons may save up to $26,000 (EUR 23,000) over the lifetime of a child.

Some have argued that proton beam is the only ethically appropriate approach for CSI, while others take a different view [14,15]. They point out that many of the radiation-induced malignancies tend to occur in the high dose volumes, and that these are similar for both photons and protons. They also make the case that the most severe growth abnormalities result from the spinal irradiation, and that the dose to this structure, the entire vertebral column, is identical regardless of therapeutic approach. Other social considerations may factor in. Is it fair to mandate that children and parents travel to a distant proton facility for lengthy treatment that may impose social and financial hardships on them for what is still a theoretical benefit?

2.4.1.2 Rhabdomyosarcoma

Rhabdomyosarcoma (RMS) is the most common soft tissue sarcoma in children. It commonly arises in the head and neck and parameningeal regions. Comparative dosimetric studies show that protons can deliver significantly lower mean doses to the retina, optic nerve, parotid, and cochlea than IMRT [16]. Clinical reports using historical controls suggest a similar tumor control rate, but with protons displaying reduced acute toxicity [17]. More recent reports treating orbital RMS with proton beam show favorable local and distant control, but a reduction in the loss of function of normal tissues [18].

2.4.1.3 Ependymoma, Craniopharyngioma, Retinoblastoma, and Glioma

Dosimetric and clinical studies have been published for patients with these three diseases again suggesting a reduced acute and long-term toxicity profile. The risk of secondary malignancy at 10 years was lower for retinoblastoma patients treated with protons, when compared to a similar group of patients treated with photons [11].

2.4.2 Adult Malignancies

While almost every treatment plan employing protons will look dosimetrically superior to an equivalent photon treatment plan due to its physical property, the potential clinical benefit in adults is not as large as it may be for children. Adult tissues are less prone to secondary malignancies and are not subject to the same growth and developmental issues. If there were no cost differential, radiation oncologists would likely use the technique they felt most comfortable with to minimize the dose to the normal tissues and best conform to the high dose. This might be protons in one situation, but it could be IMRT in another. As will be discussed below, because of the higher cost associated with proton treatment, a superior dosimetric plan is currently insufficient to justify its choice to insurers and government payers. A measurable and meaningful clinical advantage must be demonstrated. The radiation oncologist's desire to escalate radiation dose to improve tumor control is one of the opportunities for protons to outperform the photon options. Another opportunity is the desire to minimize dose to adjacent normal tissues to reduce acute effects, allowing for more strategic use of systemic agents and to reduce late effects. For every clinical scenario, however, a steep dose–response curve must first be proven, and the normal tissue toxicity that limits the ability to deliver this higher dose with photon-based treatments must exist.

We will discuss these issues for a number of adult cancers, starting with those for which proton beam is either most widely used (prostate) or most commonly recommended (ocular melanoma and chordoma/chondrosarcoma of the axial skeleton) and base this on a recent review by Mitin and Zietman [19]. We will then examine the other common malignancies in which proton beam may potentially play a role.

2.4.2.1 Prostate Cancer

It is known that prostate cancer is one of the most common diseases treated with proton beam, comprising nearly 70% of patients treated with proton beam in the United States. This, therefore, is the indication for which proton beam is receiving the most publicity and the most scrutiny. It is important to evaluate the evidence behind this strong trend. High-quality evidence has shown that higher radiation doses delivered to the prostate are associated with better cancer control [20,21]. It has also become clear that IMRT is capable of delivering doses as high as, or even higher than, can be safely delivered with protons. One recent study, for example, delivered 81 Gy with remarkably low levels of bowel and bladder toxicity [22]. A second study employing protons to deliver 82 Gy on a prospective protocol suggested that the dose "ceiling" had been reached in terms of safety [23]. Thus one of the two principal justifications for the use of protons—that a higher dose of radiation can be more safely given—appears to be incorrect. The second justification for protons is that there is lower morbidity. Talcott et al. performed a cross-sectional study using patient-reported quality of life questionnaires submitted by men with prostate cancer who were treated with either photons or protons [24]. They found no difference in any of the quality-of-life parameters between the two groups. Additionally, two SEER database analyses found no difference in the likelihood of treatment for cancer recurrence or treatment for complications when comparing photons with protons, with one unanticipated exception—the likelihood of treatment for rectal bleeding appeared to be higher among the proton-treated patients [25,26]. There are limitations to the SEER data that may weaken these conclusions; nevertheless, they did not add any evidence to suggest that there is any additional clinical benefit to prostate cancer patients from proton beam.

The American Society for Radiation Oncology (ASTRO) has recently recommended that physicians discuss these limitations in knowledge with patients prior to electing for proton treatment. ASTRO also recommends treatment be within the context of a clinical trial or registry [27]. A multi-institutional randomized phase 3 NCI trial (9368 PARTIQoL) comparing proton beam to IMRT is now in its third year and is expected to shed light on this controversial issue. It is also important to remember that prostate brachytherapy is a competitive, highly conformal, and cost-effective treatment modality for patients with prostate cancer [28]. It must be added that many patients with low-risk prostate cancer, particularly those aged 65 and above, simply do not need treatment at all [29]. Any treatment, no matter how good, is a bad treatment when given to a patient who never needed it in the first place.

This begs the question as to why no difference has emerged between protons and photons in prostate cancer? There may be physical explanations: the prostate gland sits deep within the pelvis, and the radiation beam path distance to the target is considerable. At these depths, the lateral penumbra of the proton beam is not so sharp due to Coulomb interaction. There is also end-of-range uncertainty at these depths, and complex bony anatomy must be traversed. Dosimetric comparisons have revealed a reduction in the rectal volume receiving 30 Gy or less with proton beam, but the V70 was the same [30]. Most rectal toxicity appears to be associated with the high-dose region. These similarities may separate once novel proton scanning technologies become more standard and proton image guidance improves. Indeed, recent reports now suggest that proton beam is associated with an improving safety profile. A second explanation for the current similarity in outcome may be the uncertainty surrounding the radiobiological effectiveness (RBE) of proton beam [31] (see Chapter 5). Small differences in RBE when delivering doses in the order of 80 Gy and on the steep part of the normal tissue complication curve may result in big deviations in morbidity from those anticipated. This may not be an issue for cancers that require lower doses, but it may well be a problem for those in which "the envelope is being pushed."

2.4.2.2 Uveal Melanoma

Uveal melanoma is the most common primary intraocular malignancy in adults. Surgical enucleation should always be avoided if safely possible and if the alternatives—plaque brachytherapy or external beam therapy by either protons or photons—can be delivered. A recent literature-based meta analysis appears to show a reduced rate of local recurrence when particle therapy is compared to brachytherapy, but with no significant differences in mortality or enucleation rates [32]. It is widely held that the majority of uveal melanomas can be equally well treated with either proton beam or brachytherapy, and that preference will depend upon local availability and expertize. Tumors that overlay the optic disc may be more difficult to manage by brachytherapy, whereas cataract formation is more likely to result with proton beam for large anterior lesions.

2.4.2.3 Chordoma and Chondrosarcoma

Chordomas and chondrosarcomas are locally aggressive primary bone tumors. Chordomas arise in the skull base and spine, whereas chondrosarcomas most frequently appear in the pelvis, proximal femur, and scapula. Complete surgical excision may be difficult to achieve in axial locations such as skull base, spine, and sacrum, thus radiation therapy is commonly used either as an alternative or as an adjuvant therapy. The same anatomic relationships to the cranial nerves, brain stem, and spinal cord that limit surgery may also affect the delivery of an adequate radiation dose. In this context, Schulz-Ertner and Tsujii [33] have reviewed the historical results with particle therapy. While good local control is commonly reported in the literature, it must be remembered that the evidence consists largely of single-institution series and may reflect both case selection bias and positive publication bias. As IMRT and fractionated stereotactic radiation have become more sophisticated, dose escalation has been attempted with these photon-based modalities. The early data does not seem to reproduce the rates of local control achieved with particle therapy, but again, the quality of the evidence is too poor to make real judgments [34,35]. Particle therapy has, on what might reasonably be described as a thin evidence base, established itself as the standard of care for these rare malignancies.

2.4.2.4 Breast Cancer

Adjuvant radiation therapy improves both tumor control and overall survival in women with breast cancer, yet both secondary malignancy and cardiac toxicity may adversely affect the outcomes and limit any gains achieved. Once again, dosimetric studies have shown what appear to be substantial reductions in lung, heart, and contralateral breast doses when whole breast proton plans were compared to photon plans. Protons were shown to be cost-effective for women with left-sided breast cancer, where the cost for the quality-adjusted life year was $75,000 (EUR 67,000), primarily due to a decrease in long-term cardiac toxicity. It appears that young women with a left-sided breast tumor with a long life expectancy might even benefit from proton therapy to the same extent as pediatric patients. It must again be noted that these studies used historical photon data and did not take into account advances in photon treatment, such as prone position or breath hold techniques. More recently, prospective clinical studies are beginning to report outcomes. One reports a pilot study in which 12 patients were treated with proton therapy after mastectomy to 50.4 Gy and appeared to tolerate the treatment very well. The maximum skin toxicity during treatment was just grade 2 [36]. This is important, as it had been suggested that protons would actually increase the skin dose and thus worsen the cosmetic outcomes.

As accelerated partial breast irradiation with large daily fractions (APBI) is gaining acceptance among physicians and patients, protons are being evaluated as one of several delivery methods. Kozak et al. reported significant acute skin toxicity when a single-field proton beam per fraction was used, but this has been resolved in more recent phase 2 trials employing multiple fields per day [37,38,39] and scaling modulation width. Another challenging clinical scenario is the patient with bilateral implants after mastectomy, and again protons may play a role here.

It is uncertain whether proton beam will be widely used in breast cancer. It is more likely that it will find selective use in certain clinical scenarios where the patient's anatomy poses particular cardiac or pulmonary risks with photon therapy. This "personalized" use of protons may be the model for future use of the technology throughout the body.

2.4.2.5 Lung Cancer

The use of proton therapy in patients with non-small cell lung cancer (NSCLC) has theoretical advantages in terms of sparing of organs at risk in the chest while at the same time maintaining adequate target coverage. A recent meta-analysis of dosimetric studies revealed both statistically and clinically significant decrease in lung and heart dose with proton beam plans in comparison to photon plans [40]. However, the role of dose escalation in locally advanced NSCLC is currently controversial. The standard dose is approximately 60 to 63 Gy, although local failure rates at this dose level are substantial. Dose escalation in the RTOG 0617 randomized phase 3 trial employing 3D CRT or IMRT failed to demonstrate improved survival in the arms with 74 Gy in comparison to 60 Gy [41]. This unanticipated outcome may be related to an increase in toxicity when delivering 74 Gy with the chosen photon techniques. If this is indeed the case, then protons may offer an opportunity for safer, more effective dose escalation. A phase II trial of 44 patients treated with 74 Gy proton radiation therapy with concurrent paclitaxel/carboplatin reported encouraging results, including a median survival time of 29 months with no grade 4–5 events and no local failures in nine patients [42]. A subsequent randomized phase II trial comparing protons to IMRT for 66 and 74 Gy dose levels with concurrent chemotherapy is nearing completion (www.clinicaltrials.gov NCT00915005). Proton therapy may also be useful in the setting of trimodality therapy for stage IIIA NSCLC where it is important to spare the contralateral lung, especially in patients who are pneumonectomy candidates (www.clinicaltrials.gov NCT01565772).

Photon-based stereotactic body radiation therapy (SBRT) has recently established itself as the standard of care for medically inoperable early stage NSCLC. These results may be improved by the use of protons, not in terms of local control, which is already high, but in terms of radiation-related morbidity. A literature-based meta analysis compared particle beam therapy to SBRT and found no significant differences in survival between SBRT and particle beam treatments in patients with inoperable Stage I NSCLC, but then none would have been anticipated [43]. Centrally located tumors are another challenge, and here again, the sharper lateral penumbra and the use of active scanning might allow for better sparing of the critical structures with proton-based SBRT (www.clinicaltrials.gov NCT01511081). In addition, the ability of proton beam radiation to achieve adequate target coverage with only two to three beams may be advantageous in settings involving particularly poor lung function, prior chest irradiation, or multi-focal lung cancers that require more than one treatment course [44,45].

Realizing the potential benefits of proton therapy in patients with lung cancer is currently a technical challenge, mainly due to physical problems associated with the delivery of protons to moving targets surrounded by tissues with large inhomogeneities. Proton radiation therapy for lung cancer is still in its early stages of clinical testing, particularly with regard to the development of appropriate dose algorithms, IMPT optimization, motion management, volumetric image guidance, and adaptive planning techniques [46]. Clinical trials are also in place for lung treatment.

2.4.2.6 Brain Tumors

Glioblastoma is the most feared primary brain tumor. It is currently treated by a maximal resection, followed by adjuvant radiation therapy to 60 Gy with concurrent and adjuvant temozolomide chemotherapy. Before the era of concomitant chemotherapy, proton beam was explored as a mean of dose escalation. Two small phase I/II trials suggested small gains in tumor control and survival rates, but these gains were offset by a marked increase in necrosis requiring surgical intervention and by failure of the disease ultimately outside the high-dose regions [47,48]. Dose escalation alone was simply not the optimal approach to this disease. Current investigations are using proton therapy in the management of low-grade and favorable high-grade gliomas in hopes of reducing radiation-associated adverse cognitive effects in patients who have several years of survival ahead of them.

Meningiomas sit at the other end of the brain tumor spectrum. The majority of patients achieve long-term tumor control and often enjoy normal life expectancies.

Here the advances in therapy will come not from dose escalation, but from minimizing the cerebral late effects of radiation and thus minimizing the effect of therapy on the patient's quality of life. While the evidence is currently thin, these are patients in whom clinical benefit might be anticipated [49,50].

2.4.2.7 Head and Neck Cancers

Proton therapy has been used on a clinical trial basis at several institutions for the treatment of nasopharyngeal carcinoma, oropharynx, and nasal and paranasal sinus malignancies. The value of protons for the most important head and neck sites (nasopharynx and paranasal sinuses) resides in the ability to limit dose to optic structures and brainstem and secondarily the mandible and salivary glands. Comparative dosimetric studies show the potential for a significant reduction in dose to radiosensitive structures, such as the mandible and the parotid gland, which may result in a decreased risk of mandibular osteonecrosis and xerostomia [51]. The head and neck, however, is like the lung— a considerable challenge for proton physicists due to air cavities that may be variably filled with tumor or fluid and due to the complexity and inhomogeneity of the bones. Good prospective studies are underway and early results are now being reported [52].

2.4.2.8 Gastrointestinal Malignancies

Particle therapy has a very promising future in the treatment of hepatocellular carcinoma (HCC) [53]. Clinical evidence shows encouraging local control and toxicity profiles with considerable ability to spare the liver. The integral dose benefits of proton beam may be particularly helpful in sparing this quite radiosensitive organ, and it is possible that, with evidence, it could become the preferred treatment modality for patients with Child-Pugh class B and C cirrhosis. At present, the published experience is limited to a few institutional series.

As for lung cancer, advances in photon SBRT are simultaneously narrowing the potential advantage of protons. SBRT is not only now in routine use, but it has also developed a considerable body of evidence against which protons will have to be measured. The treatment of locally advanced esophageal cancer requires either chemotherapy alone or in combination with surgery. Non-cancer deaths are common in the first year of treatment and are largely related to cardiopulmonary toxicity [54]. The ability of protons to spare the heart might decrease cardiac toxicity and death, but this requires further clinical investigation.

2.5 Assessing Technological Innovation in Medicine

New technologies are constantly being developed and brought into the practice of medicine. Their path into clinical use differs in important ways from drugs, and for good reasons. While medical technologies sometimes represent something radically novel (MRI for example), they are far more commonly representative of the steady incremental development and improvement of existing technology. In radiotherapy, there is a clear precedent for this with an evolution in treatment machines from orthovoltage x-rays to ^{60}Co gamma rays to linear accelerator-generated x-rays of increasing energy. Such changes are generally the consequence of efforts to improve the efficacy and "user-friendliness" of a device. In technology evolution, the expected changes in efficacy and the cost increases of each change tend to be modest, even when the benefits to patients are substantial. These incremental evolutionary advances in technology tend to be made frequently and rapidly, rendering the objective evaluation of cost-benefit very difficult. It is therefore unrealistic to imagine that all such advances can be tested in formal clinical trials.

Unfortunately, medical devices in the United States are regulated differently from drugs. Each drug is a distinct biological entity and is required to demonstrate medical efficacy with acceptably few side effects, while devices only have to do what they are supposed to do, and do it safely. Drugs have enormous biological unpredictability, so the FDA has established a process for drug approval that involves Phase I, II, and III testing. The third phase is a randomized controlled trial in which the novel agent is tested against standard drug therapy or no therapy at all. In contrast, medical devices—particularly when they result from technological evolution—tend to simply represent a superior tool rather than new and unpredictable biology. The FDA simply demands that such devices are safe and that they perform their advertised functions. There is no mandate that they demonstrate clinical effectiveness or superiority over existing devices.

Proton therapy certainly represents an example of technological evolution in the sense that protons are very similar to well-established x-rays, delivering ionizing radiation with well-understood biological effects, but they do so with an improved distribution of dose. As proton therapy simply represents a "sharper knife" for shallow depths in particular, it is argued that it need not be exposed to the kind of expensive and time-consuming clinical scrutiny that is placed on novel drugs or biologics.

2.5.1 The Proton Therapy Controversy

The superiority of protons over x-rays is based on two uncontroversial propositions [55]:

1. The dose distributions of protons are, in the great majority of cases, superior to those of x-rays.
2. Normal tissues are damaged by radiation, and this damage increases with higher dose and greater volumes.

Despite these incontrovertible truths, there is, however, far more controversy than might be anticipated. Some argue that the superiority of protons is "theoretical" and without any level I clinical evidence in its support [56]. They take the following lines of argument:

1. The improvements in dose distribution are often small and may be of no clinical relevance, bringing little detectable benefit to patients.
2. The technical difficulties of proton delivery—particularly across mobile and inhomogeneous tissues such as lung, sinuses, and hips—may lead to inferior dose distributions.

3. The effects of radiation are biological, like drugs, and are far less predictable than previously thought, leading to current uncertainties about the RBE of protons, which appears to vary according to beam energy, linear energy transfer (LET), tissue, point along the Bragg peak, and fraction size.
4. The medical literature is replete with examples of randomized trials with unanticipated negative outcomes, illustrative of the over-optimism and occasional hubris of investigators.

Behind all of this is the issue of cost. At present, the size of the investment required to establish a large proton therapy center with three to four treatment rooms stands in the order of up to $150 million, although single-gantry facilities are only $20 to $30 million [57]. These investments are without precedent among medical devices. In terms of the relative cost of a single treatment fraction, one study estimated that proton therapy was at present more expensive than IMRT by a factor of 2.4 [58]. This cost factor may be even higher, depending on the manpower and technical expertise needed for a particle beam therapy, by a factor of three to five times [59,60]. This ratio will rise or fall in the future due to both technological evolution of proton therapy and changes in the way treatment is reimbursed. These extraordinary setup and maintenance costs lead to substantial charges and expose this particular device to unprecedented scrutiny from payers of all stripes.

On the other side, many committed "protoneers" argue that the rationale for the superiority of protons is incontrovertible and that the technical difficulties can and will be overcome with 4D planning, gating, scanned beams, and intensity modulation. They maintain that the biologic differences are small, or in most clinical situations irrelevant, and where important, are likely to be incorporated into planning and risk estimation in the near future [55]. As a result, protons can be expected to almost always provide better treatments than x-rays (without saying how much better).

The truth probably lies somewhere between these extremes, and as long as there is a cost differential, there is an expectation on the part of those paying for the care that costly resources be wisely used [61]. In other words, there is an expectation that protons will demonstrate not just dosimetric superiority, but meaningful clinical superiority, and that this can be quantified such that estimates of "value" can be made. Once the magnitude of the clinical benefit and additional cost is known, then policymakers, not clinicians or physicists, will determine whether it is appropriate relative to the other things on which finite health care dollars must be spent.

2.5.2 The Role of the Randomized Trial

Is it reasonable that protons should be put through randomized trials when most medical devices do not have to undergo such scrutiny [62]? Does this not slow down the introduction of a technology that has been with us since the 1940s? When one considers the capital costs required for proton beam installation, is it possible that any center would even consider it without the guarantee of returns that, at the very least, amortize the investment? Many have argued that the "theoretical" arguments in favor of a device should be discounted under certain conditions. They remind us that, since there are possible downsides to protons, equipoise is assured and randomization is appropriate [56].

Four criteria have been proposed to identify when innovations should be considered significant enough to justify randomized trials [63]. Use randomized trials:

1. when there is need for retraining and re-credentialing of physicians to be allowed to perform the new procedure;
2. when the innovation provides a diagnosis or treatment for a condition for which none previously existed;
3. when an innovation, although directed toward improving the health of a sick individual, could also cause harm; and
4. when the price of an innovation is so high to the health care system that major opportunity costs are engendered.

It is clear that proton beam (as well as IMRT and SBRT) meet three of these criteria, yet still we use them routinely.

Opponents of the use of the randomized trials to assess protons argue that the difficulties faced by those currently using protons are well understood and can be managed with good confidence, and that the largely indirect evidence for the superiority of protons is strong. They therefore consider that the true equipoise necessary is unattainable and, thus, randomized trials of protons are unethical.

The pragmatic approach would state that, given the high cost of proton therapy, clinical trials are desirable wherever possible, and that phase III randomized trials can be ethical in certain well-defined clinical situations. In those situations in which there is not a comparable likelihood of downside—for example pediatric cancers—true equipoise may not be achievable, and a phase III trial would be ethically impossible. For other cancers, the situation may be far more evenly weighted. For example, if some technical aspects of proton therapy (say lung motion) or biological aspects (say RBE when giving the prostate doses close to 80 Gy) are not properly taken into account, clinical outcomes could even be worse.

It is also likely that when or if randomization proves difficult, then novel prospective, observational, case-controlled studies could serve as a valid alternative. These types of studies could assess endpoints of real interest, such as local tumor control, morbidity, quality of life, and cost-effectiveness. Bias is a risk in these studies due to possible systematic differences in the patient populations and because patients who have researched and sought out what they judge to be a superior treatment are often very satisfied with the outcome, even when it is functionally poor [56].

The main difference between protons and x-rays often lies in the larger medium- and low-dose radiation doses delivered by the latter, which will likely manifest itself primarily in late effects of low frequency that may be hard to measure and which may require long follow-up. This problem applies to both randomized clinical trials and observational studies, but the latter—which can include very large numbers of patients—may in this regard prove superior.

2.5.3 Rationalizing the Use of a Limited Resource?

Recognizing that protons are now, and will for the foreseeable future, remain more expensive to install and use than x-rays, attention must turn to the wise and selective use of an effective but limited resource.

In the spectrum of conditions currently being treated with protons, there is a wide range of opinions about their appropriateness. A great consensus exists that the use of protons in the treatment of certain, perhaps most, pediatric tumors is desirable and should be supported without further proof of effectiveness. Normal tissues in the growing child are particularly radiosensitive, and the morbidity from conventional x-ray therapy (e.g., growth abnormalities, intellectual deficits, and second malignancies) can be substantial and grave. The improvement in the dose distributions made possible by protons can be dramatic, and it is widely accepted that this will be advantageous, although the magnitude of the benefit has yet to be quantified.

The current reality is, however, very different. It has been estimated that 70% of all proton treatments delivered in the United States at the present time are not for pediatrics, but for prostate cancer [64,65]. There is considerable doubt as to whether the advantage of protons in the treatment of prostate cancer is sufficient to justify the additional cost. This phenomenon has largely been driven by the economic necessity of treating large numbers of patients to cover the installation costs of a proton facility. It might thus be argued that prostate patients are subsidizing pediatric treatments. Bearing in mind that many older men have perfectly acceptable alternatives to proton beam for their prostate cancer, often including no treatment at all, then a situation in which men may be treated unnecessarily to cover the losses engendered by more lengthy and complex pediatric treatments is a moral conundrum.

Putting economics aside and simply employing consensus informed by the available data, Zietman et al. have suggested a number of criteria to evaluate the appropriateness of protons in a given clinical situation [61]. Protons would likely be preferable when:

1. conventional x-ray treatments have serious, predictable, and frequent side effects;
2. the normal tissues receiving significant levels of conventional radiation are "particularly important," such as neural tissue or developing bone;
3. local control of the tumor needs improvement, and conventional therapy is at the limit of tolerance;
4. late morbidity is a credible (possibly model-predicted) problem with conventional therapy;
5. the target volume and the normal tissues have a complex geometric interrelationship to one another so that, with x-rays, it is difficult to deliver an adequate dose to the tumor while sparing the normal tissues; and
6. the target volume is large relative to the size of the anatomic compartment that it is desired to spare.

The treatment of pediatric tumors generally satisfies criteria 1, 2, 4, and sometimes 6. Treatment of prostate cancer with protons usually satisfies none of them. It generally fits into the category of a treatment that is expected to have only a marginally lower risk of toxicity.

In an era of personalized medicine, however, one cannot simply categorize all pediatric therapies as needing proton therapy and all prostate cancer as never needing it. For any given tumor site, there may be situations in which proton therapy is desirable and others in which it is unnecessary. For example, if small tumor-bearing portions of the prostate needed to be "boosted" to higher doses, protons may be advantageous. In addition, when chemotherapy is given concurrently with radiation therapy, the added toxicity may, in certain situations, be less with protons, making proton use advantageous. Palliative pediatric cases have no need for proton therapy.

Another potential application is in the re-treatment of recurrent tumors in critical locations when surrounding tissues have already received radiation doses close to their tolerance. The most likely scenario in which benefit will be rapidly appreciated is in the head and neck or at the skull base, but many other anatomic sites can be considered. Here well-designed trials will again be necessary as the benefits of re-irradiation are not clear. Cure is far from certain, and the morbidity from necrosis—even when little surrounding tissue is irradiated—may be intolerable or, in some cases, lethal.

2.5.4 The Financial Challenge to Proton Therapy and Policy Solutions

It is impossible to write about proton beam without a consideration of its cost. When a hospital invests in a proton therapy center, it takes a very substantial financial risk. It has likely elected to reduce its investment in other important areas of healthcare, it needs to amortize its costs rapidly, and it needs ultimately to generate a profit. Thus, the use of protons becomes as much a business decision as a clinical one; creditors and investors may drive the utilization and potentially the patient mix [64,65].

Because of the cancer's prevalence, and because of the speed and simplicity of its treatments, prostate cancer has become the economic driver for many new proton facilities. Aggressive marketing and high rates of reimbursement mean that the treatment of prostate cancer with protons can be highly profitable. The pressure to undertake such profitable treatments is exacerbated when the success of the business model requires a high throughput of patients. The more proton therapy centers with a prostate cancer concentration are established, the more likely it is that patients will be drawn away from other approaches of equal or greater value, such as a technically well-performed radical prostatectomy in younger men, brachytherapy for early-stage disease, or active surveillance in older men. The irony of the current U.S. reimbursement situation is that proton centers that focus their efforts on complex cases, such as pediatric treatments, are unable to amortize their debt at present rates, and prostate treatments are used to subsidize the more complex cases. Reimbursement rates would ideally be set so as to be financially neutral with respect to the choice of sites to be treated.

In an ideal world, independent and neutral evaluation groups would set the rational use of protons. In several European countries, such bodies have been used to evaluate the national need for proton centers, their most appropriate geographic location, and the strongest clinical indications for their use. In the United States, third-party payers have considerable power to guide technological evolution through their reimbursement policies [66]. Private insurers are beginning to question those areas where proton beam may not justify its cost, although their approach is often unselective and unscientific. There is risk that the entire technology will be "tarred" by the overuse of proton therapy in prostate cancer. If no substantial advantage is anticipated, reimbursement of that technology should be at the level of the least costly reasonable alternative. Another approach that has precedent in the United States is known as *coverage with evidence development*. This is a way for federal insurers to reimburse proton therapy's huge capital investment while at the same time forcing physicians to gather meaningful data. These data may ultimately reinforce or refute the case for the technology for specific indications.

A concern that is not unique to the United States, but which is certainly very prevalent within it, is the desire of some hospital administrations to use technologies such as proton beam for marketing. Recent *New York Times* articles have reminded us that technological advantages in medicine must only be adopted by institutions able to guarantee appropriate staffing levels and safety oversight. Otherwise, the consequences may be disastrous [67]. Likewise, physicians must curb their enthusiasm and monitor their institution's public statements, stepping in before they cross the line. Physicians should recognize which of their patients fit into the "zone of controversy" and commit to clinical studies, recognizing that data serves everyone, themselves included. In-house protocols run the risk of being lightweight covers to justify patient throughput. Far better are national randomized clinical trials or prospective registries in which large quantities of data from the new treatment and its alternatives can be accumulated to provide solid comparative, scientific and effectiveness data. Several excellent registries are now being established for just this purpose.

2.6 Summary

Proton beam is an attractive treatment for cancer. It is based on a legacy of sound theory and decades of clinical practice. The challenge for advocates of this therapy is to prove that it is indeed superior to its alternatives in terms of real patient outcomes. All the tools of modern clinical investigation—including quality of life studies and economic analyses—must be employed, in addition to the more traditional phase II and phase III trials. Clinicians must be selective in the use of this therapy and creative in its application to build a credible case for its continued use and for continued payment.

References

1. Harbron RW, Feltbowe RG, Glaser A, Lilley J, Pearce MS. Second malignant neoplasms following radiotherapy for primary cancer in children and young adults. *Pediatr Hematol Oncol* 2014,31:259–67.

2. Ginsberg JP, Goodman P, Leisenring W, et al. Long term survivors of childhood Ewing Sarcoma: report from the childhood cancer survivor study. *J Natl Cancer Inst* 2010, 102:1272–1283.
3. Davis EJ, Beebe-Dimmer JL, Yee CL, et al. Risk of second primary tumors in men diagnosed with prostate cancer: a population based study. *Cancer* 2014,120:2735–2741.
4. Suit H, Wette R. Radiation dose fractionation and tumor control probability. *Radiat Res* 1966,29:267–281.
5. Suit H.D., Zietman A.L., Tomkinson K.N., et al. Radiation response of xenografts of a human squamous cell carcinoma and a glioblastoma multiforme. *Int J Radiat Oncol Biol Phys* 1990;18:365–373.
6. Wilson RR: Radiological use of fast protons. *Radiology* 1946;47:487.
7. Bassal M, Mertens AC, Taylor L, Neglia JP, Greffe BS, Hammond S, et al. Risk of selected subsequent carcinomas in survivors of childhood cancer: a report from the childhood cancer survivor study. *J Clin Oncol* 2006;24(3):476–83.
8. Oeffinger KC, Mertens AC, Sklar CA, Kawashima T, Hudson MM, Meadows AT, et al. Chronic health conditions in adult survivors of childhood cancer. *N Engl J Med* 2006;355(15):1572–82.
9. Chung CS, Yock TI, Nelson K, Xu Y, Keating NL, Tarbell NJ. Incidence of second malignancies among patients treated with proton versus photon radiation. *Int J Radiat Oncol Biol Phys* 2013;87(1):46–52.
10. Sethi RV, Shih HA, Yeap BY, Mouw KW, Petersen R, Kim DY, et al. Second nonocular tumors among survivors of retinoblastoma treated with contemporary photon and proton radiotherapy. *Cancer* 2014,120:126–133.
11. St. Clair WH, Adams JA, Bues M, Fullerton BC, La Shell S, Kooy HM, et al. Advantage of protons compared to conventional x-ray or IMRT in the treatment of a pediatric patient with medulloblastoma. *Int J Radiat Oncol Biol Phys* 2004;58(3):727–34.
12. Lee CT, Bilton SD, Famiglietti RM, Riley BA, Mahajan A, Chang EL, et al. Treatment planning with protons for pediatric retinoblastoma, medulloblastoma, and pelvic sarcoma: how do protons compare with other conformal techniques? *Int J Radiat Oncol Biol Phys* 2005;63(2):362–72.
13. Kumar RJ, Zhai H, Both S, Tochner Z, Lustig R, Hill-Kayser C. Breast cancer screening for childhood cancer survivors after craniospinal irradiation with protons versus x-rays: a dosimetric analysis and review of the literature. *J Ped Hematol/Oncol* 2013;35(6):462–7.
14. Johnstone PA, McMullen KP, Buchsbaum JC, Douglas JG, Helft P. Pediatric CSI: are protons the only ethical approach? *Int J Radiat Oncol Biol Phys* 2013;87(2):228–30.
15. Wolden SL. Protons for craniospinal irradiation: are clinical data important? *Int J Radiat Oncol Biol Phys* 2013; 87:231–232.
16. Kozak KR, Adams J, Krejcarek SJ, Tarbell NJ, Yock TI. A dosimetric comparison of proton and intensity-modulated photon radiotherapy for pediatric parameningeal rhabdomyosarcomas. *Int J Radiat Oncol Biol Phys* 2009;74(1):179–86.
17. Timmermann B, Schuck A, Niggli F, Weiss M, Lomax AJ, Pedroni E, et al. Spot-scanning proton therapy for malignant soft tissue tumors in childhood: First experiences at the paul scherrer institute. *Int J Radiat Oncol Biol Phys* 2007;67(2):497–504.
18. Yock T, Schneider R, Friedmann A, Adams J, Fullerton B, Tarbell N. Proton radiotherapy for orbital rhabdomyosarcoma: clinical outcome and a dosimetric comparison with photons. *Int J Radiat Oncol Biol Phys* 2005;63(4):1161–8.
19. Mitin T, Zietman AL. The promise and pitfalls of heavy particle therapy. *J Clin Oncol 2014* (ePub August 11).
20. Zietman AL, Bae K, Slater JD, Shipley WU, Efstathiou JA, Coen JJ, et al. Randomized trial comparing conventional-dose with high-dose conformal radiation therapy in early-stage adenocarcinoma of the prostate: long-term results from proton radiation oncology group/american college of radiology 95–09. *J Clin Oncol* 2010;28(7):1106–11.
21. Kuban DA, Tucker SL, Dong L, Starkschall G, Huang EH, Cheung MR, et al. Long-term results of the M. D. Anderson randomized dose-escalation trial for prostate cancer. *Int J Radiat Oncol Biol Phys* 2008;70(1):67–74.
22. Zelefsky MJ, Chan H, Hunt M, Yamada Y, Shippy AM, Amols H. Long-term outcome of high dose intensity modulated radiation therapy for patients with clinically localized prostate cancer. *J Urol* 2006;176(4 Pt 1):1415–9.
23. Coen JJ, Bae K, Zietman AL, Patel B, Shipley WU, Slater JD, Rossi CJ. Acute and late toxicity after dose escalation to 82 GyE using conformal proton radiation for localized

prostate cancer: initial report of American College of Radiology phase II study 03–12. *Int J Radiat Oncol Biol Phys* 2011;81:1005–9.
24. Talcott JA, Rossi C, Shipley WU, Clark JA, Slater JD, Niemierko A, et al. Patient-reported long-term outcomes after conventional and high-dose combined proton and photon radiation for early prostate cancer. *JAMA* 2010,303:1046–53.
25. Sheets NC, Goldin GH, Meyer AM, Wu Y, Chang Y, Sturmer T, et al. Intensity-modulated radiation therapy, proton therapy, or conformal radiation therapy and morbidity and disease control in localized prostate cancer. *JAMA* 2012;307(15):1611–20.
26. Yu JB, Soulos PR, Herrin J, Cramer LD, Potosky AL, Roberts KB, et al. Proton versus intensity-modulated radiotherapy for prostate cancer: patterns of care and early toxicity. *J Nation Cancer Institute* 2013;105(1):25–32.
27. ASTRO releases list of five radiation oncology treatments to question as part of national choosing wisely campaign. https://www.astro.org/News-and-Media/News-Releases/2013/ASTRO-releases-list-of-five-radiation-oncology-treatments-to-question.aspx2013 [updated September 30, 2013; cited 2013 November 1].
28. Grimm P, Billiet I, Bostwick D, Dicker AP, Frank S, Immerzeel J, et al. Comparative analysis of prostate-specific antigen free survival outcomes for patients with low, intermediate and high risk prostate cancer treatment by radical therapy. Results from the prostate cancer results study group. *BJU international* 2012;109 Suppl 1:22–9. Epub 2012/02/18.
29. Bill-Axelson A, Holmberg L, Ruutu M, et al. Radical prostatectomy versus watchful waiting in early prostate cancer. *N Engl J Med* 2011;364(18):1708–1717.
30. Trofimov A, Nguyen PL, Coen JJ, Doppke KP, Schneider RJ, Adams JA, et al. Radiotherapy treatment of early-stage prostate cancer with IMRT and protons: a treatment planning comparison. *Int J Radiat Oncol Biol Phys* 2007;69(2):444–53.
31. Chaudhary P, Marshall TI, Francesca M., et al. Relative biological effectiveness variation along monoenergetic and modulated bragg peaks of a 62-MeV therapeutic proton beam: A preclinical assessment. *Int J Radiat Oncol Biol Phys* 2014,90:27–35.
32. Wang Z, Nabhan M, Schild SE, Stafford SL, Petersen IA, Foote RL, et al. Charged particle radiation therapy for uveal melanoma: a systematic review and meta-analysis. *Int J Radiat Oncol Biol Phys* 2013;86(1):18–26.
33. Schulz-Ertner D, Tsujii H. Particle radiation therapy using proton and heavier ion beams. *J Clin Oncol* 2007;25(8):953–64.
34. Colli B, Al-Mefty O. Chordomas of the craniocervical junction: follow-up review and prognostic factors. *J Neurosurg* 2001;95(6):933–43.
35. Debus J, Schulz-Ertner D, Schad L, Essig M, Rhein B, Thillmann CO, et al. Stereotactic fractionated radiotherapy for chordomas and chondrosarcomas of the skull base. *Int J Radiat Oncol Biol Phys* 2000;47(3):591–6.
36. MacDonald SM, Jimenez R, Paetzold P, et al. Proton radiotherapy for chest wall and regional lymphatic radition: dose comparisons and treatment delivery. *Radiat Oncol* 2013;8:71.
37. Kozak KR, Smith BL, Adams J, et al. Accelerated partial-breast irradiation using proton beams: Initial clinical experience. *Int J Radiat Oncol Biol Phys* 2006,66:691–698.
38. Galland-Girodet S, Pashtan I, MacDonald S, et al. Long-term cosmetic outcomes and toxicities of proton beam therapy compared with photon-based 3-dimensional conformal accelerated partial-breast irradiation: a phase 1 trial. *Int J Radiat Oncol Biol Phys* 90,3:493–503.
39. Bush DA, Do S, Lum S, et al. Partial breast radiation therapy with proton beam: 5-year results with cosmetic outcomes. *Int J Radiat Oncol Biol Phys* 2014,90:501–505.
40. De Ruysscher D, Chang JY. Clinical controversies: proton therapy for thoracic tumors. Sem Radiat Oncol 2013;23(2):115–9.
41. Bradley JD, Paulus R, Komaki R, et al. A randomized phase III comparison of standard-dose (60 Gy) versus high-dose (74 Gy) conformal chemoradiotherapy with or without cetuximab for stage III non-small cell lung cancer: results on radiation dose in RTOG 0617. *J Clin Oncol* 2013;31(suppl):abstr 7501.
42. Chang JY, Komaki R, Lu C, Wen HY, Allen PK, Tsao A, et al. Phase 2 study of high-dose proton therapy with concurrent chemotherapy for unresectable stage III nonsmall cell lung cancer. *Cancer* 2011;117(20):4707–13.
43. Grutters JP, Kessels AG, Pijls-Johannesma M, De Ruysscher D, Joore MA, Lambin P. Comparison of the effectiveness of radiotherapy with photons, protons and carbon-ions for non-small cell lung cancer: a meta-analysis. *Radiother Oncol* 2010;95(1):32–40.

44. Westover KD, Seco J, Adams JA, Lanuti M, Choi NC, Engelsman M, et al. Proton SBRT for medically inoperable stage I NSCLC. *J Thoracic Oncol* 2012;7(6):1021–5.
45. Hoppe BS, Huh S, Flampouri S, Nichols RC, Oliver KR, Morris CG, et al. Double-scattered proton-based stereotactic body radiotherapy for stage I lung cancer: a dosimetric comparison with photon-based stereotactic body radiotherapy. *Radiother Oncol* 2010;97(3):425–30.
46. Chang JY, Li H, Zhu XR, et al. Clinical implementation of intensity modulated proton therapy for thoracic malignancies. *Int J Radiat Oncol Biol Phys* 2014,90:809–818.
47. Fitzek MM, Thornton AF, Rabinov JD, Lev MH, Pardo FS, Munzenrider JE, et al. Accelerated fractionated proton/photon irradiation to 90 cobalt gray equivalent for glioblastoma multiforme: results of a phase II prospective trial. *J Neurosurg* 1999;91(2):251–60.
48. Mizumoto M, Tsuboi K, Igaki H, Yamamoto T, Takano S, Oshiro Y, et al. Phase I/II trial of hyperfractionated concomitant boost proton radiotherapy for supratentorial glioblastoma multiforme. *Int J Radiat Oncol Biol Phys* 2010;77(1):98–105.
49. Noel G, Bollet MA, Calugaru V, Feuvret L, Haie-Meder C, Dhermain F, et al. Functional outcome of patients with benign meningioma treated by 3D conformal irradiation with a combination of photons and protons. *Int J Radiat Oncol Biol Phys* 2005;62(5):1412–22.
50. Weber DC, Lomax AJ, Rutz HP, Stadelmann O, Egger E, Timmermann B, et al. Spot-scanning proton radiation therapy for recurrent, residual or untreated intracranial meningiomas. *Radiother Oncol* 2004;71(3):251–8.
51. Foote RL, Stafford SL, Petersen IA, Pulido JS, Clarke MJ, Schild SE, et al. The clinical case for proton beam therapy. *Radiat Oncol* 2012;7:174.
52. Holliday EB, Frank SJ. Proton radiation therapy for head and neck cancer: a review of the clinical experience to date. *Int J Radiat Oncol Biol Phys* 2014,89:292–302.
53. Skinner HD, Hong TS, Krishnan S. Charged-particle therapy for hepatocellular carcinoma. *Sem Radiat Oncol* 2011;21(4):278–86.
54. Monjazeb AM, Blackstock AW. The impact of multimodality therapy of distal esophageal and gastroesophageal junction adenocarcinomas on treatment-related toxicity and complications. *Sem Radiat Oncol* 2013;23(1):60–73.
55. Suit H, Goldberg S, Niemierko A, et al. Proton beams to replace photon beams in radical dose treatments. *Acta Oncol* 2003;42:800–808.
56. Glatstein E, Glick J, Kaiser L, et al. Should randomized clinical trials be required for proton radiotherapy? An alternative view. *J Clin Oncol* 2008;26:2438–2439.
57. Das IJ, Moskvin VP, Zhao Q, Cheng CW, Johnstone PA. Proton therapy facility planning from a clinical and operational model. *Technol Cancer Res Treat* 2014;10.7785.
58. Goitein M, Jermann M. The relative costs of proton and x-ray radiation therapy. *Clin Oncol (R Coll Radiol)* 2003;15:S37–S50.
59. Peeters A, Grutters JPC, Pijls-Johannesma M, Reimoser S, De Ruysscher D, Severens JL, et al. How costly is particle therapy? Cost analysis of external beam radiotherapy with carbon-ions, protons and photons. *Radiother Oncol* 2010;95(1):45–53.
60. Pijls-Johannesma M, Pommier P, Lievens Y. Cost-effectiveness of particle therapy: Current evidence and future needs. *Radiother Oncol* 2008;89(2):127–34.
61. Zietman A, Goitein M, Tepper JE. Technology evolution: is it survival of the fittest? *J Clin Oncol* 2010;28(27):4275–9.
62. Suit H, Kooy H, Trofimov A, et al: Should positive phase III clinical trial data be required before proton beam therapy is more widely adopted? *Radiother Oncol* 2008;86:148–153.
63. Halperin E. Personal communication. 2010.
64. Johnstone PA, Kerstiens J, Richard H. Proton facility economics: the importance of simple treatments. *J Am Coll Radiol* 2012;9(8):560–563.
65. Kerstiens J, Johnstone PA. Proton therapy expansion under current United States reimbursement models. *Int J Radiat Oncol Biol Phys* 2014;89(2):235–40.
66. Emanuel E, Tanden N, Altman S, et al. A systemic approach to containing health care spending. *N Engl J Med* 2012;367(10):949–54.
67. Bogdanich W: Radiation offers new cures, and ways to do harm. *New York Times* Jan 24, 2010; http://www.nytimes.com/2010/01/24/health/24radiation.html.

Chapter 3

Proton Beam Interactions: Basic

Hugo Palmans[1,2]

[1]EBG MedAustron GmbH
Wiener Neustadt, Austria
[2]National Physical Laboratory,
Teddington, UK

3.1 Introduction

While interactions of photons, electrons, and positrons with matter are important for some aspects of proton therapy physics, e.g., the relation of proton stopping powers with CT images, prompt gamma imaging, PET imaging, and secondary electron effects in dosimetry, we focus here on the interactions of protons with matter. For background on the interactions of other particles, we refer to other textbooks [1–3]. Protons are charged particles that are heavy compared to electrons, but light compared to most other ions. Their basic modes of interaction with matter are mainly due to Coulomb interactions (just as other charged particles, such as electrons and heavier ions), but the momentum of the proton makes that protons with ranges of radiotherapeutic interest behave to some extent between electrons and heavy ions. While on the one hand the path of the proton is much less affected by a single interaction as that of an electron, the scattering it undergoes is substantially larger than that of heavier ions, and while the probability of a nuclear interaction is much larger than that of an electron, interactions are simpler as for heavier ions since the proton as a projectile does not undergo spallation or fragmentation itself. The amount of energy the proton transfers per unit depth is also slightly higher as for electrons, but even the highest values (toward the end of the range) are still much smaller than those of heavier ions. This chapter provides definitions of quantities relevant to proton therapy physics and the basics of the interactions of therapeutic proton beams with matter. This chapter is by no means comprehensive, and more in-depth discussion of the subjects addressed in this chapter can be found in ICRU Reports [3,4–13] and basic textbooks [1,2,14,15]. Additional sources on specific topics will be given in the subsections of this chapter.

3.2 Definition of Quantities

A quantity is a property of a phenomenon, body, or substance, where the property has a magnitude that can be expressed as a number and a unit (JCGM [16]). The unit is a particular sample of the quantity which is used as a reference. The International System of Quantities and Units define seven base quantities (length, mass, time, electric current, thermodynamic temperature, amount of substance, and luminous intensity) and their respective base units (meter, kilogram, second, ampere, kelvin, mole, and candela) with unit symbols m, kg, s, A, K, mol, and cd. All other quantities are derived quantities and can be expressed as a function of the base quantities. Derived units may have a special unit, such as energy (joule, $\mathrm{J} = \mathrm{kg\ m^2\ s^{-2}}$) or absorbed dose (gray, $\mathrm{Gy} = \mathrm{J\ kg^{-1}}$). A complete description of the International System of Quantities and Units can be found at the BIPM website (www.bim.org) or (BIPM [17]). Note that the base units will be redefined in terms of fundamental constants of nature in the coming years [18], but this will have limited impact on the use of quantities and units in radiotherapy physics. The definition of radiation quantities in the following subsections conform to ICRU Report 85a [13].

3.2.1 Macroscopic Dosimetric Quantities

3.2.1.1 Quantities Describing the Radiation Field

The description of radiation fields is based on the number of particles that are emitted, transferred, or received, defined as the particle number N, or the total energy of the particles that are emitted, transferred, or received, defined as the radiant energy R. The relation between the particle number and radiant energy of a particle radiation field is described by the type of the particles, and the distribution in space, direction, and time of N and R based on the quantities defined in this section.

The fluence, Φ, at a point P in space is defined as

$$\Phi = \frac{\mathrm{d}N}{\mathrm{d}a} \tag{3.1}$$

where dN is the number of particles incident on a sphere of cross-sectional area da centered at the point P. The use of a sphere expresses the fact that one considers the area perpendicular to the direction of each particle. The main reason for defining fluence in this way is its relation to quantities such absorbed dose, which do not depend on the direction of incidence of a particle on a small sphere. The SI unit of fluence is m^{-2}.

An equivalent definition, which is for practical reasons often used to estimate fluence in a Monte Carlo simulation, is given by Chilton [19]:

$$\Phi = \frac{\sum l}{V} \tag{3.2}$$

where Σl is the expectation value of the sum of the track lengths of all particles passing the sphere and V is its volume. This definition also applies to a volume of any arbitrary shape.

The energy fluence, Ψ, at a point P in space, which is defined as the product of the fluence and the energy of the particle, is commonly used in photon beam dosimetry, but since it is not used in proton beam, it is not further discussed here.

It is seldom, if not only in theoretical cases, that all particles crossing the sphere have the same energy and direction of motion, and the radiation field at the point P is described in terms of the distributions of fluence and energy fluence with respect to energy and angle:

The fluence differential in energy is defined as:

$$\Phi_E = \frac{\mathrm{d}\Phi}{\mathrm{d}E} \tag{3.3}$$

The fluence differential in angle is defined as:

$$\Phi_\Omega = \frac{\mathrm{d}\Phi}{\mathrm{d}\Omega} \tag{3.4}$$

where Ω is the solid angle with respect to a specified direction. In spherical coordinates $\mathrm{d}\Omega = sin\theta \mathrm{d}\theta d\phi$.

The fluence rate is defined as:

$$\dot{\Phi} = \frac{\mathrm{d}\Phi}{\mathrm{d}t} \tag{3.5}$$

This quantity is sometimes referred to as particle flux density.

3.2.1.2 Interactions and Cross Sections

An interaction is a process occurring between radiation and matter. In an interaction, the energy or the direction (or both) of the incident particle is altered, or the particle is absorbed. The probability of an interaction taking place is described by the cross section, σ, defined as the mean number of interactions, N, that occur in a target per unit of particle fluence, Φ, the target is exposed to:

$$\sigma = \frac{N}{\Phi} \tag{3.6}$$

The SI unit of cross section is m^2. A special unit that is often used is the barn, b = 10^{-28} m^2 = 100 fm^2. If more than one type of interaction is possible, the total cross section is the sum of the cross sections for each of the individual interaction types.

The mass attenuation coefficient, μ/ρ, is defined as:

$$\frac{\mu}{\rho} = \frac{1}{\rho dl}\frac{dN}{N} \tag{3.7}$$

where dN / N is the expectation value of the fraction of the particles that experience interactions in traversing a distance dl in the material of density ρ. The quantity μ is the linear attenuation coefficient and is the reciprocal of the mean free path. The mass attenuation coefficient is related to the total cross section:

$$\frac{\mu}{\rho} = \frac{N_A}{M}\sigma \tag{3.8}$$

where N_A is the Avogadro constant and M the molar mass of the target material.

In the context of proton interactions and dosimetry, the relation between the cross sections and an attenuation coefficient pertains to non-elastic nuclear interactions (see Section 3.4).

For protons of a given energy, the linear stopping power, S, is defined as

$$S = \frac{dE}{dl} \tag{3.9}$$

where dE is the expectation value of the energy lost by the protons in traversing a distance dl in the material. Given that the linear stopping power is to a good approximation inverse proportional to the mass density, ρ, the mass stopping power, S/ρ, is defined as:

$$\frac{S}{\rho} = \frac{1}{\rho}\frac{dE}{dl} \tag{3.10}$$

The SI unit of stopping power is: J m^2 kg^{-1} but is in dosimetry often given in MeV g^{-1} cm^2. The total mass stopping power is constituted of a sum of components due to interactions with atomic electrons resulting in ionization or excitation, due to emission of Bremsstrahlung and due to elastic Coulomb interactions with the nucleus. These are called the mass electronic stopping power, the mass radiative stopping power, and the mass nuclear stopping power, respectively. Each of these components can be expressed in terms of interaction cross sections, e.g., for the mass electronic stopping power $\frac{S_{el}}{\rho}$:

$$\frac{S_{el}}{\rho} = \frac{N_A}{M} Z \int \varepsilon \frac{d\sigma}{d\varepsilon} d\varepsilon \tag{3.11}$$

where Z is the atomic number, $d\sigma / d\varepsilon$ the differential cross section per atomic electron, and ε the energy loss.

The lineal energy transfer or restricted linear electronic stopping power, L_Δ, of a material for protons is defined as:

$$L_\Delta = \frac{dE_\Delta}{dl} \tag{3.12}$$

where dE_Δ is the expectation value of the energy lost by the protons due to electronic interactions in traversing a distance dl, minus the mean sum of the kinetic energies in excess of Δ of all the electrons released from the material due to the action of the protons.

3.2.1.3 Dosimetric Quantities

Cema is a quantity for protons that is defined in a similar way as kerma (kinetic energy released per unit of mass) for uncharged particles. Cema is defined as

$$C_{med} = \frac{dE_{el}}{dm} \tag{3.13}$$

where dE_{el} is the mean energy lost in electronic interactions by all charged particles (i.e., primary protons and secondary charged particles produced in the beam path) except electrons in a mass dm. The SI unit of cema is J kg^{-1}, and the special name for the unit of cema is the gray (Gy). The relation between cema and the mass electronic stopping power is given by:

$$C_{med} = \Phi \left(\frac{S_{el}}{\rho} \right)_{med} \tag{3.14}$$

The quantity of absorbed dose to a medium, D_{med}, is defined as

$$D_{med} = \frac{d\bar{\varepsilon}_{med}}{dm} \tag{3.15}$$

where $d\bar{\varepsilon}_{med}$ is the mean energy imparted to a mass dm of the medium. Energy imparted is the energy incident minus the energy leaving the mass, plus any decrease in rest energy within the mass. The medium should always be specified, e.g., absorbed dose to water, absorbed dose to muscle, etc. The SI unit of absorbed dose is J kg^{-1}, and the special name for the unit of absorbed dose is the gray (Gy).

In the case of there being charged particle equilibrium, absorbed dose equals cema. In the more general case, the relation between absorbed dose and the restricted mass electronic stopping power is given by

$$D_{\text{med}} = \Phi \left(\frac{L_{\Delta}}{\rho} \right)_{\text{med}} \tag{3.16}$$

However, it is important that contributions from energetic secondary particles generated outside the volume of interest are accounted for as well. Absorbed dose, even for mono-energetic protons, should therefore be integrated over the entire charged particle energy spectrum of all particle types, *i*, crossing the cavity:

$$D_{\text{med}} = \sum_{i} \left[\int_{\Delta}^{E_{\max,i}} \Phi_{E,i} \left(\frac{L_{\Delta,i}}{\rho} \right)_{\text{med}} dE + TE_{i} \right] \tag{3.17}$$

TE_i is a track end term accounting for charged particles that originate outside the volume of interest, but that fall under the energy cutoff, Δ, within the volume of interest.

3.3.1.4 Radioactivity

Radioactivity is relevant for proton beams with respect to the production of radioisotopes by non-elastic nuclear interactions. Radioactivity is the phenomenon of the spontaneous transformation of an atomic nucleus to another nucleus or to another energy state by emission of energy under the form of radiation (e.g., alpha particles, protons, neutrons, beta rays, or gamma rays). Radioactivity is a stochastic process in which also atomic electrons are involved, since nuclear transformations can affect the atomic shell structure and cause emission or capture of electrons and the emission of photons. A nuclide is a species of atom having a specified number of protons and neutrons in its nucleus, and unstable nuclides are called radionuclides.

Activity

The activity, A, of a radionuclide is defined as:

$$A = -\frac{dN}{dt} \tag{3.18}$$

where $-dN$ is the expectation value of the number of nuclear transformations or decays in a differential time interval dt. The special name for the SI unit of activity is the becquerel ($\text{Bq} = \text{s}^{-1}$)

Decay Constant

The decay constant λ is defined as:

$$\lambda = -\frac{dN / N}{dt} \tag{3.19}$$

and thus corresponds with the expectation value of the fraction of the radionuclides in a sample that decay per unit of time, *dt*. The consequence of this is that if the initial amount of radionuclides N_0 is not replenished, the number of radionuclides as a function of time follows an exponential decay:

$$N = N_0 e^{-\lambda t} \tag{3.20}$$

The activity of a sample is thus equal to the product of the decay constant and the number of radionuclides present in the sample:

$$A = \lambda N = \lambda N_0 e^{-\lambda t} = A_0 e^{-\lambda t} \tag{3.21}$$

A quantity related to the decay constant is the half-life:

$$t_{1/2} = \frac{\ln(2)}{\lambda} \tag{3.22}$$

which is the mean time taken for half of the number of radionuclides to decay.

Parent–Daughter Relation

If a sample contains an initial number of (parent) radionuclides $(N_1)_0$ that decay into another (daughter) radionuclide, and the initial number of the daughter radionuclides is $(N_2)_0 = 0$, then the activity of the parent and daughter nuclides are

$$A_1 = \lambda_1 N_1 = \lambda_1 \left(N_1\right)_0 e^{-\lambda_1 t} \tag{3.23}$$

and

$$A_2 = \lambda_2 N_2 = \lambda_1 (N_1)_0 \frac{\lambda_2}{\lambda_2 - \lambda_1}(e^{-\lambda_1 t} - e^{-\lambda_2 t}) \tag{3.24}$$

Radioactivation

In proton beams, radionuclides are produced as the reaction products of nuclear interactions due to irradiating stable nuclei (see Section 3.4). If in a sample there are N_T target atoms exposed to a fluence rate $\dot{\Phi}$ of incoming proton radiation and the production cross section of the radionuclide of interest by that incoming radiation is σ_{RN}, then the number of radionuclides as a function of time is:

$$A_{RN} = \lambda_{RN} N_{RN} = \dot{\Phi} N_T \sigma_{RN} \left(1 - e^{-\lambda_{RN} t}\right) \tag{3.25}$$

If the half-life of the radionuclide is long compared to the irradiation time t_{irr}, then the activity at the end of the irradiation can be approximated as:

$$A_{RN} = \lambda_{RN} N_{RN} \approx \lambda_{RN} \dot{\Phi} N_T \sigma_{RN} t_{irr} = \lambda_{RN} \Phi_{irr} N_T \sigma_{RN} \tag{3.26}$$

where $\Phi_{\text{irr}} = \int_0^{t_{\text{irr}}} \dot{\Phi}\, dt$ is the total fluence delivered during the irradiation. This applies, for example, to the production of ^{11}C from irradiating ^{12}C atoms with protons for irradiations not larger than one minute given that the half-life of ^{11}C is about 20.3 min. ^{11}C is the radionuclide most commonly detected in post-treatment verification using positron-emission tomography. For ^{15}O, formed after irradiating ^{16}O with protons, on the other hand, the full expression needs to be used, given that the half-life is only about 2 min.

3.2.2 Microdosimetric Quantities

Microdosimetry concerns the determination of the spatial and temporal distribution of interactions of ionizing radiation with micrometer-sized volumes of matter. It is assumed that the distribution of ionizations on the microscale is correlated with indirect damage inflicted to the DNA via the diffusion of radiation-induced reactive species. The energy dependence of the radiobiological effectiveness of protons can be correlated to the variation of the microdosimetric properties. Microdosimetric quantities are stochastic and are, therefore, given in terms of interaction probabilities. Structural microdosimetry aims at deriving microdosimetric quantities and actions from detailed 3D distributions of energy transfer points and is not further discussed here. Relevant quantities in regional microdosimetry are [20]:

- the specific energy z in a volume V, defined as $z = \frac{\varepsilon}{\rho V}$ where ε is the energy imparted in the volume and ρ the mass density of the medium, and its probability distribution $f(z)$;
- the lineal energy y, defined as $y = \frac{\varepsilon_1}{\bar{l}}$, where ε_1 is the energy imparted in the volume in a single event and $\bar{l}$ is the mean chord length of the volume, and its probability distribution $f(y)$, usually presented as $y \times f(y)$ or $y^2 \times f(y)$ plotted against $\log(y)$;
- the first moment of $f(y)$, $\bar{y}_F = \int_0^\infty y f(y)\, dy$, also called the frequency mean lineal energy,
- the dose distribution, $d(y) = y f(y) / \bar{y}_F$, which is important for obtaining the dose components of the microdosimetric spectrum, and
- the first moment of $d(y)$, $\bar{y}_D = \int_0^\infty y d(y)\, dy$, also called the dose mean lineal energy.

For a given biological system and endpoint, an empirical relation between the radiobiological effectiveness of radiation (RBE) and the microdosimetric spectrum has been suggested [21]:

$$\text{RBE} = \int_0^{\infty} r(y)d(y)dy \tag{3.27}$$

where $r(y)$ is an empirical weighting function derived from cell survival data for a range of ion beam qualities.

3.2.3 Nanodosimetric Quantities

When ionization clustering in the vicinity or within the DNA becomes very high, the diffusion and long-term chemistry (on a time scale longer than 10^{-7} s) of reactive species becomes less important and substantial DNA damage will be more correlated with the cluster density distribution. For protons, this situation occurs mostly in the Bragg peak and is most pronounced in its distal edge. The measurement or simulation of the clustering distributions within the track structure on the nanoscale is the subject of nanodosimetry. The characterization of track structure is based on the stochastic quantity called ionization cluster size, and its frequency distribution (ionization cluster size distribution, ICSD). The ionization cluster size is defined as the number of ionizations produced by a particular particle track in a specific target volume. The size of the target volumes considered in nanodosimetry is always smaller than the lateral extension of the penumbra of the primary particle track, where interactions are due to secondary electrons. ICSD depends on the size of the target, its geometry and material composition, and on the geometrical relation between the primary particle trajectory and the target. This is accounted for by specifying the impact parameter d which is the smallest distance between the primary particle trajectory and the center of the target volume. Relevant quantities in nanodosimetry are [20]:

- ν: the ionization cluster size, defined as the number of ionizations produced in the nanometric target volume by a single primary particle track, including ionizations produced in interactions of secondary electrons within the site;

- $P_\nu(d;l)$ or $P(\nu \mid d;l)$: the probability distribution of ionization cluster size, which depends on the impact parameter of the primary particle trajectory with respect to the target d and the size of the target l;

- $F_k(d;l) = \sum_{\nu=k}^{\infty} P_\nu(d;l)$: the (complementary) cumulative probability distribution of ionization cluster size, giving the probability that an ionization cluster size of k or larger is produced in the target volume; and

- $M_k(d;l) = \sum_{\nu=0}^{\infty} \nu^k P(\nu \mid d;l)$: the kth statistical moments of $P(\nu \mid d;l)$. $M_1(d;l)$ is also called the mean ionization cluster size.

3.3 Electromagnetic Interactions

Protons interact with atomic electrons or with the nuclei along the track in a medium. Since the electrons in solid matter are mostly bound to atoms, interactions do not occur up to infinite distance from the proton track. The further the electron is from the track, the smaller the momentum that can be transferred. Proton interactions are categorized as a function of the classical impact parameter, b, defined as the closest distance between the initial trajectory of the incident proton and the nucleus, by comparing b with the atomic radius, a.

3.3.1 Interactions with Electrons

When b >> a, the incident proton interacts with the atom as a whole, and only a small amount of energy is transferred from the incident proton to the atom. These interactions are often called soft collisions. The energy transfer can result in atomic excitation (raising an orbital electron to a higher allowed orbital state) or ionization. When $b \approx a$, the incident proton can interact with a single orbital electron, resulting in a large energy transfer to that electron, termed a "knock-on electron." These interactions are often called hard collisions. The knock-on electrons ejected from the atom are also termed δ-rays. Since the electron binding energy is usually small compared to the energy transferred to the electrons, the collision can be approximated by the Rutherford cross section based on classical mechanics so that the high-energy part of the distribution of the recoil electrons, differential in energy, can be described as [22]:

$$\frac{\mathrm{d}\sigma}{\mathrm{d}E_e} = \frac{4\pi a_0^{\,2}}{E_p}\frac{m_p}{m_e}\left(\frac{R_\infty}{E_e}\right)^2 \tag{3.28}$$

where E_e is the kinetic energy of the electron, E_p the kinetic energy of the incoming proton, m_e the electron rest mass, m_p the proton rest mass, a_0 the Bohr radius and R_∞ the Rydberg energy. The maximum electron energy is $E_{e,\max} = 4\left(m_e / m_p\right)E_p$.

If the electron binding energy I is not negligible, an improved approximation is given by

$$\frac{\mathrm{d}\sigma}{\mathrm{d}E_e} = \frac{4\pi a_0^{\,2}}{E_p}\frac{m_p}{m_e}\left(\frac{R_\infty}{E_e + I}\right)^2 \tag{3.29}$$

Rudd et al. [22] provide a range of more sophisticated approaches and improvements to the Rutherford formula, among which are corrections for relativistic spin effects according to Bhabha [23]:

$$\frac{\mathrm{d}\sigma}{\mathrm{d}E_e} = \frac{4\pi a_0^{\,2}}{E_p}\frac{m_p}{m_e}\left(\frac{R_\infty}{E_e}\right)^2\left(1 - \beta^2\frac{E_e}{E_{e,\max}}\right) \tag{3.30}$$

where the maximum electron recoil energy is

$$E_{e,\max} = \frac{\beta^2}{1-\beta^2}m_p c^2 \Big/ \left(\frac{1}{\sqrt{1-\beta^2}} + \frac{1}{2}\left(\frac{m_p}{m_e} + \frac{m_e}{m_p}\right)\right) \tag{3.31}$$

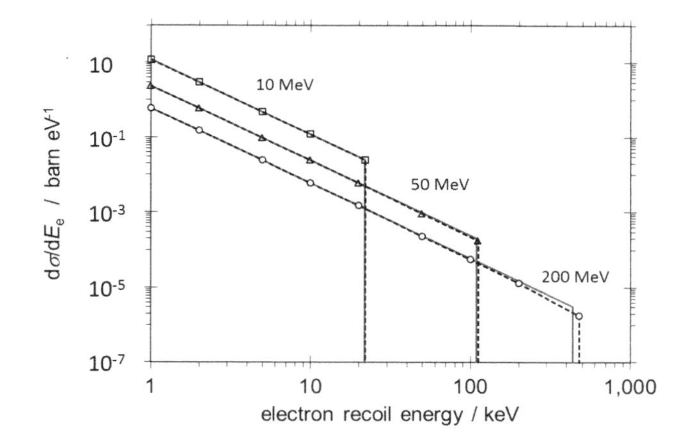

Figure 3–1 Distribution of the recoil electrons, differential in energy, using the Rutherford cross section (gray full lines) and the Bhabha cross section accounting for relativistic spin effects (symbols connected by black dashed lines).

Figure 3–1 compares the Rutherford and Bhabha distributions for 10, 50, and 200 MeV protons. It is clear that the differences are very small. In most proton therapy applications, the range of secondary electrons is so small that they are not accounted for in radiation transport. Exceptions are dosimetry using a low-density detector, such as an ionization chamber, and dose distributions in the vicinity of interfaces. In ionization chambers, the range of the most energetic secondary electrons is of the order of the detector dimensions and, even though the differences between the Rutherford and Bhabha distributions are small, the most pronounced differences occur at the highest energies where electron fluence perturbation in ionization chambers will not be negligible. Indeed, it has been demonstrated that perturbation correction for ionization chambers in clinical proton beams are of the order of 0.5% [24,25].

3.3.2 Interactions with Nucleus

When $b \ll a$, (b is approximately equal to the nuclear radius), the incident proton interacts with the nucleus via elastic and non-elastic interactions. Elastic scattering can result in large angular deflections, often with limited energy transferred to the target nucleus due to the small ratio of the mass of the incident proton to that of the target nucleus. Protons can also induce nuclear interactions, but note that the effect of nuclear interactions is not included in the stopping power. Dosimetrically, their contribution is accounted for by considering their effects on the attenuation of the primary proton beam and additions to the secondary particle fluence.

3.3.3 Stopping Power

The definition of mass stopping power is given in Section 3.2.1.2. For protons, the mass stopping power is dominated by the mass electronic stopping powers. Only at low proton energies is the nuclear stopping power an important contribution. For most clinical applications, the mass stopping power practically equals the mass electronic stopping power. The stopping power should be considered as the mean energy loss of a large number of particles of the same energy. Each individual particle will undergo a fluctuating energy loss that is described by an energy straggling distribution.

3.3.3.1 Bethe Theory

Bethe developed the basic theory for electronic stopping powers of high-energetic charged particles based on the first Born approximation [26]. The Bethe equation for the mass electronic stopping power is [8]:

$$\frac{S_{\text{el}}}{\rho} = \frac{4\pi r_e^2 m_e c^2}{\beta^2} \frac{1}{u} \frac{Z}{A} z^2 \left[\frac{1}{2} ln\left(\frac{2 m_e c^2 \beta^2 W_m}{1-\beta^2} \right) - \beta^2 - ln(I) \right] \tag{3.32}$$

where r_e is the classical electron radius, $m_e c^2$ is the electron rest energy, u is the atomic mass unit, β is the particle velocity in units of the velocity of light, Z and A are the atomic number and relative atomic mass of the target atom, z is the charge number of the projectile (unity for protons), I is the mean excitation energy of the medium, and W_m is the largest possible energy loss in a single collision with a free electron which, based on classical relativistic mechanics, is given by

$$W_m = \frac{2 m_e c^2 \beta^2}{1-\beta^2} / \left[1 + \frac{2}{\sqrt{1-\beta^2}} \frac{m_e}{M} + \left(\frac{m_e}{M} \right)^2 \right] \tag{3.33}$$

The standard value of $4\pi r_e^2 m_e c^2/u$ given in ICRU Report 49 [8] is 0.307075 MeV cm^2 g^{-1}.

3.3.3.2 Corrections

ICRU Report 49 [8] applies a number of corrections to the Bethe formula:

$$\frac{S_{\text{el}}}{\rho} = \frac{4\pi r_e^2 m_e c^2}{\beta^2} \frac{1}{u} \frac{Z}{A} z^2 \left[\frac{1}{2} ln\left(\frac{2 m_e c^2 \beta^2 W_m}{1-\beta^2} \right) - \beta^2 - ln(I) - \frac{C}{Z} - \frac{\delta}{2} + B_1 + B_2 \right] \tag{3.34}$$

where C/Z is the shell correction to account for the bound state of K and L shell electrons, $\delta/2$ is the Fermi density effect correction, B_1 the Bloch correction, and B_2 the Barkas correction. The Bloch correction accounts for deviations from the first-order Born approximation [27], and with the inclusion of this correction term, the stopping power formula is accurate when the proton velocity is larger than the atomic electron velocities. The Barkas correction accounts for another departure from the first-order

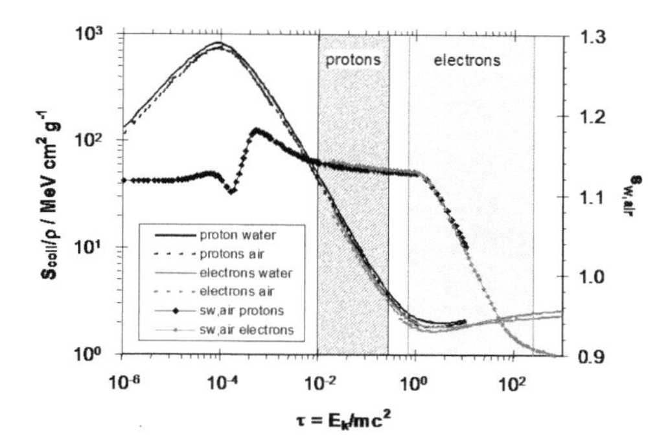

Figure 3–2 Mass electronic stopping powers and water-to-air mass electronic stopping power ratios for protons and electrons in water and air as a function of the kinetic energy over the rest mass as a measure for the relativistic speed (from ICRU Report 49 [8] for protons and ICRU Report 37 [3] for electrons). The shaded areas represent the therapeutic energy range down to the energy where the csda range of the particle is 1 mm in water.

Born approximation—making the stopping power for a negative charged particle slightly smaller than that for a positively charged particle of the same mass and velocity [28]. The shell correction accounts for the reduced energy transfer in interactions with K- and L-shell electrons at low projectile velocities [29]. The density effect accounts for the reduced stopping power due to the polarization of the medium induced by the projectile itself, and it is only important at relativistic speeds. For low-Z media, the Barkas and shell correction terms are only important for protons energies below 10 MeV, while the density effect correction is only important above 300 MeV, so none of these corrections is substantial in the clinical proton energy range.

While the basic theory is the same as for stopping powers of electrons, there are slight differences. It is nevertheless interesting to compare the stopping powers and stopping power ratios for protons and electrons/positrons as shown in Figure 3–2.

It is clear that the stopping powers are predominantly dependent on the relativistic speed of the projectile. Noticeably, however, there are also some important differences that become clear from the shaded areas marked in Figure 3–2. These cover energy ranges for protons and electrons from the maximum energy usually encountered in radiotherapy down to energies for which the particle range is reduced to 1 mm in water. A distinct difference is that in the range of clinical electron energies, the density effect correction has a large influence on the water-to-air stopping power ratios, whereas in the range of clinical proton energies, the water-to-air stopping power ratio is nearly constant. This is obviously an advantage with respect to the measurement of depth–dose distributions with an ionization chamber. On the other hand, within the clinical proton energy range, the stopping power itself, and thus the linear energy transfer (LET), varies by two orders of magnitude, leading to the characteristic Bragg peak at the low energies near the end of the range of the particle. When the remaining range is around 1 mm, the LET is about a factor 10 higher than at higher energies, so there will be a component in the Bragg peak with an even higher LET, although its significance is not a priori clear. This increased LET component will have a substantial influence on the response of detectors that exhibit a LET dependence, such as the ferrous sulphate dosimeter, alanine, solid-state detectors, thermoluminescent detectors (TLDs), gel dosimeters, film, etc. This has important practical implications in the measurement of proton depth–doses and, in particular, the distal "falloff," and influences the biological effects in tissue, as will be discussed in Chapter 5.

3.3.3.3 Restricted Stopping Powers

Dosimetry using small-volume detectors requires the calculation of restricted stopping powers, such as in the Spencer–Attix stopping power ratios. The restricted collision mass stopping power with cut-off energy Δ is obtained by replacing W_m in Equation 3.34 with Δ:

$$\frac{S_{el}}{\rho} = \frac{4\pi r_e^2 m_e c^2}{\beta^2}\frac{1}{u}\frac{Z}{A}z^2\left[\frac{1}{2}ln\left(\frac{2m_e c^2\beta^2\Delta}{1-\beta^2}\right) - \beta^2 - ln(I) - \frac{C}{Z} - \frac{\delta}{2} + B_1 + B_2\right] \tag{3.35}$$

The ratio of restricted to unrestricted collision stopping powers for some materials of dosimetric interest are shown in Figure 3–3. The ratios of restricted and unrestricted stopping power ratios show that the deviation of Spencer–Attix to Bragg–Gray stopping power ratios is dependent on the material surrounding the cavity.

3.3.3.4 Bragg Additivity Rule for Compounds

The collision stopping power for a compound can be approximated by the weighted sum of the stopping powers of the atomic constituents of the compound. For the mass electronic stopping power for the compound, the additivity rule takes the form:

$$\left(\frac{S_{el}}{\rho}\right)_{comp} = \sum_i w_i \left(\frac{S_{el}}{\rho}\right)_i \tag{3.36}$$

where w_i is the fraction by weight and the summation goes over all elemental constituents of the compound. This additivity rule is equivalent to replacing, in the mass stopping power formula, the quantities $(Z/A)_{comp}$ and $ln(I_{comp})$ as follows:

$$\left(\frac{Z}{A}\right)_{comp} = \sum_i w_i \left(\frac{Z}{A}\right)_i \tag{3.37}$$

and

$$ln(I_{comp}) = \frac{\sum_i w_i \left(\frac{Z}{A}\right)_i ln(I_i)}{\left(\frac{Z}{A}\right)_{comp}} \tag{3.38}$$

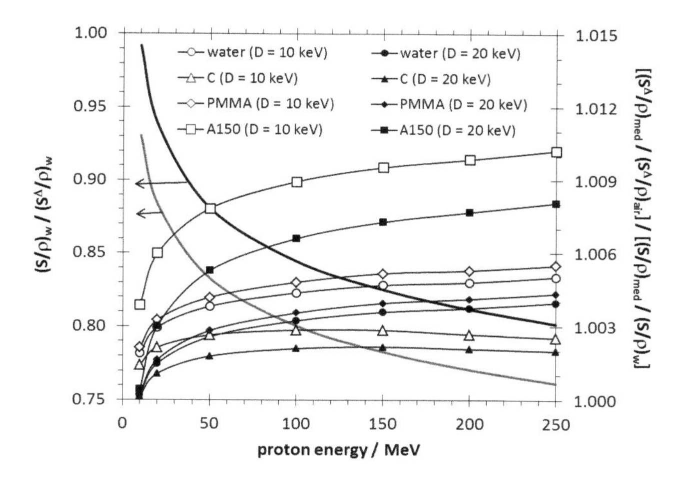

Figure 3–3 Ratio of restricted to unrestricted mass collision stopping powers of water (solid lines and values represented on left vertical axis) and medium-to-air restricted mass collision stopping power ratios relative to medium-to-air unrestricted mass collision stopping power ratios (connected symbols and values represented on right vertical axis).

where, $(Z/A)_i$ and I_i are for the ith element. It should be noted that $(Z/A)_{comp}$ is equal to the number of electrons in the molecule divided by the molecule weight [3]. *I*-values depend on physical state (gas/condensed). The latter makes that it is not sufficient to just combine stopping powers for elementary components if the latter refer to different physical states. For example, stopping powers for alanine used in the literature have been calculated from the elemental stopping powers and *I*-values. The value for alanine obtained that way is 69.3 eV. However, this rule does not account for the influence of chemical binding effects on the mean excitation energy. The mean excitation energies used in ICRU Report 49 [8] are entirely based on the extensive study of the subject for the electron and positron stopping powers of ICRU Report 37 [3]. This report recommends the use of experimental *I*-values when a sufficiently accurate value is available. If an experimental value is not available, a calculation using an adapted Bragg rule should be done, in which different *I*-values are assigned to the constituent elements, depending on the physical phase of the material. This way, the mean excitation energy calculated for alanine is 71.9 eV. An alternative presented in ICRU Report 37 uses Thompson's assignment scheme that accounts for the type of chemical bond each element exhibits in condensed organic compounds. However, due to the large uncertainties on the elemental data, it does not provide more accurate values than using the adapted Bragg rule applied above. Thompson's rule yields an *I*-value for alanine of 73.0 eV. There is another argument to be consistent with the mean excitation energies of ICRU Report 37: Medin and Andreo [30] pointed out that, since

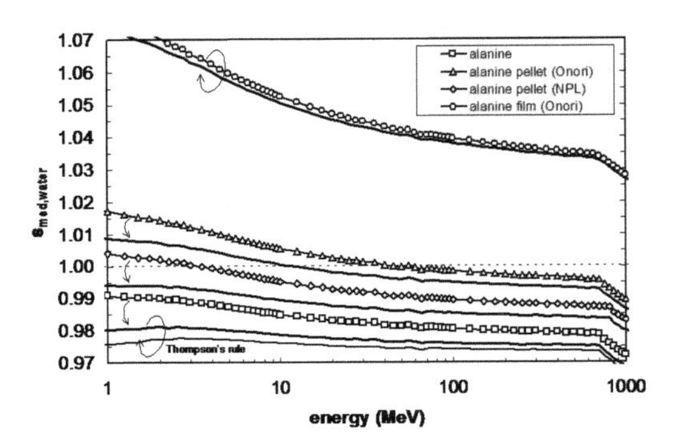

Figure 3–4 Stopping power ratios medium to water for various alanine materials as a function of proton energy. The symbols represent the calculations according to Bragg's rule using the *I*-values for the elemental media. The solid lines are those with *I*-values consistent with ICRU Report 37. (Data from Palmans et al. [31].)

calibrations of ionization chambers and other detectors are often performed in ^{60}Co gamma beams (thus involving electron stopping power data), it improves the internal consistency of dosimetry when for proton dosimetry the same *I*-values as for electrons are used (*I*-values for electrons and protons are the same and are mostly derived experimentally from data for protons). Examples of the magnitude of this is shown in Figure 3–4 for various alanine mixtures.

3.3.3.5 Energy Straggling

The stopping power should be considered as the mean energy loss of a large number of particles of the same energy. Each individual particle will undergo a fluctuating energy loss that is described by an energy straggling distribution. For long path lengths *s*, the distribution of energy losses Δ can be approximated by a Gaussian:

$$F(\Delta, s)\mathrm{d}\Delta = \frac{1}{\Omega\sqrt{2\pi}} e^{-\frac{\left(\Delta - s\frac{S}{\rho}\right)^2}{2\Omega^2}} \tag{3.39}$$

where the variance of the energy straggling distribution is given by (Fano [32]):

$$\Omega^2 = \frac{2\pi r_e^2 m_e c^2 N_A Z}{\beta^2} s W_m \left(1 - \beta^2 + \frac{2S_1}{ZW_m} ln\left(\frac{2m_e c^2 \beta^2}{I_1}\right)\right) \tag{3.40}$$

where S_1 and I_1 are the first moments of the oscillator strength and mean excitation energy distributions, respectively. For short path lengths, the expressions for the energy straggling distribution become more complicated and are given by the Landau distribution [33] or the Vavilov distribution [34].

3.3.4 Relation Stopping Power/Range

The range in the continuous slowing down approximation (csda) is the distance along its path a proton of energy E_p would travel if it is continuously losing energy according the mass stopping power:

$$r_0\left(E_p\right)=\int_{E_p}^{0}\left(-\frac{S}{\rho}\right)^{-1}\mathrm{d}E \tag{3.41}$$

The average path length protons travel (the range for individual particles fluctuates because of energy straggling) is not exactly the same as the csda range and is slightly larger [35]. In practice, we are rather interested in the depth particles reach on average, which is characterized by the range, corresponding to the depth reached by 50% of the protons that have not undergone a nuclear interaction. The ratio between the range and the csda range is expressed as a detour factor in ICRU Report 49 and is very close to unity (within 0.1%) for clinical proton energies. As rule of thumb, the range corresponds with the depth, z_{80}, on the distal edge of the Bragg peak where the dose is reduced to 80% of the maximum dose [36]. For clinical dosimetry, often a range–energy relation is established to determine the proton energy as a function of the measured range in water. The energy of the proton beam used in clinical practice will then depend on the choice of the stopping power data, and it is important to realize that there are various stopping power data tables in use [8,37].

In modern codes of practice for dosimetry of clinical proton beams, the residual range is used as a beam quality specifier. The residual range, R_{res}, for a measurement point at depth z_{ref} is defined in function of the practical range, R_p, as:

$$R_{\mathrm{res}}=R_{\mathrm{p}}-z_{\mathrm{ref}} \tag{3.42}$$

The practical range is defined as the depth distal to the Bragg peak (or spread-out Bragg peak) at which the dose is reduced to 10% of its maximum value on the depth–dose curve. Kempe and Brahme [38] provided a relation for the practical range as a function of proton energy E_p:

$$R_p=CE_p^{\,k}+R_\Delta \tag{3.43}$$

where the parameters are, for protons, given by $k = 1.770 \pm 0.023$, $C = (2.18 \pm 0.21)\ 10^{-3}$ g cm^{-2} MeV^{-k}, and $R_\Delta = 1.9\ 10^{-4}$ g cm^{-2}. From this relation it is clear that range–energy tables should be interpolated double-logarithmically.

3.3.5 Ionization of Gases

From the beginning of the history of ionizing radiation, the ionization of gases has been a key measurement [39,40]. Bragg [41] and later Gray [42] developed theories for relating the energy deposition per unit volume in a condensed medium to the ionization in a small air cavity within that medium, thus forming the basis of absolute dosimetry using ionization chambers. For ionization measurements, the ions formed in the gas are collected on electrodes of different electric potential, generating an electric field in the gas. Depending on the electric field strength, the collecting instrument works

- as an ionization chamber (low electric field strength) aiming to collect the charge of all radiation-induced ions,
- as a proportional counter (intermediate electric field strength) in which each ion pair induces a single avalanche so that the signal is proportional to the amount of energy deposited, or
- as a Geiger–Müller tube (high electric field strength) in which the avalanches overlap or produce secondary avalanches, making it sensitive to even a single ionization but with the disadvantage that the energy information is lost.

Ionization in gases by protons is quantified by the mean energy expended to produce an ion pair in a gas, W_{gas}, defined as the quotient of the kinetic energy E_p of the proton and the number of charges N formed of either sign when the initial kinetic energy of the proton is completely dissipated:

$$W_{\mathrm{gas}} = \frac{E_{\mathrm{p}}}{N} \tag{3.44}$$

The differential value, W_{gas}, of the mean energy expended to produce an ion pair in a gas is the quotient of $\mathrm{d}E_{\mathrm{p}}$, the expectation value of the energy lost by a proton of kinetic energy E_{p} in traversing a layer of gas of infinitesimal thickness, and $\mathrm{d}N$, the expectation value of the charge of either sign when $\mathrm{d}E_{\mathrm{p}}$ is completely dissipated in the gas:

$$w_{\mathrm{gas}} = \frac{\mathrm{d}E_{\mathrm{p}}}{\mathrm{d}N} \tag{3.45}$$

The theoretical calculation of W_{gas} and w_{gas} requires accurate cross section data for all inelastic collisions at all kinetic energies of all the particles in the ionization track, and is thus limited by the incompleteness, fragmentary availability, and substantial uncertainty on those cross section data. The theoretical evaluation can be done by simulating a large number of tracks using the Monte Carlo method, however, even with modern-day computing power, this is a considerable challenge given the enormous amount of interactions that need to be simulated. The advantage of a Monte Carlo simulation is that it is straightforward to incorporate contributions from all secondary, tertiary, etc. particles. This so-called "bookkeeping" is more difficult in analytical methods, but various approaches have been described in ICRU Report 31 [6]. Among those, we note methods based on continuous slowing down spectra [43] of all charged particles i generated by the initial proton with energy E_{p} and expressed as their total path length differential in energy: $y_i(E_{\mathrm{p}},E)$. The number of ionizations is then

$$N(E_p) = n_{gas} \sum_i \left[\int_{E_I}^{E_p} \sigma_i(E) y_i(E_p, E) \mathrm{d}E \right] \quad (3.46)$$

where n_{gas} is the number of gas molecules per unit volume, E_I the first ionization energy and $\sigma_i(E)$ the ionization cross section of the gas for a charged particle of kinetic energy E. Grosswendt and Baek [44] arranged the contributions differently based on the track contributions from the proton completely slowing down in the csda approximation and a function $f(E)$ to include the ionization yields of secondary electrons:

$$N(E_p) = n_{gas} \int_{E_I}^{E_p} \frac{\sigma_t(E) f(E)}{S_{el}(E)} \mathrm{d}E \quad (3.47)$$

where $\sigma_t(E)$ is the sum of the cross section for the charge-exchange cycle and of the weighted cross section for ionization by protons and neutral hydrogen projectiles. Grosswendt and Baek [44] used this to calculate W_{gas} values for tissue-equivalent (TE) gas and air. Figure 3–5 illustrates the importance of including the ionizations by all secondary electrons.

The experimental determination of W_{gas} requires the measurement of the number of electron-ion pairs produced and either the incident particle energy or the amount of energy absorbed (the latter being the only approach for the measurement of the differential value, w_{gas}). The amount of electron-ion pairs is either quantified using pulse height measurements in proportional counter mode or via a measurement of the current generated in the gas in ionization chamber mode. In the latter case, the signal has

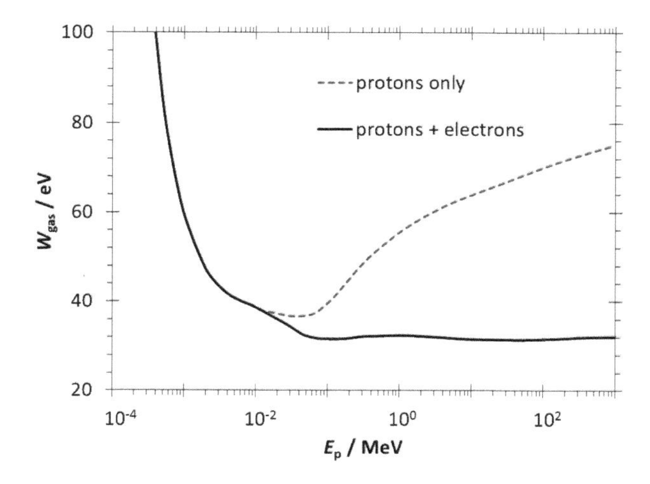

Figure 3–5 W_{gas} values for protons in TE gas calculated using the continuous slowing down approximation. (Data extracted from Grosswendt and Baek [44].)

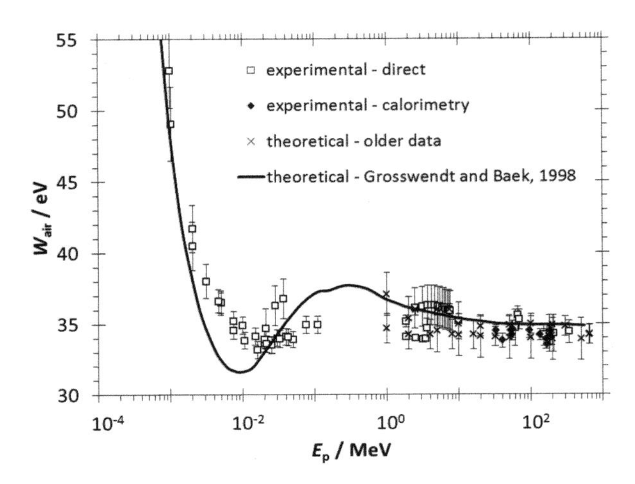

Figure 3–6 Experimental W_{air} values for protons in dry air compared with calculated values (data extracted from Grosswendt and Baek [44] and Jones [46]). The curve of theoretical values by Grosswendt and Baek [44] was obtained using Equation 3.49.

to be corrected for ion recombination losses. The determination of the incident particle energy is usually based on time of flight measurements, measurement of the deflection angles in a magnetic field, measurement of the accelerator potential, measurement of nuclear reaction products, or the use of the range–energy relation. The determination of the energy absorbed can be done using calorimetry, chemical dosimetry, calibrated proportional counters, ionization measurements in a reference gas with known W_{gas} value, or calculation from the stopping power. If the determination is based on the incident particle energy, then the number of incident particles also needs to be measured or counted, which is usually done using a Faraday cup.

Many determinations in the literature are based on comparing the ionization in one gas with that in another gas for which the value of W_{gas} is better known. This method was, for example, used by Moyers et al. [45] to determine the value for air by comparing with the values for nitrogen and argon. A comparison of experimental and calculated values of the mean energy expended to produce an ion pair in air, W_{gas}, is shown in Figure 3–6 based on experimental and calculated data from various sources quoted by Grosswendt and Baek [44] and Jones [46].

3.4 Nuclear Interactions

When speaking about nuclear interactions, a distinction has to be made between elastic and non-elastic nuclear interactions. Elastic nuclear interactions are those interactions of the proton and the nucleus in which kinetic energy is conserved. These effects are included in the stopping power and multiple scattering mechanisms and are discussed in the previous section. Non-elastic nuclear interactions refer to nuclear interactions that are not elastic, i.e., kinetic energy is not conserved, while inelastic nuclear interactions are a particular type of non-elastic interactions in which the final nucleus is the same as the original target nucleus.

This section discusses non-elastic nuclear interactions, which are important in many aspects of proton therapy. More thorough discussions of the basics can be found in ICRU Reports 28 and 63 [5,11], PTCOG Report 1 [47], and Ipe [48]. First of all, non-elastic nuclear interactions result in an attenuation process since protons are being removed from the primary beam due to non-elastic nuclear interactions. After the proton penetrates the nucleus, an intranuclear cascade takes place, which is essentially a collision avalanche with individual nucleons leading to ejection of forward-directed protons, neutrons, or pions, if the projectile's kinetic energy is high enough for pion formation. The energy not carried away by cascade particles leaves the remaining nucleus in a highly excited state and leads to potential emission of nucleons and light fragments both during the redistribution of this excitation energy (the precompound stage) and after the energy distribution over the nucleons has reached an equilibrium state (compound stage). This compound nucleus is characterized only by its mass, charge, and excitation energy, so the history of the collision and cascade is "forgotten" at that moment, and any further emission of particles is isotropic. The compound nucleus can lose energy by evaporation emitting protons, neutrons or light fragments (mainly alpha particles), and by emitting gamma rays. Eventually a stable nucleus is reached. The kinetic energy that the emitted particles carry away at all stages can contribute to the local energy deposition, which explains the contribution of the reduction in rest mass to the definition of energy imparted and, thus, to the absorbed dose. The charged particles emitted in the intranuclear cascade stage usually have high energy (of the same order of magnitude as the projectile proton) and can thus carry away energy from the projectile absorption point. Those generated in the evaporation stage, on the other hand, have low energy and will deposit all their energy close to the generation point. Though emitted neutrons make a negligible contribution to the local energy deposition, they are important to consider in shielding, radioprotection, and estimation of secondary cancer risks. Gamma rays contribute little to the local dose deposition, but they also need to be considered in shielding and, more importantly, are the key process detected in prompt gamma imaging to reconstruct the distribution of the primary proton absorption point by locating the origin of the gamma rays. The production of radionuclides leads to the possibility of treatment plan verification using positron emission tomography (PET), as will be discussed in Chapter 7.

3.4.1 Theory of Non-elastic Nuclear Interaction

Two notations are commonly used to categorize nuclear interactions:

$$x + A \rightarrow \sum_i y_i + B \tag{3.48}$$

or

$$A\left(x, \prod_i y_i\right)B \tag{3.49}$$

where x is the projectile, A is the target nucleus, y_i the emitted particles, and B the final nucleus. For example, the production of ^{11}C by proton radiation, which is used in PET dose verification, can be described as

$$p + {}^{12}C \rightarrow p + n + {}^{11}C \tag{3.50}$$

or

$$^{12}C\left(p, pn\right)^{11}C \tag{3.51}$$

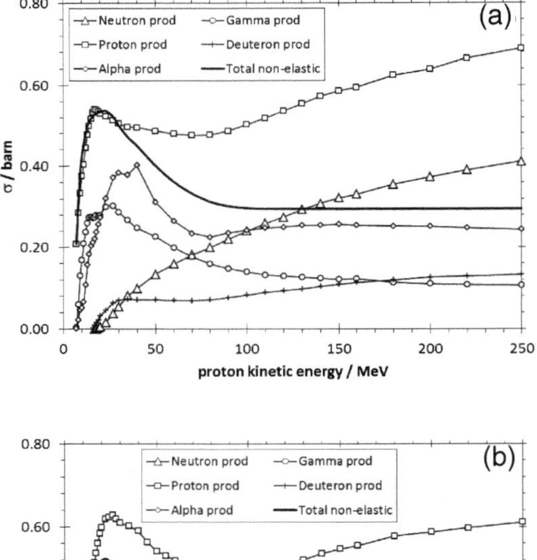

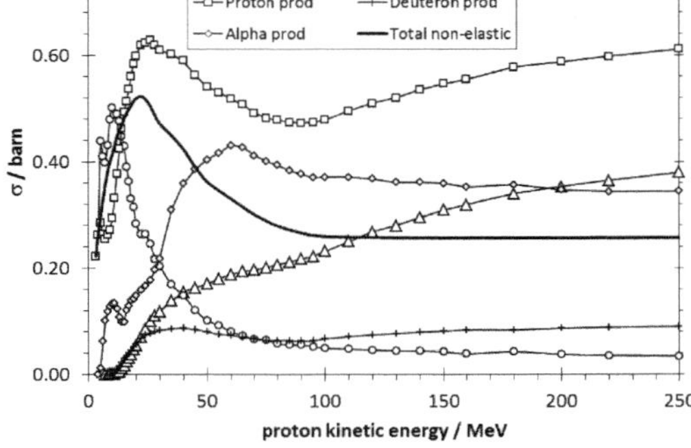

Figure 3–7 Total non-elastic nuclear interaction cross sections and production cross sections of various secondary particles (a) as a function of the kinetic energy of the incoming protons for ^{16}O as the target nucleus and (b) for ^{14}N as the target nucleus. The data are taken from ICRU Report 63 [11].

Protons in the therapeutic energy range have a substantial probability to undergo a non-elastic nuclear interaction. Figure 3–7 shows the total non-elastic nuclear interaction cross section of protons on ^{14}N and ^{16}O as a function of the proton's kinetic energy, as well as the production cross sections of neutrons, gammas, protons, deuterons, and alpha particles. It is clear that there is a resonance at low energies.

Figure 3–8 shows the kinetic energy of the emitted particles expressed as a fraction of the kinetic energy of the incoming protons for ^{16}O and ^{12}C as the target nuclei. As a rule of thumb, we can assume that 60% of the energy of the incoming particle is transferred to secondary charged particles, and thus will contribute to the dose. The total non-elastic nuclear interaction cross sections is almost independent of energy (the resonance at low energy accounts only for the last 2 cm of the range), so apart from a secondary proton buildup (or secondary proton disequilibrium) at the entrance,

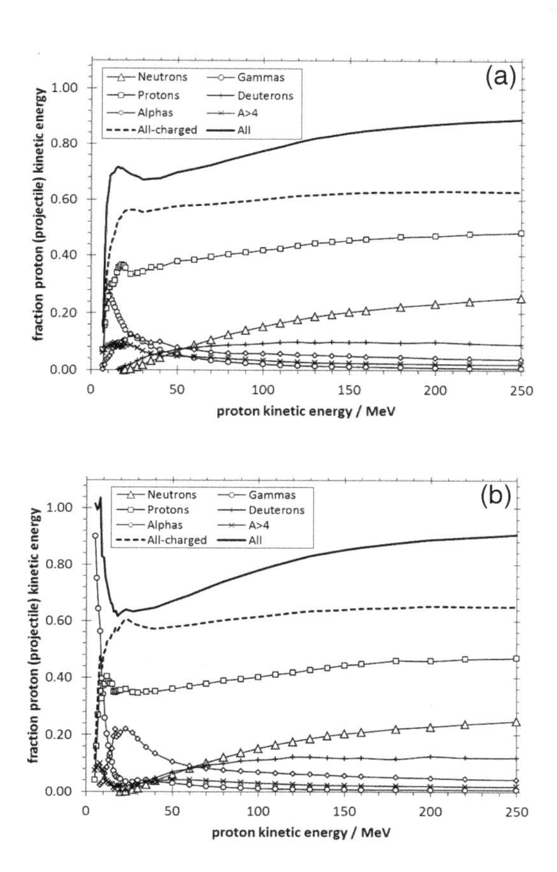

Figure 3–8 Kinetic energy of the emitted particles expressed as a fraction of the kinetic energy of the incoming protons for ^{16}O as the target nucleus (a) and for ^{12}C as the target nucleus (b). The data are taken from ICRU Report 63 [11].

the contribution of nuclear interactions to the dose is almost constant as a function of depth. It must be realized, though, that this represents a varying contribution to the total dose, given the variation of the mass electronic stopping powers as a function of depth/energy. Figure 3–9 illustrates this by showing a depth–dose distribution for a 200 MeV proton beam, as well as the separate contribution from primary and secondary protons. The dose deposited by alpha particles is about 10 times lower than that from secondary protons, consistent with the data in Figure 3–8, and is not visible in this figure (it can be visualized using a log-vertical scale, see [49]). The alpha particle contribution to the overall dose is thus small (about 0.5% in water) but is not necessarily a negligible component in all situations; e.g., when comparing doses in water and graphite (or other carbon-containing materials like plastics or tissues) given the substantially different fraction of the energy that goes to alphas for ^{16}O and ^{12}C as target nuclei (see Figure 3–8) at low energies. Figure 3–9 also shows the fractional contribution of secondary protons to the total dose, in which it can be observed that the falloff of the secondary proton contribution at large depths (close to the Bragg peak) is much steeper than the falloff of the secondary proton dose itself due to the increasing electronic stopping power at those depths.

It is clear that especially at higher energies, non-elastic nuclear interactions play an important role. Palmans et al. [50] showed that the water-equivalence of a medium is entirely determined by the differences in these nuclear interactions. When converting dose in one medium to dose in another medium (e.g., water), the fraction of the

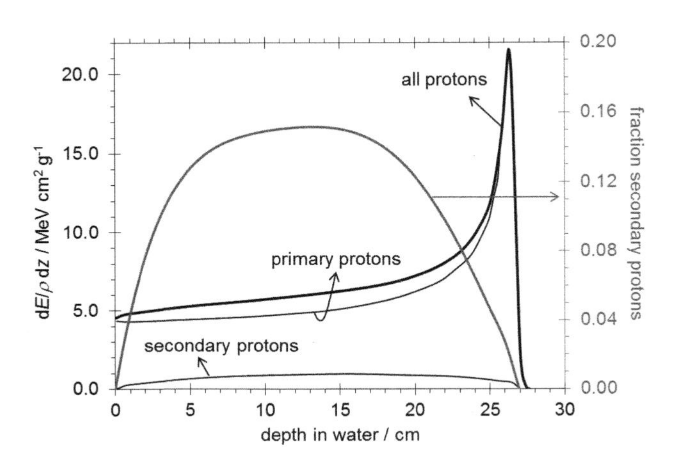

Figure 3–9 Depth–dose distribution for a broad 200 MeV proton beam with indication of the contributions from primary and secondary protons. Alpha particles contribute about a 10x lower dose as compared to the secondary protons. The grey curve (scale on the second vertical axis) represents the fractional contribution of secondary protons to the total dose.

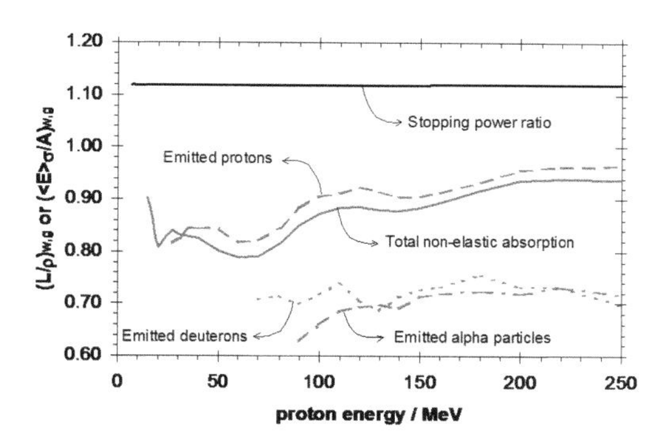

Figure 3–10 Water to graphite mass collision stopping power ratios from ICRU Report 49 and ratios of total non-elastic nuclear absorption cross sections as well as production cross sections for proton, deuterons, and alpha particles from ICRU Report 63 [11]. The ratio of non-elastic and production cross section are only plotted above the energy where they contribute more than 1% of the total energy deposition.

dose deposition due to non-elastic nuclear interactions should be converted separately from the electromagnetic contribution. The latter should be converted with the ICRU Report 49 stopping power ratios, whereas the former should be converted with the ratios of non-elastic nuclear cross sections (for generation of charged secondaries) per nucleon. The differences of the cross sections in two materials result in a different nuclear attenuation. This causes the fluence at equivalent depths to be different, making fluence correction factors necessary [50].

Figure 3–10 shows the ratio of total non-elastic interaction cross sections per atomic mass unit for protons in water and graphite, as well as the ratios of secondary proton, deuteron and alpha production cross sections.

3.4.2 Measurement of Non-elastic Nuclear Interaction Cross Sections

Two types of experiments can be distinguished for the determination of non-elastic nuclear interaction cross sections. Direct measurements of primary beam attenuation and production of particle types, energies, and angles on thick or thin targets can be performed using techniques commonly used in nuclear physics research. Indirect measurement (or validation) can also be performed via the dosimetric effects of the nuclear interactions. These methods are briefly reviewed here.

3.4.2.1 Direct Methods

Total cross sections at high energies are usually determined by attenuation measurements over thick targets with subtraction of the loss by large-angle scattered particles (determined from the elastically scattered proton distribution). The experiment requires a time-of-flight measurement to establish the coincidence of the target interaction (or beam exit) and the particle detection using scintillators. At lower proton energies, multiple Coulomb scattering angles become larger, and to include these particles one can attempt to have a wider angle detector, but then inelastically scattered protons and secondary protons from non-elastic nuclear interactions are counted as well, underestimating the number of attenuation events. This could be overcome by using a magnetic spectrometer to sweep the lower-energy inelastically scattered and secondary protons out of the beam path. The total non-elastic interaction cross sections can also be derived from integration over all partial cross sections. This method is usually only feasible when the number of reaction channels is limited. For double differential cross sections, a similar detector configuration is used under different angles with respect to the beam. To enable a better discrimination of particle type, usually a telescope of scintillators that contain one or more energy loss stages and an absorption stage is used, combined with a time of flight measurement. A comprehen-

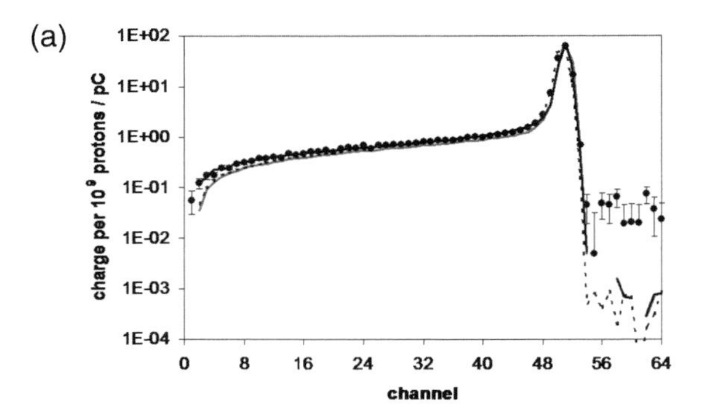

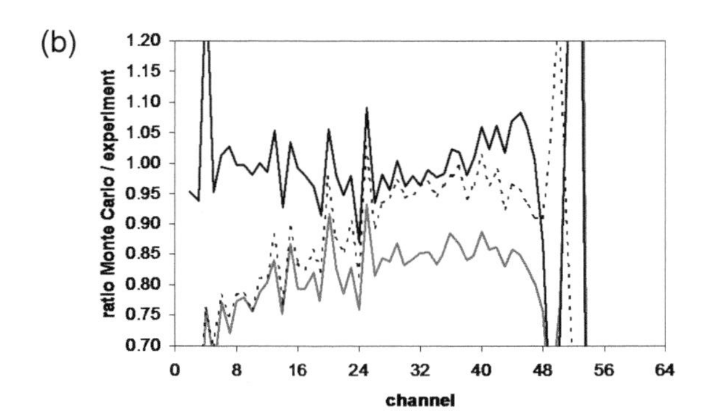

Figure 3–11 Upper graph: charge per 10^9 protons collected in each metallic plate of the multilayer Faraday cup from experiment (symbols) and from three Monte Carlo codes (dotted black line: FLUKA, full black line: Geant4, full grey line: SHIELD-HIT). Lower graph: ratios of the Monte Carlo calculated data and the experimental data. Figure reproduced from Palmans [56], with permission.

sive review of experimental data is provided in ICRU Report 63 [11]. Comprehensive evaluated data are also available in IAEA's EXFOR data base [51].

3.4.2.2 Indirect Methods

An indirect experiment to quantify the dosimetric contribution is the use of a multilayer Faraday cup (MLFC). This instrument quantifies the number of nuclear interactions taking place in slabs of nonconducting materials by using the principle of mirror charges. The charge collected in each of a series of thin metallic foils separated by the insulating absorber plates is measured, and the method determines in which slab protons stop (by nuclear interactions in the entrance plateau and at the end of their range due to electromagnetic interactions in the Bragg peak). This can serve as a test for both nuclear interaction and energy straggling models in the Monte Carlo simulation of the device, and the experimental data from such an experiment has been proposed as a benchmark data set [52,53]. In Figure 3–11, experimental results are compared with numerical results obtained using the Monte Carlo codes Geant4 [52], FLUKA [54], and SHIELD-HIT [55]. The agreement in the Bragg peak is good in all cases, but underestimations of the nuclear interactions are visible in the entrance regions for some of the codes or interaction models. The experimental data beyond the Bragg peak could be the result of measurement noise or neutron-induced secondary protons. None of the models would predict such a large effect.

3.5 Scattering

As a charged particle passes through material, it is deflected by many small-angle Coulomb scattering events with the nuclei. The combined effect of multiple events is called multiple Coulomb scattering.

3.5.1 Single Scattering

With respect to single scattering, the Rutherford cross section [57] is again the simplest approximation, but this time with the nucleus as the target (the scattering with electrons results in much smaller deflections of the proton):

$$\frac{d\sigma}{d\Omega} = \frac{Z^2 e^2}{4\pi\varepsilon_0} \frac{1}{m_p c^2 \beta^2 sin^4\left(\theta/2\right)} \tag{3.52}$$

3.5.2 Multiple Scattering

Multiple Coulomb scattering is the combined effect of many small-angle elastic and semi-elastic collisions. Theories to model multiple scattering generally assume that the number of individual scattering events is large so that the mean square of the angular deflection can be derived using a statistical approach. Applications of multiple scattering theories range from analytical modeling of scatter components in the beam (e.g., scatter foils, range modulators) to doing calculations using pencil beam algorithms and condensed history Monte Carlo transport algorithms. The angular and spatial distributions of an initially parallel pencil beam of charged particles after passing through an absorber of thickness t can be approximated to first order by a Gaussian distribution with a mean square angle of $\overline{\theta^2}(t)$, which was proposed by Fermi and Eyges [58] using the small angle approximation ($\sin\theta \approx \theta$):

$$f(\theta)\theta d\theta = \frac{2\theta}{\overline{\theta^2}(t)} e^{-\frac{\theta^2}{\overline{\theta^2}(t)}} d\theta \tag{3.53}$$

A more comprehensive theory that is widely used and which was developed also based on the small angle approximation was formulated by Molière [59]:

$$f(\theta)\theta d\theta = \frac{\theta}{\chi_c^2 B}\left[2e^{-\frac{\theta^2}{\chi_c^2 B}} + \frac{f^{(1)}\left(\frac{\theta}{\chi_c\sqrt{B}}\right)}{B} + \frac{f^{(2)}\left(\frac{\theta}{\chi_c\sqrt{B}}\right)}{B^2}\right] d\theta \tag{3.54}$$

where

$$f^{(n)}\left(\frac{\theta}{\chi_c\sqrt{B}}\right) = \frac{1}{n!}\int_0^\infty J_0\left(\frac{\theta}{\chi_c\sqrt{B}}y\right) e^{-\frac{y^2}{4}}\left[\frac{y^2}{4} ln\left(\frac{y^2}{4}\right)\right]^n ydy \tag{3.55}$$

$J_0(x)$ is Bessel's function of the first kind of order zero.

The characteristic scattering angle χ_c is for protons given by:

$$\chi_c = 0.4\sqrt{\frac{Z^2}{A}}\frac{\sqrt{t}}{pc\beta} \tag{3.56}$$

The parameter B in equations 3.54 and 3.55 can be thought of as the natural logarithm of the number interactions the protons undergoes when passing through a thickness t (in g cm^{-2}) of material and is the result of a complex calculation involving the solution of a transcendent equation, which can be summarized as:

$$B - ln(B) = b \tag{3.57}$$

where

$$b = ln\left(\frac{\chi_c^2}{1.167\chi_a^2}\right) \tag{3.58}$$

in which the characteristic screening angle is calculated as

$$\chi_a^2 = \left(7.68\times10^{-5} + 1.36\times10^{-8}\frac{Z^2}{\beta^2}\right)\frac{(m_e c^2)^2 Z^{2/3}}{(pc)^2} \tag{3.59}$$

Bethe [60] suggested various improvements to the theory, among which that for low-Z materials, replacing Z^2 in these formulas by $Z(Z+1)$ better accounts for the scatter with atomic electrons, which adds up to a non-negligible contribution. Comparing equations 3.53 and 3.54 one could assume that the first term of the Molière distribution represents the simpler Gaussian distribution, i.e. $\theta_0 = \sqrt{\overline{\theta^2}(t)} = \chi_c\sqrt{B}$, but the second term in the Molière distribution actually narrows the distribution significantly. A better approximation for the narrow angle part of the distribution was proposed by Hanson et al. [61]: $\theta_0 = \chi_c\sqrt{B-1.2}$. Highland [62] proposed a fit to the entire Molière distribution that is easier to calculate and also provides a good approximation:

$$\theta_0 = \sqrt{\overline{\theta^2}(t)} = \frac{20\,\text{MeV}}{pv}\sqrt{\frac{t}{L_R}}\left[1+\frac{1}{9}\log_{10}\left(\frac{t}{L_R}\right)\right] \tag{3.60}$$

where L_R is the radiation length which is a material-specific property. Radiation lengths as a function of Z are shown in Figure 3–12. For elements with $Z > 2$, the relation can be approximated by a simple function, which is also indicated in the figure. For a compound with elemental weight fractions w_i, the reciprocal of the radiation length is obtained as:

$$\left(\frac{1}{L_R}\right)_{\text{comp}} = \sum_i w_i \left(\frac{1}{L_R}\right)_i \tag{3.61}$$

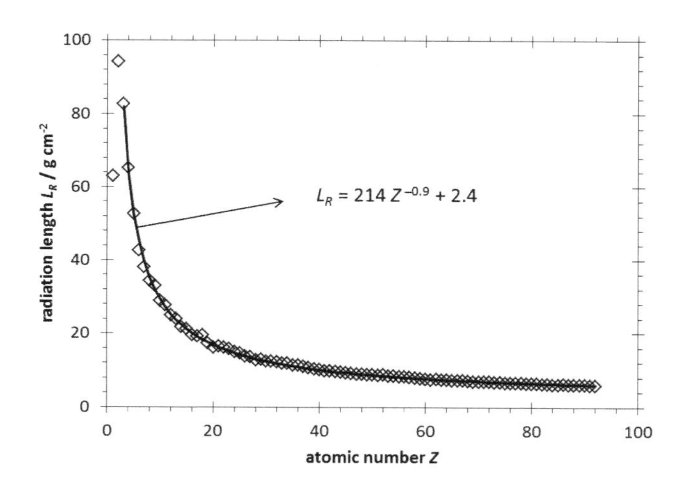

Figure 3–12 Radiation length plotted as a function of the atomic number (data from Tsai [63]). The curve presents a fit as indicated in the graph for all data points with Z>2.

A more in-depth analysis of the Molière multiple scattering theory is given by Gottschalk [64,65]. Molière's theory is widely used, but it doesn't model the large-angle tails of the multiple scattering distributions well due to the approximations made in the theory, which do not properly account for rare large-angle, single-scattering events. More exact theories were devised by Goudsmit and Saunderson [66] and Lewis [67] which are correct to higher-order statistical moments as the Molière theory, correctly modeling the large-angle scattering tails, and forming the basis of radiation transport algorithms in several modern Monte Carlo codes, e.g., Geant4 and MCNPX.

3.5.3 Scattering Power

The mass scattering power quantifies how much particles in a beam are on average scattered away from the beam direction when passing a certain distance through a material. In analogy with the mass stopping power, the mass scattering power is defined as the rate of increase of the variance of the scattering angle per unit of material thickness (in g cm^{-2}) traversed:

$$\frac{T}{\rho} = \frac{\mathrm{d}\overline{\theta^2}}{\rho \mathrm{d}z} \tag{3.62}$$

Integrated over a slab of mass thickness t, the mass scattering power gives

$$\overline{\theta^2}(t) = \overline{\theta^2}(0) + \int_0^t \frac{T}{\rho}(\rho z) d(\rho z) \tag{3.63}$$

and thus allows to account with the angular distribution at the surface of the phantom/patient which is used in pencil beam dose calculation algorithms. The mass scattering power increases approximately as Z^2/A, while the mass stopping power varies by not more than a factor two for proton energies above 50 MeV and Z>2. This means that for the same amount of energy loss, a high-Z metal is more effective in scattering than a low-Z material. It explains why scattering foils are made from heavy metals while collimator materials have to consider a compromise between effective stopping and minimized scatter (besides activation and secondary particle production in nonelastic nuclear interactions).

3.6 Aqueous Radiochemistry

Radiation chemistry, or radiochemistry, describes the effect of ionizing radiation on chemical compounds and their chemical interaction. Aqueous radiochemistry is concerned with the radiolysis of water and the reaction of the radiation-induced species with each other, with water components, with dissolved molecules, ions and gases, and with contact surfaces with another medium. Since the production of reactive species is LET dependent, this section starts with a few basic aspects of aqueous radiochemistry and then discusses the dependence on proton energy of those aspects. A comprehensive introduction to aqueous radiochemistry is given by Spinks and Woods [15].

3.6.1 Radiolysis of Water

The initiation of radiochemical effects in water is related to the physical distribution of ionization and energy deposition. The density of ion pairs has a big influence on the number of initial species formed and on the probability of recombination and formation of excited states. On a scale at which the chemical species formed can very quickly react with each other, spurs, blobs, and short tracks can be distinguished. A spur consists of about 2–3 ionizations with a maximum energy transfer of about 100 eV separated from other events so that the chemistry in this spur at short time scales develops independently of what happens in other parts of the water bulk. A blob is a larger cluster of ionization in which 100–500 eV is deposited. A short track, which can also be considered a row of overlapping spurs, is formed along the track of a low-energy delta ray and typically involves an energy transfer up to 5000 eV.

Within these spurs, blobs, and short tracks, the first reactions that take place are a direct result of ionization and excitation:

$$H_2O \overset{rad}{\rightarrow} H_2O^+ + e^- \tag{3.64}$$

$$H_2O \overset{rad}{\rightarrow} H_2O^* \tag{3.65}$$

H_20^+ is a very reactive species, and within 10^{-17} to 10^{-16} s it reacts with other water molecules:

$$H_2O^+ + H_2O \rightarrow H_3O^+ + {}^\bullet OH \tag{3.66}$$

Note that this is much faster than the time between ionizations along the track, which is of the level of 10^{-15} s, so these effects can be regarded as instantaneous, and these reactions are considered part of the physical stage involving no significant motion of molecules.

3.6.2 Production of Reactive Species

The physical stage is followed by a physico-chemical stage in which part of the deposited energy is transferred to internal energy and excited water molecules dissociate in, for example, hydrogen and hydroxyl radicals ($H^\bullet$ and ${}^\bullet OH$, respectively). Free electrons, once reduced to thermal energies that do not recombine with positive ions, are hydrated on a time scale of 10^{-11} s, corresponding to the dipole relaxation time, forming hydrated electrons (e^-_{aq}). Also all other charged species with only thermal energy become hydrated (e.g., H^+_{aq}). An increasingly complex reaction chain leads to a number of reactive species, including, besides hydrogen and hydroxyl radicals, hydroperoxyl radicals ($HO_2{}^\bullet$), hydroxide ions (OH^-), hydroperoxide ions ($HO_2{}^-$

), peroxide ions (O_2^{2-}), superoxide ions (O_2^-), hydrogen peroxide (H_2O_2), molecular hydrogen (H_2), and molecular oxygen (O_2). The diffusion of species during this stage is limited to a few molecular diameters, and all chemistry is still happening within the spur or blob. After about 10^{-7} seconds, the diffusion of the species formed in the spurs, blobs, or short-tracks becomes significant, and the chemical stage starts involving reactions of radicals with other radicals and chemical species in bulk, including dissolved ions, organic molecules, and gases.

3.6.3 Chemical Yields

The radiation chemical yield, $G(x)$, of a chemical species, x, is defined as:

$$G(x) = \frac{n(x)}{\bar{\varepsilon}} \tag{3.67}$$

where $n(x)$ is the expectation value of the amount of substance of the species x produced, destroyed, or changed in a system by the mean energy imparted $\bar{\varepsilon}$ to the matter of that system. The SI unit is mol J^{-1}, but the radiation chemical yield is more commonly expressed as the number of units of species x produced per 100 eV of energy absorbed. Note that a radiation chemical yield can be negative if an irradiated compound is converted into other chemical species as a result.

The quantification of the chemical yields of primary species is performed either by pulse radiolysis or using scavengers. In pulse radiolysis, a short intense pulse of ionizing radiation is delivered, and the fast time dependence of the absorption of light is measured immediately after the radiation pulse. This approach is mainly used to quantify the yields of e^-_{aq} and $^{\bullet}OH$ (the latter by UV absorption). Using picosecond pulse radiolysis, it is even possible to determine the development of the e^-_{aq} concentration within the spurs. The chemical yields of other reactive species are usually determined using low concentrations of scavengers, which swiftly react with a specific species. The concentration of the reaction products can be determined chemically, photospectroscopically, or with fluorescence measurements, depending on their nature. Reaction products that alter light absorption can also be employed for pulse radiolysis.

The reaction of reactive species with chemicals or scavenger species in the bulk solution can be exploited for dosimetry by quantifying the reaction products. An example is the ferrous sulphate dosimeter in which ferrous ions are converted into ferric ions in various reactions with reactive species [15]:

$$Fe^{2+} + HO_2 \rightarrow Fe^{3+} + HO_2^- \tag{3.68}$$

$$Fe^{2+} + H_2O_2 \rightarrow Fe^{3+} + {}^{\bullet}OH + OH^- \tag{3.69}$$

$$Fe^{2+} + {}^{\bullet}OH \rightarrow Fe^{3+} + OH^- \tag{3.70}$$

Ferric ions have strong absorption peaks of light, with wavelengths of 224 nm and 303 nm, thus allowing the quantification of their concentration via spectrophotometry.

Another example is polymer gel dosimetry in which radicals (generic symbol $R^\bullet$), such as $^\bullet OH$, react with monomers that are locked up in the gel matrix in a suitable aqueous environment (Baldock et al. [68]):

$$R^\bullet + M \rightarrow R + M^\bullet \tag{3.71}$$

initiating propagation reactions:

$$M_m^\bullet + M_n \rightarrow M_{m+n}^\bullet \tag{3.72}$$

leading to a polymerization process that ends with termination reactions such as:

$$M_m^\bullet + M_n^\bullet \rightarrow M_{m+n} \tag{3.73}$$

or

$$R^\bullet + M_n^\bullet \rightarrow RM_n \tag{3.74}$$

The resulting polymers form opaque clusters in the gel matrix which can be quantified in 3D by optical tomography or, since the polymers also have a different magnetic susceptibility than the original solution with the monomers, by magnetic resonance imaging.

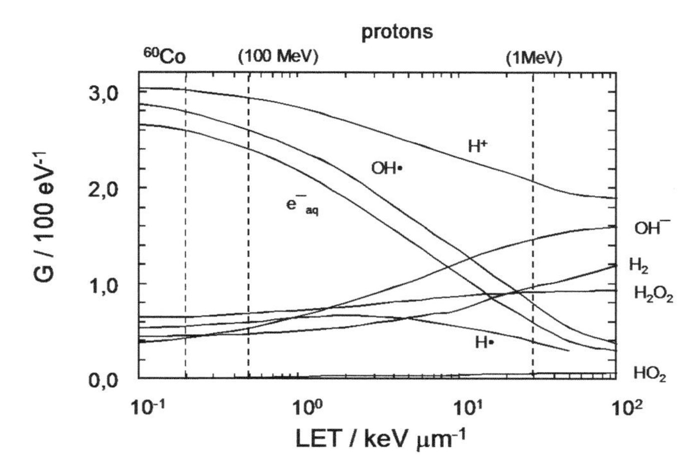

Figure 3–13 Chemical yields of primary species 10^{-7} s after the passage of an ionizing particle as a function of LET. (Adapted from Palmans and Vynckier [69] with permission of the publisher and the authors.)

The chemical yield of species within the spurs as well as in bulk after 10^{-7} s are both dependent on the pH of the solution as well as on the LET of the ionizing particles due to the effect of the ionization density on the chemistry within the spurs and blobs. Figure 3–13 shows the dependence of the chemical yields of primary species as a function of LET, determined mainly from pulse radiolysis and scavenger experiments. Effects that are dependent on the radiation chemistry, such as the chemical heat defect of water (see Chapter 12) and the response of chemical dosimeters, such as the ferrous sulphate dosimeter, polymer gels, and radiochromic films (see Chapter11), will thus in general also be LET dependent.

3.6.4 Radiolysis of DNA

DNA damage by ionizing radiation is classified as direct damage if direct ionizations of molecular bonds in the DNA structure are the result of densely clustered ionization events, or as "indirect damage" if it is due to reactive radicals formed in the aqueous solution and the subsequent chemical reaction chain. For low-LET radiation, the latter processes are responsible for up to 70% of the total DNA lesions produced by radiation exposure, while for high-LET radiation, direct action of the ionizing particles is the major reason for their increased RBE. In both cases, the DNA damage is determined by the amount of single-strand and double-strand breaks (SSB and DSB) and how they are distributed/clustered along the DNA.

Most modeling of direct DNA damage by investigating ionization clustering at the nanoscale is performed in liquid water. However, the cross sections for direct radi-

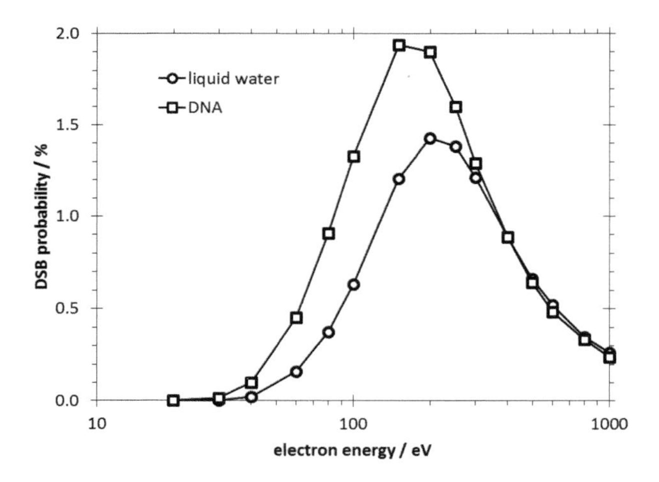

Figure 3–14 Simulated probability to produce a DNA double-strand break (DSB) as a function of kinetic energy of proton-induced secondary electrons, when cross section data of either liquid water or DNA medium are used. (Adapted from Palmans et al. [20] with permission of the publisher and the authors.)

olysis of DNA by the secondary electrons produced in the proton track are substantially different compared to those for water, as can be seen in Figure 3–14. Furthermore, to analyze the significance of ionization clustering and SSB or DSB damage clustering, it is also important to consider the structure of the DNA (the double helix is winded around histone proteins which link to form a backbone of basic elements, chromatin fibers curled up in fiber loops). Concerning indirect DNA damage, the hydroxyl radical ($^{\bullet}OH$) plays a key role in cell killing and non-lethal effects such as chromosome aberrations. So do various other reactive species.

3.6.5 Relevant Site—Relation to Cellular Geometry

Again, the relevant site on the cellular or tissue scale is determined by the biological pathway being considered. For direct DNA damage, it is generally assumed that ionization clustering at the scale of a few nanometers has to be considered. For indirect effects, it should be realized that the majority of reactive species will not diffuse by more than 10–50 nm in the biomolecular environment within the nucleus, so microdosimetric information at the scale of the nucleus to quantify the amount of radical species formed should also be complemented with clustering information at that intermediate scale. Oxidative stress caused by the production of reactive oxygen species should be considered on the microscale, but also at the scale of the entire cell, and even at the tissue scale given the influence this has on cell signaling. Indirect damage to the mitochondria can also lead to cell death, so the complexity of processes that contribute to DNA damage is, in fact, so large that it is now recognized that only the combined information at different scale levels can predict the incidence of initial biological effects. A multi-scale model was proposed for this purpose by Solov'yov et al. [70], and a generic formalism was proposed by Palmans et al. [20]. A Monte Carlo tool for simulating microdosimetric distributions in various components of the cell (cell membrane, cytoplasm, endoplasmic reticulum, nuleus, and nucleolus) was nicely illustrated by Douglass et al. [71].

3.7 Summary/Conclusions

This chapter provides definitions of macroscopic, microdosimetric, and nanodosimetric quantities relevant to proton therapy physics and a brief overview of the basics of the interactions of therapeutic proton beams with matter, including electromagnetic interactions with electrons and the nucleus, stopping power theory, ionization, nuclear interactions, single and multiple scattering, and aqueous radiation chemistry. Further information on these basic matters, specifically for protons, can be found in ICRU Reports 49, 59, 63, and 78 [8,10–12], Paganetti [14], Janni [37], Palmans et al. [72], Karger et al. [73], and Moyers and Vatnitsky [74].

References

1. Johns HE, Cunningham JR. The physics of radiology. 4th ed. Springfield (IL): Charles C Thomas Publisher Ltd, 1983.
2. Attix FH. Introduction to radiological physics and radiation dosimetry. New York (NY): John Wiley and Sons, 1986.

3. ICRU Report 37. Stopping powers for electrons and positrons. Bethesda, MD: International Commission on Radiation Units and Measurements, 1984.
4. ICRU Report 16. Linear energy transfer. Washington, DC: International Commission on Radiation Units and Measurements, 1970.
5. ICRU Report 28. Basic aspects of high energy particle interactions and radiation dosimetry. Washington DC: International Commission on Radiation Units and Measurements, 1978.
6. ICRU Report 31. Average energy required to produce an ion pair. Washington, DC: International Commission on Radiation Units and Measurements, 1979.
7. ICRU Report 36. Microdosimetry. Bethesda, MD: International Commission on Radiation Units and Measurements, 1983.
8. ICRU Report 49. Stopping powers and ranges for protons and alpha particles. Bethesda, MD, USA: International Commission on Radiation Units and Measurements; 1993.
9. ICRU Report 55. Secondary electron spectra from charged particle interactions. Bethesda MD: International Commission on Radiation Units and Measurements, 1996.
10. ICRU Report 59. Clinical proton dosimetry Part I: Beam production, beam delivery and measurement of absorbed dose. Bethesda, MD: International Commission on Radiation Units and Measurements, 1998.
11. ICRU Report 63. Nuclear data for neutron and proton radiotherapy and for radiation protection dose. Bethesda, MD: International Commission on Radiation Units and Measurements Report 63, 2000.
12. ICRU Report 78. Prescribing, recording, and reporting proton-beam therapy. Bethesda, MD: International Commission on Radiation Units and Measurements, 2008.
13. ICRU Report 85a. Fundamental quantities and units for ionizing radiation (revised). Bethesda, MD: International Commission on Radiation Units and Measurements, 2011.
14. Paganetti H, Ed. Proton therapy physics. London: CRC Press, 2011.
15. Spinks JWT, Woods RJ. An introduction to radiation chemistry. New York: Wiley, 1964.
16. JCGM Report JCGM/200. International vocabulary of metrology—basic and general concepts and associated terms (VIM). Sèvres Cedex, France: Joint Committee for Guides in Metrology, 2012.
17. BIPM. The international system of units (SI). 8th ed. Sèvres Cedex, France: Intergovernmental Organisation of the Metre Convention, 2014.
18. Newell DB. A more fundamental International System of Units. *Physics Today* 2014;67:35–40.
19. Chilton AB. A note on the fluence concept. *Health Phys* 1978;34:715–716.
20. Palmans H, Rabus H, Belchior AL, et al. Future development of biologically relevant dosimetry. *Br J Radiol* 2015;87:20140392.
21. Pihet P, Menzel HG, Schmidt R, et al. Biological weighting function for RBE specification of neutron therapy beams; intercomparison of 9 european centres. *Radiat Prot Dosim* 1990;31:437–442.
22. Rudd ME, Kim Y-K, Madison DH, Gay TJ. Electron production in proton collisions with atoms and molecules: energy distributions. *Rev Mod Phys* 1992;64:441–490.
23. Bhabha HJ. On the penetrating component of cosmic radiation. *Proc R Soc London A* 1938;164:257–294.
24. Verhaegen F, Palmans H. A systematic Monte Carlo study of secondary electron fluence perturbation in clinical proton beams (70–250 MeV) for cylindrical and spherical ion chambers. *Med Phys* 2001;28:2088–2095.
25. Palmans H. Secondary electron perturbations in Farmer type ion chambers for clinical proton beams. In: Standards, Applications and Quality Assurance in Medical Radiation Dosimetry—Proceedings of an International Symposium, Vienna 9–12 November 2010, Vol. 1. Vienna, Austria: IAEA, 2011. pp. 309–317.
26. Bethe H. Zur theorie des durchgangs schneller korpuskularstrahlen durch materie. *Ann Phys* 1930;397:325–400.
27. Bloch F. Zur Bremsung rasch bewegter teilchen beim durchgang durch die materie. *Ann Phys*1933;16:285–320.
28. Barkas WH, Dyer NJ, Heckmann HH. Resolution of the Σ^-- mass anomaly. *Phys Rev Lett* 1963;11:26–28.
29. Bichsel H. Stopping power and ranges of fast ions in heavy elements. *Phys Rev A* 1992; 46:5761–5773.

30. Medin J, Andreo P. Monte Carlo calculated stopping-power ratios water/air for clinical proton dosimetry (50–250 MeV). *Phys Med Biol* 1997;42:89–105.
31. Palmans H, Thomas R, Shipley D, Kacperek A. Light-ion beam dosimetry. NPL report DQL-RD-003, Teddington, UK: National Physical Laboratory, 2006.
32. Fano U. Penetration of protons, alpha particles, and mesons. *Ann Rev Nucl Sci* 1963;13:1–66.
33. Landau LD. On the energy loss of fast particles by ionization. *J Phys* (USSR) 1944;8:201–205.
34. Vavilov PV. Ionization losses of high-energy heavy particles. *Sov Phys JETP* 1957;5:749–751.
35. Lewis HW. Range straggling of a nonrelativistic charged particle. *Phys Rev* 1952;85:20–24.
36. Moyers MF, Coutrakon GB, Ghebremedhin A, et al. Calibration of a proton beam energy monitor. *Med Phys* 2007;34:1952–1566.
37. Janni JF. Proton range-energy tables, 1 keV–10 GeV. At Data Nucl Data Tables 1982;27:147–339.
38. Kempe J, Brahme A. Energy-range relation and mean energy variation in therapeutic particle beams. *Med Phys* 2008;35:159–170.
39. Rutherford E. On the electrification of gases exposed to Roentgen rays, and the absorption of Roentgen radiation by gases and vapours. *Philos Mag* 1897;43:241–255.
40. Thomson JJ. On the theory of the conduction of electricity through gasses exposed to Roentgen rays. *Philos Mag* 1899;47:253–268.
41. Bragg WH. Studies in Radioactivity. London, UK: MacMillan, 1912.
42. Gray LH. The absorption of penetrating radiation. *Proc R Soc London Ser. A* 1929;122:647–668.
43. Spencer LV, Fano U. Energy spectrum resulting from electron slowing down. *Phys Rev* 1954;93:1172–1181.
44. Grosswendt B, Baek WY. W values and radial dose distributions for protons in TE-gas and air at energies up to 500 MeV. *Phys Med Biol* 1998;43:325–337.
45. Moyers MF, Vatnitsky SM, Miller DW, Slater JM. Determination of the air w-value in proton beams using ionization chambers with gas flow capability. *Med Phys* 2000;27:2363–2368.
46. Jones DTL. The w-value in air for proton therapy beams. *Rad Phys Chem* 2006;75:541–550.
47. PTCOG. Shielding design and radiation safety of charged particle therapy facilities. PTCOG Report 1; Villigen, Switzerland: Particle Therapy Cooperative Group, 2000.
48. Ipe NE. Basic aspects of shielding. In: Proton therapy physics. Ed. Paganetti H. London: CRC Press; 2011, pp. 525–554.
49. Paganetti H. Nuclear interactions in proton therapy: dose and relative biological effect distributions originating from primary and secondary particles. *Phys Med Biol* 2002;47:747–764.
50. Palmans H, Al-Sulaiti L, Andreo P, et al. Fluence correction factors for graphite calorimetry in a low-energy clinical proton beam: I. Analytical and Monte Carlo simulations. *Phys Med Biol* 2013;58:3481–3499.
51. IAEA, https://www-nds.iaea.org/exfor/; accessed 27 Feb 2015.
52. Paganetti H, Gottschalk B. Test of GEANT3 and GEANT4 nuclear models for 160 MeV protons stopping in CH_2. *Med Phys* 2003;30:1926–1931.
53. Palmans H, Capote Noy R. Summary report second research coordination meeting on heavy charged-particle interaction data for radiotherapy. INDC(NDS)-0567 Distr. G+NM. Vienna, Austria: IAEA, 2010.
54. Rinaldi I, Ferrari A, Mairani A, et al. An integral test of FLUKA nuclear models with 160 MeV proton beams in multi-layer Faraday cups. *Phys Med Biol* 2011;56:4001–4011.
55. Henkner K, Sobolevsky N, Jäkel O, Paganetti H. Test of the nuclear interaction model in SHIELD-HIT and comparison to energy distributions from GEANT4. *Phys Med Biol* 2009;54:N509–517.
56. Palmans H. Monte Carlo calculations for proton and ion beam dosimetry. In: Monte Carlo applications in radiation therapy, Ed. Verhaegen F and Seco J. London: Taylor & Francis, 2013, pp. 185–199.

57. Rutherford E. The scattering of α and β particles by matter and the structure of the atom. *Philos Mag* 1911;21:669–688.
58. Eyges L. Multiple scattering with energy loss. *Phys Rev* 1948;74:1534–1535.
59. Molière G. Theorie der streuung schneller geladener teilchen II - Mehrfach- und vielfach-streuung. *Z. Naturforsch* 1948;3a:78–97.
60. Bethe H. Moliere's theory of multiple scattering. *Phys Rev* 1953;89:1256–1266.
61. Hanson AO, Lanzl LH, Lyman EM, Scott MB. Measurement of multiple scattering of 15.7-MeV electrons. *Phys Rev* 1951;84:634–637.
62. Highland VL. Some practical remarks on multiple scattering. *Nucl Instr Meth* 1975;129:497–499.
63. Tsai YS. Pair production and bremsstrahlung of charged leptons. *Rev Mod Phys* 1974;46:815–851.
64. Gottschalk B. On the scattering power of radiotherapy protons. *Med Phys* 2010;37:352–367.
65. Gottschalk B. Physics of proton interactions in matter. In: Proton therapy physics. Ed. Paganetti H. London: CRC Press, 2011, pp. 19–59.
66. Goudsmit S, Saunderson JL. Multiple scattering of electrons. *Phys Rev* 1940;57:24–29.
67. Lewis HW. Multiple scattering in infinite medium. *Phys Rev* 1950;78:526–529.
68. Baldock C, De Deene Y, Doran S, et al. Polymer gel dosimetry. *Phys Med Biol* 2010;55:R1–63.
69. Palmans H, Vynckier S. Reference dosimetry for clinical proton beams. In: Recent developments in accurate radiation dosimetry. Ed. Seuntjens JP and Mobit PN. Madison, WI: Medical Physics Publishing, 2002, pp. 157–194.
70. Solov'yov AV, Surdutovich E, Scifoni E, et al. Physics of ion beam cancer therapy: a multiscale approach. *Phys Rev E* 2009;79:011909.
71. Douglass M, Bezak E, Penfold S. Development of a randomized 3D cell model for Monte Carlo microdosimetry simulations. *Med Phys* 2012;39:3509–3519.
72. Palmans H, Kacperek A, Jäkel O. Hadron dosimetry. In: Clinical Dosimetry Measurements in Radiotherapy (AAPM 2009 Summer School), Ed. Rogers DWO and Cygler J, Madison, WI: Medical Physics Publishing, 2009, pp. 669–722.
73. Karger CP, Jäkel O, Palmans H, Kanai T. Dosimetry for Ion beam radiotherapy. *Phys Med Biol* 2010;55:R193–R234.
74. Moyers MF, Vatnitsky SM. Practical implementation of light ion beam treatments. Madison, WI: Medical Physics Publishing, 2013.

Chapter 4

Proton Beam Interactions: Clinical

Indra J Das, Ph.D.[1], Alejandro Mazal, Ph.D.[2], Ludovic De Marzi, M.Sc.[3], and Vadim P. Moskvin, Ph.D.[4]

[1]Professor & Director of Medical Physics,
Department of Radiation Oncology, Indiana University School of Medicine,
Indianapolis, IN
[2] Head of Medical Physics,
Institut Curie,
Paris, France
[3]Medical Physicist, Centre de Protonthérapie d'Orsay,
Institut Curie,
Paris, France
[4]Proton Therapy Research Physicist, Division of Radiation Oncology,
Department of Radiological Sciences,
St. Jude Children's Research Hospital,
Memphis, TN

4.1 Introduction

Charged particles such as protons interact with a medium and cause excitation, ionization, and nuclear interactions, depending on the beam energy slowing down in depth. The energy loss is associated with the stopping power (*S*) and is related to the range, which is characteristic of every charged particle. Due to their finite range, charged particles are appealing for clinical applications (see Chapter 2). The basic interactions of a proton beam have been discussed in Chapter 3. However, additional details in the clinical context are provided in this chapter dealing mainly with clinical aspects of proton physics and its practical implications for patient treatment.

Beam modification is carried out by inserting various types of materials in the beam, such as energy degraders, scatterers, range modifiers, range shifters, apertures, and compensators. Patients may also contain high-Z materials, such as prosthesis and implants. Additionally, spacers might be surgically placed in the clinical setting to reduce the range to avoid dose to critical organs. The use of hyaluronic acid and various other gels have been suggested for creating spacers between the rectum and the prostate, as described in the review article by Mok et al. [1]. In proton therapy, such attempts have been reported in the abdomen, but these materials have been taken as water equivalent [2]. However, in proton beam, water equivalent thickness (WET) needs to be known to provide suitability of such process, including surgically placed materials in the beam path [3].

The range of protons is associated with their interactions in a medium. The treatment table top and immobilization devices in the beam produce a reduction in range. This needs to be considered clinically in the context of WET during commissioning and with new devices that are implemented periodically. WET is related to range and depends on density, atomic number (*Z*), stopping power (*S*), and scattering power (*T*). Even for those materials where density is very close to water, if the *T* and *S* are different, WET values will be different, which is discussed along with other radiological parameters in this chapter.

4.2 Practical Quantities

Chapter 3 has discussed most of the relevant interaction properties of charged particles. This ability of a charged particle to release energy in a medium is expressed in stopping power. For clinical use, *S*, range, WET, and effective atomic number (Z_{eff}) are important parameters which are discussed in this section.

4.2.1 Stopping Power, S

The concept of *S* was defined by Bragg and Kleeman [4,5] for alpha particle passage in gas. It is defined as the ability for a medium to stop charged particles, which has been discussed in the previous chapter. This can be stated either in linear or mass expressions i.e., *S* (MeV/cm) or S/ρ (MeV cm^2/g) where ρ is the physical density of the medium. The proton range, i.e., the distal position of the treatment field, is directly related to stopping power. This range is often defined in a medium as WET or water equivalent depth (WED). Various publications have provided mathematical methods to provide this quantity [6–10].

A general equation for the energy loss by a charged particle was provided by Bethe and Bloch [11,12], which is a generally accepted equation, except for some small modification to Equation 4.1 [13].

$$\frac{dE}{\rho dx} = \frac{4\pi N_A r_e^2 m_e c^2}{\beta^2} \frac{Z}{A} z^2 \left[ln\left(\frac{2m_e c^2 \beta^2}{I(1-\beta^2)} \right) - \beta^2 \right] \tag{4.1}$$

where E is particle energy (MeV), N_A is Avogadro's number, r_e is the classical electron radius, m_e is mass of electron, c is velocity of light, z is charge of particle, Z is atomic number of medium, A is the atomic mass of the medium, β is relative velocity (v/c) of the particle, and I is the ionization potential of the medium. A detailed description on this topic is also shown in Chapter 3. Even though stopping power has two components (collision and radiative), for heavy charged particles, radiative losses are very small [14], i.e., the collision stopping power can be approximated with the total stopping power.

$$\left(\frac{S}{\rho} \right) \approx \left(\frac{S}{\rho} \right)_{col} \tag{4.2}$$

For clinically used tissue equivalent type media, S does vary significantly as shown in Figure 4–1(a). Furthermore, S is a strong function of proton energy rising rapidly with decreasing energy. This is important, especially toward the end of the range or near zero energy. The biological consequences of high S are reflected in the relative biological effectiveness (RBE), which will be discussed in Chapter 5. However, for dosimetry, the mass stopping power (S_m) ratio is a critical factor for ion chamber-based dosimetry where S_m(m, air) needs to be known for clinically used materials (m). ICRU-78 and IAEA TRS-398 [15,16] have provided more details on the consequences in proton beam calibration, which is also discussed in Chapter 12.

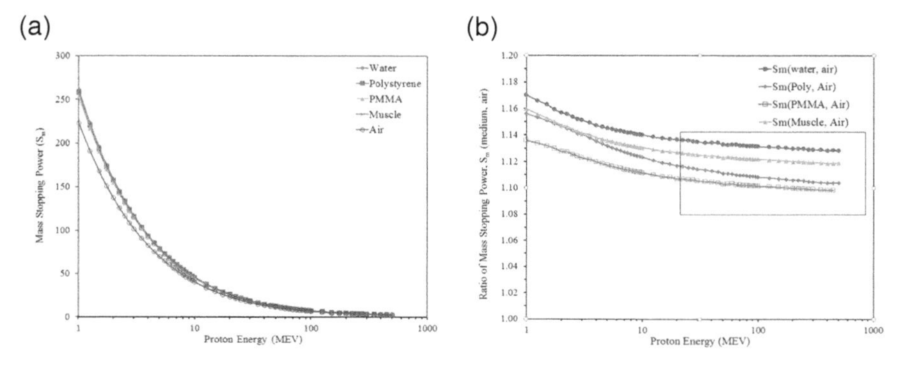

Figure 4–1 (a) Mass stopping power of proton beams for various materials and (b) ratio with respect to air versus proton beam energy.

For clinical and dosimetric use, the S_m(W, air) is relatively constant between 20–500 MeV [13,15] as shown in Figure 4–1(b) and Equation 4.3 in terms of residual range of the proton beam ($R_{res} = R_p - x_{ref}$), where x_{ref} is the point of measurement or reference depth, for example in the middle of the SOBP.

$$\left(\frac{S}{\rho}\right)_{air}^{w} = 1.137 - 4.3\times10^{-5} R_{res} + \frac{1.84\times10^{-3}}{R_{res}} \tag{4.3}$$

4.2.1.1 Stopping Power Application

Dose and Calibration

The particle fluence, ϕ, provides a direct measure of the radiation dose as shown in Equation 4.4.

$$\begin{aligned} Dose &= \ Fluence \ \times \ stopping\ power \\ &= \phi \bullet S / \rho => \# / cm^2 MeV / g\ cm^2 \\ &= MeV / g => J / kg = Gy \end{aligned} \tag{4.4}$$

Thus, it is important to know the particle fluence. In a proton beam, this can be measured through various devices, like a Faraday cup, which measures the beam current, which can then provide the fluence as shown in Equation 4.5. A typical beam current from a cyclotron-based proton facility is 10–100 nA. The number of protons (fluence in Equation 4.4) for a 100 nA proton beam current can be computed as below.

$$\begin{aligned} \text{\# of Proton/s} &= 100\times10^{-9}\left(A = \ C / s\right) / \ 1.6\ \times10^{-19} C / \ proton \\ &= 6.25\times10^{11}\text{protons/s} \ = 3.76\ \times10^{13}\ protons / min \end{aligned} \tag{4.5}$$

Obviously these protons are spread over the treatment area, thus the fluence per cm^2 is reduced significantly. For example, for a beam of 2 mm diameter spread over a 10 cm diameter area, the fluence will reduce to 4×10^{-4} #/cm^2. This has implications for large fields where the fluence falls of rapidly by the area, and so does the dose. To maintain a clinical dose rate, i.e., 200 cGy/min, beam currents needs to be adjusted accordingly. In such calculations, proton loss through scattering materials and beam-modifying devices need to be accounted for, too. Details of beam calibration and the importance of S is discussed in Chapter 12.

Treatment Planning

CT imaging is the backbone in modern radiation therapy. The CT data is correlated with electron density via CT–ρ_e curve for each scanner that is used for external beam dose planning. Most photon and electron dose calculation algorithms depend on the knowledge of ρ_e [17]. However, in proton therapy, this approach is not feasible, and the relative stopping power (RSP) is computed and correlated with CT data. The RSP versus CT number is acquired for each scanner used for treatment planning [18–21]. An innovative idea using dual-energy CT has been suggested for characterizing tissue

Table 4–1 Ionization potential (I), physical density, and electron density of materials

Material	Physical Density, ρ (g/cm^3)	Electron Density, ρ_e (electrons/g)	Ionization Potential, I (eV)
Air	0.001293	3.006×10^{23}	85.7
Water	1.00	3.343×10^{23}	75.0
Muscle	1.04	3.312×10^{23}	75.3
Fat	0.916	3.340×10^{23}	63.2
Bone	1.65	3.192×10^{23}	106.4
PMMA	1.19	3.247×10^{23}	74.0
Polystyrene	1.044	3.238×10^{23}	57.4
Lucite	1.18	3.248×10^{23}	74.0
Aluminum (Al)	2.70	2.902×10^{23}	166.0
Carbon (C)	2.25	3.008×10^{23}	78.0
Lead (Pb)	11.36	2.383×10^{23}	823

for dose calculation in proton therapy [22], but this still needs clinical evaluation. Proton tomography has also been suggested [23]. Since high-Z materials cause metal artifacts and CT data gets corrupted, the accurate estimation of RSP might be difficult. Various publications have provided methods to estimate RSP for high-Z materials [8,24–28]. Active research is continuing in this area, either to suppress artifacts or to account for them by using different imaging methods, such as dual-energy CT.

4.2.1.2 Factors Associated with Stopping Power

As shown in Equation 4.1, S and S/ρ depend on various factors. The impact of these parameters, especially ρ_e and I is discussed below.

Electron Density, ρ_e

Charged particles interact primarily with atomic structures, mainly electrons, and lose energy electromagnetically as they travel in the medium. At medium and high energies the contribution of nuclear interactions is significant as well in terms of energy loss. The interaction between charged particles and medium depends on the electron density, thus understanding of the electron density is an important consideration when determining stopping power and range of the particle. As shown in Equation 4.1, the quantity $N_A(Z/A)$ is equal to electron density, ρ_e, expressed in the number of electrons/g that influences the stopping power. It is obvious from Equation 4.1 that the stopping power decreases with increasing values of Z, e.g., it is lower with decreasing electron density. The numerical values of various materials of clinical interest are shown in Table 4–1.

Ionization Potential (I) in Medium

The ionization potential, I, is the geometric mean value of all ionization and excitation potentials of an atom of the medium, and it is computed based on Equation 4.6, where f_i is the oscillator strength for the transition with an excitation energy E_i. It depends on the chemical composition and binding energy of the compounds. This has been studied extensively. Attix [29] showed that I/Z is a slow-varying function of Z, but it increases at low Z. For compounds, it can be computed as the weighted average of I values depending on the atomic composition [29,30]. It is known that I is relatively independent of the charged particle. The stopping power is a function of ln(I) as shown in Equation 4.1, and its effect on S is relatively week. Table 4–1 shows I values for various materials.

$$ZlnI = \sum_i^n f_i \ln(E_i) \tag{4.6}$$

4.2.2 Straggling

Energy loss is a discrete phenomenon depending on the atoms or molecules with which the proton beam interacts in its path. This is a rather stochastic phenomenon. Thus, each individual proton will have a different interaction probability, so the energy at a point is statistically distributed, which is known as energy straggling. Wilson [31] provided an empirical equation for the dose and proton range relationship:

$$D = k\frac{jt}{(R-x)^{0.44}} \qquad for\ x < R \tag{4.7}$$

$$P(x)dx = \frac{R}{\alpha\sqrt{\pi}} e^{-\frac{(R-x)^2}{R^2\alpha^2}} dx \tag{4.8}$$

$$\alpha = \frac{7}{\sqrt{E_o}}\left(\frac{NZz^2R}{E_o}\right)^{-0.055} \tag{4.9}$$

where k is constant ≈ 4.8×10^{10}, j is the beam current (ampere/cm^2), t is time in sec-

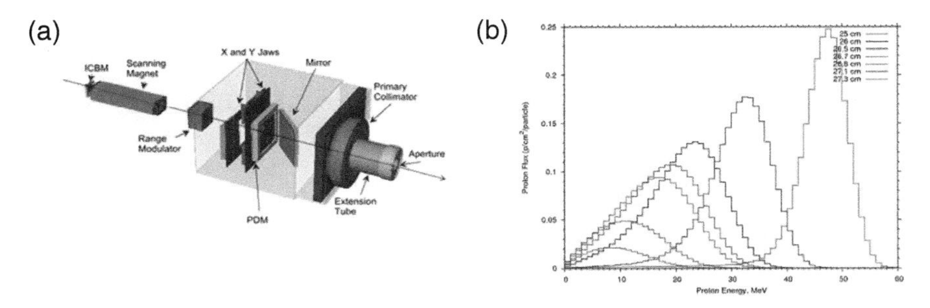

Figure 4–2 (a) The components in a Monte Carlo simulation of the IU health proton beam. (b) The energy spectra at different depths. Note that at deeper depths, e.g., lower energy, the spectrum becomes Gaussian.

onds, R is range, P is probability, N is atom/cm^3, z is charge of the particle (for proton it is 1), and E_0 is rest energy of the charge particle (MeV). This probability distribution gives rise to energy straggling, which is shown in Figure 4–2 using a Monte Carlo simulation. The IU health proton beam as shown in Figure 4–2a was modeled to simulate the energy distribution shown in Figure 4–2b) at various positions in water. Similar spectra have also been simulated by Brooks et al. [32]. Note that at deeper depths (lower energy) the spectrum becomes nearly Gausian. Also, the shape change is more pronounced at deeper depth. The energy spectrum leads to the range straggling that is discussed in the following section.

4.2.3 Range

Range is a fundamental characteristic of a charged particle which is related to the stopping power in a medium. Attix [29] stated the actual meaning of range as "range of a charged particle of a given energy in a medium is the expectation value of the path length that it follows until it comes to rest (discounting thermal motion)." Range is inversely proportional to stopping power (dE/dx) and, more importantly, given as

$$R = \int_0^E \frac{dE}{(dE/dx)} \tag{4.10}$$

Range has many definitions depending on the observation and measurements [29,30]. Range verification is performed in water using a proper detector (see Chapter 11). Various definitions of range are discussed below.

4.3.3.1 Definitions of Range

There are many types of range definitions, R_{max}, maximum range, R_{avg}, average range, R_p, practical range, R_{ex}, extrapolated range, R_{CSDA}, continuous slowing down approximation, and R_{90} and R_{80} as 90% and 80% dose, respectively. The farthest travel and the actual pathlength of a particle are different. For heavy charged particles, they are very close due to the small multiple Coulomb scattering angle. Figure 4–3

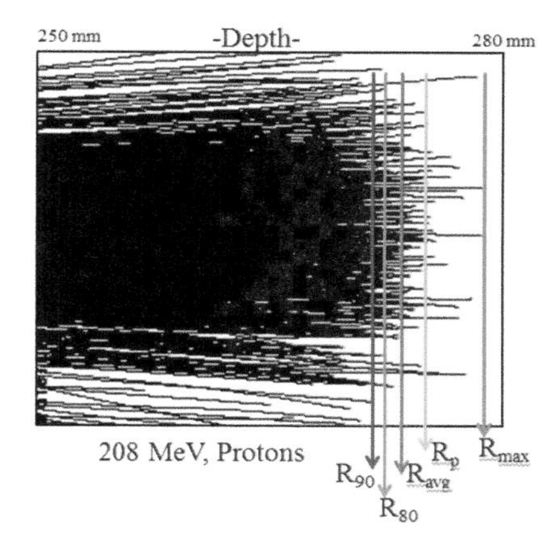

Figure 4–3 Monte Carlo simulated range of 208 MeV protons. The nominal range is 27 cm. Note various ranges due to stochastic nature of proton interaction giving different ranges. The measured ranges R_{90} and R_{80} are also shown.

shows various types of definition of range. Even though energy deposition is random and discrete, mathematically it is defined as a continuous process. If stopping power is then described in terms of continuous loss of energy, then the associated range will be called continuous slowing down approximation range (R_{CSDA}). For a monoenergetic proton energy, range in a specific medium is known and tabulated in ICRU-49 [13] calculated by integration as shown in Equation 4.11. For simplicity, it can be calculated analytically as shown in Equation 4.12 using a quadratic equation. Various other models also have been suggested that are also empirical in nature [10,29,30].

$$R_{CSDA} = \int_0^{E_{max}} \left(\left(\frac{S}{\rho} \right)^{-1} \right) dE \tag{4.11}$$

$$R = aE + bE^2 \tag{4.12}$$

where a and b are fitting parameters equal to 0.033 cm/MeV and 0.0005 cm/MeV^2, respectively, as shown in Figure 4–4.

The interaction rate of each particle is very different depending on the interaction probability with materials in the path to lose energy stochastically. This is an event-by-event loss; hence, at a point in space, energy is distributed. This is known as energy straggling, as described in the previous section. The straggling is more pronounced at a deeper depth. Figure 4–3 provides energy and range straggling. It is logical that the energy and range straggling are more pronounced in double scattering than in a scanned beam due to more energy loss in scatterers. Even though ionization measurements in a water phantom provide depth–dose curves where some of the

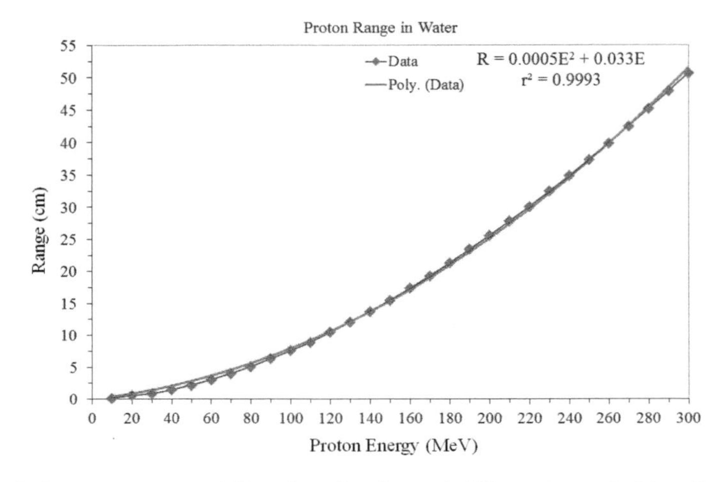

Figure 4–4 Range energy relationship. A polynomial fit as shown in Equation 4.12 is also shown, indicating a simple way to calculate range in water.

ranges described in Figure 4–3 can be measured, other innovative approaches have been proposed, such as proton-induced x-ray emissions (PIXE) by fiducial markers [33]. However, range uncertainty is more pronounced when metal artifacts are present, especially in clinical settings in CT data. Newhauser et al. [25] have attempted to provide a Monte Carlo approach to this problem.

4.2.3.2 Range Uncertainty

As discussed above, range is not a fixed number, and it depends on the definition, medium, and energy. Additionally, the calculation of range depends on the estimation of *S*. Yang et al. [34] showed that in a low-Z medium similar to tissue, the range uncertainty associated with *S* is only 3.0% to 3.4%. However, for clinical situations, this is important for patient treatment. Measurements of range and range uncertainty in patients (*in vivo*) are hard to achieve [35]. Paganetti [36] suggested that Monte Carlo simulations could reduce range uncertainties in patients considerably. A detailed discussion on range uncertainties is provided in Chapter 21.

4.3 Water Equivalent Thickness (WET)

Proton beam commissioning is performed in a water phantom, which is most convenient. This is then related to CT data for treatment planning. The proton range in water for energy (E) is usually measured. However, when compensators or range shifters are used, it is important to know the equivalent thickness of a material to place in the beam to provide a desired range. Figure 4–5 shows a Bragg peak with range R_1 in water. If a material with thickness, *t*, is placed in the beam, that shifts the range to R_2. The range shift $\Delta = R_1 - R_2$ in water is related to t and is called the WET (Figure 4–6) of the material. For a small thickness that does not change the beam spectrum significantly, the ratio Δ/t could be constant, as given in Equation 4.13:

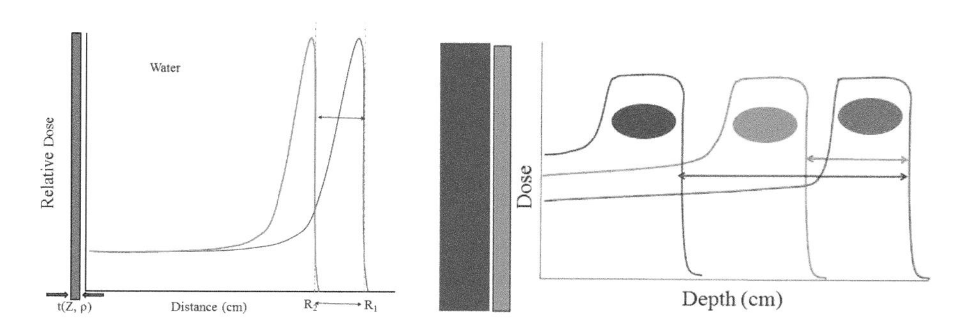

Figure 4–5 The concept of range shift for a pristine Bragg peak and an SOBP, related to the water equivalent thickness, WET, is shown. The thicknesses in the front represent the WET of the material.

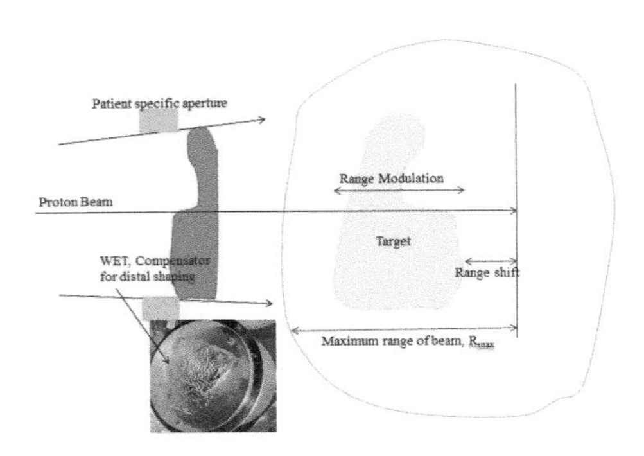

Figure 4–6 Depiction of target and associated proton range shift from maximum range. The distance is then translated into WET to create a compensator.

$$\Delta = kt \tag{4.13}$$

where k is constant for an energy and is rather complex function of E, Z, and ρ. Theoretical calculations of this equivalence have been provided by many investigators [6,9,37]. For clinical applications, WET can be simply measured as shown by Nichiporov et al. [7]. An analytical approach for WET has also been published that can be used [38]. In other words, t is not simply a property of the material, but also of specific measurement conditions. Therefore, the above equation is only accurate for thin slabs of materials and for a given beam energy. However, this simple equation can be used successfully in practice, especially for low-Z materials, for quickly estimating range shifts in water produced by the insertion of materials in the beam path.

Secondary products from inelastic nuclear interactions contribute to the depth–dose distribution of therapeutic proton beams. Different phantom materials produce different levels of inelastic reactions and different non-elastic products [39]. ICRU-46 [13] provided stopping power data for various tissues that can be used for range calculations. However, the data are based on monoenergetic beams and may not be applicable for a clinical beam. TRS-398 [16] provided some information for scaling plastic materials used for dosimetry in proton beams. The depth in a plastic material z_M (expressed in g/cm^2) should be scaled to provide corresponding depth in water z_w by

$$z_w = z_M C_{M;}\ and\ C_M \approx \frac{(R_{CSDA})_W}{(R_{CSDA})_M} \tag{4.14}$$

where C_M is a depth scaling factor and can be calculated with a good approximation, as the ratio of R_{CSDA} ranges (g/cm^2) in water and in medium. For plastic, PMMA, and clear polystyrene, it has a value of 0.974 and 0.981, respectively [13].

4.3.1 Theoretical

Many theoretical approaches have been proposed for the calculation of WET [6]. WET can be computed based on S_m as

$$t_W = t_M \frac{\rho_M}{\rho_W} \frac{S_m(E)^M}{S_m(E)^W} \tag{4.15}$$

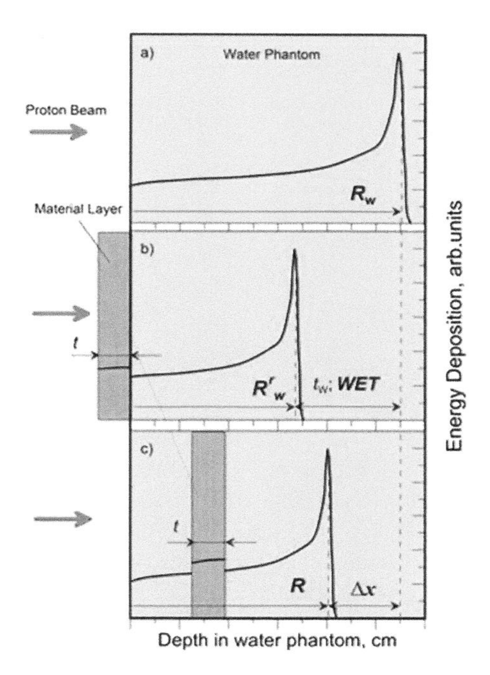

Figure 4–7 The geometry for the WET measurements. (From Moskvin et al. [38].)

where t is the thickness and the subscripts and superscripts M and W represent medium and water, respectively. If a beam of energy E has a range in water of R_W, and R_r (=R^r_w in reference [38]) is the observed range (measured) after placing the material t_M in water, then the range shift DR is calculated as

$$RW = Rr + t_w - t_M \tag{4.16a}$$

$$\Delta R = t_M \left(\frac{\rho_M}{\rho_W} \frac{S_m(E)^M}{S_m(E)^W} \right) - t_M \tag{4.16b}$$

$$\Delta R = t_M \left[\left(\frac{\rho_M}{\rho_W} \frac{S_m(E)^M}{S_m(E)^W} \right) - 1 \right] \tag{4.16c}$$

It is important to pay attention that many may ignore 1 in the bracket term in Equation 4.16 c, which is adequate if t_M is very small. However, this equation should be used for range shift in estimating WET. A detailed discussion on this issue can be found elsewhere [7]. The above equations can be further simplified assuming ρ_W =1 and possible ignoring E.

The relationship between WET and material thickness t is not always a straight line due to the changes in S_m which is a slow-varying function of energy, except at the end of range (at low energy). Additionally, spectral changes are relatively small for thin sheets of materials in the beam. In such situations, an empirical equation for WET and ΔR has been proposed by Moskvin et al. [38]. It was shown that the empirical method is energy insensitive and valid from 50–250 MeV and for Z ranging up to 82 within 3% accuracy. Additional details on analytical function is described in Section 4.5.2.

$$WET\left(t_M, Z\right) = t_M \rho_M \left(1.192 - 0.158\ln\left(Z\right)\right) (cm) \tag{4.17a}$$

$$\Delta R\left(t_M, Z\right) = t_M \left[\rho_M \left(1.192 - 0.158\ln\left(Z\right)\right) - 1.0\right] (cm) \tag{4.17b}$$

4.3.2 Measurement

The WET can be measured in water by measuring ΔR with an ionization chamber (as shown in Figure 4–7) either for a pristine Bragg peak or for an SOBP (as shown in Figure 4–5). There are two methods for measuring WET, 1) materials outside the tank, and 2) inside the tank. A shift then can be measured by placing a known thickness of material in the beam [7]. Theoretically, one could imagine a different WET value due to spectral changes and relative S_m dependence with energy in these two settings. When material is close to the surface it has more pronounced scattering compared to when it is placed inside the water phantom close to the detector. It was observed that the differences noted in WET values are minimal in both cases within the limit of measurement [7]. Using proton radiography, Hurley et al. [40] estimated WET values. A solid phantom approach using a scintillating fiber hodoscope and range telescope was used by Schneider et al. [41] to successfully estimate WET for low-Z materials.

The calculated and measured values of the range shift over a wide energy range and different materials is shown to be within <0.2 mm [7]. Other investigators [6,9] have also found very good agreement with either Monte Carlo simulation or measurements. For compounds, tissues, and materials with unknown compositions, calculations may be difficult due to poor knowledge of S_m and R_{CSDA}. In such situations, measurement using a range shift or analytical approaches (such as shown in Equation 4.17 and suggested by Moskvin et al. [38]) could be used.

4.3.3 Clinical Relevance

In a clinical setting, the WET can be measured in a tissue sample from animals that can be approximated to human tissue [20]. For example, the WET of liver, heart, lung, etc. can be performed from tissues from pigs or cows and measured in the respective beam, eventually compared to a CT image of the sample. Such approaches are more accurate than calculation as described in Section 4.3.1 for WET. The tissue samples can be acquired from butcher shops or from grocery stores. There is some speculation that these data may not reflect human tissue data due to blood flow and oxygenation, however differences are very small and are within the limit of accuracy in measurements.

4.4 Scattering Power, *T*

A theoretical review of the scattering power is provided in Chapter 3. Apart from energy loss, particle scattering is also important clinically to understand how scattering impacts the dose in patients and to optimally design a delivery system. The multiple Coulomb scattering is mainly described by Molière scattering [42], but for the charged particles used in clinical applications, it is large angle scattering. This is described by scattering power, which is the ability of a medium to scatter a charged particle to a certain angle based on columbic interactions. For materials in the beam, this can be measured in terms of full width half maximum (FWHM) which should be known for designing components for the beam line.

In the estimation of WET, scattering power also play an important role (Figure 4–8). However, for low-Z materials like beryllium and plastic, scattering power is nearly independent. This is why these materials are used to minimize the scattering for a given range modification.

The angle of scattering increases linearly with thickness of the material and Z^2. Like the stopping power, scattering power can be described in linear and mass terms as T and T/ρ. In close analogy to S_m, T/ρ is defined as below (Equation 4.18) in terms of the mean square scattering angle, $\overline{\theta^2}$ of a scattering medium.

$$\frac{T}{\rho} = \frac{1}{\rho}\frac{d\overline{\theta^2}}{dx} \tag{4.18}$$

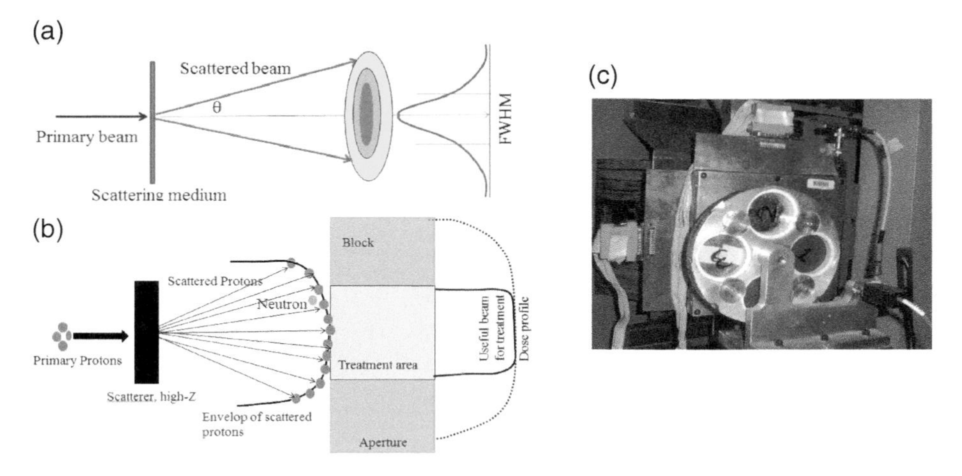

Figure 4–8 (a) View of proton scattering at a maximum angle θ, projecting various particle fluence that gives rise to Gaussian distribution. (b) Clinical implementation of scattering for beam broadening used in proton therapy. (c) Actual beam scattering device in the beam line used at Indiana University.

Gottschalk et al. [43,44] provided an extensive review of the T/ρ related to the design of scattering foils used in most scattering beams today. Scattering is also an important factor in patients when collimators are used. Titt et al. [45] showed that scattered dose contributes up to 20% at the central axis depending on the collimator size (higher for smaller collimators and increasing ranges). Additionally, significant streaking of scattered protons can be seen at the wall of collimator, especially in small fields.

4.4.1 Secondary Particles

When protons pass through media, secondary particles are generated in inelastic collisions that include several light ions and secondary electrons and delta rays (figures 4–9 and 4–10). The direction is governed by the mass and kinetic energy. If m_e and m_p are the mass of electron and particle (proton), respectively, E is kinetic energy of the secondary electron, and E_{p0} is the primary proton energy, then the direction (θ) and maximum electron energy, E_{max}, can be given as below.

$$Cos(\theta) = \frac{1}{2}\sqrt{\frac{m_p E}{m_e E_{p0}}} \tag{4.19a}$$

$$E_{max} = 4\frac{m_e}{m_p}E_{p0} \tag{4.19b}$$

Apart from secondary electrons, various other particles are emitted from interaction of proton beams with media. Using the Monte Carlo code FLUKA, it was shown that a 150 MeV proton beam produces secondary protons (42.8%), $^4He^{++}$ (23.2%),

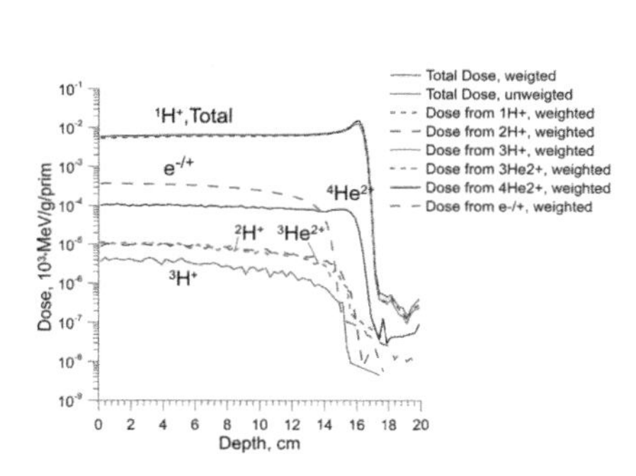

Figure 4–9 Radiobiological weighted dose from secondary particles of a 150 MeV proton beam in a water phantom for a large beam. (Calculated by coauthor, Moskvin.)

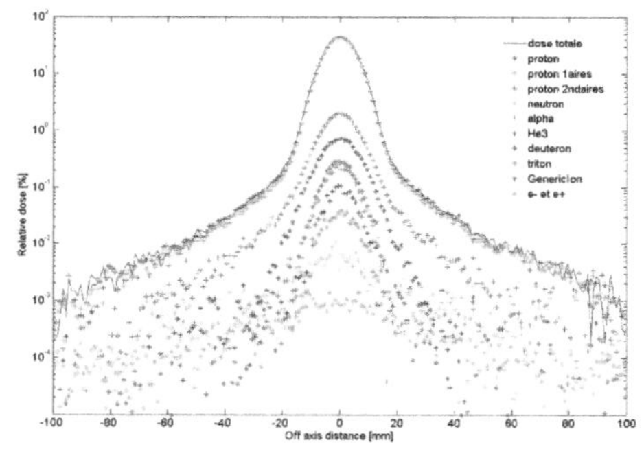

Figure 4–10 Lateral profile of a pristine 160 MeV proton beam of 5 mm width in water at mid-range (8 cm depth) showing the dose for different secondary components. (Calculated by coauthor De Marzi.)

neutrons (18%), photons (12.3%), deuterons (2.1%), $^3He^{++}$ (1.1%), and tritons (0.6%). These particles could contribute to the dose which is shown in Figure 4–9. The lateral extent of dose due to emitted particles is shown in Figure 4–10. The contribution of other particles compared to primary protons to the dose is relatively small as noted by the logarithmic scale (figures 4–9 and 4–10).

Using advanced Monte Carlo techniques, Gomà et al. [46] provided a clear idea of the effect of secondary electron on dose based on the Spencer–Attix cavity theory for S_m(W, air) in a proton pencil beam. It was found that this effect only increases S_m values to 0.1%, and insignificant effect on total dose from nuclear fragments was computed for pencil beam dosimetry.

4.4.2 Perturbation

Unlike electron beams, where scattering is more pronounced, proton beam scattering is relatively small. Usual dose perturbations as seen in electron and photon beams with high-Z medium [47–51] are not visible in proton beams. Dose perturbations and interface dosimetry in proton beams has been studied in clinical situations [52–55]. However, additional research for clinically relevant situations is needed. Experimentally, this was studied recently by Nichiporov et al. [7] in the context of WET for various high-Z materials, and they noted an increased dose at the downstream interface. There was no backscatter dose enhancement, which is usually seen in a case of photons and electrons. Figure 4–11 shows this effect on the central axis. Based on this finding, one could conclude that dose perturbations for thin, high-Z materials are relatively small (<3%). Verhaegen and Palmans [56] also studied secondary electron fluence perturbations from high-Z media and concluded that the perturbation even at gold interfaces is a mere 3%.

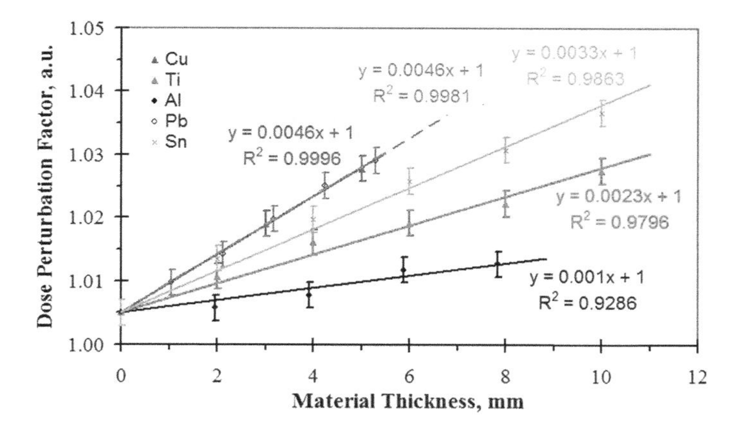

Figure 4–11 Dose perturbation as a function of thickness of various high-Z materials. Please note a linear relationship for a 208 MeV proton beam with thickness and maximum perturbation is less than 3%.

4.5 Effective Atomic Number (Z_{eff}) for Protons

Radiological parameters like S and R of charged particles are directly related to the Z of the medium based on the Bethe-Bloch equation 4.1. However, for compounds, an effective Z needs to be determined. In photon beams, this is governed by the interaction probability, e.g., photo-electric, Compton, and pair-production that has been described in detail in literature [57–60]. However for proton beams, the effective Z needs a different method which is described below.

4.5.1 Definition

An effective atomic number Z_{eff} is uniquely defined by the range shift in water $\Delta x(t,Z_{eff})$ of a proton beam, with a given energy, caused by a layer of material with the Z_{eff} of thickness t of a compound placed at the plateau part of the proton depth–dose curve and, hence, by the WET(t,Z_{eff}) of the material layer.

The Z_{eff} is commonly used as a parameter to characterize radiological properties of materials. The values of Z_{eff} for compounds are defined, conventionally, from the x-ray interactions with the material. Photon Z_{eff} could be defined from the mass attenuation coefficients [61] or through interpolation of adjacent cross section data [62,63]. The conventional method to determine Z_{eff} of a material for photon transport is based on the photoelectric coefficient per electron or total photon energy absorption cross section per electron in composition [57,59], or total photon interaction cross section per atom [58,64] for each element in the material. The estimated value of the Z_{eff} varies with the photon energy due to the weighting of the photoelectric interaction process and depends on the calculation technique used.

The interaction of protons in a medium is fundamentally different from photons. The total proton interaction is defined by collision (electronic) and nuclear stopping processes. The values of Z_{eff} for a few dosimetric materials were calculated by Prasad et al. [60] for proton energies of 1 to 200 MeV using the stopping power data. The procedure of the Z_{eff} calculation for charged particles involves the utilization of the S [60,64].

The estimation of Z_{eff} for a known elemental composition in stopping power calculations is based on the known weight factors for each element and effective nuclear density.

$$Z_{eff} = \frac{\sum \omega_i \frac{Z_i}{A_i}}{\sum \frac{\omega_i}{A_i}}, \tag{4.20}$$

where Z_i is atomic number of *i*-element in composition, A_i is atomic weight, and ω_i is the weight of the element in composition. The values of the Z_{eff} thus calculated are different from those obtained for the conventional megavoltage photons. For instance, the classical value for Z_{eff} for water for photons given by Johns and Cunningham [57] is 7.4, while it is 3.3 for protons as determined by Prasad et al. [60].

The data for dosimetry for materials and some tissues in interval atomic numbers from 1 to 14 were computed from adjacent stopping cross section data by Kurudirek [65] for proton energies 1 keV to 10 GeV based on the adjacent cross section method. The value of Z_{eff} for water varies from 2.5 for 1 KeV to 3.4 for 1 GeV protons. The adjacent stopping cross section method shows strong variation in Z_{eff} of a given material compared to water.

The methods to define Z_{eff} through direct calculation from S (nuclear density method) or cross sections assume knowledge of the exact elemental composition of compounds. However, the elemental composition of a tissue varies depending on gender, age, and life history of a patient [66]. A method for estimating Z_{eff} for an unknown elemental composition was proposed recently by Moskvin et al. [67]. This method utilizes the effect of the proton range shift and the associated definition of WET of the material layer different from water relatively to proton range in water.

4.5.2 Z_{eff} and WET of Compounds

Let's consider a slab of thickness *t* (in cm) of a material, *M*, placed in the front of a water tank (Figure 4–7). A unidirectional proton beam with energy in the range of 50 to 250 MeV is incident normally to the surface M. The amount of energy deposited in the slab corresponds to the amount of energy deposited in a slab of WET, is defined according to Zhang et al. [6,9] as for medium M and water (W):

$$WET(t, Z_{eff}) = t\frac{\rho_M}{\rho_W}\left(\frac{S}{\rho}\right)_W^M . \tag{4.21}$$

A more commonly encountered situation is that of an inhomogeneity inside a patient, such as surgical clips, spinal prosthesis, metallic breast implants, dental fillings, and hip prosthesis whose compositions are not available readily.

The proton range R_{in} in Figure 4–7(c) is defined as the range of protons in an inhomogeneous system consisting of water embedded with a layer of material. Medin and Andreo [68] showed that the S_m(w, air) changes slowly as a function of depth up to the proximal slope of the Bragg peak in a homogeneous medium where the Bragg peak starts forming (Fig 4.1 b). The analysis of the ICRU 49 data [13,69] shows that the S_m(M, air), has a weak dependence on proton energy above 50 MeV for low Z_{eff} (for example, for pure materials, about 2% for Z=13) and above 80 MeV for high-Z (about 5% for Z=82) materials also confirmed by Figure 4–1. Thus, in this semi-empirical model, the slab M is placed before the Bragg peak in water for a given energy. Additionally, the thickness of the slab *t* satisfies the condition of the material thickness less than proton range in this material.

The observed range shift, ΔR, can be defined as $R_W - R_W^r - t$, where R_W is the range of protons in water, and R_W^r is the reduced range of protons in water due to the presence of material different from water. The difference, $R_W - R_W^r$, is equal to the WET of the material (see Figure 4–7). The observed proton's range shift in a water phantom embedded with a layer t of a material M with atomic number Z_{eff} is then given by:

$$\Delta R\left(t, Z_{eff}\right) = WET\left(t, Z_{eff}\right) - t \tag{4.22}$$

The function, $WET(t,Z_{eff})$ has a monotonic dependence on atomic number Z_{eff}, that was fitted by a function as shown in Equation 4.17a.

$$WET\left(t, Z_{eff}\right) = \alpha\left(Z_{eff}\right) t \frac{\rho_M}{\rho_W} \tag{4.23}$$

where $\alpha(Z)$ is the fitting function expressing the dependence of S_m(M, W), on atomic number Z of the material [38]. It was shown above that the $WET(t,Z_{eff})$ of a thin material layer and the range shift of protons in water in the presence of material layer, is a function of the atomic number, Z, of the material. The proposed semi-empirical equation for calculation of $WET(t,Z_{eff})$ can be achieved very similar to Equation 4.17 as shown by Moskvin et al. [38] by replacing Z to Z_{eff}.

4.5.3 A Semi-empirical Model for Z_{eff} Determination

Based on invariance of the $WET(t,Z_{eff})$ with initial energy of protons in the interval from 50 to 250 MeV as suggested above, the expression for the Z_{eff} could be derived from the equation for $WET(t,Z_{eff})$ in a form

$$Z_{eff} = a \cdot \exp(-b\,\chi) \tag{4.24}$$

$$\chi = \frac{WET(t, Z_{eff}) \cdot \rho_W}{t \cdot \rho_M} \tag{4.25}$$

where χ is a dimensionless quantity defined is in Equation 4.24. To calculate the coefficients in the Equation 4.23 for Z_{eff}, the proton range shift was computed with the

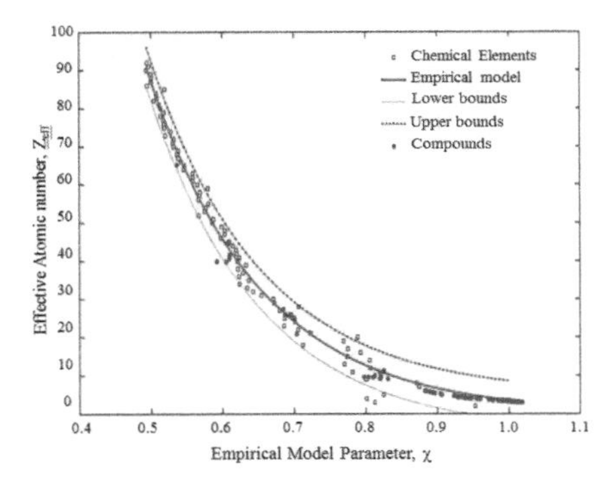

Figure 4–12 The effective atomic number semi-empirical model fit as shown in Equation 4.23.

Table 4–2 The model coefficients *a* and *b* for various initial proton energies

Energy, MeV	Model Coefficients	
	a	*b*
50	2179	6.435
100	2960	7.095
200	2917	7.227

SRIM/TRIM (the Transport of Ions in Matter) [70] code for pure chemical elements, and a set of compounds of known composition. The known elemental compositions for human tissue, dental implant materials, and some plastics commonly used in radiotherapy applications were calculated for proton energies from 50 MeV to 250 MeV [66] (see Figure 4–12). The model coefficients *a* and *b* are energy dependent as shown in Table 4–2.

A robust definition of Z_{eff} of a given material is based on proton interaction that can be based on the range shift in water due to the presence of the material layer. The proposed semi-empirical model allows calculations of Z_{eff} based on the measured range shift difference to that in water in the presence of the material of unknown composition as shown in Figure 4–7 and Equation 4.23.

4.6 Influence of Basic Interactions on Clinical Parameters

Important radiological parameters have been discussed in this chapter. However, their implication on patient treatment needs further explanation.

4.6.1 Physical Interactions of Charged Particles and Clinical Applications

Charged particles like those used in radiation therapy are affected by two types of basic interactions as discussed above and in other chapters: those with electrical and magnetic fields, and those with the traversed matter. The interaction with electric fields is the basis for the particle acceleration, while the interaction with magnetic fields bends the trajectory of particles at a given speed, as explained in Chapter 6.

Several atomic interactions occur between these accelerated charged particles while they penetrate in matter, such a patient or an element of the beam line and nozzle. We shall evaluate the influence on clinical parameters of three types of them (Figure 4–13):

1. The inelastic collisions with nuclei (nuclear interactions), losing the primary particle and producing neutrons and protons scattered with large angles.
2. The inelastic collisions with electrons, causing ionizations and excitations of the atomic targets, largely responsible for the absorbed dose in the media of interest clinically. The increase with depth of the average energy loss of the particle per unit path length (the stopping power), combined with the fluence loss due to nuclear interactions, determines the Bragg peak. When all the

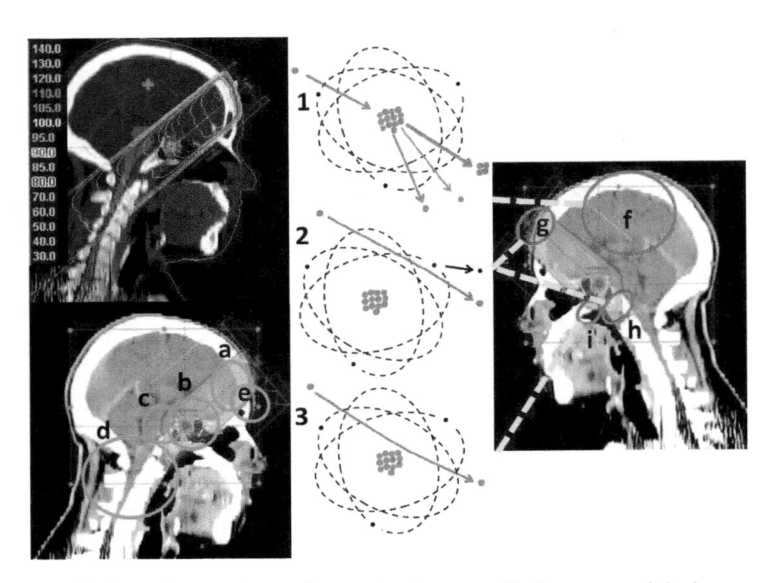

Figure 4–13 Various interactions of a proton beam with tissues and their macroscopic effects in the dose distribution into a patient.

particles are stopped, nearly no dose is deposit beyond. The primary protons scatter with a small angle.

3. The elastic collisions with nuclei, at the origin of multiple small angle deflections of the charged particle traversing the medium (Coulomb scattering), resulting in a lateral spreading of the proton beam.

A simplified representation of these three interactions is presented in Figure 4–13 with a direct link to their incidence in basic clinical parameters of the macroscopic dose distributions in matter. Let us evaluate a patient with an intracranial tumor and beam is passing through the head as shown in Figure 4–13.

When compared with a single photon beam in Figure 4–13 (upper left figure, maximal dose close to entrance and an exponential decrease over and after the target), an energy-modulated proton beam (lower left) has the advantages of (a) a flat entrance dose, (b) a homogeneous dose to the target, (c) a sharp decrease after the target, (d) no dose behind the target and, depending on the beam-shaping devices, (e) a small lateral penumbra at the entrance. The basic interactions at the origin of these effects also impose some constraints that must be dealt with:

1. The nuclear interactions deliver neutrons to the healthy tissues (f) and, for heavier ions, an important tail of fragments after the peak;
2. The ionizations by a modulated, multiple-energy proton beam increase the skin dose (g) and combined with the multiple scattering through complex inhomogeneities in tissue, imply a large uncertainty in the range (h);

3. The Coulomb scattering in the patient itself increases the lateral penumbra in depth (i), which will be dealt with later.

4.6.2 Effect of Nuclear Interactions and Ionizations

Figure 4–14 represents the fluence of a 200 MeV proton beam penetrating in water, showing an initial slope with losses of 1% per cm due to the nuclear interactions, and a final steep gradient related to energy and range straggling, as discussed in previous sections. When the energy spectrum of the particles is large, the final dose falloff is more shallow as the particles stop at different ranges. This can be due to the original beam acceleration and transport, but it can also be due to the media when protons traverse complex inhomogeneities and they mixed in space by the multiple Coulomb scattering.

Figure 4–15 presents a typical depth–dose curve in water with the Bragg peak, as the result of the product of the fluence and the stopping power as discussed earlier. At the entrance of the Bragg peak there is a small increase in dose caused by the accumulation of nuclear interaction products emitted predominantly in the forward direction, known as nuclear buildup. It follows an entrance plateau that is a compromise between the reducing number of protons by nuclear interactions and the increase of the stopping power with depth. The important increasing slope toward the peak is related to the high increase in stopping power when the protons are losing all their energy.

The final descending slope is related to the decrease of the number of particles until all of them arrive to the end of their range as a function of their energy. This distal part (beyond the Bragg peak) of the particle depth dose curve is known as "distal falloff." The depth of the 90% of the dose at the peak is known in clinical applications

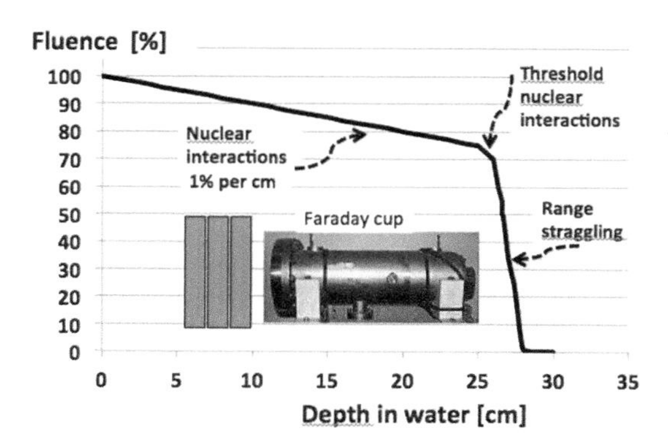

Figure 4–14 Fluence of a 200 MeV proton beam in water and associated physical phenomena affecting its shape. This fluence can be measured with a Faraday cup and slabs of WET interposed in front of it to simulate depth in water. (Data from Institut Curie, France.)

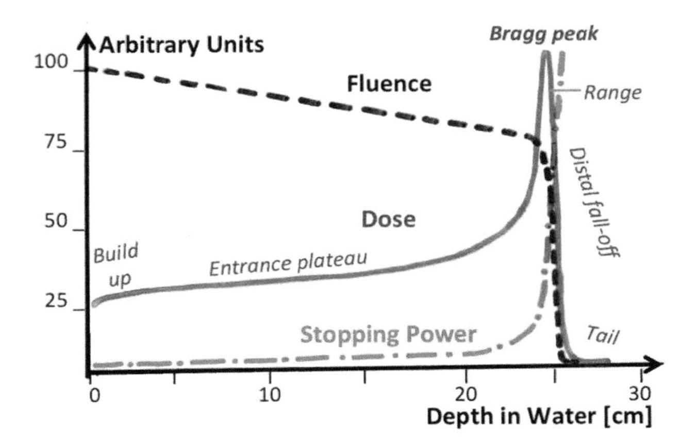

Figure 4–15 Fluence, stopping power, and depth–dose curve of a 200 MeV proton beam in water.

as the "distal range," corresponding approximately to the 50% of the fluence curve in depth (other values have been adopted, e.g., 95%, 85%,...). The distance from the depth where the dose is, for instance, 80% of the dose at the peak and the depth where the dose is, for instance, 20% of the dose at the peak, is known as the distal penumbra. For this example, the distal penumbra would be denoted by P_{80-20}. Penumbra could also be defined for different dose levels, for instance as P_{90-10} or P_{90-50}, etc. representing dose levels.

For incident ions with higher atomic number, the S increases to higher values and with a higher gradient than protons in depth, and had less range straggling and less multiple scatter, giving thinner Bragg peaks (or what is equivalent, a higher peak-to-entrance dose). But as those ions produce a larger number of fragments in nuclear interactions with the media, being lighter but with a high energy, they produce a *tail* behind the Bragg peak, limiting in some way their use in clinics. A good compromise between clinical effects at the Bragg peak and secondary effects and risks by the tail is given by the carbon ions.

4.6.2.1 Beam Penumbrae

The multiple Coulomb scattering of the charged particles in both the beam transport and in the patient increases the beam lateral penumbra. In "passive" beam lines, any element interposed to modify the range, to modulate the energy, and to monitor both the beam and the drift spaces in air increases the beam scattering and the penumbra at the entrance of the patient. For this reason, the apertures, designed to shape the section of the beam, are placed as close as possible to the patient to reduce this entrance penumbra. Additional discussion on the penumbra formation in beam line is proved by Urie et al. [71] where it is shown that the gap between the nozzle and the patient is one of the most critical factors for the enlargement of the penumbra.

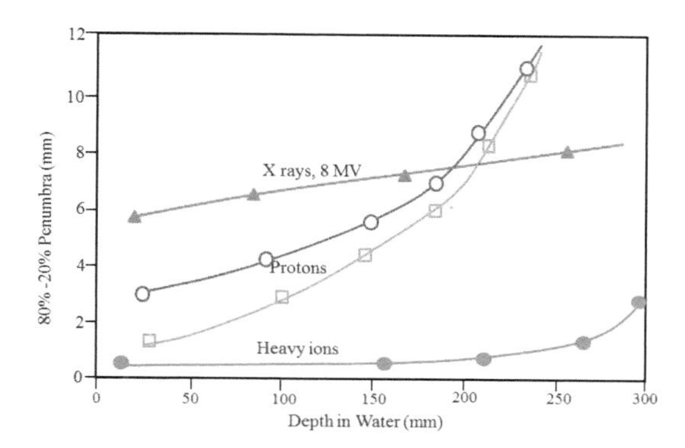

Figure 4–16 Typical lateral penumbra on beam profiles in water for photons, passive protons, and pencil beam heavy ions. The small entrance penumbra of the proton beam with white squares is related to a collimation close to the skin. It increases in depth from multiple Coulomb scattering in the media (the patient itself). At depths of around 20 cm in water, the penumbra for a 200 MeV proton beam becomes larger than the penumbra of an 8 MV photon beam. (Redrawn from A. Mazal, "Proton Beams in Radiotherapy," Chapter 46 in *Handbook of Radiotherapy Physics*, edited by P. Mayles, A. Nahum, and J-C Rosenwald. Taylor & Francis, 2007.)

For dynamically scanned pencil beams, there are in principle only monitoring devices interposed in the beam, having thin electrodes with a limited scattering. The superposition of Gaussian shapes of individual pencil beams affects the lateral penumbra. Several factors determine the steepness of this lateral penumbra (Figure 4–16) for scanned beams, including the practical considerations on the minimal spot size used to cover a target, the limited focusing properties of the beam transport system, the required air drift before arriving to the patient, and the weights assigned to every beam spot (in particular in the beam edges). These factors are evaluated in Chapter 9, Field Shaping: Scanning Beam.

For both passive and dynamic delivery systems, the multiple Coulomb scattering of charged particles into the patient itself increases the beam lateral penumbra following a law with a power of the inverse of the particle speed. Toward the end of the range, when the speed is slowing down, the particles deviate more and more from their ray trajectory. The well-known mushroom shape of the dose distribution for electron beams also exists for more heavy charged particles, with a reduced effect: the lateral penumbra of carbon beams is smaller than for protons, and this last is lower than for electrons.

The multiple scattering in the beam line and in the air before arriving to the patient are the important contributors to the lateral penumbra for rather superficial tumors treated with low energies. When treating deep tumors with higher energies, it is the multiple scatter in the patient which mainly determines the lateral penumbra in

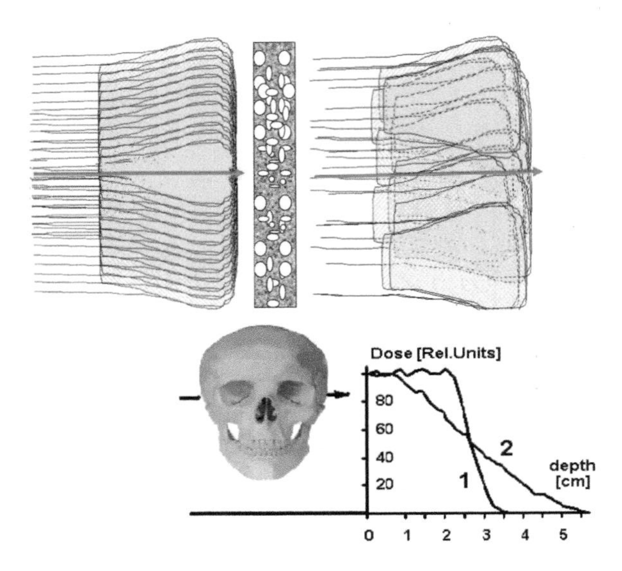

Figure 4–17 Effect of multiple scattering on dose distributions when crossing complex inhomogeneities. The sharp distal falloff is degraded (from 1 to 2) as particles cross different water equivalent paths and combine again, superposing different residual ranges. Modified from Urie et al. [71].

depth. For these reasons, even when using dynamic scanned pencil beams, it is suggested to use a patient aperture to optimize the lateral penumbra for superficial targets.

4.6.2.2 Multiple Scattering from Inhomogeneities

Several devices and scanning systems have been conceived to modify the initial pristine Bragg peak in order to cover a given target both laterally and in depth, as it will be described in chapters 6, 8, and 9. In every case, it is necessary to take into account that the basic interactions of particles with tissues will affect the theoretical dose distribution as a too simple model could calculate them.

A typical example is the incidence of multiple Coulomb scattering when the particles cross complex inhomogeneities. As the trajectories of the particles will cumulate different WET and then will mix, the final effect is equivalent to having a very large range straggling (Figure 4–17). In the same figure we include a typical case using clinical targets that has been presented for beams crossing through complex inhomogeneities at the base of the skull [54]. The distal falloff of a modulated SOBP beam is degraded compared to the original one that should have kept its original gradient if rather homogeneous tissues were crossed.

4.7 Summary/Conclusions

Particle interactions with media—especially in patients containing air, soft tissues, bone, and metals—are part of the treatment planning process that involves an understanding of *S*, range, scattering power, WET, effective Z, nuclear interactions, and other factors. We have provided here an addendum to Chapter 3 about basic interactions in the context of clinical applications. We have also tried to provide computation and analytical approaches to some of these parameters for elemental and compound materials. The parameter *S* is so important that it has to be understood in the context of treatment planning in terms of CT values, range, and RSP.

In clinical practice, these basic interactions will directly affect not only the range itself, but also the range straggling (and so the distal falloff), the homogeneity of the SOBP, as well as the lateral penumbra, the absolute dose, and the neutron dose. Knowing and reducing the uncertainties in these parameters is a critical aspect of proton beam therapy as they affect margins (e.g., on PTVs or on beam parameters) and the choice of treatment strategies and optimization based on beam angles, weights, and delivery system. Present efforts are oriented toward the management of these uncertainties using basic principles and their relationship with macroscopic effects: range verification by PET, prompt gammas, proton radiography, and others (see chapters 7 and 28). The principle of nuclear interactions used for some of these range verification systems are also at the origin of the neutron dose delivered off target, with an associated second cancer risk. The reduction of this effect is done by minimizing the material interposed in the beam and using beam scanning as the delivery system.

References

1. Mok G, Benz E, Vallee JP, Miralbell R, Zilli T. Optimization of radiation therapy techniques for prostate cancer with prostate-rectum spacers: a systematic review. *Int J Radiat Oncol Biol Phys* 2014;90:278–288.
2. Komatsu S, Hori Y, Fukumoto T, Murakami M, Hishikawa Y, Ku Y. Surgical spacer placement and proton radiotherapy for unresectable hepatocellular carcinoma. *World J Gastroenterol* 2010;16:1800–1803.
3. Akasaka H, Sasaki R, Miyawaki D, Mukumoto N, Sulaiman NS, Nagata M, Yamada S, Murakami M, Demizu Y, Fukumoto T. Preclinical evaluation of bioabsorbable polyglycolic Acid spacer for particle therapy. *Int J Radiat Oncol Biol Phys* 2014;90:1177–1185.
4. Bragg WH, Kleeman R. On the ionization curves of radium. *Phil Mag J Sci* 1904;8:726–738.
5. Bragg WH, Kleeman R. On the a particles of radium and their loss of range in passing through varios atoms and molecules. *Phil Mag J Sci* 1905;10:318–340.
6. Zhang R, Taddei PJ, Fitzek MM, Newhauser WD. Water equivalent thickness values of materials used in beams of protons, helium, carbon and iron ions. *Phys Med Biol* 2010;55:2481–2493.
7. Nichiporov D, Moskvin V, Fanelli L, Das IJ. Range shift and dose perturbation with high-density materials in pproton beam therapy. *Nucl Instr Meth Phys Res B* 2011;269:2685–2692.
8. Jäkel O. Ranges of ions in metals for use in particle treatment planning. *Phys Med Biol* 2006;51:N173–177.
9. Zhang R, Newhauser WD. Calculation of water equivalent thickness of materials of arbitrary density, elemental composition and thickness in proton beam irradiation. *Phys Med Biol* 2009;54:1383–1395.
10. Kempe J, Brahme A. Energy-range relation and mean energy variation in therapeutic particle beams. *Med Phys* 2008;35:159–170.

11. Bethe HA. Zur Theorie des Durchgangs schneller Korpuskularstrahlen durch Materie. Ann Phys (Leipzig) 1930;5:324–400.
12. Bloch F. Zur Bremsung rasch bewegter Teilchen beim Durchgang durch Materie. *Ann Phys* (Leipzig) 1933;16:285–320.
13. ICRU Report 49. Stopping Powers and Ranges for Protons and Alpha Particles. Bethesda, MD: International Commission on Radiation Units and Measurements, 1993.
14. ICRU Report 46. Photon, Electron, Proton and Neutron Interaction Data for Body Tissues. Bethesda, MD: International Commission on Radiation Units and Measurements, 1992.
15. ICRU Report 78. Prescribing, Recording, and reporting Proton Beam Therapy. Bethesda, MD: International Commission on Radiation Units and Measurements, 2007.
16. IAEA TRS 398. Absorbed Dose Determination in External Beam Radiotherapy: An International Code of Practice for Dosimetry Based on Standards of Absorbed Dose to Water. Technical Reports Series No. 398. Vienna, Austria: International Atomic Energy Agency, 2000.
17. Thomas SJ. Relative electron density calibration of CT scanners for radiotherapy treatment planning. *Br J Radiol* 1999;72:781–786.
18. Moyers MF. Comparison of x ray computed tomography number to proton relative linear stopping power conversion functions using a standard phantom. *Med Phys* 2014;41:061705.
19. Cheng CW, Zhao L, Wolanski M, Zhao Q, James J, Dikeman K, Mills M, Li M, Srivastava SP, Lu XQ, et al. Comparison of tissue characterization curves for different CT scanners: implication in proton therapy treatment planning. *Transl Cancer Res* 2012;1:236–246.
20. Schaffner B, Pedroni E. The precision of proton range calculations in proton radiotherapy treatment planning: experimental verification of the relation between CT-HU and proton stopping power. *Phys Med Biol* 1998;43:1579–1592.
21. Schneider U, Pedroni E, Lomax A. The calibration of CT Hounsfield units for radiotherapy treatment planning. *Phys Med Biol* 1996;41:111–124.
22. Hunemohr N, Paganetti H, Greilich S, Jakel O, Seco J. Tissue decomposition from dual energy CT data for MC based dose calculation in particle therapy. *Med Phys* 2014;41:061714.
23. Schulte RW, Bashkirov V, Klock MC, Li T, Wroe AJ, Evseev I, Williams DC, Satogata T. Density resolution of proton computed tomography. *Med Phys* 2005;32:1035–1046.
24. Coolens C, Childs PJ. Calibration of CT Hounsfield units for radiotherapy treatment planning of patients with metallic hip prostheses: the use of extended CT-scale. *Phys Med Biol* 2003;48:1591–1603.
25. Newhauser WD, Giebeler A, Langen KM, Mirkovic D, Mohan R. Can megavoltage computed tomography reduce proton range uncertainties in treatment plans for patients with large metal implants? *Phys Med Biol* 2008;53:2327–2344.
26. Verburg JM, Seco J. Dosimetric accuracy of proton therapy for chordoma patients with titanium implants. *Med Phys* 2013;40:071727.
27. Andersson KM, Ahnesjo A, Vallhagen Dahlgren C. Evaluation of a metal artifact reduction algorithm in CT studies used for proton radiotherapy treatment planning. *J Appl Clin Med Phys* 2014;15:112–119.
28. Jäkel O, Jacob C, Schardt D, Karger CP, Hartmann GH. Relation between carbon ion ranges and x-ray CT numbers. *Med. Phys* 2001;28:701–703.
29. Attix FH. Introduction to Radiological Physics and Radiation Dosimetry. New York: John Wiley & Sons, 1986.
30. Bichsel H. Passage of charged particles through matter, In American Institute of Physics Handbook. New York: McGraw-Hill, 1972.
31. Wilson RR. Radiological use of fast protons. *Radiology* 1946;47:487–491.
32. Brooks FD, Jones DTL, Bowley CC, Symons JE, Buffler A, Allie MS. Energy spectra in the NA proton therapy beam. *Radiat Prot Dosim* 1997;70:477–480.
33. La Rosa V, Kacperek A, Royle G, Gibson A. Range verification for eye proton therapy based on proton-induced x-ray emissions from implanted metal markers. *Phys Med Biol* 2014;59:2623–2638.
34. Yang M, Zhu XR, Park PC, Titt U, Mohan R, Virshup G, Clayton JE, Dong L. Comprehensive analysis of proton range uncertainties related to patient stopping-power-ratio estimation using the stoichiometric calibration. *Phys Med Biol* 2012;57:4095–4115.
35. Knopf AC, Lomax A. In vivo proton range verification: a review. *Phys Med Biol* 2013;58:R131–160.

36. Paganetti H. Range uncertainties in proton therapy and the role of Monte Carlo simulations. *Phys Med Biol* 2012;57:R99–117.
37. Al-Sulaiti L, Shipley D, Russell T, Kacperek A, Regan P, Palmans H. Water equivalence of various materials for clinical proton dosimetry by experiment and Monte Carlo simulation. *Nucl Instr Meth Phys Res A* 2010;619:344–347.
38. Moskvin V, Cheng CW, Fanelli L, Zhao L, Das IJ. A semi-empirical model for the therapeutic range shift estimation caused by inhomogeneities in proton beam therapy. *J Appl Clin Med Phys* 2012;13:3–12.
39. Wroe AJ, Cornelius IM, Rosenfeld AB. The role of nonelastic reactions in absorbed dose distributions from therapeutic proton beams in different medium. *Med Phys* 2005;32:37-41.
40. Hurley RF, Schulte RW, Bashkirov VA, Wroe AJ, Ghebremedhin A, Sadrozinski HFW, Rykalin V, Coutrakon G, Koss P, Patyal B. Water-equivalent path length calibration of a prototype proton CT scanner. *Med Phys* 2012;39:2438–2446.
41. Schneider U, Schaffner B, Lomax T, Pedroni E, Tourovsky A. A technique for calculating range spectra of charged particle beams distal to thick inhomogeneities. *Med Phys* 1998;25:457–463.
42. Molière G. Theorie der Streuung schneller geladener Teilchen II Mehrfach—und Vielfachstreuung-streuung. *Z. Naturforsch* 1948; B 3A:78–97.
43. Gottschalk B, Koehler AM, Schneider RJ, Sisterson JM, Wagner MS. Multiple Coulomb scattering of 160 MeV protons. *Nucl Instr Meth Phys Res B* 1993;74:467–490.
44. Gottschalk B. On the scattering power of radiotherapy protons. *Med Phys* 2010;37:352–367.
45. Titt U, Zheng Y, Vassiliev ON, Newhauser WD. Monte Carlo investigation of collimator scatter of proton-therapy beams produced using the passive scattering method. *Phys Med Biol* 2008;53:487–504.
46. Gomà C, Andreo P, Sempau J. Spencer–Attix water/medium stopping-power ratios for the dosimetry of proton pencil beams. *Phys Med Biol* 2013;58:2509–2522.
47. Zhang H, Das IJ. Dosimetric perturbations at high-Z interfaces with high dose rate Ir source. *Phys Med* 2014;30(7):782–90.
48. Das IJ, Khan FM. Backscatter dose perturbation at high atomic number interfaces in megavoltage photon beams. *Med Phys* 1989;16:367–375.
49. Das IJ. Forward dose perturbation at high atomic number interfaces in kilovoltage X-ray beams. *Med Phys* 1997;24:1781–1787.
50. Das IJ, Moskvin VP, Kassaee A, Tabata T, Verhaegen F. Dose perturbations at high-Z interfaces in kilovoltage photon beams: Comparison with Monte Carlo simulations and measurements. *Radiat Phys Chem* 2002;64:173–179.
51. Das IJ, Cheng CW, Mitra RK, Kassaee A, Tochner Z, Solin LJ. Transmission and dose perturbations with high-Z materials in clinical electron beams. *Med Phys* 2004;31:3213–3221.
52. Goitein M, Chen GT, Ting JY, Schneider RJ, Sisterson JM. Measurements and calculations of the influence of thin inhomogeneities on charged particle beams. *Med Phys* 1978;5:265–273.
53. Hong L, Goitein M, Bucciolini M, Comiskey R, Gottschalk B, Rosenthal S, Serago C, Urie M. A pencil beam algorithm for proton dose calculations. *Phys Med Biol* 1996;41:1305–1330.
54. Urie M, Goitein M, Holley WR, Chen GT. Degradation of the Bragg peak due to inhomogeneities. *Phys Med Biol* 1986;31:1–15.
55. Urie M, Goitein M, Wagner M. Compensating for heterogeneities in proton radiation therapy. *Phys Med Biol* 1984;29:553–566.
56. Verhaegen F, Palmans H. Secondary electron fluence perturbation by high-Z interfaces in clinical proton beams: a Monte Carlo study. *Phys Med Biol* 1999;44:167–183.
57. Johns HE, Cunningham JR. The Physics of Radiology. 4th ed. Springfield, IL: Charles C. Thomas, 1983.
58. Manjunathaguru V, Umesh TK. Effective atomic numbers and electron densities of some biologically important compounds containing H, C, N and O in the energy range 145–1330 keV. *J Phys B: Atomic, Molecular and Optical Physics* 2006;39:3969–3981.
59. Yang NC, Leichner PK, Hawkins WG. Effective atomic numbers for low-energy total photon interactions in human tissues. *Med Phys* 1987;14:759–766.

60. Prasad SG, Parthasaradhi K, Bloomer WD. Effective atomic numbers of composite materials for total and partial interaction processes for photons, electrons, and protons. *Med Phys* 1997;24:883–885.
61. Elmahroug Y, Tellili B, Souga C. Determination of total mass attenuation coefficients, effective atomic numbers and electron densities for different shielding materials. *Ann Nuc Energy* 2015;75:268–274.
62. Taylor ML, Smith RL, Dossing F, Franich RD. Robust calculation of effective atomic numbers: The Auto-Z(eff) software. *Med Phys* 2012;39:1769–1778.
63. Manohara SR, Hanagodimath SM, Thind KS, Gerward L. On the effective atomic number and electron density: A comprehensive set of formulas for all types of materials and energies above 1 keV. *Nuc Instr Meth Phys Res B* 2008;266:3906–3912.
64. Parthasaradhi K, Rao BM, Prasad SG. Effective atomic numbers of biological-materials in the energy region 1 to 50 Mev for photons, electrons, and He ions. *Med Phys* 1989;16:653–654.
65. Kurudirek M. Effective atomic numbers of different types of materials for proton interaction in the energy region 10keV–100GeV. *Nuc Instr Meth Phys Res B* 2014;336:130–134.
66. ICRP-89. Annals of the ICRP, Basic anatomical and physiological data for use in radiological protection: reference values. New York: Pergamon Press, 2002.
67. Moskvin V, Suga M, Cheng C, Das I. An effective atomic number of the compounds in proton beam therapy. *Med Phys* 2013;40:373.
68. Medin J, Andreo P. Monte Carlo calculated stopping-power ratios, water/air, for clinical proton dosimetry (50–250 MeV). *Phys Med Biol* 1997;42:89–105.
69. Berger MJ, Coursey JS, Zucker MA, Chang J. ESTAR, PSTAR, and ASTAR: Computer programs for calculating stopping-power and range tables for electrons, protons, and helium ions (version 1.2.3). [Online] Available: http://physics.nist.gov/Star (2010) National Institute of Standards and Technology, Gaithersburg, MD. 2005.
70. Ziegler JF, Biersack JP, Ziegler MD. SRIM, the stopping and range of ions in matter: www.SRIM.org, SRIM Company, 2008.
71. Urie MM, Sisterson JM, Koehler AM, Goitein M, Zoesman J. Proton beam penumbra: effects of separation between patient and beam modifying devices. *Med Phys* 1986;13:734–741.

Chapter 5

Proton Relative Biological Effectiveness

Harald Paganetti, Ph.D.

Professor and Director of Physics Research, Department of Radiation Oncology,
Massachusetts General Hospital and Harvard Medical School,
Boston, MA

5.1 Introduction

5.1.1 Radiation Reaction on Cells

Energy deposition events associated with radiation interactions with tissues can cause damage to cellular structures such as the DNA, which is the main mechanism of interest for radiation therapy. Protons lose their energy mainly in electromagnetic interactions, causing δ-rays (low-energy electrons). The number of energy deposition events differs between proton and photon radiation even for the same macroscopic dose to the cellular volume. The energy deposited per incident proton is much higher than the energy deposited per incident photon. In other words, for the same dose, the number of protons crossing a region of interest is typically lower than the corresponding number of photons. This, in turn, causes energy depositions in a cellular structure to be more heterogeneous in proton radiation fields compared to photon radiation. A parti-

Table 5–1 Average yield of damage in a single mammalian cell after 1 Gy delivered by photons (low LET) or low-energy a particles (high LET). Nucleus diameter = 8 mm; energy deposition per ionization = 25 eV. Based on [13,14].

Radiation	Photons	α Particles
Tracks in nucleus	1000	2
Ionizations in nucleus	10^5	10^5
Ionizations in DNA	1500	1500
DNA single strand breaks	700–1000	300–600
DNA double strand breaks (initially)	18–60	70
DNA double strand breaks (after 8 h)	6	30
Lethal lesions	0.2–0.8	1.3–3.9
Cells inactivated	10–50%	70–95%

cle track with its secondary electrons comprises the track structure. Thus, the track structure differs between proton and other ionizing radiation.

For a typical proton radiation field, ~70% of the energy lost is transferred to secondary electrons (δ-rays), ~25% is needed to overcome their binding energy, and the residual 5% produces neutral excited species [1]. Each photon or proton track is associated with δ-rays with a wide distribution of electron energies and a maximum energy of roughly 500 keV in proton beams. An electron with an energy of 100 keV will cause ~500 energy deposition events with energies larger than 10 eV in a 6 nm^2 target, but less than 10 with energies larger than 150 eV [2]. The energy of these electrons from protons, and thus their range, is typically much lower than the electrons produced in photon beams.

The key to understanding radiation effects lies in the spatial distribution of energy deposition events and the complex lesions this may cause. One of the consequences of track structure differences is that the distribution of double-strand breaks (DSB) caused by radiations can differ significantly. Damage from protons can be more complex than damage from photons [3]. The clustering of radiation damage is largely responsible for the effectiveness of high-LET (linear energy transfer) radiation [4,5]. The linear energy transfer is the average amount of energy a radiation imparts to the local medium per unit length. This affects not only the type of damage, but also the capacity of the cell to repair it [6–12].

Table 5–1 shows the approximate number of events in a mammalian cell in different radiation fields after a dose of 1 Gy [13,14]. The number of ionizations per cell and the initial yield of DSBs shows little variation with ionization density, but the number of residual breaks at eight hours after cellular repair differs substantially.

5.1.2 Definition of RBE

The optimization of radiation therapy treatment plans is typically solely based on physical dose. Since there is no simple relationship between dose and clinical endpoint, prescription doses and dose constraints are based on empirical data [15]. Biological dose optimization in radiation therapy includes three main aspects: a) predicting tumor control probability (TCP) or normal tissue complication probability (NTCP), b) considering fractionation by applying a biologically equivalent dose via a linear quadratic dose–response model, and c) considering biological differences among radiation modalities. The first aspect is applied when analyzing clinical data, but it is typically not explicitly taken into consideration when doing treatment plan optimization [16]. The second aspect is considered when applying hyper- or hypofractionation [17,18]. The third aspect is covered by the relative biological effectiveness (RBE).

Treatment planning is typically based on achieving a desired prescription dose to the target while meeting specific dose constraints to critical structures. Dose is a physical parameter, so differences in biological effect between modalities such as photons and protons need to be corrected for. Furthermore, this practice allows benefiting from the experience gained in photon treatments. Consequently, proton therapy prescriptions are based on physical dose times a factor to account for the difference in biological effect at the same dose when treating with photons, i.e., the RBE.

The proton RBE is defined as the ratio of the photon and the proton dose required to reach the same biological effect. For instance, in cell experiments this effect could be the survival level of cells. The RBE can be defined using either the dose of the reference radiation or the dose of the proton radiation. The International Commission on Radiation Units and Measurements (ICRU) has suggested that doses in proton therapy should be prescribed as Gy(RBE) [19]. Older documents still use the term CGE (Cobalt Gray Equivalent). For an RBE of 1.1 and a desired photon equivalent dose of 2 Gy, the prescription dose would be 2 Gy(RBE), corresponding to a proton dose of ~1.8 Gy.

The linear quadratic formalism (Equation 5.1) is widely used to describe a dose-response curve for cell survival endpoints [20,21]. The parameters are the number of initial cells, N_0, the number of affected cells (e.g., surviving cells), N, the dose, D, as well as the two fit parameters N and D.

$$N/_{N_0}(D) = \exp(-\alpha D - \beta D^2) \quad (5.1)$$

This simple expression neglects the dose rate and time-dependent repair [22]. Further, the linear quadratic formalism might not be valid below ~1Gy (e.g., due to hypersensitivity or adaptive response) [23,24], which could affect dose-response estimations for organs at risk typically receiving less that 1 Gy for a treatment using 2 Gy per fraction to the target. It is also problematic for doses above ~10 Gy due to transition of the dose-response to an exponential behavior [25,26], which could have implications when considering hypofractionation. These dose limits are depending on the endpoint and are largely controversial.

Assuming that the linear quadratic equation is a good approximation, one can define the RBE as a function of the α_x and β_x values for the reference photon radiation, the α_x and β_x of the proton radiation, and the proton dose per fraction, D_p, as shown in Equation 5.2.

$$RBE\left(LET_d, D_p, \left[\alpha/\beta\right]_x\right) = \frac{1}{D_p}\left(\sqrt{\frac{1}{4}\left[\alpha/\beta\right]_x^2 + \left[\alpha/\beta\right]_x \frac{\alpha(LET_d)}{\alpha_x} D_p + \frac{\beta(LET_d)}{\beta_x} D_p^2} - \frac{1}{2}\left[\alpha/\beta\right]_x\right) \tag{5.2}$$

Note that the α and β values for each modality depend on the dose-averaged linear energy transfer (LET_d). This is made explicit for protons in Equation 5.2. For α_x and β_x, this causes the RBE definition to be dependent on the specific photon reference radiation, e.g., ^{60}Co or 6 MV photons. Using the asymptotic values of RBE at doses of 0 and ∞ Gy, one can also define the relationship using RBE_{max} and RBE_{min}, respectively [27].

$$\begin{aligned} RBE_{max}\left(LET_d, \alpha, \alpha_x\right) &= \alpha(LET_d) / \alpha_x \\ RBE_{min}\left(LET_d, \beta, \beta_x\right) &= \sqrt{\beta(LET_d) / \beta_x} \end{aligned} \tag{5.3}$$

$$RBE\left(LET_d, D_p, \left[\alpha/\beta\right]_x\right) = \frac{1}{2D_p}\left(\sqrt{\left[\alpha/\beta\right]_x^2 + 4\left[\alpha/\beta\right]_x RBE_{max} D_p + 4RBE_{min}^2 D_p^2} - \left[\alpha/\beta\right]_x\right) \tag{5.4}$$

Because the RBE depends on the photon reference radiation, the reference has to be stated when reporting RBE values. Clinically, the RBE relative to 6 MV photons is of particular interest. However, cell experiments are often based on lower-energy photons. For example, some laboratory experiments use 250 kVp x-rays as a reference which, depending on the endpoint, does show an RBE of ~1.15 relative to 6 MV photons [28].

5.1.3 Low-dose RBE

The definition of RBE as the ratio of two doses is valid for all doses, but specific considerations need to be taken for low radiation doses. Equations 5.2 and 5.4 predict an increase in RBE as dose decreases. The low-dose RBE refers to RBE_{max} (Equation 5.3) and is relevant particularly for radiation protection considerations. In this domain, however, the goal is to be conservative. Consequently, a conservative estimation defined by regulatory bodies is typically applied, not RBE_{max}. These are based on either the radiation weighting factor or the radiation quality factor. Low-dose RBE values are particularly of interest because of the low dose background caused by neutrons in proton therapy (see Chapter 10 on "Shielding").

The "dose equivalent" has been defined for radiological protection purposes according to Equation 5.5 [29].

$$H = DQ \tag{5.5}$$

Here, Q is the quality factor, defined as a function of the unrestricted linear energy transfer of charged particles in water (LET_{∞}) [30]. Its maximum value is ~30 as defined by the International Commission for Radiological Protection (ICRP) [31]:

$$Q(LET_{\infty}) = \begin{cases} 1 & LET_{\infty} < 10\,keV/\mu m \\ 0.32 \times LET_{\infty} - 2.2 & 10 \leq LET_{\infty} \leq 100\,keV/\mu m \\ 300/\sqrt{LET_{\infty}} & LET_{\infty} < 10\,keV/\mu m \end{cases} \tag{5.6}$$

For a given volume, the quality factor is obtained by integrating over the dose-weighted contributions of charged particles:

$$Q = \frac{1}{D} \int_{LET_{\infty}=0}^{\infty} Q(LET_{\infty}) \frac{dD}{dLET_{\infty}} \times dLET_{\infty} \tag{5.7}$$

A different multiplier to the average dose to an organ is the "radiation weighting factor," w_R, [32]. This leads to organ "equivalent dose" instead of "dose equivalent." From the available data, a set of continuous functions was suggested by the ICRP. For example, an energy-dependent bell-shaped curve was recommended for neutron radiation, while the proton w_R is defined as a constant of 2 [33,34]. For the calculation of radiation weighting factors for neutrons, the latest recommendation shows a maximum weighting factor of ~20 at energies around 1–2 MeV [35].

The use of the equivalent dose to assess the biological effectiveness in proton therapy patients is problematic because the definition is valid for an external radiation beam, and not necessarily for neutrons or protons generated in the patient. For instance, the proton w_R of 2 is considering an external radiation source and not protons generated in the patient as originating as secondary particles from proton or neutron nuclear interactions, i.e., the proton w_R does not thus apply to protons caused internally by neutrons. Note that neutrons deposit their energy primarily via secondary protons [36,37]. This makes the scoring of equivalent doses in Monte Carlo simulations difficult because the weighting factor would depend on the particle history. This problem can be overcome by using the quality factor depending on the LET in the region of interest [36]. In contrast to neutrons generated in the patient, those neutrons that are generated in the treatment head (and thus act as an external radiation source impinging on the patient) can be weighted with w_R values for conservative radiation protection purposes.

5.2 Rationale for the Use of 1.1

Proton treatments are currently planned using a generic constant RBE of 1.1. This value was deduced as an average value of measured RBE values *in vivo* in the center of

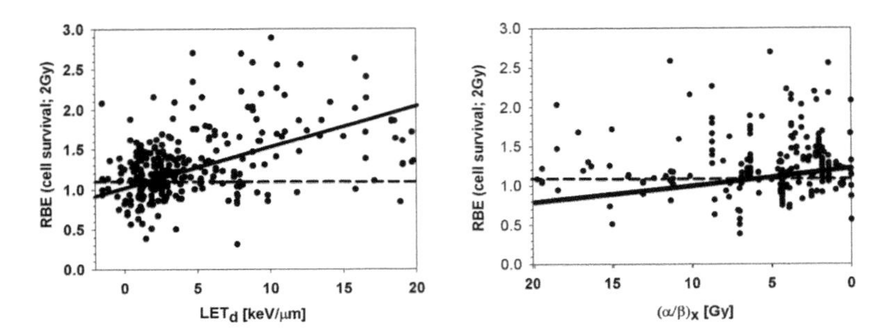

Figure 5–1 RBE as a function of LET (left) and as a function of $(a/b)_x$ (for LET_d values below 20 keV/μm) at a dose of 2 Gy (from [42]). The solid line is a fit through the data demonstrating an increasing RBE as LET_d is increasing and $(a/b)_x$ is decreasing. The dashed lines indicate the clinically used value of 1.1.

a spread-out Bragg peak radiation field done in the early days of proton therapy [38–41]. It has been shown that when averaging RBE values over all endpoints and experiments reported, the value is about 1.15 in the center of a typical SOBP [42]. Because one aims at a conservative RBE definition for tumor control, this is in line with the clinical use of 1.1, if an average RBE is to be applied. As shown in Figure 5–1, measured proton RBE values show significant variations [42]. A trend is visible as a function of LET_d as well as a function of α/β. However, the spread is too large to pinpoint these relationships for proton treatment planning.

There are several practical advantages when using a generic RBE. For example, clinical dosimetry can be based on homogeneous dose distributions in the target. On the other hand, a generic value disregards the dependencies of the RBE on physical and biological properties (e.g., proton beam energy, depth of penetration, biological end point, dose per fraction, position in the SOBP, initial beam, particular tissue, etc.). It is known that the RBE varies, but there are no conclusive clinical data indicating that these variations matter, i.e., we have not (yet) clearly identified toxicities or recurrences that were a result of RBE effects [43–45]. On the other hand, it is difficult to assess RBE effects in critical structures from clinical data because the dose distributions differ between photon and proton irradiations.

5.3 Variations of RBE

5.3.1 RBE as a Function of Dose

For most endpoints, the linear quadratic dose response curve shows a shoulder on the logarithmic scale, i.e., β is not zero. Due to the typically more pronounced shoulder in the x-ray survival curve compared to the proton survival curve, the RBE increases with decreasing dose. This is shown in most, but not all, cell survival data [42]. However, the dose dependency of the RBE is difficult to assess in the clinically relevant region because most laboratory experiments do not include detailed analysis of the dose below 2 Gy. Yet, typical prescription doses are at ~2 Gy per fraction and thus relevant doses for organs at risk are below this value. Potential limitations of the linear quadratic model below ~1–2 Gy (see above) further complicate the interpretation of experimental data. Biophysical models suggest an increase in RBE as dose decreases, and this becomes more pronounced as LET increases and $(\alpha/\beta)_x$ decreases [46]. It thus is likely that hypofractionated regimens will most likely result in lower RBE values.

5.3.2 RBE as a Function of Tissue and Biological Endpoint

Based on the validity of the linear quadratic model, one expects a dependency of the RBE on $(\alpha/\beta)_x$, i.e., a higher RBE for low $(\alpha/\beta)_x$ [47]. Figure 5–1 shows that there is indeed a weak tendency—albeit with large uncertainties due to the spread of the data and the inter-correlation with LET_d—in the measured data. This trend implies a higher RBE for late damage as compared to early-responding tissues [48]. This tendency toward an increased RBE in cells exhibiting smaller $(\alpha/\beta)_x$ ratios is apparent at lower dose levels. This might cause an underestimation of the RBE when treating, for example, prostate carcinoma [49], but it might also increase the risk for side effects for other sites. For example, the spinal cord, another tissue with a low $(\alpha/\beta)_x$, sometimes has to be partially irradiated in order to achieve sufficient tumor coverage.

Several studies have indicated that the increase of RBE with decreasing $(\alpha/\beta)_x$ is significant only at low $(\alpha/\beta)_x$ values (<~5 Gy) [50,51]. The spread in experimental data seems to be too large to confirm this based on experiments *in vitro*. One would also expect the slope to be bigger for high LET_d values, which was not shown in a compilation of experimental data due to uncertainties in the experiments [42].

5.3.3 RBE as a Function of LET

The dependency of the RBE on dose and $(\alpha/\beta)_x$ are associated with considerable uncertainties because these parameters are based on the biological characteristics of tissues. Uncertainties are expected to be smaller when considering physics, i.e., the dependency of RBE on the LET, a parameter describing the energy deposition in tissue by protons.

The LET is a macroscopic dosimetric parameter because it does not describe energy deposition in a biological target (e.g., a cell nucleus) but rather the energy deposition per path length of a particle. However, although the exact track structure and the energy deposition in defined cellular targets is neglected, the dose-averaged LET, LET_d, is a reasonable approximation to characteristics of a proton radiation field. Fortunately, other than for heavy ions, there are many proton tracks crossing a cell for typical therapeutic doses [52], so neglecting the fine structure to characterize the radiation field is reasonable.

In the LET_d region relevant for proton therapy, the RBE is expected to increase as the LET_d increases. In fact, for a given endpoint, i.e., $(\alpha/\beta)_x$, the relationship is nearly linear. This is illustrated in Figure 5–2 for an average RBE based on many independently experiments using various cell liens *in vitro* [42]. Note that the RBE also

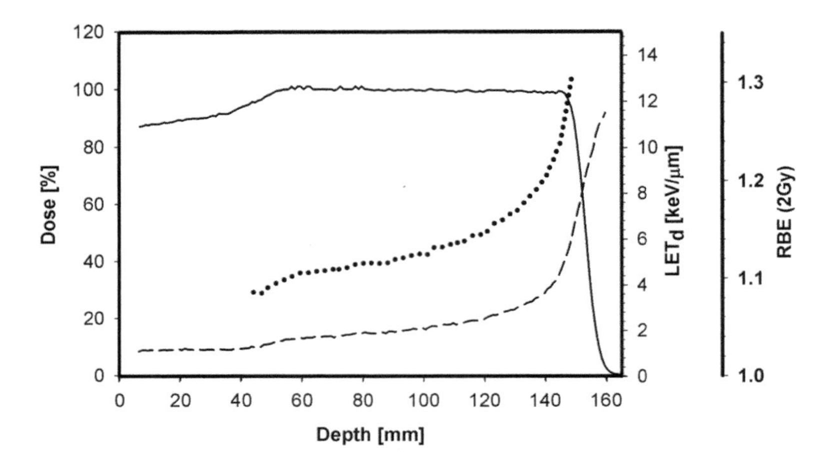

Figure 5–2 Spread-out Bragg peak with a range of 15 cm and a modulation width of 10 cm (both relative to the 90% dose level); solid line. Also shown is the dose-averaged LET_d as a function of depth (dashed line, right axis) and the RBE as a function of depth (dotted line, far right axis). (Figure from [42].)

depends on the tissue, e.g., the $(\alpha/\beta)_x$, and from model predictions one would expect a steeper slope as $(\alpha/\beta)_x$ decreases. To put the shown LET_d values into perspective, the LET_d values in the entrance region of an SOBP up to the center of the SOBP are typically between 0 and 3 keV/μm (relative to the photon LET_d), between 3 and 6 keV/μm in the downstream half of an SOBP, between 6 and 9 keV/μm in the distal edge region, and between 9 and 15 keV/μm in the dose fall-off region. The LET_d in the central region of an SOBP is typically between ~2.0 and ~3.0 keV/μm. Based on these typical LET_d values in an SOBP, it was estimated that the average RBE for 2 Gy does increase with depth from ~1.1 in the entrance region, to ~1.15 in the center, ~1.35 at the distal edge, and ~1.7 in the distal fall-off region [42].

Variation of LET distributions in a patient can be significant [53–55]. One can expect that the average LET_d across the irradiated volume is lower for a bigger prescribed range because the average proton energy is higher. Further, because the LET_d increases with depth—being more pronounced toward the end of range—its value in the center of an SOBP increases with decreasing modulation width. Most importantly, intensity-modulated proton therapy (IMPT) does not necessarily result in an SOBP dose distribution per field, which could influence the LET_d distribution significantly [55]. One can expect LET_d values to be overall more heterogeneous in IMPT, which may lead to RBE hot spots.

The increase in RBE as a function of depth in an SOBP also impacts the position of the distal dose fall-off. An increasing RBE with depth due to LET increase can lead to a shift in the dose fall-off (physical dose versus biological dose) of a few mm

[47,56,57]. This has implications when considering range uncertainties (see Chapter 21 on "Uncertainties in Planned Dose Distributions—Their Impact and Their Management").

5.4 Clinical Relevance of Experimental RBE Values

5.4.1 RBE for Tumor Control Probability

The RBE values discussed in Section 5.3 are based on laboratory experiments, mostly using the end point of cell survival *in vitro*. It would be more meaningful to measure TCP directly and compare photon and proton treatments to deduce an RBE. However, tumor variability, different margins, and patient populations hamper this approach. Animal models have been used to measure TCD_{50}, i.e., the dose for 50% local control of the tumor, in xenografts. For example, a study of tumor growth delay of human hypopharyngeal squamous cell carcinoma cells in mice resulted in an RBE between 1.1 and 1.2 at ~20 Gy relative to 6 MV photons using a 23 MeV proton beam [58].

Given the limited number of *in vivo* data and the lack of mechanistic knowledge about tumor response to radiation in patients, we do largely rely on RBE values deduced for clonogenic cell survival as summarized in Section 5.3. While cell survival might be a valid surrogate for tumor control, there are various pathways leading to cell death, some of which are not caused by radiation-induced damage itself, but by a combination of damage and apoptosis or failure to complete mitosis. A difference between *in vitro* and *in vivo* response—as well as between typically two-dimensional (cell culture) systems and three-dimensional tissue environments—is to be considered as well.

5.4.2 RBE for Normal Tissue Complication

Cell survival might be a relevant endpoint to predict the RBE in tumors, but other endpoints could be potentially more relevant for predicting RBE values for NTCP. Organ-specific effects of interest are early effects like erythema and late effects like lung fibrosis, lung function, or spinal cord injury. These are not directly correlated with cell survival. For instance, no relation has been found between fibroblast radiosensitivity and the development of late normal tissue effects such as fibrosis [59,60]. Surviving cells with unrepaired or mis-repaired damage can transmit changes to descendant cells, and malignant transformation can thus be initiated by gene mutation.

Various RBE values for endpoints other than cell survival have been measured, e.g., induction of reactive oxygen species (leading to oxidative stress and regulate a variety of response pathways), DNA strand breaks in the form of single (SSB) or double strand breaks (DSB), foci formation, repair proteins, gene expression, chromosome aberrations, mutations, micronuclei formation, apoptosis and cell cycle effects, and others. Further, many endpoints have been studied in animal models, such as skin reactions or organ weight loss. The variety of endpoints does not allow a comprehensive analysis toward a clinical RBE for NTCP considerations. The most promising endpoints related to NTCP might be the induction of inflammatory molecules such as cytokines. However, the relationship to any given tissue response is complicated, as various cytokines are typically involved in a given response, and there may be complex temporal relationships between different cytokines.

Organ effects are dependent on the dose distribution, and the mean dose is not necessarily a valid approximation. Thus, deducing RBE values for organs at risk based on clinical data would only be possible if data analysis would be done on a voxel-by-voxel basis instead of an organ-based basis because proton dose distributions in critical structures are typically more heterogeneous compared to photon therapy. The difference in dose distribution between proton and photon therapy could have even more complex consequences, e.g., when interpreting side effects, we might have to investigate physiological interactions of different organs, like the lung and heart [61]. There has been progress in identifying fundamental differences between photon- and proton-induced radiation effects [62]. There are differences on the molecular, cellular, and tissue level, as well as on proton-specific effects on gene expression, signaling, cell cycle disruption, and angiogenesis. The RBE concept might be fundamentally flawed considering these complexities. More research into these phenomena is desirable.

5.5 Clinical Impact of RBE Variations

5.5.1 Current Clinical Practice and Potential Consequences of RBE Variations

The proton RBE for clinically relevant LET_d values, $(\alpha/\beta)_x$, and dose is still associated with considerable uncertainties [42]. Furthermore, as an average value, the use of a generic clinical RBE of 1.1 at 2 Gy/fraction does not seem unreasonable for an SOBP field, keeping in mind that the RBE should be selected conservatively, i.e., on the low side with respect to tumor control.

On the other hand, using an average value could impact tumors that are at the extremes of the RBE spectrum. There have been concerns that we may underestimate or overestimate the RBE for tissues with either high or low $(\alpha/\beta)_x$, and that this could potentially impact the clinical efficacy of proton therapy [45,63,64]. Because the RBE for cell survival increases with decreasing $(\alpha/\beta)_x$, one can expect RBE values in excess of 1.1 for tumors such as prostate, possibly around 1.2 or 1.3. This could affect the interpretation of clinical trials comparing photon and proton treatments, as patients on the proton arm would effectively receive a higher dose. While some tumor types might show higher RBE values on average, there might also be tumors with lower RBE values based on their high $(\alpha/\beta)_x$. Tumors with a very high α/β could cause biologically equivalent doses that are on the order of 5% to 10% below the prescription doses [45,63]. There is currently no clear clinical evidence that this effect is significant in terms of outcome. It has been speculated that because medulloblastoma has a high $(\alpha/\beta)_x$, it could be underdosed when using protons because of an RBE below 1.1 [64]. However, in an analysis of patterns of failure in 16 out of 109 patients treated with protons, no indication was found that the RBE might have been overestimated [45].

The RBE increases with increasing LET_d, which increases with increasing depth if an SOBP is being delivered. Consequently, the RBE is expected to be significantly higher toward the distal end of an SOBP. This is of concern for critical structures immediately downstream of the target area, both because of an underestimation of the dose and because it causes a slight shift in range. Treatment planners have to keep these effects in mind. RBE variations are thus considered similar to physics-related range uncertainties.

If regions of high LET_d are within organs at risk with a low $(\alpha/\beta)_x$, we may severely underestimate the RBE. One example is the brainstem, which sometimes is located at the end of the range of an SOBP field [43,44]. Brainstem and cervical spine toxicities (e.g., necrosis) found in 4 out of 111 medulloblastoma patients were analyzed, but no clear correlation between elevated LET and regions of toxicity was found [44].

Although no clear clinical evidence for significant RBE variations has been found in patients, one has to keep potential variations in mind when comparing doses in clinical trials or when analyzing toxicities and tumor recurrences. This is hampered by the expected patient variability of tissue radiosensitivity [65]. Furthermore, delivery uncertainties in proton therapy will be reduced in the future, which may lead to a reduction in margins, which, in turn, might expose RBE variations due to elevated LET at the end of range [47,66]. Proton beam scanning and intensity-modulated proton therapy are expected to become the standard of care. There could be significant differences in RBE between scanned beams and passive scattered beams [54,55,67].

5.5.2 Biological Optimization

Because RBE increases more or less linearly with LET_d for a given dose and $(\alpha/\beta)_x$, the changes in LET_d can be used as a surrogate for RBE changes. The LET is based on physical properties and can be calculated using Monte Carlo simulations based on the treatment plan information [54,55]. It is important to emphasize that a homogeneous dose distribution does not guarantee a homogeneous distribution of LET. This allows biological dose optimization despite uncertainties in RBE values, i.e., without requiring additional biological data.

IMPT allows the delivery of inhomogeneous dose distributions for each field using beam scanning [68]. Interestingly, LET distributions can be influenced in IMPT without altering the dose constraints in treatment planning, i.e., dosimetrically equivalent plans can show significant differences in LET distributions (Figure 5–3) [55]. This can potentially be utilized to increase the efficacy of proton therapy, thus turning the disadvantage of variable RBE values into a clinical opportunity. However, LET-based treatment planning is not available in commercial treatment planning systems.

The LET-based planning concept was demonstrated in a multi-criteria optimization framework. Significant differences in LET_d distributions were observed in different base plans, in particular for organs at risk, while preserving target coverage [69]. This might be a first step toward incorporating RBE variations in proton therapy. The consideration of actual RBE values—and thus RBE-weighted doses in the treatment planning process—is still problematic because RBE variations are in the same order

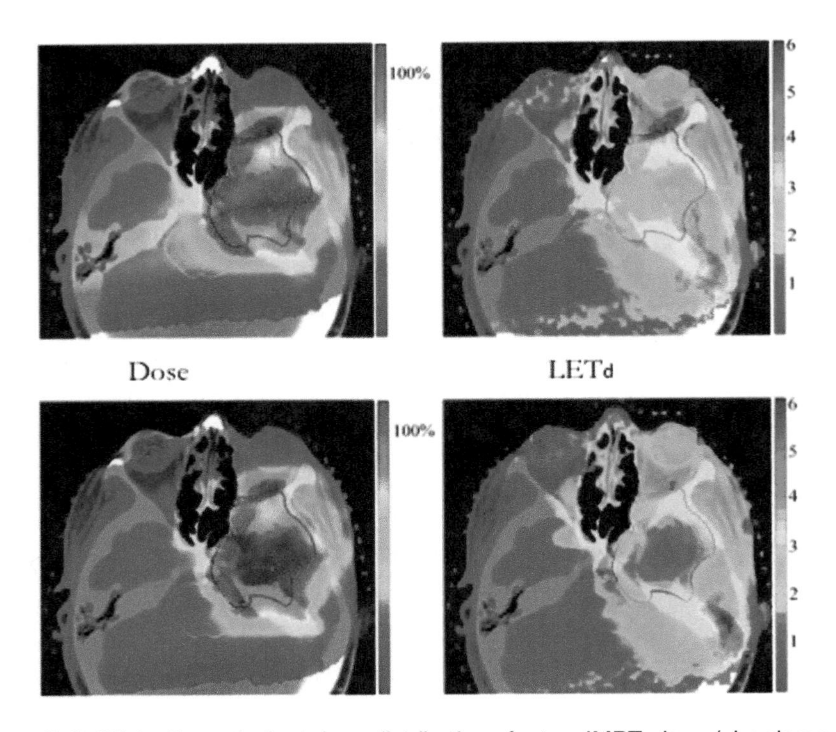

Figure 5–3 Clinically equivalent dose distributions for two IMPT plans (chordoma; dose in percent of prescribed dose; GTV in blue). The right column shows the LET_d distributions in keV/μm differing significantly. (From [55] with permission.)

as uncertainties in clinically relevant RBE values, particularly when it comes to the RBE for normal tissue complications [42].

5.6 Modeling of RBE

5.6.1 Linear Energy Transfer-based Modeling of RBE

There are various models to predict RBE values for proton therapy. A review is outside of the scope of this chapter. Notably, other than with heavy ions, where models have to include the track structure of the particle, much simpler approaches are feasible in proton therapy. There are two reasons for this. First, the typical relevant proton LET range is low enough to show a linear relationship with dose-response parameters such as α or β. Second, there is no significant contribution of heavy charged secondary particles.

Consequently, various modeling approaches have focused on parameterizing RBE as a function of dose, α, β, and LET_d. These models are typically based on equations 5.2 or 5.3. Some of the approaches assume that both α or β increase linearly with LET_d [47,70], while others consider β to be independent of LET_d [71–75]. For example, a parameterization for α/α_x as shown in Equation 5.8 was suggested with a fit value for q of 0.434 ±0.7 Gy um/keV based on 10 different cell lines and experiments *in vitro* [74].

$$\alpha/\alpha_x = 1 + \frac{q \times LET_d}{(\alpha/\beta)_x} \quad (5.8)$$

The parameterizations in these modeling approaches are based on a small subset of experimental data. The assumption that β does not depend on LET ($\beta=\beta_x$) has the advantage that it allows the RBE to be calculated as a function of LET_d and $(\alpha/\beta)_x$ without the explicit knowledge of α_x and β_x.

5.6.2 Complex Models for Predicting RBE

While LET-based models might be sufficiently accurate considering the uncertainties of underlying modeling parameters, more complex models have been suggested. They offer the advantage of being applicable not only to protons, but also to heavy ions, and they allow a more mechanistic interpretation of biological effects. A few examples are outlined below.

The repair-misrepair-fixation (RMF) model [70,76] links the particle type and kinetic energy via the microdosimetric frequency-mean specific energy ($\bar{z}_F$) to the initial formation of DNA double strand breaks (DSB induction Σ) and then to α and β. The RMF model predicts that the minimum and maximum RBE for the endpoint of reproductive cell death is as given in Equation 5.9, where RBE_{DSB} is the ratio of Σ for the proton to Σ for the ^{60}Co reference radiation. Estimates of RBE_{DSB} can be obtained from Monte Carlo simulations [77].

$$RBE_{max} \equiv \frac{\alpha}{\alpha_x} \cong RBE_{DSB}\left(1 + \frac{2\bar{z}_F RBE_{DSB}}{(\alpha/\beta)_x}\right), RBE_{min} \equiv \sqrt{\frac{\beta}{\beta_x}} = RBE_{DSB} \quad (5.9)$$

The microdosimetric-kinetic model MKM [73,78,79] combines assumptions from microdosimetry with kinetic relations for lesion repair and transformation assuming β to be constant while α is calculated from the dose mean lineal energy $\bar{y}_D$, the latter representing a radiation field by the energy deposited in a small sub-cellular volume (domain) within the framework of microdosimetry. The quantity of interest is the lineal energy, where ε is the energy deposited in the sensitive volume and $\bar{l}$ is the mean chord length of the volume. Due to large fluctuations of energy deposition in sub-micrometer volumes, the ionizing radiation is characterized by $\bar{y}_D$. The linear quadratic model parameters are given in Equation 5.10, where ρ and r_d are the density and radius of a spherical domain and α_0 being for a zero LET [73,80–82].

$$\alpha = \alpha_0 + \frac{1}{\rho \pi r_d^2} \bar{y}_D \beta, \beta = \beta_x \quad (5.10)$$

One of the most popular modeling approaches in heavy ion radiation therapy is the Local Effect Model (LEM) [83,84], which is based directly on the particle's track structure. The LEM is successfully in use in heavy ion therapy [85]. Within this model, the number of lethal events as a function of dose follows a Poisson distribution. Local doses in subcellular structures are converted to a local response via a modified linear quadratic dose response relationship. The modification ensures the validity of the linear quadratic model at very high doses, which can occur close to the particle tracks.

5.7 Out-of-field Effects

Within the radiation field, the biological effects relating to TCP and NTCP are of main concern. As described in the previous sections, there is an RBE to correlate these effects to the effects typically seen in photon therapy, where we have vast amounts of clinical data. Outside of the main radiation field—and even outside of the region of the patient considered for treatment planning or imaged—the focus is on effects from very low doses of radiation.

The dose in the high- to medium-dose region is predominantly caused by electromagnetic interactions of primary protons. However, protons undergo nuclear interactions in the treatment head and the patient itself, which can result in neutron radiation (see Chapter 10 on "Shielding"). Thus, far outside the main radiation field, the majority of the dose might be due to neutrons. Neutrons have a very long mean free path and deposit very little dose due to their low interaction probability. Consequently, the low-dose RBE has been a concern mainly for neutrons. Note, though, that neutrons deposit most of their dose via subsequent nuclear interactions, resulting in protons.

The main concern when it comes to neutron radiation is the development of radiation-induced malignancies. The RBE at very low doses (RBE_{max}) can be substantial, which is reflected in quality or weighting factor definitions (see 5.1.3). Furthermore, because of uncertainties in RBE values for carcinogenesis, one uses regulatory quantities, such as the quality factor. The neutron RBE varies considerably as a function of dose, dose rate, and biological endpoint. Data derived from human data and exposure to high-energy neutrons are sparse. Some of the high neutron RBE values (up to 59) reported in NCRP-104 were for carcinogenesis in mice involving fission neutrons that have an energy of less than 2 MeV [86], which is not relevant to proton therapy. In fact, the majority of neutron RBE values are derived from experiments using neutrons in the 1–2 MeV regions. Although the majority of neutrons impinging on organs have energies of about 1–2 MeV, the dose contribution is predominantly from high-energy neutrons up to 250 MeV [87]. Very little data are available on the biological effects of high-energy neutrons in the range of 20 MeV to 250 MeV that would be most pertinent to proton therapy. A recent study in mice using a therapeutic proton beam showed an RBE for cancer induction much lower than expected from the ICRP recommendations [88].

Despite these uncertainties in RBE, the risk for second cancers from proton and photon therapy has been compared in several publications [89–92]). Doses outside of the field are typically lower compared to photon therapy. For scanned beams, the scattered or secondary dose is likely much lower than with photon therapy even if a very conservative quality factor would be assumed. However, for passive scattering, this is not necessarily the case, particularly for fields with a small aperture opening [93–95]. It has been concluded that in-field, protons offer an advantage. However, the out-of-field risk between 6 MV photon therapy (with the dose mainly from scattered photons) and proton therapy (with the dose mainly from neutrons) might on average be comparable if passive scattered proton therapy is being used.

These low-dose considerations have to be weighed against the fact that, compared to photon therapy, proton therapy offers a substantial benefit in integral dose, i.e., the total energy deposited in the target.

5.8 Summary and Outlook

Other than assuming a 10% difference in required prescription doses and dose constraints, the biological difference between proton and photon therapy is not considered in detail clinically. With proton therapy patients increasing rapidly, one might be able to define prescription doses based on proton therapy alone, which would make the RBE concept obsolete. On the other hand, due to the typically inhomogeneous dose distribution in organs at risk, the concept of RBE or assigning constraints based on proton mean dose might be flawed, except if a voxel-by-voxel analysis is being done.

While the value of 1.1 is appropriate if a generic RBE is being applied as also recommended by ICRU report 78 [19], the proton therapy community will for sure move toward variable RBEs in the future. We know of low α/β for some tissues and we have the capability of calculating LET distributions in patients. While we may not be able to predict RBE values down to 10% accuracy, this nevertheless does allow to identify regions where adjustments to the value of 1.1 might benefit patient outcome [96].

The issue of low dose RBE and the risk for second malignancies in proton therapy is not resolved but with the move toward beam scanning will become less relevant.

References

1. Paretzke HG. Radiation track structure theory. Kinetics of nonhomogeneous processes. In: Freeman, GR Ed, John Wiley & Sons 1987:90–170.
2. Nikjoo H, Goodhead DT. Track structure analysis illustrating the prominent role of low-energy electrons in radiobiological effects of low-let radiations. *Phys Med Biol* 1991;36:229–238.
3. Goodhead DT, Thacker J, Cox R. Effects of radiations of different qualities on cells: Molecular mechanisms of damage and repair. *Int J Radiat Biol* 1993;63:543–556.
4. Rydberg B. Clusters of DNA damage induced by ionizing radiation: Formation of short DNA fragments. II. Experimental detection. *Radiat Res* 1996;145:200–209.
5. Holley WR, Chatterjee A. Clusters of DNA damage induced by ionizing radiation: Formation of short DNA fragments. I. Theoretical modeling. *Radiat Res* 1996;145:188–199.
6. Brenner DJ, Hall EJ. Commentary 2 to cox and little: Radiation-induced oncogenic transformation: The interplay between dose, dose protraction, and radiation quality. *Adv Radiat Biol* 1992;16:167–179.

7. Jenner TJ, Belli M, Goodhead DT, Ianzini F, Simone G, Tabocchini MA. Direct comparison of biological effectiveness of protons and alpha-particles of the same let. III. Initial yield of DNA double-strand breaks in v79 cells. *Int J Radiat Biol* 1992;61:631–637.
8. Goodhead DT, Nikjoo H. Track structure analysis of ultrasoft x-rays compared to high- and low-let radiations. *Int J Radiat Biol* 1989;55:513–529.
9. Jenner TJ, deLara CM, O'Neill P, Stevens DL. Induction and rejoining of DNA double-strand breaks in V79-4 mammalian cells following gamma- and alpha-irradiation. *Int J Radiat Biol* 1993;64:265–273.
10. Prise KM, Ahnstroem G, Belli M, Carlsson J, Frankenberg D, Kiefer J, Loebrich M, Michael BD, Nygren J, Simone G, Stenerloew B. A review of DSB induction data for varying quality radiations. *Int J Radiat Biol* 1998;74:173–184.
11. Frankenberg D, Brede HJ, Schrewe UJ, Steinmetz C, Frankenberg-Schwager M, Kasten G, Pralle E. Induction of DNA double-strand breaks by 1H and 4He ions in primary human skin fibroblasts in the let range of 8 to 124 kev/microm. *Radiat Res* 1999;151:540–549.
12. Pastwa E, Neumann RD, Mezhevaya K, Winters TA. Repair of radiation-induced DNA double-strand breaks is dependent upon radiation quality and the structural complexity of double-strand breaks. *Radiat Res* 2003;159:251–261.
13. Goodhead DT. Radiation effects in living cells. *Can J Phys* 1990;68:872–886.
14. Nikjoo H, Uehara S, Wilson WE, Hoshi M, Goodhead DT. Track structure in radiation biology: Theory and applications. *Int J Radiat Biol* 1998;73:355–364.
15. Bentzen SM, Constine LS, Deasy JO, Eisbruch A, Jackson A, Marks LB, Ten Haken RK, Yorke ED. Quantitative analyses of normal tissue effects in the clinic (QUANTEC): An introduction to the scientific issues. *Int J Radiat Oncol Biol Phys* 2010;76:S3–9.
16. Semenenko VA, Reitz B, Day E, Qi XS, Miften M, Li XA. Evaluation of a commercial biologically based imrt treatment planning system. *Med Phys* 2008;35:5851–5860.
17. Carabe-Fernandez A, Dale RG, Hopewell JW, Jones B, Paganetti H. Fractionation effects in particle radiotherapy: Implications for hypo-fractionation regimes. *Phys Med Biol* 2010;55:5685–5700.
18. Stuschke M, Pottgen C. Altered fractionation schemes in radiotherapy. *Front Radiat Ther Oncol* 2010;42:150–156.
19. ICRU. Prescribing, recording, and reporting proton-beam therapy. International Commission on Radiation Units and Measurements, Bethesda, MD: Report No. 78, 2007.
20. Fowler JF. Dose response curves for organ function or cell survival. *Brit J Radiol* 1983;56:497–500.
21. Fowler JF. The linear-quadratic formula and progress in fractionated radiotherapy. *Brit J Radiol* 1989;62:679–694.
22. Sachs RK, Hahnfeld P, Brenner DJ. The link between low-LET dose-response relations and the underlying kinetics of damage production/repair/misrepair. *Int J Radiat Biol* 1997;72:351–374.
23. Joiner MC, Lambin P, Malaise EP, Robson T, Arrand JE, Skov KA, Marples B. Hypersensitivity to very-low single radiation doses: Its relationship to the adaptive response and induced radioresistance. *Mutat Res* 1996;358:171–183.
24. Skarsgard LD, Wouters BG. Substructure in the cell survival response at low radiation dose: Effect of different subpopulations. *Int J Radiat Biol* 1997;71:737–749.
25. Andisheh B, Edgren M, Belkic D, Mavroidis P, Brahme A, Lind BK. A comparative analysis of radiobiological models for cell surviving fractions at high doses. *Technol Cancer Res Treat* 2013;12:183–192.
26. Kirkpatrick JP, Brenner DJ, Orton CG. Point/counterpoint. The linear-quadratic model is inappropriate to model high dose per fraction in radiosurgery. *Med Phys* 2009;36:3381–3384.
27. Carabe-Fernandez A, Dale RG, Jones B. The incorporation of the concept of minimum RBE (RBEmin) into the linear-quadratic model and the potential for improved radiobiological analysis of high-let treatments. *Int J Radiat Biol* 2007;83:27–39.
28. Sinclair WK. The relative biological effectiveness of 22-MVp x-rays, cobalt-60 gamma-rays, and 200 kVp x-rays. Vii. Summary of studies for five criteria of effect. *Radiat Res* 1962;16:394–398.
29. ICRU. Report of the RBE committee to the international commission on radiological protection and of radiological units and measurements. *Health Phys* 1963;9:357–384.

30. ICRU. Radiation quantities and units. International Commission on Radiation Units and Measurements. Bethesda, MD, 1971.
31. ICRP. Relative biological effectiveness (RBE), quality factor (Q), and radiation weighting factor (WR). International Commission on Radiological Protection. Pergamon Press, 2003.
32. ICRP. Recommendations of the international commission on radiological protection. International Commission on Radiological Protection (ICRP), 1991.
33. ICRP. Relative biological effectiveness, radiation weighting and quality factor. International Commission on Radiological Protection, 2003.
34. ICRP. The 2007 recommendations of the international commission on radiological protection. ICRP Publication 103. Ann ICRP, 2007.
35. ICRP. Recommendations of the international commission on radiological protection. International Commission on Radiological Protection (ICRP), 2007.
36. Xu XG, Paganetti H. Better radiation weighting factors for neutrons generated from proton treatment are needed. *Radiat Prot Dosim* 2010;138:291–294.
37. Zacharatou Jarlskog C, Paganetti H. Sensitivity of different dose scoring methods on organ specific neutron doses calculations in proton therapy. *Phys Med Biol* 2008;53:4523–4532.
38. Dalrymple GV, Lindsay IR, Ghidoni JJ, Hall JD, Mitchell JC, Kundel HL, Morgan IL. Some effects of 138-MeV protons on primates. *Radiat Res* 1966;28:471–488.
39. Dalrymple GV, Lindsay IR, Hall JD, Mitchell JC, Ghidoni JJ, Kundel HL, Morgan IL. The relative biological effectiveness of 138-MeV protons as compared to cobalt-60 gamma radiation. *Radiat Res* 1966;28:489–506.
40. Tepper J, Verhey L, Goitein M, Suit HD. In vivo determinations of RBE in a high energy modulated proton beam using normal tissue reactions and fractionated dose schedules. *Int J Radiat Oncol Biol Phys* 1977;2:1115–1122.
41. Urano M, Goitein M, Verhey L, Mendiondo O, Suit HD, Koehler A. Relative biological effectiveness of a high energy modulated proton beam using a spontaneous murine tumor in vivo. *Int J Radiat Oncol Biol Phys* 1980;6:1187–1193.
42. Paganetti H. Relative biological effectiveness (RBE) values for proton beam therapy. Variations as a function of biological endpoint, dose, and linear energy transfer. *Phys Med Biol* 2014;59:R419–R472.
43. Indelicato DJ, Flampouri S, Rotondo RL, Bradley JA, Morris CG, Aldana PR, Sandler E, Mendenhall NP. Incidence and dosimetric parameters of pediatric brainstem toxicity following proton therapy. *Acta Oncol* 2014;53:1298–1304.
44. Giantsoudi D, Sethi RV, Yeap B, Ebb D, Caruso P, Chen Y-L, Yock TI, Tarbell NJ, Paganetti H, MacDonald SM. Brainstem toxicity and let correlations following proton radiation for medulloblastoma. *Int J Radiat Oncol Biol Phys* 2015; under review.
45. Sethi RV, Giantsoudi D, Raiford M, Malhi I, Niemierko A, Rapalino O, Caruso P, Yock TI, Tarbell NJ, Paganetti H, Macdonald SM. Patterns of failure after proton therapy in medulloblastoma; linear energy transfer distributions and relative biological effectiveness associations for relapses. *Int J Radiat Oncol Biol Phys* 2014;88:655–663.
46. Carabe A, Espana S, Grassberger C, Paganetti H. Clinical consequences of relative biological effectiveness variations in proton radiotherapy of the prostate, brain and liver. *Phys Med Biol* 2013;58:2103–2117.
47. Carabe A, Moteabbed M, Depauw N, Schuemann J, Paganetti H. Range uncertainty in proton therapy due to variable biological effectiveness. *Phys Med Biol* 2012;57:1159–1172.
48. Withers HR, Thames Jr. HD, Hussey DH, Flow BL, K.A. M. Relative biological effectiveness (RBE) of 50 MV (be) neutrons for acute and late skin injury. *Int J Radiat Oncol Biol Phys* 1978;4:603–608.
49. Fowler J, Chappell R, Ritter M. Is α/β for prostate tumors really low? *Int J Radiat Oncol Biol Phys* 2001;50:1021–1031.
50. Gerweck L, Kozin SV. Relative biological effectiveness of proton beams in clinical therapy. *Radiother Oncol* 1999;50:135–142.
51. Paganetti H, Gerweck LE, Goitein M. The general relation between tissue response to x-radiation (α/β-values) and the relative biological effectiveness (RBE) of protons: Prediction by the Katz track-structure model. *Int J Radiat Biol* 2000;76:985–998.
52. Paganetti H. Interpretation of proton relative biological effectiveness using lesion induction, lesion repair and cellular dose distribution. *Med Phys* 2005;32:2548–2556.

53. Giantsoudi D, Grassberger C, Craft D, Niemierko A, Trofimov A, Paganetti H. LET-guided optimization in IMPT: Feasibility study and clinical potential. *Int J Radiat Oncol Biol Phys* 2013;87:216–222.
54. Grassberger C, Paganetti H. Elevated LET components in clinical proton beams. *Phys Med Biol* 2011;56:6677–6691.
55. Grassberger C, Trofimov A, Lomax A, Paganetti H. Variations in linear energy transfer within clinical proton therapy fields and the potential for biological treatment planning. *Int J Radiat Oncol Biol Phys* 2011;80:1559–1566.
56. Matsumoto Y, Matsuura T, Wada M, Egashira Y, Nishio T, Furusawa Y. Enhanced radiobiological effects at the distal end of a clinical proton beam: In vitro study. *J Rad Res* 2014;55:816–822.
57. Robertson JB, Williams JR, Schmidt RA, Little JB, Flynn DF, Suit HD. Radiobiological studies of a high-energy modulated proton beam utilizing cultured mammalian cells. *Cancer* 1975;35:1664–1677.
58. Zlobinskaya O, Siebenwirth C, Greubel C, Hable V, Hertenberger R, Humble N, Reinhardt S, Michalski D, Roper B, Multhoff G, Dollinger G, Wilkens JJ, Schmid TE. The effects of ultra-high dose rate proton irradiation on growth delay in the treatment of human tumor xenografts in nude mice. *Radiat Res* 2014;181:177–183.
59. Peacock J, Ashton A, Bliss J, Bush C, Eady J, Jackson C, Owen R, Regan J, Yarnold J. Cellular radiosensitivity and complication risk after curative radiotherapy. *Radiother Oncol* 2000;55:173–178.
60. Russell NS, Grummels A, Hart AA, Smolders IJ, Borger J, Bartelink H, Begg AC. Low predictive value of intrinsic fibroblast radiosensitivity for fibrosis development following radiotherapy for breast cancer. *Int J Radiat Biol* 1998;73:661–670.
61. Paganetti H, van Luijk P. Biological considerations when comparing proton therapy with photon therapy. *Semin Radiat Oncol* 2013;23:77–87.
62. Girdhani S, Sachs R, Hlatky L. Biological effects of proton radiation: What we know and don't know. *Radiat Res* 2013;179:257–272.
63. Jones B, Underwood TS, Dale RG. The potential impact of relative biological effectiveness uncertainty on charged particle treatment prescriptions. *Br J Radiol* 2011;84 Spec No 1:S61–69.
64. Jones B, Wilson P, Nagano A, Fenwick J, McKenna G. Dilemmas concerning dose distribution and the influence of relative biological effect in proton beam therapy of medulloblastoma. *Brit J Radiol* 2012;85:e912–e918.
65. Liu Q, Ghosh P, Magpayo N, Testa M, Tang S, Biggs P, Paganetti H, Efstathiou JA, Lu HM, Held KD, Willers H. Lung cancer cell line screen links fanconi anemia pathway defects to increased relative biological effectiveness of proton radiation. *Int J Radiat Oncol Biol Phys* 2015; in press.
66. Paganetti H. Range uncertainties in proton therapy and the role of Monte Carlo simulations. *Phys Med Biol* 2012;57:R99–R117.
67. Gridley DS, Pecaut MJ, Mao XW, Wroe AJ, Luo-Owen X. Biological effects of passive versus active scanning proton beams on human lung epithelial cells. *Technol Cancer Res Treat* 2014;(1):81–98.
68. Lomax A. Intensity modulation methods for proton radiotherapy. *Phys Med Biol* 1999;44:185–205.
69. Giantsoudi D, Grassberger C, Craft D, Niemierko A, Trofimov A, Paganetti H. Linear energy transfer-guided optimization in intensity modulated proton therapy: Feasibility study and clinical potential. *Int J Radiat Oncol Biol Phys* 2013;87:216–222.
70. Frese MC, Yu VK, Stewart RD, Carlson DJ. A mechanism-based approach to predict the relative biological effectiveness of protons and carbon ions in radiation therapy. *Int J Radiat Oncol Biol Phys* 2012;83:442–450.
71. Wilkens JJ, Oelfke U. A phenomenological model for the relative biological effectiveness in therapeutic proton beams. *Phys Med Biol* 2004;49:2811–2825.
72. Chen Y, Ahmad S. Empirical model estimation of relative biological effectiveness for proton beam therapy. *Radiat Prot Dosim* 2012;149:116–123.
73. Hawkins RB. A microdosimetric-kinetic theory of the dependence of the RBE for cell death on LET. *Med Phys* 1998;25:1157–1170.
74. Wedenberg M, Lind BK, Hardemark B. A model for the relative biological effectiveness of protons: The tissue specific parameter alpha/beta of photons is a predictor for the sensitivity to let changes. *Acta Oncol* 2013;52:580–588.

75. Tilly N, Johansson J, Isacsson U, Medin J, Blomquist E, Grusell E, Glimelius B. The influence of RBE variations in a clinical proton treatment plan for a hypopharynx cancer. *Phys Med Biol* 2005;50:2765–2777.
76. Carlson DJ, Stewart RD, Semenenko VA, Sandison GA. Combined use of Monte Carlo DNA damage simulations and deterministic repair models to examine putative mechanisms of cell killing. *Radiat Res* 2008;169:447–459.
77. Stewart RD, Yu VK, Georgakilas AG, Koumenis C, Park JH, Carlson DJ. Effects of radiation quality and oxygen on clustered DNA lesions and cell death. *Radiat Res* 2011;176:587–602.
78. Hawkins RB. A microdosimetric-kinetic model of cell death from exposure to ionizing radiation of any LET, with experimental and clinical applications. *Int J Radiat Biol* 1996;69:739–755.
79. Hawkins RB. A statistical theory of cell killing by radiation of varying linear energy transfer. *Radiat Res* 1994;140:366–374.
80. Hawkins RB. Mammalian cell killing by ultrasoft x-rays and high-energy radiation: An extension of the MK model. *Radiat Res* 2006;166:431–442.
81. Kase Y, Kanai T, Matsufuji N, Furusawa Y, Elsasser T, Scholz M. Biophysical calculation of cell survival probabilities using amorphous track structure models for heavy-ion irradiation. *Phys Med Biol* 2008;53:37–59.
82. Sato T, Watanabe R, Kase Y, Tsuruoka C, Suzuki M, Furusawa Y, Niita K. Analysis of cell-survival fractions for heavy-ion irradiations based on microdosimetric kinetic model implemented in the particle and heavy ion transport code system. *Radiat Prot Dosim* 2011;143:491–496.
83. Scholz M, Kellerer AM, Kraft-Weyrather W, Kraft G. Computation of cell survival in heavy ion beams for therapy. The model and its approximation. *Radiat Environ Biophys* 1997;36:59–66.
84. Elsasser T, Kramer M, Scholz M. Accuracy of the local effect model for the prediction of biologic effects of carbon ion beams in vitro and in vivo. *Int J Radiat Oncol Biol Phys* 2008;71:866–872.
85. Krämer M, Scholz M. Treatment planning for heavy ion radiotherapy: Calculation and optimization of biologically effective dose. *Phys Med Biol* 2000;45:3319–3330.
86. NCRP. The relative biological effectiveness of radiations of different quality. National Council on Radiation Protection and Measurements Report 1990;104.
87. Jiang H, Wang B, Xu XG, Suit HD, Paganetti H. Simulation of organ-specific patient effective dose due to secondary neutrons in proton radiation treatment. *Phys Med Biol* 2005;50:4337–4353.
88. Gerweck LE, Huang P, Lu HM, Paganetti H, Zhou Y. Lifetime increased cancer risk in mice following exposure to clinical proton beam-generated neutrons. *Int J Radiat Oncol Biol Phys* 2014;89:161–166.
89. Fontenot JD, Lee AK, Newhauser WD. Risk of secondary malignant neoplasms from proton therapy and intensity-modulated x-ray therapy for early-stage prostate cancer. *Int J Radiat Oncol Biol Phys* 2009;74:616–622.
90. Fontenot JD, Bloch C, Followill D, Titt U, Newhauser WD. Estimate of the uncertainties in the relative risk of secondary malignant neoplasms following proton therapy and intensity-modulated photon therapy. *Phys Med Biol* 2010;55:6987–6998.
91. Athar BS, Paganetti H. Comparison of the risk for developing a second cancer due to out-of-field doses after 6-mv imrt and proton therapy. *Radiother Oncol* 2011;98:87–92.
92. Athar BS, Bednarz B, Seco J, Hancox C, Paganetti H. Comparison of out-of-field photon doses in 6-MV IMRT and neutron doses in proton therapy for adult and pediatric patients. *Phys Med Biol* 2010;55:2879–2892.
93. Zacharatou Jarlskog C, Paganetti H. The risk of developing second cancer due to neutron dose in proton therapy as a function of field characteristics, organ, and patient age. *Int J Radiat Oncol Biol Phys* 2008;72:228–235.
94. Zacharatou Jarlskog C, Lee C, Bolch W, Xu XG, Paganetti H. Assessment of organ specific neutron doses in proton therapy using whole-body age-dependent voxel phantoms. *Phys Med Biol* 2008;53:693–714.
95. Mesoloras G, Sandison GA, Stewart RD, Farr JB, Hsi WC. Neutron scattered dose equivalent to a fetus from proton radiotherapy of the mother. *Med Phys* 2006;33:2479–2490.

96. Paganetti H. Relating proton treatments to photon treatments via the relative biological effectiveness (RBE)—should we revise the current clinical practice? *Int J Radiat Oncol Biol Phys* 2015; in press.

Chapter 6

Proton Beam Production and Dose Delivery Techniques

Marco Schippers

Paul Scherrer Insitute,
Villigen, Switzerland

6.1 Introduction

A proton beam from an accelerator usually does not have the right characteristics to be used for therapy directly. The modification of the beam characteristics to the desired values is strongly dependent on the dose delivery technique that is applied at the patient. The devices that match the shape of the dose distribution to the tumor volume in 3D are typically located in the *nozzle* or treatment head, which is the last part of the beam line, just before the patient. In the first section of this chapter, the two major techniques are described, followed by a discussion related to the most important beam parameters.

Then a section describing the currently used proton accelerators is given, followed by a section on new and currently investigated ideas for such accelerators. At the end of the chapter, an overview is given of the devices that rotate the beam delivery system around the patient to aim the proton beam at any direction to the tumor. It will be made clear why these devices (*gantries*) are quite big compared to the ones used in conventional photon therapy and how they can match the proton beams in the appropriate way.

6.2 Dose Delivery Techniques

6.2.1 Scattered and Scanned Beam

The dose has to be distributed over the target volume (tumor) both in the lateral direction (perpendicular to the beam direction) and in the longitudinal direction (in depth). Some techniques mostly shape the dose distribution in the lateral direction. The dose control along the beam direction (depth in the patient) is performed by adapting the beam energy. First, the two laterally working techniques will be described, followed by a description of the depth control. These techniques describe some essential specifications of the beam transport system and accelerator. In chapters 8 and 9 these dose delivery techniques will be described in more detail.

Since dimensions of the treated tumors can vary between a few cm^3 and several liters, the lateral cross section of the applied dose distribution can vary between a few cm^2 and several hundreds of cm^2. The two techniques to spread the dose over such areas in the lateral direction are the passive scattering technique—by which the beam cross section is increased [1]—and the scanning technique, which scans a relatively narrow beam over the area in a plane crossing the target volume in the lateral direction [2,3].

6.2.1.1 The Scatter Technique

The system performing the scattering is mounted approximately 1–2 m in front of the patient. Before the scattering system, the beam diameter will typically have a full width at the half minimum (FWHM) on the order of 1 cm. In the scattering system, the beam is intercepted by a foil of "heavy" material, e.g., lead or tungsten (see Figure 6–1). The protons are randomly scattered in this foil with an almost Gaussian-shaped distribution of the divergence, with a FWHM of up to 20 milliradians, depending on the foil thickness. After crossing the foil, the intensity profile has approximately a Gaussian shape in the lateral direction, having a width of several cm, depending on the distance behind the foil. By using a collimator with a small aperture (a few cm), one can select the beam fraction just around the center of this distribution, which has a rather flat intensity profile (the top of the Gaussian). This can then be aimed at the tumor. This single scattering technique has the disadvantages of losing a lot of beam at the collimator (less than a few percent of the beam intensity reaches the patient) and a dose profile that is still not flat (say, variations within 3%). The smaller the collimator aperture, the more homogeneous the lateral dose profile. An advantage, however, is the quite sharp lateral penumbra of the beam behind the collimator. In practice, this technique is typically only used for small tumors, e.g., ocular melanoma. In the double scattering technique, a second scatterer is mounted approximately 1–2 m downstream of the first foil. This is not a flat foil, however. In the most extreme design, a thick metal cylinder is mounted on the center of a flat foil. The thickness of this cylinder is such that it stops protons that hit the cylinder. In this way, the central

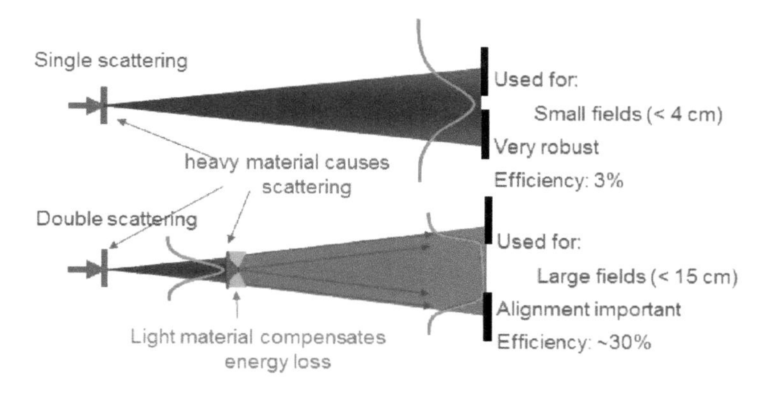

Figure 6–1 Principle of single scattering and double scattering systems.

peak in the distribution created by the first foil is cut out. Due to the scattering in the foil next to the cylinder, this gap is filled again, and the sharp edge due to the "shadow" of the cylinder is smeared out after some distance downstream of this foil. Of course, protons are also scattered to very large angles with respect to the beam axis, but these will be stopped in collimators before the patient. The intensity behind the foil region next to the cylinder is reduced by the out-scattering from the foil. At a specific distance, this intensity equals the filling of the central region (the "shadow"), so that a flat lateral intensity distribution is obtained over a diameter of typically 10 cm (field size) at the isocenter. Compared to the single scattering, more protons are scattered into the target region. But, of course, quite a few protons are lost in the cylinder. In a refinement of this method that is being used at the majority of passive scattering installations, the second foil has no cylinder, but instead has a thickness that varies as a function of the distance to the beam axis. The thickness is maximal (but doesn't stop the protons) in the center of the foil and decreases with radius (see figures 6–1 and 6–2).

The exact optimum shape of the thickness as a function of radius of this contoured scatter foil can be quite complicated and has been described by many, e.g., [4–9]. Protons hitting the thick central part of the second foil will be scattered more than those hitting the outer radius of the foil. At some distance downstream (typically 1 m), the intensity behind the foil center is increased by protons scattered into this region from larger radius and vice versa, so that a flat intensity profile with a diameter of up to several tens of cm is obtained at isocenter and, compared to the single scattering technique, a smaller fraction of the beam is lost. Typically 10% to 30% of the beam will reach the patient, who is positioned behind a patient-specific aperture.

Due to the scattering in the second foil, the dose distributions obtained by the double scattering methods have a larger penumbra than those obtained with the single scattering method. However, at larger depths in the patient, the contribution of multiple scattering in the patient tissue will also increase the lateral penumbra. At a depth of 15 cm, the 20% to 80% penumbra can typically be in the order of 7 mm when using either technique.

From the point of view at the isocenter, the protons seem to come from a source. The location of this virtual source is approximately in between the two scatter foils. The distance from the isocenter to this source (the virtual source-to-axis distance (SAD), and the divergence of the beam behind the two foils determine the number of protons per cm^2, which decreases with $1/SAD^2$. This should be taken into account in the treatment planning algorithm to take care of the dose upstream of the target volume (e.g., skin). It is preferred to have SAD as large as possible to reduce skin dose, so typically larger than 1.5 m.

The maximum achievable field size is determined by the thickness of the scatter foils and the distance to isocenter. Of course, a thick foil decreases the beam energy and the maximum depth at the patient. When the thickness variation of the second scatter foil is large, there will also be an energy (and thus range) variation over the field. To homogenize the energy loss, the second foil can be covered by a plate of light material, e.g., Lucite. The thickness of this plate is large where the metal foil thickness is small, and it is thinnest at the center, where the metal foil is thickest (see Figure 6–2). The Lucite plate mostly decreases energy and hardly contributes to the scattering. The sum of the energy loss in the metal and the loss in the Lucite is independent of the position where this second scatter foil is crossed, so protons traversing this composed second scatterer all have the same energy loss and, thus, the same range.

The proton beam coming from the beam transport system must be aimed at the center of the scattering system. This is important, especially in the case of contoured

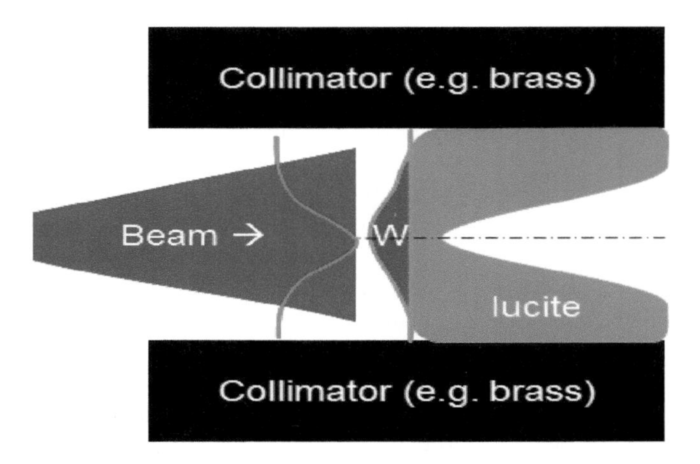

Figure 6–2 The second scatterer in a double scattering system is a combination of a heavy material like lead or tungsten (W) to scatter the central high-intensity part of the beam, and a light material such as Lucite, which reduces the energy of the protons in the outer part of the beam by the same amount as the heavy material.

scatter foils, such as in a double scattering system. Apart from some misalignment, a wrong beam centering will cause nonuniformities in the field.

The beam intensity and beam size should be matched to the field size, i.e., the collimator aperture, since the protons lost on the collimator will contribute to a neutron dose at the patient.

In order to achieve larger field sizes than can be achieved with scattering, the scattered beam can be moved by sweeper magnets located just behind the scattering system. These magnets move the beam continuously over a circular shape or along a zig-zag raster pattern so that the irradiated lateral dimensions are increased. This method is called beam wobbling or uniform scanning. Despite the term "scanning," it is basically an extension of the scattering technique.

In all cases, the effective beam size is designed to be larger than the lateral target size in the scattering technique, with or without wobbling. Therefore, for each patient a field-specific collimator is mounted just before the patient. The collimator cuts the beam to the exact lateral shape of the target volume as seen from the beam direction. This collimator is a metal (e.g., brass) disk in which a hole has been milled equal to the lateral shape of the target (see Figure 6–3). When changing a field or a patient, the collimator must also be changed.

6.2.1.2 Pencil Beam Scanning

Beam scanning is made by two fast magnets (here called sweepers), which can deflect the beam in two orthogonal planes [2,3]. The speed of field changes in these magnets is high, so that the beam shifts typically with a speed in the order of a few cm/millisecond at the isocenter. The distance between the scanning magnets and isocenter

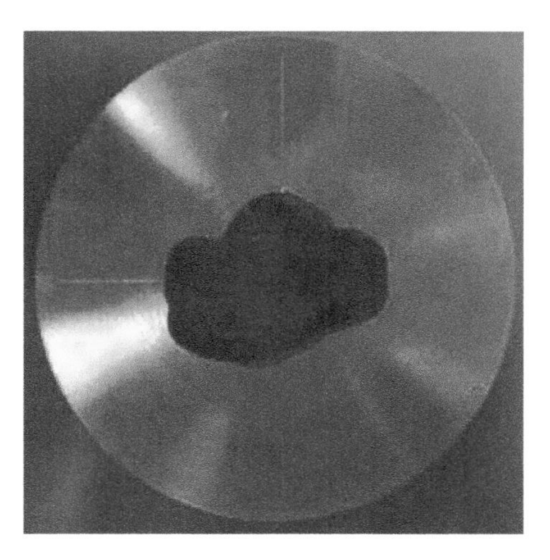

Figure 6–3 The final collimator (aperture) before the patient in a scattered beam setup. The aperture is specific for each patient and each beam direction and has the same shape as the target, as seen from the associated beam direction.

equals the SAD, and it is therefore typically chosen to be >1.5–2.5 m. Different scan techniques, such as spot scanning and continuous scanning, are in use or in development. In the spot scan (or a step and shoot [2] technique), the beam is aimed at a certain location (a certain volume element, a spot, or voxel). Then the beam is switched on, and the prescribed dose at that location is applied. When the dose has been deposited, the beam is switched off, and the sweepers aim the beam to the next voxel, etc. The beam intensity does not play a big role; one just fills with dose and takes the time it needs. The beam intensity may vary per voxel or during a voxel irradiation. In some systems, switching the beam on/off is not done when shifting to the next voxel, but, of course, this is taken into account in the treatment planning algorithm. Beam intensity, beam diameter, distance to next voxel, and sweeper speed are the important parameters in this case. But since a discrete pattern is still the basis of the scanning sequence, it is typically also called spot scanning.

In continuous scanning, the sweepers let the beam make a continuous shift over the target. The pattern of these shifts can be a series of parallel lines, a zig-zag raster, or a curved line pattern. The lengths and positions of the lines in these patterns are all within the target volume. The dose to be deposited varies along a line. Therefore, either the scan speed is adjusted at a constant beam intensity, or the beam intensity is varied and the scanning speed is kept constant. This technique requires accurate and fast control of the sweeper speed and the beam intensity. The continuous scanning method is still in development at several institutes.

The two scanning techniques are illustrated in Figure 6–4. In both scanning techniques, field sizes of up to 20 x 12 cm^2 can be achieved, a dimension that is limited by the aperture of the last magnet on the gantry. At PSI's Gantry2 [10] the scanning magnets are located just before the last bending magnet (90°) before the isocenter. The beam optics of this magnet have been made such that the pencil beam moves parallel

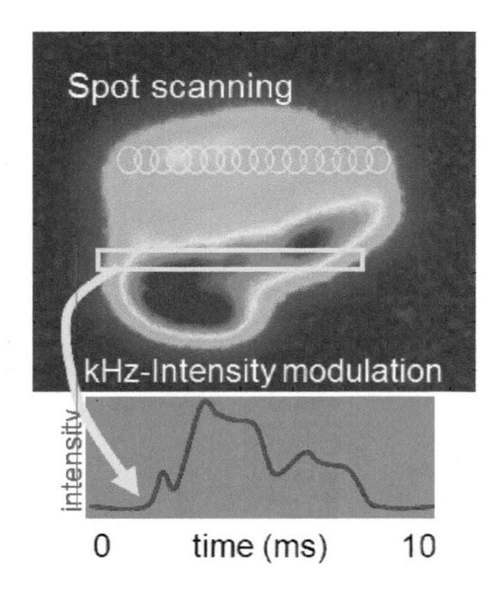

Figure 6–4 Pencil beam scanning by means of spot scanning (row of circles) and continuous line scanning (rectangle). The beam intensity (curve at bottom) varies in time during the continuous sweep made by the scanning magnet.

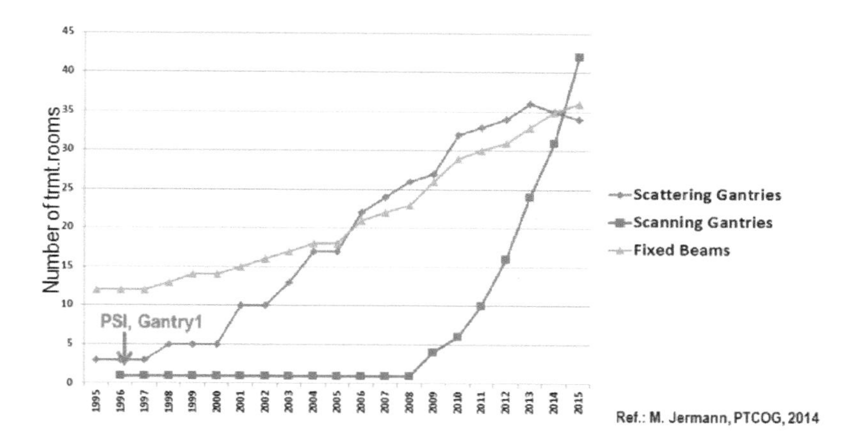

Figure 6–5 The number of treatment rooms world wide equipped with a pencil beam scanning gantry (squares), a gantry delivering a scattered beam (diamonds), and with a fixed horizontal beam line (triangles). (Courtesy of Martin Jermann, PSI.)

at isocenter, so that SAD is infinite. Since the gap in the bending magnet is limited in the direction orthogonal to the bending plane, the field is limited to 12 cm in this direction.

The scanning pattern in both the spot scanning and the continuous scanning technique is such that the beam is aimed at voxels within the target volume only. When using a narrow beam (a so-called pencil beam) having a diameter in the order of 5 mm FWHM, its contribution to the lateral penumbra at some depth is small compared to the scattering in tissue. Therefore, a collimator is usually not needed when using narrow pencil beams. Another advantage of this technique is that almost 100% of the protons are used in this technique. And, above all, this scanning technique offers possibilities for intensity-modulated proton therapy, adding an additional degree of freedom to increase dose conformality. This explains the dramatic increase in the number of treatment rooms for particle therapy with scanning beams since the start of PSI's gantry-1, as shown in Figure 6–5.

6.2.1.3 Matching the Beam Energy

In both the scattering technique and in the scanning technique, the range of the protons has to be adapted to the thickness and depth of the tumor as well. This is done by means of energy variation. Different methods will be discussed that can be used in combination with the scattering or scanning technique.

When controlling the energy, one has to distinguish between the maximum energy needed in a certain beam direction (field) and the needed variation of the beam energy. The maximum needed energy is determined by the deepest (in g/cm^2) location of the target volume. The variation of the energy causes a variation of the depth of the Bragg peak location. The needed magnitude of this variation is determined by the tar-

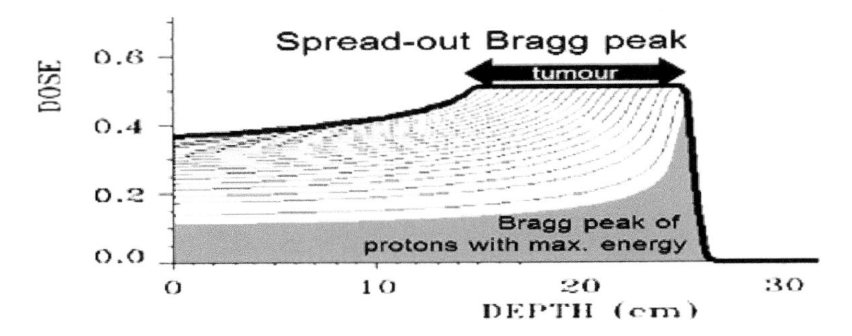

Figure 6–6 The spread-out Bragg peak is the sum of Bragg peaks of different proton energies. A homogeneous dose depth over the tumor depth is obtained in the target region.

get thickness over which the Bragg peak dose has to be applied. To obtain the desired depth–dose distribution, the different beam energies are applied with proper weight factors. When a homogeneous depth–dose distribution is required in the target volume, this is achieved by weight factors that will create the so-called spread out Bragg peak (SOBP) (see Figure 6–6).

The maximum obtainable proton energy is determined by the proton accelerator. As will be discussed later in this chapter, synchrotrons can accelerate protons to any energy between a few MeV and the machine's maximum energy. For each field, one can select the maximally needed proton energy from a synchrotron. Cyclotrons, however, cannot change the beam energy. Protons are always accelerated until a fixed energy, given by the cyclotron design. The maximum needed beam energy in a field is obtained by degrading: slowing down protons in an absorber in the beam transport system shortly after leaving the cyclotron.

Energy modulation can be done in several ways. The method mostly used in combination with the scattered beam technique is by means of a range modulation wheel in the nozzle (see Figure 6–7). This is a wheel, usually made of light material like Lucite (see Chapter 8 for more details). Along the circumference of the wheel, the thickness of the wheel varies. The beam is crossing the wheel at a certain radius. When rotating the wheel, different amounts of material will cross the beam path, each causing a certain energy loss of the protons. The thickness variation, along the circumference of the wheel, is such that the exiting protons with all of their different energies will create a spread-out Bragg peak (SOBP) in tissue. In fact, the azimuthal width of each thickness equals the weight factor coupled to the corresponding group of protons with a certain energy. In a treatment room, one usually stores a library of range modulator wheels, each wheel being specific for a certain SOBP width and incoming beam energy. Together with bolus and the collimator, the modulation wheel is mounted in the nozzle before dose application at each field.

Figure 6–7 The energy modulation to obtain the SOBP is made by rotating a, e.g., Lucite wheel in the proton beam. The different thicknesses cause the different proton energies behind the wheel.

Since the scattering technique fills the full field in the lateral direction, this energy variation is equal over the whole field. Since the chosen magnitude of the energy variation (the width of the SOBP) is determined by the thickest part of the tumor, healthy tissue in front of thinner tumor regions will thus get the same dose as the tumor gets. This is the major disadvantage of the scattering technique.

Usually the distal edge of the tumor is not equal to a flat surface at constant depth. When applying a scattering technique, a range compensator (or bolus) is mounted near the collimator. This bolus is a Lucite disk with a position-dependent varying thickness. The protons that are aimed at a target region at shallow depth will traverse more bolus material (e.g., Lucite), but the protons that need to reach deeper locations will traverse less material. Such a compensator is thus patient- and field-specific. Similar to the collimators, these are fabricated during the preparation phase of the treatment and put in place at each used gantry angle. Since energy variation can be matched perfectly in the pencil beam scanning technique (as will be described later), such a compensator is not needed when using a scanning technique.

In the scanning technique, the energy is selected per group of pencil beams. In practice, all pencil beams having their Bragg peak at a certain layer are grouped into one energy layer. As illustrated in Figure 6–8, this especially reduces the dose in the healthy tissue in front of the tumor.

In synchrotrons, the energy can be changed at the accelerator level, as will be discussed below. In cyclotron systems, the energy variation between each energy layer is done by quickly inserting a certain amount of material in the beam path. This can be done either in the nozzle or at an upstream location. The method to vary the amount of material in the beam can be done by means of range shifter plates—such as a stack of 20–30 small, 4 mm thick Lucite plates that can be quickly shifted into/out of the beam

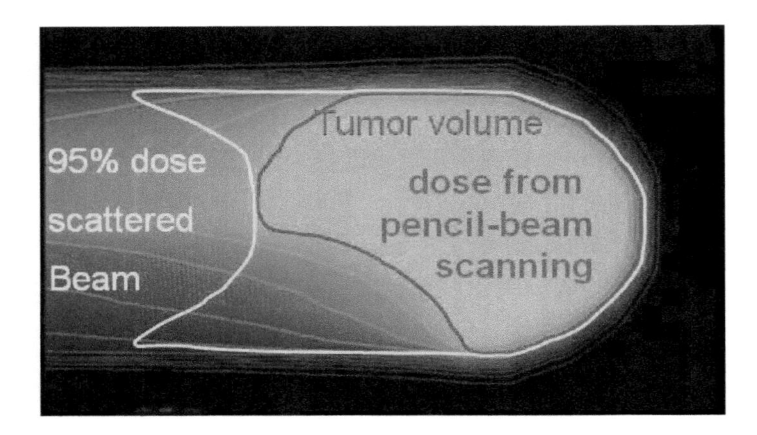

Figure 6–8 Comparison of the dose distributions for a scattered beam (light curve represents the 95% isodose line) and one obtained with pencil beam scanning (grey value; red curve is 95% isodose line and coincides with the target volume).

path—that can set the range at discrete step sizes of 5 mm water equivalent. This can also be done by a degrader consisting of a partially overlapping pair of Lucite or graphite wedges. Such a degrader can decrease the beam energy in a continuous range by any amount by adjusting the overlap (see Figure 6–9).

The location of a range-shifter can be in the nozzle. However, this location causes beam size increase due to scattering, which can be compensated for partially when it is mounted upstream in the beam line. When using an upstream location, all following beam line magnets must be set such that they are matched to the beam energy. Due to the self-inductance of the magnets, it is challenging to do this fast. However, it also gives the possibility to have the range shifter system being followed by an energy selection system consisting of a bending magnet and a slit system. Such an energy selection has an advantage because the beam, having traversed a certain amount of material, is not mono-energetic, but has some spread due to the statistical process of energy loss. When shifting the beam energy from 250 MeV to 70 MeV, the energy spread is approximately 24% or 17 MeV (FWHM), causing a range variation of 14 mm (FWHM). This is much larger than the typically required distal penumbra of several mm, and it would increase the distal penumbra dramatically. So, it is advantageous to have the range shifter system being followed by an energy selection system. When using a magnetic bending system, the trajectories of protons with high energy will be bent less than the ones with lower energies. Behind the magnet, the beam will be spread out in the bending plane, with a strong correlation between energy and position. Using a slit system at this position, a fraction of the energy spectrum (typically 1% to 2% spread) can be selected and sent onward. A further advantage of such a system is that it also acts as a check of the range shifter: only the correct beam energy can pass the slits. A disadvantage is that all magnets in the beam transport system and

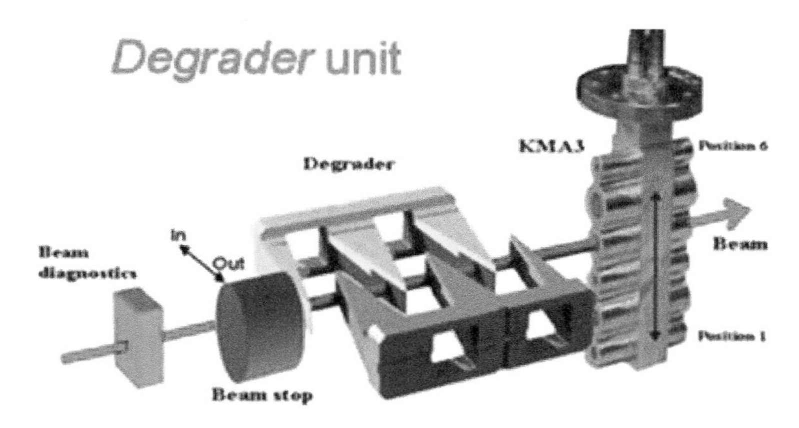

Figure 6–9 The degrader and its following collimator to define the beam line aperture for the scattered beam from the degrader at PSI.

gantry have to follow the energy changes with the same speed as the range shifter. Also the readout of all downstream monitors and beam diagnostics must be scaled with the correct energy-dependent factors.

6.2.2 Timing Considerations

The proton beam coming from the accelerator can be subject to time-dependent variations. Slow variations (>seconds) due to the temperature changes of magnets can easily be compensated or corrected for in control loops. Faster variations may interfere with the dose delivery.

Most variations that occur are those in beam intensity. These can be oscillations or random fluctuations (beam noise) in intensity. The beam is also usually pulsed. As will be described later, a synchrotron operates in a so-called spill mode. During a spill, there is beam for a period of seconds up to a fraction of a minute. The time between two spills is in the order of several seconds. During the spill, the beam intensity is pulsed at a rate of several tens of MHz. A cyclotron beam is pulsed at a frequency typically in the range of 50–100 MHz. In addition to that, the beam from a synchrocyclotron also pulses with a typical repetition rate of a few hundred Hz. So at a rate of, say, 1000 Hz, protons are coming during a period of several microseconds. Within this period, the intensity is pulsed at the cyclotron frequency rate of 50–100 MHz.

In most techniques, the MHz repetition rates, which thus occur in all cyclotrons, do not play a role since their occurrences at every 10–20 nanoseconds are much faster than the time constants in the scattering or scanning process (in milliseconds-seconds). However, when designing or studying processes with microsecond time scales in dosimetry devices, such as ionization chambers, it may be necessary to consider this "burst" nature of the proton beam intensity, as for example when dealing with space charge effects.

The scattering technique is the most robust technique with respect to timing issues in the proton beam. However, its range modulation is sensitive to timing issues. The rotation of the wheel causes energy changes at a frequency of a few hundred Hz up to a kHz. If there are intensity oscillations of similar (or related) frequencies, these may cause a distortion in the depth–dose distribution. This, for example, must be taken care of when using a synchrocyclotron.

The spot scanning technique is quite insensitive to time-dependent intensity variations since each voxel is just filled with dose, and if the intensity is low, it will only take more time to reach the required dose. However, very large intensity jumps could cause an error if the beam switch-off is too slow.

Continuous scanning is quite sensitive, however. The intensity must be controlled to be at a certain value during a fraction of a millisecond. When the applied dose is controlled by means of the scanning speed, low speed and low beam intensity are needed to prevent dose errors due to a too-slow matching of the scan speed. The spill character of a beam from a synchrotron or the kHz pulses from a synchrocyclotron are also interfering with continuous scanning. The sweep has to stop and be continued at the next pulse. Spot scanning is certainly a good possibility with a synchrotron beam, but since the pulse rate of a synchrocyclotron is in the same order as the spot scan rate, and since the intensity regulation per pulse could be not accurate enough, it is difficult to apply the prescribed spot dose with a single pulse, especially if the needed dose varies per spot. Therefore several low-dose synchrocyclotron pulses are needed per spot, which causes long treatment times.

Apart from timing issues in the beam, one must also consider the effects of organ/tumor motion during the dose delivery [11] (see chapter 25 on Motion Management). Also in this case, the scattering technique is less sensitive. In scanning, hot and cold spots could occur in the dose distribution due to the interference between organ motion and scanning. Methods considered are: fast scanning in combination with rescanning, which averages out the eventual hot and cold spots, or gating, so that the dose is only applied in the correct phase of the patient motion. A synchronous following of the motion by correcting the sweeper magnets or beam energy is a future option that is being investigated, but it is too complicated to use at present. The fast rescanning technique should average out the organ motion effects. When applying rescanning of a volume and not of an energy layer, it also requires a fast energy change. In that case, it would be optimal when the typical energy step of about 2% is done within 0.1 second. Then one could "repaint" a volume of 1 liter approximately 6–7 times in a minute, during which the dose is summed up to 2 Gy. In practice, this has only been reached at PSI at the moment, but this increase of speed in energy change is in development both at PSI and at other places.

As discussed in the section describing the beam energy setting, typical beam line magnets can change their strength only rather slowly. This is due to the electromagnetic property self-inductance (L, in Henry) of the magnet. The higher this value, the more voltage is needed to make a fast jump in magnetic field strength. This voltage is limited, however, by the insulation between the different turns in the coil of the magnet. In addition to this, the change in magnetic field is partly compensated by eddy currents, which arise in the iron yoke and pole of the magnet due to changing magnetic fields. These currents decay within a few seconds, which is too slow for fast energy variations. By making the yoke and pole out of laminated iron (iron sheets of only a few mm thickness and insulated from each other) in a certain orientation with respect to the magnetic field lines, the effect of the eddy currents can be reduced dramatically. With an energy change of 2% per 0.1 seconds, PSI's energy change system is the fastest currently in operation.

6.2.3 Dose Rate and Pulsed or Continuous Beam

Of course, the dose rate is determined by the beam intensity and beam energy at the patient. Dose rate is an important issue, since it directly affects the treatment time. A long treatment time yields a lower throughput (and therefore less income), but it also has a higher risk of motion problems. On the other hand, a high dose rate needs a more sensitive dosimetry and fast-reacting beam switch-off systems.

So first of all, the beam intensity from the accelerator needs to be sufficient. In principle, cyclotrons and synchrotrons are accelerators capable of delivering sufficiently high beam intensities. Depending on the proton source, one should even protect against too high beam intensities when using a cyclotron.

In all accelerators (see below), the acceleration is made by an electric field that oscillates with a certain frequency in the radio frequency (RF) range. In the acceleration process, only the particles that cross this field at a certain phase (voltage) will be accelerated to the desired final energy. Therefore the beam intensity is pulsed with this RF, which is quite high—typically 10–100 MHz in the RF domain—but it can be considered to be continuous in the context of biological processes. Although most biological processes do not depend on the pulse rates or beam intensities currently in use, it is important to consider the effects of the pulsed beam intensity on instruments, such as some dosimetry devices. When accelerators can provide this RF frequency only during short moments, the accelerator is said to be pulsed. This pulse frequency is much lower. The time between pulses can be a few seconds up to 0.001 sec. As will be discussed later, this has immediate consequences for the dose delivery. To obtain a sufficiently high beam intensity, the time when there is no beam between the pulses is compensated for: to obtain enough intensity averaged over time, a pulsed beam will typically have a high beam intensity during the pulse. It must be verified whether dose measurement and intensity-measuring equipment can deal with this.

Synchrocyclotrons typically deliver a pulsed beam only during 1% of the time that a normal cyclotron would provide beam, so the intensity per pulse is increased a lot using the ion source. Of course, the synchrotron design needs to deal with this high intensity, as well as with the high losses of protons that are not accepted in the acceleration process. For scattering this is not a problem, but for scanning this may lead to dose delivery inaccuracies, as discussed before.

Synchrotrons are able to give enough intensity, but the achieved values are limited by the amount of protons that can be injected in a single spill. This number can be increased by increasing the injection energy, but that requires a more powerful (and longer) injector. The spill structure of the synchrotron has implications for the types of beam delivery systems that can be used in the treatment areas. When beam gating (interruption of the beam extraction) or line scanning is applied to deal with breathing motion, for example, it is of advantage when the synchrotron has a good beam-storing quality so that the stored intensity allows spill lengths of tens of seconds.

6.3 Beam Production Machine

6.3.1 General Requirements

Acceleration of protons to 200–250 MeV energies requires powerful accelerators that typically have footprints approximately equal to that of a treatment room. The costs of an accelerator are still quite high, and since the time an actual treatment uses the beam lasts only a few minutes, and since the room occupation due to patient setup and checks typically takes 20–30 minutes, it makes sense to have one proton accelerator serving multiple (up to 3–5) treatment stations (≈rooms). For the accelerator, therefore, a very high availability is required. Also switching between treatment rooms must be possible within a faction of a minute.

The very large fraction of time that the accelerator and beam transport system are not accessible by personnel requires a very robust and service-friendly machine. Service friendliness requires extensive and easy-to-interpret machine diagnostics and a modular device with easy access to all relevant components. Some components need frequent service and should thus be optimized in that sense.

Due to inevitable activation of machine parts and activation of the air, it may be necessary that a waiting time is necessary before entering the machine areas after switching the beam off. The accelerator and beam transport design should also be designed such that the inevitable beam losses are concentrated at a few discrete and well-known locations. These places (e.g., the degrader region in a cyclotron facility) can be shielded locally so that one can pass these places without excessive exposure.

Before starting the treatments, several checks have to be done in the morning. Some of these checks, but also the development of new beam line settings, need beam diagnostics to measure beam profiles and intensities. However, measurements with interceptive beam diagnostics can influence the beam characteristics due to beam–material interactions. Therefore, one must have procedures that guarantee that no such devices are inserted during treatments, or that certain devices are always in the beam. The verification of such issues is a typical control system task.

To obtain a safe and reliable system, redundancy is often implemented, both at the measuring side and at the "beam off" command side. Parallel or sequential actions to intercept the beam and critical measurements of beam characteristics or machine status are typically to be performed with redundant methods. However, one should distinguish actions and measurements related to the treatment procedure and patient safety from those for proper machine operation. So in this respect, for example, the dose measurement has another importance than the measurement of the temperature of magnet-cooling water. A modularity to group different systems is essential to make a clear distinction between "therapy-relevant parameters" and typical technical machine parameters, as well as the different types of actions following certain measurement outcomes [12]. The control system of the machine plays an important role in the safety of the patient, but also in a high availability of the system [12].

6.3.2 Synchrotron

Although particle therapy originally started at cyclotron laboratories, the first hospital-based facility has been equipped with a synchrotron as its accelerator. This first synchrotron delivering beam for proton therapy has been installed at the Loma Linda University Medical Center (CA, USA). Since then, synchrotrons are being offered by various companies, and they have been installed in Japan and one in Houston, TX.

A synchrotron system consists of different clearly separated devices, all grouped together in the accelerator room. First of all, there is a proton source injecting particles into a pre-accelerator, from which the protons are injected into the synchrotron ring.

Current proton synchrotrons [13–15] have a typical diameter of 6–8 m (Figure 6–10). A high-intensity proton beam of 10–20 mA [14] is obtained from an external ion source. The pre-acceleration of the protons is usually done by means of two subsequent linear accelerators, an RFQ (radiofrequency quadrupole) and a DTL (drift tube linac). When filling the synchrotron, the protons must have a minimum energy of typically 4–7 MeV. The higher this energy, the more particles can be stored in the ring, since the repelling effect of the space charge forces is less destructive at higher energy. The synchrotron does not accelerate the protons during the filling process. When the filling process has been completed and the ring is filled with about 10^8–10^{11} protons, the electric RF field in a cavity in the ring is switched on, and the pro-

Figure 6–10 A typical modern synchrotron for proton therapy.

tons are accelerated to the desired energy. This acceleration phase lasts about a second. The RF cavity needs to vary its frequency from typically 1 to 16 MHz, synchronous to the velocity increase of the protons. To limit the needed power, the acceleration voltage is not so high. But since the particles cross the acceleration cavity approximately every 0.1 ms, the particles get accelerated in a short time. During the acceleration, the magnets of the ring are ramped up, synchronous to the increasing momentum of the protons. When the desired energy (which can be any between the injected energy and the maximum energy) has been reached, the acceleration is stopped and the particles are stored in the ring.

Different methods exist to extract the protons out of the synchrotron ring. The process of injection, acceleration, and extraction is called a spill, which can take several up to 10 seconds, depending on the filling of the ring and the extraction method.

Fast extraction is performed by switching on a kicker magnet that directs the beam out of the ring into the beam transport system. Then all particles from the ring are extracted within one orbit period (a fraction of a microsecond). The dose control is quite complicated in this case, since intensity cannot be controlled, but it is just set by the ring filling and eventual losses. The dose per pulse also needs to be very high, otherwise the dose delivery would take too long due to the spill's time structure. Therefore, the method of slow extraction is used.

Slow extraction from the ring is done by careful kicking of the beam with an RF kicker. The kicker amplitude is small, so that the beam shifts only a little bit in the ring and only a fraction of the stored particles are extracted by an electrostatic extraction septum and a septum magnet [14,16]. Only when particles need to be extracted is this RF perturbation switched on, so that it increases the amplitude of the oscillation of the particles around their equilibrium orbit (betatron oscillation). After several turns, the amplitude of the betatron oscillations has increased, and when protons reach the outer side of the septum, they are extracted. Thus, a sliver of the circulating stored beam is being "peeled off" and sent without further losses to the treatment area. When the necessary amount of particles has been extracted, the remaining content of the ring is deposited on a beam dump. In several synchrotron systems, the particles are decelerated prior to dumping to minimize the amount of radioactivity at the beam dump.

This extraction method has a very fast response of typically 100–200 microseconds to switching the extraction (i.e., the RF kicker) on and off, and it can be used as the primary "beam on/off" switch in the treatment control.

The extracted beam is peeled off from the circulated beam, and thus it is very sensitive to orbit fluctuations in the synchrotron. This causes fast intensity fluctuations of the extracted beam that can be up to 50% to 100% [14,17,18]. This is no problem for beam delivery systems based on scattering of spot scanning (see Section 6.3.2), but when continuous beam scanning is applied, these fluctuations compromise this method. Developments are going on to deal with these fluctuations [19].

When using a slow extraction method, the horizontal emittance and momentum spread of a beam from a synchrotron can be up to a factor of 10 smaller than the emittance of a cyclotron beam. However, the emittance is not nicely Gaussian-shaped and shows large asymmetries. As a consequence of the horizontal extraction process, the vertical profile can have a Gaussian shape, but the horizontal profile can resemble a trapezoid with rounded corners [18]. For smaller beam widths and lower energies, the effect is compensated to some extent due to scattering in the beam exit window and the treatment monitoring system. However, in case there is no such smoothing component, the preservation of the small emittance is used in advantage to keep low beam losses in the beam transport. However, when gantries are being used, it is essential to have a symmetric emittance at the gantry entrance in order to prevent a gantry angle-dependent beam shape at the patient. Different methods are being studied [17,18]. In case there is an extraction energy dependence of the beam emittance, the beam line settings must be given a beam energy-dependent correction in addition to the normal momentum scaling. Often the position and the emittance of the extracted beam vary with energy, so these must be corrected for in the beam transport system between synchrotron and gantry.

Typical advantages of the synchrotron are the relatively small size of the components and the easy access to the machine's parts for maintenance. Furthermore, the activation of machine components is very low [21]. This can be attributed to very small beam losses during acceleration and beam transport in the synchrotron and beam transport lines, and because degrading does not need to be performed.

Current developments for synchrotrons concentrate on the improvement of the beam intensity stability and energy change during extraction [22]. A proposal for a rapid-cycling synchrotron [23] is aimed at a faster spill sequence. Other ongoing efforts at existing synchrotrons are aimed at an increase of the lifetime of the stored beam and at an increase of the stability of the extracted beam. At the HIMAC synchrotron, the intensity stability of the carbon beam has been improved to approximately 10% [17].

Following a recent proposal [24], a small ring of 5 m in diameter has been developed for proton acceleration up to 330 MeV. The protons are injected by a 1 MeV linear accelerator, so the intensity in the synchrotron is rather limited. But the relatively low cost of the facility is expected to be the major advantage of these small machines.

6.3.3 Cyclotron and Synchrocyclotron

Cyclotrons in physics laboratories usually have the possibility to change parameters, such as beam energy or particle type. Modern cyclotrons dedicated for proton therapy, however, have fixed many of their parameters, since they are always operating in a limited number of modes. They accelerate protons to a fixed energy of 230 or 250 MeV. Although the development of cyclotrons for carbon ions and protons are in progress [25], the current (synchro)cyclotrons are used for therapy with protons only. Furthermore, and motivated by the increasing demand for hospital-based facilities, the new cyclotrons (see Figure 6-12) are rather compact, having magnet heights of approximately 1.5 m and typical diameters between 3.5 m and 5 m.

The major components of a typical compact cyclotron as shown in Figure 6–11 are:

- an RF system that provides strong oscillating electric fields between cavities (called "dee") by which the protons are accelerated,

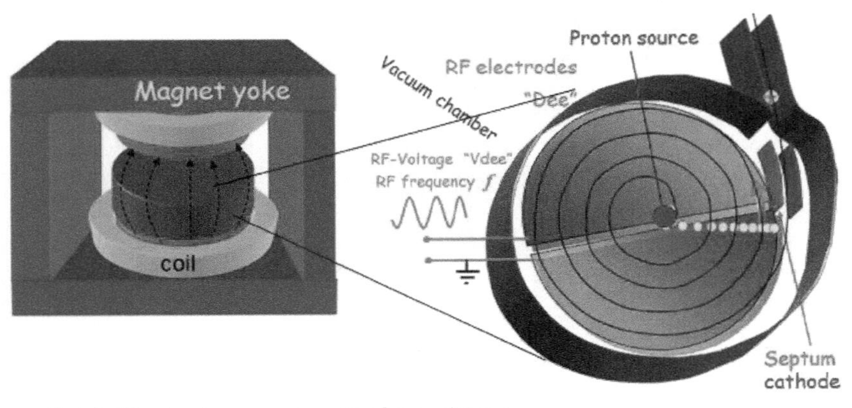

Figure 6–11 The basic components of a cyclotron.

- a strong magnet that confines the particle trajectories into a spiral-shaped orbit so that they can be accelerated repeatedly by the RF fields,
- a proton source in the center of the cyclotron in which hydrogen gas is ionized and from which the protons are extracted into the accelerating structure, and
- an extraction system consisting of a septum and cathode at the outer edge of the cyclotron. This system guides the particles that have reached their maximum energy out of the cyclotron into a beam transport system.

The protons are created in an ion source in the center of the cyclotron (see Figure 6–13). By means of an arc between some electrodes, hydrogen gas is ionized and a plasma is created in a pencil-sized vertical tube mounted at the center of the cyclotron. In the wall of this tube is a slit facing an RF electrode (the "puller"). Protons can leave the tube though this narrow slit and are pulled toward the puller when it is at the RF phase of having a negative voltage. The proton beam intensity can be controlled by changing the arc parameters controlling the ionization. The response time of an ion source to its setting can vary between milliseconds and seconds. For application of the *scatter technique* or the *spot scanning technique*, this is no problem, but for the application of the *continuous scanning technique*, controlled accurate intensity variations are needed that are much faster. To achieve a faster intensity control, a so-called vertical deflector plate has been mounted in the cyclotron at PSI. The ion source is delivering a constant supply of protons, and control of the accelerated intensity is done by regulation of a vertical electric field just behind the source exit (see Figure 6–13). This field deflects the beam in the vertical direction, and a controlled amount can thus be intercepted by a vertically limiting collimation. Beam intensity adjustment is done within 100 microseconds.

The RF system usually consists of one, two, three, or four electrodes which, due to their shape in the one-electrode system in the first cyclotrons built, are often called "*dees*." The dees are mounted as sectors in the circular beam region. The dees are connected to an RF generator driving an oscillating voltage between 30 and 100 kV at the dees with a fixed frequency. To reduce power consumption and to stabilize the frequency, the dees are parts of an oscillating system consisting of resonant *cavities* that have their resonance frequency equal to the desired RF frequency. Each dee consists of a pair of copper plates on top of each other with a few centimeters in between, in which the protons move in a plane parallel to the dees. The top and bottom plates are at the same voltage, so there is no electric field in the space between these two dee

Figure 6–12 Two examples of cyclotrons currently in use: left with super conducting coils. Left: the 250 MeV superconducting cyclotron from Varian (previously ACCEL), of which the first one has been installed at PSI, Switzerland [27,28]. Right: the 230 MeV proton cyclotron of IBA (Louvain la Neuve, Belgium), of which the first one has been installed in Boston, MA (USA) [29].

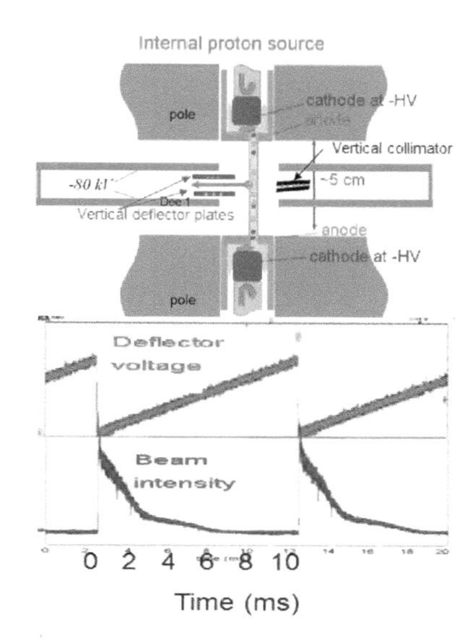

Figure 6–13 Left: Principle of an ion source in the center of a cyclotron and the vertical deflector plates controlling the intensity of the beam passing through the vertical collimator. Right: The beam intensity (lower curve) extracted from the PSI cyclotron as a function of the linearly varied voltage (top curve) on the vertical deflecting plates. This voltage deflects the beam in the vertical direction so that part of it is intercepted by the vertical collimator.

plates. The dees are placed between the magnet poles such that the magnetic field is perpendicular to the plane in which the dees are located. The magnet iron outside the dees is at ground potential. When a proton crosses the gap between the region in a dee and the grounded region at the moment the dee is at a high voltage of the appropriate polarity, it is accelerated. During its trajectory within the dee or in the region at ground potential, the proton does not experience electric fields, and at those moments, the voltages on the electrodes change sign without affecting the protons. The magnetic field forces the particle trajectory along a circular orbit so that it crosses the gaps between different dees and ground several times during one circumference. So, when having four dees, a proton gets eight accelerations in one turn. Due to the energy gain, the radius of the proton circular orbit increases, so that the orbit spirals outward. The maximum energy E_{max} (typically 230 or 250 MeV) is reached at the outer radius of the cyclotron's magnetic field. In a magnetic field of strength B, the orbit radius r scales with the particle momentum and charge q according to:

$$B.r = \frac{p}{q} \tag{6.1}$$

Due to the almost quadratic relation between kinetic energy and momentum, the constant energy gain per turn of, say, 0.5 MeV, yields a decreasing radius increase between the orbits in the acceleration process. The higher the energy, the larger the radius, but the smaller the radius difference between neighboring orbits.

The most important operational parameter of the RF system is the frequency of the RF voltage, which typically has a value between 50–100 MHz. At each dee it must be synchronous to the azimuthal frequency of the protons at all radii.

The magnetic field between the poles, having a typical value between 1 and 2 T, is created by means of a pair of coils surrounding the poles. Recently, cyclotrons have been developed with super conducting coils that allow stronger fields of, say, 3–4 T at much lower power consumption. The magnetic field is determining the beam dynamics and must be correct within a few times 10^{-5}. This specification originates from the basic specification that the field must be *isochronous*: at each radius in the cyclotron it must have the appropriate strength so that the time T a proton needs to make one turn equals the time of an integer number of RF periods. Then the proton enters and leaves the dees when they have the correct voltage phase at these moments. From the relation between momentum of a particle with mass m and field strength B, one can derive that the time T to make one turn equals:

$$T = \frac{2\pi}{q}\frac{m}{B} \tag{6.2}$$

Note that this is independent of radius, so the rotation time at all orbits (energies) equals T! That is why a cyclotron works at all. However, due to relativistic effects, this holds until approximately the radius where the proton energy approaches 30 MeV. Then the speed v of the protons becomes so large relative to the speed of light c that the particle mass m increases with respect to its rest mass m_0 due to relativity:

$$m = \frac{m_0}{\sqrt{1 - \frac{v^2}{c^2}}} \tag{6.3}$$

At 30 MeV this increase is approximately 3%, which is already enough to cause beam loss. At 250 MeV the effect becomes dramatic: the mass is increased by 27%. This would cause a longer orbit time T for the protons. But to keep a constant orbit frequency at larger radii (energies), one can design the magnet such that the field increases with radius as well to compensate the mass increase with radius. This design is applied in *isochronous cyclotrons.*

An increasing magnetic field with radius, however, would defocus the proton beam in the vertical direction, and the particles would be lost at collisions with the poles or dees. To reduce these losses, vertical focusing can be added. Then the magnet pole consists of *hills* and *valleys*. In that case, the pole of the magnet has regions with a small gap (*hill*) and large gaps (*valley*) between the poles. In the hills (valleys) the field is stronger (weaker) than the average field, respectively. The hills and valleys provide a variation of the magnetic field along the circular particle orbits. Therefore, these cyclotrons are called *AVF-cyclotrons*, (AVF = azimuthally varying field). This variation provides a focusing force to confine the protons within a horizontal plane during their acceleration. In many AVF cyclotrons, one has conveniently used the space in the valleys to mount the dees so that the gap between the upper and lower hill can be minimized to make a strong magnetic field along the particle orbit.

The increasing magnetic field with radius, in combination with the hills and valleys, was introduced in the 1950s. Until then, one swept the RF-frequency synchronous to the increasing proton mass (*synchrocyclotrons*). A bunch of particles is accelerated at each sweep of the RF so that the beam intensity is pulsed at the frequency by which this sweep is made, typically a few hundred Hz up to 1 kHz. Initially "old" and existing synchrocyclotrons of 160–200 MeV have been used for proton therapy at Harvard, Berkeley, and Uppsala. This technique has experienced a revival, however, in recent designs of very compact super conducting synchrocyclotrons (200–230 MeV). The combination of compactness (machine diameter <2m) and a very strong magnetic field (5–8 T) causes a field strength decrease with radius. In addition to the relativistic mass increase, this also requires an RF-frequency sweep synchronous to the orbit frequency, which decreases with radius. Such compact cyclotrons are enabling proton therapy facilities with a single treatment room [28] and even enable the mounting of the cyclotron on a rotating gantry [29].

At extraction, one has to separate the orbit to be extracted from the previous one that has to follow another turn. Similar to a synchrotron, this separation is made by a *septum*, a thin metal plate with a strong radial electric field on one side. There protons are deflected out of the cyclotron by this field. However, as mentioned before, the radial distance between two consecutive orbits decreases with radius. Especially at extraction radius, this would cause beam losses at the septum. Using a small but carefully made distortion in the magnetic field, a sudden shift of orbits can be obtained. This increase of orbit separation is made at the location of the septum. Using such

Table 6–1 Comparison of the most important parameters of the accelerators currently in use for proton therapy

Parameter	Synchrotron	Isochronous Cyclotron	Synchrocyclotron
Machine size (∅)	6–8 m + injection system	3.5–5 m	<2 m
Time structure beam intensity	Spill structure, dead time >10%	Continuous	Pulsed, Dead time 99%
Fast energy scanning	Wait for next spill *During extraction: in development*	Degrader	Degrader
Activation	No problem	Degrader needs shielding	Degrader needs shielding
Beam intensity	Limited in magnitude and range	"Any" Adjustable within <1ms	"Any," but low on average. Adjustable within a few ms
Intensity stability	10–20% *In development*	2–5%	20%
Scattering	Suitable	Suitable	Suitable
Spot scanning	Suitable	Suitable	Long time needed
Beam gating	Suitable	Suitable	Suitable
Fast continuous scanning	Difficult due to pulse structure	Suitable	Not possible due to pulse structure

tricks an *extraction efficiency* of 30% to 80% can be achieved. Since beam losses at extraction cause a lot of radioactivity (difficult to service) and increase failure probability (high power losses), high extraction efficiency is of advantage.

In general, the most important advantages of a cyclotron are its small size, the continuous character of the beam (in the case of isochronous cyclotrons), and that its intensity can be adjusted very quickly to virtually any desired value. Another advantage of a cyclotron-based system is its capability to change the beam energy very quickly when using the appropriate energy degrader and an energy selection system, as discussed before. The recent cyclotron developments are aiming at very compact systems that use very strong magnetic fields and exploit the synchrocyclotron concept again. Although the commercial advantage on a short term is clear, the search for a good balance between treatment quality (e.g., a low neutron dose [30] and accurate pencil beam scanning), and compactness of the facility is still a big challenge.

6.3.4 Comparison of Accelerators

Synchrotrons and (synchro)cyclotrons can be used for high-quality proton therapy. However, each machine has its pros and cons, which strongly depend on many factors, one of which is the dose application method. Table 6–1 presents a comparison of the most relevant technical parameters of the accelerator types used today: the synchrotron, the "normal" isochronous (AVF) cyclotron, and the synchrocyclotron.

Considering the strong, increasing interest in the scanning technique, the table shows that, at the moment, the isochronous cyclotrons offer the best possibilities for advanced treatments such as continuous scanning, closely followed by the synchrotrons. The interesting developments on energy change within a spill will make these two machines very comparable. Of course, each upcoming facility has to decide which parameters get high priority. Certainly many other arguments—like available space, technological and clinical environment, interest (research) in further developments, and financial aspects—will play a crucial role in the selection process.

6.4 New Accelerator Developments

In the accelerator field, different and new acceleration methods are being investigated and developed. New developments have originated from high-energy physics and military applications. Usually these techniques focus on a shorter path length to accelerate so that the accelerator size is reduced. Most of the new developments in this field are aiming at a reduction of the equipment cost due to a reduction of dimensions or due to a "simpler" system. However, it should be noted that in most of these developments one just assumes that the beam quality at the patient or the dose delivery is at least similar to equipment currently in use. But, of course, this should be investigated as well. Although these ideas are extremely interesting, at the moment all of them are in a very experimental phase and still far from a safe, reliable, and high-quality patient treatment. In many cases, the really achievable advantages of these new systems with respect to the currently used systems based on a cyclotron or a synchrotron are not clear yet. But since these techniques potentially offer extremely interesting possibilities, it is useful to give a short overview of these developments.

6.4.1 Laser Plasma Acceleration

The big potential advantage of laser acceleration is that "only" a laser beam needs to be transported to the treatment room. Such transport of light can be made within narrow tubes using lenses, mirrors, and fiber-optics. This is, of course, much simpler than the conventionally used magnets for proton transport and the necessary concrete shielding of ionizing radiation. The current idea is to focus the laser beam at a target in the treatment room. Due to the interaction of the laser beam with the target, protons will be accelerated out of the target, and they will be directed toward the patient using an energy selection and focusing system. As shown in Figure 6–14, the energy selection and focusing is planned to take place just before the patient. Also, scanning the light beam would, in principle, provide opportunities for pencil beam scanning [31]. Several methods are currently being studied to accelerate protons by means of strong laser pulses [32]. At the moment, most experience has been obtained with the *TNSA* (target normal sheet acceleration) method [33].

All techniques are based on the acceleration of protons in the extremely strong electric fields in the beams of high-power lasers. As shown in Figure 6–14, a very intense laser beam hits a target: a metal foil, sometimes doped with hydrogen. The very strong electric fields may ionize the atoms in the foil. A plasma is created due to the energy absorption in the foil, and electrons in this plasma emerge from the rear surface. The electrons are pushed forward by the electric fields in the laser light, and

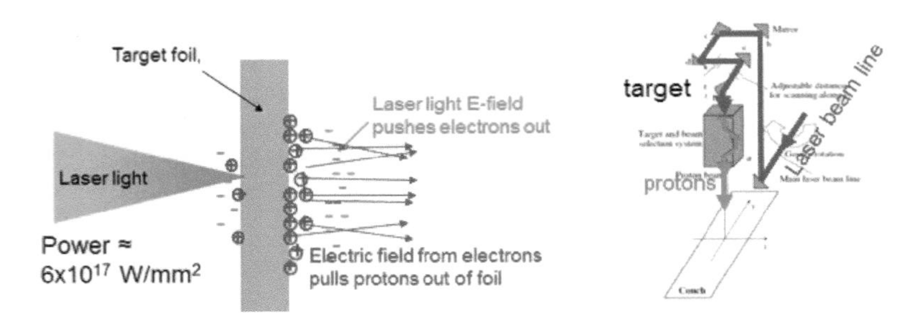

Figure 6–14 A laser-driven accelerator is based on a high-power laser pulse impinging on a target doped with hydrogen atoms. The laser light accelerates the electrons out of the foil so that an electric field is created that pulls protons out of the foil. Right: a conceptual illustration on a therapeutic use of a laser-driven proton accelerator.

the electrons may leave the foil at the rear side. When the laser pulse is sufficiently strong, the space charge created by the electrons will pull and accelerate protons out of plasma (the foil remnant). The typical energy obtained by the protons can be up to 20 MeV [34]. Of course, the proton beam needs to be focused for further transport.

There are several critical issues to be solved before such an application can be used in proton therapy. The first issue concerns the laser power. In typical experimental setups, one uses approximately 10^{18} W/mm^2. This can currently be made by a pulsed laser (with pulse length of 320 fs) with a duty cycle of 1–10 minutes. Quite some target technology also needs to be developed since the target evaporates when being hit by a laser pulse.

Above all, the amount of protons leaving the foil is still by far not enough for therapy, and it is not clear how the intensity can be controlled accurately. Also, the energy of the emitted protons is still far too low. The maximum proton energy one obtains is typically 10–30 MeV, and with a continuous energy spectrum [35]. So for treatment, an energy selection must be done, and a lot of protons can thus not be used. To save magnets of a proton beam transport system, this energy selection is proposed to take place just before the patient, which would cause a huge neutron background.

An alternative method uses radiation pressure acceleration (*RPA*) [36]. Here the light pressure of a laser pulse incident on a foil accelerates the whole foil as a plasma slab. Simulations predict that the RPA method can provide higher proton energies and less energy spread than TNSA.

However, the RPA method faces even more technological challenges. A huge amount of work still needs to be done. Before considering this method for application in proton therapy, both a power increase and a repetition rate increase of at least a factor 100 are needed. Better target technology and methods to optimize the energy spectrum are also needed [37]. Even though the field is developing very fast [38], it is expected that it will still take several decades to develop a laser-driven medical facility for proton therapy [39].

6.4.2 Linear Accelerator

Linear accelerators are the standard accelerators of electrons for photon therapy. In a linear accelerator, protons are accelerated in several RF sections located behind each other. When protons cross such a section at the appropriate moment, they cross a volume with a very strong electric field, which causes the acceleration. Due to the limitation of the electric field strengths, the maximal energy gain per meter is approximately 20 MeV/m. Therefore, linacs for >200 MeV protons need at least a length of 10–15 m, which would give a large footprint in a clinical facility. Furthermore, their pulsed operation is not convenient for several dose delivery methods. The maximum energy gain of typically 10–20 MeV/m is determined by the strength of the electric field, which is limited by the probability of creating a spark. This probability increases with field strength, but also with the time such a field is maintained. Therefore, RF with high frequencies (several GHz) is often used to reach very strong electric fields.

In a design of the Italian TERA group [40], it is proposed to mount a linear accelerator on a gantry ("TULIP"). The linac is preceded by a small 30 MeV synchrocyclotron that injects the proton beam into the linac ("cyclinac"). The linac is made of two accelerating sections. The accelerated protons are bent to the isocenter via a scanning system. Novel technological developments in this project concentrate on the linac's RF cavity design to achieve a stronger acceleration field (using technology from the CLIC project at CERN, Geneva), but also on technical issues like rotatable junctions in the RF power line between the amplifiers and the RF cavities in the linac and the design and the coupling between cyclotron and linac [41]. A potential advantage of this linac is its possibility to switch on only a controlled variable number of RF cavities during a pulse or to vary the RF power per cavity. In this way, the linac can select the beam energy per pulse, which occurs at a rate of 100–200 Hz. This would yield the fastest method to modulate the energy from an accelerator in a controlled way.

Currently one is working on the design, and many tests of components have been done successfully, but no such gantry has been built yet.

6.4.3 Dielectric Wall Acceleration

If no resonating cavity is used, but when a strong electric field is simply used between two conducting plates, acceleration can be made between these plates. A hole in the center of the plates allows the passage of the protons to the next plate. Fields over insulators sandwiched between these conducting plates will break down when the voltage difference between the conductors is too large. For conventional insulators, a shortening of the high-voltage pulse from 1000 ns to 1 ns has shown an increase of the surface break down field from 5 to 20 MV/m. If the insulators are made of a specially developed castable dielectric material called high gradient insulator (HGI dielectric), field strengths of approximately 100 MV/m have been reached. In theory, this would mean that 200 MeV proton beams can be created in a 2 m long structure, a dielectric wall accelerator (DWA). Following this possibility, the idea came up to have a linac turning around a patient like a tomograph [42]. A linac structure based on this idea has been developed [43]. This structure consists of a stack of 4000 metal electrodes at

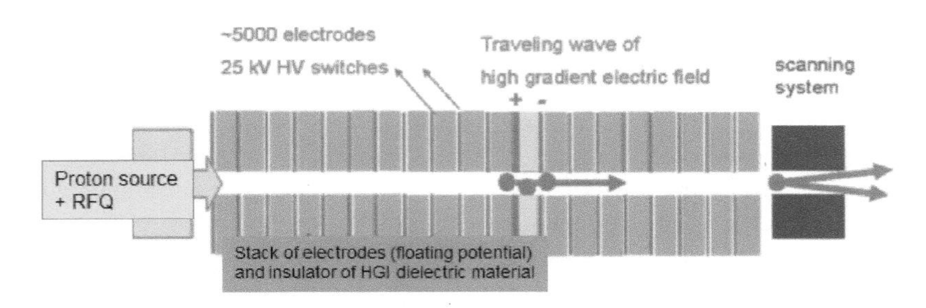

Figure 6–15 Principle of operation of the dielectric wall accelerator (DWA). It consists out of a stack of electrodes insulated by dielectric material (the blue boxes). When made of HGI, this material can stand very high voltages. A traveling high-gradient field (shown as a yellow box) is created by switching high voltages on electrodes.

a "floating" potential, with 0.25 mm thick HGI sheets in between them. One applies a 25 kV pulse during 2 ns over two neighboring electrodes and a bunch of protons is accelerated. A little bit later, synchronous to the speed of the proton bunch, one applies this to the next electrode pair, etc. To prevent electric breakdowns, the electrodes are on floating potential when they are not connected to the high voltage. This reduces the amount of secondary electrons in an eventual discharge. By switching the HV at the sequential electrodes on and off, one shifts the 25 kV pulse along the stack of 4000 electrodes. The movement of the 25 kV pulse can be used to pull a bunch of protons with it (see Figure 6-15). It is proposed that this process will accelerate a bunch of approximately 10^8 protons, and it is repeated at a rate of 10–50 Hz.

However, tests of such a DWA have shown that this ideal situation has not yet been reached [44]. On average, only 15 MV/m is obtained [45], and one also needs a few meters of length for the ion source (an RFQ injector), a beam aiming (scanning) system, and some SAD at the exit of the accelerator. Also, it is not yet clear how accurately the dose per pulse can be controlled [46]. Above all, there are several operational difficulties, such as the 200 kW cooling, the low dose rate at 10 Hz, the reliability of the 4000 25 kV switches, and the reliability of the proton energy. But work is in progress, and a 4 m prototype for 150 MeV protons is under construction.

6.5 Gantries

6.5.1 Requirements

A gantry is used to direct the particle beam from different directions to the tumor in a patient lying on a treatment table. This is a mechanical system that rotates magnets of the last part of the beam line system around the patient.

Gantries for particle therapy are of considerable size. To bend and focus the proton beam, large (and thus heavy) magnets are needed. Typical magnets bend the proton tracks with a radius of curvature of approximately 1.5 m. A distance (throw) of at least 2 m is needed to spread the beam behind a scattering system. For pencil beam scanning, a similar amount of space is needed to get sufficient lateral displacement of the scanned beam at moderate angular divergence. Also, between the exit of the last magnet and the patient, space is needed for additional equipment, such as dose monitors, a range shifter, an energy modulator, and collimation systems. So the large magnets and other heavy equipment make the gantry quite large. A diameter of 10–12 m, a length of 10–15 m, and a weight of 100–200 tons are typical (see, for example, the gantry shown in figures 6–16 and 6–17 [47]).

If one obtains an almost parallel pencil beam scanning at isocenter, one can work with a relatively short distance between the nozzle exit and the isocenter of approximately 0.5 m. With "downstream" scanning (scanning magnets between the bending magnets and the isocenter) or with a scattering system to spread the beam, one typically needs 1.5–2 m space between the scanning magnets or scatter foils and the isocenter (large enough SAD needed). The other part that contributes to the total gantry radius is the last bending magnet section. With a typical bending radius of 1.5 m for protons in conventional bending magnets, approximately 2 m should be counted for the magnet. One also needs approximately 0.5 m for nozzle equipment like dose monitors, range shifters, etc. This all yields a minimum proton gantry radius of approximately 5 m, when assuming a gantry with a fixed isocenter in the room (*isocentric* gantry). In many cases one prefers to have a fixed isocenter in the room for optimal and gantry angle and independent access to the patient.

6.5.2 Overview of Existing Systems

The first three gantries for protons originated from a design by the Harvard group [48] and were installed at the facility in Loma Linda [13]. The beam transport system layout on this type of gantry is referred to as a *corkscrew* and consists of a double, achromatic, bending system.

In general, the beam optics in a gantry design must be *achromatic*. This implies that the beam position and beam size at the isocenter are, within a certain range, not sensitive to the beam energy. A bending magnet always causes a difference in bending angle of the proton trajectory for different energies. This energy-dependent shift of the track is referred to as dispersion. If the total dispersion of a system equals zero due to compensation of the dispersions of the different bending magnets, the system is achromatic. For a gantry, this means that the position and cross section of the beam focus at isocenter is, within 0.5 percent or so, independent of beam energy. This makes the gantry optics quite robust against small energy errors. The double achromatic system of the corkscrew (first: 2 x 45°, second: 2 x 135°) is an excellent example of such a system.

The gantry nozzle of this gantry has been designed for a scattered beam. Gantries developed later also used single magnets of 135° (Figure 6–18), and designs have been made both for scattering and for scanning. In some gantries, this 135° magnet has been split up into several shorter bending magnets with quadrupoles in between to comply with the large dispersion.

Figure 6–16 The construction phase of a gantry, clearly indicating the typical dimensions.

Figure 6–17 A model of the gantry at the Loma Linda proton therapy facility. The blue arrow indicates the proton beam entering the gantry from the right. In the gantry rotation position shown, the beam is bent upwards by 45° after entering the gantry. Then it is bent another 45° into the vertical direction. From the vertical direction, the beam is bent in a horizontal direction and aimed at the patient position by using two 135° magnets.

With the increasing demand for pencil beam scanning, dedicated nozzle designs have been made for scanning. In some cases, the nozzle is also equipped with an option to perform passive scattering. In the case of pencil beam scanning, the location of the scanning magnets—which are usually mounted at the exit of the last bending magnet—is the virtual source of the pencil beams. The large distance to isocenter to obtain a 2 m SAD is in a complicated conflict with the need for small pencil beam diameters at isocenter.

At PSI, a compact proton gantry ("Gantry1") [3] has been optimized for pencil beam scanning. In this gantry, a scanning magnet has been mounted before the 90° magnet that bends the beam toward the isocenter. The beam optics of this 90° magnet has been designed such that the scanning magnet causes a parallel shift of the pencil beam at the isocenter. In this way, the virtual source of the pencil beams (SAD) is located at infinity, and an orthogonal pencil beam arrangement is obtained. The other two orthogonal displacements are performed by inserting range shifter plates in the nozzle to shift the Bragg peak in depth and by shifting the table in the direction orthogonal to the magnetic scanning direction. The thus-obtained orthogonal grid and parallel beam configuration have the advantage that there is virtually no field size limitation; by shifting the patient table, one can easily apply large fields ("field patching"). However, the total treatment time is rather long due to the slow motion of the patient table.

Although Gantry1 was the first gantry in the world applying spot scanning, the single scanning direction was limiting its possibilities. In order to have fast magnetic scanning in two directions, "Gantry2" has been built [10] and is in use since 2013

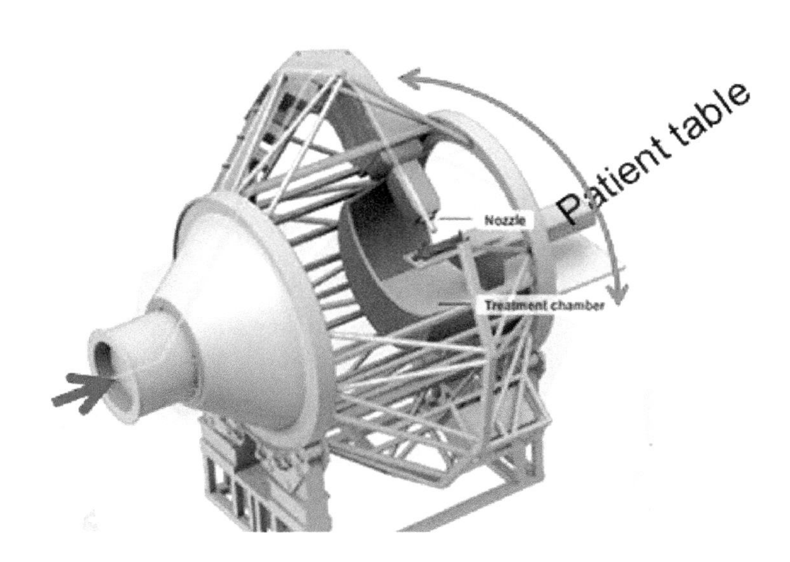

Figure 6–18 A gantry with a single 13° magnet.

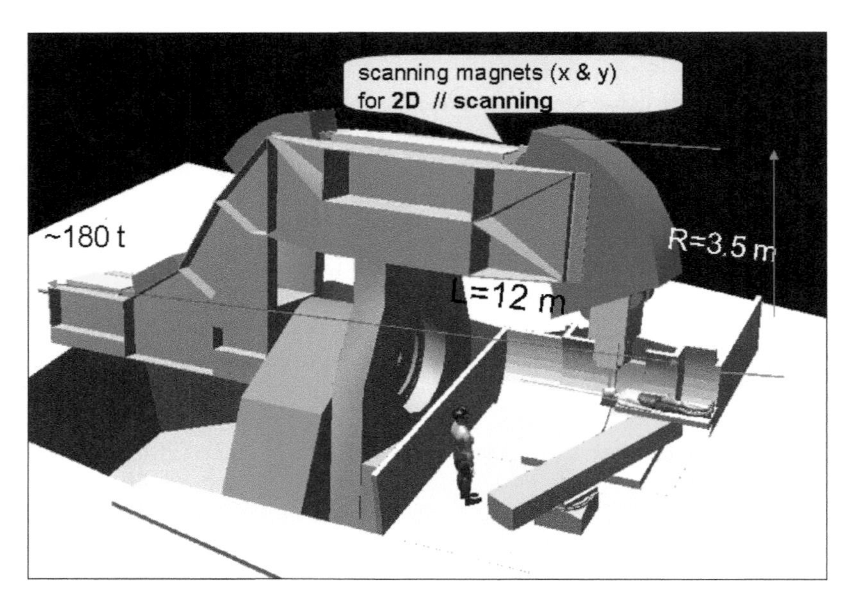

Figure 6–19 Design and realization of Gantry 2 at PSI. This gantry is designed for parallel beam scanning in two dimensions and fast energy scanning. It rotates a little bit more than 180° around the patient to have good access to the patient at any gantry angle.

(Figure 6–19). It allows for double magnetic scanning in a field of 12 x 20 cm^2 with parallel beam displacements in both directions. To allow scanning in the direction orthogonal to the bending plane, of course, a relatively large gap of the last 90° bending magnet is needed. The magnets of Gantry 2 are laminated so that a range change of 5 mm in water is achieved in <0.1 sec.

There are several important constraints on the beam optics to couple the beam into the gantry. At the rotation coupling point, it is of utmost importance that the beam is well aligned. It must be symmetric in size, as well as in divergence, and it must be dispersion free. Otherwise the beam shape, position, or intensity at the isocenter could depend on the gantry angle. The alignment of the beam position and direction at the gantry entrance are especially important for gantries that use the pencil beam scanning technique, but small position errors are easy to compensate for by using small steering magnets in the gantry. At PSI, the beam optics of both gantries is designed such that the entrance of the gantry is imaged to the isocenter with a magnification factor close to unity. A collimator has been mounted at the entrance of the gantry which thus defines the size of the pencil beam size at the isocenter. A misalignment of the fixed beam or a focusing error then only results in a loss of beam intensity at the collimator, but such errors have no effect on the position of the pencil beam position at the isocenter.

Parallel to the developments to enable pencil beam scanning, there is a strong interest to reduce the gantry size, or to include energy degrading to reduce the facility footprint. In fact, the only way to reduce the diameter of proton gantries substantially is to configure the gantry for pencil beam scanning with "upstream" scanning, placing scanning magnets before the last bending section before the isocenter.

6.5.3 Technical Issues and Developments

The need to deal with moving targets or organs requires advanced and fast methods to verify the position of the tumor and critical organs with respect to the beam. The position of the x-ray tubes in the gantry has important implications for the nozzle design, as well as for the obtained beam quality. The x-ray tubes are typically mounted in, next to, or opposite the nozzle. Recently, and in analogy to techniques used in photon treatments, one has introduced x-ray systems rotating around the patient at the treatment position. This allows the possibility to make so-called cone-beam CT images, which are very helpful in patient positioning (verification). Of course, it is no small feat to mount such systems in combination with the nozzle of a proton or ion gantry.

A gantry design employing a beam-focusing concept normally used in fixed filed alternating gradient (FFAG) accelerators allows a very large energy acceptance [10] and has a radius reduction which is mostly important for carbon ion gantries. The design of the magnet system shows very tightly packed focusing and defocusing magnets, with gradients up to 70 T/m, to be realized with superconducting magnets. Despite the very interesting beam optics, no such device has been built yet.

A gantry rotation over 180° (actually about 200°) instead of 360° reduces the footprint of a gantry by approximately 30% and allows a very good patient access (now also from the side) as well as an easy possibility of shifting the patient table into an in-room imaging device (e.g., CT or PET). A horizontal rotation of the patient table over 180° is necessary, and this combination has been applied for the first time in PSI's Gantry-2 [10]. Since then, this idea has been taken over by several companies. In the newly developed scanning gantry from IBA, the beam analysis is made behind the first 45° magnet in the gantry (see Figure 6–20). Although this gantry does not have a small radius, this combination of included energy analysis and the 180° gantry rotation saves space since it can be directly coupled to a cyclotron+degrader system.

Gantry size reduction by using super-conducting (SC) magnets is only successful in gantries for heavy ion therapy [49], where the bending radius of the carbon beam ions is approximately 5 m. Stronger magnetic fields by means of SC-magnets will significantly reduce this radius. In proton gantries, the magnet dimensions (and thus the gantry radius) could also be reduced by 0.5–1 m by using SC magnets, but the distances between nozzle and isocenter are fixed. Therefore, the most important contributions of SC magnets are in weight reduction (by a factor of 4–5) and in beam optics.

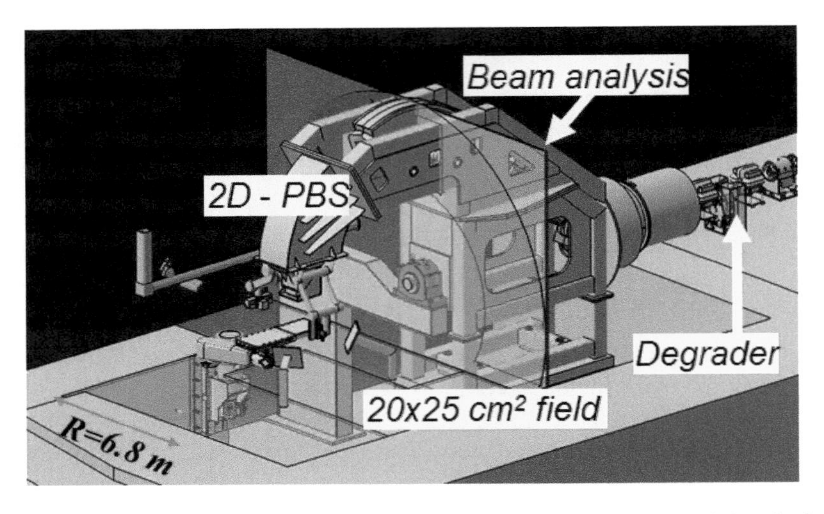

Figure 6–20 Design of a scanning gantry, rotating approximately 180°. It has included the analysis of the degraded beam into the gantry.

6.6 Conclusions and Outlook

Current developments in proton beam production and dose delivery techniques concentrate on five "highs" and one "low":

The Five Highs:

1. High reliability of the equipment to guarantee the patient throughput.
2. Higher quality: better dose distribution by using scanning techniques, and better target imaging and definition, e.g., image-guided PT.
3. Higher flexibility:
    a. Invest for a long (>25 yrs) operation period. "Future proof" by making it possible to make upgrades without too many changes or interruptions to operation.
    b. Have flexible, dynamic treatment:
        • fast and accurate intensity control and
        • fast energy modulation
4. Higher intensity to deal with organ motion.
5. Higher energy to enable proton radiography/proton-CT at all tumor sites for better range calculation.

...and one "Low":

1. Reduction of size to reduce the investment costs: smaller cyclotrons, synchrocyclotrons, and synchrotrons.

As discussed in this chapter, new accelerator concepts are being and have been developed, aiming at a reduction of the system size to reduce the costs of a proton therapy facility. Two initiatives—one based on small synchrocyclotrons and one based on a small synchrotron—have recently started in clinical use or are expected to be in operation soon. For other developments, there is either still a long way to go from the proof of principle to a clinical system, or these ideas are still in a fundamental research phase. Although the ideas are based on very interesting concepts, one should be very critical about such new developments and so-called cheap solutions. It is important for proton therapy as a whole that new developments provide *at least the same therapy quality* as we have now.

With the five "highs" and one "low" listed above, a boost is given to the possibilities and quality of proton therapy on a longer term. These and the above-mentioned developments indicate that proton therapy technology is an evolving discipline, and that there are many interesting developments in the pipeline.

References

1. Chu WT, Ludewigt BA, Renner TR. Instrumentation for treatment of cancer using proton and light-ion beams. Rev Sci Intrum 1993;64:2055–2122.
2. Pedroni E, Bacher R, Blattmann H, Böhringer T, Coray A, Lomax A, Lin S, Munkel G, Scheib S, Schneider U, et al. *Med Phys* 1995;22(1):37–53.
3. Haberer T, et al. Magnetic scanning system for heavy ion therapy. *Nucl Instrum Meth A* 1993;296–305.
4. Schippers JM, et al. PSI Scientific Report 2006. Paul Scherrer Institute, 5232 Villigen PSI, Switzerland, 62–63.
5. De Laney TH, Kooy HM, Eds. Proton and charged particle radiotherapy. Philadelphia: Lippincott Williams & Wilkins, 2007.
6. Gottschalk B, unpublished book available in pdf format at: http://huhepl.harvard.edu/~gottschalk/.
7. Koehler AM, Schneider RJ, Sisterson JM. Flattening of proton dose distributions for large-field radiotherapy. *Med Phys* 1977;4:297–301.
8. Grusell E, Montelius A, Brahme A, Rikner G, Russell K. A general solution to charged particle beam flattening using an optimized dual-scattering-foil technique, with application to proton therapy beams. *Phys Med Biol* 1994;39:2201–2216.
9. Takada Y. Optimum solution of dual-ring double-scattering system for an incident beam with given phase space for proton beam spreading. *Nucl Instrum Methods Phys Res A* 2002;485:240–261.
10. Pedroni E, Bearpark R, Böhringer T, Coray A, Duppich J, Forss S, George D, Grossmann M, Goitein G, Hilbes C, Jermann M, Lin S, Lomax A, Negrazus M, Schippers M, Kotle G. The PSI gantry 2: a second generation proton scanning gantry. *Z Med Phys* 2004;14:25–34.
11. Phillips MH, Pedroni E, Blattmann H, Boehringer T, Coray A, Scheib S. Effects of respiratory motion on dose uniformity with a charged particle scanning method. *Phys Med Biol* 1992;37:223–233.
12. Ipe NE, et al., Ed. Shielding design and radiation safety of charged particle therapy facilities. PTCOG Report 1, 2010. http://www.ptcog.ch/index.php/ptcog-publications.
13. Slater JM, Archambeau JO, Miller DW, Notarus MI, Preston W, Slater JD. The proton treatment center at Loma Linda University Medical Center: rationale for and description of its development. *Int J Radiat Oncol Biol Phys* 1992;22:383–389.
14. Hiramoto K, et al. The synchrotron and its related technology for ion beam therapy. *Nucl Instrum Methods Phys Res B* 2007;261;786–790.
15. Hirao Y, Ogawa H, Yamada S, Sato Y, Yamada T, Sato K, Itano A, Kanazawa M, Noda K, Kawachi K, Endo M, Kanai T, Kohno T, Sudou M, Minohara S, Kitagawa A, Soga F,

Takada E, Watanabe S, Endo K, Kumada M, Matsumoto S. Heavy ion synchrotron for medical use—HIMAC project at NIRS-Japan. *Nucl Phys A* 1992;538:541–550.
16. Badano L, Benedikt M, Bryant P, Crescenti M, Holy P, Knaus P, Maier A, Pullia M, Rossi S. Synchrotrons for hadron therapy I. *Nucl Instr Meth A* 430 1999;512–522.
17. Furukawa T, et al. Design of synchrotron and transport line for carbon therapy facility and related machine study at HIMAC. *Nucl Instr Meth A* 2006;5621050–1053.
18. Ondreka D, Weinrich U. The Heidelberg ion therapy (HIT) accelerator coming into operation. Proceedings of EPAC 2008 11th European Particle Accelerator Conference, Genoa, Italy, June 23–29,2008. TUOCG01–979 (abstr).
19. Schomers C, Feldmeier E, Naumann J, Panse R, Peters A, Haberer T. Patient-specific intensity-modulation of a slowly extracted beam at the HIT-synchrotron. IPAC'13: Proc. 4th Int. Particle Accelerator Conf. (Shanghai, China) 2013;2944–6.
20. Tang JY, Liu L, Yang Z, Fang SX, Guan XL. Emittance balancing technique for the resonant slow extraction from a synchrotron. *Phys Rev ST Accel Beams* 2009,050101–9.
21. Moyers MF, Lesyna DA. Exposure from residual radiation after synchrotron shutdown. *Radiat Meas* 2009;44(2):176–181.
22. Iwata Y, et al. Multiple-energy operation with quasi-DC extension of flattops at HIMAC. IPAC10, Kyoto, Japan. OPEA008,79–81.
23. Peggs S, et al. The rapid cycling medical synchrotron, RCMS. Proc. 8th European Particle Accelerator Conference, Paris, France, 2002, 2754–2756.
24. Balakin VE, Skrinsky AN, Smirnov VP, Valyaev YD. TRAPP-facility for proton therapy of cancer. First European Particle Accelerator Conference, Rome, Italy, June 7–11, 1988, p.1505.
25. Jongen Y, Beeckman W, Blondin A, Kleeven W, Vandeplassche D, Zaremba S, Aleksandrov VS, Glazov A, Gurskiy S, Karamysheva G, Kazarinov N, Kostromin S, Morozov N, Samsonov E, Shevtsov V, Shirkov G, Syresin E, Tuzikov A. IBA C400 cyclotron project for hadron therapy. In: Proc. of Cyclotrons 2007, Italy, 151–153 (abstract).
26. Schillo M, et al. Compact superconducting 250 MeV proton cyclotron for the PSI PROSCAN therapy project. Proceedings 16th International Conference. *Cycl Appl* 2001;37–39.
27. Schippers JM, et al. The SC 250 MeV cyclotron and beam lines of PSI's new proton therapy facility PROSCAN. *Nucl Instr Meth B* 2007,261:773–776.
28. IBA. http://www.iba-worldwide.com.
29. Mevion Medical Systems. www.mevion.com/.
30. Allen PD, Chaudhri MA. Photoneutron production in tissue during high energy Bremsstrahlung radiotherapy. *Phys Med Biol* 1988;33:1017–1036.
31. Ma CM. Development of a laser-driven proton accelerator for cancer therapy. *Laser Physics* 2006,16(4):639.
32. Tajima T, Habs D, Yan X. Laser acceleration of ions for radiation therapy, *Rev Acc Sci Tech* 2009;2:201–28.
33. Wilks SC, et al. Energetic proton generation in ultra-intense laser—solid interactions. Phys Plasmas 2001;8:542–9.
34. Fuchs, et al. Laser-driven proton scaling laws and new paths towards energy increase. *Nature Phys* 2006;2:48–54.
35. Malka V, Fritzler S, Lefebvre E, d;Humières E, Ferrand R, Grillon G, Albaret C, Meyroneinc S, Chambaret JP, Antonetti A, Hulin D. Practicability of proton therapy using compact laser systems. *Med Phys* 2004;31:1587–1592.
36. Robinson APL, et al. Radiation pressure acceleration of thin foils with circularly polarized laser pulses. *New J Phys* 2008;10:013021,1–13.
37. Linz U, Alonso J. What will it take for laser driven proton accelerators to be applied to tumor therapy? *Phys Rev ST Accel Beams* 2007;10:094801.
38. Ledingham K. Desktop accelerators. Going up? *Nature Phys* 2006;2:11–12.
39. Ma CM, Maughan RL. Point/counterpoint. Within the next decade conventional cyclotrons for proton radiotherapy will become obsolete and replaced by far less expensive machines using compact laser systems for the acceleration of the protons. *Med Phys* 2006;33:571–573.
40. Amaldia U, Berraa P, Crandalla K, Toeta D, Weissa M, Zennaroa R, Rossob E, Szelessb B, Vretenarb M, Cicardic C, De Martinisc C, Giovec D, Davinoe D, Masulloe MR, Vaccaroe V. LIBO—a linac-booster for proton therapy: construction and tests of a prototype. *Nucl Instr Meth A* 2004;521:512–529.

41. Degiovanni A, Amaldi U, Bergesio D, Cuccagna C, Lo Moro A, Magagnin P, Riboni P, Rizzoglio V. Design of a fast-cycling high-gradient rotating linac for proton therapy. Proceedings of the 4th International Particle Accelerator Conference, IPAC2013, THPWA008, 3642–3644 (abstract).
42. Caporaso GJ, Akana GL, Anaya R, Blackfield DT, Carroll J, Chen Y-J, Cook EG, Falabella S, Guethlein G, Harris JR, Hawkins SA, Hickman BC, Holmes C, Horner A, Nelson SD, Paul A, Poole BR, Rhodes MA, Richardson RA, Sampayan S, Sanders M, Sullivan S, Wang L, Watson JA, Pearson DW, Slenes KM, Weir JT. Status of the dielectric wall accelerator. Proceedings of the 23rd Particle Accelerator Conference, Vancouver, Canada, 2009, TH3GAI02, 3085 (abstract).
43. Caporaso GJ, et al. Compact accelerator concept for proton therapy. *Nucl Instr Meth B* 2007;261:777.
44. Caporaso GJ, Chen Y-J, Sampayan SE. The dielectric wall accelerator. *Rev Acc Sci Tech* 2009;2:253–63.
45. Zografos A, Brown T, Cohen-Jonathan C, Hettler C, Huang F, Joshkin V, Leung K, Moyers M, Parker YK, Pearson D, Rougieri M, Hamm RW. Development of the dielectric wall accelerator, Proceedings of the 4th International Particle Accelerator Conference, IPAC2013, THOAB201, 3115–3117 (abstract).
46. Chen Y-T, Akana GL, Anaya R, Anderson D, Blackfield DT, Caporaso GJ, Carroll J, Cook EG, Falabella S, Guethlein G, Harris JR, Hawkins SA, Hickman BC, Holmes C, Nelson SD, Poole BR, Richardson RA, Sampayan S, Sanders M, Stanley J, Sullivan S, Wang L, Watson JA, Pearson DW, Weir JT. Compact proton injector and first accelerator system test for compact proton dielectric wall cancer therapy accelerator. Proceedings of the 23rd Particle Accelerator Conference, Vancouver, Canada, 2009, TU6PFP094, 1516 (abstract).
47. http://medgadget.com/archives/2009/02/. europe_approves_varians_proton_therapy_system_a_cancer_zipping_cyclotron.html).
48. Koehler AM. Proc. 5th PTCOG meeting: Int. workshop on Biomedical Accelerators (Lawrence Berkeley Lab. Berkeley, CA), 1987,147–158.
49. Trbojevic D, Gupta R, Parker B, Keil E, Sessler AM. Superconducting non-scaling FFAG gantry for carbon/proton cancer therapy. Particle Accelerator Conference, 2007. PAC. IEEE, 2007,3199–3201.
50. Iwata Y, Furukawa T, Itano AI, Mizushima K, Noda K, Shirai T, Amemiya N, Obana T, Ogitsu T, Tosaka T, Watanabe I, Yoshimoto M. Design of superconducting rotating-gantry for heavy-ion therapy. IPAC2011 Proceedings. San Sebastián, Spain, 2011, IPAC 2011, 3601–3603 (Abstract).

Chapter 7

Imaging for Proton Therapy

Katja Langen, Ph.D.[1], Jerimy Polf, Ph.D.[2], and Reinhard Schulte, M.D.[3]

[1]Associate Professor, Department of Radiation Oncology,
University of Maryland,
Baltimore, MD
[2]Assistant Professor, Department of Radiation Oncology,
University of Maryland,
Baltimore, MD
[3]Professor, Division of Radiation Research,
Loma Linda University,
Loma Linda, CA

7.0 Introduction

A variety of patient-specific images are used for proton therapy planning and image guidance during treatment. Since anatomical variations have a greater impact on the proton dose distribution compared to photon dose distributions, repeat imaging frequently accompanies the course of proton therapy. In addition, new proton-specific image modalities such as proton CT are under development and other image modalities, such as PET or prompt gamma imaging, have been adapted or are under development for the purpose of proton range verification. This chapter provides an overview of the different imaging modalities that are used in proton therapy.

7.1 Imaging for Treatment Planning

An accurate representation of the patient anatomy, both in terms of geography and material composition, is a prerequisite for proton planning. Only an accurate delineation of target and organ at risk volumes, as well as an accurate tissue heterogeneity map, allows the clinician to fully exploit the physical advantages of proton therapy. While PET and MRI images are often used to aid with volume definitions and the staging of patients, a CT image is required for proton therapy planning purposes. An exception to this general rule is the treatment of ocular tumors (see Chapter 27 on Small Field Dosimetry: SRS, and Eyes). For these treatments, typically an anatomical model of the eye is used and combined with the tumor location as determined in ancillary imaging studies.

7.1.1 Established CT-based Methods for Dose and Range Computations

The planning CT image should be of the fully immobilized patient in the treatment position. Any material that is in the proton beam path, such as immobilization devices and tabletops that are used in the treatment room, should be accurately represented in the planning CT scan. Attention should be paid to ensure that the CT field of view includes all materials that are potentially in the proton beam's path. This is of importance in proton therapy planning since any material the in the beam's path influences the proton beam range.

7.1.1.1 CT Calibration

Proton depth–dose profiles are measured in water for a range of proton beam energies during beam commissioning. For materials other than water, the proton ranges are scaled by the material's stopping power ratio (SPR) relative to water. Ideally, a direct measure of the SPR as imaged with a proton CT would be used for proton dose calculations. While research into the development of proton CT is ongoing and will be discussed in detail later in this chapter, conventional x-ray CT images are typically used to deduce the SPR from the CT Hounsfield unit (HU) values. However, uncertainties in the HU to SPR calibration exist, and these translate into range uncertainties that have to be accounted for in the treatment planning process.

Hounsfield units present an image of the x-ray attenuation μ in the material relative to water, i.e., HU are calculated by

$$HU = \frac{\mu(material) - \mu(water)}{\mu(water)} \times 1000. \tag{7.1}$$

Compton scatter and the photoelectric effect attenuates x-rays in tissue, and the latter's contribution increases as x-ray energy decreases and the atomic number Z of the material increases. While the HU are dominated by the relative electron density (RED), the translation from HU to RED is not linear and requires calibration. A typical HU-to-RED curve has two separate straight-line fits—one for HUs up to water and one for higher HUs—because of the transition from predominantly Compton effect to predominantly photoelectric effect for higher-Z materials.

For photon therapy, the HU are calibrated in terms of relative electron density, while proton calculations require SPRs. The relationship between the two quantities can be derived from the Bethe–Bloch formula:

$$SPR_w^m = \frac{\rho_{e,m}}{\rho_{e,w}} \times \frac{\ln(2m_e c^2\beta^2 / I_m(1-\beta^2)) - \beta^2}{\ln(2m_e c^2\beta^2 / I_w(1-\beta^2)) - \beta^2}, \tag{7.2}$$

or

$$SPR_w^m = \frac{\rho_{e,m}}{\rho_{e,w}} \times K, \tag{7.3}$$

where $\rho_{e,m}$ is the electron density in material m, $c\beta$ is the proton velocity, m_e is the electron mass, and I_m is the mean ionization energy of material m. For a given proton energy of 219 MeV, Schneider et al. calculated K for several tissues [1]. The factor K ranged from 0.975 to 1.025, which indicates that while the SPR is closely related to the relative electron density, the ratio of SPR to the RED is not unity, nor is it constant.

Uncertainties in the HU calibration curve have several sources of errors. First, HU vary with multiple parameters, such as the imaging beam quality (e.g., tube potential), phantom size, and the material's location within the phantom. However, a single HU-to-RED curve is typically used in photon therapy for a given CT scanner. Uncertainties introduced by using a single HU calibration curve for photon therapy result in dose calculation uncertainties of about 2% in photon therapy [2]. Since these uncertainties translate into range uncertainties in proton therapy, the use of multiple calibration curves should be considered. For example, the use of different HU-to-SPR curves for anatomies that vary in size has been suggested by Ainsley and Yeager [3].

The second source of uncertainty in the HU calibration lies in the calibration procedure. A common procedure is to image tissue substitutes of known density and composition for a direct calibration of the HU. However, the elemental composition of these substitutes is not truly tissue-like, and it has been demonstrated that their use can introduce an error when the calibration curve is used to determine the electron density or SPR of actual tissues [1]. The stoichiometric calibration procedure proposed by Schneider offers a more precise calibration, and it has been widely adopted by the proton community [1].

In this method, the composition of real tissues is used to calculate their HU and relative electron density or relative stopping power, thus allowing a "virtual" HU calibration based on real tissue. Tissue substitutes are only used for a characterization of the imaging beam that is needed to calculate HU for real tissues. For a given CT scanner and tube potential, several materials of known chemical composition and density are imaged. The attenuation of the x-ray beam and, hence, the HU, is governed by three components—photoelectric effect, coherent scattering, and Compton scattering.

$$\mu = \rho\, N_g(Z,A)\,\left[\sigma^{ph} + \sigma^{coh} + \sigma^{incoh}\right], \tag{7.4}$$

where ρ is the density and N_g is the number of electrons per unit mass and σ is the cross section for the respective interaction. The cross sections can be parameterized by their dependence on Z.

$$\mu = \rho\, N_g(Z,A)\,\left[K^{ph}Z^{3.62} + K^{coh}Z^{1.86} + K^{KN}\right] \tag{7.5}$$

where K^{ph}, K^{coh}, and K^{KN} are constants that determine the contribution of the photoelectric effect, coherent scattering, and the Klein-Nishina cross section. Using a scaled Hounsfield unit definition of H = 1000 μ/μ_w and several measurements of H with materials of precise density and material composition allows a fit of the three K parameters.

Once the constants K^{ph}, K^{coh}, and K^{KN} are determined, the expected Hounsfield units for virtual tissue materials can be calculated. The stopping power for the virtual tissues can also be calculated, and this allows for the calculation of a HU-to-SPR calibration curve. A calibration curve that was derived using this method is reproduced in Figure 7–1 [4].

For biological tissues, this calibration method results in an accuracy of about 1% [4]. However, nonbiological tissues do not necessarily lie near the biological tissue calibration curve. Figure 7–2 shows that two different bolus materials that have the

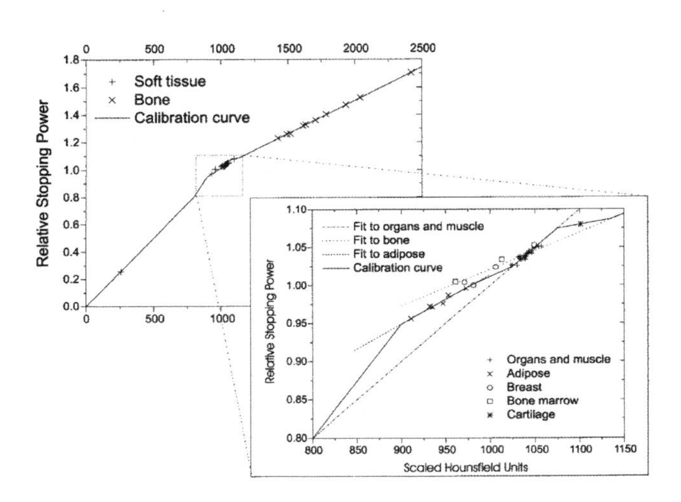

Figure 7–1 A relative stopping power to HU calibration curve determined using the stoichiometric method developed by Schneider. Separate linear fits are used for different biological tissues. (Reproduced from Schaffner and Pedroni [4] with permission from IOP Publishing.)

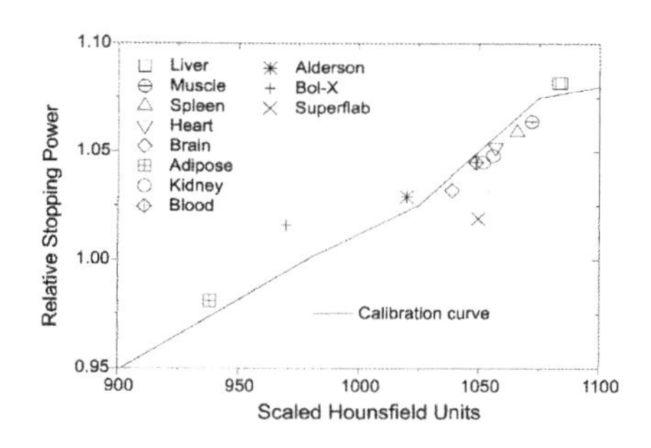

Figure 7–2 A relative stopping power to HU calibration curve determined for several biological tissues. Two bolus materials do not conform to the determined calibration curve. (Reproduced from Schaffner and Pedroni [4] with permission from IOP Publishing.)

same SPR can have different HUs. For the two examples, neither of the materials conforms to the calibration curve. The correct SPR may have to be assigned manually to these materials.

Yang et al. [5] provided a comprehensive review of range uncertainties associated with the stoichiometric calibration. Five sources of uncertainties were evaluated. These were

1. uncertainties inherent to the CT imaging process,
2. uncertainties in the determination of the K parameters,
3. uncertainties that are introduced by using standard ICRU tissues that do not include tissue variations with age or health of the patient,
4. uncertainties in the mean excitation energies, and
5. uncertainties that are introduced in the dose calculation algorithm by using a constant SPR assumption across all proton energies.

Combined uncertainties (1 σ) varied by tissue type, with lung and bone tissues having SPR uncertainties of 5% and 2.4% respectively, while SPR uncertainties were smallest (1.6%) for soft tissues. The composite range of uncertainties for various patient cases varies less (3% to 3.4% at the 95th percentile) since soft tissue is the most dominant tissue in all cases.

7.1.1.2 CT Artifacts/ Contrast CT

For some patients, contrast material is injected during the planning CT to aid volume segmentation. The presence of this contrast material influences the proton range calculation. Since the contrast material is not present during treatment, this material causes an inaccurate calculation of the proton's range in the patient. It is preferable to use a non-contrast CT for proton planning purposes. At a minimum, the density of the contrast-enhanced volumes will need to be overridden to better represent the volume's actual density during treatment. Similarly, the presence of metal artifacts results in inaccurate Hounsfield units. Metal artifacts need to be reduced or overridden.

7.1.2 Novel Methods for Dose and Range Computations

7.1.2.1 Dual-energy CT

Dual-energy CT (DECT) uses two separate scans acquired with different tube potentials or filtration to calculate the electron density $\rho_{e,m}$ and effective atomic number Z^{eff} of the imaged material. This knowledge opens the door to an alternative determination of the SPR. Please recall equation 7.2.

$$SPR_w^m = \frac{\rho_{e,m}}{\rho_{e,w}} \times \frac{\ln(2m_e c^2\beta^2 / I_m(1-\beta^2)) - \beta^2}{\ln(2m_e c^2\beta^2 / I_w(1-\beta^2)) - \beta^2}, \tag{7.2}$$

To calculate the SPR for a given proton energy requires several parameters, two of which—the electron density of material $\rho_{e,m}$ and the mean ionization energy I_m of material *m*—are unknown. If these quantities are known, the SPR can be directly calculated. DECT provide the first of two quantities by design. However, the mean ionization energy of the material is a function of the material's atomic number Z. Yang et al. discovered an empirical relationship between the effective atomic number Z^{eff} and the mean excitation energy I_m for biological tissues [6]. The empirical relationship is shown in Figure 7–3. Thus, the information provided from a DECT scanner can be used to directly calculate the SPR.

For standard biological tissues, this calibration and the stoichiometric calibration by Schneider both result in SPR determination with root mean square errors of less than 1% [6]. However, the advantage of the DECT method lies in its robustness

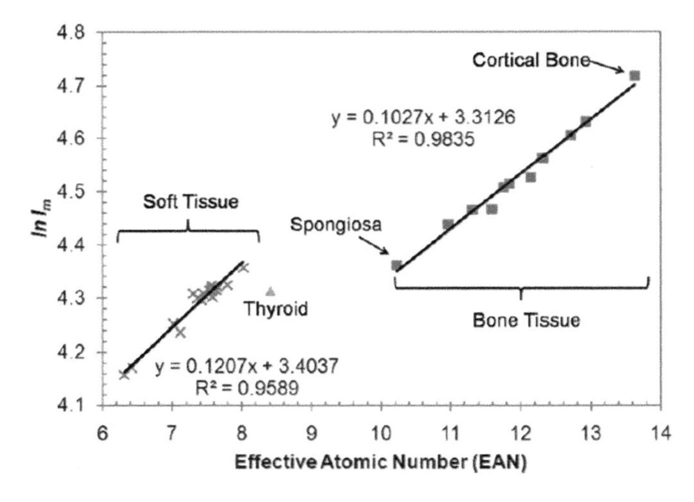

Figure 7–3 The empirical relationship between the effective atomic number and the mean ionization energy as measured by Yang et al. [6]. (Reproduced with permission from IOP Publishing.)

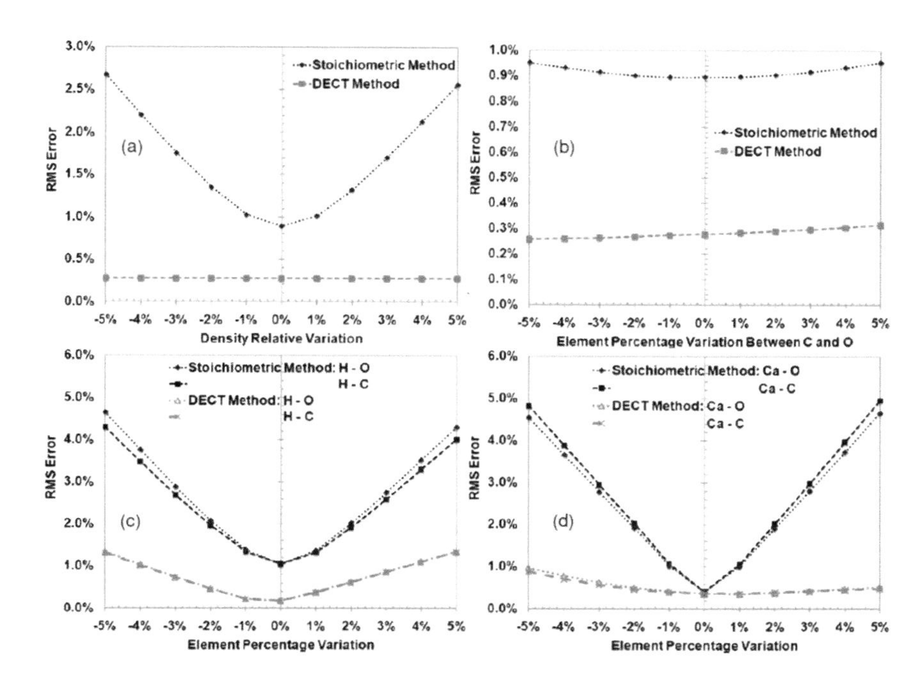

Figure 7–4 The RMS error of both calibration curves with variations in density and elemental compositions in biological materials. (Reproduced from Yang el al. [6]. (Reprinted with permission from IOP Publishing.)

against variations in tissue densities and compositions that may be encountered with variations in patient age and gender. For example, a 4% variation in density should result in an equal increase in the SPR. But the stoichiometric calibration curve may not accurately translate this change. Yang et al. [6] varied tissue density and elemental compositions and determined the error in SPR for both methods. Figure 7–4 shows that the DECT method is more robust against variations in density elemental compositions.

Huenemohr et al. [7] investigated the advantages of DECT for Monte Carlo-based proton range calculation. Mass density and elemental composition were derived from DECT scans, and this leads to an improvement in the SPR by up to 2% for selected materials.

7.1.2.2 Artifact Reduction Methods

The image quality of x-ray CT is susceptible to the presence of high-density materials, in particular compact bone and metal implants or dental fillings. High-density artifacts in CT images are primarily due to the beam hardening effect. CT images are reconstructed based on the assumption that the x-ray source is mono-energetic. In all diagnostic scanners, including those used for proton treatment planning, the energy

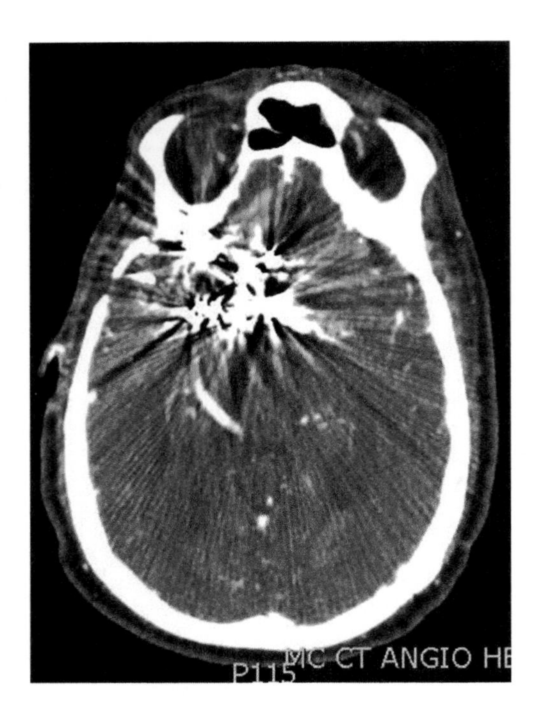

Figure 7–5 CT angiogram of a patient with large arteriovenous malformation (AVM) in the right parasellar region referred for proton radiosurgery. The presence of embolization glue enhanced with iodinated contrast and the resulting streak artifacts make the identification of the AVM nidus practically impossible. In this case, MR angiography was used to assist in the nidus delineation, and the streak artifacts, as well as the glue itself, present in the non-contrast CT were segmented and their HU values replaced by that of brain and water, respectively.

spectrum is poly-energetic, typically ranging from 20 keV to 120 keV. The faulty assumption of a monoenergetic source leads to beam hardening artifacts in normal tissues, too. Modern CT scanners apply a correction function to the attenuation data to match the attenuation data for an equivalent mono-energetic x-ray source [8]. This internal correction is, however, not accurate for metal objects. This leads to prominent streaking artifacts, both dark and bright, surrounding the metal objects. Another problem with metal implants is that the calibration of HU to relative stopping power is not optimized for nonbiological tissues, and the SPR values assigned can be inaccurate. Lastly, HU values of metal implants tend to saturate, giving an inaccurate estimate of relative stopping power.

The presence of streaking artifacts from metal implants or dental fillings, and the CT representation of the materials themselves, thus result in inaccurate HU values, as well as interference with accurate delineation of tumor volumes and organs at risk. The effect can be particularly severe when embolization glue is injected into large arteriovenous malformations prior to a course of proton radiosurgery (Figure 7–5). High-density material streak artifacts can be reduced by manually or automatically segmenting these artifacts and overriding their HU values with more realistic values based on the underlying tissue. Manual segmentation of artifacts can, however, be a tedious and time-consuming task.

Newhauser et al. [9] quantified streak-related range errors in proton treatment plans of a heterogeneous phantom and for a single prostate cancer patient with a metal hip prosthesis using corrected and uncorrected kV-CT images alone, uncorrected MV-CT images, and a hybrid approach using co-registered MV-CT and kV-CT. A TomoTherapy Hi-ART II unit (TomoTherapy, Inc., Madison, WI, USA) was used to acquire the MV-CT images. Without any correction, streak-induced range errors of 5–12 mm were present in the kV-CT-based patient plan. Correcting the streaks by manually overriding them with estimated true HU values reduced the range error to <3 mm, which could also be achieved by using the uncorrected MV-CT-based plan. The hybrid approach yielded the best result by providing reasonable delineation accuracy of soft tissues due to better contrast resolution, and its MV-CT images were less artifact-affected, providing smaller range uncertainties without requiring artifact correction. However, this method is not readily available in most proton therapy departments.

Dietlicher et al. [10] performed an anthropomorphic phantom study with an embedded cervical-spine titanium implant. To study the impact of the implant, the delivered 2D dose distribution was measured with GafChromic® films inserted between the different segments of the phantom and compared with that of three different plans comprised of four proton fields, a single-field-uniform-dose (SFUD) plan with and without artifact correction, and an intensity-modulated proton therapy IMPT plan with the artifacts corrected using manual segmentation and HU override to an average value for soft tissue. The comparison using the gamma criterion showed good agreement between prescribed and delivered dose distributions when artifacts were corrected, with >97% and 98% of points fulfilling the gamma criterion of 3%/3 mm for both SFUD and the IMPT plans, respectively. Without the artifact correction, up to 18% of measured points of the SFUD plan failed the gamma criterion of 3%/3 mm.

Alternatively, a number of artifact reduction algorithms have been developed to correct conventional kV-CT planning studies that are affected by artifacts from metal and other high-density material. These algorithms may be classified into two groups [11]. *Projection completion methods* [12] treat the projection data involving the regions causing the artifacts as ~~a~~ "missing" or corrupted data in sinogram space and fills them in using data obtained from interpolation of uncorrupted sinogram regions. These methods are usually based on fast filtered back projection (FBP) reconstruction algorithms and may result in significant metal artifact reduction in the presence of metallic implants [13]. However, the reduction of artifacts does not mean that the images are faithful. Interpolation-based methods are always associated with a loss of information caused by first removing and then interpolating the projection data, and replacing the missing data by estimated (but incorrect) data may cause additional artifacts. By removing and replacing projections, information about the artifact-causing materials and all objects contributing to the removed projections is partly lost, including the exact location of edges between high-density and regular-density (soft) tissues.

Iterative algorithms [14] provide an alternative approach to the interpolation-based artifact reduction methods. These methods utilize a more advanced, usually scanner-specific forward projection model, assuming that most artifacts arise because of incorrect or incomplete modeling of the data acquisition. FBP reconstruction algorithms use an inverse model, while iterative methods use a more complex, and thus more correct, forward model. The main obstacle for iterative methods to be imple-

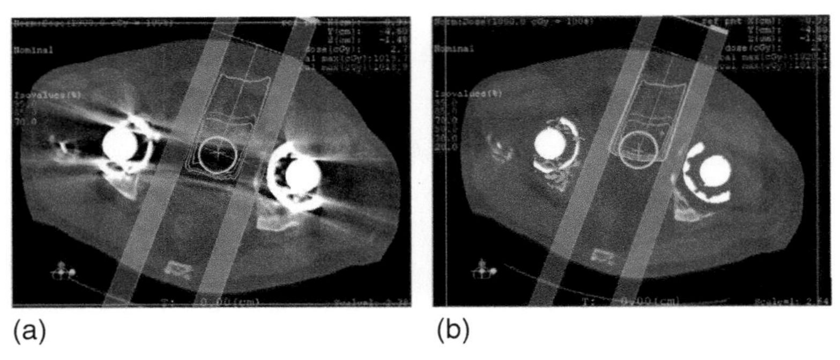

(a) (b)

Figure 7–6 (a) Dose distribution of a single 150 MeV proton beam planned to deliver a dose of 10 Gy to the centroid of a hypothetical prostate tumor (circle) based on the artifact-contaminated image. (b) Delivery of the same beam to an artifact-corrected CT image, which represents a distribution of HU and SPR values closer to the truth, results in a severe underdosage of the target. (From [15] with permission from IOP.)

mented is the increased computation time, which increases with the degree of complexity of the forward model.

Despite many publications on metal artifact reduction algorithms over the last two decades, studies of their usefulness in proton therapy treatment planning are rather scarce. Wei et al. [15] presented a case study of a patient with bilateral hip prostheses treated with a single-field proton, electron, or photon plan to a hypothetical prostate target. For all three beam modalities, the target was underdosed if metal artifact-contaminated CT images were used for dose calculations instead of images that were treated with a metal artifact suppression algorithm previously developed by the same authors [16]. The underdosage was most severe for the proton plan (Figure 7–6).

7.1.2.3 Proton CT

The ability to predict the range to 1% of the true range of protons in tissue is an important goal for accurate proton treatment planning. As discussed above, with the prevalent uncertainties in the steps of planning and delivery of proton therapy, foremost related to the inherent inaccuracy of converting HU numbers to relative stopping power, this goal has yet to be reached. Other uncertainties contributing to inaccurate proton beam range are daily variations in the patient setup, the presence of organ motion during and between treatments, and the presence of CT hardening and metal artifacts (discussed above). Ideally, the correct range of proton beams should be verified before or even during treatment, which is difficult with in-room x-ray-based imaging technologies, due to limited accuracy of range predictions based on x-ray CBCT and the relatively high dose accumulated when CBCT is used on a daily basis. Proton CT, an innovative technological development that has not yet reached clinical maturity, offers a potential solution to these problems.

The idea of proton CT is not new; it was first suggested by Cormack, the "inventor" of CT reconstruction [17, 18] as one option to create tomographic images from

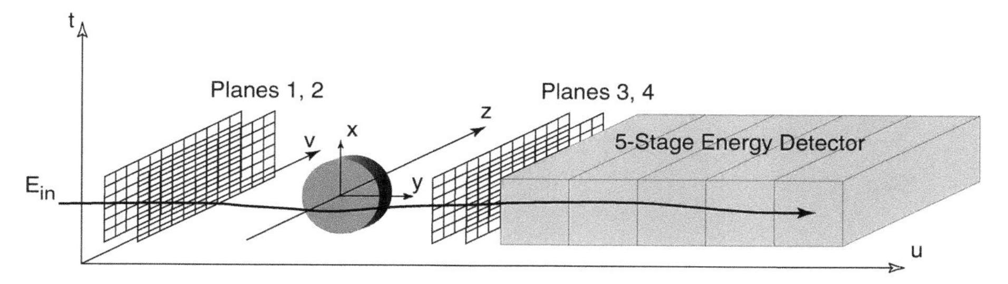

Figure 7–7 Schematic concept of a modern proton CT scanner. Four tracking detector planes are arranged to measure trajectories of individual protons of constant energy E_{in} in front of and behind the scanned object. The protons are then stopped in a multi-stage scintillating detector, which resolves their residual energy or range.

radiological measurements. Due to the wide availability of x-ray sources compared to the need for a large accelerator and a gantry to do proton CT, the latter technology remained in its infancy for a long time. Only during the last decade was interest in it renewed because of its potential to improve proton range predictions and planning for proton therapy. Detectors and fast data acquisition systems borrowed from modern high-energy physics, i.e., LHC technologies for the detection of productions from proton–proton collisions at high beam intensities have matured over the last 15 years and allow detection of individual protons at rates exceeding 1 million protons per second over a detector area of about 200 cm^2, a technology concept first proposed by the U.S. proton CT collaboration in 2004 [19]. Figure 7– 7 shows the schematic concept of a proton CT scanner with single-particle detection capability.

Proton CT has a problem with spatial resolution, i.e., the blurring of edges and washing out of small heterogeneities due to multiple Coulomb scattering, which is most pronounced in dense and high-atomic-number materials, such as bone. One possibility to reduce the effect of MCS is the "most likely path" (MLP) concept [20, 21]. In the first approximation, one can assume that the object is water-equivalent and predict the most likely lateral deviation from a straight line given by the direction of the incoming proton using a Bayesian formalism that takes into account knowledge of entry and exit directions of the protons, measured with pairs of position-sensitive detectors on the entry and exit side of the patient (Figure 7–7). The second important quantity one needs to know from each proton is its residual energy after traversing the object. Since energy loss and the traversed water-equivalent path length (WEPL) of the proton path are closely related through the Bethe formula [2], one can calibrate the instrument that measures residual energy or range so that the systems output WEPL directly [22].

The reconstruction problem of proton CT can be modeled as a so-called "feasibility problem" of finding a solution to a large and sparse linear equations system, with each linear equation representing the sum of the (unknown) SPR of intersected voxels times the intersection length, which is calculated from the MLP that is equal to the

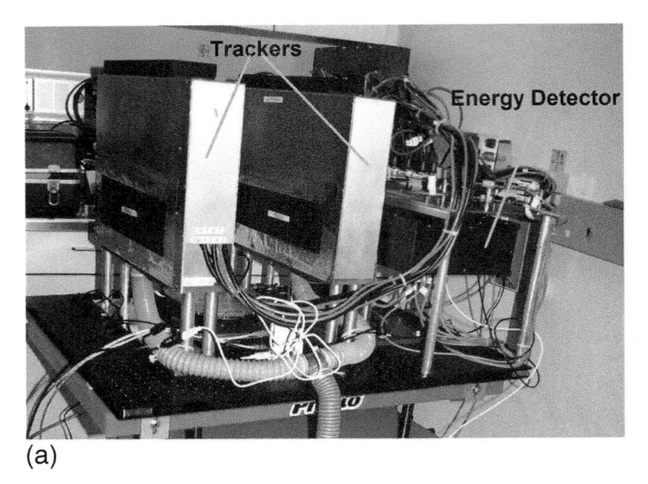

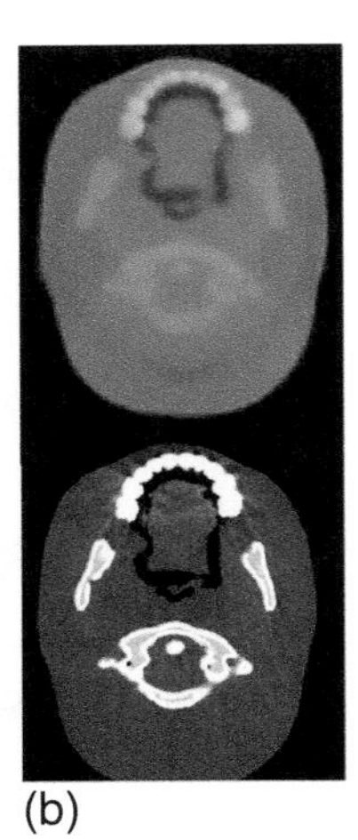

(a) (b)

Figure 7–8 (a) Phase II proton CT scanner built by the U.S. proton CT collaboration. The experimental prototype is operated on a horizontal proton research beam line and the scanned object (not shown) is rotated between the tracking detectors on a vertical-axis rotational stage. (b) Example of a proton CT head phantom image (top) compared to a diagnostic quality x-ray CT image. Note that absence of streak artifacts in the proton CT image.

measured WEPL. After initially having used the algebraic reconstruction technique (ART) [23], which is a sequential projection method, more recent approaches have utilized faster, parallelizable algorithms, such as block-iterative projection (BIP) and string-averaging projections (SAP) algorithms [24], which are suitable for implementation on fast graphics processing units (GPUs). The group at CREATIS in France has developed a different proton CT reconstruction approach, which is based on a fast FBP algorithm combined with the MLP approach [25].

The successive technology development of pCT technology has been primary driven by the Italian INFN group (PRIMA project) [26] and the U.S.-based pCT collaboration [27]. Figure 7–8 shows the Phase II scanner of the pCT collaboration and a head phantom reconstruction as an example of the state of the art of experimental pCT systems realizing the concept described in Schulte et al. [19] shown in Figure 7–7. Even more advanced, modern high-energy, physics-based technologies are likely to arise over the next few years. Development of a clinical proton CT scanner, and the first large-animal and human pilot studies in the head region, where proton CT appears most feasible at this time, seem possible within the next 5–10 years.

7.1.3 Imaging for Volume Segmentation and Treatment Planning

Proton therapy planning has basic requirements similar to photon treatment planning in the sense that images are used to identify structures of interest, to obtain an anatomical image for dose calculations, and to properly stage the patient. However, since

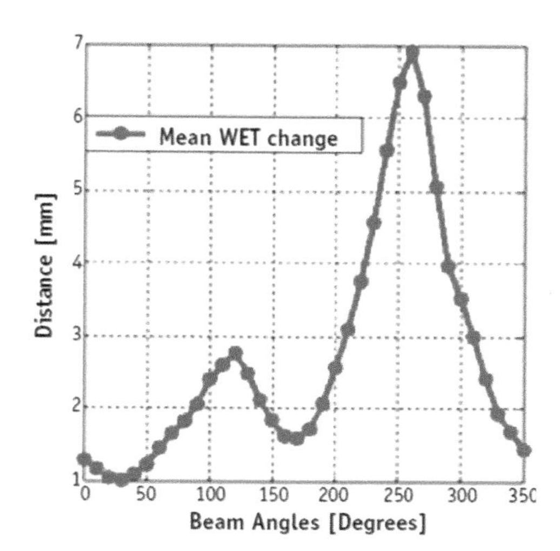

Figure 7–9 The mean water-equivalent thickness (WET) change is plotted against the beam angles. (Reproduced from Chang el al. [28] with permission from Elsevier.)

the proton range is sensitive to the material that is in the beam's path, details of how images are used in treatment planning may be different for proton therapy. In the following, these proton-specific issues are emphasized.

7.1.3.1 Use of 4DCT

First and foremost, 4DCT studies are used to assess the motion amplitude of a tumor that moves with respiration. Some proton clinics have specific motion thresholds and do not treat tumors that move beyond those thresholds. For example, a recent study from MD Anderson Cancer Center in Houston reported that patients with tumor motion larger than 5 mm were ineligible for an IMPT lung study [28]. In situations like that, a 4DCT may be used to determine the eligibility of patients for proton therapy. If it is decided to treat a moving tumor with proton therapy, the 4DCT study is used to define the ITV, similar to what is typically done for photon therapy. However, variations in tumor location and patient anatomy with breathing can change the dose distribution in the different breathing phases to a greater extent, and special care must therefore be taken.

4DCT images can be used to generate various CT images, such a maximum intensity projection (MIP) or average CT studies. It is not obvious whether a MIP, free breathing, or average CT is optimal for proton therapy planning. The use of a MIP image appears conservative since it would force the use of proton energies that are high enough such that in all phases the distal edge of the tumor is covered. However, Kang et al. found that a MIP image, if calculated in individual breathing phases, can result in an underdosage of the proximal target [29]. Protons that are aimed at the proximal edge also have a higher energy in MIP image plans, and under free breathing conditions, these protons may travel beyond the proximal tumor edge. Free breathing or average CT studies may result in an underdose of the distal target edge in phases where the radiological path length is larger than what is presented in the free breathing or average CT. Kang el al. proposed the use of an average CT with an ITV that encompasses all motion, but which has its density overridden to a uniform value of 100 HU [29]. Plans generated on this CT image were found to be most robust against breathing motion in terms of target coverage and doses to organs at risk.

In an effort to generate plans that are robust against motion, 4DCTs have been used to select optimal beam angles [28]. At different angles, breathing can cause different variations in proton range. These variations can be mapped for all beam angles, and a plot of the mean range water-equivalent thickness (WET) change with gantry angles can help to select preferred beam angles. An example of such a plot is shown in Figure 7–9. In this example, beams entering near 30 and 160 degrees should be least susceptible to breathing motion.

After a nominal dose distribution is generated, 4DCT images can be used to assess changes of the dose distribution with breathing. The nominal plan can be recalculated for each 4DCT phase to estimate tumor coverage and doses to normal tissues during delivery. These data are used to evaluate if a plan is robust enough to be acceptable for treatment.

However, to assess the effect of the interplay between spot positions and tumor motion requires synchronization of the beam delivery and tumor motion during dose calculation, i.e., a true 4D dose calculation engine. Such engines have been used for research purposes to assess the dosimetric consequence of the interplay effect [30].

7.1.3.2 Use of MRI/4DMRI

The use of MRI-based treatment planning is being debated in photon planning [31]. However, due to the need for precise electron densities, MRI-based treatment planning has not been investigated for proton therapy. On the other hand, frequent re-imaging is often used in proton clinics to monitor anatomical changes. To reduce patient dose from repeat imaging, MRI studies may have a role. A new planning CT may be triggered by observations made on repeat MRI images to accurately assess the dosimetric effect of anatomic change.

7.2 Image Guidance for Proton Therapy

Image guidance in the treatment room has rapidly evolved for photon radiation therapy over the last 20 years, but it is still in its relatively early stages for proton therapy. Here we give on overview of established and upcoming technology used to assist in aligning patients and tracking inter- and intra-fraction tumor motion.

7.2.1 Digital Radiography

Diagnostic-quality x-ray verification with 2D radiographic films, usually taken in beam direction and at an orthogonal angle, have been used for many years to align patients using skeletal features [32]. For fixed (horizontal) beam lines, additional x-ray sources were installed in order to allow for the alignment of patients who were treated with vertex or oblique proton fields. Skeletal landmarks were identified by the radiotherapist on the films, and the 2D axis of the isocentric treatment coordinate system was manually reconstructed on the films based on a comparison with digitally reconstructed radiographs (DRR). This technique made the quality of the alignment dependent upon the experience and skills of the treatment personnel, and the attending physician usually had to be present if complex field arrangements were treated at oblique angles to supervise the patient alignment verification.

In later years, x-ray films were largely replaced by digital radiographs acquired by electronic patient imaging devices (EPIDs) with one or two flat panel imagers installed in the treatment room. With this technology developed in the late 1990s, digital x-ray images could be processed to enhance skeletal features, assisting the radiotherapist in finding appropriate landmarks. Manual alignment procedures, previously practiced with x-ray films, were replaced by automated alignment algorithms, including translations and rotations. In the latest versions of these digital radiography-based alignment systems, the output from the verification algorithm has been directly coupled to the robotic patient positioner, which performs the required correction on demand. Additional images may then be captured to verify that the corrections were satisfactory. This technology is now standard in all proton treatment centers. Doses of single images are in the 10s of mGy range, and EPID-based verification can thus be performed for each fraction. Commissioning procedures for EPIDs are described in the AAPM TG-58 report [33].

7.2.2 3D and 4D CBCT

Volumetric and time-resolving image guidance techniques, now considered standard in photon therapy, are just beginning to become established in proton treatment rooms. In-room cone-beam CT (CBCT) systems with analysis of implanted fiducial markers, automated registration algorithms, and integration with robotic positioning are currently implemented in proton treatment centers. This development is eased by the fact that beam line kV x-ray sources and flat panel detectors are already commonplace in proton treatment rooms. This will allow 3D and, for lung tumors, 4D imaging-based alignment techniques that can be fully automated.

The integration of CBCT is not without challenges. The image quality of 3D CBCT images is usually not sufficient for replanning and adapting the treatment in the room. Thus, even though a redistribution of soft tissues relative to the original treatment plan may be detected, correcting for it at the time of treatment would be difficult. On the other hand, relatively large changes in tumor size or surrounding normal tissue can be visualized, prompting a rescanning of the patient and adaptation of the plan. In-room 4D CBCT images, e.g., for verifying the preplanned ITV of lung tumors, have additional challenges: image acquisition is slow, and the image quality is hampered by motion and other CT artifacts. Clinical experience and the development of proton therapy-specific protocols will inform on how to best use these new techniques in the near future.

The use of CBCT for volumetric alignment is often coupled to the analysis of the position of radiopaque fiducial markers implanted near or in the target. It should be noted that high-density fiducial markers implanted directly into the target volume can pose disturbances of the proton dose distribution [34,35].

7.2.3 CT-on-Rails

CT-on-rails is another option for image guidance in the proton treatment room. Because it is a standard CT imaging system, it offers better image quality, a lower dose per scan, and faster data acquisition than CBCT. Different from CBCT, CT-on-rails allows replanning of the patient and the possibility of adaptation of the plan if needed. The disadvantage of this system is that it must be moved in and out of the scanning position in order to not interfere with the beam delivery. This implies some delay and potential for movement of the patient during this procedure.

Using a CT-on-rail system, Zhang et al. [36] repeated CT scans that were used to study the dosimetric effect of inter-fraction motion, i.e., the day-to-day variation in the distribution of targeted and non-targeted tissues, in 10 prostate cancer patients treated with proton therapy. It was found that when the center of the prostate target volume was aligned on each of eight repeated CT scans, the PTV coverage with 98% of the target dose was adequate in all but one patient using generally accepted range and lateral uncertainties for proton therapy of prostate patients. In the one patient in which the PTV was not adequately covered, despite alignment of the prostate, the CTV included both the prostate and the seminal vesicles. It is known that in some patients the seminal vesicles move independently from the bulk of the prostate gland. This study thus demonstrated that under certain circumstances, an adaptive replanning technique may be necessary.

7.2.4 Auxiliary Positioning Systems (Ultrasound, Optical Imaging, and Electromagnetic Transponder Tracking)

There are auxiliary positioning guidance systems—including those using ultrasound, optical surface tracking, or implanted electromagnetic transponders—that do not require ionization radiation. Therefore, they can be used on a daily basis, or even for monitoring during therapy. The experience with these systems in proton therapy is relatively limited, and not much has been published. On the other hand, there is experience with these systems from conventional photon-based radiation therapy. Each system has certain specific tumor sites for which it can be used.

In-room ultrasound (or sonographic) pretreatment alignment of accessible anatomical tumor sites has been reported to decrease setup errors due to inter-fraction motion in prostate as well as in lung, abdominal, and breast tumor sites. There are several commercial systems available, as listed in the AAPM TG-154 report [37], which also gives guidelines on commissioning of such systems. While pretreatment alignment has been clinically performed in a 2D or 3D fashion, systems for 4D intra-fraction tracking of tumors during treatment are now becoming available. Ultrasound guidance has been most extensively used to position patients with prostate cancer for conformal RT or IMRT [38–41]. The performance of the technique is user-dependent, and improper pressure during probe placement may displace the prostate. Overall, the utility and clinical accuracy of this method is somewhat controversial [42]. Pretreatment ultrasound guidance has also been used in a few proton centers, but no systematic study has been reported in such setting.

In-room surface tracking with ceiling-mounted camera systems can be used to monitor patient positioning during the treatment in order to monitor and detect intra-fraction motion. This method uses rigid body transformations in combination with a least-square fit to minimize the difference between the actual surface and the expected 3D reference surface from the camera view point. At the time of this writing, two commercial systems were available for patient surface tracking during RT [43–45]: AlignRT, VisionRT (London, UK) and C-Rad Sentinel, C-RAD AB (Uppsala, Sweden). In-room alignment is based on a reference 3D model of the patient surface acquired at the time of the planning CT scan, and its geometric relationship to the room isocenter is determined during the treatment planning process. Commissioning and routine QA with an anthropomorphic phantom has been described in the AAPM TG-147 report [46].

Real-time tracking of tumor motion during treatment has become a reality with the development of electromagnetic transponders implanted within the tumor. The Calypso system (Calypso Medical Technologies, Inc., Seattle, WA) is currently the only commercially available system, and it is used for online tracking of prostate motion in patients treated for prostate cancer. The system continuously monitors the location of three implanted electromagnetic beacon transponders at the rate of 10 Hz, thus permitting 3D tracking of the prostate during treatment. The delivery of radiation can be stopped if the prostate moves outside of a user-defined threshold parameter for a set period of time. Commissioning of the system, detailed in AAPM TG-147 [46], includes system safety tests as well as phantom measurements with three transponders embedded at known distances inside the phantom to determine the accuracy and precision of the localization in the prostate [47–51] and lung [52]. Additional details of the system's operation can be found in the report of the ASTRO Emerging Technology Committee on continuous localization systems [53].

To the authors' best knowledge, there is no current use of the Calypso system in any proton center. The implantation of the transponder beacons into the tumor could lead to an unforeseen degradation of the proton dose distribution, which may be a reason not to use them, but there is no report on this subject in the literature. Tang et al. [54,55] at the University of Pennsylvania used population data of prostate motion acquired with the Calypso system in a cohort of patients treated with photon RT and applied the worst-case scenario motion to CT data sets of 10 prostate cancer patients treated with proton beam scanning techniques to study the effect of prostate motion (in the worst case) on CTV coverage. In their first study, they only considered intra-fraction motion. In the second study, the combined effect of intra- and inter-fraction motion was studied. They found that the CTV D99 (percentage dose to 99% of the CTV) varied up to 10% when compared to the initial plan in individual fractions. But over the entire treatment course of 44 fractions, the total dose degradation of D99 was only 2% to 3%, with a standard deviation of less than 2%. Thus, organ motion data obtained with auxiliary position guidance systems in larger photon RT cohorts of patients may be applied to proton therapy patient cohorts with similar characteristics to study the general adequacy of CTV margins and delivery techniques that could be sensitive to motion.

7.3 Emerging Imaging Techniques for Beam Verification and Adaptation

7.3.1 PET, Prompt Gamma, and Thermoacoustic Imaging for Range Verification

The existence of uncertainty in the *in vivo* range of the proton treatment beam has led to a large effort to improve range determination through better techniques for patient imaging, dose calculation, and patient setup/alignment, as outlined thus far in this chapter. In addition, a large effort is also underway to develop methods to image the treatment beam *in vivo* during or immediately after dose delivery, so-called "*in vivo* range verification." In fact, the only way to truly verify that the dose was delivered to the patient as intended is measure the actual *in vivo* dose distribution. Work to develop methods to measure and verify accuracy of *in vivo* dose delivery has mainly focused on imaging secondary radiation emitted from the patient during treatment delivery, such as positron annihilation gammas, characteristic prompt gammas, and thermoacoustic waves.

By far the most studied method for proton (and carbon ion) range verification is the use of positron emission tomography (PET) imaging, either during or after proton or carbon ion beam treatment [56–58]. Currently, this is the only *in vivo* imaging method that is used in clinical practice for ion beam therapy [59], while its development for clinical use with proton therapy is still under development [60–62]. In this method, unstable isotopes (^{15}O, ^{11}C, etc.) are created through inelastic proton (carbon ion) interactions with nuclei in the irradiated tissue. These radionuclides decay through $\beta+$ emission that annihilates, producing coincident 511 keV gammas which can be imaged with a PET scanner. Since the distribution of annihilation gamma emission from the patient is not directly correlated to dose delivery by the treatment beam, the measured PET image is compared to a Monte Carlo-calculated "expected" distribution of positron annihilation gamma emission in the patient during irradiation. Differences between the measured PET distribution and the Monte Carlo-calculated distribution is then used to identify mismatches between the expected and actual delivered range of the treatment beam. This method has shown the ability to detect shifts in the distal falloff of the proton Bragg peak as small as 1–2 mm in both phantom and patient studies [63,64].

Development of PET imaging for *in vivo* verification of proton and carbon ion therapy originally focused on the use of commercial PET imagers. The patient was taken to the imager immediately after the daily treatment was delivered for imaging of the induced annihilation gamma signal. However, several limitations to these post-treatment PET protocols were found, including long delay times between treatment and imaging that yield significantly reduced signal, as well as biological washout effects [65] from physiological processes in the body, such as blood perfusion. This has led to a large effort to produce PET imagers specifically designed for placement within a proton or carbon ion therapy treatment room. This includes the development of stand-alone PET scanners placed in the room adjacent to the treatment nozzle [67] and flat panel and partial ring detectors [60,61,67] attached to the treatment machine.

With these systems, PET image acquisition is performed during or within 5–10 minutes immediately after treatment delivery. This has greatly reduced the low signal and biological washout problems encountered with using PET scanners located outside of the treatment room. However, further developments are needed to overcome issues with imaging artifacts introduced due to the limited projection angles measured by flat panel and partial ring detectors in order to make online PET imaging systems viable for *in vivo* monitoring during proton or carbon ion treatment delivery.

In addition to the use of PET imaging techniques, researchers have also studied the idea of Prompt Gamma (PG) imaging as a means of verifying proton and ion beam treatment delivery. In contrast to the processes that produce positron annihilation gammas, interactions of the treatment beam with elemental nuclei in the tissue can leave behind an intact nucleus in an excited state. This excited state quickly ($< 10^{-8}$ seconds) decays to its ground state by emitting a prompt gamma with an energy that is characteristic to the element. In theory, since the PG emission is measured while the beam is on and only occurs in regions in which the treatment beam interacts, the distribution of PG creation in the patient is correlated to the distribution of dose delivery to the patient. In fact, studies by several researchers have confirmed a strong correlation between PG emission and dose delivery for both proton and carbon ion beams [68–72]. Additionally, PG emission has been shown to be proportional to the concentration of each element within the irradiated tissues [69,73,74]. This has led to an interest in placing a gamma camera (or perhaps multiple cameras) near the patient to measure PG emission during proton or ion beam treatment delivery. These measurements would then be used to perform analysis of the elemental composition of the tumor and healthy tissues irradiated during treatment delivery.

Due to the relatively high energy of PGs emitted from tissue (2–15 MeV), early testing of the detection efficiency of standard gamma detectors (SPECT imagers, standard Compton cameras, etc.) were shown to have very poor detection efficiency [75,76]. This has led to a large effort to design gamma detectors specifically for PG imaging during proton and carbon ion therapy. This includes design studies of two-stage [77–79] and three-stage Compton cameras [76,80,81], pinhole and slit cameras [82–84], multi-slit collimated array detectors [85], and time-of-flight detectors [86, 87]. Of these detectors, the slit camera has been extensively tested for proton therapy and has the capability to measure a 1D profile of PG emission as a function of depth in the patient [84]. Also, a prototype three-stage Compton camera is currently being tested with clinical proton beams which provides the capability of reconstructing 2D and 3D images of the PG emission from the patient during proton treatment delivery [88], as shown in the Figure 7–10. With these measured profiles and reconstructed images, the *in vivo* range and day-to-day variation of the *in vivo* range can be detected.

Another emerging area of research is in the use thermoacoustic imaging for range verification in proton and ion beam therapy. As the beam passes through the patient, large heat gradients are produced along the tracks of individual protons (or carbon ions) as they interact and deposit energy. These heat gradients occur on the scale of ~1–100 nanometers around the particle track within 10^{-15} to 10^{-10} seconds of the particle interaction at a given point in the tissue [89]. The thermal expansion and contrac-

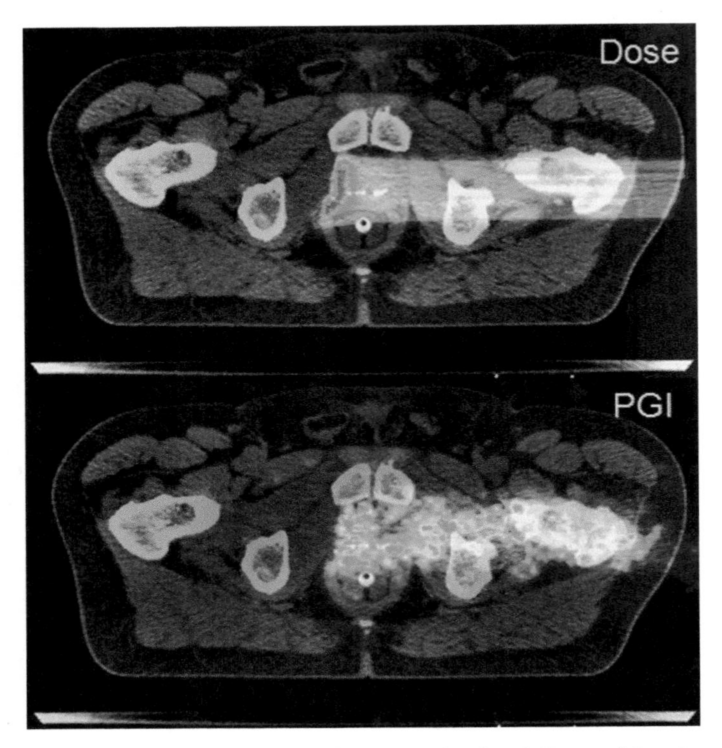

Figure 7–10 Dose and reconstructed PG images for the delivery of the deepest energy layer of a proton scanning beam prostate treatment overlaid onto the planning CT for evaluation.

tion induced by this thermal gradient produces an acoustic wave with frequencies ranging from ~10 MHz up to ~1 GHz in the tissue, which can be detected by an acoustic transducer. Since most of the energy of the beam is deposited in the Bragg peak, the greatest heat gradients will also be produced in this region, leading to an increased acoustic signal near the Bragg peak, with no signal measured beyond it, thus leading to a method to image and localize the position of the Bragg peak and its distal falloff within the patient [90,91].

7.3.2 Proton CT and Radiography

Proton CT not only improves the accuracy of range prediction in proton treatment planning, but it also bears promise to be useful for pretreatment verification of accurate tumor dose coverage. While proton CT gives the full 3D information of the tumor and surrounding tissues, it requires at least a half gantry rotation and a few minutes for image reconstruction and analysis with parallelizable reconstruction algorithms and GPU-based computer clusters. In comparison, 2D proton radiography does not require a gantry rotation and can be reconstructed in a few seconds. Radiography would be useful for an integral range verification. In the future, it may even be possible to switch from the treatment energy to a higher imaging energy for an intra-treatment integral range check before the next target layer is treated. In addition, proton radiography and CT could also be used for patient alignment verification and inter-fraction tumor tracking if markers are implanted into the tumor, taking advantage of the lower dose and absence of related high-density artifacts.

Current proton delivery systems are, however, not built for patient imaging. The required beam intensities are about three orders of magnitude smaller than those for therapy, i.e., 10^6 protons per second for imaging versus 10^9 protons for therapy. This poses challenges at two levels. First, the beam current needs to be reduced to allow low beam intensities at the level of the patient, which is difficult with high-current cyclotrons. Secondly, the low beam intensities are below the sensitivity threshold of existing beam intensity and position monitors integrated into to therapeutic proton beam lines, which makes it difficult to monitor beam intensity and position during imaging. One solution is to use the CT/radiography tracking detectors to measure the beam intensity and to provide a feedback signal to control the beam intensity for imaging at the accelerator level.

For imaging, the beam needs to be spread out into a large cone or fan beam covering the entire patient circumference. Assuming the intensity problem has been solved, this can be accomplished by modern active scanning systems. Another possibility recently proposed for the creating of a therapeutic proton fan beam [92] is to use a pair of electromagnetic quadrupole magnets to shape a single pencil beam into a large fan beam at the level of the patients.

Lastly, proton imaging systems require higher proton energies than those typically used for treatment due to the need to penetrate the patient. Current proton delivery systems (synchrotrons and cyclotrons) deliver protons of either 230 MeV or 250 MeV. Only one system, the Protom synchrotron, delivers up to 330 MeV. The range of 230 MeV, 250 MeV, and 330 MeV in water is 33, 38, and 60 cm, respectively. Thus, most existing proton facilities would be able to perform proton CT and radiography in the brain, head and neck, and chest regions (with arms extended above the shoulders). This will be sufficient to gain first clinical experience with these imaging systems.

7.3.3 Strategies for Adaptive Proton Therapy

Anatomical changes due to weight loss/gain or tumor shrinkage/growth are always of concern in radiation therapy. In photon therapy, variations in the doses received by targets or organs at risk are mainly due to geographical changes. If, for example, the parotid gland moves medically due to weight loss, it can move into high-dose regions and receive more dose than intended. The actual dose distribution itself is, however, remarkably robust against anatomical variation. In proton therapy, the concern is that in addition to geographical misplacements, the actual dose distribution itself changes in shape and location due to anatomical variations, thus making proton plans particular sensitive to anatomical variations. For patient alignment, it is not necessarily good enough to align the target position. For accurate dose assessment, density changes in the path of the proton beam need to be known. It is, hence, more common to re-image the patient during a course of proton therapy. An IMPT study for lung and mediastinal tumors conducted at MD Anderson in Houston used repeat CT during the course of 34 patients [28]. Of these, 9 patients were replanned due to treatment-related changes that led to a decrease in target coverage or overdosing of organs at risk. The mean time to adaptation was at day 10.

References

1. Schneider U, Pedroni E, Lomax A. The calibration of CT Hounsfield units for radiotherapy treatment planning. *Phys Med Biol* 1996;41:111–124.
2. Guan H, Yin FF, Kim JH. Accuracy of inhomogeneity correction in photon radiotherapy from CT scans with different settings. *Phys Med Biol* 2002;47:N223–231.
3. Ainsley CG, Yeager CM. Practical considerations in the calibration of CT scanners for proton therapy. *J Appl Clin Med Phys* (American College of Medical Physics) 2014;15:4721.
4. Schaffner B, Pedroni E. The precision of proton range calculations in proton radiotherapy treatment planning: experimental verification of the relation between CT-HU and proton stopping power. *Phys Med Biol* 1998;43:1579–1592.
5. Yang M, Zhu XR, Park P, et al. Comprehensive analysis of proton range uncertainties related to patient stopping-power-ratio estimation using the stoichiometric calibration. *Phys Med Biol* 2012;57:4095–4115.
6. Yang M, Virshup G, Clayton J, et al. Theoretical variance analysis of single- and dual-energy computed tomography methods for calculating proton stopping power ratios of biological tissues. *Phys Med Biol* 2010;55:1343–1362.
7. Huenemohr N, Paganetti H, Greilich S, et al. Tissue decomposition from dual energy CT data for MC based dose calculation in particle therapy. *Med Phys* 2014;41(6):061714.
8. Herman G. Correction for beam hardening in computed tomography. *Phys Med Biol* 1979;24:81–106.
9. Newhauser W, Giebler A, Langen KM, et al. Can megavoltage computed tomography reduce proton range uncertainties in treatment plans for patients with large metal implants? *Phys Med Biol* 2008;53:2377–2344.
10. Dietlicher I, Casiraghi M, Ares C, et al. The effect of surgical titanium rods on proton therapy delivered for cervical bone tumors: experimental validation using anthropomorphic phantom. *Phys Med Biol* 2014;59:7181–7194.
11. Xu C, Verhaegen F, Laurendeau D, et al. An algorithm for efficient metal artifact reductions in permanent seed implants. *Med Phys* 2011;47 (2011); doi: 10.1118/1.3519988.
12. Lewitt RM, and Bates RTH. Image reconstruction from projections: III. Projection completion methods. *Optics* (Stuttgart) 1978; 50:189–204.
13. Kalender WA, Hebel R, Ebersberger J. Reduction of CT artifacts caused by metallic implants. *Radiology* 1987;164:576–577.
14. Wang G, Snyder DL, O'Sullivan JA, et al. Iterative deblurring for CT metal artifact reduction. *IEEE Trans Med Imaging* 1996;15:657–664.
15. Wei J, Sandison GA, Hsi WC, et al. Dosimetric impact of a CT metal artefact suppression algorithm for proton, electron and photon therapies. *Phys Med Biol* 2006;51:5183–5197.
16. Wei J, Chen L, Sandison GA, Liang Y, et al. X-ray CT high-density artefact suppression in the presence of bones. *Phys Med Biol* 2004;49:5407–5418.
17. Cormack AM. Representation of a function by its line integrals, with some radiological applications. *Journal of Applied Physics* 1963;34:2722–2727.
18. Cormack AM. Representation of a function by its line integrals, with some radiological applications. II. *Journal of Applied Physics* 1964;35:2908–2913.
19. Schulte RW, V. Bashkirov V, T. Li T, et al. Conceptual design of a proton computed tomography system for applications in proton radiation therapy. *IEEE Trans Nucl Sci* 2004;51:866–872.
20. Williams DC. The most likely path of an energetic charged particle through a uniform medium. *Phys Med Biol* 2004;49:2899–911.

21. Schulte RW, Penfold SN, Tafas JT, et al. A maximum likelihood proton path formalism for application in proton computed tomography. *Med Phys* 2008;351:4849–4856.
22. Schulte RW, Penfold SN, Tafas JT, et al. Water-equivalent path length calibration of a prototype proton CT scanner. *Med Phys* 2012;39:2438–2446.
23. Li T, Liang Z, Singanallur JV, Satogata TJ, et al. Reconstruction for proton computed tomography by tracing proton trajectories: a Monte Carlo study. *Med Phys* 2006;33:699–706.
24. Penfold SN. Image reconstruction and Monte Carlo simulations in the development of proton computed tomography for applications in proton radiation therapy. Ph.D. thesis, University of Wollongong, 2010.
25. Rit S, Dedes G, Freud N, et al. Filtered backprojection proton CT reconstruction along most likely paths. *Med Phys* 2013;40:031103.
26. Scaringella M, Bruzzi M, Bucciolini M, et al. A proton computed tomography based medical imaging system. *JINST* 2014; 9:C12009.
27. Sadrozinski HF, Johnson RP, Macafee S, et al. Development of a head scanner for proton CT. *Nucl Instrum Methods Phys Res A*. 2013;699:205–210.
28. Chang JY, Li H, Zhu XR, et al. Clinical implementation of intensity modulated proton therapy for thoracic malignancies. *Int J Radiat Oncol Biol Phys* 2014; 90:809–818.
29. Kang Y, Zhang X, Chang JY, et al. 4D Proton treatment planning strategy for mobile lung tumors. *Int J Radiat Oncol Biol Phys* 2007;67:906–914.
30. Grassberger C, Dowdell S, Lomax A, et al. Motion interplay as a function of patient parameters and spot size in spot scanning proton therapy for lung cancer. *Int J Radiat Oncol Biol Phys* 2013;86:380–386.
31. Uh J, Merchant TE, Li Y, et al. MRI-based treatment planning with pseudo CT generated through atlas registration. *Med Phys* 2014;41:051711.
32. Schulte RW. Proton treatment room concepts for precision and efficiency. *Technol Cancer Res Treat* 2007;6:55–60.
33. Herman MG, Balter JM, Jaffray DA, et al. Clinical use of electronic portal imaging: report of AAPM radiation therapy committee task group 58. *Med Phys* 2001;28:712–737.
34. Giebeler A, Fontenot J, Balter P, et al. Dose perturbations from implanted helical gold markers in proton therapy of prostate cancer. *J Appl Clin Med Phys* 2009;10:63–70.
35. Huang JY, Newhauser WD, Zhu XR, et al. Investigation of dose perturbations and the radiographic visibility of potential fiducials for proton radiation therapy of the prostate. *Phys Med Biol* 2011;56:5287–5302.
36. Zhang X, Dong L, Lee AK, et al. Effect of anatomic motion on proton therapy dose distributions in prostate cancer treatment. *Int J Radiat Oncol Biol Phys* 2007;67:620–629.
37. Molloy JA, Chan G, Markovic A, et al. AAPM task group 154. Quality assurance of U.S.-guided external beam radiotherapy for prostate cancer: report of AAPM task group 154. *Med Phys* 2011;38:857–71.
38. Lattanzi J, McNeeley S, Pinover W, et al. A comparison of daily CT localization to a daily ultrasound-based system in prostate cancer. *Int J Radiat Oncol Biol Phys* 1999;43:719–725.
39. Serago CF, Chungbin SJ, Buskirk SJ, et al. Initial experience with ultrasound localization for positioning prostate cancer patients for external beam radiotherapy. *Int J Radiat Oncol Biol Phys* 2002;53:1130–1138.
40. Langen KM, Pouliot J, Anezinos C, et al. Evaluation of ultrasound-based prostate localization for image-guided radiotherapy. *Int J Radiat Oncol Biol Phys*. 2003;57:635–644.
41. Poli ME, Parker W, Patrocinio H, et al. An assessment of PTV margin definitions for patients undergoing conformal 3D external beam radiation therapy for prostate cancer based on an analysis of 10,327 pretreatment daily ultrasound localizations. *Int J Radiat Oncol Biol Phys* 2007;67:1430–1437.
42. Scarbrough TJ, Ting JY, Kuritzky N. Ultrasound for radiotherapy targeting. *Int J Radiat Oncol Biol Phys* 2007;68:1579; author reply 1579–1580.
43. Bert C, Metheany KG, Doppke KP, et al. Clinical experience with a 3D surface patient setup system for alignment of partial-breast irradiation patients. *Int J Radiat Oncol Biol Phys* 2006;64:1265–1274.
44. Brahme A, Nyman P, Skatt B. 4D laser camera for accurate patient positioning, collision avoidance, image fusion and adaptive approaches during diagnostic and therapeutic procedures. *Med Phys* 2008;35:1670–1681.

45. Moore C, Lilley F, Sauret V, et al. Opto-electronic sensing of body surface topology changes during radiotherapy for rectal cancer. *Int J Radiat Oncol Biol Phys* 2003;56:248–258.
46. Willoughby T, Lehmann J, Bencomo JA, et al. Quality assurance for nonradiographic radiotherapy localization and positioning systems: report of task group 147. *Med Phys* 2012;39:1728–1747.
47. Santanam L, Noel C, Willoughby TR, et al. Quality assurance for clinical implementation of an electromagnetic tracking system. *Med Phys* 2009;36:3477–3486.
48. Langen KM, Willoughby TR, Meeks SL, et al. Observations on real-time prostate gland motion using electromagnetic tracking. *Int J Radiat Oncol Biol Phys* 2008;71:1084–1090.
49. Litzenberg DW, Willoughby TR, Balter JM, et al. Positional stability of electromagnetic transponders used for prostate localization and continuous, real-time tracking. *Int J Radiat Oncol Biol Phys* 2007;68:1199–1206.
50. Willoughby TR, Kupelian PA, Pouliot J, et al. Target localization and real-time tracking using the Calypso 4D localization system in patients with localized prostate cancer. *Int J Radiat Oncol Biol Phys* 2006;65:528–534.
51. Balter JM, Wright JN, Newell LJ, et al. Accuracy of a wireless localization system for radiotherapy. *Int J Radiat Oncol Biol Phys* 2005;61:933–937.
52. Shah AP, Kupelian PA, Waghorn BJ, et al. Real-time tumor tracking in the lung using an electromagnetic tracking system. *Int J Radiat Oncol Biol Phys.* 2013;86:477–483.
53. D'Ambrosio DJ, Bayouth J, Chetty IJ, et al. Continuous localization technologies forradiotherapy delivery: Report of the American Society for Radiation Oncology Emerging Technology Committee. *Pract Radiat Oncol* 2012;2:145–150.
54. Tang S, Deville C, McDonough J, et al. Effect of intrafraction prostate motion on proton pencil beam scanning delivery: a quantitative assessment. *Int J Radiat Oncol Biol Phys* 2013;87:375–382.
55. Tang S, Deville C, Tochner Z, et al. Impact of intrafraction and residual interfraction effect on prostate proton pencil beam scanning. *Int J Radiat Oncol Biol Phys* 2014;90:1186–1194.
56. Litzenberg DW, Roberts DA, Lee MY, et al. On-Line Monitoring of Radiotherapy Beams: Experimental Results with Proton Beams. *Med. Phys.* 1999;26:992–1006.
57. Oelfke U, Lam GK, Atkins MS. Proton dose monitoring with PET: quantitative studies in Lucite. *Phys Med Biol* 1996;41:177–196.
58. Paans A, Schippers J. Proton therapy in combination with PET as monitor: a feasibility study. *IEEE Trans. Nucl. Sci.* 1993;40:1041–1044.
59. Enghardt W, Parodi K, Crespo P, et al. Dose quantification from in-beam positron emission tomography. *Radiother Oncol* 2004;73 Suppl 2:S96–98.
60. Attansi F, Belcari N, Moehrs S, et al. Characterization of an In-Beam PET prototype for proton therapy with different target compositions. *IEEE Trans Nucl Sci* 2010;57:1563–1569.
61. Nishio T, Miyatake A, Ogino T, et al. The development and clinical use of a beam ON-LINE PET system mounted on a rotating gantry port in proton therapy. *Int J Radiat Oncol Biol Phys* 2010;76:277–286.
62. Parodi K, Enghardt W, Haberer T. In-beam PET measurements of positron radioactivity induced by proton beams. *Phys Med Biol* 2002;46:21–36.
63. Parodi K, Paganetti H, Cascio E, et al. PET/CT imaging for treatment verification after proton therapy: A study with plastic phantoms and metallic implants. *Med Phys* 2007;34:419–435.
64. Parodi K, Paganetti H, Shih H, et al. Patient study of in vivo verification of beam delivery and range using positron emission tomography and computed tomography imaging after proton irradiation. *Int J Radiat Oncol Biol Phys* 2007;68:920–934.
65. Knopf A, Parodi K, Bortfeld T, et al. Systematic analysis of biological and physical limitations of proton beam range verification with offline PET/CT scans. *Phys Med Biol* 2009;54:4477–4495.
66. Zhu X, Espana S, Daartz J, et al. Monitoring proton radiation therapy with in-room PET imaging. *Phys Med Biol* 2011;56:4041–4057.
67. Surti S, Zou W, Daube-Witherspoon ME, et al. Design study of an in situ PET scanner for use in proton beam therapy. *Phys Med Biol* 2011;56:2667–2685.
68. Min CH, Kim CH, Youn M, et al. Prompt gamma measurements for locating dose falloff region in proton therapy. *App Phys Lett* 2006;89:183517.

69. Moteabbed M, Espana S, Paganetti H. Monte Carlo patient study on the comparison of prompt gamma and PET imaging for range verification in proton therapy. *Phys Med Biol* 2011;56:1063–1082.
70. Polf J, Peterson S, Ciangaru G, et al. Prompt gamma-ray emission from biological tissues during proton irradiation: a preliminary study. *Phys Med Biol* 2009;54:731–743.
71. Testa E, Bajard M, Chevallier M, et al. Dose profile monitoring with carbon ion beams by means of prompt-gamma measurements. *Nucl Intstrum Methods Phys Res. B* 2009;267:993–996.
72. Verburg JM, Riley K, Bortfeld T, et al. Energy- and time-resolved detection of prompt gamma-rays for proton range verification. *Phys Med Biol* 2013;58:L37–49.
73. Polf J, Peterson S, Roberts DA, et al. Measuring prompt gamma ray emission during proton radiotherapy for assessment of treatment delivery and patient response. *AIP Conf Proc* 2011;1336:364–367.
74. Polf JC, Panthi R, Mackin DS, et al. Measurement of characteristic prompt gamma rays emitted from oxygen and carbon in tissue-equivalent samples during proton beam irradiation. *Phys Med Biol* 2013;58:5821–5831.
75. Park MS, Lee W, Kim JM. Estimation of proton dose distribution by means of three-dimensional reconstruction of prompt gamma rays. *App Phys Lett* 2010;97:153705.
76. Richard MH, Chevillier M, Dauvergne D, et al. Design study of a Compton camera for prompt gamma imaging during ion beam therapy. IEEE Nucl Sci Sympos Conf Record. Orlando FL: IEEE, 2009. pp. 4172–4175.
77. Frandes M, Zoglauer A, Maxim V, et al. A tracking Compton-scattering imaging system for hadron therapy monitoring. *IEEE Trans Nucl Sci* 2010;57:144–150.
78. Kang BH, Kim JW. Monte Carlo design study of a gamma detector system to locate distal dose falloff in proton therapy. *IEEE Trans Nucl Sci* 2009;56:46–50.
79. Roellinghoff F, Richard MH, Chevillier M, et al. Design of a Compton camera for 3D prompt gamma imaging during ion beam therapy. *Nucl Intstrum Methods Phys Res A* 2011;648:s20–23.
80. Peterson SW, Roberts D, Polf JC. Optimizing a 3-stage Compton camera for measuring prompt gamma rays emitted during proton radiotherapy. *Phys Med Biol* 2010;55:6841–6856.
81. Robertson D, Polf JC, Peterson S, et al. Material efficiency studies for a Compton camera designed to measure characteristic prompt gamma rays emitted during proton beam radiotherapy. *Phys Med Biol* 2011;56:3047–3059.
82. Bom V, Joulaeisadeh F, Beekman F. Real-time prompt gamma monitoring in spot-scanning proton therapy using imaging through a knife-edge-shaped slit. *Phys Med Biol* 2012;57:297–308.
83. Kim D, Yim H, Kim J-W. Pinhole camera measurements of prompt gamma rays for detection of beam range in proton therapy. *J Kor Physi Soc* 2009;55:1673–1676.
84. Smeets J, Roellinghoff F, Prieels D, et al. Prompt gamma imaging with a slit camera for real-time range control in proton therapy. *Phys Med Biol* 2012;57:3371–3405.
85. Min CH, Lee HR, Kim CH, et al. Development of array-type prompt gamma measurement system for in vivo range verification in proton therapy. *Med Phys* 2012;39:2100–2107.
86. Golnik C, Hueso-Gonzalez F, Muller A, et al. Range assessment in particle therapy based on prompt gamma-ray timing measurements. *Phys Med Biol* 2014;59:5399–5422.
87. Gueth P, Dauvergne D, Freud N, et al. Machine learning-based patient specific prompt-gamma dose monitoring in proton therapy. *Phys Med Biol* 2013;58:4563–4577.
88. Mccleskey M, Kaye, W, Mackin, D, et al. Evaluation of a multistage CdZnTe Compton camera for prompt gamma imaging for proton therapy, *Nucl Instr Met Phys Res A* 2015;785:163–169.
89. Toulemonde M, Surdutovich E, Solov'yov A. Temperature and pressure spikes in ion-beams cancer therapy. *Phys Rev E Stat Nonlin Soft Matter Phys* 2009;80:031913.
90. Tada J, Hayakawa Y, Hosono K, et al. Time resolved properties of acoustic pulses generated in water and in soft tissue by pulsed proton beam irradiation—a possibility of doses distribution monitoring in proton radiation therapy. *Med Phys* 1991;18:1100–1104.
91. De Bonis G. Acoustic signals from proton beam interactions in water—comparing experimental data and Monte Carlo simulations. *Nucl Intstr Met Phys Res A* 2009;604:s199–s202.
92. Hill P, Westerly D, Mackie T. Fan-beam intensity modulated proton therapy. *Med Phys* 2013;40:111704.

Chapter 8

Field Shaping: Scattered Beam

Alejandro Mazal, Ph.D.[1], Annalisa Patriarca, M.Sc.[2], Claas Wessels, M.Sc.[2], and Indra J. Das, Ph.D.[3]

[1]Head of Medical Physics, Institut Curie, Paris, France
[2]Medical Physicist, Medical Physics Service, Centre de Protonthérapie d'Orsay. Institut Curie, Paris, France
[3]Professor & Director of Medical Physics, Department of Radiation Oncology, Indiana University School of Medicine, Indianapolis, IN

8.1 Introduction

The original proton (or ion) beam extracted from the accelerator and transported along the beam transport line has a single (and variable) energy with a small Gaussian profile, as presented in Chapter 6 (Proton Beam Production and Dose Delivery Techniques) for both fixed lines and gantries. These characteristics are not well adapted to irradiating a clinical target, imposing a field shaping process. This is performed at the end of the beam line ("nozzle"), just before arriving to the patient, with its final components placed in a telescopic "snout" supporting customized pieces, such as apertures and compensators, also known as patient-specific devices.

Two types of nozzles for beam shaping and delivery have been developed for clinical applications related to the concepts of "scattered beams" and "scanned beams." In this chapter we present the devices being used in scattered beam lines nowadays, also called "passive" lines (even if some elements may not be "passive"), while the scanned beams will be presented in Chapter 9, Field Shaping: Scanning Beam.

A draft of a passive scattered nozzle for beam shaping is presented in Figure 8–1, and typical values of specifications are presented in Table 8–1. A passive scattered nozzle includes the following basic elements:

- single or double scattering foils
- range modifiers
- range modulators
- beam monitoring devices
- apertures
- compensators
- beam light simulators
- imaging systems for patient positioning

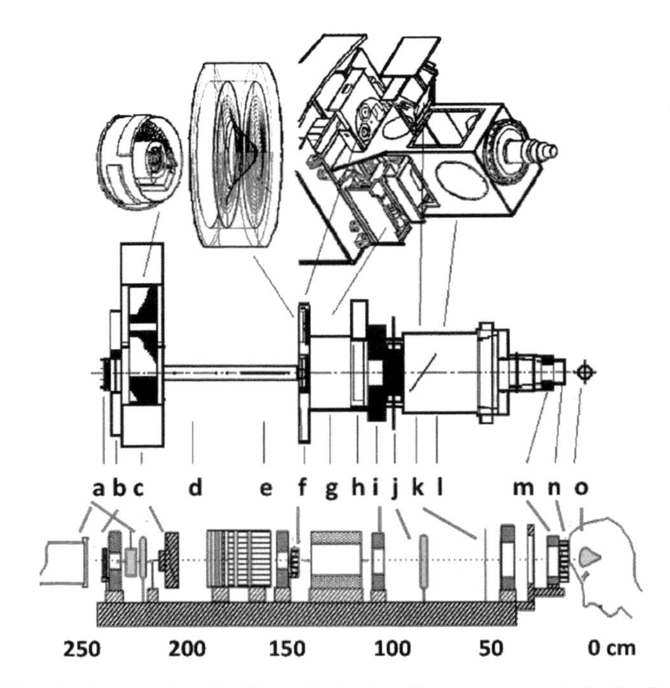

Figure 8–1 Passive beam-shaping lines, including the components in the "nozzle" and the "snout." Upper and central figures are from an industrial proposal (IBA, Belgium and courtesy of H. Paganetti). The bottom figure corresponds to a homemade line (Orsay, France, courtesy C. Nauraye). (a) Vacuum window and first monitor chamber. (b) First scatter. (c) Modulator wheel. (d–e) Space for scanning magnets. (f) Second scatter. (g) Removable x-ray tube. (h–i) Collimators. (j) Monitor chamber. (k) Mirror for light simulation. (l) Snout holder. (m) Aperture. (n) Compensator. (o) Isocenter. (Modified from Mazal [2].)

Table 8–1 Typical specifications of a passive scattered beam line devoted to all kinds of treatments as presented in Figure 8–1. (From the Centre de Protonthérapie d'Orsay, France.)

Parameter	Definition	Units	Values	Comments
Field Size	50% max diameter	cm	22	Others available. Vendor specific
Range	90% distal in SOBP	g/cm^2	4.5 to 29	Down to 0 (surface) with absorbers
Modulation	95% proximal to 98% distal	g/cm^2	0.5 (pristine) to 19	Ripple lower than ±2.5%
Distal Falloff	80% to 20%	g/cm^2	0.35 (low E) to 0.70 (high E)	Function of range and spectra
Lateral Penumbra	80% to 20% in air	mm	2 to 5	Air gap 10 cm, 6–8 mm in depth for typical SOBP
Virtual and Effective SAD	From field sizes and square of distance law respectively	cm	230 to 250	Small dependence on beam parameters and beam line devices
Original FWHM	Field size at exit vacuum window	mm	5	Beam not modulated in energy
Dose Rate	At isocenter. Middle of SOBP	Gy/min	2	Nominal value. Higher values possible

In this chapter we describe those elements modifying the beam for clinical purposes. A detailed description of devices and instrumentation for treatment of cancer using proton and light-ion beams has been presented by Chu et al. [1].

8.2 Passive Single and Double Scattering

The beam is transported from the accelerator to the treatment rooms in long pipelines with vacuum. It usually has a quasi-Gaussian cross-profile with a small diameter, on the order of a few millimeters. The energy spread is due to extraction issues from the accelerator and Coulomb interactions, but usually it is less than 0.1% or a fraction of MeV.

The tumor linear dimensions in patients vary from a few millimeters up to 30 cm. In order to cover those realistic patient targets with dimensions of several centimeters at isocenter, the first approach is to insert one single or double spreading device into the *nozzle* in what are called *"passive lines"* with *"scattered beams."* They are based on the multiple Coulomb scattering produced by high atomic number (Z) materials by the charged particle beam as described by Gottschalk et al. [3]. The spread of the beam is calculated based on scattering power of the high-Z medium for the energy in the beam line. Scattering power has been discussed in detail in Chapter 3.

A *scatterer* is a high-Z beam-spreading device with a flat or contoured profile calculated to produce a final lateral beam profile at the treatment position having a flat distribution of a required size (Figure 8–2). The approach is similar to what is done for clinical electron beams in conventional linacs.

The use of a single flat scatterer, often composed of a thin foil of high-Z material such as tungsten or lead, simply increases the width of the original Gaussian profile (Figure 8–2a). This can be used when very small sizes are required (e.g., ophthalmic and radiosurgery in field sizes up to around 3 cm in diameter). The efficiency of the system is low, as a large amount of the beam will be blocked. When only the central part of a Gaussian is kept, the final efficiency is only 3% to 5%. Special single scatterers with higher efficiency have been developed that take into account the real beam of some prototype facilities, such as elliptical foil with a central hole described by Nauraye et al. [4] which produces a 3 cm beam for ophthalmic applications with 1% flatness and losing only 0.5 MeV and 21% of intensity of the original 76 MeV proton beam, measured without the scattering foil.

A single scatter cannot provide optimum beam for large tumors, hence double scattering devices are used. The use of a *double scattering system* is typical for proton treatment heads and is based on spreading the beam laterally with a pair of specially designed scattering devices placed on the beam's central axis. It consists of a first flat scatterer and a second scatterer with either occluding rings (Figure 8–2b, as proposed by Koehler et al. [5] or with a contoured shape (Figure 8–2c, as proposed by Grusell et al. [6]. Takada [7] described an intermediate technique with two "compensated rings." In either situation, only a fraction of the beam in the center is taken out for

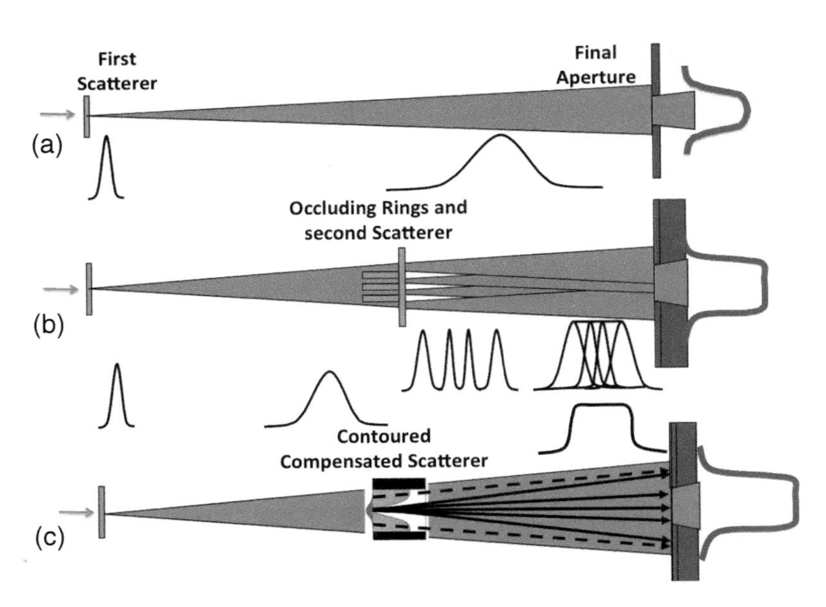

Figure 8–2 Scattering systems: (a) Single scattering. (b) Second scatter with occluding rings (courtesy N. Schreuder). (c) Second scattered contoured lead and compensated with PMMA. (From Gottschalk et al. [8,3].)

clinical use that meets the flatness criterion as described in Chapter 15. An aperture blocks the outer scattered beam.

The technique of double scattering has a higher efficiency of beam usage compared to the use of a collimated Gaussian produced by a single scatterer. The contoured scatter, in the shape of a dome, can also be "compensated" in energy: in order to avoid that the range of central rays traversing larger thicknesses of high-Z material become shorter than the rays at the edges of the filter, a "negative" of the dome is done with low-Z material. The calculation is done in such a way that the final required scattering is obtained with a constant energy loss for each ray of the beam. Efficiency of double scattering can be pretty high, e.g., 30% to 45% with a minimum energy loss (<5 MeV).

The calculations to design scattering systems are performed using the theories of Moliere, Highland, and others and implemented by Gottschalk et al. [3], providing the scattering angle as a function of the energy and the thickness of the scatterer, and optimizing the profile and the position of the scatterers. In practice, sometimes a range absorber or a range modulator can play the role of a first scatterer, but it must be taken into account that they can vary for different applications. Similarly, a carousel of scattering foils can be implemented to have the choice of optimized scatterers for different beam sizes and energies. Most of the existing scatterers produce circular homogeneous beams. All double scattering systems are sensitive to a beam misalignment on the second scatter (some centering feedback is usually installed) and not too sensitive to changes in the beam divergence.

The widening of the beam starts will determine the position of the *effective* source and the *virtual* source of the beam.

The virtual source-to-axis distance is the distance of a point on the central ray in space from which the radiation appears to originate, diverging to determine the size of the beam. The virtual source phenomenon is associated with every charged particle diverging due to Coulomb interactions in material interposed in the beam and the air up to isocenter. The virtual source position is closer to the patient than the "original" physical source position from where the beam comes from. Due to long source-to-treatment distance in proton therapy, the effect is not that pronounced compared to electron therapy. However, it is important to know this position for the calculations and the construction of patient-specific devices (apertures, compensators, etc.), and for scaling field dimensions.

The effective source-to-axis distance is the distance between a point upstream from the isocenter from which the inverse square law can be applied for a particle beam to calculate the change in dose rate as a function of distance. It is needed for dosimetry and for monitor unit calculations.

As presented later in this book, the beam size can also be selectively increased by the use of dynamic systems, e.g., either uniform scanning or pencil beam scanning.

8.3 Range Modifiers

The beam energy extracted from the accelerator is usually fixed with cyclotrons and variable in synchrotrons.

For cyclotrons, usually a beam energy degrader of rather low-Z (e.g., beryllium, carbon, or aluminum) is placed after the extraction to change the energy to a required value. A set of dipoles with slits or apertures completes the role of the degrader to constitute an energy selection system with a small transmitted energy spectrum. Details of these energy selection systems are described in Chapter 6, Proton Beam Production and Dose Delivery Techniques. Energy selection is usually performed, for example, by a pair of wedges in opposite directions in the beam line or with equivalent methods.

Independent of the method used to vary the energy upstream in the transport system, there may be some constraints to transport the lowest energies. For example, the transmission efficiency of the energy selection system is too low when degrading the original energy of a cyclotron (e.g., 200–250 MeV) to treat superficial targets. Typical minimal energy values entering the treatment rooms are in the order of 70 to 100 MeV.

In order to use lower energies for superficial targets, "range modifiers" or "absorbers" are sometimes installed in the beam-shaping nozzle, in-line with the patient. Different solutions have been provided, including variable water columns, a combination of moving wedges, and, mainly, "binary filters," where the final required range can be achieved by a combination of slabs of different thicknesses. In some extreme cases, the absorber is close to the patient, like a compensator or, if in contact with the patient, like a "bolus." Typical resolution of these kinds of energy degraders is a few tenths of millimeters, but this depends mainly on the location, composition, and control system of the absorber.

8.4 Energy Modulation

Since the depth covered by the Bragg peak of a single energy is small compared to the thickness of nearly all of the clinical targets, it is necessary to increase the width of this peak to cover the targets in the beam direction with a homogeneous dose in depth [9]. A solution has been proposed by a weighted superposition of a set of quasi-monoenergetic beams producing a region of longitudinally flat dose distribution of a finite size in depth [10].

This depth–dose distribution is known as the "*Spread-out Bragg Peak*" (SOBP). A "proximal range" is defined as the first depth where the dose is x% of the flat dose distribution (x typically defined as either 80%, 90%, or 95%). The water equivalent thickness between this proximal range of the SOBP and the distal range is known as the "*modulation width.*" The modulation width can thus be defined by two dose levels, such as 98–90 if the proximal range is defined at 98% and the distal range is defined as 90% dose. A higher dose level is sometimes used proximally because of the shallowness of the proximal dose fall-off.

The distal fall-off of the SOBP compared to the pristine peak is only slightly affected by the superposition of the distal beams. The difference between the pristine peak range and range of an SOBP with the same maximum energy is typically ~1 mm. Between the surface and the most proximal peak, a flat dose distribution is also obtained (the "entrance plateau"), in general with a dose level lower than in the region of the superposition of peaks, but higher than the flat entrance dose in the original pristine peak.

The entrance dose is proportional to (R−M)/R, where R is the range and M the modulation. The increase of the entrance dose could be a serious limiting factor in clinical applications of proton beams related to the skin tolerance, very similar to the kilovoltage beam. Proton beams give higher skin doses than those related to the build-up for lack of electronic equilibrium with photon beams.

A device used to modulate the penetration of a beam into a patient to cover the intended target during the delivery of one portal with an SOBP is called a "*range modulator*." The device may consist of a "*range modulator wheel*," a "*ridge filter*," a cone or set of cones, a set of blocks with uniform thickness programmable in a binary fashion, a variable water column, etc. The modulation can also be produced with a scanning beam, or by a combination of methods (see Figure 1–6).

A "*range modulator wheel*" or "*propeller*" consists of a stack of layers of PMMA or other low-Z material (e.g., aluminum) connected to a hub (Figure 8–3). The entire stack rotates perpendicular to the beam axis, inserting different thicknesses of material in the path of a thin beam of monoenergetic protons, thereby modulating the penetrability of the beam and producing the *SOBP*.

The use of low-Z materials minimizes the scattering produced by this device. Each displacement in water of the peak is proportional to the water equivalent thickness (WET) (as described in Chapter 4) of each sector of the stack. The required weight is represented by each relative angular sector. The values of thicknesses and weights for each sector are a function of the shape of the original pristine Bragg peak, the beam attenuation by the thickness of each sector, the position in the beam line,

Figure 8–3 Modulator wheel in PPMA. The sector in air corresponds to the maximal range. Four double "arms" give eight modulations per turn, increasing the speed of the "repainting" of the SOBP in depth up to 8 × rpm. Similar approaches are done with aluminum wheels.

and the beam energy. Please note that for each SOBP width, one should need a separate wheel. Hence, a library of wheels should be used for each combination of SOBP and energy. In practice, other solutions exist, as presented in the following paragraphs.

The displacement from the first (air) to the second sector is very critical as the highest gradients and weights are being combined. Usually the distance between peaks is in the order of the 80% proximal to the 80% distal width of the original pristine peak, but it must be finely adjusted during calculation and implementation. A too large or too short distance between consecutive peaks in an SOBP can result in nonhomogeneous doses, i.e., dips or bumps in the SOBP plateau. A weight higher than the optimal theoretical value may be adopted for the last peak, keeping a higher dose than the flat region at the end, in order to keep a very steep distal falloff. For the less weighted beams, the distance between peaks can be varied. The most proximal peak determines the proximal range, and so the modulation. The addition of layers determines the skin dose, and this is a limiting factor for superficial or thick targets requiring large modulations.

A "compensated modulator wheel" can be conceived by combining low-Z material (PMMA) and high-Z (lead) materials in order for each sector to produce the same scattering independent of the range variation. As presented previously, several modulators are needed to have at one's disposal all the required values of modulation for clinical applications, and to have them well adapted to different ranges. A time-dependent *"beam current modulation"* can be implemented, synchronized with the rotation of the modulator wheel to selectively modify the weight of each sector. This is a way to optimize the flat distribution, to use the same modulator at different energies, or to change the modulation width. The extreme use of this beam current modulation is to stop the beam for certain sectors. An equivalent approach is done in some lines manually by adding metallic inserts covering some of the thickest sectors, thereby reducing the amount of modulation using a single wheel. In both cases (covering by metal or cutting the beam) there is a reduction on the total dose rate.

A beam line with a modulator wheel is not a fully "passive" system, as the beam position is changing in depth during the treatment. A typical wheel speed in a clinical setting is on the order of 200–1000 RPM. In order to produce mechanically balanced wheels, at least two spans are calculated and built. A wheel with four symmetrical spans produces eight times the beam modulation per turn. In this way, the target is "painted" dynamically at a very high frequency, making the modulator a device that can be used with continuous and pulsed beams (in the last case, the superposition of peaks is done on an statistical base). The fast painting is also well adapted in the management of moving targets. In any case, it must be guaranteed that there is no synchronism, either with the beam pulses or with the target movement.

A *"ridge filter"* is a static range modulator that consists of several ridges and valleys presenting different thicknesses of material to an incident beam to vary the penetration into the patient. It can be designed to perform a similar function as the *"range modulator wheel,"* i.e., to construct an SOBP. The principle of a ridge filter is presented in Figure 8–4. The rays crossing the ridges and valleys will have different ranges, and they will be combined in the target by their incidences after being modified

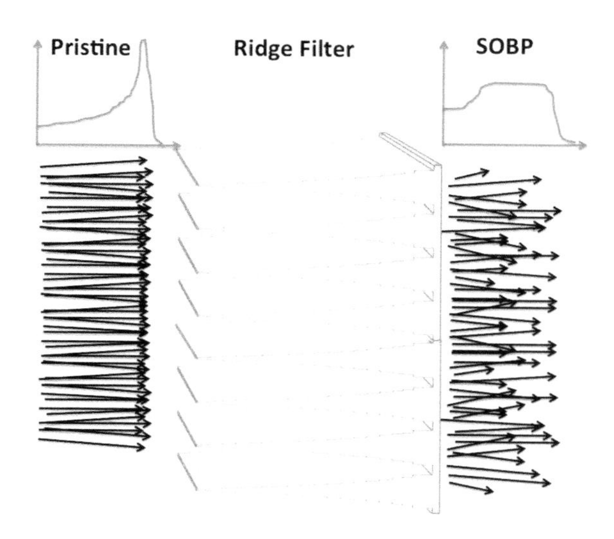

Figure 8–4 Conversion of a pristine Bragg Peak into an SOBP by a passive ridge filter. The arrows represent the particles in air, and their lengths represent their residual range in water.

by multiple scattering. One ridge filter must be built for each discrete value of modulation width required, and it will be used for a limited number of range values. A library of filters is usually created. The shape of each ridge is a function of the original shape of the pristine Bragg curve, the required modulation, the material and position of the filter in the beam line or nozzle, and the beam energy. It has the advantage to irradiate simultaneously the full depth of the target. Ridge filters are widely used in facilities in Japan for both protons and ions [11]. A rotational spiral ridge filter has also been conceived at the Lawrence Berkeley Laboratory, in particular for heavy charged particle beams [1]. The Russian proton center also uses a spiral ridge filter [12].

A thin ridge filter, called "ripple filter," produces just enough variation in light ion energies entering the patient so that a reduced number of individual accelerator energies may be used to cover the target in depth.

8.5 Apertures

The property of charged particles having a finite range in matter makes it even simpler than for photon beams to conceive personalized apertures to delimit the beam shape laterally (Figure 8–5). In general, high-Z materials produce significant amounts of neutrons and activation. A typical material is alpha-beta brass (e.g., Cu 59% Zn 39%) sometimes including a small percentage component of lead (e.g., 2% to enhance the machinability of brass). Low melting point alloys (Lipowitz, or commercial brand Cerrobend, a combination of lead, bismuth, and tin, with or without cadmium) have also been used, but the type of activation seems to be a limit on their use [13, Antoni

R, personal communication]. Due to this, brass is optimum and used universally [14]. Nevertheless, for a typical treatment field, the majority of unwanted neutron background is produced in the patient-specific aperture (see Chapter 10, Secondary Radiation Production and Shielding for Proton Therapy Facilities).

In general, a stock of cylinders of different diameters is prefabricated (e.g., diameters of 70, 100, and up to 250 mm diameter). For the smaller diameters, the thickness is adapted to the maximal energy (e.g., 200 MeV, brass cylinders of ~50 mm thickness). When the diameter is large, the weight of the aperture can be too high to manipulate. Consequently, those apertures are fabricated in 2–3 slices of about 20 mm thickness per slice. For ophthalmic applications (low energy, small field size), the apertures may be created with smaller diameters (e.g., on the order of 50–60 mm) using ~15 mm of brass. The typical density is on the order of 8370 to 8460 kg/m^3, so the usual weights vary from 0.2 (ophthalmic) to more than 7 kg. For patient-specific quality assurance, it is important to have a locking system—a hardware feature on the aperture and compensator to guarantee the correct orientation. In addition, an engraved axis and aperture identification (e.g., bar codes) for quality assurance purposes might be used. The cost of a high-resolution milling or wire-cutting machine may exceed $100,000. The basic brass cylinders to built each compensator varies between $50 and $200 apiece. The price of these machines, devices, raw material, and final personalized pieces is strongly related to vendors and local solutions. If the fabrication is outsourced, apertures may be done in 24–72 hours.

The use of apertures in passive proton beams is a key factor to getting very well-defined beam sections with a high resolution in their shape and a reduced penumbra.

Figure 8–5 Apertures for "large" fields and for ophthalmic applications.

But apertures are creating additional workflow issues. For instance, they need to be fabricated, checked, and set up prior to treatment. Furthermore, they have to be treated as radioactive material after their use. The production of neutrons from nuclear interactions in the high-Z material close to and directed toward the patient can cause secondary dose. If apertures are close to the patient, the lateral penumbra at the entrance is reduced, but some secondary doses are delivered due to edge scattering.

Protons being scattered at different thicknesses of the aperture but close to its internal surface are scattered back toward the central axis with a large spectrum of energies, including both high and low components. The low-energy components create "horns" in the lateral profile of the beam close to the skin.

This mix of energies in the beam created problems in the past when performing calculations of delivered dose using Faraday cups to measure the beam current: the conversion from cumulated charge to dose uses the stopping power. If the calculation is done with a nominal single energy, the stopping power does not reflect this mix of energies, creating uncertainties as large as 10%. These uncertainties may increase for apertures delimiting beams of small size and for non-divergent apertures.

After using the apertures for the treatments, it is necessary to evaluate, control, and follow them as active pieces, taking into account the real amount of dose delivered using each aperture. Isotopes produced by spallation on Cu, Zn, and Pb are in the region of A=60 and A=200. Isotopes of ^{62}Cu and ^{64}Cu are predominant during the first days after irradiation. After one year, the main isotopes are ^{65}Zn and ^{57}Co. Depending on the local regulations, different protocols should be used to deal with this activated material before disposal—determining how long to keep them (e.g., 1–2 years), evaluating the spectra after decay and the limit of mass activation, and measuring the global activity or hot spots, etc. A risk analysis must be performed for the staff using these apertures in clinical applications and for the staff in charge of dealing them as radioactive waste. Whenever apertures are used, radiation safety is a common concern for activation producing radiation hazard (see Chapter 10, Secondary Radiation Production and Shielding for Proton Therapy Facilities). A process similar to the one described by Walker et al. [15] should be used in such situations.

8.6 Compensators

A "*range compensator*" is a patient- and field-specific range-modifying device used primarily in passively scattered proton therapy (PSPT). It provides differential penetration laterally across a charged particle beam for each ray within the beam coming from the virtual source. The objective is to shape the distal edge of the high-dose region to conform to the distal surface of the target plus a margin in order to prevent or reduce dose to normal tissues and critical structures distal to the target. Therefore, the compensator must take into account the entrance surface, the inhomogeneities in the path, and the physical depth of the distal surface of the target. A sketch of a compensator effect in patient volume has been presented in Chapter 4. A typical compensator in PMMA is presented in Figure 8–6.

The first step in the design of a range compensator is usually done by ray tracing from the *virtual source* to the distal edge of the target. For each ray, the *water equiva-*

lent thickness (WET) is calculated. The maximal WET (with any additional margin to take into account uncertainties) will determine the required (maximal) range of the beam with a zero thickness for the compensator in the projection of this ray. For all the other rays, the difference between WETs compared to the maximal one will determine the thickness in WET of the compensator at each point. For practical reasons, a small thickness is added to all the calculations to avoid a hole at the place of the maximal range. Usually a theoretical matrix is calculated with a rather high resolution (e.g., each mm or less). These calculations are performed within the treatment planning system.

Considering that for multiple Coulomb scattering it is not possible to have compensators produce either deep isolated rays equivalent to small-sized pencil beams (the Bragg peak of the related small beam should disappear), nor very high gradients in the compensator shape (to avoid deformation effects due to multiple scattering), a "smearing process" is calculated on the theoretical matrix. The thickness of each point will be modified to take into account any smaller thickness around it at a given distance called the "*smearing radius*." This widening of the erosion of a compensator is also related to take into account anatomical misalignments due to setup uncertainties and inter- and intra-fractional anatomic variations.

Finally, the manufacturing process is simulated. For example, if the compensator will be made using a drilling machine, the size of the drill will be modeled is such a way that the depth of penetration of the drill into the compensator material will be determined by the deepest point into its diameter, which corresponds to a second smearing process.

Within the TPS, the WET matrix must be converted into thicknesses of the real material of the compensator to prepare the manufacturing file. Low-Z material such as polymethyl methacrylate (PMMA, $C_5H_8O_2$, H8%-C60%-O32%, or the commer-

Figure 8–6 Compensators in Lucite for high-energy proton beams, with the shape given by a drilling machine on a cylindrical base. One compensator is built for each field.

cial brands Plexiglas, Lucite, and others), and wax are used to conceive the compensators. In this way, the compensator modifies by ionization the range of every ray while minimizing the multiple scattering of the charged particles. While wax is recyclable, it is more fragile than PMMA.

Prefabricated cylindrical blocs can be bought/prepared in large series, e.g. with diameters of 100–300 mm diameter and thicknesses up to 100–150 mm of PMMA. The density of PMMA is 1200 g/m^3. Typical pieces have a weight of 0.5–1 kg. The final shape is either made at a local workshop with milling machines, or bought through servicing companies. Special tools and procedures are developed to speed up the compensator manufacturing, considering that in passive lines each beam for each patient requires one specific compensator.

A large milling machine with the capacity of working several pieces in parallel (including during the night) may cost more than $100,000. Typical costs of prefabricated cylinders of PPMA vary from around $10 to $100 apiece. If the fabrication is outsourced, compensators may be done in 24–72 hours.

A comprehensive quality control must be done for each compensator, either mechanically or by using some kind of tool (optical, X-rays, the proton beam, etc.) to verify the thickness of all or at least several points. Various techniques for QA have been proposed, but each institution should adopt the QA process that is economical and accurate for patient treatment in their own environment [16,17].

In practice, for most of the existing lines, it is necessary to enter in the treatment room between beams to set up the compensator corresponding to the beam to be used. Specific holders are conceived to warranty the correct positioning of the compensator related to the beam axis. A range compensator may also be used for scanning beams to deliver layers with the same shape as the distal surface of the target.

As discussed in the previous section on apertures, it is necessary to have a process of radiation protection associated with the activation of the compensators. While wax is recyclable, PPMA can be used only once. The main isotopes are ^{3}H (12.3 years, 23%) and mainly ^{7}Be (53 days, 77%), and if stored for decay, the usual storage times are a few months (e.g., 6 months).

8.7 Multileaf Collimator (MLC)

The process of manufacturing apertures [18] has radiation safety concerns [15], is relatively expensive, and does not provide any changes that may be needed after filming. Das et al. [19] has provided some estimates regarding the cost related to manufacturing.

Takahashi has discussed the early concept of MLC for the replacement of conformal blocks for radiation treatment in 1965 [20]. Later it was extended through computer control processes [21]. The lifting of heavy apertures creates occupational hazards as noted Aribisala [22] and McCullough et al. [23]. The desire for dynamic changes in blocks and the reduction in labor, manufacturing, and throughput led to the marketing of MLC in external beam therapy [24]. Every manufacturer developed and marketed various types of MLC during the 1990s, leading to the revolution in conformal therapy [25–28]. The success in the MLC technology led to the creation of nonuniform dose profiles needed for the success of IMRT [29]. The use of an MLC in proton therapy is relatively new.

The cost associated with manufacturing an aperture is significant [19] and adds to the total cost of proton treatment. Unfortunately, brass apertures are not dynamic or cannot be changed in real time for any changes needed due to tumor shape. Manufacturers also provide nozzles in diameters of 10, 20, 25, and 30 cm, each with a different aperture base size. To change a nozzle, it takes nearly 20 minutes of time as these are heavy devices. The process is usually performed robotically or manually. In such situations, it is clear that an MLC could provide advantages for dynamic field shaping, cost reduction, and throughput.

The world's first MLC for particle therapy was designed for neutron beams whose characteristics have been presented by Maughan et al. [30]. For charged particles, Mitsubishi provided an MLC to be used with protons and carbon ions at Hyogo, Japan. The characteristics of this MLC were presented by Das et al. [31]. This Mitsubishi MLC is made out of carbon-coated steel blades. Varian produced a tungsten MLC for the IBA snout, whose activation characteristics were presented by Diffenderfer et al. [32]. Moskvin et al. [14] showed that the MLC material is critical with respect to neutron production, i.e., in double scattering proton beam, a tungsten-based MLC may not be the optimum. Apart from neutron dosage from an MLC and some limitations in their resolution, these devices provide unique advantages for beam shaping and cost saving to institutions. Additional research is needed to produce MLCs made of titanium or low-Z material to reduce neutron dose and probable activation dose to therapists and engineers.

8.8 Specific Lines: Ophthalmic Treatments

Scattered beams are and have been widely used for clinical applications with a large clinical follow-up. They have been adopted for fixed lines (horizontal, vertical, oblique, etc.) and for gantries. They also have been the basis to develop lines devoted to specific treatment sites, such as ophthalmic treatments [33] and radiosurgery.

A typical beam line for treatment of ocular targets is presented in Figure 8–7. This beam line has a homemade specific single lead scatterer based on a disk diameter of 14 mm and a thickness of 0.1 mm. The entrance energy is 76 MeV (vacuum window at the end of the beam transport to the room). The complete line shown in the image is in open air. The range shifter is done with a set of binary thickness polycarbonate slabs. The energy modulator is done in PMMA (with additional brass sectors hiding the thickest sectors to reduce modulation width with a single wheel). The beam monitoring is done with a pixel ionization chamber called MOPI [34] and a flat parallel transmission monitor chamber, as in most of the centers [35]. The apertures are in brass. The gaze of the patient is oriented toward a diode mounted at the end of the line on a polar system to reproduce the virtual simulation at the treatment planning system. A system of orthogonal x-rays and detecting flat panels are used to set up the patient (with implanted fiducials in Tantalum) and an infrared camera to monitor the position during the treatment.

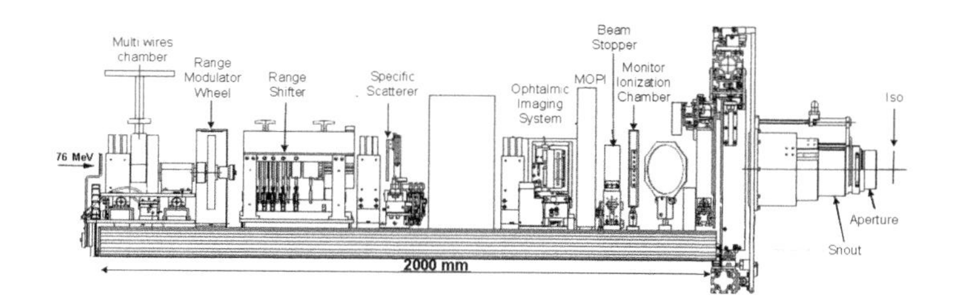

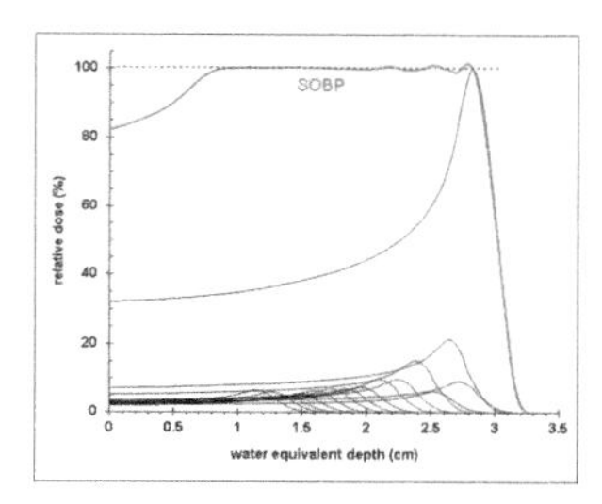

Figure 8–7 The horizontal line with a passive system for treatment of eye tumors at the Institut Curie Centre de Protonthérapie d'Orsay, France (courtesy C. Nauraye). Upper and center: scheme and view of the beam line, where components, including the last aperture, can be changed for eyes or intracranial targets. Lower left: a typical SOBP for ophthalmic treatments. Lower right: end of line for eye treatments, with aperture, light to fix the gaze, infrared diodes, and camera.

Typical data for beams produced with this beam line are presented in Table 8–2.

8.9 Summary/Conclusion

In this chapter we presented the main components for field shaping in scattered proton beams. Scattered beams have been in use since the start of proton therapy, providing today the largest clinical experience with a large follow-up.

The specifications of scattered beams are strongly determined by a small set of components placed downstream of the beamline, after the transport system (fixed line

Table 8–2 Typical beam specifications of a scattered beam line for eye treatments (Courtesy C. Nauraye, Orsay, France)

Opthalmic Line Specifications, Orsay	Pristine Bragg	Reference SOBP	Transverse Profile	
Distal falloff 90/10	2.4 mm	2.6 mm	Range	25 mm
Distal falloff 80/20	1.7 mm	1.9 mm	SOBP	18 mm
Range 90%	39.8 ±0.5 mm water	25 mm water	Depth	16 mm
FWHM = 7.6 mm	7.6 mm		P90/10	2.4 mm
SOBP 95%		18 mm	P80/20	1.6 mm
Pic-to-surface ratio = 3.5	3.5		D50%	30.8 mm
Max field diameter	30 mm	30 mm	D95%	27.5 mm
Dose rate		16 ±2 Gy/min		
Air gap	6 cm	6 cm		

or in a gantry) and close to the patient, in what is usually called a nozzle, ending in a movable "snout." These components are single- or double-scattering foils, range modifiers, energy modulators, customized apertures (or multi-leaf collimators), and compensators. They were fabricated locally by the pioneering centers, but lately they are being outsourced to specialized facilities.

Other devices are installed, e.g., monitors for the dose, the dose rate, and the beam centering, as well as tools for positioning and monitoring the patient during the treatment.

The evolution of scanning techniques (see the next chapter) makes them the ongoing tool of choice for present and future applications, including the capacity to perform intensity-modulated therapy. In spite of this tendency, the existing and some new scattered beam lines are treating and will continue to treat a large volume of patients. Dedicated lines use scattering for specific treatments, such as those providing small field sizes (e.g., ophthalmic treatments and radiosurgery). Scattered beams also provide robust solutions for the treatment of moving targets. And the scanning techniques may still use passive components in specific clinical cases, such as apertures, compensators, and ridge filters to optimize some parameters, such as the penumbra, the distal edge, and the time to cover the target in depth.

Acknowledgments

We thank Sabine Delacroix, Ph.D. and Catherine Nauraye, Ph.D., for their data from the Centre de Protonthérapie d'Orsay, Institut Curie, France, and the support from France Hadron.

References

1. Chu WT, Ludewigt BA, Renner TR. Instrumentation for treatment of cancer using proton and light-ion beams. *Rev Sci Instrum* 1993;64(8):2055–2122.
2. Mazal A. Proton beams in radiotherapy. In: *Handbook of radiotherapy physics*. Ed P. Mayles, A. Nahum, JC Rosenwald. Taylor & Francis, 2007, 1005–1032.

3. Gottschalk B, Koehler AM, Schneider RJ, Sisterson JM, Wagner MS. Multiple Coulomb scattering of 160 MeV protons. *Nucl Instr Meth B* 1993;74:467–490 and "NEU user guide" at http://gray.mgh.harvard.edu/attachments/article/212/neu.pdf.
4. Nauraye C, Mazal A, Delacroix S, Bridier A, Chavaudra J, Rosenwald JC. An experimental approach to the design of a scattering system for a proton therapy beam line dedicated to ophthalmological applications. *Int J Radiat Oncol Biol Phys* 1995;32(4):1177–83.
5. Koehler AM, Schneider RJ, Sisterson JM. Flattening of proton dose distributions for large-field radiotherapy *Med Phys* 1977;(4):297–301.
6. Grusell E, Montelius A, Brahme A, Rikner G, Russell K. A general solution to charged particle beam flattening using an optimized dual scattering foil technique, with application to proton therapy beams. *Phys Med Biol* 1994;39:2201–2216.
7. Takada Y. Dual-ring double scattering method for proton beam spreading. *Japan J Appl Phys* 1994;33:353–359.
8. Gottschalk B, Wagner MS. Contoured scatterer for proton dose flattening. *HCL Technical Note* (1989).
9. Wilson RR. Radiological use of fast protons *Radiology* 1946;47(5):487–91.
10. Koehler AM, Schneider RJ, Sisterson JM. Range modulators for protons and heavy ions. *Nucl Instr Meth* 1975;131:437–440.
11. Akagi T, Higashi A, Tsugami H, et al. Ridge filter design for proton therapy at Hyogo Ion Beam Medical Center. *Phys Med Biol* 2003;48:N301–N312.
12. Kostjuchenko V, Nichiporov D, Luckjashin V. A compact ridge filter for spread out Bragg peak production in pulsed proton clinical beams. *Med Phys* 2001;28:1427–1430.
13. Rouse I, Martz M, Glisson C, Siebers J, Elmer T. Analysis of Cerrobend activation produced by a 250-MeV medical proton accelerator. 30th midyear meeting: Health physics of radiation-generating machines. Health Physics Society. San Jose, CA, 1997.
14. Moskvin V, Cheng CW, Das IJ. Pitfalls of tungsten multileaf collimator in proton beam therapy. *Med Phys* 2011;38:6395–6406.
15. Walker PK, Edwards AC, Das IJ, Johnstone PAS. Radiation safety considerations in proton aperture disposal. *Health Phys* 2014;106:523–527.
16. Zhao Q, Wu H, Das IJ. Quality assurance of proton compensators. In: M Long, ed. Quality assurance of Proton compensators. IFMBE proceedings 2013. Springer, 1719–1722.
17. Yoon M, Kim JS, Shin D, et al. Computerized tomography-based quality assurance tool for proton range compensators. *Med Phys* 2008(35); 3511–3517.
18. Moyers MF, Benton ER, Ghebremedhin A, Coutrakon G. Leakage and scatter radiation from a double scattering based proton beamline. *Med Phys* 2008;35:128–144.
19. Das IJ, Moskvin VP, Zhao Q, Cheng CW, Johnstone PA. Proton therapy facility planning from a clinical and operational model. *Technol Cancer Res Treat* 2014;10.7785/tcrt.2012.500444 (eprint).
20. Takahashi S. Conformation radiotherapy: Rotation techniques as applied to radiography and radiotherapy of cancer. *Acta Radiologica Supplement* 1965;242:1–142.
21. Hounsell AR, Sharrock PJ, Moore CJ, Shaw AJ, Wilkinson JM, Williams PC. Computer-assisted generation of multi-leaf collimator settings for conformation therapy. *Br J Radiol* 1992;65:321–326.
22. Aribisala E. Cumulative trauma disorder: Occupational hazards of the 90s. *Radiat Therap* 1993;2:27–29.
23. McCullough EC, Buchholtz L, Hagedorn C, Hewlett T, Petry J, Schwarz T. Potential reductions in the daily lift weight of a radiation therapist. Varian CenterLine, Palo Alto, California, November, 1994.
24. Helyer SJ, Heisig S. Multileaf collimation versus conventional shielding blocks: A time and motion study of beam shaping in radiotherapy. *Radiother Oncol* 1995;37:61–64.
25. Klein EE, Harms WB, Low DA, Willcut V, Purdy JA. Clinical implementation of a commercial multileaf collimator: Dosimetry, networking, simulation, and quality assurance. *Int J Radiat Oncol Biol Phys* 1995;33:1195–1208.
26. Thompson AV, Lam KL, Balter JM, McShan DL, Martel MK, Weaver TA, Fraass BA, Ten Haken RK. Mechanical and dosimetric quality control for computer controlled radiotherapy treatment equipment. *Med Phys* 1995;22:563–566.
27. Das IJ, Desobry GE, McNeeley SW, Cheng EC, Schultheiss TS. Beam characteristics of a retrofitted double-focused multileaf collimator. *Med Phys* 1998;25:1676–1684.
28. Galvin JM, Han K, Cohen R. A comparison of multileaf collimator and alloy-block field shaping. *Int J Radiat Oncol Biol Phys* 1998;40:721–731.

29. Bortfeld T. IMRT: A review and preview. *Phys Med Biol* 2006;51:R363–379.
30. Maughan RL, Yudelev M, Aref A, Chuba PJ, Forman J, Blosser EJ, Horste T. Design considerations for a computer controlled multileaf collimator for the harper hospital fast neutron therapy facility. *Med Phys* 2002;29:499–508.
31. Das IJ, Akagi T, Kagawa K, Mayahara H, Yanou T, Suga D, Sakamoto H, Hishigawa Y, Abe M. Geometric and dosimetric characteristics of a proton beam multileaf collimator (MLC). [Abstract]. *Med Phys* 2004;31:1797.
32. Diffenderfer ES, Ainsley CG, Kirk ML, McDonough JE, Maughan RL. Comparison of secondary neutron dose in proton therapy resulting from the use of a tungsten alloy MLC or a brass collimator system. *Med Phys* 2011;38:6248–6256.
33. Bonnet DE, Kacperek A, Sheen MA, Goodall R, Saxton TE. The 62 MeV proton beam for the treatment of ocular melanoma at Clatterbridge. *Br J Radiol* 1993;66(790):907–14.
34. Bonin R, Boriano A, Bourhaleb F, Cirio R, Donetti M, Garelli E, Giordanengo S, Marchetto F, Peroni C, Sanz Freire CJ, Simonetti L. A pixel chamber to monitor the beam performances in hadron therapy. *Nucl Instr Meth Phys Res A* 2004;519:674–686.

Chapter 9

Field Shaping: Scanning Beam

X. Ronald Zhu, Ph.D., Falk Poenisch, Ph.D., Heng Li, Ph.D., Xiaodong Zhang, Ph.D., Narayan Sahoo, Ph.D., and Michael T. Gillin, Ph.D.

Department of Radiation Physics,
The University of Texas MD Anderson Cancer Center,
Houston, TX

9.1 Introduction

A relatively new form of particle therapy that has revolutionized the practice of particle therapy in recent years is delivered using scanning beam technology [1,2]. In scanning beam delivery, a particle pencil beam is magnetically scanned in the plane transverse to the beam direction, creating a large field without requiring scattering elements into the beam paths, as shown in Figure 9–1. Monoenergetic pencil beams with different energies from an accelerator and energy selection system are stacked to create the desired dose distribution along the beam direction [3,4]. Neither an aperture nor a compensator is required, although an external collimation could be beneficial for reducing the penumbra of low-energy beams. For proton therapy, either synchrotrons or cyclotrons have been used, as discussed in Chapter 6.

The desired dose distribution over the clinical target volumes can be achieved by optimizing the weights of individual pencil beams with different energies using an inverse planning process. Two general approaches are available to optimizing a scanning beam plan. The first is to simultaneously optimize all spots from all fields, called

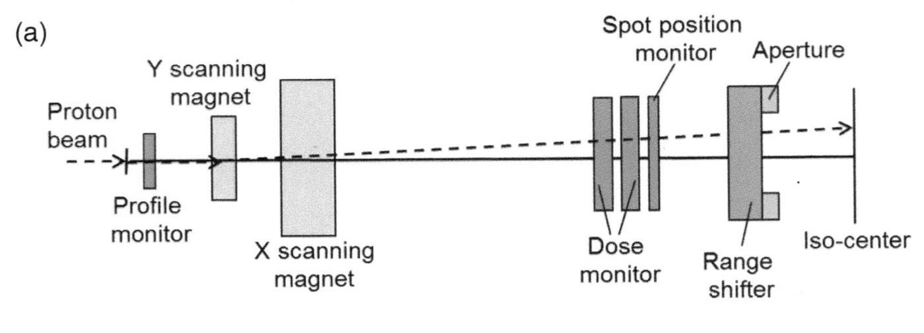

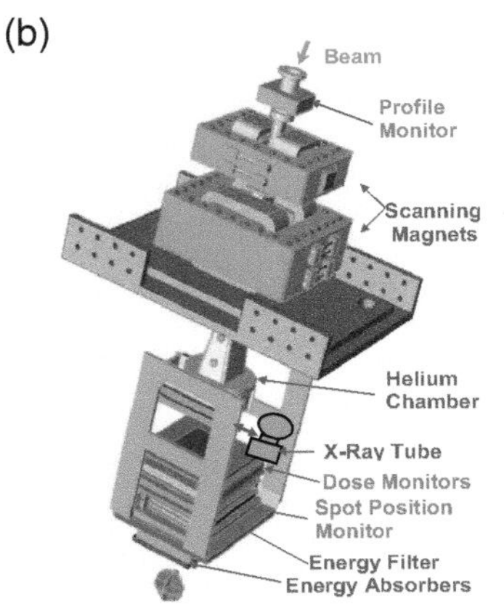

Figure 9–1 Schematic of the Hitachi scanning nozzle at the MD Anderson Cancer Center. Upper panel: major components (adapted from Figure 1 of reference [27] with permission). Bottom panel: three-dimensional rendering (from Figure 4 of reference [22] with permission.)

multi-field optimization (MFO), more commonly known as intensity-modulated proton (particle) therapy (IMPT) [5–9]. IMPT is the most advanced form of radiation therapy to date. The second approach is that each field is optimized individually to deliver a fraction of the prescribed dose to the entire target volume(s). This method is known as single-field optimization (SFO) [10,11]. The most common application of SFO is to produce a uniform dose over the entire target volume by each field, known as single-field uniform dose (SFUD) [5,10,12,13]. The major advantage of SFUD plans over passive scattering plans (only distal conformity is obtained using a compensator) is that both proximal and distal conformity in the dose distribution can be achieved. An application of the SFO technique called single-field integrated boost (SFIB) has recently been reported by Zhu et al. [11]. In SFIB planning, the SFO technique with dose constraints is used to create the desired coverage of different target volumes by different dose levels within a single plan. Clinically, MFO/IMPT and SFO techniques have been applied to various disease sites [14–21]. In this chapter, we will focus on field shaping using scanning beam for proton therapy.

9.2 Synchrotron vs. Cyclotron for Scanning Beam

One of the most important distinctions between the synchrotron and cyclotron for the scanning beam is the energy levels available from the accelerators. Numerous energy levels can be generated by the synchrotron [22,23]. On the other hand, only one energy is produced by the cyclotron [4], typically 200 to 250 MeV for proton therapy. An energy selection system—including an energy degrader, commonly made of a pair of carbon or beryllium wedges— is used with the cyclotron to produce various energies for scanning beam proton therapy.

The other important difference between the synchrotron and cyclotron for the scanning beam is the total time that it takes to change energy. The time needed for the adjustment of the beam transport line is similar between the two types of accelerators. For cyclotrons, the additional time is determined by how fast the energy selection system can be adjusted for the next energy. It has been reported that the energy change can be as fast as less than 0.1 seconds using an energy selection system [24], although most systems in clinical use would take approximately 1 to 2 seconds. For synchrotrons, the typical time required to change energy is on the order of 1 to 2 seconds [22,23]. This is the time required to de-accelerate the remaining protons with the previous energy in the synchrotrons and accelerate a new spill of protons with the new desired energy. This time is expected to be reduced significantly with a new technology called "multiple energy extraction" (MEE) [25], which is currently being developed and implemented for commercial accelerator systems.

9.3 Scanning Nozzle

An example of schematic representation of the scanning nozzle is shown in Figure 9–1. A pencil beam entering the nozzle first passes the profile monitor. Then the pencil beam goes through two scanning magnets (Y and X) followed by the sub-dose and main dose monitors and the spot position monitor. A large part of the beam path in the nozzle is either in vacuum or in a helium chamber, depending on the design and vendors, to minimize the spot size enlargement while the beam transports through the nozzle. Optional components in the nozzle include a range shifter, aperture, ridge filter, and x-ray tube [26]. The safe, accurate, and efficient treatment delivery of scanning beam is controlled by the irradiation control system and monitored by the safety system.

Unlike in the passive scattering nozzle, the range loss of the beam in the scanning nozzle is small, typically on the order of a few millimeters. On the other hand, the enlargement in pencil beam size at the isocenter caused by multiple scattering in the nozzle could be as large as 10 mm for low-energy beams [22]. The pencil beam position, size, and shape are determined by the profile monitor and spot position monitor. For the new generation of scanning nozzles, the profile monitor is an optional device. These monitors are multiwire ionization chambers [22].

Scanning magnets (X and Y) control the lateral positions of the pencil beam. The scanning magnets are normally located more than 2 meters upstream from the isocenter [22]. The scanning speed is in the range of 5 to 20 m/s, depending on designs and vendors. For a given design and vendor, in general, the low-energy beams can be scanned faster. During the treatment planning process, the treatment volume is divided into energy layers, and each layer is irradiated sequentially from highest to lowest energy. For each energy layer, there are several scanning methods available, namely, spot scanning, raster scanning, and continuous scanning, all of which are discussed in following sections.

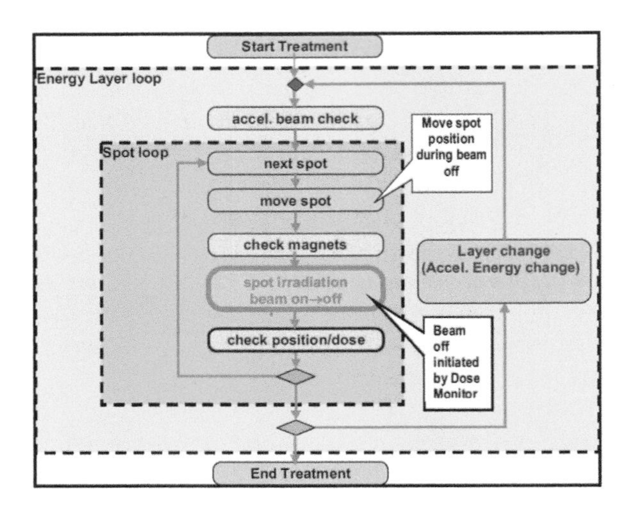

Figure 9–2 Example of flow chart for a discrete spot scanning method. The spot loop is revisited for each spot in the energy layer. When the irradiation of an energy layer has been completed, the energy layer loop cycles to the next lower energy. (From Figure 5 of Reference [22] with permission.)

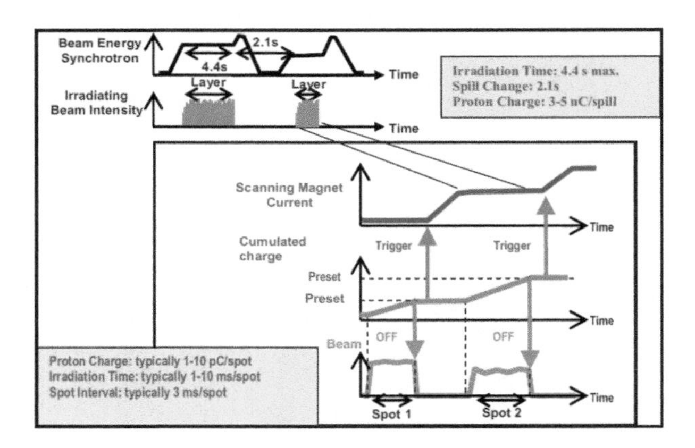

Figure 9–3 Example of a timing chart for a synchrotron-based discrete spot scanning. Two energy layer cycles are shown in the upper part. The delivery sequence between spots within an energy layer is shown in the lower part. (From Figure 6 of Reference [22] with permission.)

9.3.1 Spot Scanning

This is a discrete step-and-shot approach of delivering a scanning beam. The beam stops at the planned position and delivers a specific amount of dose controlled by the dose monitor. The beam is then turned off and moves to the next position [3], and no dose is delivered during the transition. The main advantages of this approach include: a) no modulation in beam intensity and speed scanning is required since the intensity modulation is achieved by varying the monitor units (MUs) delivered to each spot, and b) the safety of the delivering process can be easily ensured since only one spot could potentially deliver to the wrong position or incorrect amount of dose as long as the interlocks of the delivery system work correctly. The major disadvantage of this approach is that it is less efficient owing to the "dead time" between the spots.

For the Hitachi scanning nozzle at MD Anderson Cancer Center, the minimum and maximum MUs for each spot are 0.005 and 0.04, respectively. For this scanning nozzle, the flow chart and timing chart of the spot scanning beam delivery process are shown in Figure 9–2 and Figure 9–3, respectively. As shown in Figure 9–2, there are two loops involved in the process of delivering each treatment field: the energy layer loop and the spot loop. For each energy layer loop, the beam energy is first checked by measuring the revolution frequency of the proton beam and the orbit position in the synchrotron ring before extraction. Then the spot loop starts: the spot is moved to the current position while the beam is off; after the bending magnetic field strength is verified, the spot irradiation begins; the beam is turned off, initialed by the dose monitor. Once the MU to the current position reaches the spot MU given by the treatment plan; and finally the spot position and dose are checked while the beam is moving to the next spot. The spot loop is repeated until all spots are delivered within the same energy layer before moving on to the next energy layer. After all energy layers completed, the treatment delivery is completed for this field.

As shown in Figure 9–3, the timing chart of synchrotron-based discrete spot scanning delivery is viewed in two different time scales: in the scale of energy layers and within a specific energy layer. The maximum irradiation time for each energy layer is 4.4 seconds and spill change takes 2.1 seconds for this scanning nozzle. It takes typically 1 to 10 ms to deliver a spot, depending on the dose to be delivered, and typically 3 ms to verify the position, shape, and size of a spot and to move the spot to the next location [22,23]. While a spot is being delivered, the scanning magnet current is kept constant, and the accumulated charge in the dose monitor increases until the stop is completely delivered and the beam is turned off. Then the scanning beam current will change to drive the spot to the next position while the beam is off, and the accumulated charge in the dose monitor remains constant. This process is repeated for the next spot.

9.3.2 Raster Scanning

Raster scanning is similar to spot scanning except the beam is not switched off when moving to the next spot [3,28,29]. This causes a small amount of transit dose that is delivered during the move. The optimization algorithm may have to consider this transit dose in the treatment planning system (TPS) [30]. This is a hybrid method of discrete spot and continuous line scanning (as discussed in Section 9.3.3). This technique has been clinically implemented in several synchrotron-based systems for carbon ion therapy because it utilizes more efficiently the beam during spills [28,29,31,32]. This technique is faster than spot scanning for treatment delivery because the dead time is reduced to the time required to travel to the next position.

9.3.3 Continuous Line Scanning

In this approach, the beam is continuously scanned, normally along the axes in the plane of constant energy, although it is possible to have contour scanning [24,33]. The beam is only switched off when the energy is changed. The scanning speed and the beam intensity are modulated such that an arbitrary complex proton fluence can be achieved along the scanning line [24,33]. The continuous line scanning is technically more challenging to realize than either spot scanning or raster scanning [24], but it can deliver a beam very fast and can be very useful for repainting.

9.4 Energies, Range Resolutions, and Range Shifter

The proton energies normally used for scanning beam are approximately from 70 MeV up to 240 MeV, corresponding to proton beam ranges of approximately 4 to 36 g/cm^2. Some accelerators could also have higher energy, such as 330 MeV, for proton radiography or even proton computed tomography in the future. The selection of other energies between the minimum and maximum energies is determined by the ability to create a uniform SOBP using the scanning beam with superposition of pristine Bragg peaks. For example, 94 energies—72.5 to 221.8 MeV, corresponding to the proton beam ranges 4.0 to 30.6 g/cm^2 as shown in Table 9–1—are available for the Hitachi synchrotron at the MD Anderson Proton Therapy Center for spot scanning beam delivery. Figure 9–4A shows integral depth dose (IDD) curves, known as Bragg

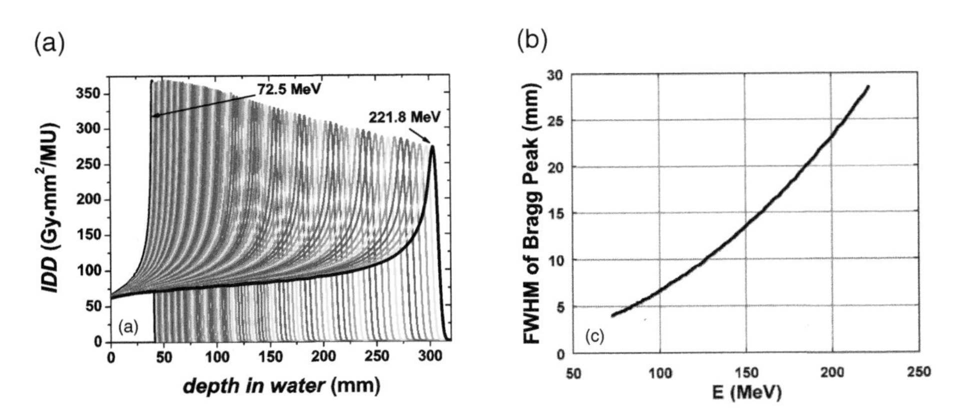

Figure 9–4 (a) Integral depth doses for 94 energies in units of Gy mm^2/MU generated using Monte Carlo simulation and (b) full-width-half-maximum (FWHM) of Bragg peak as a function of energy. (From Figure 2 of reference [27] with permission.)

Table 9–1 Energy, number of energy layers, and corresponding proton beam ranges and resolutions in water for the Hitachi scanning nozzle at the University of Texas MD Anderson Cancer Center

Energy (MeV)	72.5 –98.0	98.0 –122.5	122.5 –144.9	144.9 –161.6	161.6 –195.6	195.6 –221.8
No. Energy Layers	31	18	13	8	14	10
Range (g/cm^2)	4.0–7.1	7.1–10.7	10.7–14.5	14.5–17.6	17.6–24.6	24.6–30.6
Range Resolution (mm)	1.0	2.0	3.0	4.0	5.0	6.0

curves, for these proton energies and the full-width-half-maximum (FWHM) of Bragg peaks in the depth direction as a function of proton beam energy. As shown in Figure 9–4B, one of the most unique features of the Bragg curves is that the FWHM of Bragg peak increases with energy due to the range straggling. Thus the proton beam range resolutions vary from 0.1 to 0.6 cm as energy increases, as listed in Table 9–1. The energy levels and the corresponding ranges shown in Table 9–1 and Figure 9–4 are for a specific accelerator, and they do not apply to other accelerators, including synchrotrons and cyclotrons.

Considering the time that it takes to switch energy and to scan each energy layer, the delivery efficiency of a system with many low-energy layers using the range resolution 0.1 cm may not be desirable. To overcome this challenge, a ridge filter could be used to broaden the Bragg peak in the longitudinal direction to reduce the treatment delivery time [34].

For a given set of available energies, a proton energy optimization and reduction strategy for IMPT has been proposed, and an estimated up to 27% reduction of energy layers could be achieved using the optimization algorithm in the TPS [35]. For cyclotron-based systems, any energy could be selected by the TPS with a typical resolution of 0.1 MeV between the minimum and maximum energy, although normally only a selected number of energies are used for each field. The minimum energy of 70 MeV is used by almost all systems because it is the lowest energy at which the beam can be effectively transported from the accelerator/energy selection system to the delivery nozzle, and because the energy selection system in the case of cyclotrons is still reasonably efficient. The proton beam range of 70 MeV is approximately 4 g/cm^2, which is deeper than superficial target volumes, such as those encountered in head and neck cancers. A range shifter of 4 to 7 g/cm^2 placed at the end of the scanning nozzle is normally commissioned to bring the Bragg peak to the patient's skin. Using Monte Carlo simulation, Titt et al. [36] have demonstrated using a range shifter to effectively adjust the lateral and longitudinal size of scanning proton beam spots to optimize the penumbra and delivery efficiency. It is desirable to allow variable air gaps for the range shifter. The air gap should be kept as small as possible to minimize the spot size. Figure 9–5 shows the in-air spot size in FWHM at the isocenter for a range shifter of 6.7 g/cm^2 as a function of energy for three different air gaps. Also included in Figure 9–5 is the spot size without the range shifter. It is clearly demonstrated that the smaller air gap results in a smaller spot size.

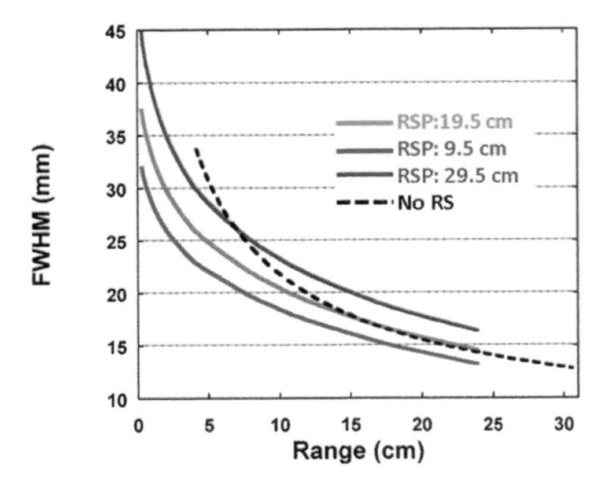

Figure 9–5 In-air spot sizes in full-width-half-maximum (FWHM) at the isocenter as a function of proton beam range with a range shifter of 6.7 g/cm^2 at different distances from the isocenter. RSP: the bottom of range shifter position from the isocenter. No RS: no range shifter. Also included is the FWHM without the range shifter.

In order to use the variable air gap range shifter, the TPS should be able to model the dosimetric variation as a function of the air gap [37]. As an alternative to range shifters, a helmet-like universal bolus around the head of a patient has recently been reported to bring the Bragg peaks to the patient skin [38]. In addition, the universal bolus keeps the air gap small, thereby maintaining the integrity of the physical properties of the pencil beam. This design is particularly useful for a fixed beam line with fixed snout.

9.5 Spot Sizes, Spot Spacing, and Spot Position Accuracy

It is generally recognized that a smaller spot size is more desirable for creating a conformal plan to allow sparing more normal tissues. It has been an industrial trend that the new generation of scanning nozzles from all vendors should have smaller spot sizes compared to their previous generation [23,39]. For example, the in-air spot sizes (1σ) at the isocenter are approximately 5 mm to 14.5 mm for the Hitachi scanning nozzle currently in clinical use at MD Anderson Cancer Center [22,23], while the new generation of nozzle from the same vendor would be reduced to 2 to 6 mm, corresponding to energies from 230 MeV to 70 MeV. Treatment planning studies have demonstrated the benefits of smaller spot sizes. A study by van de Water et al. [40] looked at potential improvements of salivary gland sparing in oropharyngeal cancer patients using IMPT with reduced spot size pencil beams. The authors found significant reduction of mean doses to the parotid glands and submandibular glands with the reduced spot sizes. Wang et al. [41] studied the impact of spot size on plan quality for radiosurgery for peripheral brain lesions and concluded that a spot size (1σ) less than 7 mm on the patient surface would result in comparable or smaller brain necrosis normal tissue control probability (NTCP) relative to photon radiosurgery.

For given spot sizes, the first factor determining the spot spacing is the ability to create a uniform dose in the plane transverse to the beam. For example, Figure 9–6 shows a simple simulation using two-dimensional Gaussian functions of lateral dose

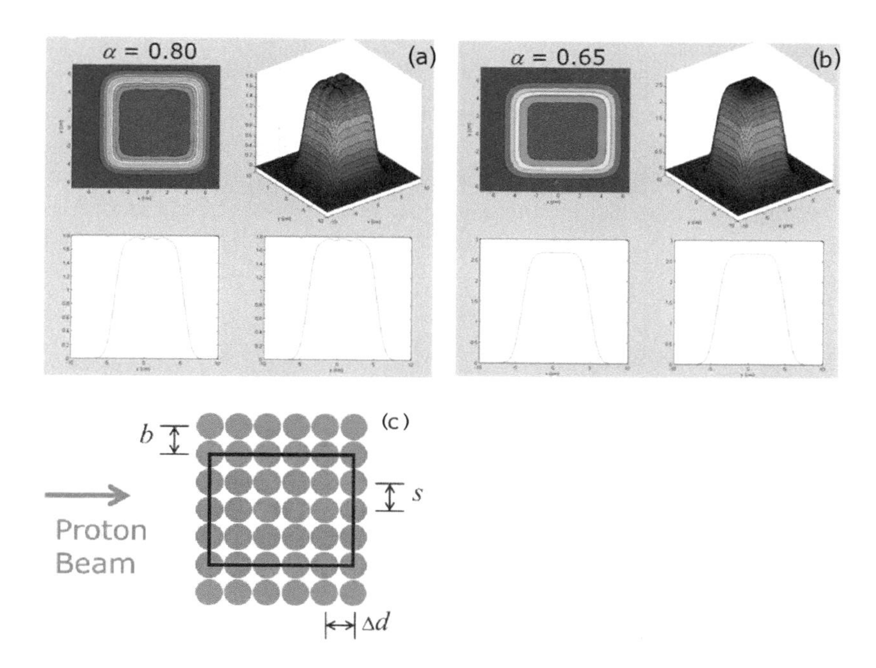

Figure 9–6 Simulation of spot spacing, placement, and effect of spot position error and lateral dose distributions. (a) The spot spacing, $s = 0.8$ *FWHM*, resulting in about 2% ripple in dose profile. (b) $s = 0.65$ *FWHM*, resulting in very uniform dose profile. (c) Spot placement with transverse spot spacing *s*, longitudinal spacing, Δd, and the distance between the target edge and the center of an outside spot, *b*.

distribution with two different spot spacings, $s = \alpha\, FWHM$, where α is a fraction of *FWHM*. When α is 0.8, there is about 2% ripple in the dose profile (Figure 9–6a); when α is 0.65, the dose profile is nearly perfectly flat (Figure 9–6b). At MD Anderson Cancer Center, the first group of prostate cancer patients were treated using a spot spacing corresponding to $\alpha = 0.65$, which resulted in a spot spacing of approximately 0.9 cm [10,12]. The second factor to be considered for spot spacing is related to penumbra when the aperture is not used. Most current commercial planning systems only allow spots to be placed on a rectilinear grid with a constant spot spacing. Normally, one would place at least one spot outside the target volume, as shown in Figure 9–6c. The distance between the target edge and the center of an outside spot, *b*, can be considered as if it was block margin, which is equal to spot spacing *s* for a constant-spacing rectilinear grid (Figure 9–6c). In general, the spot spacings and weights can be optimized to achieve sharper penumbra [42].

The delivered spot positions could deviate from the planned positions [43]. Shown in Figure 9–7 is an example of planned, measured, and recorded (in the treatment log file) spot positions for a proton beam with energy of 173.7 MeV. Figure 9–

7A shows a nine-position pattern used in this example, and Figure 9–7B is a close-up view for the planned position at (100.0 mm, 100.0 mm). The measured position using film (100.2 mm, 100.8 mm) and the mean recorded position (100.4 mm, 100.6 mm) were within 1 mm of each other. If all spots deviated from the planned positions systematically in the same direction by a small amount (e.g., <1 mm), this would have minimum impact on dose distribution in the patient. However, if some spots at certain locations deviated from the planned positions or, even worse, if neighboring spots

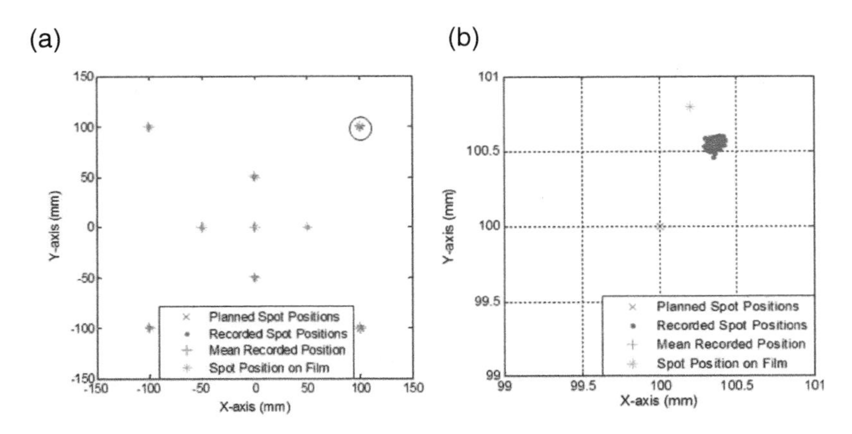

Figure 9–7 Example of planned, measured, and recorded (in the treatment log file) spot positions for a proton beam with energy of 173.7 MeV. (a) A pattern with spots at (0,0), (–50 mm, 0), (50 mm, 0), (0, –50 mm), (0, 50 mm), (–100 mm, 100 mm), (100 mm, 100 mm), (100 mm, –100 mm), and (–100 mm, –100 mm). (b) A close-up view of the spot position planned at (100.0 mm, 100.0 mm), the measured position using film (100.2 mm, 100.8 mm), and the mean recorded position (100.4 mm, 100.6 mm).

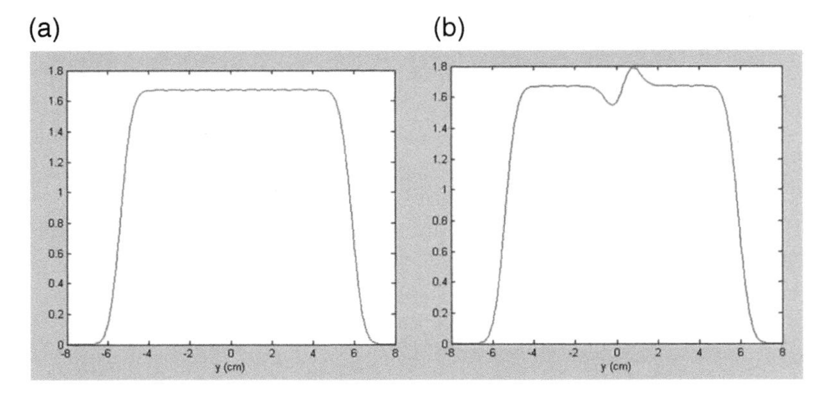

Figure 9–8 (a) Uniform dose created by evenly spaced spots in a plane and (b) 1 mm spot position error for the spots at Y = 0 mm, resulting in dose deviation about ±7%.

deviated from planned positions either toward or away from each other, there would be significant dose error. In the example shown in Figure 9–8, spots are represented by Gaussians with $\sigma = 0.5$ cm, and spot spacing equals $1.5\sigma = 0.75$ cm; a uniform dose profile is achieved without position error. However, with 1 mm error for the spots at Y = 0 mm, dose deviation of ±7% is observed.

Peterson et al. [44] attributed the delivered spot position errors to uncertainties in scanning magnetic fields in the steering magnets. The authors established an analytical relationship between magnetic field uncertainties in the steering magnets and the resulting lateral displacements of the spot positions. They studied the worst-case scenarios, in which four adjacent spots deviated either toward or away from each other, as well as a more realistic clinical scenario where all spots randomly displaced. A deviation of 5% of dose was observed for position error of merely 0.5 mm when four adjacent spots deviated either toward or away from each other, while the same amount of dose deviation would result from a 1 mm random spot position error.

In a recent study, Yu et al. [45] reported that the smaller discrete spot spacing is, the less sensitive it is to spot position error. However, if the spot spacing becomes too small, the effect of rounding error caused by the delivery constraint of the minimum monitor unit (MU) may affect the treatment plan quality if the TPS optimizer does not take the minimum MU constraint into account [10].

9.6 Deliverable MU of a Spot

9.6.1 Maximum MU Per Spot

The maximum MU is necessary for safety consideration. If the spot is in wrong position or with the wrong dose, the scanning nozzle would stop delivering the next spot before too much dose is delivered incorrectly. This is also important to achieve precise measurements of the spot position and size, as the gain setting for the spot position monitor might have limited dynamic range [10]. If a spot has higher MU than the allowed maximum, the spot can be repainted until the prescribed spot dose is reached. Repainting is designed in the treatment planning system in accordance with the constraints of the delivery system. The repainting options include individual spots, sequential spots, or all spots in an energy layer. The repainted spot(s) would have the same positions as the spot(s) already treated, but they would have a new spot number, spot weight, and spot MU. The Hitachi scanning nozzle at the MD Anderson Cancer Center has the maximum MU of 0.04. The absorbed dose at the Bragg peak irradiated by a single spot with 0.04 MU varies from 1.6 to 4.4 cGy, depending on the energy of the pencil beam. The proton beam with the energy of approximately 160 MeV has the maximum Bragg peak dose [27]. The maximum MU per spot dose may indirectly contribute to the minimum MU constraint, as shown in Section 9.6.2. However, it is possible that after splitting the original spots into one or more spots with the maximum deliverable MU, the remaining spot has an MU less than the minimum deliverable MU. This indirect minimum MU constraint can be completely eliminated if a different splitting strategy is used. For example, if, after splitting *N* times, the remaining spot has less than the minimum MU value, one can simply split the original spot into N+1 equally weighted spots. The maximum MU constraint has no significant dosimetric impact in treatment planning since the spot dose can be delivered through repainting [10]. This may not be the case for the minimum MU per spot constraint.

9.6.2 Minimum MU Per Spot

The minimum MU of a spot is imposed on by the delayed charge, which is a small amount of charge that leaks out because of the finite time required to stop the delivery of the spot. Normally, the minimum MU is set to be about twice the delayed charge. For example, the minimum MUs are 0.005 MU and 0.003 MU for the Hitachi nozzles at MD Anderson and the upcoming Mayo Clinic proton beam, respectively. Current commercial TPSs may not include this delivery constraint in their optimization algorithms. Spots with deliverable MUs are created during post-processing, in which a rounding process is performed for the spots with MU less than the minimum MU. For example, spots with MUs less than half of the minimum MU are rounded down to zero (i.e., these spots are discarded), and spots with MUs greater than half of the minimum MU and less than the minimum MU are rounded up to the minimum MU. In a recent study, Zhu et al. [10] demonstrated that these types of rounding errors could create significant distortions from ideally optimized dose distributions depending on the intended target dose, SOBP width, spot spacing, and proton range. The observed distortions could be explained by "MU starvation," that is, there are not enough MUs available to be shared by so many spots to keep the MUs larger than the minimum MU. To characterize the dose distortion, a relative height of the distorted peak of the depth–dose was defined as, $H = \Delta D/D_0 \times 100$ where ΔD is the maximum dose increase above the uniform dose D_0 of the SOBP. The change of the width of Bragg peak was also defined, $\Delta W = W_{SOBP'} - W_{SOBP}$, where $W_{SOBP'}$ and W_{SOBP} are the widths of the SOBP with and without distortion. Figure 9–9 demonstrates the dose distortion caused by the "MU starvation" effect owing to the dose per treatment field, SOBP width, proton beam range, and spot spacing, respectively:

- As the dose per treatment field decreases from 100 to 20 cGy, the distortion peak height (H) increases from 0% to almost 17% (Figure 9– 9A).
- As the width SOBP increases from 4 to 20 cm, H increases to about 8% (Figure 9–9B).
- As the proton beam range decreases from 16 to 8 cm, H increases to about 25% (Figure 9–9C).
- As the spot spacing decreases from 9 to 3 mm, the average number of spots suffering the rounding error caused by the minimum MU constraint for a group four prostate patients increases from about 10% to 75% (Figure 9–9D).

A comparison of prostate plans optimized with spot spacing of 4 and 7 mm is shown in Figure 9–10. For the plans without delivery constraint due to the minimum MU, the 4 and 7 mm spacing plans are similar in terms of target coverage and normal tissue spacing (Figure 9–10a). With a delivery constraint of a minimum 0.005 MU, the 4 mm plan lost some target coverage without consideration of this constraint during optimization (Figure 9–10b).

A two-stage linear programming approach to optimizing spot intensities and deliverable MU values simultaneously has recently been reported [46]. Thus, the post-processing procedure is eliminated and the associated optimized plan deterioration can be avoided. Howard et al. [47] performed a study similar to Zhu et al. [10] on the effect of minimum MU on the quality of the treatment plan. They used a pediatric brain case and a head and neck case to demonstrate the minimum MU effect that depends on spot size and spot spacing as well as the geometric parameters (shape, size, and location) of the target volume(s). The authors also showed that the minimum MU effect could be reduced when it was considered during the optimization process.

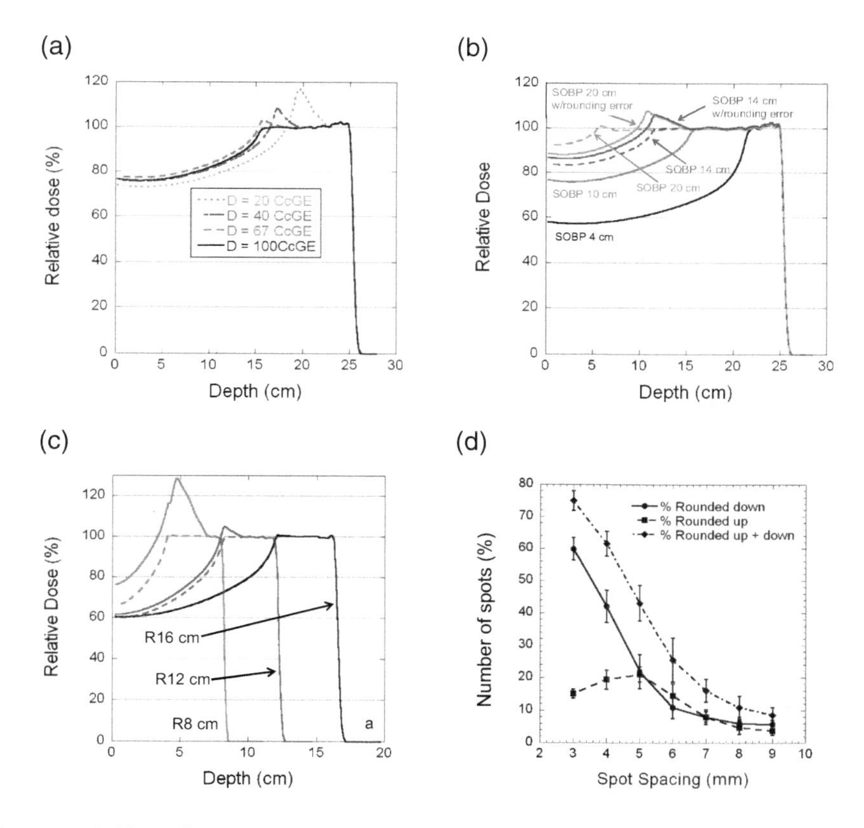

Figure 9–9 The effect of minimum MU constraint on the dose distribution when TPS does not incorporate this constraint during optimization. (a) The distortion peak height (*H*) increases as dose per field decreases. (b) *H* increases as the width of SOBP increases. (c) *H* increases as the proton beam range decreases. (d) the percent spots suffers from rounding error increases as the spot spacing decreases for four prostate patients. (From Figure 2, 3, 5, and 6 of reference [10] with permission.)

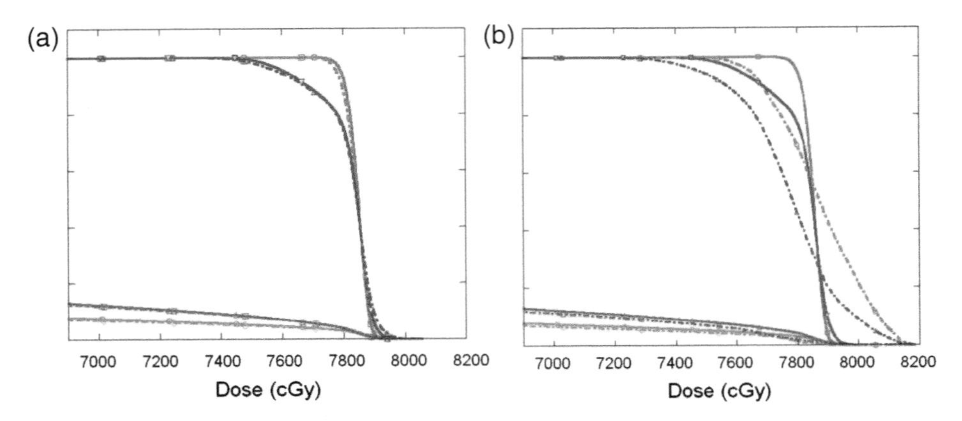

Figure 9–10 Comparison of dose volume histograms (DVHs) between treatment plans generated without (a) and with (b) a minimum MU constraint using spot spacings of 4 and 7 mm for a prostate cancer patient. The solid lines are for the 7 mm plan, and the dash-dotted lines are for the 4 mm plan. (From Figure 8 of reference [10] with permission.)

The effect of the delayed charge on the delivered treatment has recently been simulated [48]. The clinically available method to account for the delayed charge is by adjusting the MU of the current spot based on the cumulative MU. With this correction method, the effect is mainly on the dose delivered to the first spot in each energy layer. If the first spot was randomly selected in each energy layer, the delayed charge effect would be reduced. An alternative method using a pre-calculated value of delayed charge is proposed in this study; that method appears to be superior to the cumulative MU model for minimizing the effects of delayed charge [48].

9.7 Future Directions for Scanning Beam Field Shaping

9.7.1 Collimation for Scanning Beam

While collimation is not required to deliver scanning beam particle therapy, the benefit of using aperture [49] and multi-leaf collimator (MLC) [50,51] has been recognized for sharpening the penumbra and reducing the out-of-field dose for shallow target volumes, especially when a range shifter is used. The advantages and disadvantages of using an MLC versus aperture for field shaping were the subject of a recent debate [52]. The concern with an MLC is that the lateral penumbra may not be as good as aperture due to bulky size, necessitating a larger air gap [53]. In addition, the 3D collimator effect (the leaves always have their maximum thickness) and the scalloping effect (the leaves only approximate the ideal shape) are also concerns [53]. For aperture-based collimation, however, the aperture could only be optimal for the energy layer with the largest area in the beam's-eye view, which is also not ideal.

Recently, Hyer et al. [54,55] reported the concept of a dynamic collimation system (DCS) for shaping the lateral penumbra reduction in spot scanning proton therapy. The DCS consists of two pairs of parallel nickel trimmer blades of 2 cm thickness perpendicular to each other. The blades move to the target edge before the scanning spot arrives. The DCS has the potential to reduce the penumbra for each energy layer independently. One of the weaknesses of the DCS is it may increase the treatment delivery time [54].

9.7.2 Scanning Beam for Moving Targets

The management of the interplay effect between the motion of the scanning beam and respiratory motion is a subject of intensive research in the last few years [56]. This topic is discussed in detail in Chapter 24, Motion Management. Here we will only discuss the evaluation of dosimetric impact of the interplay effect over a course of treatment delivery [57–59].

At MD Anderson Cancer Center, this was accomplished by a dynamic dose simulator using an analytic pencil beam dose algorithm [60,61] with realistic breathing patterns and patient four-dimensional computed tomography (4DCT) images. The simulator first randomly selects a delivery starting point among the respiratory phases. The spots for an entire treatment field were then grouped into different respiratory phases based on the realistic machine delivery time sequence and the breathing pattern. The doses contributed by each group of spots were fully calculated on the CT images corresponding to the respiratory phase in which the spots were delivered. The doses were then accumulated to the reference phases of the 4DCT using deformable registration to generate the dynamic dose that provides an estimation of what might be delivered under the influence of the motion interplay effect. This process was then repeated for each field to calculate the dynamic dose for the current fraction [58]. It has been shown that the accumulated dynamic dose over the course of treatment delivery would converge to the corresponding 4D dose, which is the equally weighted average of the doses over the respiratory phases [62]. In other words, the motion interplay effect on the accumulated dynamic dose will be averaged out over a normal course of the fractionated treatment delivery, although the biological effect of fraction-to-fraction variation of delivered dose is unknown. In a follow-up study, Kardar et al. [59] used a single-fraction dynamic dose as a metric to determine the magnitude of dose degradation caused by the motion interplay effect and concluded that the interplay effect could potentially be mitigated by increasing the amount of isolayered repainting in each fraction of delivery. A similar study was performed by Grassburger et al. [57] using Monte Carlo simulation for dose calculation. The additional finding was that the interplay effect was smaller for larger spot sizes. For the single-fraction dynamic dose, Li et al. [63] have recently reported a novel technique of optimizing the spot delivery sequence to reduce the motion interplay effect. This optimizing strategy was based on maximizing the effective delivery time for each spot.

9.7.3 Other Improvements of Scanning Beam Delivery

In the last few years, significant efforts have been devoted to developing the new generation of scanning nozzles with much smaller spot sizes than the previous ones to sharpen the penumbra. However, a smaller spot size may increase the treatment delivery time and reduce the robustness of the treatment with respect to range and dose calculation uncertainties [8], the spot position error [45], and the interplay effect between respiratory motion and scanning motion [57]. One suggestion is to have mixed spot sizes for a given energy of the proton pencil beam, with larger spots placed in the central portion of the field and smaller spots positioned at the peripheral part of the field [8,40]. To make this idea practical, the scanning nozzle has to be able to switch to different spot sizes quickly during the delivery of each energy layer without any human intervention.

With reduced spot sizes, the delivery efficiency may be a concern, especially for spot scanning with synchrotron-based systems with sharp Bragg peaks of the low-energy proton pencil beams. Ridge filters have been developed to broaden the Bragg peaks of low-energy beams [34] to reduce the number of energies required for a treatment field. For synchrotron-based systems, a more general approach could be incorporated in designing the scanning nozzle, including the use of a ridge filter with optimized energy selections for the most efficient beam delivery. Another way to improve treatment delivery efficiency is to implement raster scanning and continuous line scanning.

One of the most active research areas for scanning beam delivery has been mitigating the interplay effect between respiratory motion and scanning beam motion. Using repainting to average out the interplay effect has drawn significant attention in recent years. The most important requirements of performing volumetric repainting is to have a faster energy switch. It has been reported that the energy change can be as fast as less than 0.1 second for a cyclotron-based scanning beam delivery system [24]. For synchrotron-based systems, techniques of multiple energy extraction are currently being developed [25].

9.8 Summary

Particle therapy with scanning beam delivery technology has created a new era of radiation therapy. Field shaping with scanning beams is achieved through scanning magnets, placing each pencil beam with proper energy at the planned transverse position. The weight of each pencil beam is determined by an optimization process in the TPS. In this regard, all scanning beam treatment delivery techniques can be considered intensity-modulated therapy. IMPT has only been practically implemented with the scanning pencil beams. Mitigating interplay effect of respiratory motion and scanning beam motion remains one of most important challenges for scanning beam delivery. 3D collimation for low-energy proton pencil beams is yet to be developed and implemented. Delivery efficiency is critical to the success of any busy clinic, especially for treating large target volumes with a modern scanning nozzle with very small spot sizes.

References

1. Delaney TF, Kooy HM. Proton and charged particle radiotherapy. Philadelphia: Wolters Kluwer Lippincott Williams & Wilkins, 2008.
2. ICRU. Prescribing, recording, and reporting proton-beam therapy. Washington DC: International Commission on Radiation Units and Measurements, 2007.
3. Haberer T, Becher W, Schardt D, Kraft G. Magnetic scanning system for heavy ion therapy. *Nucl Instrum Methods Phys Res A* 1993;330:296–305.
4. Pedroni E, Bacher R, Blattmann H, Bohringer T, Coray A, Lomax A, Lin S, Munkel G, Scheib S, Schneider U, et al. The 200-MeV proton therapy project at the Paul Scherrer Institute: Conceptual design and practical realization. *Med Phys* 1995;22:37–53.
5. Lomax A. Intensity modulation methods for proton radiotherapy. *Phys Med Biol* 1999;44:185–205.
6. Lomax AJ, Boehringer T, Coray A, Egger E, Goitein G, Grossmann M, Juelke P, Lin S, Pedroni E, Rohrer B, Roser W, Rossi B, Siegenthaler B, Stadelmann O, Stauble H, Vetter C, Wisser L. Intensity modulated proton therapy: A clinical example. *Med Phys* 2001;28:317–324.
7. Lomax AJ, Bohringer T, Bolsi A, Coray D, Emert F, Goitein G, Jermann M, Lin S, Pedroni E, Rutz H, Stadelmann O, Timmermann B, Verwey J, Weber DC. Treatment planning and verification of proton therapy using spot scanning: Initial experiences. *Med Phys* 2004;31:3150–3157.
8. Lomax AJ. Intensity modulated proton therapy and its sensitivity to treatment uncertainties 1: The potential effects of calculational uncertainties. *Phys Med Biol* 2008;53:1027–1042.
9. Lomax AJ. Intensity modulated proton therapy and its sensitivity to treatment uncertainties 2: The potential effects of inter-fraction and inter-field motions. *Phys Med Biol* 2008;53:1043–1056.
10. Zhu XR, Sahoo N, Zhang X, Robertson D, Li H, Choi S, Lee AK, Gillin MT. Intensity modulated proton therapy treatment planning using single-field optimization: The impact of monitor unit constraints on plan quality. *Med Phys* 2010;37:1210–1219.
11. Zhu XR, Poenisch F, Li H, Zhang X, Sahoo N, Wu RY, Li X, Lee AK, Chang EL, Choi S, Pugh T, Frank SJ, Gillin MT, Mahajan A, Grosshans DR. A single-field integrated boost treatment planning technique for spot scanning proton therapy. *Radiat Oncol* 2014;9:202.
12. Zhu XR, Poenisch F, Song X, Johnson JL, Ciangaru G, Taylor MB, Lii M, Martin C, Arjomandy B, Lee AK, Choi S, Nguyen QN, Gillin MT, Sahoo N. Patient-specific quality assurance for prostate cancer patients receiving spot scanning proton therapy using single-field uniform dose. *Int J Radiat Oncol Biol Phys* 2011;81:552–559.
13. Fredriksson A, Forsgren A, Hardemark B. Minimax optimization for handling range and setup uncertainties in proton therapy. *Med Phys* 2011;38:1672–1684.
14. Frank SJ, Cox JD, Gillin M, Mohan R, Garden AS, Rosenthal DI, Gunn GB, Weber RS, Kies MS, Lewin JS, Munsell MF, Palmer MB, Sahoo N, Zhang X, Liu W, Zhu XR. Multifield optimization intensity modulated proton therapy for head and neck tumors: A translation to practice. *Int J Radiat Oncol Biol Phys* 2014;89:846–853.
15. Chang JY, Li H, Zhu XR, Liao Z, Zhao L, Liu A, Li Y, Sahoo N, Poenisch F, Gomez DR, Wu R, Gillin M, Zhang X. Clinical implementation of intensity modulated proton therapy for thoracic malignancies. *Int J Radiat Oncol Biol Phys* 2014;90:809–818.
16. Grosshans DR, Zhu XR, Melancon A, Allen PK, Poenisch F, Palmer M, McAleer MF, McGovern SL, Gillin M, DeMonte F, Chang EL, Brown PD, Mahajan A. Spot scanning proton therapy for malignancies of the base of skull: Treatment planning, acute toxicities, and preliminary clinical outcomes. *Int J Radiat Oncol Biol Phys* 2014;90:540–546.
17. Pugh TJ, Munsell MF, Choi S, Nguyen QN, Mathai B, Zhu XR, Sahoo N, Gillin M, Johnson JL, Amos RA, Dong L, Mahmood U, Kuban DA, Frank SJ, Hoffman KE, McGuire SE, Lee AK. Quality of life and toxicity from passively scattered and spot-scanning proton beam therapy for localized prostate cancer. *Int J Radiat Oncol Biol Phys* 2013;87:946–953.
18. Weber DC, Rutz HP, Pedroni ES, Bolsi A, Timmermann B, Verwey J, Lomax AJ, Goitein G. Results of spot-scanning proton radiation therapy for chordoma and chondrosarcoma of the skull base: The Paul Scherrer Institut experience. *Int J Radiat Oncol Biol Phys* 2005;63:401–409.

19. Weber DC, Rutz HP, Bolsi A, Pedroni E, Coray A, Jermann M, Lomax AJ, Hug EB, Goitein G. Spot scanning proton therapy in the curative treatment of adult patients with sarcoma: The paul scherrer institute experience. *Int J Radiat Oncol Biol Phys* 2007;69:865–871.
20. Rutz HP, Weber DC, Sugahara S, Timmermann B, Lomax AJ, Bolsi A, Pedroni E, Coray A, Jermann M, Goitein G. Extracranial chordoma: Outcome in patients treated with function-preserving surgery followed by spot-scanning proton beam irradiation. *Int J Radiat Oncol Biol Phys* 2007;67:512–520.
21. Munier FL, Verwey J, Pica A, Balmer A, Zografos L, Abouzeid H, Timmerman B, Goitein G, Moeckli R. New developments in external beam radiotherapy for retinoblastoma: From lens to normal tissue-sparing techniques. *Clin Experiment Ophthalmol* 2008;36:78–89.
22. Smith A, Gillin M, Bues M, Zhu XR, Suzuki K, Mohan R, Woo S, Lee A, Komaki R, Cox J, Hiramoto K, Akiyama H, Ishida T, Sasaki T, Matsuda K. The M. D. Anderson proton therapy system. *Med Phys* 2009;36:4068–4083.
23. Gillin MT, Sahoo N, Bues M, Ciangaru G, Sawakuchi G, Poenisch F, Arjomandy B, Martin C, Titt U, Suzuki K, Smith AR, Zhu XR. Commissioning of the discrete spot scanning proton beam delivery system at the University of Texas M.D. Anderson Cancer Center, proton therapy center, Houston. *Med Phys* 2010;37:154–163.
24. Zenklusen SM, Pedroni E, Meer D. A study on repainting strategies for treating moderately moving targets with proton pencil beam scanning at the new gantry 2 at PSI. *Phys Med Biol* 2010;55:5103–5121.
25. Iwata Y, Kadowaki T, Uchiyama H, Fujimoto T, Takada E, Shirai T, Furukawa T, Mizushima K, Takeshita E, Katagiri K, Sato S, Sano Y, Noda K. Multiple-energy operation with extended flattops at HIMAC. *Nucl Instrum Meth A* 2010;624:33–38.
26. Smith AR. Vision 20/20: Proton therapy. *Med Phys* 2009;36:556–568.
27. Zhu XR, Poenisch F, Lii M, Sawakuchi GO, Titt U, Bues M, Song X, Zhang X, Li Y, Ciangaru G, Li H, Taylor MB, Suzuki K, Mohan R, Gillin MT, Sahoo N. Commissioning dose computation models for spot scanning proton beams in water for a commercially available treatment planning system. *Med Phys* 2013;40:041723.
28. Furukawa T, Inaniwa T, Sato S, Tomitani T, Minohara S, Noda K, Kanai T. Design study of a raster scanning system for moving target irradiation in heavy-ion radiotherapy. *Med Phys* 2007;34:1085–1097.
29. Noda K, Furukawa T, Fujimoto T, Inaniwa T, Iwata Y, Kanai T, Kanazawa M, Minohara S, Miyoshi T, Murakami T, Sano Y, Sato S, Takada E, Takei Y, Torikai K, Torikoshi M. New treatment facility for heavy-ion cancer therapy at HIMAC. *Nucl Instrum Meth B* 2008;266:2182–2185.
30. Inaniwa T, Furukawa T, Tomitani T, Sato S, Noda K, Kanai T. Optimization for fast-scanning irradiation in particle therapy. *Med Phys* 2007;34:3302–3311.
31. Rieken S, Habermehl D, Nikoghosyan A, Jensen A, Haberer T, Jakel O, Munter MW, Welzel T, Debus J, Combs SE. Assessment of early toxicity and response in patients treated with proton and carbon ion therapy at the Heidelberg ion therapy center using the raster scanning technique. *Int J Radiat Oncol Biol Phys* 2011;81:e793–801.
32. Rieken S, Habermehl D, Haberer T, Jaekel O, Debus J, Combs SE. Proton and carbon ion radiotherapy for primary brain tumors delivered with active raster scanning at the Heidelberg ion therapy center (HIT): Early treatment results and study concepts. *Radiat Oncol* 2012;7:41.
33. Schatti A, Meer D, Lomax AJ. First experimental results of motion mitigation by continuous line scanning of protons. *Phys Med Biol* 2014;59:5707–5723.
34. Courneyea L, Beltran C, Tseung HS, Yu J, Herman MG. Optimizing mini-ridge filter thickness to reduce proton treatment times in a spot-scanning synchrotron system. *Med Phys* 2014;41:061713.
35. Cao W, Lim G, Liao L, Li Y, Jiang S, Li X, Li H, Suzuki K, Zhu XR, Gomez D, Zhang X. Proton energy optimization and reduction for intensity-modulated proton therapy. *Phys Med Biol* 2014;59:6341-6354.
36. Titt U, Mirkovic D, Sawakuchi GO, Perles LA, Newhauser WD, Taddei PJ, Mohan R. Adjustment of the lateral and longitudinal size of scanned proton beam spots using a pre-absorber to optimize penumbrae and delivery efficiency. *Phys Med Biol* 2010;55:7097–7106.
37. Schaffner B. Proton dose calculation based on in-air fluence measurements. *Phys Med Biol* 2008;53:1545–1562.

38. Both S, Shen J, Kirk M, Lin L, Tang S, Alonso-Basanta M, Lustig R, Lin H, Deville C, Hill-Kayser C, Tochner Z, McDonough J. Development and clinical implementation of a universal bolus to maintain spot size during delivery of base of skull pencil beam scanning proton therapy. *Int J Radiat Oncol Biol Phys* 2014;90:79–84.
39. Kooy HM, Clasie BM, Lu HM, Madden TM, Bentefour H, Depauw N, Adams JA, Trofimov AV, Demaret D, Delaney TF, Flanz JB. A case study in proton pencil-beam scanning delivery. *Int J Radiat Oncol Biol Phys* 2010;76:624–630.
40. van de Water TA, Lomax AJ, Bijl HP, Schilstra C, Hug EB, Langendijk JA. Using a reduced spot size for intensity-modulated proton therapy potentially improves salivary gland-sparing in oropharyngeal cancer. *Int J Radiat Oncol Biol Phys* 2012;82:e313–319.
41. Wang D, Dirksen B, Hyer DE, Buatti JM, Sheybani A, Dinges E, Felderman N, TenNapel M, Bayouth JE, Flynn RT. Impact of spot size on plan quality of spot scanning proton radiosurgery for peripheral brain lesions. *Med Phys* 2014;41:121705.
42. Pedroni E. Active scanning beams: 1. Modulation delivery. Paul Scherrer Institute - Proton Therapy Winter School 2010.
43. Li H, Sahoo N, Poenisch F, Suzuki K, Li Y, Li X, Zhang X, Lee AK, Gillin MT, Zhu XR. Use of treatment log files in spot scanning proton therapy as part of patient-specific quality assurance. *Med Phys* 2013;40:021703.
44. Peterson S, Polf J, Ciangaru G, Frank SJ, Bues M, Smith A. Variations in proton scanned beam dose delivery due to uncertainties in magnetic beam steering. *Med Phys* 2009;36:3693–3702.
45. Yu J, Beltran CJ, Herman MG. Implication of spot position error on plan quality and patient safety in pencil-beam-scanning proton therapy. *Med Phys* 2014;41:081706.
46. Cao W, Lim G, Li X, Li Y, Zhu XR, Zhang X. Incorporating deliverable monitor unit constraints into spot intensity optimization in intensity-modulated proton therapy treatment planning. *Phys Med Biol* 2013;58:5113–5125.
47. Howard M, Beltran C, Mayo CS, Herman MG. Effects of minimum monitor unit threshold on spot scanning proton plan quality. *Med Phys* 2014;41:091703.
48. Whitaker TJ, Beltran C, Tryggestad E, Bues M, Kruse JJ, Remmes NB, Tasson A, Herman MG. Comparison of two methods for minimizing the effect of delayed charge on the dose delivered with a synchrotron based discrete spot scanning proton beam. *Med Phys* 2014;41:081703.
49. Dowdell SJ, Clasie B, Depauw N, Metcalfe P, Rosenfeld AB, Kooy HM, Flanz JB, Paganetti H. Monte Carlo study of the potential reduction in out-of-field dose using a patient-specific aperture in pencil beam scanning proton therapy. *Phys Med Biol* 2012;57:2829–2842.
50. Bues M, Newhauser WD, Titt U, Smith AR. Therapeutic step and shoot proton beam spot-scanning with a multi-leaf collimator: A Monte Carlo study. *Radiat Prot Dosimetry* 2005;115:164–169.
51. Daartz J, Bangert M, Bussiere MR, Engelsman M, Kooy HM. Characterization of a mini-multileaf collimator in a proton beamline. *Med Phys* 2009;36:1886–1894.
52. Daartz J, Maughan RL, Orton CG. Point/counterpoint: The disadvantages of a multileaf collimator for proton radiotherapy outweigh its advantages. *Med Phys* 2014;41:020601.
53. Gottschalk B. Multileaf collimators, air gap, lateral penumbra, and range compensation in proton radiotherapy. *Med Phys* 2011;38:i–ii.
54. Hyer DE, Hill PM, Wang D, Smith BR, Flynn RT. A dynamic collimation system for penumbra reduction in spot-scanning proton therapy: Proof of concept. *Med Phys* 2014;41:091701.
55. Hyer DE, Hill PM, Wang D, Smith BR, Flynn RT. Effects of spot size and spot spacing on lateral penumbra reduction when using a dynamic collimation system for spot scanning proton therapy. *Phys Med Biol* 2014;59:N187–196.
56. Rietzel E, Bert C. Respiratory motion management in particle therapy. *Med Phys* 2010;37:449–460.
57. Grassberger C, Dowdell S, Lomax A, Sharp G, Shackleford J, Choi N, Willers H, Paganetti H. Motion interplay as a function of patient parameters and spot size in spot scanning proton therapy for lung cancer. *Int J Radiat Oncol Biol Phys* 2013;86:380–386.
58. Li Y, Kardar L, Li X, Li H, Cao W, Chang JY, Liao L, Zhu RX, Sahoo N, Gillin M, Liao Z, Komaki R, Cox JD, Lim G, Zhang X. On the interplay effects with proton scanning beams in stage III lung cancer. *Med Phys* 2014;41:021721.

59. Kardar L, Li Y, Li X, Li H, Cao W, Chang JY, Liao L, Zhu RX, Sahoo N, Gillin M, Liao Z, Komaki R, Cox JD, Lim G, Zhang X. Evaluation and mitigation of the interplay effects of intensity modulated proton therapy for lung cancer in a clinical setting. *Pract Radiat Oncol* 2014;4:e259–268.
60. Li Y, Zhu RX, Sahoo N, Anand A, Zhang X. Beyond gaussians: A study of single-spot modeling for scanning proton dose calculation. *Phys Med Biol* 2012;57:983–997.
61. Zhang X, Liu W, Li Y, Li X, Quan M, Mohan R, Anand A, Sahoo N, Gillin M, Zhu XR. Parameterization of multiple Bragg curves for scanning proton beams using simultaneous fitting of multiple curves. *Phys Med Biol* 2011;56:7725–7735.
62. Li H, Li Y, Zhang X, Li X, Liu W, Gillin MT, Zhu XR. Dynamically accumulated dose and 4d accumulated dose for moving tumors. *Med Phys* 2012;39:7359–7367.
63. Li H, Zhang X, Zhu XR. Minimizing dose uncertainty for spot scanning proton therapy of moving tumors by optimization of the spot delivery sequence. *Int J Radiat Oncol Biol Phys* 2014;90:S926.

Chapter 10

Secondary Radiation Production and Shielding for Proton Therapy Facilities

Nisy Elizabeth Ipe, Ph.D.[1] and C. Sunil, Ph.D.[2]

[1]Consultant, Shielding Design, Dosimetry, and Radiation Protection, San Carlos, CA

[2]Health Physics Division, Bhabha Atomic Research Centre, Mumbai, India

10.1 Introduction

During the past 80 years, a reasonable understanding of secondary radiation production, as well the physical processes that govern radiation transport through accelerator shielding has been gained from the operation of proton accelerators. For the purposes of this chapter, the proton energies are classified as follows (where E_P is the energy of the incident proton):

Very Low Energy:	$E_p < 10$ MeV
Low Energy:	10 MeV $\leq E_p <$ 50 MeV
Intermediate Energy:	50 M eV $\leq E_p \leq$ 1000 MeV (1 GeV)
High Energy:	$E_p > 1$ GeV

The prompt radiation field produced by protons in the therapeutic energy range of interest, i.e., 67 MeV to 330 MeV (intermediate-energy range), is comprised of a mixture of charged and neutral particles, as well as photons. Neutrons form the dominant component and have energies as high as the incident proton energy. The goal of structural shielding is to attenuate the prompt secondary radiation to dose levels that are within regulatory or design limits for individual exposure. Structural shielding can be supplemented with localized shielding (hereafter referred to as "local shielding") on an as-needed basis.

Residual radiation from induced activation can also contribute to individual exposure, either through external exposure or internal exposure, i.e., inhalation and ingestion. In proton therapy facilities, local shielding is used to minimize external exposure from activation and radiation damage to equipment, while ventilation is used to minimize internal exposure from activation. Finally, during treatment with the primary proton beam, the patient is also exposed to secondary radiation from proton interactions with beam-shaping devices and nearby components. Secondary dose to the patient depends upon the beam delivery system and is higher with passive scattering than with active beam scanning systems. Therefore, the risk of secondary cancers must be considered. The biological effects of secondary dose to patients are discussed in Chapter 5 (Proton Relative Biological Effectiveness).

Shielding is the primary focus of this chapter. Activation and secondary dose to patients will only be described briefly. The reader is referred to the Particle Therapy Cooperative Group (PTCOG) Report 1 [1] and Chapter 17 of *Proton Therapy Physics* [2] for an extensive coverage of shielding design and radiation safety for charged particle therapy facilities. The references in this chapter are not meant to be extensive and, therefore, only include references that are deemed relevant to the topics of discussion.

10.2 Secondary Radiation

Secondary radiation is produced by the interaction of protons with any material in their path. Thus, as protons are being delivered to the treatment room during the normal operation of proton therapy facilities, there can be 1) beam losses (i.e., losses of primary proton beam intensity) occurring along the beam line, 2) full beam incident on intentional targets, such as the degrader, beam stops, etc., and 3) partial beam interception by devices such as collimators, apertures, and slits. All the above are broadly referred to as "beam losses" in this chapter. Locations of beam loss, beam incidence on targets, or beam interception by devices occur in the synchrotron and cyclotron during injection; during energy degradation in the cyclotron; during beam transport to the treatment room; in passive scattering systems, range degraders and modulators;

and in beam-shaping devices such as collimators, apertures, and range compensators placed near the patient [1]. Secondary radiation is also produced in the patient and the dosimetric phantom. Thus, the proton therapy facility requires structural shielding.

10.2.1 Prompt and Residual Radiation

Secondary radiation consists of both prompt and residual radiation. Prompt radiation is produced only while the machine is on. Residual radiation is produced by activated materials, i.e., materials that have become radioactive during beam operation. Thus, residual radiation remains for a time period that is determined by the half-life of the activated material, even after the machine is turned off. The large structural shielding thicknesses for the facility are determined by the prompt radiation, while the residual radiation requires considerably less local shielding. Therefore, it is important to understand the physics behind secondary radiation production.

10.2.1.1 Physics of Secondary Radiation Production

Proton interactions have been described in Chapter 3 (Particle Beam Interactions: Basic). Nuclear evaporation and the intra-nuclear cascade are the two nuclear processes that are important in the determination of particle yields from proton nuclear interactions [1,3–4].

The interaction of very low-energy protons ($E_P < 10$ MeV) with a nucleus can be described by the compound nucleus model [4]. A compound nucleus is formed when the incident particle is absorbed by the target nucleus. The compound nucleus is in an excited state and has a number of allowed decay channels, the preferred decay channel being the entrance channel. The number of levels available to the incident channel increases considerably as the energy of the incident particle increases. Instead of discrete levels in the quasi-stationary states of the compound nucleus, there is a complete overlapping of levels inside the nucleus. Under these conditions, the emission of particles can be described by an evaporation process that is similar to the evaporation of a molecule from the surface of a liquid.

The interaction of intermediate-energy protons with matter results in the production of an intra-nuclear cascade (spray of particles) in which neutrons have energies as high as the incident proton. Thus, the intra-nuclear cascade plays an important part for therapeutic protons. There are five distinct and independent stages to be considered, as shown in Figure 10–1 [3].

1. intra-nuclear cascade
2. production of muons
3. electromagnetic cascade
4. evaporation process
5. activation

10.2.1.2 Intra-nuclear Cascade

A qualitative description of the intra-nuclear cascade is provided. An intra-nuclear cascade is produced when an incoming hadron (proton, neutron, etc.) with energy less

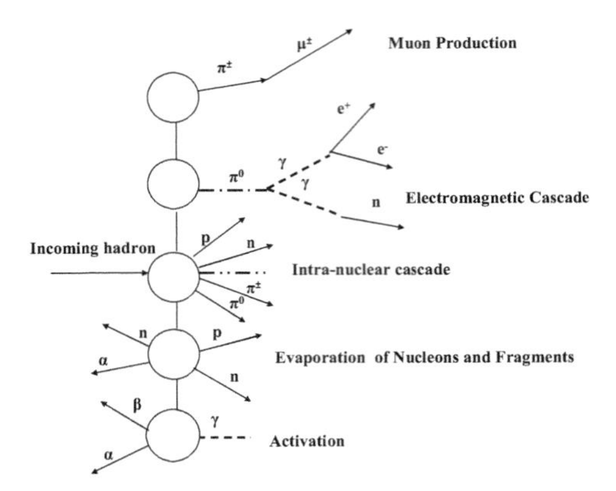

Figure 10–1 Schematic representation of various stages of intra-nuclear cascade. (Adapted with permission of the International Commission on Radiation Units and Measurements from [6].)

than ~2 GeV interacts with individual nucleons in a nucleus, producing a spray of particles, such as protons, neutrons and π mesons (pions). At low particle energies (~100 MeV), all interactions occur just between nucleons, and the process is called nuclear cascade [5]. As the incident particle energy increases, the threshold energies for particle production in nucleon-nucleon collisions are exceeded. Pions can be charged ($\pi^{\pm}$) or neutral (π). Charged pions have a rest mass of 139.6 MeV/c^2 (where c is the velocity of light) and neutral pions have a mass of about 135 MeV/c^2. Therefore, the threshold for pion production in the center of mass system is 135 MeV. Conservation of momentum requires the proton to have energy of ~290 MeV for pion production.

The scattered and recoiling nucleons from the interaction proceed through the nucleus. Each of these nucleons may, in turn, interact with other nucleons in the nucleus, leading to the development of a cascade. Some of the cascade particles that have sufficiently high energy escape from the nucleus, while others do not. These particles are emitted in the direction of the incident particle. A large fraction of the energy in the cascade is transferred to a single nucleon. This nucleon, with energy greater than 150 MeV, propagates the cascade. The cascade neutrons which arise from individual nuclear interactions are forward-peaked and have longer attenuation lengths than evaporation neutrons. The attenuation length is defined later in this section. The rest of the energy is equally distributed among nucleons in the nucleus, which is left in a highly excited state [5]. During the creation of a compound nucleus, the energy which is initially concentrated on a few nucleons spreads through the composite nucleus, which evolves toward a state of statistical equilibrium. During this pre-compound stage, nucleons or fragments having considerable energy may be ejected. These particles are referred to as pre-equilibrium or "pre-compound" particles. These particles may be emitted after each interaction between the incident parti-

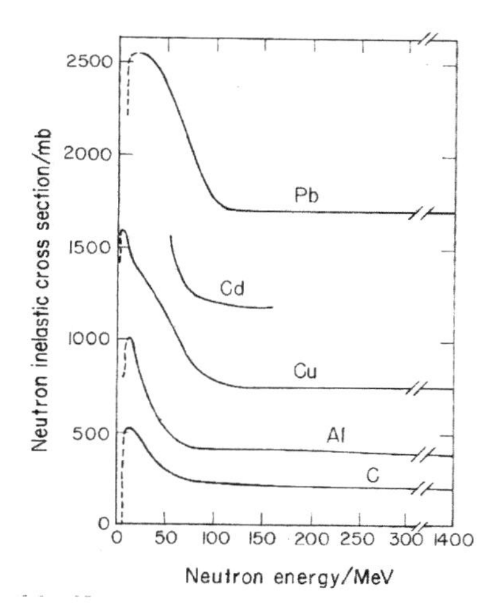

Figure 10–2 Neutron inelastic cross sections as a function of neutron energy for various materials. (Reprinted with permission of the International Commission on Radiation Units and Measurements from [6].)

cle (or another cascade particle) and a nucleon inside the nucleus. The energies of pre-equilibrium particles are greater than the energies of particles emitted during the equilibrium decay.

The residual nucleus evaporates particles such as alpha particles and other nucleons. These "evaporation nucleons" do not contribute to the cascade, but they do contribute to local energy absorption. In the third stage, after particle emission is no longer energetically possible, the remaining excitation energy is emitted in the form of gamma rays. Figure 10–2 shows the neutron inelastic cross sections for various materials [6].

The inelastic cross sections can be represented by the following equation:

$$\sigma_{in} = 43.1 A_N^{0.7} mb \tag{10.1}$$

for $A_N \geq 3$ and $E_p \geq 150$ MeV, regardless of whether the particle is a proton or neutron. A millibarn (mb) = 10^{-24} cm^2.

The cross sections increase with increasing mass number. They decrease with increasing energy to constant values above ~150 MeV. Thus, cascade neutrons with energy >150 MeV will eventually control the radiation field for protons with energy above 150 MeV. The macroscopic inelastic cross section Σ_{in} is given by:

$$\sum_{in} = \frac{\sigma_{in} N_A}{A} cm^{-1} \tag{10.2}$$

where, N_A is the Avogadro constant and A is the molar mass of the medium.

The neutron absorption length, λ_{in}, which is the reciprocal of the macroscopic inelastic cross section is given by:

$$\lambda_{in} = \frac{1}{\sum_{in}} cm \tag{10.3}$$

Radiation transmission can be approximated by an exponential function over a limited range of thickness [4]. The attenuation length is the distance traveled in the medium through which the intensity of the radiation is reduced to 37% of its original value. The neutron attenuation length, λ, is given by:

$$\lambda = \frac{1}{\sum} cm \tag{10.4}$$

where Σ is the total neutron macroscopic cross section.

It is important to note that while the high-energy neutrons transport the cascade, the lower-energy neutrons deposit a major fraction of the absorbed dose, even outside thick shields, except in the forward direction.

Nucleons with energies between 20 and 150 MeV transfer energy to several nucleons. Therefore, on average, each nucleon receives about 10 MeV. Charged particles at these energies are quickly stopped by ionization. Thus, neutrons predominate at low energies.

10.2.1.3 Production of Muons and Electromagnetic Cascade

Charged pions decay into muons and neutrinos. Muons are charged particles and deposit their energy by ionization. Protons and pions with energy less than 450 MeV have a high rate of energy loss. Thus, neutrons are the principal propagators of the cascade with increasing depth in the shielding.

Neutral pions decay into two energetic gamma rays, which initiate electromagnetic cascades. The photons that are produced interact through pair production or Compton collisions, resulting in the production of electrons. These electrons radiate high-energy photons (Bremsstrahlung) which, in turn, interact to produce more electrons. At each step in the cascade, the number of particles increases and the average energy decreases. This process continues until the electrons fall into the energy range where collision losses dominate over radiative losses, and the energy of the primary electron is completely dissipated in excitation and ionization of the atoms, resulting in heat production. This entire process—resulting in a cascade of photons, electrons, and positrons—is called an electromagnetic cascade. A very small fraction of the Bremsstrahlung energy in the cascade goes into the production of hadrons, such as neutrons, protons, and pions. The energy that is transferred is mostly deposited by ionization within a few radiation lengths. However, the attenuation length of these cascades is much shorter than the absorption length of the neutrons. Thus, the electromagnetic cascade does not contribute significantly to the energy transport. It is important to note that the intra-nuclear cascade dominates for protons in the therapeutic energy range of interest.

10.2.1.4 Evaporation Process and Activation

The energy of those particles that do not escape is assumed to be distributed among the remaining nucleons in the nucleus, leaving it in an excited state. It then de-excites by emitting particles, mainly neutrons and protons which are referred to as evaporation nucleons, alpha particles and some fragments. The evaporation nucleons are so called because they can be considered as boiling off a nucleus which is heated by the absorption of energy from the incident particle. The energy distribution of emitted neutrons can be described by the following equation:

$$n(E)dE = aEe^{-E/\tau} \tag{10.5}$$

where a is a constant, E is the energy of the neutron, and τ is the nuclear temperature which has the dimensions of energy with a value that lies between 0.5 and 5 MeV. The evaporated particles are emitted isotropically in the laboratory system, and the energy of the evaporation neutrons extends to ~8 MeV in shielding materials such as concrete [3,4]. Similar equations may be used to describe the emission of charged particles, but the emission of low-energy charged particles is suppressed by the Coulomb barrier. Therefore, charged particles produced by evaporation are unimportant in shielding considerations.

If low-energy particles are emitted, they are stopped near their point of emission [3]. These particles do not contribute to the cascade, but they contribute to local energy deposition. Therefore, charged particles produced by evaporation do not impact the determination of shielding thickness. The evaporation neutrons travel long distances, continuously depositing energy. Evaporation neutrons produced by interactions near the source contribute to dose inside the shield and to leakage dose through doors and openings. However, because they are strongly attenuated in the shield, they do not contribute to dose outside the shield. The dose outside the shield is dominated by evaporation neutrons produced near the outer surface of the shield. The remaining excitation energy may be emitted in the form of gammas. The de-excited nucleus may be radioactive, thus leading to residual radiation.

10.2.2 Spallation

The interaction of a high-energy proton or neutron with kinetic energies from ~100 MeV to several GeV with a target nucleus results in the emission of a large number of nucleons and fragments. This process is referred to as spallation, and it includes the intranuclear cascade and de-excitation. Spallation plays an important role in activation.

10.3 Prompt Radiation

The prompt radiation field produced by protons of energies up to 330 MeV encountered in proton therapy is quite complex, consisting of a mixture of charged and neutral particles, as well as photons, including:

1. neutrons, charged particles (like pions, kaons, and ions), and nuclear fragments emitted in inelastic hadronic interactions;
2. prompt gamma radiation from the interaction of neutrons or protons with matter;
3. muons and other particles;
4. characteristic x-rays due to transfer of energy from the proton to an electron in the bound state and the subsequent emission of a photon from the decay of the excited state; and
5. Bremsstrahlung radiation.

Neutrons dominate the prompt radiation field. For structural shielding, neutrons are the dominant component. For mazes and penetrations, low-energy neutrons and capture gamma rays contribute to dose. Therefore, it is important to understand how neutrons interact.

10.3.1 Neutron Energy Classification and Interactions

Neutrons are classified according to their energy as follows:

Thermal:	$\bar{E}_n = 0.025$ eV at 20°C. Typically $E_n \leq 0.5$ eV
Intermediate:	$0.5 \text{ eV} < E_n \leq 10$ keV
Fast:	$10 \text{ keV} < E_n \leq 20$ MeV
Relativistic:	$E_n > 20$ MeV
High-energy Neutrons:	$E_n > 100$ MeV

where E_n is the energy of the neutron and $\bar{E}_n$ is the average energy of the neutron. Neutrons can travel significant distances in matter without undergoing interactions because they are uncharged. Neutron collisions with atoms can result in elastic or inelastic reactions [7]. In an elastic reaction, the total kinetic energy of the incoming particle is conserved, whereas in an inelastic reaction, the nucleus absorbs some energy and is left in an excited state. Inelastic scattering can occur only at energies above the inelastic scattering threshold, i.e., the lowest excited state of the material.

Thermal neutrons (n_{th}) are in approximate thermal equilibrium with their surroundings. They gain and lose only small amounts of energy through elastic scattering. However, they diffuse about until captured by atomic nuclei. Thermal neutrons can undergo radiative capture, i.e., the absorption of a neutron leads to the emission of a gamma ray, such as in the $^{1}H(n_{th},\gamma)\,^{2}H$ reaction. The capture cross section for this reaction in hydrogen is 0.33×10^{-24} cm^2, and the energy of the emitted gamma ray is 2.22 MeV. This reaction occurs in hydrogenous shielding materials such as polyethylene and concrete. For shielding purposes, borated polyethylene is used instead of polyethylene because the cross section for capture in boron is much higher (3480×10^{-24} cm^2) than for hydrogen. Additionally, the subsequent capture gamma ray from the $^{10}B(n_{th},\alpha)\,^{7}Li$ has a much lower energy of 0.478 MeV. The capture cross sections for low-energy neutrons (<1 keV) decrease with increasing neutron energy. Intermediate-energy neutrons lose energy by scattering and are absorbed.

Neutrons can also be captured or absorbed by a nucleus in reactions such as (n,2n), (n,p), (n,α) or (n,γ). Fast neutrons include evaporation neutrons and neutrons from specific nuclear reactions such as (p,n), etc. They interact with matter mainly through a series of elastic and inelastic scattering. They are finally absorbed after giving up their energy [7]. Approximately 7 MeV is given up to gamma rays, on an average, during the slowing down and capture process. Therefore, the shielding calculations must take capture gamma rays into account for both structural shielding (if it is thick enough) as well as scattering through mazes and penetrations. Inelastic scattering is the dominant process in all materials at neutron energies above 10 MeV. Elastic scattering dominates at lower energies. Below 1 MeV, elastic scattering is the principal process by which neutrons interact in hydrogenous materials. When high-Z material is used for shielding, it must always be followed by hydrogenous material. A useful rule of thumb is that the hydrogenous material should have a thickness of at least one high-energy inelastic interaction mean free path [7]. The mean free path is defined as the average distance traveled by the particle in the material between two interactions. The reason for the latter requirement is because the energy of the neutrons may be reduced by inelastic scattering to a lower energy where they may be transparent to the non-hydrogenous material. For example, lead is virtually transparent to neutrons with energy below 0.57 MeV [8].

The sum of the inelastic and (n, 2n) cross sections for neutrons with energy <20 MeV is called the non-elastic cross section [8]. The inelastic scattering dominates at lower energies, while the (n, 2n) reactions dominate at higher energies. In an inelastic collision, there is a minimum energy loss which equals the energy of the lowest excited state; however, the energy loss in any inelastic collision cannot be determined exactly. The binding energy of a nucleus is the energy that would be needed to take it apart into its individual protons and neutrons. Typically, there is a large energy loss in a single collision, resulting in the excitation of energy states above the ground state, followed by the emission of gamma rays. The minimum energy loss is equal to the binding energy of the neutron in the (n, 2n) reaction, which produces a large number of lower-energy neutrons because the energies of the two neutrons that are produced are similar. In high-Z materials, a large amount of elastic scattering takes place, but it results in negligible energy loss. However, the mean free path or path length of the neutrons in the shielding material increases, thus providing more opportunities for inelastic and (n, 2n) reactions to occur.

Relativistic neutrons arise from cascade processes in proton accelerators. They are important in propagating the radiation field. The high-energy component of the cascade with neutron energies above 100 MeV (cascade neutron) propagates the neutrons through the shielding, and it continuously regenerates lower-energy neutrons and charged particles at all depths in the shield via inelastic reactions with the shielding material [9]. Neutrons with energy above 100 MeV undergo spallation reactions, thus producing copious amounts of neutrons.

The reactions occur in three stages for neutrons with energies between 50 and 100 MeV [10]. In the first stage, an intra-nuclear cascade develops, where the incident high-energy neutron interacts with an individual nucleon in the nucleus. In the second stage, the residual nucleus is left in an excited state and evaporates particles such as alpha particles and other nucleons. In the third stage, after particle emission is no longer energetically possible, the remaining excitation energy is emitted in the form of gamma rays. The de-excited nucleus may be radioactive. For neutrons with energy below 50 MeV, only the second and third stages are assumed to be operative.

10.3.2 Unshielded and Shielded Neutron Spectra

The characteristics of the unshielded and shielded neutron spectra are discussed in this section. Figure 10–3 shows the calculated unshielded neutron spectra for neutrons at various emission angles produced by 250 MeV protons incident on a thick iron target [11].

Lethargy plots are plots in which the differential neutron spectra are multiplied by the neutron energy, E, to account for the logarithmic energy scale. Neutron lethargy, or logarithmic energy decrement, u, is a dimensionless logarithm of the ratio of the energy of source neutrons (E_0) to the energy of neutrons (E) after a collision:

$$u = \ln \frac{E_0}{E} \tag{10.6}$$

$$E = E_0 \exp(-u) \tag{10.7}$$

A plot of E vs. u will show an exponential decay of energy per unit collision, indicating that the greatest changes of energy (ΔE) result from the early collisions.

The energy distributions in these figures are typically characterized by two peaks: a high-energy peak (produced by the scattered primary beam particle) and an evapora-

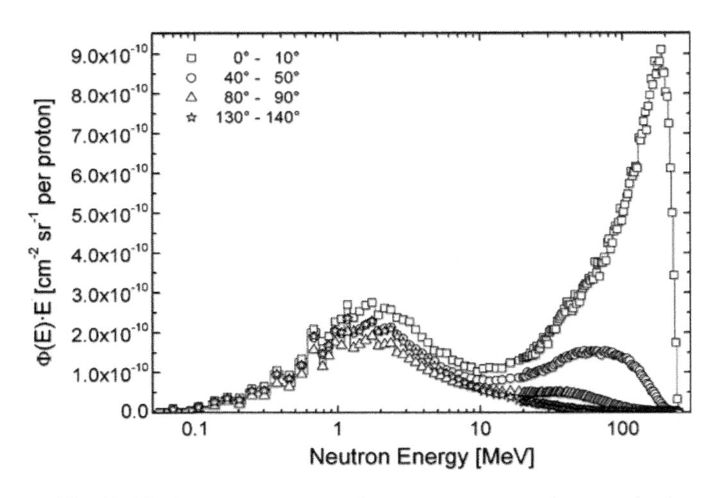

Figure 10–3 Unshielded neutron spectra for neutrons at various emission angles produced by 250 MeV protons incident on a thick iron target (without any concrete shielding). (Reprinted from [11] with permission from Elsevier.)

tion peak at ~2 MeV. The high-energy peaks shift to higher energies with increasing proton energies, which are particularly evident in the forward direction (0° to 10°). The high-energy peak for the unshielded target is not the usual 100 MeV peak that is observed outside thick concrete shielding, which will be discussed in the next section.

This high-energy component of the cascade with neutron energies (E_n) above 100 MeV propagates the neutrons through the shielding and continuously regenerates lower-energy neutrons and charged particles at all depths in the shield via inelastic reactions with the shielding material [9]. However, the greater yield of low-energy neutrons is more than compensated for by greater attenuation in the shield due to a higher cross section at low energy. Shielding studies indicate that the radiation field reaches an equilibrium condition beyond a few mean-free paths within the shield. Neutrons with energies greater than 150 MeV regenerate the cascade, even though they are present in relatively small numbers. They are accompanied by numerous low-energy neutrons produced in the interactions. The typical neutron spectrum observed outside a thick concrete shield consists of peaks at a few MeV and at ~100 MeV.

Figure 10–4 shows the normalized neutron spectra in the transverse direction at the surface of the concrete and at various depths in the concrete, for 250 MeV protons incident on a thick iron target [11].

The low-energy neutron component is attenuated up to about a depth of 100 cm with a short attenuation length, thus giving rise to a less intense but more penetrating spectrum with a longer attenuation length. Beyond 100 cm, the spectrum reaches equilibrium.

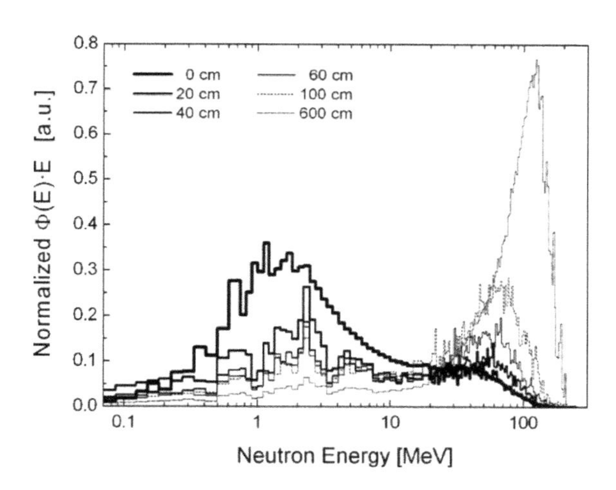

Figure 10–4 Normalized neutron spectra in transverse direction at surface of a concrete shield, and at various depths, produced by 250 MeV protons incident on a thick iron target. (Reprinted from [11] with permission from Elsevier.)

Thus, the neutron energy distribution consists of two components—high-energy neutrons produced by the cascade and evaporation neutrons with energy peaked at ~2 MeV. As previously mentioned, the high-energy neutrons are anisotropic and forward peaked, but the evaporation neutrons are isotropic. The highest-energy neutrons detected outside the shielding are those that arrive without interaction, or that have undergone only elastic scattering or direct inelastic scattering with little loss of energy and a small change in direction. Low-energy neutrons and charged particles detected outside the shielding are those that have been generated at the outer surface of the shield. Thus, the yield of high-energy neutrons ($E_n > 100$ MeV) in the primary collision of the protons with the target material determines the magnitude of the prompt radiation field outside the shield in the therapeutic proton energy range of interest.

The charged particles produced by the protons will be absorbed in shielding that is sufficiently thick to protect against neutrons. Thus, neutrons dominate the radiation field outside the shielding. Degraded neutrons might undergo capture reactions in the shielding, giving rise to neutron-capture gamma rays.

10.3.3 Shielding Calculations

Shielding calculations are performed to ensure that the facility is designed so that exposures of personnel and the public are well within regulatory limits. The purpose of radiation monitoring is to demonstrate compliance with design or regulatory limits [3]. This implies that calculations and measurements must be expressed in terms of quantities in which the limits are defined.

10.3.3.1 Dose Quantities

The International Commission on Radiological Protection (ICRP) defines dose limits which are expressed in terms of protection quantities, such as the effective dose (E) which is measured in the human body [12,13]. However, E is not directly measurable. For external individual exposure, the operational quantity ambient dose equivalent $H^*(d)$, as defined by the International Commission on Radiological Units (ICRU), can be used.

The protection quantities and operational quantities can be related to the particle fluence and, in turn, by conversion coefficients to each other. ICRP Publication 60 [12] introduced the concept of equivalent dose. ICRP Publication 103 [13] modified the weighting factors. The definitions of effective dose, equivalent dose, and ambient dose equivalent taken from ICRP Publication 60, ICRP Publication 103, and ICRU Report 51 [14] are as follows:

The **effective dose**, E, is given by $E = \sum_T w_T H_T$, where H_T is the equivalent dose in the tissue or organ, T, and w_T is the corresponding tissue weighting factor. The effective dose is expressed in Sv.

The **equivalent dose**, H, in a tissue or organ is given by $H = \sum_R w_R D_{T,R}$, where $D_{T,R}$ is the mean absorbed dose in the tissue or organ, T, due to radiation, R, and w_R is

Table 10–1 Radiation weighting factors recommended by ICRP Publication 103

Radiation Type	Energy Range	Radiation Weighting Factor
Photons, electrons and muons	All energies	1
Neutrons	<1 MeV	$W_R = 2.5 + 18.2\exp\left[-\frac{(\ln(E))^2}{6}\right]$
Neutrons	1 MeV to 50 MeV	$W_R = 5 + 17\exp\left[-\frac{(\ln(2E))^2}{6}\right]$
Neutrons	>50 MeV	$W_R = 2.5 + 3.5\exp\left[-\frac{(\ln(0.04E))^2}{6}\right]$
Protons, other than recoil protons	>2 MeV	2
Alpha particles, fission fragments, and heavy nuclei	All energies	20

the corresponding radiation weighting factor. The unit of equivalent dose is the sievert (Sv).

The weighting factor, w_R, for the protection quantities recommended by ICRP Publication 103 [13] is shown in Table 10–1. In the case of neutrons, w_R varies with energy and, therefore, the computation for the protection quantities is made by integration over the entire energy spectrum.

For radiation protection, the operational dose quantities provide a good and conservative estimate of effective dose under most exposure conditions, when the w_R values from Table 10–1 are used for neutrons.

The **ambient dose equivalent**, *H*(d)*, at a point in a radiation field is the dose equivalent that would be produced by the corresponding expanded and aligned field, in the ICRU sphere (diameter = 30 cm, composition: 76.2% O, 10.1% H, 11.1% C, and 2.6% N) at a depth, *d*, on the radius opposing the direction of the aligned field [19]. The ambient dose equivalent is measured in Sv. For strongly penetrating radiation, a depth of 10 mm is recommended. For weakly penetrating radiation, a depth of 0.07 mm is recommended. In the expanded and aligned field, the fluence and its energy distribution have the same values throughout the volume of interest as in the actual field at the point of reference, but the fluence is unidirectional.

10.3.3.2 Conversion Coefficients

Conversion coefficients are used to relate the protection and operational quantities to physical quantities characterizing the radiation field [14]. Radiation fields are frequently characterized in terms of absorbed dose or fluence. The fluence, Φ, is the quotient of $d\underline{N}$ by $d\underline{a}$ where $d\underline{N}$ is the number of particles incident on a sphere of cross-sectional area $d\underline{a}$. The unit is m^{-2} or cm^{-2}. Thus, for example, the effective dose, *E*, can be obtained by multiplying the fluence with the fluence-to-effective dose conversion coefficient. The ambient dose equivalent, *H*(d)*, can be obtained by multiplying the fluence with the fluence-to-ambient dose equivalent conversion coefficient. Conversion coefficients have been calculated by various authors using Monte Carlo transport codes including FLUKA [15,16] for many types of radiation (photons, electrons, positrons, protons, neutrons, muons, charged pions, and kaons) and incident energies up to 10 TeV. Pelliccioni [17] has summarized most of the data. Because the conversion coefficient for *H*(10)* for neutrons becomes smaller than that for *E*(AP) (where AP refers to anterior-posterior) above 50 MeV, use of *E*(AP) may be considered more conservative for high-energy neutrons. The conversion coefficient for *E*(AP) becomes smaller than that for posterior-anterior (PA) irradiation geometry, *E*(PA), at neutron energies above 50 MeV. However, the integrated dose from thermal neutrons to high-energy neutrons is highest for AP geometry, and therefore the choice of *E*(AP) is more conservative than the choice of *E*(PA). The term "dose" or "dose equivalent" is used in a generic sense for the effective dose equivalent or ambient dose equivalent in this chapter. For photons and low-energy neutrons, such as those encountered at the maze exit, the ambient dose equivalent is more conservative than *E*(AP).

10.3.4 Calculational Methods

Shielding calculations can be performed using analytical methods, Monte Carlo codes, or computational models.

10.3.4.1 Analytical Methods

Most analytical models consist of line-of-sight or "point kernel" models. They are limited in their use because they are based on simplistic assumptions and geometry. Many of the models are restricted to transverse shielding. Further, they do not account for changes in energy, angle of production, target material and dimensions, and concrete material composition and density.

Attenuation Length

Attenuation length was defined in Section 10.2.1.2. The attenuation length, λ, is usually expressed in cm (or m) and in g cm^{-2} (or kg m^{-2}) when multiplied by the density (ρ). The value of λ changes with increasing depth in the shield for thicknesses (ρd) that are less than ~100 g cm^{-2} because the "softer" radiations are more easily attenuated, and the neutron spectrum "hardens." The attenuation length may eventually reach an equilibrium value after the spectrum hardens, which is referred to as the effective attenuation length, λ_{eff}.

A radiation source can be considered a point source if the distance at which the dose is determined is at least five times the dimension of the radiation source. For a point source, the dose decreases as the square of the distance. Because protons have a defined range in the target material, the dimension of the source can be considered to be the range of the proton in the target material. For example, the range of a 230 MeV proton in tissue is 32 cm, while the distance from the target to the point of interest outside the shielding walls is typically greater than 1.6 m (5 x 32 cm).

At a given proton energy, the dose equivalent, $H(d, \theta)$, at a depth d and an angle θ in the shield, is approximately given by the following equation over a limited range of shielding thicknesses [4]:

$$H(d,\theta) = \frac{H_0}{r^2}\exp\left[-\frac{d(\theta)}{\lambda_{eff}}\right] \tag{10.8}$$

where H_0 is the dose equivalent extrapolated to zero depth in the shield and at a corresponding angle θ, at a distance of 1 m from the source; r is the distance from the source to the point of interest outside the shield; and λ_{eff} is the effective attenuation length for dose equivalent through the shield.

Figure 10–5 shows the calculated attenuation length ($\rho\lambda$) for broad beams of monoenergetic unidirectional neutrons perpendicularly incident on concrete at a depth greater than 1 m as a function of neutron energy.

The attenuation length increases with increasing neutron energy at energies greater than ~20 MeV. The increase in attenuation length is indicative of the change from the energy region—in which neutrons interact mainly by elastic scattering with the target nuclei as a whole—to the region where interaction occurs more likely with individual nucleons in a target, thus leading to an intra-nuclear cascade. In the past, it has typically been assumed that the attenuation length reaches a high-energy limiting value of about 120 g cm^{-2}, even though the data in Figure 10–5 shows a slightly increasing trend above 200 MeV for monoenergetic neutrons.

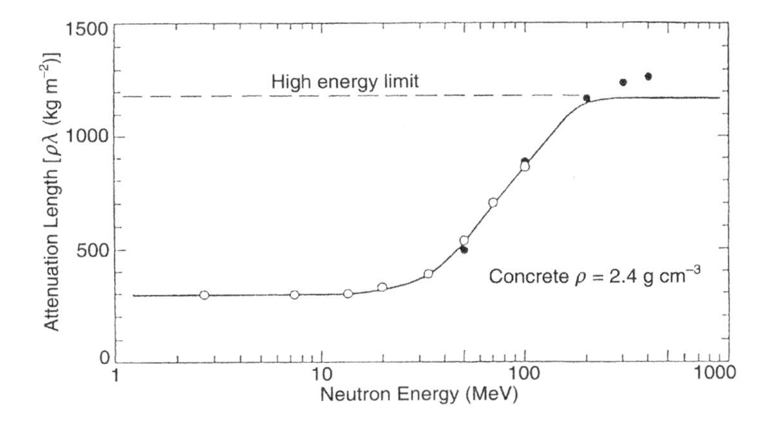

Figure 10–5 Attenuation length ($\rho\lambda$) for monoenergetic neutrons in concrete as a function of energy. (Reprinted with permission from National Council on Radiation Protection and Measurements, http://NCRPpublications.org, [3].)

Comparison of neutron dose attenuation lengths measured at various facilities, for concrete and iron, respectively, as a function of the effective maximum energy (E_{max}) of the source neutrons, for neutrons with energies from thermal to maximum has been made by Nakamura [18]. The data include measurements for E_{max} ranging from 22 MeV to 700 MeV, and various production angles for a variety of neutron sources. According to Nakamura, the measured neutron dose attenuation length (thermal to maximum energy) for concrete at 22 MeV is about 30 g cm^{-2} in the forward direction. The attenuation length gradually increases above 100 MeV to a maximum value of about 130 g cm^{-2}, which may be considered the high-energy limit. Work performed by Chen et al. [19] for thick targets indicate that for 10 MeV $\leq E_p <$ 30 MeV, the attenuation length can be treated as a constant with a value of 29.57 cm^{-2}, irrespective of the neutron emission angles and the target materials. The target materials that were investigated in Chen's study were carbon, nitrogen, aluminum, iron, copper, and tungsten.

The effective attenuation length in concrete as a function of proton energy for thin and thick copper targets is reported in the literature [4]. Thick targets are targets in which the protons are completely stopped, i.e., the thickness is greater than or equal to the particle range. By contrast, thin targets are targets with thicknesses that are significantly less than the particle range. Thus, for example, the protons lose an insignificant amount of energy in the target, and the kinetic energy available for neutron production in the target is the full incident proton energy [4]. The values of λ_{eff} in the forward direction for neutrons from a thin target will be higher than those from a thick target because of the softer spectrum emitted from the thick target compared to the thin target.

It is important to note that, in addition to particle type and energy, λ also depends upon the production angle (θ), material composition, and density.

Moyer Model

A semi-empirical method for the shield design of the 6 GeV proton Bevatron [3] was developed by Burton Moyer in 1961. This model is only applicable to the transverse shielding for a high-energy proton accelerator (GeV range) and is restricted to the determination of neutron dose equivalent produced at an angle between 60° to 120°. The simple form of the Moyer model [20] is given by:

$$H = \frac{H_0}{r^2}\left[\frac{E_P}{E_0}\right]^{\alpha} \exp\left[-\frac{d}{\lambda}\right] \tag{10.9}$$

where H = maximum dose equivalent rate at a given radial distance (r) from the target, d = shield thickness, E_P = proton energy, E_0 = 1 GeV, $H_0 = 2.6 \times 10^{-14}$ Sv m^2, and α is about 0.8.

The Moyer model is effective in the GeV region because the neutron dose attenuation length (λ) is nearly constant, regardless of energy. At proton energies in the therapeutic range of interest, the neutron attenuation length increases considerably with energy. Therefore, the Moyer model is inappropriate for use in proton therapy shielding.

Kato and Nakamura have developed a modified version of the Moyer model which includes changes in attenuation length with shield thickness, and also includes a correction for oblique penetration through the shield [21].

In the past, high-energy accelerators were shielded using analytical methods. However, sophisticated Monte Carlo codes have superseded analytical methods with the advent of powerful computers.

10.3.4.2 Monte Carlo Calculations

The Monte Carlo codes FLUKA and MCNP [22] have been used extensively in shielding calculations for particle accelerators of all energies. An extensive coverage of other codes used in shielding can be found in PTCOG Report 1 [1]. These codes can be used to perform a full simulation, modeling the accelerator and room geometry in its entirety. They can also be used to derive computational models as discussed in the next section. Monte Carlo codes have been used in the shielding design of several particle therapy facilities [23–27].

In the early stages of design, a facility typically undergoes several iterations of changes in layout. Therefore, a full Monte Carlo simulation is not practical or cost effective. The use of Monte Carlo codes is time consuming. Full simulations should be performed only after the layout has been finalized. Monte Carlo simulations are especially effective for special issues such as maze design, penetration shielding, skyshine, and groundshine (described in Section 10.5).

10.3.4.3 Computational Models

Computational models (derived using Monte Carlo codes) that are independent of geometry typically consist of a source term and an exponential term that describes the attenuation of the radiation. Both the source term and the attenuation length are dependent on particle type, incident particle energy, and production angle. Shielding can be estimated over a wide range of thicknesses by the following equation for a point source which combines the inverse square law and an exponential attenuation through the shield, and is independent of geometry. A more universal form of Equation 10.8 is shown below [28]:

$$H\left(E_p,\theta,d/\lambda(\theta)\right)=\frac{H_0(E_p,\theta)}{r^2}\exp\left[-\frac{d}{\lambda(\theta)g(\theta)}\right] \tag{10.10}$$

where H is the dose equivalent outside the shielding; H_0 is source term at a production angle θ with respect to the incident beam and is assumed to be geometry independent; E_p is the energy of the incident particle; r is the distance between the target and the point at which the dose equivalent is scored; d is the thickness of the shield; $d/g(\theta)$ is the slant thickness of the shield at an angle θ; $\lambda(\theta)$ is the attenuation length for dose equivalent at an angle θ and is defined as the penetration distance in which the intensity of the radiation is attenuated by a factor of e; $g(\theta) = \cos\theta$ for forward shielding; $g(\theta) = \sin\theta$ for lateral shielding; and $g(\theta) = 1$ for spherical geometry. Computational models are useful especially during the schematic phase of the facility design, when the design undergoes several changes to determine the barrier shielding [29]. The entire room geometry is not modeled, but usually spherical shells of shielding material are placed around the target, and Monte Carlo codes are used to score dose at given angular intervals and in each shell of shielding material. Plots of dose vs. shielding thickness can be fitted to obtain source terms and attenuation lengths as a function of angle, and at the energies of interest, with the appropriate target. For thick targets, the data may be fitted with a single or double exponential, as discussed in Section 10.3.5.

Most of the published computational models provide source terms and attenuation lengths for neutrons only. For shielding purposes, the total dose from all particles should be considered instead of only neutron dose, as first proposed by Ipe [30]. As stated previously, the source terms and attenuation lengths will depend upon the combination of composition and density of the shielding material. Ray traces can be performed at various angles, and the source terms and attenuation lengths can be used for dose calculations. These models are also useful in identifying thin shielding and to facilitate improved shield design.

A stopping target can be used to determine dose from the beam incident on the patient, but the use of a stopping target is not necessarily conservative in all cases. The dose equivalent (Sv-m^2/p) in the forward direction as a function of depth in concrete for protons is typically higher for a thin target than for a thick target. Further, as previously stated, the effective attenuation length for a thin target is higher than for a thicker target. In Section 10.3.6, it is shown that the forward-directed neutrons are more penetrating for a thin target than for a thick target in the case of a graphite target. Thus, neutrons from a thin target may propagate the intra-nuclear cascade in the downstream shielding, thus requiring more shielding than in the case of a thick target.

Several authors have published computational models. Published computational models are only of academic interest and should not be used for calculations since they are typically based on stopping targets and some theoretical concrete composition and density, which can be significantly different from the site-specific concrete composition and density. Computational models for proton energies, target material, and dimensions—as well as concrete composition and density that are facility-specific—should be derived on a case-by-case basis.

10.3.5 Neutron Yield and Angular Distribution

The neutron yield of a target is defined as the number of neutrons emitted per incident primary particle. The neutron yield from a target depends upon the target material and dimensions, and the particle energy. The neutron yield from a thin target will be proportional to the target thickness. The protons in the beam will be scattered, so some particles may strike the accelerator or beam-line components. The neutrons from a thin target are more penetrating than those from a thick target, so the shielding in the forward direction (0–10°) will be thicker for a thin target when compared to a thick target, as illustrated in Section 10.3.6.

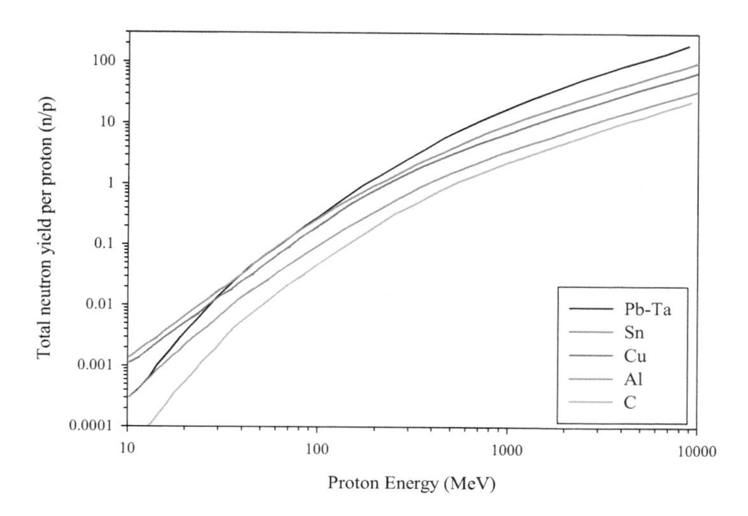

Figure 10–6 Total neutron yield per proton for different target materials. (Data from [11].)

As the target thickness increases, the proton interactions increase, resulting in an increase in the secondary neutron yield. At first, the yield is dominated by the high-energy neutrons, but as the thickness is further increased, the high-energy neutrons interact, producing more low-energy neutrons. Therefore, the high-energy neutron yield decreases and the low-energy neutron yield increases, while the overall neutron yield increases. The low-energy neutrons get attenuated in the target as the thickness is further increased. The net result is an increase in total neutron yield with increasing target thickness until it reaches a maximum, and then a decrease occurs due to the attenuation of low-energy neutrons in the target material. Therefore, the dimensions of the target play an important role in the determination of shielding thickness.

As the proton energy increases, the threshold for nuclear reactions is exceeded, and more nuclear interactions can occur. At energies above 50 MeV, the intra-nuclear cascade process becomes important. Neutron yields increase with increasing proton energy because the thresholds for neutron production are exceeded. Figure 10–6 shows the measured total neutron yield per proton for thick targets (except at high energies) of different materials [31].

According to Tesch, between proton energies of 50 and 500 MeV the neutron yields for thick targets increase as approximately E_P^2 for all target materials, where E_P is the energy of the incident proton. The ratio of neutron yields from different target materials for thick targets is independent of E_P in the energy range of 20 MeV to 1 GeV (which covers the therapeutic energy range of interest). The data are given relative to medium mass number (copper/iron) by:

C:Al:Cu-Fe:Sn:Ta-Pb = (0.3±0.1):(0.6±0.2):(1.0):(1.5±0.4):(1.7±0.2)

Calculations and measurements of neutron yields, energy spectra, and angular distributions for protons of various energies incident on different types of materials have been reported in the literature [4,11,32–36]. As the proton energy increases, the average neutron energy in the forward direction (0° to 10°) increases, thus resulting in the hardening of the spectra [11]. However, at very large angles (130° to 140°) the average energy does not change significantly with increasing proton energies, so the spectra does not change much.

The angular neutron yield and spectra (usually expressed as double differential neutron yield), are more meaningful than the total neutron yield because targets encountered in proton shielding are not necessarily thick targets. For example, the thickness of the degrader (usually graphite) in the cyclotron varies depending upon the energy required. The degrader would never be a stopping target, because if all the protons are stopped, no protons would be transported to the treatment room. Doses from thick and thin graphite targets are reported in Section 10.3.6.

10.3.6 Angular Dose Profile and Dependence on Shielding Thickness and Target Dimensions

Figure 10–7 shows the angular dose profile from thick unshielded ICRU tissue targets in the forward direction for various proton energies calculated using the Monte Carlo code, FLUKA (Ipe, this work). The total ambient dose equivalent from all particles is normalized to a distance of 1 m and expressed in pSv per proton. For a given production angle, as the energy increases, the dose increases because the thresholds for nuclear interactions are exceeded. At a given energy, the dose decreases significantly

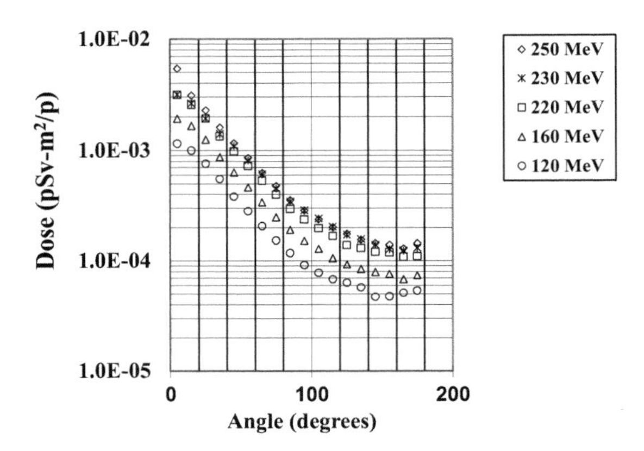

Figure 10–7 Angular dose profile from unshielded thick ICRU tissue targets for various proton energies calculated using the Monte Carlo code, FLUKA (Ipe, this work).

Table 10–2 ICRU tissue dimensions as a function of energy (Ipe, this work)

Energy (MeV)	250	230	220	160	120
Target Length (cm)	38	34	31	18	11
Target Radius (cm)	20	17	16.5	10	6.5

with increasing angle initially, and then it levels off at the larger angles. The decrease is due to the softer spectra at the larger angles. Therefore, more shielding will be required in the forward direction.

Table 10–2 shows the target dimensions as a function of energy.

Figures 10–8 and 10–9 show the total ambient dose equivalent attenuation curves in concrete in the 0–10° direction and in the 80–90° direction, respectively, from a pencil beam of protons incident on thick ICRU tissue targets for various proton energies (Ipe, this work).

Figure 10–8 shows that there is a dose buildup in the forward direction at small depths. For thicknesses greater than 75 cm, the source term and attenuation length in

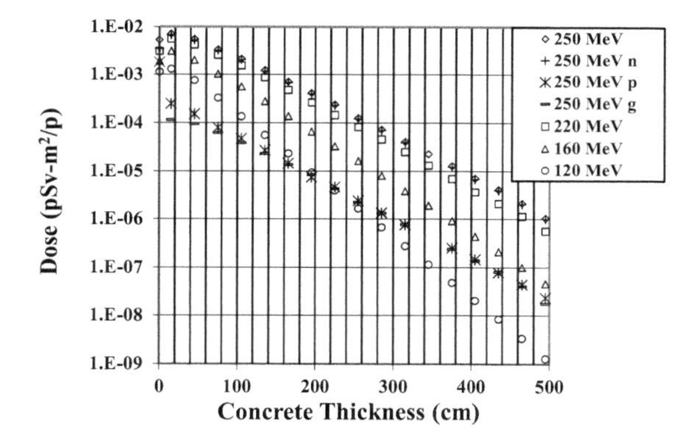

Figure 10–8 Dose as a function of shielding thickness in the 0–10 ° direction for protons incident on ICRU tissue target. Total dose is shown for all energies. At 250 MeV, neutron (n), proton (p), and gamma (g) dose contributions are also shown. (Ipe, this work.)

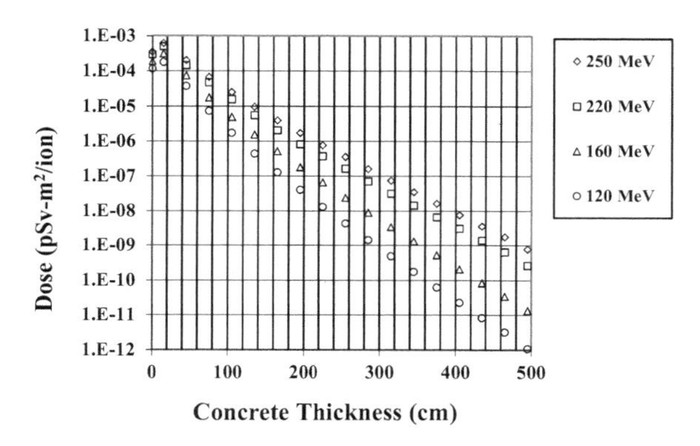

Figure 10–9 Dose as a function of shielding thickness in the 80–90° direction for protons incident on ICRU tissue target (Ipe, this work).

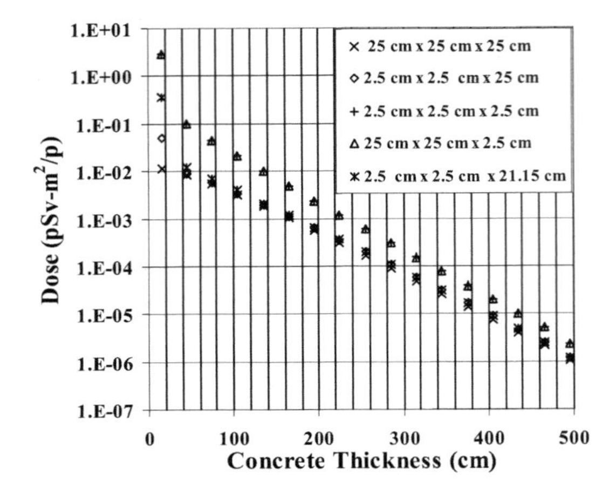

Figure 10–10 Dose in the 0–10° direction as a function of shielding thickness for 250 MeV proton incident on graphite targets of various dimensions. (Ipe, this work.)

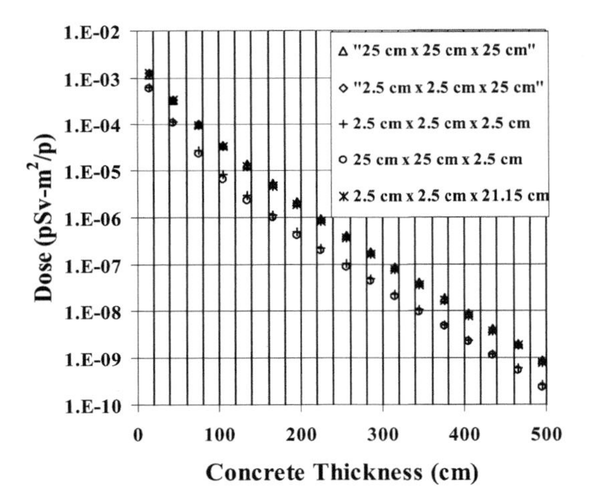

Figure 10–11 Dose in the 80–90° direction as a function of shielding thickness for 250 MeV proton incident on graphite targets of various dimensions. (Ipe, this work.)

the forward direction (0–10°) produced by 250 MeV protons can be fitted as follows: $H_0 = 1.66 \times 10^{-14}$ Sv m^2 per proton and $\lambda_{tot} = 122.4$ g cm^{-2}, respectively. For thin targets, multi-component fits may be required.

Also shown are the neutron (n), proton (p), and gamma (g) components of dose at 250 MeV in the forward direction. In the forward direction, more than 95% of the dose is from neutrons. The proton dose is higher than the photon dose for concrete thicknesses up to 165 cm. Beyond 165 cm, the photon dose is greater than the proton dose. At large angles, the ratio of neutron dose to total dose decreases. Therefore, it is important to use total dose for all shielding calculations.

Figure 10–9 shows that, at large angles, a two-component fit may be required for thick targets. Plots of dose equivalent vs. shielding indicate that there is a dose buildup in the forward direction at small depths. However, at large angles, the low-energy component of the radiation is attenuated quickly in small thicknesses. This is because the neutron spectrum changes with depth in the shield. The spectrum hardens with depth and reaches equilibrium after a depth greater than about 100 cm. Thus, there are two attenuation lengths. The second attenuation length in this case is referred to as the effective attenuation length, and it is valid for thicknesses greater than 100 cm of concrete.

Figures 10–10 and 10–11 show the total effective dose, E(AP), from all particles as a function of concrete shielding thickness for 250 MeV protons incident on a graphite target with varying dimensions in the 0–10° direction and the 40–50° direction, respectively. The 25 cm x 25 cm x 25 cm target represents a thick target for 250 MeV protons (Ipe, this work).

Figure 10–10 shows that in the forward direction (0–10°) direction, the graphite with thickness of 2.5 cm results in a higher dose than the graphite with thickness of 25 cm. The dose increases as the target thickness decreases. Therefore, the use of a stopping target is not conservative in the forward direction for graphite. For example, if the thick graphite target requires a shielding thickness of 405 cm, the thin graphite target will require a thickness of about 445 cm. Thus, the shielding in the 0–10° direction will be underestimated by about 40 cm if the graphite target is a thin target. Change in the lateral dimensions does not seem to have a significant effect in the forward direction for the thin graphite target. A slight change is observed for the thick graphite target.

Figure 10–11 shows that in the 80–90° direction, the thick graphite target results in a higher dose than the thin graphite target. Therefore, the use of a thick target for graphite overestimates the dose in the 80–90° direction. Change in lateral dimensions does not appear to have a significant effect. Thus, it is important to consider both the target material and dimensions in shielding calculations, instead of always assuming a stopping target. Note that the error bars on the figures in this report generated by Ipe are within a few percentage points in the 0–10° direction and within 10% in the 80–90° direction. Hence, these error bars are too small to be seen on the figures.

10.4 Shielding Design Considerations

The primary purpose of shielding is to attenuate secondary radiation to levels that are within regulatory or design limits for individual exposure. Shielding design requires knowledge of the various parameters, such as beam parameters and losses (including location, target material, and dimensions), treatment parameters, beam delivery system, facility layout, adjacent occupancies, shielding material composition and density, and regulatory dose limits [1–3, 30].

10.4.1 Beam Parameters and Losses

For proton therapy facilities, the concrete shielding thicknesses for various parts of the facility may range from about 60 cm to about 7 m of concrete. Cyclotron rooms have the thickest shielding. Beam losses or interception of beam by any material (including the patient) result in the production of secondary radiation. Therefore, in order to design effective shielding, the beam parameters (energy, intensity, beam dimensions) losses and sources of radiation for proton therapy facilities must be well understood. This requires knowledge of how the accelerators operate and deliver beam to the treatment rooms. Specific details of beam losses, duration, frequency, targets (material and dimensions), and locations should be provided by the equipment vendor. Higher beam losses will occur during start-up and commissioning as the beam is tuned and delivered to the final destination, and this should be anticipated.

10.4.1.1 Synchrotron- and Cyclotron-based Systems

Synchrotrons are designed to accelerate protons to the exact energy required for therapy. Since synchrotrons do not have energy degraders, less structural and local shielding is required when compared to cyclotrons. Further, there is less activation of beamline components. The sources of radiation that need to be considered include x-rays from the ion source, x-rays produced by back-streaming electrons striking the linac (linear accelerator) structure, and neutrons produced by the interaction of the protons with the linac structure toward the end of the linac. The target material is typically copper or iron. The production of x-rays from back-streaming electrons will depend upon the vacuum conditions and the design of the accelerator [10]. The use of a Faraday cup or beam stop to intercept the beam downstream of the linac must also be considered. Distributed or local beam losses can occur typically during the injection process, during RF capture and acceleration, and during extraction. Since losses are machine-specific, the equipment vendor should provide this information. Particles that are not used in a spill may be discarded by deflection onto a beam dump or stop. These discarded particles have sufficient energy to produce energetic secondary radiation in the beam stop or dump. Therefore, they should to be considered in the shielding design and activation analysis. Particles that are decelerated before being discarded are not of concern in the shielding design or activation analysis. X-rays may be produced at locations such as the injection and extraction septa due to the voltage applied across electrostatic deflectors. During acceleration, about 20% to 50% of the beam particles can be lost continuously in the cyclotron. The steel in the magnet yoke provides considerable shielding, except in regions where there are holes or cutouts through the yoke. These holes should be considered in the shielding design. Losses at very low proton energies can contribute to activation of the cyclotron, but they are not of concern for structural shielding. The structural shielding is determined mainly by beam losses that occur at higher energies, close to the extraction energy (230 MeV to 250 MeV depending upon the cyclotron type). These beam losses may occur in the dees and the extraction septum, which are typically made of copper, and also cause activation.

The Energy Selection System (ESS) in a cyclotron consists of an energy degrader, collimators, spectrometer, energy slits, and a beam stop. The ESS allows the proton energy to be lowered after extraction. The intensity from the cyclotron is increased as the degraded energy is decreased in order to maintain the same dose rate at the patient. Thus, large amounts of neutrons are produced in the degrader, especially at the lower energies, resulting in thicker local shielding requirements in this area because the of the higher incident proton intensity. The degrader scatters the protons and increases the energy spread. A collimator is used to reduce the beam emittance. A magnetic spectrometer and energy slits are used to reduce the energy spread. Beam stops and Faraday cups are used to tune the beam. Neutrons are also produced in the collimator, slits, and beam stop. Losses in the ESS are large, and they also result in activation.

Losses occur in the beam transport line for synchrotron- and cyclotron-based systems. Although these losses are usually very low (~1%) and distributed along the beam line, they need to be considered for shielding design. The target material is typically copper or iron. During operation, the beam is steered onto Faraday cups and beam stops. Beam incident on these components should also be considered in the shielding design.

10.4.1.2 Treatment Rooms

The radiation produced from the beam impinging on the patient (or phantom) is a dominant source for the treatment rooms. Thus, a thick tissue or water target or should be assumed in computer simulations for shielding calculations. The range of treatment depths should be defined for the specific energy of interest, as well as the treatment field size. In addition, losses in the nozzle and in beam-shaping and range-shifting devices must also be considered in the shielding design. The contributions from adjacent areas, such as the beam transport and other treatment rooms, should also be taken into account. Typically, a large facility's treatment rooms do not have shielded doors, and therefore the effectiveness of the maze design is critical. The smaller single-room facilities have shielded doors. In such cases, a full computer simulation for the maze is recommended. Treatment rooms either have fixed beam rooms or gantries.

10.4.1.3 Fixed Beam and Gantry Rooms

For proton therapy facilities, the use factor (U) may be defined as the fraction of beam operation time during which the primary proton beam is directed toward a primary barrier [1]. Either a single horizontal fixed beam or dual (horizontal and vertical or oblique) beams are used in fixed-beam rooms. Shielding walls in the forward direction are much thicker than the lateral walls and the walls in the backward direction. For a horizontal fixed-beam room, the primary beam direction is fixed, and the U is 1 for the barrier toward which the primary beam is directed for rooms with dual beams. The use factor for the wall in the forward (0°) direction for each beam should be considered. For example, this may be either 1/2 for both beams or 2/3 for one beam and 1/3 for the other. For a single beam, U is 1 for the wall in the forward direction.

The beam is rotated about the patient in gantry rooms. In some cases, the gantries may rotate completely (360°), while in other cases only partial rotation is possible (~180°). For full rotation, on average, it can be assumed that the use factor for each of the four barriers (two walls, floor, and ceiling) is 0.25. In some designs, the gantry counterweight (made of large thicknesses of steel) acts as a stopper in the forward direction. However, it usually covers a small angle and is asymmetric. The ceiling, lateral walls, and floor are exposed to the forward-directed radiation. The walls in the forward direction can be thinner than for fixed beams because of the lower use factor.

10.4.1.4 Beam Delivery and Beam-shaping Techniques

The various techniques used to shape and deliver the beam to the patient can be divided into two categories: passive scattering and active beam scanning. Passive scattering (PS) includes single and double scattering. In passive scattering, lateral spread of the beam is achieved by scatterers, and a spread-out Bragg peak (SOBP) is produced by a range modulation wheel or a ridge filter located in the nozzle [37]. Typically, for small fields, a single scatterer is used, while for large fields, a double scatterer is used. A collimator (specific to the treatment field) located between the nozzle exit and the patient is used to shape the field laterally. A range compensator is used to correct for the shape of the patient surface, inhomogeneities in the tissues traversed by the beam, and the shape of the distal target volume. A much higher beam current is required at the nozzle entrance when compared to the other delivery techniques because there are losses due to the incidence of the primary beam on the various beam delivery and shaping devices (see Table 10–3). The typical maximum efficiency of a passive scattering system with a patient field is about 45%.

Active scanning includes uniform scanning (US) and pencil beam scanning (PBS). In the IBA US system, two perpendicular dipoles are used to scan a large spot along a fixed pattern. A fixed scatterer, range modulator, and patient aperture and compensator are also used.

In PBS, a proton pencil beam is magnetically scanned throughout the target volume without the need for scattering, flattening, or compensating devices. Therefore, for PBS, the intensity of secondary radiation in the treatment room is much lower than for PS and US because there is little material in the beam, and almost 100% of the beam at the nozzle entrance reaches the patient. The depth of penetration of the Bragg peak is varied by adjusting the energy of the beam before it enters the treatment room.

Nozzle entrance is typically defined as the entrance position of the beam into the treatment room, i.e., where the beam enters beam monitoring (for the IBA system, this is the location of ion chamber IC1 where the beam current is measured before entering the nozzle) or beam-shaping devices. Nozzle exit is defined as the position upstream of the patient-specific aperture and compensator. One may, however, also define it as including aperture and compensator, as IBA does.

Table 10–3 shows the typical currents required at the cyclotron exit (Icyclo) and the treatment nozzle entrance (Inozzle) to deliver a dose rate of 2 Gy/min to 1 liter of water at distal energies of 160 MeV and 220 MeV for three delivery techniques: passive beam scattering, uniform scanning, and pencil beam scanning. In each case, the current loss (Icyclo-Inozzle) is also shown. The extracted proton energy from the IBA cyclotron is 230 MeV. The proton energy incident on the degrader for exit energies of 160 and 220 MeV is 230 MeV.

From Table 10–3, one notes that PS requires the highest proton currents (Icyclo and Inozzle) when compared to US and PBS to deliver a dose rate of 2Gy/liter at 160 MeV and 220 MeV. Therefore, more secondary radiation is produced in the cyclotron, the beam transport line, and the treatment room in PS when compared to active scan-

Table 10–3 Typical beam currents needed to deliver a dose rate of 2 Gy/min in a water target of 1 liter for various beam delivery techniques at 220 MeV and 160 MeV. (Adapted with permission from IBA [38,39].)

	Passive Scattering (PS)			Uniform Scanning (US)			Pencil Beam Scanning (PBS)		
Energy (MeV)	Icyclo (nA)	Inozzle (nA)	Loss (nA)	Icyclo (nA)	Inozzle (nA)	Loss (nA)	Icyclo (nA)	Inozzle (nA)	Loss (nA)
220	12	6.5	5.5	6.5	4	2.5	9.28	0.595	8.68
160	130	11	119	22.5	2.6	19.6	17.36	0.493	16.87

ning (US or PBS) at both energies. Thus, these areas require more shielding for PS compared to US and PBS. Secondary dose to patient also increases with PS compared to active scanning. Less current at the cyclotron exit is required at 220 MeV for PS when compared to 160 MeV. The current loss for 160 MeV is greater than for 220 MeV. Therefore, more shielding is required for the cyclotron at 160 MeV compared to 220 MeV.

At 160 MeV, the proton currents at the cyclotron exit and the nozzle entrance are higher for US than for PBS. Therefore, more shielding is required in the cyclotron and treatment room for US compared to PBS. However, at 220 MeV the proton current at the cyclotron exit is higher for PBS when compared to US. According to IBA, in PBS the efficiency of the ESS is reduced by closing the divergence-limiting slits for the following reasons:

1. Because of noise in instrumentation and because of the ion source characteristics, the cyclotron beam current regulation is not as good at very low current, so the slits are closed to operate the cyclotron at a somewhat higher current.
2. Closing the slits also has the effect of decoupling the beam in the beam line and at isocenter from small angular deviations of the beam at the degrader caused by cyclotron fluctuations.

Therefore, for PBS at 220 MeV most of the beam losses occur at the divergence-limiting slits, thus impacting the shielding of the cyclotron room in the vicinity of the slits. For the reasons mentioned above, activation is also higher in the cyclotron room and the treatment room for PS compared to active scanning.

10.4.2 Workload

The term "workload" is used in a generic sense to include for each treatment room, the proton energy of interest, the beam-shaping method, the number of fractions per week, the time per fraction, the dose per fraction, and the proton current/charge required to deliver a specific dose rate/dose. Once the workload for the treatment room has been established, one must work backwards to determine the corresponding energies and currents from the cyclotron or the synchrotron. The workload for each facility will be facility-specific and equipment-specific. Therefore, the workload will vary from facility to facility and from one equipment vendor to the other. For cyclotrons, the degrader is the dominant source of radiation. The proton intensities incident on the degrader at the maximum energy of the cyclotron (typically 250 or 230 MeV) for lower exit energies (thicker degrader) are much higher than the proton intensities at the higher exit energies (thinner degrader) because of the low transmission efficiencies at the lower energies.

One school of thought prescribes use of the lowest proton treatment energy for shielding calculations of the cyclotron and the highest proton treatment energy for the treatment rooms and the beam transport line. This is an overly conservative approach and results in much more shielding than is necessary for the cyclotron room, beam transport line, and treatment rooms. Excessive shielding has a detrimental impact on the cost of the facility. For example, at 250 MeV, the effective half value layer (thickness required to reduce the dose by a factor of two) could range from about 30–35 cm of concrete. Our opinion is that a reasonable approach would be to reduce all treatments to three energies for shielding calculations—one at the lower end, one at the higher end, and one in the middle. An example of a workload can be found in PTCOG Report 1 [1].

10.4.3 Regulatory Dose Limits

The use of protons for therapy purposes is associated with the generation of secondary radiation. Therefore, protection of the occupationally exposed workers and members of the public must be considered. Most of the national radiation protection regulations are based on international guidelines or standards. In the United States, medical facilities are subject to state regulations. These regulations are based on standards of protection issued by the U.S. Nuclear Regulatory Commission [40]. The U.S. dose limits for occupational exposure are as follows.

Dose Limits for Occupational Exposure

An annual limit, which is the more limiting of:

1. the total effective dose equivalent being equal to 0.05 Sv or
2. the sum of the deep-dose equivalent and the committed dose equivalent to any individual organ or tissue other than the lens of the eye being equal to 0.5 Sv

In the interests of ALARA (as low as reasonably achievable), it is prudent to use an annual dose constraint of 5 mSv.

Dose Limits for the Individual Members of the Public

1. the total effective dose equivalent to individual members of the public does not exceed 1 mSv in a year
2. the dose in any unrestricted area from external sources 0.02 millisievert in any one hour

Since this is a dose to an area, the occupancy factor is assumed to be 1. The occupancy factor is defined below.

The occupancy factor (T) for an area is the average fraction of the time that the maximally exposed individual is present in the area while the beam is on [41]. If the use of the machine is spread out uniformly during the week, the occupancy factor is the fraction of the working hours in the week during which the individual occupies the area. For instance, corridors, stairways, bathrooms, or outside areas have lower occupancy factors than offices, nurse stations, wards, staff rooms, or control rooms. The occupancy factor for controlled areas is typically assumed to be 1, and this is based on the premise that a radiation worker works 100% of the time in one controlled area or another. In the United States, the regulatory agencies allow the use of occupancy factors; however, some countries do not allow the use of occupancy factors.

The United States does not have an instantaneous dose rate limit, but some countries have limits as low as 1 μSv/h. Therefore, a one-size-fits-all approach could potentially underestimate or overestimate the shielding in some areas for a proton therapy facility in another country, assuming similar patient workload, usage, beam parameters, equipment, and shielding material.

10.4.4 Shielding Materials

Earth, concrete, and steel are typically used for particle accelerator shielding [4]. Other materials, such as polyethylene and lead, are used to a limited extent. When using steel, a layer of hydrogenous material must be used in conjunction with the steel because neutrons are the dominant component.

10.4.4.1 Earth

Earth or soil is often used as shielding material at underground accelerator facilities and provides an alternative to more costly shields. It is suitable for shielding of both photons and neutrons, but it must be compacted to minimize cracks and voids and to attain a consistent density [4,41]. Shielding of photons is not sensitive to the water content. Water is present in soil as bound water, hygroscopic water, and gravitational water (i.e., rainwater prior to its flow to a dryer soil where it can be held as capillary or hygroscopic water). The capillary water content should not be included for shielding calculations because it evaporates easily. The water content can vary from 0% to 30%. The density of earth typically ranges from 1.7 g/cm^3 to 2.2 g/cm^3 and depends upon the soil type, water content, and degree of compaction. The site-specific earth composition and oven-dry density should be used for all shielding calculations. The potential for activation of the groundwater must also be considered for underground facilities.

10.4.4.2 Concrete and Heavy Concretes

Concrete is the most widely used shielding material, and it is relatively inexpensive. Concrete is made from a mixture of cement, water, and aggregates [41]. The density of concrete depends on the amount and density of the aggregate, the amount of air that is entrapped or purposely entrained, and the water and cement contents. Typically, ordinary concrete has a density that varies between 2.2 and 2.4 g cm^3. The hydrogen content of concrete is important for its effectiveness in neutron shielding. Almost all of the hydrogen in concrete is in the form of water, which is present as bound water (i.e., water of hydration in the cement and aggregate) as well as free water in the pores of the cement. Both forms of water may be lost at elevated temperatures, thus reducing the neutron attenuation of concrete. Free water may be lost over time by diffusion and evaporation. Typically, the initial free water content is about 3% by weight. However, this water is lost by curing of the concrete. Over a period of 20–30 years, about 50% of the bound water may be lost. This loss is more rapid at elevated temperatures. The water content of the concrete is important in the shielding of neutrons with energies between 1 and 15 MeV [42]. However, the concrete absorbs moisture from the surrounding environment until it reaches some equilibrium. Therefore, all shielding calculations should be performed using the equilibrium density, not the wet density. The use of oven-dry density is a conservative approximation, since the equilibrium density is usually not known beforehand. Brandl et al. [43] showed than an increase in water content of 50 cm-thick ICRU concrete from –4 to +4% resulted in an increased neutron ambient dose equivalent attenuation by a factor of 18.

Figure 10–12 shows the dose attenuation in three poured concretes with different compositions but the same density, in the 0–10° direction for 250 MeV protons incident on a thick ICRU tissue target [Ipe, this work].

The dose attenuation for concrete A is higher than the dose attenuation for concretes B and C. The silicon content of concrete A is about 1.7 times higher than concrete B and about 2.2 times higher than concrete C. In order to get the same dose attenuation as 405 cm of concrete C, a thickness of 448 cm of concrete A would be required. Thus, the difference in composition can lead to differences in concrete thickness of about 43 cm, which is significant. For the same thickness and the same water content, siliceous types of concretes are not as effective as carbonaceous types of concretes. Brandl et al. also studied the dose attenuation a total of 33 ordinary and shielding concretes (each 50 cm thick) with varying compositions and densities rang-

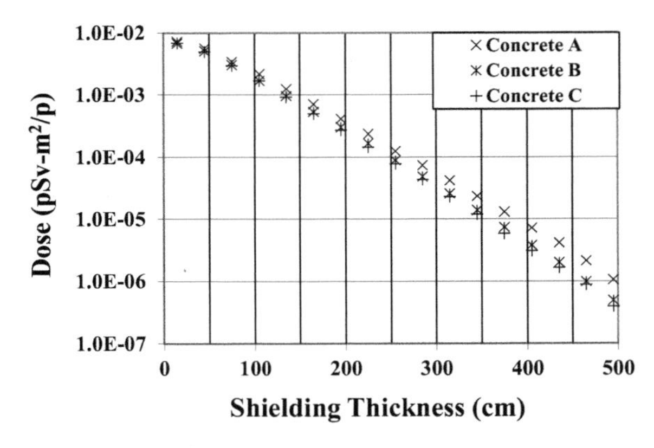

Figure 10–12 Dose Attenuation for three concrete types with same density but different compositions. (From Ipe, this work.)

ing from 1.6–5.38 g cm^3. The variations in neutron ambient dose equivalent at a proton energy of 250 MeV ranged from −50 to +30%. The combination of composition and density of the shielding material affects the dose attenuation. Therefore, all shielding calculations should be performed using the density and composition of the site-specific concrete and not some theoretical composition.

Heavy concretes contain high-Z aggregates or small pieces of scrap steel or iron which increase its density and effective Z, and they are available with densities as high as 4.8 g cm^3. Concrete enhanced with iron ore is particularly effective for the shielding of relativistic neutrons. The pouring of such high-Z-enhanced concrete is a special skill because the high-Z aggregates could sink to the bottom, resulting in a non-uniform composition and density. The high-Z aggregate-enhanced concrete is also sold in the form of prefabricated interlocking or non-interlocking modular blocks. The interlocking blocks are more effective than ordinary blocks in minimizing the streaming of radiation through gaps between the blocks. One important consideration in the choice of shielding materials is their susceptibility to radioactivation by neutrons, which can last for decades. Dose attenuation by 250 MeV protons on a tissue target for heavy concrete and composite shielding (concrete and steel) have been reported by Ipe [2,44].

10.4.4.3 Steel and Iron

Steel is an alloy of iron and is used for shielding photons and high-energy neutrons. The attenuation of steel can be considered equivalent to the attenuation of iron for photon shielding, but not for neutron shielding, Steel is often used in conjunction with concrete (composite shielding) when space is at a premium. Iron has an important deficiency in shielding neutrons because it contains no hydrogen. Natural iron is composed of 91.7% ^{56}Fe, 2.2% ^{57}Fe, and 0.3% ^{58}Fe. The lowest inelastic energy level of ^{56}Fe is 0.847 MeV [4]. Neutrons with energy above 0.847 MeV will lose energy by inelastic scattering in ^{56}Fe, but below this energy, neutrons can only lose energy by elastic scattering, which is a very inefficient process. Therefore, there is a buildup of neutrons below this energy. Furthermore, the radiation weighting factor is close to the maximum in this energy region. In addition, natural iron has two regions where the total cross section is very low because of resonances in ^{56}Fe. There is one resonance at 27.7 keV (minimum cross section = 0.5 barn) and another at 73.9 keV (minimum cross section = 0.6 barn). The net result is an increased attenuation length. Thus, large fluxes of low-energy neutrons are found outside steel or iron shielding. If steel is used for the shielding of high-energy neutrons, it must be followed by a hydrogenous material for shielding the low-energy neutrons that are generated. Dose attenuation by 200 MeV protons on a thick iron target for iron shielding have been reported by Sheu et al. [35]. Due to the large variety of nuclear processes, including neutron capture reactions of thermalized neutrons, steel can be highly activated [1].

10.4.4.4 Polyethylene

Polyethylene $(CH_2)_n$ is used for neutron shielding. Attenuation curves in polyethylene of neutrons from 72 MeV protons incident on a thick iron target have been published by Teichmann [45]. The thermal neutron capture in polyethylene yields a 2.2 MeV gamma ray, which is quite penetrating. Therefore, boron-loaded polyethylene can be used. Thermal neutron capture in boron yields a 0.478 MeV gamma ray. Borated polyethylene can be used for shielding of doors and ducts and other penetrations.

10.4.4.5 Lead

Lead is used primarily for the shielding of photons. It has a very high density (11.35 g cm^{-3}) and is available in bricks, sheets, and plates. Lead is malleable [4] and therefore cannot support its own weight when stacked to large heights. Therefore, it will require a secondary support system. Lead is transparent to fast neutrons and, therefore, it should not be used for door sills or thresholds for proton therapy facilities where secondary neutrons dominate the radiation field. However, it does decrease the energy of higher-energy neutrons by inelastic scattering down to about 5 MeV, making the hydrogenous material following it more effective. Below 5 MeV, the inelastic cross section for neutrons drops sharply.

10.5 Special Topics

The characteristics of prompt radiation in the immediate vicinity of the radiation source depend strongly upon the proton energy. The radiation field consists of two components at larger distances—direct and scattered [3]. If the roof shielding is thin compared to the lateral walls, secondary radiation may escape the roof and then be scattered down by the atmosphere, or it may be scattered by the roof to the ground level. This process is loosely referred to as "skyshine."

Similarly, "groundshine" refers to radiation escaping the floor slab, reaching the earth, and scattering upward, or radiation scattered upward from the floor slab. Roof and floor slab shielding should be thick enough to prevent skyshine and groundshine. Neutrons are not easily absorbed above thermal energies, so neutron scattering is more important than photon scattering. Furthermore, the neutron component is much higher than the photon component at proton therapy facilities. It is also important to consider the presence of adjacent elevated floors and multi-storied buildings.

Cyclotron and synchrotron rooms can either have direct-shielded doors or mazes. Direct-shielded doors are used when space is at a premium, and they must provide the same attenuation as the shield wall. The thickness of the direct-shielded door can be minimized using pre-fabricated high-Z aggregate concrete blocks or a combination of steel, lead, and borated polyethylene. All gaps around the door should be minimized because they provide a path for scattered radiation. An overlap of shielding at least 10 times the gap with should be used.

Most facilities use mazes without shielded doors. The radiation at the maze entrance consists of neutrons that scatter through the maze and capture gamma rays. The forward-directed radiation from the target should never be aimed toward the maze opening. The dose at the maze entrance can be reduced by reducing the cross-sectional area of the maze. As the number of legs increases, the attenuation increases. The legs should be perpendicular to each other to increase maze effectiveness. If the two legs are not perpendicular to each other, a single scatter could still reach the maze entrance. The radiation should scatter at least twice before reaching the maze entrance. The maze walls should be thick enough to reduce the direct radiation at the maze entrance to negligible levels. The sum of the maze shield wall thicknesses between the source of radiation and the maze entrance should be at least equal to the direct shield wall thickness that would be required if there was no maze [3]. For facilities with direct-shielded doors, ducts and other penetrations above the door would also require shielding.

Penetrations in the shielding wall are required for the routing of various utilities, such as air conditioning, RF, cooling water, electrical conduits, etc. Various options can be used to reduce the amount of radiation scattered through a penetration. They include path extension, use of bends, and covering the penetration with a shadow shield. However, one must also ensure that there is no direct external radiation from the source pointing in the direction of the penetration or a portion of the penetration, resulting in an unshielded path.

10.6 Activation

Activation in proton therapy facilities is much less of a problem than at proton accelerators because of the low proton currents that are used in proton therapy. Activation is the transformation of the atomic nucleus by its interaction with ionizing radiation. An isotope of the same chemical species—or of a nucleus with completely different chemical properties—can be produced. These new atoms are unstable and undergo radioactive decay. In principle, all of the nuclides that have atomic mass and atomic number equal to or less than the sum of numbers of the target plus the projectile nuclei can be produced. Personnel working with activated materials during operation and maintenance are exposed to the emitted radiation. In order to activate a nucleus, the incident radiation should be able to overcome the Coulomb repulsion of the positively charged nucleus. This requirement favors neutrons because they are electrically neutral. Energetic charged particles—such as protons, deuterons, alpha particles, and heavy ions—can also cause various types of nuclear reactions, resulting in residual radioactive nuclides. Activation in a proton accelerator is thus caused by both the protons and secondary neutrons. The accelerator shielding, air, cooling water, and groundwater in the vicinity of the accelerator can also become activated. Activation in proton accelerators has been described in the literature [1,3,4,46–51].

10.6.1 Nuclear Reactions Leading to Activation

The mechanisms for the production of activation at high-energy accelerators have been summarized by Barbier [46]. Thus, activation is produced during various nuclear reactions, which are described below.

Since neutrons are uncharged, they will not be repelled by the charge of the atomic nucleus. Unlike charged particles, they are not affected by the Coulomb barrier of the nuclei. Thus, neutrons of any energy can interact with nuclei. Thermal and intermediate-energy neutrons undergo radiative capture by the nucleus, i.e., the capture results in the emission of a gamma ray. The resulting atom is of the same kind, but heavier by one neutron. The capture cross sections are generally proportional to $1/v$ (v

is the neutron velocity) or $1/\sqrt{E}$, where E is the energy. They fluctuate at the resonance energy region according to the characteristics of the nuclide [1]. Thermal neutron capture is the fundamental reaction for thermal neutron activation. Thermal neutrons can produce fission fragments when interacting with a small number of heavy elements. Thermal neutrons also undergo (n, p) and (n, α) reactions with light elements, for example, ^{4}N (n, p) ^{14}C and ^{6}Li (n, α) ^{3}H.

Neutrons with energy higher than the excited level of the target nucleus induce (n, n') reactions. The excited nucleus de-excites accompanied by gamma ray emission. When the neutron energy is sufficiently high enough to cause particle emission, many types of activation reactions occur. Neutrons or protons with energies from a few MeV to about 50 MeV are capable of knocking out one or more nucleons, or even a fragment of the target nucleus. Proton reactions of the type (p, n), (p, pn), (p, 2n) (p, p2n), (p, α), etc. have a threshold energy for the incident proton, above which these reactions take place. The probability of a particular reaction is measured in units of barns (10^{-24} cm^2).

A similar situation exists for the neutron-induced reactions (n, p), (n, pn), (n, 2p), (n, n2p), (n, α), etc. In each of these cases, it is assumed that the incident particle is absorbed by the target nucleus and distributes its energy in a random manner to various nucleons. This results in the formation of the compound nucleus in an excited state and the subsequent evaporation of neutrons, as previously described.

Radioactive nuclides are also produced by secondary neutrons, the energies of which extend up to the primary proton energy. As the energy of the neutrons increases, the number of reaction channels, as well as the number of radionuclides, increases. High-energy neutrons cause spallation reactions that emit any type of nuclide lighter than the target nucleus. The neutrons that are produced can escape from the nucleus or be captured and give up their energy to excite the entire nucleus. The spallation neutrons cause activation at threshold energies above 20 MeV. A much wider variety of new atoms are formed than previously discussed. The excited residual nucleus can de-excite by emitting evaporation nucleons or whole groups of nucleons in a manner similar to the evaporation of particles from compound nuclei. De-excitation of the nucleus can also occur by fission. The de-excited nucleus is radioactive.

10.6.2 Activation Calculations

Once radionuclides have been produced, they decay. The rate of decay of a radionuclide is described by its activity, i.e., by the number of atoms that disintegrate per unit time. The activity A of a radionuclide at any given time t is given by:

$$A = A_0 e^{-\lambda_D \tau} \tag{10.11}$$

where A_0 = initial activity and λ_D is the decay constant

$$\lambda_D = \frac{\ln(2)}{T_{1/2}} \tag{10.12}$$

where $T_{1/2}$ is the half-life.

Half-lives of radionuclides can range from fractions of seconds to thousands of years. Half-lives for radionuclides are shown in parentheses in this section. The activity is measured in becquerels (Bq).

The activity of the radionuclide builds up during irradiation; however, there is the competing process of radionuclide decay. Eventually, the radioactivity reaches saturation. The buildup of activation after an irradiation time, t_i, is given by:

$$A_t = A_S(1 - e^{-\lambda_D t_i}) \tag{10.13}$$

where A_S is the saturation activity, i.e., the activity after a long irradiation period, and $t_i >> T_{1/2}$.

If there is continuous ventilation during irradiation, λ_D in Equation 10.13 can be replaced by an effective removal constant, λ_E

$$\lambda_E = \lambda_D + \lambda_v \tag{10.14}$$

where λ_v is the ventilation constant (ventilation rate/volume).

Activation calculations for proton accelerators using analytical methods are described in the literature [3,4]. Such methods require the knowledge of the number of parent nuclei, the neutron fluence and energy distribution, and the cross sections. In such calculations, an average cross section and a flat neutron spectrum are typically assumed [3]. However, these calculations are not accurate because neutron fluence and cross sections are a function of energy. Further, many capture reactions have several resonance peaks in the epithermal energy range (eV to keV) which cannot be neglected. For example, Figure 10–13 shows the ^{40}Ar capture cross section (right ordinate). Also shown are the neutron spectra that are emitted from a thick Cu target bombarded by 230 MeV protons (blue line) and when they are reflected from a concrete shield wall. Both were estimated using the FLUKA Monte Carlo code. In the reflected spectrum, the thermal neutron component is visibly enhanced, while the epithermal neutrons are also significant. There are also many resonance peaks between 10 keV and 1 MeV. The total activation of air will be caused by the emitted and the reflected components. The activity (A) should, therefore, be calculated as a convolution of the total number of target nuclei (N) and the energy-dependent cross section, $\phi(E)$, and neutron fluence, $f(E)$, as given by the following equation:

$$A = N\int_E \phi(E)\sigma(E)dE \tag{10.15}$$

The best approach for activation calculations is the use of Monte Carlo radiation transport codes such as FLUKA or MCNP. Actual beam parameters and targets are included with the complete modeling of the room geometry. The energy of the emitted neutron is a function of the angle of emission with respect to the beam direction, and this, in turn, influences the reflected neutron energy. Neutrons emitted in the backward direction will have much less energy than those in the forward direction, and they are easily moderated with fewer scatters. Monte Carlo codes can take these

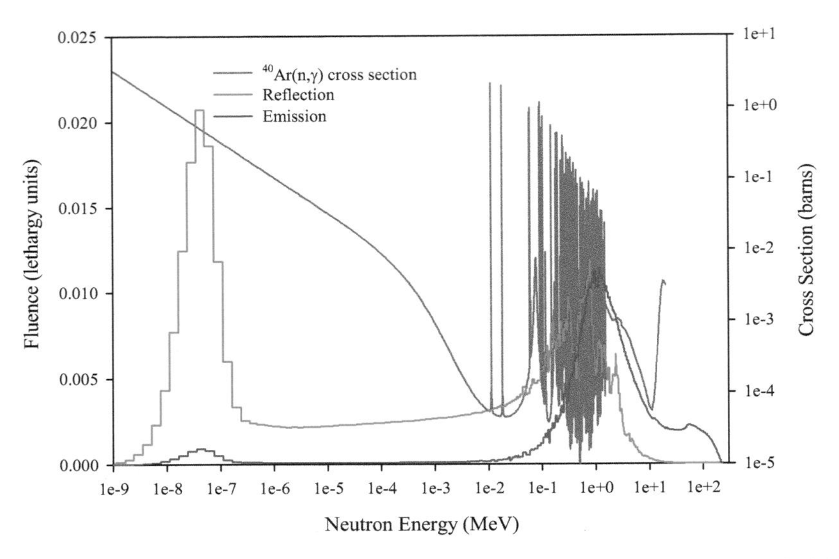

Figure 10–13 Incident and reflected neutron spectra from 230 MeV protons incident on thick copper target and ^{40}Ar capture cross section. (Sunil, this work.)

effects into account accurately. Biju et al. have performed analytical and Monte Carlo calculations for ^{41}Ar produced in a 15 m x 15 m x 15 m concrete room from a pencil beam of 230 MeV protons incident on a thick copper target [47]. Their results indicate that analytical methods under predict the ^{41}Ar activity compared to Monte Carlo calculations.

10.6.3 Activation of Accelerator Components and Other Solid Materials

Radioactive nuclides are mainly produced in the accelerator and beam-line components (including beam-shaping and delivery devices) and the energy selection system (ESS). The materials used in the construction of the accelerator and beam-line components are typically aluminum, steel, stainless steel (nickel, chromium and iron), iron, and copper. The radioactivity is usually distributed throughout the irradiated materials, even in large components such as magnet yokes, etc. [3]. Surface dose rate measurements are often adequate to indicate the presence of radioactivity, and they can be used to determine the locations or vicinity of beam losses. However, they are not sufficient to quantify the amount of radioactivity because of geometrical effects.

Steen et al. performed a systematic study of the activation of the 250 MeV SC cyclotron for proton therapy at Paul Scherrer Institute (PSI) [48]. Figure 10–14 shows the measured dose rates from activation at different locations in the cyclotron, 10 cm from the surface for a 230 MeV proton beam, 50 minutes after beam termination. The

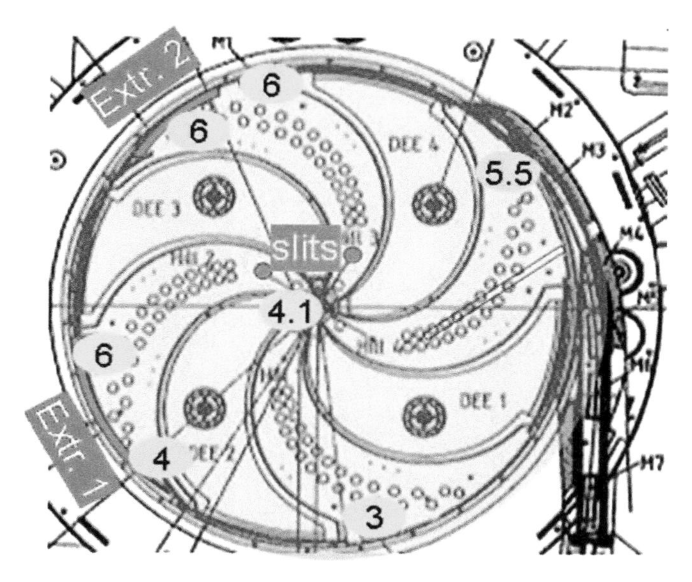

Figure 10–14 Measured dose rates in mSv/h from activation at different locations in the PSI 230 MeV cyclotron, 10 cm from the surface measured 50 min after beam termination, for an extracted charge of 805 μA-h. (Adapted from [49].)

extracted charge is 805 nA-h. The major radionuclides in copper are ^{64}Cu (12.8 h) and ^{61}Cu (3.32 h), and the dominant radionuclides produced in iron are ^{52}Mn (5.6 d), ^{54}Mn (303 d), and ^{56}Mn (2.576 h). Iron can be activated by thermal neutrons, fast neutrons, and protons.

The cross sections for neutron capture, reactions by proton and (n, p) reaction, are shown for ^{56}Fe in Figure 10–15. The data is obtained from ENDF database [49]. The reactions by protons and the (n, p) reaction have a certain threshold (typically about a few MeV). The thermal neutron capture process has the highest cross section and is inversely dependent on the energy. The ^{56}Fe (n, p) ^{56}Mn fast neutron reaction occurs only if the incident fast neutron energy is about 5 MeV. In general, cross sections for threshold reactions rapidly increase beyond the threshold energy and have a peak. They decrease beyond the peak energy since other reaction channels open with the increase of energy. The ^{59}Co (n, γ)^{60}Co reaction is important for the activation of stainless steel by thermal neutrons.

In cyclotron-based facilities, the highest activity is encountered in the ESS, i.e., primarily in the degrader and associated collimators. Since these systems need to be accessed only for maintenance or repairs, they can be shielded locally.

In the passive scattering system, collimators, ridge filters, and range modulators—which are located in the treatment nozzle—are significantly activated [1]. However, the patient collimator for each patient is irradiated for only a short time, and the

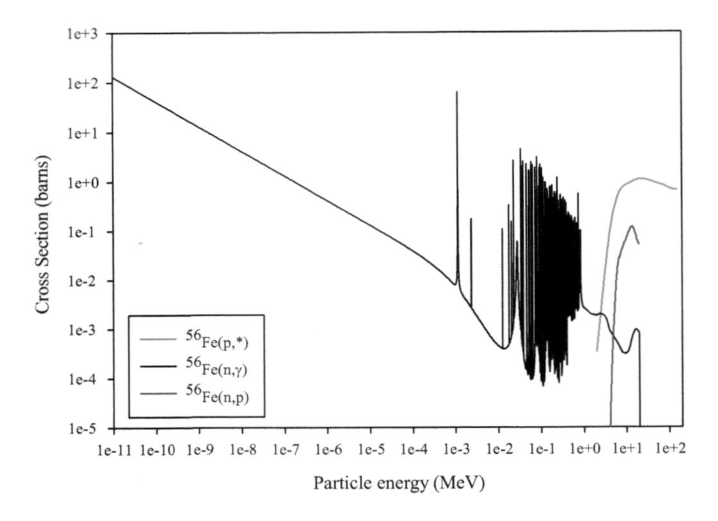

Figure 10–15 Cross sections for neutron capture, reactions by proton and (n,p) reaction for ^{56}Fe. (Sunil, this work.)

residual activities last only for a relatively short period after irradiation because of the short half-lives ($T_{1/2}$) of the induced radioactive isotopes. Collimators are typically stored for a period of time before they are shipped out of the facility. Therefore, radiation exposure to those handling the collimators is low.

Lee et al. studied the activation in a brass collimator and a polymethyl methacrylate (PMMA) phantom after irradiation with 230 MeV protons [49]. The highest dose rates for the brass aperture resulted from ^{62}Cu (9.76 min) and ^{63}Zn (38.33 min), both of which have short half lives. The brass aperture had a higher dose rate than the PMMA phantom. The major radionuclides in PMMA were ^{10}C (19.25 s), ^{11}C (20.39 min), ^{13}N (9.96 min), ^{14}O (1.18 min), and ^{15}O (2.04 min). They found that the dose rate decreased rapidly with time because the dominant radionuclides were all short-lived.

The amount of induced radioactivity and activity concentration in concrete used for shielding is smaller than that in the accelerator components that are directly irradiated by the primary accelerator beams. Thermal neutron capture in ^{23}Na leads to the production of ^{24}Na (15 h), which is a gamma emitter. Exposure by gamma rays from ^{24}Na in a concrete room can be significant during the first few hours after beam termination. Thermal neutron capture reactions with Co and Eu impurities in concrete result in ^{60}Co and ^{152}Eu. The amounts of these impurities are small, but the ^{59}Co (n, γ) and ^{151}Eu (n, γ) cross sections are large. The long-lived radioactive nuclides of concern in the decommissioning of concrete are ^{22}Na (2.62 y), ^{60}Co (5.26 y), and ^{152}Eu (13.52 y) [1]. Depending upon the concrete composition, other radionuclides may also be of concern.

Table 10–4 Radionuclides produced in water by spallation

Nuclide	Half-life	Decay Mode, γ-ray Energy, and Emission Probability
^{3}H	12.3 years	β^-
^{7}Be	53.3 days	EC, 0.478 MeV γ, 10.5%
^{11}C	20.4 min	β^+
^{13}N	9.97 min	β^+
^{14}O	1.18 min	β^+, 2.3 MeV γ, 99.4%
^{15}O	2.04 min	β^+

10.6.4 Water and Earth Activation

Cooling water for accelerator and beam-line components are activated by secondary neutrons produced by beam losses of the accelerated particles [1]. However, in the cyclotron, there may be specific locations—such as slits and the extraction deflector—where the protons directly strike and activate the cooling water. Thus, both protons and neutrons may produce activation. Therefore, all cooling water circuits should be confined to the primary loop. Care should be taken to ensure external dose rates from activation of water are low in the event that cooling water pipes pass through uncontrolled areas [4]. Rupture of these water pipes could lead to local contamination, as well as a release of the activated water to the drainage system, and eventually to the environment. The radionuclides produced in the accelerator components and the concrete shielding are fairly immobile. However, radionuclides produced in the earth below or in the groundwater are free to migrate. High-energy secondary neutrons produced by beam losses may penetrate the shielding in the cyclotron and treatment room to activate the groundwater.

High-energy neutrons produce ^{14}O, ^{15}O, ^{13}N, ^{11}C, ^{7}Be, and ^{3}H through spallation reactions of oxygen [1]. The significant radionuclides produced in water by spallation with half lives greater than 1 min are shown in Table 10–4. The dominant short-lived radionuclide after 1–5 h of irradiation is ^{11}C [4]. The only radionuclide which is a gamma emitter is ^{7}Be. The radioactivity of ^{14}O, ^{15}O, ^{13}N, and ^{11}C, all of which have short half-lives, reaches saturation in a short irradiation time. The annihilation photons produced by these positron-emitting nuclides and the gamma rays from ^{7}Be increase the dose rate around cooling water pipes and heat exchangers.

Only a few of the radionuclides produced in earth have been observed in groundwater. The radionuclides of concern in groundwater are ^{3}H and ^{22}Na (2.6 y). Though ^{3}H is produced directly in water, ^{22}Na is produced in earth through high-energy neutron interactions. The radionuclides that are produced in groundwater might enter the general groundwater system and potentially contaminate public water supplies. Further, the possibility that the radionuclides produced in the earth may leach into the groundwater should also be considered.

Table 10–5 Radionuclides from neutron activation of air

Radionuclide	Half-life	Emission	Parent Element	% of Parent Element in Air
^{3}H	12.2 y	β^-	^{12}C ^{14}N ^{16}O	0.012 75.5 23
^{7}Be	53 d	γ, EC	^{12}C ^{14}N ^{16}O	0.012 75.5 23
^{11}C	20.5 min	β^+	^{12}C ^{14}N ^{16}O	0.012 75.5 23
^{13}N	10 min	β^+	^{14}N ^{16}O	75.5 23
^{15}O	2.1 min	β^+	^{16}O	23
^{41}Ar	1.8 h	β^-, γ	^{40}Ar	0.013

10.6.5 Air Activation

Interaction of primary protons and secondary radiation with nitrogen, oxygen, argon, and carbon in the air contained in the accelerator rooms and the treatment rooms results in the production of radioactive gases [6,4,3]. Radionuclides with short half-lives (<1 min) quickly decay away, while radionuclides with long half-lives have low production rates. Entry into the rooms can be delayed to allow for decay of the short-lived radionuclides. If the air is confined to the room, the concentrations of the radioactive gases can be high. However, air circulation is used in order to reduce the concentration of these radionuclides to acceptable levels. The radionuclides of concern are ^{3}H, ^{7}Be, ^{11}C, ^{13}N, ^{15}O, and ^{41}Ar [3]. Table 10–5 summarizes the relevant parameters for these radionuclides. Most of these radionuclides are produced by high-energy neutrons [4] except for ^{41}Ar, which is produced mainly by thermal neutrons. Activation by high-energy neutrons is largely due to spallation and occurs at production thresholds above 20 MeV. Measurements and calculations performed at various proton accelerator facilities indicate that the only ^{15}O, ^{13}N, ^{11}C, and ^{41}Ar need to be considered based on activities produced and allowable concentrations.

10.7 Secondary Dose to Patient

Secondary dose to patient is only addressed briefly. The literature is replete with data on secondary dose. The reader is referred to the literature [1,53–55] for an overview of secondary dose to patient. Any additional dose deposited outside the treatment volume can contribute to the risk of secondary cancer induction. The physics of secondary radiation production was previously discussed. The secondary dose (dose outside the treatment field) is dominated by secondary neutrons produced by the interaction of protons with beam-shaping and modifying devices in the treatment head or nozzle and the patient. The beam-shaping devices are typically made of materials such as lead, brass, steel, carbon, or nickel. In some cases, range shifting is used to change the energy of the beam. Devices that are close to the patient—such as the brass aperture and range compensator—scatter the primary beam and produce secondary neutrons [56]. The neutrons produced in the forward direction (0–10°) are more penetrating than the neutrons produced at larger angles. The maximum energy of the neutrons is equal to the incident proton energy. The neutron fluence is also higher in the forward direction.

The main contribution to the dose in the patient outside the primary proton field is from protons that have scattered at large angles and from neutrons produced in the nozzle and the patient. Secondary photons should also be considered.

Cyclotrons typically use energy degraders of variable thicknesses and collimators, slits, etc. to change the beam energy. In multi-room facilities, the energy selection system is far away from the patient treatment room, while in compact proton therapy systems, the energy selection system can be fairly close to the patient. Secondary neutrons produced by the interaction of protons with these beam-modifying systems also contribute to the patient secondary dose.

The secondary neutron dose also decreases with increasing distance from the target, as well as increasing lateral distance from the beam axis. Thus, the closer the patient is to the beam-shaping device, the higher the secondary neutron dose. For example, the patient collimator is usually just upstream of the patient.

The secondary neutrons produced can scatter from the treatment room concrete walls and any additional material that they encounter, such as the gantry counterweight. The scattered neutrons in the room will reach an equilibrium condition in which the thermal and scattered fast-neutron fluences are constant in the concrete rooms [8] Thus, the neutron fluence inside the room will be comprised of:

1. a direct neutron component consisting of relativistic, fast, and thermal neutrons from the isocenter that falls off with inverse square distance;
2. scattered fast neutrons from the walls with a spectrum which remains constant in the room; and
3. a thermal neutron component which is constant throughout the room.

Thus, the neutron fluence measured as a function of the distance from the isocenter will drop off more slowly than predicted by the inverse square law [8]. However, the dose will drop off more rapidly than the fluence because the thermal and scattered contributions are lower in energy than the direct fast neutron component, and they contribute less to the dose. Inside the room, the dose will still drop off more slowly than predicted by the inverse square law. Howell and Burgett measured the secondary neutron spectrum from a 250 MeV passively scattered proton beam in air at a distance of 100 cm laterally from the isocenter. Measurements were performed with a medium snout (18 x 18 cm two-aperture opening) and closed brass aperture using an extended-range Bonner sphere measurement system [57]. The ambient dose equivalent H*(10)

Table 10–6 Effective dose and lifetime risk of secondary cancer for three treatment modalities [58]

Treatment Mode	Effective Dose ($mSv\ Gy^{-1}$)	Lifetime Risk
PBS	1.900	1.037
ProteusOne	1.910	1.052
RS (Range Shifting)	4.901	3.262

was calculated using measured fluence and fluence-to-ambient dose equivalent conversion coefficients. The neutron fluence spectrum was characterized by a high-energy neutron peak, an evaporation peak, a thermal peak, and an intermediate energy continuum between the thermal and evaporation peaks. The ambient dose was dominated by the neutrons in the evaporation peak because of both their large numbers and the high fluence-to-dose equivalent conversion coefficients in that energy interval. The ambient dose equivalent at 100 cm laterally from isocenter was 1.6 mSv per proton Gy (at isocenter). Neutrons with energies ≥20MeV contributed to ~35% of the total dose equivalent. Secondary photons resulting from neutron capture should also be considered.

The highest secondary dose to the patient is received from passive beam scattering when compared to US and PBS because passive beam scanning typically requires beam shaping and beam-modifying devices, unlike in active beam scanning. The patient aperture, typically made of high-Z material, is a major source of neutrons. As the aperture opening decreases, neutron fluence increases; hence the dose increases because the proton beam intercepts more of the aperture material. Therefore, beam scanning reduces the secondary neutron dose from the treatment head significantly, especially for small apertures.

Intensity-modulated proton therapy (IMPT) requires active beam delivery methods such as PBS. The neutron doses from PBS are significantly lower than from passive scattering because PBS requires less material in the nozzle [55]. Therefore, the majority of the neutrons are produced in the patient. In some cases, patient-specific apertures and compensators may be required, thus increasing the secondary dose. However, the dose in such cases will still be lower than passive scattering because the patient-specific aperture is only used to stop protons that are scattered at large angles.

In compact proton therapy systems, the cyclotron or synchrotron is much closer to the patient. A significant amount of secondary radiation is produced in the energy selection system, primarily in the degrader. The radiation also contributes to secondary patient dose. The IBA ProteusOne uses an energy selection system, but the secondary radiation in the forward direction (the direction toward patient) is relatively well shielded [58]. For a cyclotron-based system which does not have a degrader, the spread-out Bragg peak can be obtained by using a range shifter made of low-Z material, such as Lexan. Stichelbaut has analyzed the secondary neutron dose to patients treated in two different proton therapy systems, comparing it to pure PBS. Table 10–6 summarizes the effective dose and lifetime risk of secondary cancer for three treatment modalities. The results indicate that the ProteusOne system does not result in any increase in lifetime risk. However, a compact proton therapy system using range shifting (RS) results in an increase in effective dose by a factor greater than 2.6, and this increases the lifetime risk by a factor of 3.14 when compared to pure PBS.

10.8 Conclusions

Secondary radiation production and the physical processes that govern radiation transport through accelerator shielding have been described. Shielding design considerations, activation, and secondary dose to patient have been discussed.

References

1. PTCOG. Ipe NE. Task Group Leader. Shielding design and radiation safety of charged particle therapy facilities. PTCOG Report 1, Particle Therapy Cooperative Group, 2010. http://www.ptcog.ch/archive/Software_and_Docs/Shielding_radiation_protection.pdf. Accessed on 20 November 2014.
2. Ipe NE. Basic aspects of shielding. In: Paganetti H, editor. Proton Therapy Physics. New York: CRC Press, 2011, pp. 525–554.
3. NCRP. Radiation protection for particle accelerator facilities. NCRP Report 144. Maryland: National Council on Radiation Protection and Measurements, 2003.
4. IAEA. Radiological safety aspects of the operation of proton accelerators. IAEA Technical Reports Series No. 283. Vienna: International Atomic Energy Agency, 1988.
5. Krasa A. Spallation reaction physics. Lecture notes for students of the faculty of nuclear sciences and physical engineering, Czech Technical University in Prague. May 2010. http://ojs.ujf.cas.cz/~krasa/ZNTT/SpallationReactions-text.pdf. Accessed 29 September 2014.
6. ICRU. Basic aspects of high energy particle interactions and radiation dosimetry. ICRU Report 28. Maryland: International Commission on Radiation Measurements and Units, 1978.
7. Turner JE. Atoms, radiation, and radiation protection. New York: Pergamon Press, 1986.
8. NCRP. Neutron contamination from medical electron accelerators. NCRP Report 79. Maryland: National Council on Radiation Protection and Measurements Report, 1984.
9. Moritz LE. Radiation protection at low energy proton accelerators. *Radiat Prot Dosim* 2001;96(4):297–309.
10. NCRP. Protection against neutron radiation. NCRP Report 38. Maryland: National Council on Radiation Protection and Measurements, 1971.
11. Agosteo S, Magistris M, Mereghetti A, Silari M, Zajacova Z. Shielding data for 100 to 250 MeV proton accelerators: Double differential neutron distributions and attenuation in concrete. *Nucl Instrum Methods Phys Res B* 2007;265:581–589.
12. ICRP. Recommendations of the International Commission on Radiological Protection. ICRP Publication 60. Annals of ICRP 21(1–3) UK: Pergamon Press, 1991.
13. ICRP. Recommendations of the International Council on Radiological Protection, ICRP Publication 103. Annals of the ICRP. UK: Elsevier Science, 2007
14. ICRU. Conversion coefficients for use in radiological protection against external radiation. ICRU Report 57 Maryland: International Commission on Radiation Units and Measurements, 1998.
15. Ferrari A, Sala PR, Fasso A, Ranft J. FLUKA: a multi-particle transport code. CERN yellow report CERN 2005-10; INFN/TC 05/11, SLAC-R-773. CERN, Geneva, Switzerland, 2005.
16. Battistoni G, Muraro S, Sala PR, Cerutti F, Ferrari A, Roesler S, et al. The FLUKA code: Description and benchmarking. In: Albrow M, Raja R, Ed. Proceedings of the Hadronic Shower Simulation Workshop 2006; Sep 6–8; Fermilab, Battavia, Illinois. AIP Conference Proceedings 2007; pp. 31–49.

17. Pelliccioni M. Overview of fluence-to-effective dose and fluence-to-ambient dose equivalent conversion coefficients for high energy radiation calculated using the FLUKA Code. *Radiat Prot Dosim* 2000; 88(4):277–97.
18. Nakamura T. Summarized experimental results of neutron shielding and attenuation length. In: SATIF7 Proc. Shielding Aspects of Accelerators, Targets and Irradiation Facilities, 17–18 May 2004, Portugal, Nuclear Energy Agency. Paris: Nuclear Energy Agency, Organization for Economic Co-operation and Development, 2004. pp. 129–146.
19. Chen CC., Sheu RJ, Jiang SH. Calculations of neutron shielding data for 10–100 MeV proton accelerator. *Radiat Prot Dosim* 2005;116: 245–251.
20. Thomas RH. Practical aspects of shielding high-energy particle accelerators. Washington DC: U.S. Department of Energy. Report UCRL-JC-115068, 1993.
21. Kato T, Nakamura T. Analytical method for calculating neutron bulk shielding in a medium-energy facility. *NIM Phys Res B* 2001;174:482–90.
22. Goorley JT. MCNP6.1.1-Beta Release Notes. LA-UR-14-24680, 2014.
23. Agosteo S, Arduini G, Bodei G, Monti S, Padoani F, Silari M, et al. Shielding calculations for a 250 MeV hospital-based proton accelerator. *NIM Phys Res A* 1996; 374:254–68.
24. Dittrich W, Hansmann T. Radiation measurements at the RPTC in Munich for verification of shielding measurements around the cyclotron area. In: SATIF8 Proc of Shielding Aspects of Accelerators, Targets and Irradiation Facilities; 2006 May 22–24; Gyongbuk, Republic of Korea; Paris: Nuclear Energy Agency, Organization for Economic Co-operation and Development, 2007. pp. 345–349.
25. Hofmann W, Dittrich W. Use of isodose rate pictures for the shielding design of a proton therapy centre. In: SATIF7 Proc of SATIF7 Shielding Aspects of Accelerators, Targets and Irradiation Facilities; 2004 May 17–18; Portugal, Paris: Nuclear Energy Agency, Organization for Economic Co-operation and Development, 2005 pp. 181–187.
26. Kim J. Proton therapy facility project in national cancer center, Republic of Korea. *Journal of the Republic of Korean Physical Society* 2003; 43:50–54.
27. Porta A, Agosteo S, Campi F. Monte Carlo simulations for the design of the treatment rooms and synchrotron access mazes in the CNAO hadron therapy facility. *Radiat Prot Dosim* 2005;113(3):266–274.
28. Agosteo S, Fasso A, Ferrari A, Sala, P R, Silari M, Tabarelli de Fatis P. Double differential distributions and attenuation in concrete for neutrons produced by 100–400 MeV protons on iron and tissue targets. *NIM Phys Res B* 1996; 114:70–80.
29. Ipe NE. Particle accelerators in particle therapy: The new wave. In: Proc of the 2008 Mid-Year Meeting of the Health Phys Society on Radiation Generating Devices; Oakland, CA. VA:Health Phys Society, 2008.
30. Ipe NE, Fasso A. Preliminary computational models for shielding design of particle therapy facilities. In: Proceedings of SATIF8 Shielding Aspects of Accelerators, Targets and Irradiation Facilities; 2006 May 22–24; Gyongbuk, Republic of Korea. Paris: Nuclear Energy Agency, Organization for Economic Co-operation and Development; 2007, pp. 351–359.
31. Tesch K. A simple estimation of the lateral shielding for proton accelerators in the energy range 50 to 1000 MeV. *Radiat Prot Dosim* 1985;11(3):165–72.
32. Kato T, Kurosawa K, Nakamura T. Systematic analysis of neutron yields from thick targets bombarded by heavy ions and protons with moving source mode. *NIM Phys Res A* 2002;480:571–90.
33. Nakashima H, Takada H, Meigo S, Maekawa F, Fukahori T, Chiba S, et al. Accelerator shielding benchmark experiment analyses. In: Proceedings of SATIF-2 Shielding Aspects of Accelerators, Targets and Irradiation Facilities; 1995 Oct 12–13; Geneva, Switzerland. Paris: Nuclear Energy Agency, Organization for Economic Co-operation and Development, 1996 pp. 115–145.
34. Tayama R., Handa, H., Hayashi, H., Nakano, H., Sasmoto, N., Nakashima, H., Masukawa, F. Benchmark calculations of neutron yields and dose equivalent from thick iron target for 52–256 MeV protons. *Nucl Eng and Design* 2002;213:119–31.
35. Sheu RJ, Chen YF, Lin UT, Jiang SH. Deep penetration calculations in concrete and iron for shielding of proton therapy accelerators. *NIM Phys Res B* 2012;280:10–17.
36. Oh J, Lee HS, Park S, Kim M, Sukmo H, Ko S, Cho WK. Comparison of the FLUKA, MCNPX, and PHITS codes in yield calculation of secondary particles produced by intermediate energy proton beam. *Prog Nucl Sci Tech* 2011;1:85–88.
37. Smith AR. Vision 20/20: Proton therapy. *Med Phys* 2009;36(2):556–68.

38. IBA. Radiation Sources in the Proteus 235 System Internal Document 35597 Rev. A. 2012.
39. Jongen Y. Private communication, 11/2014.
40. USNRC. United States Nuclear Regulatory Commission. Standards for Protection Against Radiation 10CFR20, Code of Federal Regulations. http:// www.nrc.gov/reading-rm/doc-collections/cfr/part020/. Accessed 17 September 2014.
41. NCRP. Structural shielding design and evaluation for megavoltage X- and gamma-ray radiotherapy facilities. NCRP Report 151. Maryland: National Council on Radiation Protection and Measurements, 2005.
42. Chilton AB, Schultis JK, Faw RE. Principles of radiation shielding. New Jersey: Prentice Hall, 1984.
43. Brandl A, Hranitzky C, Rollet S. Shielding variation effects for 250 MeV protons on tissue targets. *Radiat Prot Dosim* 2005;115(1–4):195–199.
44. Ipe NE. Transmission of shielding materials for particle therapy facilities. *Nucl Tech* 2009;168(2):559–63.
45. Teichmann S. Shielding parameters of concrete and polyethylene for the PSI proton accelerator facilities In: Proc of SATIF8 Shielding Aspects of Accelerators, Targets and Irradiation Facilities; 2006 May 22–24, Gyongbuk, Republic of Korea; Paris: Nuclear Energy Agency, Organization for Economic Co-operation and Development: 2007, pp. 45–54.
46. Barbier M. Induced radioactivity. New York: Elsevier Science, 1969.
47. Biju K, Sunil C, Sarkar PK. Estimation of 41Ar production in 0.1–1.0-GeV proton accelerator vaults using FLUKA Monte Carlo code. *Radiat Prot Dosim* 2013;157:437–441.
48. Thomadsen B, Nath R, Bateman FB, Farr J, Glisson C., Islam MK, Moore ME, Xu G, Yudelev K. Potential hazard due to induced radioactivity secondary to radiotherapy: The report of task group 136 of the American Association of Physicists in Medicine. *Health Phys* 2014;107(5):442–460.
49. Steen G, Amrein B, Frey N, Frey P, Kiselev D, Kostezer M, Lüscher R, Mohr D, Morath O, Schmidt A, Wohlmuther M, Schippers JM. Activation of a 250 SC-cyclotron for proton therapy. In Proc of CYCLOTRONS 2010. The 19th international conference on cyclotrons and their applications. Lanzhou, China. September 6–10, 2010, p 72.
50. Lee SH, Cho S, Kwak J, Kim DH, Kim S, Chan HK, Dongho S, Park SY, Lee SB. Study of the induced radioactivity by therapeutic proton beam in National Cancer Center, Korea. Poster presented at PTCOG 50, 2011.
51. Hsu YC, Sheu RJ. Comparing the predicted neutron yields and radioactivities of two accelerators in Taiwan: 235-MeV proton cyclotron and 3-GeV electron synchrotron. ANS RPSD 2014 18th Topical Meeting of the Radiation Protection & Shielding Division of ANS.
52. http://www.nndc.bnl.gov/exfor/endf00.jsp. Last accessed 22 September 2014.
53. Jarlskog CZ, Lee C, Bolch WE, Xu XG, Paganetti H. Assessment of organ-specific neutron equivalent doses in proton therapy using computational whole-body age- dependent voxel phantoms. *Phys Med Biol* 2008;53:693–717.
54. Paganetti H, Late effects from scattered and secondary radiation. In Proton Therapy Physics, Paganetti H ed. New York: CRC Press, 2011.
55. Dowdell SJ. Pencil beam scanning proton therapy: the significance of secondary particles. Doctor of Philosophy Thesis, University of Wollongong, Center for Radiation Physics, 2011.
56. Classie B. Assessment of out-of-field absorbed dose and equivalent dose in proton fields. *Med Phys* 2010;37(1):311–32.
57. Howell RM, Burgett EA. Secondary neutron spectrum from 250-MeV passively scattered proton therapy: Measurement with an extended-range Bonner sphere system. *Med Phys* 2014;41:92–104.
58. Stichelbaut F, Closset M, Jongen Y. Secondary neutron doses in a compact proton therapy system. *Radiat Prot Dosim* 2014;161(1–4):368–372.

Chapter 11

Detector Systems

Stanislav Vatnitsky, Ph.D.[1] **and Hugo Palmans, Ph.D.**[1,2]

[1]MedAustron GmbH,
Wiener Neustadt, Austria
[2]National Physical Laboratory
Teddington, United Kingdom

11.1 Introduction

The instrumentation that is used in proton therapy dosimetry is similar to that used in dosimetry of high-energy photon and electron beams, with only minor exceptions. Therefore, medical physicists coming to the proton facility from conventional clinics are familiar with the majority of detector systems needed for dosimetry of proton beams. However, the understanding of detector performance and the conversion of the signal into a dose in proton beams requires specific knowledge of proton physics. For example, the ideal detector for depth–dose measurements is defined not only by dose and dose rate linearity, but mostly by its linear energy transfer (LET) independence. As ionization chambers are the reference instruments for proton depth–dose measurements, all other detectors should be carefully checked against data measured with ionization chambers. This chapter is written to guide medical physicists through the practical implementation of different detectors in performing dosimetry tasks at a proton therapy facility. In each case practical recommendations are given to justify the selection of the detector system to perform the specific task. There are very few detector systems that are developed explicitly for proton dosimetry, and in this case, more details are given to provide the user with information needed. Typical dosimetry tasks can be associated with acceptance testing and commissioning of beam delivery systems, reference calibration of clinical beams, acceptance testing and commissioning of treatment planning system (TPS), periodic quality assurance (QA) checks, and verification of dose delivery. These tasks are completed by measuring absorbed dose, while at other times these tasks are simply related to measuring relative doses. The determination of absorbed dose requires either a dosimeter that can independently measure quantities related to energy deposition per unit mass, a so-called *absolute dosimeter*, or a dosimeter that has been cross-calibrated against an absolute dosimeter. Absolute dosimeters that can be used in proton beams include calorimeters, Faraday cups, Fricke dosimeters, carbon activation dosimeters, and ionization chambers. This chapter describes the use of so-called *relative dosimeters* that shall be calibrated against one of the absolute dosimeters in conditions that are as close as possible to those in the clinical applications.

Generally, a distinction should be made between what are often called active and passive dosimeters (Figure 11–1). The first group of instruments (active dosimeters) provides a direct display of the momentary dose rate or the accumulated dose. Examples of such devices are ionization chambers or solid state detectors (diodes or diamond detectors) in connection with a suitable electrometer. The second group of instruments (passive dosimeters) uses a probe that accumulates the dose during the irradiation. The value of the dose is obtained after the irradiation by means of a suit-

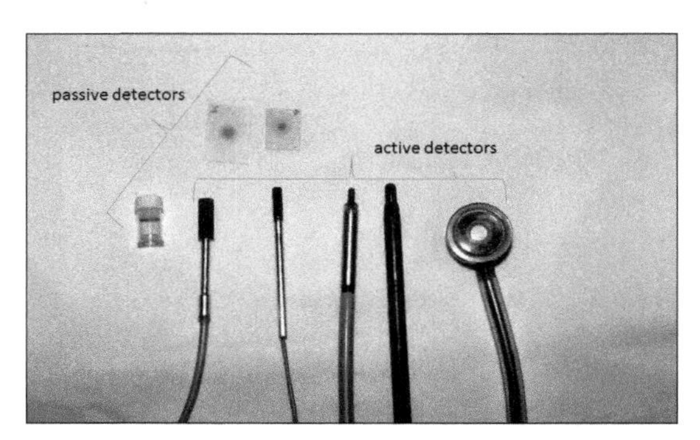

Figure 11–1 Passive detector probes (TLD, films) and active detector probes (diamond, diode, ionization chambers).

able read-out device. The read-out process may either be destructive (e.g., in thermoluminescence dosimetry, or TLD), or non-destructive (e.g., alanine, radiochromic film). For the latter group of instruments, the dosimeter package consists of the combination of probe detector, read-out device, software, and additional devices, such as an annealing oven for TLDs.

While proton dosimetry protocols (see Chapter 12, Dosimetry and Beam Calibration) describe how adsorbed dose is to be measured under reference conditions, i.e., in water phantoms, most dosimetric tasks in clinical practices refer to non-reference conditions. These tasks may use tissue equivalent rather than water phantoms for measurements during checks of monitor calibration, acceptance, and commissioning of beam delivery systems, TPS, and patient-specific QA checks. The following sections describe the use of different detectors and corresponding phantoms, with an emphasis on choosing specific dosimetry systems and the practical considerations for their use.

11.2 Active and Passive Detector Systems

Relative dose measurements use active or passive detectors systems that require no detector calibration, other than the verification of the linearity of response, energy independence, LET dependence, and spatial resolution within the assumed dynamic range of the measurement conditions. The time structure of the beam must also be considered. The measurements of the dose distribution delivered with scanning beams differ from those with scattered beams only from the point of view of the longer integration time needed for the measurements. For this reason, depth–dose or profile measurements based on dose rates using moving detectors cannot be used in scanned beams. Systems with multiple detectors in 1D, 2D, or 3D arrays can save time in measurements of dose distributions, particularly in scanned beams. Dosimetric data required for the commissioning of beam delivery systems and TPS are measured in water. In this case, the probe of the active dosimeter is immersed in water and required depth–dose data or cross-sectional profiles are obtained with computer-controlled water tanks by scanning the probe. If motion of the probe is not required, then it is often faster, easier, and more reproducible to place probes of passive or active dosimeters into a tissue or water-equivalent plastic slab phantom. Tissue-equivalent materials are also employed for manufacturing sectional anthropomorphic phantoms used for QA measurements and end-to-end tests. The sections of these phantoms can accommodate films or are equipped with cavities that are customized for placement of different probes.

11.2.1 Ionization Chambers

Applications in relative measurements: active probe

- standard probe for verification/calibration of other detectors in the clinic,
- probe for depth–dose and profile measurements, output measurements, and verification of dose delivery,
- probe element of computer-controlled water tank,

- probe element of multi-detector array for 2D QA measurements and verification of dose delivery in scattered and scanned beams.

Ionization chambers have been adopted as standard dosimetry instrumentation for relative measurements. Selection of an ionization chamber for a specific measurement is a compromise between the sensitive volume and spatial resolution. Parallel-plate chambers are recommended for depth–dose measurements, while smaller-volume cylindrical chambers are recommended for lateral profile measurements. Depth–dose measurements for pencil beams and beam calibration in terms of dose–area product should be performed with large area parallel-plate ionization chambers. The charge measured at a given depth and normalized to the applied monitor units is thus essentially proportional to the lateral two-dimensional dose distribution integrated over the whole plane at that depth. The calibration coefficient to convert the ionization chamber response to dose can be obtained by performing cross-calibration of the large area parallel-plate ionization chamber against a reference thimble chamber using the reference conditions as indicated in Chapter 12.

Some practical considerations should be accounted with when working with different ionization chambers:

- Parallel-plate ionization chambers are recommended for depth–dose measurements. The diameter of the parallel-plate ionization chamber used for depth–dose measurements in scanned proton beam should large enough to intercept all primary protons and secondary products in the beam. Alternatively, a method to apply necessary corrections should be used, which is described in [1].

- Cylindrical ionization chambers with volumes of about 0.1 cm^3 that give a relative dose precision of a few percent can be used for lateral profile measurements. Smaller chambers exhibit a low signal-to-noise ratio over the duration of the scan and require special treatment for measurements in pulsed beams [2,3]. Small volume ionization chambers are also recommended for patient-specific or portal dose verification.

- The scanning of ionization chambers with existing commercial water tanks provide accuracy for the calibration of the beam energy or range to within 100 μm. A more accurate solution is to use a water-filled bellows with an ionization chamber on each end to measure the range carefully. The commercial device (PTW Peak Finder, see Figure 11–2) uses a water column that is especially designed for high-precision Bragg peak position detection of proton and heavier ion beams. This closed water column allows scans up to 35 cm depth with increments of 10 μm. Because of its sealed construction, it can be used in any spatial orientation with special adapters.

11.2.2 Diodes

Applications in relative measurements: active probe]

- probe for depth–dose and profile measurements,

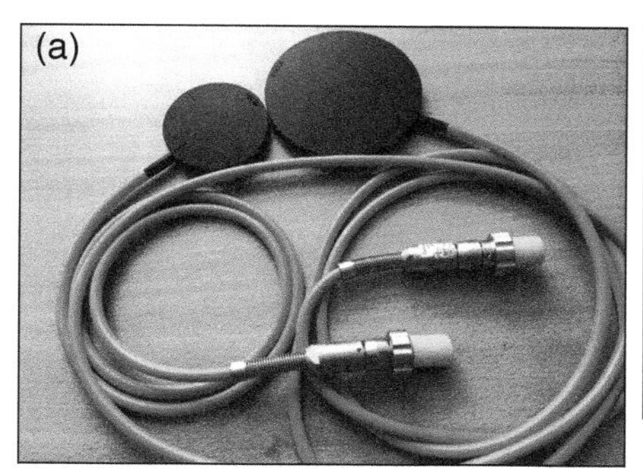

Figure 11–2 (a) PTW Plane—parallel chambers for depth–dose measurements. (b) PTW Peak Finder device designed for high-precision Bragg peak position detection of scanned proton beams. (Courtesy of PTW.)

- probe element of multi-detector array for 2D QA measurements and verification of dose delivery in scattered and scanned beams.

Silicon diodes have been used in proton dosimetry mainly because they have high sensitivity within a small volume, good spatial resolution, and real-time response. However, a discrepancy between responses of ionization chambers and diodes near the Bragg peak region was consistently observed with both monoenergetic and range-modulated proton beams [4,5]. For this reason ICRU 78 [6] advised caution when detectors based on n-type silicon or on low-resistivity p-type silicon are used for depth–dose measurements. Silicon diodes exhibit an increase of response with temperature because of an increase of the energy of the charge carriers that escape from recombination [7].

Commercially available diodes are generally made of a small quantity of silicon doped with phosphorous (n-type diodes) or boron (p-type diodes). They can yield accurate measurements, provided that their response is regularly monitored and corrected by factors taking into account both the aging of the diode and the irradiation damage [8]. In most cases, Si diodes are operated with the zero bias across the diode junction. In contrary to n-type diodes [9], the depth ionization curve determined by p-type detectors made of highly doped p-type material (Hi-p Si diodes) does not change as much with accumulated dose after pre-irradiation [10]. Using Hi-p Si diodes, the depth–dose distributions in proton beams were found to correspond closely to the distributions obtained with parallel-plate ionization chambers [10,11,12]. However, the sensitivity of this type of diode decreases rapidly with accumulated dose. Nevertheless, as the relative change in sensitivity per unit dose decreases with the total accumulated dose, this effect can be reduced to manageable levels by pre-irradiation of the detector to a dose of 20 Gy. A comparison of depth–dose curves in the Bragg peak region, determined in a 126 MeV non-modulated beam with a Hi-p-type silicon diode, with two PTW ionization chambers, and with the diamond detector [12] is

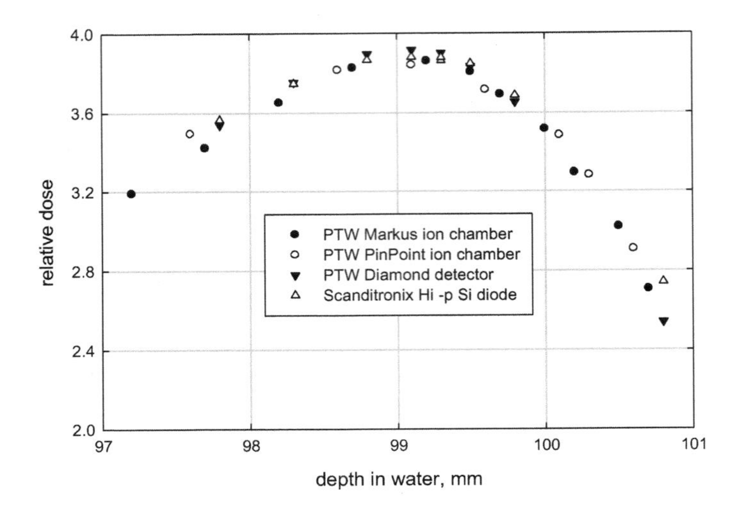

Figure 11–3 A part of depth–dose curves near Bragg peak for a 126 MeV non-modulated beam produced with a 30 mm collimator measured with different detectors. All curves are normalized to the maximum at the depth of 99.2 mm. Data taken from [12].

shown in Figure 11–3. It can be seen that the relative responses from all detectors are identical with water depth near the Bragg peak and that the differences are within the positioning accuracy of each detector in a water phantom. Hi-p-type silicon detectors are probably the most reliable diode detectors, and they are preferably the type used for depth–dose measurements in proton beams. This type of detector gives a signal which is proportional to the ionization density in the silicon crystal in all parts of the Bragg curve and for all levels of accumulated dose to the detector [10,12]. The agreement of depth–dose measurements between the diode and ionization chamber, combined with the small size of the diode's sensitive volume, make the Hi-p Si diodes especially attractive for depth–dose studies in radiosurgery beams. When the detector is used for lateral profile measurements, it is moving perpendicularly to the beam axis, thus the proton energy and the stopping powers are fairly constant, and only the dose rate and the mean ionization density vary by a large factor. In this case, no deviation is expected between lateral profiles measured with a diode and other small-volume detectors [6], so the diodes can safely be used for profile measurements.

When performing measurements in regions of high dose gradient, such as in the lateral penumbra or the distal edge, it is important to arrange the small dimension of the active volume, typically 50 mm, with the direction of greatest gradient. If the construction details of the diode are unknown, then x-ray pictures of the device from multiple directions should be taken to determine the most appropriate arrangement.

Some practical considerations should be accounted for when working with diode probes:

Only use diodes whose characteristics in proton beams have been published.

- Any practical use of diodes should be carefully checked by comparing results with other detectors—for instance, a parallel-plate ionization chamber for depth–dose measurements and a small-volume ionization chamber for profile measurements.
- When using diodes with the stem perpendicular to the beam axis for profile measurements, potential asymmetry of the profile due to asymmetry of the detector's construction should be checked.
- Silicon diodes exhibit an increase of response with temperature. In practice, it is recommended that the increase in sensitivity with temperature is determined for each new diode and checked periodically.

11.2.3 Diamond Detectors

Application in relative dosimetry: active probe

- Probe for profile and depth–dose measurements in water and plastic phantoms.
- Probe for absolute dose determination and output factor measurements.

The potential advantages of natural diamond detectors for relative dosimetry in proton beams were reported in several publications [13,14,15,16,17]. These detectors were supplied by PTW, Freiburg within two decades. PTW recently limited the production of natural diamond detectors and switched to the detectors with the synthetic single crystal microdiamond T 60019 [18,19,20,21,22]. Since natural diamond detectors are still in use at different institutions, this section describes the features of both types of detectors with natural and synthetic diamonds (Table 11–1).

11.2.3.1 Natural Diamond Detector

The natural diamond probes are radiation resistant, practically tissue-equivalent, have high sensitivity and stability, are relatively small, and are sealed in a waterproof polystyrene housing. Before each dose measurement, a pre-irradiation of the detector is

Table 11–1 Characteristics of typical PTW diamond detector T60003 and microdiamond detector T60019

	T60003	T60019
Thickness of sensitive volume, mm	0.30	1 μm
Sensitive area, mm^2	4.0	3.8
Outer probe diameter, mm	7.3	6.9
Sensitivity to ^{60}Co radiation (nC Gy^{-2})	60.0	1.0
Pre-irradiation dose, Gy Bias	5.0–15.0 +100 V	5.0–12 0

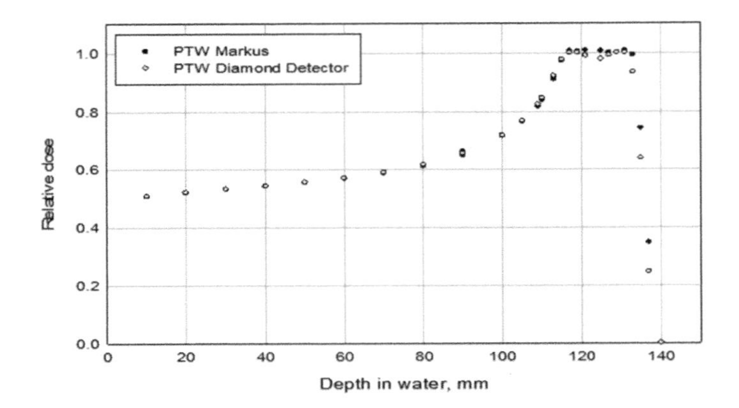

Figure 11–4 Comparison of depth–dose curves for a 155 MeV range-modulated proton beam shaped with a 30 mm circular aperture. Open circles = PTW Markus ionization chamber. Closed squares = PTW diamond detector T60003. Data taken from [12].

required to stabilize the response. The characteristics of a typical PTW natural diamond detector T60003 are given in the middle column of Table 11–1. The detector can be calibrated in terms of absorbed dose-to-water and used for dose measurements in proton beams as described in [23].

The middle of the sensitive volume is usually assumed to be the effective point of measurement. Relative dose measurement comparison results have shown good agreement with those of ionization chamber measurements in high-energy (≥100 MeV) proton beams [5,12,13].

Figure 11–4 shows a comparison of depth–dose curves measured with a PTW Markus ionization chamber and a PTW natural diamond detector in a 155 MeV range-modulated beam. Several studies researched an application of a diamond detector for quality control and the dosimetry in low-energy small proton beams delivered by eye-beam lines [5,24]. The reported differences between depth–dose curves measured with natural diamond detectors and parallel-plate ionization chambers in low-energy proton beams are explained by perturbation effects of the finite detector thickness [25] and LET dependence of the response. A practical procedure to correct the response of a diamond detector in low-energy proton beams was suggested in [24]. The procedure accounts for dependence of detector response on dose rate and includes correction of the absorbed dose to water calibration coefficient at reference depth on LET dependence. Figure 11–5 shows that, after application of the developed calibration procedure, depth–dose measurements obtained with the PTW natural diamond detector and parallel plate ionization chamber in a 62 MeV modulated proton beam show close agreement.

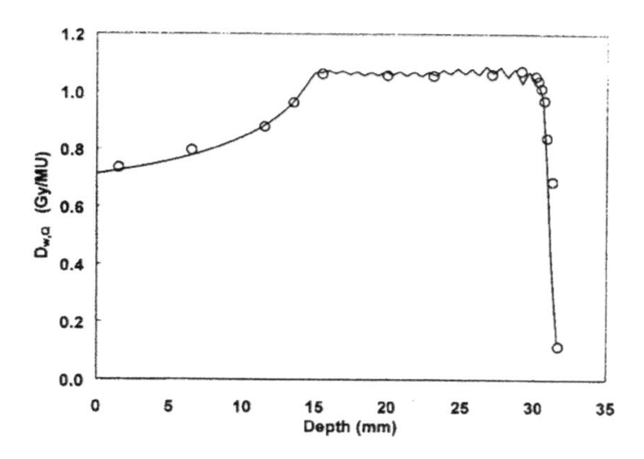

Figure 11–5 Effect of calibration procedure of PTW natural diamond detector on depth–dose measurement results in a 62 MeV modulated proton beam. Open circles = PTW Markus ionization chamber. Continuous line = corrected data for a PTW diamond detector. (Reprinted from [24] with permission from American Association of Physicists in Medicine.)

11.2.3.2 Synthetic Crystal Diamond Detector

A comparison of the technical characteristics of both types of diamond detector probes—the older model T60003 and the new model T60019—is given in Table 11–1. A performance of the prototype of synthetic single crystal diamond probe in relative dose measurements in clinical proton beams has been studied [18,19]. These papers reported that the observed dosimetric properties of the synthetic single crystal diamond detector indicate that its behavior is proton energy independent and dose rate independent in the investigated energy and dose rate range, and it is suitable for accurate relative dosimetric measurements in large-field and small-field, high-energy clinical proton beams. A synthetic single crystal diamond-based Schottky photodiode was tested in the proton beam line (62 MeV) dedicated to the radiation treatment of ocular disease [20]. Depth–dose curves were measured for the 62 MeV pristine proton beam and for three unmodulated range-shifted proton beams. Furthermore, the spread-out Bragg peak was measured for a modulated therapeutic proton beam. Measured dose distributions were compared with the corresponding dose distributions acquired with reference plane-parallel ionization chambers. Field size dependence of the output factor (dose per monitor unit) in a therapeutic modulated proton beam was measured with the diamond detector over the range of ocular proton therapy collimator diameters (5 to 30 mm). Output factors measured with the diamond detector were compared to the ones made by a Markus ionization chamber, a Scanditronix Hi-p Si stereotactic diode, and a radiochromic EBT2 film. Signal stability within 0.5% was demonstrated for the diamond detector with no need of any pre-irradiation dose. Dose and dose rate dependence of the diamond response was measured. Deviations from linearity resulted to be within ±0.5% over the investigated ranges of 0.5 to 40.0 Gy and 0.3 to 30.0 Gy/min, respectively. Output factors from diamond detector measured with the smallest collimator (5 mm in diameter) showed a maximum deviation of about 3% with respect to the high-resolution radiochromic EBT2 film. Depth–dose curves measured by diamond for unmodulated and modulated beams were in good agreement with those from the reference plane-parallel Markus chamber, with relative differences lower than ±1% in peak-to-plateau ratios, well within experimental uncertainties. A 2.5% variation in diamond detector response was observed in angular dependence measurements carried out by varying the proton beam incidence angle in the polar direction. A recent study [21] shows the performance of the commercial PTW microdiamond detector model T60019 in proton beam. Before characterizing this detector in proton beam, the authors performed an extensive testing in photon beam and obtained the data that were very similar to the data presented in [22] in terms of linearity, leakage, pre-irradiation, radiation damage, and reproducibility.

The comparison of depth–doses (Figure 11–6a) shows that microdiamond provided identical results to that of an ion chamber in three different proton beam energies without LET dependence. The small field profiles (Figure 11–6b) taken with a microdiamond detector have a slightly larger spread. However, being tissue equivalent and having small size, the detector does not create any perturbation and is most suited in small fields where other detectors fail. The dosimetric characterization of the tested synthetic single crystal diamond detector clearly indicates its suitability for relative dosimetry in proton beams.

Some practical considerations should be accounted when working with natural and synthetic diamond detector probes:

- The operating bias of the natural diamond detector is +100 V, whereas the PTW synthetic diamond detectors are used unbiased (0 V).

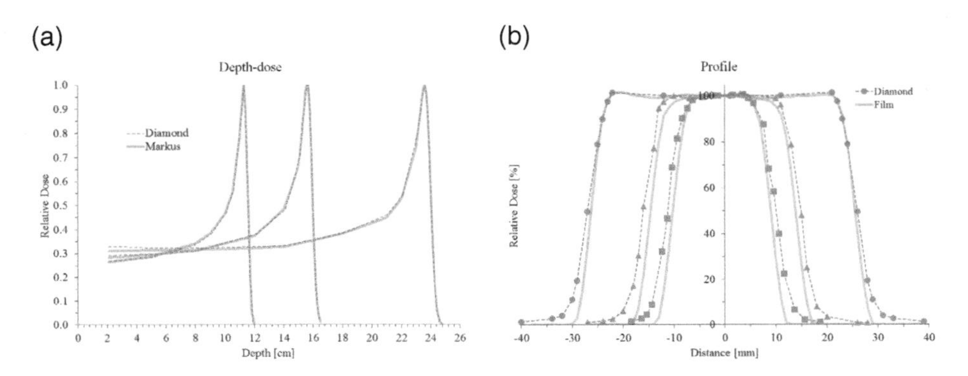

Figure 11–6 (a) Comparison of depth–dose measurement results in monoenergetic proton beams with PTW microdiamond detector and PTW Markus ionization chamber. (b) Comparison of profiles of SOBPs measured with a PTW microdiamond detector and EBT films. (Courtesy of I. Das.)

- Diamond detectors can be calibrated in terms of absorbed dose to water and can be used for absorbed dose determination in proton beams.
- A pre-irradiation of the detector to 5 to 15 Gy (natural diamond probe) and 5 to 12 Gy (synthetic diamond probe) is required to stabilize the response. The pre-irradiation should be repeated each time the bias is turned off and back on.

The use of natural diamond detectors for absorbed dose determinations in low-energy proton beams requires determination of dose rate and LET correction factors, while synthetic diamond detectors do not need any correction factors accounting for dose rate and linear energy transfer dependence. Further studies are needed to support the reported advantages of the synthetic diamond probe for measurements in clinical proton beams.

11.2.4 Films

11.2.4.1 Radiographic Films

Application in relative dosimetry: passive detector

- Detector for measurements of 2D dose distributions in QA applications.
- Profile measurements in beam characterization.
- Radiation leakage measurements.

Films are usually used for measurements in the plane perpendicular to the beam axis being placed between plastic slabs. The response or darkening of the film is dependent upon the number of small individual silver halide grains that are given enough energy by the radiation beam to create a latent image. A detailed description of film performance in photon beams can be found in [26]. However, for proton beam, very limited references are available [27].

There are two effects that influence film response for protons:

- The stopping power of protons varies tremendously with primary and secondary particles and with energy, so the relationship between fluence and dose is not constant.
- The energy transferred to the grain from protons can be greater than that required to produce the latent image, therefore some of the energy is not used in producing a response.

A combination of these two effects results in a film response that is variable with depth and limits the use of films in irradiations along beam axis. To obtain accurate results with the dosimetric film XV-2, it is usually pressed between slabs of plastic in order to eliminate air pockets near the film. After exposure, the film is developed and may be digitized with charge coupled devices (CCD) or laser scanners. The variation of the overall film response is slightly dependent on the type of scanner used and on the film batch (typical variations being 5 to 15%). However, the main variation comes from the processing conditions, which can yield differences greater than 30% [28, 29]. The energy dependence of radiographic film in proton beams can be corrected if calibration curves produced at the entrance region and at the depth of the SOPB are

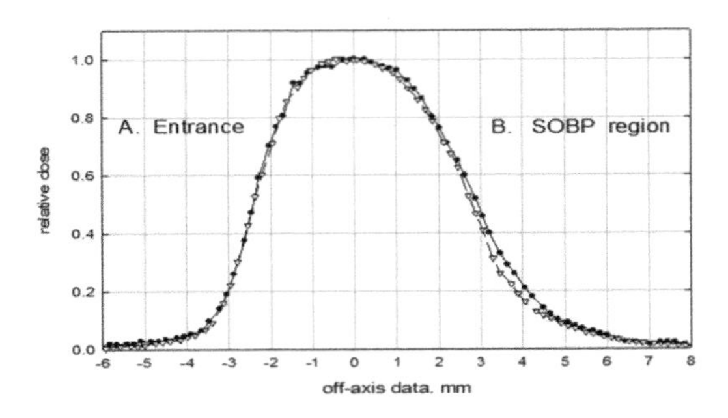

Figure 11–7 Lateral beam profiles for a 126 MeV range-modulated proton beam. Full circles = radiochromic film MD-55. Open triangles = Kodak XV-2 film. (Adapted from [12] with permission from IOP Publishing, Bristol, UK.)

used for beam profile characterization at the corresponding depths [12]. An application of this calibration technique showing close agreement of beam profiles obtained with a radiochromic film and with a radiographic film processed with two different calibration curves is illustrated in Figure 11–7.

EDR-2 film, formerly known as ready pack ECL, contains cubic crystals of AgBr about 0.2 μm on a side, resulting in a crystal volume about 1/8 of that found for the crystals in the XV-2 film [30]. Dose response calibration curves for EDR-2 film obtained for different charged particle beams of multiple energies were reported in [31], where it was also indicated that the low sensitivity of the EDR-2 film to protons can be advantageous for quality control measurements and constancy checks. The processing temperature and condition of the chemicals for EDR-2 film, as for other radiographic films, should be carefully controlled to obtain accurate results. A third silver halide film that is useful for applications in proton dosimetry is Kodak PPL-2. Due to a high dose sensitivity as shown in [31], this film can be quite useful for determining radiation leakage from the treatment head.

Practical considerations:

- Radiographic films should be pressed between slabs of phantom in order to eliminate dose- distorting air pockets.
- The response of films exposed at different depths should be converted to dose using film calibrations performed at the corresponding depths using the same beam type and residual range.
- Dose range for protons: Kodak EDR-2 film's range is 0.2 to 0.4 Gy; Kodak XV-2 film's range is 0.05 to 0.8 Gy; and Kodak PPL film's range is 0.003 to 0.06 Gy.
- The user should carefully control developing conditions to have reproducible results.

11.2.4.2 Radiochromic Film

Application in relative dosimetry:

- Detector for measurements of 2D dose distributions for multiple beam combinations, suitable for scanned beams.
- Point detector for absolute dose determination and output factor measurements.

Radiochromic film consists of a thin polymer sensor layer coated on the side of a transparent polyester base. Recent versions of radiochromic films have sensor layers on both sides of the base. The film is colorless or has an initial distinct color for type identification, and a radiation-induced color change (a polymerization process) is induced without the need for any chemical processing [32]. Radiochromic films are stable at temperatures up to 60°C. The optical density of the blue tint depends on the amount of the radiation dose. The film is usually placed, for irradiation, in a plastic phantom, however, with special precautions the irradiation may be performed in a water phantom [33,34]. An attractive feature of radiochromic film is that the emulsion sensor and film base are made of low-atomic-number constituents, hence the stopping power ratios of the film material for protons are very similar to those of water, making The film tissue equivalent and much less sensitive to the spectral hardening observed with silver halide films. The main performance features of the radiochromic films are given for Gafchromic MD-55 types, as these features hold for modern versions of the films. Radiochromic film exhibits post-irradiation development (Figure 11–8a), and an increase of optical density (OD) up to 15% may be observed in the first 10 days [32,33,35,36]. Studies of Gafchromic MD-55 film and its more sensitive version,

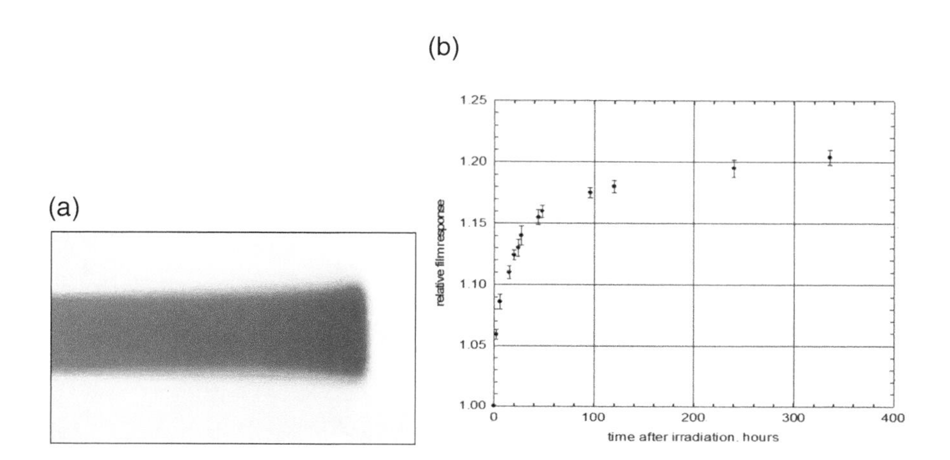

Figure 11–8 (a) Image of range-modulated proton beam obtained with a radiochromic film. (b) Post-irradiation development of MD-55 film. (Reprinted from [35] with permission from Elsevier.)

MD-55-2, have shown a similar linear response within the dose region from 2 to 100 Gy for radiotherapy beams, except when the films are irradiated in a monoenergetic proton beam at the Bragg peak location [33,35]. The use of radiochromic-type film for 2D dosimetry may be complicated by variations in optical density (OD) observed on different parts of the film, which ranged from 6 to 15% due to the nonuniform dispersal of the sensor medium [33,36]. These irregularities can be corrected by using a double exposure technique [37] or multi-color channel technique [38]. With this precaution and the selection of the proper digitized resolution [35], a precision level of ±3% in absorbed dose determination with Gafchromic films can be achieved at doses of 10 to 30 Gy. Figure 10–9 shows a comparison of proton depth dose curves for a 100 MeV monoenergetic proton beam and for a 155 MeV range-modulated beam obtained with MD-55 films positioned perpendicular to the beam axis and a parallel-plate ionization chamber [35]. This comparison of depth–dose distributions clearly demonstrates the LET dependence of the film and, correspondingly, the suppression of the dose (5 to 10%) in the Bragg peak region (at the distal part of SOBP for range-modulated beam). However, the difference in penetration of the beam—as given by the depths of the distal 80% and 50% doses measured with the film and the ionization chamber—is within 0.2 mm. To reduce artifacts that might be produced by any air gap between the film and the phantom, the films may be positioned almost parallel to the central axis but tilted slightly away from the beam axis by a small angle (~10°). Orientation of the film in this manner allows the measurement of two-dimensional dose distributions throughout a target volume on one film. The performance of the next version of radiochromic film—Gafchromic EBT—in proton beams is similar to MD-55 films, but EBT film is more practical since the saturation level of this film is low-

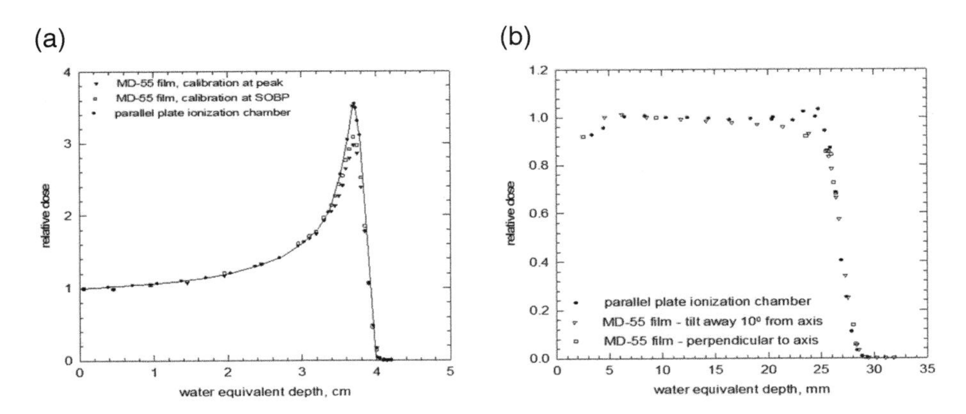

Figure 11–9 A comparison of the proton depth–dose curves obtained with MD-55 film and parallel-plate ionization chamber. (a) 100 MeV proton beam, no modulation, aperture diameter 3 cm. (b) 155 MeV range-modulated proton beam, modulation 3 cm, aperture diameter 3 cm. (Adapted from [36] with permission from Elsevier.)

ered to 2 Gy [39]. The dose response of EBT film was studied in terms of absolute dose for a scanned proton beam at HIT [40]. They have found that films calibrated at any clinical proton beam energy can be used for dose measurements at the entrance region.

In this case, the reported overall uncertainty in dose determination for proton beams was 4.6%, or one standard deviation. The dose response of the latest version of Gafchromic EBT 2 film was investigated in proton spread-out Bragg peaks created with passive beam delivery system [41,42], and the differences within 4.5% were found for beam energies of 100 and 250 MeV. Since the EBT and EBT 2 films contain the same media, the performance features of EBT 2 films can be associated with the features of EBT films.

The energy dependence of the new generation EBT, EBT2, and EBT3 films has been extensively documented in [41, 42, 43, 44]. The preferable scanner recommended in recent publications was a flatbed scanner (EPSON Expression 10000XL) employing transmission unit. Depth–dose curve can be also obtained in a single measurement by exposing EBT film parallel to the beam axis [42]. The authors of study [42] determined the relative effectiveness of EBT films and corrected under-response at the Bragg peak region for films exposed parallel to the beam axis with small tilt. With these corrections, EBT films can be considered as a useful tool for depth–dose verification in scanning proton beams. The latest version of EBT3 film has been released recently. As claimed by the manufacturer, EBT3 films are constructed similarly to EBT2 films, with similar expected performances. EBT3 films have the added features of a symmetric construction (the effective point of measurement is now at center) and anti-Newton ring artifacts coating. The symmetry of the film should allow a safer procedure for scanning films in clinical routine. The study [45] compared the work performed by other authors on earlier versions of EBT films to evaluate the performance of EBT3 in terms of measurement uncertainty in various clinical radiation beams, including protons. It was reported that the global uncertainty on acquired optical densities is within 0.55% and could be reduced to 0.1% by placing films consistently at the center of the scanner and using red channel. The total uncertainties on calibration curve due to film reading and fitting were within 1.5% for proton beams, which makes EBT3 a promising candidate for proton dosimetry. Based on the available data, radiochromic film dosimetry is more convenient than silver halide film dosimetry in terms of calibration, as films may be exposed at multiple depths and the response converted to dose using a single calibration curve. Depth–dose characterization of very narrow proton beams is very sensitive to detector alignment with respect to the beam axis, therefore the radiochromic film technique can serve as a valid reference. Nevertheless, caution should be taken when analyzing the very distal part of the Bragg peak curve because of the LET dependence of the response [41,42]. With their high spatial resolution and quasi water-equivalent composition, radiochromic films may be used to measure complex dose distributions in an irradiated phantom, thus enabling verification of dose delivery of proton Bragg peak stereotactic radiosurgery with multiple non-coplanar beams [46, 47]. With appropriate care, radiochromic film dosimetry can validate the prescribed dose delivery to within ±5% (one standard deviation) and detect possible misalignments greater than 2 mm [12,39,27].

Practical considerations:

- Radiochromic film should be calibrated using a well-characterized uniform radiation field. The dose response curve and film sensitivity should be obtained in the dose range and conditions of interest. Films should preferably be scanned with a flatbed scanner (EPSON Expression 10000XL) and proper frequency should be used for calibration and measurement.
- Profiles, depth–dose, and absorbed dose values for small radiosurgery beams may be obtained from a single exposure of several films placed in the phantom at different depths along the beam axis.
- For depth–dose measurements, films may also be positioned along the central axis but tilted away from the beam axis over a small angle to check distal edge boundaries precisely.
- Since the film response changes with time, especially during the first 24 h after irradiation, the exposure time and readout time for all the films should be documented and, if necessary, appropriate correction factors for instabilities should be applied.
- Older Gafchromic film (MD55): range 10 to 100 Gy. More recent Gafchromic MD-55-2: range 10 to 30 Gy; EBT and EBT2 and DBT3 films: range 0 to 6 Gy.
- LET dependence should be clearly understood and accounted for.

11.2.5 Luminescent Dosimeters

11.2.5.1 Thermoluminescent Dosimeters (TLDs)

Application in relative dosimetry: passive probe

- Point detector for dose verification and output factor measurements.
- Foil detector for measurements of 2D dose distributions, suitable for scanned beams.

TLDs have been used for *in vivo* and in phantom applications with photon and electron beams primarily because of their small volume, but sometimes because they are easily transportable. A disadvantage of TLDs is the long preparation and readout times required for their use. They have been used quite often in phantoms for patient-specific dose verifications. The TLD material most often used for patient dosimetry is LiF:Mg, Ti-type TLD 100. One study [23] described a calibration method for these dosimeters for use with proton beams. Results of TLD measurements to determine dose/MU have shown good agreement with ionization chamber data, particularly in range-modulated beams in which the LET variation is diluted [12,48]. However, getting more precise results by accounting for effects of an inhomogeneous energy loss and a finite track length of the projectiles in the sensitive detector volume of the TLD dosimeters is difficult because the energy dependence follows complex models [49]. The application of TLDs for remote dose auditing in proton therapy was presented in [50]. Novel TLD foils, manufactured as a hot-pressed mixture of LiF:Mg,Cu,P powder and ethylene-tetrafluoroethylene (ETFE) copolymer, may be used for 2D dosimetry of proton beams to allow fast and dynamic imaging of the proton beam parameters [51]. To achieve this goal, it is necessary to use more sophisticated recording and data transfer techniques compared to those currently used.

The new system of CCD camera with scintillating screen to read TLD foils uses an efficient scintillating screen that converts proton radiation to visible light, which is then reflected by a 45-degree glass mirror. Reflected light is focused optically and then viewed and recorded by a high-resolution CCD camera. To achieve relatively high time resolution, a USB-based data acquisition system is employed. The images are recorded and analyzed by dedicated software programmed in the LabView environment. To avoid radiation damage to the CCD camera sensor from protons in the beam, the camera is positioned perpendicularly to the beam axis, and light from the scintillator is reflected by a 45-degree glass mirror. The results from this dosimeter are generally consistent with measurements performed using radiochromic films and results of Monte Carlo transport calculations using the MCNPX code. Due to detector reusability, good linearity, and adequate spatial resolution, the 2D planar TL dosimetry system can potentially complement existing 2D techniques, such as radiochromic films or scintillation screens. Figure 11–10 shows the results of TL foil analysis irradiated for determination of spot parameters at different proton gantry angles. The relatively short time required to evaluate 2D TLD foils makes them very suitable for quality assurance in radiotherapy.

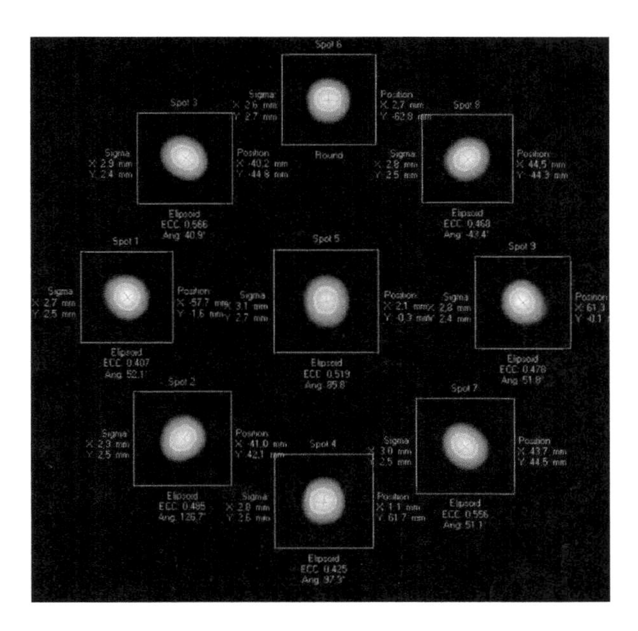

Figure 11–10 Results of analysis of TL foil irradiated for determination of spot parameters at different proton gantry angles. (Courtesy of P. Olko.)

Advantages of this technique are TL foil reusability and rapid processing of the results which are directly available in digital form (e.g., DICOM-compatible if needed).

11.2.5.2 Optically Stimulated Luminescence Dosimeters (OSLDs)

Application in relative dosimetry: passive probe

- Point detector for dose determination and output factor measurements.

OSL is the luminescence emitted from an irradiated carbon-doped aluminum oxide (Al_2O_3:C) crystal as a result of light stimulation. The difference between TL and OSL is that optical stimulation is used instead of thermal stimulation to release the trapped charges. The important point, however, is that, as in the case of TL, stimulation of the dosimeter depletes the concentration of trapped charges, thereby zeroing the luminescence signal. Since the OSL dosimeter is continuously exposed to the radiation field, during the measurements one has access to the radioluminescence (RL) in addition to the OSL signal produced by optical stimulation. In this regard, the RL signal may be used for real-time monitoring of the dose rates during irradiation, while the OSL signal is read only after the irradiation is over to obtain the total absorbed dose during the treatment [52,53]. Such optical dosimetry system was tested in a 175 MeV clinical proton beam for depth–dose measurements [54] and a significant decrease in luminescence efficiency with LET was observed. Even the RL technique cannot fully resolve the peak–plateau ratio. The study implies that the RL-signal from Al_2O_3:C could potentially be suitable for medical proton dosimetry in the 0 to 0.3 Gy range even without any LET-dependent correction factors. The similar results for the proton irradiations at the different facilities are reported in [55]. This study demonstrated the potential for use of OSL dosimeters as postal dosimeters for intercomparison of the output of therapy machines.

11.2.5.3 Radio-photoluminescence Glass Dosimeter (RPLGD)

Application in relative dosimetry: passive probe

- Point detector for dose determination and output factor measurements.
- Dosimeter for postal proton dosimetry intercomparison.

RPLGD uses glass compound as the luminescent material and applies different excitation methods, along with different readout techniques, compared to OSL or TLD. The basic principle of RPLGD is that the color centers are formed when the luminescent material inside the glass compound exposed to radiation and fluorescence are emitted from the color centers after being irradiated with ultra-violet light. RPLGDs have been used in different applications of radiation dosimetry, but only recently, when the GD-301 dosimeter became commercially available, was the study of its applications in proton dosimetry initiated. GD-301 has a length of 8.5 mm and a diameter of 1.5 mm. These dosimeters have small differences in individual sensitivity and small fading effect. An important feature is that the luminescence signal does not disappear after readout; therefore, repeated readout for a single exposure is possible. One study [56] investigated the dosimetric characteristics of the GD-301 dosimeter (reproducibility, linearity, dose rate, fading, angular dependence, and depth–dose distribution) in high-energy proton beams and its feasibility to be used as an *in vivo* dosimeter. The obtained results have shown the advantages of the glass dosimeter compared with TLD-100 and in the phantom study; the difference of isocenter dose between the delivery dose calculated by the TPS and that measured by the glass dosimeter was within ±5%. The obtained results indicated that the glass dosimeter has considerable potential for use with *in vivo* patient proton dosimetry. The feasibility study of the glass dosimeter GD-301 for postal dose intercomparison of high-energy proton therapy beams was performed in a cooperative study involving several institutions [57]. The authors developed the methodology of the absorbed dose determination with the glass dosimeter system, establishing the calibration coefficient and various correction factors (nonlinearity, fading, energy, and holder). The participating proton therapy centers were asked to irradiate the glass dosimeter to 2 Gy with similar setup and conditions. The difference in measured absorbed dose values between the glass dosimeter and an ionization chamber was within ±2% as a function of proton beam quality; residual ranges determined from measured depth–dose curves were between 2.1 and 9.0 cm. The influence of the holder material in absorbed doses of the proton beams is less than 1%. In the accuracy evaluation of the glass dosimeter system established in blind test, agreement within ±2.5% with ionization chamber dosimetry was obtained for the proton beam. In this feasibility study, the results on the proton beam output check were within ±6% for all participating centers, showing considerable potential for the glass dosimeter to be used for a postal dose audit program in proton beams. Use of such detectors is still limited to a few centers, and more studies are required for its implementation in clinical practice.

11.2.6 Alanine

Application in relative dosimetry: passive probe

- Point detector for dose determination and output factor measurements.
- Dosimeter for postal proton dosimetry intercomparison.

The use of alanine for dosimetry in clinical proton beams was first proposed in [58,59]. Recently, interest in its use as a proton dosimeter has been renewed [60,61]. Alanine detectors were also successfully used for end-to-end tests in scanned proton beams [62]. Usually crystalline alanine is used in powder form, film, or as pellets when the alanine is compressed with a binding agent like paraffin wax, polyethylene, or cellulose. When crystalline alanine is bombarded by ionizing radiation, free radical products are produced and locked in the crystal lattice that can subsequently be quantified by electron spin resonance (ESR) spectroscopy. Advantages of alanine as an absolute dosimeter are: preservation of the signal after read-out with long-time stability, nearly tissue equivalent composition and stopping power, and linear response over a wide dose range. Also, the energy dependence of alanine can be well understood and described by a simple single-hit model [63]. Disadvantages of alanine dosimetry include the need for specialized readout equipment, cumbersome readout procedures (given that the response is dependent on the orientation of crystalline lattice with the magnetic field), and its relatively low sensitivity: pellets of 2 mm thickness and 5 mm diameter require doses of 10 Gy or more to achieve a standard deviation of readout of 1% [64,65]. Smaller pellets are available (e.g., 0.5 mm thickness and 5 mm diameter or 2.5 mm thickness and 2.5 mm diameter) but these require doses of at least 30 Gy for the same standard deviation. Use of the alanine dosimeter for absorbed-dose determinations in proton and other light ion beams requires consideration of the particle energy fluence spectrum and control of the environment during and after irradiation [66]. The energy dependence has been widely reported, and a summary of experimental findings and its impact on the measurement of depth–dose distributions is given in [61]. Recent work demonstrated that the beam quality can be extracted from the different spin relaxation modes, providing a means to correct for the energy dependence without knowing the fluence differential in energy [67]. The study [68] investigated the dose response of alanine pellets in a 169 MeV proton beam and reported a linear response (±2%) in the dose range from 50 to 300 Gy and a relative inter-detector response of ±1.5%. Depth–dose measurements at residual proton energy of 50 MeV were indirectly compared with ionization chamber measurements and found to be in good agreement. The studies [48, 65] used alanine pellets in a 62 MeV clinical proton beam and reported good agreement between depth–doses measured with alanine and parallel-plate ionization chambers. When measuring depth–dose curves using a stack of pellets in a plastic phantom, dose tails beyond the Bragg peak were observed [65]. Monte Carlo simulations of the depth–dose curves in a stack of alanine pellets centered on the 60 MeV clinical proton beam axis and surrounded by phantom material were studied in [61]. The results showed that the observed dose tails may be explained by inscatter from the surrounding phantom material when the range in the material is larger than in the alanine pellet material. To perform accurate measurements and avoid dose tails, it was advised to use alanine mixed with a binding agent that closely matches the electron density and mass density of the phantom material. Also shown was that protons "tunnelling" in the air gap between the pellet and the phantom could contribute to the observed dose tails. This effect can be substantial when measuring a depth–dose curve with a stack of pellets, and it can be avoided by orienting the stack under a small angle with respect to the beam axis. With careful adjustment of the ESR spectrometer and by using the appropriate dosimeters dose, values within the clinical dose range can be determined with an overall uncertainty of 3.5%, at a confidence level of 95%. Alanine dosimeters are small, compact, and easy to handle. They are characterized by low influence of temperature, humidity, and dose rate. They also have a wide measuring range, which makes them applicable for radiation therapy, in blood components irradiators, and in industrial irradiation facilities. Because of their high quality and low costs, this dosimetry system can be utilized both for reference and routine dosimetry.

The future use of alanine detectors also depends on improving its sensitivity. It must be recalled, though, that the precision of alanine pellets is decreased at low doses, limiting their use to over 5 Gy if better than 2% precision is required. Films of alanine would seem very attractive because a 0.25 mm spatial resolution could be achieved, but unfortunately, their current low sensitivity to ionizing radiation prevents them from being used clinically in this manner. Measurements of alanine pellets during one-year storage at room conditions did not give any evidence of a long-term alanine response decrease [60]. This, in addition to the mailing suitability and ruggedness of the dosimeters, supports the possible use of alanine for transfer proton dosimetry [62].

11.2.7 Plastic Nuclear Track Detector (PTND)

The most common PTND is CR-39. CR-39, or allyl diglycol carbonate (ADC), is a plastic polymer commonly used in the manufacture of eyeglass lenses. The abbreviation CR-39 stands for "Columbia Resin #39" because it was the 39th formula of a thermosetting plastic developed by the Columbia Resins project in 1940. As a radiation detector, it is sensitive to charged particles with a LET in water greater than 5 keV/mm [69]. When a charged particle traverses the detector, it breaks the chemical bonds of the polymer, producing what is referred to as a latent damage trail along the particle's trajectory [70]. Following irradiation, the detectors are chemically processed in a sodium hydroxide solution. This processing preferentially etches the trails of latent damage produced by the charged particles. These etchings take the form of conical pits in the surface of the plastic that can be measured using an optical or atomic force microscope. The dimensions of the elliptical opening of each track are proportional to the LET of the charged particle that produced it. By measuring a statistically significant number of tracks within a given area on the detector surface, one can thus generate a plot of the LET spectrum. This spectrum can then be used to both determine dose and dose equivalent. Neutrons can also be detected indirectly by CR-39 PTND by placing a layer of plastic just upstream of the PTND. Secondary charged particles produced within the upstream plastic layer pass into the PTND, creating latent damage trails. One of the most common applications of PTND in proton treatments is the monitoring of neutron dose in and around the treatment rooms and the accelerator [71]. Several investigators have used PTND within proton beam treatment fields [72, 73]. This application is primarily to determine the fraction of dose within the field due to the contaminating high-LET component because the LET of protons with energies greater than 20 MeV is lower than the track registration threshold of CR-39 PTND. Except in the region very close to the Bragg peak, PTNDs cannot be used to measure dose from the primary protons. Another light ion beam application is determining the dose and dose equivalent outside of the field, much of which is transported there by neutrons. The major disadvantage of PTNDs for specialized applications is the large amount of effort required in reading the detectors and analyzing the results.

11.2.8 Scintillation Detectors

Application in relative dosimetry:

- Probe for relative depth–dose and profile measurements.
- Detector for measurements of 2D dose distributions for QA, suitable for scanned beams.

Small-volume light emission probes based on material scintillation or gas electron multiplier technology are attractive for small-field dosimetry due to their high spatial

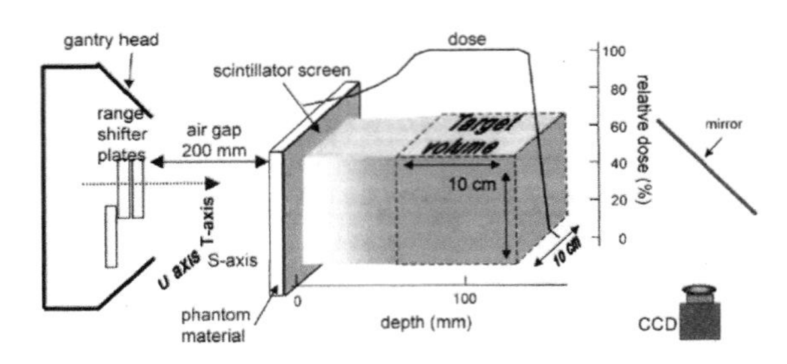

Figure 11–11 The setup for fluorescence measurements with a CCD camera. (Reprinted from [79] with permission from American Association of Physicists in Medicine.)

resolution, water equivalence, and instantaneous readout. The main disadvantage of plastic scintillators when irradiated by protons is the quenching effect, which reduces the amount of scintillation and results in dose underestimation by close to 30% at the Bragg peak for beams of 150 MeV or more [74,75]. This effect can be accounted for in the calculation of the dose response [76,77], however, small-volume light emission probes are not yet implemented in routine practice of proton dosimetry. It should be noted that the quenching is not a problem if the detector is used as a QA device, where the only purpose is to show consistency in the measured parameter.

A solution for two-dimensional scintillation dosimetry with high spatial resolution for QA purposes was proposed in [78,79]. The measurements of dose distribution can be performed by placing a scintillation screen perpendicular to the proton beam axis and viewing it through a 45° mirror by a CCD camera (Figure 11–11). The most used scintillating material was polystyrene with a small concentration of an aromatic dye to shift the intrinsic emitted light spectrum produced by the polystyrene to one with a higher efficiency of detection by the CCD. The studies performed at PSI [80, 81] have shown that CCD pictures of a scintillator plate placed at several depths underneath a PMMA stack can verify dose delivery with scanned proton pencil beams. The system used at PSI allows visualization of the lateral dose distribution in two dimensions at each depth with a sub-millimeter position resolution and a dose reproducibility of a few tenths of a percent. The use of a fluorescence screen with a CCD camera for 2D QA measurements was implemented by IBA Dosimetry into a commercial device that was used at several facilities for QA and commissioning measurements [82,83]. The system Lynx (IBA Dosimetry) has an active surface of 30 x 30 cm^2 and provides effective resolution of 0.5 mm for single shot and movie mode measurements (Figure 11–12).

Another device that can measure a 3D distribution with a CCD camera is the large multi-array detector system developed at PSI [84] that employs 400 scintillator/fibers. The scintillators (sensitive volume of 5 mm^3) are optically coupled to the fiber light guides (2 mm diameter by 1 m long). The detectors are embedded at different

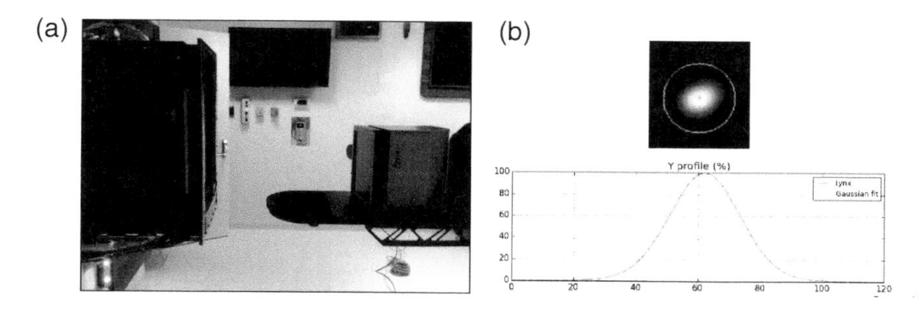

Figure 11–12 (a) The Lynx device (IBA Dosimetry) for fluorescence measurements with a CCD camera. (b) Measured shape of the pencil beam.

depths in a polyethylene phantom. The light signals from the scintillators are carried by the bundle of fibers and are mapped on a CCD by means of a lens system. The device allows measuring the location and shape of beam spots as a function of position off-axis and measure 2D dose distributions at different depth by placing phantom material in front of the screen [80].

Practical considerations:

- A scintillating screen viewed by a CCD system may be considered as very useful tool to visualize the delivered dose in both static and dynamic beam deliveries. It is possible to measure the relative dose in two dimensions at different depths with a position resolution of a fraction of a millimeter and with a reproducibility of the data of about 1%.
- In addition to dosimetric characteristics of the screen CCD camera system, the immediate availability of the digital data (DICOM format) also makes it very useful in the QA procedures of scanning beams.

11.2.9 Activation Detectors

Radioactivation of carbon has been used for absolute dosimetry for very high dose rate radiosurgery proton beams where significant recombination in ion chambers can occur [85]. The technique is based on the activation of ^{11}C (half-life = 20.4 minutes) via the $^{12}C(p,pn)^{11}C$ reaction as described in detail in several publications [86,87]. Carbon activation (CA) detectors (Figure 11–13) used for the dose measurements are usually made of scintillation material and are thus not pure carbon. They are generally irradiated on the front surface of a phantom. Measurement of the radioactivation is essentially an off-line method; that is, irradiation of the carbon sample is performed in the beam and then the sample is removed and subsequently moved to a 4π β-γ coincidence scintillator counter.

To count only the positron decays from the ^{11}C and not emissions resulting from other reactions, the counting system requires the detection of the positron in coincidence with the two 0.511 MeV γ-rays. The uncertainty of the carbon activation

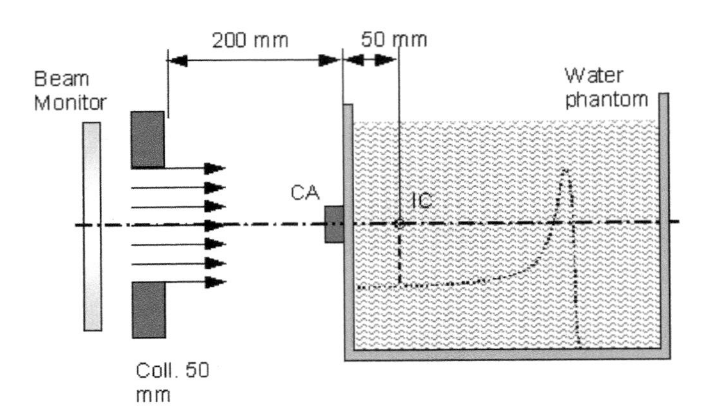

Figure 11–13 Schematic geometry for radioactivation dosimetry in light ion beam. Reprinted from [86] with permission from American Association of Physicists in Medicine.

method results from a combination of a number of sources, the most significant of which is the $^{12}C(p,pn)\ ^{11}C$ reaction cross section. A recent study showed relative agreement of 1.7% between doses measured using carbon activation and ionization chambers [86]. The radioactivation technique is not suitable for routine applications, as the experimental setup is quite complicated, but it can be used for independent verification of ionization chamber dosimetry. Such devices require a special lab and may not be suitable in a clinical setting.

11.2.10 3D Gel and Plastic Dosimeters

Application in relative dosimetry:

- Detector for measurements of 3D dose distributions for QA, suitable for scanned beams.

The common interest in gel dosimetry systems is in their ability to measure complex three-dimensional dose distributions with good spatial resolution and to integrate the dose delivered from all beams in a treatment session. The oldest gel system is based upon the ferrous sulphate or Fricke chemical reaction. This system is generally used to measure doses of between 25 and 30 Gy. With Fricke gels, the change in paramagnetic properties due to the oxidation of ferrous ions to ferric ions can be measured using MRI, and as a result, three-dimensional images of the dose distributions can be provided [88]. Image quality and accuracy of measured dose values depend upon a good understanding and selection of the MRI sequencing parameters and techniques, as well as the characteristics of the local scanner [89]. An alternative analyzing approach uses optical CT scanning [90] to produce maps of radiation-induced changes in the local optical absorption coefficient.

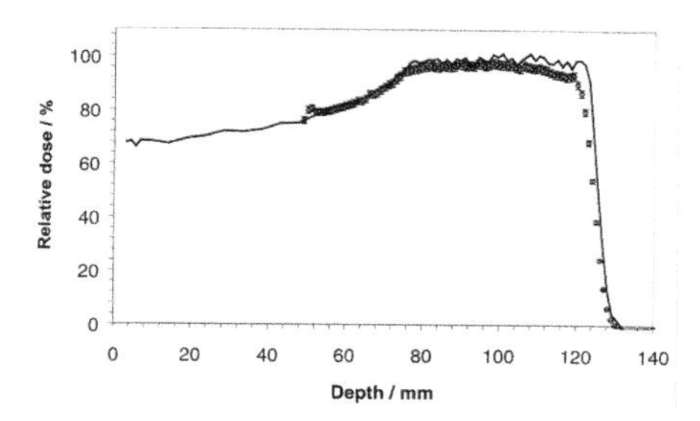

Figure 11–14 Depth–dose distributions for a range-modulated 132 MeV proton beam measured with FeMRI (open circles) and with an ionization chamber (solid line). The curves were normalized to yield the same relative dose at the depth of 60 mm (Reprinted from [91] and [88] with permission from IOP Publishing, Bristol, UK.)

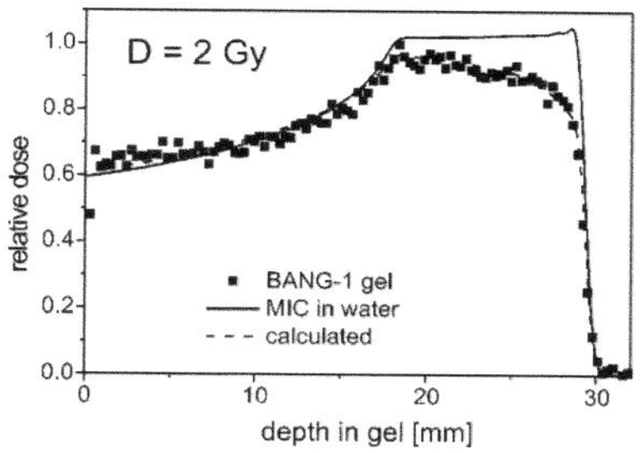

Figure 11–15 Depth–dose distribution for a modulated beam measured with BANG-1 gel. The solid line indicates the distribution measured with a Markus chamber in water. The dashed line shows a SOBP calculation from the modulator parameters using the fitted quenched Bragg peak data for a monoenergetic beam. (Reprinted from [96] with permission from AAPM.)

The study of Fricke gel systems in proton beams [88, 91] has demonstrated an underestimation of the dose close to the distal edge of the Bragg peak (Figure 11–14); this change in sensitivity was explained by the LET dependence of the chemical yield. Another disadvantage of Fricke gels is the fast diffusion of ferric ions in the gel matrix.

A different gel system, PAG, was based on the radiation-induced polymerization of polyacrylamide monomers infused in a gelatin matrix. Free radicals created during irradiation induce polymerization exactly in the irradiated region. The number of these free radicals and, correspondingly, the polymer yield is dependent upon the dose absorbed in the gel. The study [92] showed that a commercial version of the PAG system, BANG™, has a higher sensitivity (lower dose requirement) and reduced diffusion of radiation products compared to the Fricke gel and can be effectively analyzed with a conventional MR scanner.

In addition to MR scanning, other methods, e.g., optical scanning [93] or x-ray CT [94], have been tested to analyze the changes in gel structure.

The few studies of BANG gel dosimetry applied to proton beams [95, 96] have all indicated that the detector response as for other gel systems strongly depends on LET (Figure 11–15). There has been renewed interest in gel dosimetry with the advent of new products such as PRESAGE™, a three-dimensional polyurethane dosimeter with block polymers containing radiochromic dyes. Upon irradiation, PRESAGE generates a color change and hence OD change [97]. The handling of the dosimeter has become easier, and optical tomography measurements on such detectors may be done with a CCD-based optical system. Preliminary studies of the PRES-

AGE dosimeter in a monoenergetic proton beams showed that the measured ratio of the response at the peak and entrance is approximately 2:1, whereas the true value is approximately 5:1 [98]. For range-modulated proton beams, the response close to the end of the proton range, i.e., at the Bragg peak, is underestimated by approximately 20% compared to the corresponding diode measurement. Several studies focused on a new formulation of PRESAGE that has minimal under-response in a Bragg peak area of the proton beam. The study [99] described a modified formulation of PRESAGE that replaced the reporter molecule used in the original formulation, leucomalachite green (LMG), with a new LMG derivative. These dosimeters were irradiated in the plateau region of an unmodulated proton beam, and the response of the dosimeter irradiated in the water phantom was converted to dose using the calibration curve. The PRESAGE measured the absolute dose delivered within 3%. However, the comparison of depth–doses measured in water with an ionization chamber and the PRESAGE dosimeter showed good agreement between the PRESAGE and ion chamber data in the Bragg Peak region, but the PRESAGE overestimated the dose in the plateau region by as much as a 7.4%. Similar results were reported in [100] for the so-called "proton formula" PRESAGE dosimeter. Their results suggest that this proton formula PRESAGE dosimeter has the potential for 3D dosimetry of small fields in proton therapy, but further investigation is needed to improve the dose under response of the PRESAGE in the Bragg peak region [100,101].

Practical considerations:

- Enthusiasm about the use of gel detectors is based on their possibility of measuring three-dimensional dose distributions, but further investigations and a revised formulation are necessary before PRESAGE can be used routinely for proton beam therapy dosimetry.
- The proton formula PRESAGE dosimeter could be used for relative dosimetry of small proton fields.

11.3 Multi-detector Dosimetry Systems

The need to accurately and quickly measure beam parameters during commissioning or to perform periodic verification of dosimetric data and radiation beam quality makes these tasks difficult and demanding for proton beams. The time structure of the beam must also be considered. The measurements of the dose distribution delivered with scanning beams differ from those with scattered beams only from the point of view of the longer integration time needed for the measurements. For this reason, depth–dose or profile measurements based on dose rates using moving detectors cannot be used in scanned beams. Systems with multiple detectors in 1D, 2D, or 3D arrays can save time in measurements of dose distributions, particularly in scanned beams. Thus the development and implementation of custom made and commercial multi-detector instruments for measurements in the beam direction or in the plane perpendicular to the beam direction is becoming extremely important for proton dosimetry. As small-volume ionization chambers provide good reproducibility, high special resolution, and proton energy-independent signal, the multi-detector systems based on small-volume ionization chambers are of primary interest.

11.3.1 Multilayer Ionization chambers (MLIC)

The custom instrument for radiation quality assurance depth–dose measurements in uniform scanning and energy stacking proton beams—a transportable multilayer ionization chamber (MLIC) suitable in uniform scanned beam—was described in [102]. The MLIC contains 122 small-volume ionization chambers stacked at 1.82 mm water-equivalent steps to enable depth–dose profile measurements (Figure 11–16). The MLIC detector can measure dose distributions up to 20 cm in depth, can determine the 80% distal dose fall-off with about 0.1 mm precision, and is connected to a data acquisition system. A comparison of depth–dose curves for a 10 cm range-modulated proton beam measured with the MLIC detector and a Markus ionization chamber in water is shown in Figure 11–16. For most of the points, the maximum difference between the MLIC and the water phantom measurements was reported to be less than 1%, with only a few points showing a difference of 1.5%. The use of such detectors offers significant time savings during measurements in actively delivered beams compared with traditional measurements using a water phantom.

Very similar to the MLIC as described above, IBA Dosimetry introduced a commercial multi-layer ionization chamber (MLIC) system, the Zebra, consisting of 180 independent vented parallel-plate ionization chambers with 2 mm detector spacing and a circular-shaped collecting electrode with a diameter of 2.5 cm, covering up to 33 cm of water-equivalent thickness. This MLIC can measure the depth–dose curves much faster than the computerized water tank with ion chamber with a very short setup time. Another device from IBA Dosimetry—the MLIC system named Giraffe—also has 180 parallel-plate ionization chambers with 2 mm detector spacing, but the circular-shaped collecting electrode of this device has a diameter of 12 cm. This allows using the device for scanned beam measurements. The Zebra and Giraffe devices are shown in Figure 11–17. An evaluation of these detectors was performed by several groups, and the data obtained by the group from MD Andersen Cancer

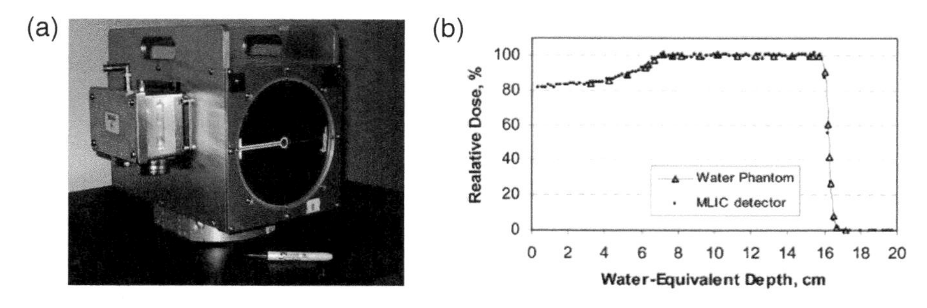

Figure 11–16 (a) Multilayer ionization chamber. (b) Comparison of depth–dose curve for a 10 cm range-modulated proton beam measured with the MLIC and a Markus ionization chamber in water. (Reprinted from [102] with permission from American Association of Physicists in Medicine.)

Center [103] is shown in Figure 11–18. The differences between percentage depth–dose values in the plateau region measured by Zebra and those measured by the Markus chamber in a PTW computerized water tank are within 2%. However, in the high-dose distal gradient regions, the difference is more than 2%, an amount that can be explained by the limited spatial resolution due to the 2 mm inter chamber spacing of the Zebra.

Nevertheless, the analysis of a large set of measurements shows that the commonly assessed beam quality parameters obtained with the Zebra are within the acceptable variations specified by the manufacturer of the delivery system, and the device can be successfully used for checking the constancy of data. The initial calibration of the Zebra and Giraffe systems should be performed by the manufacturer using ion chamber water phantom measurement results obtained on site. The drawback of the systems from the user's point of view is that a uniformity recalibration for the chambers should be performed before each use, and periodic calibration is also needed.

(a) 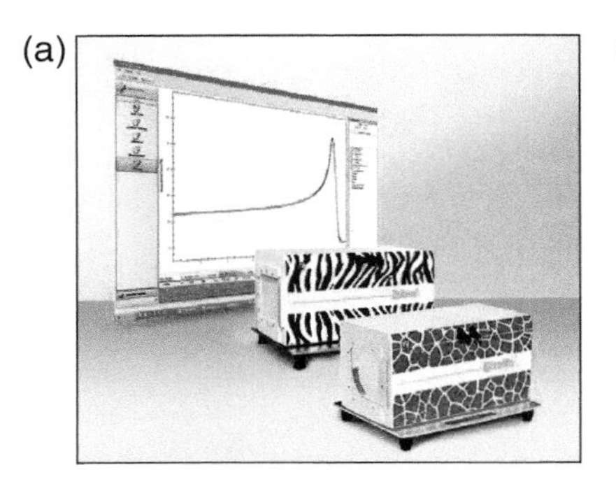

(b) 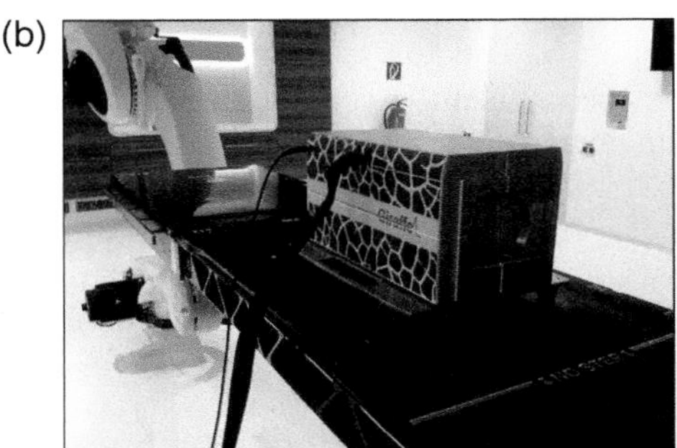

Figure 11–17 (a) Multilayer ionization chamber detectors Zebra and Giraffe. (Courtesy of IBA Dosimetry). (b) The Giraffe installed on a robot table for depth–dose measurements.

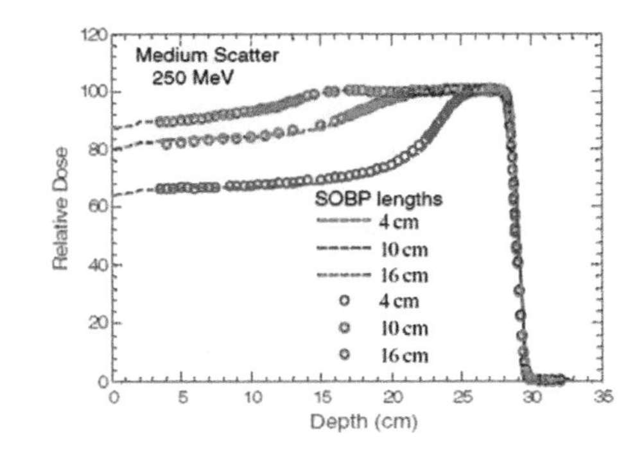

Figure 11–18 Comparison of the depth–dose distributions measured by the Zebra (dashed lines) and by the Markus chamber in a PTW water tank (open circles) for 250 MeV nominal SOBP widths of 4, 10, and 16 cm. (Reprinted from [103] with permission from American Association of Physicists in Medicine.)

For this purpose, the depth–dose measured with the MLIC in a monoenergetic beam should be compared to a reference depth dose distribution that was measured with the ionization chamber in the water tank under the same conditions. Based on this comparison, the correction factor for each of the chambers is determined, and this chamber calibration is used for subsequent measurements.

11.3.2 2D Arrays for Relative Dosimetry in Cross-sectional Plates

2D ionization chamber arrays (Figure 11–19) were primarily designed for dosimetry of high-energy photon and electron beams; however, they may be also used for 2D dosimetry in proton quality assurance dosimetry. The two-dimensional array of ionization chambers referred to as MatriXX, manufactured by IBA Dosimetry, was used in [104] for QA measurements in proton beam. This device is equipped with 32 x 32 parallel-plate ionization chambers, each chamber having a diameter of 4.5 mm and a 7.62 mm center-to-center separation.

The reproducibility of dose/MU and the consistency in the beam flatness and symmetry measured by MatriXX were reported to be similar to the characteristics determined with film dosimetry and with a standard ionization chamber data in water. This device can be immersed in water with a specially designed holder. For scanning beam applications, the manufacturers have produced special versions with higher collection voltages. Another example of tools for two-dimensional and quasi-three-dimensional QA dose measurements involves the use of stacks of two-dimensional arrays of ionization chambers interleaved with plastic slabs [105]. The "Magic Cube" detector (Figure 11–20) has a sensitive area of 24 x 24 cm^2 and consists of 64 planes of parallel plate ionization chambers. Each plane is divided into 24 individual readout ionization chamber strips, each 0.375 cm wide and 24 cm long. A closely related detector that was developed by the same group is the Pixel Ionization Chamber (PXC). This detector uses pixel anode segmentation to yield a matrix of 1024 cylin-

(a)

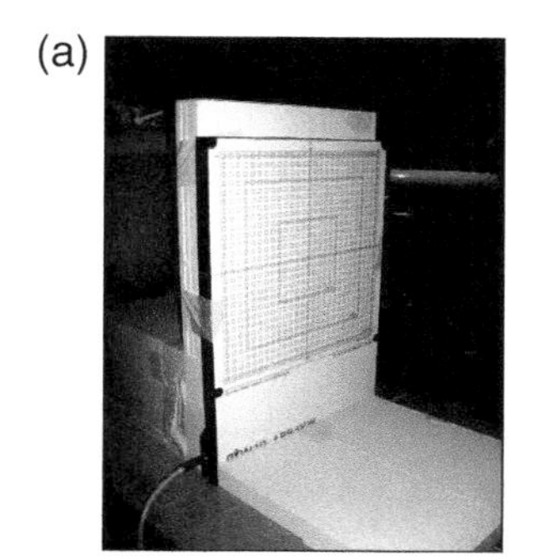

(b)

Figure 11–19 Two-dimensional arrays of ionization chambers. (a) 2D array from PTW (courtesy of PTW). (b) ImRT MatriXX from IBA (courtesy of IBA).

Figure 11–20 View of the chambers in the Magic Cube.

drical ionization cells arranged in a square having a 24 x 24 cm^2 area. Each cell has a diameter of 0.4 cm and a height of 0.55 cm, with a pitch of 0.75 cm separating the centers of adjacent cells. The sensitive volume of each single ionization cell is 0.07 cm^3.

Both the Magic Cube and PXC detectors are read out using custom-designed, front-end microelectronics and a PC-based data acquisition system. The testing results reported in [105] demonstrated the capability of the detectors for dose measurements in both scattered and scanned beam delivery systems. It was also demonstrated that the Magic Cube can be used as a fast scanning beam monitor exploiting the high readout speed allowed by the microelectronics chips. On the other hand, these devices are bulky, and the practical setup of the systems for periodic QA measurements is problematic. The most efficient application of devices similar to the Magic Cube is fast on-line measurements during installation and alignment of beam delivery systems.

11.4 Relative Dosimetry for Scattered Beams

Relative dosimetry or measurements in non-reference conditions in scattered proton beams are similar to those that are made in conventional radiotherapy with high-energy photon or electron beams. The measurements are usually performed as a part of the commissioning or acceptance testing procedures or as a part of dosimetric QA procedures [27] and include the following:

- Constancy checks of dose monitor calibration.
- Measurements of output factors.

- Relative dose measurements in the plane perpendicular to the beam axis
- Relative dose measurements in beam direction
- Patient-specific portal calibration

For scattered beams where the radiation fields are similar to that in conventional radiotherapy, the majority of the measurements can be done with stationary or moving ionization chambers. Depending on the dosimetry task, other detector systems (see Section 11.2) can also be used. However, in addition to evaluation of angular dependence, spatial resolution, linearity of the response, radiation hardness, and long-term time stability within the assumed dynamic range of the measurements, the user should also check LET dependence of the detector response against the reference instrument, an ionization chamber. Typical basic requirements to the detectors that are sufficient for the majority of proton dosimetry point measurements in scattered proton beams are as follows:

- Spatial resolution of 1–2 mm
- Linearity of response: dose 0.05 to 30 Gy; 0.05 to 100 Gy/min.

11.4.1 Constancy Checks of Dose Monitor Calibration

The calibration of the beam dose monitor in dose per MU for scattered beams is performed by placing an ionization chamber in the center of reference SOBP for a field size of 10 x 10 cm^2. Reference conditions for absorbed dose to water determination in scattered beams will be described in Chapter 12 (Dosimetry and Beam Calibration), and an example of instrumentation is shown in Figure 11–21. The purpose of constancy checks that are performed as a part of daily QA procedure is to verify that the dose per MU value is within established tolerances. The checks are usually done with the same type of ionization chambers that were used for reference beam calibration. To make the daily QA procedure more practical, the water phantom used in reference monitor calibration is immediately replaced with the plastic phantom, and baseline data for constancy check are determined. More details of QA procedures are given in Chapter 15 (Machine Quality Assurance).

(a)

(b)

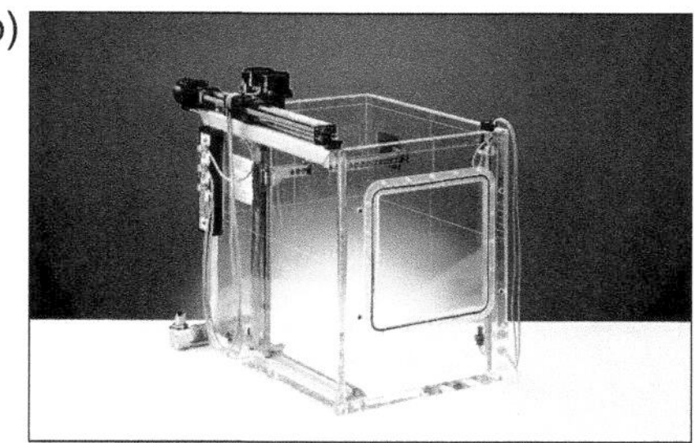

Figure 11–21 Equipment for constancy checks of dose monitor calibration (courtesy of PTW).

11.4.2 Measurements of Output Factors

Different probes of active and passive dosimeters that can be used for point dose measurements in a water or plastic phantom were described in section 11.2. The measurement of output factors or dependence of dose per MU on field size can be performed with ionization chambers or solid state detectors (diodes, diamond, or TLD) if their linearity of response was checked against an ionization chamber within the used clinical range of proton beams. The measurement of output factors for field sizes below 5 x 5 cm^2 requires detectors with high spatial resolution, therefore small volume (pinpoint) ionization chambers or solid state detectors should be used. Small-field dosimetry is discussed in Chapter 26 (SRS) and is an evolving field.

11.4.3 Relative Dose Measurements in the Plane Perpendicular to the Beam Axis

The purpose of these measurements is to determine the profile of an irradiation field. The main requirement of the probe detector for profile measurements is high spatial resolution. The measurements of profiles for small radiosurgery beams should be performed with radiochromic films placed in plastic slabs perpendicular to the beam axis. As long as the energy is constant over the irradiation field, the response of the used probes is either directly proportional to the dose, or it may be converted into the dose by a calibration curve, which has to be determined for the respective beam quality. One-dimensional dose distributions may be measured by a computer-controlled water phantom that allows positioning of a probe detector within 1 mm or better. The software of these phantoms can be programmed to acquire specific sets of scans in an appropriate sequence, display the results in real time, and store the results in digital form. However, the accurate setup of the computer-controlled tank requires substantial effort and time to achieve precision of horizontal and vertical alignment with beam axes to within 1 mm.

Some practical considerations should be accounted for when working with a computer-controlled water phantom for profile measurements. The most common factors are the same as described in AAPM Report TG-106 [106]:

- The orientation of the detector should be appropriate relative to the direction of scanning to provide proper spatial resolution.
- The user should check that the detector's axes of motion are parallel, in depth, to the CAX and, transversally, to the water surface, to a precision of 1 mm over the range of travel.

To make the measurements more efficient, it is possible to extract beam profiles from two-dimensional dose distributions obtained with film, a scintillating screen, or TLD foils as described in Section 11.2 or from 2D measurements with multi-detector arrays as described in Section 11.3.

11.4.4 Relative Depth–Dose Measurements for Scattered Beam

In passive scattering, beam depth–dose distribution may be measured by a computer-controlled water tank. The recommended probe is a plane-parallel ionization chamber. In some regions of the depth–dose curve, the dose changes rapidly with depth; therefore, the depth must be determined very accurately. To achieve this, the depth–dose distributions input into the treatment planning system are usually performed with a horizontal beam traversing the wall of the water tank rather than through the top surface of the water. This avoids the problems associated with water waves and allows a very accurate determination of depth. On the other hand, this can make measuring the dose close to the surface difficult, which can be important for capturing aperture scatter and the small amount of buildup seen with high-energy proton beams. It is therefore recommended that one wall of the water tank have a section with a thin window, as seen in Figure 11–22.

Some practical considerations should be accounted for when working with computer-controlled water tank for depth dose measurements:

- For measurements in vertical beams, the speed of scanning should be adjusted to avoid both water ripples and surface tension on the water surface.
- Mechanical sag of the tank support system and evaporation of water should be controlled to verify stability of the reference point and the coordinate system relative to the isocenter. The most convenient way is to use a robotic positioner with specific immobilization tools to perform the reproducible setup of the water tanks.

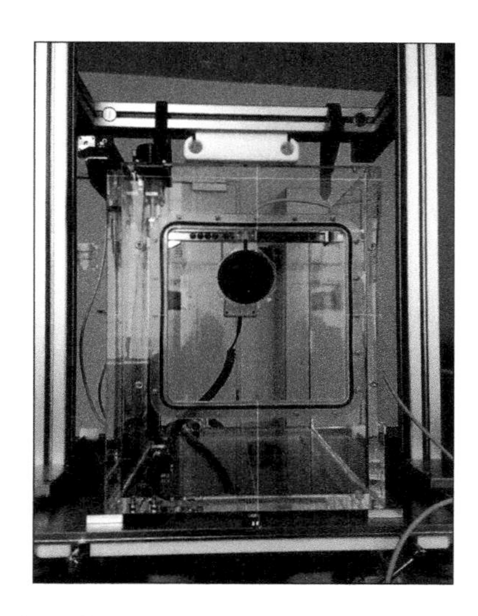

Figure 11–22 A view of the front window of a computerized water tank.

- The orientation of the detector should be appropriate relative to the direction of scanning.
- The user should check that the detector's axes of motion are parallel, in depth, to the CAX and, transversally, to the water surface, to a precision of 1 mm over the range of travel. This requirement is extremely important for depth–dose measurements in small beams.

11.4.5 Patient-specific Portal Calibration

The measurement of portal-specific calibration coefficients in terms of dose per MU is usually performed with the same instrumentation as for constancy checks of dose monitor calibration (Section 11.4.1). The selection of ionization chambers for field sizes below 5 x 5 cm^2 require a detector with high spatial resolution, therefore small-volume ionization chambers should be used. More details are given in Chapter 16 (Patient QA).

11.5 Relative Dosimetry for Scanned Beams

The major requirements of the detectors for measurements in scanned beams are similar to those used in dosimetry of scattered beams; however, due to the changes of dose deposition in time in both directions, laterally and in depth, the used detectors require additional dose rate linearity, positional accuracy, and the ability to capture the whole beam profile at once. The scanned beam delivery system uses a superposition of many pencil beams with different energies and quality. To select the proper dosimetry system, the user should take into account the following considerations:

- The dose to the single slice is delivered with very high dose rate.
- The dose delivered to each voxel is a superposition of many spots, and to achieve high accuracy in dose measurements (2 to 3%) the accuracy of measurements for a single spot should be better than 1%.

11.5.1 Relative Depth–Dose Measurements

Since in a narrow beam it is difficult to measure dose at a point on the central axis, it is more practical to measure the dose over the whole lateral plane at the depth of interest, so essentially dose-area-product is measured. The depth dose-area-product distribution for a pencil beam is typically measured using a large-diameter, parallel-plate ionization chamber immersed in a water tank. Compared to the measurements in scattered beams that use a small diameter (4 to 20 mm) parallel-plate ionization chamber, for these measurements a chamber with a diameter of 80 to 120 mm should be used—in other words, a chamber large enough to intercept not only all the primary protons, but also the secondary charged particles generated within the beam. The charge produced in the chamber at a given depth and normalized to the applied monitor units is thus essentially proportional to the lateral dose distribution integrated over the whole plane at that depth. The central axis depth–dose distributions for the pencil beams without range shifting might also be measured using a small-volume solid state detector (diode, diamond, etc.). The accuracy of this type of measurement is dependent upon the mechanical accuracy of the tank and of the positioning of the detector.

11.5.2 Relative Dose Measurements in the Plane Perpendicular to the Beam Axis

The input data requirements for modulated scanning dose delivery with pencil beams require lateral profiles for several of the available energies and spot sizes. The lateral profiles of the beam spots are usually needed to fit the beam model for dose distribution calculations. During commissioning of a beam delivery system, the variation of spot shape and size in air at different distances from the isocenter, different off-axis positions, and different depths within a phantom needs to be measured. These measurements are performed with films, TLD foils, and scintillation screens with CCD cameras as described in Section 11.2.

11.5.3 Verification of Dose Delivery

The scanning beam delivery with proton beams allows the buildup of the dose as a superposition of many thousands of individually placed and weighted pencil beams. Therefore, a simultaneous verification of absorbed dose at many points is required. Such practice is established in conventional intensity-modulated radiotherapy (IMRT), where measured and calculated dose distributions are compared on a matrix of points using multi-detector systems. The verification of treatment plans for scanned beam delivery can be done with radiochromic films and with the planar arrays of ionization chambers, as described in Section 11.3.2. The arrays can be placed at different depths using plastic slab phantoms, and the measured results are compared with dose distributions from TPS using software similar to that used in the verification of IMRT. In addition to stacked or planar arrays of ionization chambers, ion chambers can also be clustered to give a sampling of a 3D dose distribution. A device developed at GSI [107] and commercially available through PTW Freiburg uses 24 small-volume pinpoint ionization chambers individually calibrated to the dose absorbed in water and connected to two multichannel electrometers. The ionization chambers are fixed in bore holes of a PMMA mounting, which is attached to the motorized arm of the water phantom (Figure 11–23). The chambers are arranged in such a way that they do not cover each other in beam's eye view. By rotating the mounting 90°, dose distributions in the planes perpendicular to the beam direction can be measured. Using this system, chambers can be positioned under visual control to any point in the phantom to measure any part of the 3D dose distribution. In addition to cluster arrangement, verification of dose delivery can be performed with a linear array that is attached to the motorized arm of the water phantom (Figure 11–23b). Twenty-four small-volume calibrated pinpoint ionization chambers are installed in an array connected to two multichannel electrometers.

All of the ionization chamber arrays discussed above have advantages and drawbacks as QA tools; unfortunately, none of them can successfully address the whole

(a)

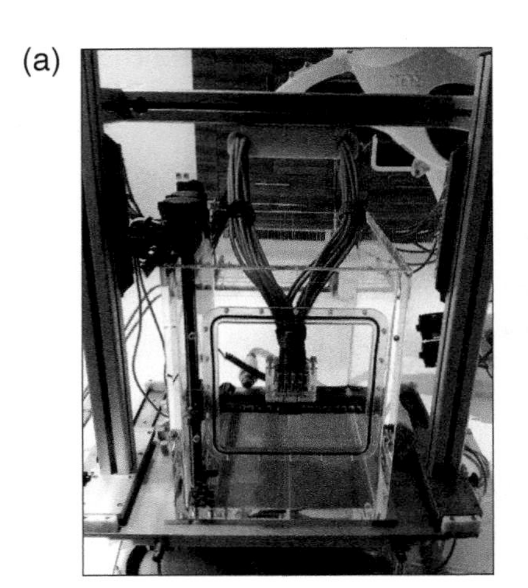

(b)

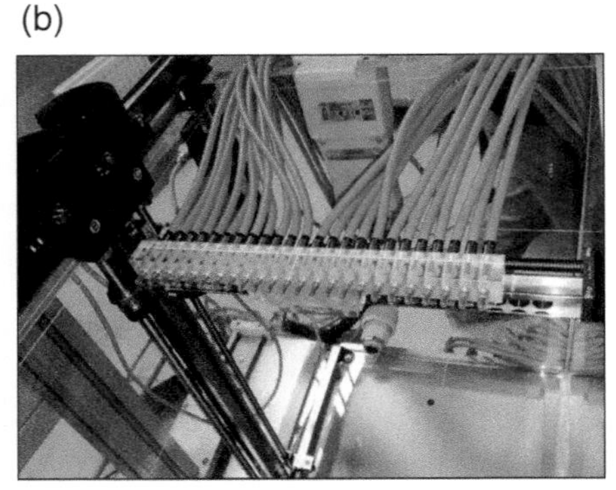

Figure 11–23 (a) The PTW multi-detector system with 24 ionization chambers densely packed into a PMMA mounting and installed in a water tank. (b) The 24 ionization chambers grouped into linear array, which is attached to the motorized arm of a water phantom.

spectrum of requirements that a detector system must meet in order to characterize a dose field created by scanning beam delivery.

The available software allows control position of the cluster or linear array and registers the measured responses; however, the whole package for comparison of measured and calculated data from treatment plan is not available commercially and should be customized by the user.

Practical considerations:

- The radiation detectors used most commonly in relative dosimetry are air-filled ionization chambers. Ionization chambers have long-term stability, high sensitivity, and very small energy dependence in response to proton beams, and they should be used as reference detectors for calibration or cross-calibration of other detectors.
- Stacked or planar arrays of ionization chambers used for QA measurements should periodically be checked against cylindrical or parallel plate chambers with small active volumes to verify calibration of elements in the multi-detector array.

11.6 Summary

In this chapter we have outlined the main features of instrumentation for dosimetry in proton therapy and have guided medical physicists through the practical implementation of different detectors and systems in performing dosimetry tasks at the proton therapy facility. Dosimetry is a critical foundation of the radiotherapy process, and the more advanced the beam delivery sequence is, the more complicated the instrumentation becomes to support acceptance testing and commissioning of beam delivery systems, reference calibration of clinical beams, acceptance testing and commissioning of TPS, periodic QA checks, and verification of dose delivery. It is emphasized that the user should know the limits of detectors in terms of dose and dose rate linearity, energy, angular, temporal and spatial response, LET dependence, quenching, and other limitations. We hope that the guidance provided will help medical physicists select the proper equipment and successfully use it to complete their dosimetry tasks.

References

1. Gillin MT, Sahoo N, Bues M, Ciangaru G, Sawakuchi G, Poenisch F, Arjomandy B, Martin C, Titt U, Suzuki K, Smith AR, Zhu XR. Commissioning of the discrete spot scanning proton beam delivery system at the University of Texas M. D. Anderson Cancer Center, Proton Therapy Center, Houston. *Med Phys* 2010;37:154–163.
2. Nohtomi A, Sakae T, Tsunashima Y, Kohno R. Dosimetry of pulsed clinical proton beams by a small ionization chamber. *Med Phys* 2001;28:1431–1435.
3. Schreuder AN, Jones DTL, Kiefer A. A small ionization chamber for dose distribution measurements in a clinical proton beam. In: Amaldi U, Larsson B, Lemoigne Y., Eds. Advances in Hadrontherapy Amsterdam: Elsevier Science B.V; 1997; pp.284–289.
4. Koehler AM. Dosimetry of proton beams using small silicon diodes. *Radiat Res* 1967; (Suppl 7):53–63.
5. Onori S, De Angelis C, Fattibene P, Pacilio M, Petetti E, Azario L, Miceli R, Piermattei A, Barone L, Tonghi B, Cuttone G, Nigro SL. Dosimetric characterization of silicon and diamond detectors in low-energy proton beams. *Phys Med Biol* 2000;45:3045–3058.
6. International Commission on Radiation Units and Measurements (ICRU) Prescribing, Recording, and Reporting Proton Beam Therapy. ICRU Report 78 *Journal of ICRU* 2007.
7. Coutinho L, Das I, Würfel J, Cheng CW. Characteristics of diode detectors in proton beam therapy. *Med Phys* 2012;39:3886.
8. Newhauser WD, Burns J, Smith AR. Dosimetry for ocular proton beam therapy at the Harvard Cyclotron Laboratory based on the ICRU Report 59. *Med Phys* 2002;29:1953–1961.
9. Nichiporov D, Kostjuchenko V, Symons J, Khrunov V. On the properties of n-type silicon diode detectors for clinical proton dosimetry. *Radiat Meas* 2011;46:1628–1633.
10. Grusell E, Medin J. General characteristics of the use of silicon diode detectors for clinical dosimetry in proton beams. *Phys Med Biol* 2000;45:2573–2582.
11. Pacilio M, De Angelis C, Onori S, Azario L, Fidanzio A, Miceli R, Piermattei A, Kacperek A. Characteristics of silicon and diamond detectors in a 60 MeV proton beam. *Phys Med Biol* 2002;47:N107–N112.
12. Vatnitsky SM, Miller DW, Moyers MF, Levy RP, Schulte RW, Slater JD, Slater JM. Dosimetry techniques for narrow proton-beam radiosurgery. *Phys Med Biol* 1999;44:2789–2801.
13. Vatnitsky SM, Khrunov VS, Fominych VI, Schuele E. Diamond detector dosimetry for medical applications. *Radiat Prot Dosim* 1993;47:515–518.
14. Vatnitsky S, Järvinen H. Application of a natural diamond detector for the measurement of relative dose distributions in radiotherapy. *Phys Med Biol* 1993;38:173–184.
15. Khrunov VS, Martynov SS, Vatnitsky SM, Ermakov IA, Chervjakov AM, Karlin DL, Fominych VI, Tarbeyev YV. Diamond detectors in relative dosimetry of photon, electron and proton radiation fields. *Radiat Prot Dosim* 1990;33:155–157.
16. De Angelis C, Onori S, Pacilio M, Cirrone GAP, Cuttone G, Raffaele L, Bucciolini M, Mazzocchi S. An investigation of the operating characteristics of two PTW diamond detectors in photon and electron beams. *Med Phys* 2002;29:248–254.
17. Bucciolini MF, Buonamici FB, Mazzocchi S, De Angelis C, Onori S, Cirrone GAP. Diamond detector versus silicon diode and ion chamber in photon beams of different energy and field size. *Med Phys* 2003;30:2149–2154.

18. Mandapaka AK, Ghebremedhin A, Patyal B, Marinelli M, Prestopino G, Verona C, Verona-Rinati G. Evaluation of the dosimetric properties of a synthetic single crystal diamond detector in high energy clinical proton beams. *Med Phys* 2013;40:121702.
19. Akino Y, Gautam A, Coutinho L, Würfel J, Das IJ. Characterization of a new commercial single crystal diamond detector for radiation dosimetry. *J Rad Res* 2015 (in press).
20. Marinelli M, Prestopino G, Verona C, Verona-Rinati G, Mandapaka AK, Ghebremedhin A, Patyal B. Cirrone GAP, Cuttone G, La Rosa S, Raffaele L. Dosimetric evaluation of a microdiamond prototype in clinical proton and carbon-ion beams. *Physica Medica* 2014; 30 (Suppl 1):e34.
21. Das I, Akino Y, Reed R, Gautam AS, Jan U, Würfel JU. Characteristics of a commercial micro-diamond detector for proton beam. Poster presentation 54th meeting of the PTCOG, May 15–23, 2014 Shanghai, China.
22. Ciancaglioni I, Marinelli M, Milani E, Prestopino G, Verona C, Verona-Rinati G, Consorti R, Petrucci A, De Notaristefani F. Dosimetric characterization of a synthetic single crystal diamond detector in clinical radiation therapy small photon beams. *Med Phys* 2012;39:4493.
23. Vatnitsky S, Miller D, Siebers J, Moyers M. Application of solid state detectors for dosimetry of therapeutic proton beams. *Med Phys* 1995;22:469–473.
24. Fidanzio A, Azario L, De Angelis C, Pacilio M, Onori S, Kacperek A, Piermattei A. A correction method for diamond detector signal dependence with proton energy. *Med Phys* 2002;29:669–675.
25. Bichsel H. Calculated Bragg curves for ionization chambers of different shapes. *Med Phys* 1995;22:1721–1726.
26. Pai S, Das IJ, Dempsey JF, Lam KL, LoSasso TJ, Olch AJ, Palta JR, Reinstein LE, Ritt D, Wilcox EE. Radiographic film for megavoltage beam dosimetry. *Med Phys* 2007;34:2228–2258.
27. Moyers MF, Vatnitsky S. Practical implementation of light ion beam treatments. Madison, WI: Medical Physics Publishing, 2012.
28. Georg D, Kroupa B, Winkler P, Poetter R. Normalized sensitometric curves for the verification of hybrid IMRT treatment plans with multiple energies. *Med Phys* 2003;30:1142–1150.
29. Srivastava SP, Das IJ. Effect of processor temperature on film dosimetry. *Med Dosim* 2012;37:138–139.
30. Zhu XR, Yoo S, Jursinic PA, Grimm DF, Lopez F, Rownd JJ, Gillin MT. Characteristics of sensitometric curves of radiographic films. *Med Phys* 2003;30:912–919.
31. Moyers MF. EDR-2 film response to charged particles. *Phys Med Biol* 2008;53(10):N165–N173.
32. Zhao L, Coutinho L, Cao N, Cheng CW, Das IJ. Temporal response of Gafchromic EBT2 radiochromic film in proton beam irradiation. In: Long M, Ed. World congress proceeding of med phys. Beijing, China: Springer, 2012; pp. 1164–1167.
33. Piermattei A, Miceli R, Azario L, Fidanzio A, Delle Canne S, De Angelis C, Onori S, Pacilio M, Petetti E, Raffaele L, Sabini MG. Radiochromic film dosimetry of a low energy proton beam. *Med Phys* 2000;27:1655–1660.
34. Butson M, Cheung JT, Yu PKN. Radiochromic film dosimetry in water phantoms. *Phys Med Biol* 2001;46:N27–N31.
35. Vatnitsky SM. Radiochromic film dosimetry for clinical proton beams. *Appl Radiat Isot* 1997;45:641–653.
36. Daftari I, Castenadas C, Petti PL, Singh RP, Verhey LJ. An application of GafChromic MD-55 film for 67.5 MeV clinical proton beam dosimetry. *Phys Med Biol* 1999;44:2735–2745.
37. Zhu Y, Kirov AS, Mishra V, Meigooni AS, Williamson JF. Quantitative evaluation of radiochromic film response for two-dimensional dosimetry. *Med Phys* 1997;24:223–31.
38. Micke A, Lewis DF, Yu X. Multichannel film dosimetry with non-uniformity correction. *Med Phys* 2011;38:2523–2534.
39. Mumot M, Mytsin G, Luchin YI, Molokanov A. A comparison of dose distributions measured with two types of radiochromic film dosimeter MD55 and EBT for proton beam of energy 175 MeV. 46th meeting of the PTCOG, May 18–23, 2007; Wanjie Proton Therapy Center, China.
40. Martiskova M, Ackermann B, Klemm S, Jaekel O. Use of Gafchromics EBT films in heavy ion therapy. *Nucl Instr Meth* 2008;A591:171–173.

41. Arjomandy B, Tailor R, Anand A, Sahoo N, Gillin M, Prado K, Vicic M. Energy dependence and dose response of Gafchromic EBT2 film over a wide range of photon, electron, and proton beam energies. *Med Phys* 2010;37:1942–1947.
42. Zhao L, Das IJ. Gafchromic EBT film dosimetry in proton beams. *Phys Med Biol* 2010;55: N291–N301.
43. Kirby D, Green S, Palmans H, Hugtenburg R, Wojnecki C, Parker D. LET dependence of GafChromic films and an ion chamber in low-energy proton dosimetry. *Phys Med Biol* 2010;55:417.
44. Fiorini F, Kirby D, Thompson J, Green S, Parker DJ, Jones B, Hill MA. Under-response correction for EBT3 films in the presence of proton spread out Bragg peaks. *Physica Medica* 2014;30:454–461.
45. Sorriaux J, Kacperek A, Rossomme S, Lee JA, Bertrand D, Vynckier S, Sterpin E. Evaluation of Gafchromic EBT3 films characteristics in therapy photon, electron and proton beams. *Physica Medica* 2013;29:461–469.
46. Luchin YI, Vatnitsky SM, Miller DW, Kostyuchenko VI, Nichiporov DF, Slater JD, Slater JM. The use of radiochromic film in treatment verification of proton radiosurgery. *J Radiosurg* 2000;3:69–75.
47. Vatnitsky SM, Schulte RWM, Galindo R, Meinass HJ, Miller DW. Radiochromic film dosimetry for verification of dose distributions delivered with proton beam radiosurgery. *Phys Med Biol* 1997;42:1887–1898.
48. Fattibene P, Calicchia A, d'Errico F, De Angelis C, Egger E, Onori S. Preliminary assessment of LiF and alanine detectors for the dosimetry of proton therapy beams. *Radiat Prot Dosim* 1996;66:305–309.
49. Besserer J, Bilski P, de Boer J, Kwiecien T, Moosburger M, Olko P, Quicken P. Dosimetry of low-energy protons and light ions. *Phys Med Biol* 2001;46:473–485.
50. Ibbott G. The radiological physics center TLD proton dosimetry credentialing program. 2008, private communication.
51. Czopyk L, Klosowski M, Olko P, Swakon J, Waligorski MPR, Kajdrowicz T, Cuttone G, Cirrone GAP, d Di Rosa F. Two-dimensional dosimetry of radiotherapeutical proton beams using thermoluminiscence foils. *Radiat Prot Dosim* 2007;126:185–189.
52. McKeever SWS, Blair MW, Bulur E, Gaza R, Kalchgruber R, Klein DM, Yukihara EG. Recent advances in dosimetry using the optically stimulated luminescence of Al2O3:C. *Radiat Prot Dosim* 2004;109:269–276.
53. Karsch L, Beyreuther E, Burris-Mog T, Kraft S, Richter C, Zeil K, Pawelke J. Dose rate dependence for different dosimeters and detectors: TLD, OSL, EBT films, and diamond detectors. *Med Phys* 2012;39:2447–2455.
54. Andersen CE, Edmund JM, Medin J, Grusell E, Jain M, Mattsson S. Medical proton dosimetry using radioluminescence from aluminium oxide crystals attached to optical-fiber cables. *Nucl Instr Meth* 2007;A580:466–468.
55. Reft C. The energy dependence and dose response of a commercial optically stimulated luminescent detector for kilovoltage photon, megavoltage photon, and electron, proton, and carbon beams. *Med Phys* 2009;36:1690–1699.
56. Rah JE, Oh do H, Kim JW, Kim DH, Suh TS, Ji YH, Shin D, Lee SB, Kim DY, Park SY. Feasibility study of glass dosimeter for in vivo measurement: dosimetric characterization and clinical application in proton beams. *Int J Radiat Oncol Biol Phys* 2012;84(2):e251–6.
57. Rah JE, Oh do H, Shin D, Lee SB, Kim TH, Kim JY, Kase Y, Li Z, Ibbott GS, Koss PJ, Lin L, McDonough J, Arjomandy B, Park SY. Feasibility study of glass dosimeter for postal dose intercomparison of high-energy proton therapy beams. *Radiat Meas* 2013;59:66–72.
58. Zink S. Proton dosimetry at ITEP accelerator using alanine and radiochromic detectors: a preliminary study. XVII PTCOG meeting 1992; Loma Linda, CA.
59. Gall K, Siebers J, Moyers M, Miller D, Coursey B, Desrosiers M, Dick C, Puhl J. Proton dosimetry intercomparison: potential for ESR. *Particles* 1993;11:A2.
60. Onori S, De Angelis C, Aragno D, Mattacchioni A. Possible applications of alanine-based dosimetry in radiotherapy. *Radiother Oncol* 2001;61(S1):106–107.
61. Palmans H. Effect of alanine energy response and phantom material on depth dose measurements in ocular proton beams. *Technol Cancer Res Treatment* 2003;2:579–586.
62. Ableitinger A, Vatnitsky S, Herrmann R, Bassler N, Palmans H, Sharpe P, Ecker S, Chaudhri N, Jäkel O, Georg D. Dosimetry auditing procedure with alanine dosimeters for light ion beam therapy. *Radiother Oncol* 2013;108(1):99–106.

63. Hansen JW, Olsen KJ. Theoretical and experimental radiation effectiveness of the free radical dosimeter alanine to irradiation with heavy charged particles. *Radiat Res* 1985;104:15–27.
64. Bartolotta A, Fattibene P, Onori S, Pantaloni M, Petetti E. Sources of uncertainty in therapy level alanine dosimetry. *Appl Radiat Isot* 1993;44:13–17.
65. Onori S, d'Errico F, De Angelis C, Egger E, Fattibene P, Janovsky I. Alanine dosimetry of proton therapy beams. *Med Phys* 1997;24:447–453.
66. Waligórski MPR, Danialy G, Loh KS, Katz R. The response of the alanine detector after charged-particle and neutron irradiations. *Appl Radiat Isot* 1989;40:923–933.
67. Marrale M, Longo A, Brai M, Barbon A, Brustolon M. Discrimination of radiation quality through second harmonic out-of-phase cw-ESR detection. *Radiat Res* 2014;181(2):184–192.
68. Nichiporov D, Kostjuchenko V, Puhl JM, Bensen DL, Desrosiers MF, Dick CE, McLaughlin WL, Kojima T, Coursey BM, Zink S. Investigation of applicability of alanine and radiochromic detectors to dosimetry of proton clinical beams. *Appl Radiat Isot* 1995;46:1355–1362.
69. Benton EV, Ogura K, Frank AL, Atallah T, Rowe V. Response of different types of CR-39 to energetic ions. *Nucl Tracks Radiat Meas* 1986;12:79–82.
70. Benton EV. On latent track formation in organic nuclear charged particle track detectors. *Radiation Effects and Defects in Solids* 1970;2(4):273–280.
71. Schneider U, Agosteo S, Pedroni E, Besserer J. Secondary neutron dose during proton therapy using spot scanning. *Int. J. Radiat Oncol Biol Phys* 2002;53:244–251.
72. Molokanov AG, Begusova M, Spurny FB, Vlcek B. Microdosimetric characteristics of the clinical proton beams at JINR, Dubna. *Radiat Prot Dosim* 2002;99:433–434.
73. Pachnerová Brabcová K, Ambrožova I, Kubančak J, Puchalska M, Vondraček V, Molokanov AG, Sihver L, Davídková M. Dose distribution outside the target volume for 170-MeV proton beam. *Radiat Prot Dosimetry* 2014;161:410–6.
74. Archambault L, Polf JC, Beaulieu L, Beddar S. Characterizing the response of miniature scintillation detectors when irradiated with proton beams. *Phys Med Biol* 2008;53:1865–1876.
75. Torrisi, L. Plastic scintillator investigations for relative dosimetry in proton-therapy. *Nucl Instrum and Meth* 2000;B170:523–530.
76. Seravalli E, de Boer MR, Geurink F, Huizenga J, Kreuger R, Schippers JM, van Eijk CWE. 2D dosimetry in a proton beam with a scintillating GEM detector. *Phys Med Biol* 2009;54:3755–65.
77. Wang LLW, Perles LA, Archambault L, Sahoo N, Mirkovic D, Beddar S. Determination of the quenching correction factors for plastic scintillation detectors in therapeutic high-energy proton beams. *Phys Med Biol 2012*;57(23):7767–82/
78. Boon SN. Dosimetry and quality control of scanning proton beams. Doctoral thesis in mathematics and natural sciences (Rijksuniversteit Groningen, The Netherlands), 1998.
79. Boon SN, van Luijk P, Böhringer T, Coray A, Lomax A, Pedroni E, Schaffner B, Schippers JM. Performance of a fluorescent screen and CCD camera as a two-dimensional dosimetry system for dynamic treatment techniques. *Med Phys* 2000;27:2198–2208.
80. Lomax A, Böhringer T, Bolsi D, Coray A. Emert F, Goitein G, Jermann M, Lin S, Pedroni E, Rutz H, Stadelmann O, Timmermann B, Verwey J, Weber DC. Treatment planning and verification of proton therapy using spot scanning: Initial experiences. *Med Phys* 2004;31:3150–3157.
81. Pedroni E, Scheib S, Boehringer T, Coray A, Grossmann M, Lin S, Lomax A. Experimental characterization and physical modelling of the dose distribution of scanned proton pencil beams. *Phys Med Biol* 2005;50:541–561.
82. Grevillot L, Bertrand D, Dessy F, Freud N, Sarrut D. A Monte Carlo pencil beam scanning model for proton treatment plan simulation using GATE/GEANT4. *Phys Med Biol* 2011; 56:5203–5219.
83. Farr JB, Mascia AE, Hsi WC, Allgower CE, Jesseph F, Schreuder AN, Wolanski M. Clinical characterization of a proton beam continuous uniform scanning system with dose layer stacking. *Med Phys* 2008;35:4945–4954.
84. Safai S, Lin S, Pedroni E. Development of an inorganic scintillating mixture for proton beam verification dosimetry. *Phys Med Biol* 2004;49:4637–4655.
85. Kostjuchenko V, Nichiporov D. Measurement of the 12C(p,pn)11C reaction from 95 to 200 MeV. *Int J Appl Radiat Isot* 1993;44:1173–1175.

86. Nichiporov D. Verification of absolute ionization chamber dosimetry in a proton beam using carbon activation measurements. *Med Phys* 2003;30:972–978.
87. Nichiporov D, Luckjashin V, Kostjuchenko V. Measurement of the activity of ^{11}C and ^{22}Na sources using 4pi-beta-gamma coincidence system. *Appl Radiat Isot* 2004;60:703–716.
88. Johansson S, Magnusson AP, Medin J, Grusell E, Olsson LE, Olsson P. Clinical proton dosimetry using two different dosimeter gels and MRI. *Med Biol Eng Comput* 1997;35:961.
89. Oldham M, Siewerdsen JH, Shetty A, Jaffray DA. High resolution gel-dosimetry by optical-CT and MR scanning. *Med Phys* 2001;28:1436–45.
90. Kelly RG, Jordan KJ, Battista JJ. Optical CT reconstruction of 3D dose distributions using the ferrous-benzoic-xylenol (FBX) gel dosimeter. *Med Phys* 1998;25:1741–1750.
91. Bäck S, Medin J, Magnusson P, Olsson P, Grussel E, Olsson E. Ferrous sulphate gel dosimetry and MRI for proton beam dose measurements. *Phys Med Biol* 1999;44:1983–1996.
92. Maryanski MJ, Schulz RJ, Ibbott GS, Gatenby JC, Xie J, Horton D, Gore JC. Magnetic resonance imaging of radiation dose distributions using a polymer-gel dosimeter. *Phys Med Biol* 1994;39:1437–1455.
93. Doran SJ, Koerkamp KK, Bero MA, Jenneson P, Morton EJ, Gilboy WB. A CCD-based optical CT scanner for high-resolution 3D imaging of radiation dose distributions: equipment specifications, optical simulations and preliminary results. *Phys Med Biol* 2001;46:3191–213.
94. Hilts M, Jirasek A, Audet C, Duzenli C. X-ray CT polymer gel dosimetry: applications in stereotactic radiosurgery and proton therapy. *Radiother Oncol* 2000;56:S1–S80.
95. Gustavsson H, Bäck SA, Medin J, Grusell E, Olsson LE. Linear energy transfer dependence of a normoxic polymer gel dosimeter investigated using proton beam absorbed dose measurements. *Phys Med Biol* 2004;49:3847–3855.
96. Heufelder J, Stiefel S, Pfaender M, Luedemann L, Grebe G, Heese J. Use of BANG® polymer gel for dose measurements in a 68 MeV proton beam. *Med Phys* 2003;30:1235–1240.
97. Adamovics J, Maryanski M. A new approach to radiochromic three-dimensional dosimetry-polyurethane. *Journal of Physics: Conference Series* 2004;3:172–175.
98. Al-Nowais S, Doran S, Kacperek A, Krstajic N, Adamovics J, Bradley D. A preliminary analysis of LET effects in the dosimetry of proton beams using PRESAGE™ and optical CT. *Appl Radiat Isot* 2009;67:415–418.
99. Grant R, Ibbott G, Zhu X, Carroll M, Adamovics J, Oldham M, Followill D. Investigation of PRESAGE® Dosimeters for Proton Therapy. *Med Phys* 2011;38:3571.
100. Zhao L, Das IJ, Zhao Q, Thomas A, Adamovics J, Oldman M. Determination of the depth dose distribution of proton beam using PRESAGE™ dosimeter. *J Phys Conf Ser* 2010;250:012035
101. Zhao L, Newton J, Oldham M, Das IJ, Cheng CW, Adamovics J. Feasibility of using PRESAGE® for relative 3D dosimetry of small proton fields. *Phys Med Biol* 2012;57(22): 431–444.
102. Nichiporov D, Solberg K, His W, Wolanski M, Mascia A, Farr J, Schreuder A. Multichannel detectors for profile measurements in clinical proton fields. *Med Phys* 2007;34:2683–2690.
103. Dhanesar S, Sahoo N, Kerr M, Taylor MB, Summers P, Zhu XR, Poenisch F, Gillin M. Quality assurance of proton beams using a multilayer ionization chamber system. *Med Phys* 2013;40(9):092102 1–9.
104. Arjomandy B, Sahoo N, Ding X, Gillin M. A 2D ion chamber array detector as a QA device for spot scanning proton beams. *Med Phys* 2008;35:2779–2784.
105. Amerio S, Boriano A, Bourhale F, Cirio R, Donetti M, Fidanzio A, Garelli E, Giordanengo S, Madon E, Marchetto F, Nastasi U, Peroni C, Piermattei A, Freire CJS, Sardo A, Trevisiol E. Dosimetric characterization of a large area pixel-segmented ionization chamber. *Med Phys* 2004;31:414–420.
106. Das IJ, Cheng C-W, Watts RJ, Ahnesjö A, Gibbons Li JXA, Lowenstein Mitra RK, Simon WE, Zhu TC. Accelerator beam data commissioning equipment and procedures: Report of the TG-106 of the therapy physics committee of the AAPM. *Med Phys* 2008;35:4186–4210.

107. Karger CP, Jäkel O, Hartmann GH, Heeg P. A system for three-dimensional dosimetric verification of treatment plans in intensity-modulated radiotherapy with heavy ions. *Med Phys* 1999;26:2125–2132.

Chapter 12

Dosimetry and Beam Calibration

Hugo Palmans[1,2] and Stanislav Vatnitsky[1]

[1]EBG MedAustron GmbH,
Wiener Neustadt, Austria
[2]National Physical Laboratory,
Teddington, UK

12.1 Introduction

Accurate dosimetry of clinical proton beams is an essential step in achieving tumor control while minimizing side effects and, thus, for exploiting proton therapy to its full clinical potential. Given typical uncertainty requirements on the dose delivered to the target volume of 3 to 7% [1–3], the uncertainty on reference and relative dosimetry, constituting only one step in the process of dose delivery, should be well below that, i.e., ideally lower than 1%. Concepts of proton dosimetry are, in general, very similar to those used in conventional radiotherapy using high-energy photon or electron beams. Nevertheless, there are a number of important distinctions, especially when performing dosimetry of scanned proton beams. This chapter provides an overview of the methods used for reference dosimetry, the physics and procedures of current dosimetry protocols, and how consistency in dosimetry is verified using traceable detector calibrations, auditing, and integration in end-to-end tests. Dosimetry of special systems using small fields and scanned delivery modes is treated in a separate section. The last section covers micro and nanodosimetry.

12.2 Primary Standards

Primary standards are instruments of the highest metrological quality to determine the unit of a quantity from its definition and of which the accuracy has been verified by comparison with standards of other institutions at the same level [4]. In most cases, only national metrology institutes, such as the National Institute of Science and Technology (NIST) in the United States, can afford the resources to maintain instruments at the required level and engage in the required comparisons. For the quantity of absorbed dose, the only instruments that measure this quantity directly according to its definition are calorimeters. Nevertheless, absorbed dose could be derived from the measurement of a different quantity if the relation between the measured quantity and absorbed dose is considered a fundamental constant of nature. For example, it can be derived from a measurement of ionization if the mean energy required to produce an ion pair in the medium, W_{med}, is accurately known of from a measurement of fluence using a Faraday cup if the mass stopping power, S_{med}/ρ of the medium is accurately known.

12.2.1 Concepts, Traceability, etc.

"Traceability" is the concept that describes the chain that links a measurement of a quantity in the field to a standard. Traceability is obtained by the process of "calibration," and the entire path between a clinical measurement and a standard is referred to as the calibration chain. User instruments can be directly calibrated against these primary standards, but more often they are calibrated against secondary standards, calibrated against a primary standard, that are maintained in accredited dosimetry calibration laboratories (ADCL) (or more generically in the international context: secondary standard laboratories, or SSDLs). The exact role and position of PSDLs, SSDLs, the Bureau International des Poids et Mésures (BIPM), and the role of key comparisons is explained at the BIPM web pages (http://www.bipm.org) and in particular for dosimetry also in IAEA TRS-398 [1]. Quality assurance programs must be in place to ensure the quality of the calibration chain.

12.2.2 Faraday Cup

The Faraday cup is the oldest instrument that has been used for absolute dosimetry of proton beams based on its ability to measure the number of protons in a beam accurately, provided that it is well-designed. The number of protons is derived from the charge collected on the collecting electrode (CE) measured by an electrometer. The

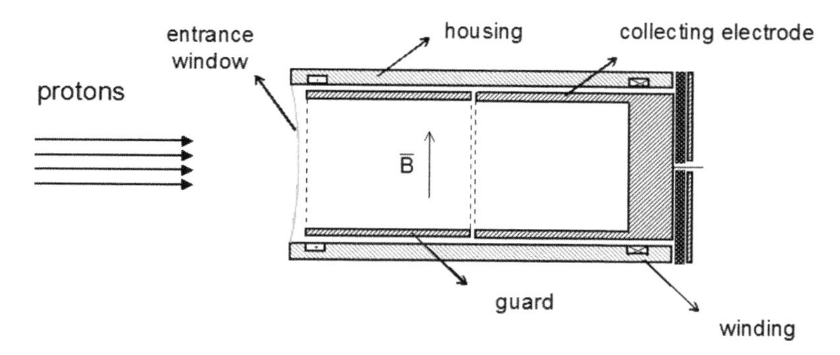

Figure 12–1 Schematic diagram of a reference dosimetry level Faraday cup with internal vacuum. Shown are the collecting electrode, the guard electrode (which is at negative potential with respect to the collecting electrode), the entrance window, and the windings creating a magnetic field, B, to suppress the loss of electrons generated in the collecting electrode. (Replotted from Palmans and Vynckier [5], with permission of the publisher and the authors.)

CE should be sufficiently thick to stop all incident charged particles, as well as those generated within it. For broad beams, a major uncertainty contribution comes from the determination of the field area, from which particles are collected. For pencil beams, this uncertainty vanishes, provided the diameter of the CE is large enough to capture the entire beam.

A major concern is the influence of secondary electrons generated by ionization in the entrance window that reach the collecting electrode (and thus reduce the signal) as well as electrons liberated in and escaping from the collecting electrode (which enhance the signal). Both sources of perturbation to the measurement are usually suppressed by a guard electrode with a negative potential with respect to the electrode and casing. Further improvements of the efficiency can be achieved by adding windings in the constructions that generate a magnetic field perpendicular to the beam direction, improving the collecting efficiency by trapping electrons and forcing them to be deposited close to the generation point. (Figure 12–1).

12.2.3 Calorimeters

The most direct way to measure absorbed dose is calorimetry; all the energy deposited by the radiation will appear as heat if no other changes of internal energy take place in the medium, such as chemical, phase, or lattice energy changes. The energy deposition in calorimeters is measured by quantifying the electrical energy dissipation that is needed to realize the same temperature rise in the medium as the radiation does. If the specific heat capacity, c_{med}, of the medium is known, absorbed dose to the medium can also be derived from a measurement of the temperature rise DT as:

$$D_{med} = c_{med} \Delta T \quad (12.1)$$

While primary standards of absorbed dose in high-energy photon and electron beams based on calorimeters have been well established [6,7], no such primary standards exist for proton beams. Nevertheless, calorimeters made of water, graphite, and tissue-equivalent plastic have been used in experimental settings [8], and their feasibility to serve as primary standard instruments for proton beams has been demonstrated.

12.2.3.1 Water Calorimeters

Water calorimeters have the main advantage of measuring the quantity of interest in radiotherapy, i.e., absorbed dose to water, directly. The signal-to-noise ratio is, however, small, and while the low thermal diffusivity of water is an advantage in the sense that is allows sufficient time to measure a temperature increase before heat conduction results in a large perturbation of the signal, it has the disadvantage of resulting in instruments that equilibrate very slowly.

In addition to the simple calorimetry equation given above, a number of correction factors k_i are required for any heat transported away or toward the measurement point, field non-uniformity and fluence perturbations due to the presence of non water-equivalent materials. In addition, the radiolysis of the medium results in free radicals triggering a chain of reactions, which can be endothermic or exothermic, resulting overall in a net heat defect (h) and a correction factor $\frac{1}{1-h}$. This leads to the overall equation:

$$D_w = c_w \Delta T \frac{1}{1-h} \prod k_i \tag{12.2}$$

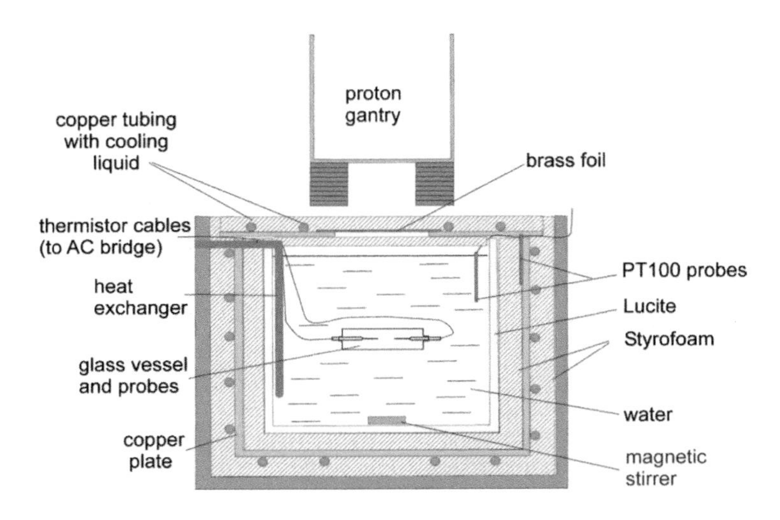

Figure 12–2 Schematic cross section of a sealed water calorimeter. (Reproduced from Sarfehnia et al. [9] with permission of the publisher and the authors.)

Figure 12–2 shows a typical example of a modern sealed water calorimeter that has been used in passively scattered as well as in scanned proton beams. It consists of a water phantom which is stabilized at a temperature of approximately 4 °C where the density of water reaches a local maximum and, consequently, the onset of convection due to density differences in minimized. The radiation-induced temperature rise is measured using thermistors fixed in thin glass probes. These glass probes are located in a thin-walled glass container containing high-purity water to control the chemical heat defect. The location of the measurement point is essentially the position of the thermistors.

The main source of uncertainty in water calorimetry is due to chemical heat defect. Since the primary yield of the initial radicals is linear energy transfer (LET) dependent, the resulting heat defect will be LET dependent (see Chapter 3).

From experiments and simulations, the following is known about the chemical heat defect in protons:

1. For high-energy (low-LET) protons, pure water saturated with a chemically inert gas like argon or nitrogen exhibits a small (<0.1%) initial heat defect, reaching a steady state after irradiation [10]. For high-LET protons, however, simulations show a steady increase in the chemical energy in the aqueous system due to a higher production of hydrogen peroxide than what is decomposed, resulting in a non-zero endothermic heat defect [11]. This was confirmed experimentally by comparing the heat defect of water with that of aluminum by measuring the heating of a dual-component water/aluminum absorber [12] in which a beam of incident protons is either completely absorbed in water or in aluminum. The following fit can be derived for the relative chemical heat defect h, as introduced in Equation 12.2, of pure water as a function of LET in keV μm^{-1} [13]:

$$h = \left(0.041 \pm 0.004\right)\left(e^{-(0.035\pm0.010)LET} - \left(1.000 \pm 0.001\right)\right) \qquad (12.3)$$

where the uncertainties given are 1 SD.

2. For pure water saturated with hydrogen, both simulations and relative comparison of the heat defect with other systems indicate that it exhibits a zero heat defect over the entire LET range, which can be explained by an enhanced decomposition of hydrogen peroxide compared to the nitrogen system. When initial oxygen concentrations are present, the hydrogen system exhibits an initial exothermic heat defect which increases until depletion of oxygen, after which the heat defect drops abruptly to zero. This way it can be monitored when the steady-state, zero-heat-defect condition is reached.
3. For water with a known quantity of sodium formate (which serves as a deliberately added organic impurity) saturated with oxygen, the exothermic heat defect in a modulated beam is only about half of that in a ^{60}Co beam with the same dose rate. This was explained by a combination of the lower chemical

yields for certain species at high LET and the time structure of the formation of chemical species due to the beam modulation [10].

12.2.3.2 Graphite Calorimeters

Graphite calorimeters have the advantage of a lower specific heat capacity than water, leading to a six times higher signal for the same dose. The higher thermal diffusivity of graphite as compared to water, however, means that a sample of graphite has to be isolated from the phantom to prevent the radiation-induced heat from flowing away from the measurement point too quickly. This sample is called the core, and it is usually separated from the phantom environment by one or more vacuum gaps. A portable graphite calorimeter for proton beam dosimetry is shown in Figure 12–3.

Two modes of operation are possible:

1. Quasi-adiabatic mode in which all components are left to stabilize passively until temperature drifts are sufficiently small to measure the radiation-induced temperature rise. The operational principle is similar as that of the water calorimeter, and Equation 12.1 applies again with a number of correction factors for heat transfer by radiation over the vacuum gaps and conduction through the thermistor wires, fluence perturbation due to the presence of vacuum gaps, volume averaging over the core, beam uniformity, impurities of the thermistors, etc.

2. Isothermal mode in which the temperature of every component of the calorimeter is actively stabilized to a constant temperature (above the environmental temperature). The energy deposited due to irradiation is then determined by integrating the electrical power reduction that is needed to keep the core at constant temperature.

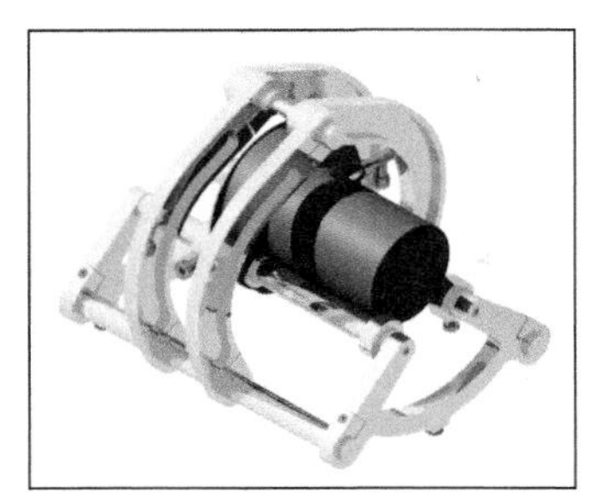

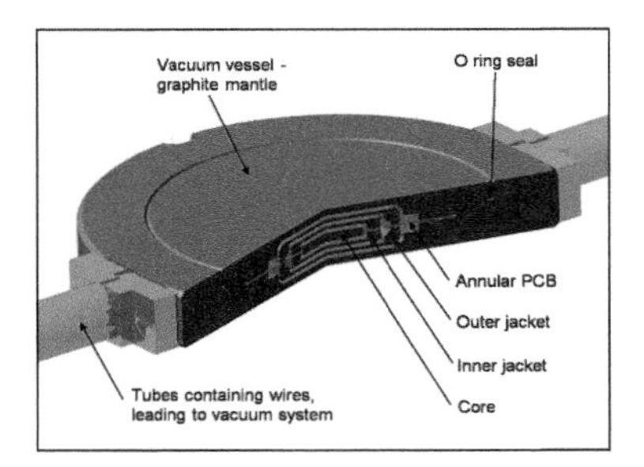

Figure 12–3 Schematic drawing of a portable graphite calorimeter for light ion beam dosimetry built at the National Physical Laboratory, UK. Left: assembly of the frame, phantom, and central calorimeter parts. Right: central calorimeter parts containing the core, two jackets, a printed circuit board to organize wiring of thermistor leads, and the vacuum vessel, which is positioned in the phantom.

It is generally assumed that there is no chemical heat defect in graphite calorimeters, although it has been suggested that if dissolved oxygen is present in the graphite matrix between the grain boundaries, about 5 kGy of pre-irradiation dose is needed to remove an initial heat defect of about 2% [14]. Apart from that, investigation of graphite used in nuclear reactors has revealed that a small amount of the energy deposited in graphite by ionizing radiation is stored in lattice defects and is released upon heating the graphite [15]. Schulz et al. [16] used a double graphite/aluminum absorber experiment, similar in concept to the experiment of Brede et al. [12] mentioned earlier, finding an endothermic heat defect of 0.4% with a standard uncertainty of 0.3%.

The largest uncertainty contribution in graphite calorimetry for radiotherapy, however, comes from the need to convert dose-to-graphite, D_g, to dose-to-water, D_w. The ratio of dose to graphite at the measurement depth in a graphite phantom and dose to water at the water-equivalent-depth (WED) in water (scaled by the ranges, see Chapter 4) can be derived from charged particle spectra differential in energy, $\Phi_{E,g,i}$ and $\Phi_{E,w,i}$, at those equivalent depths in both phantoms as:

$$\frac{D_w(\text{WED})}{D_g} = \frac{\sum_i \left[\int_0^{E_{\max,i}} \Phi_{E,w,i} \left(\frac{S_{\text{el},i}}{\rho} \right)_w dE \right]}{\sum_i \left[\int_0^{E_{\max,i}} \Phi_{E,g,i} \left(\frac{S_{\text{el},i}}{\rho} \right)_g dE \right]} \tag{12.4}$$

where i refers to the charged particle type (the summations are over all charged particle types), $\left(\frac{S_{\text{el},i}}{\rho} \right)_{\text{med}}$ is the electronic mass collision stopping power, and $E_{\max,i}$ the maximum energy of particle type i in the fluence distribution.

If the charged particle spectra are identical at the equivalent depths in both phantoms, i.e., $\Phi_{E,g,i} = \Phi_{E,w,i}$, then the dose conversion is adequately described by the mass stopping power ratio water to graphite for the charged particle spectrum in either phantom:

$$\frac{D_w(\text{WED})}{D_g} = \frac{\sum_i \left[\int_0^{E_{\max,i}} \Phi_{E,w,i} \left(\frac{S_{\text{el},i}}{\rho} \right)_w dE \right]}{\sum_i \left[\int_0^{E_{\max,i}} \Phi_{E,w,i} \left(\frac{S_{\text{el},i}}{\rho} \right)_g dE \right]} = \frac{\sum_i \left[\int_0^{E_{\max,i}} \Phi_{E,g,i} \left(\frac{S_{\text{el},i}}{\rho} \right)_w dE \right]}{\sum_i \left[\int_0^{E_{\max,i}} \Phi_{E,g,i} \left(\frac{S_{\text{el},i}}{\rho} \right)_g dE \right]} = S_{w,g} \tag{12.5}$$

where $S_{w,g}$ is the Bragg-Gray stopping power ratio water-to-graphite. If the charged particle spectra at the measurement depth in graphite and at the WED in water are not the same, then a fluence correction factor has to be introduced, which can be derived by multiplying and dividing Equation 12.4 with $\sum_i \left[\int_0^{E_{\max,i}} \Phi_{E,w,i} \left(\frac{S_{\text{el},i}}{\rho} \right)_w dE \right]$:

$$\frac{D_w(\text{WED})}{D_g} = \frac{\sum_i \left[\int_0^{E_{\max,i}} \Phi_{E,w,i} \left(\frac{S_{\text{el},i}}{\rho} \right)_w dE \right] \sum_i \left[\int_0^{E_{\max,i}} \Phi_{E,g,i} \left(\frac{S_{\text{el},i}}{\rho} \right)_w dE \right]}{\sum_i \left[\int_0^{E_{\max,i}} \Phi_{E,g,i} \left(\frac{S_{\text{el},i}}{\rho} \right)_g dE \right] \sum_i \left[\int_0^{E_{\max,i}} \Phi_{E,g,i} \left(\frac{S_{\text{el},i}}{\rho} \right)_w dE \right]} = s_{w,g} k_{fl} \tag{12.6}$$

Comparing equations 12.5 and 12.6 shows that

$$k_{fl} = \frac{\sum_i \left[\int_0^{E_{\max,i}} \Phi_{E,w,i} \left(\frac{S_{\text{el},i}}{\rho} \right)_w dE \right]}{\sum_i \left[\int_0^{E_{\max,i}} \Phi_{E,g,i} \left(\frac{S_{\text{el},i}}{\rho} \right)_w dE \right]} \tag{12.7}$$

demonstrating that this correction factor accounts purely with difference in charged particle fluence, since the stopping power is the same in numerator and denominator. The differences in fluence are due to the differences in the absorption of primary protons in non-elastic nuclear interactions and differences in the production of secondary charged particles in water and graphite. A simple analytical calculation using nuclear interaction data from ICRU Report 63 [17] shows that the number of primary protons absorbed over the entire range in graphite is different from that in water by 2% for 60 MeV protons and 8% for 200 MeV protons [18]. Monte Carlo simulations, on the other hand, indicate that this is, to some extent, compensated by the difference in secondary particle production, leading to dose conversion corrections limited to 0.4% at 60 MeV and 0.6% at 200 MeV.

Water to graphite stopping power ratios and fluence correction factors calculated by Monte Carlo using Geant4 are shown in figures 12–4 and 12–5, respectively.

12.2.3.3 Tissue-equivalent Calorimeters

Solid tissue-equivalent calorimeters have been developed and used in proton beams, and their operational principles are the same as for graphite calorimetry. The main difference is that the polymer-based plastics used in these calorimeters can exhibit a substantial chemical heat defect due to break-up of the polymer structure. Fleming and Glass [23], McDonald and Goodman [24], and Schulz et al. [16] used a similar double absorber as described above to determine the heat defect of A-150 tissue-equivalent plastic. The results of the three studies show an endothermic heat defect close to 4%. Because of this, both AAPM Report 16 of AAPM TG-20 [25] and the European

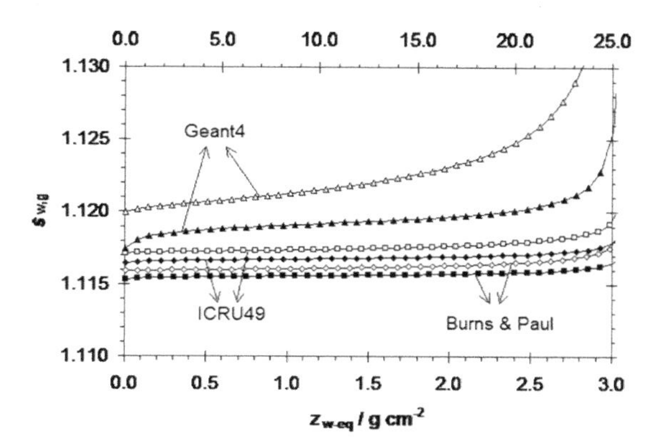

Figure 12–4 Water to graphite stopping power ratios as a function of depth for 60 MeV protons (hollow symbols, lower horizontal axis) and 200 MeV protons (solid symbols, upper horizontal axis) based on the stopping powers from Geant4, those from ICRU Report 49 [19], and those from Burns [20] and Paul [21]. (Figure reproduced from Palmans [22], with permission.)

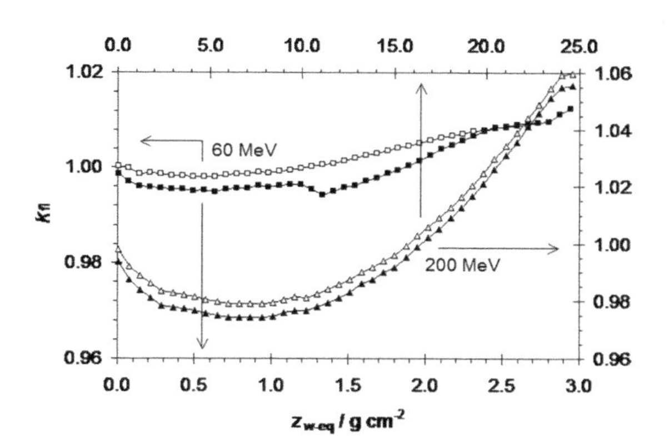

Figure 12–5 Monte Carlo-calculated fluence correction factors as a function of depth for 60 MeV incident protons based on Equation 12.7 using only the proton fluence (hollow squares) and using all charged particles (solid squares) and for 200 MeV protons using only the proton fluence (hollow triangles) and using all charged particles (solid triangles). (Figure reproduced from Palmans [22], with permission.)

Charged Heavy Particle Dosimetry group ECHED [26,27] recommend a correction factor of 1.040 ±0.015 for the heat defect of an A-150 calorimeter.

12.2.4 Ionization Chambers

Ionization chambers, described in the previous chapter, can in principle offer a direct determination of absorbed dose to water without calibration. The relation between dose to water in the phantom and the average dose to air in the cavity is given by Bragg-Gray cavity theory, provided the charged particle fluence in the air cavity is the same as at the measurement point in water:

$$D_w = \bar{D}_{\mathrm{air}}\, s_{w,\mathrm{air}}\, p \qquad (12.8)$$

where $s_{w,air}$ is the Bragg–Gray or Spencer–Attix mass collision stopping power ratio water over air for the charged particle spectrum at the measurement point in water and

p is a correction factor to account for any deviation from the conditions under which Bragg–Gray cavity theory is valid, such as the presence of the wall and central electrode. The average dose to air in the cavity can in principle be obtained from the measured ionization, M, knowledge of the mass of air in the cavity ($m_{air} = r_{air}\ V_{cav}$) and the mean energy required to produce an ion pair in air, W_{air}.

$$\bar{D}_{air} = \frac{M\,W_{air}}{\rho_{air}\,V_{cav}} \tag{12.9}$$

In practice, however, the volume of commercial ionization chambers is not known with the required precision, and one has to rely on a calibration in a reference beam to estimate the volume, or bypass the requirement of knowing the volume. It is worthwhile to mention, however, that if an accurate estimate of the volume is available independently of a calibration, e.g., from a manufacturer's blueprints or measured dimensions (as in primary standards of air kerma), dosimetry using an ionization chamber could be based on first principles using equations 12.8 and 12.9. It has been shown that this can provide reliable monitor unit calibration for a transmission ionization chamber to link with a Monte Carlo-based planning system [28].

12.3 Reference Dosimetry Using Ionization Chambers

Ionization chambers are the preferred reference dosimeter in the clinical environment because of their ease of operation, robustness, and long-term stability. An ionometric dosimeter system for reference dosimetry in proton beams should contain the following components:

- one or more ionization chamber assemblies, which include the electrical fitting and any permanently attached cable, intended for different purposes;
- a measuring assembly (electrometer), often separately calibrated in terms of charge or current per scale division;
- one or more phantoms with waterproof sleeves; and
- one or more stability check devices.

12.3.1 Formalisms

All formalisms for absorbed dose to water start from the fundamental relation between fluence and dose:

$$D_{med} = \sum_i \left[\int_0^{E_{max,i}} \Phi_{E,med,i} \left(\frac{S_{el,i}}{\rho} \right)_{med} dE \right] = \Phi_{med} \overline{\left(\frac{S_{el}}{\rho} \right)}_{med} \tag{12.10}$$

where i refers to the charged particle type, $\Phi E_{med,i}$, is the charged particle fluence differential in energy, $\left(\frac{S_{el,i}}{\rho} \right)_{med}$ is the electronic mass collision stopping power, and

$E_{\max,i}$ the maximum energy of particle type i in the fluence distribution, $\overline{\left(\frac{S_{\text{el}}}{\rho}\right)}_{\text{med}}$ is the mean stopping power given by

$$\overline{\left(\frac{S_{\text{el}}}{\rho}\right)}_{\text{med}} = \frac{\sum_i \left[\int_0^{E_{\max,i}} \Phi_{E,\text{med},i} \left(\frac{S_{\text{el},i}}{\rho}\right)_{\text{med}} dE\right]}{\sum_i \left[\int_0^{E_{\max,i}} \Phi_{E,\text{med},i} dE\right]} \tag{12.11}$$

This can be either directly from the fluence in the water (replace "med" with "w" in above equations) or $D_{w,Q}$, in a proton beam of beam quality Q (representing the entire charged particle fluence differential in energy for all charged particles):

$$D_{w,Q} = \bar{D}_{air,Q} \left(s_{w,air}\right)_Q p_Q \tag{12.12}$$

In order to facilitate later the discussion of contributions to differences in dosimetry protocols, the overall expression to derive dose to water from the ionization measurements can be split in three factors as follows:

$$D_{w,Q} = \left[M_Q\right]\left[\frac{1}{\rho_{air} V_{cav}}\right]\left[\left(W_{air}\right)_Q \left(s_{w,air}\right)_Q p_Q\right] \tag{12.13}$$

All protocols for dosimetry of proton beams using ionization chambers and calculated data can be reduced to this factorization in which the second factor, representing an estimate of the ionization chamber volume, is solely related to the calibration conditions, whereas the third factor is solely related to the proton beam.

12.3.2 Fluence-based Approach

While not related to ionization chamber measurements, it is worth mentioning the fluence-based approach to dosimetry in the formalism since it is the most direct way to link dose and fluence, and because it enables making the formal link between these approaches and ionization chamber-based codes of practice. In a broad proton beam, dose to water at a shallow depth, z, can in principle be derived from the proton fluence at the surface, Φ, as:

$$D_w(z) = \Phi \overline{\left(\frac{S_{\text{el}}}{\rho}\right)}_w \prod k_i \tag{12.14}$$

where $\prod k_i$ is the product of correction factors for beam divergence, scatter, field non-uniformity, beam contamination, and secondary particle buildup. It is obvious that this method relies on accurate values of the proton stopping power in water, for which the uncertainty is estimated to be 1% to 2% according to ICRU Report 49 [19]. The Faraday cup is the most common way to measure the fluence, but it can also be derived from the activation of a sample if the production cross section of the radioisotope being quantified is known [29].

12.3.3 Air Kerma-based Approach

Before the advent of absorbed dose-based protocols for reference dosimetry in light ion beams, most recommendations for reference dosimetry using ionization chambers were based on calibrations in terms of exposure or air kerma. ICRU Report 59 [30] was the last international recommendation of this type, and we take this as the basis of discussing the factorization of Equation 12.13. The general equation for absorbed dose determination is

$$D_{w,p} = M_p^{corr} \frac{N_K(1-g)A_{wall}A_{ion}}{s_{wall,g}\left(\mu_{en}/\rho\right)_{air,wall} K_{hum}} \left(s_{w,air}\right)_p \frac{\left(w_{air}\right)_p}{\left(W_{air}\right)_c} \tag{12.15}$$

the second factor in Equation 12.13 then becomes:

$$\left[\frac{1}{\rho_{\text{air}} V_{\text{cav}}}\right] = \frac{N_K(1-g)A_{\text{wall}}A_{\text{ion}}}{\left(W_{\text{air}}\right)_c s_{\text{wall},g}\left(\mu_{\text{en}}/\rho\right)_{\text{air,wall}} K_{\text{hum}}} \tag{12.16}$$

where N_K is the air kerma calibration coefficient in ^{60}Co, g is the correction for radiative losses, A_{wall} is the correction factor for absorption and scatter in the wall and build-up cap in ^{60}Co, A_{ion} is the ion recombination correction factor in the calibration beam, $(W_{air})_c$ is the mean energy required to produce an ion pair in dry air in the calibration beam quality, and K_{hum} is the humidity correction factor in the calibration beam. While it is interesting and necessary to study the influence of this factor on dosimetry using different dosimetry formalisms, it is not related to the proton beam.

In spite of several deficiencies of the air kerma approach [31], the adoption of ICRU Report 59 in multiple proton centers showed a considerable improvement in dosimetry homogeneity, and a comparison among various institutions resulted in agreements within 1.5% for all participants in the calibration of a common proton beam using their own instrumentation [32]. For this reason, the air kerma calibration approach is still in use in several institutions [33].

12.3.4 Dose to Water-based Approach

Dose to water-based formalisms are the simplest since they rely on an absorbed dose to water calibration coefficient $N_{D,w,Q0}$ in a calibration beam with beam quality Q_0 and only require one overall correction factor to account for the difference of the detector response in the calibration beam and the users beam. If we take IAEA TRS-398 [1] as an example, dose to water is determined as:

$$D_{w,Q} = M_Q N_{D,w,Q_0} k_{Q,Q_0} \tag{12.17}$$

The beam quality correction factors $k_{Q,Q0}$ can be measured as a ratio of calibration coefficients:

$$k_{Q,Q_0} = \frac{N_{D,w,Q}}{N_{D,w,Q_0}} \quad (12.18)$$

or calculated as

$$k_{Q,Q_0} = \frac{(W_{air})_Q (s_{w,air})_Q \, p_Q}{(W_{air})_{Q_0} (s_{w,air})_{Q_0} \, p_{Q_0}} \quad (12.19)$$

With this formalism, the second factor in Equation 12.13 is given by

$$\left[\frac{1}{\rho_{air} V_{cav}}\right] = \frac{N_{D,w,Q_0}}{(W_{air})_{Q_0} (s_{w,air})_{Q_0} \, p_{Q_0}} \quad (12.20)$$

For this approach to work, a suitable proton beam quality index has to be used that is able to characterize the charged particle spectrum at the measurement point in the proton field uniquely. IAEA TRS-398 uses the residual range, R_{res}, defined as the distance from the measurement point to the depth at which the absorbed dose beyond the Bragg peak or SOBP falls to 10% of its maximum value. It was shown that for beams with $R_{res} > 1$ g cm^{-2} (corresponding to an effective mono-energetic proton energy of 30 MeV), this beam quality index enables the selection of water to air stopping power ratios with an error smaller than 0.1%, and differences between unmodulated and modulated beams are smaller than 0.4%.

Interesting to note is that there may not be a great difference in uncertainty on the chamber volume as derived by this factor from air kerma or absorbed dose calibrations. However, in the absorbed dose-based protocols, the quantities occurring in the denominator of Equation 12.20 and those in the third factor in Equation 12.3 are the same, except for the difference in beam quality and, thus, they are expected to be more strongly correlated than those in the air kerma-based approach where the factors A_{wall} and k_m refer to very different conditions, as well as to the build-up cap, which is not present with in-phantom measurements. This shows that this factorization is only for illustrative purposes, but that the factors two and three cannot be considered independently.

12.3.5 Relations between Different Calibration Methods

The above equations allow easy linking of the different calibration methods. To give an example of this, take an air kerma-based formalism consistent in notation with IAEA TRS-277 [34]:

$$D_w = M_{corr} N_K (1-g) k_{att} k_m \frac{(W_{air})_Q}{(W_{air})_{Q_0}} (s_{w,air})_Q \, p_Q \quad (12.21)$$

The second term in Equation 12.13 thus becomes

$$\left[\frac{1}{\rho_{\text{air}} V_{\text{cav}}}\right] = \frac{N_K (1-g) k_{\text{att}} k_{\text{m}}}{\left(W_{\text{air}}\right)_{Q_0}} \tag{12.22}$$

Comparing equations 12.20 and 12.22 gives us a relation between $N_{D,w,Q0}$ and N_K:

$$N_{D,w,Q_0} = N_K (1-g) k_{\text{att}} k_{\text{m}} \left(s_{w,\text{air}}\right)_{Q_0} p_{Q_0} \tag{12.23}$$

This shows that the difference between absorbed dose and air kerma-based dosimetry is purely due to differences in the calibration beam and has no relation with quantities in the proton beam. Similar relations can be established for any pair of absorbed dose/air kerma-based formalisms and can reveal inconsistencies between experimental values of $N_{D,w,Q0}$ and N_K and theoretical data used in the protocols.

Comparing the fluence-based approach of Equation 12.13 with the absorbed dose-based approach of equations 12.17 to 12.19 shows that this can provide a value of the mean energy required to produce an ion pair in dry air:

$$\left(W_{\text{air}}\right)_Q \approx \left(W_{\text{air}}\right)_Q p_Q = \Phi \overline{\left(\frac{S_{\text{el}}}{\rho}\right)}_{\text{air}} \prod k_i \frac{\left(W_{\text{air}}\right)_{Q_0} \left(s_{w,\text{air}}\right)_{Q_0} p_{Q_0}}{M_{\text{corr}} N_{D,w,Q_0}} \tag{12.24}$$

where the only interaction data related quantity on the right-hand side of the equation is the mass electronic stopping power in air.

12.3.6 Calibration Routes

Ionization chambers are the preferred instruments for reference dosimetry in the clinic. Since the volume of commercial ionization chambers is normally not known with the required accuracy for dosimetry in radiotherapy, they are calibrated in terms of air kerma or in terms of absorbed dose to water in a calibration beam. If we only consider the latter for proton beams, this calibration beam would be, in order of preference:

1. the clinical proton beam itself,
2. another proton beam with different characteristics, or
3. a beam which is not a proton beam like a high-energy electron beam or a photon beam.

The first two options are at present unavailable, and it is unlikely that any calibration laboratory will have availability of proton beams soon. But those options may become available if standards laboratories would have permanent access to one clinical facility. Both approaches for calibration can be based on a calorimetric determination of absorbed dose to water in a proton beam of quality Q_{p0} and the use of a so-called dummy calorimeter, a replica of the calorimeter geometry except for the sensing thermistors, with an ionization chamber insert [16]. From the dose to water

$\left[D_{w,Q_{p0}}\right]_{\text{CAL}}$ measured by the calorimeter and the charge reading of the ionization chamber in the dummy calorimeter, M_{Qp0}, the absorbed dose to water calibration coefficient for the proton calibration beam quality Q_{p0} is then

$$N_{D,w,Q_{p0}} = \frac{\left[D_{w,Q_{p0}}\right]_{\text{CAL}}}{M_{Q_{p0}}} \tag{12.25}$$

The absorbed dose to water in a clinical beam of the same quality can then be derived directly from the corrected ionization chamber reading, M_{Qp0}, as:

$$D_{w,Q_{p0}} = M_{Q_{p0}} N_{D,w,Q_{p0}} \tag{12.26}$$

While this option, corresponding to the first calibration route, would be the ideal, in general it is impractical and inaccessible to provide these types of calibrations in every single clinical beam used (this is the case for all radiotherapy beam, but it is even more so for protons, given the higher cost of beam time). More likely would be to have a calibration coefficient available in a proton calibration beam quality Q_{p0} and use a beam quality correction factor for using the calibrated ionization chamber in a proton beam of a different quality, Q_p. This corresponds with the second calibration route, for which the absorbed dose to water in the proton beam with quality Q_p is given by

$$D_{w,Q_p} = M_{Q_p} N_{D,w,Q_{p0}} k_{Q_p,Q_{p0}} \tag{12.27}$$

where M_{Qp} is the charge collected in the ionization chamber at the reference point in water, corrected for influence quantities, and the factor $k_{Qp,Qp0}$ corrects the ionization chamber's response for the difference between the proton beam qualities Q_p and Q_{p0}. According to the recommendations of IAEA TRS-398 [1] and ICRU Report 78 [3], assuming constant values of $(W_{air})_Q$ and $p_Q \approx 1$ for proton beams, Equation 12.19 that is used to calculate the proton beam quality correction factor is then just simplified to a ratio of the stopping power ratios of water to air in the proton beam qualities Q_p and Q_{p0}:

$$k_{Q_p,Q_{p0}} = \frac{\left(s_{w,\text{air}}\right)_{Q_p}}{\left(s_{w,\text{air}}\right)_{Q_{p0}}} \tag{12.28}$$

The advantage of this approach becomes clear from Figure 12–6 showing that $k_{Qp,Qp0}$ values are close to unity and do not vary much with energy. For beam quality indices $R_{res} > 1$ g cm^{-2} the values are not more than 0.2% different from unity for any ionization chamber and any clinical proton beam quality. This picture does not take into account recent information on perturbation correction factors for ionization chambers, which will be discussed in Section 12.4.3. The plateau region of a monoen-

Table 12–1 Type-B uncertainties of the ionization chamber dosimetry with a Farmer-type chamber in a clinical proton beam (one standard deviation). The estimates are from IAEA TRS-398 [1], Vatnitsky et al. [35], and Palmans et al. [36].

Method	Contribution	Type-B Uncertainty (%)
$D_{w,Q_p} < N_{D,w,{}^{60}Co}$		
	$N_{D,w,{}^{60}Co}$	0.6
	$k_{Q_p,{}^{60}Co}$	1.7
	Total	2.0*
$D_{w,Q_p} < N_{D,w,Q_{p0}}$		
	$N_{D,w,Q_{p0}}$	1.0
	$k_{Q_p,Q_{p0}}$	0.6
	Total	1.2*

* The combined uncertainty on *Dw,Qp* also includes uncertainties for the ionization chamber reading and correction for influence quantities which are not listed in the table.

ergetic beam is the preferable location for the measurements at the reference beam quality due to the lower uncertainty in the determination of the stopping power ratio. A comparison of the type B uncertainties for dose measurements using ionization chambers with equations 12.19 and 12.28 in a clinical proton beam is given in Table

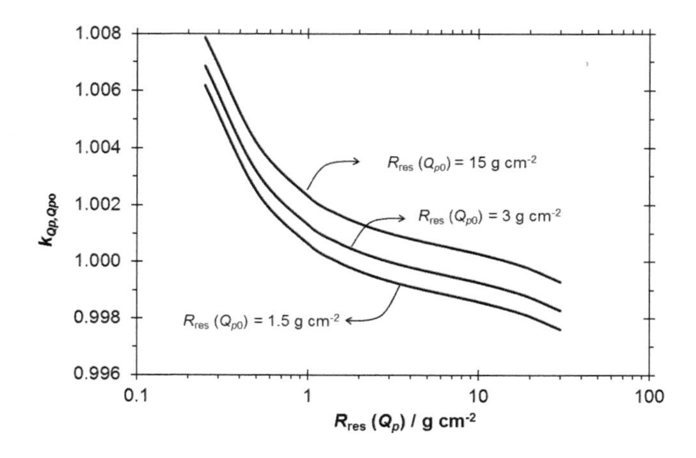

Figure 12–6 $k_{Qp,Qp0}$ values with reference to three different proton beam calibration qualities Q_{p0} (R_{res} = 15 g cm^{-2}, 3 g cm^{-2}, and 1.5 g cm^{-2}) as derived from the data of IAEA TRS-398 for any ionization chamber.

12–1. The table shows that the advantage of using an ion chamber with a calorimetry-based calibration coefficient obtained directly in a light ion beam compared to a calibration coefficient obtained in a ^{60}Co beam is a reduction in the uncertainty.

The major sources of uncertainty typically associated with ion chamber dosimetry, namely the proton W_{air} values, do not participate in dose determination using equations 12.27 and 12.28. The results shown in Table 12–1 are encouraging the development of calorimetry-based ion chamber calibrations.

A similar approach, using a cross calibration of a plane-parallel chamber against a cylindrical ionization chamber in a high-energy proton beam, has been proposed and can serve as a basis. Consistent with IAEA TRS-398 and ICRU Report 78, the formalism for the measurement of absorbed dose to water in the proton beam with beam quality Q_p can be written as:

$$D_{w,Q_p} = M_{Q_p} N_{D,w,Q_{p-\text{cross}}} k_{Q_p,Q_{p-\text{cross}}} \tag{12.29}$$

where M_{Qp} is the ionization chamber reading corrected for influence quantities, $N_{D,w,Qp\text{–cross}}$ the calibration coefficient in terms of absorbed dose to water in a calibration proton beam with beam quality $Q_{p\text{–cross}}$ and $k_{Qp,Qp\text{–cross}}$ and the beam quality correction factor to correct the calibration coefficient for use in the beam with beam quality Q_p. While theoretical values of these beam quality correction factors are not provided in IAEA TRS-398, they can be derived from the values with ^{60}Co as a reference quality as $k_{Q_p,Q_{p-\text{cross}}} = k_{Q_p,{}^{60}Co} / k_{Q_{p-\text{cross}},{}^{60}Co}$ or in IAEA TRS-398's short notation $k_{Q_p,Q_{p-\text{cross}}} = k_{Q_p} / k_{Q_{p-\text{cross}}}$. The advantage of this approach is that $k_{Q_p,Q_{p-\text{cross}}}$ values are close to unity and do not vary much with energy; as shown in Figure 12–6 for beam quality indices $R_{res} > 1$ g cm^{-2} the values are not more than 0.2% different from unity for any ionization chamber and any clinical proton beam quality. This picture does not take into account recent information on perturbation correction factors for ionization chambers which will be discussed in Section 12.4.3.3. The third approach is the one most commonly applied nowadays with ^{60}Co as the calibration reference. The k_Q values can therefore be directly found in IAEA TRS-398. However, some investigators have also explored the possibility of cross calibrating ionization chambers, in particular plane-parallel chambers, in electron beams, for which the same formulas as above can be used. There are indications that when using plane-parallel chambers, the consistency of proton dosimetry is improved when using cross calibrations in electrons as compared to ^{60}Co calibrations.

12.4 Dosimetry Protocols

Current dosimetry protocols or codes of practice based upon ionization chamber dosimetry include several major components:

- basic equations, referring to a formalism, to convert charge to absorbed dose for well characterized reference conditions,

Table 12–2 Reference conditions for the determination of proton beam quality R_{res} (from IAEA TRS-398)

Influence Quantity	Reference Value or Reference Characteristics
Phantom Material	Water
Chamber Type	Cylindrical and plane-parallel
Reference Point of Chamber	For plane-parallel chambers, on the inner surface of the window at its center. For cylindrical chambers, on the central axis at the center of the cavity volume
Position of Reference Point of Chamber	For plane-parallel and cylindrical chambers, at the point of interest
SSD	Clinical treatment distance
Field Size at the Phantom Surface	10 cm x 10 cm For small field applications (i.e., eye treatments), 10 cm x 10 cm or the largest field clinically available

- guidance on the selection of the necessary physics data and correction factors for the application of the protocol, and
- guidance on the correction of the response of the ionization chamber for different influence quantities.

These three aspects are described below, with particular emphasis on the needs for scattered proton beam dosimetry. Those aspects where scanned proton beam dosimetry deviate from these recommendations are discussed in Section 12.5.

12.4.1 Reference Conditions

As mentioned before, IAEA TRS-398 [1] and ICRU Report 78 [3] use the residual range, R_{res}, as a beam quality index to characterize the proton energy spectrum at the reference depth. Older protocols use different indices, such as the effective energy, but these can be related to R_{res} via the stopping power and range tables that were used in those protocols to make the conversion. The residual range R_{res} should be derived from a measured depth–dose distribution, obtained using the reference conditions given in Table 12–2. The preferred choice of detector for the measurement of central axis depth dose distributions is a plane-parallel chamber. The reference conditions for the determination of absorbed dose according to Equation 12.17 using an ionization chamber calibrated in a beam quality Q_0 are given in Table 12–3.

12.4.2 Influence Quantities

Most corrections for influence quantities do not require considerations that are specific for proton beams. The correction factor for deviation of the air density from the normal conditions for which the calibration is valid is given by:

$$k_{Tp} = \frac{(T + 273.15)}{(T_0 + 273.15)} \frac{p_0}{p} \tag{12.30}$$

Table 12–3 Reference conditions for the determination of absorbed dose in proton beams (from IAEA TRS-398)

Influence Quantity	Reference Value or Reference Characteristics
Phantom Material	Water
Chamber Type	For R_{res} 0.5 g cm^{-2}, cylindrical and plane-parallel. For $R_{res} < 0.5$ g cm^{-2}, plane-parallel.
Measurement of depth z_{ref}	Middle of the SOBP*.
Reference Point of Chamber	For plane-parallel chambers, on the inner surface of the window at its center. For cylindrical chambers, on the central axis at the center of the cavity volume
Position of Reference Point of Chamber	For plane-parallel and cylindrical chambers, at the measurement depth z_{ref}.
SSD	Clinical treatment distance
Field Size at the Phantom Surface	10 cm x 10 cm, or that used for normalization of the output factors, whichever is larger. For small field applications (i.e.m eye treatments), 10 cm x 10 cm or the largest field clinically available.

* The reference depth can be chosen in the "plateau region" at a depth of 3 g cm^{-2}, for clinical applications with a mono-energetic proton beam (e.g., for plateau irradiations).

where T is the temperature in °C and p the pressure in kPa of the air in the cavity of the ionization chamber and T_0 and p_0 are the reference conditions for temperature and pressure for which the calibration coefficient of the ionization chamber is valid and which amount to 22 °C in the United States (20 °C in most other countries in the world) and 101.325 kPa, respectively. No correction is made for the relative humidity if the ionization chamber is used in a range of 20% to 80% relative humidity and has a calibration coefficient valid at a relative humidity of 50%. If the ionization chamber and electrometer are calibrated separately, the calibration coefficient for the ionization chamber is given in units Gy/C. The calibration coefficient k_{elec} obtained for the electrometer converts the electrometer reading to charge and is expressed in unit rdg/C. If the reading of the electrometer is in terms of charge, the electrometer calibration coefficient is dimensionless. If the ionization chamber and the electrometer are calibrated together as one measurement assembly, no separate electrometer calibration coefficient has to be applied. The correction factor for polarity in a given radiation beam should be measured as:

$$k_{pol} = \frac{|M_+| + |M_-|}{2M} \tag{12.31}$$

where M_+ and M_- are the electrometer readings obtained at positive and negative polarity, respectively, and M is the electrometer reading taken at polarity used routinely.

Ion recombination effects, however, need some special attention. Recombination takes place because ion pairs formed in the air cavity may recombine before being collected at the electrodes, resulting in an underestimation of the ionization signal. There are two types of recombination processes: initial recombination within a single ionizing track and volume recombination involving ions formed in different particle tracks. In particular, the initial recombination has long been known to depend strongly on the ionization density distribution within the track and, thus, could potentially be higher in a proton beam as compared to photon or electron beams. However, an experimental investigation at the Clatterbridge Cancer Centre, UK, revealed that both types of recombination behaved in the same way as in photon beams [37]. Only very near the distal edge of the SOBP was there an indication of a slightly increased initial recombination effect. Another aspect is that the time-dependent structure of a proton dose delivery in a SOBP is important for recombination. This is of importance in modulated beams, as well as in scanned proton beams, as will be discussed further on.

Codes of practice such as IAEA provide formulas to apply the two-voltage method in pulsed and continuous beams. It is important to consider what a pulsed beam means, since it could be assumed that proton beams generated by accelerators such as cyclotrons and synchrotrons are inherently all pulsed. However, with respect to ion recombination, the definition of a pulsed beam is one for which the dose in a pulsed beam is delivered in a short time interval compared with the ion collection time, and in the same time the period between pulses is large compared to that collection time. The latter condition is not fulfilled in proton beams since the pulse repetition frequency is very high. Many pulses are delivered within an ion collection time interval so that with respect to recombination, proton beams should be treated as continuous beams [37]. The two-voltage method can then be used and a correction factor derived using the relation

$$k_s = \frac{\left(\frac{V_1}{V_2}\right)^2 - 1}{\left(\frac{V_1}{V_2}\right)^2 - \left(\frac{M_1}{M_2}\right)^2} \qquad (12.32)$$

This relation is based on a linear dependence of $1/M$ on $1/V^2$, which describes the effect of general recombination in continuous beams. Initial recombination is usually small. However, if initial recombination is significant, it may perturb this linearity and a modified version of Equation 12.32 should be used. It should be noted that the reference conditions for the calibration of ionization chambers in standards laboratories recommend that the calibration certificate states whether or not a recombination correction has been applied.

12.4.3 Data

Only data related to the proton beam will be discussed here, i.e., those quantities occurring in the third term in Equation 12.13: $(W_{air})_Q$, $(s_{w,air})_Q$, and p_Q. For the quantities occurring in the second term related to a ^{60}Co calibration beam quality we refer to dosimetry codes of practice [1,3,25,30,31,38–40].

12.4.3.1 Mean Energy Required to Produce an Ion Pair in Dry Air

The value of $(W_{air})_Q$ for protons has been the subject of controversy in the past. At the end of the 1980s, ECHED [26,27] adopted the value of 35.18 J C^{-1} from ICRU Report 31 [41], while AAPM TG-20 [25] previously recommended a value of 34.3 J C^{-1}. This discrepancy of 2.6% remained the source of differences in dosimetry recommendations until the publication of ICRU Report 78 [3], which adopted the same value of 34.2 J C^{-1}as was recommended in IAEA TRS-398 [1]. Experimental data on this quantity have come from two types of experiments:

1. simultaneous measurement of the energy loss over an air column and the ionization produced per proton and
2. comparison of the response of an ionization chamber and a calorimeter.

The first method is cumbersome and requires a correction for electron losses, which is difficult to determine. The second method has the disadvantage that it does not provide a direct measurement of $(W_{air})_Q$, but the pragmatic advantage is that if ionization chamber dosimetry is based on a value derived from calorimetry, it provides consistency with the dosimetry in high-energy photon beams. It is also worth noting that $(W_{air})_Q$ depends on protons energy, but that this dependence has not been very well characterized in the clinical energy range.

12.4.3.2 Water to Air Mass Collision Stopping Power Ratio

$(s_{w,air})_Q$ is governed by the same physics as for electrons and positrons, and the theoretical models for calculations at high energies are based on the same Bethe-Bloch formulas with a series of correction terms. For consistency with photon and electron dosimetry, ICRU Report 49 [19] recommended proton stopping powers using the same values for the mean excitation energy, I, as used in the electron and positron stopping power tables of ICRU Report 37 [42]. An important difference is that, compared to electrons, the clinically relevant range of proton energies is at much lower (non-relativistic) velocities where the density effect is of no importance, but where there is a strong energy (v^{-2}) dependence of the stopping powers [19]. The result is that the water to air collision stopping power ratios are fairly constant over the entire clinical energy range. The most recent recommended values of $(s_{w,air})_Q$ are Spencer-Attix stopping power ratios, including the contributions of secondary protons and electrons, and are given as a function of the residual range R_{res} by [1]:

$$\left(s_{w,air}\right)_Q = a + bR_{res} + \frac{c}{R_{res}} \tag{12.33}$$

where $a = 1.137$, $b = -4.3\ 10^{-5}$ and $c = 1.84\ 10^{-3}$.

12.4.3.3 Ionization Chamber Perturbation Factor

The ionization chamber perturbation factor, p_Q, corrects for deviations from Bragg-Gray conditions and can, according to IAEA TRS-398 [1], be described as the product of four factors:

1. the displacement correction factor, p_{dis}, for the deviation of the effective point of measurement from reference point of the ionization chamber,
2. the cavity perturbation correction factor, p_{cav}, for the perturbation of the charged particle fluence distribution due to the presence of air cavity,
3. the wall perturbation factor, p_{wall}, for the non-water equivalence of the ionization chamber's wall, and
4. the central electrode correction factor, p_{cel}, for the presence of the central electrode.

The first one can alternatively be dealt with by positioning the effective point of measurement at the required measurement depth. For proton beams, it is slightly easier to determine an effective point of measurement than for photon beams given the small lateral deflections that protons undergo. A reasonable approximation is thus to regard protons as traveling along straight lines once they enter the ionization chamber geometry and integrate their dose contributions over the cavity volume. For a cylindrical air cavity with radius R_{cav} in water, it is easy to show that this approach results in an effective point of measurement which is relative the center of the cavity positioned a distance $Dz_{cav} = 8R_{cav}/3\pi \approx 0.85R_{cav}$ closer to the phantom surface [43]. For Farmer-type chambers, the higher density of the wall and central electrode materials brings this slightly toward the center of the chamber and closer to the value of $Dz_{cav} = 0.75R_{cav}$ recommended in IAEA TRS-398 for ion beams, but this is not the case for other cylindrical chambers with a thick wall or central electrode, for which substantial deviations from this rule may occur. Regarding the other perturbation factors, all the

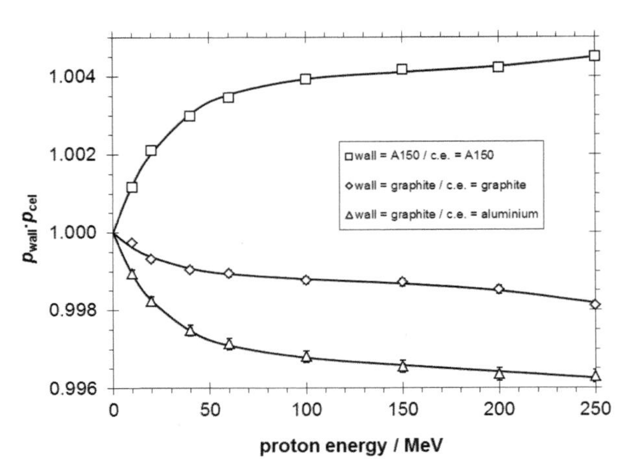

Figure 12–7 Secondary electron perturbation correction factors (i.e., the product of the wall and central electrode correction factors) for Farmer-type ionization chamber with different wall/central electrode material combinations obtained from Monte Carlo simulations. The data are derived from [36].

evidence points to corrections of less than 1% [36,44], and relative correction factors of 1.005 have been demonstrated both experimentally, by cavity theory, and by Monte Carlo simulation for A150-walled ionization chambers as compared to graphite-walled chambers [36]. For the central electrode effect, Monte Carlo simulations indicate that an aluminum central electrode in a Farmer-type chamber results in correction factors around 0.998 [36], while experimentally no values deviating significantly from unity have been demonstrated [44]. Figure 12.7 illustrates the relative importance of wall and central electrode correction factors in graphite-walled and A150-walled ionization chambers.

In the calculated values for protons ($k_Q = k_{Q,QCo-60}$) in IAEA TRS-398, the following assumptions are made: $(W_{air})_Q$ is constant as a function of proton energy or proton beam quality, and the perturbation correction factor in the proton beam is unity ($p_Q = 1$). The first assumption is consistent with the best compilation of calorimetric data available, with an adequate uncertainty assigned to it. The second assumption has been investigated to some extent since the publication of IAEA TRS-398. As shown in the previous paragraph, for cylindrical ionization chambers, gradient perturbation correction factors can be modeled analytically and can be represented by an effective depth in water, displaced from the reference point of the ionization chamber. Studies of secondary electron perturbations demonstrate that perturbations for ionization chambers in proton beams are substantial enough to be accounted for in dosimetry protocols, but until the present, comprehensive data are lacking.

12.5 Reference Dosimetry of Small and Scanned Beams

While scanned proton beams are mentioned in IAEA TRS-398 [1], no specific guidance was provided, leaving several aspects unclear. One of the issues is that the monitor chambers are usually calibrated in terms of the number of protons N in the narrow beam which is being scanned, or its equivalent in terms of energy deposition, which is the dose-area-product to water, DAP_w. This calibration has to be performed as a function of beam energy at a shallow point, but deep enough to establish secondary proton equilibrium. The quantities N and DAP_w are related by the following equation under the condition that the spectrum of the particles doesn't change as a function of the lateral position in the beam:

$$N \approx \frac{DAP_w}{(S/\rho)_w} \tag{12.34}$$

The approximation sign is there because the relation between dose and number of particles actually involves the fluence, which is only approximately the same as the number of protons incident per unit area. This equation also establishes the link between Faraday cup measurements and the determination of dose-area-product to water as explained below. Since IAEA TRS-398 only provides procedures for the dosimetry of broad fields, either an alternative method has to be used to determine DAP_w or a link has to be established between the absorbed dose to water determined at a point in a broad beam and the quantity of interest. A paper by Alfonso et al. [45]

addressed reference dosimetry for non-conventional photon beams by introducing machine-specific reference fields and mentioned that scanned proton beams could be regarded as examples of such fields. Here, in analogy, three types of machine-specific reference fields are considered, but more research in this area is needed:

1. A single static pencil beam in which dose-area-product to water $(\mathrm{DAP}_{w,Q}^{\infty})$ for the proton beam with beam quality Q is determined at a shallow depth using a large-area plane-parallel ionization chamber. The infinity sign in superscript indicates that the lateral integration of dose should be over the entire lateral plane. In practice, the dose-area-product $(\mathrm{DAP}_{w,Q}^{A})$ integrated over the area A of the ionization chamber's collecting electrode is determined. Provided the diameter of the collecting electrode is sufficiently large, it can be assumed that $(\mathrm{DAP}_{w,Q}^{\infty}) \approx (\mathrm{DAP}_{w,Q}^{A})$. In practice, this assumption is not fulfilled, as the size of the collecting electrode of commercially available chambers is not large enough. Therefore, the measured value of $(\mathrm{DAP}_{w,Q}^{A})$ should be corrected using film measurements or Monte Carlo simulation to account for the part of $(\mathrm{DAP}_{w,Q}^{\infty})$ that is deposited outside of the collecting diameter of the large-area plane-parallel ionization chamber. This method has been used by Gillin et al. [46].
2. A single layer scanned field of sufficient size (establishing lateral charged particle equilibrium at the central axis in a time-averaged sense) with a constant number of monitor units delivered per spot and equidistant lateral spot spacing in two dimensions in which $D_{w,Q}$ at shallow depth is determined at a point using a code of practice for broad beam dosimetry. Dose-area-product to water is then derived as $\mathrm{DAP}_{w,Q}^{\infty} = D_{w,Q}\Delta x \Delta y$, where D$x$ and Dy are the constant spot spacings in both orthogonal directions perpendicular to the beam axis (Hartmann et al. [47], Jäkel et al. [48]).
3. A modulated scanned field of sufficient size resulting in a uniform spread-out Bragg peak with a constant number of monitor units delivered per spot in each layer and equidistant lateral spot spacing in two dimensions in which $D_{w,Q}$ at a point in the spread-out Bragg peak is determined. In principle, a dose-area-product could also be derived, but it is less meaningful given that the total number of particles is composed of contributions with different energies. This option for reference dosimetry and monitor calibration is rather useful as a reference for fields with the same modulation.

For the determination of $\mathrm{DAP}_{w,Q}^{\infty}$ using a large-area plane-parallel ionization chamber, alternative equations to those given in Section 12.3 need to be introduced, keeping nevertheless to similar notations and concepts. $(\mathrm{DAP}_{w,Q}^{A})$ is determined as

$$\mathrm{DAP}_{w,Q}^{A} = M_Q N_{\mathrm{DAP},w,Q_0} \kappa_{Q,Q_0} \tag{12.35}$$

The symbol $\mathrm{k}_{Q,Q0}$ is used here for the beam quality correction factor to indicate that it corrects a different quantity than the conventional beam quality correction factor $k_{Q,Q0}$. In this case $k_{Q,Q0}$ is defined as a ratio of dose-area-product to water calibrations coefficients in the narrow proton beams with beam qualities Q and the calibration beam with beam quality Q_0:

$$\kappa_{Q,Q_0} = \frac{N_{\text{DAP},w,Q}}{N_{\text{DAP},w,Q_0}} \tag{12.36}$$

At present, the problem with the application of these equations is that calibrations coefficients in terms of DAP_w are not provided by standards laboratories. Additionally, even if such calibrations would become available, then it remains a problem that $k_{Q,Q0}$ values are not known for commercially available large-area plane-parallel ionization chambers. The only option is then to cross calibrate the plane-parallel ionization chamber (indicated by "PP") against a conventional ionization chamber (indicated by "REF") on the central axis of a broad proton beam cross calibration beam quality Q_{cross}:

$$N_{\text{DAP},w,Q_{\text{cross}}} = \frac{\left[M_{Q_{\text{cross}}} N_{D,w,Q_0} k_{Q_{\text{cross}},Q_0}\right]_{\text{REF}}}{\left[M_{Q_{\text{cross}}}\right]_{\text{PP}}} \times \iint_A \text{OAR}(x,y)\,dxdy \tag{12.37}$$

where OAR(x,y) is the off-axis ratio at the reference depth and A the area of the collecting electrode of the large-area plane-parallel ionization chamber. If the lateral beam profile is uniform over the area of the ionization chamber, this expression reduces to

$$N_{\text{DAP},w,Q_{\text{cross}}} = \frac{\left[M_{Q_{\text{cross}}} N_{D,w,Q_0} k_{Q_{\text{cross}},Q_0}\right]_{\text{REF}}}{\left[M_{Q_{\text{cross}}}\right]_{\text{PP}}} \times A \tag{12.38}$$

Dose area product to water is then obtained as

$$\text{DAP}^A_{w,Q} = M_Q N_{\text{DAP},w,Q_{\text{cross}}} \kappa_{Q,Q_{\text{cross}}} \tag{12.39}$$

where $k_{Q,Qcross}$ can be assumed to be unity since both Q and Q_{cross} refer to proton beam qualities.

Equations 12.38 and 12.39 correspond with the approach described by Gillin et al. [46]. The relation with the number of particles in the pencil beam via the mass stopping power of water is given by Equation 12.34 and links Faraday cup-based dosimetry (as used by Pedroni et al. [49] and Lorin et al. [50]) with the DAP-based dosimetry method.

Recombination in scanned beams deserves separate attention because of two effects. The first one is that, similarly as in IMRT, the instantaneously partial irradiation conditions affect volume recombination. The volume recombination for a Gaussian pencil beam with FWHM of 1 cm in water hitting the base of a Farmer-type chamber is found to be 10% to 15% greater than when it hits the center of the cavity

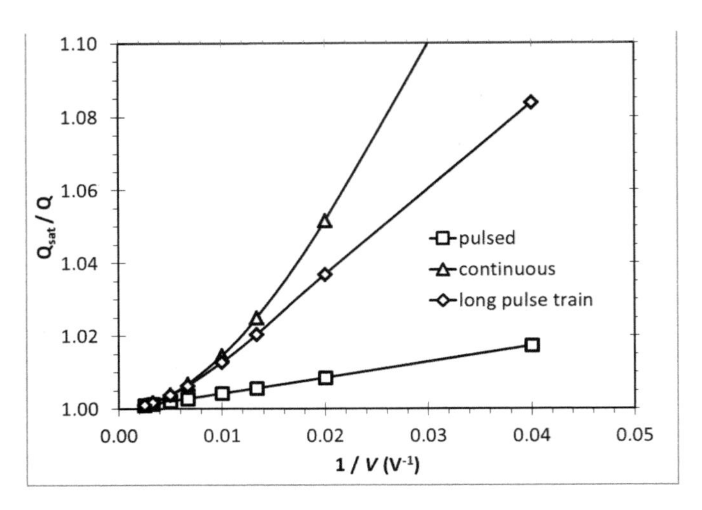

Figure 12–8 Jaffé plots for pulsed, continuous, and intermediate regime (long pulse train) proton radiation (from Palmans [51]). The recombination correction at 400 V is the same for each curve.

[51]. Depending on the position of the scan lines with respect to the chamber, the integrated recombination can vary by more than 10%, and given that volume recombination can amount to 10% in a spot-scanned beam, this can lead to dose errors of more than 1%. The second effect is caused by the deviation of the ionization current from continuous or pulsed irradiation conditions. A dose pulse at one point in a scanned proton field can be as short as 250 μs, and given that the ion collection time is of the order of hundreds of μs, a pulse train observed by an ionization chamber may be neither short nor long in comparison with the ion collection time. Consequently, the Boag theory for both extremes is not correct. Figure 12–8 shows a Jaffé plot for different irradiation regimes [51]. Using the two-voltage method for either pulsed or continuous radiation, as recommended in codes of practice, will thus result in erroneous estimates of the volume recombination.

12.6 QA of Dissemination

While accurate dosimetry of clinical proton beams is ensured by traceable calibration of ionization chambers and internationally accepted dosimetry protocols, there are several steps involved where errors can occur. To capture such potential errors, redundant quality assurance measurements are implemented. In the days when a limited number of proton therapy centers were clinically operating and the approach to reference dosimetry taken by various institutes was very nonuniform—using different methods (Faraday cups, ionization chambers) and different protocols (AAPM TG-20 [25], ECHED [26,27]) or in-house procedures—the most efficient way of testing the consistency in dosimetry was by direct intercomparisons. With the recent growth in the number of proton therapy centers, however, it has become impractical to involve all facilities in such intercomparison exercises, and it is more realistic for auditing services to provide additional quality assurance tools [33].

12.6.1 Dose Intercomparisons

Early examples of dose intercomparison studies for low-energy proton beams were reported by Kacperek et al. [52] and Jones et al. [53] showing differences in reference dosimetry on the order of 3%, within the uncertainties of the different methods and dosimetry protocols used. The first example of a similar comparison in a high-energy proton beam was reported by Jones et al. [54], followed by two larger-scale comparisons by Vatnitsky et al., [32,55]. The earlier of the latter two studies [55] was based on calibrations in terms of exposure or air kerma using the AAPM TG-20 [25] and ECHED [26,27] protocols, and it showed the importance of using a common dosimetry protocol with uniform data to improve dosimetric consistency. Using institution-specific procedures and data, the maximum deviation of reference dose from the average was 3%, which could be improved to 1.5% when using a common dosimetry protocol. The second study [32] demonstrated that by using ICRU Report 59 [30], similar consistency could be reached. This study also showed that by calibrating all involved ionization chambers in a common ^{60}Co beam, the observed differences could not be explained only by differences in traceability of ^{60}Co calibration coefficients. This indicated that chamber-to-chamber specific perturbation factors in the proton beam may contribute to the difference (confirmed later in an experimental Monte Carlo study by Palmans et al. [56]). Contributions to the differences could also arise from the chamber-specific perturbation factors in ^{60}Co used in ICRU Report 59. (As pointed out by Medin et al. [31], these are even omitted applying ICRU Report 59 using absorbed dose to water calibration coefficients.)

12.6.2 Traceability Audits

Moyers et al. [33] described a traceability audit of reference dosimetry at eight proton therapy centers in North America in collaboration with the Radiological Physics Center (M.D. Anderson Cancer Center, Houston, Texas) by performing on-site dosimetry using either an Exradin P11 parallel-plate ionization chamber or an Exradin A12 Farmer-type chamber in the reference fields and the cross-calibrated four central ionization chambers of a MatriXX Evolution array for clinical fields. The dosimetry was performed by the audit group using three different procedures (ICRU Report 59 [30] using either an air kerma or an absorbed dose to water calibration coefficient or ICRU Report 78 [3]), and it was compared with the dose reported by the local hospital. Using ICRU Report 78, the mean difference between the audit dose and reported dose was 0.3% with a standard deviation of 1.6%, confirming the adequate traceability and implementation of dosimetric procedures since most centers currently use ICRU Report 78 for their reference dosimetry. Also using ICRU Report 59 with air kerma calibration coefficients (correcting, however, for the errors in this procedure reported by Medin et al. [31]), a similar level of agreement was found. On the other hand, using ICRU Report 59 with absorbed dose to water calibration coefficients, the mean deviation was 3.4%, suggesting that this protocol should not be used in combination with the two others for clinical trials. Overall, the study of Moyers et al. (2014) shows that dosimetric auditing is feasible for proton therapy and can aid in ensuring dosimetric consistency for multi-institutional clinical trials.

12.6.3 End-to-end Tests

The most efficient solution for dosimetry intercomparison of proton beam delivery systems is to use end-to-end test-based auditing procedures as they allow verifying that the whole logistical chain of treatment delivery is operable and leads to the desired results with sufficient accuracy. During the testing, the phantom is moving through the workflow as a real patient to simulate the clinical procedure. Such intercomparisons would also contribute to a dosimetric harmonization among proton beam therapy centers, which is currently challenged by the lack of international and national primary dose standards for protons and heavier ion beams. But compared to dosimetric auditing procedures for conventional photon and electron beam therapy, additional dose rate and LET effects might be present, and this needs to be taken into account when analyzing the detector response in scanning proton beams.

An example of the end-to-end test for proton (and carbon ion beams) with alanine detectors is given by Ableitinger et al. [57] who reported a trial end-to-end dosimetry audit for scanned particle beams. In this study, alanine pellets traceable to the National Physical Laboratory's primary standards were used as an absolute dosimeter next to a Farmer ionization chamber. The results showed that for scanned proton beams, the correction for the relative effectiveness of the alanine dosimeters was limited to 5% for all measurement points distributed over the irradiated box-volume, and the agreement with planned doses was better than 3.5% for all points. The mean difference was 2.4%, with a standard deviation of 0.9%, which was in good agreement with the deviation of the Farmer measurement from the planned dose, which was 1.7% in the same direction, demonstrating that alanine can be used for such end-to-end dosimetry tests or in dosimetric credentialing procedures for clinical trials. Even for carbon ions, despite quenching corrections being up to 25%, a similar level of agreement as for protons was observed in the study of Ableitinger et al. [57]. Thermoluminescent dosimeters (TLDs) were used for absolute dosimetry in another dosimetric end-to-end test in an anthropomorphic spine phantom by Cho et al. [58]. Due to the more complex and more pronounced energy dependence of TLDs for low-energy protons, however, these detectors can only be used for accurate dosimetry in the entrance region, as was done in the study of Cho et al. [58].

12.7 Microdosimetry and Nanodosimetry

Radiotherapy prescription is based on the quantity absorbed dose to water multiplied with a weighting factor to quantity the effect of a given radiation quality on a given biological system under given conditions. The radiation quality is entirely described by the local charged particle spectrum. The biological system can be an organ, a tissue, an individual cell, a cell component, DNA, etc. The conditions can, for example, refer to the length and time structure of irradiation, the fractionation, and the environmental temperature. It is understood that for a given biological system and given conditions, the biological effect of radiation is dependent on the distribution of ionization and energy deposition on the length scales of the cellular and tissue components that are involved in the processes leading to the biological effect or endpoint [59]. These distributions are characterized by microdosimetry and nanodosimetry.

12.7.1 Microdosimetry

Microdosimetry concerns the determination of the spatial and temporal distribution of interactions of ionizing radiation with micrometer-sized volumes of matter. It is assumed that the distribution of ionizations on the microscale are correlated with indirect damage inflicted to the DNA. Microdosimetric quantities are stochastic and, therefore, given in terms of interaction probabilities. Structural microdosimetry aims at deriving microdosimetric quantities and actions from detailed 3D distributions of energy transfer points, and is not further discussed here. Relevant quantities in regional microdosimetry are described in Chapter 3.

Regional microdosimetry is most commonly performed using tissue-equivalent proportional counters (TEPCs). In its most simple form, it is an ionization chamber operated at low pressure with a high voltage applied such that an avalanche takes place. The number of ion pairs produced is assumed to be proportional to the energy transferred. The pressure is adapted such that the mean imparted energies are equal, i.e. $\bar{\varepsilon}_T = \bar{\varepsilon}_G$, where the subscript "T" stands for "Tissue" and "G" for "Gas." The same equation can be written as:

$$\left(\frac{S}{\rho}\right)_T \bullet \rho_T \bullet \bar{l}_T = \left(\frac{S}{\rho}\right)_G \bullet \rho_G \bullet \bar{l}_G \tag{12.40}$$

where $\frac{S}{\rho}$ are the mass stopping powers, ρ the mass densities, and $\bar{l}$ the mean chord lengths of the sites. If the gas has the same atomic composition of the tissue, i.e. $\left(\frac{S}{\rho}\right)_T = \left(\frac{S}{\rho}\right)_G$, then

$$\rho_T \bullet \bar{l}_T = \rho_G \bullet \bar{l}_G \tag{12.41}$$

If, furthermore, the wall material has the same atomic composition, then the fluence of secondary charged particles is independent of density variations (Fano's theorem). For proton beams, the energy deposition in a single event can be overestimated due to

- the delta-ray effect (a secondary electron that would normally not cross the same site as the primary proton does enter the volume because of the larger dimensions),

- the V-effect (a secondary proton or heavier particle from a nuclear interaction that would normally not cross the same site as the primary proton does enter the volume because of the larger dimensions), and
- the re-entry effect (after leaving the collecting volume, a secondary electron following a strongly curved path that would normally not re-enter the measurement volume does re-enter because of the larger dimensions) [60].

More sophisticated TEPCs are wall-less, either by using special electrodes to shape the field and that way define the measurement volume without the presence of a wall or by using a grid wall, thus minimizing the amount of wall material. The size of TEPCs limits the spatial resolution, which is especially a problem in the Bragg peak. Pile-up is another issue at radiotherapy level dose rates. Mini-TEPCs, which have much smaller volumes than conventional TEPCs, reduce the magnitude of various correction factors and allow for the gas pressure to be much higher [61]. Also gas microstrip detectors are used to overcome the problems of corrections and resolution for proton microdosimetry [62].

Calibration of a TEPC consists of establishing a relation between pulse height and energy deposition, and it is done using either a source of particles with known energy which are completely absorbed in the detector volume or by making use of the proton edge. The latter exploits the fact that near the end of the range, the proton energy deposition is highest and the lineal energy distribution has a sharp edge (which is, however, blurred by energy and range straggling and by the contaminant presence of heavier recoils). The edge lineal energy for protons was determined as $y_{edge} = 136$ keV μm^{-1} for a water sphere of 2 μm in diameter (and should be scaled for the size and the operational gas density of the TEPC). Knowledge of the mean energy required to produce an ion pair in the gas, W_{gas}, is another source of uncertainty. W_{gas} is known to decrease with energy for a given particle mass and increase with particle mass for a given energy, and it is assumed to add about 5% uncertainty to the energy absorption measured with a proportional counter.

Silicon-based devices for microdosimetry also measure ionization via electron-hole pair creation in a depletion layer. Their main advantage compared to a TEPC is that their size is much closer to the site of interest in water or tissue. For radiotherapy this results in both a higher spatial resolution and a reduction of the corrections mentioned above. Disadvantages are that the collecting volume is poorly defined, that they are prone to radiation damage, and that they are not exactly tissue equivalent, so that a conversion is needed. The mean energy required to produce an electron-hole pair, W_{Si}, is about a factor 10 lower than W_{gas} in TEPCs and depends on the particle energy and type.

Application of an array of p-n junctions with a pixel area of 0.04 cm^2 to two therapeutic proton beams revealed that in large beams, even with a silicon diode, pile-up effects can occur, but silicon devices can be further miniaturized. Another interesting feature of silicon technology is that a construction integrating more than one detector with different functions is possible. For example, a ΔE–E silicon telescope has been described [63] consisting of two layers of silicon detectors sharing the same p^+ electrode, the upper one being very thin (1 μm) and the lower one being 500 μm thick.

When a particle passes through both layers, the signal from the upper layer provides the energy loss over 1 μm of silicon while the sum of both signals provides the total energy of the particle (under the condition that it is stopped within the thick lower layer). This coincidence measurement allows resolving the particle type that hits the detector, since different particle types will occupy different regions in a ΔE–E map, and is thus of interest in mixed particle fields. Application of the detector to protons showed that contributions of other ion species in the radiation field are marginal.

12.7.2 Nanodosimetry

Nanodosimetry is concerned with measuring track structure down to nanometer resolution, the scale of DNA base pairs. The characterization of track structure is based on the stochastic quantity called ionization cluster size, which is defined as the number of ionizations produced by a particular particle track in a specific target volume and its frequency distribution (ionization cluster size distribution, ICSD). The size of the target volumes considered in nanodosimetry is always smaller than the lateral extension of the penumbra of the primary particle track, where interactions are due to secondary electrons. ICSD depends on the size of the target, its geometry and material composition, and on the geometrical relation between the primary particle trajectory and the target. This is accounted for by specifying the impact parameter d, which is the smallest distance between the primary particle trajectory and the center of the target volume. Relevant quantities in nanodosimetry are described in Chapter 3.

The most common method for nanodosimetry relies also on the measurement of ionization in a small volume filled with a low-density gas in order to simulate the size of a nanometer structure, such as the DNA. This is also based on a density scaling principle that allow equivalent ICSDs in target volumes of different size and material composition to be obtained. Two sites A and B are said to be equivalent when the mean ionization cluster sizes obtained in the two sites for the same radiation quality are equal, i.e., $M_1^{(A)} = M_1^{(B)}$. It has been demonstrated that for targets that fulfill this equality, ICSD obtained in propane are similar to those produced in nitrogen or in liquid water. The mean ionization cluster size M_1 is related to the diameter l of the sensitive volume and the mean free path for ionizing interactions of the primary particle with the medium λ_{ion}, such that $M_1 \propto \frac{l}{\lambda_{ion}}$ where the proportionality factor depends on the impact parameter d.

Three types of nanodosimeter devices (Jet Counter, StarTrack, and Ion Counter) have been developed that are capable of measuring the frequency distribution of ionization cluster size in a gas [59]. The Jet Counter at Narodowe Centrum Badán Jądrowych (NCBJ) detects positive ions produced by primary particles of electrons or ions in a jet of nitrogen gas propagating inside a cylindrical tube, where the number density of molecules can be adjusted to obtain biological target sizes in the range of 2 nm to 20 nm. The measured ICSDs relate to a central passage of the primary particle through the target.

In the StarTrack detector at the Legnaro National Laboratories of the Italian Nuclear Research Institute (INFN), the target volume consists of an almost wall-less cylinder 3.7 mm in diameter and height defined by electrode wires. Filled with pure propane and operated at a gas pressure of 3 mbar, the effective water length of the target is 20 nm. The target volume can be moved perpendicularly to the particle beam with an accuracy of 0.1 mm, enabling measurements for different impact parameters. Since the StarTrack device can detect electrons generated by an ion traversing the target volume or passing close by, it is able to distinguish ionizations generated in the core and the penumbra of the track, respectively.

The Ion Counter at the Physikalisch-Technische Bundesanstalt (PTB) detects positive ions produced in a wall-less gas volume. When operated with propane at a gas pressure of 1.2 mbar, it simulates a liquid water cylinder of about 3 nm in diameter. When operated with nitrogen at the same gas pressure, the diameter of the simulated water volume reduces to about 0.5 nm. The possibility of using other operating gases such as water vapor or gas mixtures of DNA ingredients has also been demonstrated. The Ion Counter is equipped with a position-sensitive trigger detector to record the position of the primary ion impinging on the detector surface. This position-sensitive detector enables the reconstruction of the primary particle's path, thus allowing the extraction of ICSDs with different impact parameters in order to discriminate between ICSDs originating from the core and the penumbra of the primary particle's track.

12.8 Summary and Conclusions

This chapter reviewed established methods for reference dosimetry using calorimeters and calibrated ionization chambers. The chemical heat defect and the conversion from dose to graphite to dose to water were identified as the main issues in water calorimetry and graphite calorimetry, respectively, and current data on these effects were discussed. Dosimetry formalisms using calibrated ionization chambers were summarized, with attention to reference conditions and available data, and a factorization was introduced that allows easy comparison of the influence of data and their uncertainty contributions in different codes of practice. The role of detector calibrations, auditing, and end-to-end tests using ionization chambers, alanine, and TLDs were discussed in the process of ensuring traceable dosimetry and dosimetric consistency across proton therapy centers. A section was devoted to methods and instruments used for micro- and nanodosimetry, which are essential quantities in establishing the beam quality of proton beams in relation to their biological and therapeutic effects. Overall, this chapter gave a picture of the challenges to perform reference dosimetry in proton beam with the same level of accuracy as in conventional high-energy photon beams, a prerequisite to exploiting proton therapy to its full potential.

References

1. IAEA. Absorbed dose determination in external beam radiotherapy: an international code of practice for dosimetry based on standards of absorbed dose to water. IAEA Technical Report Series 398. Vienna, Austria: IAEA, 2000.
2. Wambersie, A. What accuracy is required and can be achieved in radiation therapy (review of radiobiological and clinical data). *Radiochim. Acta* 2001;89:255–64.
3. ICRU. Prescribing, recording, and reporting proton-beam therapy. International Commission on Radiation Units and Measurements Report 78. Bethesda MD: ICRU, 2007.
4. IAEA. Calibration of dosimeters used in radiotherapy. IAEA technical report series 374. Vienna, Austria: IAEA, 1994.
5. Palmans H, Vynckier S. Reference dosimetry for clinical proton beams. In: Seuntjens JP, Mobit PN, Eds. Recent developments in accurate radiation dosimetry. Madison, WI: Medical Physics Publishing, 2002, p. 157–194.
6. Seuntjens J, Duane S. Photon absorbed dose standards. *Metrologia* 2009;46:S39–S58.
7. McEwen MR, DuSautoy AR. Primary standards of absorbed dose for electron beams. *Metrologia* 2009;46:S59–79.
8. Karger CP, Jäkel O, Palmans H, Kanai T. Dosimetry for ion beam radiotherapy. *Phys Med Biol* 2010;55:R193–234.
9. Sarfehnia A, Clasie B, Chung E, Lu HM, Flanz J, Cascio E, Engelsman M, Paganetti H, Seuntjens J. Direct absorbed dose to water determination based on water calorimetry in scanning proton beam delivery. *Med Phys* 2010;37:3541–50.
10. Palmans H, Seuntjens J, Verhaegen F, Denis JM, Vynckier S, Thierens H. Water calorimetry and ionization chamber dosimetry in an 85-MeV clinical proton beam. *Med Phys* 1996;23:643–50.
11. Sassowsky M, Pedroni E. On the feasibility of water calorimetry with scanned proton radiation. *Phys Med Biol* 2005;50:5381–400.
12. Brede HJ, Hecker O, Hollnagel R. Measurement of the heat defect in water and A-150 plastic for high-energy protons, deuterons and α-particles. *Radiat Prot Dosim* 199770:505–508.
13. Palmans H. Dosimetry. In: Proton Therapy Physics, Ed. H. Paganetti. London: Taylor & Francis, 2011, pp. 191–219.
14. Bewley DK, Page BC. Heat defect in carbon calorimeters for radiation dosimetry. *Phys Med Biol* 1972;17:584–85.
15. IAEA. Stored energy and the thermo-physical properties of graphite. In: Irradiation damage in graphite due to fast neutrons in fission and fusion systems. IAEA Technical Document 1154. Vienna, Austria: IAEA, 2000.
16. Schulz RJ, Venkataramanan N, Huq MS. The thermal defect of A-150 plastic and graphite for low-energy protons. *Phys Med Biol* 1990;35:1563–74.
17. ICRU. Nuclear data for neutron and proton radiotherapy and for radiation protection dose. International Commission on Radiation Units and Measurements Report 63. Bethesda MD, USA: ICRU, 2000.
18. Palmans H, Al-Sulaiti L, Andreo P, Thomas RAS, Shipley DR, Martinkovič J, Kacperek A. Conversion of dose-to-graphite to dose-to-water in clinical proton beams. In: Standards, Applications and Quality Assurance in Medical Radiation Dosimetry–Proceedings of an International Symposium, Vienna 9–12 November 2010–Vol. 1. Vienna, Austria: IAEA, 2011, pp. 343–355.
19. ICRU. Stopping powers and ranges for protons and alpha particles. International Commission on Radiation Units and Measurements Report 49. Bethesda MD: ICRU, 1993.
20. Burns DT. A re-evaluation of the I-value for graphite based on an analysis of recent work on W, S(c,a) and cavity perturbation corrections. *Metrologia* 2009;46:585–90.
21. Paul H. A comparison of recent stopping power tables for light and medium-heavy ions with experimental data, and applications to radiotherapy dosimetry. *Nucl Instr Meth B* 2006;247:166–72.
22. Palmans H. Monte Carlo calculations for proton and ion beam dosimetry." In: Monte Carlo Applications in Radiation Therapy, Ed. F. Verhaegen and J Seco. London: Taylor & Francis, 2013, pp. 185–99.
23. Fleming DM, Glass WA. Endothermic processes in tissue-equivalent plastic." *Radiat Res* 1969;37:316–22.

24. McDonald JC, Goodman LJ. Measurements of the thermal defect for A-150 plastic. *Phys Med Biol* 1982;27:229–33.
25. AAPM Protocol for heavy charged-particle therapy beam dosimetry. A report of task group 20 radiation therapy committee. American Association of Physicists in Medicine Report 16. New York: AIP,1986.
26. Vynckier S, Bonnett DE, Jones DTL. Code of practice for clinical proton dosimetry. *Radiother Oncol* 1991;20:53–63.
27. Vynckier S, Bonnett DE, Jones DTL. Supplement to the code of practice for clinical proton dosimetry. *Radiother Oncol* 1994;32:174–9.
28. Paganetti H. Monte Carlo calculations for absolute dosimetry to determine machine outputs for proton therapy fields. *Phys Med Biol* 2006;51:2801–12.
29. Nichiporov D. Verification of absolute ionization chamber dosimetry in a proton beam using carbon activation measurements. *Med Phys* 2003;30:972–78.
30. ICRU. Clinical proton dosimetry Part I: Beam production, beam delivery and measurement of absorbed dose. International Commission on Radiation Units and Measurements Report 59. Bethesda MD: ICRU, 1998.
31. Medin J, Andreo P, Vynckier S. Comparison of dosimetry recommendations for clinical proton beams. *Phys Med Biol* 2000;45:3195–211.
32. Vatnitsky SM, Moyers M, Miller D, Abell G, Slater JM, Pedroni E, Coray A, Mazal A, Newhauser W, Jaekel O, Heese J, Fukumura A, Futami Y, Verhey L, Daftari I, Grusell E, Molokanov A, Bloch C. Proton dosimetry intercomparison based on the ICRU report 59 protocol. *Radiother Oncol* 1999;51:273–279.
33. Moyers MF, Ibbott GS, Grant RL, Summers PA, Followill DS. Independent dose per monitor unit review of eight U.S.A. proton treatment facilities. *Med Phys* 2014;41:012103.
34. IAEA. Absorbed dose determination in photon and electron beams: an international code of practice. IAEA Technical Report Series 277. Vienna, Austria: IAEA,1987.
35. Vatnitsky SM, Siebers JV, Miller DW. k_Q factors for ionization chamber dosimetry in clinical proton beams. *Med Phys* 1996;23:25–31.
36. Palmans H. Secondary electron perturbations in Farmer type ion chambers for clinical proton beams. In: standards, applications and quality assurance in medical radiation dosimetry–proceedings of an international symposium, Vienna 9–12 November 2010–Vol. 1. Vienna, Austria: IAEA, 2011, pp. 309–317.
37. Palmans H, Thomas R, Kacperek A. Ion recombination correction in the Clatterbridge Centre of Oncology clinical proton beam. *Phys Med Biol* 2006;51:903–17.
38. AAPM. Task group 21: A protocol for the determination of absorbed dose from high-energy photon and electron beams. *Med Phys* 1983;10:741–71.
39. AAPM. Task group 51: protocol for clinical reference dosimetry of high-energy photon and electron beams. *Med Phys* 1999;26:1847–70.
40. IAEA. The use of plane-parallel ionization chambers in high-energy electron and photon beams: an international code of practice for dosimetry. IAEA Technical Report Series 381. Vienna, Austria: IAEA, 1997.
41. ICRU. Average energy required to produce an ion pair. International Commission on Radiation Units and Measurements Report 31. Washington, DC: ICRU,1979.
42. ICRU. Stopping powers for electrons and positrons. International Commission on Radiation Units and Measurements Report 37. Bethesda MD: ICRU,1984.
43. Palmans H. Perturbation factors for cylindrical ionization chambers in proton beams. Part I: corrections for gradients. *Phys Med Biol* 2006;51:3483–501.
44. Palmans H, Kacperek A, Jäkel O. Hadron dosimetry. In: Rogers DWO, Cygler JE, Eds. Clinical dosimetry measurements in radiotherapy. Madison WI: Medical Physics Publishing, 2009, pp. 669–722.
45. Alfonso R, Andreo P, Capote R, Huq MS, Kilby W, Kjäll P, Mackie TR, Palmans H, Rosser K, Seuntjens J, Ullrich W, Vatnitsky S. A new formalism for reference dosimetry of small and nonstandard fields. *Med Phys* 2008;35:5179–86.
46. Gillin MT, Sahoo N, Bues M, Ciangaru G, Sawakuchi G, Poenisch F, Arjomandy B, Martin C, Titt U, Suzuki K, Smith AR, Zhu X. Commissioning of the discrete spot scanning proton beam delivery system at the University of Texas M.D. Anderson Cancer Center, Proton Therapy Center, Houston. *Med Phys* 2010;37:154–63.

47. Hartmann GH, Jäkel O, Heeg P, Karger CP, Kriessbach A. Determination of water absorbed dose in a carbon ion beam using thimble ionization chambers. *Phys Med Biol* 1999;44:1193–206.
48. Jäkel O, Hartmann GH, Karger CP, Heeg P, Vatnitsky S. A calibration procedure for beam monitors in a scanned beam of heavy charged particles. *Med Phys* 2004;31:1009–13.
49. Pedroni E, Scheib S, Böhringer T, Coray A, Grossmann M, Lin S, Lomax A. Experimental characterization and physical modelling of the dose distribution of scanned proton pencil beams. *Phys Med Biol* 2005;50:541–61.
50. Lorin S, Grusell E, Tilly N, Medin J, Kimstrand P, Glimelius B. Reference dosimetry in a scanned pulsed proton beam using ionisation chambers and a Faraday cup. *Phys Med Biol* 2008;53:3519–29.
51. Palmans H. Theoretical models for volume recombination in scanned proton beams. *Radiother Oncol* 2014;111(Suppl. 1):315.
52. Kacperek A, Vynckier S, Bridier A, Herault J, Bonnet DE. "A small scale European proton dosimetry intercomparison. Proc. of the proton therapy workshop at PSI (Feb 28–March 1, 1991, Villigen, Switzerland). PSI bericht no 11176, 1991, pp. 76–79.
53. Jones DTL, Kacperek A, Vynckier S, Mazal A, Delacroix S, Nauraye C. A European proton dosimetry intercomparison. NAC annual report NAC/AR/92-01,1992, pp. 61–63.
54. Jones DTL, Schreuder AN, Symons JE, Vynckier S, Hayakawa Y, Maruhashi A. Proton dosimetry intercomparison at NAC. NAC Annual report NAC/AR/94-01, 1994, pp. 94–95.
55. Vatnitsky S, Siebers J, Miller D, Moyers M, Schaefer M, Jones D, Vynckier S, Hayakawa Y, Delacroix S, Isacsson U, Medin J, Kacperek A, Lomax A, Coray A, Kluge H, Heese J, Verhey L, Daftari I, Gall K, Lam G, Beck T, Hartmann G. Proton dosimetry intercomparison. *Radiother Oncol* 1996;41:169–177.
56. Palmans H, Verhaegen F, Denis J-M, Vynckier S, Thierens H. Experimental p_{wall} and p_{cel} correction factors for ionization chambers in low-energy clinical proton beams. *Phys Med Biol* 2001;46:1187–204.
57. Ableitinger A, Vatnitsky S, Herrmann R, Bassler N, Palmans H, Sharpe P, Ecker S, Chaudhri N, Jäkel O, Georg D. Dosimetry auditing procedure with alanine dosimeters for light ion beam therapy: results of a pilot study. *Radiother Oncol* 2013;108:99–106.
58. Cho J, Summers PA, Ibbott GS. An anthropomorphic spine phantom for proton beam approval in NCI-funded trials. *J Appl Clin Med Phys* 2014;15:252–65.
59. Palmans H, Rabus H, Belchior AL, Bug MU, Galer S, Giesen U, Gonon G, Gruel G, Hilgers G, Moro D, Nettelbeck H, Pinto M, Pola A, Pszona S, Schettino G, Sharpe PH, Teles P, Villagrasa C, Wilkens JJ. Future development of biologically relevant dosimetry. *Br J Radiol* 2015;87:20140392.
60. Bradley PD. The development of a novel silicon microdosimeter for high LET radiation therapy. PhD thesis, University of Wollongong, Australia, 2000.
61. De Nardo L, Cesari V, Donà G, Colautti P, Conte V, Tornielli G. Mini TEPCs for radiation therapy. *Radiat Prot Dosim* 2004;108:345–352.
62. Waker AJ, Dubeau J, Surette RA. The application of micro-patterned devices for radiation protection dosimetry and monitoring. *Nucl Technol* 2009;168:202–206.
63. Wroe A, Schulte R, Fazzi A, Pola A, Agosteo S, Rosenfeld A. RBE estimation of proton radiation fields using a ΔE–E telescope. *Med Phys* 2009;36:4486–4494.

Chapter 13

Acceptance Testing of Proton Therapy Systems

Jonathan B. Farr, Ph.D., D.Sc.[1] and Jon J. Kruse, Ph.D.[2]

[1]Chief of Radiation Physics and Associate Member,
Department of Radiation Oncology,
St. Jude Children's Research Hospital,
Memphis, TN

[2]Assistant Professor, Department of Radiation Oncology,
Mayo Clinic,
Rochester, MN

13.1 Acceptance Testing Overview

Acceptance testing usually represents a comprehensive body of acceptance tests from specific to system wide that are defined contractually between the vendor(s) or integrator and the customer (user). The goals of acceptance testing are focused on validating contractual specifications represented by testable utility conditions, but they may include ancillary activities, such as providing first experiences with the system for the user. Acceptance testing is also critical to ensure patient and operator safety and for generating baseline data for future comparisons [1].

Acceptance testing is purely vendor-specific operational testing, whereas commissioning is a set of tests and measurements to collect data for patient treatment, such as described in TG-106 [2]. The specifications, tolerances, test methods, and methods of analysis for the acceptance tests need to be agreed upon between the vendor and user. In most cases, they are contractual and will reside in an acceptance testing plan. The process is largely vendor driven with user input and participation. Acceptance testing is not a substitute for clinical commissioning. During commissioning all options of energy, field size, beam angle, calibration, range verification, output, etc. must be validated. It is not practical to perform such a high number of tests during acceptance testing, which is usually contemporaneous with vendor project integration. Instead, acceptance testing should seek to validate the operation across the possible range of control values, thereby ensuring the capability for commissioning. When well conceived and well written, acceptance testing provides the ideal precursor to clinical commissioning such that, at the completion of acceptance testing, the user has a safe, stable, operational system with which to prepare for clinical use. Prior to such a handover, the vendor usually directs acceptance testing with the user functioning in an observing, sub-operating, or test record review capacity.

Acceptance tests should cover all relevant safety, physical, radiologic, and control aspects of the proton therapy system. The tests range from very specific unit or component testing to broad-use case scenario tests (end-to-end). After proton therapy system installation and during vendor integration, acceptance testing usually follows the same schedule (i.e., as soon as a subsystem has been completed, enabling a particular set of acceptance tests, the tests are performed in a piecemeal manner). This naturally follows physical testing, basic irradiation alignment (usually planar or several parallel planes), testing of control systems, volumetric irradiations, and end-to-end testing. Depending on the facility layout and contractual schedule, acceptance testing may follow parallel tests in multiple treatment rooms, or it may test the rooms individually or in phases.

The nature of acceptance testing can vary from direct, on-site user participation to factory inspections or simple record reviews of vendor-completed tests. Acceptance testing usually comprises portions of each. For example, if the vendor's fabrication is in a distant location, then occasional factory visits can be backed up with test reviews.

Each test described here will include discussion of specification, method of measurement, and method of analysis. The specifications reported are from the authors' experience and are not necessarily representative of specific contracts or vendor acceptance testing schemes. As specifications continue to evolve, site-specific acceptance tests will need to be synchronized. If the acceptance testing results do not satisfy the specifications, then the user should consider the delivery of the system to be incomplete and work with the vendor(s) to resolve related issues until mutually acceptable. Because of the complicated nature of proton therapy system equipment, this testing may occur over an extended timeline, so one should be proactive in specifying the time requirement for acceptance testing and commissioning as shown for external beam [2]. If the system is modified, upgraded, or repaired, then the acceptance tests can serve as a basis upon which to establish suitability for resuming clinical use.

13.2 Timing and Staff Planning

Once a proton therapy system has been contracted for, component fabrication usually begins within months. Typically, these will be accelerator components, including magnets and large structure items, such as the gantry superstructure. Inspection of these components in the factory is recommended and has the additional benefit of building the vendor relationship that becomes even more important during on-site testing. Because factory visits are fairly infrequent, the user's project principals usually meet the staffing requirements. After fabrication, the proton therapy system components are shipped to the user location and the installation is performed. Prior to acceptance testing and to the extent possible, the staff should visit other centers with the same or similar equipment in operation to familiarize themselves with their proton therapy system model and to understand the process, challenges, and solutions associated with it. The staff should also read all available literature about their specific system.

Installation consists of placing all the components (including the large accelerator and gantry pieces), performing an accurate alignment, and achieving vacuum and basic integration of service-mode type control systems. At this point, the proton therapy system can accelerate a proton beam and perform basic mechanical motions, enabling basic acceptance testing. During the remaining period of vendor integration and calibration, the proton therapy system becomes more capable of higher-level acceptance testing. A suitable level of user staffing for the acceptance testing according to the testing schedule should be agreed upon with the vendor. Figure 13–1 provides an illustrative example of a schedule of the on-site proton therapy system

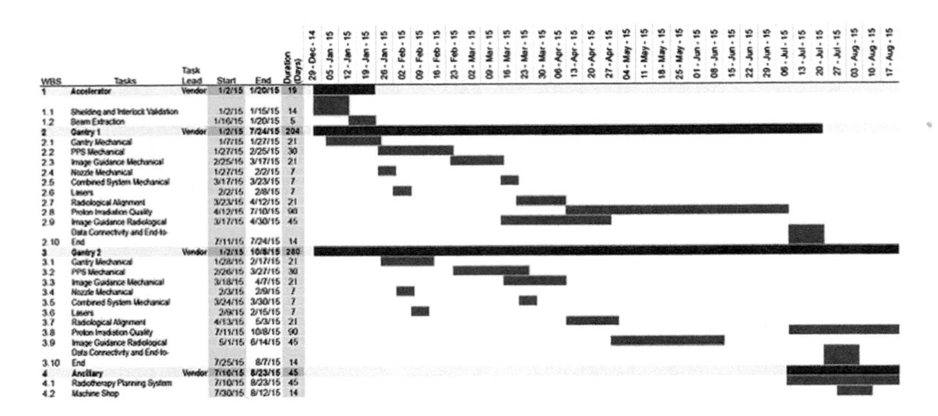

Figure 13–1 Example acceptance testing schedule for a two-gantry proton therapy facility.

acceptance testing for a typical two-gantry installation. The example is extensible to higher room numbers and is intended to show multi-room schedule dependencies. In this example, two full-time equivalent physicists are expected to cover the acceptance testing on a one shift per day basis (the estimate is only for technical acceptance testing, not facility operation). The increased personnel costs of bringing a proton facility online should be calculated and allowed for. Please refer to Chapter 17 on Clinical Workflow for details on staffing requirements. With regard to vendor staffing, the user should request an experienced staff from the vendor side in the acceptance testing process rather than relying on a novice staff. Such fore-planning could smooth the acceptance testing process and save valuable time.

The initial accelerator run-up requires several weeks of concurrent testing to validate the center radiation shielding (Figure 13–1). Several weeks of beam tuning along the beam transport line usually ensue, during which mechanical testing of the first gantry can occur. Because acceptance tests are identical among gantry treatment rooms, the project is usually organized into two to three gantry blocks for testing. Once an acceptance test is completed in the first room, then the test is repeated in the subsequent rooms soon after. This process leads to the pattern of acceptance testing timing in Figure 13–1, with similar acceptance tests staggered between the gantry rooms. After beam transport calibration, the proton beam comes into the gantry rooms, and central beam(let) alignment can be tested. (This is also true for scattering systems, although they are less sensitive in this regard than scanning systems.) The remaining acceptance testing schedule is dedicated to therapeutic field production tests, image guidance, and system-wide acceptance tests. Acceptance testing of the radiotherapy planning system may or may not be directly included in the proton therapy system acceptance testing, depending on contractual arrangements. Some connectivity is usually required for end-to-end testing.

13.3 Acceptance Testing Equipment Use

Measurement equipment used during acceptance testing can be a combination of vendor-specific proton therapy system tools as well as available commercial equipment. The timeline for procurement and availability parallels that of acceptance testing. The earlier acceptance tests require physical measurement tools and simple radiation detectors, whereas later, more advanced, testing may require sophisticated multi-element detectors and anthropomorphic phantoms. It may make sense to cooperate in tool-sharing with the vendor because many of the tools can serve common acceptance testing and clinical commissioning needs. In this way, the cost can be shared, and the user may gain valuable tool-use experience with their proton therapy system before the handover. Another common need is the ability to mount measurement equipment to the proton therapy system in a rigid manner. This can be accommodated by specialty "jigs" adapting the measurement equipment to hard mount points on the proton therapy system [3].

13.4 Radiation Safety Acceptance Testing

Facility shielding verification and radiation safety systems validation are types of acceptance testing. They can include third-party facility radiation safety systems, such as area monitors and door interlocks, but also include integration with proton therapy system safety interlocks for the accelerator, beam delivery systems, and kilovoltage image guidance systems. Before the first accelerating potential is applied, usually to the proton source, it is important to have a radiation safety testing plan in place with appropriate measurement equipment and qualified staff. The importance of safety awareness during this period cannot be overstated because staff will initially be unfamiliar with the facility installation, some integrated systems may not be operating as designed, and some visitors or workers may be unaware of the potential for harm.

Radiation levels in the proton accelerator and treatment areas can vary significantly—from being the same as background, to being elevated from activated components, to being very high during beam operations. Neutron and gamma radiation detectors within high-radiation areas monitor area radiation levels for staff awareness. The output of the facility radiation monitors, usually neutron and gamma, should be connected to optical and audible reporting devices. The safety system integration should be verified by testing and should be in accordance with state/national radiation protection guidelines.

13.4.1 Validation of Radiation Protection Systems and Safe Use Policies

No person is permitted in high-radiation (restricted) areas during beam operations. A facility safety assessment analysis should describe safety and interlock systems that promote personnel safety. The safety and interlock systems include the search process, door interlocks, emergency power off (EPO) buttons, in situ neutron and gamma radiation detectors, cameras, motion sensors, and intercom systems.

13.4.1.1 Area Search

The search process ensures that no person, except the patient, remains in the area in preparation for beam operation. The process requires a trained person to observe each area to ensure that they are vacated. The person may press a button or turn a key to acknowledge the observation. The search is completed by closing an interlocked door. Not completing the search within a specified time requires the process to be repeated. The search normally initiates audible or visual warnings that the area is being cleared to alert remaining workers in the area to leave. The area search function should be validated for each searchable zone.

13.4.1.2 Door Interlocks

Access from an unrestricted area to a restricted, high-radiation area must be interlocked to prevent or terminate beam operation. All interlocked doors must be tested for functionality prior to further activities with proton radiation.

13.4.1.3 Emergency Power Off Interlocks

Pressing an emergency power off (EPO) button removes power from the facility. This button is different from the PTS enable button described in Section 13.6. The EPO button must have dual electrical circuitry to enhance reliability. Once pressed, a positive action must occur, such as turning the button or unlocking it with a key, to release the interlock. At least one facility EPO test and recovery should be performed as part of acceptance testing. It is debatable if the EPO should be tested annually as it shuts down the whole facility, but its operation should be verified at least once during acceptance testing.

13.4.1.4 In Situ Neutron and Gamma Radiation Detectors

Neutron and gamma radiation levels inside restricted areas must be continuously monitored. Neutron detectors should be able to detect neutrons from thermal to maximum energy with good sensitivity. The Wide Energy Neutron Detector Instrument (WENDI-2) detectors from Thermo Fischer Scientific (Model FHT 762 Wendi-2 Wide-Energy Neutron Detector) or similar detectors exhibit excellent neutron measurement sensitivity for energies from thermal to 5 GeV. Neutron and gamma radiation detectors should accurately measure radiation dose levels to the maximum dose rate. Detector placement should be carefully considered to avoid obstructions and be close to high-radiation sources for accurate indications of radiation levels in the area, yet not so close that persistent, high-radiation levels damage the detector. A recording system of radiation readings would be helpful in the event of a radiological incident. Instruments in pulsed radiation facilities, such as synchrotrons, should have acceptable radiation integration in terms of dead time and detection characteristics for accurate measurements.

13.4.1.5 Cameras

Clinical cameras must be installed to monitor the patient from many angles. Other cameras can be placed in the facility for safety purposes. These cameras should have zoom functions to visualize the close-in view of irradiation.

13.4.1.6 Motion Detectors

Once an area has been searched and vacated, there should be no motion. Motion detectors can be placed in the cameras' blind spots. When a movement is detected, it should prevent or terminate proton operations. Motion detector testing can consist of electrical contact switching into the safety system.

13.4.1.7 Intercom Systems

An ability to communicate throughout a facility is essential for facility and patient safety. Radiation therapist or operators must be able to audibly monitor the patient or phantom irradiation. The system's functionality should be verified. A signal repeater system that allows radio and cellular phone communications within the facility can be a practical replacement for non-patient communications.

13.4.1.8 Safety and Interlock System Testing and Verification

Safety and interlock systems should be tested and verified in full during acceptance testing. The testing should ensure that a "beam ready" condition cannot be met when interlocks are engaged. Each interlock must be tested at least once and, perhaps, in multiple combinations and conditions of readiness. When the interlocked systems are engaged, the beam must not be initiated. Upon termination, the facility's warning lights and audible indicators must accurately reflect the interlock conditions. Interlock and "beam ready" statuses must be observable at the control stations.

13.4.1.9 Signage Verification

Proton treatment rooms are restricted areas that must be posted with one of two warning signs depending upon the state/national guidelines—either a "caution, high radiation area" (greater than 0.1 mSv/h) sign or a "grave danger, very high radiation area" sign (greater than 5 Gy/h). A waiver to use lesser warning terms, such as "caution," can be obtained from state regulatory authorities, especially for patient care areas. Other signs pertaining to magnetic fields, authorized access, or other restrictions should also be posted, if relevant. A walk-through should be performed, noting the appropriate signage for the areas prior to use.

13.4.1.10 Radiation Safety Detection System

In situ radiation detector readings must be present at the entrance of restricted areas and for operators at control stations. Elevated readings should trigger a visual warning indicator for facility staff. A lower alarm level may correlate to 20 mSv/h, and a second alarm triggers at 1 mSv/h. The actual alarm dose levels can be much lower than these values, depending on the beam direction and the detector's distance and placement from the radiation source. The warning indicator lights (Figure 13–2a) correlate the radiation levels with advisory information for entry. A green light might indicate that the radiation detectors read below the first alarm level, which can be interpreted as safe. A yellow light might indicate that elevated radiation levels are present. A red light illuminates when the beam is on, forbidding entry.

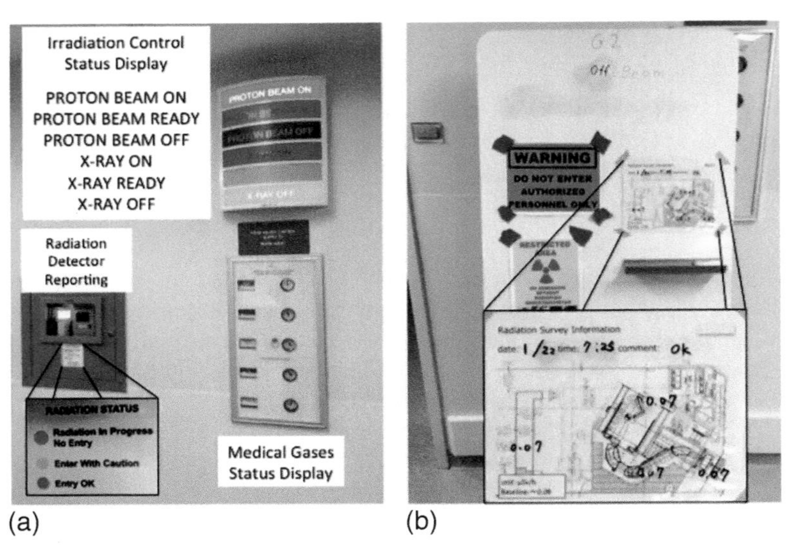

Figure 13–2 Example of appropriate treatment room radiation safety condition reporting directly outside the interlocked entrance door to a treatment room. (a) The irradiation control status display illuminates according to the proton therapy control system output, whereas the radiation detector reporting display illuminates in response to direct radiation measurements in the treatment room. (b) The physical note board reporting to the workers highlighting the most recent in-room activation survey and permission to work in the room.

13.4.1.11 Beam Status Display

A beam status display (Figure 13–2 a, Irradiation Control Status Display) shows the operational status of the proton beam in the respective area. It should show that the beam is off, that the areas are clear and ready for the beam, or that the beam is on. The "beam on" status for the radiation and beam status displays must agree for an effective warning indication.

13.4.1.12 Warning Systems within Restricted Areas

Proton facilities can cover large areas and numerous compartments. Each area where high radiation levels can be present must be searched to ensure that no one is present in the room. A radiation warning tone and visible indicator should alert workers that the area is being secured for beam operation. A red warning light within the restricted area should flash or strobe when the beam is on.

13.4.2 Shielding Verification Testing

Neutron and gamma radiation measurements must be made in areas adjacent to shielded spaces to ensure that levels are within state/national regulatory limits and consistent with the philosophy of as low as reasonably achievable (ALARA). The shielding acceptance testing should be performed according to a developed plan. The plan should include shielding confirmation under the conditions used during the shielding calculations, including energies, gantry angles, and calculation (test) positions. Monitored locations should be selected based on worst-case scenarios, such as in-line with the beam isocenter, at conduit penetrations, or at calculated points where the shield integrity might be questionable. A low-Z beam stop must be at the beam isocenter during treatment room operations.

Physicists should note the neutron quality or weighting factors used by calibration facilities for the instruments used and shielding calculation methods. Differences between these neutron dose correction factors can be significant. This information is useful when comparing readings between neutron detection systems, calculations, or other proton therapy center measurements.

Personal badge-type dosimeters should also be placed on the outside of selected shielding wall locations and monitored on a monthly basis, beginning with acceptance testing.

13.4.3 Activation Levels after Use

Facility activation is a significant issue for proton treatment centers [4]. Handling activated components can increase radiation doses to physicists, therapists, engineers, and other workers. Control and proper disposition of activated materials is also essential for regulatory compliance. Each facility should have a sufficiently shielded, secure room in which to store components. The room shielding should be validated, and radiation safety procedures should delineate processes for tracking, assessing, and disposing of or releasing components. Some information and guidance is available in the literature for proton therapy-related radiation safety aspects [4,5]. These policies and procedures should be considered as precursors to beam testing. Prior to each in-room acceptance test, a survey should be performed to verify a safe working condition.

13.5 Gantry Acceptance Testing

13.5.1 Gantry Testing Equipment

The proton therapy system is specified to small geometric values on the order of millimeters, plus a tolerance value for operation. Part of acceptance testing involves verifying that the delivered equipment conforms to the overall specification. Practical metrology suggests that the measuring equipment should have an accuracy 4–10 times greater than the unit specification, which requires that the geometrical test tools possess sub-millimeter accuracy.

13.5.1.1 Theodolites and Laser Interferometers

Theodolites are sometimes used to align and verify geometric accuracy of the proton therapy system. They are actually angular measuring devices in horizontal and vertical planes, but when combined with reference linear target scales, they can be used in a linear manner over small angles, providing sub-millimeter accuracy. Use of a theodolite in such a manner is shown in Figure 13–3 in a proton gantry. Laser interferometry-based systems are commercially available and are becoming the standard for geometrical testing in proton centers [6]. The systems are commonly termed "laser trackers." A typical laser tracker system consists of a base unit, control computer, and multi-faceted mirror target ball (Figure 13–4). The target ball can have different mounting possibilities, such as magnetic or threaded. Because a laser tracker can measure over a wide range of angles and distances, many challenging geometries can be measured accurately and efficiently [7].

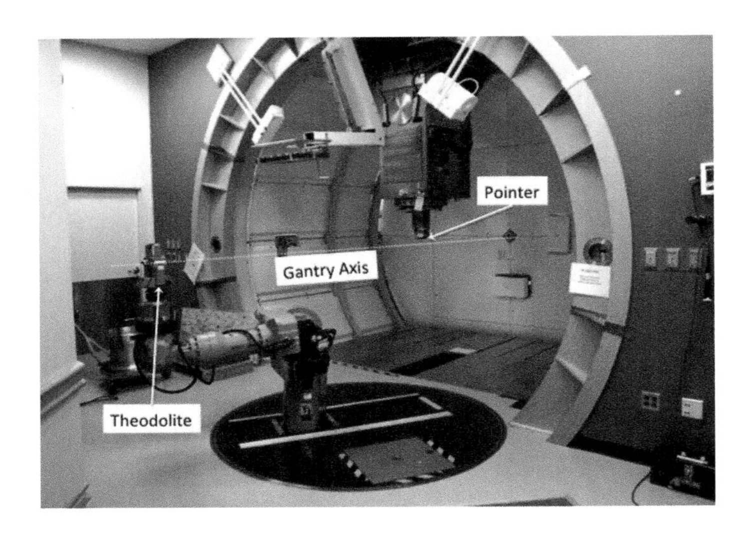

Figure 13–3 Method of proton gantry alignment verification with theodolite and pointer, with line-of-sight shown as the dashed line from the theodolite through the pointer and into the gantry axis point on the rear wall where an optical cross hair was placed.

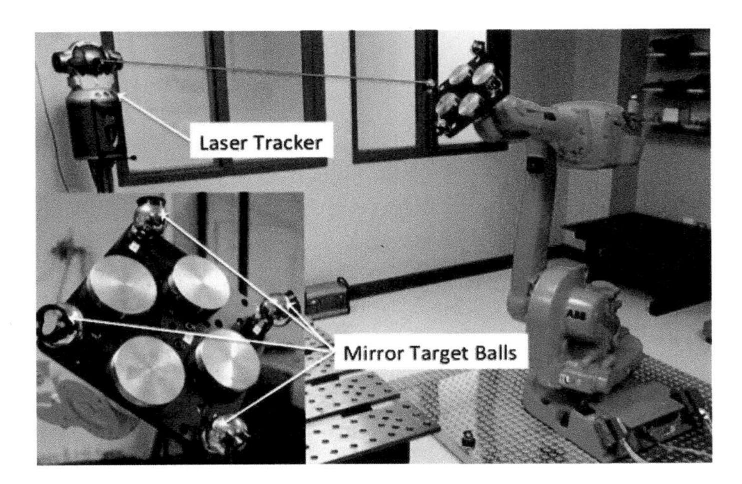

Figure 13–4 Experimental setup of the industrial robot (ABB Group, Zurich; model: IRB 1600) and the a laser tracker (Faro, Inc. Lake Mary, FL; model: Laser tracker) [8] for geometrical alignment.

13.5.1.2 Mechanical Dial Indicators, Spheres, Photo-electric Gates, and Force Gauges

Another type of geometric test apparatus can be fashioned from precision interacting spheres attached to mechanical dial indicators (Figure 13–5). Specifically, this apparatus is for geometric accuracy of a gantry and the patient-positioning system (PPS). As the gantry rotates around the spheres attached to the PPS, the dial indicators attached to it record sub-millimeter positional data [9]. A similar type of test capable of radiation isocentric testing shown in Figure 13–6.

Photoelectric gates are sometimes used for temporal measurements of translational and rotational speeds. They can be more accurate tests than conducted by eye with the use of a stopwatch.

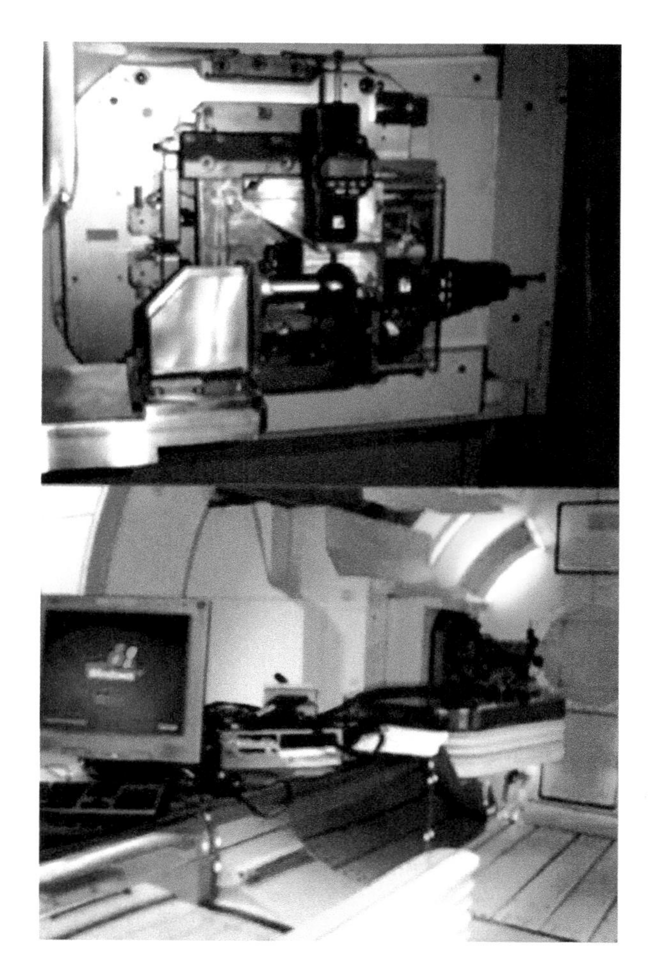

Figure 13–5 Gantry and PPS alignment test apparatus consisting of two dial indicators and a precision sphere.

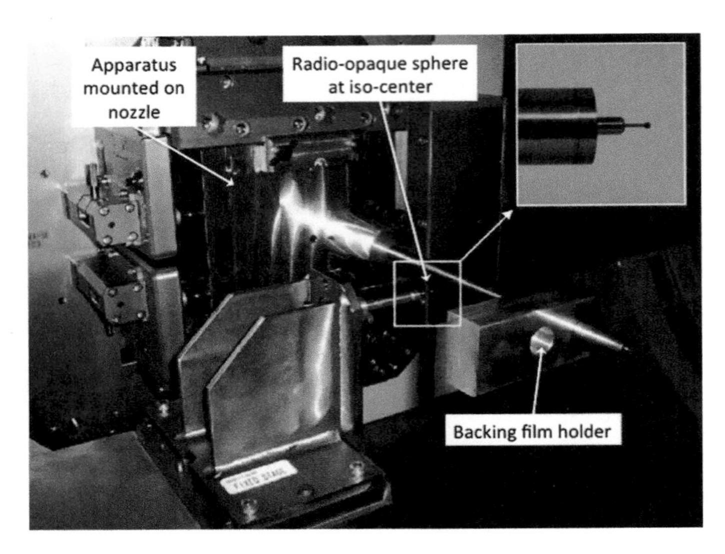

Figure 13–6 Test apparatus for gantry isocentricity with a 2 mm steel ball mounted on the patient positioner rigid plate and film holder extension from gantry nozzle. The brass plate mounts rigidly into the nozzle providing a small aperture beam block irradiation through the radio-opaque sphere into film contained in the backing film holder.

A simple push-pull force gauge (example: Extech Instruments, Nashua, New Hampshire, USA; model: 475044-SD High Capacity Force Gauge/Datalogger) is needed to apply known forces. It is used for applying 150 N force to test crush limits between physical devices.

13.5.2 Gantry Testing

Gantry-related physical acceptance tests validate performance and ensure safe use of a machine with extremely high rotational inertia.

13.5.2.1 Isocentricity

There are two types of methods for validating mechanical gantry isocentricity: mechanical and radiological. In some systems, patient positioner system (PPS) corrections are applied to correct known gantry isocentric deviations that are gantry angle-dependent. Hence, the gantry isocentricity performance may be evaluated either with or without PPS corrections. When using PPS corrections, it is still good practice to validate the gantry isocentric performance initially without the corrections to understand the magnitude and direction of the corrections that will be applied. The corrected gantry isocentricity should be <1 mm tolerance. Various methods as described below could be used to perform the testing.

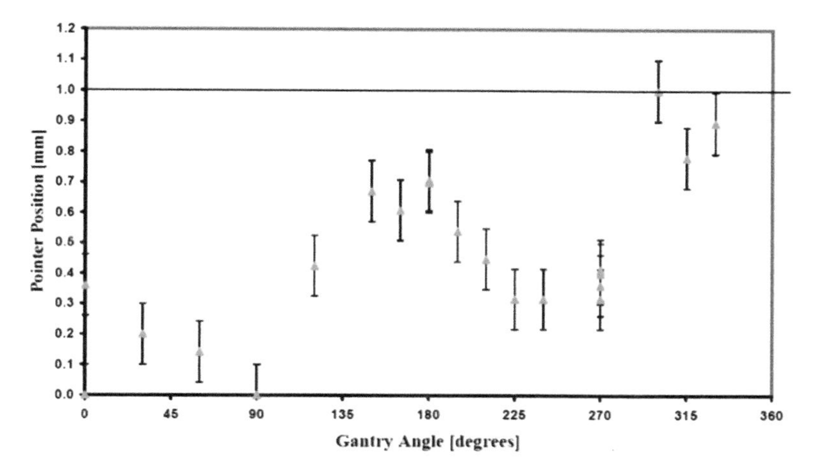

Figure 13–7 Isocentricity acceptance test results from a 360-degree gantry rotation. Error bars reflect the measurement accuracy. The pattern of measurements is sometimes observed where the gantry has certain "jumps" at under-couch angles, for example. These are likely due to the gravity statics geometry of all the gantry components acting together.

Method 1: Physical

Fit a ridged mechanical pointer into the nozzle/snout as shown in the measurement setup in Figure 13–3. Adjust the setup so that the pointer tip is located at the isocenter. Rotate the gantry through discrete degree positions, and record the deflection of the pointer from its initial position in mm. The magnitude of maximum deviation among the measurements defines the in-plane isocentricity accuracy, as shown in Figure 13–7. The repeated measurements determine the precision. A variation of this method uses mechanical dial indicators and a high-accuracy sphere positioned at the isocenter (Figure 13–5) and described in reference [9]. The methods are similar because both use mechanical means to determine the physical isocentricity performance of the gantry. A further variation of this method is to use a laser tracker device and place the optical target onto the nozzle frame. Align the device to the room coordinate system by using the permanent survey benchmarks that are usually placed during facility construction (assuming room and gantry isocenter are referenced from these points, as is the normal practice). Place the measurement reflector on the nozzle frame. Rotate the gantry continuously between the angle limits, and acquire data with the tracker. Software fitting can determine the centroid of rotation and the circularity thereof, representing isocenter accuracy and isocentricity, respectively.

Method 2: Radiological

This method uses the proton beam to evaluate gantry isocentricity. Fix a film at the isocenter, perpendicular to the gantry axis. Prepare a 1 cm or smaller central axis,

Figure 13–8 Proton gantry alignment star shot result.

high-energy, pristine proton beam by using an aperture, if necessary. By successfully irradiating from various gantry angles, perform a star shot through the film. After the film is ready, trace through the centroids (Figure 13–8) of the beamlet tracks. The circle diameter defines the gantry isocentricity magnitude. A variation of this method is to place a radio-opaque 2 mm sphere at the isocenter and shoot the central proton beam through it onto a backing film (Figure 13–6) [3]. It is useful to use proper software to analyze the star shot rather than simply using hand drawing to estimate deviation.

13.5.2.2 Angular Accuracy

It is important to verify gantry angular accuracy because it could affect geometric sparing of contralateral structures, and it is especially important when using a range compensator aligned to anatomy [10]. The gantry's angular accuracy should be verified within ±0.5° accuracy and ±0.1° precision. The test can be performed by setting up a large protractor on the gantry axis and verifying angular rotations with room lasers. More attractive current methods include the use of a calibrated digital protractor or a laser tracker.

13.5.2.3 Rotation Speed and Gantry Stopping

Proton gantries have several speed modes (low, medium, and high) and should be tested. The high-speed mode is limited to 5–6 degrees per second. Safety standards dictate that the gantry must decelerate and come to a complete stop within 5 degrees in an emergency. Both of these specifications should be verified by stopwatch or an optical gate system in both clockwise and counter-clockwise directions. The emergency stop verification must be performed in high-speed mode.

13.6 Patient Positioning System Acceptance Testing

Physical acceptance tests of geometric precision and accuracy form the basis of guaranteeing that a proton therapy system can easily and reliably be aligned within radiological guidelines. Motion safety acceptance tests are also required. Physical acceptance tests are typically the initial ones performed (Figure 13–1).

PPS systems (couches and robots) require acceptance testing for geometrical accuracy (six degrees of freedom), collisions, weight limits, and other interference. The types of tests are the same for both systems. If two or more treatment rooms are intended to share patients between them without regard for treatment re-planning, care should be taken during PPS acceptance testing to verify that the alignment of the systems is universal between the rooms.

13.6.1 Patient Positioning System Motion Safety

The PPS must not move unless enabled to do so. An emergency stop button depress must be capable of stopping PPS motion. Through a sequence of system enable and emergency stop actions, the system must respond appropriately. A test of the PPS collision force sensor stop must be made by exerting a force with a force gauge (Section 13.2) of 150 N. When the force reaches 150 N, the PPS must stop.

13.6.2 Patient Positioning System Motion Range

The specified 6-degree range of motion should be verified in all speed ranges. Use a ruler or protractor to check the lateral, longitudinal, and vertical stroke and the maximum pitch, roll, and yaw. Vertical stroke is shorter for therapy couches and ~50 cm and greater for robotic PPSs. The combination of pitch and roll is typically less than 10 degrees for the robotic systems and 3 degrees for therapy couches. The lowest position from the floor should also be verified to ensure ease of patient loading.

13.6.3 Patient Positioning System Translational and Rotational Accuracy and Precision

There are several approaches to PPS accuracy and precision testing. Either physical or radiological methods are used. The approaches differ in their objectives. Mechanical measurements test solely the PPS performance relative to that of either other physical components, such as the nozzle or snout, or defined benchmarks, such as room isocenter, whereas radiological testing uses the image guidance system, testing PPS performance relative to it. The significant difference between the methods is that the physical method requires traceability to a separate image guidance test, whereas the radiological implicitly includes it. In both cases, the PPS's performance in comparison to a chosen reference is made by calculating Euclidian distance (ED) as follows:

$$d((x_1,x_2),(y_1,y_2),(z_1,z_2)) = \sqrt{(x_2 - x_1)^2 + (y_2 - y_1)^2 + (z_2 - z_1)^2} \quad (13.1)$$

It is also important to test the movement and rotational accuracy and precision under a variety of loading conditions: maximum load (typically 180 kg), low load (10 kg), and differential loading.

All motion directions and angular rotations should be verified. Place the selected load on the PPS. Move the PPS to the start position at isocenter. Record this position as (x1, y1, z1). In successive directions ±x, ±y, ±z, either move to the test coordinate or move to the test coordinate and back to the start position. Then record the new position (x2, y2, z2). Repeat the test with individual rotations in pitch, roll, and yaw within the specification limits. All calculated EDs should be within specification, typically 1 mm total (±0.5 mm). The special case of eye treatment and intra-cranial radiosurgery positioners are tested to greater accuracy, i.e., in the range 0.2–0.3 mm.

13.6.4 Patient Positioning System Isocentric Rotations

The manner in which isocentric rotations are accomplished differs between traditional axial positioners and robotic PPS. Axial positioners rotate about physical axes, resulting in isocentric rotations intrinsically. Robotic systems require multiple joint movements to produce isocentric movement. Excursions from isocentric movement specifications and the need for supplementary corrections for robotic PPSs have been reported [11], resulting in the need for careful acceptance testing in these cases. By using one of the methods below, determine the ED (Equation 13.1) at each test position away from isocenter: PPS test positions 0, 45, 90, 180, and 270 degrees. Agreement should be within ±0.5 mm.

Method 1

Use the planar image guidance system to radiograph the spherical target at each test position. As observed with the image guidance system, the spherical target should not deviate from its start position by more than ±0.5 mm through the rotations tested.

Method 2

Position a laser tracker target on the PPS at 10 cm away from the isocenter. Record the target position at the test positions, and use the laser tracker software to fit a circle through them. The reported center of the circle is the absolute PPS isocenter, and the eccentricity represents the isocentricity of the PPS.

Method 3

Position a radiographic or radiochromic film flat on the PPS surface, centered at the approximate isocenter. Position the proton beam horizontally. Perform a star shot through the test positions with a narrow central axis proton beam (collimated or pencil beam). The start shot fit to intersecting exposure lines gives the isocentricity directly (similar to Figure 13–8, but for the PPS instead of gantry).

13.7 Laser Testing

Laser performance of systems intended for patient positioning within the treatment room may or may not be specified contractually between the vendor and user. In-room laser systems are a combination of line lasers and cross-lasers. A specification of ±1 mm for the laser system at the isocenter and within the patient treatment space is typical, although current systems may perform even better. Modern laser systems are ultra-fine and might be tested to the sub-millimeter level. Because of the high accuracy of the PPS, its use as a test apparatus is a reasonable choice for laser acceptance tests.

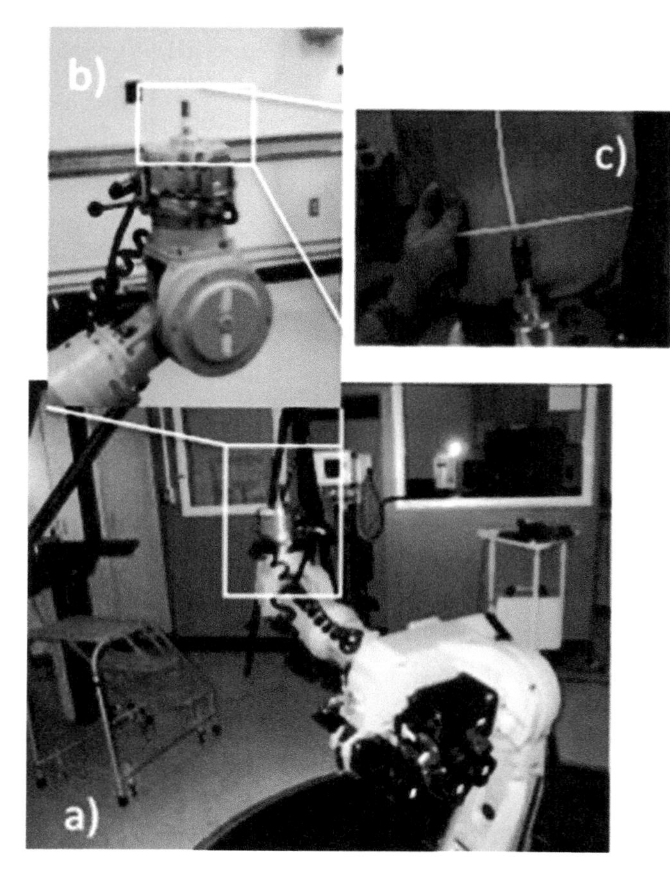

Figure 13–9 Positioning-laser acceptance testing showing (a) a robotic patient positioner modified to hold a pointer device, (b) a physical pointer device, and (c) the use of the device to confirm laser accuracy at a point. The positioner moves to confirm multiple points along a laser's projection line within the patient's space.

Method 1

Use the 2 mm opaque sphere of setup from Figure 13–6. Place a viewing screen distal and perpendicular to the laser source(s), and use the PPS to move the sphere into the laser path, fully occluding it. Jog the PPS along the laser path and ensure that the sphere occludes the laser beam over the length of the volume intended for patients' use.

Method 2

Remove the couch and mount a pointer to a robotic PPS arm or on the couch. The pointer needs to be pre-calibrated with a laser tracker at points along the orthogonal directions of laser projections. Details of the method are given in Figure 13–9.

13.8 Nozzle Testing

The primary performance characteristics of the proton therapy system indicate the properties of the dose distribution delivered by the system to a phantom or patient. Starting with the accelerator itself, the maximum energy of the machine will dictate the greatest range of the proton system, and the number of energy levels available will determine the resolution in depth with which the distal edge of a Bragg peak may be placed in a patient. Synchrotron-based proton systems will have specific accelerator settings for each clinical beam energy, and cyclotron systems are designed to always extract the beam at maximum energy. That maximum energy is then reduced to the desired clinical setting by using an energy selection system. Energy degraders, typically wedge-shaped blocks of material, are driven into the beam to reduce the beam energy to a desired level. Because of the statistical nature of proton energy loss in matter (see Chapter 4 on Energy and Range Straggling), the width of the energy distribution of the beam emerging from the energy degrader can be significantly greater than the incoming beam, and this requires an achromatic magnetic spectrograph and a set of aperture slits to trim the tails from this distribution. The width of the energy distribution of the beam emerging from an energy selection system or a synchrotron will affect the steepness of the dose falloff at the distal end of a proton beam.

Scanning-beam proton systems use a pair of dipole magnets to deflect the beam laterally over the extent of the target volume, and weighted stacking of individual energy levels controls the shape of a scanning beam's dose distribution in the proximal–distal direction. As such, the performance of a scanning beam system can be almost completely characterized by parameters that describe the energy, energy distribution, and lateral width of pristine pencil beams. Acceptance testing of a scanning beam system, therefore, requires very few measurements beyond depth–dose measurements in water and profile measurements in air for each energy level as described by Farr et al. [12]. Even so, uniform irradiation fields are usually tested for IMPT systems, too.

Scattered proton systems, however, generally use a pair of scatterers in the nozzle to spread the beam laterally over the target volume. An inherent trade-off exists between the scattering power of the devices—their ability to spread the beam over a large lateral area—and the energy degradation suffered by the beam as it passes through the scattering material. For that reason, most scattered proton systems use an assortment of scatterers, which can be used to provide the lateral scattering power and maximum residual range of the beam in a variety of clinical situations. Each of these scatterer combinations, generally referred to as options, must be tested individually over a range of SOBP widths and field sizes at the time of acceptance testing for beam uniformity. For additional details on the design and function of these proton therapy systems, please refer to Chapter 3 for double scattering (DS) and uniform scanning (US) systems and Chapter 4 for intensity-modulated proton therapy (IMPT) systems.

Proton beam acceptance testing consists of testing irradiation produced by the accelerator, energy selection or tuning, beam transport, and irradiation nozzle. Although the nozzle is the distal component that produces and delivers radiation fields, its beam testing is also representative of all the proton therapy system proximal components as well.

The nozzle is physically mounted to the gantry or, in the case of a fixed-beam installation, a rigid structure. In some systems, a snout is connected to the nozzle and serves as a telescoping aperture (block) and compensator-mounting (bolus) mechanism that allows those devices to close the distance between the nozzle and the patient. Snouts are used with DS and US scanning systems. Some IMPT systems may contain a telescoping snout to close the distance to the patient surface, and others may not. IMPT systems without a true snout may include a slot for inserting a range shifter and, possibly, an aperture. Basic nozzle testing includes physical, radiation leakage, and simple proton beam tests accomplished with a single beamlet (IMPT and US) or a simple open field (DS). After the basic nozzle tests have been passed, the acceptance testing can move on to advanced nozzle testing comprised of testing all the therapeutic field-producing aspects: longitudinal, transverse, and dose delivery.

13.8.1 Proton Beam Testing Equipment

A range of testing tools is needed for the proton beam acceptance testing. The tools represent a subset of those required for clinical commissioning. The suggested proton beam test tools are presented in Table 13–1. The tools listed in Table 13–1 are indicated for use with DS, US, and IMPT systems. The most significant difference among the DS, US, and IMPT measurements is the scanned nature of the US and IMPT systems. Scanned beams cause detectors operating in continuous acquisition to lose signal as the scanned beam moves off the detector. For this reason, accommodation must be provided when performing scanned tests by using an integrating sensor, an integrating detector mode, or multi-element detectors capable of simultaneous acquisition. Care must also be taken for selection and use in pulsed beam systems. Even DS tests in pulsed beam systems require a triggering signal and functionality for the detector control system. Two specific testing apparatus are presented here.

Figure 13–6 depicts a rigid rig for DS isocentric tests [3]. It has been adapted for IMPT testing as well, without the small aperture seen on the brass mounting plate. The apparatus shown in Figure 13–10 is a combined scintillator and camera system based on a design developed earlier for dynamic photon multileaf collimator testing [13].

Table 13–1 Proton beam acceptance testing-related tools

No.	Device	Delivery Type	Comments
1	Electrometer for absolute dose measurement	DS, US, IMPT*	Capable of supporting connected instruments. Sufficient bias capability for measurements in scanning beams.
2	Water phantom for 3D dose profile measurement	DS, US, IMPT	Trigger functionality needed for pulsed beam systems; limited to point-by-point integrated measurements for IMPT and US in scanning mode; for horizontal beam measurements, a thin window is attractive but care should be taken to avoid window deflection.
3	Software for water phantom control/analysis	DS, US, IMPT	Should include proton curve–specific analyses functions.
4	Waterproof Markus-type ionization chamber for depth dose measurements	DS, US, IMPT	Limited to point-by-point integrated measurements for US and IMPT in scanning mode; also handy for measurements in solid phantoms.
5	Ionization chambers for 3D/lateral dose measurement and output	DS, US, IMPT	PinPoint Ionization Chamber ~0.015 cm^3, flex chamber ~0.015 cm^3; limited to point-by-point integrated measurements for US and IMPT.
6	Ionization chamber for Bragg Peak measurement	IMPT	Larger diameter chambers may be preferred depending on beamlet size.
7	Multi-element ionization planar detector or scintillator for 2D dose profile measurement	IMPT	Needed for scanned lateral profile measurements.
8	Multi-layer Ionization Chamber	US, IMPT	Increases measurement efficiency in comparison to point-by-point depth dose measurements in scanned beams.
9	Film	DS, US, IMPT	Gafchromic or radiochromic.
10	Film scanner and analysis	DS, US, IMPT	Transmission type.
11	Cables	DS, US, IMPT	An assortment of tri-axial, co-axial, and Ethernet cables that is needed between the measurement equipment location and the power supply and control system. For control systems remaining in the treatment room, cables should be long enough to provide a standoff distance for the controller in the shielding maze.
12	Water-equivalent solid phantom sets	DS, US, IMPT	Needed for beam stopping and range loss material measurements. Proton-validated materials are preferred.
13	Control/analysis laptop PC	DS, US, IMPT	A centralized data acquisition computer is recommended.
14	Standard thermometer with probe	DS, US, IMPT	Calibrated; air and water temperature measurement capability.
15	Standard barometer	DS, US, IMPT	Calibrated.

*Double Scattering (DS), Uniform Scanning (US), Intensity-modulated Proton Therapy (IMPT)

13.8.2 Basic Nozzle Testing

13.8.2.1 Nozzle Mechanical

The nozzle frame and snout extension should be evaluated for flexure under gantry rotation. This testing is usually implicit in the gantry testing of Section 13.2 when tested as a combined unit. Snout extension accuracy should be tested. The snout has an indicator of its extension position in cm. The indicated extension can be verified

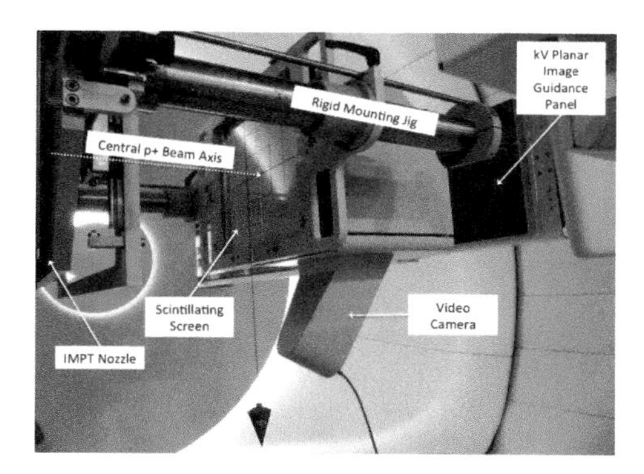

Figure 13–10 Test apparatus for IMPT spot alignment and fluence map scanning. The rigid mounting jig positions the scintillating screen in the axial plane at isocenter. The positioning is confirmed by beamline kV irradiation onto the imaging panel. Subsequent to setup, the proton beam is delivered statically or scanned onto the scintillating screen, producing visible light that is captured by the attached video camera in a light-proof housing.

within 1 mm by using a tape measure. The accessory slot mounting physical tolerance should be confirmed so that movement "slop," mostly attributable to gravity sag, is less than 1 mm at all gantry angles.

13.8.2.2 Nozzle Leakage

Radiological testing for nozzles and snouts consists of safety verification against direct proton transmission through holes or gaps. Finding "radiation leaks" is more common than it might be supposed; therefore, this testing is critical for patient safety. An appropriately designed and built proton therapy system guards against leaks. It is important that the patient never receive any direct proton beam other than the intended field of irradiation. With beam blocking, prompt gamma emission and neutron radiation will be produced. For systems using these beam-limiting devices, the absorbed dose within the patient plane behind the interchangeable beam-limiting device must be less than 0.5% of the dose when the device is not present [14].

Typically, out-of-field neutron-dose testing is out of the scope of acceptance testing. The user is advised to perform it as part of clinical commissioning. Testing of beam-limiting device radiation blocking and nozzle/snout leakage testing can be performed by using the proton beam and radiographic or radiochromic film. A good first step is to check for obvious radiation leaks by placing a light source in the nozzle and determining whether any light can be seen shining out of it in the darkened treatment room. The radiological acceptance testing method consists of wrapping the nozzle and snout with film and placing an array of films in the patient treatment plane, as is performed in external beam radiation survey. Figure 13–11 shows an example of radiation leakage testing with film.

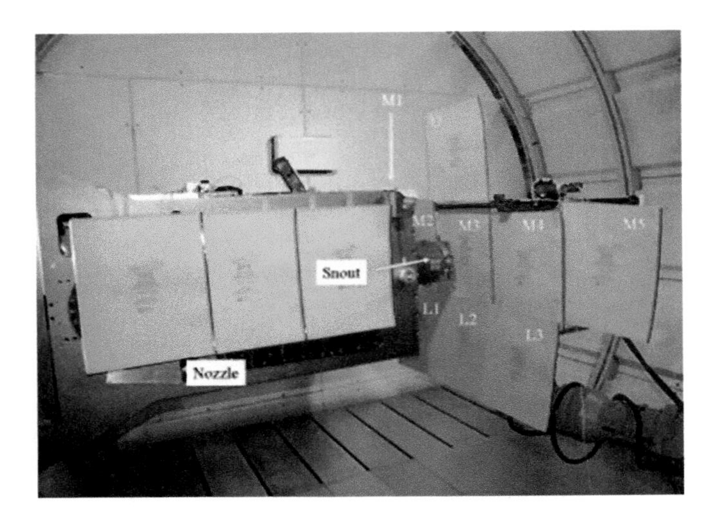

Figure 13–11 Experimental setup for snout leakage testing. Five films (M1–M5) were placed along the patient's position. One film was positioned above the patient's position (U), and three films below (L1–L3). An additional three films were placed directly on one face of the nozzle. This setup is not representative of a nozzle leakage test, in which the entire nozzle (all surfaces) should be covered with film. The setup shown here is specifically for evaluating dose in the patient's place. The additional films placed on the nozzle side were to confirm the patching of a previously found radiation leak through the nozzle side. The test is conducted with highest energy.

The general test condition for this type of test is to deliver a large enough dose to observe optical density on the films, with 100 Gy usually sufficient. The dose may be subdivided between a couple of clinically related scatter conditions—snout extended and retracted, for instance. The monitor units can be determined from simple dosimetric factors from the planned surface dose (no bolus) to the in-field portion of the film. An example result is presented in Figure 13–12. The acceptance testing for IMPT is similar. The scan conditions should be set to maximum field size. If apertures or range shifters are used, then they should be included in the test conditions.

13.8.2.3 Beam Alignment

Position accuracy of the proton beam is crucial for clinical system use, and it is verified by acceptance tests. Beam alignment may refer to either the centroid of the beam delivery system or the centroid of a broader therapeutic field. In this sense, the two may be the same for DS systems, but different for US and IMPT systems. Primarily therapeutic field alignment must be compared to the patient treatment position.

Central Beam Alignment

The beam alignment to central axis specification may vary depending on the system type. Both DS and US types define the clinical beam by physical collimation. This

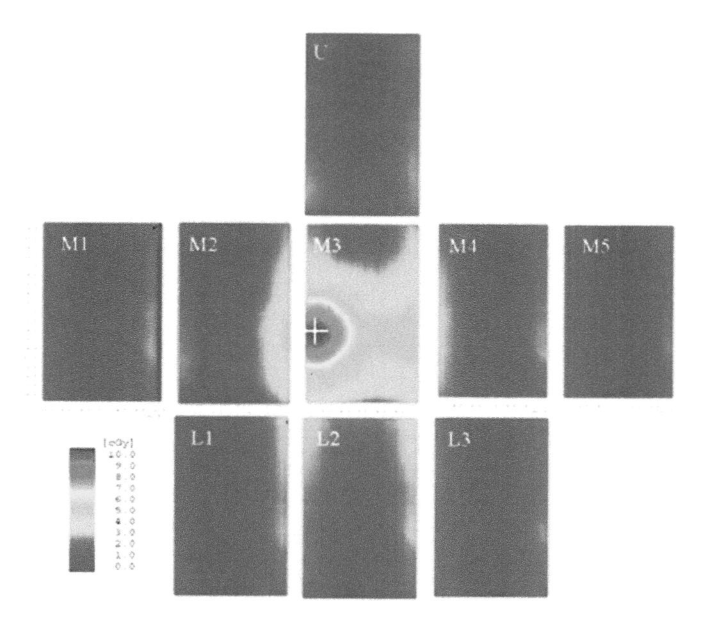

Figure 13–12 Film scan results from the snout leakage test. The dose scale is bottom left. The cross represents the approximate beam axis. The scanned background film (base + fog) did not differ from M1 or M5. 10 cGy/10,000 cGy corresponds to ~0.1% leakage.

makes magnetic alignment of the pre-collimated, post-scattered beam less critical for these systems than for an IMPT system of small spot size.

For all system types, the intent is to deliver the central beam or beamlet to isocenter. Such delivery is tested by placing a radio-opaque sphere at isocenter and irradiating through it onto a backing film (Figure 13–6) or scintillating detector [15]. The DS test uses a beam-defining aperture. Both small-diameter (8 mm) and larger-diameter (60 mm) apertures have been used. Because of the use of a beam-defining aperture, DS and US systems can be less sensitive to beam alignment than the IMPT. However, central beam alignment does play a role in field flatness of DS and US scanning systems. US systems scan beyond the maximum aperture diameter, providing tolerance of slight beam off-centering. The DS systems are more critical in this area because of the need for high-accuracy centering on the second scatterer. In both cases, the field flatness acceptance test is another test to review in terms of beam centering. Because IMPT systems usually deliver therapeutic fields without collimation, acceptance testing the magnetic beam alignment is critical. Testing US and IMPT beam centering precedes the same as for DS, except for the requirement of a single central axis beamlet for the test. When using a gantry, the isocentric beam alignment should be tested at several gantry angles, covering all quadrants. Example scanning test results are pre-

Table 13–2 Central beam alignment test results for an IMPT system and pass/fail calculation

Gantry Angle, °	0	0	0	90	180	–90
PPS Angle, °	0	20	–20	0	0	0
Measurements Position of the center of the field on the image						
X–, mm	–37.7	–42.7	–27.37	–22.2	14.9	–23.3
X+, mm	26.0	24.65	40.31	43.4	81.6	43.1
Y–, mm	–31.3	–45.9	–30.49	–32.2	–34.5	–29.9
Y+, mm	32.4	20.3	37.61	33.1	31.6	36.1
Center coordinates (calculated from above)						
X, mm	–5.85	–9.025	6.47	10.6	48.25	9.9
Y, mm	0.55	–12.8	3.56	0.45	–1.45	3.1
Position of the central sphere on the image						
X, mm	–6.30	–9.10	5.90	10.50	48.20	9.90
Y, mm	0.90	–12.80	3.10	–0.10	–1.50	3.90
Analysis Euclidian distance						
D, mm	0.57	0.07	0.73	0.56	0.07	0.80
Test succeeds if 100% of the values are below 1 mm						
				D_{mean} =	0.47	
				SD =	0.32	
					PASS	

sented in Table 13–2. In case two or more treatment rooms are intended to be dosimetrically "twinned," the central beam alignment tests should agree between the two rooms with the single room tolerances stated here.

The off-axis beamlet's placement accuracy must be verified for IMPT systems.

Off-axis Beam Alignment—Method 1

Use lasers to set up the film at isocenter. Then irradiate a central spot (beamlet) and additional pre-programmed spots in off-axis positions. Repeat this process at several gantry angles. Take care to measure spot positions out to the maximum field size: fringe magnetic field effects can make scanning magnet tuning more error-prone in this region. Measure the off-axis spot centers relative to the central spot center and confirm their alignment within ±0.5 mm.

Off-axis Beam Alignment—Method 2

Mount a scintillating detector system (Figure 13–10) to the nozzle or PPS. Follow the generalities of Method 1. Use applied fiducials or internal benchmarks to determine the deviation from isocenter from an aligned image guidance system, as shown in Figure 13–13. This test requires accurate spot placement within ±0.5 mm with respect to the image guidance center, coincident with isocenter.

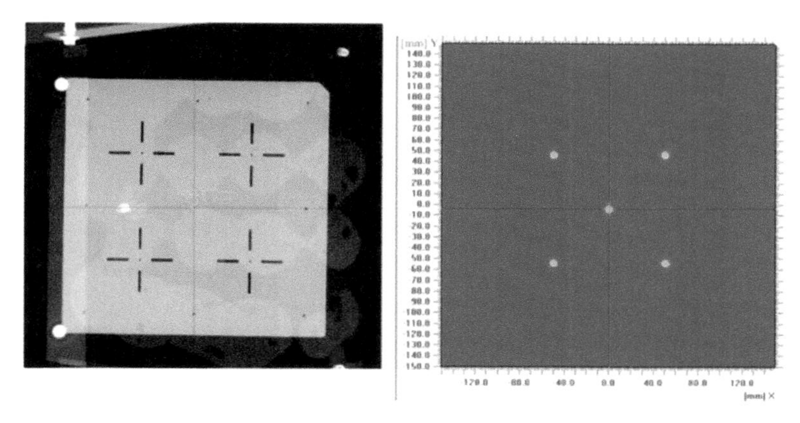

Figure 13–13 (a) Scintillator detector alignment to the isocenter using a kilovoltage image-guidance system. The green vector and magnitude indicate the position away from the isocenter. (b) Test spot irradiations at preset positions with respect to the isocenter.

The results of Method 2 represent true alignment of the proton beamlets to the treatment geometry as defined by the image guidance system. In contrast, Method 1 requires additional traceability to the treatment isocenter. The traceability can be established by combining the central beam alignment test with the image guidance isocenter verification testing described in Section 13.2.

13.8.2.4 Spot Size and Shape

Besides range, beam spot size and shape are the fundamental specifications for a scanning proton beam (IMPT) system. The size of the spot is usually given at various energies in air at isocenter. The specification can be in either sigma (σ) of the Gaussian distribution (assuming it is Gaussian) or full width half maximum (FWHM). One FWHM is equivalent to 2.35σ for the Gaussian distribution. Specification and verification of the beam shape is critical because the lateral size of the pristine proton beam will determine the lateral penumbra of a scanning beam's dose distribution. The shape of a pristine proton beam is generally well described by a Gaussian distribution, although for many delivery systems, the spot actually exhibits an elliptical distribution that requires fitting both the major and minor axes of the spot shape with Gaussian functions. There is evidence in many proton systems of a second Gaussian contribution to the spot shape. This component, arising from nuclear interactions of the proton beam with material in the nozzle, has a much larger sigma than does the primary component, although the relative area under this second curve is approximately 5% that of the primary. The broad width of this second Gaussian, combined with its small weight relative to that of the primary component, renders it difficult to parameterize with normal beam measurements. Indirect measurement of a second Gaussian can be performed via a field size measurement in which the dose measured in the center of a uniform layer of beam spots increases with the lateral growth of that layer because of the contribution of increasing numbers of secondary protons [16].

The width of the Gaussian spot shape is energy-dependent because lower-energy proton beams are more readily scattered by material in the nozzle and in the column of air between the vacuum window and the isocenter. Users may specify not only the size of the Gaussian beam spot, but also that the shape of the beam spot be invariant with respect to its position within the scanning area and that the beam sizes are matched between treatment rooms within a facility. The major and minor axes of an elliptical beam spot will precess with changing gantry angle, but the ratio of the major axis' length to the minor axis' length should be independent of gantry position.

Pinpoint ionization chambers, film or scintillators are commonly used to make proton spot measurements. When using an ion chamber, consideration must include the convolution effect of the chamber diameter with the spot size to be measured. Users should verify, in this case, that the spatial resolution of the pinpoint chamber is sufficient to avoid smearing out the true lateral profile of the proton beam. Another potential complication in measuring lateral beam profiles with a pinpoint chamber is that in systems with elliptical beam spots, the major and minor axes of the beam spot precess about the beam axis with gantry rotation. If the spot profile is measured at a gantry angle for which the major and minor axes of the beam spot do not align with the cardinal scanning angles of the water phantom, then users will underestimate the size of the beam spot. Some of the complications described above may be alleviated by measuring the lateral beam dimensions with a two-dimensional detector. Radiochromic film is an obvious choice for this task, but the dose response of the film must be carefully parameterized before its use. Commercially available scintillator screen detectors may also provide a two-dimensional measurement of the lateral beam profile. An advantage of both film and scintillator screen detectors (Figure 13–10) is that they may be affixed to the end of the nozzle and rotated with the gantry, facilitating measurement of the spot size and shape as a function of the gantry angle.

In addition to spot profile measurements performed at the isocenter for various gantry angles, the beam spot profiles should also be sampled near the extremes of the available scanning area. Assuming that the width of the scanning area is small compared to the distance between the scanning magnets and isocenter, the divergence of the beam should be small near the edges of the maximum field size, and the beam spot profiles should be independent of position.

Data from the spot profile measurements are fit with a one- or two-dimensional Gaussian profile, as needed. Modern scanning systems produce proton beam spread with σ ranging from 2 to 4 mm for high-energy beams and 5 to 8 mm for low-energy beams in air. The measured beam profiles should match the energy-dependent size specification to within a fraction of a millimeter, although the matching criteria may be energy-dependent as well. Some scanning systems may specify a maximum ratio between the length of the major and minor axes of an elliptical spot, and this ratio should be verified. Additionally, users should verify that the size and shape of the beam spot does not change with the gantry angle.

The beam shape parameter that is of the most interest to clinical users of a proton system is the lateral penumbra width of a treatment field. Without application of an aperture, the penumbra of a spot-scanning proton field is largely determined by the size of the individual beam spots, although the lateral falloff may ultimately be affected by the spot-spacing and spot-weighting of the beam spots within a field. Because the lateral penumbra is affected by treatment planning parameters, it is not generally considered a technical specification of a spot scanning system, and only the width of individual beam spots is measured at acceptance. For twinned treatment rooms, the beam spot shape parameters should be comparable within 10%.

13.8.2.5 Beam Energy and Energy Selection

Beam energy as determined by water-equivalent range measurements are a basic performance characteristic for all types of proton therapy systems. A typical acceptance test plan includes the minimum energy, maximum energy, and number of energy levels available that should be tested. Usually proton therapy systems have minimal energy levels near 70 MeV, corresponding to a water-equivalent thickness (WET) of 4 cm, and maximum energy levels near 230 MeV, corresponding to a range of 32 cm in water. Some proton systems fill the entire therapeutic range of energy levels with Bragg peaks spaced 1 mm apart, but other systems offer fewer energy levels in the deeper range of the spectrum because these Bragg peaks exhibit shallower distal dose falloffs beyond the peak. The range of a pristine proton beam is usually described as the depth of the distal 90% or 80% dose level. The slope of the distal falloff of a proton Bragg peak is partially dictated by the statistical nature of the protons' energy loss in matter, but the width of the energy distribution of the incoming beam will also contribute to the distal falloff. An advantage of describing the range of a pristine proton beam as the depth of distal 80% dose level is that this depth is fairly independent of the width of the beam's incoming energy distribution [17]. The distal falloff of the Bragg peak is also an important specification of a scanning beam system because it affects the ability of a user to spare critical structures downstream of a target.

The range of a pristine proton beam is typically measured by scanning a large-area parallel plate chamber, a Bragg peak chamber, through a water phantom. Ideally, this chamber is large enough in diameter to capture the entire lateral dimension of the beam as it scatters in the water phantom. This measurement is known as an integrated depth dose (IDD). Gillin et al. have shown, however, that even an 8 cm diameter Bragg peak chamber will lose protons to lateral scatter at therapeutic depths, necessitating Monte Carlo-based corrections to the IDD data. Because of the complex time structure of many proton beams, some delivery systems may require that IDD curves be measured point-by-point, stepping the Bragg peak chamber through the water phantom. To avoid having to compensate for the energy loss of the beam by the plastic wall of the water phantom, as many of these measurements as possible should be performed with the gantry at zero degrees.

An alternative to the Bragg peak chamber/water phantom method of IDD measurement involves the use of a multi-layer ion chamber [18,19]. This device is a long stack of parallel plate ionization chambers in which the ionization level at many points along the depth of the beam can be recorded simultaneously. The obvious advantage of this system lies in its efficiency— a single irradiation of the device produces an entire IDD curve, in contrast to the point-by-point data collection of the

Bragg peak chamber. Performance of this device must be benchmarked against either a Bragg peak chamber or a Monte Carlo calculation for a selection of energies to gauge the water-equivalent depth of the various ionization planes and to parameterize the loss of protons to lateral scatter with depth.

Each of the measured IDD curves should be normalized to 100% at the Bragg peak. The depth of the distal 80% dose level then specifies the range of pristine proton energy. Because proton delivery systems usually offer 1 mm range spacing over at least a portion of the therapeutic range, typical acceptance criteria may be that the depth of the 80% dose level matches the specification to within 0.5 mm.

In addition to the range of each proton beam's energy, the IDD data should be evaluated to determine the distal falloff of the Bragg peak. Range-straggling of the proton beam—the statistical spread in the range of individual protons—increases with depth and will cause longer-range Bragg peaks to exhibit shallower distal falloff. The range-straggling of a proton beam, σ_{RS}, can be estimated by Equation 13.2:

$$\sigma_{RS} = 0.0143 * R[mm] \tag{13.2}$$

where R is the depth in mm of the distal 80% dose level. The distal falloff of each Bragg peak, DF_m, described as the distance between distal 80% and 20% dose levels, is due to a combination of range-straggling, σ_{RS}, and the width of the incoming proton beam's energy distribution, DF_w. DF_w, calculated as shown in Equation 13.3:

$$DF_w = DF_m - \sigma_{RS} \tag{13.3}$$

Synchrotron-based proton systems should exhibit a DF_W of less than 1 mm, but the energy selection system in a cyclotron system will produce broader energy distributions.

Just as for the scanning systems described in the previous section, energy levels of a scattered system may be measured in a phantom with a Bragg peak chamber, in water, or with an MLIC. These measurements are used to verify the minimum and maximum range of the unmodulated proton energies, the range adjustment resolution of the system (number of energy levels), and the distal falloff of the monoenergetic beams.

The ability of the proton therapy system to adjust its range should be tested to the 1 mm level. This is usually performed around one or more energies and confirmed as water equivalent depth in a water phantom. Figure 13–14 provides an example test result. For twinned treatment rooms, the energy, energy selection, and distal falloff testing results should compare between the rooms within the single room tolerances stated here.

13.8.3 Advanced Nozzle Testing—Lateral

13.8.3.1 Uniform Field Flatness and Symmetry

In the same way that various range modulator wheels are used to produce SOBPs of differing modulation widths, DS proton systems will also use an array of hardware

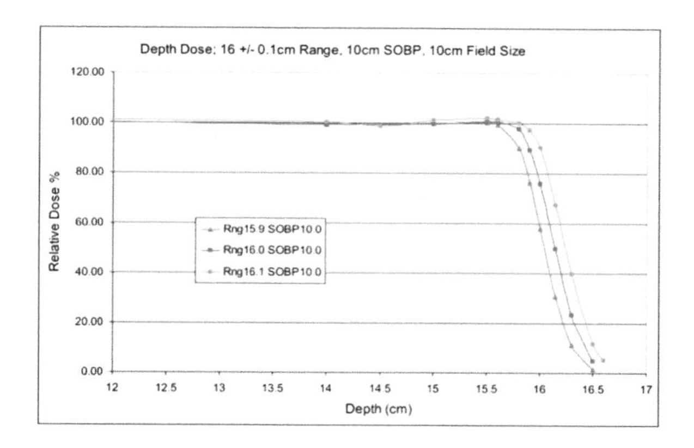

Figure 13–14 This test was designed to evaluate the ability of the energy selection system to vary water equivalent range by ±1 mm from a nominal setting (16 cm). The tolerance for range delivery was ±0.5 mm.

options to achieve differing field sizes. Lateral dose profiles should be measured to verify the performance and alignment of the scattering hardware. The lateral profiles may be measured with an ion chamber in a water phantom, or two-dimensional dose maps may be obtained in a lateral plane by using film or an ion chamber array.

Each set of primary and secondary scatterers is optimized for a range of field sizes and depths. Lateral profiles should be measured to determine the minimum and maximum field size for each scattering option. In addition to producing uniform proton fluence across the designated field size, the materials of a scattering system are meant to have constant WET as a function of distance off-axis [21]. This property of the scattering system should be verified for each option by measuring the range of the beam at various points off-axis.

The flatness and symmetry of a DS proton field are very sensitive to alignment of the beam with the scattering system. Two-dimensional lateral dose measurements from film or an ion chamber array should be analyzed at various gantry angles to verify mechanical rigidity of the beam optics and scattering system as the system rotates.

The US delivery technique shares similarities with both spot-scanning and DS proton systems. Like a spot scanning system, US uses a pair of magnets to distribute the protons laterally over a treatment field and apply various energy levels one layer at a time. Like DS systems, the final field shaping is performed with apertures. Hence, the primary specifications for the lateral dimension of the beam—such as symmetry, flatness, and penumbra—are defined in a similar manner for macroscopic fields shaped by apertures.

An example therapeutic transverse radiation field test result is presented in Figure 13–15. The results are for a US system using point-by-point measurements in a water phantom. Similar continuous detector scanning results could be obtained for a DS system. In acceptance testing, the intent is to confirm the capability of the system

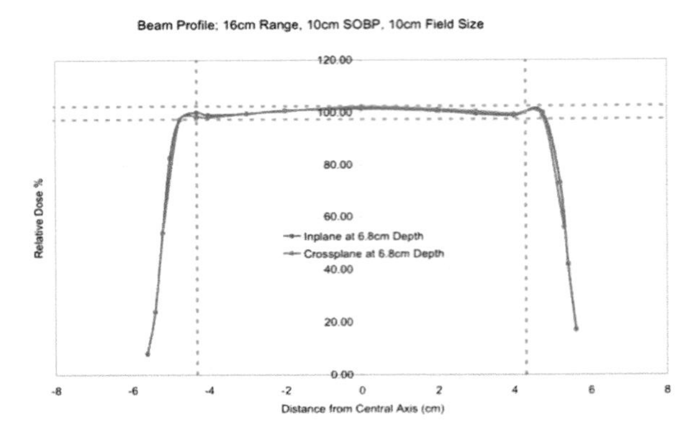

Figure 13–15 This test and others like it was designed to evaluate the lateral beam production capability of the system at a variety of field sizes and depths. The dotted lines represent the acceptance criteria, ±2.5% within 1.5 lateral 80/20% penumbra widths.

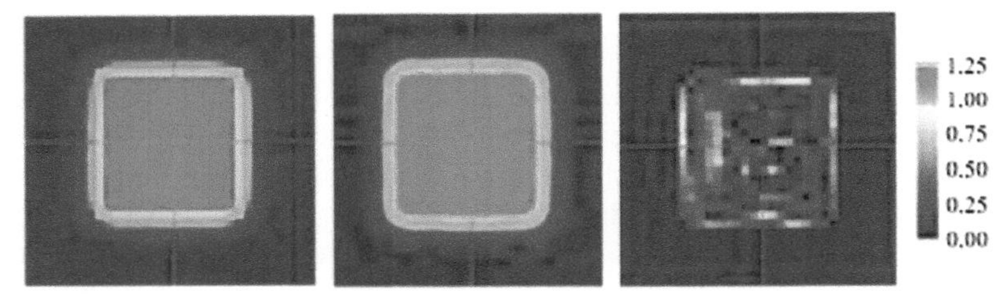

Figure 13–16 One liter uniform dose cube scanned test result showing (a) planned volume of uniform dose export from radiotherapy planning system, (b) measurement result from MatriXX detector, and (c) gamma criteria result at 3%/3mm.

to produce acceptable fields over the specified range of operation, and this should be verified for each option. IMPT should also be tested for uniform field output. This can be performed by delivering a 1 liter "dose cube," usually representing a 10 x 10 x 10 cm^2 irradiation volume. This is commonly generated by evenly spaced beamlets over enough energy layers to provide flatness and symmetry in the longitudinal direction. An example dose uniformity test result from a scanning system is provided in Figure 13–16. A multi-element ionization chamber-based detector or film is required to be placed in the center of the water equivalent thickness, midway within the 10 cm cube. The analysis depends on specification either ±2.5% to 3% as with DS and US or 3%/3mm as shown here. It is also common to test uniform dose cubes across the proton therapy system energy span of low, medium, and high.

13.8.3.2 Penumbra

Lateral penumbra in a proton treatment field is typically defined as the distance between the 80% and 20% dose levels, and it may be recorded from the measurement

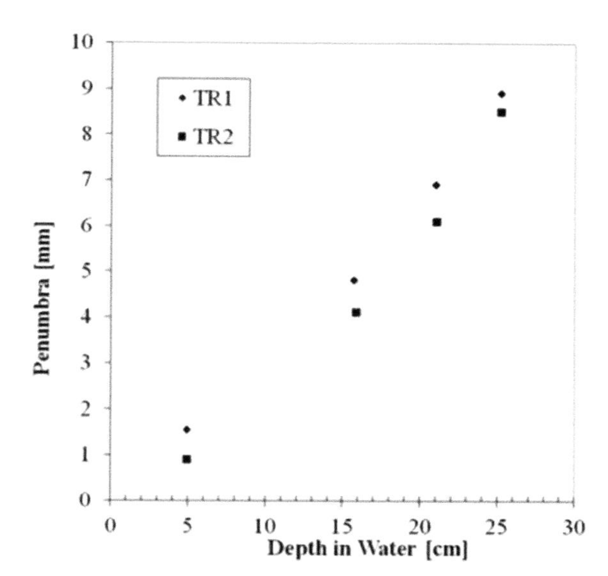

Figure 13–17 Lateral beam penumbra for two different treatment nozzles. Double scattering = TR1 and uniform scanning = TR2, as a function of depth in water. Reproduced from [12] with permission.

of lateral beam profiles by using a pinpoint chamber or film. The width of the penumbra is largely dictated by scattering of the beam within the patient, so it varies strongly with depth. However, the scattering and angular difference of the beam as it emerges from the nozzle also affects the penumbra width, so they should be specified and measured for a variety of field sizes and modulation widths. Figure 13–17 shows the penumbra width as a function of depth for two different nozzle configurations at the Midwest Proton Radiotherapy Institute, Indiana. The disparity in penumbra width between the two nozzles can be attributed to differences between DS (TR1 in the figure) and US (TR2 in the figure) with regard to penumbra [22].

13.8.3.3 Field Size

The size and shape of the area in the plane of isocenter that can be reached with the scanning beam is an important clinical specification that should be verified at the time of acceptance. Large-format radiochromic film may be used to measure the location and shape of beam spots at the limits of the scanning field area. Modern spot scanning systems typically feature field sizes that are 30 to 40 cm on a side. The field size dimension is significantly smaller than the distance between the isocenter and the scanning magnets (2 to 3 m), so the spot shape should be nearly independent of its location in the field. The most challenging case is maximum energy at maximum field size due to the largest scanning magnet current required to deflect the highest-energy protons. Field sizes are smaller for DS systems due to the inefficiency of large field production, which requires very thick scattering material. The testing method is the same as for scanning systems.

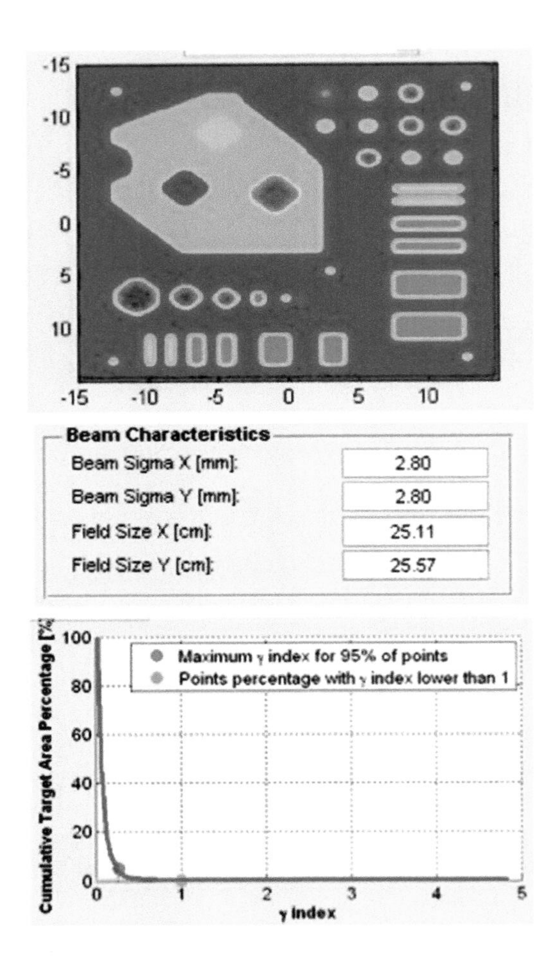

Figure 13–18 Gamma analysis results from an IMPT fluence irradiation for 230 MeV. The upper panel is the scintillator measurement from proton scanning irradiation of a planned test target. Software analysis tools calculated the field size, beamlet size (mm sigma) and the gamma index for every voxel in the image in comparison to the reference target (not shown).

13.8.3.4 Scanned Fluence Map

The spot acceptance tests described thus far use the scanning system only to make single spot dwell position irradiations. For additional confidence in the ability of an IMPT system to scan dynamically, fluence map irradiation acceptance tests can be performed [15] by using the setup shown in Figure 13–10. The combination of spot size, spot position accuracy, and MU chamber accuracy can be demonstrated by delivering geometric patterns of beam spots to a planar detector. Farr et al. [23] measured the performance of two types of scanning delivery systems with the assistance of a scintillating screen detector and software. Idealized beam spot patterns were convolved with measured spot size for a given energy to produce a predicted measurement pattern using fluoroscopic system as shown in Figure 13–18. These predicted fluence planes were compared with measurements via a two-dimensional gamma index to verify accuracy of the spot location and MU chambers.

13.8.4 Advanced Nozzle Testing—Longitudinal

13.8.4.1 Uniform Field Flatness and Symmetry

The proton therapy system should be tested for its ability to deliver flat and symmetric uniform dose in the depth (longitudinal) direction. The uniformity of the SOBP in the beam direction must meet specifications over the entire modulation width for each setting, typically 2.5% to 3% [24]. For DS and US systems, this usually involves testing a range of spread-out Bragg peak (SOBP) deliveries. Figure 13–19 presents an example of four different SOBP irradiations from a US system performed point-by-point in a water phantom with a parallel plate ionization chamber. The testing would be similar for DS, except the ionization chamber could be scanned. An MLIC might also be used for testing if it is known to provide water-equivalent measurements. SOBP extent testing results should be self-consistent for twinned treatment rooms as well.

In addition to IDD measurements for IMPT systems, SOBP type testing can be performed on the 1 liter dose cubes as well. An MLIC is usually used in this case, with the field scanned onto the detector.

13.8.4.2 Low-energy Range Absorber

Spot scanning proton systems typically produce a minimum beam energy with a Bragg peak at a depth of 4 to 7 cm in water. Treatment of targets shallower than this minimum range requires the use of an energy absorber (EA) or range shifter. In this sense, energy absorbers on the nozzle are IMPT-specific because on the DS and US systems they are simply applied to compensator (bolus) thickness. The EA is typically

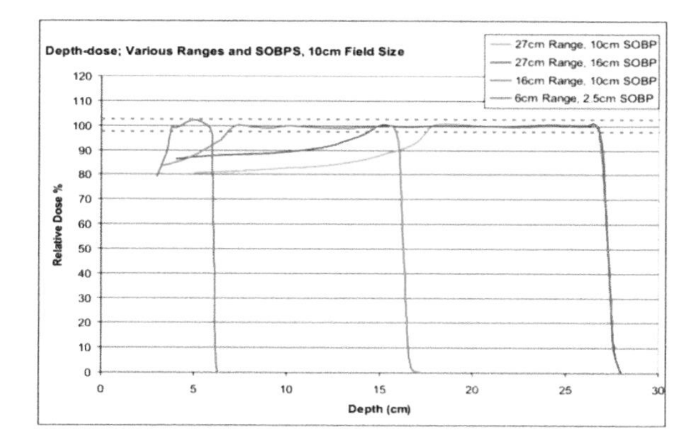

Figure 13–19 These depth–dose measurements where designed to test the system's ability to deliver a range of SOBPs from narrow to wide at several test energies.

a block of plastic or wax with stopping power and scattering properties similar to those of water. The radiologic thickness of the EA should be similar to the range of the minimum beam energy from the spot scanning system so that Bragg peaks may be placed at the surface of the patient. EAs are usually mounted in a metallic frame that can be placed in a slot in the scanning nozzle. The distance between the EA slot and isocenter may be variable in some systems so that the EA may be brought as close as possible to the patient surface, effectively reducing the size of the spot in the patient.

Acceptance testing of the EA requires that the user verify the WET of the device (see Chapter 4), its homogeneity, and, if applicable, the accuracy of the variable positioning system. The WET of the EA should be verified by using a Bragg peak chamber in a water phantom or by using a multi-layer ion chamber. IDDs for several moderate- to high-energy proton beams should be measured with the detector system, and the range of the beams is parameterized as the depth of the distal 90% dose depth. The EA is inserted in the nozzle, and the range of the same beam energies is measured to determine the WET of the device. All measurements should be taken with the beam on the central axis because off-axis beam spots will traverse the EA at a slight angle, influencing the effective thickness of the EA. The WET of the EA on the central axis should match the specification to within 1 mm WET for all beam energies sampled.

The most efficient method of verifying homogeneity of the EA is to acquire a CT scan of the device. The EA must be removed from its metallic frame before the CT scan because artifacts from the mounting hardware will cause artifacts in the images. The CT image should be inspected carefully to ensure that there are no cracks, voids, or heterogeneities in the material. Mean and standard deviation Hounsfield units (HU) should be recorded for regions of interest placed over the extent of the EA. There should be no regional variation in mean HU greater than the standard deviation of HU within any of the measurement regions.

Finally, if the EA is mounted in a movable slot in a telescoping nozzle, then the distance between the EA and isocenter should be measured at several points over its range of motion. The downstream face of the EA should be within 1 mm of its prescribed location over the entire range.

13.8.4.3 Ridge Filter

A ridge filter is a device with narrow variations in thickness that is placed in the proton beam to produce spatial differences in the proton beam's energy [25]. The ridges are narrow enough that the spatial correlation between the protons and their energy is quickly washed out by scattering within the patient. Historically, ridge filters have been used in scattering systems as an alternative to modulator wheels to produce spread-out Bragg peaks. However, mini ridge filters, which reduce the proton range by a maximum of only a few millimeters, may find application in modern spot scanning systems. Synchrotron-based proton facilities do not rely on energy selection systems to produce low-energy beams, so the energy distribution of these beams is very narrow. The narrow energy distribution, paired with a minimal amount of range-straggling, means that pristine Bragg peaks for low-energy synchrotron beams are quite

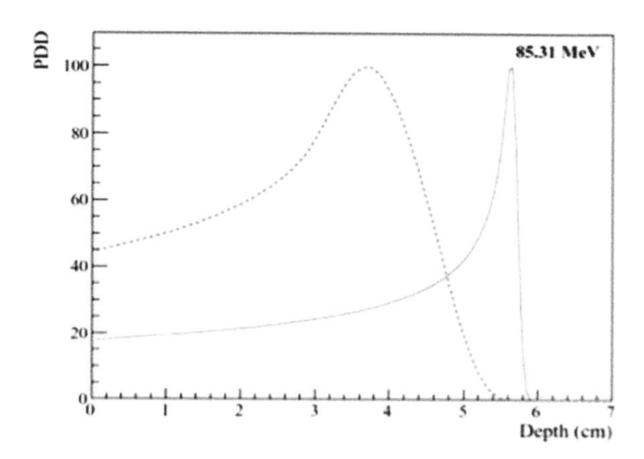

Figure 13–20 Monte Carlo-simulated PDD for 85.31 MeV proton pristine Bragg peak (solid line) compared to the same beam energy passing through a mini ridge filter (dashed line). From Courneyea et al. [27] with permission.

sharp, so producing a smooth dose distribution in the beam direction requires the summation of many energy levels. Use of a mini ridge filter can change the distal fall-off of low-energy pristine Bragg peaks from less than one millimeter without the filter to several millimeters with the device. Fewer energy levels are subsequently required to produce a uniform dose distribution, and treatment time can be dramatically reduced for superficial targets [26]. Figure 13–20 shows the change in the shape of a low-energy Bragg peak as it passes through a mini ridge filter using Monte Carlo simulation.

Acceptance testing of a ridge filter requires measurement of IDD curves for low-energy beams passing through the device on the central axis. Measurements may be performed with a Bragg peak chamber in a water phantom or with an MLIC. Users should verify that the filter produces the specified broadening in depth of low-energy Bragg peaks. A CT scan of the filter should be performed to check for localized voids or heterogeneities. Finally, a lateral dose distribution should be measured with film in a solid water phantom for a plan designed to deliver a uniform dose to a superficial rectangular target. The film should be at a shallow depth in the phantom (1 to 2 cm) to verify that the spatial correlation of beam energy is washed out by scattering in the phantom and that the pattern of the ridge filter is not visible in the measured dose distribution.

13.8.5 Advanced Nozzle Testing—Output

The proton therapy system output should be tested. Machine output testing is not calibrated dose output testing. Dose calibration is part of user commissioning. Some-

Table 13–3 Machine output adjustability test results from an IMPT system

Relative Output	Measured nC	Temp °C	Pressure mBars	Temp K	Corrected nC	Dose (Gy)	Variation
1.00	41.52	23.00	993.00	296.15	42.80	2.04	
1.00	41.52	23.00	993.00	296.15	42.80	2.04	
1.00	41.55	23.00	993.00	296.15	42.83	2.05	
				AVERAGE	**42.81**	**2.04**	–
1.03	42.72	23.00	993.00	296.15	44.04	2.10	
1.03	42.75	23.00	993.00	296.15	44.07	2.10	
1.03	42.74	23.00	993.00	296.15	44.06	2.10	
				AVERAGE	**44.05**	**2.10**	**2.9%**
0.97	40.24	23.00	993.00	296.15	41.48	1.98	
0.97	40.27	23.00	993.00	296.15	41.51	1.98	
0.97	40.27	23.00	993.00	296.15	41.51	1.98	
				AVERAGE	**41.50**	**1.98**	**–3.1%**

times a rough dose calibration factor is used in acceptance testing, but only for the convenience of doing so. It should not be regarded as system calibration.

13.8.5.1 Output Adjustability

The output adjustability acceptance test verifies that the proton therapy system dose output can be verified within a range suitable for performing machine radiation output calibration during clinical commissioning. The output range is machine-dependent and may be in the range ±3% to 7%. Meant here is only the range of adjustability for the dose monitor system, not its accuracy of calibration. The accuracy and precision specification is an order of magnitude less than the percent output change tested. Temperature and pressure should be monitored during the test. Sample data from such a test are presented in Table 13–3.

13.8.5.2 Dose Rate Linearity

Technically, it is increasingly challenging to produce a dose-monitoring system with acceptable linearity as the spot size decreases. This phenomenon is a result of the localized high beamlet flux rates required to deliver reasonable dose rates over field sizes used in clinical treatment.

Because dose rate is a limitation for IMPT systems, they perform at an optimized dose rate for all spot deliveries. Therefore, dose rate uniformity testing is implicit in MU linearity testing. Example output linearity acceptance testing results are presented here for a DS system (Figure 13–21) and a US system (Figure 13–22).

13.8.5.3 Monitor Unit Linearity

Machine monitor unit (MU) output linearity should be verified over a wide range for clinical use. Lower MU settings may be indicative of multiple clinical fields contributing to uniform dose or low spot-weighting (IMPT) as a result of the planning process. High MU settings can represent radiosurgery or hypofractionated treatments. It

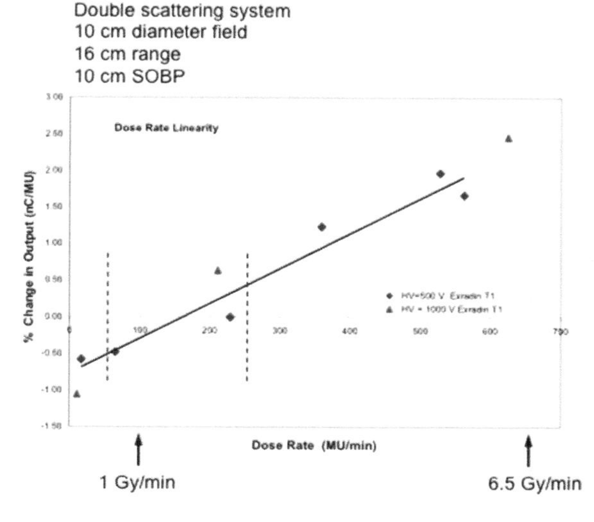

Figure 13–21 The variation in the beam output measured with a calibrated ionization chamber as a function of dose rate for two different polarizing voltages to the transmission ionization chambers in the irradiation nozzle. The vertical dashed lines indicate the operational dose rate boundaries used in clinic.

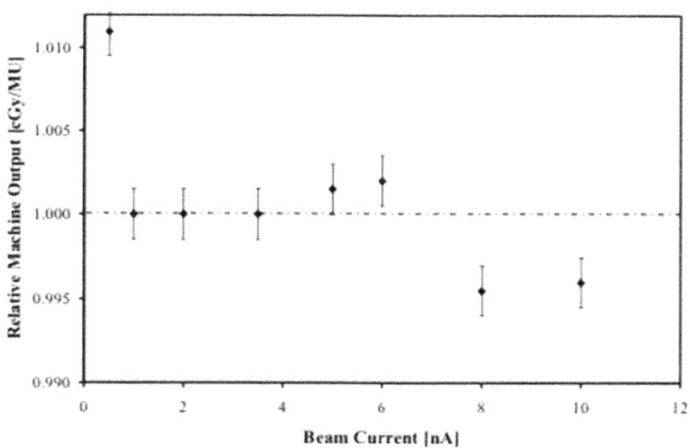

Figure 13–22 Dose rate uniformity results from a uniform scanning system. In the case of acceptance testing, Pion corrections were not performed. The clinical use of this system was between 0.5 and 5.0 Na, so the test was judged to be satisfactory.

is important to check both ends of the MU range. Testing is accomplished by setting up a reference geometry within an irradiation phantom, usually at mid-depth of a 10 cm^3 uniform field. Monitor unit settings are made in the proton therapy system before irradiation and ionization charge collection. The measurement should be repeated several times to verify repeatability. Example results of a US system MU linearity test are presented in Figure 13–23. At the time of testing, the minimum deliverable MU (a weakness of scanning systems) was 50. Subsequent improvement of the dose monitor system was later validated at a lower MU. This is an example of when it is appropriate to retest a system after significant modification. The extra-cameral effects in elongated fields should also be tested as shown by Zhao et al. [28]. Similarly, IMPT beam dose uniformity as shown in Figure 13–24 should be tested. Note that for some systems, low MU may show nonlinearity, as has been observed in photon and electron beams [29,30].

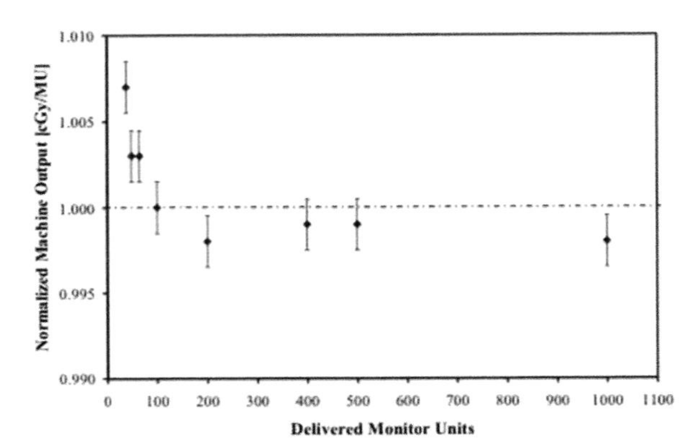

Figure 13–23 Output linearity with set monitor units.

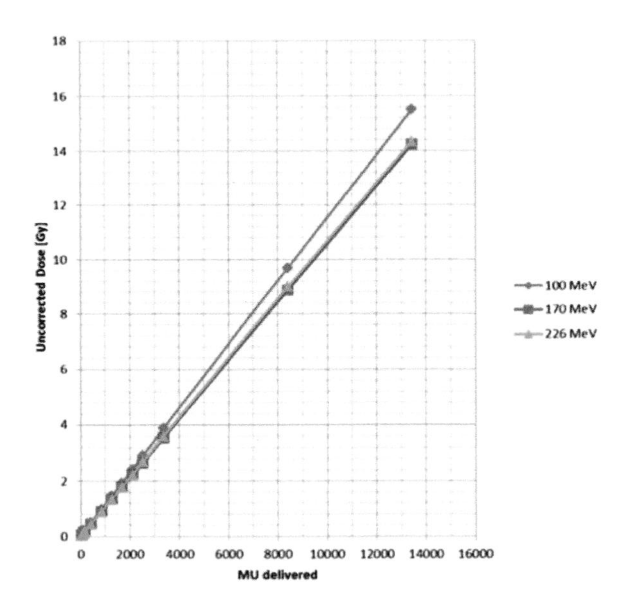

Figure 13–24 IMPT linearity.

13.8.5.4 Output Reproducibility and Stability

It is important to verify the output's repeatability, invariance to gantry angle, and applied temperature and pressure corrections (Ct, p). Repeatability testing (Table 13–4) calls for pre-setting the same MU on the PTS and repeating irradiations to validate the variance, which should be within ±0.3%. The test setup is with an irradiation phantom, ionization chamber, and electrometer. The same setup is used to determine gantry angle output dependency (Table 13–5) and multi-day output stability, including Ct, p corrections (Table 13–6).

Table 13–4 An example of reproducibility testing for an IMPT system

Range: 32 g/cm^2 Map: 10 x 10 cm Spot Distance: 2,5 mm Requested MU: 1800.00			
Repetition	Collected Charge (nC)	Delivered MUs	Dose (Gy)
1	42.6	1802	2.04
2	42.6	1802	2.03
3	42.8	1800	2.03
4	42.6	1802	2.03
5	42.6	1796	2.03
6	42.6	1795	2.03
7	42.6	1795	2.03
8	42.6	1796	2.03
9	42.6	1795	2.03
10	42.6	1795	2.03
	Average Dose		**2.03**
	Reproducibility		**0.05**
		Temp (°C)	**Pressure (hPa)**
Start Time	14:00	23.5	993
End Time	14:00	23.5	993

Table 13–5 Gantry angle output dependency results

Gantry Angle (degrees)	Run 1 (C × 10^{-8})	Run 2 (C × 10^{-8})	Run 3 (C × 10^{-8})	Average (C × 10^{-8})
0	3.39	3.41	3.38	3.39
45	3.39	3.39	3.40	3.39
90	3.40	3.39	3.40	3.40
135	3.40	3.39	3.41	3.40
180	3.39	3.41	3.37	3.39
225	3.40	3.40	3.40	3.40
270	3.40	3.39	3.40	3.40
315	3.39	3.39	3.39	3.39
	Average	3.39	C × 10^{-8}	
	Max	3.41	C × 10^{-8}	
	Min	3.37	C × 10^{-8}	
	Error Max	0.4%	Pass	
	Error Min	1.0%	Pass	

Table 13–6 Multi-day output stability testing with applied temperature and pressure corrections

	Date	Measured E-9 C	Temp °C	Pressure mBars	Temp K	Corrected E-9 C	Normalized Dose (cGy)
1	21-Dec	41.52	23.0	993	296.2	42.80	204.4
2	21-Dec	41.52	23.0	993	296.2	42.80	204.4
3	21-Dec	41.55	23.0	993	296.2	42.83	204.6
4	22-Dec	42.58	23.5	993	296.7	43.97	210.0
5	22-Dec	42.59	23.5	993	296.7	43.98	210.0
6	22-Dec	42.57	23.5	993	296.7	43.96	209.9
7	23-Dec	41.34	23.4	992	296.6	42.72	204.0
8	23-Dec	41.37	23.4	992	296.6	42.75	204.2
9	23-Dec	41.37	23.4	992	296.6	42.75	204.2
10	18-Jan	43.47	23.3	1012	296.5	44.01	210.2
11	18-Jan	43.47	23.3	1012	296.5	44.01	210.2
12	18-Jan	43.48	23.3	1012	296.5	44.02	210.3
13	20-Jan	43.17	23.4	1020	296.6	43.36	207.1
14	20-Jan	43.13	23.4	1020	296.6	43.32	206.9
15	20-Jan	43.13	23.4	1020	296.6	43.32	206.9
16	21-Jan	43.14	23.4	1025	296.6	43.12	206.0
17	21-Jan	43.14	23.4	1025	296.6	43.12	206.0
18	21-Jan	43.13	23.4	1025	296.6	43.11	205.9
	Average	**43.54**				**43.33**	**207.0**
	StdDev					**0.52**	**0.02**
	in %					**1.2%**	**1.2%**

13.9 Image Guidance System Testing

Careful, precise image-guidance radiation therapy (IGRT) is absolutely essential to effectively leverage the benefits of proton therapy. Exact knowledge of the tumor and normal tissue locations at the time of treatment allows clinicians to reliably treat the intended target with a minimal margin expansion to account for spatial uncertainty. This approach, in turn, minimizes the volume of normal tissue receiving a substantial radiation dose.

Every modern proton therapy system is sold with at least one image-guidance modality integrated into the delivery equipment. These systems are discussed in Chapter 7 (Imaging for Proton Therapy), but acceptance testing for the most common systems—radiographic flat panels or CT systems—is addressed here. Amorphous, silicon flat-panel imagers are ubiquitous in proton therapy treatment rooms. These detectors, paired with kV diagnostic x-ray tubes, are an efficient, effective means of imaging and locating radiographically evident features in the patient, such as bony anatomy or fiducial markers. Many treatment rooms are outfitted with a pair of x-ray tubes and panels, either mounted to fixed locations in the room or on the gantry so that orthogonal images can be obtained simultaneously. The downside to straight

Table 13–7 Image guidance acceptance testing-related tools

No.	Device	Imaging Type	Comments
15	X-ray test object	2D	Radiographic image quality
16	2D alignment test object	2D	Planar geometric alignment
17	3D volumetric test object	3D	Volumetric image quality
18	3D alignment test object	3D	Volumetric geometric alignment
19	CT ion chamber	3D	CT integral dose (CTDI)
20	Anthropomorphic phantom	2D or 3D	Suitable for proton therapy use

radiographic localization is that they give very little information about the patient's soft tissue anatomy. For this reason, volumetric localization, either via cone-beam CT (CBCT) or CT on rails, has started to enter the proton therapy domain. The broad details of IGRT system acceptance are common between 2D and 3D systems.

13.9.1 Image Guidance Testing Equipment

The end-to-end test (system test) and image guidance-related testing tools are presented in Table 13–7. Depending on the configuration of the user's system, planar (2D) or volumetric (3D) acceptance tests will be indicated. As is customary in image quality testing, the test objects are multi-function devices.

13.9.2 Image Guidance System Mechanical Testing

Image guidance systems are either rigidly fixed (rigid type) to the treatment room/gantry superstructure or have moving parts (motion type) such as swing arm detector panels, robotic cone-beam computed tomography (CBCT) systems, and rail-track mounted computed tomography (CT) systems (CT-on-rails). Because an accurate and precise mechanical foundation translates directly to image guidance quality, the aspects related to spatial reproducibility need to be verified by acceptance tests. These components essentially consist of established technologies from general radiation therapy practice. The AAPM Report No. 104 serves as a comprehensive guide [31]. Both planar and volumetric types of imaging systems require accurate and precise mechanical performance. The measurements for mechanical acceptance tests are quite similar, with the difference being that the planar imaging mechanical tests are usually discrete, whereas the volumetric imaging mechanical tests are continuous.

Rigid systems can be verified as part of room surveying. Some care might be warranted from gantry-induced vibrations if mounted on the superstructure, although the situation is static during imaging. Moving planar imaging panel arms should be verified for motion accuracy and precision. A panel imaging position location feedback sensor should be included and tested by moving the panel 0.5 mm out of position and ensuring an imaging interlock is given. The panel transit system should be run in 50 times to ensure repeatability.

13.9.2.1 Gantry-mounted Systems

Gantry-mounted image guidance systems usually consist of extendable image panel arms. Depending on the length of extension of the transit arms and their rigidity, a gravity-induced sag effect of some magnitude will be observed at the imaging panel. In addition, because of the transit arm rotation with the gantry, the relative sag will vary according to angle with respect to vertical. The sag should be measured and accounted for, if possible, in the reconstruction process. Several approaches exist to characterize the imaging panel sag.

Method 1

Using a beam line x-ray (IEC Z-axis), rigidly fix a central cross-wire to the nozzle and another one on the flat panel, and perform successive exposures at varying gantry angles. The resulting offset correlation between the two cross-wires is the angle-dependent panel sag with respect to the nozzle. The nozzle can also experience angle-dependent sag to a lesser extent. However, this possibility is a limitation of this method. Another limitation of this method is that it cannot account for gantry isocentric corrections performed by the PPS. Such PPS corrections are common practice. Therefore, this method is best applied when interrogating the uncorrected geometry or in fixed-beam rooms.

Method 2

Using a laser tracker, determine the gantry and panel sag performance independently by making full gantry rotations while collecting continuous data. The laser tracker software gives the circularity and deviation from isocenter iso-arcs directly. The lack of PPS integration in the test is the same as that for Method 1.

Method 3

Using a fixed cylindrical kV detector/phantom (Medcom, Darmstadt, Germany) placed at isocenter (Figure 13–25), map the panel sag behavior with the gantry angle every 45°. The resulting values are used as input for software correction in the reconstruction process. The reconstructed radius should be 0.5 mm or less. This method has the advantage of including the PPS in the process when gantry isocentric corrections are applied. A similar apparatus and method has been developed for megavoltage imaging on a linear accelerator. Four radiopaque markers are mounted on the surface of a cylindrical phantom, and users measure the distance between these markers in the reconstructed images [32].

13.9.2.2 C-arm Systems

Systems with C-arms use a mounted kV source/detector panel pair to generate planar imaging, fluoroscopy, or volumetric imaging. The C-arm might be mounted on a robotic positioning system or the patient couch. C-arms represent mechanical dipoles. Rigid restraint is required to achieve adequate circular rotation around the imaging center. Because of their closed and smaller form, C-arm mechanical tests are usually performed radiologically. Robotic C-arm systems may correct for the dipole-related imaging center deviations due to gravity. For testing, a fixed or variable collimator

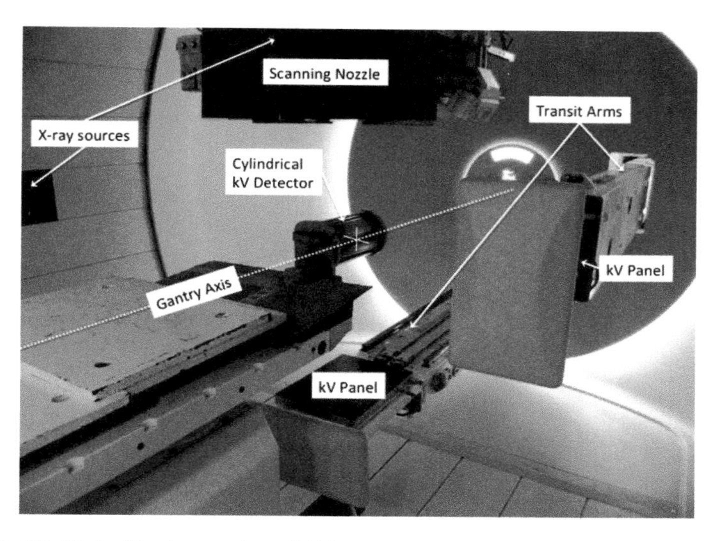

Figure 13–25 Cylindrical mapping of kV panel sag gantry angle dependence. The nozzle (beam line) x-ray source exposes the lower kV panel and the x-ray source on the left (gantry 270°) exposes the kV panel on the right (gantry 90°) in the image. Both kV panels are shown extended out on their transit arms from the gantry structure. The cylindrical kV detector is placed at gantry iso-center co-axial with the gantry axis.

may be used by selecting the smallest possible field size of less than 1 cm. By fitting centroids of fixed angle exposures, the discrete sag values can be determined, whereas a continuous exposure rotation can measure the sag by excursion from the pinhole, with divergence included. Laser trackers have also been used to determine C-arm system mechanical performance. By acquiring x-ray source and imaging panel position data during rotation and fitting this to the travel path, the imaging system's center and isocentricity can be determined.

13.9.2.3 Rail Track-mounted Systems

Acceptance tests for rail track-mounted systems include the CT scanner gantry, the PPS, and the coincidence of CT and PPS isocenters [33–35]. CT geometric accuracy can be validated by using a cross-wire phantom and imaging [33–36]. The PPS geometric accuracy acceptance tests described in Section 13.6 can be applied, with additional care taken over the long transit distance from CT location to isocenter. The absolute coincidence between CT and treatment isocenter must be verified within 0.5 mm. This verification may be performed by using a radio-opaque sphere and assessing the CT scan's center position with respect to the treatment isocenter. The correlation between imaging scan center and treatment isocenter can be established by either 1) kV planar imaging at treatment isocenter and traceability to proton beam isocentric alignment or 2) directly to the proton beam alignment by axial proton irradiation through the sphere and backing film.

13.9.3 Image Guidance System Radiological Testing

13.9.3.1 Image Quality

The primary specifications for a radiographic localization system are spatial and contrast resolution. Any number of commercially available phantoms may be used to verify the image quality performance of a kV radiographic or CT imaging system, but some rely on a user's judgment to determine the visibility of high-frequency or low-contrast features. A preferable approach is to use a system that includes automated image analysis software to calculate performance metrics. These quantitative measures can then be used as baseline data to monitor imaging performance after the system is deployed clinically.

13.9.3.2 CT Number Accuracy

Besides imaging the patient to determine tumor location immediately prior to treatment, volumetric IGRT systems may be used to detect changes in patient anatomy that would require a revision of the treatment plan. Ideally, the CT images used for tumor localization could be fed back to the treatment planning system to recalculate the patient's plan based on their anatomy of the day. Unfortunately, the cone angle used in CBCT imaging means that the scatter-to-primary ratio in these images is quite high, leading to cupping artifacts and low CT number accuracy [37]. If the users intend to use localization images from a CT-on-rails for dose assessment or replanning, then they must establish and verify a calibration curve for CT number to proton stopping power [38].

13.9.3.3 Imaging Dose

Acceptance testing of a radiographic IGRT system should include measuring beam hardness with an ionization chamber. Typical localization systems have x-ray tubes with inherent filtration equivalent to 2 to 3 mm of Al [39]. This level is necessary to ensure that a patient's internal anatomy can be imaged without delivering an excessive surface x-ray dose.

13.9.4 Registration Time

A variety of registration techniques may be used to compare daily localization images to reference images from the treatment plan. Two-dimensional radiographs may be aligned to digitally reconstructed radiographs (DRRs) manually or by using an automated algorithm, such as a mutual information calculation. Registration of 2D radiographs to 2D DRRs (2D-2D matching) results in a three- to four-element correction vector in which translations and, possibly, a yaw rotation may be sent to the patients' couch.

Alternatively, the IGRT system may use the treatment planning CT as a reference image and repeatedly calculate DRRs with rotations and translations until an image is found that matches the orientation of the acquired 2D radiographs. The result of this process (2D-3D matching) may be a full six degree-of-freedom (6 DOF) correction vector in which three translations and three rotations are sent to the patients' couch. A

6 DOF correction vector may also be obtained in volumetric localization processes in which the treatment planning CT is registered to a CBCT or CT-on-rails image acquired in the treatment room (3D-3D matching).

Even carefully immobilized patients do not stay in a single position for a very long time. Either voluntary or involuntary motion may cause the position of the target tissue to move after imaging with the IGRT system. For this reason, as well as general clinical efficiency, it is important that automated registration algorithms return a valid result within 30 to 60 seconds. This specification may be validated in conjunction with registration accuracy tests, described below.

13.9.5 Registration Accuracy

The accuracy of each registration technique must be tested by using geometric and anthropomorphic phantoms. These tests are trivially performed after the patient-alignment lasers have been accepted in the treatment room. A phantom with radiographically evident internal features is scanned in the CT simulator, and the treatment planning system is used to generate reference DRRs or CT data sets for a test alignment plan. The isocenter of the plan should coincide with external marks on the surface of the phantom. The phantom is placed on the treatment couch with either translational or translational and rotational errors relative to its planned orientation. The phantom is imaged and re-aligned by the IGRT system. The user may then verify that the treatment room lasers point to the external marks on the surface of the phantom within a tolerance of 0.5 mm. Phantoms with straight geometric features are well-suited for testing localization capabilities with manual matching algorithms, but anthropomorphic objects should be used to test automated 2D-3D or 3D-3D registration algorithms. Tests that combine IGRT localization with beam delivery accuracy are described in the section on end-to-end testing.

13.10 Treatment Planning System Testing

Users must verify proper functionality and communication between their treatment planning system and the proton delivery system. Starting with a CT scan of a phantom with known geometry, the CT image should be sent to the treatment planning system as if it was a patient's image for clinical operation, most likely through a DICOM protocol. The geometric integrity and image quality of the phantom CT scan should be checked after import into the planning system. Proton treatment plans should be generated for the phantom, with at least one plan corresponding to every delivery modality in the facility (spot-scanning, double-scattering, etc.). For spot-scanning proton plans, the optimizer must generate a reasonable dose distribution that respects the delivery constraints of the treatment machine, including minimum and maximum MU per spot and maximum field size. The DS plans should similarly obey the machine's delivery constraints and generate aperture and compensator files to be sent to the machine shop for fabrication. The AAPM Task Group 53 provides comprehensive guidance for commissioning the remaining, non-proton-specific functionalities of a treatment planning system [40]. Another resource is the TRS-430 report [41].

13.11 Gating Testing

Treatment of a moving tumor is particularly problematic with proton beams because the radiologic depth of the tumor may vary in addition to its position relative to the beam axis. Spot scanning proton plans may be particularly sensitive to tumor motion due to the interplay effect, in which portions of the moving tumor may be over- or under-treated by the scanning beam. Several strategies have been proposed to mitigate the interplay effect, including breath-held treatments, rescanning, tracking, and gating [42]. Respiratory gating is a proven method of reducing irradiated volumes in x-ray treatments of moving tumors, so it is readily applicable to proton therapy. For a detail of motion management, see Chapter 25.

An important specification in a respiratory-gated delivery system is the latency period between activation or termination of the "beam on" signal and the corresponding response of the treatment beam. Many x-ray and proton delivery systems require several tenths of a second to produce a beam after the respiratory gate is activated, but the beam should be terminated within a millisecond of the gate closure [43]. These characteristics should be tested by using a digital storage oscilloscope. One channel of the scope will display the 5-volt gating enable signal, and a second channel will measure the current collected in the dose monitor in the proton nozzle. The latency in both beam delivery and termination relative to the gating window should be measured and compared to specifications. Dosimetric accuracy of a gated delivery should be verified for a variety of regular and irregular gating patterns. A test pattern is fed into the gating signal input so that the delivery of the 1-liter dose cube is gated on and off. The dose plane measured by the ion chamber array should match the ungated baseline delivery to within 2% at all points within the 1-liter cube.

13.12 Machine Shop Testing

Most DS proton centers manufacture their own apertures and compensators. Communication of aperture shapes and compensator drilling patterns should be verified between the treatment planning system and the fabrication machines. Test apertures should be cut, and their orientation and dimensional accuracy should be verified. Users should confirm that the aperture can only be inserted in the treatment nozzle in the correct orientation. Trial compensators should also be manufactured to verify that they are cut with the designated drill diameter and taper. QA of manufactured compensators may be performed via CT scanning or other methods [44,45].

13.13 Data Connectivity and End-to-End Testing

End-to-end testing is an opportunity for the user to verify that all components of a proton treatment system—from simulation to patient treatment chart work—are in concert and as they were intended. It is also an effective means for the therapy staff to gain valuable experience in delivering daily treatments to patients before a facility is clinically operational. Various end-to-end tests may be developed to demonstrate functionality of a single attribute of the proton system, but they should mimic the actual treatment flow as closely as possible, with simulation and treatment performed by therapists, treatment plans calculated by dosimetrists, and verification of delivered dose and electronic chart accuracy provided by physicists.

13.13.1 Radiotherapy Planning System-oncologic Information System-treatment Control System Testing

The basic end-to-end test sends a phantom through the entire simulation, planning, and treatment process. The phantom should have external marks for visual alignment and internal features to facilitate IGRT registration. Some portion of the testing should be performed on a phantom that contains an ion chamber array so that absolute delivered dose can be verified after treatment. The basic plan may deliver a clinically likely dose, such as a uniform 2 Gy dose to a 1-liter cube, and this measured dose plane should be stored as a comparison baseline for other delivery simulations. The plan may be delivered later, with various operational perturbations, and each of these delivered dose planes may be compared to the baseline dose distribution.
Each of the available IGRT registration methods should be tested, and the system should deliver the correct dose to the embedded detection system within 2% and 1 mm. After completion of a treatment fraction, the patient's chart in the oncology information system should reflect the delivery of the correct dose. Image data and registration vectors from the pre-treatment IGRT should also be available for review and accurately represent the pre-treatment registration process.

13.13.2 Operational Perturbations

Any number of factors may interrupt a routine proton treatment fraction. Patients may become ill or need assistance during a proton treatment fraction. This possibility should be simulated by interrupting the beam manually during delivery and resuming treatment after a brief interval. The dose plane measured by the ion chamber array should be compared to the plane measured during the uninterrupted delivery, and all points within the 1-liter cube should agree to within 2%. The patient's chart should reflect successful delivery of the prescribed treatment fraction.
Patients who become more ill during a treatment may need a longer interval before resuming treatment. In this case, the user should stop delivery part-way through the 1-liter dose cube and close the treatment on the delivery console. A partial treatment record should be sent back to the patient's chart. A new appointment is scheduled for this patient, and the remaining portion of the treatment plan is delivered. The ion chamber array in the phantom can be used to measure the combined dose from the two partial deliveries, and the total dose should agree with the uninterrupted baseline delivery to within 2%. This is especially important for US and PBS.

Failure of the delivery system during a treatment fraction should also be simulated and tested. A delivery interlock is manually asserted during delivery of the 1-liter cube, forcing the recording of a partial treatment record in the patient's chart. A new appointment is scheduled to deliver the remaining portion of the treatment. The total dose delivered between the two partial treatments should be compared to the baseline delivery and should agree to within 2%. In addition to testing with a geometric phantom and detector array, some portion of the end-to-end testing should be performed with anthropomorphic phantoms. Users should verify the calculation and transmission of DRRs from the planning system to the IGRT console and the functionality of the various registration methods.

13.13.3 Throughput Timing

Efficient throughput of patients in a proton center is critical for operational success of a facility and for the patients' comfort and convenience. An acceptance test measuring the total delivery time for treatment of a phantom ensures that all components of the system are operating efficiently. A timer is started when a phantom is brought into the room and placed on a table. IGRT alignment of the phantom is performed, and the proton plan is loaded and delivered. The plan should be clinically realistic, including two to three treatment fields with gantry rotations and, possibly, couch movements. The treatment record is stored in the patient's electronic chart. Many proton centers may plan on 20-minute time slots for basic patient treatments, which should then serve as the criteria for this test.

13.13.4 Treatment Modalities

For systems with multiple options, like DS, US, and IMPT, each of the expected treatment modalities should be represented in the end-to-end testing, including spot scanning with single-field uniform dose plans, IMPT plans, US, and DS. The test doses should be clinically realistic, ranging from 1.8 Gy up to 10 or 12 Gy if hypofractionated plans will be delivered in the facility.

13.13.5 Electronic Medical Record, EMR

The EMR should accurately reflect delivery of all test plans during end-to-end testing. Tolerance tables for setup variations should be defined in the EMR, and users should verify that they are enforced during delivery of test plans. Any parameter overrides should be reflected in the patient's chart, along with the credentials of the user who performed the override. Image data from IGRT processes should be available for review in the EMR, along with registration vectors representing the action that was performed by the therapists during treatment. Finally, once a test patient receives the prescribed dose, the EMR should prevent further treatment to that prescription site.

13.14 Summary/Conclusions

Acceptance testing represents the fulfillment of a contract between the vendor and the user. It is wide ranging, encompassing all aspects of system performance. Clearly written specifications with testable conditions result in less confusion at the time of testing. The acceptance testing of a proton therapy system is a significant commitment in time and effort for all parties, typically extending beyond one year for multiroom facilities. The acceptance testing performance is an indication of what issues will be faced during clinical commissioning. The vendor and user should work cooperatively to resolve the issues until all acceptance tests can be satisfied. In case of system repair or upgrade-specific issues, acceptance tests can serve to validate a return to useable conditions.

Acknowledgments

We thank Tom Mohaupt, Radiation Safety Officer, St. Jude Children's Research Hospital, for helpful discussions on the subjects of initial shielding verification and safety conditions prior to acceptance testing. We also thank Rick Jesseph, who performed some of the measurements for the US system used as examples. Also we acknowledge the sources of the generalized acceptance testing concepts and examples from the former Indiana University Health Proton Therapy Center, Indiana, Ion Beam Applications (Leuvan-la-Neuve, Belgium) and Hitachi Ltd. (Hitachi City, Japan).

References

1. Nath R, Biggs PJ, Bova FJ, Ling CC, Purdy JA, van de Geijn J, Weinhous MS. AAPM code of practice for radiotherapy accelerators: Report of AAPM radiation therapy task group no. 45. *Med Phys* 1994;21:1093–1121.
2. Das IJ, Cheng CW, Watts RJ, Ahnesjo A, Gibbons J, Li XA, Lowenstein J, Mitra RK, Simon WE, Zhu TC. Accelerator beam data commissioning equipment and procedures: Report of the TG-106 of the therapy physics committee of the AAPM. *Med Phys* 2008;35:4186–4215.
3. Ciangaru G, Yang JN, Oliver PJ, Bues M, Zhu M, Nakagawa F, Chiba H, Nakamura S, Yoshino H, Umezawa M, Smith AR. Verification procedure for isocentric alignment of proton beams. *J Appl Clin Med Phys* 2007;8:65–75.
4. Thomadsen B, Nath R, Bateman FB, Farr J, Glisson C, Islam MK, LaFrance T, Moore ME, Xu XG, Yudelev M. Potential hazard due to induced radioactivity secondary to radiotherapy: The report of task group 136 of the American Association of Physicists in Medicine. *Health Phys* 2014;107:442–460.
5. Walker PK, Edwards AC, Das IJ, Johnstone PAS. Radiation safety considerations in proton aperture disposal. *Health Phys* 2014;106:523–527.
6. Pella A, Riboldi M, Tagaste B, Bianculli D, Desplanques M, Fontana G, Cerveri P, Seregni M, Fattori G, Orecchia R, Baroni G. Commissioning and quality assurance of an integrated system for patient positioning and setup verification in particle therapy. *Technol Cancer Res T* 2014;13:303–314.
7. Burge JH, Su P, Zhao C, Zobrist T. Use of a commercial laser tracker for optical alignment - art. No. 66760e. *P Soc Photo-Opt Ins* 2007;6676:E6760–E6760.
8. Nubiola A, Bonev IA. Absolute calibration of an abb irb 1600 robot using a laser tracker. *Robot Cim-Int Manuf* 2013;29:236–245.
9. Arjomandy B, Sahoo N, Zhu XR, Zullo JR, Wu RY, Zhu M, Ding X, Martin C, Ciangaru G, Gillin MT. An overview of the comprehensive proton therapy machine quality assurance procedures implemented at the University of Texas M. D. Anderson cancer center proton therapy center-Houston. *Med Phys* 2009;36:2269–2282.
10. Sejpal SV, Amos RA, Bluett JB, Levy LB, Kudchadker RJ, Johnson J, Choi S, Lee AK. Dosimetric changes resulting from patient rotational setup errors in proton therapy prostate plans. *Int J Radiat Oncol Biol Phys* 2009;75:40–48.
11. Hsi WC, Law A, Schreuder AN, Zeidan OA. Utilization of optical tracking to validate a software-driven isocentric approach to robotic couch movements for proton radiotherapy. *Med Phys* 2014;41:179–188.
12. Farr JB, Mascia AE, Hsi WC, Allgower CE, Jesseph F, Schreuder AN, Wolanski M, Nichiporov DF, Anferov V. Clinical characterization of a proton beam continuous uniform scanning system with dose layer stacking. *Med Phys* 2008;35:4945–4954.
13. Ma L, Geis PB, Boyer AL. Quality assurance for dynamic multileaf collimator modulated fields using a fast beam imaging system. *Med Phys* 1997;24:1213–1220.

14. Moyers MF, Benton ER, Ghebremedhin A, Coutrakon G. Leakage and scatter radiation from a double scattering based proton beamline. *Med Phys* 2008;35:128–144.
15. Farr JB, Dessy F, De Wilde O, Bietzer O, Schonenberg D. Fundamental radiological and geometric performance of two types of proton beam modulated discrete scanning systems. *Med Phys* 2013;40:072101.
16. Li Y, Zhu RX, Sahoo N, Anand A, Zhang X. Beyond Gaussians: A study of single-spot modeling for scanning proton dose calculation. *Phys Med Biol* 2012;57:983–997.
17. Bortfeld T. An analytical approximation of the Bragg curve for therapeutic proton beams. *Med Phys* 1997;24:2024–2033.
18. Dhanesar S, Sahoo N, Kerr M, Taylor MB, Summers P, Zhu XR, Poenisch F, Gillin M. Quality assurance of proton beams using a multilayer ionization chamber system. *Med Phys* 2013;40:092102.
19. Nichiporov D, Solberg K, Hsi W, Wolanski M, Mascia A, Farr J, Schreuder A. Multichannel detectors for profile measurements in clinical proton fields. *Med Phys* 2007;34:2683–2690.
20. Janni JF. Energy loss, range, path length, time-of-flight, straggling, multiple scattering, and nuclear interaction probability: In two parts. Part 1. For 63 compounds part 2. For elements $1 \leq z \leq 92$. Atomic Data and Nuclear Data Tables 1982;27:147–339.
21. Koehler AM, Schneider RJ, Sisterson JM. Flattening of proton dose distributions for large-field radiotherapy. *Med Phys* 1977;4:297–301.
22. Nichiporov D, Hsi W, Farr J. Beam characteristics in two different proton uniform scanning systems: A side-by-side comparison. *Med Phys* 2012;39:2559–2568.
23. Farr JB, Dessy F, De Wilde O, Bietzer O, Schonenberg D. Fundamental radiological and geometric performance of two types of proton beam modulated discrete scanning systems. *Med Phys* 2013;40:072101.
24. ICRU. Report no. 78: Prescribing, recording, and reporting proton-beam therapy. International Commission on Radiation Units and Measurements, 2007.
25. Kostjuchenko V, Nichiporov D, Luckjashin V. A compact ridge filter for spread out Bragg peak production in pulsed proton clinical beams. *Med Phys* 2001;28:1427–1430.
26. Courneyea L, Beltran C, Tseung HS, Yu J, Herman MG. Optimizing mini-ridge filter thickness to reduce proton treatment times in a spot-scanning synchrotron system. *Med Phys* 2014;41:061713.
27. Courneyea L, Beltran C, Tseung HSWC, Yu J, Herman MG. Optimizing mini-ridge filter thickness to reduce proton treatment times in a spot-scanning synchrotron system. *Med Phys* 2014;41.
28. Zhao QY, Wu HM, Cheng CW, Das IJ. Dose monitoring and output correction for the effects of scanning field changes with uniform scanning proton beam. *Med Phys* 2011;38:4655–4661.
29. Das IJ, Kase KR, Tello VM. Dosimetric accuracy at low monitor unit settings. *Br J Radiol* 1991;64:808–811.
30. Das IJ, Harrington JC, Akber SF, Tomer AF, Murray JC, Cheng CW. Dosimetric problems at low monitor unit settings for scanned and scattering foil electron-beams. *Med Phys* 1994;21:821–826.
31. Yin F, Wong J., et al. Report no. 104: The role of in-room kV x-ray imaging for patient setup and target localization. American Association of Physicists in Medicine, 2009.
32. Gayou O, Miften M. Commissioning and clinical implementation of a mega-voltage cone beam CT system for treatment localization. *Med Phys* 2007;34:3183–3192.
33. Court L, Rosen I, Mohan R, Dong L. Evaluation of mechanical precision and alignment uncertainties for an integrated ct/linac system. *Med Phys* 2003;30:1198–1210.
34. Kuriyama K, Onishi H, Sano N, Komiyama T, Aikawa Y, Tateda Y, Araki T, Uematsu M. A new irradiation unit constructed of self-moving gantry-CT and linac. *Int J Radiat Oncol Biol Phys* 2003;55:428–435.
35. Charlie CM, Paskalev K. In-room CT techniques for image-guided radiation therapy. *Med Dosim* 2006;31:30–39.
36. Cheng CW, Wong JR. Commissioning and clinical implementation of a CT scanner installed in an existing treatment room for precise tumor localization. *Cancer J* 2001;7:546–547.
37. Bissonnette JP, Balter PA, Dong L, Langen KM, Lovelock DM, Miften M, Moseley DJ, Pouliot J, Sonke JJ, Yoo S. Quality assurance for image-guided radiation therapy utilizing ct-based technologies: A report of the AAPM TG-179. *Med Phys* 2012;39:1946–1963.

38. Moyers MF. Comparison of x ray computed tomography number to proton relative linear stopping power conversion functions using a standard phantom. *Med Phys* 2014;41(6):061705.
39. Murphy MJ, Balter J, Balter S, BenComo JA, Jr, Das IJ, Jiang SB, Ma CM, Olivera GH, Rodebaugh RF, Ruchala KJ, Shirato H, Yin FF. The management of imaging dose during image-guided radiotherapy: Report of the AAPM task group 75. *Med Phys* 2007;34:4041–4063.
40. Fraass B, Doppke K, Hunt M, Kutcher G, Starkschall G, Stern R, Van Dyke J. American Association of Physicists in Medicine radiation therapy committee task group 53: Quality assurance for clinical radiotherapy treatment planning. *Med Phys* 1998;25:1773–1829.
41. Van Dyke J, Trs 430: Commissioning and quality assurance of computerized planning systems for radiation treatment of cancer. International Atomic Energy Agency, 2004.
42. Knopf A, Bert C, Heath E, Nill S, Kraus K, Richter D, Hug E, Pedroni E, Safai S, Albertini F, Zenklusen S, Boye D, Sohn M, Soukup M, Sobotta B, Lomax A. Special report: Workshop on 4d-treatment planning in actively scanned particle therapy—recommendations, technical challenges, and future research directions. *Med Phys* 200;37:4608–4614.
43. Keall PJ, Mageras GS, Balter JM, Emery RS, Forster KM, Jiang SB, Kapatoes JM, Low DA, Murphy MJ, Murray BR, Ramsey CR, Van Herk MB, Vedam SS, Wong JW, Yorke E. The management of respiratory motion in radiation oncology report of AAPM task group 76. *Med Phys* 2006;33:3874–3900.
44. Yoon M, Kim JS, Shin D, Park SY, Lee SB, Kim DY, Kim T, Shin KH, Cho KH. Computerized tomography-based quality assurance tool for proton range compensators. *Med Phys* 2008;35:3511–3517.
45. Park S, Jeong C, Kang DY, Shin JI, Cho S, Park JH, Shin D, Lim YK, Kim JY, Min BJ, Kwak J, Lee J, Cho S, Kim DH, Park SY, Lee SB. Proton-radiography-based quality assurance of proton range compensator. *Phys Med Biol* 2013;58:6511–6523.

Chapter 14

Clinical Commissioning of Proton Beam

Lei Dong, Ph.D.

Scripps Proton Therapy Center,
San Diego, CA

14.1 Introduction

"Commissioning" refers to the work following the acceptance of a machine's installation but prior to treating the first patient. Although the primary task for commissioning is related to beam measurement and commissioning of the treatment planning system, there are many other tasks that should be planned to get the center ready for the first patient treatment. For this reason, the commissioning work includes, but is not limited to, 1) beam data acquisition, 2) modeling and validating the beam model in the treatment planning system, 3) establishing a patient-specific quality assurance

(QA) program, 4) establishing machine QA baselines, 5) ensuring regulatory compliance and radiation safety, 6) developing operational procedures, and 7) training all personnel for various clinical procedures. Most of these topics will be discussed in this chapter.

14.1.1 Acceptance Testing vs. Commissioning

The goal of acceptance testing is to ensure that the machine meets its performance specifications. However, these specifications may not tell you how to treat a patient. Although the goals are different between these two tasks, the timing of the acceptance test is so close to the beam data acquisition that sometimes it is possible to share some of these test procedures to save commissioning time. However, unless the acceptance test procedure was designed to meet the beam acquisition task for the treatment planning system to be used, it is rare that these two processes can be combined entirely. In addition, the vendors would like to complete the acceptance test as quickly as possible so that a partial payment can be secured. Acceptance testing is a major milestone for the vendor, while the start of commissioning is also a key milestone for the center in preparation for patient treatment. After passing the acceptance test, the equipment (the accepted portion) is officially transferred to the center. The staff should take advantage of the available beam time and execute the commissioning plan. It is important to work with the vendor to ensure adequate beam access and utilization of the equipment.

14.1.2 Goal Setting

It is well known that proton therapy systems are expensive. Facilities usually borrow heavily for equipment purchasing and to prepare for the lengthy installation and complicated commissioning processes. There is high pressure to initiate clinical treatments as soon as possible to save money on bank loans. Therefore, it is critical to set up clinical goals and carefully plan timelines. Management and execution of commissioning tasks are highly time-sensitive.

Although the actual start of the commissioning usually follows the completion of the acceptance test, planning for commissioning began long before hiring new staff. The administrative managers should create a business plan that is well integrated with the planned clinical activities and the recruitment of appropriate technical staff. The following tasks and questions need to be carefully considered:

1. What are the unique clinical services that could establish the center in the existing care giving environment? What is the clinical focus by the new proton therapy center?
2. What kind of staff and expertise need to be recruited to achieve these clinical goals? One of the common mistakes is to hire overwhelmingly junior staff. Junior staff needs mentorship and clinical knowhow for some of these very complicated clinical flow and measurement tasks. On the other hand, it may be difficult to find experienced staff. Sometimes, it is advantageous to pair staff that can complement each other. The most critical skill sets of the new

staff for commissioning work are measurement expertise, beam modeling experience for treatment planning systems, clinical workflow experience, and successful training experience.

3. What types of equipment that can work for the clinical system? Based on the beam specifications provided by the vendor, there is a careful selection process to purchase the right equipment, quality assurance devices, etc.

Most importantly, what are the timelines for various tasks, including staff learning and training? Unlike other matured radiotherapy equipment, proton therapy is still evolving and changing rapidly. A staff's ability to learn and adapt to new procedures plays a critical role in a center's success. Adequate timelines should be proposed so that the staff has the opportunity to prepare in advance. The lead medical physicist is a key in this commissioning process because most technical and clinical duties to set up the center fall on the physics team.

It is also important to realize that most centers will have a ramp-up plan, which means that the center is willing to do a partial commissioning (with limited capacity or limited options for patient treatment), or may treat only a portion of the eligible patients while the center is continuing the commissioning task (for additional treatment rooms, releasing new treatment options, etc.). This approach may have some logistic and financial advantages; however, it may create conflicts for beam time for patient treatment and remaining commissioning measurements. In addition, supporting both patient treatments and commissioning measurement would put staff into long working hours. Therefore, it is important to staff adequately or make shift arrangements to take advantage of the available beam time.

It is also important to know that there are many clinical issues and logistics when treating patients. Some of these clinical questions should be addressed early, before the beam is becoming available for measurement.

It is worthwhile negotiating with vendors on beam time support. Unlike linear accelerators, the operation of a proton therapy system requires a team of vendor-trained engineers who can support clinical treatment as well as maintenance of the machine. Because beam time is so critical in the commissioning phase, a facility may need to negotiate separately with the vendor to balance short-term needs for beam acquisition with the machine installation needs for the remaining treatment rooms. This is particularly true for multi-room installations that require careful planning with vendors to schedule both clinical time and available measurement time for commissioning.

14.1.3 The Big Picture

It is useful to realize that new clinical equipment is usually associated with other clinical activities, especially for a new center. There are numerous administrative tasks that will require the physics team's input at various levels. A short summary of key activities for a new center is listed in Table 14–1. Physics is involved in many activities, such as assisting accreditation, identifying patients' immobilization needs (supplies), assisting in marketing contents etc. In particular, the physics team plays a key

Table 14–1 Various tasks involved with a new proton therapy center. The physics team plays an important role in this process.

Department	Activities	Department	Activities
Human Resources	Organization structure	**Supply Chain**	Office supplies
	Compensation plan		Patient treatment supplies
	Competencies		Pharmaceutical supplies
	Position descriptions	**Marketing**	Marketing strategies
	Recruitment		Community connection
Regulation	Joint commissioning accreditation		Marketing content development
	Licensing requirements		Advertising contracts
	Risk management		Tours
	Radiation safety	**Education**	HR training
	Waste management		User training
Business and Legal	Contract negotiation		Patient orientation
	Reimbursement		Patient satisfaction
	Local cost determination	**Information Services**	Network connectivity
	Patient intake process		Equipment
	Property management		Software applications
	Partnership agreements		Interface needed
	Business associate agreements		Emergency procedure
	Charity care agreements		Backup and storage strategies
	Insurance and malpractice coverage		Access control
Equipment commissioning	TPS commissioning		Hospital information system
	Calibration procedure		PACS
	QA procedures		Meaningful use implementations
	Treatment protocol development		
	Special procedures		

role in radiation safety, IT configuration, user training, and, of course, equipment commissioning.

14.1.4 State/Government Requirements

Before the start of commissioning, it may be useful to collect various requirements by the state or regulators that grant the facility permission to begin patient treatment. To reduce the time for starting the first patient treatment, it is necessary to optimize the commissioning work so that a part of the paperwork can be completed before the TPS commissioning starts.

For example, in the State of California, the paperwork to apply for authorization for treatment includes the following: 1) a registration form (which includes the x-ray tube information used for IGRT), 2) acceptance test report, 3) physics calibration report, 4) room survey report (shielding), 5) outline of QA procedures, and 6) major policies and procedures used by the center. Most of these policies are related to operational procedures (emergency procedures, IT policies, etc.), but some of them may be related to the physics—for example, the RBE value used in the center. Due to current limitations of commercial treatment planning systems, the relative biological equivalent (RBE) dose is usually not clearly labeled. In addition, it may be premature to use variable RBE in the treatment planning process. Therefore, it may be useful to have a standing policy to state that all dose displayed in the treatment planning system or treatment plan report assumes an RBE value of 1.1. The policy will clarify all dose units used in the center if they are not labeled with RBE. (The detailed discussion about RBE is in Chapter 5.)

In order to apply for state approval (for example, in California), it becomes advantageous to perform beam calibration at the beginning of the commissioning process so that the application to the state can be requested while other commissioning tasks are still underway. Beam calibration also establishes the monitor unit (MU) setting, which is useful to define at the beginning, instead of at the end of a commissioning process.

14.2 Commissioning of Imaging Devices

Imaging devices—such as CT scanners, MR scanners, PET/CT scanners, or even in-room imaging devices—can be commissioned prior to the availability of proton beams. A CT scanner, for example, is perhaps required to be commissioned as the first step in TPS commissioning. The general principle and application for various imaging systems are described in Chapter 7, which will not be repeated here.

14.2.1 CT Scanner

Most centers use the stoichiometric calibration method described by Schneider et al. [1]. This procedure requires a physical phantom with known density and material compositions. Therefore, it is important to purchase the phantom from a vendor who can provide such elemental composition data. Otherwise, the stoichiometric calibration cannot be performed.

After performing the conversion from tissue substitutes to ICRU standard human tissues (ICRU Report No. 44 [2]), the relative stopping power vs. HU curve will look like Figure 14–1a. It is important to review if the curve is reasonable and increasing monotonically with HU because most treatment planning systems can't handle multiple stopping power values for the same HU value. Another important aspect is to create an extended CT table, which may handle some higher HU materials used in the phantom measurement or to accommodate some special materials of known stopping power ratios (for density override). Due to potential CT imaging artifacts (especially in the presence of metals), some of the higher HU values may not be accurate. One possible solution is to create a "saturated" stopping power table at high HU values, as shown in Figure 14–1b. This will minimize the risks of using unexpectedly high HU values (due to imaging artifacts, for example). For a large area of (known) high HU values, it is always a good idea to review the area and perform a manual density override, instead of relying on the extrapolated stopping power table.

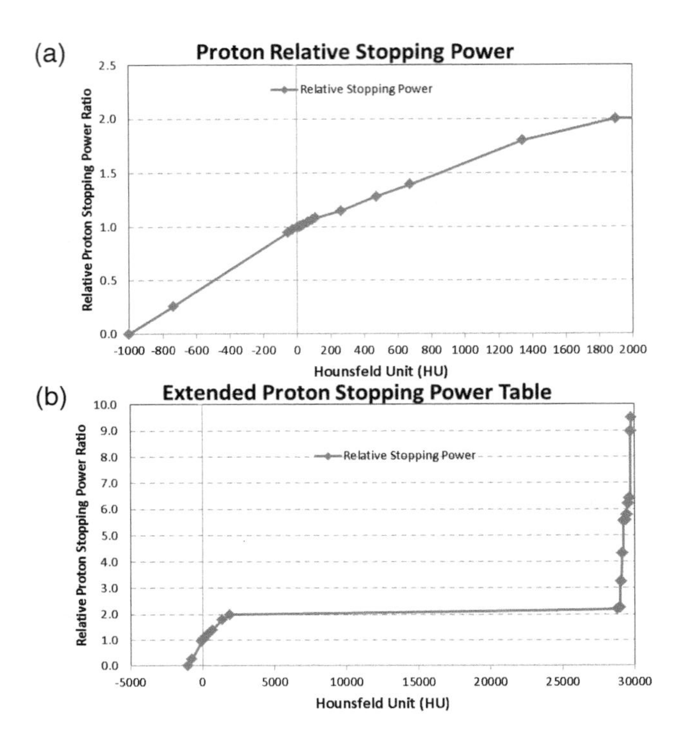

Figure 14–1 Construction of the relative proton stopping power to HU conversion table based on the stoichiometric calibration method. (a) An example of a typical table. (b) An intentionally flattened table at extended HU range (>2000) when such calibration may not be accurate. In addition, some of special materials can be assigned in the table so that the treatment planning system can locate the stopping power for dose calculation.

There are a limited number of CT scanner vendors. Most CT scanners from the same manufacturer may have a very similar HU density/stopping power table. It may be a good idea to compare these stopping power ratio curves with other proton therapy centers with the same model of CT scanner. If significant differences are found, the phantom or CT stoichiometric calibration process should be investigated.

One addition task in the commissioning of the CT scanner is to create a special material table. Some of these materials may be encountered during patient treatment planning. Instead of using the measured HU table for the stopping power conversion, these materials should be used manually using the assigned material properties. One example of such a special material table is shown in Table 14–2.

14.2.2 Other Imaging Devices

Other imaging devices may include 1) diagnostic imaging devices, such as an MR scanner or PET/CT imaging devices and 2) devices that related to in-room image

Table 14–2 Examples of special material table for CT HU override. (Please note: these values are for reference only.)

Examples of Special Materials	SPR	Relative Electron Density	Mass Density (g/cm^3)
Lucite	1.157	1.147	1.190
Teflon	1.839	1.904	2.200
Aluminum	2.260	2.345	2.699
Pyrex glass	2.200	1.999	2.230
Titanium	3.240	3.762	4.540
Tin	4.317	5.551	7.310
Lipowitz metal	5.812	7.078	9.741
Iron	5.541	6.610	7.874
Stainless steel	5.560	6.726	8.000
Brass	5.786	6.891	8.360
Silver	6.409	8.249	10.500
Nickel	6.210	7.657	8.902
Tantalum	8.980	12.114	16.654
Gold	9.490	13.972	19.320

guidance, such as x-ray imaging systems, CBCT, surfacing imaging systems, or other patient-positioning systems. While acceptance testing is necessary to verify the performance specifications of these devices, additional commissioning is required to integrate these devices into the radiotherapy workflow.

This chapter will not deal with the specifics of these devices, but for commissioning methodology, it may be useful to focus on key functions of these devices. For example, network connectivity of diagnostic imaging devices should be integrated with the hospital PACS system and the treatment planning system. Imaging protocol development should be another key commissioning task for these devices. On the other hand, patient imaging and positioning systems should be tested for workflow (using the use case scenarios, see later of this chapter) and logistics. Physicians and therapists should begin using these systems both at console and remotely. Site-specific guidelines should be developed for all of these systems, including re-imaging/shift policies. The end-to-end tests will be described in a later section.

14.3 Commissioning of Treatment Planning Systems (TPS)

Commissioning of TPS is the focus of the entire commissioning process. The overall goal of this process is to commission a treatment planning system that can accurately design dose distributions for a patient to be treated for a particular treatment unit. Therefore, the dose calculation from the TPS should match the measured dose for the same configuration. The TPS is also required to simulate various beam-shaping options (range shifters, compensators, or collimators/apertures, etc.). These treatment options should be implemented and verified as well.

14.3.1 Using a "Runplan" to Organize Commissioning Activities

As previously mentioned, when the beam time is available, the commissioning should start as soon as possible to save overall startup costs for the proton therapy center. It is almost mandatory to set up a commissioning plan to deal with various time-sensitive activities. It may be useful to list these tasks first, assign a time for completion, and then re-order these tasks in the most logical and efficient sequence. Some of these tasks have to be done in a certain order (for example, equipment testing should be completed before beam measurement), while other tasks can be done in parallel to save time. An example of a run-plan (prior to the actual execution of the plan) is shown in Figure 14–2 for a proton therapy facility that only has the pencil beam scanning (PBS) mode.

14.3.2 First Week's Activities at the Machine

If a runplan is important for the overall commissioning, the first week is perhaps the most important week when the beam is becoming available. The main goal for the first week is to ensure that the measurement equipment is functional and initial measurements are within expectations. If not, a change in the plan or looking for alternative solutions may be necessary. The worst thing would be to find out that measurements were incorrect half way through the commissioning process. The first week is also the time to get familiar with the machine and equipment, which may answer some of the questions that may arise from planning the commissioning activities.

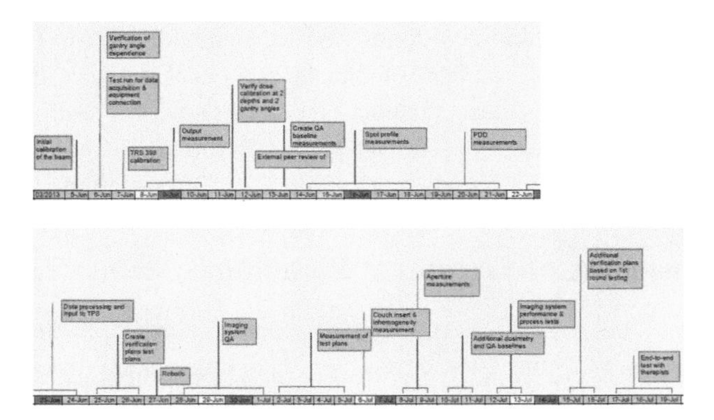

Figure 14–2 An example of a "runplan" which assigns various tasks with different durations at different time points of the commissioning process.

Although the main goal for the first week is to evaluate feasibilities for commissioning, it would be useful to complete a few commissioning tasks on the way. The following tasks will be very useful to be completed early in the commissioning process:

- electrometer and ion chamber evaluation (leakage, linearity, end-effect, ion collection efficiency, etc.)
- dose calibration
- water tank and scanning hardware and software tests
- detector tests for measuring beam profiles
- gantry angle dependence (to ensure that measurement can be done, and measurements at one gantry angle would be adequate)
- availability of accessories and other functions of the treatment unit

14.3.3 Machine Calibration and MU Definition

The first step in commissioning a TPS is to perform dose calibration and set machine monitor unit (MU) for a reference condition. The goal of this step is to establish the dosimetric relationship between the absolute dose (Gy) and the machine beam-on time (MU). The charge collected in the monitor chamber in the nozzle is an internal indicator for dose. However, this internal indicator needs to be calibrated (linked) to reflect the absolute dose for a given treatment condition. This process involves the use of an ionization chamber to measure the dose delivered by the treatment machine under a specifically selected reference condition. The TPS should be able to calculate the dose under the same reference condition and match the dose for the treatment condition. MU definition connects the treatment delivery system and the dose calculation system. Therefore, this is an important process. The system should be able to reproduce the same condition under various software upgrades or machine hardware upgrades. For this reason, the selection of this reference calibration condition is not entirely arbitrary.

The dosimetric protocol that is used for calibration is already described in more detail in Chapter 12 (Dosimetry and Beam Calibration). The most popular protocol in current practice is the IAEA TRS-398 [3]. To use this protocol, an ionization chamber, a reference class electrometer, and a water phantom are required. In particular, the ion chamber should meet the TRS-398 specifications, and perhaps have the (chamber-specific) K_Q factor available for various beam quality factors [3].

The selection of the reference condition is important. This reference condition defines the calibration condition: the radiation field to be used, the MU meter set, and the dose measurement point for the ionization chamber.

Traditionally, passive scatter and uniform scanning proton beams were commonly calibrated using an SOBP field. This is because of the availability of built-in SOBP settings in these delivery modes. The ion chamber is usually placed in the middle of the SOBP, as illustrated in Figure 14–3a.

However, for pencil-beam scanning, the SOBP is usually created by an inverse planning system. For this calibration, the inverse planning system may not have been commissioned yet (for example, the spot profiles may not have been entered into the system yet). In addition, the deliverables for an SOBP may not be unique when designed by an inverse planning system. For all these reasons, it will be more convenient—and perhaps arguably more accurate—to select the reference calibration point near the entrance of a uniformly spaced single-layer spot pattern. Other than these reasons, the selection of the reference field is theoretically arbitrary. Even in the pencil beam mode (modulated scanning with energy stacking technique for range modulation), the center of the SOBP could be used as the calibration point [4]. Figure 14–3b shows an example of the central axis depth–dose distribution from a uniformly spaced pencil beam spot pattern with a constant MU per spot at a fixed energy (160 MeV). Although this single layer uniform spot pattern does not have the characteristic of a flat SOBP, it has a relatively smooth entrance dose distribution, which is suitable for dose calibration.

The use of IAEA TRS-398 protocol for dose/MU calibration will not be repeated here (see Chapter 12 under the "Dose to Water Based Approach"). TRS-398 uses the residual range, R_{res}, defined as the distance from the measurement point to the depth at which the absorbed dose beyond the Bragg peak or SOBP falls to 10% of its maximum value. The chamber-specific perturbation factor is a function of R_{res}, and the factor becomes stabilized when $R_{res} > 5$ cm [3]. This is another advantage when calibrating the chamber at a shallower depth (for a greater R_{res} value). The corresponding R_{res} has been marked on Figure 14–3a and b.

It may be useful to remind here that the ion chamber and electrometer need to be calibrated by ADCL to ensure accuracy and regulatory compliance. Although both thimble or parallel-plate chambers are acceptable by TRS-398 protocol, parallel-plate chambers are preferred for high dose rate beam delivery because their ion collection

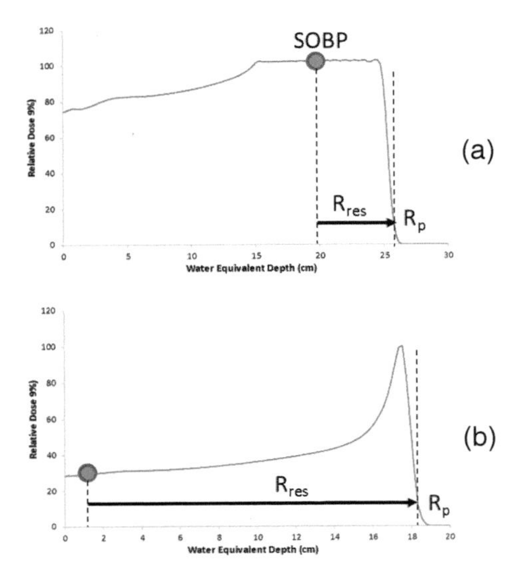

Figure 14–3 Two typical dose calibration conditions for MU definition. (a) The calibration point is in the middle of the SOBP, which is most convenient for passive scatter or uniform scanning delivery modes. (b) For pencil beam scanning, it is advantageous to select the calibration point near the entrance of a uniformly spaced 2D spot matrix with a constant MU per spot at a fixed energy.

efficiency is usually better than thimble-type chambers, and the average energy at the chamber position is easier to determine.

Before starting the actual calibration measurement, it is important to test your equipment:

1. The ion chamber needs to be measured for ionic recombination effect using the common two-voltage technique [5]. For modern pencil beam scanning systems, the instantaneous dose rate can be very high. It should be alerted if the ion recombination correction factor (pion) is greater than 1%, which usually indicates that the dose rate may be too high as a reference condition. Sometimes it may be possible to change dose rate in the selected reference irradiation condition.
2. Polarity correction, in which the bias can be reversed to get an average reading.
3. Monitor chamber linearity test and end effect. Because the dose monitor plays an important role in integrating doses, the linearity test should be performed to ensure the impact of monitor chamber nonlinearity is minimum. Beam output should be measured for a set of MU settings, which covers the full range of typical dose settings. A linear regression analysis can show any significant deviation from linearity, and the intercept on the abscissa from the regression will provide the monitor system's end effect.
4. Temperature and pressure correction, which is a standard when doing a calibration with an ionization chamber.

14.3.4 Peer Review of Calibration

Because beam calibration is such an important step in the commissioning process, it is usually a good practice to have an external peer review. This is typically recommended by practical guidelines (ACR-AAPM Technical Standard for Performance of Proton Beam Radiation Therapy, 2013). In order to make this process more independent, the external reviewer should be familiar with proton therapy but not directly involved in the commissioning process for the institution. The external reviewer should bring his/her own electrometer and ionization chamber and perform an independent measurement to verify the dose for the reference calibration condition. The process should be documented as a part of commissioning steps. If there is a difference, the external reviewer and the physicists from the institution should work together to resolve the difference and provide an explanation.

14.3.5 Collecting Machine Data for TPS Commissioning

The measurement requirements for commissioning a treatment planning system (TPS) depend heavily on the specific dose calculation algorithm used in a TPS. In general, when the commercial treatment planning system is selected, the vendor should provide detailed information on measurement data requirements for commissioning. This book chapter will not deal with a specific TPS. We will only provide examples of such data required for common treatment planning systems.

14.3.5.1 Passive Scatter/Uniform Scanning

The unique aspect of both passive scatter and uniform scanning is to create various SOBP with different widths (laterally) and lengths (proximal and distally in the beam's direction). In order to create very large fields, the beamline components may need to be modified by hardware, while smaller fields may require less scatter and are more efficient in producing higher dose rate for treatment. Similarly, the modulation length may require different hardware in the system in order to generate broad or narrow SOBPs. Sometimes the system may use different hardware components when treating in different beam energy ranges. Due to these different hardware configurations, there are many beam delivery "options." Some of these options may overlap in energy or modulation range, etc. In order to treat in one of the options, measurement data are required for the TPS for each option in order to model or verify beam characteristics accurately. Therefore, there are usually many (sometimes more than 50) options and various combinations. An example of the depth–dose measurement on a Hitachi system using the rotating modulation wheel (RMW) at 250 MeV/25 cm maximum range is shown in Figure 14–4.

The beam measurements for passive scatter and uniform scanning systems are usually categorized as:

- pristine Bragg peaks (for various energies),
- snout sizes (for various lateral field sizes),

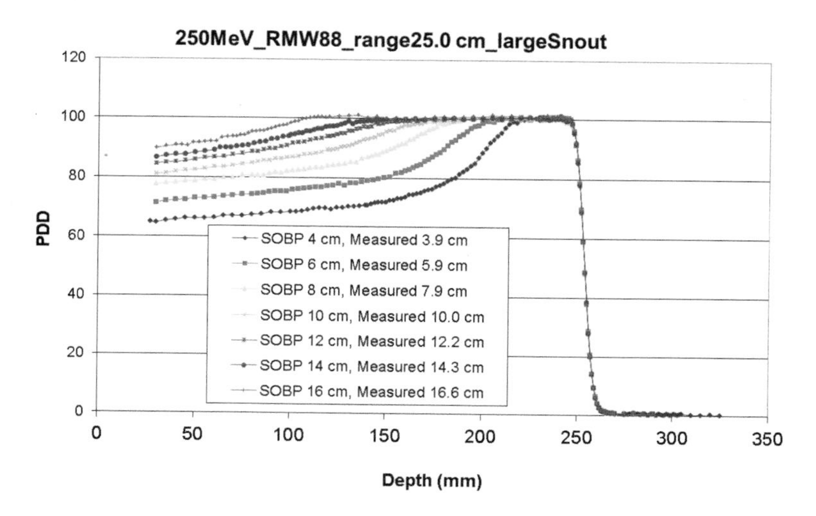

Figure 14–4 SOBP depth–dose measurements for the Hitachi system using a large snout and maximum range setting.

- modulation depths (for combinations of energies/range shifter settings and width of the modulation wheels),
- modulation ranges (for combinations of energies/range shifter settings and width of the modulation wheels), and
- relative output factors (for establishing dosimetric relationship among field size and modulation width).

Dose measurements for various accessories and snout positions are also needed, which include, but not limit to, 1) compensators, 2) range shifter/energy absorbers, 3) aperture/collimator, 4) air gap factors, etc.

Depending on beam delivery modes (pulsed beam from a synchrotron or quasi-continuous beam from a cyclotron), sometimes dose measurement must be conducted in the integration mode (rather than the "ratio" mode when using a reference ion chamber). These measurements could be very time-consuming. Sometimes it may be advantageous to use Monte Carlo methods to model some of the relative beam characteristics, which allow for fewer measurements without losing accuracy [6–9].

14.3.5.2 Pencil Beam Scanning

Compared to passive scatter and uniform scanning, the beam data requirement for pencil beam scanning is usually straightforward because there is almost no additional beam-shaping hardware for pencil beam delivery. Typically, the beam measurements for TPS are

- integrated depth–dose for a set of pencil beams at different energies,
- in-air spot profiles at several distances from the isocenter,
- relative output factor at a reference depth of calibration, and
- in-air spot profiles for each range shifter.

Integrated depth–dose measurement requires a large parallel plate chamber, which needs to capture the entire pencil beam profile at various depths. Unfortunately, some of the beam profiles at depth may be too broad to be captured by the size of the ion chamber. Corrections are needed to ensure the integrated dose measurement is accurate. This has been described by Anand et al. [10] in more detail. The large scatter component due to nuclear interaction (the "Halo" effect) is usually responsible for this large low-dose penumbra [11–13].

The in-air spot profile measurement for small pencil beam could be challenging as well. Some of the high-energy pencil beams have very small spot size (2–3 mm). It may be difficult to use an ion chamber for such measurements. Therefore, most institutions use imaging techniques, such as CCD cameras or radiographic films. Although the measurement is easier with CCD cameras or films, the data processing and calibration could be tricky. Because of potential nonlinear response to dose measurement, CCD camera data may require a good validation before proceeding for all beam measurements. It is also noted that the beam profile measurement may need to be measured with more accuracy for pencil beams, because when thousands of pencils added together to deliver a full treatment, such small differences in beam penumbra may contribute to noticeable large dose errors. This will be discussed later.

14.3.5.3 Creating Machine QA Baselines

One important step during the beam measurement process is to create machine QA baselines. Once beam data are acquired for the treatment planning system, they become the gold standard in designing patient treatment plans. For this reason, it will be extremely useful to irradiate the QA devices immediately after beam calibration. Extended delay in creating QA baselines may run into potential risks that the machine beam characteristics may be changed. Daily, weekly, and monthly machine QA procedures are discussed in Chapter 15.

14.3.6 Modeling in TPS and Validating Beam Model

Once the commissioning data sets are acquired, they need to be imported into the TPS according to instruction. It is important to pay attention to the unit of data that need to be scaled or normalized properly. Sometimes it is useful to build RBE values directly into the dose output, which makes the display of dose in units of RBE in the TPS. For depth–dose measurements, sometimes it is necessary to make range shift corrections to account for the thickness of the tank wall or chamber wall. Incorrect water-equivalent thickness (WET) will result in error in the measured energies.

Each TPS should have fitting tools to extract parameters used in analytical dose calculation. The TPS should also have dose comparison tools to evaluate the differences between measured data and calculated/fitted beam data. Sometimes it may be necessary to manually adjust beam parameters to get a better compromise between the measured data and calculated data.

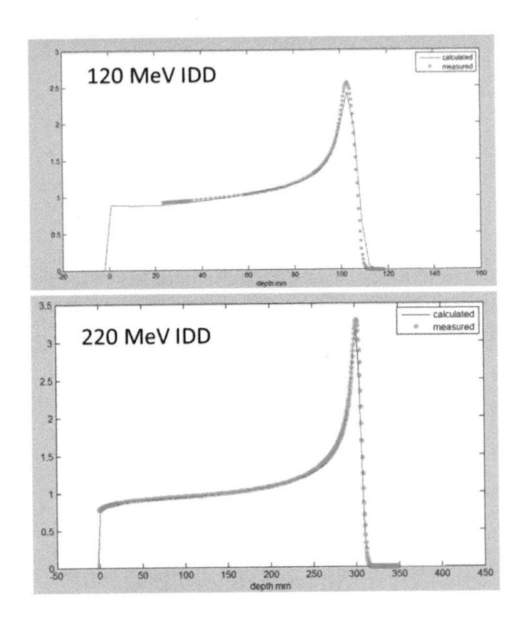

Figure 14–5 Comparison of the measured and calculated integrated depth–dose curves for monoenergetic pencil beam 120 MeV (top) and 220 MeV (bottom). The calculation grid was limited to 2.5 mm.

Once the calculation model is satisfactory to the measured data, it may be useful to create additional validation plans outside the commissioning tools. For example, treatment plans to simulate the single pencil beam integrated depth–dose curve measurement could be created and compared to the calculation (summing all doses in the transverse plane along the depth direction). An example of such a comparison is shown in Figure 14–5, in which the measured IDDs of 120 MeV (top) and 220 MeV (bottom) are compared to the calculated ones. The difference shown in 120 MeV was mainly due to the limited spatial resolution (2.5 mm grid size) of the treatment plan. The measurement of the depth–dose curve of a single pencil beam is not recommended even using a small volume ionization chamber due to alignment challenges. Positioning and scanning angle uncertainties could impact the accuracy of depth–dose measurement for such a small field.

Fitting calculated curves with measurements is only the first step in beam model validation. Physicists should not spend too much time on this because specifically designed test plans may be a better way to validate or tune beam parameters (see below).

14.3.7 Specific Validation Plans for Commissioning and Parameter Optimization

Vendor-specified commissioning data are necessary to create the beam model in the TPS. However, in order to assess the accuracy of the dose calculation algorithm, it is almost mandatory to create additional validation plans. The main goal of these validation plans is to evaluate if the dose calculation algorithm can apply to a variety of clinical cases. The test results can also provide a fine-tuning opportunity to change the parameters used in the dose calculation algorithm so that they can match better with dose measurements. Therefore, this is an important part of the commissioning process, and arguably the most time-consuming part of the commissioning.

Because SOBP from a single proton field plays an important role in treatment planning (all delivery modalities, i.e., passive scatter, uniform scanning, or pencil beam scanning), the first assessment is to evaluate dose calculation accuracy along the central axis of an SOBP plan. Measurements near the entrance, proximal side, middle, and both sides of the SOBP are important. In addition, the field size effect (which is strongly impacted by the lateral scatter of the beam), modulation range, and depth of penetration (maximum energy of the beam) are typical clinical scenarios. By designing a set of SOBP plans, such as the ones shown in Figure 14–6, trend analysis can illustrate the weakness of the dose calculation algorithm and provide opportunities for fine-tuning fitting parameters. In general, the validation plans should cover the following treatment scenarios:

- small, medium, and large tumor volumes,
- at shallow and deep locations,
- use of treatment accessories (range shifter, aperture, compensator, etc.), and
- variable air gaps.

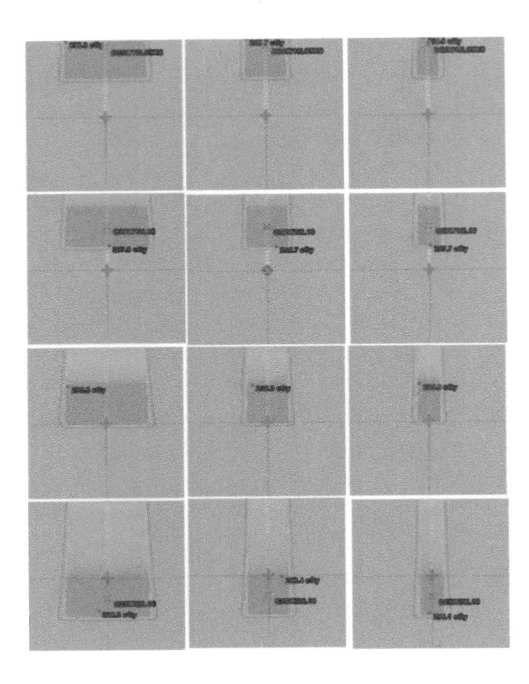

Figure 14–6 SOBP test plans to evaluate dose calculation accuracy for different field sizes, modulation ranges, and maximum energies. These specifically designed test plans vary each parameter at a time and will provide useful information in trend analysis for tuning dose calculation parameters to match with measurements.

One of the major tasks in commissioning is to understand the dose calculation algorithm used in the TPS. The full discussion of dose calculation algorithms used in proton therapy is described in Chapter 19, which will not be repeated here. Ideally, the dose calculation algorithm determines the types of plans that should be created for testing and parameter optimization. Although understanding of dose calculation algorithms can and should be prepared well in advance (before the start of beam commissioning), sometimes it is necessary to create additional tests based on observed disagreements between calculations and measurements. Therefore, it is important to analyze data in parallel with other commissioning activities to make the overall progress manageable.

Commissioning the pencil beam-based dose calculation algorithm (see Chapter 12.3.3) can serve as a good example. Zhu et al. described the commissioning of this pencil beam dose calculation algorithm in much detail using the spot scanning delivery mode on a Hitachi machine [13]. The basic beam measurements for the algorithm only require in-air lateral profiles, in-water integrated depth–dose curves, and dose/MU at a reference point for individual pencil beams. However, in order to model the Gaussian parameters, the field size effect plays an important role. The tails of the Gaussian model will not only contribute to beam penumbra, but they will also have a significant impact for output factor calculations (or the field size effect). Sometimes it is more effective using the field size effect to tune the beam model. In this study, the single Gaussian model was proved inadequate to model the beam satisfactorily.

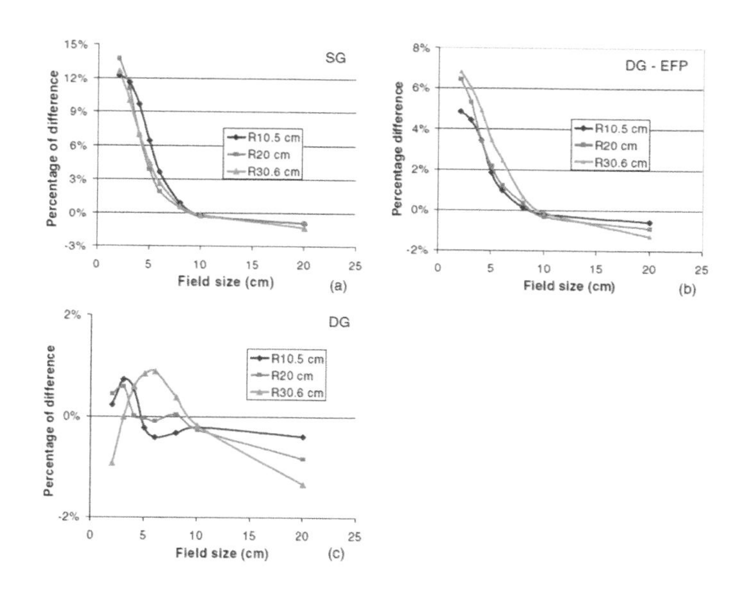

Figure 14.7 Field size effects when comparing different Gaussian models in pencil beam commissioning: Single-Gaussian (SG) model produced large calculation errors compared to the measurements; using vendor provided Double-Gaussian model (DG-EFP) improved the beam model; however; the Double-Gaussian (DG) model with user-empirical parameters produced the best results. From Zhu et al. (2013) with permission.

Therefore, it is critical to plan additional commissioning measurements that will address the need for parameter tuning.

Figure 14–7 illustrates the effectiveness of various beam models (single Gaussian vs. double Gaussian) when modeling the field-size effect. The shape of the low-dose tail is usually difficult to measure directly due to limited dynamic range of the measurement device; the field-size effect becomes an effective tuning tool to model the collective effect of many pencil beams. The relative output factor as a function of field size can be tested by using different calculation models and compared to measurements (Figure 14–7). If the spot size is big or behaves unlike a Gaussian function, the single Gaussian model will not be able to model the field size effectively (Figure 14–7 SG). On the other hand, the parameters used in a vendor-provided “double Gaussian” model may need to be optimized further to provide the best fitting for the field size effect (Figure 14–7 DG).

14.3.8 Alternative Commissioning Data

Pencil beam scanning is usually delivered in a dynamic mode, which means that the entire treatment field is delivered in a time sequence, during which the pencil beam will vary its position and energy to deliver the cumulative dose at the end of the sequence. This behavior creates some measurement challenges. Point dose measurement has to wait until the entire sequence of spot delivery to be completed, which is very inefficient to measure a large amount of data. For this reason, Monte Carlo simulation techniques were actively sought to provide supplemental commissioning data [8,13–16]. Combined with measurement validation, Monte Carlo data speeds up the entire commissioning process and provides some valuable validation data that might be difficult to measure as well.

Monte Carlo simulation can also provide a validation data set when tissue heterogeneity is concerned for the calculation model. Most commissioning measurements were performed in water or solid uniform slab geometries. It will be difficult to perform accurate dose measurement directly in heterogeneous media.

14.3.9 Validation using Patient QA Plans

The last step in validating the beam model is to use the treatment planning system and newly created beam model to plan a variety of clinical cases. The previous steps in the commissioning process focus on "single factor" effect (for example, dose accuracy near the entrance, field size effect, accuracies at high or low energies, etc.). These validation data will help to understand various inaccuracies and assist in the optimization of the beam model parameters. However, it is also important to test the beam model in realistic clinical applications, which may reveal further issues.

The strategy is to combine the protocol development for treating various disease sites and deliver near realistic clinical plans in phantom geometry. This is also a good time to test the patient-specific QA procedure. The details of patient-specific QA are covered in Chapter 16. Briefly, a patient treatment plan will be copied to a phantom geometry (either using water tank or slab geometry). Ion chamber measurements are typically used in patient-specific QA. However, 2D planar dose evaluation may reveal more information about relative or absolute dose distribution.

An ion chamber or 2D array chamber detector can be used to measure the dose calculated for the phantom geometry. Dose calculated in the phantom will be compared with the measurement at the same detector location. Percent of dose differences will be reported. If 2D dose measurement was performed, the gamma passing rate will be computed to account for both the absolute dose difference and the distance of agreement [17,18]. If 2D measurement is performed, it may be useful to compare the dose profile differences at different locations. An example of such comparison is shown in Figure 14–8, which shows a head and neck case planned with IMPT technique. The locations of the ion chambers are plotted as square overlay for the measured dose plane (Figure 14–8(a)). A profile was taken from the inferior to the superior direction, and the resultant dose comparison is shown in Figure 14–8(b). As can be seen, the measured doses follow the calculated dose very well in this case. The measurement location of the ion chamber is informative if a large dose difference is shown near a gradient region. In this case, the spacing between any two rows of ion chamber is 1 cm. The 1 cm spacing seems reasonable for this highly modulated case. However, if the treatment field is smaller, the 1 cm spatial resolution may not be

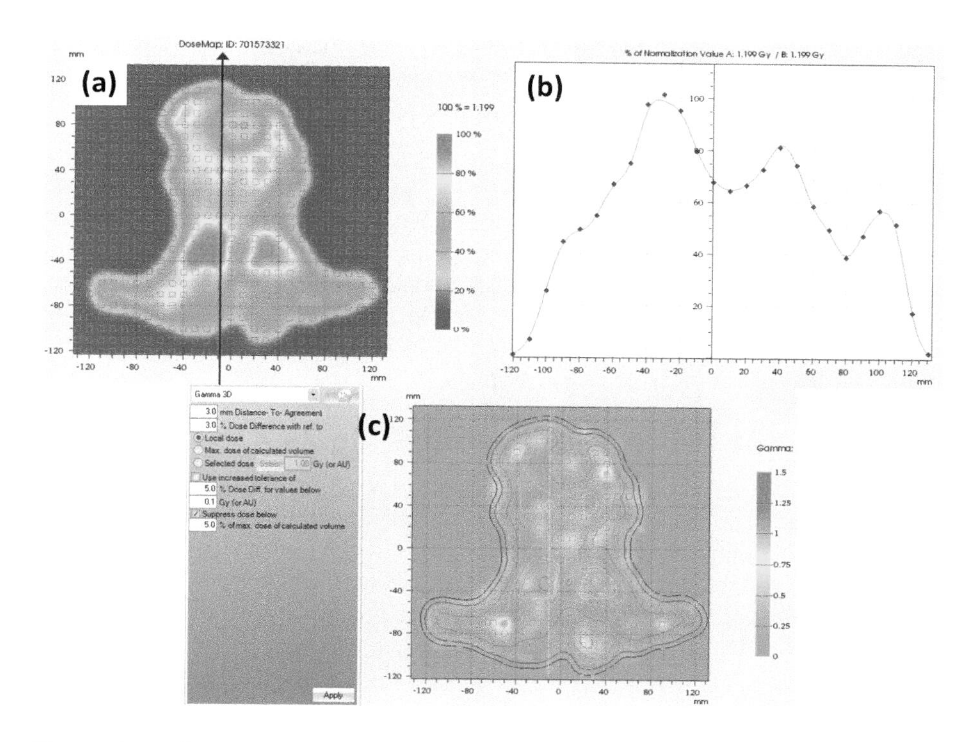

Figure 14–8 Realistic patient QA cases should be analyzed to evaluate any suspicious dose deviations from TPS calculations. In this example, a highly modulated head and neck plan was compared to TPS dose calculation in the same measurement geometry. (a) The original dose distribution for this posterior treatment field with 2D ion chamber array positions is shown. (b) A dose profile tool shows point-by-point dose deviations. (c) A 2D gamma display shows suspicious dose or distance deviations at various locations.

enough and may require two or four measurements with 5 mm shifts to improve the spatial resolution (if necessary).

A suggested set of patient QA plans include 1) prostate cancer, 2) brain cancer (large and small target volumes), 3) head and neck cancers, 4) a CSI case with two isocenters, 5) a large sarcoma case, 6) breast cancer cases with both the whole breast and APBI cases, and 7) a lung cancer case and a liver cancer case measured with a motion phantom. These test cases should cover a variety of clinical scenarios and would be ideal if the same cases are used in the treatment protocol development. Sometimes, it may be advantageous to reuse these cases when software upgrade and new commissioning become necessary.

Although it may be difficult to perform patient QA measurement at the same treatment angle as planned, it may be useful to select a few representative cases and perform such measurements during the commissioning process. This is perhaps a good opportunity to evaluate the gantry dependence. It will add confidence to the commissioning data and machine delivery system.

14.3.10 Secondary MU Calculation

The discussion on how to develop a secondary MU calculation program is described in more detail in Chapter 26, which will not be repeated here. However, secondary MU calculation is also an integral part of the commissioning process. It would be extremely useful to perform TPS-independent MU verification based on separate sources of machine data. The test cases used in the commissioning could be used to evaluate the secondary MU calculation program.

14.3.11 Beam Matching Considerations

For operational reasons, it would be a big advantage if the proton beam is "matched" in different treatment rooms within the same facility or at different proton therapy centers. The definition of a "matched" beam means that the same treatment plan can be delivered to different treatment rooms with the same result (within a clinically acceptable tolerance). To validate beam matching, the institution at least needs to compare the following:

1. The beam energy is matched with the same range by measuring the integrated depth–dose curves in a water phantom (to within typical measurement uncertainties or <1 mm).
2. The dose output at the reference point should be identical (or within the measurement uncertainties).
3. The same patient QA results are obtained when delivering the same patient plan in two different treatment rooms (within the uncertainty of QA measurement).

If the beam is confirmed to be "matched," there is no need to commission the beam model from scratch, which could be a significant time saver for the institution.

14.3.12 Reports and Summary

The final step in TPS commissioning is to summarize all performance tests for the beam model. It is important to see if there is any trend in SOBP measurements as well as in patient-specific QA. Regions that are commonly reviewed include 1) superficial region at shallow depth, 2) the proximal and distal shapes of SOBP plans, 3) field size effects, 4) when range shifter is used, 5) large air gaps, 6) junction region of two abutting fields, etc.

It is also important to create a list of tables and figures to assist for planning, chart checking, and monthly or annual QA baselines. Typical tables and figures include, but are not limited to, 1) output factors (which may be defined differently for passive scatter and pencil beam scanning), 2) energy vs. range (Figure 14–9), 3) spot size vs. beam energy, 4) a list of beamline delivery modes/options, 5) special material

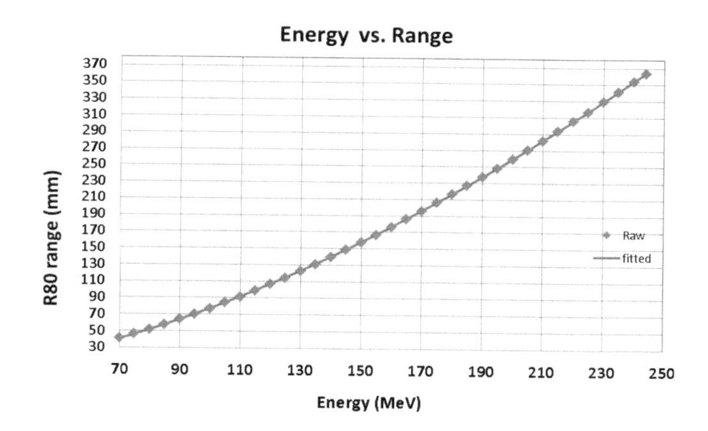

Figure 14–9 Pencil beam energy vs. range without a range shifter. The range is defined by the depth of the 80% of the distal falloff for a pencil beam integrated depth–dose curve. One potential use of this figure is to estimate buildup thickness when performing QA measurement.

table for HU/density override, 6) available range shifters and WETs, 7) recommended pencil beam spot spacing for different energies, etc.

14.4 End-to-End Tests

Because commissioning includes the task to verify various treatment procedures, it is important to develop clinical scenarios for the purpose of testing and organizing test procedures. In addition, some subsystems (such as the couch top and image guidance) may need specific implementations.

14.4.1 "Use Case" Scenarios

In software and systems engineering terminology, a use case is a list of steps to achieve a goal. It usually consists of a written description of how users will perform certain tasks in order to achieve the goal. Use case scenario can be perfectly used in the commissioning process to develop the steps and identify obstacles in performing certain clinical procedures. Here are common use cases in proton therapy that can be used to test and develop procedures for various situations:

- Ideal workflow: from CT simulation to treatment delivery. Treat a phantom by going through CT imaging, setting isocenter in treatment planning, loading the phantom on couchtop, performing in-room imaging, and treatment delivery using vendor supplied R&V system, then verify treatment records. The goal of this test is to ensure data transfer in various steps and to identify procedure difficulties and accuracy.

- Orientation tests: perform combinations of head first/feet first and supine/prone CT scans of a phantom; mark orientations, and verify if the same orientation is consistent through treatment planning and treatment delivery.
- Non-coplanar setup and imaging procedures: design a treatment plan containing a couch-kick or a non-coplanar couch angle. Test imaging procedure; verify shift direction and accuracy, etc.
- Two-iso setup: design a treatment plan containing two isocenters. Perform treatment setup and imaging, and verify if the procedure needs additional implementation.
- Partial treatment scenarios: interrupt a treatment by intentionally opening the door or pushing beam-off button, etc. Observe system response and system error message logging. Resume treatment and see if there is any difficulty in delivering the remaining treatment. Check and understand treatment records regarding treatment interruption.

14.4.2 Image Guidance Systems

Although in-room imaging performance tests have been previously described, the use of image guidance systems in various clinical scenarios still needs further evaluation. Therapists should work with physicians to develop guidelines for patient alignment based on disease-specific bony landmarks or soft tissue structures (if cone-beam CT imaging is available). Due to anatomical deformation, rigid-body based alignment may not perform a perfect setup correction, the site-specific guidelines help therapists to resolve alignment dilemmas.

The second important task is to develop shift protocols. For example, when a large shift (for example, more than 1 cm) is performed, it is required to perform post-shift imaging to evaluate the residual setup errors. If the shift is small, perhaps there is no need to re-image. This type of shift policy will eventually save setup time while maintaining reasonable accuracy.

The patient positioning system is also an integral part of the image guidance system. It is important to verify the orientation and angular setting of the coordinate systems to match with TPS settings. It is also important to verify shift accuracy, particularly for small shifts, such as 2–3 mm setup corrections. Couch sag is a reality. Couch sag may not be reproducible and can vary at different couch extensions, rotation angles, and load conditions. It may be important to develop a policy—for example, re-imaging is necessary when the couch is rotated more than 20 degrees from the previously imaged angle. It may be challenging to image when the couch is rotated to certain angles.

14.4.3 Modeling the Couchtop

To minimize inter-fractional variations in patient's anatomy, sometimes it is necessary to angle the beam through the couchtop. For this reason, it is very important to measure the water-equivalent thickness of the couch or the couch material. In addition, the shape of the couch should be carefully handled during the treatment planning process.

Most CT couches are different from the treatment couches. This makes the modeling of the treatment couch in the TPS an important task. Techniques such as the digital couch solution—replacing the CT couch with a digitally reconstructed treatment couch in the CT images—have been developed to assist the modeling of the treatment couch during the treatment planning process. Alternatively, the contour of the treatment couch can be used with appropriate density override to represent the treatment couch. A QA procedure should be established to ensure the couchtop was accurately modeled with appropriate water-equivalent thickness. A planning procedure should be developed to incorporate the couchtop into the patient planning CT image.

14.4.4 Development of Checklists

Checklists are good practices to minimizing operational errors. Physicists should assist therapists to develop a simple checklist to ensure correct treatment fields are delivered to the right patient at the correct angle. Treatment accessories and immobilization devices need to be verified. Sometimes even the rail positions in some of the couch designs need to be verified. Proton therapy is sensitive to materials in the beam path.

14.4.5 User Training

The final step in the commissioning process is user training. At this time, various treatment procedures have been finalized, and every task is near its final stage for patient treatment. Sometimes the user training has been provided earlier in order to get familiar with the vendor's equipment. However, when it becomes closer to patient treatment, it is a good idea to review the operation of equipment, and practice various treatment procedures prior to treating the first patient. The lead therapists should organize practice sessions with the physicist's assistance. Different use cases should be practiced with a phantom.

Therapists should be also trained by physicists to perform daily QA tasks. These tasks include, but are not limited to, 1) verify audio/video equipment is functional in the morning, 2) verify imaging system and laser alignment, and 3) perform daily output measurement. Depending on system complexity, these daily QA tasks can vary by different implementations. Some institutions use physicists or physicist assistants to perform daily QA duties.

14.5 Summary of Commissioning

Commissioning is a length process to prepare everything necessary prior to treating the first patient. In this chapter, we reviewed typical tasks in the commissioning of a proton therapy center. Planning for commissioning is perhaps as important as the commissioning itself. Most tasks can be planned in parallel, instead of serial, which saves on overall time. The key task in the commissioning is to set up the treatment planning system. Depending on the system requirements, both beam data acquisition and beam model validation take considerable time. The knowledge learned during the preparation for commissioning (prior to the availability of beam time) may prove to be the best investment in this process. Sometimes, the quality of such preparation, for example, fully understanding the dose calculation algorithm, plays a critical role in speeding up the commissioning process or making definitive decisions. It is fair to say that an effective commissioning should improve the quality of patient care and improve the financial aspect of the clinical operation.

References

1. Schneider U, Pedroni E, Lomax A. The calibration of CT hounsfield units for radiotherapy treatment planning. *Phys Med Biol* 1996;41:111–124.
2. ICRU. Tissue substitutes in radiation dosimetry and measurement. ICRU Report 44. 1989.
3. Andreo P, Burns DT, Hohlfeld K, Huq MS, Kanai T, Laitano F, Smythe VG, Vynckier S. Absorbed dose determination in external beam radiotherapy: An international code of practice for dosimetry based on standards of absorbed dose to water. IAEA Technical Report Series 398. Vienna: IAEA, 2000.
4. Moyers MF, Ibbott GS, Grant RL, Summers PA, Followill DS. Independent dose per monitor unit review of eight U.S.A. Proton treatment facilities. *Phys Med Biol* 2014;41:012103.
5. Weinhous MS, Meli JA. Determining pion, the correction factor for recombination losses in an ionization chamber. *Med Phys* 1984;11:846–849.
6. Peterson SW, Polf J, Bues M, Ciangaru G, Archambault L, Beddar S, Smith A. Experimental validation of a Monte Carlo proton therapy nozzle model incorporating magnetically steered protons. *Phys Med Biol* 2009;54:3217–3229.
7. Titt U, Sahoo N, Ding X, Zheng Y, Newhauser WD, Zhu XR, Polf JC, Gillin MT, Mohan R. Assessment of the accuracy of an MCNPX-based Monte Carlo simulation model for predicting three-dimensional absorbed dose distributions. *Phys Med Biol* 2008;53:4455–4470.
8. Flanz J, Paganetti H. Monte Carlo calculations in support of the commissioning of the northeast proton therapy center. *Australas Phys Eng Sci Med* 2003;26:156–161.
9. Paganetti H, Jiang H, Lee SY, Kooy HM. Accurate Monte Carlo simulations for nozzle design, commissioning and quality assurance for a proton radiation therapy facility. *Med Phys* 2004;31:2107–2118.
10. Anand A, Sahoo N, Zhu XR, Sawakuchi GO, Poenisch F, Amos RA, Ciangaru G, Titt U, Suzuki K, Mohan R, Gillin MT. A procedure to determine the planar integral spot dose values of proton pencil beam spots. *Med Phys* 2012;39:891–900.
11. Li Y, Zhu RX, Sahoo N, Anand A, Zhang X. Beyond gaussians: A study of single-spot modeling for scanning proton dose calculation. *Phys Med Biol* 2012;57:983–997.
12. Pedroni E, Scheib S, Bohringer T, Coray A, Grossmann M, Lin S, Lomax A. Experimental characterization and physical modelling of the dose distribution of scanned proton pencil beams. *Phys Med Biol* 2005;50:541–561.
13. Zhu XR, Poenisch F, Lii M, Sawakuchi GO, Titt U, Bues M, Song X, Zhang X, Li Y, Ciangaru G, Li H, Taylor MB, Suzuki K, Mohan R, Gillin MT, Sahoo N. Commissioning dose computation models for spot scanning proton beams in water for a commercially available treatment planning system. *Med Phys* 2013;40:041723.
14. Sawakuchi GO, Titt U, Mirkovic D, Ciangaru G, Zhu XR, Sahoo N, Gillin MT, Mohan R. Monte Carlo investigation of the low-dose envelope from scanned proton pencil beams. *Phys Med Biol* 2010;55:711–721.
15. Tourovsky A, Lomax AJ, Schneider U, Pedroni E. Monte Carlo dose calculations for spot scanned proton therapy. *Phys Med Biol* 2005;50:971–981.
16. Koch N, Newhauser W. Virtual commissioning of a treatment planning system for proton therapy of ocular cancers. *Radiat Prot Dosimetry* 2005;115:159–163.
17. Low DA, Morele D, Chow P, Dou TH, Ju T. Does the gamma dose distribution comparison technique default to the distance to agreement test in clinical dose distributions? *Med Phys* 2013;40:071722.
18. Low DA, Dempsey JF. Evaluation of the gamma dose distribution comparison method. *Med Phys* 2003;30:2455–2464.

Chapter 15

Quality Assurance, Part 1: Machine Quality Assurance

Bijan Arjomandy, Ph.D.

McLaren Proton Therapy Center,
Flint, MI

15.1 Introduction

In recent years, interest in using proton beams as an alternative modality in radiation treatment of cancer patients has grown, and it is expected that this awareness will increase rapidly as relevant treatment sites are extended beyond those already established [1–4]. As the technology of the proton therapy machine advances and the cost to acquire these technologies reduces, the popularity is expected to increase, and it will be more accessible for the treatment of cancer [5–7]. As the number of proton therapy centers surges, it is important and critical to provide broad guidelines and procedures for quality assurance (QA) of proton therapy machines in order to make sure that the patients are treated safely and accurately.

The QA procedures for proton therapy are complex and time demanding because of the complexity of equipment and control systems that are used for beam delivery systems as well as the beams' physical characteristics [8]. In the past, there have been only limited publications that address some of the QA procedures for proton therapy [9–11].

In the recent years, the American Association of Physicists in Medicine (AAPM) has established additional task groups (TG) to address some of the needs for this new modality. For that purpose, TG-224 (which is currently under review) was established to provide guidelines and recommendations for the comprehensive quality assurance procedures for proton machine QA. The guidelines and recommendations in this report are for systems with double scattering (DS), uniform scanning (US), and pencil beam scanning (PBS), as well as the other equipment that may be used in conjunction with these machines, e.g., cone beam CT (CBCT), portable CTs, CT on-rail, and, in some instances, multi-leaf collimators (MLCs).

A robust QA program should consider the failure modes of the system and implement the necessary measures to prevent and detect these possible failures. The collection of data (such as daily, weekly, and monthly QA) could help in foreseeing some of these failure modes, and it is highly recommended that a database be established for analysis and documentations. This chapter provides the general recommendations for proton therapy equipment QA checks. Since there are many different proton therapy manufacturers, the reader is encouraged to follow the vendors' recommendations in many instances. Specific QA limits and procedures not only depend on the design of the machine, but also on the delivery techniques. The AAPM task groups recommend the specific tolerances for many QA procedures.

15.2 Background and Purpose

15.2.1 Methodology of Quality Assurance

Quality assurance is a method to prevent mistakes or defects, avoid problems, and to predict mishaps when delivering a service to a customer. QA is a necessary procedure to provide confidence that a radiotherapy machine delivers radiation and is functioning as safely and accurately as it was when commissioned.

In radiotherapy, the underlying principal for QA is based on ICRU reports [12–14], and specifically *Report 24*, where it was concluded that for radiation therapy to be effective for certain tumors, the absorbed dose needs to be accurate to within –5% to +7% of prescribed dose. Although this report was written for radiotherapy using photon and electron beams (the modality mostly used at the time) the recommendations are certainly true for heavy ions, such as proton and carbon ion beams.

There are two key branches in QA procedures: 1) general equipment functionality that includes dosimetry, imaging, and machine QA procedures and 2) patient treatment field dose delivery-related QA procedures.

In the case of equipment, it is important to verify the beam parameters that are obtained during the commissioning procedures for configuration of the treatment planning system (TPS). In the case of patient QA, it is necessary to ensure that the dose calculated by the TPS is deliverable by the equipment (Chapter 16: Patient-specific QA). As discussed in Chapter 14 (Clinical Commissioning), the clinical commission of a beam delivery system involves appropriate measurements of specific and correct beam parameters for configuration of the TPS. Given that the calculation of patient dose distributions depends on the validity of these parameters, their consistency must be checked.

Equipment parameters and clinical data used for dose distributions and monitor unit (MU) calculations must be checked periodically to ensure proper dose delivery. There are two categories of QA procedures. One category involves the validation of dosimetry parameters that monitor a) absolute target dose that is measured with a calibrated device (such as an ionization chamber) and b) relative dose distribution that is measured with a suitable detector, such as film, ionization chamber, or two-dimensional (2D) detector array. The other category relates to accuracy of equipment, such as mechanical checks and imaging checks, which ensure the precise position of the target receiving the dose [15] as well as safety checks that monitor proper functionality of the equipment to ensure patient safety.

It is highly recommended that facilities record all the measured data during QA procedures for statistical analysis. This will help in differentiating between systematic and random errors and setting action levels to prevent mishaps.

15.2.2 Clinical Machine Characteristics and Parameters

There are different types of particle beam accelerators that provide beams of protons for clinical applications as discussed in Chapter 2 (Clinical Perspective of Proton Beam Therapy). However, there are three nozzle designs that make these proton beams useful for clinical applications, namely scattered, including passive scattering (PS) or double scattering (DS), uniform scanning (US), and pencil-beam scanning (PBS) or modulated scanning [16]. The typical beam energies are in rage of 70–250 MeV, which correspond to a range of ~4–37.5 cm in water. These beams are modified for clinical use by implementing different techniques to spread and make the beams uniform transversely and longitudinally (discussed in details in Chapter 6, Beam Production and Delivery Technology). In the case of the cyclotron and synchrocyclotron, a range shifter (called a degrader) is used to change the range of protons, impacting the beam quality by increasing its angular dispersion and emittance. A series of magnetic dipoles with a combination of spatially and spectrally collimating slits restrict unsuitable portions of beams and pass along the rest of the beam. On the other hand, a synchrotron produces beams of protons with variable energies by splitting the protons in intervals of several seconds. In either case, a narrow beam of nearly monoenergetic proton beams with a Gaussian shape is transported to the nozzle where it encounters different beam shaping components. These components spread the beam transversely either physically with scatterers or magnetically throughout the target. In the case of double scattering, the beams are spread longitudinally, either using modulation wheels or ridge filters to produce a uniform dose distally and proximally to the target. To produce a uniform spread-out Bragg peak (SOBP), the intensity of the proton beams are varied at different modulation wheel thicknesses. Any of these components can malfunction at any time and produce an undesirable outcome. The control system for these accelerators is responsible to monitor and deliver the parameters, which are calculated and provided by the TPS for a specific treatment. There are many well-considered interlocks that are implemented by manufacturers to prevent any malfunction, and it is the responsibility of a qualified medical physicist to ensure these control systems and the corresponding interlocks are functioning properly and are in good working conditions prior to beam delivery for treatment.

These requirements are met by implementing suitable QA procedures on a daily, weekly, monthly, and annual basis. Based on the delivery systems and accelerator designs, some of these procedures may be modified to suit the requirements of the institution's or manufacturer's recommendations.

It is important to understand the beam parameters that need to be measured and how these parameters are related with the physical devices that control or modify these beam parameters. Therefore, identifying the vital beam parameters and relating them to equipment that influences their dosimetric properties is critical. Some of the key beam properties for scattering and scanning beams are shown in Figure 15–1.

To monitor these parameters, one needs to make sure that 1) the beam flatness and symmetry stays within acceptable limits and 2) the monitoring chambers that measure these parameters are functioning properly and terminate the beams if the limits are exceeded. The other relevant parameters for scattered beams are the energy, energy spread, initial beam size, and beam current. It is important to monitor these parameters to ensure consistency and accuracy of beam delivery. The manufacturers monitor some of these parameters on-line prior to treatment, e.g., by introducing a beam profiler in the beam path to measure the beam cross-section and current. The monitoring chambers in the nozzle are responsible for measuring some of these parameters and stopping the beams if the values of these parameters exceed the specific limits. The reproducibility of these on-line devices needs to be checked because they are intertwined with the control system. However, some external devices are

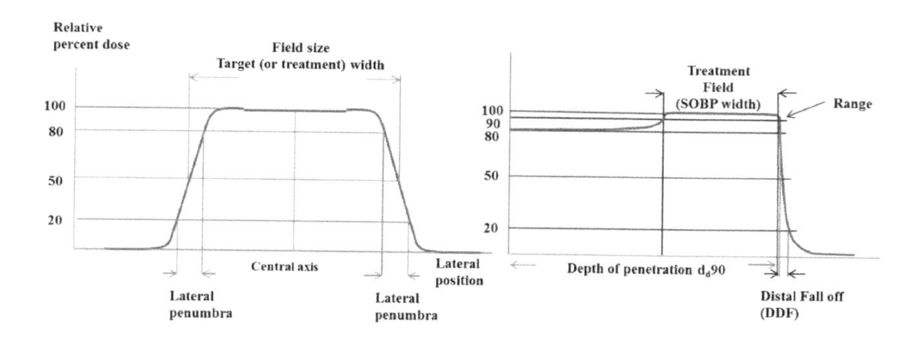

Figure 15–1 Dosimetric parameters for scattering and scanning proton beams.

required to perform routine checks on these parameters as a part of daily, weekly, monthly, and annual QA procedures.

The ICRU 78 [17] has identified some parameters for broad proton beams that pertain to DS, US, and PBS delivery systems as shown in Figure 15–1.

15.2.3 Multileaf Collimators in Proton Therapy

Multileaf collimators are being used extensively in photon radiotherapy, but multileaf collimation in proton therapy has only been adopted in one center in the United States in conjunction with scattering and uniform scanning delivery systems as an alternative to apertures. Whether this device will be utilized in other proton therapy centers is not clear at this time because most newer proton therapy machines use a PBS system. In special cases (i.e., brain tumors, head and neck) PBS may require additional aperture to reduce dose to organs at risk (OAR). However, PBS could mostly achieve dose conformality without apertures or multileaf collimators (see Chapter 8).

In any case, if such a device is implemented in a proton therapy clinic, it requires additional QA procedures to ensure it is reproducible and safe to operate. These procedures include mechanical and radiation safety QA of the device.

15.2.4 Imaging in Proton Therapy

Imaging is an important aspect in proton therapy for the success of treatment. Many existing imaging technologies are being implemented in proton therapy. Most imaging devices have been discussed in AAPM task group reports. Readers are expected to be familiar with the QA procedures for all existing imaging tools.

The most commonly used image guidance system in proton therapy is kilovoltage x-rays acquired with digital imaging panels. Planar x-ray images are obtained and registered to digitally reconstructed radiographs (DRRs) generated from the TPS. The x-ray tubes may be mounted in the nozzle, in orthogonal planes, or in the floor/ceiling as long as stereoscopic alignment can be obtained. Evaluation of geometric accuracy of this imaging system relative to the proton beam is essential. In recent years, many proton therapy centers acquired and implemented volumetric imaging for accurate position of the beam onto the target and evaluation of reproducibility of the anatomic position. Suitable image guidance for routine patient setup is an essential clinical instrument and, as such, the imagining devices must function as expected and reproduce an accurate patient positioning to have a successful proton therapy treatment [18–24].

There are substantial recommendations provided by TG-142 [25] and TG-104 [26] on image guidance systems that can be used for imagining systems within proton therapy centers.

15.3 Daily Quality Assurance Procedures

Parameters that could influence the dose distributions and cause inaccuracies in dose delivery —resulting in harming patients or staff who work around the treatment area—need to be checked on daily basis. All these parameters need to be verified prior to patient treatment to ensure safe treatment conditions. If there are devices that are used on-line prior to beam delivery (such as a multi-layer Faraday cup), the integrity of such devices needs to be checked at least annually or more frequently if there is any concern about their functionality.

There are different types of ionization chambers, devices, and phantoms that are available for daily checks. It is prudent to use phantoms and devices that are easy to set up and give reproducible results. Plastic Water® (Computerized Imaging Reference System, Inc., Virginia, USA) is a good choice of material to be used for daily setup as well as other QA measurements. However, care should be taken in determining their relative linear stopping power (RLSP), the energy dependence of RLSP, and the effects of nuclear interaction [27–29]. Table 15–1 lists the parameters that are important to check on a daily basis prior to patient treatment for each delivery system. The recommended tolerance values will be outlined in the upcoming AAPM TG-224. A selection of phantom and detector description can be found in Chapter 11 (Detector Systems).

15.3.1 Dose Per Monitor Unit (D/MU)

Dose per monitor unit is one of the important parameters that needs to be verified prior to patient treatment on a daily basis. This parameter is associated with proper functionality of the ionization chambers that monitor the delivered dose (charge) and proton beam fluence consistency. In essence, this test should verify the accumulated charges for a specific delivery as well as the accurate functionality of temperature–pressure monitoring sensors, which are located in the nozzle for correction of air density. The timer set, which is directly related with MU set for each treatment field, should also be verified during this procedure.

It is recommended to institute a standard condition that relates to reference proton beam calibration conditions that are normally established during commissioning procedures. Almost all proton therapy centers have implemented the International Atomic Energy Agency Technical Report Series (IAEA TRS) 398 protocol as the

Table 15–1 Suggested daily QA parameters and their possible tolerances for proton therapy beams. (Refer to TG-224 for the appropriate recommendations of tolerance limits and details of these procedures.)

Parameters	Tolerances*
Dosimetry Parameters Checks:	
Output	±2%
Depth verification	
Distal	±2 mm
Proximal	±2 mm
SOBP width	2%
Sigma	10.0%
Spot position	±2/±1 mm
Range uniformity	±0.5 mm
Mechanical Checks:	
Couch translation motion	±1 mm
Lasers position accuracy	±1 mm
Imaging System Functionality:	
x-ray isocenter v. laser isocenter	±1 mm
x-ray and proton beam isocenter coincidence	±2 mm
Image acquisition and communication	Functional
Safety Checks:	
Door interlock	Functional
Audiovisual monitor	Functional
Beam on indicator	Functional
X-ray on indicator	Functional
Search/clear button	Functional
Pause beam button	Functional
Emergency beam stop button	Functional
Monitor units interlocks	Functional
Collision interlocks	Functional
Radiation monitor (neutron and X-ray)	Functional
Optional:	
Range modulation wheel timing	±2%
Field light	Functional
Field width	±2 mm
Field symmetry and flatness	±1% and ±2%
Dose rate	±2%
Gantry angle readout accuracy	±1°
Interlock test therapy delivery system	Functional
Interlock test therapy verification system	Functional

*These are possible tolerances. Readers should refer to AAPM-224 for appropriate recommendations.

standard proton beam calibration [30]. Clinical beams of different characteristics (i.e., range, SOBP width, field size, etc.) may be calibrated relative to a standard beam to obtain a relative D/MU factor. The relative D/MU of different beams may be checked on alternative days as a part of daily QA. However, it is not necessary to measure multiple beam energies because the objective is to check the consistency of the dose monitor functionality.

15.3.2 Range

Daily proton beam range measurements are highly recommended to ensure the integrity of the accelerator control system. For a synchrotron machine, a selection of different beams (to include all secondary scatterers) with different ranges (also SOBP widths) may be measured at a reasonable time interval. Even though a cyclotron machine uses a degrader for energy selection, it is important to check the proper energy selection by an in-room QA measurement. A parallel plate ionization chamber is a well-suited measuring device for this purpose when a large field is applied because it has higher special resolution for distal fall off compared to a Farmer type cylindrical chamber. A multi-layer ionization chamber (MLIC) device may be utilized to measure the range of proton beams efficiently, as described in Chapter 11 (Detector Systems). Range shifter integrity checks should be part of daily QA checks to ensure the appropriate thicknesses of inserts are implemented for a specific pull-back range.

15.3.3 SOBP Width Measurement

It is highly recommended to measure the SOBP (or range modulation) width on a daily basis for delivery systems that employ either energy stacking, gating, or current modulation used with PBS and US or in combination with a rotating range modulation wheel (RMW) in DS. In some proton accelerators for DS, the beam intensity and the extractions are synchronized with modulation wheel rotation to produce uniform dose distributions in a homogenous media [31,32]. Systems that use fixed RMW or ridge filters can be exempt from this QA check procedure. The modulation width is commonly defined between the proximal and distal 90% dose levels (Figure 15–1). However, different centers may adopt different dose levels for defining the SOBP widths, and this definition may be in collaboration with the system manufacturer. In any case, the SOBP width definition has to be consistent with the TPS commission data. This is an important parameter that is used in designing clinical treatment planning with specific widths that include some additional margins [33]. In the case of a scanning delivery system, a combination of precise energy stacking and fluence results in uniform SOBP width dose distributions [34]. Any change in SOBP width may result in an overdose to critical structures or an underdose to the target.

SOBP width can easily be verified by measuring the doses at predetermined depths at the proximal and distal bounds of the SOBP in a phantom. Cross-calibrated MLIC is an ideal tool for this test. Precise daily range and D/MU measurements could be an alternative to SOBP checks [35]. It is recommended to measure different combinations of range and SOBP width on alternative days.

15.3.4 Range Uniformity

The range uniformity is referred to as the distal layer of broad field dose uniformity. The range uniformity is an important quantity that is required to be measured on a daily basis to validate the proton beam ranges that correspond to depth of distance 90% dose in water. The tolerance in range uniformity should be restricted to ±0.5 mm. The uniformity in proton ranges is crucial in sparing the critical organs that are located distally and adjacent to the target in patients. This is especially critical in patch field combinations, where a change in range could result in a large over- or underdose of 20 to 50% more or less than the prescription for healthy tissue [36]. It is also crucial in intensity-modulated proton therapy (IMPT) where multiple fields of different gantry angles with specific beam spots are designed to produce a uniform dose distribution in the target. It is recommended to check proton beams of different ranges on different days. One method to facilitate this task is to use a daily QA device with multiple ionization chambers that can measure D/MU, range uniformity, as well as flatness and symmetry of broad proton beams at the same time [37].

15.3.5 Spot Profile Monitor Constancy

Proton beam scanning nozzles are equipped with a monitoring chamber (see Chapter 6, Beam Production and Delivery Technology) that measures the spot profiles during or prior to irradiation. The functionality of these monitoring devices needs to be checked frequently to ensure a consistent beam delivery. The beam profiles may be measured by film dosimetry, an ion chamber array, strip chambers, or CCD detecting systems, e.g., Lynx® from IBA dosimetry [38].

15.3.6 PBS Spot Position and Width

The TPS designates and uses specific beams with known widths during dose calculation and allocates these beams at specific spot positions in the target. Therefore, any deviation in shape of these profiles relative to the commissioned data—as well as their position and number of protons at any given time—could result in incorrect dose delivery. A beam is characterized with a Gaussian shape with specific width (FWHM $= 2.35\sigma$) as shown in Figure 15–2.

During treatment, the characteristics and location of beams is monitored by the nozzle beam monitoring systems. Therefore, it is a function of QA procedures to measure and confirm these parameters with an external device to ensure these monitoring devices function adequately.

There are a variety of ways to deliver 3D dose distributions. For example, one can have a fixed spot spacing for all layers. Alternatively, one can vary the spot spacing as a function of different layers since the beam spots are changing for different beam energies. Furthermore, the spot size and spacing can be adjusted for a given energy or layer. Whether or not a spot pattern is affected by incorrect beam position or beam width depends on the amount these parameters could influence the desired dose distribution, which leads to QA tests of these parameters.

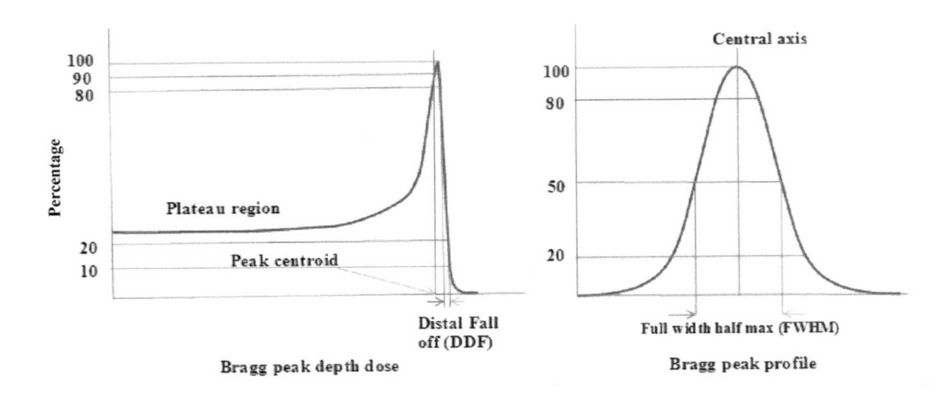

Figure 15–2 Depth–dose and lateral profiles parameters to characterize a narrow monoenergetic beam (pristine Bragg peak).

It is recommended to measure a minimum set of spots, including the central spots (undeflected beam) and spots in the far corner of the maximum deliverable field size. However, a pattern of well-designed spots that covers these criteria may be used to enable automated analysis of the desired beam parameters [39]. The analysis should include the centroid of the individual spots, beam width of the spots, or the penumbra of geometric figures. If complex patterns are used, then a gamma index analysis [40] could be used to determine the pass/fail of these patterns with respect to a standard pattern.

There are two effects that could potentially influence the dose distributions: 1) a systemic effect and 2) a random or isolated effect. Figure 15–3 represents the case of three different spot positional separations of broad beams with separation of 0.5σ, 1.0σ, and 2.0σ relative to the beam sigma (σ). As shown, the uniformity of the dose depends on the spot separation for a given spot size. For a continuous line scanning, the number of spots indicates the time required for an irradiation; therefore, one tries to minimize this number, resulting in spot spacing of 1.5σ. If the beam size changes (i.e., it is too small), then the resulting dose distribution will be improper because the spot spacing seems to be too large. A different spacing will result in different dose distribution and, consequently, requires different tolerance.

There are instances when an isolated positional error could occur. As shown in Figure 15–4, a local nonuniformity of 9% could result when one spot (with spot spacing of 1.0σ) is shifted by 20% of beam σ. The positional tolerances depend on the beam σ for a given relative spot spacing. Thus, the local shift must remain below 13% of beam σ in order to obtain a uniformity of ±3%. For example, for a 3 mm or 10 mm σ beam, this means a positional tolerance of about ±0.4 mm or ±1.3 mm, respectively. Therefore, determination of the absolute tolerance depends on a particular facility's installation.

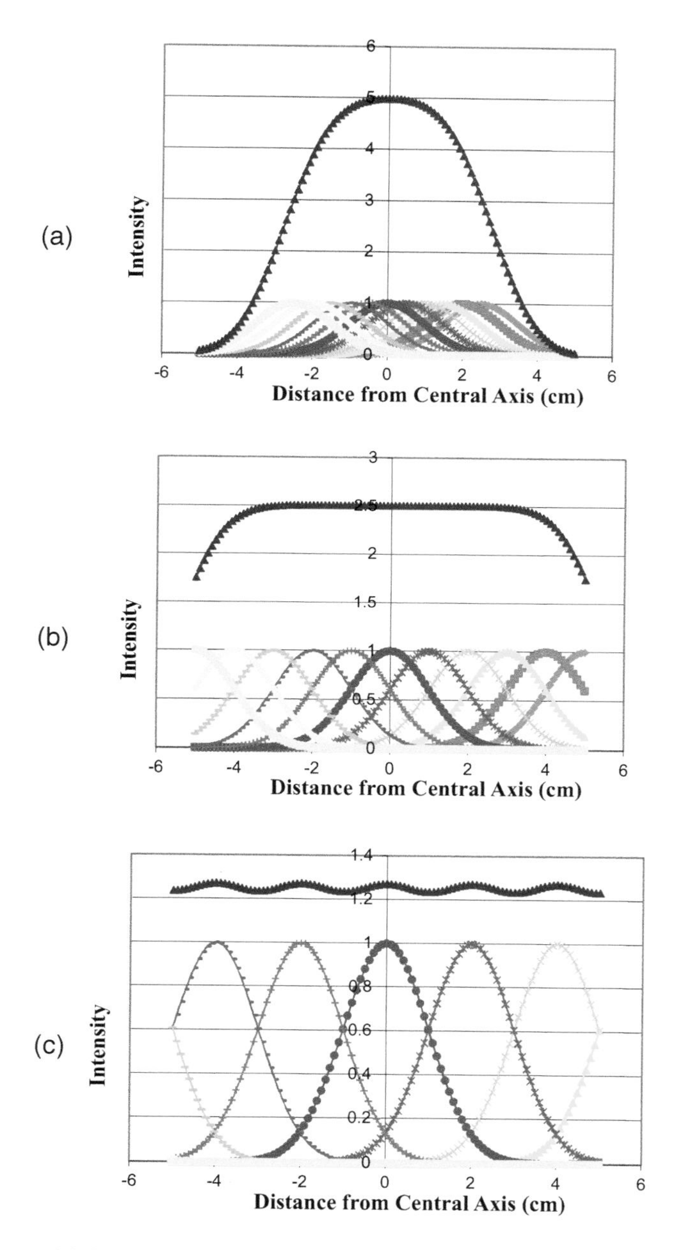

Figure 15–3 (a) Schematic representation of a broad beam profile from the sum of the proton beam spots with a Gaussian shape and with 0.5σ spot positional separation. (b) Schematic representation of a broad beam profile from the sum of the proton beam spots with a Gaussian shape and with 1.0σ spot positional separation. (c) Schematic representation of a broad beam profile from the sum of the proton beam spots with a Gaussian shape and with 2.0σ spot positional separation.

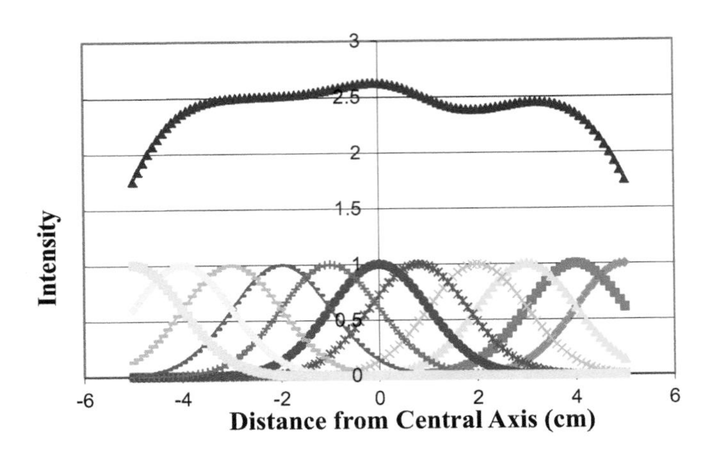

Figure 15–4 Schematic representation of a broad beam profile from the sum of the proton beam spots with a Gaussian shape and with 1.0σ positional separation. In this case, one spot is shifted by 20% of the beam, resulting in 9% nonuniformity of the beam profile.

15.3.7 Patient Setup Verification

15.3.7.1 Lasers

The main purposes of the lasers are to project marks at the location of gantry isocenters. If lasers are used for initial setup of the patient, then daily verification of their accuracy is required. A few different commercial QA devices are available for daily QA of lasers. It is recommended to use a device that contains fiducial markers that could be used for imaging the device to verify the accuracy of the phantom positioning via lasers.

It is also possible to verify the laser alignment against the mechanical device that has been provided by the manufacturer. Marking the laser lines on opposing walls in the treatment room or gantry enclosure may facilitate a quick check of laser alignment.

15.3.7.2 Imaging Systems

Proton therapy is image-guided radiotherapy (IGRT). Prior to patient treatment, patient setup is verified using either orthogonal planar imaging or volumetric imaging. The verification of imaging alignment systems can be accomplished using a phantom (as indicated in the previous section) in conjunction with room lasers, or by the cross-wires, which could be embedded either in the imaging system or on the detector panel. The fusion of acquired images with baseline images is an elaborate method to verify the imaging alignment accuracy. The IGRT QA phantom can be used to check the accuracy of IGRT and the radiation system by placing radiographic

Table 15–2 Suggested weekly quality assurance parameters and tolerances for proton therapy beams

Parameters	Tolerances*
Review daily QA checks	Review for systematic problems
Mechanical checks for all delivery systems:	
Gantry angle	±1°
Snout extension	±1 mm
Optional:	
Couch positional accuracy	±1 mm/1°
Range consistency	±0.5 mm
Image quality of imaging system	See TG-142

*These are possible tolerances. Readers should refer to AAPM TG-224 for appropriate recommended tolerances.

films on the phantom and irradiating with proton beams after imaging shifts have been applied [41–43].

15.3.8 Communication

A simple image acquisition and file swapping between the record and verification (R&V) system and the imaging registration software needs to be performed prior to patient treatment to detect any communication problem. The control and delivery systems in proton therapy could be more complex than those in conventional linac equipment, and in many instances the machine, TPS, and R&V system are not from the same vendor, which could cause intermittent problems. Thus, testing is required to ensure proper electronic communication among systems. A simple QA patient/phantom residing within the R&V system could aid in the setup of the measurements, verification, and recording.

15.3.9 Safety

The functionality of many safety interlocks that are mandated by state regulations and provided by the manufacturer for safe operation of equipment needs to be verified on a daily basis prior to patient treatment. The mechanical safety checks can be verified based on the following procedures. Produce an interlock or warning messages by pressing an emergency stop button. This should trigger a proper response, i.e., movement is stopped. The recovery from this stall state should be within a reasonable period.

The proper operation of beam pause needs to be verified on a daily basis, and collision sensors that are installed on the moving equipment (i.e., nozzle, couch, x-ray components, etc.) need to be checked to ensure the safety of the patients and staff.

15.4 Weekly Quality Assurance Procedures

Weekly QA checks should be limited to the parameters that have less potential to impact patient safety and a lower probability of occurrence than tests implemented on a daily basis. If it is determined that some of these tests are for QA items that have a lower risk of occurrence, then these tests may be allocated along with monthly tests. However, it is highly recommended that review of daily QA results should be part of weekly QA to evaluate any systematic drift or inconsistencies in the acquired daily tests. These tests should be conducted by a qualified proton medical physicist or trained personnel, and any out-of-tolerance parameters should be reported to the supervising physicist. Table 15–2 lists the parameters that are recommended by AAPM TG-224.

15.4.1 Gantry Angles vs. Gantry Indicators

Any small deviation in the actual gantry angles that were intended during the clinical treatment planning may cause the proton beams to pass through the more or less dense media, resulting in different dose distribution than was intended for the patient [44]. Thus, verification of gantry angle indicators or digital readout is critical in accurate dose delivery. The verification of gantry angles should at least be done for cardinal angles, and a simple mechanical or digital leveling device or lasers may be used for this purpose.

15.4.2 Snout or Applicator Extension

A snout is used to hold apertures, compensators, and, in some cases, range-shifters. Apertures are used for limiting the dose to critical structures by defining the treatment field and sharpening the penumbra in cases of scattering, uniform scanning, and, in some instances, for PBS. In addition, range-shifters are sometimes introduced in the beam path to treat shallow tumors. To reduce the scattering and sharpening of the penumbra, the snout is extended to minimize the air gap between the aperture and patient surface to about 2 cm. Therefore, it is important to make sure that the snout extension is accurately reflecting the requirement of the clinical treatment plan. The extension accuracy of snout can be verified using a tape or a ruler with respect to isocenter location.

15.4.3 Imaging Systems

In some imaging systems, the acquired images for positional verification of the patients are performed outside of the physical space where the treatment of the patient is done (such as in-room CT-on-rails). When imaging for patient positioning is performed outside of the treatment space, weekly checks of the patient positioning system (PPS) are required to verify the accuracy and reproducibility.

15.5 Monthly Quality Assurance Procedures

Monthly QA procedures are designated to parameters that have a lower probability of drifting or causing error on a short time scale or which have less severe consequences if they are out of tolerances. Table 15–3 specifies the parameters for monthly QA procedures for proton therapy beams.

Table 15–3 Suggested monthly quality assurance parameters and tolerances for proton therapy beams

Parameters	Tolerances*
Dosimetry parameters checks:	
Output	2%
Field symmetry and flatness	1% and 2%
Range check	1 mm
Uniformity of spot shapes	2% and 2 mm
Mechanical checks for all delivery systems:	
Gantry isocentricity	≤2 mm
Couch isocentricity	≤2 mm
Couch translational accuracy	≤1 mm
Couch rotational accuracy	≤1 mm
Couch trueness	≤1 mm
Snout trueness	≤1 mm
MLC:	
Light/radiation field coincidence (symmetric)	2 mm or 1%
Light/radiation field coincidence (asymmetric)	1 mm or 1%
Collimator angle indicator (four cardinal angles)	1°
Leaf position accuracy for two designated patterns	2 mm
Compensator placement accuracy	2 mm
Imaging and treatment coordinate coincidence (four cardinal angles)	2 mm
Congruence of proton and x-ray field	2 mm
Safety checks:	
Emergency push buttons	Functional
All interlocks	Functional
Exposure from long-term activation	<0.02 mSv/hr
Respiratory gating	Refer to TG-76

*These are possible tolerances. Readers should refer to AAPM TG-224 for appropriate recommended tolerances.

15.5.1 Beam Dosimeter Parameters

Beam qualities such as flatness, symmetry, intensity, and range are influenced by many components that are used to shape the beam in a scattering system. In the case of PBS, the beam shaping components have less impact on these parameters because scanning magnets and the spot size affect the beam quality instead. However, other factors—such as beam spread and energy resolution—could influence the beam quality in scanning beams.

Although the nozzle components for different manufacturers could be different in design for the same delivery system, the quality of the beam delivery requirements share the same criteria. Such criteria could be beam centering on primary and secondary scatterers for double scattering systems [45] or spot size and shape reproducibility in the PBS systems [46,47]. The monitoring chambers' function is to check these clinical parameters for accurate dose delivery. Therefore, it is important to routinely verify the proper operation of the monitoring chambers as well as their control system. There are selected devices that could be implemented to measure these parameters and verify that the monitoring chambers are operating properly. For example, a two-dimensional ionization chamber array could be used to measure the flatness and symmetry of a broad beam while simultaneously measuring dose per monitoring unit (D/MU) at different gantry angles [48]. Higher-resolution dosimeters such as conductive wire profiler monitors or gas emission multiplication detectors could also be used for this purpose [49]. The gamma index analysis [50] could be used in conjunction with evaluation of the data for uniform broad field PBS spots. The choice of QA device can be a factor in determining the extent to which monthly QA tests need to be performed. For example, strip chambers can be used to efficiently measure the beam width and its position in the air at the isocenter plane and other planes for a large number of spots on a daily basis. Different gantry angles and energies can be verified on different weekdays, whereby a monthly check could be superfluous.

15.5.2 Dose Per Monitor Unit (D/MU)

A monthly calibration check of the dose monitor should be conducted using a different dosimeter system (ionization chamber and electrometer) than the one used for daily checks. Since the fluence of the proton beam can change over time and at different gantry angles, it is recommended to measure D/MU at multiple gantry angles. There are different techniques that may be used to verify the fluence at different gantry angles. This could be done in a phantom at a fixed depth along the central axis of a broad field or by utilizing a jig that holds the chamber at the nozzle for different gantry angles. Dose per monitor unit should be verified for all options (i.e., for all range modulations in the case of scattering) for reference SOBP widths using a calibrated chamber in a solid phantom or water. All measurements should be compared with the base lines that were established during commissioning procedures or annual QA checks.

15.5.3 Range

The range of proton beams for scanning, uniform, or double scattering systems generated by any type of accelerators should be verified. These could be a single, central beamlet for PBS distribution that spans the entire range of available energies for patient treatment. These can be verified either using an MLIC or daily QA devices [51].

15.5.4 Flatness and Symmetry of Broad Fields

The lateral flatness and symmetry should be measured for the largest available fields. The lateral flatness and symmetry should be checked for different gantry angles and should be within ±2% and ±1%, respectively, with respect to the commissioning data in the TPS.

15.5.5 Mechanical Checks

15.5.5.1 Gantry Isocentricity

A slight variation in the location of the isocenter could affect the proton beam radiological path, which influences the range and lateral position of the field and, consequently, the dose distributions. Gantry isocentricity should be verified to make sure the correct dose is delivered at the intended target location. The verification can be done using a star shot technique with the proton beam or any apparatus used in conjunction with CBCT isocentricity checks [52].

15.5.5.2 Couch Translational and Vertical Accuracy

The couch translational movement should be checked not only during daily QA procedures over the range of distances used for the patient setup, (i.e., ±10 cm) but also for maximum possible range of motion during monthly QA. This test verifies the linear motion of the couch, which may not be detectable during daily QA for a short range of motions. The test should be done for all axis of motion. The tolerances for couch translation motion should be ±1 mm for the most extreme movements. The tolerances for newer robotic patient positioners should be even smaller [53,54]. However, similar checks should be performed on robotic couch as well.

The "trueness" of the couch positioning is defined as the motion of an object moving in a straight line without any deviation from the straight line. The couch vertical axis trueness should be verified because the couch can be subject to accidental collision or to malfunction resulting from wear and tear.

15.5.5.3 Snout or Applicator Longitudinal Accuracy

The snout or the applicator is the moving part of the nozzle that is constantly used during patient treatment and QA procedures. Wear and tear and possible collisions may have an impact on its motion accuracy. Therefore, the snout's longitudinal motion, as well as its trueness, should be verified on a monthly basis or when there has been a collision.

15.5.6 Emergency Stop

All emergency stop buttons located inside or outside of the room for stopping the mechanical components in case of an emergency need to be checked for proper operation. In addition to mechanical avoidance, some of these emergency buttons will stop the x-ray radiation as well as proton beam delivery. Therefore, these checks must test the functionality of the emergency button for both mechanical and radiation safety.

15.5.7 Coincidence of X-ray and Proton Radiation Field

For most treatment conditions, the x-ray images are used to align the patient for proton field treatment. Hence, the congruence of the field edges of the x-ray vs. proton fields should be verified to ensure correct patient alignment for the treatment fields. However, one needs to be aware that the x-ray source and proton effective source are often at different distances. The tolerance limits for x-ray and proton field coincidences should be within 2 mm.

15.5.8 Multileaf Collimator

The monthly QA procedures for proton machines equipped with MLCs are associated with alignment, leaf positioning, activation, and interlock functionality of the MLCs. Along with the common QA process for photon beam MLC, proton MLC needs additional QA that is addressed below.

15.5.8.1 Alignment

The heavy weight of the nozzle assembly and MLCs could influence the positional reproducibility and movement of the leaves at different gantry angles. Hence, verification of the coincidence of the proton treatment field and the x-ray image guidance system used for patient localization needs to be performed on a monthly basis. These tests need to be performed at multiple gantry angles (e.g., the four cardinal angles for a full 360° rotating gantry) to ensure there are no gravitational deflections. The collimator gantry readout must be verified to within 1° at four cardinal angles.

15.5.8.2 Leaf Positioning

The positional leaf accuracy should be performed on a monthly basis at a single gantry and collimator angle. Depending on the intended usage, the tolerance limits could be ±2 mm or ±1 mm if the MLCs are employed for field matching or patch techniques. Diamond-shape and X-shape field tests have been used and are suggested as a suitable assessment for MLC positional accuracy [55].

15.5.8.3 Activation

Radionuclide activation could occur when there is a nuclear interaction between high-energy proton beams and the material of an MLC [56,57]. Therefore, regular radiation level measurements should be conducted from MLC components on monthly basis at the times when there has been no beam delivery to the room for several hours (several hours prior to treatment). The national regulations [58] as well as any local regulation limits must be followed.

15.5.8.4 Interlock Functionality

All interlocks based on the manufacturer's recommendations and specifications need to be tested on a monthly basis. For example, a plan should be excluded when the leaf position is out of limits or beyond the maximum permissible extended or retracted position. Also, the MLC leaves should be parked in their retractable positions when there is no intention to use MLCs in a plan or when a multipurpose nozzle capable of delivering double scattering, uniform scanning, and PBS is used. Specifications of QA tests could vary for different manufacturers of the equipment.

15.5.9 Respiratory Gating Equipment

Respiratory motion can influence the proton dose distribution significantly due to changes in the radiological path lengths for treatment sites such as the lung, liver, and mediastinum. Respiratory gating has been used in proton therapy to reduce the interplay of the motion in such cases [59].

The criteria of QA checks for the respiratory gating system in proton therapy are the same as the recommendations outlined in AAPM TG-76 [60].

15.5.10 Imaging

The quality of an imaging system can influence the accuracy of patient positioning. It is recommended that the procedures outlined in TG-142 [25] for image quality testing be followed on a monthly basis for imaging systems used for proton therapy patient alignment.

15.6 Annual Quality Assurance Procedures

Annual QA procedures require more time to accomplish and are more comprehensive than monthly, weekly, and daily QA tests. This also may include state-based regulatory QA. Annual QA includes the checking of all mechanical functionality, evaluation of the quality and the accuracy of operation of the imaging devices, all the safety checks and interlocks, and verification of all dosimetric data measured during commission procedures and of the proton beam dose output calibration. Table 15–4 specifies the parameters for annual QA procedures. These recommendations may be modified based on the institutional requirements or depending on the particular parameters and recommendations of the manufacturer and functionality of the equipment. Readers should refer to TG-224 and the manufacturer's suggested recommendations for the appropriate tolerance limits and details of these QA procedures.

15.6.1 Beam Dosimeter Parameters

15.6.1.1 D/MU Constancy

It is recommended that the proton beam calibration be verified annually using the standard calibration protocol, the IAEA TRS Report 398 [61]. The calibration of the proton beams should be performed using a dosimeter system that has been calibrated by an accredited dosimeter calibration laboratory (ADCL). All the dosimeter devices used for daily and monthly QA tests should be cross calibrated at the time of annual calibration against the institution's ADCL calibration reference dosimetry system.

Reproducibility, linearity, and end effect of the beam D/MU should be checked. Any clinical dosimeter systems that are used for the calculation of monitor units and dose distributions should be verified against the commissioning data. In the case of any deviation from commissioning data, the cause should be investigated immediately and appropriate action should be taken to resolve the issue. For PBS, the consistency of output as a function of beam flux, as well as the beam position for different gantry angles and energies, should be performed. All the beam data, including the depth–dose

Table 15–4 Suggested annual quality assurance parameters and tolerances for proton therapy beams

Parameters	Tolerances*
Dosimetry parameters checks:	
Standard output calibration	±2%
Range verification	±1 mm
Depth–doses verification	±2%
Lateral profile	±2 mm
Field symmetry	±2%
Field flatness	±2%
Spot position	1 mm/0.5 mm
Inverse square correction	±1%
Monitoring chambers:	
Linearity	±1%
Reproducibility	±2%
Minimum/maximum dose/spot	Functional
End effect	1 MU
SOBP factors	±2%
Range shifter factors	±2%
Relative output factors	±2%
Verification of daily QA equipment	±1% or ±1 mm
Cross calibration of field chambers	±2%
Congruence of proton and x-ray field	±1 mm
Congruence of proton and light field	±1 mm
MLC leakage:	
Interleaves	Same values as measured during commissioning
Leaf end	
Shielding support	
Mechanical checks for all delivery systems:	
Gantry angle accuracy	1°
Gantry isocentricity (mechanically)	2 mm
Gantry x-ray isocentricity vs. lasers	2 mm
Couch sagging	1 mm
Snout extension accuracy	1 mm
Snout rotational accuracy	1°
CBCT isocentricity	2 mm
MLC:	
Leaf position accuracy for two designated patterns	2 mm
Leaf position reproducibility	1 mm
Spoke shot	2 mm
Imaging system functionality:	
Image quality checks	Consistence with baseline
CBCT	TG-179 and TG-142
Standard annual x-ray system checks	State regulations

Table 15–4 continued Suggested annual quality assurance parameters and tolerances for proton therapy beams

Parameters	Tolerances*
Safety checks:	
MLC activation test	<0.02 mSv/hr
Collision protection interlock tests	Functional
Dead man switch	Functional
Radiation warning sign	Functional
Door interlock	Functional
Beam pause	Functional
Room beam stop	Functional
Facility beam stop	Functional
Beam delivery indicator	Functional
Radiation monitors	Functional
Audio and visual monitoring	Functional
Gantry rotation sensor	Functional
Room clearance push button	Functional
Room sensor	Functional
Visual inspections:	
Modulation wheels	Wear and cracks checks
Block and compensator doors	Functional/wear and cracks checks

*These are possible tolerances. Readers should refer to AAPM TG-224 for appropriate recommended tolerances.

data, pencil beam lateral beam profile width, SOBP factors, range shift factors, and relative output factors for different energies should be checked and compared with commissioning data. Flatness and symmetry of broad field and D/MU of beam at different gantry angles other than those tested during monthly procedures should be checked (i.e., at every 45°). State requirements may govern the quality and the type of QA tests and the number of gantry angles for these tests. It is recommended that four gantry angles be tested for a fully rotating (360°) gantry and three for a half (180°) gantry.

15.6.1.2 Range

The range consistency of the proton field should be checked on an annual basis for detection of wear or cracks of the scatterers or modulation wheel, which could result in a different proton beam range than when they were commissioned [17,53]. For PBS, the range of individual spots or a broad field should be verified on the central axis as well as off-axis for different gantry angles. The range verification is normally tested with depth–dose measurements for the number of beams at a depth corresponding to 90% depth–dose. The accuracy of range loss and integrity of the range shifter needs to be verified as well as their encoder, if provided.

15.6.1.3 SOBP Width

The SOBP width should be verified for clinical treatment delivery combinations. These tests should include various snout sizes and a combination of range and SOBP widths. For scattering systems, a water tank is the ideal method for taking measurements. However, for scanning methods (PBS, US), water tank measurement is not practical, and multi-chamber devices have been shown to greatly increase the efficiency of data collection [48,62].

15.6.1.4 Integral Depth–Dose Distribution (PBS)

As mentioned in the commissioning chapter, the measurement of spot integral depth–dose distributions (IDDDs) is a quantity that is used in the configuration of the TPS for dose calculation in a PBS system [63]. The IDDD is the area integral dose of PPB at different depths and is usually measured using a Bragg peak chamber in a water tank. It is recommended to verify a wide selection of IDDD data collected during commissioning on an annual basis. The IDDD should be checked after any repair of scanning magnets or change in beam optics that could influence the spot dose distribution or when there is a significant difference that has been observed between the measured versus dose distribution during patient QA.

It is also recommended to measure and compare the 3D volumetric dose distribution. This can be accomplished by measuring the 2D dose distributions at different depths and depth dose curves, and comparing these measurements with the baseline that was established during the commissioning.

15.6.1.5 PBS: Spot Angular-spatial Distributions and Lateral Profiles

Spot dose profiles are important parameters used for TPS configurations (see Chapter 23, Treatment Planning). The TPS dose takes the shape and the width of the profile into account when the dose distributions are calculated for a patient. These quantities need to be verified on an annual basis, especially if there is a repair of the nozzle components or tuning of the beams. Dose profiles should be checked in air at different locations from isocenter for a range of energies and at different gantry angles. There are a variety of tools and equipment that may be used for this purpose, as described in Chapter 11. For example, a film dosimeter or scintillating foil [64,65], or a combination of a scintillating screen and camera base system [39], or gas emission multiplication detectors [49] may be used to measure these profiles with high accuracy.

15.6.1.6 PBS: Spot Position

A complete set of clinical data of spot alignment measurements at the isocenter plane and other clinical transverse planes should be performed to ensure that the machine is capable of delivering a pattern of spots that has been calculated by a TPS. The accuracy of the spot position in the treatment volume according to its position in the spot position monitor (SPM) system should be verified annually. More intricate tests need to be conducted to ensure that the SPM interlock prevents the dose delivery when the beam position is out of tolerance. Such a test also needs to be performed after any repair of SPM in addition to annual QA procedures. The manufacturer's recommendations should be followed for the required test to confirm the accuracy and validity of the spot position that could lead to aborting of the dose delivery of the spots when the interlock system of SPM is functioning accordingly.

15.6.1.7 Inverse Square Correction Test

An inverse square correction is normally applied to correct the dose due to change in the distance when the point of measurement is different from the effective source distance for a broad field. The validity of this correction factor should be tested on an annual basis [66].

15.6.1.8 Monitoring Chamber Proper Functionality

The monitoring chamber's dose linearity and reproducibility should be verified on an annual basis. In the case of PBS, the monitoring chambers have predetermined minimum and maximum dose criteria per spot that need to be checked. This can be accomplished by changing the intensity of the beam in order to enforce the tolerance on these limits. In some systems (US), the monitoring system may have cameral effect that needs to be verified [67].

15.6.1.9 Monitor Chamber End Effect

Monitoring chamber dose "end effect"—which is the difference in monitor units for a given dose (amount of charge) for a series of short radiation exposures versus a single radiation exposure—should be checked to ensure the consistency of the response of electronic equipment used for measurements. For a PBS system, this may require special spot patterns where the MU of every spot in the field are equal and can be linearly increased or decreased without exceeding the minimum or maximum MU/spot limit of the system.

15.6.1.10 Dosimeter Factor

The factors that are used for monitor unit calculation or verification of dose [68] should be tested. These factors include, but are not limited to, range shifter factors, SOBP factors, relative output factors for all available energies and modulations, and compensator factors [10,69,70]. The majority of these factors may be spot-checked to ensure their consistency.

15.6.2 MLC Activation

Radiation levels from the activation process in the treatment nozzle must be evaluated during commissioning and checked annually to ensure the safety of staff [57,71]. All the federal and state regulatory restrictions must be followed to keep the level as low as achievable [72].

15.6.3 Mechanical

All the mechanical translational and rotational movements for the couch, gantry, imaging systems, and, if applicable, the MLC leaves' accuracy need to be checked on an annual basis. The tests should verify all ranges of motions for clinical setup as well as the maximum permissible ranges of motion for all equipment. Since these tests are used as a baseline for monthly QA and for validating the commissioning and accepting tests (chapters 13 and 14), most of these tests should be performed with high-precision devices that can measure accurately in fractions of millimeters or degrees of angles. For example, the gantry isocentricity should be checked with high-precision tools such as theodolite [53,73] or precision digital front pointers [11,74] or films [75].

At some proton centers, light fields may be used for field verification or during physics QA procedures. In such cases, it is recommended to check the light field versus kV x-ray field and proton fields. Verification of kV x-ray versus proton beam irradiation isocentricity should be conducted to ensure the correct coincidence of target alignment and irradiation for all cardinal angles.

15.6.3.1 MLC Leaf Positioning

The monthly leaf position accuracy test should be performed over a range of gantry and collimator angles to ensure there is no sagging of the components due to the force of gravity. The MLC spoke shot should be used to check the integrity of the center of rotation of the MLC with respect to the delivered radiation.

15.6.4 Imaging Systems

As required by state and federal law, the accuracy and quality of the imaging system should be checked on an annual basis [25]. Image quality checks that include low- and high-contrast resolution, HU, as well as operating tests (kV, mAs), and the exposure test limits should be checked against baseline values. If treatment rooms are equipped with CBCT, the recommendation and the required QA outlined in TG-179 [76] and TG-142 [25] should be followed. All the safety interlocks on the imaging devices should be checked, which include, but are not limited to, "dead-man" switches, fluoroscopy timer warning, radiation warning lights, and door interlocks.

15.6.5 Safety

In addition to safety checks that are done during daily and monthly QA, there are other safety checks that are either recommended by the manufacturer or institutional protocols that need to be checked. There are safety checks that may require assistance from other staff, such as engineers, accelerator operators, and therapists. These could include the accelerator emergency shutdown that could be linked with the facility shutdown. All the collision avoidance sensors need to be tested for proper functionality.

Annual radiation safety checks of the RMWs, beam collimators, and apertures should be conducted to ensure that radiation levels are the same as when the equipment was accepted and commissioned.

15.6.6 Visual Inspection

The visual inspection of the gantry and proton beam delivery accessories/devices should be performed to detect wear and tear of components. For example, RMW, apertures, compensator doors that interlock on the nozzle, and any other moving mechanical components that may wear down due to normal usage should be inspected.

Acknowledgments

Most of the materials presented here are associated with the work accomplished from the contributions by members of the AAPM TG-224 report. I would like to thank members for their diligent and exhausting work, particularly Mrs. Paige Summers and Drs. Christopher Ainsley, Sairos Safai, Narayan Sahoo, Mark Pankuch, Eric Klein, Sung Yong Park, Johnathan Farr, Yuki Kase, and Jacob Flanz for their contribution to this chapter. In addition, I would like to thank our consultant team for AAPM TG-224, Drs. Ellen D. Yorke and David Followill.

References

1. Maughan RL, Van den Heuvel F, Orton CG. Point/counterpoint. Within the next 10–15 years protons will likely replace photons as the most common type of radiation for curative radiotherapy. *Med Phys* 2008;35:4285–4288.
2. De Laney TF, Kooy HM. Proton and charged particle radiotherapy. Philadelphia: Wolters Kluwer Health/Lippincott Williams & Wilkins, 2008.
3. Jakel O, Karger CP, Debus J. The future of heavy ion radiotherapy. *Med Phys* 2008;35:5653–5663.
4. Mohan R, Bortfeld T. Proton therapy: Clinical gains through current and future treatment programs. *Front Radiat Ther Oncol* 2011;43:440–464.
5. Smith AR. Proton therapy. *Phys Med Biol* 2006;51:R491–504.
6. Smith AR. Vision 20/20: Proton therapy. *Med Phys* 2009;36:556–568.
7. Ma CM, Maughan RL. Within the next decade conventional cyclotrons for proton radiotherapy will become obsolete and replaced by far less expensive machines using compact laser systems for the acceleration of the protons. *Med Phys* 2006;33:571–573.
8. International Commission on Radiation Units and Measurements. Clinical proton dosimetry. International Commission on Radiation Units and Measurements, 1998.
9. DeLaney TF; Kooy HM, Ed. Proton and charged particle radiotherapy. Philadelphia: Wolters Kluwer Health/Lippincott Williams & Wilkins, 2008.
10. Moyers MF, Miller DW. The modern technology of radiation oncology : A compendium for medical physicists and radiation oncologists. Madison, WI: Medical Physics Publishing, 1999.
11. Arjomandy B, Sahoo N, Zhu XR, Zullo JR, Wu RY, Zhu M, Ding X, Martin C, Ciangaru G, Gillin MT. An overview of the comprehensive proton therapy machine quality assurance procedures implemented at the university of Texas M. D. Anderson cancer center proton therapy center–Houston. *Med Phys* 2009;36:2269.
12. International Commission on Radiation Units and Measurements. Prescribing, recording, and reporting photon beam therapy. ICRU, 1993.
13. International Commission on Radiation Units and Measurements. Prescribing, recording, and reporting photon beam therapy. Bethesda, MD: ICRU, 1999.
14. International Commission on Radiation Units and Measurements. Determination of absorbed dose in a patient irradiated by beams of x or gamma rays in radiotherapy procedures.Washington: ICRU, 1976.
15. Ford EC, Gaudette R, Myers L, Vanderver B, Engineer L, Zellars R, Song DY, Wong J, Deweese TL. Evaluation of safety in a radiation oncology setting using failure mode and effects analysis. *Int J Radiat Oncol Biol Phys* 2009;74:852–858.
16. Paganetti H, Fontenot J. Proton therapy physics. *Med Phys* 2013;40:017301.
17. International Commission on Radiation Units and Measurements. Prescribing, recording, and reporting proton-beam therapy. Oxford, UK: Oxford University Press, 2007.
18. Arjomandy B. Evaluation of patient residual deviation and its impact on dose distribution for proton radiotherapy. *Med Dosim* 2011;36:321–329.
19. Mendenhall NP, Li Z, Hoppe BS, Marcus RB, Jr., Mendenhall WM, Nichols RC, Morris CG, Williams CR, Costa J, Henderson R. Early outcomes from three prospective trials of image-guided proton therapy for prostate cancer. *Int J Radiat Oncol Biol Phys* 2012;82:213–221.

20. Kudchadker RJ, Lee AK, Yu ZH, Johnson JL, Zhang L, Zhang Y, Amos RA, Nakanishi H, Ochiai A, Dong L. Effectiveness of using fewer implanted fiducial markers for prostate target alignment. *Int J Radiat Oncol Biol Phys* 2009;74:1283–1289.
21. Lim YK, Kwak J, Kim DW, Shin D, Yoon M, Park S, Kim JS, Ahn SH, Shin J, Lee SB, Park SY, Pyo HR, Kim DY, Cho KH. Microscopic gold particle-based fiducial markers for proton therapy of prostate cancer. *Int J Radiat Oncol Biol Phys* 2009;74:1609–1616.
22. Oshiro Y, Okumura T, Ishida M, Sugahara S, Mizumoto M, Hashimoto T, Yasuoka K, Tsuboi K, Sakae T, Sakurai H. Displacement of hepatic tumor at time to exposure in end-expiratory-triggered-pulse proton therapy. *Radiother Oncol* 2011;99:124–130.
23. Chen Y, O'Connell JJ, Ko CJ, Mayer RR, Belard A, McDonough JE. Fiducial markers in prostate for kv imaging: Quantification of visibility and optimization of imaging conditions. *Phys Med Biol* 2012;57:155–172.
24. Handsfield LL, Yue NJ, Zhou J, Chen T, Goyal S. Determination of optimal fiducial marker across image-guided radiation therapy (IGRT) modalities: Visibility and artifact analysis of gold, carbon, and polymer fiducial markers. *J Appl Clinic Med Phys* 2012;13:181–189.
25. Klein EE, Hanley J, Bayouth J, Yin FF, Simon W, Dresser S, Serago C, Aguirre F, Ma L, Arjomandy B, Liu C, Sandin C, Holmes T. AAPM task group 142 report: Quality assurance of medical accelerators. *Med Phys* 2009;36:4197–4212.
26. Yin F-F, Balter J, Benedict, Timothy C, Dong L, Jaffray D, Jiang S, Kim S, Ma C, Murphy M, Munro P, Solberg T, Wu QJ. The role of in-room kV x-ray imaging for patient setup and target localization. American Association of Physicists in Medicine, 2009.
27. Zhang R, Newhauser WD. Calculation of water equivalent thickness of materials of arbitrary density, elemental composition and thickness in proton beam irradiation. *Phys Med Biol* 2009;54:1383–1395.
28. Zhang R, Taddei PJ, Fitzek MM, Newhauser WD. Water equivalent thickness values of materials used in beams of protons, helium, carbon and iron ions. *Phys Med Biol* 2010;55:2481–2493.
29. Moyers MF, Vatnitsky AS, Vatnitsky SM. Factors for converting dose measured in polystyrene phantoms to dose reported in water phantoms for incident proton beams. *Med Phys* 2011;38:5799–5806.
30. IAEA. Absorbed dose determination in external beam radiotherapy. Technical report series 398. Vienna: IAEA, 2000.
31. Smith A, Gillin M, Bues M, Zhu XR, Suzuki K, Mohan R, Woo S, Lee A, Komaki R, Cox J, Hiramoto K, Akiyama H, Ishida T, Sasaki T, Matsuda K. The M. D. Anderson proton therapy system. *Med Phys* 2009;36:4068–4083.
32. Lu HM, Kooy H. Optimization of current modulation function for proton spread-out Bragg peak fields. *Med Phys* 2006;33:1281–1287.
33. Moyers MF, Miller DW, Bush DA, Slater JD. Methodologies and tools for proton beam design for lung tumors. *Int J Radiat Oncol Biol Phys* 2001;49:1429–1438.
34. Pedroni E, Bacher R, Blattmann H, Bohringer T, Coray A, Lomax A, Lin S, Munkel G, Scheib S, Schneider U, et al. The 200-MeV proton therapy project at the Paul Scherrer Institute: Conceptual design and practical realization. *Med Phys* 1995;22:37–53.
35. Dhanesar S, Sahoo N, Kerr M, Taylor MB, Summers P, Zhu XR, Poenisch F, Gillin M. Quality assurance of proton beams using a multilayer ionization chamber system. *Med Phys* 2013;40:092102.
36. Hill PM, Klein EE, Bloch C. Optimizing field patching in passively scattered proton therapy with the use of beam current modulation. *Phys Med Biol* 2013;58:5527–5539.
37. Ding X, Zheng Y, Zeidan O, Mascia A, Hsi W, Kang Y, Ramirez E, Schreuder N, Harris B. A novel daily QA system for proton therapy. *J Appl Clin Med Phys* 2013;14:4058.
38. Grevillot L, Bertrand D, Dessy F, Freud N, Sarrut D. A Monte Carlo pencil beam scanning model for proton treatment plan simulation using GATE/GEANT4. *Phys Med Biol* 2011;56:5203–5219.
39. Farr JB, Dessy F, De Wilde O, Bietzer O, Schonenberg D. Fundamental radiological and geometric performance of two types of proton beam modulated discrete scanning systems. *Med Phys* 2013;40:072101.
40. Low DA, Moran JM, Dempsey JF, Dong L, Oldham M. Dosimetry tools and techniques for imrt. *Med Phys* 2011;38:1313–1338.
41. Saw CB, Chen H, Beatty RE, Wagner H, Jr. Multimodality image fusion and planning and dose delivery for radiation therapy. *Med Dosim* 08;33:149–155.

42. Kessler ML. Image registration and data fusion in radiation therapy. *Br J Radiol* 2006;79 Spec No 1:S99–108.
43. Krengli M, Gaiano S, Mones E, Ballare A, Beldi D, Bolchini C, Loi G. Reproducibility of patient setup by surface image registration system in conformal radiotherapy of prostate cancer. *Radiat Oncol* 2009;4:9.
44. Sejpal SV, Amos RA, Bluett JB, Levy LB, Kudchadker RJ, Johnson J, Choi S, Lee AK. Dosimetric changes resulting from patient rotational setup errors in proton therapy prostate plans. *Int J Radiat Oncol Biol Phys* 2009;75:40–48.
45. Nishio T, Kataoka S, Tachibana M, Matsumura K, Uzawa N, Saito H, Sasano T, Yamaguchi M, Ogino T. Development of a simple control system for uniform proton dose distribution in a dual-ring double scattering method. *Phys Med Biol* 2006;51:1249–1260.
46. Phillips MH, Pedroni E, Blattmann H, Boehringer T, Coray A, Scheib S. Effects of respiratory motion on dose uniformity with a charged particle scanning method. *Phys Med Biol* 1992;37:223–234.
47. Widesott L, Lomax AJ, Schwarz M. Is there a single spot size and grid for intensity modulated proton therapy? Simulation of head and neck, prostate and mesothelioma cases. *Med Phys* 2012;39:1298–1308.
48. Arjomandy B, Sahoo N, Ding X, Gillin M. Use of a two-dimensional ionization chamber array for proton therapy beam quality assurance. *Med Phys* 2008;35:3889.
49. Klyachko AV, Friesel DL, Kline C, Liechty J, Nichiporov DF, Solberg KA. Dose imaging detectors for radiotherapy based on gas electron multipliers. *Nucl Instrum Methods Phys Res A* 2011;628:434–439.
50. Low DA, Harms WB, Mutic S, Purdy JA. A technique for the quantitative evaluation of dose distributions. *Med Phys* 1998;25:656–661.
51. Moyers MF, Ghebremedhin A. Spill-to-spill and daily proton energy consistency with a new accelerator control system. *Med Phys* 2008;35:1901–1905.
52. Bissonnette JP, Moseley D, White E, Sharpe M, Purdie T, Jaffray DA. Quality assurance for the geometric accuracy of cone-beam CT guidance in radiation therapy. *Int J Radiat Oncol Biol Phys* 2008;71:S57–61.
53. Allgower CE, Schreuder AN, Farr JB, Mascia AE. Experiences with an application of industrial robotics for accurate patient positioning in proton radiotherapy. *Int J Med Robot* 2007;3:72–81.
54. Farr JB, O'Ryan-Blair A, Jesseph F, Hsi WC, Allgower CE, Mascia AE, Thornton AF, Schreuder AN. Validation of dosimetric field matching accuracy from proton therapy using a robotic patient positioning system. *J Appl Clin Med Phys* 2010;11:3015.
55. Hounsell AR, Jordan TJ. Quality control aspects of the Philips multileaf collimator. *Radiother Oncol* 1997;45:225–233.
56. Daartz J, Bangert M, Bussiere MR, Engelsman M, Kooy HM. Characterization of a mini-multileaf collimator in a proton beamline. *Med Phys* 2009;36:1886–1894.
57. Moskvin V, Cheng CW, Das IJ. Pitfalls of tungsten multileaf collimator in proton beam therapy. *Med Phys* 2011;38:6395–6406.
58. US Nuclear Regulatory Commission. Standards for protection against radiation. Washington: U.S. Government, 2011.
59. Lu HM, Brett R, Sharp G, Safai S, Jiang S, Flanz J, Kooy H. A respiratory-gated treatment system for proton therapy. *Med Phys* 2007;34:3273–3278.

Chapter 16

Quality Assurance, Part 2: Patient-specific Quality Assurance

Bijan Arjomandy, Ph.D.[1], Mark Pankuch, Ph.D.[2], and Brian Winey, Ph.D.[3]

[1]Senior Proton Medical Physicist, McLaren Proton Therapy Center, Flint, MI

[2]Director of Medical Physics, Cadence Health Proton Center, Warrenville, IL

[3]Assistant Professor, Department of Radiation Oncology, Massachusetts General Hospital Medical School, Boston, MA

16.1 Introduction

Patient-specific quality assurance (QA) procedures play an important role to ensure that the clinical plan generated by the treatment planning system (TPS) is deliverable to the patient. The underlying principal for QA is based on ICRU reports, where it states that delivered dose needs to be accurate to within –5% to +7% of prescribed dose in order to be effective for certain tumors [1–3]. Patient-specific QA can be divided into two separate categories: 1) the procedures that involve inspecting the clinical plans to be deliverable based on equipment restrictions and criteria set by clinical protocols and 2) the procedures to evaluate if the dose distribution of the clinical plan delivered by the equipment is within acceptable tolerances compared to the intended dose distribution form the TPS. Proton beams have finite range characteristics and depth doses that are different from photon beams. In addition, the proton beam characteristics are extremely sensitive to the change in density of the material that they are traversing. Errors in the range calculation, or such things as the omission or incorrect characterization of objects within the beam path, can create significant deviations between the planed dose and the dose actually delivered to the patient [4–7]. Although the same careful evaluations are necessary for photons, the magnitude of the resultant errors can be much larger when treating with protons [8–11]. We have divided patient-specific QA in two parts: 1) planning QA and requirements to produce a deliverable clinical plan and 2) evaluation of dose and dose distributions of the plan performed by measurements.

16.2 Plan Quality Assurance

16.2.1 Simulation

Patient-specific QA begins at the time of simulation. Treatment site-specific simulation procedures must be in place to assure that the patient is adequately immobilized to reproducibly position the patient while maintaining appropriate clearance for the treatment snouts. The position of the body, including the extremities, must be taken into consideration to avoid potential collisions with the treatment snout, particularly for thorax and torso patients with larger immobilization devices of wing-boards or vacuum bags (see Chapter 18). During treatment, it is often desirable to move the snout assembly as close to the patient as possible to improve the geometric penumbra when using field apertures. If a range shifter is used when treating with pencil beam scanning (PBS), positioning the range shifter as close to the patient as possible is necessary to reduce the proton spot size. It is important that the entire simulation process results in a treatment device that is comfortable enough for the patient to maintain the simulation position for the entire length of the daily treatment fraction, plus some additional time due to potential beam availability delays.

Since pediatric patients are becoming a common proton therapy patient cohort, the immobilization devices for pediatrics also need to consider clearance for anesthesia equipment, along with the needs of the anesthesiologist. The pitch of the patient's head is often a compromise between the desires of the treatment planning team and the requirements of the anesthesia team to access the airway. If the treated area is near the intubation devices, reproducibility of the setup should be discussed with the anesthesiologist. Supine or prone treatments with or without masks can increase or decrease the complexity of the setup. Rapid and unrestricted access by the anesthesiologist must remain available at all times in case of an emergency.

The shapes and thicknesses of proton immobilization devices can differ significantly from those used for photons. Unlike photons, protons do not have a dose buildup region, and the reduced skin sparing effect for protons is minimally affected by additional material at the skin surface. Ideal proton immobilization devices are uniform in thickness with smooth, contoured, curved edges. Edges that have the potential to align in a direction parallel to the proton beam should be avoided, as should discrete edges. While the planning systems may model the dose propagation through the density variations of the immobilization devices well, patient setup errors within the immobilization devices themselves can introduce range errors. Immobilization devices should have no high-density objects, such as metallic supports. Figure 16–1 displays two examples of immobilization devices specifically designed for proton treatments—the BoS Headframe from Qfix™ and a homemade PVC pod used at Loma Linda Medical Center—but similar devices are used everywhere.

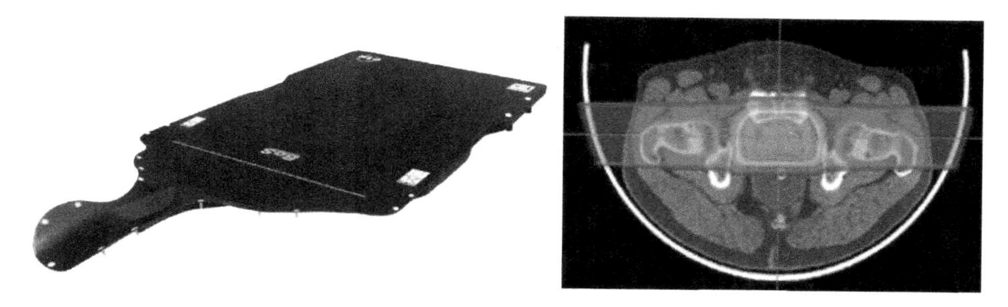

Figure 16–1 Immobilization devices used for proton therapy treatment: (a) Qfix™ Bos Headframe, (b) homemade PVC pod used for treatment of patients at Loma Linda Medical Center.

Materials used in the simulation process for immobilization devices must be well known to the physics team [12]. Relative stopping power values of non tissue-equivalent materials may not be properly characterized in the conversion process from Hounsfield units (HU) to relative stopping power [4,12–20]. New immobilization devices introduced to the center need to be evaluated by the medical physicist prior to clinical implementation. Water-equivalent characteristics of the materials of the device should be directly measured before use [17,22], as described in Chapter 4. These measured values can be used to evaluate the accuracy of the CT to relative stopping power (RSP) conversion and should be evaluated on a case by case-by-case basis for all custom-made devices [22–24]. In cases where the measured water equivalence does not match that predicted by the planning system, overriding of HU to the directly measured values is necessary.

Special care should be taken in the case of water-activated devices, such as MoldCare® head cushions that contain polystyrene beads coated in a moisture-rich polyurethane resin. These devices begin as a soft, formable pillow of beads that hardens into shape after dampening with water. CT scans obtained immediately after formation can express areas of higher CT number due to an uneven distribution of water used in the formation process. These higher concentrations of water are not present at the time of treatment due to complete evaporation over time. Range evaluations through these cushions should be carefully performed [25,26].

The planning CT scan must acquire all the essential information about the patient, including the positional extents of all immobilization devices and the treatment couch. The dosimetric effects caused by the couch top and immobilization devices were recently published in a report of AAPM Task Group 176 [27]. Proton-specific concerns were included in this publication. If the treatment device or treatment couch is not captured in the planning CT, the device must be accurately modeled and included in the dose calculation. Some modern treatment planning systems allow for the addition of a treatment couch into the planning CT. Ideally, the couch will be uniform and have minimal effects on the scattering (penumbra broadening) and energy straggling (distal broadening) of the proton beam.

The accuracy and reproducibility of the information obtained from the CT simulation is essential to build a good treatment plan. Recommendations for CT simulator QA have been clearly defined in the report of Task Group 66 [28]. Following the TG-66 guidelines and recommendations is important to ensure accurate dose calculation and patient safety. Of high importance are the guidelines recommended to maintain image performance consistency for the CT to RSP conversions for each energy (kVp) utilized for planning CT acquisition (see a the detailed discussion in Section 16.2.2.2). Daily, monthly, and annual QA results should be reviewed carefully, with attention paid to even slight drifts in consistency.

16.2.2 Treatment Planning

16.2.2.1 Patient-specific Treatment Planning QA

Acquisition of the CT simulation of the patient provides a three-dimensional model of the patient utilized for treatment planning purposes. The target volumes and the critical structures are defined on this "virtual" patient, and proton planning can begin. Target volume, MR/CT fusion, and geometric quality assurance methods are not detailed in this chapter, but they are well documented in Chapter 7.

For centers with multiple delivery systems (passive scatter/uniform scanning/pencil beam) [29,30], the choice of which delivery system to use must be determined early in the planning process based on the current site-specific protocols in place and the clinical expertise of the entire treatment team. The size of the target will determine the minimum snout size required to cover the entire target in a single field. For multi-room centers, individual rooms may be dedicated to specific snout sizes or beam delivery modalities for daily operational efficiencies. The process of changing a snout can take from 5 to 10 minutes, and it is much more involved when compared to the changing of an electron cone on a linear accelerator. The choice of treatment room and the availability of treatment time slots are often determined by the availability of the planned snout size.

When the plan has been completed, a thorough evaluation of the plan must be performed by the entire treatment planning team. All members of the planning team must find the plan acceptable based on their specific areas of expertise. The treatment planning team includes the treatment planner, the radiation oncologist, and the medical physicist.

The planner (often a dosimetrist) is tasked with evaluating the patient's unique anatomy and then using the treatment planning tools to meet or exceed the dose constraints of the given protocol for the target prescription [31]. The planner optimizes beam angles, designs apertures and compensators, and adjusts range and modulation, all in an effort to maximize the therapeutic ratio for the patient while maximizing the robustness of the plan [32–37]. If dose constraints for the target structure or organs at risk cannot be met, close communication with the physicist and radiation oncologist are necessary to quantify and prioritize potential compromises.

The radiation oncologist must determine if the presented dose distributions are acceptable and optimal for the patient. The radiation oncologist will establish criteria based on his or her expert knowledge of past clinical data, on knowledge of the patient's current condition of health, and the anticipation of the treatment established with fundamental patient participation. When evaluating proton plans, the radiation oncologist must be aware of the potential consequences of the proton treatment uncertainties and recognize the effect that the range uncertainties may have on target coverage and critical structure doses [7,37–43]. The radiation oncologist must consider "worse case" scenarios of dose distributions and must accept the finite probability of all potential dose distributions.

The medical physicist's role is to independently evaluate the plan by verifying dose protocol parameters, evaluating the physical limitations of the delivery system, evaluating range and setup uncertainties, and assessing the plan's "robustness" [35,36,44–46]. The procedures used in these evaluations can differ significantly, depending on the delivery method of the proton beam. Excellent communication with the radiation oncologist is essential to ensure the physician has all the knowledge necessary to make the best decisions for the patient.

The medical physicist's review of the treatment plan is an important portion of the patient-specific QA. The physicist is responsible for ensuring that the treatment plan is well optimized for the patient and can be delivered accurately and safely, including all appropriate margins [47–52]. The physicist must have specific knowledge of potential differences between planned [53–55] and delivered dose and must be able to quantify these differences and communicate them to the treating physician [56].

Many centers will use site-specific and delivery method-specific checklists or automated software checks for the physicist to complete as part of plan evaluations. A checklist also provides excellent documentation of the specific tasks performed by the physicist, and completed checklists should become a permanent part of the patient's medical record.

Several items should be checked as part of the plan review, and treatment centers may have very different requirements based of their specific workflow and treatment delivery system. Some examples of items to be checked, independent of the delivery method of the plan, are listed in the following section.

16.2.2.2 General Plan Review Items

Prescription

The prescribed treatment site must match the planned treatment site and provide a clear, unambiguous description of the intended target. In the treatment plan, the prescription dose must match the planned dose with clear understanding of calculation points and normalization methods.

Dose Algorithm

The appropriate dose algorithm must be chosen for the treatment site, with a suitable calculation grid to calculate dose to the target areas and completely encompass all organs at risk. Calculation point spacing should be chosen to provide the desired accuracy for the given anatomy, while maintaining reasonable calculation times.

In some cases, the capability of the dose algorithms should be verified for specific sites. For example, patient-specific measurements or Monte Carlo evaluations should be obtained when treatment plans are designed for sites with extreme heterogeneity.

Field Delivery Schema

Most proton fields deliver a uniform dose to the target area (except for multi-field-optimization in case of PBS). Since the delivered dose from each field is uniform, the sums of individual field combinations are also uniform. Not every proton field of the treatment plan will need to be treated on a daily basis to achieve excellent dose uniformity. For example, a three-field plan of fields A, B, and C may be delivered on day 1 with fields A and B, on Day 2 with fields B and C, and on Day 3 with fields A and C with equal target uniformity if the monitor units are calculated correctly. These complex treatment schemes are preferred for patient treatment efficiency, but they can cause confusion in the treatment room if not properly communicated to the treating therapists. In the cases where through/patch field combinations are used, distinct knowledge of paired fields must be presented. For matched fields, clearly defined shifts from one isocenter to the next must be outlined. A distinct understanding of the schema and unambiguous documentation of the schema are necessary to avoid a misadministration.

CT Conversion

In centers with multiple CT scanners, one needs to ensure that the correct CT scanned image data set is transferred to TPS and used for dose calculation. Many CT scanners are equipped with options of scanning the patients with different kVp to produce higher contrast and resolutions for dose calculations.

Verification that the correct CT-RSP conversion file is used in the plan must be performed. Conversion files can be obtained under several different CT scanning conditions [23]. The CT-RSP conversion curves are dependent on the kVp of the CT x-ray generator [41,57,58]. Unique conversion curves for the size of the patient and the reconstructed field of view could be acquired in an effort to more accurately model the conversion curve. Values for the RSP have been published for human tissues such as muscle, fat, brain, breast, lens of eye, heart, kidney, liver, lung, and urine in the bladder, just to name a few [17,59]. Sampling of the RSP of these easily identifiable organs and comparing them to published values within appropriate ranges provides confidence that the conversion curve has been applied appropriately to the planning CT.

The CT-RSP conversions of nonhuman tissues or implanted objects may also convert incorrectly [4,59]. Using a stoichiometric conversion curve for silicon breast implants has been shown to incorrectly predict the water-equivalent thickness of these permanent implants [60]. Other examples include cases including treatment sites that include surgical devices, such as embolization materials in AVMs or titanium hardware in spine stabilizations. In addition to incorrect prediction of stopping power ratios, these materials may introduce CT artifacts that affect the integrity of surrounding tissues [61,62].

Field Review

Each field should be reviewed to ensure it meets the criteria of a deliverable beam and that all parameters—such as range, modulation, gantry angle, and couch angle combinations—fall within equipment specifications. Such details as verifying that the aperture edges do not extend beyond the snout limits and evaluating if a beam is treating through a treatment device, couch top, or couch top edge may require attention if the planning system does not properly warn the planner. Potential for collisions for each field should be evaluated, including assessment of extents of the snout into the treatment position and clearance about the imaging guidance system.

Field Robustness

A term commonly used when describing proton plans is "robustness." Robustness is a multi-purpose term referring to how resilient the planned proton dose distribution will be when considering random setup errors, potential anatomical changes, calculation algorithm accuracy, range uncertainties, etc. [6,9,26,34–37,44–46,63–65]. Some characteristics of a robust beam may include such considerations as:

- beam approach angles with perpendicular incidence to the tissue surface,
- traversing inhomogeneities with a perpendicular incidence rather than parallel,
- evaluation of potential HU unit conversion errors along the beam path,
- selecting proton beam paths that do not cross inter-fraction variable-filled cavities such as the large bowel, sinuses, or stomach,
- assessing beams that have a distal edges stopping near a critical structure,
- careful evaluation of target and tissue motion throughout the respiratory or cardiac cycle, and
- assessment of the effects of setup error and proton range uncertainties.

Fields that are less than optimally robust are unavoidable for some treatment locations. In such cases, multiple treatment fields can substantially reduce the uncertainties of a single treatment field and increase the overall plan robustness.

Some examples of evaluations of individual field robustness for aperture- or compensator-based planning are presented in Figure 16–2. Figure 16–2(a) shows and axial image of a brain target with a proton beam entering from the right lateral direction. The field direction is en face to the patient surface and has perpendicular incidence to the skull/brain interface. A metallic clip can be noted at the center of the field. The presence of this clip has created an artifact in the muscular tissue between the skull and skin. These artifacts and the metallic clip will require careful contouring and overriding of the HU units to the appropriate values to ensure range calculation accuracy.

Figure 16–2b shows a coronal image of a target located within the left maxillary space. The beam is entering from left anterior superior direction. The oblique inci-

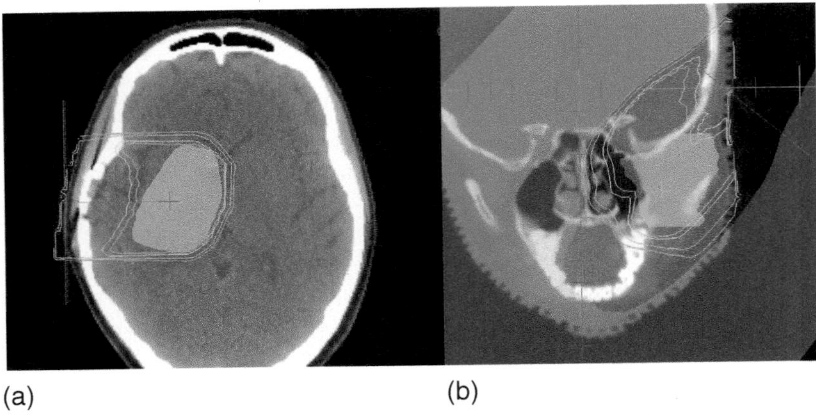

(a) (b)

Figure 16–2 Design and evaluation of a plan's robustness based on the geometric setup, material used for immobilization, and compensator and aperture designs. (a) Isodose distribution of the proton beam entering the right side of brain. (b) Isodose distribution of the proton beam entering the left maxillary space. (Images courtesy of the Northwestern Health Chicago Proton Center.)

dence of the beam relative to the patient surface may create a situation where the dose distribution is sensitive to patient alignment. The beam direction parallel to the temporal bone of the skull may be suspect to dose distribution calculation errors depending on the calculation algorithm chosen. The maxillary sinus and nasal cavity are clear in the treatment planning CT, but they could have variable levels of filling throughout the course of treatment. All potential conditions of sinus fill may require careful evaluation.

In addition to the items described above, more items should be checked that are specific to the proton beam delivery method (see Chapter 6, Proton Beam Production and Dose Delivery Technique). These additional checks can be categorized either as using aperture/compensator delivery (such as double scattering or uniform scanning) or pencil-beam delivery.

16.2.2.3 Aperture and Compensator Planning-specific Items

Items that should be checked specific to passive or double scattering delivery methods (and rarely PBS) may include the following.

- Aperture/compensator delivery systems make use of field combination techniques such as through/patch field combinations. There might be two or more pairs of through/patch field combinations to treat a target. These pairs may be delivered alternatively on different days to reduce edema or dose limits to OAR. Individual pairs should be evaluated along with evaluations of daily field combinations to determine if appropriate dose distributions are delivered on a day-to-day basis. All planned daily field arrangements should be reviewed, in addition to the cumulative sum. Extensive hot or cold spots should be approved by the physician for the cases where strict protocol definitions are not defined.

- Because of the physical size limitations of the proton beams, larger fields (in case of craniospinal irradiation [33]) can be treated by matching field edges from one field's lateral edge to the lateral edge of another field. Substantial hot or cold areas of dose can occur at the matchlines of such fields. To minimize this effect, planners will often include multiple matchlines where the spatial position of the matchline is varied (feathering). The dosimetric distribution of individual daily matches should be evaluated and approved by the physician if doses fall outside of protocol tolerances, along with the cumulative dose distribution. It should be noted that there is no perfect solution for matching and feathering, as described by Cheng et al. [66].
- Different proton vendors use different methods of delivering and monitoring dose output. These differences have led to difficulties in developing a uniform proton beam output model. Because of the lack of a widely accepted model, not all proton planning systems calculate monitor units for aperture/compensator delivery. Monitor units can be calculated from the in-depth knowledge of the planning system's weight point normalization, or they can be calculated from the dose to a point reported from the treatment plan. Careful evaluations of an institution's procedure for obtaining monitor units must be performed [66–70].
- Aperture shapes should be reviewed to ensure that extents of the edges fall within the limits of the assigned snout. Milled apertures are fabricated with a finite drill diameter that may not be capable of sharp, discrete corners. Smoothing of the field outline may be needed to ensure that the planned aperture shape will match the fabricated one to within the acceptable tolerance. Complex aperture shapes that generate small openings through the aperture may lead to incorrect dose calculations because many commercial TPS algorithms do not model the scattering factors accurately [71–74]. In these cases, careful evaluation of the predicted versus measured dose should be made. For small apertures, a correction factor to the open field output factor may have to be used to ensure that monitor units reflect the prescription dose [75].
- Compensators are designed to provide distal shaping to the target volume (see Chapter 23, Treatment Planning for Passive Scatter or Uniform Scanning). To account for potential patient setup errors and internal motions within the patient, compensator smearing must be applied [75]. Compensator smearing helps to ensure target coverage and plan robustness at the cost of degraded distal conformity. Formulas for the magnitude of the smearing radius have been proposed. A review should be performed to ensure that the smearing radius is appropriate for the given treatment site. The topography of each compensator must be evaluated to confirm that there are no steep steps, ridges, or pylons [76]. Large, discrete steps can generate unintentional

perturbation of dose that may not be properly modeled by the planning software. Some milling tools used in manufacturing compensators have an inherent taper to their design. These physical limitations of the milling tools may also make the manufacturing of discrete steps impossible.

- Range uncertainty must be considered in every proton plan and for every proton field individually. The magnitude of these expansions in the direction parallel to the proton beam has been investigated by several groups. The standard PTV expansion concept is not appropriate because these distal expansions must be extended in units of water equivalent thicknesses and are a function of the incident range of the proton field. Scaled beam-specific PTVs have been suggested [6,34]. Systematic methods of adding uncertainties to individual fields have been used [75]. Some planning systems also have integrated calculations of required uncertainties. Regardless of the method used at the specific clinic for the specific treatment site, an evaluation of the correct implementation of the site's protocol for range uncertainties should be completed. A detailed discussion on uncertainty can be found in Chapter 21.
- Additional range and modulation is added to each field to ensure appropriate coverage to the target. This will spread the dose that will actually be delivered to deeper or shallower locations than the target distal and proximal locations. Margins were added to ensure target coverage, but they will not include "worst case" scenarios for organs at risk [78]. These potential worst case doses to critical structures are most important for organs at risk that are located near the distal edge of a beam. Evaluations of these worst case OAR doses must be evaluated and presented to the physician, especially for serial-type organs, such as the spinal cord or optic nerves.

16.2.2.4 Pencil Beam-specific Items

Items that should be checked specific to pencil-beam delivery methods may include the following.

Inverse Planning Technique

Pencil-beam delivery treatments are planned using an inverse planning technique that attempts to minimize the differences between desired doses and deliverable doses within an iterative cost function. A review of this optimization should be performed to verify that penalties have been applied correctly, with attention given to over- and under-penalized areas. New techniques using multi-criteria optimization [45,79] methods may offer the physician a tool to converge to their most optimal solution in a much quicker time frame. Inaccuracies in the dose calculation algorithm, particularly in heterogeneous tissues, can negatively affect the spot weight optimization process and should be carefully evaluated.

Spot and Layer Spacing

Planning systems allow variations in spot spacing lateral to the beam direction, along with variation in energy layer spacing. Smaller spacing spots and closer layering of the range layers can increase the robustness [52,80] and uniformity of the plan at the cost of increased treatment times. For most systems, increasing the spots spacing within a layer has smaller time consequences when compared to increasing the numbers of layers. This is a consequence of the relatively long layer switching times (typically in the order of one to six seconds) of most current delivery systems. Optimal spacing should be defined with treatment site-specific recommendations that will consider

- desired uniformity,
- optimization method, such as single field uniform dose (SFUD) or multi-field optimization (MFO),
- overall treatment time, and
- expected internal motions of the targets.

Spot spacing and layer spacing parameters should be verified to be consistent with the planned treatment site's protocol.

Range Shifter

The PBS modes have minimum ranges, generally greater than those of passive scattering systems that use aperture/compensator. In PBS, low-range protons to treat targets near the surface are not an option because compensators are not used. To degrade the proton energy low enough to treat more superficial targets, range shifters are used. A range shifter influences the spot size, which is a function of the incident proton energy, the range shifter thickness, the minimum desired treatment depth, and the air gap of the range shifter [81–83]. Centers may often use more than one range shifter thickness to address this effect. The choice of the most suitable range shifter for the patient case under plan review should be confirmed.

Intensity Map (Spots Outside of Target)

The intensity maps of each field should be reviewed to ensure the spot pattern planned for treatment appears logical. A quick review to see that there are no spots placed outside of the target region can be performed, along with a check on the minimum and maximum MU per spot. The intensity pattern can also be examined to evaluate areas of extensive inhomogeneity within the patterns. Excessively large variations in spot intensities should be identified and rationalized by correlating them with anticipated regions of large gradients. Some treatment planning systems offer automated methods to minimize the potential for unacceptable spot patterns.

Repainting Parameters

To minimize the dosimetric effects of target motion, repainting within an energy layer and volumetric repainting of energy layers can be done [84] (see Chapter 25). If repainting is considered for treatment delivery, the potential consequences of the repainting should be evaluated. If repainting was not considered in the optimization, an individual spot just above the minimum MU per spot threshold can fall below deliverable limits if equally divided in all repaintings. Some treatment delivery systems will simply omit the division of such spots from the repainted layers to maintain the delivery of the originally prescribed integrated pattern. An understanding of all potential effects of repainting is important.

Robustness Analysis

Pencil-beam delivery does not make use of compensators and, therefore, the mechanical methods of adding smearing to account for setup and internal motion are not inherently included. In addition, PBS fields can possess highly inhomogeneous spot pattern intensities, and the effects of each beamlet's range uncertainty or setup uncertainty to the integrated dose is not easily accommodated. Quantifying the potential deviations from the planed dose can be performed using a robustness analysis technique. Several methods have been described for performing a robustness analysis [6,9,35–37,43,85,86]. Typical evaluations include recalculations of the approved plan's spot pattern back onto the treatment planning CT with modifications that include:

1. adding anticipated positional offsets of the isocenter to mimic setup errors,
2. adjusting target contours in conjunction with isocenter position to mimic internal motions, and
3. modifying the CT conversion curve to simulate effects of range uncertainties.

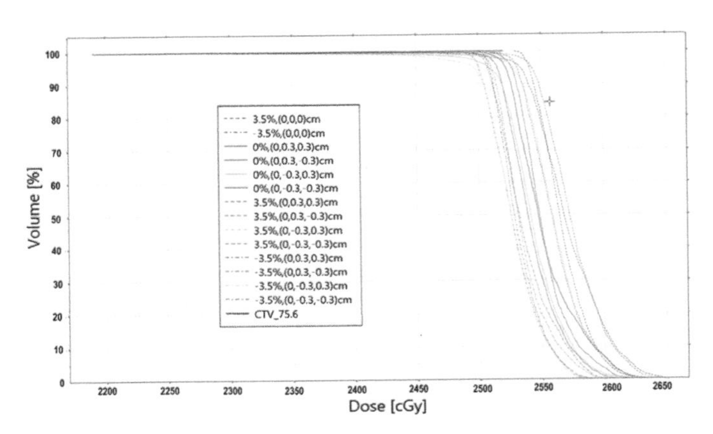

Figure 16–3 Dose volume histogram showing a PBS plan robustness evaluation for the CTV. The legend defines the various conditions tested in the evaluations, which include several combinations of translational positional errors of 0.3 cm and HU conversion errors of ±3.5%. The scale of the X-axis was expanded to emphasize the differences in the plans.

Each of these tests can be performed individually or in combinations with each other. Evaluation of all potential scenarios can result in a considerable number of plans to calculate and then evaluate. Automation of this process can be done using programming scripts, if necessary, if these tests are not included within the planning system's toolset.

The DVH results of robustness evaluation of a PBS plan are displayed in Figure 16–3. Both target structures and organs at risk are displayed. The DVHs show a distribution of potential dose distributions with the various combinations of positional offsets and CT conversion curves. These DVH plots should be used to demonstrate to the physician the anticipated range of potential differences in the planned dose distribution that could occur. In reality, the combination of systematic and random errors that are actually encountered during treatment cannot be quantified absolutely for any patient, but information from these curves can be used by the planning team to determine the magnitude of potential deviations from the idealized treatment plan.

16.2.3 Post-planning QA

Once the treatment plan has been approved by the planning team, it is exported to the record and verify system. This DICOM format contains all the mechanical positional information of the treatment delivery system, including such information as the gantry angle, treatment couch type and position, snout size and position, associated planning CT set, etc. The DICOM file also contains all the information of the treatment devices used, including all parameters to fabricate the aperture and compensator, along with the field range and modulation. For PBS plans, the spot map, nominal spot size, MU per spot, number of repaints, and energy for each layer is distributed through this file.

A review to verify correct transfer from the planning system to the record and verify system should be performed, and treatment parameters should be checked. In many systems, the record and verify system is not from the same manufacturer as the TPS or the delivery system, and additional checks for validity of the transfer are necessary.

Image guidance transfer integrity must also be included as part of the transfer reviews to verify that the DRRs are accurately generated to the imaging guidance systems. Many systems will directly use the reference CT information as a baseline for image guidance. Image transfer integrity, along with correct structure set association, should be reviewed. Accurate transfer of the treatment isocenter coordinate is essential for safe treatment delivery.

Editable parameters of the treatment fields in the record and verify system should be locked with an electronic approval by the reviewer (qualified medical physicist, QMP) once they are reviewed for correctness. Any subsequent changes in parameters should invalidate the approval, forcing the parameters to be reverified, reapproved, and relocked by the QMP.

16.2.4 MU Calculations for Aperture- and Compensator-based Systems

Different proton manufacturers use different scattering or wobbling methods to spread the protons across the area of treatment. The output response of these beams are measured using monitor chambers, which are calibrated using a standard calibration protocol [87] and provide a generic monitor unit to the user. The number of monitor units required to deliver a specific dose to a treatment volume is a function of several things, including the incident proton energy, the SOBP modulation, the type of range shifter (double scatting), the area of the wobble pattern (uniform scanning), the irradiated field size, the snout position, and other factors [29,66,68]. These numerous differences have led to struggles in developing a uniform proton beam output model that would be applicable to aperture/compensator-based systems in the treatment planning systems. Many planning systems do not offer the option to calculate field MU. If monitor units are not calculated directly by the planning system, alternative methods must be used to derive the MU required for the desired dose. There are several methods employed to generate MUs for aperture/compensator-based systems, including direct measurement, developing empirical or analytical models that are delivery system-specific, and Monte Carlo methods [66,67,88,89].

16.2.5 Patient-specific Dose Distribution

Many institutions have begun their programs measuring the prescribed dose and dose distributions of each treatment field prior to first treatment to verify the dose delivery of the beam in phantoms. Ideally, confirmation of dose distributions in three-dimensions (3D) would be optimal; however, 3D dosimeter systems are not yet commercially available. Safai et al. [90] used experimental devices to capture the light signal using a charge-coupled device (CCD) camera with inorganic scintillating mixed blocks. However, this device is not used for routine clinical use. There have been many attempts to measure the two-dimensional dose distributions in conjunction with depth–dose to verify the dose in 3D. Some noncommercial 2D devices are CCD cameras [91,92], which suffer from the quenching effect, and light-emitting detectors, such as gas electron multipliers and scintillating gas detectors [93,94], with their application as dosimeters not yet established. Several 2D array detectors are available and are used routinely to verify the dose distributions in radiation therapy departments [95–97]. In the past, many investigators have used film to verify dose distributions and perform QA checks in proton therapy [98–104]. Although films have high special resolution, they do not provide immediate results, and they require developing, scanning, calibrating, and off-line analysis, which is very time consuming. However, the recent advancement in Gafchromic films, such as EBTs, have shown that these films could facilitate the QA process and are very useful for dose and dose distribution evaluation with reasonable results [105–111]. Recently, Arjomandy et al. [112] have shown that the use of a 2D ion chamber array MatriXX (IBA Dosimeter, Schwarzenbruck, Germany) is a very useful device for patient-specific QA in proton therapy.

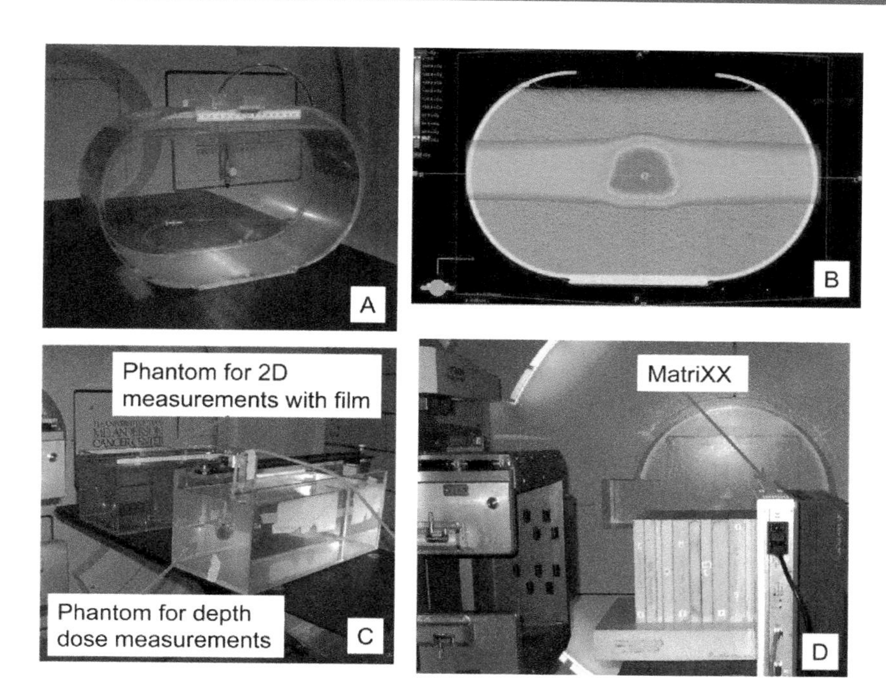

Figures 16–4 Measurements of dose and dose distributions technique using a single ionization chamber and 2D array MatriXX in different phantom materials. A) An elliptical water tank ("fish tank") used to measure absolute dose at the prescription point for prostate patients. B) CT image of fish tank used for generating patient verification plans. C) Water tank used to measurement patient's depth–doses along central axis. D) 2D depth–dose measurements in a solid water using MatriXX. (Reprinted with permission from [113]).

16.2.5.1 Patient Dose Measurement and MU Calculation

These measurements of dose can be done using a single ionization chamber or an array of chambers, such as the MartiXX (IBA Dosimetry, Germany) which is calibrated using a standard calibrated ionization chamber [112,113]. Figure 16–4 shows a typical setup for patient-specific QA with a single ionization chamber in a water phantom, as well as the use of a 2D ion chamber array MatriXX. Choice of phantoms may vary for different measurements, and care must be taken to verify the WET of the material other than water.

Dose measurement verification is generally obtained at the center and along the spread-out Bragg peak. Not all of the TPS used in proton therapy dose calculations are capable of providing the MU required for patients' treatment fields. Normally, MU is calculated from measuring the prescribed dose for each field using an ionization chamber and relating the measuring dose to relevant clinical data for that specific treatment field [66]. A few investigators have developed either empirical models or analytical techniques to calculate the MU and prescribed dose, which are compared with the measurements [66,68,114,115] either as primary or secondary MU calculations. If the calculated dose point from the planning system does not reside at the isocenter, additional inverse square corrections may be necessary. For small apertures, a correction factor to the open field output factor may have to be used to ensure that monitor units reflect the prescription dose [75].

It is commendable to calculate or measure MU on at least two independent systems. There have been efforts to generate analytic models based upon the specific properties of the beam line devices of double scattering systems, including the modulator wheel and range shifter, along with specific beam parameters, such as field size and off-axis ratios [72,116]. Analytical models for uniform scanning systems were developed that included effects for dose rate, scanned field size, aperture field size, and monitor chamber stem effects [115].

More recently, Monte Carlo has been suggested as a possible second check for patient field dosimetry verification [89,117]. The greatest advantage of Monte Carlo over the empirical or analytical model and measurement approaches is the possibility of performing the second check with the patient CT, including local heterogeneities and compensator effects beyond the more simple check performed in water.

16.2.5.2 Patient Dose Distribution Verification

As previously discussed, the two-dimensional planar dose distributions could be obtained using a film dosimeter or 2D array ionization chamber at different depths. The measured planar isodose distributions need to be compared to those calculated using TPS. This is accomplished by using a TPS "verification" module that can calculate the dose distributions using different computer tomography data sets or user-defined phantoms for a given combination of unmodified compensator, apertures, proton fluences, and spot position (in case of PBS) [112,113,118]. The gamma index is used to evaluate the agreement between planar dose distributions calculated by the TPS and those measured using a 2D detector. Low et al. [119] have shown that this technique can provide a set of criteria for distance-to-dose agreement and dose tolerances that can be set by the users for specific passing criteria, leading to gamma indices less than or equal to 1 in the 2D plane. Figure 16–5 shows the evaluation of planar dose distribution using gamma index analysis for a field used to treat lung cancer with passively scattered proton. Figure 16–6 shows the evaluation of planar dose distribution using the same tool for treating prostate cancer with pencil-beam scanning.

16.2.5.3 Aperture QA

Apertures that are used in proton therapy are manufactured mainly using two methods. They can be milled from high-density materials such as brass, or they can be cast using low-temperature metals such as Lipowitz's alloy, commercially marked as Cerrobend©. Production of the apertures can be performed on-site if the suitable equipment and mechanical expertise is available. Modern-day computer-controlled milling tools, such as the Leadwell CNC (Computer Numerical Control) machines (Figure 16–7) can be used to produce highly accurate devices in an expedited and automated manner. The typical milling time to generate a single aperture on these devices is approximately 15 minutes. There are several commercial machine shops that can manufacture these devices directly from the DICOM RTION plan and then ship the completed devices to the center for use.

The raw materials from milled devices are only used once. Then they are shipped offsite for recycling after all activation products have decayed. Onsite recycling of materials can be achieved when casting techniques using materials such as Lipowitz's metal are used. Casting techniques may not provide the manufacturing accuracy of the milling methods without considerable quality control methods, but castings offer a much lower operating expense over time.

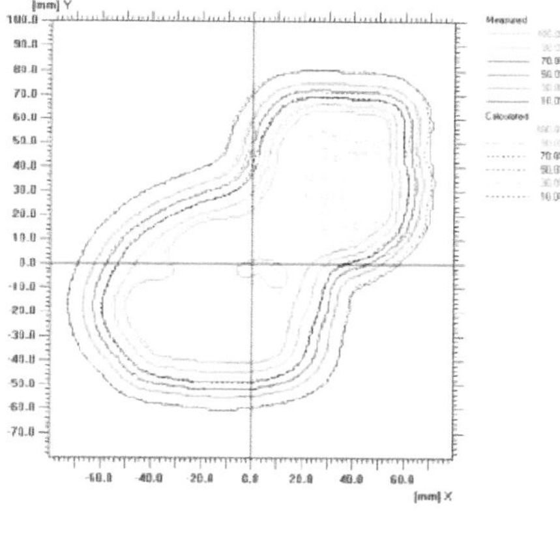

Figure 16–5 Measured (solid lines) and calculated (dashed lines) of planar isodose distributions at the center of SOBP for a field used to treat lung cancer with double scattered proton beam. Fully 100% of the pixels met the 3% dose and 3 mm distance agreement criteria. (Reprinted with permission from Arjomandy et al. 2010.)

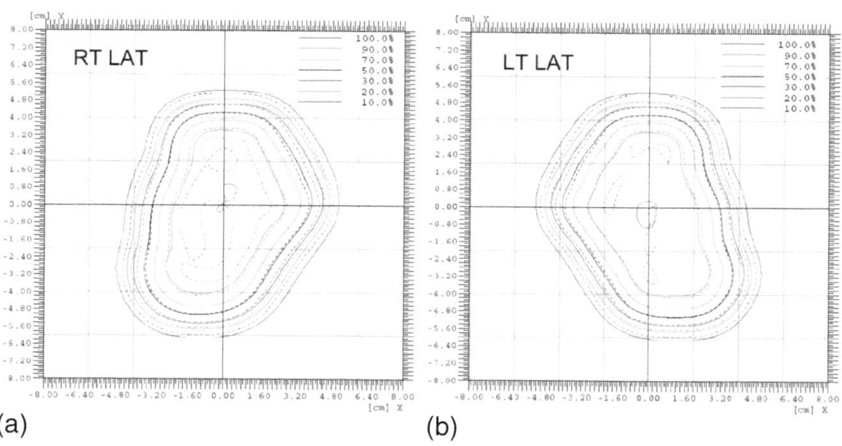

Figure 16–6 Measured (solid lines) and calculated (dashed lines) of planar isodose distributions at a depth of 23.4 cm for a field used to treat prostate cancer with pencil-beam protons. (a) Right lateral field: 95.9% of the pixels met the 3% dose/3 mm distance criteria. (b) Left lateral field: 98.9% of the pixels met the 3% dose/3 mm distance criteria. (Reprinted with permission from [113].)

Figure 16–7 The Leadwell CNC computer-controlled milling machine that is used to produce apertures from DICOM RTION transferred from TPS. (Courtesy of the CDH Proton Center.)

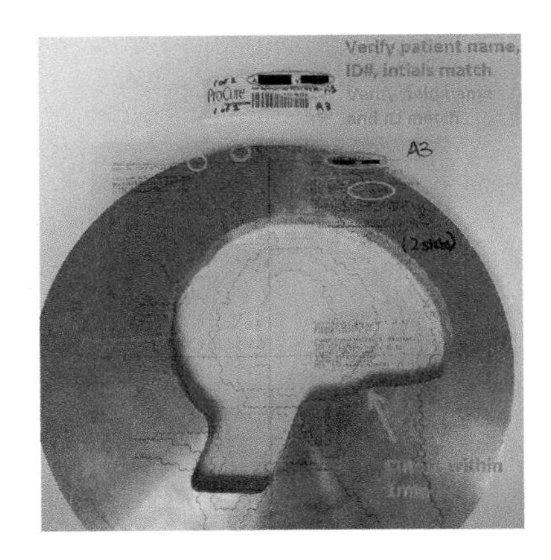

Figure 16–8 Evaluation of an aperture cut-out using a transplant overlay printed from TPS that is scaled to the planned shape.

Regardless of the manufacturing process, the device must be independently verified for accuracy. Verification can be performed physically using simple methods, such as comparing the aperture shape to a printout of the planned shape on a transparent overlay (Figure 16–8) that is accurately scaled to the planned shape. Some centers measure the proton beam through the aperture as part of the verification process, comparing the calculated distribution through the aperture to the measured dose distribution through the aperture using an ionization chamber array in a setup similar to figures 16–5 and 16–6. With proper chamber calibrations, the dosimetric comparison with a chamber array can also be used as a part of the output measurement required for calculating MUs.

Comparison of the manufactured device to the planned device must fall within a tolerance specified by the center. Consideration of the limitations of the manufacturing process and verification techniques must be considered when setting acceptable tolerance levels. These potential errors must be well thought out when setting geometric margins at the time of planning.

Apertures must be thick enough to completely stop the incident proton energies. This can lead to brass apertures that need to be greater than 10 cm think to stop a 200+ MeV proton beam. For larger snout sizes (in the order of 25 x 25 cm, the weight of such an aperture would be impractical and unsafe to lift into the snout mount. Instead, several aperture layers can be used, as long as the sum of the aperture thickness is enough to stop the incident beam. Extreme care must be taken to ensure sufficient aperture thickness is present for each field.

Most record and verify systems make use of a barcode system to assist the therapist in verifying that the correct aperture is in place prior to treatment. The barcodes are often printed stickers that are placed onto the apertures as part of the quality control process. Before attaching the sticker, a through verification of the aperture's identity must be confirmed. Extreme care must be made to ensure all apertures are labeled clearly and correctly.

16.2.5.4 Compensator QA

Compensators are manufactured in milling devices (Figure 16–7) identical to those used to mill apertures. The material used for compensator fabrication is usually acrylic or millable wax sold under the trade name of Freeman Wax. The acrylic offers the benefits of well-known material consistency with the option to easily inspect the device for internal imperfections due to the material's transparency. A disadvantage of acrylic is that the material can be brittle and may chip, crack, or break. Blue wax has the benefit that it is easily millable, which can allow for faster cutting tool speeds with quicker plunge milling movements. Wax has the additional benefit of a smaller scattering cross section than acrylics, resulting in sharper penumbras. The wax is not brittle, but it is soft, making it more susceptible to deformation if physically dropped or crushed. The chemical composition of wax may alter through melting when it is recycled repeatedly. The consistency in the chemical composition of this recycled wax can be checked by measuring the Hounsfield units of images obtained from randomly selected wax blocks by CT scanner or by direct measurements of water equivalence in the proton beam.

The design of the compensator, including a verification measurement of the thickness of the steps, should be performed on a system independent of the manufacturing device. Simple methods can be used that make use of depth gauges or calipers to verify the thickness at several points on the device. Points measured often include the maximum thickness, the minimum thickness, and several easily definable points within the compensator structure. This manual process can be tedious, and it may not be a thorough check for complex range compensators. Several publications have provided QA processes for compensators [121,122]. More complex methods have been proposed, including volumetric verification using computed tomography scans [120] and three-dimensional surface scanners.

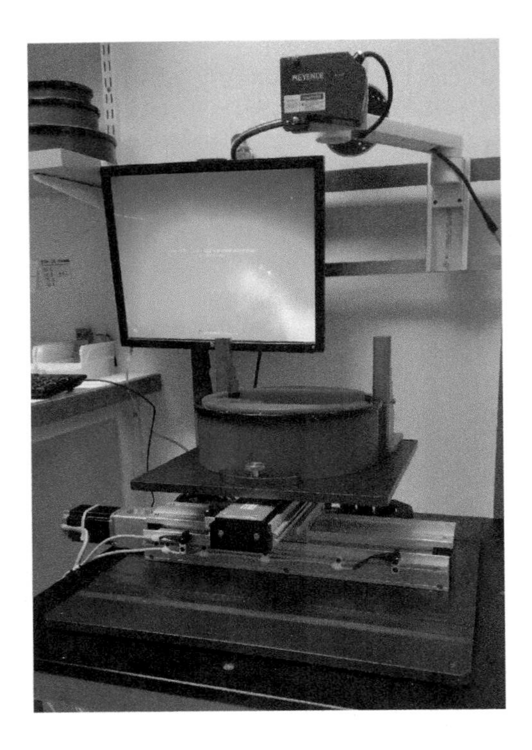

Figure 16–9 The 3D surface scanner used for verification of compensator manufacturing by measuring the depth at 100 unique points.

Figure 16–9 shows a surface scanning system in use at several centers to verify the depth at many or all points of the compensator in an automated manner. A depth-determining laser is positioned above the test compensator. The compensator is locked into a defined reference position. Two motorized arms move the device in a direction perpendicular to the laser direction. A software program loads the planned compensator pattern directly from the DICOM RTION file obtained from the treatment planning system. A gamma-test methodology is use for evaluation with a 0.5 mm thickness specification and a 1.0 mm distance to agreement. The DTA test must be used for cases of higher gradient, where portions of a step may be removed by neighboring plunges. Results often show a 100% passing rate with these criteria. This system can verify up to 100 unique points within one minute, providing higher confidence in the manufacturing process.

Many record and verify proton systems are also capable of verifying the compensator via a barcode placed on the device after verification. Processes and workflows must be in place to ensure the correct label is attached to the corresponding device.

16.2.6 Pencil-beam Scanning QA

Patient-specific PBS QA is a relatively new, but developing topic. PBS allows for more customized dose delivery to the patient and provides greater variables for the dose delivery, specifically the spot positions and weights. Whether optimized using SFUD or MFO methods, the doses for each field cannot be reduced to a simple model of range and modulation variables. The more complex dose patterns necessitate 3D dose verifications. Many 3D dose verification methods have been suggested in the literature, but the most common methods employ a 1D or 2D ion chamber array placed at several depths in water or solid water [112,113,121]. Unlike photons, the proton intensity patterns of PBS delivery are entirely depth dependent and can change very quickly with depth. Knowledge of the fluence pattern at a single depth cannot be used to quantify the fluence patterns at alternative depths. Figure 16–10 shows planned and measured PBS distributions at two different depths for a pelvis patient. Very different patterns of dose can be observed. The determination of how many depths to check and what is an acceptable evaluation criterion for each depth is, therefore, very difficult to

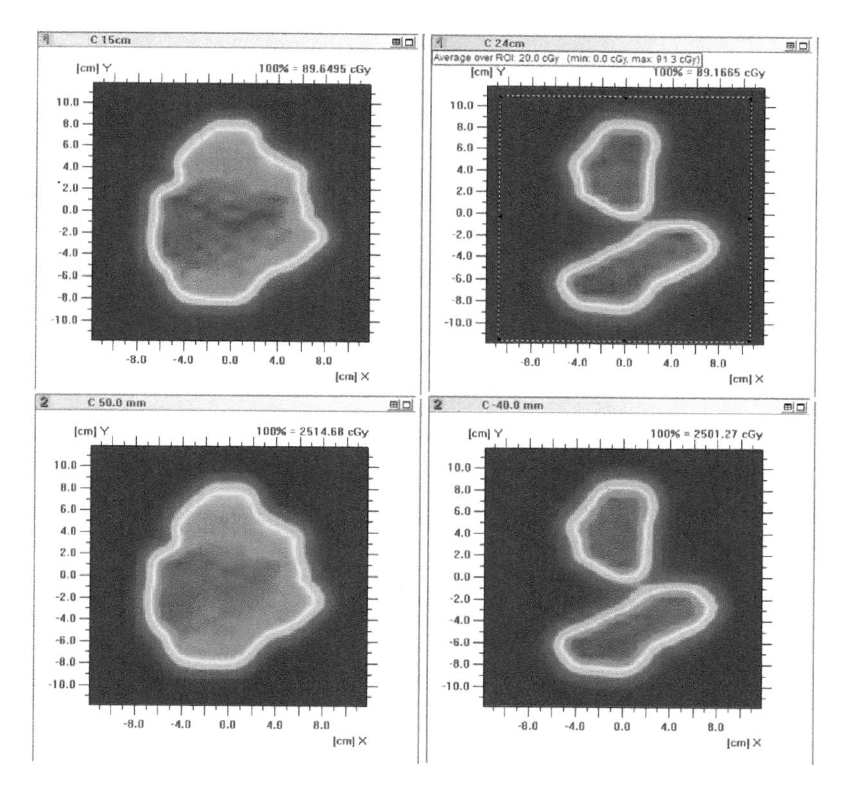

Figure 16–10 Measured (top row) and calculated (bottom row) dose intensity patterns of a single PBS field treating pelvic nodes. The left column was measured at a depth of 15 cm in a water-equivalent phantom. The right column was measured at a depth of 24 cm in a water-equivalent phantom. Dose intensity patterns are strongly dependent on depth.

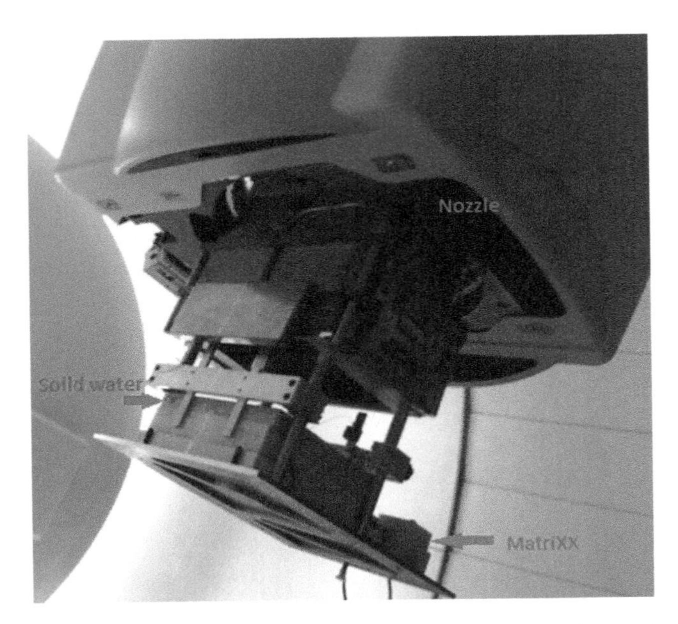

Figure 16–11 A measurement construction for PBS delivery at the gantry treatment angle with a 2D ion chamber array and solid water.

quantify. Efforts within the proton community to obtain a consensus or professional guidelines are currently in progress.

Some institutions also perform a single point dose measurement in addition to the chamber array [113]. Due to the many variables of the scanned delivery and treatment planning delivery, there can be variability in the beam delivery with gantry angle, and some institutions attempt to perform verification measurements with solid water at the gantry treatment angle (Figure 16–11).

Pencil-beam patient-specific QA methods can be very time consuming. More efficient methods have been proposed, but not fully deployed into clinical use. One option is to use the spot position and weights used for the patient delivery as input into a secondary dose calculation [121]. The primary limitation is the unknown response of the beam delivery system to the requested spots and weights, which has reduced the adoption of this method. Another suggested option is to deliver the doses to 3D dosimeters such as gels [122–124], but gel dosimeters require infrastructure support, and there are unknown effects of linear energy transfer (LET) and energy dependencies on the dose response of the gel materials.

Monte Carlo has also been explored as a secondary dose verification method for scanning delivery modalities [125]. Like scattered delivery verification methods, Monte Carlo verification for PBS allows for dose calculations in the patient geometries. Similar drawbacks include the limitation of measuring the actual beam delivery, including the inherent dose delivery variables of the gantry angle, electronic feedback, spot weight sensitivity, and other machine and nozzle-specific variables.

Summary and Conclusions

Patient-specific quality assurance is an extremely important step to ensuring that each patient's treatment will be delivered safely and with the required accuracy. A large majority of the QA procedures used for patients treated with photons are directly applicable to patients treated with protons. The paragraphs above have described several additional steps that should be considered when treating patients specifically with protons.

The finite range of the particles, along with the fact that the treatment delivery systems vary considerably from standard medical linacs, adds several layers of supplementary concerns that must be addressed by the planning team. Because of the lack of maturity of the technology, optimal and automated methods are still in the process of development. At the current time, many of these very important tasks require a considerable amount of physical resources to accomplish. Improvements offered by commercial vendors are expected to become more widely available as proton therapy technology becomes more mainstreamed into clinical practice.

References

1. International Commission on Radiation Units and Measurements. Determination of absorbed dose in a patient irradiated by beams of x or gamma rays in radiotherapy procedures. Washington: ICRU, 1976.
2. International Commission on Radiation Units and Measurements. Prescribing, recording, and reporting photon beam therapy. Bethesda, MD: ICRU, 1993.
3. International Commission on Radiation Units and Measurements. Prescribing, recording, and reporting photon beam therapy. Bethesda, MD: ICRU, 1999.
4. Schaffner B, Pedroni E. The precision of proton range calculations in proton radiotherapy treatment planning: Experimental verification of the relation between ct-hu and proton stopping power. *Phys Med Biol* 1998;43:1579–1592.
5. Verburg JM, Seco J. Dosimetric accuracy of proton therapy for chordoma patients with titanium implants. *Med Phys* 2013;40:071727.
6. Park PC, Zhu XR, Lee AK, Sahoo N, Melancon AD, Zhang L, Dong L. A beam-specific planning target volume (ptv) design for proton therapy to account for setup and range uncertainties. *Int J Radiat Oncol Biol Phys* 2012;82:e329–336.
7. Lim YK, Hwang UJ, Shin D, Kim DW, Kwak J, Yoon M, Lee DH, Lee SB, Lee SY, Park SY, Pyo HR. Proton range uncertainty due to bone cement injected into the vertebra in radiation therapy planning. *Med Dosim* 2011;36:299–305.
8. Arjomandy B. Evaluation of patient residual deviation and its impact on dose distribution for proton radiotherapy. *Med Dosim* 2011;36:321–329.
9. Park PC, Cheung JP, Zhu XR, Lee AK, Sahoo N, Tucker SL, Liu W, Li H, Mohan R, Court LE, Dong L. Statistical assessment of proton treatment plans under setup and range uncertainties. *Int J Radiat Oncol Biol Phys* 2013;86:1007–1013.
10. Moskvin V, Cheng CW, Fanelli L, Zhao L, Das IJ. A semi-empirical model for the therapeutic range shift estimation caused by inhomogeneities in proton beam therapy. *J Appl Clin Med Phys* 2012;13:3631.
11. Paganetti H, Goitein M. Radiobiological significance of beamline dependent proton energy distributions in a spread-out Bragg peak. *Med Phys* 2000;27:1119–1126.
12. Akkermana A, Breskina A, Chechika R, Lifshitz Y. Calculation of proton stopping power in the region of its maximum value for several organic materials and water. 8th Interna-

tional Symposium on Radiation Physics. In radiation physics and chemistry 2001;61:333–335.
13. Schaffner B, Pedroni E. The precision of proton range calculations in proton radiotherapy treatment planning: Experimental verification of the relation between ct-hu and proton stopping power. *Phys Med Biol* 1998;43:1579–1592.
14. International Commission on Radiation Units and Measurements. Stopping powers and ranges for protons and alpha particles. Bethesda, MD: ICRU, 1993.
15. Laitano RF, Rosetti M. Proton stopping powers averaged over beam energy spectra. *Phys Med Biol* 2000;45:3025–3043.
16. Moyers MF, Sardesai M, Sun S, Miller DW. Ion stopping powers and CT numbers. *Med Dosim* 2010;35:179–194.
17. Palmans H, Verhaegen F. Assigning nonelastic nuclear interaction cross sections to Hounsfield units for Monte Carlo treatment planning of proton beams. *Phys Med Biol* 2005;50:991–1000.
18. Parodi K, Ferrari A, Sommerer F, Paganetti H. Clinical CT-based calculations of dose and positron emitter distributions in proton therapy using the FLUKA Monte Carlo code. *Phys Med Biol* 2007;52:3369–3387.
19. Sharada KS. Proton stopping powers in some low-z elements. *Radiat Res* 1993;136:335–340.
20. Waibel E, Willems G. Stopping power and ranges of low-energy protons in tissue-equivalent gas. *Phys Med Biol* 1987;32:365–370.
21. Hurley RF, Schulte RW, Bashkirov VA, Wroe AJ, Ghebremedhin A, Sadrozinski HF, Rykalin V, Coutrakon G, Koss P, Patyal B. Water-equivalent path length calibration of a prototype proton ct scanner. *Med Phys* 2012;39:2438–2446.
22. Schneider U, Pedroni E, Lomax A. The calibration of CT Hounsfield units for radiotherapy treatment planning. *Phys Med Biol* 1996;41:111–124.
23. Schneider U, Pemler P, Besserer J, Pedroni E, Lomax A, Kaser-Hotz B. Patient specific optimization of the relation between CT-Hounsfield units and proton stopping power with proton radiography. *Med Phys* 2005;32:195–199.
24. Bentefour el H, Shikui T, Prieels D, Lu HM. Effect of tissue heterogeneity on an in vivo range verification technique for proton therapy. *Phys Med Biol* 2012;57:5473–5484.
25. Fredriksson A, Forsgren A, Hardemark B. Minimax optimization for handling range and setup uncertainties in proton therapy. *Med Phys* 2011;38:1672–1684.
26. Olch AJ, Gerig L, Li H, Mihaylov I, Morgan A. Dosimetric effects caused by couch tops and immobilization devices: Report of AAPM task group 176. *Med Phys* 2014;41:061501.
27. Mutic S, Palta JR, Butker EK, Das IJ, Huq MS, Loo LN, Salter BJ, McCollough CH, Van Dyk J., AAPM radiation therapy committee task group No. 66. Quality assurance for computed-tomography simulators and the computed-tomography-simulation process: Report of the AAPM radiation therapy committee task group no. 66. *Med Phys* 2003;30:2762–2792.
28. Smith A, Gillin M, Bues M, Zhu XR, Suzuki K, Mohan R, Woo S, Lee A, Komaki R, Cox J, Hiramoto K, Akiyama H, Ishida T, Sasaki T, Matsuda K. The m. D. Anderson proton therapy system. *Med Phys* 2009;36:4068–4083.
29. Flanz J, Smith A. Technology for proton therapy. *Cancer J* 2009;15:292–297.
30. Giebeler A, Newhauser WD, Amos RA, Mahajan A, Homann K, Howell RM. Standardized treatment planning methodology for passively scattered proton craniospinal irradiation. *Radiat Oncol* 2013;8:32.
31. Kooy HM, Clasie BM, Lu HM, Madden TM, Bentefour H, Depauw N, Adams JA, Trofimov AV, Demaret D, Delaney TF, Flanz JB. A case study in proton pencil-beam scanning delivery. *Int J Radiat Oncol Biol Phys* 2010;76:624–630.
32. DeLaney TF. Clinical proton radiation therapy research at the Francis H. Burr proton therapy center. *Tech Canc Res Treat* 2007;6:61–66.
33. Liu W, Frank SJ, Li X, Li Y, Zhu RX, Mohan R. PTV-based IMPT optimization incorporating planning risk volumes vs robust optimization. *Med Phys* 2013;40:021709.
34. Liu W, Li Y, Li X, Cao W, Zhang X. Influence of robust optimization in intensity-modulated proton therapy with different dose delivery techniques. *Med Phys* 2012;39:3089–3101.
35. Liu W, Zhang X, Li Y, Mohan R. Robust optimization of intensity modulated proton therapy. *Med Phys* 2012;39:1079–1091.

36. McGowan SE, Burnet NG, Lomax AJ. Treatment planning optimisation in proton therapy. *Br J Radiol* 2013;86:20120288.
37. Lu HM. A point dose method for in vivo range verification in proton therapy. *Phys Med Biol* 2008;53:N415–422.
38. Paganetti H. Range uncertainties in proton therapy and the role of Monte Carlo simulations. *Phys Med Biol* 2012;57:R99–117.
39. Schneider U, Pedroni E. Proton radiography as a tool for quality control in proton therapy. *Med Phys* 1995;22:353–363.
40. Yang M, Zhu XR, Park PC, Titt U, Mohan R, Virshup G, Clayton JE, Dong L. Comprehensive analysis of proton range uncertainties related to patient stopping-power-ratio estimation using the stoichiometric calibration. *Phys Med Biol* 2012;57:4095–4115.
41. Wang Y, Efstathiou JA, Sharp GC, Lu HM, Ciernik IF, Trofimov AV. Evaluation of the dosimetric impact of interfractional anatomical variations on prostate proton therapy using daily in-room CT images. *Med Phys* 2011;38:4623–4633.
42. Munck AF, Rosenschold P, Aznar MC, Nygaard DE, Persson GF, Korreman SS, Engelholm SA, Nystrom H. A treatment planning study of the potential of geometrical tracking for intensity modulated proton therapy of lung cancer. *Acta Oncol* 2010;49:1141–1148.
43. Lassen-Ramshad Y, Vestergaard A, Muren LP, Hoyer M, Petersen JB. Plan robustness in proton beam therapy of a childhood brain tumour. *Acta Oncol* 2011;50:791–796.
44. Chen W, Unkelbach J, Trofimov A, Madden T, Kooy H, Bortfeld T, Craft D. Including robustness in multi-criteria optimization for intensity-modulated proton therapy. *Phys Med Biol* 2012;57:591–608.
45. Lomax AJ. Intensity modulated proton therapy and its sensitivity to treatment uncertainties 2: The potential effects of inter-fraction and inter-field motions. *Phys Med Biol* 2008;53:1043–1056.
46. Kang JH, Wilkens JJ, Oelfke U. Demonstration of scan path optimization in proton therapy. *Med Phys* 2007;34:3457–3464.
47. Li Y, Zhang X, Mohan R. An efficient dose calculation strategy for intensity modulated proton therapy. *Phys Med Biol* 2011;56:N71–84.
48. Pflugfelder D, Wilkens JJ, Oelfke U. Worst case optimization: A method to account for uncertainties in the optimization of intensity modulated proton therapy. *Phys Med Biol* 2008;53:1689–1700.
49. Trofimov A, Bortfeld T. Optimization of beam parameters and treatment planning for intensity modulated proton therapy. *Tech Canc Res Treat* 2003;2:437–444.
50. Yoda K, Saito Y, Sakamoto H. Dose optimization of proton and heavy ion therapy using generalized sampled pattern matching. *Phys Med Biol* 1997;42:2411–2420.
51. Zhu XR, Sahoo N, Zhang X, Robertson D, Li H, Choi S, Lee AK, Gillin MT. Intensity modulated proton therapy treatment planning using single-field optimization: The impact of monitor unit constraints on plan quality. *Med Phys* 2010;37:1210–1219.
52. Li HS, Romeijn HE, Fox C, Palta JR, Dempsey JF. A computational implementation and comparison of several intensity modulated proton therapy treatment planning algorithms. *Med Phys* 2008;35:1103–1112.
53. Schell S, Wilkens JJ. Advanced treatment planning methods for efficient radiation therapy with laser accelerated proton and ion beams. *Med Phys* 2010;37:5330–5340.
54. Soukup M, Alber M. Influence of dose engine accuracy on the optimum dose distribution in intensity-modulated proton therapy treatment plans. *Phys Med Biol* 2007;52:725–740.
55. Lomax AJ, Boehringer T, Coray A, Egger E, Goitein G, Grossmann M, Juelke P, Lin S, Pedroni E, Rohrer B, Roser W, Rossi B, Siegenthaler B, Stadelmann O, Stauble H, Vetter C, Wisser L. Intensity modulated proton therapy: A clinical example. *Med Phys* 2001;28:317–324.
56. Yang M, Virshup G, Clayton J, Zhu XR, Mohan R, Dong L. Does kV-MV dual-energy computed tomography have an advantage in determining proton stopping power ratios in patients? *Phys Med Biol* 2011;56:4499–4515.
57. Cheng CW, Zhao L, Wolanski M, Zhao Q, James J, Dikeman K, Mills M, Li M, Srivastava SP, Lu XQ, Das IJ. Comparison of tissue characterization curves for different ct scanners: Implication in proton therapy treatment planning. *Transl Cancer Res* 2012;1:236–246.
58. Zhang R, Newhauser WD. Calculation of water equivalent thickness of materials of arbitrary density, elemental composition and thickness in proton beam irradiation. *Phys Med Biol* 2009;54:1383–1395.

59. Moyers MF, Mah D, Boyer SP, Chang C, Pankuch M. Use of proton beams with breast prostheses and tissue expanders. *Med Dosim* 2014;39:98–101.
60. De Marzi L, Lesven C, Ferrand R, Sage J, Boule T, Mazal A. Calibration of CT Hounsfield units for proton therapy treatment planning: Use of kilovoltage and megavoltage images and comparison of parameterized methods. *Phys Med Biol* 2013;58:4255–4276.
61. Espana S, Paganetti H. The impact of uncertainties in the ct conversion algorithm when predicting proton beam ranges in patients from dose and PET-activity distributions. *Phys Med Biol* 2010;55:7557–7571.
62. Albertini F, Hug EB, Lomax AJ. The influence of the optimization starting conditions on the robustness of intensity-modulated proton therapy plans. *Phys Med Biol* 2010;55:2863–2878.
63. Albertini F, Hug EB, Lomax AJ. Is it necessary to plan with safety margins for actively scanned proton therapy? *Phys Med Biol* 2011;56:4399–4413.
64. Cao W, Lim GJ, Lee A, Li Y, Liu W, Ronald Zhu X, Zhang X. Uncertainty incorporated beam angle optimization for IMPT treatment planning. *Med Phys* 2012;39:5248–5256.
65. Sahoo N, Zhu XR, Arjomandy B, Ciangaru G, Lii M, Amos R, Wu R, Gillin MT. A procedure for calculation of monitor units for passively scattered proton radiotherapy beams. *Med Phys* 2008;35:5088–5097.
66. Hsi WC, Schreuder AN, Moyers MF, Allgower CE, Farr JB, Mascia AE. Range and modulation dependencies for proton beam dose per monitor unit calculations. *Med Phys* 2009;36:634–641.
67. Kooy HM, Rosenthal SJ, Engelsman M, Mazal A, Slopsema RL, Paganetti H, Flanz JB. The prediction of output factors for spread-out proton Bragg peak fields in clinical practice. *Phys Med Biol* 2005;50:5847–5856.
68. Koch N, Newhauser WD, Titt U, Gombos D, Coombes K, Starkschall G. Monte Carlo calculations and measurements of absorbed dose per monitor unit for the treatment of uveal melanoma with proton therapy. *Phys Med Biol* 2008;53:1581–1594.
69. Zheng Y, Ramirez E, Mascia A, Ding X, Okoth B, Zeidan O, Hsi W, Harris B, Schreuder AN, Keole S. Commissioning of output factors for uniform scanning proton beams. *Med Phys* 2011;38:2299–2306.
70. Akagi T, Kanematsu N, Takatani Y, Sakamoto H, Hishikawa Y, Abe M. Scatter factors in proton therapy with a broad beam. *Phys Med Biol* 2006;51:1919–1928.
71. Koch NC, Newhauser WD. Development and verification of an analytical algorithm to predict absorbed dose distributions in ocular proton therapy using Monte Carlo simulations. *Phys Med Biol* 2010;55:833–853.
72. Bednarz B, Daartz J, Paganetti H. Dosimetric accuracy of planning and delivering small proton therapy fields. *Phys Med Biol* 2010;55:7425–7438.
73. Deasy JO. A proton dose calculation algorithm for conformal therapy simulations based on Moliere's theory of lateral deflections. *Med Phys* 1998;25:476–483.
74. Moyers MF, Miller DW, Bush DA, Slater JD. Methodologies and tools for proton beam design for lung tumors. *Int J Radiat Oncol Biol Phys* 2001;49:1429–1438.
75. Jursinic PA, Nelms BE. A 2-D diode array and analysis software for verification of intensity modulated radiation therapy delivery. *Med Phys* 2003;30:870–879.
76. Wroe AJ, Schulte RW, Barnes S, McAuley G, Slater JD, Slater JM. Proton beam scattering system optimization for clinical and research applications. *Med Phys* 2013;40:041702.
77. Fredriksson A, Bokrantz R. A critical evaluation of worst case optimization methods for robust intensity-modulated proton therapy planning. *Med Phys* 2014;41:081701.
78. Bokrantz R. Multicriteria optimization for volumetric-modulated arc therapy by decomposition into a fluence-based relaxation and a segment weight-based restriction. *Med Phys* 2012;39:6712–6725.
79. Dowdell S, Grassberger C, Sharp GC, Paganetti H. Interplay effects in proton scanning for lung: A 4D Monte Carlo study assessing the impact of tumor and beam delivery parameters. *Phys Med Biol* 2013;58:4137–4156.
80. Kimstrand P, Traneus E, Ahnesjo A, Grusell E, Glimelius B, Tilly N. A beam source model for scanned proton beams. *Phys Med Biol* 2007;52:3151–3168.
81. Paganetti H. Monte Carlo method to study the proton fluence for treatment planning. *Med Phys* 1998;25:2370–2375.

82. Widesott L, Lomax AJ, Schwarz M. Is there a single spot size and grid for intensity modulated proton therapy? Simulation of head and neck, prostate and mesothelioma cases. *Med Phys* 2012;39:1298–1308.
83. Matsuura T, Miyamoto N, Shimizu S, Fujii Y, Umezawa M, Takao S, Nihongi H, Toramatsu C, Sutherland K, Suzuki R, Ishikawa M, Kinoshita R, Maeda K, Umegaki K, Shirato H. Integration of a real-time tumor monitoring system into gated proton spot-scanning beam therapy: An initial phantom study using patient tumor trajectory data. *Med Phys* 2013;40:071729.
84. Moravek Z, Rickhey M, Hartmann M, Bogner L. Uncertainty reduction in intensity modulated proton therapy by inverse Monte Carlo treatment planning. *Phys Med Biol* 2009;54:4803–4819.
85. Seregni M, Cerveri P, Riboldi M, Pella A, Baroni G. Robustness of external/internal correlation models for real-time tumor tracking to breathing motion variations. *Phys Med Biol* 2012;57:7053–7074.
86. IAEA. Absorbed dose determination in external beam radiotherapy: TRS 398, 2000.
87. Jakel O, Hartmann GH, Karger CP, Heeg P, Vatnitsky S. A calibration procedure for beam monitors in a scanned beam of heavy charged particles. *Med Phys* 2004;31:1009–1013.
88. Paganetti H. Monte Carlo calculations for absolute dosimetry to determine machine outputs for proton therapy fields. *Phys Med Biol* 2006;51:2801–2812.
89. Safai S, Lin S, Pedroni E. Development of an inorganic scintillating mixture for proton beam verification dosimetry. *Phys Med Biol* 2004;49:4637–4655.
90. Pedroni E, Scheib S, Bohringer T, Coray A, Grossmann M, Lin S, Lomax A. Experimental characterization and physical modelling of the dose distribution of scanned proton pencil beams. *Phys Med Biol* 2005;50:541–561.
91. Boon SN, van Luijk P, Schippers JM, Meertens H, Denis JM, Vynckier S, Medin J, Grusell E. Fast 2D phantom dosimetry for scanning proton beams. *Med Phys* 1998;25:464–475.
92. Fetal S, van Eijk CWE, Fraga F, de Haas J, Kreuger R, van Vuure TL, Schippers JM. Dose imaging in radiotherapy with an ar–cf4 filled scintillating gem. *Nucl Instrum Methods Phys Res A*. 2003;513:42–46.
93. Seravalli E, de Boer M, Geurink F, Huizenga J, Kreuger R, Schippers JM, van Eijk CW, Voss B. A scintillating gas detector for 2D dose measurements in clinical carbon beams. *Phys Med Biol* 2008;53:4651–4665.
94. Herzen J, Todorovic M, Cremers F, Platz V, Albers D, Bartels A, Schmidt R. Dosimetric evaluation of a 2D pixel ionization chamber for implementation in clinical routine. *Phys Med Biol* 2007;52:1197–1208.
95. Poppe B, Blechschmidt A, Djouguela A, Kollhoff R, Rubach A, Willborn KC, Harder D. Two-dimensional ionization chamber arrays for imrt plan verification. *Med Phys* 2006;33:1005–1015.
96. Stasi M, Giordanengo S, Cirio R, Boriano A, Bourhaleb F, Cornelius I, Donetti M, Garelli E, Gomola I, Marchetto F, Porzio M, Sanz Freire CJ, Sardo A, Peroni C. D-IMRT verification with a 2D pixel ionization chamber: Dosimetric and clinical results in head and neck cancer. *Phys Med Biol* 2005;50:4681–4694.
97. Gueli AM, De Vincolis R, Kacperek A, Troja SO. An approach to 3D dose mapping using GafChromic film. *Radiat Prot Dosimetry* 2005;115:616–622.
98. Troja SO, Egger E, Francescon P, Gueli AM, Kacperek A, Coco M, Musmeci R, Pedalino A. 2D and 3D dose distribution determination in proton beam radiotherapy with GafChromic film detectors. *Technol Health Care* 2000;8:155–164.
99. Daftari I, Castenadas C, Petti PL, Singh RP, Verhey LJ. An application of GafChromic md-55 film for 67.5 MeV clinical proton beam dosimetry. *Phys Med Biol* 1999;44:2735–2745.
100. Nichiporov D, Kostjuchenko V, Puhl JM, Bensen DL, Desrosiers MF, Dick CE, McLaughlin WL, Kojima T, Coursey BM, Zink S. Investigation of applicability of alanine and radiochromic detectors to dosimetry of proton clinical beams. *Appl Radiat Isot* 1995;46:1355–1362.
101. Onori S, d'Errico F, De Angelis C, Egger E, Fattibene P, Janovsky I. Alanine dosimetry of proton therapy beams. *Med Phys* 1997;24:447–453.
102. Vatnitsky SM, Schulte RW, Galindo R, Meinass HJ, Miller DW. Radiochromic film dosimetry for verification of dose distributions delivered with proton-beam radiosurgery. *Phys Med Biol* 1997;42:1887–1898.

103. Vatnitsky SM. Radiochromic film dosimetry for clinical proton beams. *Appl Radiat Isot* 1997;48:643–651.
104. Arjomandy B, Tailor R, Zhao L, Devic S. Ebt2 film as a depth-dose measurement tool for radiotherapy beams over a wide range of energies and modalities. *Med Phys* 2012;39:912–921.
105. Arjomandy B, Tailor R, Anand A, Sahoo N, Gillin M, Prado K, Vicic M. Energy dependence and dose response of GafChromic ebt2 film over a wide range of photon, electron, and proton beam energies. *Med Phys* 2010;37:1942–1947.
106. Martisikova M, Jakel O. Dosimetric properties of GafChromic ebt films in monoenergetic medical ion beams. *Phys Med Biol* 2010;55:3741–3751.
107. Reinhardt S, Hillbrand M, Wilkens JJ, Assmann W. Comparison of GafChromic ebt2 and ebt3 films for clinical photon and proton beams. *Med Phys* 2012;39:5257–5262.
108. Richley L, John AC, Coomber H, Fletcher S. Evaluation and optimization of the new ebt2 radiochromic film dosimetry system for patient dose verification in radiotherapy. *Phys Med Biol* 2010;55:2601–2617.
109. Sorriaux J, Kacperek A, Rossomme S, Lee JA, Bertrand D, Vynckier S, Sterpin E. Evaluation of GafChromic ebt3 films characteristics in therapy photon, electron and proton beams. *Phys Med* 2012;(6):599–606.
110. Zhao L, Das IJ. GafChromic ebt film dosimetry in proton beams. *Phys Med Biol* 2010;55:N291–301.
111. Arjomandy B, Sahoo N, Ciangaru G, Zhu R, Song X, Gillin M. Verification of patient-specific dose distributions in proton therapy using a commercial two-dimensional ion chamber array. *Med Phys* 2010;37:5831.
112. Zhu XR, Poenisch F, Song X, Johnson JL, Ciangaru G, Taylor MB, Lii M, Martin C, Arjomandy B, Lee AK, Choi S, Nguyen QN, Gillin MT, Sahoo N. Patient-specific quality assurance for prostate cancer patients receiving spot scanning proton therapy using single-field uniform dose. *Int J Radiat Oncol Biol Phys* 2011;81:552–559.
113. Kooy HM, Schaefer M, Rosenthal S, Bortfeld T. Monitor unit calculations for range-modulated spread-out Bragg peak fields. *Phys Med Biol* 2003;48:2797–2808.
114. Zhao Q, Wu H, Cheng CW, Das IJ. Dose monitoring and output correction for the effects of scanning field changes with uniform scanning proton beam. *Med Phys* 2011;38:4655–4661.
115. Newhauser W, Fontenot J, Koch N, Dong L, Lee A, Zheng Y, Waters L, Mohan R. Monte Carlo simulations of the dosimetric impact of radiopaque fiducial markers for proton radiotherapy of the prostate. *Phys Med Biol* 2007;52:2937–2952.
116. Herault J, Iborra N, Serrano B, Chauvel P. Monte Carlo simulation of a proton therapy platform devoted to ocular melanoma. *Med Phys* 2005;32:910–919.
117. Ciangaru G, Yang JN, Oliver PJ, Bues M, Zhu M, Nakagawa F, Chiba H, Nakamura S, Yoshino H, Umezawa M, Smith AR. Verification procedure for isocentric alignment of proton beams. *J Appl Clin Med Phys* 2007;8:2671.
118. Low DA, Harms WB, Mutic S, Purdy JA. A technique for the quantitative evaluation of dose distributions. *Med Phys* 1998;25:656–661.
119. Li H, Zhang L, Dong L, Sahoo N, Gillin MT, Zhu XR. A CT-based software tool for evaluating compensator quality in passively scattered proton therapy. *Phys Med Biol* 2010;55:6759–6771.
120. Walker PK, Edwards AC, Das IJ, Johnstone PA. Radiation safety considerations in proton aperture disposal. *Health Phys* 2014;106:523–527.
121. Zhao Q, Wu H, Das IJ. Quality assurance of proton compensators. In: Long M, Ed. World Congress on Medical Physics and Biomedical Engineering. China. Berlin: Springer, 2012, pp. 1719–1722.
122. Yoon M, Kim JS, Shin D, Park SY, Lee SB, Kim DY, Kim T, Shin KH, Cho KH. Computerized tomography-based quality assurance tool for proton range compensators. *Med Phys* 2008;35:3511–3517.
123. Li, H, Zhang L, Dong L, Sahoo N, Gillin MT, Zhu XR. A CT-based software tool for evaluating compensator quality in passively scattered proton therapy. *Phys Med Biol* 2010;55:6759–6771.
124. Lomax AJ, Bohringer T, Bolsi A, Coray D, Emert F, Goitein G, Jermann M, Lin S, Pedroni E, Rutz H, Stadelmann O, Timmermann B, Verwey J, Weber DC. Treatment planning and verification of proton therapy using spot scanning: Initial experiences. *Med Phys* 2004;31:3150–3157.

125. Archambault L, Poenisch F, Sahoo N, Robertson D, Lee A, Gillin MT, Mohan R, Beddar S. Verification of proton range, position, and intensity in IMPT with a 3D liquid scintillator detector system. *Med Phys* 2012;39:1239–1246.
126. Lopatiuk-Tirpak O, Su Z, Li Z, Hsi W, Meeks S, Zeidan O. Spatial correlation of proton irradiation-induced activity and dose in polymer gel phantoms for PET/CT delivery verification studies. *Med Phys* 2011;38:6483–6488.
127. Zeidan OA, Sriprisan SI, Lopatiuk-Tirpak O, Kupelian PA, Meeks SL, Hsi WC, Li Z, Palta JR, Maryanski MJ. Dosimetric evaluation of a novel polymer gel dosimeter for proton therapy. *Med Phys* 2010;37:2145–2152.
128. Grevillot L, Bertrand D, Dessy F, Freud N, Sarrut D. Gate as a GEANT4-based Monte Carlo platform for the evaluation of proton pencil beam scanning treatment plans. *Phys Med Biol* 2012;57:4223–4244.

Chapter 17

Proton Therapy Workflow

Martijn Engelsman, Ph.D.[1], Juliane Daartz, Ph.D.[2], and James E. McDonough, Ph.D.[3]

[1]HollandPTC and Delft University of Technology, Delft, The Netherlands
[2]Department of Radiation Oncology, Francis H. Burr Proton Therapy Center, Massachusetts General Hospital, Boston, MA
[3]Department of Radiation Oncology, University of Pennsylvania, Philadelphia, PA

17.1 Introduction

It has been more than 60 years since the first proton therapy treatment of a cancer patient, and the field has seen many and marked advances since computed tomography for treatment planning, image-guided patient setup, new modalities such as uniform scanning and pencil beam scanning (PBS), to name a few. From the pioneering days of building, maintaining, and operating your own beam line, there are now more than 10 vendors, with some offering the whole integrated package of a treatment planning system, oncology information system (OIS), and proton therapy equipment. Pencil beam scanning is gaining momentum, and volumetric imaging has just entered the proton therapy treatment room. Proton therapy is a mature radiotherapy treatment modality.

It is, however, also fair to say that the field of proton therapy has not been able to keep up with many of the advances in the field of photon therapy. Even now, with more than a dozen proton therapy centers operational within the United States and an even larger number of centers in varying stages of development, proton therapy remains a niche treatment. It has, therefore, been lacking the benefits of a large user base. Until recently, this has translated into limited interest of companies to invest in this market. It also means that there was only a relatively small group of medical professionals continuously improving on existing technology and methods by means of a research and development program.

Now that the interest in proton therapy has significantly increased, both users and vendors are aiming for a competitive advantage. These vendors realize that proton therapy equipment alone is not sufficient to allow an accurate and efficient delivery of proton therapy treatments. Potential customers expect superior dose distributions to be prepared, delivered, and adapted at least as efficiently as they are used to from their photon therapy experience. Vendors and users are collaborating on further improving proton therapy solutions, but there still is a long way to go.

Let us briefly outline the proton therapy workflow challenge by means of Figure 17–1. Compared to photon therapy, the typical patient mix at a proton therapy institute is more challenging. Proton therapy, therefore, also has an increased need for applying advanced techniques such as 3D image guidance and adaptive radiotherapy. The overriding reason for this, as is explained in Chapter 21 (Uncertainty in Planned Dose Distributions), is the finite range of protons. These two features would pose an interesting challenge even with technology and software integration as available in photon therapy. Proton therapy technology is, however, not nearly at that level (see Figure 17–1). It is our professional duty to deliver the best possible proton therapy treatments to as many patients as possible in a safe and efficient manner. To achieve this, improvements in hardware, software, and integration are needed. This holds for all aspects of the treatment chain, e.g., proton production and delivery technology, treatment planning, image guidance, *in vivo* dosimetry, quality assurance, and integration with the oncology information system. The current proton therapy workflow is very labor-intensive and time-consuming. This burdens the clinic and hampers intensification of adaptive therapy. With further improvements, it is possible to serve the patients entering our clinic today, while also allowing us to use our experience to institute further improvements for future patients.

Proton therapy can deliver very good treatments to select indications with the technology as it is today, and we can foresee many improvements on the horizon. This chapter aims to provide an overview of the current state of the art in proton therapy workflow, and its limitations.

Before this chapter is continued, the authors would like to express their gratitude to 12 of the clinical operational proton therapy facilities in the United States for having participated in a survey to help paint the current state of the art in proton therapy. On average, these centers have 7.5 years of experience in applying proton therapy. Five of these centers started patient treatments in or before 2006, and the other seven

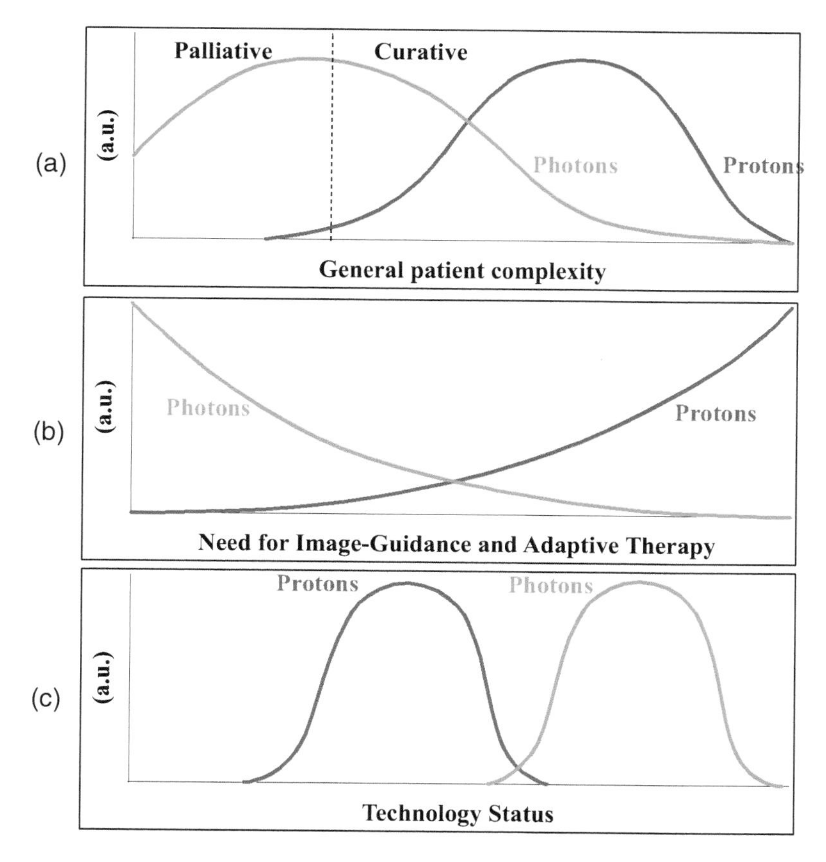

Figure 17–1 Comparing photon therapy to proton therapy regarding (a) relative patient complexity, (b) the relative need for image guidance, and (c) the current state of the technology.

centers began operating in 2010 or later. Results of this survey will be used throughout this chapter, where applicable.

A few aspects will not be covered or be briefly touched upon in this chapter. The treatment of intra-ocular tumors is one of the success stories of proton therapy. It is, however, a very specialized application. In order to not convolute the flow of this chapter, the focus will be on those aspects of proton therapy related to the treatment of deep-seated tumors. Only commercially available platforms will be discussed, as this covers the vast majority of current and future operational centers.

17.2 Patient Referral, Intake, and Complexity

In the very early days of proton therapy, the maximum proton range of a beam line was restricted. This limited the application of proton therapy to intra-cranial indications, or at least to tumors that were located at a relatively shallow depth. Another reason for limiting the choice of indications at that time was that volumetric (CT) imaging and treatment planning based on these images were not yet available to allow a high enough level of confidence in where the dose would actually be deposited, i.e., where the protons would stop. Nowadays all beam lines, whether fixed or on a gantry, have a maximum proton range in water of at least 30 cm, allowing the physician to treat virtually any tumor at any location.

It is, however, not only what protons *can* treat that determines the patient mix. The expected patient mix of a proton therapy center is mainly driven by reimbursement policies [1] and the referral patterns that have been set up. Section 17.10 (facility startup) discusses how the patient mix can vary over time as a proton therapy facility is ramping up from first patient treatment to maximum capacity and maximum experience.

17.2.1 Reimbursement

Judging by international reports on indications for proton therapy, one notices nearly worldwide consensus on three indications for which proton therapy is considered the standard of care. These are intra-ocular tumors, tumors in pediatric patients, and base-of-skull chordomas/chondrosarcomas. Although very little clinical data has been published, many other indications have already been treated at one or more institutes.

The finite range of protons is a powerful tool to reduce the integral dose and the volume of healthy tissue receiving intermediate doses. These benefits are weighed against the higher cost of proton therapy. Simply delivering "less dose" is not a sufficient criterion for recommending proton therapy. At a minimum, the choice of protons rather than photons should result in a clinically relevant benefit. Sufficient clinical evidence of this benefit has only been acquired for a few indications. This is reflected in the reimbursement policy of some of the larger healthcare insurance companies (Table 17–1). The data in this table is only a snapshot in time covering a limited number of insurance companies, and it is only meant to provide a rough indication of reimbursements. Readers are encouraged to carefully analyze their own reimbursement situation.

Table 17–1 Summary of reimbursement policies for a variety of healthcare insurers. The table is based on the Proton Beam Therapy report by the Institute for Clinical and Economic Review [2] which appeared in 2014. A green background means the indication will be reimbursed. Orange indicates limited or conditional reimbursement.

Indication	CMS	Regence Group	Premera Blue Cross	Blue Shield California	Aetna	Anthem Blue Cross Blue Shield	Humana	United Health Care
Abdominal	investigational		investigational	investigational	investigational	investigational		no
Acoustic Neuromas		investigational	investigational	investigational	investigational	investigational		no
AVM		investigational	investigational	investigational	investigational	investigational	yes	yes
Bone Metastasis			investigational	investigational	investigational	investigational		no
Breast	investigational	investigational	investigational	investigational	investigational	investigational		no
Central Nervous System	yes	children	children	children	yes	yes	yes	yes
Cervix			investigational	investigational	investigational	investigational		no
Chordoma/ Chondrosarcoma	yes	yes	yes	yes	yes	yes	yes	yes
Head and Neck	yes	investigational	investigational	investigational	investigational	investigational		no
Hodgkin			investigational	investigational	investigational	investigational		no
Liver	investigational	investigational	investigational	investigational	investigational	investigational		no
Lung			investigational	investigational	investigational	investigational		no
Lymphoma			investigational	investigational	investigational	investigational		no
Meningioma		investigational	investigational	investigational	investigational	investigational	yes	no
Non-Hodgkin			investigational	investigational	investigational	investigational		no
Ocular	yes	yes	yes	yes	yes	investigational	some	yes
Paranasal Sinus	yes		investigational	investigational	investigational	investigational		no
Pediatric	yes		investigational	investigational	yes	yes		no
Pituitary	yes		investigational	investigational	yes	yes	yes	no
Prostate	very few	no	no	no	no	some	some	no
Rectum	investigational		investigational	investigational	investigational	investigational		no
Retroperitoneal Sarcoma	yes		investigational	investigational	yes	yes		no
Sinus Tumors	yes		investigational	investigational	investigational	investigational		no
Skin Cancer	investigational		investigational	investigational	investigational	investigational		no

Nearly all insurers will reimburse the standard indications, but nearly all other indications are "under investigation." This may mean that a treatment will only be reimbursed if it is part of a clinical trial. The proton center will need the resources and ability to set up or participate in such trials. This may require additional effort in, for example, credentialing by Imaging and Radiation Oncology Core (IROC); imaging before, during, and after the treatment course; quality assurance; and follow-up. Obtaining insurance approval for proton therapy itself is a time-consuming and resource-intense process. Some insurance companies may reimburse without question. Others may require a peer-to-peer conversation between physicians, and they may even require a comparative planning study between protons and photons on the level of the individual patient. This has a huge impact on workflow. The facility needs to have the means to create these plans, as related to both personnel and treatment planning software. For stand-alone proton centers that do not offer routine photon treatments as well, this means that basic expertise in photon treatment planning has to be maintained.

Acquiring approval of reimbursement may lead to a delay in treatment start or to a reduction in time available to create a proper proton therapy treatment plan. Especially for patients on clinical trials, the protocol may only allow limited time between, for example, removal of the bulk of a tumor by means of surgical intervention and the start of any form of radiotherapy treatment. Some proton therapy centers do report that patients start with photon treatment because of this need for a quick start, or because insurance approval only comes in after the patient has started their photon treatment.

17.2.2 Referral Patterns

The number of referrals per indication also determines the patient mix. There are few proton therapy facilities that can fill up their capacity by internal referral from their directly affiliated main hospital. External referrals, therefore, partly determine the patient mix. For many indications, it is not obvious even to experienced proton therapy centers which subset of patients has the most, or at least sufficient, clinical advantage from being treated with protons. Once proper triaging is established, the entire radiotherapy community needs to be educated. Without proper understanding of the medical benefits of proton therapy, there may be a barrier for physicians in peripheral centers to transfer responsibility of their patients to a proton therapy center. Moreover, patient referral has an economic impact on the referring center.

In the United States, there is the interesting situation that some proton therapy centers have far too many referrals, while others have too few. Neither situation is ideal. Too few referrals means that equipment and expertise are not used to their full extent, and this could eventually result in the closure of a center. Too many referrals distracts from offering the best possible treatment to those patients who have been accepted for treatment. All patients who are referred to a center need to be triaged as to whether they will benefit from proton therapy. Patients who do not qualify for proton therapy will have to be referred back to photon therapy and may have a delay in their treatment start.

17.2.3 Hospital-based or Not

A proton therapy center can be a completely stand-alone facility, it can be located on the main campus but in a separate building, or it can be physically embedded into the photon radiotherapy department. The distribution of these operational models within the United States is shown in Figure 17–2.

The previous section discussed how a facility's operational model can affect patient referral and intake. Depending on the design of a facility, it will also affect other aspects of the workflow. The Skandion Kliniken in Uppsala will operate on a basis of "distributed competence," where patient intake and treatment preparation will take place in a number of regional medical centers throughout Sweden [3]. Subsequently, the patient, the immobilization, and the treatment data will be transferred to the proton therapy center which, effectively, only performs the irradiation part of the workflow. No U.S. centers follow this approach, but depending on the medical equipment available at a proton therapy center, it may still be necessary to resort to off-site

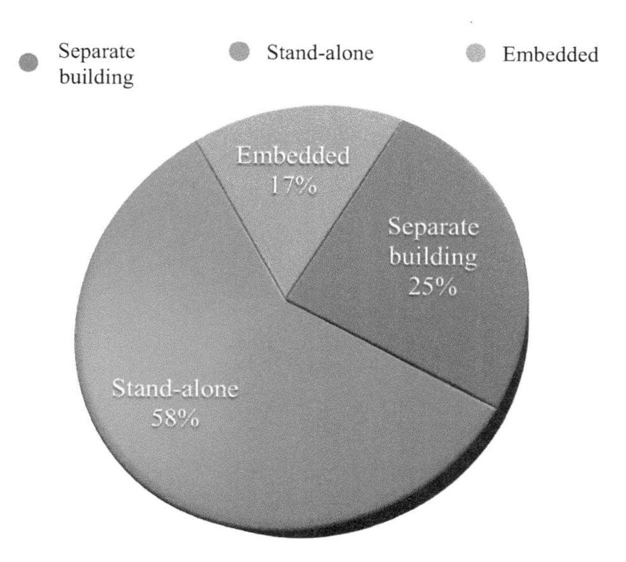

Figure 17–2 Distribution of proton therapy centers in the United States for the 12 centers that participated in the survey. "Embedded" means the center is embedded into the photon radiotherapy department. "Separate" means the center is located on the main campus, but in a separate building. "Stand-alone" means the center is physically and organizationally a separate institute.

imaging for treatment preparation (MRI, PET/CT), and perhaps even for treatment adaptation. Even for a center on the main campus, the acquisition of a treatment planning CT and the preparation of related patient immobilization may take place at some distance from the proton therapy center.

Regarding staffing, centers can choose between having dedicated staffing or sharing personnel between the proton and photon departments. Given the current labor-intensive and time-consuming workflow, it certainly is advantageous for the quality of treatments if staff is dedicated to proton therapy alone. Dedicated staff means that professionals are always on site and able to discuss ongoing patients and issues, even in an ad-hoc manner. This is especially important in the first years of starting up a new proton therapy facility, when the technology is still new and somewhat unknown. It also facilitates the transition to a workflow in which many indications treated with proton therapy will see an increase in treatment adaptation.

Sharing staff will not make the proton therapy workflow easier. It may, however, be beneficial for internal patient referral, and it may allow both proton and photon modalities to learn more effectively from each other's know-how, developments, and research.

The results of the survey show no unified personnel approach based on the physical location of the proton center, although some tendencies are apparent. Physicians and physicists tend toward being dedicated personnel in stand-alone facilities, but they are shared personnel with the photon department otherwise. Dosimetrists are typically dedicated to a stand-alone facility and shared otherwise. Therapists have a strong tendency to have dedicated personnel for proton therapy. Personnel related to information technology (IT) are dedicated in case of a stand-alone facility and shared otherwise.

17.2.4 Case and Plan Complexity

Whereas photon therapy is frequently applied with palliative intent, nearly 100% of proton therapy patients will be treated with curative intent, where the aim is maximum care in preventing unacceptable side effects, both short-term and long-term. Obviously, very complex cases are also treated in any photon therapy department and require additional effort, such as daily cone-beam CT or volumetric dose guidance. But the proportion of patients requiring such extensive care in photon therapy is, at the moment, relatively small, except perhaps for some of those radiotherapy departments embedded in a university medical center.

The reimbursement policies of the healthcare insurers, as well as the established referral patterns, have a direct impact on the plan complexity of those patients referred to proton therapy (Figure 17–1a). If a photon treatment plan shows a realistic chance of cure in combination with an acceptable probability of complications, it is not straightforward to still refer the patient for proton therapy to possibly receive an even better treatment. A fraction of the patients who are referred to proton therapy will, therefore, consist of patients who are simply too difficult for photon therapy, i.e., the challenging cases. Such challenging cases could include a tumor located very close to multiple organs at risk (OAR) or patients with local recurrences or second primaries where there is the need to minimize dose to a previously irradiated region. It will take a number of years for proton centers (and those referring patients to them) to fully understand the limits of proton therapy and to know which patients can be treated successfully with photons, but even better with protons.

The professionals working in proton therapy also expect miracles. Even if the patient geometry is not overly challenging, the tendency in proton therapy is still to make an all-out effort to ensure that maximum benefit is obtained. This may result in more challenging dose constraints, more boost volumes, and more dose prescription levels.

The typical proton therapy patient has a sizable tumor located in the head-and-neck region with difficult-to-meet constraints for the nearby OAR. Pediatric patients under anesthesia, especially when treated for medulloblastoma, are, by far, the most complex patients to treat, requiring the most care and personnel involvement. Lung tumors are probably the next most difficult to treat due to respiratory motion and tumor location in a region of substantial density heterogeneities.

At the moment, there are also patients for whom treatment preparation and treatment delivery are relatively simple. Patients with localized prostate cancer, for example, are typically treated with two lateral treatment fields applying generous safety margins both in the lateral and distal/proximal directions. Creating such a treatment plan and its related paperwork requires only a few hours. In-room time, including patient alignment and dose delivery, may require only 10 minutes per treatment fraction. The clinical benefit of proton therapy for these indications, or at least of the currently applied proton therapy techniques, is questioned by insurance companies. It is also under investigation, for example in a phase III randomized clinical trial by the Massachusetts General Hospital and the University of Pennsylvania [4]. Depending on the outcome of this trial, reimbursement for these treatments may be withdrawn or the treatment approach may drastically alter. This would imply the use of more advantageous beam angles in combination with intensity-modulated proton therapy (IMPT) and, as a consequence, much more complex imaging and on-line dose verification procedures.

Complex cases put high demands on every aspect of the treatment chain, from target and OAR delineation to creating multiple treatment plans for the physician to choose from. Complex cases may require creating patient-specific devices, or they may need a quick start with a "temporary" treatment plan just to get the treatment started, followed by additional treatment plans. Steep dose gradients needed near OAR require image guidance and adaptive therapy to be the standard of care (Figure 17–1b).

17.3 Treatment Planning

Treatment planning methodologies for passive-scattering, uniform scanning, and intensity-modulated particle therapy (IMPT) are discussed in Chapter 23 (Treatment Planning) and Chapter 24 (Intensity-modulated Proton Therapy). This current section will only discuss the workflow aspects of basic treatment planning, with treatment adaptation subsequently discussed in Section 17.8.

17.3.1 Technical Capabilities

Just as in photon therapy, but perhaps to a lesser extent, proton therapy has seen its share of home-grown treatment planning systems (TPS). Home-grown systems came about mostly because there were no commercial packages available, but they also allowed the institute to directly control the capabilities of the treatment planning system such that it better conformed to the needs of the clinic. Ten of the proton therapy institutes surveyed report they are using a commercial software package for treatment planning. Sometimes they have a second system on site, presumably for testing purposes, or it may have been acquired with the aim of near-future transition. One institute performs the bulk of its treatment planning on a commercial system, while also having developed their own software package. Only one institute reports using their own treatment planning system exclusively.

There are a few reasons why a proton therapy treatment planning system may not be as complete and efficient a tool as it should be. For one, it is unrealistic to expect companies to put together a very large development team to serve the relatively small market that proton therapy still is. But perhaps more importantly, there hasn't really been a consensus in the proton therapy community as to what tools should truly be part of a treatment planning package, and what these ideal treatment planning tools should look like. To put this somewhat bluntly, there have been more scientific papers written about repainting strategies than there have been patients actually treated with repainting. So, which strategy should a vendor implement into their TPS or their treatment delivery system (TDS)? Given the number of vendors of proton therapy planning software has increased from two to four in recent years, the additional competition will lead to an increased pace of improvements.

Features that can reduce the workload of treatment planning are auto-patching and auto matching for patched and matched field combinations, automated beam angle optimization, multi-criteria optimization (see Chapter 22, Plan Optimization), and scripting. Scripting will be a time-saver, both for treatment plan design and for the export of plan paperwork.

An important technical feature of proton therapy nozzles is that they typically allow translation along the beam axis of field-specific apertures or range shifters. This is in order to minimize the so-called air gap, thereby keeping the lateral beam penumbra small. The "extension" of the snout supporting these devices is set during the treatment planning phase. The complete geometry of the patient and, more importantly, of the snout are not exactly known to the treatment planning system, and it requires significant expertise of the dosimetrist, or an in-room simulation, to accurately select this snout extension. As far as collisions are concerned, the problem is somewhat similar to when performing a couch rotation during a photon treatment. But in proton therapy, the difficulty also holds for the majority of fields that have no couch rotation. None of the treatment planning systems yet offers a fool-proof solution. From a practical point of view, a single treatment fraction can be delivered with the hardware at the incorrect distance, but a redesign of the treatment field is then indicated.

Another tool that is missing, and which is important in the case of single-field uniform dose (SFUD) treatments, has to do with the field-set of the day. A head and neck treatment may consist from 6 to 15 fields, of which typically only two fields are treated on any given day. What combination of fields is treated on which day is up to the dosimetrist to determine and to typically enter manually into the oncology information system. This requires extensive bookkeeping, especially when a plan adaptation is required, and it makes a proper physics check of the treatment plan very cumbersome.

Figure 17–3 shows the proton therapy treatment planning packages currently in use in the United States. Both Elekta XiO and Varian Eclipse have been in clinical use for many years and have the largest user base. Philips Pinnacle and Raystation by Raysearch are relative newcomers.

17.3.2 Treatment Planning Efficiency

In a photon therapy department, the number of patients under treatment per year per fulltime equivalent (FTE) dosimetrist may be 200 up to even 300. This is at least partially explained by the proportion of palliative treatments. But it is also true that the

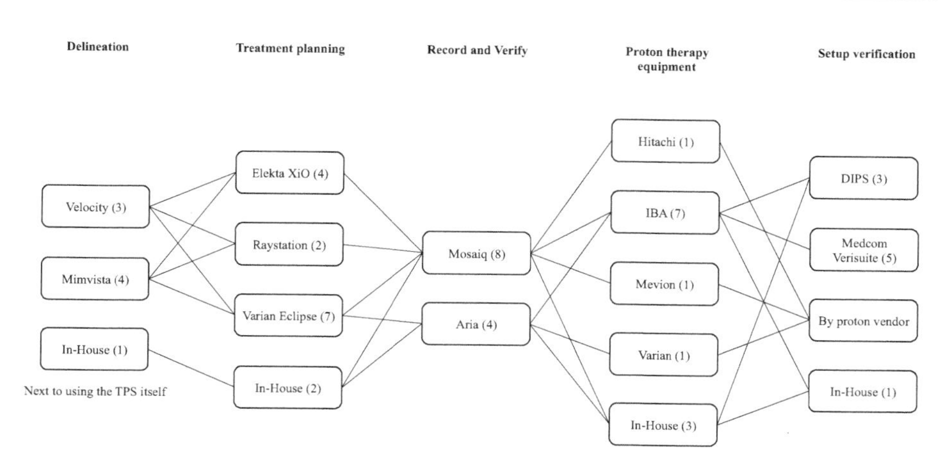

Figure 17–3 Software packages used in the various phases of treatment preparation and treatment delivery, as well as suppliers of proton therapy delivery equipment used at current operational centers in the United States. The connecting lines indicate the various combinations that are used in clinical practice. Some facilities use more than one treatment planning system or have a system that is partially commercial and partially home grown.

treatment planning software is better tailored to the typical photon therapy workflow. In proton therapy, typical patient numbers per dosimetrist are substantially lower [5].

The survey results show numbers from around 60 to 110 patients per year per FTE proton dosimetrist. On average, a proton therapy dosimetrist will complete less than two treatment courses per week! Without technological software improvements, this number will be even lower given the expected increase in treatment adaptation in proton therapy. There are a number of contributing factors. Patient complexity and the desire to squeeze the last possible benefit out of a proton therapy treatment have already been mentioned. At the moment, the majority of patients are still treated with passive-scattering or uniform scanning. Typically this requires laborious manual tweaking of apertures and range compensators. The advent of pencil beam scanning may reduce this workload significantly by automating the treatment planning process, but it may add manpower for quality assurance (QA) purposes. However, in photon therapy the introduction of IMRT did not so much decrease the workload as lead to an increase in plan quality and complexity.

Figure 17–4a shows the ideal workflow for treatment planning and delivery—a patient is referred, all required anatomical structures are delineated, and a definitive prescription is handed over by the physician to the dosimetrist. An initial treatment plan and two cone-downs are created. These are delivered during seven weeks of treatment delivery, with no adaptations needed as the patient anatomy doesn't change. In this scenario, physicians, dosimetrists, and even physicists focus their full attention

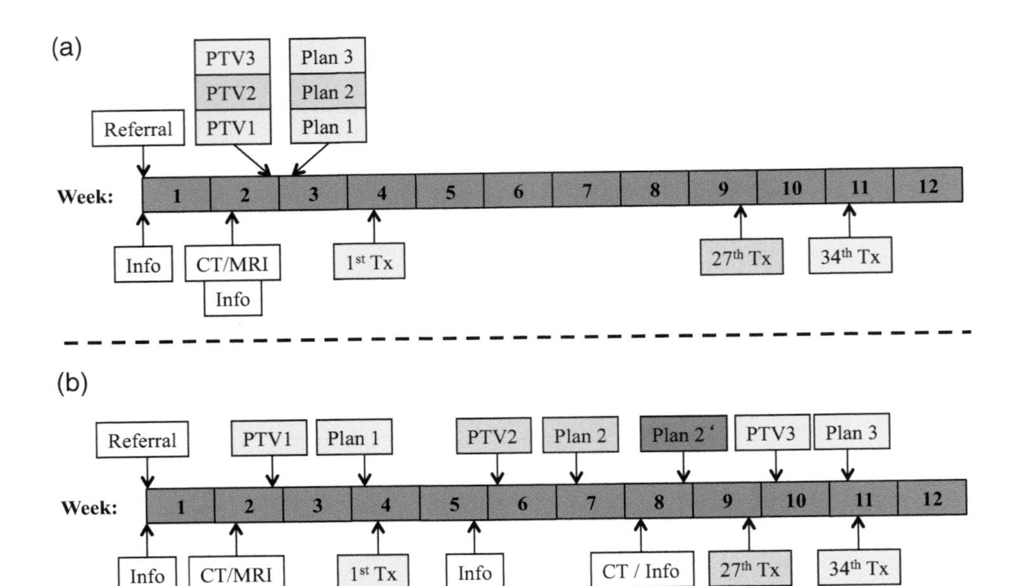

Figure 17–4 Two scenarios of treatment planning and treatment delivery workflow. (a) An ideal scenario. (b) An actual proton therapy scenario (with or without treatment adaptation).

on this patient during a brief period of time, subsequently moving their attention to the next patient.

In proton therapy, the workflow—even with no or very little treatment adaptation—tends to be represented by Figure 17–4b. Sometimes a "temporary" treatment plan (photons, or protons with a reduced number of fields) is designed and delivered for a few fractions to make up for time lost in patient referral. Sometimes the complexity of the patient may require additional imaging or other medical assessments, which will only arrive after the patient has already started treatment. Extension of the treatment preparation over a prolonged time period means that physicians and dosimetrists are continuously switching their attention back and forth between patients. From a workflow perspective, this is undesirable. A plan that is under continuous development also makes a proper physics check and peer-to-peer physician review difficult. Faster treatment planning software and procedures mitigate this.

17.3.3 Photon Back-up Planning Due to Facility Down Time

Depending on the achieved up-time for a proton facility, it may be advisable to create photon backup plans. In the very early years of clinical operation at the Francis H. Burr proton therapy center, for example, the treatment approach for a substantial portion of patients included up to 10 fractions of photon therapy. Some of these fractions were scheduled to be delivered at the end of their treatment schedule, to be moved forward in time in case of prolonged machine downtime. A second reason for these photon fractions at that time was to allow the start of radiotherapy treatment while waiting for available treatment slots on the proton gantries. A third reason for including photon treatment fractions was skin sparing for superficial lesions-for head and neck cancers and sarcomas, for example. Passive scattering delivers full dose to the skin and may even result in hotspots because of scattering effects of the range compensator and the patient surface. A fourth reason was to balance uncertainties in proton dose calculation due to the presence of titanium hardware (see Chapter 21, Uncertainty in Planned Dose Distributions).

With the uptime of proton therapy machines at 95% or higher [6], it appears to be an institution-specific choice whether or not to have photon backup plans available, and for which patients. For some cancer indications, the attending physician may be more willing to accept a brief treatment interruption than for other patients. A choice in favor of backup planning requires there to be the personnel and expertise to make these plans, either a priori or on the fly. It also requires access to photon therapy linear accelerators to treat the patient, which may be a challenge for a stand-alone facility. And, though certainly not easy or always possible, it is very useful to have as accurate an estimate as possible of how long the downtime is expected to last.

17.4 The Oncology Information System

Just as in photon therapy, the oncology information system (OIS) is the central hub between treatment preparation and treatment delivery (see Figure 17–3). The most important task of the OIS is to verify if the treatment delivery system has set the treatment parameters (beam angle, beam energy, monitor units, etc.) correctly and to track overall progress over the course of treatment (correct fields used on the correct day to the correct dose). It can also contain patient or fraction-specific site setup information, such as reference setup images and immobilization devices. It may furthermore be used for patient scheduling, workflow tracking, and quality assurance workflow support.

In photon therapy, there are only two main vendors of linear accelerators and many thousands of users. The communication between OIS and TDS has been tried and tested, and the TPS vendors make sure they fully support both of these vendors. Although a photon therapy department typically will have developed a few software routines to manipulate some aspects of the treatment data one way or another (e.g., for automated plan checking), the existing treatment management chain of TPS, OIS, and TDS allows a smooth workflow [7–9]. In contrast, proton therapy's digital connectivity that replaces tedious manual data entry has only become a priority in very recent years.

The quality, user friendliness, safety, and ease of use of the treatment management chain are best illustrated by quoting the experience of some current users. When asked, "Are you happy about the electronic integration between TPS, OIS, and TDS?", they replied:

> "We do a lot of in-house manipulations to make our OIS work with protons."

"There is no integration whatsoever."

"No way this system can be integrated."

"The integration between TPS and OIS is acceptable. The challenge is mainly in the communication between OIS and TDS."

One facility reports no connection whatsoever between OIS and TDS, but the difficulty could be in any part of the treatment management chain. Out of the 12 centers surveyed, only three responded that integration was acceptable. This is partially explained by the large variation in systems available for proton therapy (Figure 17–3) and, again, by the small user base. Even with DICOM-RT-ION as a standard for digital communication, successful integration means that both the OIS and the TPS are aware of all of the intricacies of the treatment delivery system. This requires a substantial effort from and clear communication between vendors. A few examples of what may not be supported by a typical proton therapy treatment management chain are:

- Handling of minimum and maximum spot weights, which in a worst case can prevent treatment only at the moment the data is sent to the TDS on the day of treatment.
- Field-set of the day, which requires manual bookkeeping.
- Handling of field-specific hardware. Both passive-scattering and uniform scanning-and perhaps some pencil beam scanning-require a field-specific aperture and range compensator to be attached to the treatment nozzle. Ideally, these are interlocked such that the OIS can prevent treatment. Next best is a system with barcode readers, a practice which has been successfully implemented at various proton therapy facilities. Ideally, these unique barcodes would be generated automatically by the TPS, rather than requiring manual interaction with a home-grown software system.
- Use of varying spot sizes within a single field. Several research papers discuss the benefit of being able to switch PBS spot sizes within a range layer. This feature is not covered by DICOM-RT-ION, except in a roundabout way.

The software packages in use in an affiliated photon department may determine the combination of packages that makes up the proton therapy treatment management chain. Although there is no guarantee that a desired combination of software and hardware will allow the desired smooth workflow, a large variety of treatment management chains have been tested and made to work in clinical practice (Figure 17–3). We would like to stress the safety-related aspect. Despite the efforts by the proton therapy users and vendors, the prevailing atmosphere at the moment is one of "making it work." This and the small user base are two very good reasons to pay close attention to the upcoming TG-100 on quality assurance needs [10] and TG-201 on data transfer QA [11]. Section 17.9 discusses hazard analysis.

17.4.1 The Paperless Chart

Even though the actual integration between TPS, OIS, and TDS is not ideal, 9 out of 12 proton therapy departments report having a fully paperless chart. This is perhaps even a higher proportion than currently achieved in photon therapy. Just as in photon therapy, however, it may mean that a portion of the information is printed to PDF and subsequently imported, manually or automatically, into a "general document repository." This means that the PDF of a treatment plan (with screen grabs, DVHs, etc.) to the OIS has the same intrinsic meaning as the PDF of a monitor unit (MU) sheet. Figuring out the relevant paperwork for a specific treatment fraction, especially after some treatment adaptations, requires an advanced degree.

A paperless department does not mean that all data transfer is electronic, let alone automatic, or that data validation is performed automatically. A typical exception to automatic data transfer is the manual entering of monitor units in the case of fields to be delivered using passive-scattering. Although a handful of the surveyed facilities report automated electronic data transfer, all of the facilities report the use of human verification by one, two, or even more independent people. Non-automated data transfer and human verification will become even more burdensome with the expected increase in proton treatment adaptation. Again, the reader is directed toward the upcoming TG-201 on data transfer QA [11].

The operational centers were also surveyed as to who is required to validate and sign which parts of the treatment plan paperwork (Figure 17–5).

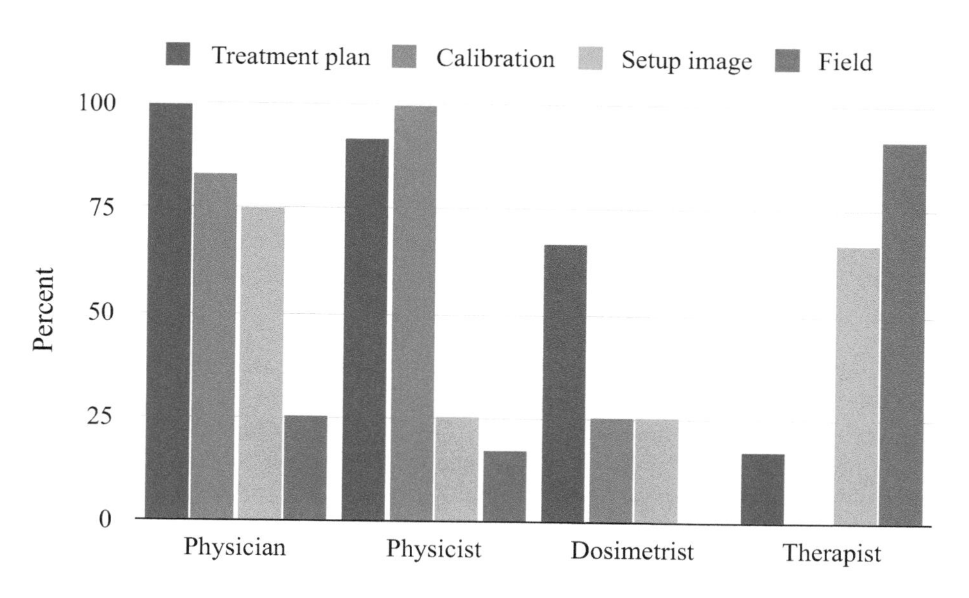

Figure 17–5 An overview of signatures required for various aspects of proton treatment preparation and delivery. The numbers indicate the percentage out of the 12 surveyed centers. "Treatment plan" denotes each individual treatment plan or cone-down. "Calibration" is related to the number of monitor units delivered for each field (either the calibration or the validation thereof). "Setup image" means the final images acquired prior to delivery beam-on. "Field" denotes a signature for each delivered treatment field.

17.5 Image Guidance

With the technical aspects of image guidance covered in Chapter 7 (Proton-specific Imaging), this section will only cover the workflow aspects. Orthogonal radiographic imaging for setup verification is available at every currently operational proton therapy center. Volumetric imaging for the same purpose has, however, only very recently successfully been introduced into the clinic. All proton therapy vendors are actively developing it or are at a point where clinical implementation is possible. This mostly pertains to isocentric cone-beam computed tomography (CBCT). Some vendors are by now also offering in-room CT, though none have actually implemented it.

Up to now, volumetric imaging was only available as part of the proton therapy treatment delivery workflow using a trolley system. Such a system allows the patient to be imaged outside the treatment bunker and then treated using the same treatment couch. There are claims that a trolley system allows a substantial increase in patient throughput by minimizing in-room setup time, but there is only limited evidence to support this [12]. Moving certain activities out of the treatment room reduces time pressure in preparing for treatment delivery and allows more time for interaction with the patient. The increased time between imaging and treatment is, however, burdensome for the patient and may increase patient motion between imaging and treatment. Only one U.S. facility has a trolley system, but it is not yet in clinical use.

Volumetric imaging certainly has to become an integral part of the proton therapy workflow sooner rather than later. It is a prerequisite for meaningful adaptive proton therapy and allows us to compete with the rapidly changing image guidance developments in photon therapy. Next to the introduction of this technology, most vendors now also offer the possibility of remote operation of the in-room imaging capabilities. This allows for a smoother workflow and time savings.

The surveyed centers provided typical in-room setup times, including initial setup verification, for a variety of indications-pediatric cranio-spinal irradiation (with and without anesthesia), intra-cranial, gastro-intestinal, prostate, head and neck, thorax, and stereotactic body radiotherapy. Not all facilities treat all of these indications. The reported data are plotted in Figure 17–6. Only for special cases, such as an SBRT treatment or treatment of a child under anesthesia, will patient immobilization and setup verification take 25 minutes or more. Setup is typically achieved in between 10 and 15 minutes, which is not much slower than the 7.5 minutes achieved in state-of-the-art photon therapy with patient alignment based on volumetric imaging [13].

On average, the reported time per treatment fraction after initial setup and initial setup verification is 18 minutes. This brings the total average total in-room time per fraction to about 30 minutes (see Figure 17–6b), which is longer than in photon therapy. Reasons for the extended in-room time per patient include the necessity to swap the field-specific apertures and range compensators, the sharing of the beam with other treatment rooms, or the imaging procedure itself. The improvements toward remote control of gantry rotation, couch rotation, and setup verification imaging will bring the average time for a treatment fraction closer to what is achieved in photon therapy. So will the continuous effort to reduce layer switching times in PBS.

Another reason for the extended fraction times can be that just over half of the centers verify the patient position prior to every field in a fraction, regardless the indication (Figure 17–7). This is very different from photon therapy experience. Three centers perform setup verification only at the start of a treatment fraction for all indications treated. None of the proton therapy centers uses a setup protocol [14,15] to

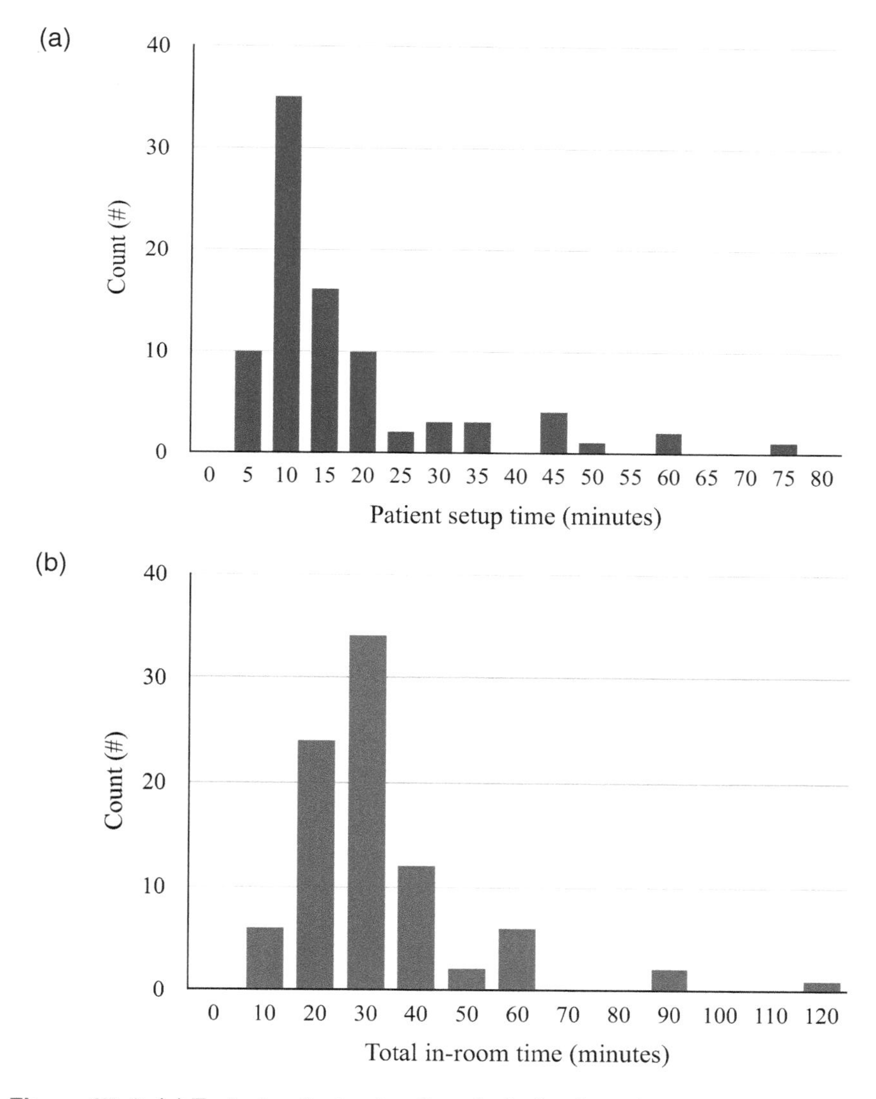

Figure 17–6 (a) Typical patient setup time, including immobilization and setup verification, for a variety of cancer indications (see text). (b) Typical total in-room time per fraction for the same indications.

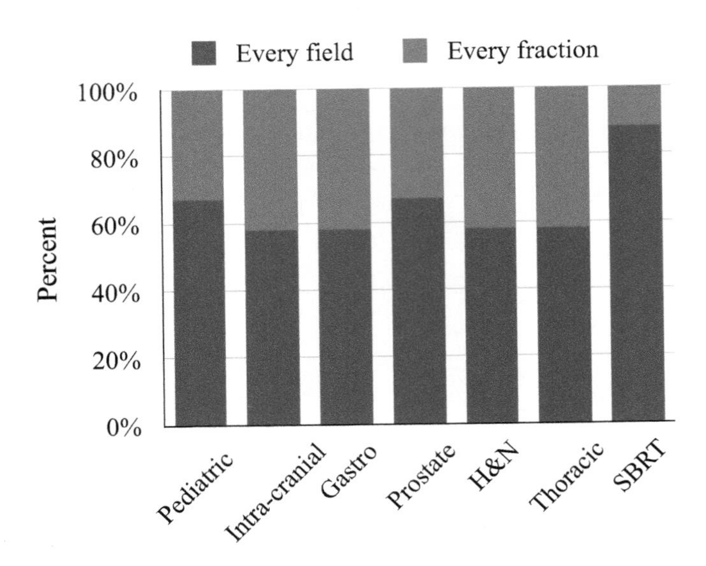

Figure 17–7 Frequency of imaging for setup verification for a variety of indications, summed over all surveyed proton therapy institutes.

reduce imaging frequency even further. Re-imaging prior to every field may be driven by the desire to maximize treatment accuracy by always using the most recent setup information. In the first few years of operation, it may be prudent to perform additional imaging while learning the ins and outs of the new proton therapy equipment. Perhaps it is even the first clinical use of the proton therapy solution offered by this specific vendor. In this respect, proton therapy is quite different from adding or replacing a photon linear accelerator.

17.6 Treatment Delivery

It is the goal of each proton therapy center to treat a high number of patients with the best possible quality of treatment. These two aspects are not mutually exclusive, but they certainly influence each other. This could be as basic as accepting a clinically non-significant increase in dose to normal tissues by treating only a subset of treatment fields each day [16]. Or a facility could decide not to wait for the development of high-quality volumetric in-room imaging and opt to use a trolley system. The efficiency of treatment delivery in patients per year or, more to the point, in treatment fractions per year, is determined by the facility layout, by technical aspects of the proton therapy equipment, and by clinical aspects of the patient mix. The number of fractions per year is a more accurate number to state, as the reduced dose to normal tissues when using proton therapy may allow an increase in hypo-fractionation. Applying photon therapy for a number of fractions may further cloud efficiency numbers.

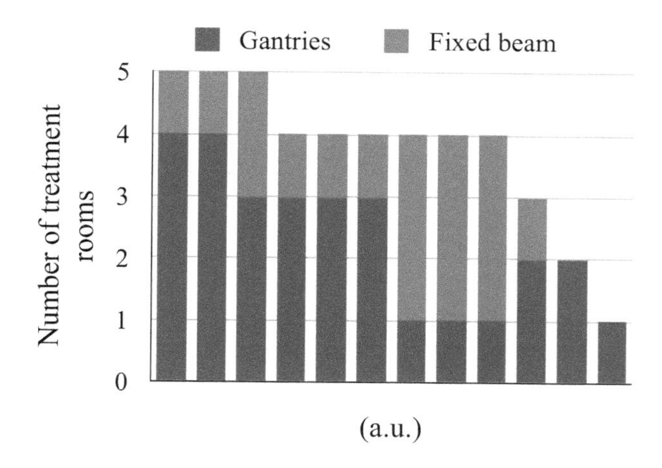

Figure 17–8 Distribution of the number of gantries and fixed beam treatments over the operational U.S. proton therapy centers. The data is ordered by the number of treatment rooms.

17.6.1 Facility Layout

The number of treatment rooms in a facility directly impacts the number of fractions that can be treated on any day or in any year. Figure 17–8 shows the number of treatment rooms for all operational proton therapy centers in the United States. At some facilities, a fixed-beam treatment room contains more than one beam line, or it can be an "inclined beam room" where a single nozzle can be positioned at the end of both a horizontal beam line and a beam line at a 30-degree angle [12]. Four of the 12 facilities surveyed have a fixed horizontal beam line optimized for and dedicated to the treatment of intra-ocular tumors. The number of fractions that can be delivered each year does not scale directly with the number of rooms, however. See Section 17.6.5 on throughput simulation.

The number of operating hours also directly determines patient throughput. Many centers operate on double shifts or at least for extended hours. On average, the U.S.-based proton therapy centers have 80 hours of user primary access time per week, with a minimum of 58 and a maximum of 100. This includes time needed for quality assurance (see Section 17.7, Quality Assurance).

17.6.2 Facility Up-time

The typically achieved up-time per our survey averages 95% to 96% [6]. This relates only to the proton therapy delivery equipment and thus does not include downtime of, for example, the OIS or the TPS. Eight out of the 12 institutes state this to be sufficient for a smooth clinical workflow. The other institutes report some inconvenience for their treatment staff and their patients. In general, the achieved up-time can be expected to be somewhat lower in the first year(s) of operation, especially if the machine is one of the first of a particular equipment vendor. It is important to realize that different institutes use different metrics to determine and report up-time [17–19], so the surveyed up-time is only a rough indication. For example, an institute may decide not to label a short break in treatment as downtime if it only delays the actual patient treatment but does not lead to any treatment cancellations. Others report any break in treatment room "readiness," even as brief as 10 minutes, as downtime.

Proton therapy equipment does not reach the typical up-time of 98% or more of a state-of-the art photon therapy machine. Prolonged downtime of a few hours or more is, however, rare. Most downtime affects only a single room, and treatment delays can typically be remedied over the remainder of the day, with or without the need for an extension of treatment hours on that day. Overall, down-time has only a very small effect on patient throughput.

17.6.3 Technical Aspects Affecting Throughput

The United States harbors only one facility where the accelerator is dedicated to a single treatment room which, regarding workflow, means that the treatment beam is always immediately available. In all other facilities, the proton accelerator is shared between multiple treatment rooms. This may lead to possible "room-interplay" effects between treatment rooms. This means that a room with a patient immobilized and aligned may have to wait for the treatment in another treatment room to finish. Only one surveyed facility with more than two treatment rooms stated they did not notice this room-interplay effect. All other centers said they experience mild to moderate room-interplay effects. In case of room-interplay, rooms that are treating or about to treat a pediatric patient or an SBRT patient typically get priority access to the proton beam. At the moment, the technical solution to prioritizing beam requests is the use of the intercom between rooms, but proton therapy vendors are developing more advanced systems. Should room-interplay effects become severe, it may also affect the imaging workflow since a prolonged wait between setup verification and the start of treatment increases the probability of patient motion.

Dose delivery of a single field by means of passive-scattering is typically the fastest, requiring about 30 to 40 seconds regardless of the field size. Next comes uniform scanning with a typical field delivery time of about one minute. The most time-consuming treatments are large PBS fields on a slow scanning system. It is not the beam current that is limiting, but the time needed to move from spot to spot and from layer to layer.

The vendors have been continuously improving the timing aspects of their systems such that room switching takes less than 30 seconds and PBS delivery of 1 Gy to 1 liter takes less than 60 seconds. Dealing with intra-fractional patient motion by means of volumetric repainting may, however, still prove a challenge as layer switching typically takes one second or longer.

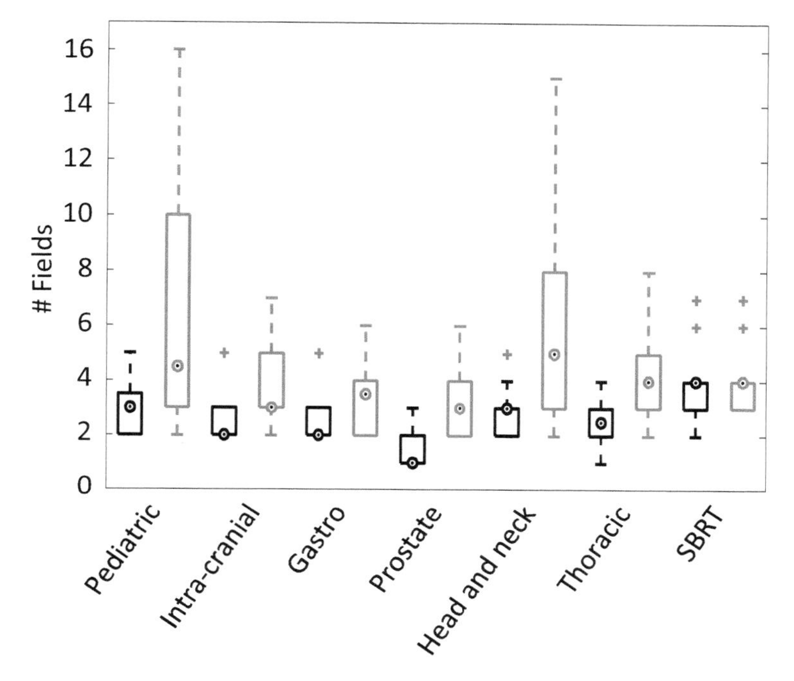

Figure 17–9 Box plot of the number of fields delivered per fraction (black) and over the entire treatment course (magenta) for a variety of indications. The small circles indicate the median, and the red crosses indicate outliers.

In some facilities the treatment nozzle is equipped with a "drawer" that allows remote switching of two sets of field-specific apertures or range compensators. The reduced need for room entry speeds up the treatment delivery.

17.6.4 Clinical Aspects Affecting Throughput

The patient mix also has a direct influence on the number of fractions that can be treated in a facility, as different indications require a different treatment approach. Patient setup has already been discussed in Section 17.5 on image guidance. This section presents data regarding the number of fields treated per fraction and, for completeness, also the number of fields for an entire treatment course (Figure 17–9). Most patients are currently treated with SFUD, mostly by means of passive-scattering or uniform scanning. In SFUD, each field delivers a homogeneous dose to the target. This allows the delivery of only a subset of fields in a treatment plan. The expectation is that this field-set of the day approach will disappear with the increased use of pencil-beam scanning. For IMPT, all fields in a treatment plan will obviously need to be delivered each day. Based on current experience of those institutes delivering PBS treatments, including the Paul Scherrer Institut (PSI) in Switzerland, the typical PBS treatment plan will consist of two or three fields. The use of four or even more fields is an exception. Looking at Figure 17–9, the use of PBS may actually lead to a small increase in the number of fields treated per day, especially if patient mix is currently biased toward prostate treatments. But the reduction in handling field-specific apertures and range compensators will likely still lead to a net time savings.

Next to the number of fields, treatment delivery time and patient in-room time will also increase if patients require gating or repainting to improve accuracy in dose delivery. Children are a category of their own, as they typically have cranio-spinal irradiations requiring the use of 4 to 5 fields per fraction in order to treat the entire brain and spinal cord. The use of abutting fields extends the time needed for accurate patient alignment, and typical in-room times stated for pediatric patients are 60 to 90 minutes. The exact time needed depends on many factors, amongst which are the choice of starting the application of anesthesia inside or outside of the treatment room and the experience of the team of radiotherapists and anesthesiologists. It is also very important to actively involve the anesthesiology department far in advance of the first treatment of a patient under anesthesia. Their choices not only affect the current patient, but also the patients waiting to be treated next.

17.6.5 Throughput Simulation

Experience is the best simulation. As such, only one proton center in the United States has been able to treat just over 1000 patients per year. The surveyed centers provided their patient numbers, as these are typically well known and readily available. They were not asked for detailed information as to how many patients of which indication were treated and what the corresponding fractionation schedules were. It is, therefore, only possible to report on patient numbers per year, rather than the number of treatment fractions per year. When ignoring the number of clinical operational hours and the time needed for quality assurance, the average number of patients per treatment room for the 12 centers is 195 patients per year. The minimum and maximum are 125 and 250 patients per year, respectively.

A throughput simulation is very helpful as input for the business case, for facility design, for the number of personnel required, etc. When estimating potential throughput, it is important to take into account all aspects of the treatment. This means not only patient mix and room switching times, but also the typical extent of treatment volumes (to best estimate time for dose delivery of a field) and the expected patient setup times. One must also be realistic about expected technical and clinical improvements. Lastly, it will have to be an actual simulation, as average numbers do not take into account room-interplay effects. At least a few proton therapy vendors have models to perform such a simulation.

A simple calculation may provide some insight. Assume that dose delivery of a single field takes 60 seconds, and that it takes another 30 seconds to rotate the gantry and patient for the next field or to swap the beam to a different treatment room. In this case, it is possible at the very best to deliver 40 fields per hour when having only a single accelerator. This may sound like a lot, but for an average of three fields per patient, this means 13 fractions per hour. For a facility with four treatment rooms, this translates into three patients per hour per treatment room. In such a case, this would limit the average in-room time per patient to 20 minutes, which is a number not yet clinically achieved. And this is without taking into account the effect of room-interplay.

The prime reason for increasing the number of treatment rooms is the cost of the single accelerator. But an extra treatment bunker and gantry are also costly. More importantly, one needs to factor in the cost of personnel waiting in a treatment room for the beam to arrive, as well as their unhappiness to have to wait.

Typical total in-room times are currently 30 minutes. There is no reason why patient setup times and total in-room times will not come down toward what is achieved in photon therapy, i.e., less than 20 minutes, including volumetric image-guided patient setup and a selection of the plan-of-the-day [20]. Both these improvements will increase the number of fractions delivered per treatment room. It will also increase the probability of room-interplay.

17.6.6 Intra-fractional Patient Motion

Chapter 25 (Motion Management) discusses the various techniques to mitigate intra-fractional patient motion or the effects thereof on the dose distribution. All centers participating in the survey reported an increase in lateral, distal, and proximal safety margins as the prime technique applied to limit these effects. On top of that, six centers limit breathing-induced motion of the tumor and changes in patient anatomy by voluntary or assisted breath-hold or by using a compression belt. Only one center reports the use of gating, and only one center is currently applying repainting.

17.7 Quality Assurance

The technical details of quality assurance are discussed in chapters 15 and 16. Here only the workflow aspects of QA are discussed. Perhaps the most striking feature is that the rapid increase in the use of proton therapy is too recent for QA procedures to have become standardized. The various institutes have developed their own QA procedures. These procedures tend to be more complex than in photon therapy, and they require more knowledge and understanding in case of an unexpected result. This means that QA tasks are typically performed by a medical physics group rather than a therapist.

17.7.1 Field-specific Hardware QA

The expectation is that field-specific apertures and range compensators will be required only rarely or not at all when applying PBS treatments. Those centers employing passive-scattering or uniform scanning have to take the time constraints of fabrication and validation of these devices into account. Depending on the incidence of individual field calibrations, these devices have to be fabricated at least one day before their first clinical use. The fabrication process itself is also time-consuming, especially for very large range compensators. A treatment plan therefore needs to be ready even more than one day before the start of treatment.

From a personnel point of view, validating the correctness of the individual hardware pieces is a laborious process requiring from a few hours per day up to a full FTE if more than a few hundred patients per year are treated. As an alternative, it is possible to have these devices fabricated and QA performed off-site before transporting them to the proton center. Regardless, the use of apertures and range compensators puts limits on the workflow, especially on a workflow aimed at adaptive therapy.

17.7.2 Daily, Weekly, and Monthly QA

Although QA procedures have not yet been standardized, the past decade has seen quite a decrease in the time needed to perform these tasks. This is partly because better measurement tools have became available and partly because extended experience has led to an increase in comfort with the proton therapy equipment. The proton centers surveyed provided information as to how much time is currently spent on QA.

Daily imaging QA typically requires 5 to 15 minutes per treatment room, which is similar to photon therapy. Daily dosimetric QA averages at about 15 minutes per treatment room, but could be as low as 7 minutes or as high as 30 minutes. Depending on the number of treatment rooms connected to the accelerator and on the available personnel for quality assurance, this means that daily QA blocks each treatment room for more than an hour each day in a facility with multiple rooms connected to a single proton accelerator. Of course, while the last room finishes QA measurements, all other rooms can start setting up the first patient to be treated, saving some time.

Some facilities have beam lines with a nozzle offering multiple modalities of proton therapy, e.g., passive-scattering and pencil-beam scanning. Although switching modalities can be fast from a technical perspective, it does require additional QA after the switch.

Not all centers have a weekly QA program, opting for monthly QA only. Based on the survey results, a rough estimate of the total time needed for an entire proton therapy center (i.e., all treatment rooms) is five hours per week for weekly QA. For monthly QA, this varies between 5 and 25 hours per month. These numbers are put in further perspective in Section 17.7.4.

17.7.3 Individual Field Measurements

Measurements of individual fields can be aimed at calibration or at verification. Calibration means that the measurement directly determines the number of MU of the field. Verification means that treatment planning software determines the MU to be used, unless the measurement shows too large a deviation. The latter case can necessitate the redesign of the treatment plan.

In case of passive-scattering or uniform scanning, the TPS typically does not provide the MUs for the fields. A measurement-with or without the use of the field-specific aperture and range compensator-is necessary, or one may revert to building and using a prediction model. This prediction model still requires regular validation, but it may significantly reduce the time spent on individual field measurements. For pencil-beam scanning, the TPS does provide the MUs, and these need to be validated. At the moment, this means planar dose measurements at multiple depths. Generally, dosimetric validation of a PBS field takes much longer than for a passive-scattering field.

The surveyed institutes state an average of about 10 hours per week for field measurements, ranging from 5 to 15 hours. Current practice is that 100% of all PBS fields are validated by means of a measurement, while for passive-scattering and uniform scanning, the average is about 60% of all fields. On average, currently only about a quarter of the delivered fields in a proton therapy facility are delivered by means of pencil-beam scanning, with the vast majority being SFUD rather than IMPT.

17.7.4 Overall In-room QA Time

For a photon linear accelerator, a good estimate is that just under 10% of all clinical operational hours are used to perform in-room quality assurance tasks. This includes all daily, weekly, monthly, and yearly imaging, and it covers dosimetric QA, imaging QA, and individual plan validations. The continued development of portal imaging devices for dosimetric quality assurance has been an especially huge time saver [21]. Using such a tool, dosimetric quality assurance of a treatment plan takes only a few minutes. In contrast, the dosimetric validation of a single treatment plan for proton therapy will typically take up to one hour (a good two treatment fractions), and it will to some extent block other treatment rooms from using the proton beam. It is probably because of this that the total amount of time spent on in-room QA, as reported in the survey, is around 20% of the clinical operational hours of a treatment room. This is twice as much as for photon therapy. This value of 20% does include performing some of the tasks in parallel over the various treatment rooms.

With the use of PBS and treatment adaptation expected to increase, the amount of time and personnel needed for individual field measurements will increase, reducing the hours available for patient treatments even further. And more field measurement does not fully guarantee treatment quality. A single measurement prior to clinical use of a field certainly catches a number of data transfer errors, but it is not nearly as complete a QA tool as the daily on-line *in vivo* dosimetry that is possible with EPIDs in photon clinics. Most importantly, these field measurements are an obstacle for adaptive therapy. The proton therapy community, both users and vendors, has to address this. Faster dosimetry tools and scripting of phantom setup and of beam delivery will help. But in the long run, increased confidence in treatment delivery has to ensure that each and every field no longer needs to be validated by means of a set of measurements. *In vivo* dosimetry (Chapter 28, *In Vivo* Dosimetry for Patient Dose Verification) will play an important role.

17.8 Treatment Adaptation

The finite range of protons makes the delivered dose distribution, especially in case of IMPT, very sensitive to any change in patient anatomy. In the best case, the patient anatomy and the treatment plans are robust over the entire treatment course, such that treatment adaptation is not necessary. Adaptive therapy is, however, not simply a buzzword, not even for the very robust treatment plans as delivered with photon therapy. Even if a clinically significant benefit is achieved for only a fraction of the proton therapy patients, the department needs to have a framework prepared to:

1. Determine which patients will benefit from a treatment adaptation.
2. Efficiently adapt and validate the treatment plan.

The impact of the former on the workflow could be as little as monitoring patient weight or as extensive as daily volumetric in-room imaging by means of cone-beam CT. Plan adaptation could, in the future, perhaps consist of a library of plans with varying safety margins or robustness, or a methodology of on-line re-optimization. Currently, however, it is a manual off-line process.

How often to assess the need to adapt a treatment plan, and how often to adapt it, may differ between patient groups. But no matter how advanced the framework, once it has been implemented, it will likely be applied for all patients, even if the clinical benefit will only be marginal. With in-room volumetric imaging available, the future of treatment adaptation in proton therapy looks quite different from what it looked like just a few years ago. The workflow needs to be adapted accordingly.

The "typical" treatment planning and treatment delivery workflow of proton therapy as shown in Figure 17–4b probably underestimates the number of required plan adaptations. More importantly, there are two intrinsic objections to it. First, these interaction moments currently are too ad hoc and should be part of a well-designed, indication-specific, adaptive therapy workflow. Second, these interaction moments are very time- and resource-consuming. A simple recalculation not only requires time on the CT scanner, it also requires someone to verify the contours, recalculate the dose, and present the results to the treating physician. It is perhaps for this second reason that adaptive therapy is only applied somewhat sporadically at the moment. In Section 17.4 it was concluded that even basic software integration needs serious improvement. Obviously, an adaptive therapy workflow increases the demand on software integration even further.

The proton therapy centers were surveyed as to the frequency of recalculations and replans. A recalculation in the survey means dose calculation of an existing plan on new CT data. Similarly, a replan means that a new treatment plan is created based on new CT data. This could mean that the plan for the initial prescription, as well as the plan for a subsequent cone-down, needs to be adapted.

As shown in Figure 17–10, when averaged over all indications, 50% of the institutes state to never do a recalculation, and another 25% of institutes do this only once per treatment course. For replans, these numbers are 60% and 25%, respectively. One of the institutes that does do a lot of recalculations and replans reported on their experience. They stated that overall they have three times as many plans approved (including photon plans) than actually delivered. Prostate treatments required the fewest plan adaptations, while lung and head and neck treatments required the most.

It is fair to state that the field of proton therapy is in the process of fine tuning the balance between incorporated margins and plan robustness versus the frequency of recalculation and replans. Assuming all institutes treat the same number of patients of the same indication, the data in Figure 17–10 shows that for lung and SBRT treatments, the number of replans is about half the number of recalculations. For the other

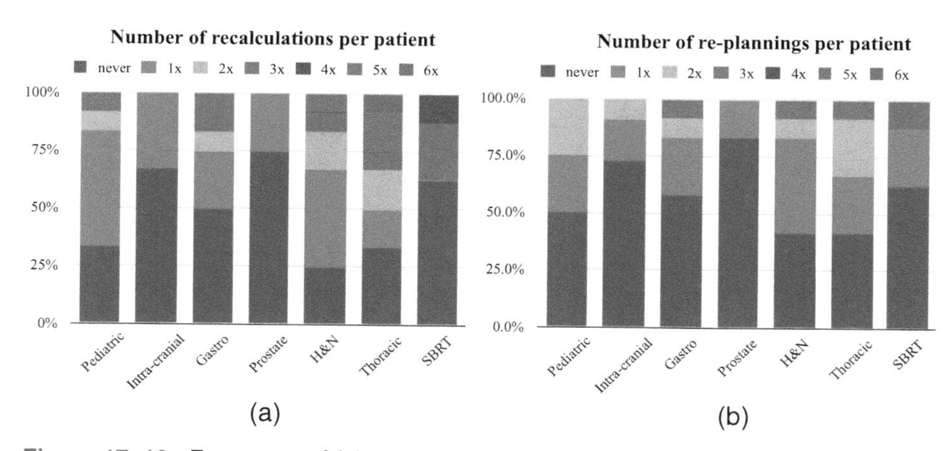

Figure 17–10 Frequency of (a) recalculations and (b) replanning per patient for a variety of cancer indications. Data should be interpreted as "per institution." For example, 75% of institutes report they never perform a recalculation in case of a prostate cancer treatment. SBRT is currently only performed in 8 out of 12 surveyed institutes.

indications, the number of replans is at least two thirds of the number of recalculations.

When asked about the tools available for recalculation and replanning, 8 out of 12 facilities stated their tools were sufficient, although some facilities indicated that "the tools could be better and more efficient." Ten out of 12 institutes use only commercial software for their adaptive therapy workflow, meaning that a large computing and software development group may not be a prerequisite for applying adaptive therapy. But single-button-press dose accumulation over multiple deformed CTs, and subsequent plan re-optimization based on the accumulated dose, has not yet found its way to the clinic.

17.9 Hazard Analysis

Every piece of new equipment added to improve the quality of our radiotherapy treatments adds more complexity to an already massively complicated system. For the most part, these additions and modifications in our technology and workflow are made in an ad hoc manner, addressing only the immediate environment of the expected change. Proton therapy could be treated as "just another machine" used for the actual delivery of a treatment fraction. When disregarding the size of the proton accelerator and beam transport line, a pencil-beam scanning treatment delivery is indeed not more complex than IMRT with a multi-leaf collimator. But if the effects of adding proton therapy to a department were indeed that limited, this chapter needn't have been written. The introduction of proton therapy puts pressure on nearly every aspect of the radiotherapy workflow. More importantly, given the still limited number of proton therapy end users, an individual facility will very likely be one of only a few users of certain commercial products, or a combination thereof. This requires the design of unique processes, considering a site-specific combination of hardware and software, and taking into account the historical perspective and future expectations of the radiotherapy department. Complex technology and a limited number of users ask for increased risk-awareness and risk-mitigation.

The professional groups have long recognized the need for a comprehensive, systematic investigation of the radiotherapy process, and they recommend formal hazard analysis as a tool to analyze past errors and to preempt new ones (see AAPM's TG-100 [10], ASTRO's "Safety is no Accident" [22], and ESTRO's AcciRad [23]). It is recognized that each institution has a different implementation of the treatment process, and therefore each institution needs to perform this analysis for their own setup.

In industry, safety engineering—the elimination or mitigation of risks by proper consideration during the design phase—is a common concept. Industry uses advanced methods of hazard analysis to ensure the safety of their systems. But industry can't test for all unique implementations of their products in a complex and continuously evolving workflow. As an end user, copying best practices in photon therapy is a good starting point. But given the relative novelty of proton therapy and the aforementioned unique product combinations, applying safety engineering principles in the development of proton therapy procedures may prove especially important.

There are various methods for hazard analysis. ASTRO and AAPM recommend failure mode and effect analysis (FMEA) [24]. This method, developed in 1949, is based on first identifying a system's parts and their potential failure modes. These are then assigned scores for probability of occurrence, severity of the consequences, and probability of detection, to be summarized into a risk level. All failure modes with a cumulative score above a user-defined limit need to be addressed. There are various forms of usage of FMEA. In new, complex systems of hardware, software, and human interaction, it often becomes infeasible to assign solid numbers of probability to a failure mode. Some hazard analysis systems therefore remove numerical scores and work with severity matrices instead.

Other methods of hazard analysis that have been applied in radiotherapy include fault tree analysis (FTA), hazard and operability analysis (HAZOP), failure mode effects and criticality analysis (FMECA), and hazard identification (HAZID). Most of these originated in the middle of the last century, and an argument can be made that these methods have not been able to keep up with the rapid development and complexity of today's socio-technical applications.

System theoretic process analysis (STPA) provides a framework for hazard analysis that specifically aims at reaching into the areas of interactions between various system components and humans, in addition to component failure assessment. STPA is applied by a rapidly growing number of users across all industries for both process design and analysis.

Whichever method is chosen, the authors highly recommend systematic hazard analysis as a tool to preemptively address safety concerns in process design.

17.10 Facility Startup

17.10.1 Patient Numbers

We would have liked to present details as to the ramp-up in patient numbers over the first four years after the first proton patient treatment, but it is difficult to interpret the data provided by the surveyed facilities. There is a large variation in the number of treatment rooms per facility and in clinical operational hours. Especially for the earlier centers, there could be a few years between acceptance of the first treatment room (and the first patient treatment) and other treatment rooms. With increasing experience of the vendors, the speed of handover of the different treatment rooms has increased. Even for a four-room facility, it is nowadays possible to have all rooms acceptance tested within six to nine months after the first room's acceptance, though there will likely be a punch-list of small issues still to be resolved for final acceptance. During these months of room acceptance, user primary access will likely be limited to a single eight-hour shift per work day, with the other hours used by the equipment vendor for maintenance and finalizing acceptance of the other rooms.

Six facilities with at least four years of clinical operational experience shared their patient numbers over these years. On average, the numbers of patients treated were 180, 440, 530, and 600 in years one to four of clinical operation. Please keep in mind that this is likely to be an underestimate as to what is achievable today, for the reasons mentioned above. On the other hand, if the patient mix eligible for reimbursement changes significantly (e.g., prostate treatments) this could also affect these numbers negatively.

17.10.2 Indications Treated

The surveyed centers were asked to describe their ramp-up in patient indications. This data is shown in Figure 17–11. Only five centers responded to this question, so the data has limited accuracy. It is not surprising that prostates are one of the first indications to be treated. That this is also true for pediatric treatments comes, perhaps, as somewhat of a surprise. Given the complexity of these treatments, they require much more training and resources. On the other hand, the medical benefit for these patients is clearly recognized, and additional effort is made to add this indication to the patient mix. Existing proton therapy centers are very willing to share their expertise, allowing a timely introduction of this indication in new centers.

Lung treatments are challenging because of intra-fractional breathing motion and tissue heterogeneities. Nevertheless, lung treatments are typically part of the patient mix within the first two years of operation. Next to the data shown, there are various other indications that are treated in the first two years-ocular tumors, brain tumors, base-of-skull chordomas, chondrosarcomas, and Hodgkin's disease. But not all centers treat these, or at least not in the first four years. The perhaps only slightly optimistic conclusion is that the available experience in the proton therapy community allows

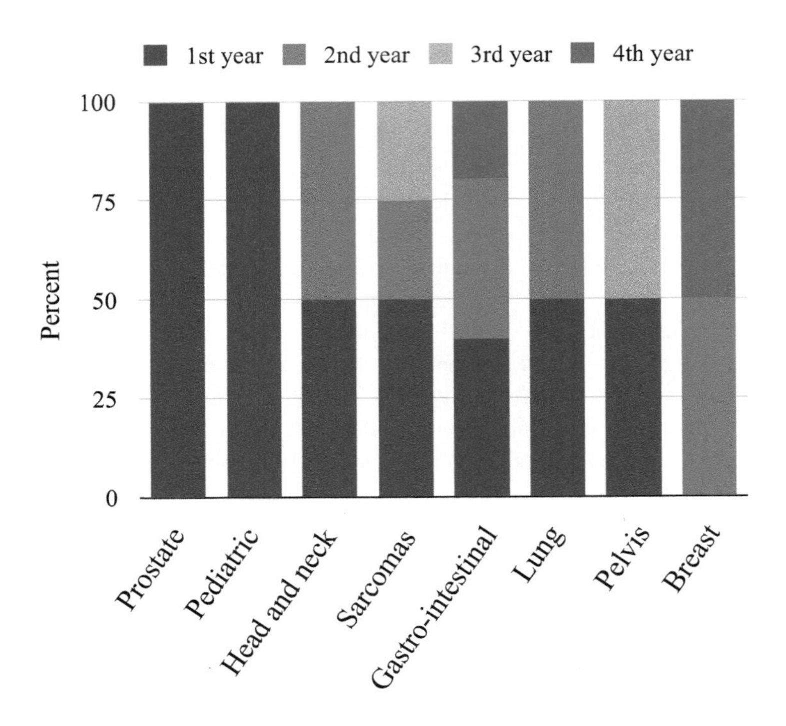

Figure 17–11 Distribution of the year a certain indication was first treated. For all indications shown, at least four centers reported their first year of treatment, with the exception of pelvis and lung treatments, for which there were only two responses.

an almost complete mix of patient treatments after only two years—provided, of course, that personnel can be trained fast enough.

17.10.3 Personnel Requirements

It may take some prolonged exposure to really have the intricacies of proton therapy ingrained. It can probably be compared to brachytherapy. Being proficient or even an expert at external beam photon therapy, it will take six to 12 months to reach a level of expertise feeling comfortable being the sole physician/physicist/therapist responsible for these treatments.

There are courses by ESTRO and PTCOG, as well as by the PSI and by the Penn Medicine Department of Radiation Oncology, that provide an overview of what to expect in and from proton therapy, both from a physics and a clinical perspective. But at least a core team of specialists from all medical professions should get some months of hands-on training. Official training programs in proton therapy are limited. Having good contacts with operational facilities is another way to arrange for training.

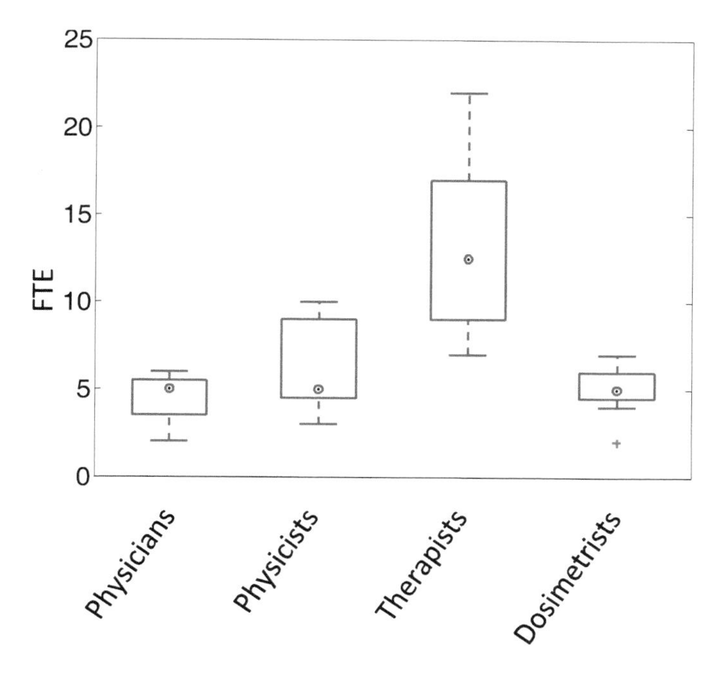

Figure 17–12 Personnel needed to treat 500 patients per year, as estimated by eight of the currently operational proton therapy facilities in the United States. The circles indicate the median, and the red crosses indicate outliers.

We hope to have provided information that will assist in determining expected ramp-up in patients and indications, and thereby in timing the employment and training of personnel. For reasons mentioned earlier in this chapter, e.g., complexity of patients and less-than-optimal software and software connectivity, proton therapy is more personnel-intensive than photon therapy. The one missing piece of data may be exactly how many personnel are needed for a certain number of patients. The operational proton therapy centers were asked to provide their estimate of "personnel required to treat 500 patients per year." A scenario as to the number of treatment rooms, clinical operational hours, or proton therapy equipment used (i.e., treatment modalities) was not provided. This may explain the relatively large variation in the number of personnel stated. The data in Figure 17–12 should furthermore be taken to reflect current clinical practice. The number of physicists may or may not include medical physics assistants. The same holds for the number of therapists, although only two institutes said they use therapist assistants. Their role is then limited to bringing the patients in and out of the room or to selecting the hardware for the treatment fraction about to be delivered. Averaged over all institutes, there are 2.75 therapists present at any time per treatment room. Four institutes have stated two therapists per room, and one institute states to have four.

Peeters et al. [5] stated personnel numbers for a three-room facility expecting to treat 1600 to 1800 patients per year. They expected to need 8 physicians, 15.5 physicists, and 35 therapists (this includes dosimetrists). The required number of physicians scales almost linearly with the number of patients. So does the number of therapists, provided the patient mix and fractionation schemes are similar. As such, current clinical practice appears to require about 100% more physicians and 50% more therapists per 500 patients than estimated by Peeters et al. [5]. Physicists spend a significant amount of their time on implementing new technology and new quality assurance procedures. The required number of physicists, therefore, is only weakly linked to the number of patients treated. A total of 15.5 physicists to treat 1600 patients, therefore, seems an overestimate.

17.11 Summary and Conclusion

From a technical point of view, the current state of proton therapy already allows very good treatments for select patients over a wide variety of indications. From a workflow perspective, proton therapy needs to become more mature. There are many opportunities to reduce the resources needed for treatment preparation and treatment delivery, while at the same time increasing treatment quality even further. Proton therapy needs to prepare itself for extensive use of in-room image guidance and adaptive therapy. This need is recognized by both users and vendors, and the increased interest in proton therapy will accelerate this transition.

Acknowledgments

We would like to thank all proton therapy institutes for participating in the survey and in answering follow-up questions. In particular, we thank, in order of the date of first patient treatment:

- Dr. Baldev Patyal of the James M. Slater, M.D. Proton Treatment and Research Center in Loma Linda, CA.
- Dr. Juliane Daartz of the Francis H. Burr Proton Therapy Center in Boston, MA.
- Dr. Indra J. Das of the Indiana University Health Proton Therapy Center in Bloomington, IN.
- Dr. Xiaorong Ronald Zu of the MD Anderson Proton Therapy Center in Houston, TX.
- Dr. Zuofeng Li of the University of Florida Proton Therapy Institute in Jacksonville, FL.
- Dr. Mark Pankuch of the CDH Proton Center in Warrenville, IL.
- Dr. Jim McDonough of the Roberts Proton Therapy Center in Philadelphia, PA.
- Dr. Ping Wong of the Hampton University Proton Therapy Institute in Hampton, VA.

- Dr. Dennis Mah of the New Jersey / Metro New York Procure Proton Therapy Center in Somerset, NJ.
- Dr. Eric E. Klein of the S. Lee Kling Proton Therapy Center in St. Louis, MO.
- Dr. Lei Dong of the Scripps Proton Therapy Center in San Diego, CA.

References

1. Das IJ, Moskvin VP, Zhao Q, et al. Proton therapy facility planning from a clinical and operational model. *Technol Cancer Res Treat* 2014; Jun 30.
2. Ollendorf DA, Colby JA, Pearson SD. Proton beam therapy. www.icer-review.org. 2014.
3. Karlsson M, Björk-Eriksson T, Mattsson O, et al. Distributed proton radiation therapy—A new concept for advanced competence support. *Acta Oncol* 2006;45:1094–1101.
4. Efstathiou J. Proton therapy vs. IMRT for low or intermediate risk prostate cancer (PARTIQoL). In: ClinicalTrials.gov. Bethesda, MD: National Library of Medicine, 2012. Available from: http://clinicaltrials.gov/ct2/show/NCT01617161.
5. Peeters A, Grutters JPC, Pijls-Johannesma M, et al. How costly is particle therapy? Cost analysis of external beam radiotherapy with carbon-ions, protons and photons. *Radiother Oncol* 2010;95:45–53.
6. Miller ED, Derenchuk V, Das IJ, et al. Impact of proton beam availability on patient treatment schedule in radiation oncology. *J Appl Clin Med Phys* 2012;13:3968–3968.
7. Mohan R, Podmaniczky KC, Caley R, et al. A computerized record and verify system for radiation treatments. *Int J Radiat Oncol Biol Phys* 1984;10:1975–1985.
8. Klein EE, Drzymala RE, Williams R, et al. A change in treatment process with a modern record and verify system. *Int J Radiat Oncol Biol Phys*. 1998;42:1163–1168.
9. Klein EE, Drzymala RE, Purdy JA, et al. Errors in radiation oncology: a study in pathways and dosimetric impact. *J Appl Clin Med Phys* 2005;6:81–94.
10. Huq MS, Fraass BA, Dunscombe PB, et al. A method for evaluating quality assurance needs in radiation therapy. *Int J Radiat Oncol Biol Phys* 2008;71:S170–3.
11. Siochi RA, Balter P, Bloch CD, et al. A rapid communication from the AAPM Task Group 201: Recommendations for the QA of external beam radiotherapy data transfer. AAPM TG 201: Quality assurance of external beam radiotherapy data transfer. *J Appl Clin Med Phys* 2010;12.
12. Fava G, Widesott L, Fellin F, et al. In-gantry or remote patient positioning? Monte Carlo simulations for proton therapy centers of different sizes. *Radiother Oncol* 2012;103:18–24.
13. Heijkoop ST, Langerak TR, Quint S, et al. Clinical implementation of an online adaptive plan-of-the-day protocol for nonrigid motion management in locally advanced cervical cancer IMRT. *Int J Radiat Oncol Biol Phys* 2014;90:673–679.
14. de Boer HCJ, Heijmen BJM. eNAL: an extension of the NAL setup correction protocol for effective use of weekly follow-up measurements. *Int J Radiat Oncol Biol Phys* 2007;67:1586–1595.
15. Gangsaas A, Astreinidou E, Quint S, et al. Cone-beam computed tomography-guided positioning of laryngeal cancer patients with large interfraction time trends in setup and nonrigid anatomy variations. *Int J Radiat Oncol Biol Phys* 2013;87:401–406.
16. Engelsman M, Delaney TF, Hong TS. Proton radiotherapy: The biological effect of treating alternating subsets of fields for different treatment fractions. *Int J Radiat Oncol Biol Phys* 2010;79(2):616–22.
17. Miller ED, Derenchuk V, Das IJ, et al. Impact of proton beam availability on patient treatment schedule in radiation oncology. *J Appl Clinl Med Phys* 2012;13(6):3968.
18. Patyal B. Maintenance and logistics experience at Loma Linda proton treatment facility. www.ptcog.ch.
19. Suzuki KK, Gillin MTM, Sahoo NN, et al. Quantitative analysis of beam delivery parameters and treatment process time for proton beam therapy. *Med Phys* 2011;38:4329–4337.
20. Heijkoop ST, Langerak TR, Quint S, Bondar L, Mens JW, Heijmen BJ, Hoogeman MS. Clinical implementation of an online adaptive plan-of-the-day protocol for nonrigid motion management in locally advanced cervical cancer IMRT. *Int J Radiat Oncol Biol Phys* 2014:1–7.

21. Podesta M, Nijsten SMJJG, Persoon LCGG, et al. Time dependent pre-treatment EPID dosimetry for standard and FFF VMAT. *Phys Med Biol* 2014;59:4749–4768.
22. Zietman AL, Palta JR, Steinberg ML. Safety is no accident. American Society for Radiation Oncology (ASTRO), 2012. Available from: https://www.astro.org/uploadedFiles/Main_Site/Clinical_Practice/Patient_Safety/Blue_Book/SafetyisnoAccident.pdf.
23. Malicki J, Bly R, Bulot M, Godet JL, Jahnen A, Krengli M, Maingon P, Martin CP, Przybylska K, Skrobala A, Valero M, Jarvinen H. Patient safety in external beam radiotherapy—Guidelines on risk assessment and analysis of adverse error-events and near misses: Introducing the ACCIRAD project. *Radiother Oncol* 2014;112(2):194–8.
24. Cantone MC, Ciocca M, Dionisi F, et al. Application of failure mode and effects analysis to treatment planning in scanned proton beam radiotherapy. *Radiat Oncol* 2013;8:127.

Chapter 18

Immobilization and Simulation

Jon J. Kruse, Ph.D.

Assistant Professor of Medical Physics,
Mayo Clinic Department of Radiation Oncology,
Rochester, MN

18.1 Introduction

Compared to photon-based external beam radiotherapy, proton radiotherapy offers two potential advantages. Depending on the delivery system and depth of treatment, the lateral penumbra of a proton beam may be sharper than that of an x-ray beam. More importantly, the finite range of proton beams allows for an additional degree of freedom in shaping a dose distribution to the target volume while sparing healthy tissues downstream of the target. Successful application of these proton beam characteristics ultimately requires special care in immobilization during the simulation and treatment processes.

Proton dose distributions have been shown to be especially sensitive to both positioning and range uncertainties. Liebl et al. [1] showed that patient positioning errors could lead to proton range changes of up to several millimeters, while Park et al. [2] observed significant changes to organs at risk (OAR) doses in the presence of setup and range errors. To take advantage of the potential advantages of proton therapy—and perhaps, more importantly, to avoid harming a patient by misplacing the steep gradient of a proton beam dose distribution—exceptional care must be exercised in the selection and application of immobilization equipment and in the treatment simulation process.

18.2 External Immobilization

Precise, safe application of a proton beam treatment plan requires that the radiological depth of the target tissue be consistent from day to day. One potential variable in the daily radiologic target depth is the patient pose—the relative position of the patient's anatomy. Changes in the neck flexion of a head and neck patient, for example, can lead to deformation of the target volume relative to the simulation image, a situation that cannot be fully corrected by three or six degree of freedom image-guided radiotherapy (IGRT) processes. For this reason, proton centers may employ immobilization processes that are more complex and comprehensive than those found in photon clinics. Owing to the importance of consistent neck flexion in head and neck treatments, for example, vacuum-assisted bite blocks may be employed in proton treatments of this site. Similarly, rectal balloons are routinely employed in the treatment of prostate cancer to try to reduce the variability of the target tissue relative to the surrounding bony pelvic anatomy [3–7].

In addition to carefully limiting daily variations in the positions of both target and healthy anatomy within the patient, it is critical to consider the impact of the immobilization devices themselves on the range of the proton beam. Treatment machine couch tops and immobilization devices produce fairly minimal changes to delivered dose distributions from an x-ray linac. In an x-ray treatment, immobilization equipment will attenuate the intensity of the therapy beam by a few percent and may also increase the dose to the patient's surface through a bolusing effect [8]. In proton treatments, however, because of the flat shape of a depth–dose profile near the surface of a patient, immobilization devices will have a minimal impact on surface dose. The range of a proton beam, though, will be altered by immobilization devices. Most modern couch tops and immobilization devices are constructed of low-density materials, yet even a lightweight appliance may have a water equivalent thickness (WET) of several millimeters, which could adversely affect the dose distribution from a proton treatment field if it is not carefully accounted for.

18.2.1 Device Characterization

The best way to account for the impact of couch tops and immobilization devices on a proton treatment plan is to include them in the treatment plan dose calculation. Typically this is done by outfitting the CT simulator with a couch top that matches the treatment machine couch top and including the couch in the CT scan. Proton dose calculations, however, use calibration curves that convert CT Hounsfield units (HU) to relative proton stopping power (RSP). These curves are constructed for a specific CT scanner and imaging technique, and they are generally only valid for biological tissues [9]. Rather than rely upon the HU-to-RSP conversion for the characterization of an immobilization device, Wroe et al. [10] described a technique for measuring the WET of immobilization equipment. A variable thickness range shifter was mounted in a treatment nozzle, and the location of the distal edge of a spread-out Bragg peak

(SOBP) was measured by a parallel plate ionization chamber. An immobilization device was placed in the beam path, and the thickness of the nozzle-mounted range shifter was adjusted until the position of the SOBP distal edge was restored. Several locations were tested on each device to check for consistency. For comparison, they also acquired CT scans of devices and used the HU-to-RSP curves within their treatment planning system to estimate the WET of each device. For each of the devices studied, their TPS predicted the actual WET to within 0.7 mm. In general, however, it may be that a TPS will do a poor job of predicting the WET of a device. This could be due to either x-ray beam hardening in the CT scanner, or because the device is made of a material that does not fall on the HU-to-stopping-power curve that is stored in the treatment planning system (TPS). In that case, the device should be contoured in a treatment plan, and the HU value for the material within the contour should be overwritten such that the WET of the device is accurately represented in the plan [8].

In addition to studying the WET prediction from the TPS, CT scans of immobilization equipment should be used to verify the homogeneity of each device that will be used within a proton center. Most modern devices are constructed of low-density materials, such as a foam core wrapped by carbon fiber and coated in epoxy. The total WET of these devices may lie in the range of only 2 to 5 mm, yet it is important to CT scan each appliance that will be used in a center and examine the scan for localized voids or high-density anomalies that could cause unexpected variations in the WET of the device within a proton treatment beam. Similarly, each of the CT scans of a given type of immobilization device should be compared to look for variation in WET between multiple copies of a given piece of equipment. Variations in manufacturing processes or materials may introduce differing WET effects among seemingly identical, interchangeable parts.

18.2.2 Edge Effects

When immobilization devices are properly characterized in a TPS, their impact on a proton treatment plan can generally be safely and reliably accounted for. This is only true, however, for treatment beams that cross smooth, homogeneous regions of an immobilization device. Proton beams that approach or traverse treatment device edges should be avoided. A common example of this is a posterior oblique treatment field that crosses the edge of a treatment couch upstream of the patient. The WET of a typical treatment couch is on the order of 2 to 5 mm, and if a portion of a treatment field traverses the couch while another portion does not, the daily variation in the distal edge of a proton treatment beam near this edge may exceed the standard safety margin used to create the patient's planning target volume (PTV). Indexing the patient's immobilization equipment to the treatment couch can reduce the daily variation in the relative position of the patient and couch edge. However, studies at MD Anderson Cancer Center showed that even indexed treatment devices may still allow a 1 cm daily variation in the position of the couch edge relative to the patient [11]. The best approach, where possible, is to avoid treatment beam angles that may include the couch edge within the field aperture or spot map.

In some cases, the immobilization equipment must be modified or specially designed to accommodate the most advantageous field arrangements for a treatment site. Patients receiving cranio-spinal irradiation at M.D. Anderson Cancer Center are treated with a pair of cranial fields angled 15 degrees below the horizontal plane to spare the cribiform plate and lenses [12]. The edge of a normal treatment couch would interfere with these beam angles, and so a 10 cm thick polystyrene foam slab is placed between the couch top and the patient for simulation and treatment. Unlike the edge of a carbon fiber couch top, the foam slab will not appreciably affect the range of the proton treatment fields.

18.2.3 Proton Treatment Couch Design

Treatment couches used in x-ray therapy are typically a simple rectangular slab of low-density material. Traditionally these couches were made of a "tennis racket" type construction that produced a low average density suitable for x-ray beams, but the localized heterogeneities of a tennis racket couch top would be unacceptable for proton therapy. The era of image-guided x-ray radiotherapy has ushered in a new generation of treatment couch tops that are constructed of a foam core wrapped with carbon fiber and epoxy. The low density and homogeneous construction of these couch tops is readily applicable to proton therapy, although the traditional rectangular shape of an x-ray treatment couch is not optimal for some proton treatment applications.

The treatment area covered by a scattered proton treatment field is defined by a customized brass aperture, and the lateral penumbra of that beam's dose distribution can be optimized by minimizing the air gap between the aperture and the patient surface. Similarly, spot scanning proton treatments to superficial targets often require use of a slab of range shifter material in the nozzle to reduce the beam energy and bring the Bragg peak toward the patient's surface. In addition to decreasing the energy of the proton beam, a range shifter will also scatter the proton beam, so any drift space between the range shifter and the patient surface will allow the beam spot to grow and increase the lateral penumbra of a spot scanning proton field. In both scattered and scanned proton treatments, then, there are instances in which the treatment nozzle will ideally be placed as close as possible to the patient.

In x-ray treatments of the brain or head and neck, an overlay is typically placed on top of the rectangular couch top, and a fixation mask attaches to the overlay close to the patient's head. The portion of the rectangular couch top on either side of a head and neck overlay, then, prevents a proton treatment nozzle from being brought close to the patient surface for lateral treatment fields. The desire to minimize the air gap between the nozzle and a patient's head has led to the development of proton-specific couch tops, such as the BoS insert (Qfix, Avondale, PA, USA), in which the superior end of the couch top is contoured to eliminate any material lateral to the patient (Figure 18–1).

Clearly this couch design is not optimal for all patient treatment sites, so many modern proton couch tops now feature a two-piece design. The inferior end of the couch is a simple rectangular slab that is mounted on the treatment machine, while the superior end features a number of interchangeable extensions that are optimized for

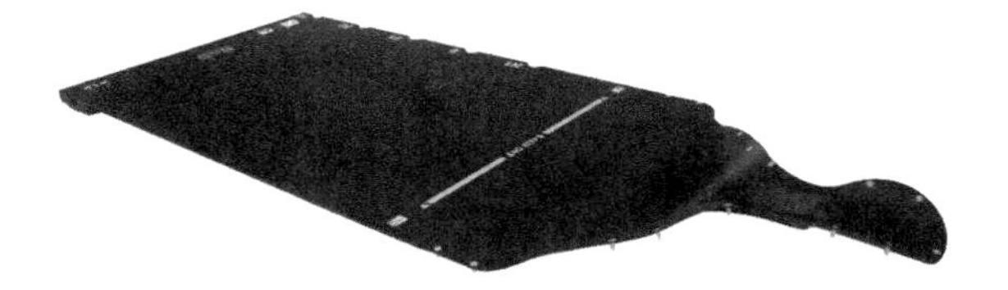

Figure 18–1 Base of skull (BoS) couch extension manufactured by Qfix.

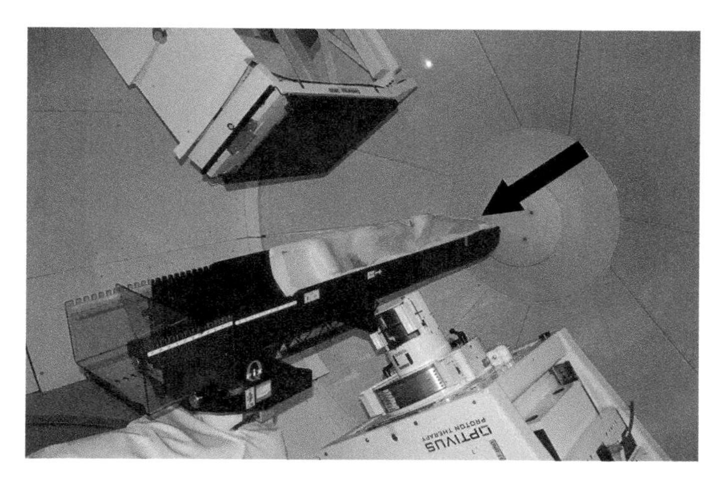

Figure 18–2 Pod immobilization device used for proton therapy at James M. Slater Proton Treatment and Research Center (JMSPTRC). (Photo courtesy of Andrew Wroe, JMSPTRC.)

various disease sites. In addition to the BoS extension, for example, the QFix couch is available with a standard rectangular extension, as well as parts that are optimized for breast and for stereotactic body treatments. This type of modular treatment couch is now offered from a variety of manufacturers.

An advantage of a modular-design treatment couch is that it can be readily applied throughout a radiotherapy clinic. Mounting a proton treatment couch in the CT simulator allows for the treatment planner to visualize the couch in a beam's eye view and avoid beam angles that traverse any sharp edges, and to include the effect of the treatment couch in the dose calculation. Modular treatment couches allow for site-specific treatment extensions to be imaged in the patient simulation, but a standard-purpose rectangular extension can be installed for more general-purpose treatment simulation. Additionally, many proton treatment centers operate in conjunction with an x-ray facility, and modular treatment couches can also be installed on x-ray linacs. By installing the same treatment couches on proton and x-ray treatment machines, radiotherapy clinics can plan multi-modality proton and x-ray treatment courses from a single simulation session. Planning both modalities from a single simulation eliminates the need to manufacture two sets of patient-specific immobilization devices, and it allows the use of complex deformable registration tools to combine dose distributions from two different simulation images. Finally, some proton therapy centers develop intensity-modulated x-ray treatment plans for each of their proton therapy patients as a contingency strategy in the instance that the proton therapy system would suffer an extended breakdown. If the proton and x-ray treatment machines use the same couch tops, patients can be moved seamlessly between the two modalities.

To maintain consistent radiological depth of the target from day to day over a course of radiotherapy, it is necessary that the external contour of the patient be reproduced for each treatment fraction, as well as its relationship to internal target tissue. This may be especially challenging to maintain for an obese patient with an excess of mobile adipose tissue. This problem was addressed at the James M. Slater Proton Treatment and Research Center at Loma Linda University through the development of a pod immobilization device [13]. Patients receiving abdominal or thoracic proton radiotherapy are immobilized in a rigid semi-circular pod, and the space between the patient and the pod is filled with expanding foam (Figure 18–2). The customized pod structure provides a reproducible patient pose over the course of treatment, as well as a repeatable external contour, minimizing intra- and inter-fractional variations in radiologic target depth.

18.2.4 MRI Compatibility

For all the advantages of carbon fiber immobilization devices in proton therapy applications, there remains a single important downside of this material in modern radiation therapy. While the stopping power data gleaned from a CT scan makes that imaging modality essential for proton therapy planning, magnetic resonance imaging (MRI) has emerged as an important imaging modality. The superior soft tissue resolution from MRI allows for better target and critical structure definition in treatment planning [14]. Traditionally, MRI scanners have been fitted with curved treatment couches, and so fusion of MRI scans with CT scans for target definition and treatment planning has required some type of deformable registration algorithm to account for differences in patient pose between the two image sets. Deformable registration has been shown to present a challenge for multiple-modality imaging studies [15], yet the accuracy of these fusions may be improved by imaging the patient in the same position in both scanners. The apparent solution would be to outfit the MRI scanner with a proton treatment couch, the same way a CT simulator is equipped. Unfortunately, however, carbon fiber is conducting, so it may heat up in an MRI scanner, causing image artifacts or even patient injury.

Two commercial solutions to this problem have emerged in the marketplace. The first is an MRI-compatible version of a couch top or immobilization device made of solid acrylonitrile butadiene styrene (ABS) plastic. These devices—with the same form factor and shape as the carbon fiber treatment versions—are only meant for MRI imaging. The patient pose is reproduced using these devices in the MRI scanner, but the patients return to the carbon fiber versions for CT imaging or treatment. The second solution is to replace carbon fiber altogether and build treatment couches and devices from some other fiber mat wrapped around a low-density foam core. Kevlar and fiberglass are the most common alternatives to carbon fiber appliances. With the increasing importance of MRI in radiotherapy, it is recommended to select immobilization devices that are either MR compatible or available in an MRI-compatible version.

18.2.5 Active Immobilization Devices

All of the immobilization and planning strategies discussed to this point have attempted to minimize the impact of the couch tops and other devices on the range of the proton beam. In some instances, however, especially for spot scanning proton treatments, devices on the couch, or even the couch itself, may be used to substantially alter the proton beam energy. Spot scanning delivery systems typically offer a minimum therapeutic range of 4–7 cm. Treatment of lesions shallower than the minimum beam range, then, require the use of a range shifter—a slab of energy-absorbing material inserted in the treatment nozzle. Unfortunately, a range shifter scatters the proton beam in addition to reducing its energy. Any air gap between the range shifter and the patient acts as a drift space and can substantially increase the proton spot size [16]. To minimize the spot size of a range-shifted spot scanning beam, Both et al. [17] developed a universal bolus to be mounted on the treatment couch for spot scanning proton treatment of cranial targets. Figure 18–3 shows the universal bolus mounted on the end of a Qfix BoS proton couch.

As opposed to the manufacturer-supplied range shifter with a WET of 7.5 cm mounted 34.5 cm from isocenter, the universal bolus is a U-shaped piece of plastic with a WET of 5.5 cm. In addition to substantially reducing the air gap between range shifter and the patient, the reduced thickness of the bolus was found to be adequate

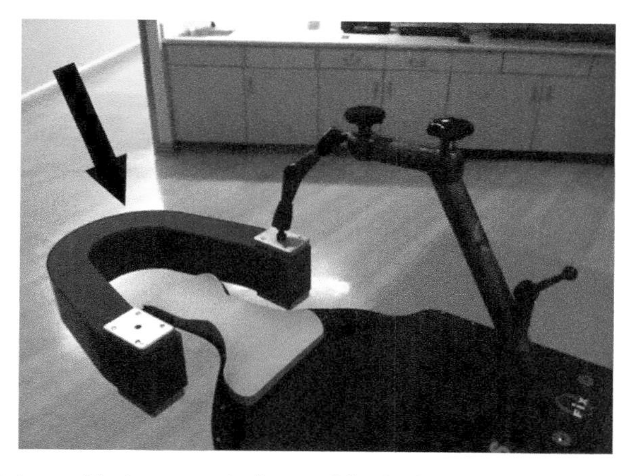

Figure 18–3 Universal bolus mounted on a Qfix BoS proton treatment couch. (From [17].)

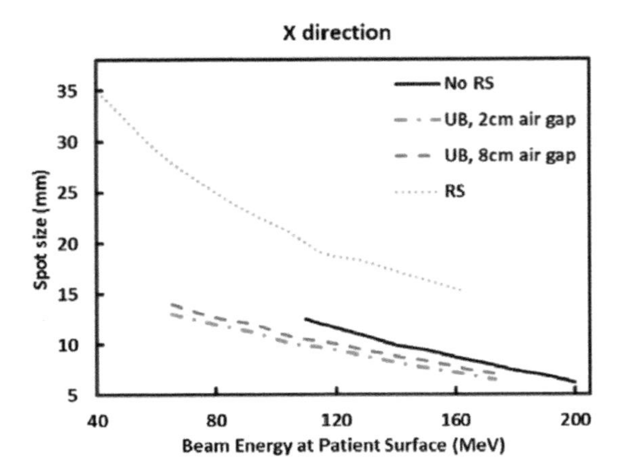

Figure 18–4 Spot size (full width at half maximum) as a function of beam energy for open beams, the standard range shifter (RS), and the universal bolus (UB). (Adapted from [17].)

Figure 18–5 Orfit Range shifting couch top with carbon fiber immobilization overlay, handholds, and face mask. (Image courtesy of Orfit Industries, Antwerp, Belgium.)

for most patient treatments. The result of the thinner device mounted much closer to the patient is that spot sizes from the universal bolus are dramatically smaller than those from the standard range shifter. Figure 18–4 depicts the impact on spot size of an air gap between a range shifter and the patient surface.

The idea of range-shifting couch components is now being applied to treatment sites outside the head. Orfit (Orfit Industries, Antwerp, Belgium) for example, has developed a rectangular extension to a modular couch design in which the entire couch surface is a 4.5 cm WET range shifter. As shown in Figure 18–5, an immobilization overlay can be placed directly on this range-shifting couch top, resulting in zero air gap between the range shifter and patient surface for any posterior treatment

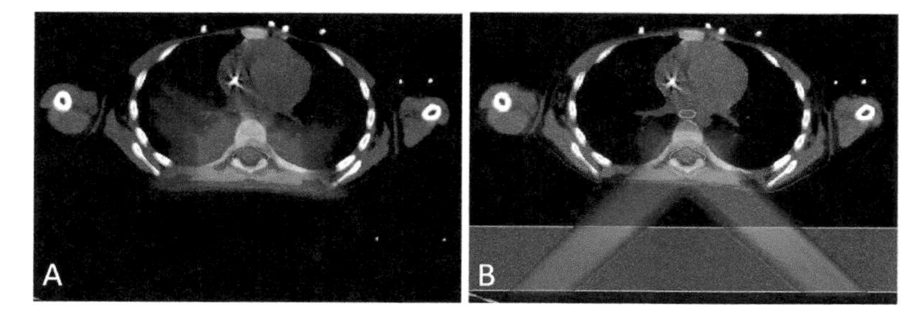

Figure 18–6 Panel A shows a cranio-spinal irradiation calculated with a 4.5 cm range shifter mounted in the treatment nozzle, approximately 35 cm upstream of isocenter. Panel B shows a plan with the patient placed directly on a 4.5 cm WET range shifting surface. (Image courtesy of Ali Tasson, Mayo Clinic.)

field. This design will be useful in treating superficial, posterior targets such as lung or supine cranio-spinal irradiation. Placing the patient directly on a range shifter will minimize the beam spot size for superficial spine irradiations. Figure 18–6 demonstrates the resulting improvement in dose conformity between cranio-spinal plans calculated with a range shifter in the treatment nozzle (Figure 18–6a) and integrated into the couch top (Fig 18–6b).

18.2.6 Immobilization Quality Assurance

Section 18.2.1 discussed the importance of imaging every couch top and immobilization device with a CT scanner prior to its deployment in the proton clinic. Imaging of these devices during their initial commissioning is meant to characterize their WET so they may be accurately accounted for in a treatment plan, but also to look for unexpected voids or heterogeneities in each device that may arise from the manufacturing process. It is equally important that immobilization equipment be re-imaged on a regular basis as part of a periodic quality assurance program. Devices that have been dropped or mishandled may develop internal cracks that are not evident from a visual inspection, but which would produce unintended localized variations in the WET of the device. This is especially important for active immobilization components, such as couch-mounted range shifters, discussed in Section 18.2.5.

Not all immobilization devices are made of rigid plastics or composite materials. Cushioned knee rests or head rests may be routinely applied in radiotherapy treatments. Because many of these devices are general application and not patient specific, radiotherapy departments may store one set of these devices in the simulation suite, and use another set in each of the treatment rooms. Immobilization cushions will often see a higher usage rate in a treatment room than the devices stored in a simulation suite, leading to a greater rate of breakdown and softening of this equipment in the treatment rooms. Even if proton treatment plans are developed to avoid sending beams through an immobilization cushion, a difference in the rigidity of a cushion between simulation and treatment could lead to a systematic change in the patient's pose over a course of treatment. For this reason, general-purpose, non-customized immobilization devices should be rotated between simulation and treatment rooms on a regular basis.

18.3 Internal Immobilization

All of the equipment and strategies discussed to this point are meant to pose and immobilize the patient's external anatomy in a repeatable fashion that closely matches the treatment simulation image. However, intra- and inter-fractional changes in the patient's internal anatomy can also adversely affect the location and radiologic depth of the target tissues. A number of strategies have been developed to deal with daily variations of internal anatomy over a course of treatment. Some devices, like rectal balloons used for prostate treatment, are true immobilization mechanisms meant to press and hold the prostate gland against the bony pelvic anatomy [3–7]. Other tools to be discussed here, such as passive or active fiducial markers, are not strictly immobilization devices, but are meant to address the problem of variable anatomy.

18.3.1 Passive Fiducial Markers

Fiducial markers have been used to assist in IGRT procedures since the early 1990s. Initially, radiopaque objects were implanted in the patient's skull, and computerized analysis of two-dimensional projection radiographs of the markers were used to calculate translational and rotational correction vectors for patient positioning [18,19]. Not long after, Balter et al. [20] showed that fiducial markers placed in a patient's prostate could be used to monitor variations in the position of internal target tissues with respect to radiographically evident bony anatomy. Routine clinical use of gold fiducials for daily pretreatment visualization and positioning of target tissues did not commence, however, until amorphous silicon portal imaging devices were readily available in x-ray therapy clinics. The prostate was the first treatment site to see widespread use of gold fiducial markers [21,22], but they have subsequently been applied to treatments of the liver [23,24], pancreas [25], and lung [26,27]. The use of fiducial markers for efficient, accurate localization of target tissues that would otherwise not be visible in a two-dimensional radiograph has allowed radiotherapy practitioners to adopt smaller target margins and reduce the volume of healthy tissue receiving prescription dose levels. There are, however, a number of potential complications associated with using fiducial markers as surrogates for target tissue location.

Stability of the markers' location in the tissue is a potential concern. In general, the position of the markers within a tumor has been shown to be relatively stable [28], although allowing for several days to a week between marker placement and treatment simulation will reduce swelling or inflammation from marker placement and decrease the chances of marker migration during treatment. Additionally, the placement of three or more markers will make the drift of a single marker within the patient readily evident during daily IGRT. A larger concern with the use of fiducial markers is

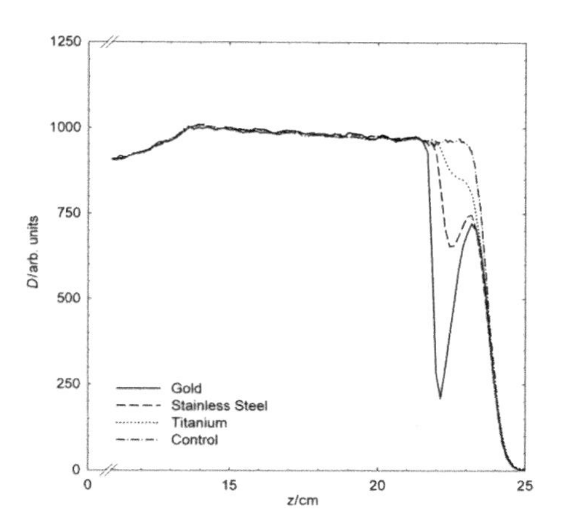

Figure 18–7 Simulated relative absorbed dose (D) as a function of depth (z) in a slab phantom. (From [31].)

the fact that the location of the markers may not alert users to the deformation of either target or healthy tissues in the treatment area. The shape of the prostate [29], and especially the seminal vesicles, has been shown to vary in ways that are not illustrated by marker configurations.

Finally, there are two potential complications with the use of fiducial markers for proton therapy especially that are not generally important in x-ray therapy applications. The first is that while gold markers are typically quite small—on the order of a few millimeters—their high density can lead to significant localized CT artifact in the simulation images and may impact the accuracy of a proton dose calculation. The second proton-specific concern is that high-density fiducial markers may cause clinically significant dose shadows downstream of the marker, especially near the distal end of the beam path. Figure 18–7 shows the impact of several different maker materials on a simulated dose distribution. Newhauser et al. [30] determined that while gold markers may cause excessive defects in a proton dose distribution, stainless steel markers represented a reasonable tradeoff between dose shadow and marker visibility in radiographic localization images.

While fiducial marker-based IGRT enjoyed widespread use in the x-ray therapy community for many years, its ubiquity has declined somewhat with the advance of in-room volumetric imaging capability, such as cone-beam CT, CT on rails, or more recently, MR-guided radiotherapy. Volumetric localization is not widely available in many current-generation proton therapy rooms, however, so there remains a viable role for fiducial markers in IGRT for proton treatments. The most advanced recent development in marker-based IGRT may be found in the real-time tumor-tracking system developed at Hokkaido University [32]. A pair of x-ray tubes and flat panel imagers mounted in a spot scanning gantry is used to track the position of a tumor under respiratory motion and develop a signal for respiratory-gated proton therapy.

18.3.2 Active Fiducial Markers

A drawback to many IGRT localization methods is that, with the exception of fluoroscopic tracking technologies, most systems allow the user to position the target tissue at the beginning of a treatment fraction, but they do not provide any data on the location of the target once the therapeutic beam has been turned on. The Calypso 4D Localization System (Varian Medical Systems, Palo Alto, CA), however, provides real-time tracking data from implanted electromagnetic transponders. Several transponders are placed in the target volume before simulation, and during treatment the beacons are activated by an electromagnetic panel which detects the location of each transponder with 10 Hz frequency. The use of Calypso to track the prostate gland during x-ray radiotherapy is widely reported in the literature [33,34]. There are challenges associated with the application of the Calypso to proton therapy—specifically the possibility for dose shadowing downstream of the transponders, as well as potential interference between the electromagnetic array panel with proton beam angles. Tang et al. [35] have used 4D data sets recorded from actual x-ray patient treatments to assess the impact of the observed intra-fractional motion on simulated scanned proton treatments. They found that in the worst-case scenario among the patient traces studied, the total CTV dose delivered over a course of treatment could be compromised by approximately 2% from intra-fractional motion alone. With the beam arrangements used (opposed laterals), they found that target coverage degradation was similar between the single-field uniform dose (SFUD) plans and the intensity-modulated proton therapy (IMPT) plans.

18.3.3 Endorectal Balloons

In order to reduce the inter- and intra-fractional variation in the position of the prostate relative to bony anatomy, patients are sometimes simulated and treated with an endorectal balloon (ERB) [36,37]. It has also been proposed that ERBs may reduce rectal toxicity by limiting the volume of rectal wall within the high-dose region of a treatment plan [38]. ERBs have been more frequently applied to prostate irradiation in proton therapy centers than they have in x-ray clinics, largely because inter-fractional variation in the relative position of the prostate and pelvic bony anatomy can have a larger effect on the dose delivered by a proton plan.

Recently, the effect of ERBs on intra-fractional motion has been studied through the use of Calypso tracking during x-ray radiotherapy [39]. Wang et al. [39] found that intra-fractional motion exceeding 1 cm was only observed in the patient cohort being treated without ERBs. Additionally, the time-dependent displacements of the prostate were smaller for the ERB patient cohort. While a 5 mm 3D internal margin would be required to cover patients without ERBs for a 6-minute treatment fraction, only 3 mm would be necessary if an ERB is used. The stabilizing effect of ERBs may be limited, however, in larger prostate treatment volumes. Wachter et al. [36] found that ERBs were not particularly useful in immobilizing the seminal vesicles.

While ERBs in x-ray treatments are generally inflated with air, there is an advantage in proton therapy to filling the ERB with water. A water-filled ERB provides consistent density near the target volume border and eliminates heterogeneous material borders along the beam path for lateral treatment fields. Vargas et al. [40] studied the benefit of a water-filled ERB vs. injecting saline directly into the rectum before treatment. While a cohort of patients benefited from the ERB, primarily because of pelvic anatomy, most patients received a similar benefit from water alone.

Finally, ERB placement technique may also play a role in the effectiveness of the device in reducing inter-fractional anatomy variations. The ERB is designed to press against the anterior rectal wall near the prostate, but stool or gas in the rectum may lead to improper positioning of the ERB. Placing a catheter in the rectum before the balloon is inserted will help evacuate excess gas near the balloon, but gas may continue to accumulate at the superior end of the ERB during treatment. Wootton et al. [41] reported that an ERB with a central gas release conduit was effective in eliminating gas pockets that would have developed during treatment.

18.3.4 Rectal Spacers

Proton treatments for prostate cancer are generally quite conformal, and the associated rectal dose volumes are well tolerated by most patients. A novel method of further reducing dose to the rectum has emerged, however, in which hyaluronic acid is injected into the perirectal fat, forming a spacer between the anterior rectal wall and the target volume. This approach was first reported in 2007, to a cohort of patients receiving high dose rate (HDR) brachytherapy boosts to a course of external beam radiotherapy [42]. Before the second of two HDR treatments, 3–7 ml of hyaluronic acid was injected into the patient, displacing the rectum 2 cm away from the prostate. Rectal doses were shown to drop significantly in the treatment after hyaluronic acid injection.

A cadaver study was performed to examine a polymer gel as a spacer between the prostate and rectum for external beam treatments. Each of the specimens received a CT simulation before and after injection with the spacer, and intensity-modulated x-ray plans were developed for pre- and post-injection CT images. It was found that a separation of 10–15 mm between prostate and rectum was sufficient to achieve an 80% reduction in rectal V70 [43]. A clinical series of 10 patients receiving HDR brachytherapy followed by intensity-modulated x-ray treatment found that injection of 9 ml of a hyaluronan gel before treatment led to a significant reduction in rectal toxicity [44]. A subsequent study of 35 patients with hyaluronic acid application before treatment found that the use of a rectal spacer was also correlated with an improved acute quality of life [45].

Rectal spacers have not yet been used clinically in proton treatments for prostate cancer. However, a cadaver study has been performed in which a rectal spacer hydrogel was injected into a specimen to explore the application of anteriorly angled beams to achieve lower doses to the rectum and femoral heads. Anterior beams are typically avoided in proton treatments to the prostate because of the danger of ranging out in the rectum if the radiological depth of the prostate is less than expected. It was found that the prostate–rectum separation afforded by the spacer may reduce rectal toxicity for uniform scanning deliveries, but not necessarily for pencil-beam scanning treatments [46].

18.4 Treatment Simulation

The treatment simulation appointment serves three purposes—to build any patient-specific immobilization devices, to acquire a CT scan for dose calculation, and to provide reference images for daily treatment alignment. The simulation process is a uniquely significant point in the radiation therapy process, since any anomalous features in patient anatomy or position may result in a systematic error over the treatment course. The patient must be immobilized in a repeatable, comfortable position that is consistent with the intended treatment technique and beam arrangement. The patient pose, such as arm position in a thoracic treatment, may impact the suitability of various field angles, so the process should begin with a discussion between simulation therapists, dosimetrists, and physicians. With knowledge of the target location, characteristics, and expected treatment technique, the simulation staff can choose the optimal patient position, determine the length of the patient that must be scanned for dose calculation, and decide whether any respiratory motion management such as 4DCT or breath hold may be required.

18.4.1 Immobilization Construction

A wide variety of patient-specific immobilization devices, such as alpha cradles, thermoplastic masks, or vacuum bags, may be used to comfortably and reproducibly immobilize the patient. Ideally, these devices will be used throughout a course of treatment, since modification of one of them would generally require a new simulation appointment and treatment plan. There are a number of reasons why an immobilization device may need to be replaced before a treatment course is complete. Patients may routinely lose weight or experience tumor shrinkage, rendering a mask or head mold too loose to reproducibly hold a patient. In this case, a subsequent simulation may not be avoidable.

A number of other common problems can be prevented, however, by separating patients' immobilization construction sessions from their CT imaging appointments. It has been shown repeatedly that patients' positions within their immobilization devices are typically much different in the first few fractions than they are over the remaining course of treatment [47,48]. Since the simulation session is the first experience in a radiotherapy course, it should be expected that patients' stress levels and general discomfort may lead to them being immobilized in a way that is inconsistent with how they may appear later in the treatment. In breast treatment, Das et al. [49] showed that patients relax significantly in second sessions. As a minimum, after any patient-specific devices are constructed, the simulation staff should ask patients to get out of the immobilization equipment, walk for a bit, and then re-enter the immobilization for CT imaging. Patient poses within a device will be much different if they are entering a pre-constructed device than if that equipment has just been formed around them. Getting patients out of their immobilization before imaging will also afford the simulation staff an opportunity to examine the freshly created devices for any problems that would interfere with proton treatment planning, such as a wrinkle in a vacuum bag that may cast a heterogeneity in the proton beam path.

Ideally, the interval between immobilization creation and CT simulation should be several hours or more as observed by Das et al. [49]. Some thermoplastic masks may shrink in the hours after they are created, resulting in a mask that is too tight to comfortably place over patients when they return for treatment. Sometimes this is handled in x-ray clinics by shimming the patient's head under the mask for simulation and removing the shim for treatment. This approach is not advised in proton therapy, however, since a change in immobilization thickness will alter the radiographic depth of the target tissue. A shim may be used to create the mask, however, if the patient is sent away after the immobilization appointment. Patients may return when the mask has shrunk to its ultimate size, and then they can be comfortably scanned and ultimately treated without a shim. In this way, patients are imaged with exactly the same immobilization equipment they'll be treated with, and the fit of the mask will be very close to treatment conditions.

18.4.2 CT Scan

The final stage in the simulation process is the CT scan. The CT scan provides a 3D model of the patient for dose calculation, as well as proton stopping power data for determination of the radiological depth of the target tissue. Calibration of CT Hounsfield Units (HU) to proton is discussed in Chapter 7. It is critical that patients are scanned with a CT technique for which an HU-to-stopping-power curve exists. in the interest of minimizing radiation dose during the imaging process, modern CT scanners may offer adaptive imaging techniques in which either the kVp or the mA are varied during a scan. HU values are typically not affected by the mA setting, but variation of the kVp during a CT simulation could lead to inaccuracies in the stopping power determination.

A number of factors can impact the accuracy of the HU, and ultimately the proton range determination, in the reconstructed CT scan. High-density materials in the patient, such as dental work or spinal stabilization hardware, may cause significant CT artifact through beam hardening, leading to incorrect stopping power determination, not just in the high-density material, but in the area around the high-density material. It has been shown that pencil-beam dose calculations exhibit considerable inaccuracies near titanium spinal implants [50], and these inaccuracies may be partially responsible for decreased local control rate for patients who present with surgical stabilization of the spine [51]. High-density immobilization equipment, such as bite block holders, may cast similar artifacts in CT images, and they should not be placed near the target volume or critical organs for simulation.

Changes in the CT x-ray spectrum with depth in the patient may also cause a variation in CT number with patient size, composition, and location in the scanner. Figure 18–8 shows a heterogeneity phantom scanned twice with the same imaging technique, but in the second scan the edge of the phantom is placed closer to the boundary of the scan field of view. High-density features in the phantom cast considerably larger artifacts in the scan when they are near the border of the image. The location of heterogeneous features in a patient may not be controllable, but users

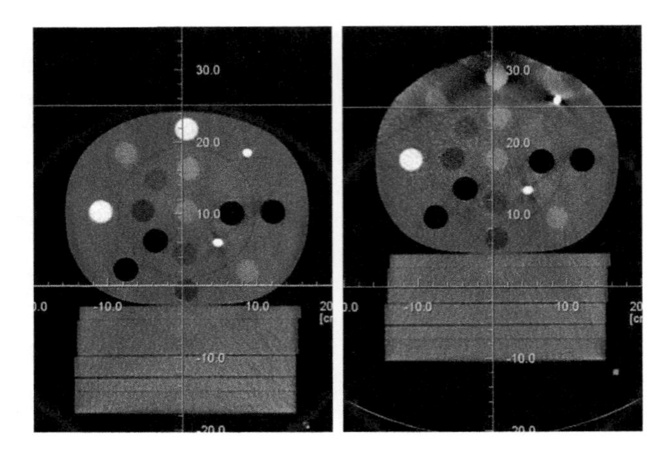

Figure 18–8 Heterogeneous features in an electron density phantom cast large image artifacts when they are situated near the border of the reconstruction field of view. (Image courtesy of Debra Brinkmann, Mayo Clinic.)

should be aware of the resulting uncertainties in stopping power and design target volumes and treatment fields accordingly.

Because of the importance of the CT simulation image on the success of a course of proton therapy, a CT study should be examined for anomalous features before the patient is sent away. A study of prostate patients who received x-ray therapy at M.D. Anderson Cancer Center showed that patients whose rectum was distended in the CT planning study had a higher rate of biochemical failure after treatment [52]. These patients were treated in the era before IGRT techniques, which could visualize and localize soft tissue targets, and so there may be a lower correlation between rectal cross-sectional area and relapse today. However, abnormal anatomical features may be associated not just with atypical location of the target tissue, but distortions of the target volume which cannot be corrected by IGRT techniques [29]. If a CT simulation shows the patient's rectum is filled with gas or stool, for example, it may be advisable to repeat the scan at a time when these features are no longer present.

18.4.3 4D Imaging

Finally, the treatment of moving targets presents a special challenge in proton therapy. The variation in position and range of a target with respiratory motion requires extensive knowledge of the anatomy throughout a respiratory cycle. Motion management in proton therapy is discussed extensively in Chapter 25, but a brief mention of the simulation process is warranted here. Depending on the range of respiratory motion of the tumor relative to its size, any number of strategies may be employed to robustly treat a moving tumor with a proton beam. Scattered beams may be used to treat an internal target volume (ITV) that contains the full range of motion of a tumor, but this approach may not be advisable with a scanning beam due to the interplay effect. Respiratory gating, breath held treatments, repainting, or tracking are all approaches that have been proposed for proton treatments of moving anatomy. Each of these strategies, even the ITV approach, requires a 4DCT scan to image deformation and range variation of anatomy over the respiratory cycle. The exception may be for breath-held treatments that could conceivably be planned from a full-exhale or full-inhale phase of a 4DCT. However, a patient's anatomy may look quite different in the inhale phase of a free-breathing 4DCT than it does in a breath-held state. For this reason, the anticipated treatment strategy should be identified before simulation so that the CT scan may most accurately model the state of the patient's anatomy during treatment.

18.5 Conclusions

The simulation process is a uniquely important step in a course of proton radiotherapy. The ability of the simulation staff to immobilize the patient in a comfortable, reproducible fashion and then to generate a representative CT image of the patient will have significant impact on the success of the treatment. Proton treatments are particularly sensitive to variations in patient pose, internal anatomy, and interference from external immobilization devices. Wider application of proton therapy as a radiation oncology modality has led to commercial development of treatment couches, immobilization hardware, and IGRT strategies that can maximize the likelihood of successful treatment. It is incumbent upon clinical users, however, to understand the relationship between immobilization, imaging, and the accuracy of proton dose calculation and delivery. It is also important to create immobilization devices that provide accuracy in parameters and comfort for the patients. Due to CT imaging and range perturbation, it is advised that such devices are made of near-tissue-equivalent media without any metal.

References

1. Liebl J, Paganetti H, Zhu M, Winey BA. The influence of patient positioning uncertainties in proton radiotherapy on proton range and dose distributions. *Med Phys* 2014;35(11):5088-97.
2. Park PC, Cheung JP, Zhu XR, Lee AK, Sahoo N, Tucker SL, Liu W, Li H, Mohan R, Court LE, Dong L. Statistical assessment of proton treatment plans under setup and range uncertainties. *Int J Radiat Oncol Biol Phys* 2013;86(5):1007–13.
3. Both S, Wang KK-H, Plastaras JP, Deville C, Bar Ad V, Tochner Z, Vapiwala N. Real-time study of prostate intrafraction motion during external beam radiotherapy with daily endorectal balloon. *Int J Radiat Oncol Biol Phys* 2011;81(5):1302–309.
4. Srivastava SP, Das IJ, Kumar A, Johnstone PA, Cheng CW. Impact of rectal balloon-filling materials on the dosimetry of prostate and organs at risk in photon beam therapy. *J Appl Clin Med Phys* 2013;14:3993.
5. Smeenk RJ, Teh BS, Butler EB, van Lin EN, Kaanders JH. Is there a role for endorectal balloons in prostate radiotherapy? A systematic review. *Radiother Oncol* 2010;95:277–282.
6. Patel RR, Orton N, Tome WA, Chappell R, Ritter MA. Rectal dose sparing with a balloon catheter and ultrasound localization in conformal radiation therapy for prostate cancer. *Radiother Oncol* 2003;67:285–294.

7. Vargas C, Saito AI, Hsi WC, Indelicato D, Falchook A, Zengm Q, Oliver K, Keole S, Dempsey J. Cine-magnetic resonance imaging assessment of intrafraction motion for prostate cancer patients supine or prone with and without a rectal balloon. *Am J Clin Oncol* 2010;33:11–16.
8. Olch AJ, Gerig L, Li H, Mihaylov I, Morgan A. Dosimetric effects caused by couch tops and immobilization devices: Report of AAPM task group 176. *Med Phys* 2014;41:061501.
9. Schneider U, Pedroni E, Lomax A. The calibration of CT Hounsfield units for radiotherapy treatment planning. *Phys Med Biol* 1996;41(1):111–24.
10. Wroe AJ, Ghebremedhin A, Gordon IR, Schulte RW, Slater JD. Water equivalent thickness analysis of immobilization devices for clinical implementation in proton therapy. *Technol Cancer Res Treat* 2014;13:415–420.
11. Ma C-MC, Lomax T. Proton and carbon ion therapy. Boca Raton, FL: Taylor & Francis, 2013.
12. Giebeler A, Newhauser WD, Amos RA, Mahajan A, Homann K, Howell RM. Standardized treatment planning methodology for passively scattered proton craniospinal irradiation. *Radiat* 2013;8:32.
13. Wroe AJ, Bush DA, Slater JD. Immobilization considerations for proton radiation therapy. *Technol Cancer Res Treat* 2014;13:217–226.
14. Devic S. MRI simulation for radiotherapy treatment planning. *Med Phys 2012;*35(11):6701–6711.
15. Brock KK. Deformable registration accuracy C. Results of a multi-institution deformable registration accuracy study (MIDRAS). *Int J Radiat Oncol Biol Phys* 2010;68(3):892–7.
16. Titt U, Mirkovic D, Sawakuchi GO, Perles LA, Newhauser WD, Taddei PJ, Mohan R. Adjustment of the lateral and longitudinal size of scanned proton beam spots using a pre-absorber to optimize penumbrae and delivery efficiency. *Phys Med Biol 2010;*55:7097–7106.
17. Both S, Shen J, Kirk M, Lin L, Tang S, Alonso-Basanta M, Lustig R, Lin H, Deville C, Hill-Kayser C, Tochner Z, McDonough J. Development and clinical implementation of a universal bolus to maintain spot size during delivery of base of skull pencil beam scanning proton therapy. *Int J Radiat Oncol Biol Phys* 2014;90:79–84.
18. Lam KL, Ten Haken RK, McShan DL, Thornton AF, Jr. Automated determination of patient setup errors in radiation therapy using spherical radio-opaque markers. *Med Phys* 1993;20:1145–1152.
19. Gall KP, Verhey LJ, Wagner M. Computer-assisted positioning of radiotherapy patients using implanted radiopaque fiducials. *Med Phys* 1993;20:1153–1159.
20. Balter JM, Sandler HM, Lam K, Bree RL, Lichter AS, Ten Haken RK. Measurement of prostate movement over the course of routine radiotherapy using implanted markers. *Int J Radiat Oncol Biol Phys* 1995;31:113–118.
21. Herman M, Pisansky TM, Kruse JJ, Prisciandaro JI, Davis BJ, King BF. Technical aspects of daily on-line positioning of the prostate for three-dimensional conformal radiotherapy using an electronic portal imaging device. *Int J Radiat Oncol Biol Phys* 2003;57:1131–1140.
22. Litzenberg D, Dawson LA, Sandler H, Sanda MG, McShan DL, Ten Haken RK, Lam KL, Brock KK, Balter JM. Daily prostate targeting using implanted radiopaque markers. *Int J Radiat Oncol Biol Phys* 2002;52:699–703.
23. Kitamura K, Shirato H, Shimizu S, Shinohara N, Harabayashi T, Shimizu T, Kodama Y, Endo H, Onimaru R, Nishioka S, Aoyama H, Tsuchiya K, Miyasaka K. Registration accuracy and possible migration of internal fiducial gold marker implanted in prostate and liver treated with real-time tumor-tracking radiation therapy (RTRT). *Radiother Oncol* 2002;62:275–281.
24. Shirato H, Seppenwoolde Y, Kitamura K, Onimura R, Shimizu S. Intrafractional tumor motion: Lung and liver. *Semin Radiat Oncol* 2004;14:10–18.
25. Ahn YC, Shimizu S, Shirato H, Hashimoto T, Osaka Y, Zhang XQ, Abe T, Hosokawa M, Miyasaka K. Application of real-time tumor-tracking and gated radiotherapy system for unresectable pancreatic cancer. *Yonsei Med J* 2004;45:584–590.
26. Harada T, Shirato H, Ogura S, Oizumi S, Yamazaki K, Shimizu S, Onimaru R, Miyasaka K, Nishimura M, Dosaka-Akita H. Real-time tumor-tracking radiation therapy for lung carcinoma by the aid of insertion of a gold marker using bronchofiberscopy. *Cancer* 2002;95:1720–1727.

27. Imura M, Yamazaki K, Shirato H, Onimaru R, Fujino M, Shimizu S, Harada T, Ogura S, Dosaka-Akita H, Miyasaka K, Nishimura M. Insertion and fixation of fiducial markers for setup and tracking of lung tumors in radiotherapy. *Int J Radiat Oncol Biol Phys* 2005;63:1442–1447.
28. Schallenkamp JM, Herman MG, Kruse JJ, Pisansky TM. Prostate position relative to pelvic bony anatomy based on intraprostatic gold markers and electronic portal imaging. *Int J Radiat Oncol Biol Phys* 2005;63:800–811.
29. Nichol AM, Brock KK, Lockwood GA, Moseley DJ, Rosewall T, Warde PR, Catton CN, Jaffray DA. A magnetic resonance imaging study of prostate deformation relative to implanted gold fiducial markers. *Int J Radiat Oncol Biol Phys* 2007;67:48–56.
30. Newhauser W, Fontenot J, Koch N, Dong L, Lee A, Zheng Y, Waters L, Mohan R. Monte Carlo simulations of the dosimetric impact of radiopaque fiducial markers for proton radiotherapy of the prostate. *Phys Med Biol* 2007;52:2937–2952.
31. Newhauser W, Fontenot J, Zheng Y, Polf J, Titt U, Koch N, Zhang X, Mohan R. Monte Carlo simulations for configuring and testing an analytical proton dose-calculation algorithm. *Phys Med Biol* 2007;52:4569–4584.
32. Matsuura T, Miyamoto N, Shimizu S, Fujii Y, Umezawa M, Takao S, Nihongi H, Toramatsu C, Sutherland K, Suzuki R, Ishikawa M, Kinoshita R, Maeda K, Umegaki K, Shirato H. Integration of a real-time tumor monitoring system into gated proton spot-scanning beam therapy: An initial phantom study using patient tumor trajectory data. *Med Phys* 2013;40:071729.
33. Willoughby TR, Kupelian PA, Pouliot J, Shinohara K, Aubin M, Roach M, 3rd, Skrumeda LL, Balter JM, Litzenberg DW, Hadley SW, Wei JT, Sandler HM. Target localization and real-time tracking using the calypso 4D localization system in patients with localized prostate cancer. *Int J Radiat Oncol Biol Phys* 2006;65:528–534.
34. Kupelian P, Willoughby T, Mahadevan A, Djemil T, Weinstein G, Jani S, Enke C, Solberg T, Flores N, Liu D, Beyer D, Levine L. Multi-institutional clinical experience with the calypso system in localization and continuous, real-time monitoring of the prostate gland during external radiotherapy. *Int J Radiat Oncol Biol Phys* 2007;67:1088–1098.
35. Tang S, Deville C, McDonough J, Tochner Z, Wang KK-H, Vapiwala N, Both S. Effect of intrafraction prostate motion on proton pencil beam scanning delivery: A quantitative assessment. *Int J Radiat Oncol Biol Phys* 2013;87:375–382.
36. Wachter S, Gerstner N, Dorner D, Goldner G, Colotto A, Wambersie A, Potter R. The influence of a rectal balloon tube as internal immobilization device on variations of volumes and dose-volume histograms during treatment course of conformal radiotherapy for prostate cancer. *Int J Radiat Oncol Biol Phys* 2002;52:91–100.
37. D'Amico AV, Manola J, Loffredo M, Lopes L, Nissen K, O'Farrell DA, Gordon L, Tempany CM, Cormack RA. A practical method to achieve prostate gland immobilization and target verification for daily treatment. *Int J Radiat Oncol Biol Phys* 2001;51:1431–1436.
38. van Lin ENJT, Hoffmann AL, van Kollenburg P, Leer JW, Visser AG. Rectal wall sparing effect of three different endorectal balloons in 3D conformal and IMRT prostate radiotherapy. *Int J Radiat Oncol Biol Phys* 2005;63:565–576.
39. Wang KK-H, Vapiwala N, Deville C, Plastaras JP, Scheuermann R, Lin H, Bar Ad V, Tochner Z, Both S. A study to quantify the effectiveness of daily endorectal balloon for prostate intrafraction motion management. *Int J Radiat Oncol Biol Phys* 2012;83:1055–1063.
40. Vargas C, Mahajan C, Fryer A, Indelicato D, Henderson RH, McKenzie C, Horne D, Chellini A, Lawlor P, Li Z, Oliver K, Keole S. Rectal dose-volume differences using proton radiotherapy and a rectal balloon or water alone for the treatment of prostate cancer. *Int J Radiat Oncol Biol Phys* 2007;69:1110–1116.
41. Wootton LS, Kudchadker RJ, Beddar AS, Lee AK. Effectiveness of a novel gas-release endorectal balloon in the removal of rectal gas for prostate proton radiation therapy. *J Appl Clin Med Phys* 2012;13(5):3945.
42. Prada PJ, Fernandez J, Martinez AA, de la Rua A, Gonzalez JM, Fernandez JM, Juan G. Transperineal injection of hyaluronic acid in anterior perirectal fat to decrease rectal toxicity from radiation delivered with intensity modulated brachytherapy or EBRT for prostate cancer patients. *Int J Radiat Oncol Biol Phys* 2007;69:95–102.
43. Susil RC, McNutt TR, DeWeese TL, Song D. Effects of prostate-rectum separation on rectal dose from external beam radiotherapy. *Int J Radiat Oncol Biol Phys* 2010;76:1251–1258.

44. Wilder RB, Barme GA, Gilbert RF, Holevas RE, Kobashi LI, Reed RR, Solomon RS, Walter NL, Chittenden L, Mesa AV, Agustin J, Lizarde J, Macedo J, Ravera J, Tokita KM. Cross-linked hyaluronan gel reduces the acute rectal toxicity of radiotherapy for prostate cancer. *Int J Radiat Oncol Biol Phys* 2010;77:824–830.
45. Wilder RB, Barme GA, Gilbert RF, Holevas RE, Kobashi LI, Reed RR, Solomon RS, Walter NL, Chittenden L, Mesa AV, Agustin JK, Lizarde J, Macedo JC, Ravera J, Tokita KM. Cross-linked hyaluronan gel improves the quality of life of prostate cancer patients undergoing radiotherapy. *Brachytherapy* 2011;10:44–50.
46. Christodouleas JP, Tang S, Susil RC, McNutt TR, Song DY, Bekelman J, Deville C, Vapiwala N, Deweese TL, Lu HM, Both S. The effect of anterior proton beams in the setting of a prostate-rectum spacer. *Med Dosim* 2013;38:315–319.
47. de Boer HC, Heijmen BJ. A protocol for the reduction of systematic patient setup errors with minimal portal imaging workload. *Int J Radiat Oncol Biol Phys* 2001;50:1350–1365.
48. Bel A, van Herk M, Bartelink H, Lebesque JV. A verification procedure to improve patient set-up accuracy using portal images. *Radiother Oncol* 1993;29:253–260.
49. Das IJ, Cheng CW, Fosmire H, Kase KR, Fitzgerald TJ. Tolerances in setup and dosimetric errors in the radiation treatment of breast cancer. *Int J Radiat Oncol Biol Phys* 1993;26:883–890.
50. Verburg JM, Seco J. Dosimetric accuracy of proton therapy for chordoma patients with titanium implants. *Med Phys* 2013;40:071727.
51. Staab A, Rutz HP, Ares C, Timmermann B, Schneider R, Bolsi A, Albertini F, Lomax A, Goitein G, Hug E. Spot-scanning-based proton therapy for extracranial chordoma. *Int J Radiat Oncol Biol Phys* 2011;81:e489–496.
52. de Crevoisier R, Tucker SL, Dong L, Mohan R, Cheung R, Cox JD, Kuban DA. Increased risk of biochemical and local failure in patients with distended rectum on the planning CT for prostate cancer radiotherapy. [see comment]. *Int J Radiat Oncol Biol Phys* 2005;62:965–973.

Chapter 19

Dose Calculations for Proton Beam Therapy: Semi-empirical Analytical Methods

Radhe Mohan, Ph.D.[1], X. Ronald Zhu, Ph.D.,[1] and Harald Paganetti, Ph.D.[2]

[1]MD Anderson Cancer Center,
Houston, TX
[2]Massachusetts General Hospital,
Boston, MA

19.1 Introduction

Computed dose distributions and the displays and indices derived from them are the basis for making treatment decisions. In order for these decisions to be judicious, the distributions need to be accurate and reflect, to a good approximation, the dose distribution that a patient actually receives in a treatment fraction and over the course of radiotherapy. To that end, dose distributions may need to be computed multiple times over the course of radiotherapy to adapt to changing anatomy, preferably taking into account dose distributions already delivered.

Best approximations of the dose distributions actually delivered at the end of a treatment course, calculated for each of a large number of patients in a study, may also be used for assessing correlations of dose with response to treatments. Such dose distributions may be calculated using multiple (e.g., weekly) images acquired over the course of radiotherapy and accumulated using deformable registration. For anatomies affected by respiratory motion, dose distributions would need to be calculated for each phase of a 4D CT.

In addition, computed dose distributions are used for optimizing treatment plans, including beam configuration optimization and robust optimization to make dose distributions resilient in the face of uncertainties. The competing requirements of speed and accuracy, especially considering the magnitude of the data involved for many of the applications mentioned here, often lead to compromises in methods of dose computation. While these needs and challenges also exist for photon therapy, they are particularly critical for proton therapy.

Although protons have been used for radiotherapy for many decades, the semi-empirical analytical (SEA) models (i.e., formalisms and algorithms) for dose calculations have not advanced to the same degree as those for photons. Because of the charged nature, heavier mass, finite range, and scattering properties of protons, the models developed for photons and electrons are not extensible to protons. In proton dose calculations, it is also necessary to deal with another dimension—that of energy.

While the most accurate method of computing dose distributions is Monte Carlo (MC) simulations (see Chapter 20), for practical reasons, namely computational efficiency, proton dose distributions in the current state of the art are carried out using SEA models. These models out of necessity make numerous assumptions and approximations that have an impact on accuracy. As different models make different assumptions and approximations, each of these models may have different deficiencies in accuracy under different situations. It is not practical to describe all available models in this chapter. Of the three models discussed, one is the common basis for many of the models in use, the second is implemented on a commonly used commercial system, and the third is arguably the most advanced. These models will be described briefly with minimal mathematical detail and with some general comments about the nature of the approximations made and their consequences.

Note that the dose in this chapter is in units of Gy (RBE) unless otherwise explicitly stated. This is valid even for photons where RBE=1.

19.2 Data Required for Dose Computations

19.2.1 CT Numbers and Stopping Power Ratios for Human Tissues

Radiation dose computations utilize 3D CT images represented as Digital Imaging And Communications in Medicine (DICOM) streams [1]. Each image voxel is characterized by a Hounsfield unit (HU). Dose calculation algorithms in photon therapy use electron density ratios (EDRs) with respect to water because the dominant energy loss process is interactions of photons and secondary electrons with atomic electrons. Protons, on the other hand, lose energy by ionizations, multiple Coulomb scattering, and non-elastic nuclear reactions. Because each interaction type has a different relationship with the materials, whose characteristics are obtained from the CT scan [2,3], stopping power ratios (SPRs) relative to water are used to define water-equivalent tissue properties. Each proton treatment planning system uses a HUs-to-SPRs conversion curve (also called the "CT calibration curve") obtained during the commissioning of the planning system.

The accuracy of dose calculations may be affected significantly by the ability to characterize tissues accurately based on CT scans [4,5]. CT numbers reflect the attenuation coefficients of human tissues for diagnostic x-rays and may be identical for different combinations of elemental compositions, elemental weights, and mass densities [6]. The HUs-to-SPRs conversion curve is a function of the kVp and the energy spectra. Therefore, calibration must be performed for each CT scanner and imaging technique. The uncertainty in CT conversion algorithms is a part of the overall range uncertainty in proton therapy [7,8]. While there are additional uncertainties in HUs due to statistical fluctuations, beam hardening, volume averaging, etc., such uncertainties will be ignored in this chapter.

Traditionally, for photon therapy, the conversion of HUs to EDRs has employed tissue substitute materials commonly used for radiation dosimetry and radiobiology. However, the chemical composition of such tissue substitutes is often quite different from that of actual human tissues. Nevertheless, the effect on computed photon dose distributions is negligible. That is not the case for protons and heavier particles, for which the so-called "stoichiometric calibration" methods have been suggested. In these methods, the measured HUs of tissue substitutes and the chemical composition of actual human tissues are used to predict HUs for the actual tissues [6,9,10]. A robust division of most soft tissues and skeletal tissues is feasible, with the exception of the CT number range between 0 and 100, which is difficult to characterize accurately. The following step-by-step process to convert HUs to SPRs has been adapted from Schneider et al. [10].

1. Several tissue substitute materials of known chemical composition and physical density are chosen.
2. CT scans of phantoms with inserts made of the tissue substitutes are obtained with the scanner using the imaging technique to be used for acquiring images for proton therapy planning.
3. The measured HUs are then fitted to the following equations:

$$HU = 1000 \times \mu / \mu_w \qquad (19.1)$$

$$\mu = \sigma \times N_g(Z, A) \times [K^{ph} \times \tilde{Z}^{3.62} + K^{coh} \times Z^{1.86} + K^{KN}] \qquad (19.2)$$

where N_g is the number of electrons per unit volume of the material being scanned. Coefficients K^{ph}, K^{coh} and K^{KN} correspond to the photo-electric, coherent, and Compton (Klein-Nishina) scattering processes, respectively, and are obtained by the fitting process. It is important to reemphasize that these coefficients must be determined for the x-ray spectrum unique to the CT scanner and the imaging technique. Zs are the averages over a mixture of elements comprising the materials and are computed using

$$\tilde{Z} = \left[\sum \lambda_i \times Z_i^{3.62}\right]^{1/3.62}, \quad \tilde{\tilde{Z}} = \left[\sum \lambda_i \times Z_i^{1.86}\right]^{1/1.86} \qquad (19.3)$$

$$and\ \lambda_i = N_g^i / N_g$$

for the i'th element of the material. While they are available in a 1989 ICRU publication [11], the chemical composition and densities of various tissue substitutes should be confirmed with the manufacturer of the substitutes.

4. The HUs of selected ICRP human tissues (ICRP 1975) [12] are then computed by inserting their chemical compositions into the equations above for the CT scanner and imaging technique.
5. With the knowledge of the chemical composition of the same selected ICRP human tissues, the relative proton stopping powers using the Bethe-Block formula are calculated.

$$\rho_s = \frac{\rho_e \left\{ log_e \left[\dfrac{2m_e c^2 \beta^2}{I_m \left(1-\beta^2\right)} \right] - \beta^2 \right\}}{\left\{ log_e \left[2m_e c^2 \beta^2 / I_{water} \left(1-\beta^2\right) \right] - \beta^2 \right\}} = \rho_e K \tag{19.4}$$

where I_m and I_{water} are the mean ionization energies of the material and water respectively, m_e is the electron mass and βc is the proton velocity.

6. A fit through the data points is made to produce the final SPRs vs. HUs calibration curve.

An example of SPR vs. HUs curve is shown in Figure 19–1. For computation of dose distributions, the CT image is transformed into a 3D matrix of SPRs.

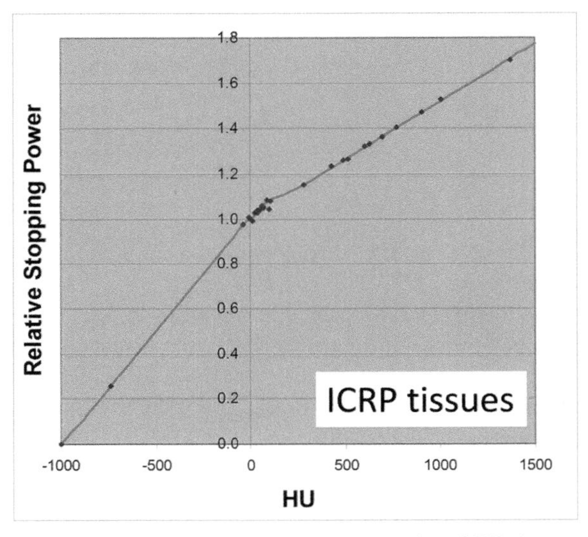

Figure 19–1 A typical HUs-to-SPRs calibration curve for ICRP tissues. This example is for the data being used at the MD Anderson Cancer Center Proton Therapy Center.

19.2.2 Machine Characteristics

For passively scattered proton therapy (PSPT), a narrow proton beam of a given energy entering the nozzle (treatment head) is spread laterally with scatterers and modulated longitudinally with a range modulator wheel (RMW) to create a spread-out Bragg peak (SOBP) of desired width. The high-dose region of the resulting dose distribution is cuboid in shape, which is modified appropriately to conform laterally and distally to the target volume, plus appropriate margins using apertures and compensators. Typical semi-empirical dose distribution calculation algorithms need the following information (see Figure 19–2). Exact requirements may vary from system to system:

1. Nominal source-to-axis (i.e., isocenter) distance (SAD): This, along with the source-to-block (aperture) distance (which may be variable), is used to scale the "effective source size," and, thus, the penumbra.
2. Virtual SAD and virtual source position: Virtual SAD is obtained from in-air beam profiles measured at multiple distances and projected to converge to a

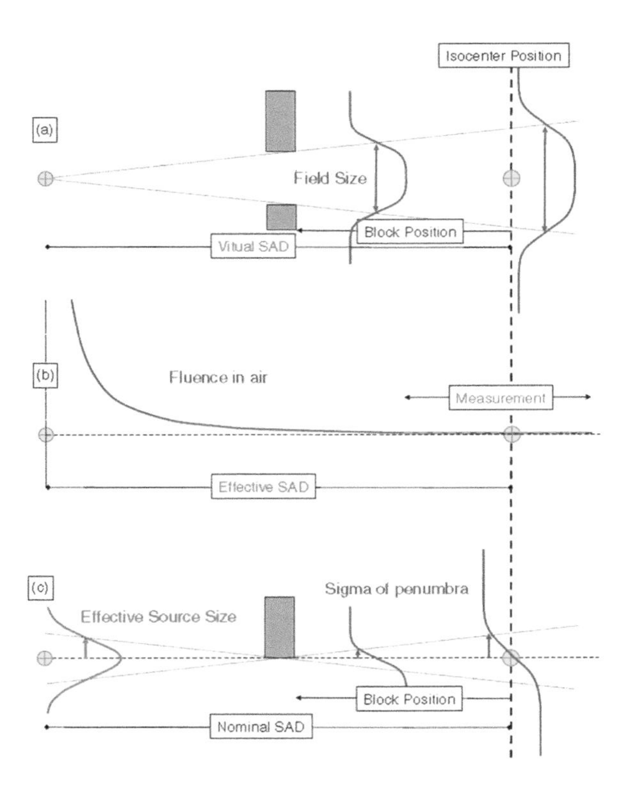

Figure 19–2 Schematic display of the geometries of measurements required for the configuration of typical passively scattered or uniform scanning dose calculation systems. The measured quantities are shown as blue lines, and the extracted parameters are in red. (a) Measurement of cross profiles of a large field. A fit to the field sizes at different positions yields the Virtual SAD. (b) The intensity distribution as a function of distance from the isocenter. (c) Measurements of the penumbra, which are fitted with a Gaussian function. The extracted penumbra parameter σ is projected back to the nominal source position. (Reproduced from Schaffner et al. [13].)

virtual source. These quantities together define the divergence characteristics of a PSPT beam.

3. Effective SAD: It is obtained from the measured dose in air as a function of distance from the source along the central ray of the beam. It is used for computing intensity as a function of distance (i.e., to apply the inverse square law).
4. Effective source size: The effective sources size is used for the computation of the shape of the beam penumbra. The source is typically assumed to be represented by a Gaussian function. The parameters of the Gaussian are obtained from half-blocked, in-air, cross-beam profiles measured at multiple positions relative to the isocenter.
5. Open-field depth–dose curves for unmodulated proton beams for each energy, i.e., the pristine Bragg curves, are obtained. Parameters of the dose computation model are extracted by fitting computed depth doses with the measured data. (NOTE: these data are needed for each combination of energy and RMW.)
6. Energy layer weights are determined from the widths of the RMW steps.
7. Apertures are specified as one or more contours. The assumption is that transmission through the aperture opening is unity and is zero in the blocked region.
8. Compensator is specified in the form of a matrix of thicknesses of the material (e.g., Lucite) used to fabricate compensators, the material density, and its SPR.

In general, the requirements for scanning beam and IMPT dose calculations are simpler, though additional steps might be necessary, including apertures close to the patient in order to improve the penumbra for large spot sizes. In general, the requirements for IMPT include the following:

1. Integral depth–dose (IDD) curves for the full range of energies used. An IDD is the integral of dose at a given depth distributed over an infinite (or a sufficiently large) plane in water for the thin scanning beam (i.e., "beamlet") of protons. It may be obtained by measurements with a large parallel plate chamber (typically about 8 cm in diameter). Considering that contributions to beamlet dose outside the chamber, especially for higher-energy beamlets and larger depths, cannot be ignored, depth-dependent correction factors, computed with MC simulations, need to be applied. Alternatively, IDDs may be calculated entirely with MC simulations. Figure 19–3 shows IDDs for the Hitachi proton therapy machine at MD Anderson Cancer Center.
2. IDDs are multiplied with beamlet profiles to compute dose distributions. Past algorithms assumed the profiles in water to be representable by Gaussians to a good approximation. In reality, they deviate from Gaussians due to large angle scattering, secondary nuclear particles, and scattering from monitoring devices (see Section 19.3.3 for examples.) It has been shown that a

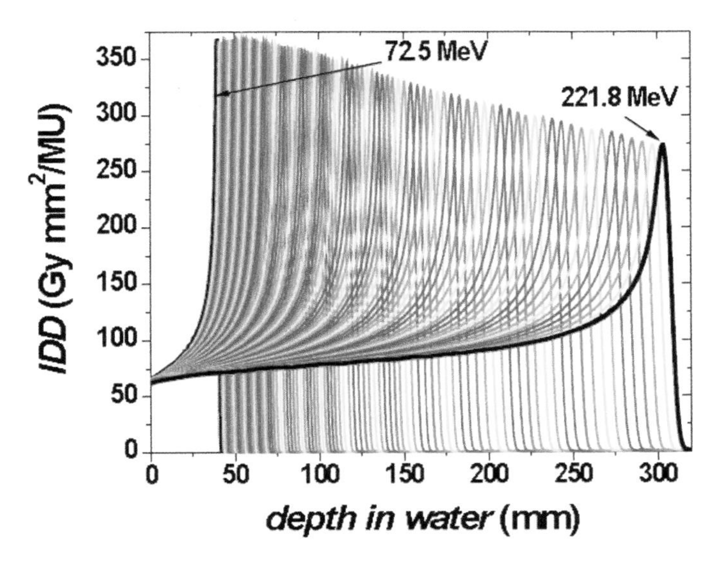

Figure 19–3 Integral Depth Dose (IDD) data for 92 energies from 72.5 MeV to 221.8 MeV in units of Gy mm^2 per MU. These data for the Hitachi proton therapy machine at MD Anderson Cancer Center were produced using a calibrated Monte Carlo system. By "calibrated" we mean that the number of protons required for MC simulation to produce the same dose under reference conditions per MU for each energy were determined empirically.

Gaussian representation is inadequate, even in air, because of scattering from the profile monitor, if one is present during treatment delivery. Increasingly, profiles are assumed to be represented by combinations of Gaussians. Sometimes functions other than Gaussians are also used [14]. The parameters of these functions are depth dependent.

19.2.3 Secondary Particles from Nuclear Interactions

Some SEA algorithms approximate contributions from the secondary nuclear particles as constant offset corrections to the dose distributions computed with the SEA algorithms, while others assume them to be embedded in the measured dose distribution data used for commissioning. Some other models attempt to explicitly, and thus more accurately, account for nuclear secondary particles. An example of such an algorithm is given in Section 19.3.3.

19.3 Dose Computation Algorithms

As mentioned, for practical reasons analytical semi-empirical formalisms and algorithms make numerous assumptions and approximations compromising their accuracy. Different developers of treatment planning systems (institutional or commercial) have different models, each making different assumptions and approximations. This means that the accuracy of a given treatment planning system may be adequate for some clinical situations but not for others. The most common approximation is the manner in which the passage of particle beams through complex media (e.g., compensators) and heterogeneous patient anatomy is managed. Most, if not all, SEA algorithms consider only the variations in densities (or SPRs) along the path of protons. These algorithms ignore, or only approximately account for, changes in the characteristics of proton beams due to scattering from neighboring regions. Other examples include ignoring the scattering from aperture boundaries and nuclear secondary particles. Illustrative examples are given in Section 19.5. The impact of such approximations and assumptions on clinical outcomes is often difficult to asses; however, considering many other sources of uncertainty, the accuracy of the current systems is assumed adequate. Nevertheless, it is important to understand the algorithms used in one's clinic and their limitations.

Most semi-empirical analytical algorithms for proton dose calculations in use today are variations of the "pencil beam" algorithm of Hong et al. [15]. This algorithm divides an arbitrary broad incident beam into pencil beams. The pencil beam dose distribution for a given energy is represented by the product of a pencil beam central axis depth dose (CADD) term and a lateral profile term. The CADD term is obtained from measured broad beam data in water, and it is essentially equivalent to the integral depth dose (IDD) of the pencil beam over an infinite plane. The lateral spreading of proton pencils, caused mostly by multiple Coulomb scattering, is assumed describable by a Gaussian. However, on theoretical grounds and for many applications and situations, Gaussian characterization is inadequate. This has been found to be particularly important for scanned beams and, therefore, combinations of multiple Gaussians or other functions are required [14,16,17]. Each profile must be normalized so that its integral over an infinite plane is unity.

For the purpose of describing the Hong algorithm [15], we will assume that the pencil beam profile is represented by a Gaussian whose σ (standard deviation, i.e., lateral spreading parameter) is obtained by fitting computed broad beam profiles with the corresponding measured data in water. To take into account beam-modifying devices and patient inhomogeneities, the σ of a given pencil beam at the depth of interest is obtained by summing in quadrature the multiple Coulomb scattering contributions from each of the beam-modifying devices and from the patient. The spreading in the intervening air gaps is also taken into account.

For passive scattering, the geometry of the treatment head relevant for dose calculations typically consists of a beam-spreading system (one or two scatterers and a range modulator), range shifters of various thicknesses for fine adjustment of the range, aperture and collimators to conform the dose distribution laterally to the shape of the target, and a range compensator to conform the distal edge of the dose distribution to the target volume plus the margins. Figure 19–4 is a schematic of a typical geometry. Beam monitoring devices are assumed to have a negligible effect on dose distributions. The beam limiting devices (apertures, collimators, blocks, etc.) are considered to be binary in nature; that is, the pencil beams in the portion of the broad pro-

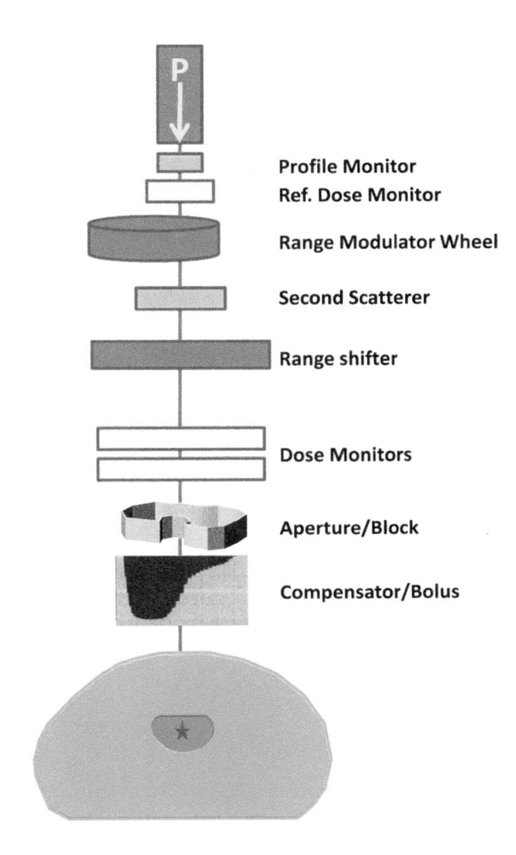

Figure 19–4 Typical geometry for dose calculations for passively scattered proton therapy. Some systems have a first scatterer built into the range modulator, while others have a separate first scatterer. Also, some systems do not use range shifters.

ton beam intercepted by the beam-limiting device are assumed to be completely absorbed.

The transmitted pencil beams pass through a range shifter of selectable thickness (if present), the compensator, and the patient. In the process, they lose energy and penetrability, and scatter and spread laterally. When passing through air gaps, they continue to spread laterally and scatter negligibly. These processes affect penumbra and dose distributions within and outside the beam boundaries, especially when complex heterogeneities (e.g., lung and H&N) are encountered.

19.3.1 The Hong Algorithm

The following description is adapted from Hong et al. [15]. Their algorithm, designed originally for PSPT, incorporates scattering in devices upstream of the beam-limiting device(s) into an effective source, which is assumed to be represented by a Gaussian. In principle, the effective source Gaussian's σ should be a function of range shifter thickness. Scattering in devices downstream of the beam-limiting device(s), as well as scattering in the patient, lead to an increase in angular confusion and radial emittance (spatial standard deviation of the distribution).

The dose $d(x', y', z')$ at a point in the patient due to a pencil beam is the product of central axis dose $C(z')$ and the lateral profile term $O(x', y', z')$. The central axis term is given by

$$C(z') = IDD(d_{eff})\ [(ssd_0 + d_{eff}) / z']^2 \tag{19.5}$$

where *IDD* is the central axis depth–dose distribution of the open broad beam in a water phantom, ssd_0 is the source-to-surface distance when the CADD data were measured, and d_{eff} is the effective depth, i.e., the water-equivalent pathlength along the pencil axis to the depth of the point (x', y', z') and includes the effect of all beam modifying devices as well as the patient. The origin of the system is assumed to be located at the source. Its effective pathlength component through the patient is given by

$$\tau_{patient} = \int_{surface}^{Zp} dz' \times SPR\left(HU\left(z'\right)\right) \tag{19.6}$$

That is, $\tau_{patient}$ is computed by voxel-by-voxel integration through the 3D CT. The Hounsfield units of the CT voxels are pre-converted into SPRs as discussed in Section 19.2. The term z_p is the position of the point of dose computation in the pencil beam coordinate system.

The off-axis pencil beam term is expressed as:

$$O\left(x', y', z'\right) = \left\{\frac{1}{2\pi\left[\sigma_{tot}(z')\right]^2}\right\} exp\left(-\frac{x'^2 + y'^2}{2\left[\sigma_{tot}(z')\right]^2}\right) \tag{19.7}$$

where $\sigma_{tot}(z')$ is the radial spread standard deviation calculated by adding in quadrature the contributions from the effective source, range compensator. and the patient.

Contributions from the effective source include the effect of increase in angular emittance by the beam-spreading components (scatterers, range modulators, and range shifters). The source parameter (σ) is typically obtained by measuring in-air penumbra of a half beam block (see Section 19.2.2). The resulting spatial spread is further scaled linearly to the patient as a function of distance to the point of interest. This is an approximation that may affect accuracy for cases where the air gaps encountered deviate significantly from those used for measuring commissioning data, and when large, complex heterogeneities are encountered. Section 19.5 gives an illustrative example. In theory, the scaling should be done on a layer-by-layer basis as proposed by Soukop et al. [18].

Hong's algorithm incorporates the effects of nuclear interactions, assuming that their effect is included in the measured central axis depth–dose data. The influence of nuclear interactions on the angular emittance of the beam is ignored, assuming that such events are relatively infrequent.

19.3.2 The Schaffner Algorithm

Schaffner et al. generalized the concept of the in-air, fluence-based method of dose calculation developed at the Paul Scherer Institute for modulated scanning beams to passively scattered and uniform scanning (wobbling) proton delivery techniques [13,19]. The aim of their approach was to reuse the software developed for one treatment machine and technique for other machines and techniques to the extent possible. The model separates in-air fluence calculation from the dose deposition calculation in the patient.

For PSPT, for instance, the calculation for a given SOBP is split into multiple "layers," i.e., component Bragg curves. Water-equivalent ranges for each grid point on the 3D dose matrix are computed and used to scale pencil beam dose distributions along the paths of the pencil beams. The fluence for each layer is convolved with the scaled pencil beam dose distributions.

The in-air fluence calculation utilizes virtual and effective SADs and the effective source size defined in Section 19.2. In principle, these parameters are functions of beam line characteristics and settings (e.g., energy of the layer, snout position, and range shifter thickness and position); however, simplifying approximations are made to reduce the complexity and computation time. For instance, the virtual SAD for the highest energy is assumed to be valid for all energy layers, and the effect of block thickness and scattering on computed penumbra is assumed to be incorporated in virtual source parameters obtained by fitting with measured beam profiles. The effect of range shifter thickness on effective source size is ignored. For each layer, the fluence is calculated using products of the following terms:

- The inverse square correction relative to the effective SAD
- The sum of error functions describing the penumbra (computed from using the effective source sigma parameter and the aperture position) at the distance of the point of interest and the distances to the beam boundary along ±x directions
- A similar term along ±y directions.
- A normalization factor N that depends on the weight of the layer. If the initial intensity of the beam entering the nozzle is constant, it is proportional to the width (in degrees) of the range modulator wheel angular step.

The above quantities combined lead to the following expression for fluence for an energy layer l:

$$\Phi_l = N\left(\frac{SAD_{eff,l} - z}{SAD_{eff,l}}\right)^2 \left[\text{erf}\left(\frac{\Delta x + (z)}{\sqrt{2}\sigma_{pl}(z)}\right) + \text{erf}\left(\frac{\Delta x - (z)}{\sqrt{2}\sigma_{pl}(z)}\right)\right] \times \left[\text{erf}\left(\frac{\Delta y + (z)}{\sqrt{2}\sigma_{pl}(z)}\right) + \text{erf}\left(\frac{\Delta y - (z)}{\sqrt{2}\sigma_{pl}(z)}\right)\right] \quad (19.8)$$

This is an excellent approximation for rectangular fields, but it may be questionable for complex, irregular fields. For improved accuracy, the Gaussian source function should be convolved with the aperture boundary.

The Schaffner algorithm models a patient-specific compensator as a perturbation of each layer's fluence. In doing so, it takes into consideration the scattering of protons and the air gap between the compensator and the patient, and thus increases the spreading of pencil beams passing through different thicknesses of the compensator. However, it ignores changes in the energy of the protons scattered into the neighboring region, which has consequences as illustrated in Section 19.5.1.1.

For calculating dose for each layer in an arbitrary inhomogeneous medium, each pencil beam is scaled according to the water-equivalent depth in the medium, similar to the manner described in Section 19.3.1. Parameters for the calculation of the pencil beam depth–dose are based on the theoretical model considerations as formulated by Ulmer et al. [20,21]. The total dose for each layer is the sum of contributions of all pencil beams multiplied by the fluence at the position of the pencil beam.

The model does not explicitly take into account the loss of protons in the compensator or the patient, secondary particles, and nuclear interactions. However, since the depth–dose, profile, and other parameters of pencil beams are determined from measured dose distributions, these effects are approximately accounted for. The Schaffner model is implemented in a widely used commercial clinical proton treatment planning system for PSPT, uniform scanning proton therapy, and IMPT. It has been found to be reasonably accurate in most clinical situations encountered and, in some ways, advances the original concepts of Hong et al. However, as in other SEA models, approximations and assumptions limit its accuracy in some situations.

19.3.3 The Soukop Algorithm

Soukup et al. have developed an algorithm that attempts to improve upon many of the limitations of the previous models [18]. Their algorithm is aimed primarily at the special accuracy and speed requirements of dose calculations for beam scanning and IMPT. In IMPT, dose is delivered with a sequence of monoenergetic scanning beamlets of varying intensities. The delivery of each energy layer may be in the form of a continuous scan pattern or in the form of a set of discrete beamlets. The terminal high-dose end of each beamlet is called a "spot." For dose calculations, it is assumed that the delivery is in the form of discrete spots. Considering that a given IMPT dose distribution may contain contributions from many thousands of beamlets (spots), small errors in single spot dose distributions may accumulate into significant errors in the final dose distribution.

In most models, such as those described above, inhomogeneities are incorporated by computing water-equivalent pathlength along the pencil beam paths, which ignores the fact that the order of inhomogeneities affects the spatial spread. For instance, spread caused by a high-density inhomogeneity followed by a low-density inhomogeneity is not the same as the spread caused by a low-density inhomogeneity followed by a high-density inhomogeneity (see the illustration in Section 19.5.1.2). Furthermore, a pencil beam perceives an inhomogeneity in its path to be a slab of infinite lateral extent. Previous models also assume that the changes in nuclear secondary particles caused by inhomogeneities are negligible. Soukup's model attempts to overcome these and other limitations. Though developed for IMPT, most of its concepts can be applied to improve the accuracy of passive scattering and uniform scanning as well. The following abbreviated description is adapted from Soukup et al. [18].

The model defines beamlets (spots) as a function of initial energy E_0; position x_0, y_0; lateral spread; divergence; and direction. Divergence characterizes spot widening as a function of distance in vacuum or air. In this model

$$\Phi(x,y) = \left(2\pi\sigma_{x_0}\sigma_{y_0}\right)^{-1} exp\left[-\frac{(x-x_0)^2}{\sigma_{x_0}{}^2} - \frac{(y-y_0)^2}{\sigma_{y_0}{}^2}\right] \tag{19.9}$$

where Φ is the relative fluence distribution of the spot in a plane across the beamlet at a point (x_0, y_0, z) on the beamlet central axis in the beamlet coordinate system. The sigmas define the spatial spread of the spot in air.

19.3.3.1 Consideration of Finite Lateral Extent of Heterogeneities

To better take into account the variation of heterogeneity across the beamlet cross-section, the spot (beamlet) is decomposed into multiple sub-spots. The origins of the sub-spots lie on the spot plane and cover the original spot. They are assigned weights equal to the fluence of the spot at the sub-spot positions. (Soukup et al. adopted the concept of spot decomposition from the original idea proposed by Schaffner et al. [19]) Another advantage of the decomposition of spots is that this approach can improve the precision of accounting for apertures and blocks. The larger the number of spots is, the greater the accuracy and precision of correcting for heterogeneities and beam boundaries is, but at the cost of computation speed.

19.3.3.2 Accounting for Multiple Coulomb Scattering in Voxel-sized Steps

To improve the accuracy of computed beamlet spread, the Soukup algorithm calculates multiple scattering $\sigma_{MS}{}^2$ as a function of depth by first computing the characteristic angle $\theta_0{}^2$ in voxel-sized steps, or layer by layer, as the beamlet (or sub-beamlet) traverses the medium. The characteristic angle may be computed using, for instance, Highland or Rossi formulae [15,18] and depends on the step size, radiation length, and particle momentum at the entry into the voxel or layer. (NOTE: radiation length characterizes the energy loss of high-energy particles traversing a medium.) To compute the depth-dependent spatial spread $\sigma_{MS}(Z,E_0)^2$ due to multiple scattering, all previous angular spreads must be taken into account according to their corresponding geometrical distances at the actual position z_n as follows.

$$\sigma_{MS}\left(Z,E_0\right)^2 = \sum_{n=1}^{N}\theta_0\left(z_n,E_n\right)^2 \times z_n{}^2 \tag{19.10}$$

The same approach can be adopted for improving the accuracy of computed dose when taking into account range shifters and air gaps for compensators for PSPT.

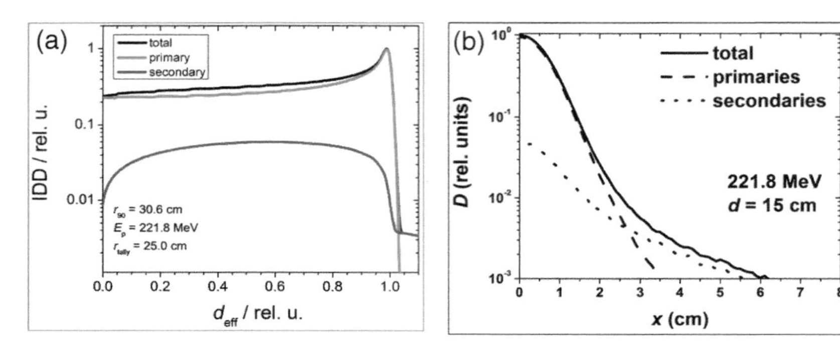

Figure 19–5 Primary and secondary components of a proton scanning beamlet. Panel (a) shows integral depth–dose data and its primary, secondary components for a 222 MeV beam. Panel (b) shows the lateral profile of the same beamlet at a depth of 15 cm (~50% range) on a semi-log plot. The secondary component is small and not discernible on a linear plot. However, the combined contributions of thousands of beamlets comprising an IMPT treatment can be significant. (Courtesy of U. Titt et al., MDACC, private communication.)

19.3.3.3 Incorporating Nuclear Interactions

As mentioned above, in typical analytical algorithms, the dose contributions of secondary nuclear products are either neglected or taken into account as a constant offset. This may be an acceptable approximation for passive scattering and for lower-energy protons, but not necessarily for beam scanning and IMPT, particularly for higher-energy protons as illustrated in Figure 19–5a for a 222 MeV scanned beam. At higher incident energy, nuclear interactions produce secondary products (mostly secondary protons). Their contributions, and the deviations of beamlet profiles from Gaussians, are undetectable on a linear plot, but they become quite apparent on semi-log plots (Figure 19–5b). The combined effect of large numbers of beamlets in an IMPT plan can be substantial, as illustrated in Section 19.5.2.2.

Soukup et al. model nuclear interactions as an additional component, i.e., as a "nuclear beamlet" [18]. The relative contribution of nuclear beamlets and their lateral spreading parameter (σ_{nuc}) are determined using Monte Carlo simulations. Since dose contributions due to the nuclear component are small, they can be dealt with in a less rigorous manner. For instance, nuclear beamlets are not decomposed into sub-beamlets. The central axis depth–dose of each beamlet, i.e., the IDD, is subdivided into nuclear and multiple Coulomb scattering components,

$$IDD(z,E_0) = W_{nuc} \times IDD_{nuc}(z_{eq},E_0) + (1 - W_{nuc}) \times IDD_{MCS}(z_{eq},E_0) \quad (19.11)$$

Sample IDD_{nuc} and IDD_{MCS} are depicted as primary and secondary components of a 222 MeV beamlet in Figure 19.5 (a). The MCS part is divided into component

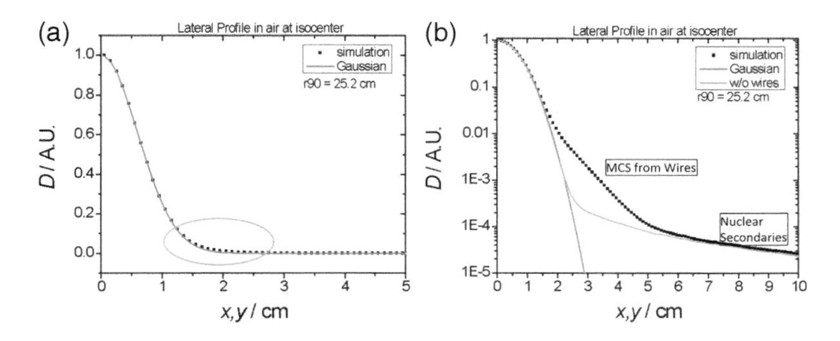

Figure 19–6 Monte Carlo simulations of dose in air at the isocenter for a beamlet of 200 MeV. Seemingly small deviation from the Gaussian—not seen on a linear scale, but obvious on a semi-log plot—can have a significant cumulative effect on a scanned beam dose distribution. (Courtesy of U. Titt et al., MDACC, private communication.)

sub-beamlets. The sub-beamlet profiles have the same functional form as the primary MCS beamlet.

It should be noted that although the incident beamlet (or pencil beam) profiles in air are assumed to be Gaussian in shape, they may deviate due to scattering from the wires of the profile monitor, air, and other monitors in the beam path. Figure 19–6(a) and (b) illustrate such deviation for a 200 MeV beamlet profile in air. As in the case of the nuclear component, the deviation is often not obvious when plotted on a linear scale (panel (a)) but becomes evident when plotted on a semi-log scale (panel (b)). The discovery of such deviations prompted the use of multiple Gaussians and other functions to characterize in-air beamlet profiles [14].

The dose at a point x, y, z due to a beamlet in Soukup's algorithm is then computed using the expression:

$$D(x, y, z, E_0) = D_{nuc}(x, y, z, E_0) + \sum_{k=1}^{N_{sub}} D_{MCS,k}(x, y, z, E_0) \tag{19.12}$$

D_{nuc} and $D_{MCS,k}$ are calculated using expressions similar to those in the Hong algorithm, except that σ_{nuc} and σ_{MCS} are computed based on data obtained with Monte Carlo simulations, and their values at the point of computation are determined in voxel-sized steps.

19.4 Relating Computed Dose to MU Settings on the Machine

A monitor unit (MU) corresponds to a specific amount of charge collected in the monitor chamber for a specific amount of dose delivered (typically 1 cGy) under calibration conditions. The computed dose distribution prescribed for delivery to the patient must be linked to the monitor unit (MU) settings of the treatment machine. The processes are different for PSPT and IMPT. For the former, the MUs are defined for the beam as a whole, and the correspondence between MUs and dose is intuitively apparent. However, for the latter, the MUs are defined for each beamlet and, just like for IMRT, the relationship between MUs and dose is highly variable and obscure. (Refer to Chapter 26 for details on monitor unit calculations.)

19.4.1 Passive Scattering

In general, PSPT delivery requires the selection of an initial energy and multiple beam-shaping devices. Their choice is normally determined by the treatment planning system (TPS) based on the patient and tumor characteristics. The TPS also normalizes the planned dose distribution so that the target is covered by the prescribed dose level. The factor for the conversion of the prescribed dose distribution to the MU setting of the treatment machine (i.e., the "output factor" or OF) is a function of the initial energy, range modulator-scatterer combination, SOBP width, range shifter, compensator, aperture, and the distance to the reference point relative to the calibration point. Empirical models for computing OFs have been proposed that require a number of empirically determined variables [22–25].

For many cases, the MUs per field may be calculated with an appropriately validated theoretical model. However, partly due to the limited confidence in such models and partly for reasons of safety—especially for small and complex apertures and complex compensators—MUs are sometimes determined for individual fields via measurements. These measurements are carried out by irradiating a water phantom with each treatment field for a specified number of MUs and measuring the dose at a point chosen to be the reference. Some commercial treatment planning systems provide a function to transfer the settings of each beam intended for treatment to another image of the same patient or to a water phantom "image" (typically, a digitally constructed 3D matrix of voxels of water HUs). One purpose of such an exercise is to verify if an originally designed treatment plan is valid for an image acquired on a subsequent day. For this reason, the second plan is called the "verification plan." Another purpose is to determine the MUs for the treatment field. The dose distribution computed for the treatment beam in the water phantom image is normalized to the measured dose to determine MUs.

The following sections give examples of approaches to determine MUs. These approaches typically cover fields only down to a certain field size. For small fields such as those used in radiosurgery, aperture scattering and loss of equilibrium require introduction of correction factors [26].

19.4.1.1 MU Calculations—the Sahoo Approach

In the approach followed by Sahoo et al. [24] for PSPT, the OF is assumed to be a function of energy, scatterers, RMW, SOBP width, thickness of the range shifter (if any), the depth–dose value relative to the normalization point in the SOBP, and scattering in both the compensator and the patient. In analogy with the procedure used for photons and electrons, Sahoo et al. introduced proton-specific factors that relate dose per MU to a reference condition. These factors include the relative output factor (ROF), SOBP factor (SOBPF), range shifter factor (RSF), SOBP off-center factor (SOBPOCF), off-center ratio (OCR), inverse square factor (ISF), field size factor (FSF), and compensator and patient scatter factor (CPSF). The ROF, SOBPF, and RSF—which are the main contributors to dose/MU—are obtained by measurements in a water phantom as a part of the clinical commissioning process of each beam and are tabulated. The following formula is used to determine the MUs for delivering the prescribed dose D at the point of interest in the patient:

$$MU = D \;/ \begin{bmatrix} ROF \times SOBPF \times RSF \times SOBPOCF \\ \times\, OCR \times FSF \times ISF \times CPSF \end{bmatrix} \tag{19.13}$$

In this expression, the term CPSF is the product of two factors: the compensator scatter factor (CSF) and the patient scatter factor (PSF). The CSF and PSF depend on the water-equivalent thicknesses of the compensator material and of the patient to the point of interest. They and the composite CPSF can be calculated using the treatment planning system applied to a verification plan.

Sahoo et al. verified the accuracy and robustness of their method by comparing calculations and measurements in water for a large number of different combinations of beam parameters. This procedure has been implemented for clinical use and found to be accurate to within 2% for 99% of the PSPT fields.

19.4.1.2 MU Calculations—the Kooy Approach

Kooy et al. proposed a theoretical model for predicting MUs for PSPT [22,23]. The model utilizes idealized pristine Bragg curves derived using the Bortfeld and Schlegel method of computing SOBP based on such idealized curves [27]. The model relies only on the ratio between the entrance dose and the dose in the SOBP plateau. Kooy et al. reduce the dependence of the output factor to just a single variable $r = (R - M) / M$, where R and M are the range and modulation width respectively. They have implemented the model into clinical practice and found that the measurements of output agree with model predictions to within 2.9%.

The method used by Kooy et al. is based on measurements performed in open fields, i.e., without patient-specific apertures or compensators. In their approach, each treatment field, excluding the compensator and aperture, is transferred to a water phantom image ("verification plan") positioned in the calibration conditions geometry. The dose per MU at the reference point is then obtained using either an empirical model or measurements. The underlying assumption is that the TPS model is able to accurately account for the aperture and range compensator. This is justified on the grounds that we, after all, make this assumption implicitly when we use the TPS-computed dose distribution to make treatment decisions. This approach does, however, require corrections based on experiments for small-aperture openings.

19.4.1.3 Alternate Approaches to Determining MUs

Ideally, dose distribution computed with a TPS should be verified in three dimensions, not just verifying the MUs at a point. This can be done using a calibrated and validated Monte Carlo system, a process which can be quite computation resource-intensive and not currently suitable for routine practice. Nevertheless, it should be used to assess the strengths and weaknesses of TPS models and for selected clinical treatment plans involving complex patient or field geometries. Monte Carlo simulations of treatment head geometries, including the ionization chamber used for output factor measurements, have been used to predict MUs in proton beams [28–30].

19.4.2 Scanning Beams

For scanning (IMPT) beams, the integral depth doses (IDDs) are tabulated in units of dose per MU for each energy. Such IDDs may be calculated using calibrated Monte Carlo systems (see Figure 19–3) [31]. At the end of optimization and dose distribution calculation, the TPS provides intensities of beamlets in terms of MUs. The treatment delivery system utilizes these settings to deliver IMPT. The IDDs per beamlet may also be expressed directly in terms of the number of protons per unit prescribed dose [32].

19.5 Accuracy of Semi-empirical Analytic Models

The numerous approximations and assumptions made in the semi-empirical analytic (SEA) models of dose calculations and their effect on the accuracy of dose distributions are generally not explicitly stated by developers and vendors. Often, such models evolve over the tenure of multiple developers and the assumptions and approximations are not even fully known. This is particularly true of commercial systems. Furthermore, for reasons of confidentiality, company reputation, and legal exposure, commercial vendors are, in general, unwilling to acknowledge the limitations and share such information with users. Generally, it is very difficult, perhaps impossible, for a user to guess *a priori* the limitations of a model in any given situation or to verify sufficiently wide variety of dose distributions in 3D. This is especially true when the details of the algorithms are not fully described in publications or documentation.

This section gives some examples of comparisons of dose distributions computed using SEA models with measured or Monte Carlo-generated dose distributions. Attempts to explain the observed discrepancies are also given. Additional examples are given to elucidate the consequences of certain approximations.

19.5.1 Passive Scattering

19.5.1.1 Compensators

Figure 19–7 shows an example of a "step compensator" experiment designed to test the accuracy of a commercial SEA algorithm to account for compensators. The geometry of the experiment and the beam characteristic are shown in panel (A). The central axis of the 225 MeV beam (range ~20 cm) is coincident with the step. Panels B1 through B4 show profiles at a sequence of depths. Agreement among SEA, MC, and measurements is excellent for the open field (no compensator) and quite reason-

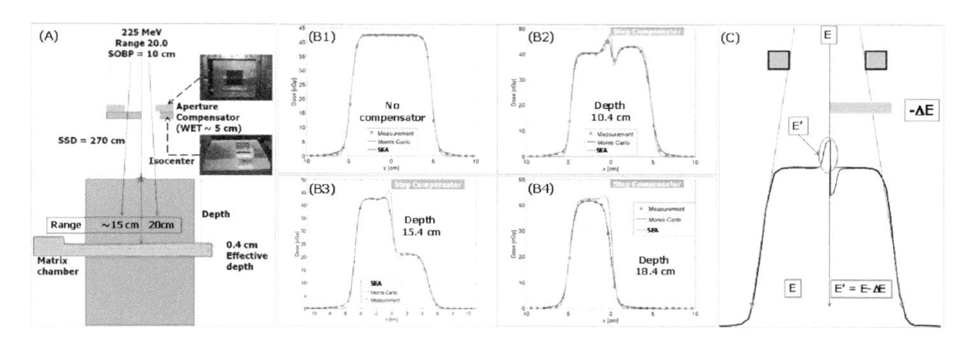

Figure 19–7 A simple experiment to evaluate the accuracy of a commercial semi-empirical analytical model to account for the compensators. Panel (A) schematically shows the experimental setup involving a square 10 cm x 10 cm aperture and a "step compensator" with its step in the middle of the beam. Panels B1–B4 compare measured, Monte Carlo, and semi-empirical analytical (SEA) model cross beam profiles normal to the compensator edge. Agreement between SEA and the other two is good in the absence of the compensator and at shallow depths, but agreement begins to breakdown at larger depths, especially near the end of the range of the protons. Panel (C) explains the reason, which is elaborated in the text.

able at the depth of 10.4 cm (near the proximal edge of the SOBP) with the compensator. The characteristic fluctuation in dose at the compensator edge is due to the scattering of the proton beam from the edge.

At larger depths near and beyond the range of protons through the thick portion of the compensator (i.e., at depths of 15.4 and 19.4 cm), discrepancies between the SEA model relative to measurements and MC calculations begin to emerge. The cause of discrepancies, explained in panel (C), is the manner in which the algorithm separates the calculation into fluence and dose and manages the scattered protons. While the algorithm takes into account the lateral scattering and spreading of protons by the compensator, it ignores the fact that the protons scattered into the open region have lower energy, and therefore higher linear energy transfer (LET), and consequently deposit higher dose at shallower depths.

Another example—that of a large, complex compensator used for an actual lung treatment—is shown in Figure 19–8. Again, the agreement within the field boundaries is good in shallower regions, but it begins to break down near the end of the range. The discrepancies between SEA vs. MC and measurements were also observed just outside the irradiated region and increase as a function of depth. In addition to the approximation discussed above related to the energy of the laterally scattered protons, the approximate manner in which lateral spreading is accounted for may also be responsible for the discrepancies observed for this complex compensator case.

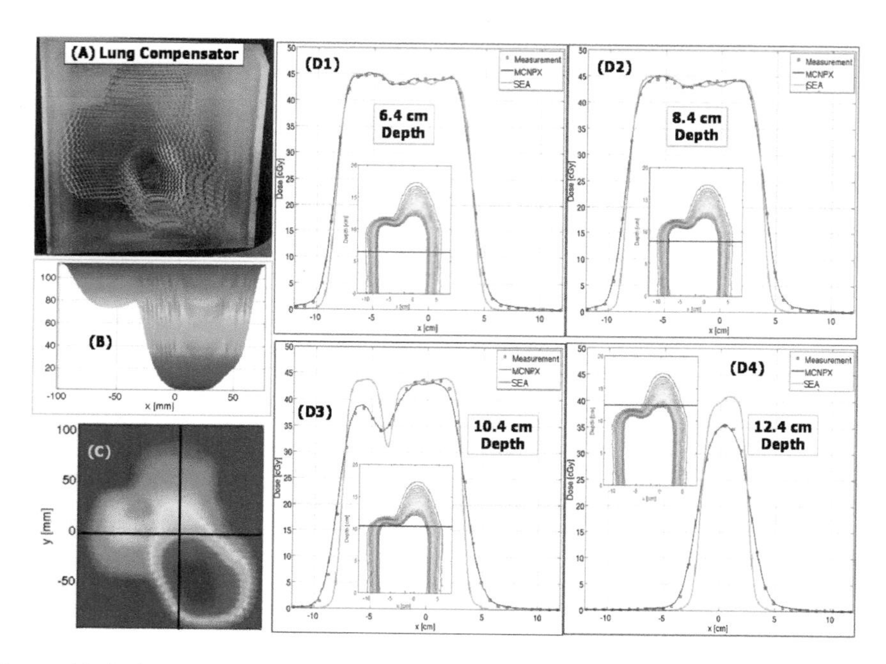

Figure 19–8 An experiment to evaluate the accuracy of a commercial semi-empirical analytical model to account for compensators. The discrepancies between the SEA model and the measurements and Monte Carlo calculations increase with depth and are not confined to only within the field, but also occur at the field boundaries.

19.5.1.2 Order of Layers in Correcting for Inhomogeneities in the Paths of Protons

Figure 19–9 is a simple, perhaps an extreme, example illustrating the importance of the order of inhomogeneities in determining the lateral spreading of proton beams. Most SEA algorithms consider only the cumulative water-equivalent pathlength to the point of interest in computing the spreading parameters. An exception is Soukup's algorithm, which uses an approach that computes lateral spreading in voxel-sized steps (or layer by layer).

Panel (a) of Figure 19–9 shows a water-equivalent phantom with two blocks. The left block is movable, but the right block, with a detector (a matrix chamber) embedded in it, is fixed relative to the beam. A passively scattered 250 MeV beam (range ~25 cm), SOBP 10 cm is incident on the phantom. Profiles perpendicular to the incident beam direction were calculated at a water-equivalent depth of 22 cm for various gaps (0, 5, 10, and 15 cm) between the right and left blocks. As expected for the SEA model, all profiles are identical. However, Monte Carlo calculations show that the discrepancy at a depth of 3 cm after the gap is ~7% on the central axis for a 15 cm gap and is significantly larger near the boundaries. Panel (d) shows the Monte Carlo-generated depth–dose data for the 15 cm air gap and indicates that at the surface of the

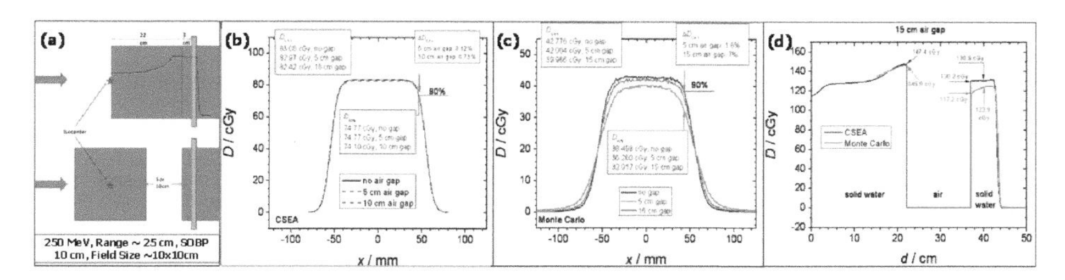

Figure 19–9 Monte Carlo simulation to assess the consequences of not accounting for the order and location of heterogeneities along the path of protons. (See text for explanation.)

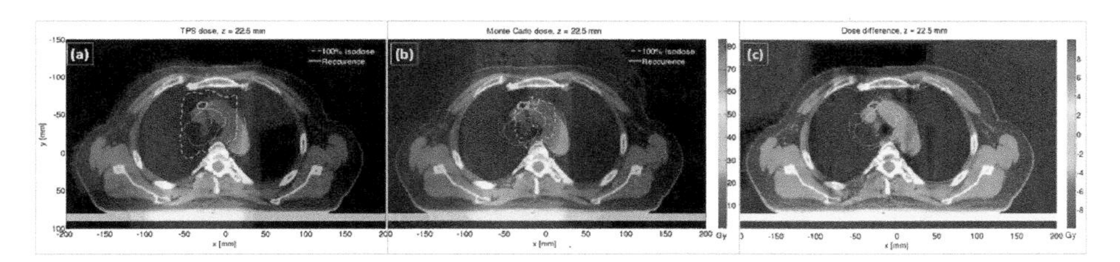

Figure 19–10 Panels (a) and (b) are the dose distributions for a locally advanced NSCLC patient calculated with an SEA model and Monte Carlo simulations, respectively. The dashed lines depict the prescription isodoses. The thick green contour identifies the region of recurrence. Panel (c) shows the difference (SEA–MC) dose distribution, i.e., positive differences indicate underdosing and negative differences show overdosing predicted by the SEA model. From [36].

second block on the central axis the error may be as high as 11%. Monte Carlo calculations were confirmed with measurements, but the results are not shown.

19.5.1.3 Patient Dose Distributions Computed Using Semi-empirical Models vs. Monte Carlo Techniques

Figure 19–10 compares dose distributions for a lung case computed with a semi-empirical analytic model (a) and Monte Carlo (b). Panel (c) shows the difference. These calculations were carried out as a part of the analyses of recurrences among lung patients treated with protons [36]. Monte Carlo calculations show the underdosing near the target boundary and may be among the factors contributing to the recurrence (the thick green outline). The 100% of the prescription isodose volume region was found to be significantly smaller on the Monte Carlo simulations compared to the SEA algorithm, and this may have resulted in a marginal miss. There are additional such examples in Chapter 20 (Dose Calculation: Monte Carlo).

19.5.1.4 Apertures

In most dose computation models, apertures are assumed be infinitesimally thin, and scattering from them is neglected. However, Slopsema et al. [33] have developed an extension of their SEA model to account for the finite thickness of apertures. They validated their model using measurements of the lateral penumbra and found that the overall effect depends on aperture thickness, the location of the point of interest relative to the aperture edge, the source size, and SAD, but the model results were insensitive to the gap between the aperture and the patient. Interestingly, in contrast with the results of Titt et al. [34] presented below, their model results were found to be insensitive to the depth in the patient.

Figure 19–11 is an example produced with Monte Carlo simulations for a passively scattered beam. It illustrates that scattering from the collimator can contribute significant dose in the neighborhood of an aperture, even at large depths [34]. Currently, apertures are not commonly used for IMPT beams, though they are gradually being introduced to sharpen beam boundaries. A similar impact on accuracy can be expected.

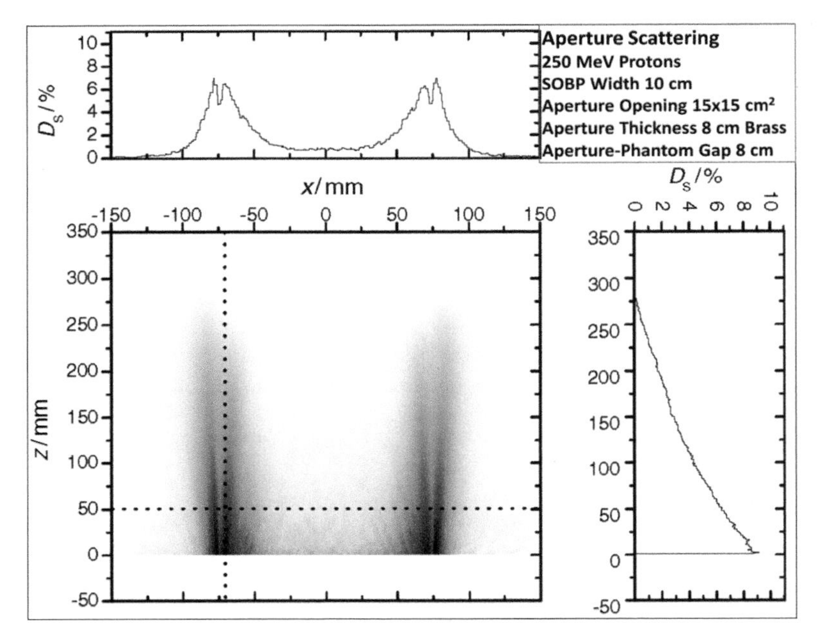

Figure 19–11 Dose (D_S) scattered from 15 cm x 15 cm aperture for a 250 MeV passively scattered beam with a 10 cm SOBP width and an 8 cm air gap between the collimator and phantom. Shown are the differences between the simulations including and excluding protons scattered from the collimator. The dose was normalized to the dose at the center of the SOBP. (From Titt et al. [34].)

19.5.2 Scanning Beams and IMPT

Because of the absence of beam-modifying devices, the accuracy of computed IMPT dose distributions is expected to be affected mainly by internal inhomogeneities. However, there are factors that may contribute to small errors in the profiles used for IMPT calculations. These include the deviation of beamlet profiles from Gaussian, the dependence of such deviation on air gaps and depth, and the approximate handling of nuclear interactions. Under certain circumstances, these small errors may add up to a significant fraction of the dose delivered.

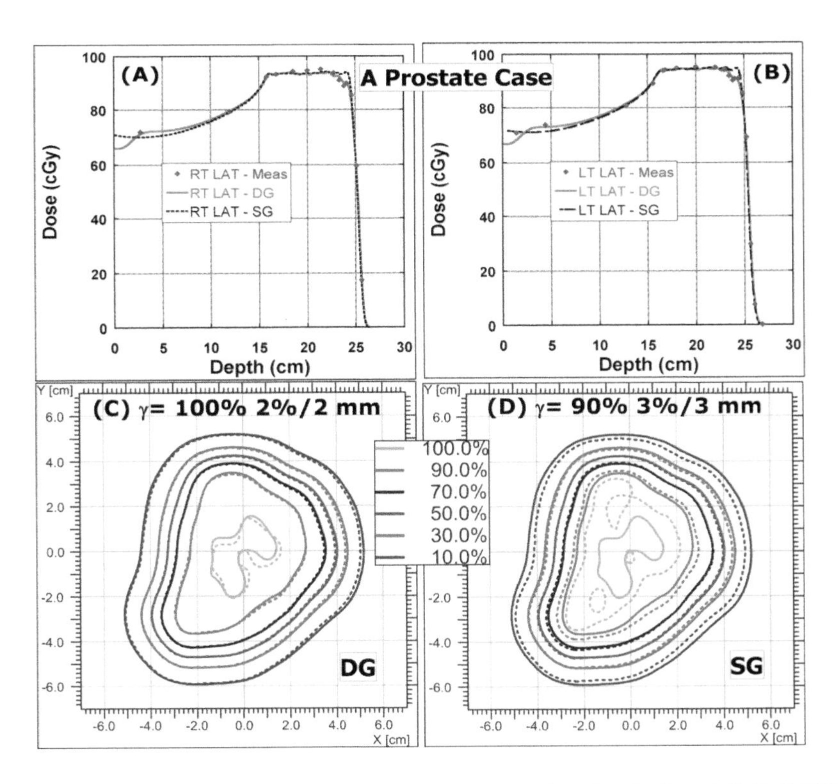

Figure 19–12 A clinical example comparing measured and calculated SFO-IMPT dose distributions for a prostate cancer patient treated with right and left lateral fields. The fields were applied to a water-equivalent phantom, and calculations were done with both double and single Gaussian (DG and SG) fluence models. Panels (A) and (B) are depth–doses (blue diamonds = measured, red solid lines = calculated by the DG fluence model, and black dashed lines = calculated by the SG fluence model). Panels (C) and (D) are isodose distributions for one of the fields in a plane transverse to the beam direction (solid lines = measured and dashed lines = calculated). [31]

19.5.2.1 IMPT Patient Dose Distributions Computed with an SEA Model vs. MC and Measurements

Figure 19–12 compares single-field optimized (SFO)-IMPT calculated and measured dose distributions in a water phantom for the left and right lateral beams for a prostate case. It demonstrates that the use of a second Gaussian to account for the long, low dose tails of the beamlet profiles is important to improve the accuracy of computed IMPT dose distributions. Figure 19–13 shows similar data for a base of skull chordoma patient. Figure 19–14 compares measured and calculated lateral profiles for a head and neck multi-field optimized (MFO)-IMPT case.

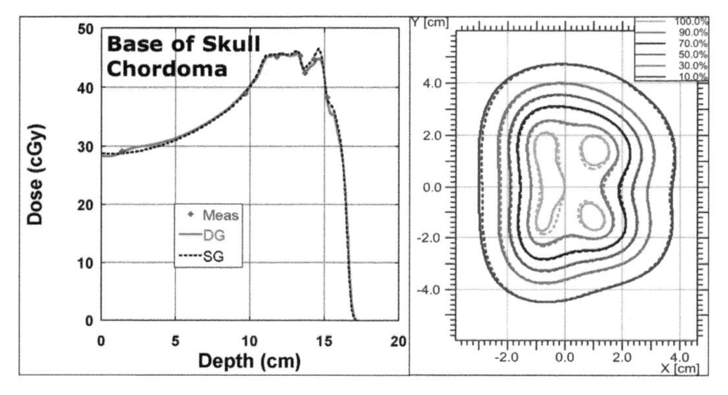

Figure 19–13 Similar to Figure 19–12, depth–dose and isodose distribution for a pediatric patient with chordoma of the base of skull [31].

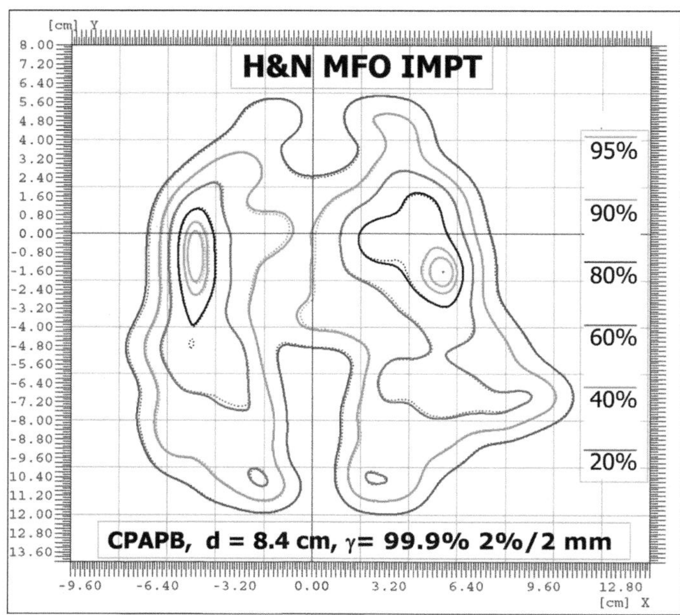

Figure 19–14 Measured (solid lines) and calculated (dotted lines) isodose distributions for one of the proton beams in a water-equivalent phantom for a head and neck case. Measurements were performed with a matrix chamber. (Courtesy of X.R. Zhu from a 2014 AAPM presentation [35].)

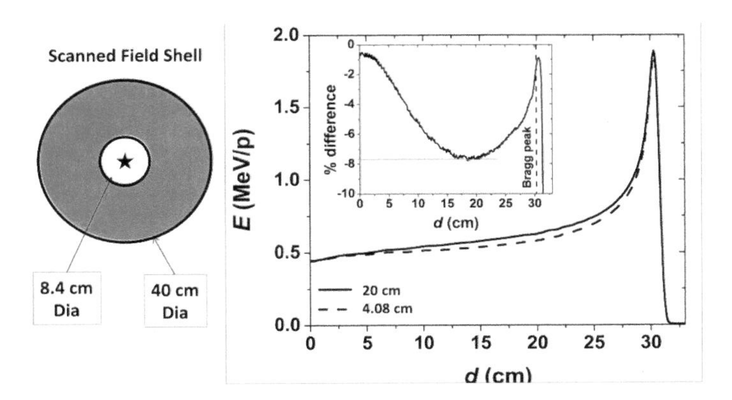

Figure 19–15 Monte Carlo simulation of a 222 MeV beamlet dose distribution illustrating the significance of the long-range contribution of secondary particles. The graph in the inset implies that beamlets in the red region between the radii of 4.2 and 20 cm may contribute as much as 8% of the dose deposited.

The data shown in these figures demonstrate excellent agreement between measured and calculated dose distributions in homogeneous phantoms. However, accurate estimation of the effect of complex heterogeneities in patients, irregular patient surface, and other factors (e.g., varying air gap) on lateral spreading and distal range would require Monte Carlo simulations. Examples of SEA vs. MC in patients are given in Chapter 20.

19.5.2.2 Approximate Handling of Nuclear Secondary Particles in SEA Models for Computing IMPT Dose Distributions

As mentioned above, most models try to account for nuclear secondary particles either as a constant offset after the calculations neglecting them have been performed, or they assume them to be included in the measured data. For PSPT models, such approximations may not be consequential. However, as the following example illustrates, large angle scattering and secondary nuclear particles can contribute significantly to dose at large distances from the beamlets or pencil beams and can affect accuracy of IMPT dose distributions. Such contributions are complex functions of beamlet energy, residual range, and heterogeneities of materials in the path of the beamlets. Some SEA models (e.g., Soukup's model) attempt to explicitly account for these factors to improve accuracy. Other models employ multiple Gaussians or combinations of Gaussians and a Lorentzian to achieve modest improvement.

The Monte Carlo calculation shown in Figure 19–15 was performed to understand the consequences of the approximations in consideration of large angle-scattered particles and the use of a "Bragg peak chamber" for measuring IDDs. It shows the IDD over a circle of 8.4 cm and 40 cm diameter circles for a 222 MeV beamlet. Dose distributions are equivalent to uniformly scanned field depth–doses for 8.4 cm and 40 cm diameters, implying that contributions of nuclear secondary particles from beamlets from the radial shell between 4.2 to 20 cm radii may be as high as ~8% of the dose. The key point is that in an arbitrary IMPT dose distribution, beamlets quite far from the point of interest may contribute significantly. Considering that the intensities of beamlets are, in general, highly nonuniform, it may be important to explicitly account for nuclear interactions in the dose computation model rather than in an average approximate manner.

19.6 Summary

Calculated dose distributions are used for prescribing radiation treatments and evaluating response. Almost universally, semi-empirical analytical (SEA) models of dose computations are used in clinical practice. In contrast with Monte Carlo techniques, which simulate radiation transport particle by particle based on basic laws of physics, the SEA models make numerous simplifying approximations and assumptions to accelerate computations and to facilitate development and implementation. For photons, while Monte Carlo techniques remain the ultimate standard of accuracy, significant advances have been made in improving the accuracy of SEA dose calculation models (e.g., the superposition-convolution model). However, the accuracy of proton SEA models for many situations is still far from adequate.

It is common to accept the dose distribution seen on a treatment plan as if it was perfect. In reality, it may be significantly different from what is delivered to the patient. The magnitude of the difference and its potential clinical impact are difficult to discern in any given situation, especially for complex geometries and in the presence of many other sources of uncertainty. Questions about how accurate a particular dose distribution is and how accurate it needs to be remain unanswered. One can argue, however, that there are no data explicitly suggesting that inaccuracies of the current models by themselves affect outcomes. It would be interesting and desirable to conduct treatment response analyses for patients on clinical trials to assess the clinical impact of uncertainty in computed dose distributions.

It should be noted that rarely, if ever, are the details of the assumptions and approximations of dose computation models and their dosimetric consequences included in the documentation or in publications. Not surprisingly, vendors and authors are inclined to put only their best feet forward. At times, there are claims of proprietary models and algorithms; at other times, the information is poorly documented or lost because the previous developers are no longer with the team. However, to appreciate the limitations of models and the scope of their validity, it is essential that their users understand them, assess their strengths and weaknesses in a wide range of clinically relevant situations, and publish their findings. In this regard, vendors must place their formalisms and algorithms in open forums. Furthermore, vendors and users need to collaborate and create compilations of comparisons of dose distributions computed with SEA models with measurements and Monte Carlo simulations. Such data would serve not only the user community, but also the vendors by helping identify the causes of the limitations of the models, leading to their improvements.

Lastly, in needs to be stated that while Monte Carlo methods continue to advance, it is important to continue to improve SEA models also. Even when MC methods are widely available, there will be many applications requiring large numbers of 3D and 4D dose distributions for each patient, necessitating fast SEA models, at least for preliminary calculations. Examples of such applications include on-line adaptive replanning, cumulative dose distributions delivered over the course of proton therapy, 3D and 4D IMPT, robust optimization, etc.

References

1. ACR-NEMA. Digital Imaging and Communications. ACR-NEMA Standards Publication, National Electrical Manufacturer's Association. Washington, D.C., 1985;300.
2. Matsufuji N, Tomura H, Futami Y, Yamashita H, Higashi A, Minohara S, Endo M, Kanai T. Relationship between CT number and electron density, scatter angle and nuclear reaction for hadron-therapy treatment planning. *Phys Med Biol* 1998;43:3261–3275.
3. Palmans H, Verhaegen F. Assigning nonelastic nuclear interaction cross sections to Hounsfield units for Monte Carlo treatment planning of proton beams. *Phys Med Biol* 2005;50:991–1000.
4. Jiang H, Seco J, Paganetti H. Effects of Hounsfield number conversions on patient CT based Monte Carlo proton dose calculation. *Med Phys* 2007;34:1439–1449.
5. Schaffner B, Pedroni E. The precision of proton range calculations in proton radiotherapy treatment planning: experimental verification of the relation between CT-HU and proton stopping power. *Phys Med Biol* 1998;43:1579–1592. http://www.ncbi.nlm.nih.gov/entrez/query.fcgi?cmd=Retrieve&db=PubMed&dopt=Citation&list_uids=9651027.
6. Schneider W, Bortfeld T, Schlegel W. Correlation between CT numbers and tissue parameters needed for Monte Carlo simulations of clinical dose distributions. *Phys Med Biol* 2000;45:459–478.
7. Paganetti H. Range uncertainties in proton therapy and the role of Monte Carlo simulations. *Phys Med Biol* 2012;57:R99–R117.
8. Yang M, Zhu XR, Park PC, Titt U, Mohan R, Virshup G, Clayton JE, Dong L. Comprehensive analysis of proton range uncertainties related to patient stopping-power-ratio estimation using the stoichiometric calibration. *Phys Med Biol* 2012;57:4095–4115. http://www.ncbi.nlm.nih.gov/pubmed/22678123.
9. du Plessis FCP, Willemse CA, Loetter MG, Goedhals L. The indirect use of CT numbers to establish material properties needed for Monte Carlo calculation of dose distributions in patients. *Med Phys* 1998;25:1195–1201.
10. Schneider U, Pedroni E, Lomax A. The calibration of CT Hounsfield units for radiotherapy treatment planing. *Phys Med Biol* 1996;41:111–124.
11. ICRU-44. Tissue Substitutes in Radiation Dosimetry and Measurement, 1989.
12. ICRP-23. Report of the Task Group on Reference Man, 1975.
13. Schaffner B. Proton dose calculation based on in-air fluence measurements. *Phys Med Biol* 2008;53:1545–1562. http://www.ncbi.nlm.nih.gov/entrez//query.fcgi?cmd=Retrieve&db=PubMed&dopt=Citation&list_uids=18367787
14. Li Y, Zhu RX, Sahoo N, Anand A, Zhang X. Beyond Gaussians: a study of single-spot modeling for scanning proton dose calculation. *Phys Med Biol* 2012;57:983–997. http://www.ncbi.nlm.nih.gov/pubmed/22297324.
15. Hong L, Goitein M, Bucciolini M, Comiskey R, Gottschalk B, Rosenthal S, Serago C, Urie M. A pencil beam algorithm for proton dose calculations. *Phys Med Biol* 1996;41:1305–1330. <Go to ISI>://A1996VA70100005.
16. Li Y, Zhang X, Lii M, Sahoo N, Zhu RX, Gillin M, Mohan R. Incorporating partial shining effects in proton pencil-beam dose calculation. *Phys Med Biol* 2008;53:605–616. http://www.ncbi.nlm.nih.gov/pubmed/18199905.
17. Li Y, Zhang X, Mohan R. An efficient dose calculation strategy for intensity modulated proton therapy. *Phys Med Biol* 2011;56:N71–84. http://www.ncbi.nlm.nih.gov/pubmed/21263173.
18. Soukup M, Fippel M, Alber M. A pencil beam algorithm for intensity modulated proton therapy derived from Monte Carlo simulations. *Phys Med Biol* 2005;50:5089–5104. http://

www.ncbi.nlm.nih.gov/entrez/query.fcgi?cmd=Retrieve&db=PubMed&dopt=Citation&list_uids=16237243.
19. Schaffner B, Pedroni E, Lomax A. Dose calculation models for proton treatment planning using a dynamic beam delivery system: an attempt to include density heterogeneity effects in the analytical dose calculation. *Phys Med Biol* 1999;44:27–41. <Go to ISI>://000078449600004.
20. Ulmer W, Schaffner B. Foundation of an analytical proton beamlet model for inclusion in a general proton dose calculation system. *Radiat Phys Chem* 2011;80:378–389. <Go to ISI>://WOS:000287292100013.
21. Ulmer W, Matsinos E. Theoretical methods for the calculation of Bragg curves and 3D distributions of proton beams. *Eur Phys J-Spec Top* 2010;190:1–81. <Go to ISI>://WOS:000286466700001.
22. Kooy HM, Schaefer M, Rosenthal S, Bortfeld T. Monitor unit calculations for range-modulated spread-out Bragg peak fields. *Phys Med Biol* 2003;48:2797–2808. http://www.ncbi.nlm.nih.gov/entrez/query.fcgi?cmd=Retrieve&db=PubMed&dopt=Citation&list_uids=14516102.
23. Kooy HM, Rosenthal SJ, Engelsman M, Mazal A, Slopsema RL, Paganetti H, Flanz JB. The prediction of output factors for spread-out proton Bragg peak fields in clinical practice. *Phys Med Biol* 2005;50:5847–5856. http://www.ncbi.nlm.nih.gov/pubmed/16333159.
24. Sahoo N, Zhu XR, Arjomandy B, Ciangaru G, Lii M, Amos R, Wu R, Gillin MT. A procedure for calculation of monitor units for passively scattered proton radiotherapy beams. *Med Phys* 2008;35:5088–5097. http://www.ncbi.nlm.nih.gov/pubmed/19070243.
25. Zhao Q, Wu H, Wolanski M, Pack D, Johnstone PA, Das IJ. A sector-integration method for dose/MU calculation in a uniform scanning proton beam. *Phys Med Biol* 2010;55:N87–95. http://www.ncbi.nlm.nih.gov/pubmed/20057011.
26. Daartz J, Engelsman M, Paganetti H, Bussiere MR. Field size dependence of the output factor in passively scattered proton therapy: influence of range, modulation, air gap, and machine settings. *Med Phys* 2009;36:3205–3210. http://www.ncbi.nlm.nih.gov/pubmed/19673219.
27. Bortfeld T, Schlegel W. An analytical approximation of depth-dose distributions for therapeutic proton beams. *Phys Med Biol* 1996;41:1331–1339. http://www.ncbi.nlm.nih.gov/pubmed/8858723.
28. Fontenot JD, Newhauser WD, Bloch C, White RA, Titt U, Starkschall G. Determination of output factors for small proton therapy fields. *Med Phys* 2007;34:489–498. http://www.ncbi.nlm.nih.gov/entrez/query.fcgi?cmd=Retrieve&db=PubMed&dopt=Citation&list_uids=17388166.
29. Koch N, Newhauser WD, Titt U, Gombos D, Coombes K, Starkschall G. Monte Carlo calculations and measurements of absorbed dose per monitor unit for the treatment of uveal melanoma with proton therapy. *Phys Med Biol* 2008;53:1581–1594. http://www.ncbi.nlm.nih.gov/entrez/query.fcgi?cmd=Retrieve&db=PubMed&dopt=Citation&list_uids=18367789.
30. Paganetti H. Monte Carlo calculations for absolute dosimetry to determine machine outputs for proton therapy fields. *Phys Med Biol* 2006;51:2801–2812. http://www.ncbi.nlm.nih.gov/pubmed/16723767.
31. Zhu XR, Poenisch F, Lii M, Sawakuchi GO, Titt U, Bues M, Song X, Zhang X, Li Y, et al. Commissioning dose computation models for spot scanning proton beams in water for a commercially available treatment planning system. *Med Phys* 2013;40:041723. http://www.ncbi.nlm.nih.gov/pubmed/23556893.
32. Clasie B, Depauw N, Fransen M, Goma C, Panahandeh HR, Seco J, Flanz JB, Kooy HM. Golden beam data for proton pencil-beam scanning. *Phys Med Biol* 2012;57:1147–1158. http://www.ncbi.nlm.nih.gov/pubmed/22330090.
33. Slopsema RL, Kooy HM. Incorporation of the aperture thickness in proton pencil-beam dose calculations. *Phys Med Biol* 2006;51:5441–5453. http://www.ncbi.nlm.nih.gov/pubmed/17047262.
34. Titt U, Zheng Y, Vassiliev ON, Newhauser WD. Monte Carlo investigation of collimator scatter of proton-therapy beams produced using the passive scattering method. *Phys Med Biol* 2008;53:487–504. http://www.ncbi.nlm.nih.gov/entrez/query.fcgi?cmd=Retrieve&db=PubMed&dopt=Citation&list_uids=18185001.

35. Zhu XR. The status of intensity modulated proton and ion therapy—dose calculational algorithms and commissioning. 2014. http://www.aapm.org/meetings/2014AM/PRAbs.asp?mid=90&aid=23810.
36. Mirkovic D, MDACC, private communication.

Chapter 20

Dose Calculations for Proton Beam Therapy: Monte Carlo

Harald Paganetti, Ph.D.[1], Jan Schuemann, Ph.D.[1], and Radhe Mohan, Ph.D.[2]

[1]Massachusetts General Hospital,
Boston, MA
[2]MD Anderson Cancer Center,
Houston, TX

20.1 The Monte Carlo Method

The Monte Carlo method employs random numbers to create mathematical scenarios based on probability density functions. Monte Carlo techniques have been used for solving many mathematical and physical problems, such as differential equations. In the context of dose calculation, Monte Carlo methods simulate how particles propagate through different materials in the delivery system or a patient. As such, Monte Carlo mimics what is happening in reality when a beam of particles penetrates a geometrical object. At each step of the simulation, the occurrence of a particular physics event is determined randomly based on probabilities which, in turn, determines the next step of the particle. A particle history is defined as the description of the trajectory of one particle, including its secondary particles created by physics interactions. Many particles need to be simulated in order to achieve a given precision, i.e., the statistical uncertainty of a quantity simulated with Monte Carlo depends on the number of random histories.

The simulation of a particle history typically starts by sampling and selecting a particle from a starting source distribution. This sampling can be based on a mathematical function describing the distribution of particle characteristics, or it can be from a list of particles stored in a phase space that characterizes a radiation field in a certain region, e.g., the entrance plane of a treatment head. Monte Carlo simulations often use a phase space, which is a data file containing characteristics of a large number of particles with defined type, energy, direction, and other user-defined identifiers such as a flag providing some information about the history of the particle. Phase spaces are typically defined at a specific plane perpendicular to the central axis of a beam. They can be quite useful if parts of the simulation are generic, allowing a phase space to be reused for subsequent simulations.

Subsequently, each particle is tracked step-by-step through a well-defined geometry, such as a treatment head or patient. Each particle travels a certain distance until it undergoes (determined by interaction probabilities) a physics process, e.g., a collision, that will lead into energy loss or inelastic scattering, from which the particle can emerge with a different energy or direction. Secondary particles may also be produced, which then have to be tracked as well. The step size is determined by the probability of certain physics interactions, but a maximum allowed step size can be defined by the user to ensure a certain geometrical resolution. Each material in the geometry is characterized by its physical properties, e.g., elemental composition, electron density, mass density, and mean excitation energy.

Charged particles like protons interact continuously (in discrete events) so that this microscopic process cannot be explicitly simulated with reasonable efficiency. Monte Carlo simulations of charged particle interactions typically use the so-called Class II condensed history algorithms, in which several energy losses and directional changes are "condensed" (or summed) into a single step [1]. For instance, because the scattering angle of a single proton scattering event is very small, multiple scattering theories provide probability density functions that represent the net result of several scattering incidents.

The Monte Carlo code itself, or its user, typically sets a maximum step size, up to which continuous energy loss and a certain multiple scattering angle is assumed, unless a so-called catastrophic event occurs. Each such event requires explicit modeling in order for the simulation results to be sufficiently accurate and include high-energy δ-electron emission or interactions with the atomic nucleus that may produce secondary particles. Thus, Monte Carlo codes based on condensed history methods use a combination of continuous and discrete processes. In proton therapy, discrete processes are typically nuclear interactions, secondary particle production (including δ-electrons), and large-angle Coulomb scattering. A large step size decreases the computing time, but may add uncertainties. Monte Carlo codes use various methods to ensure proper step size selection, particularly near boundaries where the material and, thus, the underlying physical dose changes [2–4].

Monte Carlo simulations can predict physics phenomena more accurately compared to analytical algorithms because the underlying physics is explicitly modeled on a particle-by-particle basis. This is particularly important near tissue heterogeneities, where analytical algorithms show weaknesses in considering scattering effects. The physics used in Monte Carlo codes is defined using theoretical models, parameterizations, or experimental cross section data for electromagnetic and nuclear interactions. Nevertheless, even Monte Carlo simulation results have uncertainties because the physics might not be known with sufficient precision depending on the type of interaction and material. Particularly, nuclear interaction cross sections are often not measured in the energy region of interest to radiation therapy. In these cases, models, parameterizations, or a combination of models, parameterizations, and experimental data are applied. Furthermore, some of the Monte Carlo codes were originally developed for high-energy physics applications and, therefore, span a wide range of particles and energy domains. Consequently, a Monte Carlo code might allow different physics settings for different energy domains, from which the user can choose. However, those settings and physics models might not be tailored for proton therapy [5–7].

Monte Carlo simulations typically have several components. First, there is a generic code, such as the software provided by the developer. Second, there is a user-defined code that, for instance, defines the geometry of the simulation (e.g., the patient or a treatment head). Third, there is an input file that defines certain parameters, such as the exact location of certain devices in the geometry or the conditions for particle tracking.

There are various Monte Carlo codes for use in proton therapy, e.g., FLUKA [8,9], Geant4 [10,11], MCNPX [12,13], VMCpro [14], Shield-Hit [15], and others. Most of them were originally developed in physics laboratories and then later adopted by medical physics. There are also user-friendly interfaces to some of these codes that are specifically tailored to users in proton therapy, e.g., the TOPAS framework, which is based on Geant4 [16].

20.2 Computational Efficiency of Monte Carlo Dose Calculation

20.2.1 Speed, Precision, and Accuracy

In contrast with analytical methods, the precision of Monte Carlo dose calculation depends on the number of histories. Each simulated history decreases the statistical uncertainty. The required number of particles for a given precision, e.g., a desired statistical uncertainty of the dose within the target of 2%, depends on the field parameters, such as the beam energy and treatment volume. A certain statistical precision in the target volume does not guarantee the same precision in organs at risk (OAR) because of the lower dose and, thus, potentially fewer particles involved. On the other hand, the impact of statistical imprecision is less for dose–volume analysis in OAR because the dose distribution is less homogeneous, and the dose–volume histograms (DVH) are less steep [17,18].

The speed of the Monte Carlo simulation not only depends on the number of histories, but also on the number of steps each particle takes. For dose calculations, the maximum step size is limited by the size of the CT voxels, unless several neighboring voxels are combined into one. This is because the physics settings are typically different for different voxels due to their difference in Hounsfield units and, thus, material composition or material density. Combining voxels is not straightforward because the averaging of material compositions is not exactly defined. On the other hand, it is feasible to first calculate the dose to a lower precision in each CT voxel and then average the dose to a larger grid after the simulation for dose analyzes.

One consequence of the lack of statistical precision is the inhomogeneity of the dose distributions. It has been suggested that Monte Carlo-generated dose distributions can be smoothed [19–23]. This smoothing or de-noising is a technique well known in imaging. These methods need to be applied with caution in proton therapy. For instance, some de-noising techniques tend to artificially soften dose falloffs and assume constant proportionality between fluence and dose. Both aspects are particularly problematic in proton therapy.

Further improvements in efficiency can be achieved by limiting the tracking of particles. For example, if a secondary electron has a very small energy, it may be sufficiently accurate to have its energy absorbed at the point of emission from the atom instead of having it tracked to a very small distance. Thus, typically, above a certain energy threshold, δ electrons are produced explicitly, but below the threshold continuous energy loss of the primary particle is assumed. Production cuts for secondary particles can influence energy loss and, thus, the simulation results [4].

For proton dose calculation, depending on the beam energy, primary and secondary protons account for roughly 98% of the dose [24]. This includes the energy lost via secondary electrons created by ionizations. The range of most electrons in a clinical proton beam is typically less than 1 mm in water. Thus, explicit tracking of electrons is not necessarily required for dose calculations on a typical CT grid.

The tracking of secondary electrons might be required for applications other than dose calculation, e.g., for microdosimetry, for absolute dose simulations [25], or for ion chamber simulations [25–27]. Secondary particles from nuclear interactions typically include protons, neutrons, and heavier fragments. The latter have a very short range and don't need to be tracked explicitly.

Other techniques to improve the computational efficiency of Monte Carlo dose calculation fall under the term of variance reduction techniques [28]. They have been utilized in proton therapy simulations as well [29], such as the method of splitting of particles, or the Russian roulette technique [30,31]. The particle-splitting technique splits protons N times (N protons are generated from the incident proton) at specific locations (e.g., planes perpendicular to the central beam axis) in the treatment head. These protons are then assigned reduced weights. By splitting at specific locations, it is feasible to predominantly consider particles that have a high likelihood of contributing in regions of interest. Protons with low probability of contributing to a certain scoring region can be subject to the Russian roulette technique with a probability of discarding the particle equal to $1-1/N$.

There is a relationship between computation time and the statistical uncertainty that needs to be optimized, depending on the aim of the simulation. This relationship is specified by the computational efficiency ε, which is a function of the statistical uncertainty of the quantity of interest, s, and the central processing unit (CPU) time T as shown in Equation 20.1 [32]:

$$\varepsilon = \frac{1}{(s^2 T)} \tag{20.1}$$

For dose calculation one may want to restrict the region of interest by defining the average statistical uncertainty s_Y as a measure of the overall statistical uncertainty s according to Equation 20.2 [32]:

$$s_{Y\%}^2 = \frac{1}{N} \sum_{i=1}^{N} \left(\frac{\delta D_i}{D_i} \right)_{Y\%}^2 \tag{20.2}$$

Here, D_i is the dose in voxel i having a statistical uncertainty of δD_i. Only those voxels with a value larger than $Y\%$ of the maximum value are typically considered [33].

Another way of improving computational efficiency is to minimize the scope of a Monte Carlo code so that it does focus on one specific task only, e.g., dose calculation on a CT grid, instead of being capable to also simulate other properties such as, for instance, the scattering properties in a specific treatment head device or certain secondary radiation effects. A dedicated Monte Carlo code, VMC_{pro}, was introduced solely optimized for dose calculation in human tissues for proton therapy [14]. A significant speed improvement compared to established Monte Carlo was achieved by introducing approximations in the multiple scattering algorithm and using density-scaling functions instead of actual material compositions. Furthermore, nuclear interactions are considered explicitly, but in parameterizations relying on distribution sampling. Nevertheless, the agreement between VMC_{pro} and other Monte Carlo codes was found to be excellent. Other codes have been optimized for speed by only tracking primary protons through the treatment head and the patient and treating contributions from nuclear interactions in analytical approximations without the tracking of secondary particles [34].

20.2.2 Hybrid Monte Carlo and Particle Track Repeating Algorithms

There is no clear boundary between what one would consider a full-blown Monte Carlo and an analytical algorithm. Some fast Monte Carlo codes utilize analytical approaches for certain physics phenomena, such as nuclear interactions (see above).

Kohno et al. [35] developed a simplified Monte Carlo method for proton dose calculation. The algorithm uses the depth–dose distributions in water measured in a broad proton beam to calculate the energy loss in materials based on a water-equivalent model. At each voxel, the proton's residual range is reduced according to the local material property, and a corresponding amount of energy is deposited. Multiple Coulomb scattering (MCS) is modeled by sampling scattering angles from a normal distribution parameterized by theoretical considerations. The method has been shown to be more accurate compared to common analytical algorithms [35,36].

A different strategy is applied in track-repeating algorithms [37,38]. Here, proton tracks and their interactions in water (or different materials) are tabulated after they have been simulated using Monte Carlo. The changes in location, angle, and energy for every transport step and the energy deposition along the track are recorded for all primary and secondary particles and reused in subsequent Monte Carlo calculations. Each particle trajectory stored consists of a set of steps, and for each step, the direction, step length, and energy loss. During the simulation, at each step there is a track segment loaded from the database that resembles the scenario for the tracked particle. Because complicated physics model calculations are avoided, this method can be more efficient compared to standard Monte Carlo calculations and typically cause very little compromise in accuracy.

20.2.3 Monte Carlo on Graphics Processing Units (GPU)

A graphics processing unit (GPU) is a dedicated hardware designed for accelerated processing of graphics information. Compared to a conventional CPU, it has a much higher number of processing units sharing a common memory space. A GPU executes a program in groups of parallel threads. They have long been used to accelerate not only graphics, but also other scientific computational tasks. Whether a task can be significantly accelerated using GPU hardware depends on the mathematical formalism. Analytical dose calculations are very well suited because they can be formulated as vector operations, which take advantage of multi-threading. Some of the Monte Carlo algorithms mentioned above have also been implemented on GPU cards [35,37,39]. Particularly, track-repeating algorithms can benefit from the multi-threading environment of a GPU [37,39]. The application of proton Monte Carlo dose calculation on GPUs has been reviewed by Jia et al. [40].

In addition to implementing existing algorithms on a GPU, there have been efforts to develop proton Monte Carlo dose calculation algorithms that are tailored to GPU use and are, thus, very efficient [41,42]. While particle tracking with respect to electromagnetic interactions can be done explicitly in a GPU-based code, approximations may be necessary for nuclear interactions because of required data tables that would exceed the memory capability of GPUs. For instance, the gPMC code [41] considers proton propagation by a Class II condensed history simulation scheme using the continuous slowing down approximation; however, nuclear interactions are modeled using an analytical approach [14]. While other secondary particles are neglected, the emitted secondary protons are tracked in the same way as primaries. This code has been compared with CPU-based full Monte Carlo codes for patient dose calculation and found to be in excellent agreement, with a gamma index passing rate of ~99% using a 2%/2 mm gamma index criteria. The gamma index is often used to compare distributions, and it combines features of both dose difference and distance to agreement [43–45].

The efficiency improvement using GPU codes for Monte Carlo is excellent, allowing dose calculation down to 1% statistical accuracy based on an existing treatment plan to be done in seconds instead of hours.

20.3 Dose Calculation for Patient Treatment

20.3.1 Passive Scattering

A Monte Carlo algorithm starts tracking each particle at a well-defined location. For instance, if all protons would be tracked starting at the patient surface, the distributions of proton energies, angles, etc. would need to be known at that position so that the Monte Carlo routine could randomly sample from these distributions. Typically, Monte Carlo simulations are either started with an analytical description of such distributions, or the particles are sampled from a phase space file (see above). The distribution of protons at the treatment head exit (i.e., the patient entrance) is very complex in passively scattered proton therapy because protons exiting the treatment head underwent complex scattering and interaction events in the beam-shaping devices in the treatment head. Thus, for dose calculation in passively scattered proton therapy, the Monte Carlo simulation typically starts at the treatment head entrance (i.e., the treatment room) because an analytical definition of a phase space is not feasible.

A parameterization of the phase space at the treatment head entrance can be obtained by fitting measured data [46] and then selecting a parameterization of the beam at the entrance of the treatment head based on beam energy, energy spread, beam spot size, and beam angular distribution [47,48]. Accurately measuring some of these parameters directly can be difficult, and a user might have to rely on the manufacturer's information.

For the particle tracking through the treatment head, most Monte Carlo approaches model machine-specific components in the treatment head using manufacturer's blueprints [46,48,49]. Each geometrical structure has to be defined in a Monte Carlo code via a mathematical formalism typically based on a predefined set of structures, such as disks, boxes, etc. Important beam-shaping devices are the passive scattering system, as well as apertures and compensators.

Treatment heads in passive scattering proton therapy can be complex, and the position of certain devices can change depending upon the specified field. For example, scattering foils might be inserted or different modulator wheels evoked for a certain combination of range and modulation width. A generic Monte Carlo model of a treatment head has to accommodate all possible geometrical variations [46]. The parameters defining the field-specific setting of the treatment head are provided either by a treatment planning system (if it prescribes the treatment head settings for a patient field) or by the treatment control system (if the treatment head settings are defined by an interface to the planning system or manually by an operator). An interface between the Monte Carlo code and these parameter settings is thus needed, unless parameters are edited in the Monte Carlo input file manually.

Some components in a treatment head for passively scattered proton therapy are moving, such as a rotating modulator wheel. To accurately consider moving parts, the four-dimensional Monte Carlo technique has been suggested [50]. The technique allows geometry changes after each particle history. Another time-dependent parameter is the modulation of the beam current in passive scattering systems [51]. In order to deliver a certain modulation width, the beam current might be regulated at the accelerator (beam source) level [51]. A sophisticated and user-friendly way to handle time-dependent components in a simulation has been introduced by Shin et al. [52] in the TOPAS code.

Particles at the treatment head exit are often stored in a phase space so that the patient CT-based simulation is done in a separate step. Depending on the complexity of the treatment head, it may be feasible to compute and pre-store phase spaces for all SOBP options and then sample from them for each combination of aperture, compensator, and patient.

20.3.2 Scanning Beams

To simulate a scanning beam, four-dimensional Monte Carlo techniques can be used to constantly update the magnetic field strength [50,52,53]. This allows studying beam scanning delivery parameters [53–55]. For patient dose calculation, this is typically not necessary. While it may be difficult to parameterize a beam at the treatment head exit in passive scattering, it is, in most cases, feasible for beam scanning because of the limited amount of scattering material in the beam path. Thus, the tracking of particles through the treatment head for each case can typically be avoided when simulating scanned beam delivery.

This has important consequences because it typically improves the computational efficiency substantially. The reason is the relatively complex geometry of a passive scattering treatment head compared to the regular patient CT grid. For simulations, the majority of calculation time is typically spent tracking particles through the treatment head. The efficiency of proton therapy treatment heads for passive scattering is typically between 2% and 40%, depending on the field, and thus most tracked protons do not reach the patient and contribute to the dose information. The use of beam models allows for the potential to routinely use fast Monte Carlo in the clinic for beam scanning.

It has been demonstrated that beam characterization for scanned delivery can be done based on measured depth–dose curves in water alone, without consideration of the exact treatment head system [56–58]. For example, the energy spread of the beam as a function of energy can be deduced by fitting the widths of measured pristine Bragg curves. Secondary particles, other than protons generated in the treatment head, can typically be neglected [57]. A planning system will provide a matrix of energies, weights, and positions of beam spots, which can be translated into Monte Carlo input settings (see Chapter 9). Beamlets are typically not simulated, considering the beamline and treatment head components. Thus, one has to be cautious characterizing a beam spot by a simple Gaussian fluence distribution. As shown in Figure 20–1, this leads to an overestimation of the fluence in the center of a spot due to a halo caused by protons scattered at a large angle from nuclear interactions [57,59,60], so small

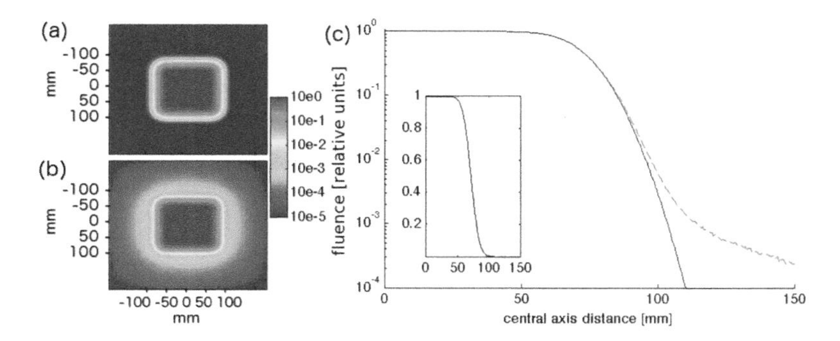

Figure 20–1 Logarithmic fluence map of 10 x10 cm^2 field consisting of a 91 MeV proton energy layer using a Gaussian spot parameterization at treatment head exit (a) and using a full treatment head simulation (b). (c) shows a comparison of the two fluence distributions along the central axis (black: Gaussian approximation; dotted red: treatment head simulation). (From [57] with permission.)

corrections might be needed, such as the consideration of a second Gaussian in the formalism.

20.3.3 Dose in Patients

In order to simulate the dose in a patient and to recalculate a treatment planning result using Monte Carlo, the patient geometry (CT) and plan information needs to be imported into the Monte Carlo code. The patient's CT image can be imported using either a DICOM stream [61,62] or by importing CT information based on a planning system's specific format [46,63]. Alternatively, the entire plan information can be transferred to the Monte Carlo code using a DICOM-RTion interface. In any case, the information provided to the Monte Carlo dose calculation engine will be translated into specific settings in the Monte Carlo code, such as the gantry angle, the patient couch angle, the isocenter position of the CT in the coordinate system of the planning program, the number of voxels and slice dimensions in the CT coordinate system, the size of the air gap between the treatment head and patient, and the prescribed dose [46]. This can either be done internally within the Monte Carlo, or by using a dedicated program that uses the plan information and translates it into an input file readable by the Monte Carlo code.

Chapter 19 describes the conversion of CT numbers to relative stopping powers (RSP) for analytical dose calculation. Instead of RSP, Monte Carlo dose calculations are based on tissue (material) compositions and mass densities for better accuracy. It is not necessary to consider each CT number as a separate tissue. Instead, different CT numbers can share common material properties, i.e., elemental composition and ionization potential, and may, for instance, only differ in density. Typically, between 5 and 30 different tissues are defined that cover the materials for the entire range of CT numbers. Each tissue spans up to a few hundred CT numbers using the same material composition with a varying density [64–66]. A relationship between a certain CT number and a combination of materials is not unique, causing some uncertainty in the process [67,68]. Also needed are mean excitation energies for each tissue, which can be interpolated based on the atomic weight of the tissue elements using Bragg's rule. Mean excitation energies for various elements [69] and averaged values for tissues [70,71] have been tabulated. Mean excitation energies are subject to uncertainties on the order of 5% to 15% for tissues. This is reflected in uncertainties in predicting the proton range [72–74].

In order to produce correct results, the CT conversion used by the Monte Carlo needs to be normalized to the departmental CT scanner. This can be done by either doing a separate stoichiometric calibration or, as an approximation, by simulating RSP values in the Monte Carlo based on an existing CT conversion and then comparing the results with the planning system conversion curve (which had been validated during the commissioning process) with subsequent fine-tuning [46,63,64].

Analytical dose calculation engines calculate dose by modeling physics relative to water (using the RSP) and reporting dose-to-water. It has been debated whether doses in radiation therapy should be reported as dose-to-water or dose-to-tissue, the latter being based on the actual density and material composition of the tissue [75]. Monte Carlo dose calculation results are naturally in terms of dose-to-tissue. However, one typically re-normalizes the Monte Carlo results to dose-to-water because clinical experience is based on dose-to-water, quality assurance and absolute dose measurements are done in water, and only dose-to-water allows proper comparison with treatment planning systems based on analytical algorithms. This re-normalization can be done based on proton energy-dependent RSP relationships and on a nuclear interaction parameterization [76]. Dose-to-water can be higher by as much as ~10% to 15% compared to dose to tissue in bony anatomy, while for soft tissues the differences are typically on the order of 2% [76].

20.3.4 Monte Carlo-based Optimization

Calculating dose using Monte Carlo and Monte Carlo treatment planning are two different issues. The discussion above deals with the former. The use of Monte Carlo in an inverse planning framework has stricter requirements in terms of computational efficiency because the dose has to be calculated more than once in the optimization loop. For example, for intensity-modulated proton therapy, each potentially desired pencil would have to be pre-calculated before evoking the optimization algorithm. Each of these would have to be calculated to high statistical precision because the weight of that particular pencil (i.e., its contribution to the final dose distribution) would not be known a priori. If the difference between a Monte Carlo and an analytical algorithm is small, one might base the optimization on the analytical algorithm and use Monte Carlo simulations for fine-tuning only. The feasibility of this strategy depends on the complexity of the patient geometry because this determines the difference between the two algorithms.

20.3.5 Commercial Monte Carlo Dose Calculation or Treatment Planning Approaches

The fact that the higher accuracy of Monte Carlo dose calculation compared to analytical methods can be clinically significant has not gone unnoticed by vendors of treatment planning software. Furthermore, improvements in both the computational efficiency of Monte Carlo software and computer hardware have made it feasible to use Monte Carlo dose calculation routinely. Consequently, several vendors are working on the implementation of proton Monte Carlo into their planning systems. Whether these implementations will use Monte Carlo dose calculation at every single optimization iteration step or only at certain checkpoints remains to be seen. Furthermore, some implementation will for sure utilize some of the efficiency improvement techniques described above, which may cause some algorithms to be less accurate compared to full Monte Carlo simulation algorithms.

20.4 Differences between Monte Carlo and Analytical Dose Calculation

As discussed in Chapter 19 and above, Monte Carlo and analytical dose calculation algorithms are two fundamentally different approaches of calculating the same physical property, i.e., the dose distribution in a patient. The two approaches were developed for different purposes. While analytical algorithms were developed to provide fast dose calculations in patients with reasonable accuracy for treatment planning, Monte Carlo codes were developed to obtain an accurate representation of the physical processes in the patient, leading to more accurate dose distributions, albeit with significantly longer computation times. In addition, Monte Carlo codes inherently provide the opportunity to record other physical properties, since every step of every particle (within defined limitations) is calculated.

The main dosimetric differences between the two algorithms is the handling of MCS. The differences in scattering events become particularly important at high-density gradients in the beam direction, such as bone/air or soft tissue/lung interfaces in a patient. The disequilibrium of scattering events causes protons to be preferentially scattered out of the higher-density material into the lower-density material. Coulomb scattering causes small-angle deflections of protons. The number of scattering events a single proton experiences is proportional to the amount of material in its path—protons are scattered more frequently in high-density materials than in low-density materials. The design of analytical algorithms makes it nearly impossible to accurately describe MCS along the proton trajectories.

Differences in modeling the proton propagation between the two calculation algorithms can cause two types of discrepancies in the resulting dose distributions: differences in the range of the protons, i.e., the distance that protons travel inside a patient, phantom, or beam delivery device, and differences in the amount of dose deposited in a voxel by changing the proton path or direction.

20.4.1 Differences of Monte Carlo and Analytical Algorithms in the Prediction of Range

The range of a proton beam for clinical use is generally defined as its range in water. The position of the dose distribution that receives 80% of the prescribed dose in the distal falloff of the proton field is defined as R_{80}. For mono-energetic proton beams with clinically used energies, R_{80} is the position where ~50% of initial protons have stopped. As a result, R_{80} is independent of the initial proton energy spread. However, the range of a proton field in clinical use is historically defined as R_{90}, i.e., the position of the 90% dose level in the distal falloff region of a spread-out Bragg peak. While the range in water is well defined, patient-specific beam-shaping devices, such as the range compensator, modulate the fields in a patient. In addition, density variations in the patient will additionally modulate the physical range. Thus, there is generally no clearly defined range of a proton field in a patient, rather each small beamlet or pencil has its own range in the patient. Analytical algorithms convert tissue into water-equivalent path lengths (WEPL) to estimate the range of each pencil, and calculations are, hence, performed assuming water (of varying density) as the target.

It is well known that analytical algorithms have deficiencies when calculating dose distributions. In clinical practice at the Massachusetts General Hospital and many other centers, range margins of the order of 3.5% +1 mm are assigned to proton fields to cover various uncertainties when predicting the range in patients. One of those uncertainties is due to deficiencies in analytical dose calculation. It has been estimated that range margins could be reduced to 2.4% +1.2 mm for all patients if Monte Carlo techniques were to be used routinely for dose calculations [77]. When using Monte Carlo simulations, three sources of range uncertainties will be reduced: the range degradation from local lateral inhomogeneities (2.5% to 0.1%), the range degradation from complex inhomogeneities (0.7% to 0.1%), and the uncertainties from the conversion of CT to tissue (0.5% to 0.2%) [73]. Using Monte Carlo simulations, Schuemann et al. [78] investigated the range margins currently used due to analytical dose calculation and demonstrated that range uncertainties due to dose calculation depend on the patient geometry and, thus, on the treatment site. Figure 20–2 shows the root mean square of the difference in R_{90} between the two algorithms across the treatment field in the high-dose region. All range differences for homogeneous patient geometries (liver and prostate) are well within the 3.5% of the range. However, a large fraction of range differences for heterogeneous patient geometries is outside the recommended range margin.

In order to cover 95% of the treatment fields with a generic margin, a range margin of 6.3% +1.2 mm for head and neck, lung and breast patients would have to be used [78]. This, however, assumes the requirement of full target coverage for each field. Thus, while single-field coverage cannot be guaranteed for such patient geometries, the use of multiple fields generally mitigates the effects in the combined dose distributions because the distal falloff regions for treatment fields with varying beam angles are located in different parts of the target. Head and neck patients have the largest range uncertainties, but their treatment plans are generally also composed of the highest number of treatment fields.

An example with large discrepancies between the two dose calculation algorithms is shown in Figure 20–3, where the delivered proton field experiences multiple high-density gradients along the beam direction at lung/tumor, lung/spine, ribs/soft tissue, and rib/lung interfaces. The effects of MCS is illustrated in Figure 20–3a. The anterior part (dotted arrow) of the field first passes a muscle/soft tissue interface and a bone/lung interface at the distal end of the field. In both instances, the field propagates along the density gradient, and the higher density is on the posterior side. At both

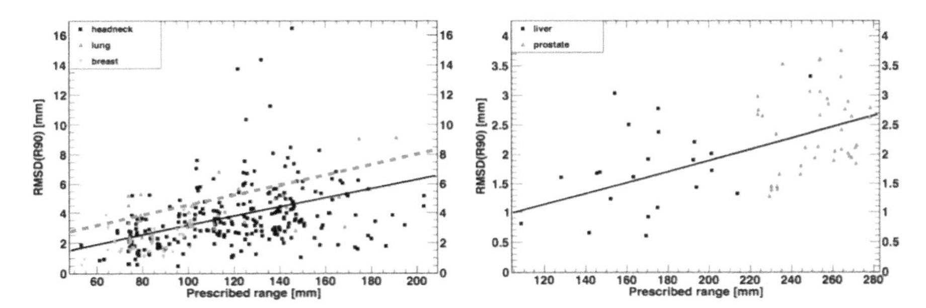

Figure 20–2 Root mean square differences for R_{90} between analytical-calculated and Monte Carlo-calculated dose distributions for single fields for five treatment sites. The solid line indicates the best fit through the data points, the red-dashed line indicates the clinically used range margin of 3.5% +1 mm. Left: range differences for heterogeneous patient geometries such as head&neck, lung and breast. Right: range differences for patient geometries with only a few heterogeneities parallel to the beam direction, such as liver and prostate (here the red dashed line is outside of the displayed area). (From reference [78].)

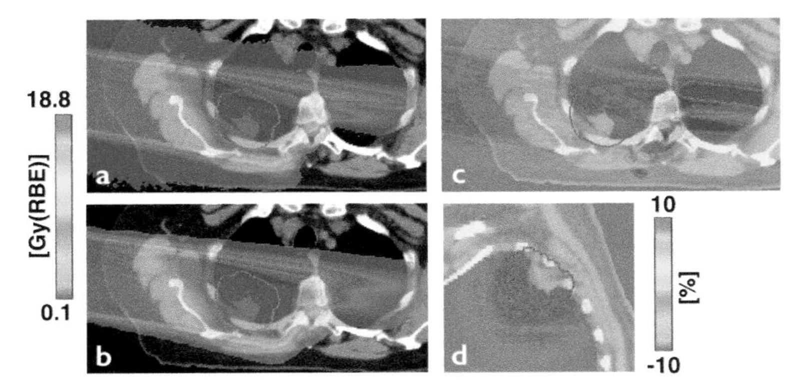

Figure 20–3 Example of a lung patient illustrating the differences in the dose distributions caused by the different calculation algorithms: a) Monte Carlo simulation, b) analytical calculation, c) the dose difference between a and b in axial view and, d) the dose difference in coronal view.

locations, protons are preferentially scattered out of the higher-density material (i.e., toward the anterior side of the patient), resulting in large regions in the distal dose falloff where analytical and Monte Carlo dose calculations differ by over 10% of the prescribed dose distal to the spine (red region in Figure 20–3c). Similarly, posterior to the high-dose region is a region of lower dose due to the preferential scattering out of the higher-density regions. The same effect can be seen from scattering along the rib and tumor volume in the center of the proton field (dashed arrow). The combination of these two scattering regions results in the pattern of alternating higher and lower doses in the right lung in Figure 20–3c. These range fluctuations cause large root mean square range differences, but they do not affect the target volume (pink contour). However, protons scattered out of the beam may have consequences with regard to normal tissue toxicities.

Another factor affecting R_{90} is the sharpness of the distal dose falloff. The slope of the dose gradient depends on the initial beam energy spread, the heterogeneity of the patient, and the range (due to the statistical nature of proton interactions, so-called range straggling). Sawakuchi et al. [79] studied the effects of phantom heterogeneity on the distal dose falloff of monoenergetic proton beams and found that the slope is largely defined by range straggling. Even in highly heterogeneous geometries, range straggling was responsible for over 80% of the slope. Note that the distal falloff along each ray, as predicted by analytical algorithms, can be considerably sharper than with Monte Carlo simulations.

Differences in the predicted range can contribute to the overall dose difference in the target between the two calculation algorithms when the range differences are larger than the safety margin around the target volume.

20.4.2 Differences in Absorbed Dose in Target

In addition to differences in range between the two types of dose calculation methods, there can be differences in the predicted dose. Soukup and Alber reported a combined systematic error of more than 6% for patients with high occurrences of lateral heterogeneities [80]. If a spot decomposition method is applied to the analytical algorithm, the error can be reduced to 2.5%. In this method, the original spot was decomposed into 121 sub-spots.

Schuemann et al. [81] compared analytical-based and Monte Carlo-based dose calculations and the effects of the dose calculation algorithm on the dose delivered to the target volume. Figure 20–4 summarizes their findings for the mean target dose and the maximum dose covering 95% (2%) of the target volume, D_{95} (D_{02}). Composite dose distributions of all proton fields in the treatment plans were analyzed for this study. The results show that Monte Carlo-predicted doses are generally lower than those predicted by analytical algorithms for all treatment sites. Differences in the mean target dose of up to 4% for head and neck and lung patients, or as low as 2% for breast and liver patients, were observed.

The systematically lower dose when using Monte Carlo simulations is caused not only by differences in the manner in which the patient geometry is considered, but it can also be due to the modeling of the proton field before the protons reach the

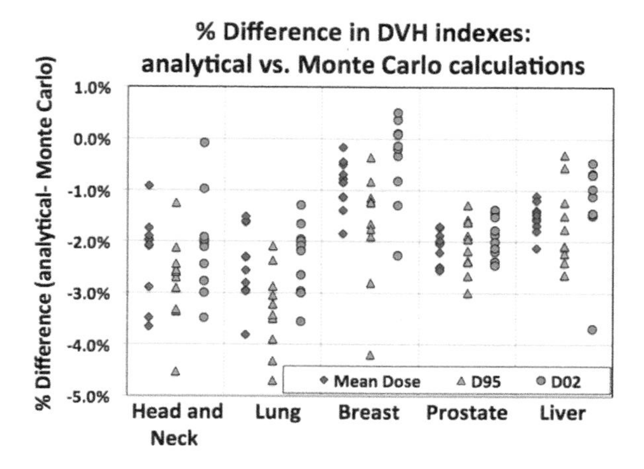

Figure 20–4 Differences in the target volume from dose calculation algorithms for five treatment sites between analytical-based and Monte Carlo-based dose calculations. Shown are the mean dose and the maximum dose that covers 95% (2%) of the target volume, D_{95} (D_{02}). (From [81]).

patient. Experimental [82] and simulation studies [83,84] have demonstrated that small apertures (less than ~5 cm in diameter) can cause a significantly lower dose than that predicted by analytical algorithms at the isocenter depending on the range in the patient. Small fields lose the equilibrium state, even in the central part of the field, resulting in non-flat transverse profiles of the dose distributions. In particular for head and neck patients—where tumor volumes are often irregularly shaped and multiple fields of different size and shapes are combined—the small-field effect can contribute to the systematically lower dose in the target region.

In general, analytical algorithms underestimate the lateral extent of the scattering. This can readily be seen by the larger extent of the dark blue (low dose) region of the predicted Monte Carlo results in figures 20–3 and 20–5 as compared to the analytical calculations. The excess dose outside the main treatment field direction is mostly caused by large angle scattering of the incoming protons and secondary electrons. These large-angle scattering events are poorly described by analytical calculations, in part because large-angle scattering inside the patient can only be accounted for via such approaches as the beam decomposition method [80] or the pencil beam redefinition approach [85]. However, both of these methods result in significant computing time penalties. Another approach that has been applied with some success is the use of an additional Gaussian or an analytical function with a longer tail to accommodate large angle scattering [86].

While MCS is the dominating source of discrepancies between the two dose calculation algorithms, the use of water-equivalent path lengths in the analytical dose calculations also introduces uncertainties when the density in the proton beam path is sufficiently different from water, such as in the lung. Lung tissue has a density of less than 1/3 the density of water. In lung treatments, protons are scattered in the chest wall, which causes the beam to diverge with distance. Since analytical algorithms calculate the spread of the beam for the WEPL, not the physical distance a proton propa-

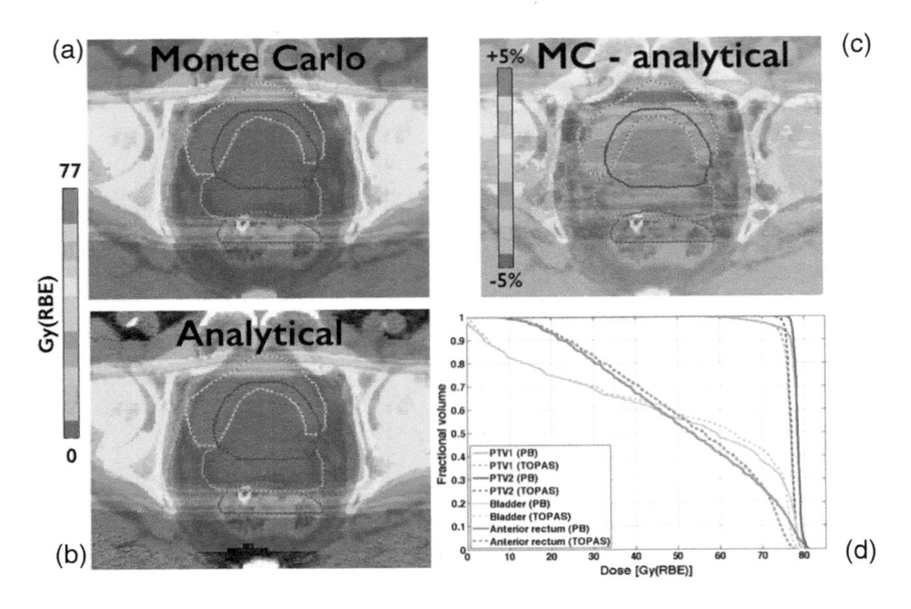

Figure 20–5 Dose calculations for a prostate patient comparing Monte Carlo (a) and analytical dose (b) and the difference, Monte Carlo minus analytical dose (c). Also shown is (d) the dose–volume histogram for the two target volumes (PTV1 includes the prostate, PTV2 includes prostate and seminal vesicles and has a lower prescribed dose than PTV1) and two organs at risk (bladder and anterior rectum) for analytical (solid line) and Monte Carlo calculation (dashed line).

gates, the beam spread is also only calculated for less than ~1/3 of the actual distance that protons travel in the patient. Accordingly, analytical algorithms generally underestimate the lateral beam spread in lung patients. The mean target dose calculated with Monte Carlo for typical lung tumors was found to be between 1.4% and 4% lower than predicted by analytical algorithms [81,87].

In general, the uncertainties in dose calculations are related to how much the irradiated geometry differs from a water phantom. A heterogeneity index was introduced by Pflugfelder et al. [88] to quantify lateral inhomogeneities. Several studies have investigated the correlation between patient heterogeneity and the difference in target dose between analytical and Monte Carlo dose calculations [80,84,88].

20.4.3 Considerations for Organs at Risk

The main concern with respect to uncertainties in dose calculation is often focused mainly on tumor coverage, e.g., a potential overestimation of the range as discussed above. However, OARs can be affected as well. For instance, analytical algorithms might underestimate the range of a field pointing toward a critical structure, such as the brainstem, or they might underestimate lateral scattering into the normal lung.

Figure 20–5 displays a typical prostate treatment plan as calculated analytically and with Monte Carlo. Prostate treatment fields generally have a long range in patients, thus small differences in the modeling of scattering can result in large differences at the isocenter. Thus, Monte Carlo simulations for prostate fields predict lower doses in the target region and higher doses in the lateral penumbra, as shown by the red regions in Figure 20–5c. Some regions in the organs at risk (rectum and bladder) receive doses up to 5% higher compared to those predicted by the analytical algorithm due to increased lateral beam spread (Figure 20–5d).

20.5 Other Monte Carlo Applications Relevant to Treatment Planning

20.5.1 Monte Carlo to Support Analytical Algorithms

As discussed in the previous chapter, Monte Carlo simulations can also be applied to support analytical methods. Spot decomposition techniques have been applied in analytical methods where multiple ray-tracings are performed for a single pencil [80,89]. One thus might perform Monte Carlo calculations only on a subset of rays that go through a highly heterogeneous geometry. Monte Carlo-generated kernels have been generated to be used as look-up tables in an analytical dose calculation framework [80]. Further, pre-calculated Monte Carlo depth–dose curves in water or other materials can serve as input for pencil-beam algorithms. The pencil-beam algorithm by Soukup et al. [89] was designed with the help of Monte Carlo, which was used to model nuclear interactions.

20.5.2 4D Dose Calculations

The capability of simulating time-dependent geometrical structures in a treatment head was discussed above. Monte Carlo simulations have also been used to simulate time-dependent patient geometries for studies of breathing motion in proton therapy of the lung [90–94]. The use of Monte Carlo for these studies is particularly important in proton therapy because small uncertainties in scattering and range can have a large impact in low-density lung tissue [87]. Monte Carlo simulations have been used to understand the interplay between breathing motion and beam motion in proton beam scanning, and they have shown that the effect is biggest for delivery systems with a small spot size [90,92,95]. The impact of beam delivery parameters—such as scanning speed or the time it takes to switch energies—have been studied using Monte Carlo simulations [94].

20.5.3 LET and RBE-weighted Dose Distributions

Prescription doses to cancerous tissue, as well as dose constraints to organs at risk, are based on clinical experience with photon beams. Proton doses are related to photon doses using the relative biological effectiveness (RBE) (see Chapter 5). Biophysical modeling is a long way from being able to simulate all radiation effects in sub-cellular structures. The physics, however, can be simulated reasonably well using Monte

Carlo. In order to interpret biological experiments, one needs to know the characteristics of the radiation beam. One parameter necessary to interpret biological effectiveness of proton beams is the linear energy transfer (LET). It has been demonstrated that analytical methods have shortcomings when estimating LET and that Monte Carlo methods are needed to accurately predict the distribution of LET_d (the dose-averaged LET) in patients [96].

Using Monte Carlo simulations, it was estimated that in the center of the SOBP, the average LET_d, is typically between 2 and 3 keV/μm. The maximum expected LET_d value in the target (doses >90% of the target dose) turned out to be ~8 keV/μm, while for the region outside the target, values up to ~12 keV/μm were predicted (for doses >2% of the target dose) [97]. For low-energy beams and small modulation widths, LET values could be significantly higher. One can simulate dose-averaged LET distributions in a patient geometry to identify potential hot spots of biological effectiveness [96,98]. Simulated LET distributions in patients have been used to identify correlations between toxicities and LET hot spots [99] or recurrences and LET cold spots [100].

As discussed in Chapter 5, there is interest in changing the treatment paradigm in proton therapy toward the use of variable RBE values. However, more research is needed to accomplish this, experimentally as well as theoretically. Monte Carlo simulations will play a role in this process as they allow simulating experimental conditions to interpret or design radiobiological experiments. Furthermore, Monte Carlo simulations can be used to provide physics input into biophysical model calculations that can then calculate dose-averaged α and β parameters for the linear quadratic dose–response model. The TOPAS Monte Carlo framework [16], widely used in proton therapy, allows exporting 3D distributions of α and β parameters based on a treatment plan that incorporates a variety of different models [101].

Monte Carlo simulations have also been applied to predict more fundamental quantities, such as DNA damage based on particle energy spectra (e.g., [102,103]).

20.5.4 Neutron Doses

Monte Carlo simulations play a huge role in radiation protection (see Chapter 10). Particularly, simulations have been used to predict the neutron radiation fields in patients. Monte Carlo techniques are necessary because dosimetry for organs outside the portion of the body that is imaged for treatment planning will need to be reconstructed using whole-body patient models and because dose calculation algorithms for treatment planning purposes are not commissioned for predicting low-dose radiation background in radiation therapy.

Comparisons of measured and Monte Carlo-simulated neutron doses have shown very good agreement [105]. For studies focusing on organs outside of the area typically imaged for treatment planning, whole-body computational phantoms have been developed and implemented into Monte Carlo codes [106]. These are being used in radiation protection as well as radiation therapy, and they are typically age and gender specific. Some of them are voxelized (e.g., [107]), while others use a mesh [104] to resemble organ shapes as close as possible (see Figure 20–6).

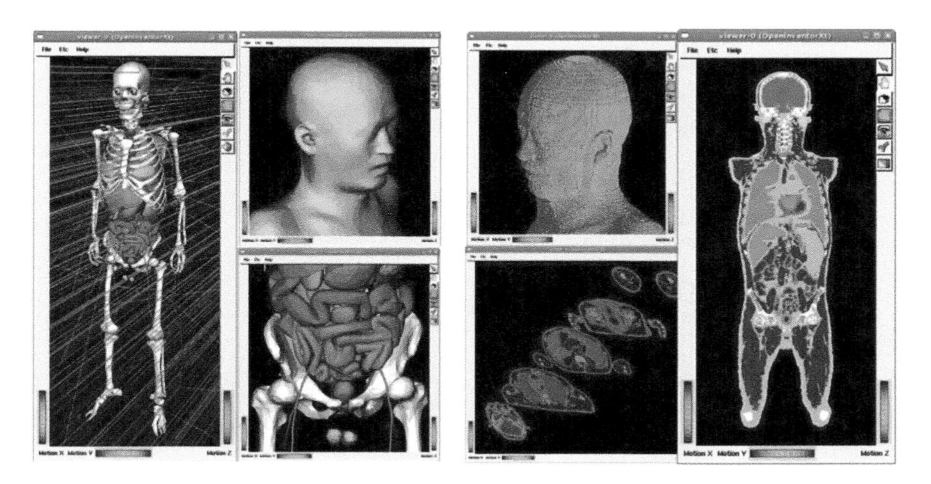

Figure 20–6 Whole-body computational phantom based on a mesh structure as implemented into a Monte Carlo code. (From [104] with permission.)

While measurements of neutron radiation should be the gold standard, Monte Carlo methods are able to differentiate between neutrons generated in the accelerator and the patient, and they can track those neutrons inside the patient, so they can be used to design delivery techniques that reduce neutron exposure. There are many Monte Carlo studies on secondary neutron radiation in proton therapy [108].

20.6 Summary

Monte Carlo simulations are the most accurate method to calculate dose in patients. Their use in proton therapy will increase in the near future because of the development of dedicated codes for proton therapy dose calculation and due to commercial interest in implementing Monte Carlo in planning systems. Eventually, Monte Carlo-simulated dose distributions will contribute to the reduction of margins as they reduce the uncertainty originating from dose calculation uncertainties. Both lateral and range uncertainty margins could be reduced by the use of Monte Carlo and, thus, could lead to a change in the way we treat proton therapy patients [109].

References

1. Berger MJ. Monte Carlo calculation of the penetration and diffusion of fast charged particles. In: Methods in computational, physics. Alder B, Fernbach S, Rotenberg M, Eds. New York: Academic, 1963.
2. Kimstrand P, Tilly N, Ahnesjo A, Traneus E. Experimental test of Monte Carlo proton transport at grazing incidence in GEANT4, FLUKA and MCNPX. *Phys Med Biol* 2008;53:1115–1129.
3. Poon E, Seuntjens J, Verhaegen F. Consistency test of the electron transport algorithm in the GEANT4 Monte Carlo code. *Phys Med Biol* 2005;50:681–694.

4. van Goethem MJ, van der Meer R, Reist HW, Schippers JM. GEANT4 simulations of proton beam transport through a carbon or beryllium degrader and following a beam line. *Phys Med Biol* 2009;54:5831–5846.
5. Herault J, Iborra N, Serrano B, Chauvel P. Monte Carlo simulation of a protontherapy platform devoted to ocular melanoma. *Med Phys* 2005;32:910–919.
6. Stankovskiy A, Kerhoas-Cavata S, Ferrand R, Nauraye C, Demarzi L. Monte Carlo modelling of the treatment line of the proton therapy center in Orsay. *Phys Med Biol* 2009;54:2377–2394.
7. Pia MG, Begalli M, Lechner A, Quintieri L, Saracco P. Physics-related epistemic uncertainties in proton depth dose simulation. *IEEE Trans Nucl Sci* 2010;57:2805–2830.
8. Battistoni G, Muraro S, Sala PR, Cerutti F, Ferrari A, Roesler S, Fasso A, Ranft J. The fluka code: Description and benchmarking. Proceedings of the hadronic shower simulation workshop 2006, Fermilab 6–8 September 2006. Albrow M, Raja R, Eds. AIP Conference Proceeding 2007;896:31–49.
9. Ferrari A, Sala PR, Fasso A, Ranft J. Fluka: A multi-particle transport code. CERN Yellow Report CERN 2005–10; INFN/TC 05/11, SLAC-R-773. Geneva: CERN. 2005.
10. Agostinelli S, Allison J, Amako K, Apostolakis J, Araujo H, Arce P, Asai M, Axen D, Banerjee S, Barrand G, Behner F, Bellagamba L, Boudreau J, Broglia L, Brunengo A, Burkhardt H, Chauvie S, Chuma J, Chytracek R, Cooperman G, Cosmo G, Degtyarenko P, Dell'Acqua A, Depaola G, Dietrich D, Enami R, Feliciello A, Ferguson C, Fesefeldt H, Folger G, Foppiano F, Forti A, Garelli S, Giani S, Giannitrapani R, Gibin D, Gomez Cadenas JJ, Gonzalez I, Gracia Abril G, Greeniaus G, Greiner W, Grichine V, Grossheim A, Guatelli S, Gumplinger P, Hamatsu R, Hashimoto K, Hasui H, Heikkinen A, Howard A, Ivanchenko V, Johnson A, Jones FW, Kallenbach J, Kanaya N, Kawabata M, Kawabata Y, Kawaguti M, Kelner S, Kent P, Kimura A, Kodama T, Kokoulin R, Kossov M, Kurashige H, Lamanna E, Lampen T, Lara V, Lefebure V, Lei F, Liendl M, Lockman W, Longo F, Magni S, Maire M, Medernach E, Minamimoto K, Mora de Freitas P, Morita Y, Murakami K, Nagamatu M, Nartallo R, Nieminen P, Nishimura T, Ohtsubo K, Okamura M, O'Neale S, Oohata Y, Paech K, Perl J, Pfeiffer A, Pia MG, Ranjard F, Rybin A, Sadilov S, Di Salvo E, Santin G, Sasaki T, Savvas N, Sawada Y, Scherer S, Sei S, Sirotenko V, Smith D, Starkov N, Stoecker H, Sulkimo J, Takahata M, Tanaka S, Tcherniaev E, Safai Tehrani E, Tropeano M, Truscott P, Uno H, Urban L, Urban P, Verderi M, Walkden A, Wander W, Weber H, Wellisch JP, Wenaus T, Williams DC, Wright D, Yamada T, Yoshida H, Zschiesche D. GEANT4—a simulation toolkit. *Nucl Instrum Methods Phys Res A* 2003;506:250–303.
11. Allison J, Amako K, Apostolakis J, Araujo H, Arce Dubois P, Asai M, Barrand G, Capra R, Chauvie S, Chytracek R, Cirrone GAP, Cooperman G, Cosmo G, Cuttone G, Daquino GG, Donszelmann M, Dressel M, Folger G, Foppiano F, Generowicz J, Grichine V, Guatelli S, Gumplinger P, Heikkinen A, Hrivnacova I, Howard A, Incerti S, Ivanchenko V, Johnson T, Jones F, Koi T, Kokoulin R, Kossov M, Kurashige H, Lara V, Larsson S, Lei F, Link O, Longo F, Maire M, Mantero A, Mascialino B, McLaren I, Mendez Lorenzo P, Minamimoto K, Murakami K, Nieminen P, Pandola L, Parlati S, Peralta L, Perl J, Pfeiffer A, Pia MG, Ribon A, Rodrigues P, Russo G, Sadilov S, Santin G, Sasaki T, Smith D, Starkov N, Tanaka S, Tcherniaev E, Tomé B, Trindade A, Truscott P, Urban L, Verderi M, Walkden A, Wellisch JP, Williams DC, Wright D, Yoshida H. GEANT4 developments and applications. *IEEE Trans Nucl Sci* 2006;53:270–278.
12. Pelowitz DBE. MCNPX user's manual, version 2.5.0. Los Alamos National Laboratory, 2005;LA-CP-05-0369.
13. Waters L. MCNPX user's manual. Los Alamos National Laboratory, 2002.
14. Fippel M, Soukup M. A Monte Carlo dose calculation algorithm for proton therapy. *Med Phys* 2004;31:2263–2273.
15. Dementyev AV, Sobolevsky NM. Shield-universal Monte Carlo hadron transport code: Scope and applications. *Radiation Measurements* 1999;30:553–557.
16. Perl J, Shin J, Schumann J, Faddegon B, Paganetti H. Topas: An innovative proton Monte Carlo platform for research and clinical applications. *Med Phys* 2012;39:6818–6837.
17. Jiang SB, Pawlicki T, Ma C-M. Removing the effect of statistical uncertainty on dose-volume histograms from Monte Carlo dose calculations. *Phys Med Biol* 2000;45:2151–2162.
18. Keall PJ, Siebers JV, Jeraj R, Mohan R. The effect of dose calculation uncertainty on the evaluation of radiotherapy plans. *Med Phys* 2000;27:478–484.
19. Deasy JO. Denoising of electron beam Monte Carlo dose distributions using digital filtering techniques. *Phys Med Biol* 2000;45:1765–1779.

20. Deasy JO, Wickerhauser M, Picard M. Accelerating Monte Carlo simulations of radiation therapy dose distributions using wavelet threshold de-noising. *Med Phys* 2002;29:2366–2373.
21. Fippel M, Nuesslin F. Smoothing Monte Carlo calculated dose distributions by iterative reduction of noise. *Phys Med Biol* 2003;48:1289-1304.
22. Kawrakow I. On the de-noising of Monte Carlo calculated dose distributions. *Phys Med Biol* 2002;47:3087–3103.
23. De Smedt B, Vanderstraeten B, Reynaert N, De Neve W, Thierens H. Investigation of geometrical and scoring grid resolution for Monte Carlo dose calculations for IMRT. *Phys Med Biol* 2005;50:4005–4019.
24. Paganetti H. Nuclear interactions in proton therapy: Dose and relative biological effect distributions originating from primary and secondary particles. *Phys Med Biol* 2002;47:747–764.
25. Paganetti H. Monte Carlo calculations for absolute dosimetry to determine output factors for proton therapy treatments. *Phys Med Biol* 2006;51:2801–2812.
26. Verhaegen F, Palmans H. Secondary electron fluence perturbation by high-z interfaces in clinical proton beams: A Monte Carlo study. *Phys Med Biol* 1999;44:167–183.
27. Verhaegen F, Palmans H. A systematic Monte Carlo study of secondary electron fluence perturbation in clinical proton beams (70–250 MeV) for cylindrical and spherical ion chambers. *Med Phys* 2001;28:2088–2095.
28. Cygler JE, Lochrin C, Daskalov GM, Howard M, Zohr R, Esche B, Eapen L, Grimard L, Caudrelier JM. Clinical use of a commercial Monte Carlo treatment planning system for electron beams. *Phys Med Biol* 2005;50:1029–1034.
29. Ramos-Mendez JA, Perl J, Faddegon B, Schuemann J, Paganetti H. Geometrical splitting technique to improve the computational efficiency in Monte Carlo calculations for proton therapy. *Med Phys* 2013;40:041718.
30. Kawrakow I. The effect of Monte Carlo statistical uncertainties on the evaluation of dose distributions in radiation treatment planning. *Phys Med Biol* 2004;49:1549–1556.
31. Sempau J, Bielajew AF. Towards the elimination of Monte Carlo statistical fluctuation from dose volume histograms for radiotherapy treatment planning. *Phys Med Biol* 2000;45:131–157.
32. Rogers DWO, Mohan R. Questions for comparisons of clinical Monte Carlo codes. Proceedings of the 13th International Conference on Computers in Radiotherapy (ICCR) 2000:120–122.
33. Fragoso M, Kawrakow I, Faddegon BA, Solberg TD, Chetty IJ. Fast, accurate photon beam accelerator modeling using BEAMnrc: A systematic investigation of efficiency enhancing methods and cross-section data. *Med Phys* 2009;36:5451–5466.
34. Tourovsky A, Lomax AJ, Schneider U, Pedroni E. Monte Carlo dose calculations for spot scanned proton therapy. *Phys Med Biol* 2005;50:971–981.
35. Kohno R, Sakae T, Takada Y, Matsumoto K, Matsuda H, Nohtomi A, Terunuma T, Tsunashima Y. Simplified Monte Carlo dose calculation for therapeutic proton beams. *Japanese Journal of Applied Phyics* 2002;41:L294–L297.
36. Kohno R, Takada Y, Sakae T, Terunuma T, Matsumoto K, Nohtomi A, Matsuda H. Experimental evaluation of validity of simplified Monte Carlo method in proton dose calculations. *Phys Med Biol* 2003;48:1277–1288.
37. Li JS, Shahine B, Fourkal E, Ma CM. A particle track-repeating algorithm for proton beam dose calculation. *Phys Med Biol* 2005;50:1001–1010.
38. Yepes P, Randeniya S, Taddei PJ, Newhauser WD. A track-repeating algorithm for fast Monte Carlo dose calculations of proton radiotherapy. *Nucl Technol* 2009;168:736–740.
39. Yepes PP, Mirkovic D, Taddei PJ. A GPU implementation of a track-repeating algorithm for proton radiotherapy dose calculations. *Phys Med Biol* 2010;55:7107–7120.
40. Jia X, Ziegenhein P, Jiang SB. GPU-based high-performance computing for radiation therapy. *Phys Med Biol* 2014;59:R151–182.
41. Jia X, Schumann J, Paganetti H, Jiang SB. GPU-based fast Monte Carlo dose calculation for proton therapy. *Phys Med Biol* 2012;57:7783–7797.
42. Ma J, Beltran C, Seum Wan Chan Tseung H, Herman MG. A GPU-accelerated and Monte Carlo-based intensity modulated proton therapy optimization system. *Med Phys* 2014;41:121707.
43. Low DA, Harms WB, Mutic S, Purdy JA. A technique for the quantitative evaluation of dose distributions. *Med Phys* 1998;25:656–661.

44. Clasie BM, Sharp GC, Seco J, Flanz JB, Kooy HM. Numerical solutions of the gamma-index in two and three dimensions. *Phys Med Biol* 2012;57:6981–6997.
45. Jiang SB, Sharp GC, Neicu T, Berbeco RI, Flampouri S, Bortfeld T. On dose distribution comparison. *Phys Med Biol* 2006;51:759–776.
46. Paganetti H, Jiang H, Parodi K, Slopsema R, Engelsman M. Clinical implementation of full Monte Carlo dose calculation in proton beam therapy. *Phys Med Biol* 2008;53:4825–4853.
47. Hsi WC, Moyers MF, Nichiporov D, Anferov V, Wolanski M, Allgower CE, Farr JB, Mascia AE, Schreuder AN. Energy spectrum control for modulated proton beams. *Med Phys* 2009;36:2297–2308.
48. Paganetti H, Jiang H, Lee S-Y, Kooy H. Accurate Monte Carlo for nozzle design, commissioning, and quality assurance in proton therapy. *Med Phys* 2004;31:2107–2118.
49. Titt U, Sahoo N, Ding X, Zheng Y, Newhauser WD, Zhu XR, Polf JC, Gillin MT, Mohan R. Assessment of the accuracy of an MCNPX-based Monte Carlo simulation model for predicting three-dimensional absorbed dose distributions. *Phys Med Biol* 2008;53:4455–4470.
50. Paganetti H. Four-dimensional Monte Carlo simulation of time dependent geometries. *Phys Med Biol* 2004;49:N75–N81.
51. Lu HM, Kooy H. Optimization of current modulation function for proton spread-out Bragg peak fields. *Med Phys* 2006;33:1281–1287.
52. Shin J, Perl J, Schumann J, Paganetti H, Faddegon BA. A modular method to handle multiple time-dependent quantities in Monte Carlo simulations. *Phys Med Biol* 2012;57:3295–3308.
53. Paganetti H, Jiang H, Trofimov A. 4D Monte Carlo simulation of proton beam scanning: Modeling of variations in time and space to study the interplay between scanning pattern and time-dependent patient geometry. *Phys Med Biol* 2005;50:983–990.
54. Peterson S, Polf J, Ciangaru G, Frank SJ, Bues M, Smith A. Variations in proton scanned beam dose delivery due to uncertainties in magnetic beam steering. *Med Phys* 2009;36:3693–3702.
55. Peterson SW, Polf J, Bues M, Ciangaru G, Archambault L, Beddar S, Smith A. Experimental validation of a Monte Carlo proton therapy nozzle model incorporating magnetically steered protons. *Phys Med Biol* 2009;54:3217–3229.
56. Kimstrand P, Traneus E, Ahnesjo A, Grusell E, Glimelius B, Tilly N. A beam source model for scanned proton beams. *Phys Med Biol* 2007;52:3151–3168.
57. Grassberger C, Lomax A, Paganetti H. Phase space based Monte Carlo simulations for active scanning proton therapy. *Phys Med Biol* 2014; under review.
58. Grevillot L, Bertrand D, Dessy F, Freud N, Sarrut D. A Monte Carlo pencil beam scanning model for proton treatment plan simulation using GATE/GEANT4. *Phys Med Biol* 2011;56:5203–5219.
59. Sawakuchi GO, Titt U, Mirkovic D, Ciangaru G, Zhu XR, Sahoo N, Gillin MT, Mohan R. Monte carlo investigation of the low-dose envelope from scanned proton pencil beams. *Phys Med Biol* 2010;55:711–721.
60. Sawakuchi GO, Zhu XR, Poenisch F, Suzuki K, Ciangaru G, Titt U, Anand A, Mohan R, Gillin MT, Sahoo N. Experimental characterization of the low-dose envelope of spot scanning proton beams. *Phys Med Biol* 2010;55:3467–3478.
61. Kimura A, Aso T, Yoshida H, Kanematsu N, Tanaka S, Sasaki T. DICOM data handling for GEANT4-based medical physics application. *IEEE Nuclear Science Symposium Conference Record* 2004;4:2124-2127.
62. Kimura A, Tanaka S, Aso T, Yoshida H, Kanematsu N, Asai M, Sasaki T. DICOM interface and visualization tool for GEANT4-based dose calculation. *IEEE Nuclear Science Symposium Conference Record* 2005;2:981–984.
63. Parodi K, Ferrari A, Sommerer F, Paganetti H. Clinical CT-based calculations of dose and positron emitter distributions in proton therapy using the FLUKA Monte Carlo code. *Phys Med Biol* 2007;52:3369–3387.
64. Jiang H, Paganetti H. Adaptation of GEANT4 to Monte Carlo dose calculations based on CT data. *Med Phys* 2004;31:2811–2818.
65. Jiang H, Seco J, Paganetti H. Effects of Hounsfield number conversions on patient CT based Monte Carlo proton dose calculation. *Med Phys* 2007;34:1439–1449.

66. Parodi K, Paganetti H, Cascio E, Flanz JB, Bonab AA, Alpert NM, Lohmann K, Bortfeld T. PET/CT imaging for treatment verification after proton therapy: A study with plastic phantoms and metallic implants. *Med Phys* 2007;34:419–435.
67. Espana S, Paganetti H. The impact of uncertainties in the CT conversion algorithm when predicting proton beam ranges in patients from dose and PET-activity distributions. *Phys Med Biol* 2010;55:7557–7572.
68. Schneider W, Bortfeld T, Schlegel W. Correlation between CT numbers and tissue parameters needed for Monte Carlo simulations of clinical dose distributions. *Phys Med Biol*2000;45:459–478.
69. ICRU. Stopping powers and ranges for protons and alpha particles. Report No. 49. International Commission on Radiation Units and Measurements. Bethesda, MD: 1993.
70. ICRU. Tissue substitutes in radiation dosimetry and measurement. Report No. 44. International Commission on Radiation Units and Measurements. Bethesda, MD: 1989.
71. ICRU. Photon, electron, proton and neutron interaction data for body tissues. Report No. 46. International Commission on Radiation Units and Measurements. Bethesda, MD: 1992.
72. Andreo P. On the clinical spatial resolution achievable with protons and heavier charged particle radiotherapy beams. *Phys Med Biol* 2009;54:N205–215.
73. Paganetti H. Range uncertainties in proton therapy and the role of Monte Carlo simulations. *Phys Med Biol* 2012;57:R99–R117.
74. Yang M, Virshup G, Clayton J, Zhu XR, Mohan R, Dong L. Theoretical variance analysis of single- and dual-energy computed tomography methods for calculating proton stopping power ratios of biological tissues. *Phys Med Biol* 2010;55:1343–1362.
75. Liu HH, Keall P. DM rather than DW should be used in Monte Carlo treatment planning. *Med Phys* 2002;29:922–924.
76. Paganetti H. Dose to water versus dose to medium in proton beam therapy. *Phys Med Biol* 2009;54:4399–4421.
77. Paganetti H. Range uncertainties in proton therapy and the role of Monte Carlo simulations. *Phys Med Biol* 2012;57:R99–117.
78. Schuemann J, Dowdell S, Grassberger C, Min CH, Paganetti H. Site-specific range uncertainties caused by dose calculation algorithms for proton therapy. *Phys Med Biol* 2014;59:4007–4031.
79. Sawakuchi GO, Titt U, Mirkovic D, Mohan R. Density heterogeneities and the influence of multiple Coulomb and nuclear scatterings on the Bragg peak distal edge of proton therapy beams. *Phys Med Biol* 2008;53:4605–4619.
80. Soukup M, Alber M. Influence of dose engine accuracy on the optimum dose distribution in intensity-modulated proton therapy treatment plans. *Phys Med Biol* 2007;52:725–740.
81. Schuemann J, Giantsoudi D, Grassberger C, Moteabbed M, Min CH, Paganetti H. Assessing the clinical impact of approximations in analytical dose calculations for proton therapy. *Int J Radiat Oncol Biol Phys* 2015; under review.
82. Daartz J, Engelsman M, Paganetti H, Bussiere MR. Field size dependence of the output factor in passively scattered proton therapy: Influence of range, modulation, air gap, and machine settings. *Med Phys* 2009;36:3205–3210.
83. Bednarz B, Daartz J, Paganetti H. Dosimetric accuracy of planning and delivering small proton therapy fields. *Phys Med Biol* 2010;55:7425–7438.
84. Bueno M, Paganetti H, Duch MA, Schuemann J. An algorithm to assess the need for clinical Monte Carlo dose calculation for small proton therapy fields based on quantification of tissue heterogeneity. *Med Phys* 2013;40:081704.
85. Shiu AS, Hogstrom KR. Pencil-beam redefinition algorithm for electron dose distributions. *Med Phys* 1991;18:7–18.
86. Li Y, Zhu RX, Sahoo N, Anand A, Zhang X. Beyond Gaussians: A study of single-spot modeling for scanning proton dose calculation. *Phys Med Biol* 2012;57:983–997.
87. Grassberger C, Daartz J, Dowdell S, Ruggieri T, Sharp G, Paganetti H. Quantification of proton dose calculation accuracy in the lung. *Int J Radiat Oncol Biol Phys* 2014;89:424–430.
88. Pflugfelder D, Wilkens JJ, Szymanowski H, Oelfke U. Quantifying lateral tissue heterogeneities in hadron therapy. *Med Phys* 2007;34:1506–1513.
89. Soukup M, Fippel M, Alber M. A pencil beam algorithm for intensity modulated proton therapy derived from Monte Carlo simulations. *Phys Med Biol* 2005;50:5089–5104.

90. Grassberger C, Dowdell SJ, Shackleford J, Sharp GC, Choi N, Willers H, Paganetti H. Effects of motion interplay as a function of patient characteristics and proton spot size when treating lung cancer with scanning proton therapy: A 4D Monte Carlo study. *Int J Radiat Oncol Biol Phys* 2013; available ahead of print.
91. Paganetti H, Jiang H, Adams JA, Chen GT, Rietzel E. Monte Carlo simulations with time-dependent geometries to investigate effects of organ motion with high temporal resolution. *Int J Radiat Oncol Biol Phys* 2004;60:942–950.
92. Grassberger C, Shackleford J, Sharp G, Paganetti H. Four-dimensional Monte Carlo simulations of lung cancer patients treated with proton beam scanning to assess interplay effects. *Med Phys* 2012;39:3998.
93. Dowdell S, Grassberger C, Paganetti H. Four-dimensional Monte Carlo simulations demonstrating how the extent of intensity-modulation impacts motion effects in proton therapy lung treatments. *Med Phys* 2013;40:121713.
94. Dowdell S, Grassberger C, Sharp GC, Paganetti H. Interplay effects in proton scanning for lung: A 4D Monte Carlo study assessing the impact of tumor and beam delivery parameters. *Phys Med Biol* 2013;58:4137–4156.
95. Grassberger C, Dowdell S, Lomax A, Sharp G, Shackleford J, Choi N, Willers H, Paganetti H. Motion interplay as a function of patient parameters and spot size in spot scanning proton therapy for lung cancer. *Int J Radiat Oncol Biol Phys* 2013;86:380–386.
96. Grassberger C, Paganetti H. Elevated LET components in clinical proton beams. *Phys Med Biol* 2011;56:6677–6691.
97. Paganetti H. Relative biological effectiveness (RBE) values for proton beam therapy. Variations as a function of biological endpoint, dose, and linear energy transfer. *Phys Med Biol* 2014;59:R419–R472.
98. Grassberger C, Trofimov A, Lomax A, Paganetti H. Variations in linear energy transfer within clinical proton therapy fields and the potential for biological treatment planning. *Int J Radiat Oncol Biol Phys* 2011;80:1559–1566.
99. Giantsoudi D, Sethi RV, Yeap B, Ebb D, Caruso P, Chen Y-L, Yock TI, Tarbell NJ, Paganetti H, MacDonald SM. Brainstem toxicity and let correlations following proton radiation for medulloblastoma. *Neuro-oncology* 2014; submitted.
100. Sethi RV, Giantsoudi D, Raiford M, Malhi I, Niemierko A, Rapalino O, Caruso P, Yock TI, Tarbell NJ, Paganetti H, MacDonald SM. Patterns of failure after proton therapy in medulloblastoma; linear energy transfer distributions and relative biological effectiveness associations for relapses. *Int J Radiat Oncol Biol Phys* 2014;88:655–663.
101. Polster L, Schuemann J, Rinaldi I, Burigo L, McNamara AL, Stewart RD, Attili A, Carlson DJ, Sato T, Faddegon B, Perl J, Paganetti H. Extension of TOPAS for the simulation of proton radiation effects considering molecular and cellular endpoints. *Phys Med Biol* 2014; submitted.
102. Semenenko VA, Stewart RD. Fast Monte Carlo simulation of DNA damage formed by electrons and light ions. *Phys Med Biol* 2006;51:1693–1706.
103. Nikjoo H, O'Neill P, Terrissol M, Goodhead DT. Quantitative modelling of DNA damage using Monte Carlo track structure method. *Radiat Environ Biophys* 1999;38:31–38.
104. Kim CH, Jeong JH, Bolch WE, Cho KW, Hwang SB. A polygon-surface reference korean male phantom (psrk-man) and its direct implementation in GEANT4 Monte Carlo simulation. *Phys Med Biol* 2011;56:3137–3161.
105. Clasie B, Wroe A, Kooy H, Depauw N, Flanz J, Paganetti H, Rosenfeld A. Assessment of out-of-field absorbed dose and equivalent dose in proton fields. *Med Phys* 2010;37:311–321.
106. Xu XG. An exponential growth of computational phantom research in radiation protection, imaging, and radiotherapy: A review of the fifty-year history. *Phys Med Biol* 2014;59:R233–302.
107. Lee C, Lodwick D, Hurtado J, Pafundi D, Williams JL, Bolch WE. The UF family of reference hybrid phantoms for computational radiation dosimetry. *Phys Med Biol* 2010;55:339–363.
108. Xu XG, Bednarz B, Paganetti H. A review of dosimetry studies on external-beam radiation treatment with respect to second cancer induction. *Phys Med Biol* 2008;53:R193–R241.
109. Paganetti H. Monte Carlo simulations will change the way we treat patients with proton beams today. *Br J Radiol* 2014;87:20140293.

Chapter 21

Uncertainties in Proton Therapy: Their Impact and Management

Radhe Mohan, Ph.D. and Narayan Sahoo, Ph.D.

MD Anderson Cancer Center,
Houston, TX

21.1 Introduction

If there is one thing that is certain in life, it is uncertainty.[1] Radiotherapy is no exception—there are uncertainties in just about every link of the radiotherapy chain.

Arguably, the main role and responsibility of medical physicists has been, is, and will continue to be the management of uncertainties—minimizing them and managing residual uncertainties. Some of this goes under the name of quality assurance (QA). The goal is to deliver treatments safely and as accurately as feasible while delivering an appropriately high tumor dose and sparing normal tissues explicitly or implicitly, considering uncertainties. Uncertainties can be minimized but cannot be eliminated. Medical physicists attempt to achieve these goals through practices and procedures developed based on many decades of experience and with the aid of tools, systems, and methods developed through research and development.

Uncertainties may be divided into several categories. Some of them are due to patient-to-patient variability. Others arise from the inconsistencies in practices and perceptions, for instance in estimating and delineating the extent of the disease, which may, in turn, be attributable to the limitations of imaging modalities. Uncertainties

1. Except for death and taxes, as they say, but they are uncertain too in terms of when and how much.

related to imaging data also contribute to uncertainties in dose distributions computed for the planning of treatments. Assumptions and approximations in dose computation algorithms further add to the uncertainties. Dose distributions are generally optimized and prescribed based on the limited knowledge of the dose–volume response of tumors of various sites and histologies, and of the tolerances of normal tissues based on prior studies and trials. Dose–volume response relationships are established assuming that the dose distributions seen on treatment plans are actually delivered. However, due to intra- and inter-fractional anatomic variations in anatomy and in response, as well as uncertainties in radiation beam incident on the patient, the actual delivered dose distributions may be significantly different from those shown in the plan. This may lead to unanticipated outcomes and affect the reliability of dose–response data and models deduced from the presumed dose distributions. Furthermore, the dose–volume histograms (DVHs) or dose–volume indices (as well as tumor control and normal tissue complication probabilities (TCP and NTCPs) computed therefrom), presumed to represent dose response, in reality summarize complex dose distributions and ignore numerous factors, such as the location of the dose and the irradiation of other tissues. The variation of chemotherapy and biological agents across the patient populations are additional confounding factors.

While uncertainties affect dose distributions delivered with photons as well as with protons and heavier charged particles (HCPs), particle dose distributions are considerably more susceptible to many of the uncertainties. The underlying reasons include the sharp distal falloff of particle dose distributions and their scattering characteristics. Furthermore, the biological response of cells and tissues to particle irradiation (represented by the relative biological effectiveness, or RBE, the ratio of photon dose to particle dose required to achieve the same biological effect) is quite different and substantially more complex than for photons, and is a major source of uncertainty. Since the focus of this chapter is proton therapy, the rest of the chapter will refer to protons only. However, much of the discussion is applicable to therapy with heavier charged particles as well.

The reasons for uncertainties are manifold. They include the limited knowledge of the highly complex processes involved and the current state of the art of equipment and systems. They are also due to the assumptions, approximations, and compromises sometimes made necessary for reasons of practicality, expediency, and to account for gaps in knowledge.

The magnitudes and impact of individual uncertainties may vary substantially depending on the type of uncertainty, clinical situation, and institutional practices. For photons, through research and through decades of experience, the consequences of various uncertainties have been estimated and solutions developed to minimize uncertainties and to incorporate residual uncertainties into the design of treatments to make dose distributions more resilient. However, these solutions sometimes do not apply to protons because many of the uncertainties affect the ranges of protons and their lateral scattering characteristics. Alternative solutions for protons have been developed and proposed and continue to evolve.

The numerous sources of uncertainties and the complexity of response to irradiation may appear to be daunting and seem discouraging. At the same time, it is worth noting that over time, even in the face of uncertainties, there has been steady progress in the improvement of the safety, effectiveness, and quality of treatments and outcomes. One can, therefore, argue that there is a potential for significant further improvement in outcomes and quality of life with the reduction and appropriate management of uncertainties, particularly for proton therapy.

21.2 Sources of Uncertainty

The following subsections briefly discuss the most important sources of uncertainty in radiotherapy. Most of these uncertainties exist for both photons and protons, except that proton dose distributions are more sensitive to many of them, and their consequences may be more serious.

21.2.1 Patient-specific Sources

In a way, radiotherapy is a form of personalized cancer therapy. In the current parlance, however, such terminology is used for customizing treatments based on a person's genetic makeup, which could unravel the biology of his or her tumor, indicate the specific dose required for tumor control, and suggest the tolerance limits of normal tissues. The use of such information in radiotherapy is currently rather limited. In practice, tumor control doses and normal tissue tolerances are based on averages over large numbers of patients treated and tend to have large uncertainty. There is clinical research ongoing to customize radiotherapy based on biomarkers. An example is a randomized NRG oropharynx trial to determine if tumor dose can be reduced for HPV-positive patients.

Another patient-specific source is the uncertainties in the ability of a patient to follow instructions and to comply with the treatment requirements.

21.2.2 Tumor Heterogeneity

In the traditional practice, a tumor is assumed to be a homogeneous mass of tissue. PET and MR imaging is now used increasingly to identify the gross tumor volume (GTV) and to identify sub-regions of resistance, proliferative activity, and tumor burden, and to tailor prescription dose accordingly. Ling et al. [1] elegantly explained these concepts and introduced the term "biological target volume" (BTV) in their 2000 article (see Figure 21–1).

In general, images are unable to reveal the extent of sub-clinical disease (microscopic extensions), which is estimated based on broad accumulated clinical experience. Furthermore, appropriate dose levels required to eradicate tumor sub-regions are not yet established. Response characteristics of sub-regions may be of particular importance to proton therapy since the way the protons ionize leads to different biological effects. For example, laboratory experiments have shown that protons and heavier particles are more effective in overcoming radiation resistance than photons. However, because of the various uncertainties associated with the determination of the biological target volume and the doses required for achieving clinical goals, bio-

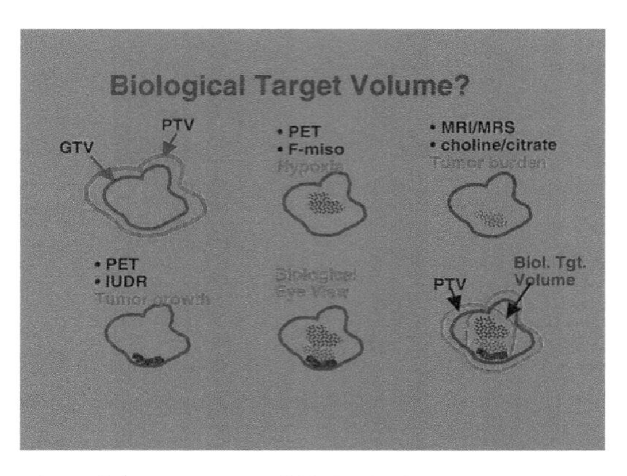

Figure 21–1 Illustration of the concept of biological target volume. The conventional target volume is defined in terms of GTV, CTV, and PTV. In contrast, the biological target volume may derive information from biological images to improve dose targeting to certain regions of the target volume ("dose painting"). For instance, regions of low pO2 level may be derived from PET-18F-misonidazole study, high tumor burden from MRI/MRS data of choline/citrate ratio, and high proliferation from PET-124IUdR measurements, etc. From Ling, et al. [1].

logical conformal radiotherapy is currently not generally practiced either for photons or for protons and other particles.

21.2.3 Variations in Individual Practices and Perceptions

Significant inconsistencies often exist in the delineation of the target volumes between one individual and another, and even more so between one institution and another. To a lesser extent, such variations occur even for the same individual attempting to outline the target at two different times on the same image. Figure 21–2 shows a head and neck example from Reigel, et al. [2]. They concluded that, even with the use of fused PET/CT images, significant variability remained, and they emphasized the need for well-defined protocols and training. In fact, ASTRO has been establishing guidelines and conducting workshops and training sessions at its various meetings to improve consistency in target delineation. This topic is still a work in progress.

In addition to target volumes, there is variability in the delineation of normal critical structures as well. This is generally due to the limited contrast of images (mainly CT), especially when artifacts resulting from the presence of high-Z materials, e.g., dental fillings, obscure the boundaries. Figure 21–3 is an illustrative example.

21.2.4 Uncertainties in Computed Dose Distributions

Dose distributions computed using CT data are used for the planning of treatments. Dose distribution displays, dose–volume histograms (DVHs), and dose–volume indices derived from dose distributions are used to make treatment decisions and to opti-

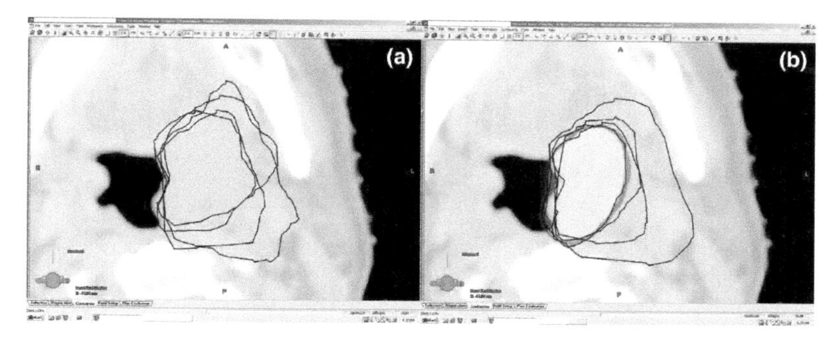

Figure 21–2 Example of inter-observer variation among four physicians in the delineation of the GTV for a head and neck patient (a) using CT and (b) using fused PET/CT. From Reigel, et al. [2].

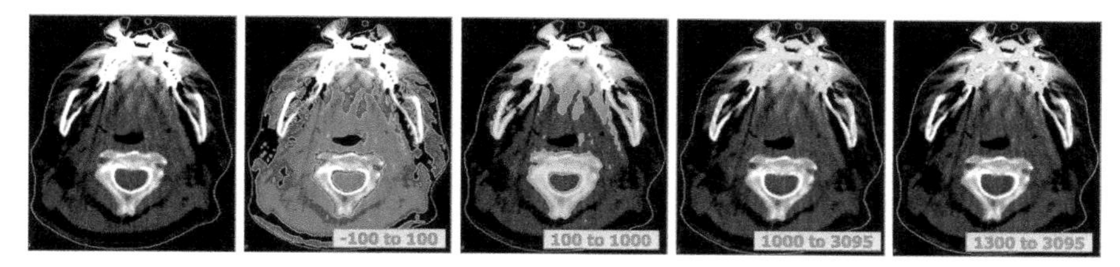

Figure 21–3 Metal artifacts obscure boundaries and introduce uncertainty in the delineation of targets and normal structures. The high-Z artifacts make CT numbers inaccurate, not just in the regions of the implants, but also in large areas in their vicinity on the CT section, thus affecting results of dose computations, especially for protons. The red text on each panel indicates the range of the CT numbers highlighted.

mize dose distributions (manually or with the aid of automated methods) to achieve an appropriate balance among target volume and critical normal tissue doses.

It is widely assumed that the dose distribution seen on a treatment plan designed at the beginning of the treatment course represents the dose distribution actually delivered. Such dose distributions for cohorts of patients are also used to determine associations between dose and dose–volume indices and dose–response models vs. observed tumor and normal tissue responses. However, dose distributions computed by treatment planning systems currently being used in the clinic may differ significantly compared to those possible with the most accurate methods (e.g., Monte Carlo techniques). Some contributing factors are described below briefly. More details can be found in chapters 19 and 20.

21.2.4.1 CT Numbers and Stopping Power Ratios

There are statistical uncertainties (random noise) in CT numbers resulting from the limits of detectors to detect sufficient numbers of quanta (photons). These uncertain-

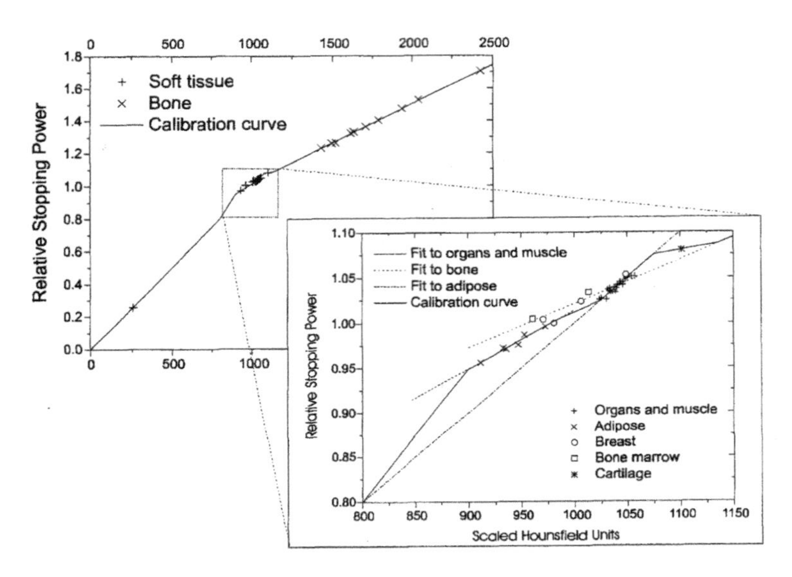

Figure 21–4 Calibration curves to convert CT numbers to proton relative stopping power ratios based on the stoichiometric method of Schneider et al [4].

ties increase with increasing image resolution and decrease with increasing x-ray dose. Statistical uncertainties have minimal impact on photon dose distributions as well as on proton ranges, but increase proton range straggling.

In addition, there are systematic uncertainties in CT numbers. They result from the limitations of the imaging systems to detect the transmitted signals, beam hardening artifacts, and the limitations of reconstruction algorithms. An example of systematic uncertainty is the dependence of CT numbers for the same material on the location of the material in the volume being imaged. Artifacts caused by high-Z materials may introduce large errors in CT numbers (Figure 21–3). Systematic uncertainties have minimal impact on photon dose distributions, but can have a significant impact on proton dose distributions. Modern dual-energy CT scanners can greatly reduce many of the systematic errors compared to those in images from the commonly used scanners.

While photon dose computations use electron density ratios derived from CT numbers, proton beam dose distributions are based on relative stopping powers (RSPs or stopping power ratios relative to water, i.e., SPRs) of various media encountered in the patient. The CT data must be converted into RSPs, and there are uncertainties in the conversion process. A major contributor is the uncertainties in the I-values of water and biological materials in the human body needed for the determination of RSP, as discussed by Andreo [3]. Furthermore, RSPs depend on proton energy, but this dependence is generally ignored. There are ongoing efforts to reduce these uncertainties. Publications by Schneider, et al. and Yang, et al. in this regard are noteworthy

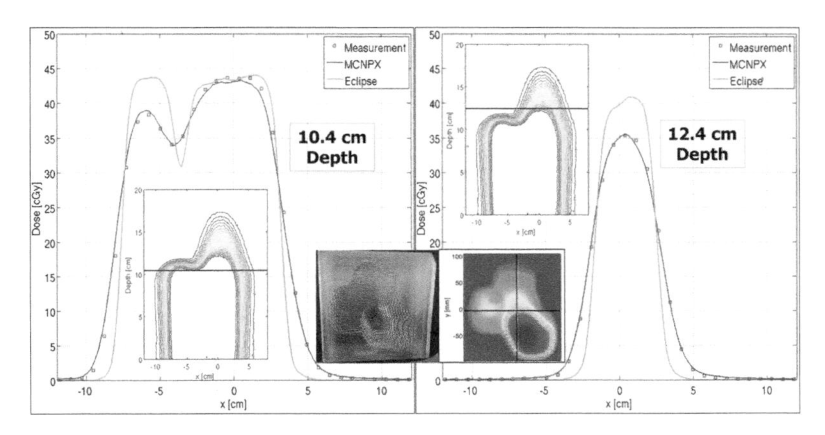

Figure 21–5 Measured and calculated dose distribution profiles in a water phantom for a passively scattered beam through a complex compensator (shown in the inset) at two different depths. Calculations were done using the MCNPX Monte Carlo code and with a semi-empirical model implemented on a commercial treatment planning system.

[4–6]. Figure 21–4 shows the results of Schneider's stoichiometric calibration method for CT scanners to more accurately estimate RSPs for biological tissues. However, as pointed out by Andreo [3], limitations of the current knowledge and methods, especially the variability in the tissue compositions from patient to patient, may not allow any further reduction in range uncertainties than the 2% to 3% determined by the currently used formalism and the values of physical constants. Thus, excluding high-Z artifacts, the current magnitude of uncertainty attributable to CT numbers and their conversion to RSPs is estimated to be between 2% to 3%.

Although it has been argued by Andreo [3] and Moyers [7] that CT data and its conversion to RSPs represents the largest source of physical uncertainty, we believe that many other factors (e.g., dose computation algorithms, inter- and intra-fractional anatomic variations, biological effectiveness) are far more important for proton therapy.

21.2.4.2 Approximation and Assumptions of Dose Computation Algorithms

For practical reasons, for instance to achieve adequate speed for the planning of treatments on affordable computers, the analytic semi-empirical models of proton dose computations used clinically make numerous assumptions and approximations. Different models make different assumptions and approximations, thus have different limitations, which manifest themselves differently in dose computations for passively scattered and scanning beam therapy.

Examples of assumptions and approximations include using ray tracing to correct for tissue heterogeneities, assuming slab geometry to estimate the lateral spreading of proton pencil beams passing through heterogeneous media, neglecting scattering from

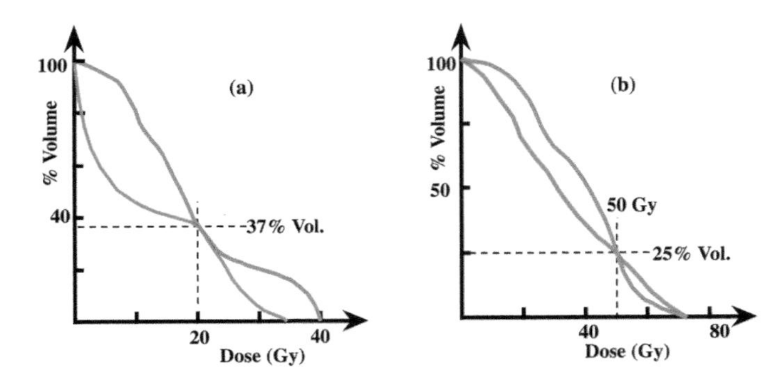

Figure 21–6 Panel (a) illustrates that a specified dose–volume constraint (e.g., V20 ≤ 37% for normal lung) can lead to an infinite number of DVHs and responses. Similarly, panel (b) illustrates that an infinite number of dose–volume combinations may lead to the same response.

apertures, assuming that the nuclear component of beams can be accounted for by a simple offset at the end of pencil beam-based calculation, neglecting scattering from beam line components such as a profile monitor,[1] etc. More details are given in Chapter 20, but we show an illustrative example here. Figure 21–5 illustrates the limitations of a dose computation model for passively scattered protons that uses the ray tracing technique and neglects the fact that protons passing through different thicknesses of a compensator lose different amounts of energy and, therefore, spread laterally differently.

Ultimately, Monte Carlo techniques (or their abridged or accelerated variations) will be necessary to overcome the limitations of the current models.

21.2.4.3 Limitations of Treatment Response Models

Minimum and maximum dose to a structure, dose–volume indices, equivalent uniform dose (EUD), and TCP and NTCP models are used to design treatment plans and as components of objective functions for the optimization of IMRT and IMPT plans. While these indices are crude surrogates of treatment response, they are useful for making radiotherapy systematic and consistent. They have numerous limitations. For instance, DVHs summarize dose distributions and hide positional information of dose deposited—two identical DVHs may correspond to very different dose distributions and may, therefore, lead to totally different responses. Similarly, the use of mean lung dose or constraints on lung V20 (volume 20 Gy or higher), for instance, used to design, optimize, and evaluate lung plans, may arise from very different DVHs (Figure 21–6) and lead to different toxicities.

1. The purpose of a profile monitor is to ensure the constancy of the shape of the cross-sectional profile of the proton beam entering the nozzle. It is a part of the nozzle hardware.

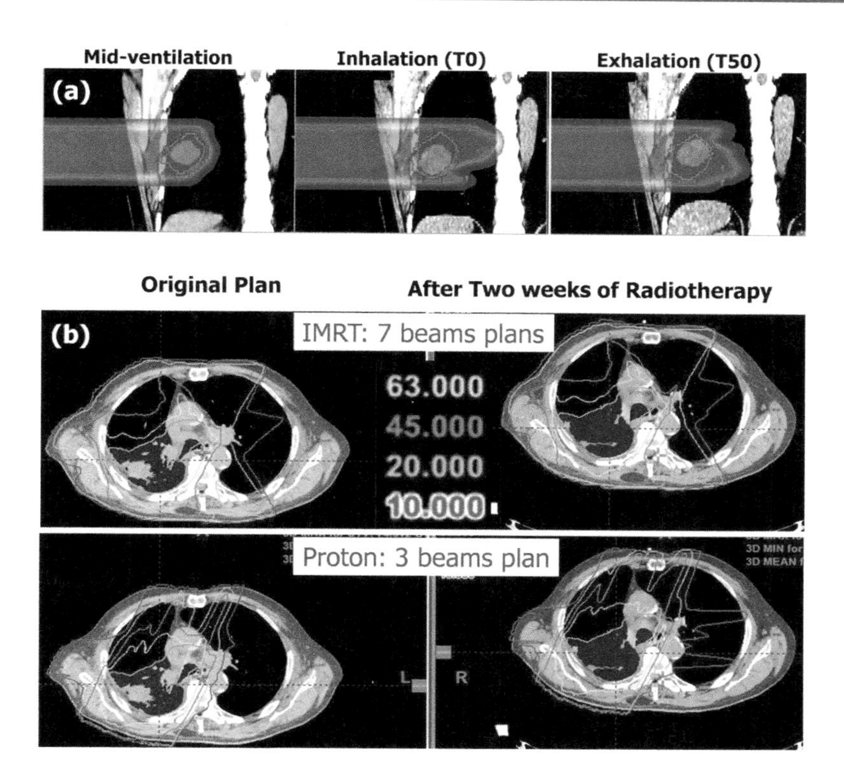

Figure 21–7 (a) Shows the impact of the respiratory motion of a lung tumor on the penetration of a proton beam. (b) Shows that, for a 7-beam IMRT plan, the effect of significant tumor shrinkage after two weeks to treatments is negligible, but it is considerable for a 3-beam passively scattered proton plan (lower panel).

One can argue that the same is the case for photon therapy. However, proton dose distribution patterns, especially for normal tissues, are very different compared to photons, so that even the limited dose–response knowledge we have gained to date from experience with the latter may not be extensible to protons. For instance, a large low-to-intermediate dose bath for photon treatments may affect the response of a particular normal tissue differently, even when the dose–volume indices for this tissue are the same for both protons and photons.

21.2.4.4 Intra- and Inter-Fractional Anatomic Variations

Photon dose distributions are minimally perturbed by setup variations, respiratory and other intra-fractional motion, and inter-fractional anatomic variations, such as those caused by tumor shrinkage or weight loss. They are considered to be a "static cloud" in space and, as long as appropriate margins are assigned to the target volume and normal structures, safe and effective dose distributions can be assumed to be delivered. In contrast, proton dose distributions are considerably more vulnerable to the factors mentioned earlier due to their finite range and sharp distal falloff. These factors are a major source of uncertainty in proton therapy. Figure 21–7 illustrates the impact of anatomic variations on proton therapy.

It is often assumed that, for photons, if the respiration-induced tumor motion is small (e.g., ≤5 mm end to end), the impact of such motion on dose distribution can be neglected. For protons, however, as demonstrated by Matney, et al. [8] it is not just the motion of the tumor but also the anatomy in the path of the proton beam that can have a significant effect on dose distributions.

21.2.4.5 Biological Effectiveness of Protons

As mentioned above, the biologic effectiveness of protons relative to photons (i.e., "relative biologic effectiveness" or RBE) has simplistically been assumed to have a generic fixed value of 1.1 [9, 10]. This value is based on an average of the results of numerous *in vitro* and *in vivo* experiments conducted under limited conditions. Most commonly, these experiments were conducted at high doses per fraction (e.g., 6–8 Gy) and in the middle of the spread-out Bragg peak where the RBE is relatively constant and close to the average value of 1.1. Furthermore, the RBE data have been acquired for only a limited number of cell lines, tissues, and endpoints. The resulting data have large error bars [10]. To justify the current assumption, it is argued that, clinically, no adverse responses have been demonstrated with the use of an RBE of 1.1. On the other hand, the large uncertainties in the treatment processes may have obscured the effect of not using a variable RBE to calculate biologically effective dose distributions. Furthermore, a case can be made that an improved knowledge of RBE could lead to improvements in treatment planning which, in turn, may enhance the effectiveness of proton therapy.

For particles heavier than protons, the RBE is not assumed to be a fixed generic value. Nevertheless, considerable uncertainties exist in RBE data for all particles, which can have a profound impact on outcomes. Chapter 5 provides more details about particle RBE.

21.3 Consequences

Because of uncertainties, the biologically effective dose distributions actually delivered to patients may be substantially different from the ones predicted by the treatment planning systems and used to make treatment decisions, evaluate outcomes, and model treatment response. The magnitude of each uncertainty and its effect are not well quantified and may depend on the treatment site, treatment modality, and many other unknown factors. The impact on clinical outcomes is even less clear, as studies directly linking one or more sources of uncertainty to outcomes are uncommon (especially for proton therapy) and almost impossible to conduct prospectively. Moreover, whatever evidence exists is not very strong.

A confounding factor is the uncertainty in the level of desired or required accuracy. Herring and Compton [11], based on their analysis of laryngeal cancer data, indicated that a deviation of >5% in delivered dose can affect outcomes. Since then,

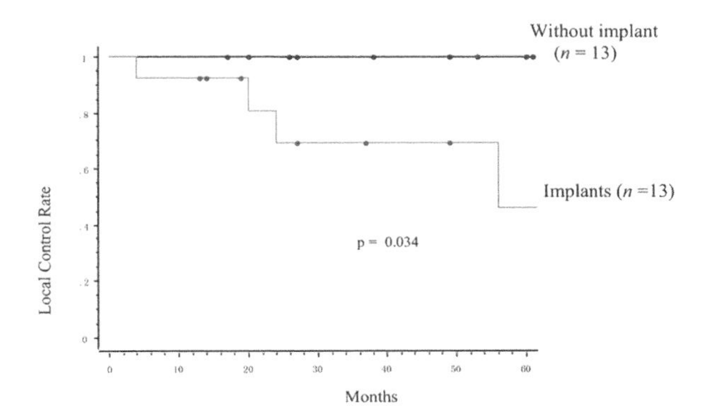

Figure 21–8 Local control in patients with an implant as compared to without implant in patients who underwent resection of chordoma followed by proton therapy. From Rutz, et al. [15].

5% accuracy in dose has come to be accepted as standard. The IROC component of NRG (known previously as the Radiological Physics Center or RPC) requires variable levels of accuracy for NRG and other cooperative group trials. These requirements may vary from ±4% to ±7% differences between measured and computed dose values in water and anthropomorphic phantoms for 80% to 95% of the points tested. These requirements depend on the disease site and are often dictated by the levels of accuracy achievable by a sufficiently large number of institutions in order that they can qualify to participate and enroll patients on trials. In all likelihood, the dose delivered to the patient would have considerably larger differences compared to those achieved by the IROC criteria.

21.3.1 Clinical Evidence

From the perspective of clinical evidence, Peters, et al. [12] reviewed outcomes of a large international phase III advanced head and neck trial to assess the impact of radiotherapy quality, and implicitly of the uncertainties, and found that centers treating only a few patients are a major source of quality deficiency.

Soares et al. [13,14] performed a meta-analysis of data from all (57) completed phase III RTOG randomized controlled trials conducted from 1968 through 2002 to test the hypothesized superiority of experimental innovative treatments vs. standard treatments. They "found no predictable pattern of treatment successes in oncology: sometimes innovative treatments are better than the standard ones and vice versa; in most cases (88%) there were no substantive differences between experimental and conventional treatments." They concluded that, taking into account the uncertainties in the trials, the "result strongly suggests that the RTOG investigators did not violate equipoise when they designed these trials." However, the large variability permitted among institutions in phantom dose distributions, and the even greater uncertainties

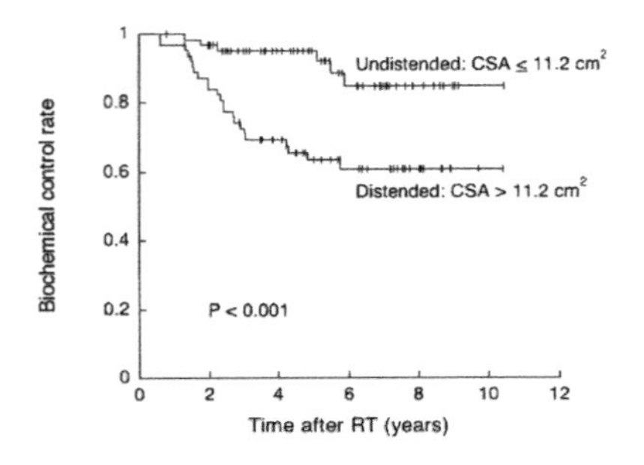

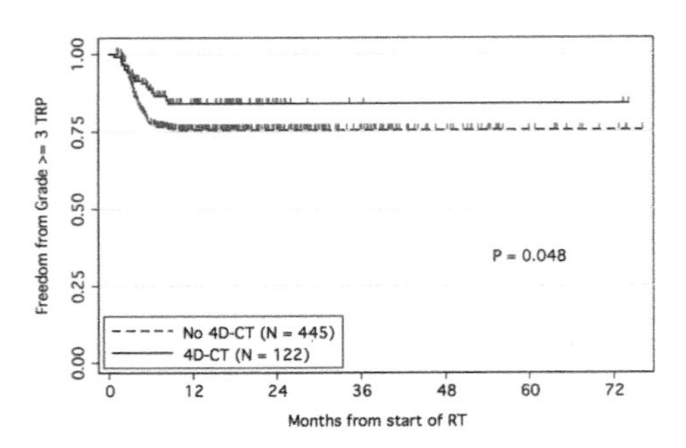

Figure 21–9 The left panel (from de Crevoisier, et al. [16]) shows the impact of uncertainty introduced by ignoring the consequences of rectal distension at the time of acquisition of CT images for treatment planning. The right panel indicates the value of reducing uncertainty in the definition of ITV using 4D CT as opposed to assigning one margin fits all for lung treatments. (S. L. Tucker, MD Anderson Cancer Center, private communication.)

occurring in actual dose distributions delivered, may have been a factor in obscuring the hypothesized superiority of innovative techniques.

An important proton therapy-specific example (Figure 21–8) of the impact of uncertainties is from a publication by Rutz, et al. [15], who found that in extracranial chordoma patients treated with proton therapy, local failures were significantly higher for those for whom post-surgical titanium implants were used for stabilization of anatomy after extensive resections. Although the titanium implants were implicated as the cause of failures, there may be multiple unknown factors, including the extent of the disease, its surgical management, and uncertainties in images and dose distributions caused by implants contributing to the observed outcomes.

An example of the consequences of uncertainties from the photon realm (Figure 21–9, left panel) is the retrospective analysis by de Crevosier, et al. [16] to determine the correlation of biochemical failure vs. rectal distension for prostate treatments. Another example is the unpublished work of Tucker, et al. which showed that customized determination of ITV for each lung patient reduces the probability of radiation-induced pneumonitis compared to the conventional practice of using a one-size-fits-all margin for intra-fractional tumor motion. Considering the greater sensitivity of protons to uncertainties, such effects are expected to be even greater.

21.3.2 Approximations and Assumptions of Conventional Dose Computation Models

An example of approximations and assumptions in conventional dose computation algorithms on dose distributions is given in Figures 21–6 above. Figure 21–10 shows another example, that of the scattering from apertures, which is commonly ignored in conventional algorithms [17]. It shows that the scatter from a square field aperture may be as high as 9% of the incident dose.

Figure 21–11 further illustrates the impact of ignoring collimator scatter in the analytical models used for clinical treatment planning. The example shown is for 160 MeV protons. The effect is greater for higher energies and at positions closer to the surface.

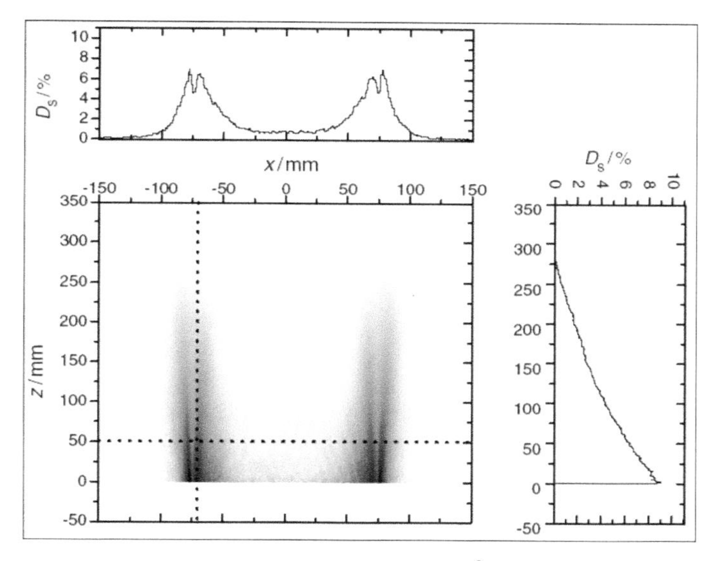

Figure 21–10 Scattered dose (D_S) from a 15 x 15 cm^2 250 MeV passively scattered field with a 10 cm SOBP width. Shown here is a plot of the difference between Monte Carlo simulations including and excluding protons scattered from the collimator. The collimator was positioned at a distance of 8 cm upstream from the phantom surface. The dose was normalized to 100% at the center of the SOBP. (From Titt et al. [17].)

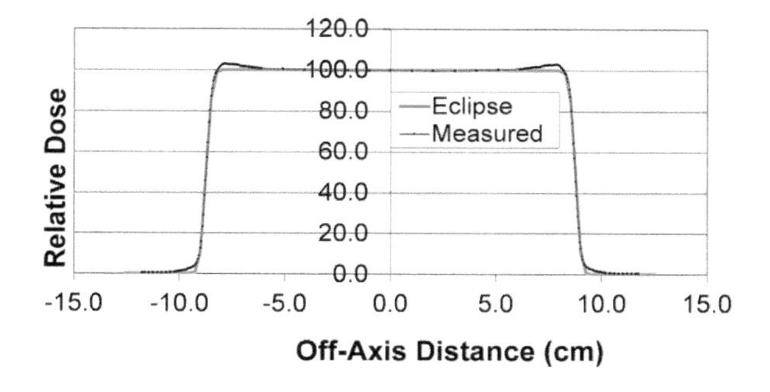

Figure 21–11 Comparison of a measured transverse profile for a 160 MeV proton beam with 4 cm SOBP width at 6 cm depth with the profile calculated with the analytical model implemented on a commercial system. The differences near the field edge are presumably due to collimator scatter.

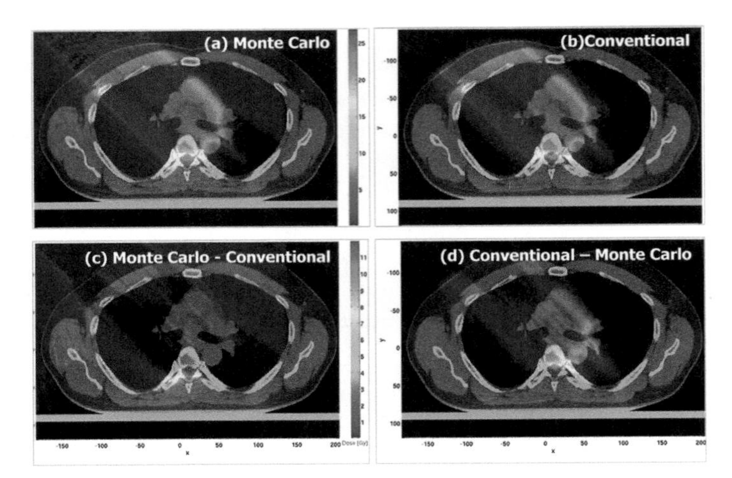

Figure 21–12 Passively scattered proton therapy dose distributions for an NSCLC patient computed with (a) Monte Carlo simulations and (b) a conventional commercial system. Panel (c) shows regions of under-dosing in conventional model predictions and panel (d) shows regions of over-dosing. (Mirkovic, et al. Unpublished, private communication.)

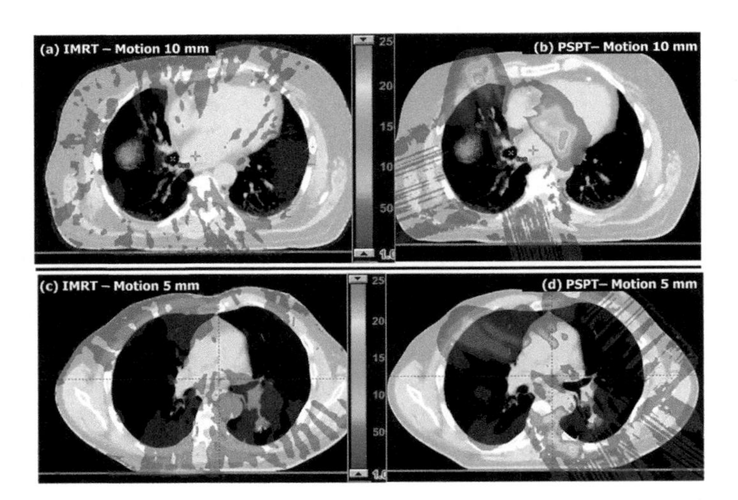

Figure 21–13 Panels (a) and (b) provide an example of a patient with >10 mm tumor motion showing the four dimensional dose distributions for IMRT and passively scattered protons (PSPT). These dose distributions were calculated by the accumulation of deformably registered dose distributions on all 10 phases of the 4D CT. They differ by less than 5 Gy (RBE) for IMRT, but differ by as much as 30 Gy for PSPT. Panels (c) and (d) are for a patient with <5 mm tumor motion showing that, even for small tumor motion, proton dose distributions are affected by as much as 10 Gy (RBE) in the contralateral lung and by 15 Gy (RBE) near the spine and ribs. (From Matney, et al. [8].)

Figure 21–12 compares dose distributions for one of the beams for a lung patient calculated with a conventional commercial planning system and with Monte Carlo simulations and shows over- dosing with the use of the former of as much as 10 Gy and underdosing of as much as 6 Gy.

21.3.3 Impact of Inter- and Intra-Fractional Anatomic Variations

Dosimetric impact of tumor respiratory motion and tumor shrinkage is illustrated above in Figure 21–7. Figure 21–13 illustrates the greater sensitivity of protons to intra-fractional motion and also makes a point that it is not just the motion of the tumor, but the motion of any part of the anatomy in the path of protons that can perturb the planned dose distribution.

21.3.4 Relative Biological Effectiveness

While clinical evidence to date may not have indicated that proton RBE deviates significantly from the currently assumed value of 1.1, substantial *in vitro* and some *in vivo* evidence exists, and more evidence is accumulating indicating that proton RBE around the Bragg peak and, especially in the distal edge, may be considerably higher. The magnitudes depend on the tissue type, cell type, and end points of interest.

It is sometimes argued that, though the RBE may be much greater than 1.1 in these regions, it is confined to very small regions because of the rapid falloff of dose and, therefore, of negligible consequence other than effectively increasing the range by a couple of mm. It should be pointed out that the distal edge is often not nearly as steep as it is shown to be in a water phantom. The passage of beams through complex heterogeneities can considerably widen the distal edge [18,19], especially when the distal edge is in a low-density medium such as the lung. To our knowledge, the biological effectiveness in the degraded edge has not been studied; it is likely that the increase in RBE in regions in and distal to the Bragg peak may not be negligible. It has been argued by Wedenberg and Toma-Dasu [20] that a fair comparison between the clinical outcomes with proton and photon therapy is not possible without consideration of the variable RBE at different depths of the proton dose deposition. Figure 21–14 shows RBE (relative to ^{60}Co) derived from measurements as a function of dose

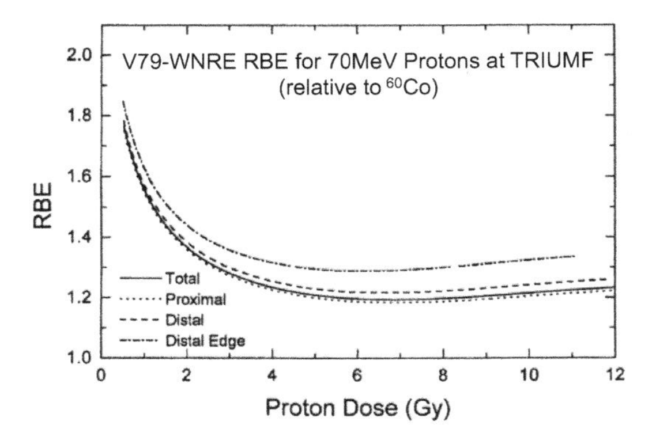

Figure 21–14 The RBE in the 70 MeV SOBP compared to ^{60}Co gamma rays as a function of proton dose illustrating that the RBE increases with increasing depth and is largest at the distal edge and that the RBE rises at low doses of radiation in all regions of the SOBP. (Adapted from Wouters, et al. [21].)

at different positions along a 70 MeV proton depth dose curve of an SOBP of 2 cm [21]. Many similar experiments have been reported.

Analyses of clinical outcomes data based on a variable RBE distribution are needed to validate the use of variable RBE in treatment planning [22]. A combination of experimental determination of the RBE for different type of tissues, both at different dose levels and at locations in the beam path, and the study of its usefulness in interpreting the clinical outcomes data, will help to reduce the RBE-related uncertainties.

21.4 Management of Uncertainties

The examples and the experience with the national clinical trials cited above stress the importance of reducing uncertainties and making quality assurance more stringent. The examples also point to the need for incorporating residual uncertainties in the design and optimization of treatment plans.

Ideally, to manage an uncertainty, it is helpful to have a reasonable estimate of its magnitude and impact. Reliable quantitative information is often not available even for some of the most consequential uncertainties, such as those related to the extent of the disease or the response of the disease and normal tissues to complex dose patterns, especially when combined with drugs and other agents. Commonly, we, the radiation oncology community, have to rely on conventional wisdom, experience, and observed and published patterns of failures and successes. Stated differently, there is uncertainty in uncertainty. Nevertheless, even in the face of uncertainties, radiotherapy is a successful modality in treating large numbers of patients (nearly one million per year in the United States), curing many of them while prolonging the lives of many others. However, failures do occur in curing or controlling the disease. Furthermore, radiotherapy may lead to acute and chronic toxicity and reduced quality of life. It is, therefore, a fair assumption that a reduction in uncertainties and the mitigation of their consequences will improve the effectiveness of radiotherapy.

It should be obvious that uncertainties and their consequences can be reduced by improving our understanding of the underlying causes and our knowledge of various facets of radiotherapy. This requires response data from carefully and accurately conducted studies and clinical trials and their analyses. Uncertainties can be minimized with improved knowledge and more sophisticated tools and processes, but they cannot be eliminated altogether. Residual uncertainties must be managed—incorporated into planning, plan evaluation, and optimization processes.

With decades of experience with radiation therapy with photons, many techniques have been developed. They will be mentioned here, but not described in detail. Some of these are not extensible to proton therapy, and alternative strategies have been developed and continue to be developed. One should keep in mind that the need to reduce uncertainties and manage residual uncertainties is greater for protons than for photons. It is not clear whether ions heavier than protons are more susceptible to uncertainties than protons.

21.4.1 Minimization of Uncertainties

21.4.1.1 Immobilization, Imaging, and Image-guided Radiotherapy (IGRT)

Common traditional ways of reducing uncertainties include immobilization, imaging, and image-guided setup for radiation treatments. In most instances, the methods used for proton therapy are the same as for photon therapy, but, in some cases, adjustments are needed. For example, immobilization devices with sharp density gradients in the path of the proton beam can perturb the planned dose distribution to a larger degree compared to photons due to inter-fractional misalignment and should be avoided. As another example, the alignment of KV setup images with planned DRRs to bony structures may not be adequate for target localization in proton therapy as the small differences in the amount of soft tissue in the beam path between the treatment position and the position for imaging for planning may not be quantifiable with this approach. As mentioned earlier, small changes in materials in the beam path can lead to large, sometimes unacceptable, perturbations in the dose distributions in proton therapy compared to photons.

It is interesting to note that image-guided setup using multiple kV x-rays was first used for proton therapy well before on-board imaging became available for photon therapy. However, the state of the art of IGRT for photons has advanced with the availability of on-board cone-beam CTs and in-room CT-on-rails for volumetric imaging, whereas image guidance in proton therapy is still almost entirely based on orthogonal kV x-rays. This has limited the use of protons in the treatment of certain cancers (e.g., SBRT of early stage lung), and it is an impediment to improving the accuracy of treatments. The good news is that there is a realization of the need to make online or in-room volumetric imaging available in proton treatment rooms.

21.4.1.2 Implanted Fiducials

Another approach to reducing uncertainties is the use of implanted fiducials for more accurate image-guided setup [23] and even for respiratory gating with real-time fluoroscopic tracking [24]. Fiducials are especially important for targets that move with respect to external marks and bony structures. Organs such as prostate and liver can be much more accurately targeted using implanted fiducials.

Furthermore, transponder-based implanted fiducials may be tracked electromagnetically (e.g., the wireless beacon transponder system by Calypso) and used for accurate setup and gating [25,26]. The undesirable aspects of implanted fiducials are the invasiveness of the implantation procedure, potential for complications, instability of marker positions, loss of markers, prolongation of the treatment course, and the extra cost of the surgical procedure.

For proton therapy, there is an additional consideration—that of perturbation of dose distributions by the markers, which may be significant depending on the size, material, and the location of the marker relative to the beam [27]. Caution is recommended when using fiducial markers for proton therapy. Fiducials fabricated with low-Z material should be preferred both to reduce the artifacts on the planning CT images, as well as to minimize the perturbation of proton dose distributions.

Table 21–1 Frequency of adaptive replanning for passively scattered proton therapy vs. intensity-modulated radiotherapy for patients enrolled in a randomized NSCLC trial

	Randomized to IMRT	Randomized to PSPT
Total enrolled	53	40
Number requiring adaptive plans	10 (18.9%)	21 (52.5%)

Treatment doses were 74 Gy (RBE) in both arms. Higher frequency in the PSPT arm suggest greater sensitivity of protons to inter-fractional changes during the course of radiotherapy.

21.4.1.3 Repeat Imaging and Adaptive Replanning

Daily volumetric IGRT is, of course, the preferred technique for daily setup. However, the potential for improved accuracy needs to be balanced against the extra imaging dose. For SRS, SRT, and SBRT, the extra imaging dose would be sufficiently small and not of significant concern.

Uncertainty in dose distributions due to inter-fractional anatomic variations can be further minimized by adaptive replanning. Periodically (perhaps weekly), a new CT image (called the "verification image") may be acquired and used to verify whether the current beam configuration is still satisfactory in terms of the original treatment objectives. If not, a new treatment plan is generated for the remainder of the treatments. Optimization for adaptive replanning should ideally take into account the dose delivered to date. The cumulative dose-to-date may be deformably registered to the latest image. As one would expect, there is a greater need for adaptive replanning for protons than for photons. Table 21–1 shows the adaptive replanning frequency for 93 locally advanced NSCLC patients treated in a randomized IMRT vs. passively scattered proton therapy (PSPT) trial. The frequency of adaptive replanning required for proton plans is nearly three times that for photons, indicating the relative resilience of IMRT dose distributions to inter-fractional anatomic changes.

21.4.1.4 Improvements in Semi-empirical Algorithms and the Acceleration of Monte Carlo Techniques for Dose Calculations

As mentioned above, the approximations and assumptions in clinical dose calculation algorithms are important sources of uncertainty in radiotherapy. While photon dose computation methods have improved steadily over the decades, proton methods still lag behind. However, over the last decade or so, investigators have published considerably improved methods.

The most accurate way of computing dose distributions is with Monte Carlo techniques. However, Monte Carlo techniques are still not fast enough, especially for proton therapy where another dimension, that of energy, increases the CPU time requirements by an order of magnitude. The CPU time requirements can be many times greater for the computation of IMPT influence matrices, for 4D (incorporating respiratory motion) and 5D (incorporating 4D and inter-fractional variations) IMPT optimization, and for robustness evaluation and robust optimization. To date, the use of Monte Carlo techniques for photons or protons has been limited to a very small number of institutions and situations. It is hoped that, with the increase in the power of computers, including parallel processing clusters and graphics processing units (GPUs), Monte Carlo techniques will be embedded widely into clinics. However, it may be necessary to continue to improve and use analytic semi-empirical methods in combination with MC techniques for applications such as robust optimization. The former, however, may be supplanted by novel accelerated Monte Carlo methods, such as the proton track repeating algorithm of Yepes, et al. [28], Macro Monte Carlo of Fix, et al. [29], or Voxel Monte Carlo for Protons of Fippel, et al. [30]. Details of conventional and Monte Carlo dose calculation methods can be found in Chapters 19 and 20, respectively.

21.4.2 Incorporation of Residual Uncertainties

In spite of our best efforts, residual uncertainties will remain. It is important to account for them in treatment planning so that the target is covered by the prescribed dose and normal tissues are adequately spared with high probability. Furthermore, in current routine practice, after appropriate measures have been taken to minimize and account for uncertainties, it is commonly assumed that the treatment plan reflects exactly what the patient will receive. Preferably, the treatment decisions should be made based on plans that reflect the range of possible dose distributions the patient is likely to receive. In other words, they should convey the confidence limits on the dose distributions being used to make treatment decisions.

The following subsections describe approaches to account for uncertainties and to evaluate dose distributions in the presence of uncertainties.

21.4.2.1 Margins

In photons, the traditional approach to account for uncertainties is to assign margins to the clinical target volume (CTV) to obtain a planning target volume (PTV). Details can be found in ICRU reports 50, 62, and 78 on prescribing, recording, and reporting dose distributions [31–33]. The margins are normally determined empirically for each treatment site and may depend on tumor location and size and on image-guidance practices at each institution. A simple recipe, provided by van Herk, et al. [34] is commonly used to estimate CTV-to-PTV margins. The margins are based on a combination of systematic and random inter-fractional uncertainties. They may not be isotropic if inter-fractional variations are expected to be non-isotropic. They ensure, for example, that a minimum of 95% of the prescribed dose to the CTV will be achieved for 90% of the patients. Intra-fractional motion may also be incorporated into such margins by first creating an internal target volume (ITV) that incorporates motion based on 4DCT and then expanding it to the PTV.

Expansion of the CTV (or ITV) to the PTV is the current standard of practice in photon therapy. It effectively ensures that there is high probability that the worst that can happen to the CTV is that it will receive a certain minimum dose (e.g., 95% minimum dose for 90% of the patients) over the course of radiotherapy. The inherent

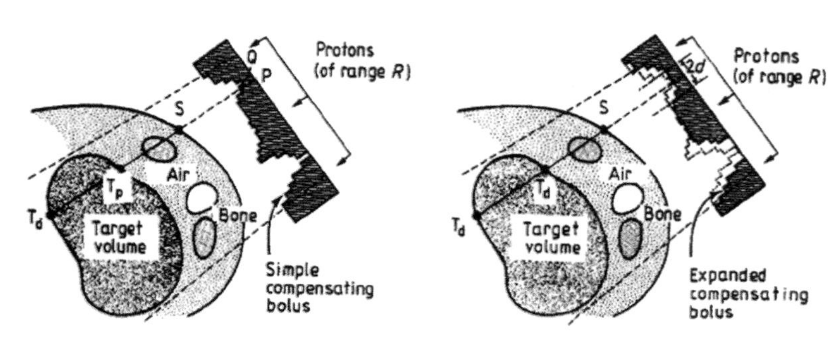

Figure 21–15 Illustration of compensator smearing to account for misalignment from planning CT to treatment fraction of bony structures and target volume. (From Urie, et al. [37].)

assumption is that the dose distribution is a "static cloud" that is not perturbed by anatomic and setup variations.

Similar to assigning margins to CTVs to obtain PTVs, organs at risk may be assigned margins for inter- and intra-fractional variations to obtain organ-at-risk volumes, or ORVs. Adoption of PTV and ORV concepts is strongly recommended, especially in cooperative group trials, but has been slow in coming. Some practitioners do not use the PTV and ORV concepts but draw CTV and organs at risk volumes that implicitly include uncertainties based on their experience.

It is important to realize that the PTV and ORV concepts are not strictly valid for protons for two reasons. One is the uncertainty in the range of protons, which has been estimated to be two to three percent of the depth of penetration (plus a constant offset), and would, therefore, depend on the direction of the beam and size of the target. The second is that anatomic variations in the path of protons may perturb dose distributions in the target and normal tissues to a considerably greater degree than for photons. Therefore, for the planning of proton treatments, beam-specific proximal and distal margins are assigned to the CTV for each beam. Note that these margins are essentially independent of the change in position of the anatomy along the beam direction. A number of investigators have devised and published strategies to estimate margins for range uncertainties. These strategies continue to be refined. The reader is referred to publications by Moyers, et al. [35], Yang, et al. [5], and Schuemann, et al. [36] and the references cited therein.

The lateral margins assigned for beams in proton therapy are the same as the lateral CTV to PTV margins in photon therapy. In addition, margins for the lateral spreading (penumbra) of the beam must also be included. For protons, the depth and energy dependence of these margins may also need to be considered.

Relative shifts or misalignments of anatomic structures of different densities in the path of protons may significantly perturb dose distributions. To ensure target coverage in the presence of such shifts for PSPT, compensators are "smeared" (expanded) and the SOBP is appropriately widened. The concept of smearing, first reported by Urie, et al. [37], involves widening the peaks and valleys of a compensator to desensitize the dose distribution to anatomic shifts and misalignments (see Figure 21–15).

Note that the beam-specific margins and smearing are applicable for designing individual beams of PSPT plans. Such margins are not appropriate for beamlets of IMPT, and alternate strategies have been developed (see subsections on robust optimization and robustness evaluation below).

21.4.2.2 Proton Plan Evaluation and Plan Comparison in the Face of Uncertainties

While beam-specific margins and smearing are appropriate for designing individual beams for passively scattered and uniform scanning proton therapy planning, they cannot be integrated in the evaluation and comparison of composite dose distributions of all beams. Nevertheless, for lack of alternatives, displays of composite dose distributions and DVHs, TCPs, NTCPs, EUDs, etc. derived from them are used.

A simple strategy for proton therapy evaluation, which is applicable to both PSPT and IMPT, is to examine individual dose distributions and derived indices for each of a set of uncertainty scenarios. These scenarios may, for instance, include shifts along the orthogonal axes, range uncertainty, end-inhale, and end-exhale phases, etc. The magnitudes of shifts may be chosen to be the same as the margins for the PTV, and the magnitudes for range uncertainty may the same as those used for designing planning beams. Such a review should reveal deficiencies in a dose distribution in one or more scenarios, and steps may be taken to rectify them. However, the process is cumbersome and time consuming.

21.4.2.3 Plan Robustness Evaluation

The goal of robustness evaluation is to assess the resilience of a dose distribution to uncertainties. A simple strategy is to compute dose distributions for each of a sufficiently large set of uncertainty scenarios and plot DVHs for each anatomic structure of interest. The band of DVHs for a given structure represents the range of possible dose distributions. A narrow band means high robustness and vice versa. In analogy with the DVH of the PTV being the "worst case" representation of the DVH of the CTV in the photon domain, one may select the DVHs that correspond to the overall worst scenario for robustness evaluation.

Band width at critical dose–volume points on the DVH (e.g., V20 for lung) may be used as a quantitative measure of robustness. Other ways of quantifying robustness have been proposed. For instance, one could compute the area of the DVH band as a robustness measure. Another possibility, proposed by Liu, et al. [38], is the root-mean-square dose–volume histogram (RMSD-VH) in which dose spread in each voxel is represented by the root-mean-square under a number of uncertainty scenarios. The area under the curve of the RMSD-VH may be used as a quantitative index of robustness.

In addition to DVH bands, one can also calculate the ranges of EUDs, TCPs, and NTCPs, etc. and use the ones corresponding to the worst-case plan for evaluation and inter-comparison of competing plans. Note that a worst case plan may be superior for some structures and inferior for some others.

21.4.2.4 Robustness Improvement and Robust Optimization

The robustness of proton dose distributions depends on many factors. Plans with larger numbers of beams tend to be more robust. The passage of beams through highly heterogeneous anatomy increases uncertainty and leads to reduced robustness. In general, IMPT dose distributions are less robust than PSPT dose distributions, which are less robust than IMRT dose distributions. For PSPT, the dose distribution due to each incident beam is designed independently of other beams to cover the target adequately. It is relatively uniform (in water) and is terminated beyond the distal edge of the target plus a margin. Therefore, the composite dose distribution due to all beams is relatively less sensitive to perturbations. It should also be noted that smearing implicitly improves the robustness of PSPT dose distributions.

For IMPT, in contrast, the dose distributions due to individual fields tend to be highly complex and have high gradients which, when combined, fit like a jigsaw puzzle to produce the desired pattern of homogeneous dose distribution in the target and sparing of normal tissues (see an example Figure 21–16). However, in the face of uncertainties (e.g., in range), the fit is lost, creating hot and cold regions.

Figure 21–17 shows an example of locally advanced non-small cell lung cancer treatment plans. IMRT dose distributions were optimized with standard techniques using PTV margins. PSPT dose distributions were designed by assigning distal and proximal margins for range uncertainties and by smearing the compensators appropriately. IMPT dose distributions were optimized in two different ways—using the conventional approach of assigning PTV margins, as for IMRT, and using robust optimization methodology.

Robust optimization considered nine different uncertainty scenarios: the nominal dose distribution, six dose distributions obtained by shifting the patient image by ±5

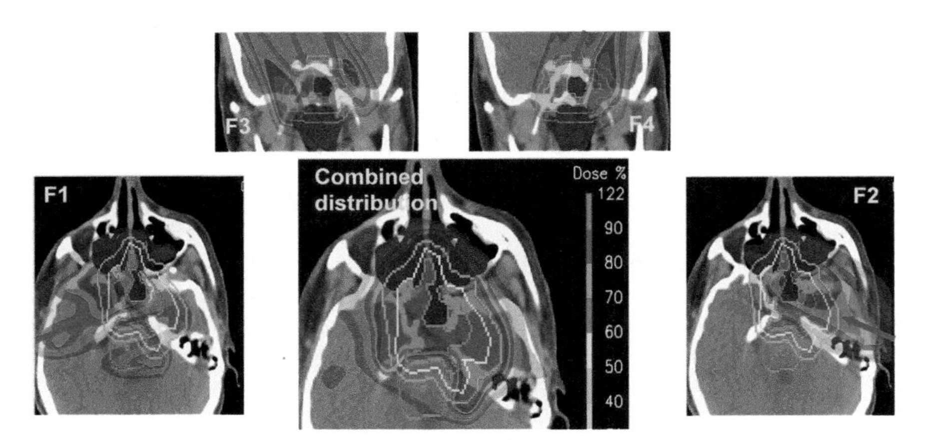

Figure 21–16 Inhomogeneous individual field IMPT target dose distributions (F1, F2, F3, F4) and apparently homogenous combined dose distribution for a head and neck case. (Adapted from a figure provided by A. Lomax, PSI, private communication.)

mm (equal to the CTV-to-PTV margin) along three orthogonal directions, and two additional dose distributions incorporating uncertainty in range of ±3%. The optimization algorithm computes the score (the value of the objective function to be minimized) in each iteration by selecting the worst dose in each voxel from among the nine scenarios. For the target voxels, the worst would be the minimum value, and for normal tissues it would be the maximum value. This is the so-called "voxel-by-voxel" worst-case approach [38,39]. Alternate worst-case approaches have been proposed and have different strengths. Fredrickson, et al., for example, proposed the "minimax" worst-case approach, in which the dose distribution corresponding to the worst score of the plan as a whole is selected in each step of the process of minimizing the objective function [40–42].

It is worth pointing out that, although robust optimization is new to our field, it has been applied for decades in other fields, including statistics, operations research, finance, engineering, etc., wherever uncertainties play an important role.

The DVHs shown in Figure 21–17 are for the nine dose distributions for the PSPT and IMPT plans and the seven dose distributions for the IMRT plan (i.e., no range uncertainty). It is important to note that, when comparing robustness, the DVHs of the CTV and OARs under various uncertainty scenarios are considered, but not the DVHs of PTV and ORVs.

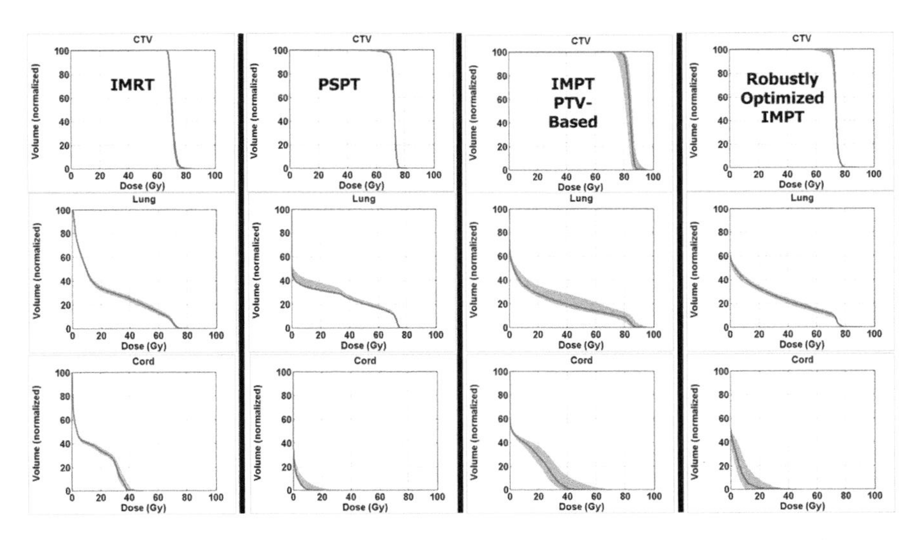

Figure 21–17 Inter-comparison of IMRT, PSPT, and IMPT dose distributions for a locally advanced NSCLC case. IMPT dose distributions were optimized in two different ways: the traditional approach of using CTV-to-PTV margins (third column) and using robust optimization. Bands of DVHs for CTV, normal lung, and spinal cord under nine different uncertainty scenarios are shown. As expected, IMRT dose distribution is the most robust, followed by PSPT. PTV margins-based optimized IMPT is the least robust, but with robust optimization, robustness improves significantly.

Based on the widths of DVHs, it is apparent that IMRT dose distributions are the most robust, followed by PSPT. As expected, IMPT dose distributions optimized based on conventional PTV-based approach are the least robust.

An interesting result of robust optimization in comparison with PTV-based optimization is that the plan optimality (e.g., sparing of normal tissues) also improves. This does not mean that robustness improves plan optimality. In fact, if a plan is to be more robust, it will likely be at the expense of optimality. However, the seemingly anomalous result seen in Figure 21–17 is presumably due to the constraint that each voxel must comply with the criteria under the worst-case scenario. Generally, there is substantial flexibility available in the IMPT optimization process due to the degeneracy of intensity distributions. As a result, improvement in the optimality of plans can be achieved, while at the same time making them robust compared to PTV-based optimization. This type of behavior has been seen commonly.

Another fortuitous result of robust optimization that has been observed is that robust optimization may make dose distributions resilient not only to uncertainties that were included in the optimization process, but also to some that were not explicitly considered. For instance, robust optimization that considers only setup and range uncertainties may lead to dose distributions that also have improved robustness in the face of inter- and intra-fractional anatomic variations. The reason appears to be that robust optimization implicitly reduces the gradients in dose distribution per field as well as for the composite, which means that the dose distribution becomes less sensitive to variations in anatomy in general.

Robust optimization is an important tool to significantly improve the confidence in proton dose distributions and to improve their optimality. There is considerable additional work needed to improve our understanding of the currently implemented methods, for the development of new ones, and for the extension of such methods to uncertainties other than setup and range.

21.4.2.5 Management of Uncertainties in RBE

As stated previously, the assumption of a constant RBE of 1.1 for protons may introduce considerable uncertainty in the biologically effective dose distributions and lead to suboptimal treatments and unforeseen consequences. While one may be tempted to use RBE models to design and evaluate treatment plans, the current knowledge of RBE is rather limited, and it is essential to proceed with caution. Realizing that there are large gaps in our understanding of the biological effectiveness of protons, an increasing number of investigators are undertaking *in vitro* and *in vivo* experiments to accurately map RBE as a function of dose, LET, cell and tissue type, and for various endpoints. Efforts are also being initiated to estimate RBE based on observed changes in images before and after treatment and on clinical response. These comments are meant mainly for proton therapy. While RBE used for heavier ions is assumed to be variable, uncertainties there are even greater.

For more on RBE uncertainties and their management, see Chapter 5.

21.5 Summary

Uncertainties are present in every step associated with radiation therapy, e.g., in the identification of the extent and response characteristic of the disease and surrounding tissues, delineation of the target and normal structure volumes, computed dose distributions, the dose delivery process, etc. Due to the sharp distal falloff of proton dose distributions and the scattering characteristics of protons, uncertainties have a greater impact on proton therapy compared with photons. The highly heterogeneous per-beam dose distributions of IMPT make it even more vulnerable to uncertainties than PSPT.

Many of the sources of uncertainties, such as target and organ delineation and setup precision, exist in both photon and proton therapies. The methods of reducing uncertainties, e.g., the use of multi-modality imaging, immobilization, and IGRT, may also be the same for both modalities. However, in general, the need for such solutions is greater for proton therapy. Issues unique to proton therapy include the greater impact of inter- and intra-fractional anatomic variations, the greater susceptibility to uncertainties in CT numbers and in factors for converting them to stopping power ratios, and the assumptions and approximations of dose computation models. Furthermore, the biological effectiveness of protons relative to photons is not well understood. Such uncertainties may lead to clinically significant differences between planned dose distributions and those actually delivered over the course of proton therapy and may negatively affect clinical outcomes.

It is not possible to eliminate uncertainties; however, it is important to minimize them to the extent reasonably practical. It is also important to estimate the residual uncertainties and their consequences and to incorporate them in the design, optimization and evaluation of treatment plans.

Uncertainties in dose delivered can be reduced by the routine use of such techniques as external and internal immobilization, tracking, in-room volumetric image-guidance, intra-fractional motion management, etc. The accuracy of computed dose distributions can be improved with Monte Carlo or equivalently accurate methods. Furthermore, inter-fractional anatomic changes, such as tumor shrinkage and weight gain or loss, can be accounted for via adaptive replanning, most effectively by taking into consideration the dose distribution already delivered.

There are other proton therapy-specific steps to mitigate uncertainties or their effects. For PSPT, beam-specific PTVs that include both range and setup uncertainties is an important way to assure that the CTV remains covered with the prescribed dose in the face of many of the physical uncertainties. Furthermore, compensator smearing is used to make dose distributions for heterogeneous anatomy resilient to positioning changes. Smearing also partially accounts for the limitations of conventional dose computation algorithms to account adequately for lateral scattering.

For IMPT dose distributions, robust optimization methods are important to account for the range, setup, and motion-related uncertainties to ensure target coverage and OAR sparing in the face of various uncertainties. For both PSPT and IMPT, robustness evaluation to quantify the resilience of dose distributions to uncertainties is essential.

Improving our understanding of RBE and the development of more accurate models of predicting RBE is crucial. For IMPT, it is important that optimization be based on RBE-weighted dose distributions. While RBE variability for particles heavier than protons is well recognized, it has essentially been ignored for protons. However, there is increasing interest in the variability of proton RBE, and it is being further scrutinized experimentally, theoretically, and clinically. Attempts are also being made to take a closer look at clinical response data to explore the possible effects of variable RBE.

Accurate dose calculation methods, improvements in positioning and motion management techniques, robust optimization, and robustness evaluation can make the delivered dose distributions more similar to the planned dose distributions and improve confidence in them. Furthermore, improved understanding of RBE variability and its clinical relevance, and the development of reliable RBE predictive models, should allow us to better correlate dose distributions with treatment response. These advances should lead to improved and more consistent clinical outcomes. A reduction in uncertainties will also enable us to reliably inter-compare competing treatment modalities in terms of their clinical and cost effectiveness.

References

1. Ling CC, Humm J, Larson S, Amols H, Fuks Z, Koutcher JA. Towards multi-dimensional radiotherapy (MD-CRT): Biological imaging and biological conformality. *Int J Radiat Oncol Biol Phys* 2000;47:551–560.
2. Riegel AC, Berson AM, Destian S, Ng T, Tena LB, Mitnick RJ, Wong PS. Variability of gross tumor volume delineation in head-and-neck cancer using CT and PET/CT fusion. *Int J Radiat Oncol Biol Phys* 2006;65:726–732.
3. Andreo P. On the clinical spatial resolution achievable with protons and heavier charged particle radiotherapy beams. *Phys Med Biol* 2009;54:N205–215.
4. Schneider U, Pedroni E, Lomax A. The calibration of CT Hounsfield units for radiotherapy treatment planning. *Phys Med Biol* 1996;41:111–124.
5. Yang M, Zhu XR, Park PC, Titt U, Mohan R, Virshup G, Clayton JE, Dong L. Comprehensive analysis of proton range uncertainties related to patient stopping-power-ratio estimation using the stoichiometric calibration. *Phys Med Biol* 2012;57:4095–4115.
6. Yang M, Virshup G, Clayton J, Zhu XR, Mohan R, Dong L. Theoretical variance analysis of single- and dual-energy computed tomography methods for calculating proton stopping power ratios of biological tissues. *Phys Med Biol* 2010;55:1343–1362.
7. Moyers M. Physical uncertainties in the planning and delivery of light ion beam treatments. Report of the AAPM Task Group 202 (unpublished).
8. Matney J, Park PC, Bluett J, Chen YP, Liu W, Court LE, Liao Z, Li H, Mohan R. Effects of respiratory motion on passively scattered proton therapy versus intensity modulated photon therapy for stage III lung cancer: are proton plans more sensitive to breathing motion? *Int J Radiat Oncol Biol Phys* 2013;87:576–582.
9. Paganetti H, Niemierko A, Ancukiewicz M, Gerweck LE, Goitein M, Loeffler JS, Suit HD. Relative biological effectiveness (RBE) values for proton beam therapy. *Int J Radiat Oncol Biol Phys* 2002;53:407–421.
10. Paganetti H. Relative biological effectiveness (RBE) values for proton beam therapy. Variations as a function of biological endpoint, dose, and linear energy transfer. *Phys Med Biol* 2014;59:R419–472.
11. Herring DF, Compton DMJ. The degree of precision required in the radiation dose delivered in cancer radiotherapy. *Br J Radiol* 1970; special issue 5:51–58.
12. Peters LJ, O'Sullivan B, Giralt J, Fitzgerald TJ, Trotti A, Bernier J, Bourhis J, Yuen K, Fisher R, Rischin D. Critical impact of radiotherapy protocol compliance and quality in

the treatment of advanced head and neck cancer: results from TROG 02.02. *J Clin Oncol* 2010;28:2996–3001.

13. Soares HP, Kumar A, Daniels S, Swann S, Cantor A, Hozo I, Clark M, Serdarevic F, Gwede C, Trotti A, et al. Evaluation of new treatments in radiation oncology: are they better than standard treatments? *JAMA* 2005;293:970–978.
14. Djulbegovic B, Kumar A, Glasziou PP, Perera R, Reljic T, Dent L, Raftery J, Johansen M, Di Tanna GL, Miladinovic B, et al. New treatments compared to established treatments in randomized trials. *Cochrane Database Syst Rev* 2012;10:MR000024.
15. Rutz HP, Weber DC, Sugahara S, Timmermann B, Lomax AJ, Bolsi A, Pedroni E, Coray A, Jermann M, Goitein G. Extracranial chordoma: Outcome in patients treated with function-preserving surgery followed by spot-scanning proton beam irradiation. *Int J Radiat Oncol Biol Phys* 2007;67:512–520.
16. de Crevoisier R, Tucker SL, Dong L, Mohan R, Cheung R, Cox JD, Kuban DA. Increased risk of biochemical and local failure in patients with distended rectum on the planning CT for prostate cancer radiotherapy. *Int J Radiat Oncol Biol Phys* 2005;62:965–973.
17. Titt U, Zheng Y, Vassiliev ON, Newhauser WD. Monte Carlo investigation of collimator scatter of proton-therapy beams produced using the passive scattering method. *Phys Med Biol* 2008;53:487–504.
18. Urie M, Goitein M, Holley WR, Chen GT. Degradation of the Bragg peak due to inhomogeneities. *Phys Med Biol* 1986;31:1–15.
19. Sawakuchi GO, Titt U, Mirkovic D, Mohan R. Density heterogeneities and the influence of multiple Coulomb and nuclear scatterings on the Bragg peak distal edge of proton therapy beams. *Phys Med Biol* 2008;53:4605–4619.
20. Wedenberg M, Toma-Dasu I. Disregarding RBE variation in treatment plan comparison may lead to bias in favor of proton plans. *Med Phys* 2014;41:091706.
21. Wouters BG, Lam GK, Oelfke U, Gardey K, Durand RE, Skarsgard LD. Measurements of relative biological effectiveness of the 70 MeV proton beam at TRIUMF using Chinese hamster V79 cells and the high-precision cell sorter assay. *Radiat Res* 1996;146:159–170.
22. Jones B. Patterns of failure after proton therapy in medulloblastoma. *Int J Radiat Oncol Biol Phys* 2014;90:25–26.
23. Kothary N, Heit JJ, Louie JD, Kuo WT, Loo BW, Jr., Koong A, Chang DT, Hovsepian D, Sze DY, Hofmann LV. Safety and efficacy of percutaneous fiducial marker implantation for image-guided radiation therapy. *J Vasc Interv Radiol* 2009;20:235–239.
24. Shirato H, Shimizu S, Kitamura K, Nishioka T, Kagei K, Hashimoto S, Aoyama H, Kunieda T, Shinohara N, Dosaka-Akita H, et al. Four-dimensional treatment planning and fluoroscopic real-time tumor tracking radiotherapy for moving tumor. *Int J Radiat Oncol Biol Phys* 2000;48:435–442.
25. Ogunleye T, Rossi PJ, Jani AB, Fox T, Elder E. Performance evaluation of Calypso 4D localization and kilovoltage image guidance systems for interfraction motion management of prostate patients. *Scientific World Journal* 2009;9:449–458.
26. Willoughby TR, Kupelian PA, Pouliot J, Shinohara K, Aubin M, Roach M, 3rd, Skrumeda LL, Balter JM, Litzenberg DW, Hadley SW, et al. Target localization and real-time tracking using the Calypso 4D localization system in patients with localized prostate cancer. *Int J Radiat Oncol Biol Phys* 2006;65:528–534.
27. Newhauser W, Fontenot J, Koch N, Dong L, Lee A, Zheng Y, Waters L, Mohan R. Monte Carlo simulations of the dosimetric impact of radiopaque fiducial markers for proton radiotherapy of the prostate. *Phys Med Biol* 2007;52:2937–2952.
28. Yepes P, Randeniya S, Taddei PJ, Newhauser WD. Monte Carlo fast dose calculator for proton radiotherapy: application to a voxelized geometry representing a patient with prostate cancer. *Phys Med Biol* 2009;54:N21–28.
29. Fix MK, Frei D, Volken W, Born EJ, Aebersold DM, Manser P. Macro Monte Carlo for dose calculation of proton beams. *Phys Med Biol* 2013;58:2027–2044.
30. Fippel M, Soukup M. A Monte Carlo dose calculation algorithm for proton therapy. *Med Phys* 2004;31:2263–2273.
31. ICRU. ICRU Report 62: Prescribing, recording and reporting photon beam therapy (supplement to ICRU Report 50). Bethesda, MD: International Commission on Radiation Units and Measurements, 1999.
32. ICRU. ICRU Report 50: Prescribing, recording, and reporting photon beam therapy. Bethesda, MD: International Commission on Radiation Units and Measurements, 1993.

33. ICRU. ICRU Report 78: Prescribing, Recording, and Reporting Proton Beam Therapy. *J ICRU* 2007;7.
34. van Herk M, Remeijer P, Rasch C, Lebesque JV. The probability of correct target dosage: dose-population histograms for deriving treatment margins in radiotherapy. *Int J Radiat Oncol Biol Phys* 2000;47:1121–1135.
35. Moyers MF, Miller DW, Bush DA, Slater JD. Methodologies and tools for proton beam design for lung tumors. *Int J Radiat Oncol Biol Phys* 2001;49:1429–1438.
36. Schuemann J, Dowdell S, Grassberger C, Min CH, Paganetti H. Site-specific range uncertainties caused by dose calculation algorithms for proton therapy. *Phys Med Biol* 2014;59:4007–4031.
37. Urie M, Goitein M, Wagner M. Compensating for heterogeneities in proton radiation therapy. *Phys Med Biol* 1984;29:553–566.
38. Liu W, Frank SJ, Li X, Li Y, Park PC, Dong L, Ronald Zhu X, Mohan R. Effectiveness of robust optimization in intensity-modulated proton therapy planning for head and neck cancers. *Med Phys* 2013;40:051711.
39. Liu W, Zhang X, Li Y, Mohan R. Robust optimization of intensity modulated proton therapy. *Med Phys* 2012;39:1079–1091.
40. Fredriksson A, Forsgren A, Hardemark B. Minimax optimization for handling range and setup uncertainties in proton therapy. *Med Phys* 2011;38:1672–1684.
41. Fredriksson A. A characterization of robust radiation therapy treatment planning methods-from expected value to worst case optimization. *Med Phys* 2012;39:5169–5181.
42. Fredriksson A, Bokrantz R. A critical evaluation of worst case optimization methods for robust intensity-modulated proton therapy planning. *Med Phys* 2014;41:081701.

Chapter 22

Treatment Plan Optimization in Proton Therapy

Jan Unkelbach, Ph.D., David Craft, Ph.D.,
Bram L. Gorissen, Ph.D., and Thomas Bortfeld, Ph.D.

Department of Radiation Oncology,
Massachusetts General Hospital,
Boston, MA

22.1 Introduction

In *treatment plan optimization* we try to find the best possible treatment plan for our patients. Note that this definition of optimization is different from the colloquial use of optimization, which is often understood as iterative improvement, but not necessarily improvement to optimality. In the mathematical literature, the term optimization has a well-defined meaning, referring to the maximization or minimization of an objective function while satisfying a set of constraints. In this chapter, we discuss the application of mathematical optimization techniques to treatment planning in proton therapy. We focus exclusively on treatment planning for intensity-modulated proton therapy (IMPT) with spot scanning, i.e., passive scattering delivery techniques are not considered.

Optimization methods have found wide-spread applications in radiation therapy since the introduction of intensity-modulated radiation therapy (IMRT) [1,2]. As far as the underlying optimization problems are concerned, there are strong similarities between optimizing x-ray treatments and proton therapy. Formally, the basic IMPT treatment plan optimization problem is virtually identical to the fluence map optimization problem in IMRT. This general optimization concept is introduced in Section 22.2.1. Radiation treatment planning typically involves several conflicting objectives, and this applies to proton therapy as well. How to deal with several objectives through multi-criteria optimization (MCO) will be described in Section 22.2.3. An overview of the solvers, i.e., the computational algorithms that find the solutions of the optimization problems, is given in Section 22.4.

On a more advanced level, there are also important differences between IMRT and IMPT planning. For example, IMRT planning systems incorporate specialized algorithms such as Direct Aperture Optimization (DAO) to optimize the leaf positions of the multi-leaf collimator. In this chapter, we emphasize some of the problems in plan optimization that are specific to proton therapy. First of all, proton therapy's finite beam range is more strongly affected by uncertainties than x-rays. This is due largely to the validity of the *dose cloud approximation* in x-ray therapy, which is not valid in proton therapy. X-ray beams produce a dose cloud in space, which is hardly affected by motion. The patient and the inner organs move within this dose cloud without changing the shape of the cloud significantly. Therefore, adding a planning target volume (PTV) margin is a legitimate way to account for motion, as long as the margin encompasses the range of motion. This approach is not generally sufficient in proton therapy, where the shape of the dose cloud is affected strongly by motion. Simple PTV margins do not work here. Addressing the problem of plan robustness against uncertainties is therefore more critically important, and it deserves more attention in proton therapy than for x-rays (see Section 22.2.2).

Another important difference between plan optimization for x-rays and protons is that protons exhibit a different relative biological effect (RBE), which is normally taken into account through a constant factor of 1.1. Plan optimization based on a more realistic RBE model, which is particularly relevant for highly modulated IMPT, is described in Section 22.3.1. The RBE effect is relevant in proton therapy, but it is much more pronounced in treatments with heavier ions. Biological effects may affect not only the optimal spatial distribution of dose, but also the temporal distribution, i.e., dose fractionation (see Section 22.3.2). Finally, the potential of IMPT to deliver dose painting based on biological variations in the tumor will be discussed.

22.2 IMPT Optimization

In this section, we discuss treatment planning for proton therapy using spot scanning as the delivery technique. The treatment planning problem is formulated as a mathematical optimization problem that determines the intensities of individual pencil beams in order to realize a desired dose distribution. The treatment planning methods described here apply to single-field uniform dose (SFUD) treatments as well as IMPT, but they are not applicable to passive scattering techniques. This section is based in parts on the chapter "Robust Optimization" from the 2011 AAPM summer school book *Uncertainties in External Beam Radiation Therapy* [3].

22.2.1 Treatment Planning as an Optimization Problem

In spot scanning, proton therapy is delivered using a narrow monoenergetic proton beam of variable energy that is magnetically scanned across the treatment volume. An individual pencil beam is described by its energy (which determines its range in the patient) and its lateral position relative to the isocenter. For treatment planning, a set of potentially used pencil beams is determined. Typically, this set of pencil beams is defined such that the Bragg peak locations uniformly cover the treatment volume. A dose calculation algorithm then calculates the dose distributions of each individual pencil beam inside the patient. The dose calculation data can be stored in a large matrix D, which we refer to as the dose-influence matrix. More specifically, a matrix element D_{ij} stores the dose contribution of pencil beam j to voxel i in the patient for unit intensity. A natural unit of the pencil beam intensity is *gigaprotons*, i.e., the number of protons delivered. In that case, the natural unit of the dose-influence matrix is Gy per gigaproton. The index j is an abbreviation for the energy and lateral position that defines the pencil beam. The total dose distribution d in the patient is given by the superposition of all pencil beam contributions

$$d_i(\boldsymbol{x}) = \sum_j D_{ij} x_j \tag{22.1}$$

where x_j is the intensity of pencil beam j. The goal of treatment planning is to find a set of pencil beam intensities $\boldsymbol{x}$ that yields a dose distribution that is as close as possible to the desired dose distribution. This task is formulated as a mathematical optimization problem. To that end, the treatment goals are formulated in terms of an objective function $f(\boldsymbol{d}(\boldsymbol{x}))$ and constraint functions $c_k(\boldsymbol{d}(\boldsymbol{x}))$, which are mathematical functions of the dose distribution. In general terms, an IMPT optimization problem can be stated as follows:

$$\begin{aligned} &\underset{\boldsymbol{x}}{\text{minimize}} \quad f(\boldsymbol{d}(\boldsymbol{x})) \\ &\text{subject to} \\ &c_k(\boldsymbol{d}(\boldsymbol{x})) \le u_k \qquad \forall k \\ &x_j \ge 0 \qquad \forall j. \end{aligned}$$

We illustrate this concept based on a concrete example. We consider a spinal metastasis treatment for the patient illustrated in Figure 22–1(a). The tumor entirely surrounds the spinal cord which is to be spared. Additional organs at risk are the esophagus, the lungs and the remaining unclassified tissue. We choose three beam directions: a posterior beam at 0° and two oblique beams at ±45°. Proton pencil

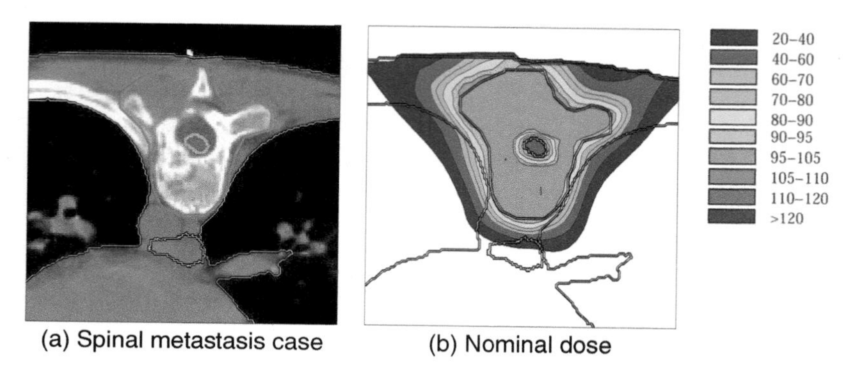

(a) Spinal metastasis case (b) Nominal dose

Figure 22–1 Illustration of a complex IMPT treatment plan for a spinal metastasis treatment (a). The target volume (red) surrounds the spinal cord (orange) which is to be spared. Using IMPT, a highly conformal homogeneous dose distribution can be delivered to the target while avoiding the spinal cord (b). The dose distribution is shown as a percentage of the prescribed dose.

beams of 5 mm sigma at the patient surface are used, which are placed on a regular grid of 5 mm resolution.

For treatment planning, we specify the following objective function:

$$f(\boldsymbol{d}) = w_T \sum_{i \in T} (d_i - D^{pres})^2 + w_C \sum_{i \in C} (d_i)^2 + w_H \sum_{i \in H} d_i . \tag{22.2}$$

The first term represents a quadratic objective function that penalizes deviation of the tumor dose from the prescription $D^{pres} = 60$ Gy, and thus aims at delivering a homogeneous dose to the target volume. The second term quadratically penalizes dose to the spinal cord, and the third term minimizes the mean dose to all adjacent normal tissues. The coefficients w_T, w_C, and w_H are weighting factors that control the relative importance of the objectives. In addition, we specify constraints on the maximum spinal cord dose, which is to be bounded by $D^{max} = 54$ Gy, i.e., the general constraints $c_k(\boldsymbol{d}(\boldsymbol{x})) \leq u_k$ become

$$d_i \leq D^{\max} \quad \forall i \in C. \tag{22.3}$$

Solving the IMPT optimization problem yields optimal pencil beam intensities $\boldsymbol{x}$ and a corresponding dose distribution, which is shown in Figure 22–1(b). With IMPT it is possible to deliver a very homogeneous and conformal dose distribution to the clinical target volume (CTV) while achieving good sparing of the spinal cord. It is illustrative to inspect how this dose distribution is created using external proton beams. Figure 22–2 shows the dose contributions of the three incident beam directions. It is apparent that each individual beam direction delivers a complex and highly nonuniform dose

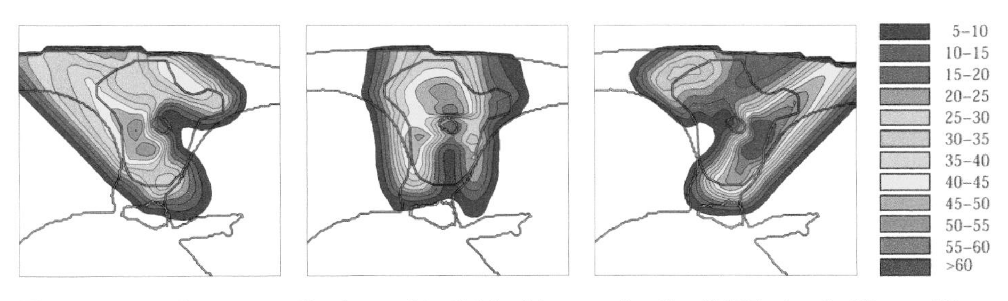

Figure 22–2 Dose contributions of individual beams for the IMPT plan in Figure 22–1(b) as a percentage of prescribed dose.

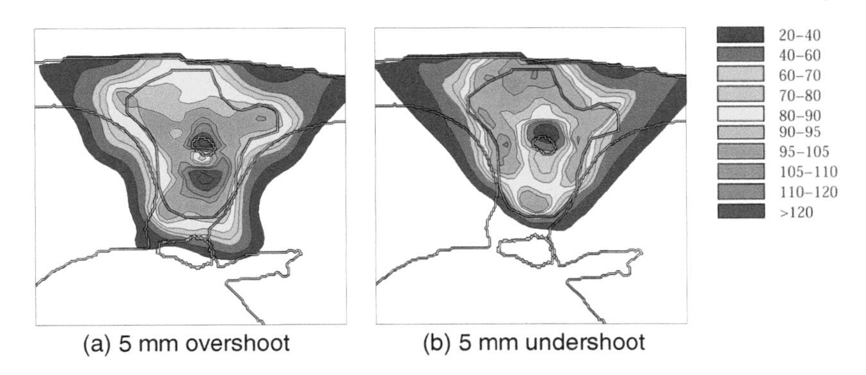

Figure 22–3 Sensitivity analysis for the conventional plan.

distribution, and only the superposition of all three beams yields the desired cumulative target dose.

22.2.2 Handling Uncertainty through Robust Optimization

The main physical advantage of proton therapy is the finite range of protons. However, the location of the Bragg peak with respect to anatomical structures is associated with uncertainty. The quality of an IMPT plan may highly degrade if the true range differs from the assumed range or if a setup error occurs. For that purpose, robust optimization methods have been developed that directly incorporate uncertainty in treatment plan optimization.

22.2.2.1 Sensitivity of IMPT Plans

We first perform a sensitivity analysis of the IMPT plan shown in figures 22–1(b) and 22–2. Figures 22–3(a) and 22–3(b) show the resulting dose distribution for a 5 mm overshoot (i.e., the Bragg peak positions of all pencil beams are shifted farther into the patient) and a 5 mm undershoot, respectively. We observe that the dose distribution is degraded. If the range of the proton beams is larger than expected, high doses are delivered to the spinal cord. Both undershoot and overshoot of protons yield highly inhomogeneous dose distributions in the target volume.

This sensitivity can be explained by analyzing the dose contributions of individual beam directions (shown in Figure 22–2). The optimization assigns a high weight to Bragg peaks placed in front of the spinal cord. This allows for the best dose sparing of the critical structure since the steep distal falloff of the Bragg peak is utilized. As a consequence, a high dose is delivered to the spinal cord if the range of a pencil beam is larger than expected (Figure 22–3(a)). Generally, dose gradients in beam direction make the plan sensitive to range variations because a range error results in a relative shift of the dose contributions of individual beams, which thus do not add up to the desired dose as planned.

22.2.2.2 Robust Optimization Methods

Although the treatment plan above does not apply a margin to the CTV, it is evident that the margin concept cannot solve the problem sufficiently. A margin around the CTV could only reduce the underdosage of the CTV at the boundary due to geometric shifts of the dose distribution. It cannot account for dose inhomogeneity in the CTV and overdosage of the spinal cord due to misalignments of beams. Adding margins has no influence on the occurrence of steep dose gradients (along the direction of a single beam) inside the CTV, which have to be reduced to improve the robustness of a treatment plan. For that purpose, robust optimization methods have been developed that incorporate a model of the uncertainty directly into the IMPT optimization problem.

In IMPT planning, uncertainty is typically modeled through a finite number of discrete error scenarios. A simple model of range uncertainty may consist of three scenarios:

Scenario 1: nominal scenario (no range error)

Scenario 2: overshoot (all pencil beams synchronously overshoot by a specified amount)

Scenario 3: undershoot (all pencil beams undershoot by a specified amount)

Similarly, a simple model of setup uncertainty may consist of seven scenarios, i.e., the nominal scenario plus six specified patient shifts in anterior, posterior, left, right, superior, and inferior direction. In each error scenario, the dose distribution realized by pencil beam intensities $\boldsymbol{x}$ is different, and given by

$$d_i^s(\boldsymbol{x}) = \sum_j D_{ij}^s x_j, \tag{22.4}$$

where s is an index to the error scenario and D_{ij}^s is the corresponding dose-influence matrix. The overall goal in robust planning is to obtain a treatment plan that is of high quality under all assumed error scenarios. In optimization theory, there are mainly two paradigms to translate this notion into mathematical terms: stochastic programming and minimax optimization.

In the stochastic programming approach, a probability distribution is assigned to the uncertain parameters. The basic idea in stochastic programming is to optimize the expected value of the objective function f. The optimization problem to be solved is given by

$$\underset{x \geq 0}{\text{minimize}} \quad \sum_{s \in S} p_s f(\boldsymbol{d}^s(\boldsymbol{x})), \tag{22.5}$$

where p_s is the probability of occurrence for error scenario s and S is the set of scenarios. The composite objective 22.5 is a weighted sum of objective values for every error scenario. Thus, we aim at a treatment plan that is good for every error that can occur, but we assign a higher weight to those errors that are likely to occur.

In minimax optimization, the goal is to obtain a treatment plan that is as good as possible for the worst case scenario that is accounted for. Mathematically, this can be expressed as

$$\underset{x \geq 0}{\text{minimize}} \quad \left[\max_{s \in S} f(\boldsymbol{d}^s(\boldsymbol{x})) \right]. \tag{22.6}$$

In other words, we consider the maximum of the objective function f taken over the error scenarios s, and we aim at determining the pencil beam intensities $\boldsymbol{x}$ such that the maximum objective value is minimized.

A typical way of handling dose constraints in robust optimization consists in requiring that the constraints are satisfied for every realization of the error. For example, if we consider a maximum dose constraint for an organ at risk as in constraint 22.3, then robust optimization means that this constraint is to be fulfilled for every realization of the error, i.e.,

$$d_i^s \leq D^{\max} \qquad \forall i \in C \ \ \forall s \in S. \tag{22.7}$$

22.2.2.3 Robust Optimization Example

To illustrate the use of robust optimization, we consider the simple uncertainty model above using only three scenarios: the nominal scenario, range overshoot, and range undershoot. To apply the probabilistic planning paradigm, we assign probabilities to these three scenarios: we assume a probability $p_1 = 0.5$ for the nominal scenario and $p_2 = p_3 = 0.25$ for the overshoot and undershoot scenario. To calculate dose for the overshoot and undershoot scenarios, we precalculate dose-influence matrices D_{ij}^s for these three scenarios. For that purpose, we need to make the model of a range error explicit. Here we model a 5 mm range overshoot by assuming that the initial proton energy of all beam spots is higher by an amount that corresponds to 5 mm water equivalent range. Alternatively, one could uniformly scale down stopping power ratios obtained from the CT. Given D_{ij}^s, we can calculate dose for every scenario as $d_i^s = \sum_j D_{ij}^s x_j$ and solve the following optimization problem:

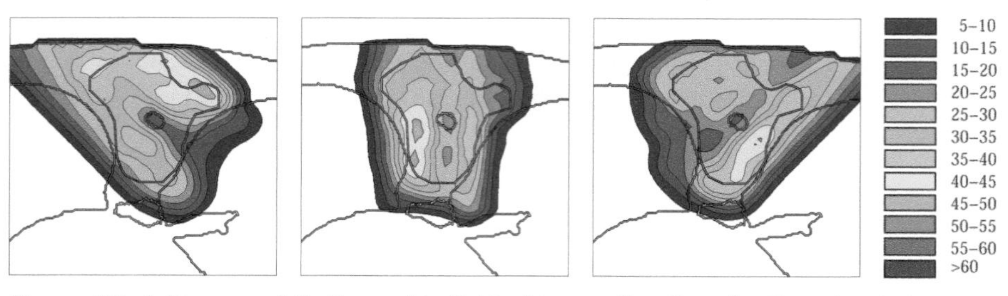

Figure 22–4 Dose contributions of individual beam directions for the robust plan.

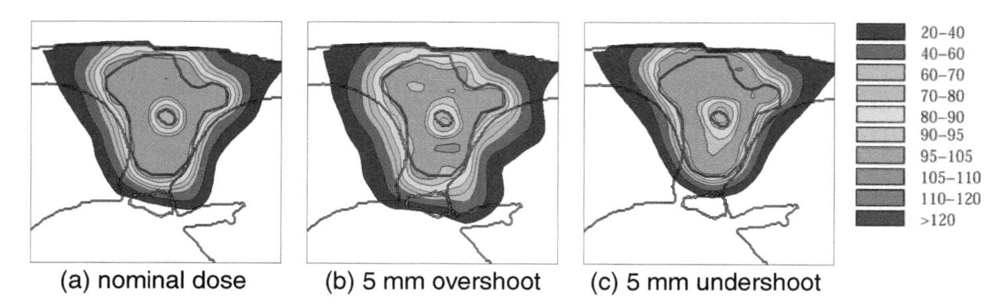

(a) nominal dose (b) 5 mm overshoot (c) 5 mm undershoot

Figure 22–5 Sensitivity analysis for the robust plan.

$$\underset{x \geq 0}{\text{minimize}} \quad \sum_{s=1}^{3} p_s \left[w_T \sum_{i \in T} (d_i^s - D^{pres})^2 + w_S \sum_{i \in C} (d_i^s)^2 \right] + w_H \sum_{i \in H} d_i^1 \qquad (22.8)$$

For the example patient, robustness primarily refers to ensuring target coverage and spinal cord sparing under range errors. In the interest of computational efficiency, it is sufficient to incorporate the remaining healthy tissue only via the nominal scenario.

Figures 22–4 and 22–5 show the treatment plan obtained. Figure 22–5(a) shows the dose distribution for the nominal scenario, and figures 22–5(b) and 22–5(c) show the overshoot and undershoot scenario, respectively. Despite a range error, the dose in the tumor remains widely homogeneous, and the sparing of the spinal cord is preserved. Figure 22–4 shows the dose contributions of the three beams for the nominal case, which explains how robustness of the plan is achieved. Protection against range uncertainty is achieved by avoiding steep dose gradients in beam direction in the dose contributions of individual beams. This includes avoiding the placement of the distal edge of a Bragg peak directly in front of the spinal cord.

22.2.2.4 Visualization Tools to Evaluate Robustness

Evaluating the robustness of treatment plans requires appropriate tools for visualization. The most obvious approach consists in evaluating the dose distribution for each error scenario of interest, as illustrated in Figure 22–5. However, to make robustness

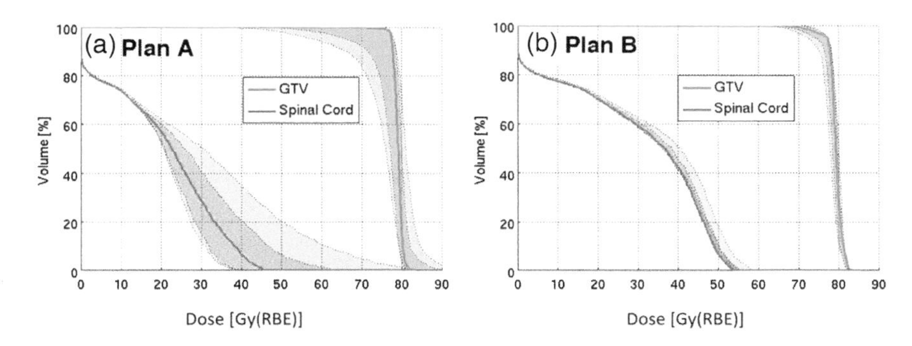

Figure 22–6 A conventional proton therapy plan (a) and a robustly optimized plan (b) that are affected differently by uncertainties. Plan A is better in the nominal case (solid lines) but the DVH may vary widely under uncertainty. Plan A is, therefore, less robust than plan B.

evaluation more efficient, it is desirable to visualize plan quality for the different error scenarios in a more condensed manner. One such approach is the display of DVH bands, which allows the treatment planner to assess the possible degradation of the DVH. An example is shown in Figure 22–6. Here we see two plans, plan A and plan B. The solid lines represent the dose volume histograms in the nominal case, and the shaded areas reflect the magnitude of degradation under range and setup uncertainties. Plan A is better in the nominal case, but it is much more affected by uncertainties. The figure illustrates both the effect of uncertainties and the so-called price of robustness. For further details on the calculation of DVH bands, we refer the reader to Trofimov et al. [4].

22.2.2.5 Status of Robust Optimization Methods

For further details on robust optimization, we refer to the scientific literature. This includes the work by Unkelbach et al. on the stochastic programming approach [5] and the work by Fredriksson et al. [6] on minimax optimization. The work by Pflugfelder [7] describes a variation of the minimax approach where the worst case is determined on a voxel-by-voxel basis. The later works by Fredriksson [8] compare different robust optimization methods.

Robust optimization methods are currently being implemented into the first commercial treatment planning systems. To our knowledge, the RayStation planning system (version 4.5) is the first system that released robust optimization functionality and uses a minimax type approach. An IMPT version of the Pinnacle planning system (not yet released) is expected to feature an implementation of the stochastic programming approach. With first commercial implementations, robust IMPT planning methodology will become available to a larger number of researchers, which may facilitate a broader evaluation and characterization of robust optimization for different treatment sites.

22.2.3 Multi-criteria Optimization

A difficulty that is often encountered in treatment planning is choosing a set of objective function weights w that leads to a desirable treatment plan. In the example in Equation 22.2, the weights w_T and w_C determine the trade-off between target coverage and spinal cord sparing. The mapping from weights w to the final treatment plan is not obvious, and this typically causes a human iteration loop where several time-consuming optimizations need to be performed until a satisfactory plan is obtained.

A radiation therapy plan involves unavoidable tradeoffs: if one were to list the goals of a radiation therapy plan, some of them would be in conflict, notably full target coverage and healthy organ sparing. The compromise between the conflicting goals that a physician or planner will ultimately select is dependent on the patient geometry and health status, and is not easily articulated prior to the optimization. This motivates the use of multi-criteria optimization (MCO). We discuss three MCO approaches below: goal programming, prioritized optimization, and Pareto surface navigation.

22.2.3.1 Goal Programming

Goal programming casts the treatment planning problem as a set of goals. Consider by way of example the use of tumor control probability (TCP) and normal tissue complication probabilities (NTCP) as the functions used to judge plan quality. Perhaps the doctor specifies that the TCP $\geq$ 95% and all NTCP functions $\leq$ 10%. Goal programming tries to find a plan that satisfies a set of such constraints. Two things can happen: either a plan can be found that satisfies these goals, or not. In the first case, one could naturally ask: can one do better, and in the second case one would ask, what goals should I sacrifice in order to achieve an acceptable plan? These two scenarios indicate a deficiency of the goal programming approach for radiation therapy. However, given the clinical reality that physicians often specify their desires in such language, the goal programming framework is relevant to modern treatment planning.

Regarding the mathematical implementation of goal programming, assume we are trying to minimize given functions f_k. For example, these could be dose-volume functions to reduce the number of voxels that violate a dose level, or f_k could be the NTCP for structure k, or they could be the quadratic penalty functions used in Equation 22.2. Let g_k denote the goal that we are trying to reach. The goal programming radiation treatment planning problem can then be written as:

$$\begin{aligned} \underset{x}{\text{minimize}} \quad & \sum\nolimits_k h_k \\ & \boldsymbol{d} = \boldsymbol{D}\boldsymbol{x} \\ & h_k \geq f_k(\boldsymbol{d}) - g_k \\ & h_k \geq 0 \\ & \boldsymbol{d} \in C, \boldsymbol{x} \geq \boldsymbol{0} \end{aligned} \tag{22.9}$$

where $\boldsymbol{d} \in C$ is a simplified notation for the constraints on the dose distribution. If $f_k(\boldsymbol{d}) \geq g_k$, that is, the goal has not been met, h_k will take on the value of the difference, or slack. If the goal has been met, h_k will take on the value 0, hence no penalty in the objective function is incurred. If all goals are met, the value of the objective function will be 0, and there is no incentive built into this optimization formulation to lower the functions f_k any further.

22.2.3.2 Prioritized Optimization

Some physicians, rather than giving a general set of goals, will prioritize the goals. This prescription style leads naturally to prioritized optimization, also called lexicographic optimization, and it is implemented by running a sequence of optimizations [9–13]. Constraints are nonnegotiable: they are the highest priority. Objectives have a sense (either maximize or minimize), but they do not have a hard level they are supposed to reach. A prioritized optimization approach to designing a treatment plan with a maximum spinal cord dose of 45 Gy as a nonnegotiable constraint might look as follows. In step 1, we maximize TCP subject to spinal cord dose ≤45 Gy. Let us assume that the TCP obtained from this optimization is 98%. This value is turned into a constraint in subsequent steps, including a relaxation factor, also known as a slip factor. Step 2 could then be: minimize mean lung dose subject to spinal cord dose ≤45 Gy and TCP ≥97%. Here an absolute slip factor of 1% has been used. Lower-priority goals would be handled in subsequent steps, each one of which will gather the previously obtained objective values and use them, with slip factors, as constraints. Formally, prioritized optimization proceeds as follows. Let $\boldsymbol{d} \in C$ denote the nonnegotiable constraints, and let f_1 be the first priority objective, f_2 be the second, etc.

Step 1:

$$\begin{aligned} &\underset{x}{\text{minimize}} \qquad f_1(\boldsymbol{d}) \\ &\boldsymbol{d} = D\boldsymbol{x}, \;\; \boldsymbol{d} \in C, \;\; \boldsymbol{x} \geq \boldsymbol{0} \end{aligned} \tag{22.10}$$

Then, letting f_1^* denote the optimal objective value for the first step, the next optimization proceeds as:

$$\begin{aligned} &\underset{x}{\text{minimize}} \qquad f_2(\boldsymbol{d}) \\ &f_1(\boldsymbol{d}) \leq f_1^* + \varepsilon \\ &\boldsymbol{d} = D\boldsymbol{x}, \;\; \boldsymbol{d} \in C, \;\; \boldsymbol{x} \geq \boldsymbol{0} \end{aligned} \tag{22.11}$$

Assuming a total of three priority levels, the optimization which yields the final treatment plan would be:

$$\begin{aligned} &\underset{x}{\text{minimize}} \qquad f_3(\boldsymbol{d}) \\ &f_1(\boldsymbol{d}) \leq f_1^* + \varepsilon \\ &f_2(\boldsymbol{d}) \leq f_2^* + \varepsilon \\ &\boldsymbol{d} = D\boldsymbol{x}, \;\; \boldsymbol{d} \in C, \;\; \boldsymbol{x} \geq \boldsymbol{0} \end{aligned} \tag{22.12}$$

22.2.3.3 Pareto Surface Navigation

Both goal programming and prioritized optimization provide only a single plan at the end of the process, without a natural way to explore tradeoffs. Pareto surface navigation is an MCO technique that puts the interactive exploration of the dosimetric tradeoffs at the front and center. For a given set of objectives and constraints, a treatment plan is Pareto optimal if it satisfies all of the constraints and none of the objective values can be improved upon without worsening at least one of the other objectives. These are the plans that should be considered for patient treatment. The set of Pareto-optimal treatment plans (termed the Pareto surface) for continuous optimization problems is typically an infinite set. Finding efficient ways to approximate it and intuitive ways to explore it are important research and development areas. The Pareto optimization problem is written as:

$$\begin{aligned} &\underset{x}{\text{minimize}} \quad (f_1(\boldsymbol{d}), f_2(\boldsymbol{d}), \ldots f_N(\boldsymbol{d})) \\ &\boldsymbol{d} = D\boldsymbol{x}, \;\; \boldsymbol{d} \in C, \;\; \boldsymbol{x} \geq \boldsymbol{0} \end{aligned} \tag{22.13}$$

where N is the number of objective functions. In a robust optimization setting, some or all of the functions f_k may be either expected values across scenarios or minimax functions. The constraint set may involve robust constraints (i.e., they are to hold for all scenarios) or nominal constraints.

Constraints are chosen to limit the range of the Pareto surface. If constraints are chosen too aggressively, the result might be that there are no feasible plans, and thus no Pareto surface. On the other hand, if constraints are chosen too loosely, the extent of the Pareto surface might be too large, thus requiring too many plans to approximate it. Ideally the Pareto surface contains a variety of plans that are all potential treatment candidates. Constraints chosen should be nonnegotiable (this is, in fact, the definition of a constraint). Negotiable requests should be framed as objectives. The number of objectives chosen should be such that all relevant trade-offs are exposed, but no more, since computational burden rises with the number of objectives, although perhaps not as fast as one might anticipate from a purely geometrical argument [14,15].

Pareto surfaces are typically generated by solving several weighted sum optimization problems where the weights w are varied to yield treatment plans in different regions of the Pareto surface. Weighted sum methods are attractive because they work well with sandwich algorithms, which produce inner and outer approximations to the Pareto surface, allowing the approximation error to be controlled [16,17]. However, these techniques require convex hulling algorithms that are practically infeasible for dimensions (number of objectives) above around 10. Therefore, especially in a robust setting where planners are interested in exploring the price of robustness, simpler methods of determining weight vectors to run need to be employed.

Once the Pareto surface is approximated by a discrete set of computed plans, the planner navigates across the surface by taking convex combinations of the pre-computed plans. IMPT is well-suited for such navigation since IMPT plans can be averaged by simply averaging the beamlet intensities, as opposed to step-and-shoot IMRT, where averaging two plans may double the number of apertures used in the averaged plan, leading to an inefficiently delivered plan.

22.3 Biological Optimization and Dose Painting

The previous sections discussed methods for optimizing the physical dose distribution in the patient that can be realized with external proton beams. Ultimately, treatment planning has to consider clinical outcome and the effectiveness of a treatment as a whole. In that regard, proton therapy shares the difficulties that arise in integrating radiotherapy planning, tumor imaging, outcome modeling, and radiobiology. In this section, we outline a few research topics on biological-based and outcome-based treatment planning that are specific to proton therapy.

22.3.1 LET-based Optimization

To this day, proton therapy planning is based on physical dose. This applies to both the passive scattering delivery techniques and spot scanning. The observation that protons show a higher relative biological effectiveness (RBE) compared to photons is taken into account through a generic scaling factor of 1.1, i.e., it is assumed that protons are 10% more effective than photons. While this approach was considered sufficient in passive scattering treatments, the emerging era of IMPT suggests we revisit the topic of RBE in proton therapy planning.

In vitro cell survival assays suggest that the biological effect is higher at the end of range than in the entrance region of the proton beam. Although RBE estimates vary over cell lines, it appears that the RBE in the entrance region is close to 1.0, while values of 1.2 and higher are commonly observed near the Bragg peak. This suggests that energy deposition alone (measured in terms of physical dose) is not sufficient to predict biological response. In addition, the ionization density along the track of a charged particle is important, i.e., how dose is deposited on a microscopic scale. Thus, a second physical quantity is needed to characterize the radiation field. Most approaches to RBE-based IMPT planning utilize the concept of Linear Energy Transfer (LET) for that purpose.

22.3.1.1 Treatment Plan Optimization Using LET-based RBE Models

Accounting for variations in RBE along the proton depth–dose curve requires a model of biological dose or RBE. To highlight basic problems in RBE-based IMPT planning, we consider a parsimonious biological dose model, in which the biological dose b depends on physical dose d and linear energy transfer L. We first consider a single pencil beam. In analogy to the biologically equivalent dose (BED) model, we can motivate the biological dose b via an exponential cell kill model. We assume that the surviving fraction is

$$S = \exp(-\alpha d), \tag{22.14}$$

and further that the radiosensitivity parameter α depends on LET L. For simplicity, we assume a linear dependence of α on LET:

$$\alpha = \alpha_0 + \lambda L. \tag{22.15}$$

In analogy to the BED model, we can define b via

$$b = -\frac{\log S}{\alpha_0} = \left(1 + \frac{\lambda}{\alpha_0} L\right) d. \tag{22.16}$$

In this expression, we can interpret $1+(\lambda/\alpha_0)L$ as a LET-dependent RBE. Thus, b is an RBE-weighted dose. In order to generalize the model to multiple pencil beams, we assume that different pencil beams have independent contributions to the overall cell kill S_i in some voxel i:

$$S_i = \prod_j \exp(-(\alpha_0 + \lambda L_{ij}) d_{ij}), \tag{22.17}$$

where L_{ij} is the LET of pencil beam j at voxel i, and d_{ij} is the dose deposited by pencil beam j in voxel i. We have

$$d_{ij} = D_{ij} x_j, \tag{22.18}$$

and the total dose in voxel i is

$$d_i = \sum_j d_{ij} = \sum_j D_{ij} x_j. \tag{22.19}$$

Here, D_{ij} is the familiar dose-influence matrix; correspondingly L_{ij} is the LET-matrix. The L_{ij} matrix can, for example, be calculated through Monte Carlo simulations or be based on analytical models of LET [18]. We can define the biological dose b via

$$b_i = -\frac{\log S_i}{\alpha_0} = \sum_j \left(1 + \frac{\lambda}{\alpha_0} L_{ij}\right) d_{ij}. \tag{22.20}$$

Given that the dose-averaged LET in voxel i is given by

$$LET_i = \frac{\sum_j L_{ij} d_{ij}}{d_i}, \tag{22.21}$$

we see that the biological dose can be written as

$$b_i = \left(1 + \frac{\lambda}{\alpha_0} LET_i\right) d_i. \tag{22.22}$$

Treatment plan optimization for IMPT will involve objective and constraint functions evaluated for LET, physical dose, and biological dose. Thus, we need to consider how these quantities depend on the decision variables, which are the pencil beam intensities x_j. The physical dose d_i is a linear function of x_j, which is desired. In contrast, LET_i is a more complex function of the beamlet intensities,

$$LET_i = \left(\frac{\sum_j L_{ij} D_{ij} x_j}{\sum_j D_{ij} x_j} \right), \tag{22.23}$$

and would be more difficult to optimize. However, the biological dose model only contains the product of physical dose and LET, which is again a linear function of the pencil beam intensities:

$$LET_i d_i = \sum_j L_{ij} D_{ij} x_j . \tag{22.24}$$

We further see that the biological dose according to the parsimonious model (22.16) remains a linear function of the beamlet intensities. If we define the biological dose-influence matrix as

$$B_{ij} = D_{ij} \left(1 + \frac{\lambda}{\alpha_0} L_{ij} \right), \tag{22.25}$$

the biological dose is given by

$$b_i = \sum_j B_{ij} x_j . \tag{22.26}$$

The biological dose model can be extended to include the quadratic term $-\beta d^2$ of the LQ model [19]. In addition, LET dependence of β can be introduced [20]. These extensions of the biological dose model are particularly important if multiple fractionation schemes are considered.

22.3.1.2 Status of RBE-based IMPT Planning

So far, accounting for variable RBE in IMPT planning remains a research topic and has not found its way into clinical use. One main reason for this is that the LET dependence of the biological effect *in vivo* is widely unknown. While tumor cell survival assays show increased RBE with higher LET, the effect cannot be quantified for normal tissues and relevant clinical endpoints. Furthermore, clinical experience is consistent with an RBE of approximately 1.1, which indicates that the overall magnitude of RBE effects is in the range of 10% of the prescription dose. This is in contrast to carbon ion therapy, where the RBE is in the order of 2–3, so that RBE models are a necessity in treatment planning.

Different approaches to RBE-based IMPT planning can be pursued. The most obvious approach is to perform plan optimization based on a biological dose model, i.e., evaluate the commonly used objective and constraint functions for biological dose rather than physical dose (see, e.g., the work by Wilkens et al. [19]). The main concern regarding this approach is the uncertainty in the LET dependence of the biological dose. An alternative approach consist in using treatment planning objectives for physical dose and LET, which may provide a hybrid approach that lies between the current physical dose planning and full RBE-based planning. In such an approach, the main planning goals can be formulated in terms of physical dose, but additional objectives can aim at avoiding high LET in adjacent critical structures or concentrating high LET within the gross tumor volume (GTV) [21].

22.3.2 Optimization of Fractionation Schemes

Treatment planning typically refers to the optimization of the spatial dose distribution in the patient, realized through external radiation beams. In addition to the spatial dose distribution, outcome is determined by the fractionation scheme, i.e., how radiation is administered over time. Broadly speaking, proton therapy as it is practiced today uses the fractionation schemes that were established in x-ray therapy. However, there is an interdependence between the spatial dose distribution and fractionation decisions as discussed below. Therefore, the depth–dose characteristics of a proton beam may potentially give rise to altered fractionation schemes—even though this is not clinical practice today.

22.3.2.1 Hypofractionation in Proton Therapy

The biologically equivalent dose (BED) model is the most common concept to compare the effectiveness of different fractionation regimens. The BED b is given by

$$b = nd\left(1 + \frac{d}{(\alpha / \beta)}\right) \tag{22.27}$$

where d is the physical dose per fraction, n is the number of fractions, and α/β is an endpoint-specific constant that quantifies a tissue's sensitivity to fractionation. One approach to determining optimal fractionation schemes consists in determining the number of fractions n that minimizes the BED in the dose-limiting normal tissue for a fixed prescribed BED delivered to the tumor. This problem has been studied by Mizuta [22] for an idealized situation in which the tumor receives a uniform dose d_T and the normal tissue receives a uniform dose d_N. In this case, the BED model provides a simple guideline for fractionation decisions: If

$$\frac{(\alpha / \beta)_N}{(\alpha / \beta)_T} > \frac{d_N}{d_T} \tag{22.28}$$

the BED model suggest hypofractionation, otherwise standard fractionation is indicated. Typically, values for the α/β ratios in the tumor and normal tissue are $(\alpha/\beta)_T = 10$ and $(\alpha/\beta)_N = 4$. Thus, treatment sites where the normal tissue dose is much lower than the tumor dose bear potential for hypofractionation, even for unfavorable α/β ratios. Proton therapy has the potential to reduce the dose to normal tissue d_N. Hence, the above model suggests that proton therapy may give rise to hypofractionated schedules in situations were standard fractionation should be used for x-ray therapy.

However, a more detailed analysis of this problem questions the increased potential for hypofractionation in proton therapy. For many treatment sites, normal tissues within or directly adjacent to the treatment volume are dose limiting and determine the fractionation scheme. In these situations, proton therapy cannot reduce the dose to the relevant structures compared to photon therapy. Thus, increased potential for hypofractionation with protons is limited to treatment sites in which the reduction in integral dose has impact on fractionation decisions. This suggest that tumors embedded in a dose-limiting parallel organ, such as lung or liver tumors, may be eligible for hypofractionation in proton therapy more than in x-ray therapy. This question is investigated by Unkelbach et al. [23] (and also Keller et al. [24] and Gay et al. [25]). In these works, the fractionation model above is generalized to an arbitrary inhomogeneous dose distribution in a parallel OAR, such as the lungs. It is concluded that the BED model does not necessarily support increased use of hypofractionation in proton therapy since proton therapy tends to reduce the volume of normal tissue exposed to low doses, but not the amount of normal tissue in the high-dose region.

22.3.2.2 Nonuniform Spatiotemporal Fractionation Schemes

The rationale for fractionation is the observation that most healthy tissues can tolerate a much higher total dose if the radiation is split into small fractions. On the other hand, fractionation typically requires that a higher total dose is delivered to the tumor in order to achieve the same level of response. Fractionation decisions, therefore, face the trade-off between increasing the number of fractions to spare normal tissues and increasing the total dose to maintain the same level of tumor control. In that regard, the ideal treatment would fractionate in normal tissues, and at the same time hypofractionate in the tumor. This appears to be impossible at first glance because the dose to normal tissues is an unavoidable consequence of delivering dose to the tumor. Generally, increasing the dose to the tumor in a given fraction will increase the dose to healthy tissues in that fraction. However, interestingly, it is possible to achieve some degree of hypofractionation in parts of the tumor while exploiting the fractionation effect in normal tissues. The latter can be achieved by delivering distinct dose distributions in different fractions. The fractions have to be designed such that their dose distributions are similar in normal tissues, but different fractions deliver high single fraction doses to different regions of the tumor. Proton therapy provides one mechanism to realize such dose distributions. The dose in the entrance region of a proton beam is almost independent of its range. This provides some freedom to modify the dose in the tumor without affecting the dose in the entrance region substantially. A proof-of-principle of this concept has been presented in Unkelbach et al. [26]. Further research is needed to identify treatment sites that may clinically benefit from spatiotemporal fractionation schemes.

22.3.3 Dose Painting with IMPT

It has been proposed to modulate the dose to the tumor target volume based on biological variations due to oxygen supply and tumor cell density [27]. These biological variations lead to a variation of the prescription dose. For example, hypoxic areas

Table 22–1 Dose-influence matrix D_{ij} (Gy/gigaproton)

	Voxel *i*			
	1	2	3	4
Beam spot *j* = 1	1.0	0.8	0.7	1.5
Beam spot *j* = 2	0.9	1.0	0.4	1.2

need more dose than well-oxygenated ones. To deliver dose distributions that are matched to the biologically motivated nonuniform prescription is called dose painting. Intuitively, one would expect that IMPT, with its greater flexibility, should be better suited for dose painting delivery than IMRT with x-rays. However, several studies using dose prescriptions based on functional imaging (measuring hypoxia) have found that the dose painting capability of IMPT is actually comparable to that of advanced x-ray therapy, such as tomotherapy [28,29]. This finding can be understood because IMRT is fundamentally limited by only two things: the limited lateral sharpness of the x-ray pencil beam and the inability to deliver negative beam intensities. The former (lateral sharpness) is similar for protons and x-rays, or even slightly better for x-rays. The non-negativity normally affects x-ray therapy more than proton therapy because the proton finite range helps. However, when modulating a dose distribution on top of a uniform "background," as is the case with most dose painting studies, the "background" dose level provides a positive offset of the intensity profiles, and the negativity constraint is typically not active within the projection of the target [30]. The situation could change if one goes to stronger dose painting modulation, with variations in the prescribed dose going down to almost zero.

22.4 Optimization Algorithms

After formulating a treatment plan optimization problem, a mathematical optimization algorithm (also called a solver) has to be applied to find its optimal solution. To facilitate a good understanding of different solvers, we consider a toy optimization problem with just two pencil beams and four voxels. The dose-influence matrix D_{ij} is given in Table 22–1, where voxels 1 and 2 are in an OAR and voxels 3 and 4 are in the target. The objective will be to maximize the mean dose to the target while the maximum dose to the OAR is limited to 1 Gy. Using the general notation introduced in Section 22.1.1, this can be written as

$$\begin{aligned} \underset{x}{\text{maximize}} \quad & \frac{1}{|T|}\sum_{i \in T} d_i(\boldsymbol{x}) \\ \text{subject to} \quad & d_i(\boldsymbol{x}) \le 1 \quad \forall i \in \text{OAR} \\ & \boldsymbol{x} \ge \boldsymbol{0}, \end{aligned}$$

or, using the data from Table 22–1:

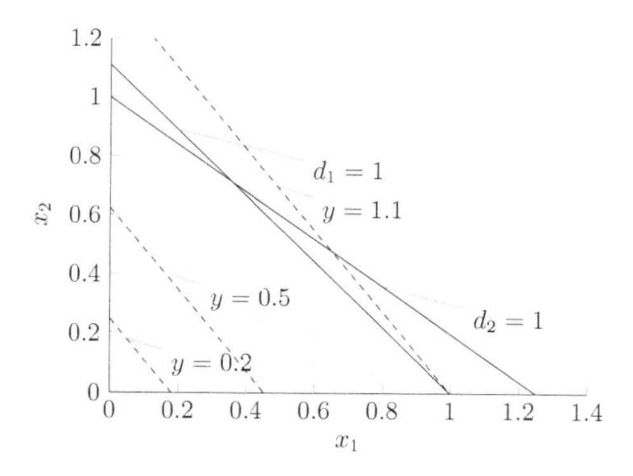

Figure 22–7 The shaded area is the feasible set of optimization problem 22.29, which is enclosed by the axes and the constraints $d_1 \leq 1$ and $d_2 \leq 1$. The dashed lines indicate isolines on which the objective y is constant.

$$\begin{aligned} \max_x \quad & y = 1.1x_1 + 0.8x_2 \\ s.t. \quad & x_1 + 0.9x_2 \leq 1 \\ & 0.8x_1 + x_2 \leq 1 \\ & x_1 \geq 0,\ x_2 \geq 0, \end{aligned} \tag{22.29}$$

The set of treatment plans that satisfy all constraints is called the *feasible region* and is drawn in Figure 22–7. All constraints are linear inequalities, which means that the treatment plans (x_1, x_2) that satisfy the constraints to equality are straight lines in the two-dimensional space. The figure additionally shows the objective function in the form of isolines, i.e., lines where the objective y is constant. Since the objective function is linear in the beamlet intensities, the isolines of the objective function are also straight lines. From the picture it is apparent that $(x_1 = 1, x_2 = 0)$ is the optimal solution with an objective value of 1.1.

Next, we will show how to solve this problem with a projection method and an interior point method. The choice for these particular algorithms is motivated by their relatively low complexity. For further information on gradient descent-based techniques and sequential quadratic programming, which are also used in practice, we refer the interested reader to [31–33]. Although the objective and the constraints of 22.29 are linear in the optimization variables, the methods described here can easily be adapted to the nonlinear case.

22.4.1 Projection Method

The projection method is based on a concept called projections onto convex sets (POCS). In essence, POCS can be used to find a feasible point by iteratively project-

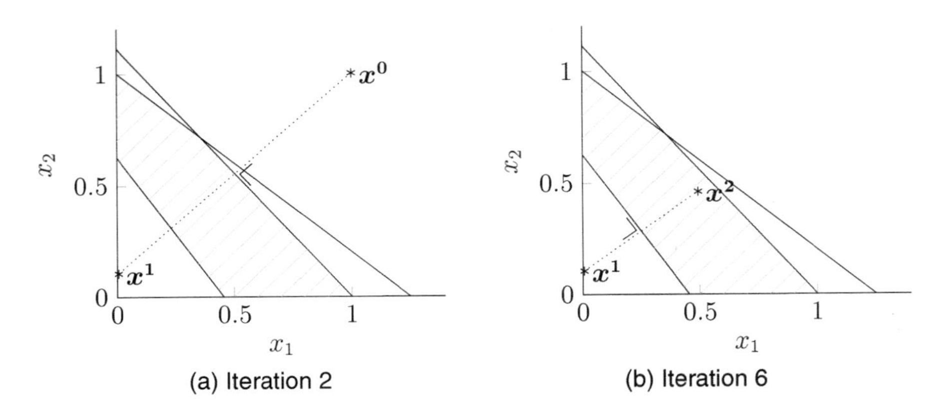

Figure 22–8 Two iterations of a projection method.

ing onto the constraints until all constraints are satisfied [34]. Let us demonstrate the projection method to find a treatment plan for which the mean target dose is at least 0.5 Gy starting from the point $\boldsymbol{x}^0 = (1, 1)$. We iteratively go over the constraint set $\{1.1x_1 + 0.8x_2 \geq 0.5, x_1 + 0.9x_2 \leq 1, 0.8x_1 + x_2 \leq 1, x_1 \geq 0, x_2 \geq 0\}$, one by one. Our initial point $\boldsymbol{x}^0$ satisfies $1.1x_1 + 0.8x_2 \geq 0.5$, so we move on to the next constraint. It does not satisfy $x_1 + 0.9x_2 \leq 1$, and hence, we make a projection to obtain a point that satisfies this constraint. In practice, it turns out to be beneficial to mirror the point in the constraint to obtain the new iterate x_1 (Figure 22–8(a)). Since $\boldsymbol{x}^1$ satisfies $0.8x_1 + x_2 \leq 1$, $x_1 \geq 0$ and $x_2 \geq 0$, we are done iterating over all of the constraints. However, since we have performed at least one projection step, we have to go over all constraints once more to verify that none of them is violated. It turns out that $\boldsymbol{x}^1$ does not satisfy $1.1x_1 + 0.8x_2 \geq 0.5$, so we have to do another projection (see Figure 22–8(b)). The new iterate $\boldsymbol{x}^2$ satisfies all constraints, so we have found a treatment plan for which the mean target dose is at least 0.5 Gy.

To use the projection method for optimization, we do a binary search. Suppose we know that the optimal objective value of 22.29 is somewhere between 0 and 2 Gy, we can invoke the projection method to figure out if the optimal objective value is in [0, 1] or in [1,2] Gy. We do this by verifying if a treatment plan exists for which the mean dose in the target is at least 1 Gy. Such a plan indeed exists and is found by the projection method, so we know that the optimal value is at least 1 Gy, and we therefore continue our search on the interval [1, 2] Gy. We keep narrowing down the search interval until it is sufficiently small. A downside of projection methods is that if they do not find a feasible point after a large number of iterations, it is not possible to tell whether the feasible set is empty or if more projections are necessary.

22.4.2 Interior Point Method

Interior point methods [35] got their name because they iteratively generate a sequence of points in the interior of the feasible region. The optimization problem 22.29 is first transformed into an unconstrained optimization problem with a logarithmic barrier and a positive parameter μ:

$$\max_{x} 1.1x_1 + 0.8x_2 + \mu\{\log(1 - x_1 - 0.9x_2) + \log(1 - 0.8x_1 - x_2) + \log(x_1) + \log(x_2)\}. \tag{22.30}$$

Note that at least one of the logarithmic terms goes to $-\infty$ near the boundary of the feasible set, so indeed these terms serve as a barrier. For each μ, Equation 22.30 has a unique optimal solution $\boldsymbol{x}(\mu)$. The set of $\boldsymbol{x}(\mu)$ is called the central path, which converges to an optimal solution as μ goes to zero (Figure 22–9(a)). Since it is numerically difficult to directly solve Equation 22.30 for small values of μ, typically it is first solved for a large value of μ, and a near optimal solution is used as a starting point for a slightly smaller μ. Since Equation 22.30 does not have constraints, it can be solved with Newton's method, which iteratively maximizes a quadratic Taylor approximation. Recall that the Taylor approximation around a point $\boldsymbol{a}$ is given by:

$$g(\boldsymbol{x}) = f(\boldsymbol{a}) + \nabla f(\boldsymbol{a})(\boldsymbol{x} - \boldsymbol{a}) + \frac{1}{2}(\boldsymbol{x} - \boldsymbol{a})^{\mathrm{T}} \nabla^2 f(\boldsymbol{a})(\boldsymbol{x} - \boldsymbol{a}). \tag{22.31}$$

Since g is a strictly concave function, its maximum is found by setting its derivative to 0: $\nabla f(\boldsymbol{a}) + \nabla^2 f(\boldsymbol{a})(\boldsymbol{x} - \boldsymbol{a}) = 0$. This is a system of linear equalities, and hence, the

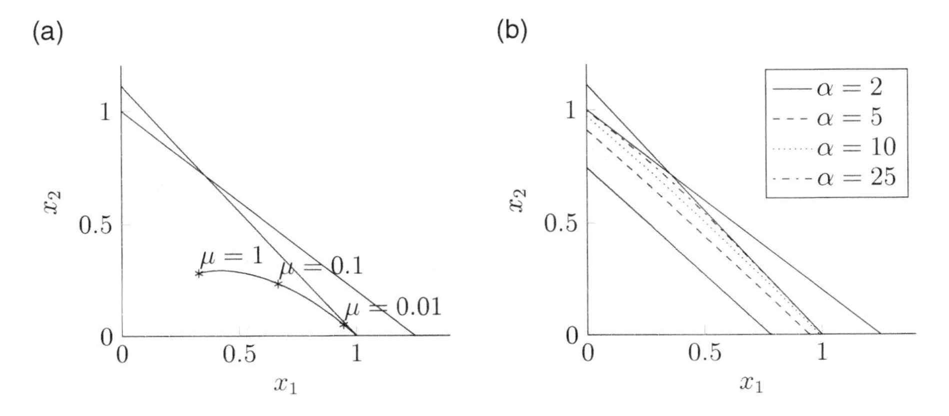

Figure 22–9 (a) The central path runs toward an optimal solution as μ goes to 0. (b) The EUD can be used to approximate a constraint on the maximum dose. The larger the EUD parameter α, the better the approximation.

maximizer $\boldsymbol{x}$ can be found with straightforward linear algebra. Suppose we start with $\boldsymbol{x}^0$ = (0.1, 0.1) and μ = 1, then the maximizer of the Taylor approximation is $\boldsymbol{x}^1$ = (0.1847, 0.1803). Repeating Newton's method three more times yields $\boldsymbol{x}^4$ = (0.3281, 0.2819). By alternatingly reducing μ and making a few steps with Newton's method, the solution converges to (1, 0). Interior point methods can find an optimal solution in around 50 iterations of Newton's method. Problems with 10,000 pencil beams can be solved in less than an hour, which makes the method suitable for larger spot sizes. When the number of pencil beams is larger, solving the linear system becomes a bottleneck.

22.4.3 Convex Optimization

The formulation 22.29 contains a linear objective and linear constraints. Since there is one linear constraint for each voxel in the OAR to restrict the maximum dose, the number of constraints is typically of the order of 10^4–10^5. To reduce the number of constraints and to allow for more flexibility in the problem formulation, one could use nonlinear functions to formulate a problem. For example, the maximum dose constraint can be formulated as a single constraint on the EUD with a large parameter α:

$$((x_1 + 0.9x_2)^\alpha + (0.8x_1 + x_2)^\alpha)^{1/\alpha} - 1 \le 0. \tag{22.32}$$

This constraint is depicted in Figure 22–9(b) for different values of α and approaches the original two maximum dose constraints as $\alpha \rightarrow \infty$. As long as the objective is to minimize a convex function (or to maximize a concave function) and the constraints can be written as $f_i(\boldsymbol{x}) \le 0$, where f_i is convex, there is a good amount of theory on finding an optimal solution. Solvers for convex optimization are complex, but the general underlying ideas can be understood by generalizing the discussed algorithms.

For the projection method, in each iteration we can linearize the corresponding constraint [36]. The remainder of the algorithm is then the same as in the linear case.

For the interior point method, we have to solve the unconstrained problem 22.30, which is already nonlinear due to the logarithmic barrier terms. We can therefore still solve optimization problem 22.30 with Newton's method when the objective and constraint functions are nonlinear. However, the scalability issue when the number of pencil beams is large still exists. To overcome this, extensions have been developed that rely on advanced linear algebra and low-cost approximations of second derivatives.

22.4.4 Future Challenges

The scale of the optimization problems increases continuously over time due to the use of higher resolution voxel grids, the inclusion of multiple scenarios for robust optimization, and the commissioning of proton beams with smaller spot sizes. Advances in computing power alone may not meet the expectations, and methods that efficiently use parallel computing still have to be developed.

References

1. Shepard DM, Ferris MC, Olivera GH, Mackie TR. Optimizing the delivery of radiation therapy to cancer patients. *SIAM Review* 1999;41:721–744.
2. Bortfeld T. IMRT: a review and preview. *Phys Med Biol* 2006;13:R363–397.
3. McQuaid D, Unkelbach J, Trofimov A, Bortfeld T. Robust optimization. In Palta J and Mackie TR, Eds. Uncertainties in external beam radiation therapy. Madison, WI: Medical Physics Publishing, 2011.
4. Trofimov A, Unkelbach J, DeLaney TF, Bortfeld T. Visualization of a variety of possible dosimetric outcomes in radiation therapy using dose-volume histogram bands. *Pract Radia Oncol* 2012;3:164–171.
5. Unkelbach J, Bortfeld T, Martin BC, Soukup M. Reducing the sensitivity of IMPT treatment plans to setup errors and range uncertainties via probabilistic treatment planning. *Med Phys* 2009;36(1):149–163, 2009.
6. Fredriksson A, Forsgren A, Härdemark B. Minimax optimization for handling range and setup uncertainties in proton therapy. *Med Phys* 2011;38(3):1672–1684.
7. Pflugfelder D, Wilkens JJ, Oelfke U. Worst case optimization: a method to account for uncertainties in the optimization of intensity modulated proton therapy. *Phys Med Biol* 2008;53(6)1689–1700.
8. Fredriksson A. A characterization of robust radiation therapy treatment planning methods—from expected value to worst case optimization. *Med Phys* 2012;39(8):5169–5181.
9. Wilkens J, Alaly J, Zakarian K, Thorstad W, et al. IMRT treatment planning based on prioritizing prescription goals. *Phys Med Biol* 2007;52(6):1675–1692.
10. Falkinger M, Schell S, Müller J, Wilkens J. Prioritized optimization in intensity modulated proton therapy. *Zeitschrift für Medizinische Physik* 2011;22(1):21–28.
11. Jee KW, McShan D, Fraass B. Lexicographic ordering: intuitive multicriteria optimization for IMRT. *Phys Med Biol* 2007;52(7):1845–1861.
12. Clark V, Chen Y, Wilkens JJ, Alaly J, et al. IMRT treatment planning for prostate cancer using prioritized prescription optimization and mean-tail-dose functions. *Linear Algebra Appl* 2008;428(5):1345–1364.
13. Breedveld S, Storchi P, Voet P, Heijmen B. iCycle: integrated, multicriterial beam angle, and profile optimization for generation of coplanar and noncoplanar IMRT plans. *Med Phys* 2012;39:951.
14. Craft D, Bortfeld T. How many plans are needed in an IMRT multi-objective plan database? *Phys Med Biol* 2008;53(11):2785–2796.
15. Spalke T, Craft D, Bortfeld T. Analyzing the main trade-offs in multiobjective radiation therapy treatment planning databases. *Phys Med Biol* 2009;54(12)3741–3754.
16. Rennen G, Van Dam ER, Den Hertog D. Enhancement of sandwich algorithms for approximating higher dimensional convex Pareto sets. *INFORMS J Computing* 2011;23(4):493–517.
17. Bokrantz R, Forsgren A. An algorithm for approximating convex Pareto surfaces based on dual techniques. *INFORMS J Computing* 2012;25(2):377–393.
18. Wilkens JJ, Oelfke U. Analytical linear energy transfer calculations for proton therapy. *Med Phys* 2003;30(5)806–815.
19. Wilkens JJ, Oelfke U. Optimization of radiobiological effects in intensity modulated proton therapy. *Med Phys* 2005;32(2)455–465.
20. Carabe-Fernandez A, Dale RG, Hopewell JW, Jones B, et al. Fractionation effects in particle radiotherapy: implications for hypo-fractionation regimes. *Phys Med Biol* 2010; 55;(19)5685–5700.
21. Giantsoudi D, Grassberger C, Craft D, Niemierko A, et al. Linear energy transfer-guided optimization in intensity modulated proton therapy: Feasibility study and clinical potential. *Int J Radiat Oncol Biol Phys* 2013;87(1):216–222.
22. Mizuta M, Takao S, Date H, Kishimoto N, et al. A mathematical study to select fractionation regimen based on physical dose distribution and the linear-quadratic model. *Int J Radiat Oncol Biol Phys* 2012;84(3):829–833.
23. Unkelbach J, Craft D, Salari E, Ramakrishnan J, et al. The dependence of optimal fractionation schemes on the spatial dose distribution. *Phys Med Biol* 2013;58(1):159–167.

24. Keller H, Hope A, Meier G, Davison M. A novel dose-volume metric for optimizing therapeutic ratio through fractionation: Retrospective analysis of lung cancer treatments. *Med Phys* 2013;40(8)084101.
25. Gay HA, Jin JY, Chang AJ, Ten Haken RK. Utility of normal tissue-to-tumor α/β ratio when evaluating isodoses of isoeffective radiation therapy treatment plans. *Int J Radiat Oncol Biol Phys* 2013;85(1):e81–87.
26. Unkelbach J, Zeng C, Engelsman M. Simultaneous optimization of dose distributions and fractionation schemes in particle radiotherapy. *Med Phys* 2013;40(9):091702.
27. Ling CC, Humm J, Larson S, Amols H, et al. Towards multidimensional radiotherapy (MD-CRT): biological imaging and biological conformality. *Int J Radiat Oncol Biol Phys* 2000;47(3):551–560.
28. Thorwarth D, Soukup M, Alber M. Dose painting with IMPT, helical tomotherapy and IMXT: A dosimetric comparison. *Radiother Oncol* 2008;86(1):30–34.
29. Flynn RT, Bowen SR, Bentzen SM, Mackie TR, et al. Intensity-modulated x-ray (IMXT) versus proton (IMPT) therapy for theragnostic hypoxia-based dose painting. *Phys Med Biol* 2008;53(15):4153–4167.
30. Bortfeld TR, Boyer AL. The exponential radon transform and projection filtering in radiotherapy planning. *Int J Radiat Oncol Biol Phys* 1995;6(1):62–70.
31. Hedar AR, Fukushima M. Derivative-free filter simulated annealing method for constrained continuous global optimization. *J Global Optimization* 2006;35(4):521–549.
32. Lasdon LS, Waren AD, Jain A, Ratner M. Design and testing of a generalized reduced gradient code for nonlinear programming. *ACM Trans Mathematical Software* 1978;4(1):34–50.
33. Nocedal J, Wright SJ. Numerical optimization. Springer, 2006.
34. Chen W, Craft D, Madden TM, Zhang K, et al. A fast optimization algorithm for multicriteria intensity modulated proton therapy planning. *Med Phys* 2010;37(9):4938–4945.
35. Boyd SP, Vandenberghe L. Convex optimization. Cambridge University Press, 2004.
36. Gibali A, Küfer KH, Süss P. A generalized projection-based scheme for solving convex constrained optimization problems. 2014. Working paper.

Chapter 23

Treatment Planning for Passive Scattering Proton Therapy

Heng Li, Ph.D.[1], Annelise Giebeler, Ph.D.[2], Lei Dong, Ph.D.[2], Xiaodong Zhang, Ph.D.[1], Falk Poenisch, Ph.D.[1], Narayan Sahoo, Ph.D.[1], Michael T. Gillin, Ph.D.[1], and X. Ronald Zhu, Ph.D.[1]

[1]The University of Texas MD Anderson Cancer Center
Houston, TX
[2]Scripps Proton Therapy Center
San Diego, CA

23.1 Introduction

The typical prescription in proton beam therapy is to deliver a uniform radiation dose to the target volume. To achieve that goal, a narrow monoenergetic proton beam from the accelerator is spread uniformly over a large volume. There are two different techniques implemented to spread the proton beam: passive scattering [1–3] and uniform scanning [4]. Usually no distinction is made between them for treatment planning purposes [5], as they both produce uniform dose distributions that can be characterized by the proton range, field size, and modulation width of the Spread-out Bragg Peak (SOBP).

It is usually impossible to generate an acceptable dose distribution using a single beam in photon radiotherapy. However, this is not the case for proton radiotherapy. One of the unique aspects of proton beam is that a single beam can be used to cover the entire target with a uniform dose distribution by designing the proton beam with beam-specific parameters and apparatus, including apertures and compensators. When multiple beams are used, each beam is also independently designed to cover the target volume entirely. This is because passive scattering and uniform scanning are a form of three-dimensional conformal therapy using particle beams. The exception only occurs when no single beam direction can be used to cover the entire target volume, in which case patch fields (Section 23.7.1), where two uniform fields are matched using lateral and distal penumbras, respectively, could be considered. Regardless, in all cases, it is highly recommended to review the dose distribution for individual beam, as each beam should contribute uniform dose, even when multiple beams or patch fields are used.

This chapter presents the treatment planning considerations for a spread-out proton beam. In this respect, the chapter first discusses some basic concepts and considerations for proton beam treatment planning. Next, it covers topics such as margin considerations and beam-shaping parameters related to apertures and compensators. Then it discusses the robustness and potential limitations of passive scattering/uniform scanning plans, followed by a final section that presents case studies and special considerations. The details of passive scattering/uniform scanning delivery systems are described in chapters 6 and 8, and monitor unit calculation of treatment plans is covered in Chapter 26.

23.2 Treatment Planning Considerations for Passive Scattering or Uniform Scanning Systems

23.2.1 Target Definition

In clinical practice, the goal of radiotherapy is to kill tumor cells by delivering a uniform therapeutic dose to the target volume, as prescribed by the treating radiation oncologist, while limiting dose to normal tissue using dose–volume constraints, which are established during treatment planning. Treatment planning also takes into consideration internal variation and setup uncertainties that may occur during the treatment.

Several International Commission on Radiation Units and Measurements (ICRU) reports [6-8] defined important anatomic volumes—including the gross target volume (GTV), clinical target volume (CTV), internal target volume (ITV), and planning target volume (PTV) for tumors—and defined OAR and planning organs at risk volume (PRV) for normal tissues. Except for PTV and PRV, the delineation of these volumes is independent of specific treatment modality, and these volumes have been used across all modalities, including proton therapy. ICRU 78 emphasizes that the radiation modality should not affect the delineation of these volumes since "(i) their definitions are modality independent; and (ii) one might wish to combine or compare or retrospectively analyze treatment plans for more than one modality" [8].

PTV and PRV were initially introduced for photon radiotherapy treatment planning to account for internal variation and setup uncertainties and to design lateral beam margins to cover the target volume. Depth–dose distributions do not change significantly, owing to the setup uncertainty in the beam direction for all external beam modalities, including photon, electron, and proton [9,10]. However, for proton radiotherapy, in addition to the setup uncertainty, there is also range uncertainty in the depth direction that should be considered on a beam-by-beam (BEV) basis. Therefore, in principle, one would need a separate beam-specific PTV to design the proton beam margins for the lateral and depth directions.

23.2.2 Patient Support Devices

Patient support devices, including treatment couches and immobilization devices, are an important component in all forms of radiotherapy. These devices ensure that the patient stays in the same position throughout the course of treatment, and they allow for accurate and precise delivery of radiation dose to the patient. These devices are particularly important in proton radiotherapy, as dose distributions in proton radiotherapy are usually considered more sensitive to inter-fractional changes in patient position or to intra-fractional patient motion than photon radiotherapy, and effective immobilization devices could reduce these changes (see Chapter 18). Additionally, for proton radiotherapy, the patient supporting devices in the beam path act as range shifters; if these devices are not included in the dose calculation or are not modeled correctly for a proton beam, significant dose error could occur at the end of the proton range for that particular beam [11].

To calculate patient dose correctly with beams that pass through a patient support device, the treatment planning system (TPS) has to identify the radiologic path length of the device. Since the material composition, and thus relative stopping power (RSP), of the device is usually different from the tissue materials used in the CT number calibration process, potential errors could occur in the stopping power calculation if the same calibration curve is used. To address this problem in the TPS, the device structure in the planning CT is digitally replaced by a full CT scan of the device with assigned CT numbers corresponding to the correct RSP of the device. The TPS then calculates dose using the proton stopping power inferred from the assigned CT numbers. The calculation should be validated with measurements during the commissioning of the TPS/supporting device. The entire supporting device should be included in the planning CT scan to calculate the radiologic path length and dose correctly.

When each beam passes through the treatment couch or patient immobilization devices, the device water-equivalent thickness (WET) should be taken into account for beam line calculation and compensator design [12,13]. In addition, the uncertainty of WET of the devices, especially near sharp edges, owing to CT number calibration and patient setup, should also be included in the calculations. This practice could minimize the dosimetric effects of such uncertainties on the target coverage [14]. Rounded edges are preferred for the couch top to minimize these effects.

23.2.3 Beam Angle Selection

For passive scattering and uniform scanning proton radiotherapy, the beam angle selection plays a vital role in treatment planning. Although proton radiotherapy shares many treatment-planning considerations with photon conformal radiotherapy, the finite range of proton beams offers another dimension of freedom in terms of geometrically sparing OARs. In addition, to minimize low-dose exposure to normal tissues, proton therapy typically uses fewer beams than photon therapy; therefore, it is important to select beam angles more carefully. In principle, fewer proton beams lead to less conformal dose distribution, less robustness, and less integral dose to patient, whereas more beams lead to better high-dose conformity and improved robustness, but more integral dose to patient. In practice, usually one to four beams are used per isocenter.

Both geometrical and physical factors should be considered during beam angle selection. An important and obvious consideration is geometrical OAR sparing. For proton radiotherapy, both the steep distal dose falloff and the sharp penumbra from the aperture could be used to shape the proton beam and spare normal tissues. Although both the distal dose falloff and the lateral penumbra are subject to the setup uncertainty, the distal dose falloff of the proton beam is also subject to range uncertainty, which increases with the proton range. Therefore, although it is reasonable to protect normal tissues using a proton beam that stops before reaching the critical organ, it is usually not recommended to design a proton beam that stops just in front of the critical structure. In other words, there should be enough distal separation between the target and the critical structure to allow for potential range uncertainties. Details on range uncertainties have been provided in Chapter 21.

For proton therapy, it is usually considered advantageous to use beam angles that take the shortest radiologic path length from the surface of the patient to the target to minimize both the integral dose to the patient and the range uncertainty. The beam direction should avoid abrupt changes in the proton path length, as these changes could lead to less robust treatment plans that are sensitive to setup uncertainties. Examples of such abrupt changes are beams that run parallel the skin surface, that travel through high-density materials in the patient (such as titanium rods), or that travel through the edges of immobilization devices. When abrupt path length changes are unavoidable, multiple beam directions should be considered to mitigate possible dosimetric uncertainties. The BEV features in the TPS are often helpful for visualizing the geometrical features of the tumor and OAR or other structures, such as high-density inserts.

The combination of gantries and robotic treatment couches allows almost any incident beam direction for a given patient. However, in order to reduce beam penumbra, the treatment snout is placed as close as possible to the patient's skin in proton treatment, which poses a challenge for collision checks between the treatment snout and the patient. Considering the size of state-of-the-art proton gantries and the dosimetric consequences of last-minute changes to treatment parameters, it is highly recommended to incorporate some sort of collision check in the treatment planning process.

23.3 Margin-based Treatment Planning

This section presents the classic technique for proton treatment planning, where the proton beamline, aperture, and compensator are designed for individual beams to generate a uniform dose distribution that conforms to the target. Conformity of the proton beam in the lateral direction is achieved using custom-milled apertures or multileaf collimators, whereas conformity in the depth direction is achieved using the range of the proton beam and custom-made compensators. To account for physical and patient-specific uncertainties (discussed in detail in Chapter 21), margins are included in the beamline, aperture, and compensator designs.

23.3.1 Lateral Conformity with Apertures

Apertures are used to avoid undesired exposure to normal tissues from the proton beam. In the TPS, apertures are shaped to follow the target in the BEV, with setup margins (SM) to account for the setup uncertainty and dosimetric margins to account for the physical and geometrical penumbra. The dosimetric margin could be established from the distance between the 50% isodose level (the field edge) and the desired isodose coverage (typically 90% or 95%), and it is a function of proton energy and snout position (aperture to isocenter distance). Internal margin and SM depend upon patients and treatment guidance techniques and can be determined using the same techniques used in photon beam therapy (for example, using 4DCT and the margin formula proposed by van Herk [15]). Currently, two approaches are implemented in different centers:

1. using the CTV directly (without using a PTV as an intermediary step), where the lateral margin from the CTV includes both the SM and the dosimetric margin (Figure 23–1a); and
2. using the PTV, which is the uniform expansion of the CTV with SM. In this case the lateral margins are equal to the dosimetric margin (Figure 23–1b).

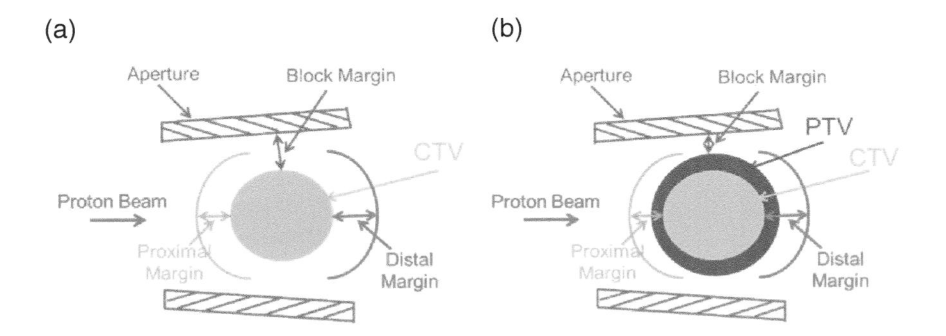

Figure 23–1 Different approaches for proton beam margin design. (a) Using CTV directly without PTV and (b) using PTV for lateral margin design. In both approaches, CTV is used to calculate distal and proximal margins.

In both approaches, the distal margins (DM) and proximal margins (PM) are calculated from the CTV. These two methods are equivalent if the proper margins are used in each approach. Advantages of using the PTV include the visual guidance to assess the lateral coverage of each field and record and report dose distribution (ICRU-78), especially for the purposes of comparison with photon plans. To spare OARs for a given beam angle, the shape of the aperture sometimes needs to be edited in addition to the simple geometric expansion of the target in the BEV. One of the limitations of both approaches is that the DM and PM are calculated for the maximum ranges. For disease sites with large inhomogeneity, this "one size fits all" approach may use a larger margin than necessary for part of the target volume. The beam-specific PTV (bsPTV) approach [16], which has not yet been implemented in most commercial planning systems, was proposed to overcome this limitation. It is discussed in detail in Section 23.3.4.

23.3.2 Beamline Design

The finite range of proton radiotherapy offers the unique opportunity to spare OAR. However, many uncertainties are associated with calculating the proton beamline, which is essentially the combination of the range and SOBP of the proton beam. Therefore, PM and DM are required to cover the target during treatment planning and delivery. The DM is particularly important because an error in range, or the distal end of the proton beam, could result in a dose error of up to 100% of the beam dose to the target. The distal range of the proton beam is chosen to ensure distal target coverage with the prescribed dose for that particular beam, and it is typically determined by calculating the WET or accumulated relative linear stopping power of the beam contributed from everything in the beam path to the most distal point of the CTV. The maximum range needed to cover the distal edge of the target is chosen as the proton beam range (R_d) and is used to determine the DM, and the minimum range needed to cover the proximal edge of the target (R_p) is used to determine the PM. Both DM and PM are range-dependent, and currently 3.5% of the range plus 1–3 mm (this number differs at different institutions; see Chapter 21) are used [5,13]. The 1–3 mm additional margin is to cover the uncertainties in energy/range of the proton beam, the WET of the manufactured compensators, the estimation of WET for immobilization, and, in some degree, for the uncertainties of dose calculation algorithms. Therefore, the range required for a given beam in centimeters is

$$R = R_d + DM = 1.035\, R_d + 0.3\ (cm) \tag{23.1}$$

The SOBP required for the beam in centimeters is

$$\begin{aligned} SOBP &= \left(R_d - R_p\right) + PM + DM \\ &= 1.035\, R_d - 0.965\, R_p + 0.3\ (cm) \end{aligned} \tag{23.2}$$

The SOBP covers from the most distal part to the most proximal part of the target. Normally, conformity in the proximal end of the target cannot be achieved with passive scattering. Additionally, if the range compensator has a minimum thickness (for milling purposes, usually 1–2 mm), this value should be added to the ranges of the beamline calculation. When considering range uncertainty, note that the proton range is increased beyond the target in the distal end, and the SOBP is pulled toward the surface of the patient to ensure target coverage. Also note that the proton beam could overshoot or undershoot compared with the calculated range in the TPS, and the DM is designed to ensure tumor coverage even if the proton beam undershoots in the patient (the PM protects against overshooting). However, if the proton beam does overshoot or undershoot, using the prescribed DM or PM, respectively, would irradiate more normal tissue. Therefore, efforts should be made to minimize the range uncertainty and, hence, the DM and PM to fully take advantage of the proton beam.

23.3.3 Distal Conformity with Compensators

Properly designed beamlines ensure the distal and proximal coverage of the CTV, even with range uncertainties. However, with a spread-out proton beam, such as in passive scattering and uniform scanning delivery, the proton dose distribution does not conform to the distal end of the CTV. To further protect OAR distal to the target, compensators are used to conform the distal end of the proton beam to the CTV with DM by pulling back the distal end of the proton beam [17].

The TPS creates compensators by making a three-dimensional model of the range compensator to calculate the dose distribution in the patient. The compensator is beam specific and is easier to define in the BEV coordinate system, where the thickness of the range compensator of each ray is calculated. A simple method to calculate the thickness is to calculate the difference of the water-equivalent thicknesses between the prescribed proton beam range (R) and the proton beam range required to cover the distal surface of the target volume along a given proton ray line by using a ray-tracing algorithm:

$$RC_i = R - R_i\,(cm) \tag{23.3}$$

Smearing (Figure 23–2b) accounts for setup uncertainty and is an important aspect of compensator design. The smearing process could be described as follows. After the desired thickness of the range compensator at each point across the calculation grid is calculated with a ray-tracing algorithm, circles with a predefined smearing radius (SR) are superimposed over the calculation grid and centered at each point. The thickness of the compensator is reduced within each circle to the minimum thickness at any point encompassed by the circle. The SR could be calculated as described in [13]:

$$SR = \sqrt{(IM + SM)^2 + (0.03R)^2} \tag{23.4}$$

where the first term accounts for internal motion and setup uncertainty and the second term ensures lateral proton scatter is achieved [13,17,18].

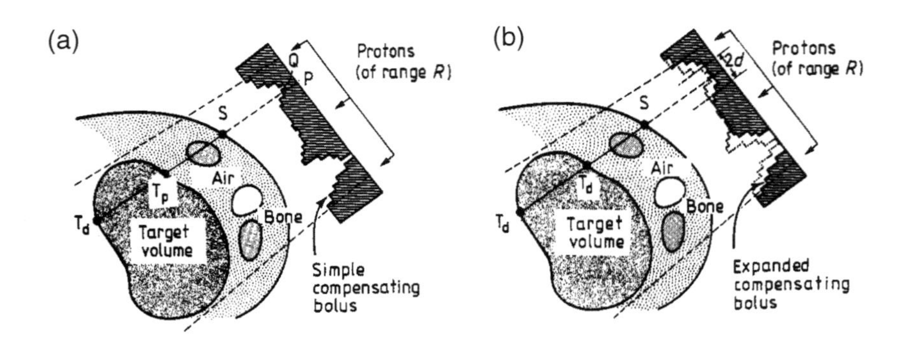

Figure 23–2 Design of a compensator for protons of range R to compensate for surface and target irregularities and tissue heterogeneities. (a) Sample line ray along which a line integral is performed to obtain the water-equivalent areal density between the skin (S) and the proximal (T_p) and distal (T_d) target volume surfaces. (b) Expansion of the compensator to ensure target volume treatment within positioning and motion uncertainties of distance d. (From Figure 1 of reference [17] with permission.)

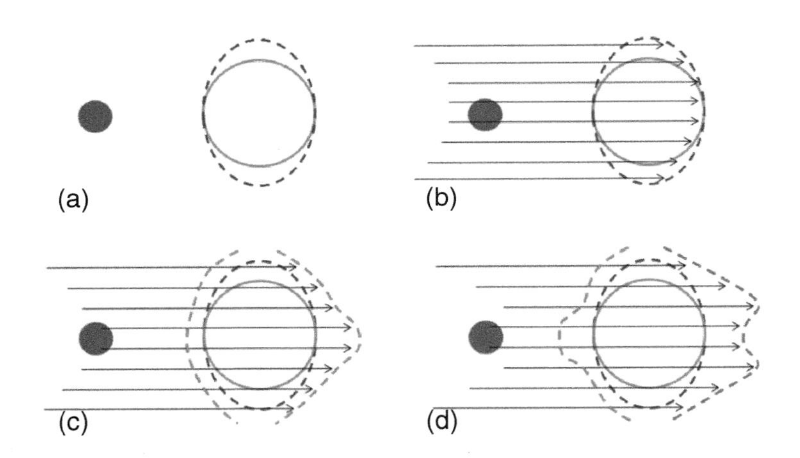

Figure 23–3 An illustration of the four essential steps in creating the bsPTV (red contour) from a CTV (green contour) with a dense object (gray sphere) along the beam path. (a) The CTV is expanded laterally away from the beam axis using the expected motion margin (IM) and setup margin (SM). (b) From a given beam angle, ray tracing is performed to calculate the radiologic path length of each ray from the source to both the distal and proximal surface of the laterally expanded CTV (blue contour). (c) The fraction of the total radiologic range calculated in the previous step is used to determine the distal margins per ray. (d) The interplay effect of setup and range error is corrected by applying the correction kernel, and the radiologic path length margins are converted to physical depth margins. (From Figure 2 of reference [16] with permission.)

23.3.4 Smearing and Smearing Radius

Smearing is an effective technique to maintain dose coverage to the target with setup uncertainty (shifts of the entire patient with magnitude up to SR). However, the smearing process reduces the thickness of the compensator except at the thinnest thickness and results in dose overshooting in the final dose distribution. Improved patient alignment minimizes the need of smearing and reduces undesired overshooting in the healthy tissue.

Conventionally, a computerized milling machine is used to drill the compensator, which has an inherent smearing radius due to the finite size of the drill bit, and which is not usually taken into account by the TPS. A different manufacturing technique has been reported to produce compensators with smooth surfaces that might improve accuracy and generate more conformal plans than drilled compensators [19].

23.3.5 Beam-specific PTV

The bsPTV approach [16,20] generates a beam-specific volume using a ray-tracing technique to account for setup and range uncertainties, and it is more effective than the conventional planning technique (described above) when inhomogeneity is present in the beam path. Figure 23–3 illustrates the essential steps for creating the bsPTV. It is then used as the planning target for the specific beam angle.

23.4 Robustness of Passive Scattering Plans

Robustness in proton radiotherapy can be loosely viewed as the sensitivity of dose distribution to various uncertainties, including setup uncertainty, range uncertainty, and patient anatomy changes. Robustness can usually be quantified by modeling the change of dose distribution owing to these variations. Passive scattering and uniform scanning deliveries are usually considered more robust than intensity-modulated proton therapy (IMPT) because most uncertainties are explicitly included in the beam design. Indeed, robustness evaluation demonstrates that typical passively scattered proton therapy (PSPT) plans are highly insensitive to setup and range uncertainties. However, PSPT treatment plans do not guarantee robustness, for example, when a patient's anatomy is not accurately reflected in the CT images used for the plan design or dose calculation.

Figure 23–4 shows an example of a lung patient with a prescription for 45 Gy to CTV (blue contour). Figure 23–4a and 4b show a three-field and a single-field passive scattering plan, respectively. Figure 23–4c and 4d show the DVH presentation of the worst-case robustness evaluation results for the two plans, respectively, using 3 mm setup and 3.5% range uncertainties. The single-field plan (Figure 23–4b and 4d) shoots through the humeral head and stops close to the spinal cord, which results in poor robustness in both target coverage (blue lines, uncertainty curves show a potential reduction of 20% in prescription coverage) and spinal cord dose (red lines, large spread of the uncertainty curves). The three-field plan, on the other hand, is very robust. Patch fields, or aperture/compensator editing to spare critical structures, could also result in less robust plans.

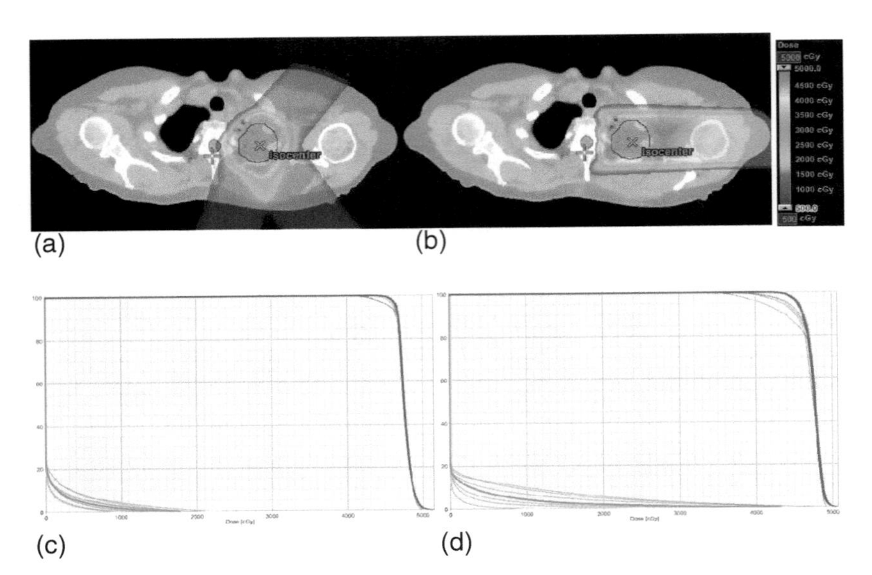

Figure 23–4 Robustness of different passive scattering plans. Color wash dose distribution of two treatment plans using the techniques described in this chapter. (a) A three-beam plan and (b) a single lateral beam plan are shown for the same patient. The DVH presentation of the worst-case robustness evaluation of the two plans using 3 mm setup uncertainty and 3.5% range uncertainty are shown in (c) and (d), respectively. The blue and red solid line are the nominal DVH of the ITV and the spinal cord, respectively. The dash lines are the DVHs with uncertainties.

An alternative planning technique has therefore been proposed in which the robustness of the treatment plan is used to guide the treatment planning parameters, such as setup margin, PM, DM, and SR [21]. The treatment planning parameters are tuned to minimize dose to normal tissues while maintaining acceptable robustness.

23.5 Plan Evaluation and Dose Reporting

Treatment plans, which estimate the delivered dose to the patient, are important decision-making tools for radiation oncologists. While the process of evaluating a proton plan is similar to evaluating a photon treatment plan, it is critical to understand the uncertainties associated with proton therapy and incorporate these considerations in the planning and decision-making process.

For classic PSPT treatment planning, the beamline, aperture, and compensator design processes are field-specific and, in theory, lead to uniform dose distribution for each field in the plan. Therefore, as mentioned above, field dose evaluation is an important component for plan evaluation. Checklists should include:

1. checking the entrance path of the beam or objects in the beam path between the snout and the patient (e.g., treatment couch or immobilization devices),

2. finding density heterogeneities along the beam path and, if possible, avoiding them, and
3. verifying the proximal, distal, and lateral coverage of each beam.

As mentioned above, although the robustness of PSPT plans is usually very good, robustness evaluation may be important for some patients for whom aperture blocks or compensator edits (e.g., distal blocks) are used in the treatment planning process, and when patch fields are used.

23.6 Limitations of Passive Scattering or Uniform Scanning Techniques

One of the major limitations of passive scattering or uniform scanning techniques is the inability to conform proximal dose. In passive or uniform scanning beams, the proton beam is spread and modulated to a fixed width (SOBP) over the entire field, so the proximal contribution from each contributing Bragg peak additively increases proximal dose. Therefore, if any portion of the tumor is deep within the patient (tumor size can vary in the depth direction across the field), a large SOBP would have to be used to cover the entire tumor, and some normal tissue along the beam path could receive an unnecessarily high dose (Figure 23–4a, posterior region of the patient). Passive and uniform scanning beams also require patient-specific hardware (apertures and compensators) and are less flexible than intensity-modulated proton therapy.

23.7 Examples

23.7.1 Patch Fields

Patch field planning is a technique that can be used when a target structure wraps or folds around a dose-limiting structure (OAR), such that any approach to the target results in an over-tolerance dose to the OAR. As shown in Figure 23–5 from reference

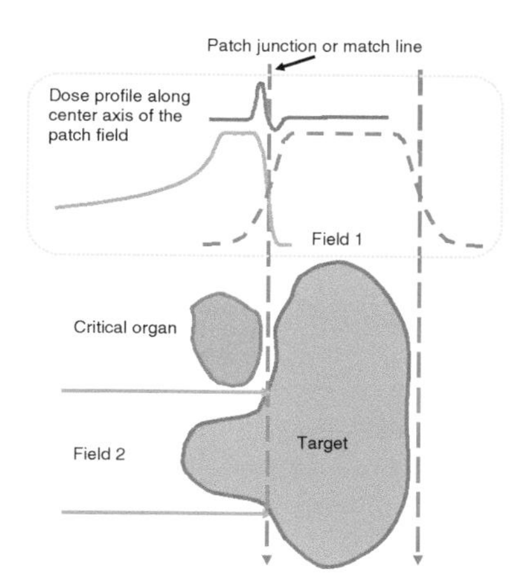

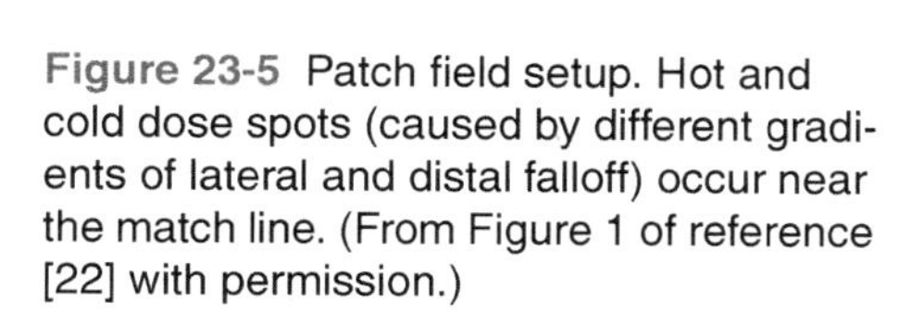
Figure 23-5 Patch field setup. Hot and cold dose spots (caused by different gradients of lateral and distal falloff) occur near the match line. (From Figure 1 of reference [22] with permission.)

[22], this technique typically involves two treatment fields called the "shoot-through" field (field 1) and the "patch" field (field 2). These fields cover the target in a piecewise fashion and use aperture edges as physical barriers to reduce dose to nearby OAR. The junction between the treatment fields is defined by the lateral penumbra (50%) from the shoot-through field and by the distal penumbra (50%) from the patch field. However, the respective distal and lateral penumbras do not always immediately match, as they can represent very different ranges/energies depending on their path to the target. As a result, cold and hot spots can occur near the match line, as shown in Figure 23–5. Thus, to reduce dosimetric uncertainty at the junction location, patch plans are designed with multiple alternative junctions to feather or smear out the uncertainty.

The patch field planning process is complicated and often highly iterative. The first step is to map beam angles and the corresponding locations of the patch junctions, which are always located within the target volume. The angle between the patch and shoot-through fields should be kept equal or greater than 90 degrees in order to avoid cold regions.

Once the appropriate angles are set, apertures and compensators for the two fields are then independently designed to avoid the OAR. Typically, the shoot-through field is designed first, beginning by editing its aperture edge off the OAR. Then, it is optimized as a stand-alone field to provide homogeneous coverage of the portion of the target in its path. Once the shoot-through field is optimized, its lateral 50% isodose line is used to create a field-specific target volume for the patch field. The distal boundary of the patch field's target corresponds to the lateral 50% of the shoot-through field. Compensator edits are often used to conform the distal shape of the patch field to reduce any hot or cold regions in the combined plan. As mentioned above, multiple junctions are used to reduce dosimetric uncertainties in the plan because individual junction regions can receive up to 115% to 120% of the prescription dose, depending on the prescribed coverage.

Figure 23–6 shows an example of treating a C-shaped chordoma tumor (blue contour) close to the spinal cord (green contour) using two pairs of patch fields. Using two junctions reduces the hot junction lines and improves the robustness of the plan. Note that the two junction pairs use the same beam angles in this example.

Because patch fields are designed with piecewise coverage, they are not as robust as other passive scattering plans. Therefore, when patch fields are used, robust evaluations are recommended to understand the possible dosimetric consequences of uncertainties.

23.7.2 Craniospinal Irradiation

Craniospinal irradiation (CSI) is a radiation technique used to treat central nervous system malignancies that have a substantial risk of subarachnoid spread [23]. CSI is part of the standard of care for medulloblastoma, the most common brain cancer in children, and could be used for other diagnoses, including primitive neuroectodermal tumors, atypical teratoid rhabdoid tumors, central nervous system germ cell tumors,

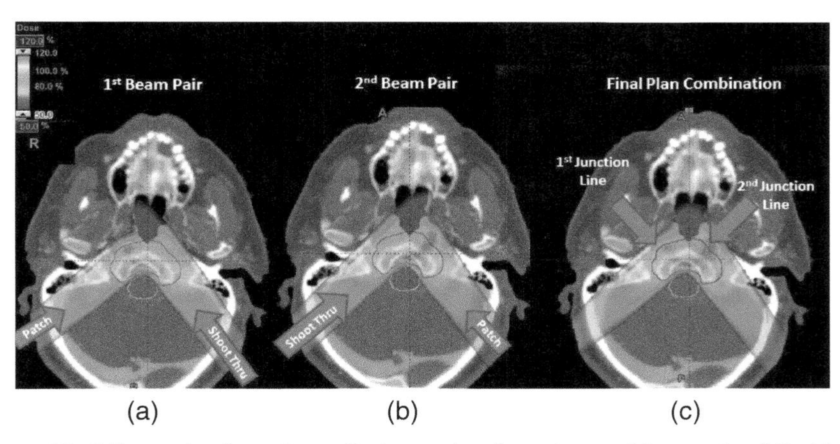

Figure 23–6 Example of treating a C-shape chordoma tumor (blue contour) that is close to the spinal cord (green contour) using the patch field technique. Two pairs of patch fields are designed using the same beam angles, and the individual and total dose distributions are shown in (a), (b), and (c), respectively.

or any tumor that has evidence of cerebrospinal fluid spread diagnosed by a lumbar puncture or a magnetic resonance myelogram.

However, many toxicities are associated with conventional x-ray CSI (xCSI) [24–28]. Proton-based CSI (pCSI) has been studied and increasingly used over the past 15 years, as the lack of an exit dose of the proton beam could reduce normal tissue dose and toxicities. Several studies of dosimetric evaluations and clinical outcomes have reported that pCSI provides equal tumor control and is associated with fewer radiotherapy-related toxicities [29–31].

Giebeler et al. [32] reported a detailed standardized treatment planning methodology using passive scattering proton therapy to optimize the dose to the thyroid gland, esophagus, and lenses, and their method will be followed in this chapter.

For all patients undergoing CSI, the CTV contour includes the entire craniospinal fluid space (including the brain and the spinal canal through the cauda equina to the level of the S2/S3 vertebral junction), as shown in Figure 23–7, the anterior edges of the target are defined differently on the basis of the maturity of the patient's skeletal development. For patients younger than 15 years, an additional normal tissue target volume, which includes the entire vertebral bodies, is included as part of the CTV to avoid asymmetric growth of the vertebral body in these patients. For older patients, the CTV includes the spinal canal and extends no more than 2–3 mm into the vertebral bodies to reduce the dose to the bone marrow, which may allow for better tolerance of the required chemotherapy.

Supine setup, which allows for sedation and patient comfort, is often used for pCSI. A typical pCSI plan using supine setup includes two cranial fields that are angled 15 degrees from the horizontal plane to reduce dose to the lens and improve

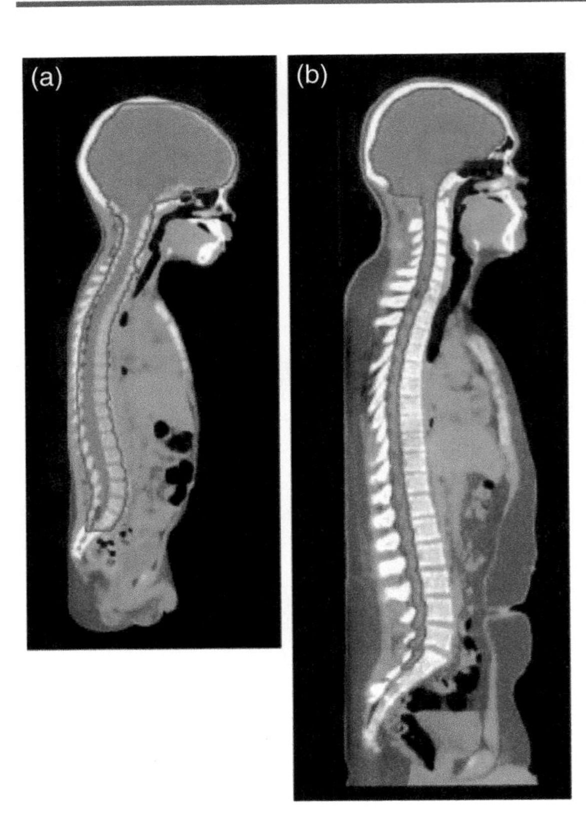

Figure 23–7 Age-specific target volumes for two patients. (a) The target volume for a 4-year-old patient, including the spinal canal and the entire vertebral body, and (b) the target volume for a 15-year-old patient, including only the spinal canal. Both volumes also include the brain. (From Figure 3 from reference [32] with permission.)

dosimetric coverage of the cribriform plate and includes 2–3 posterior spinal fields, as shown in Figure 23–8. Field-specific considerations in treatment planning, including isocenter selection, aperture design, and junction considerations, are described in detail in Giebeler et al. [32]. As in the approach used for traditional xCSI, the spine/spine and spine/cranial junctions are shifted once a week in pCSI. Cheng et al. [33,34] evaluated the dosimetry at the junction between spine fields by comparing a moving gap approach with proton therapy and x-rays of the spine and concluded that xCSI and pCSI shared similar dosimetric characteristics between the abutting spine fields.

Typical pCSI and xCSI dose distributions for a patient are shown in Figure 23–9 [35]. Normal tissues anterior to the spine are completely spared with pCSI.

23.7.3 Lung

The methodology of using proton therapy to treat lung tumors was first reported by Moyers et al. in 2001 [13] and was further developed by Kang et al. [36], who incorporated 4DCT images. In essence, 4DCT is used to delineate internal gross tumor volume (IGTV), which is defined by the combined volume of the GTVs at all 4DCT phases and ITV by the uniform expansion of IGTV. A modified averaged CT (of all

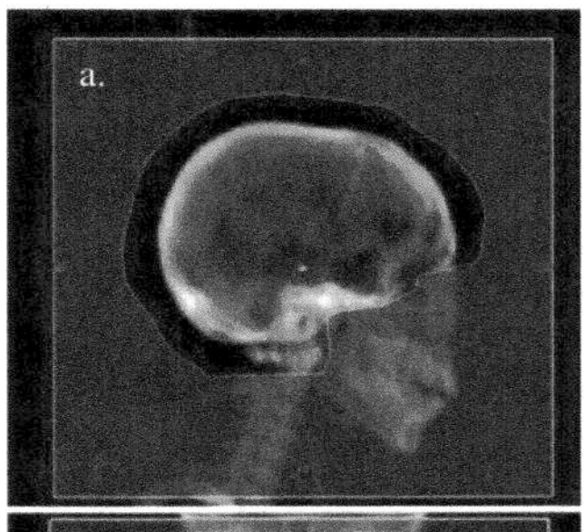

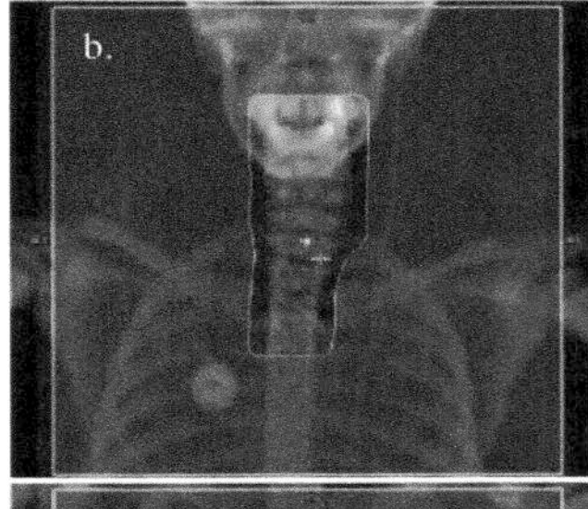

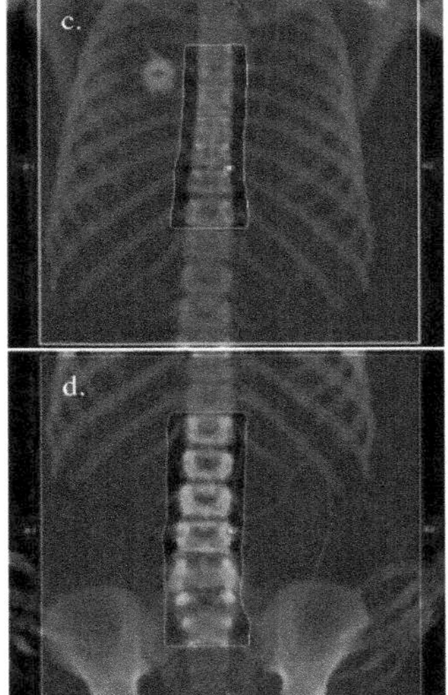

Figure 23–8 Proton therapy cranial and spinal field borders for craniospinal irradiation (CSI) (a representative patient) in the (a) cranial field, (b) upper spinal field, (c) middle spinal field, and (d) lower spinal field. Field isocenters are indicated with red circles. (From Figure 4 of reference [32] with permission.)

Figure 23–9 A dosimetric comparison of proton and photon treatment plans for a child treated with craniospinal irradiation. Sagittal images of proton and photon plans are depicted, with the excess dose deposited from photon treatment highlighted on the right. (From Figure 1 of reference [33] with permission.)

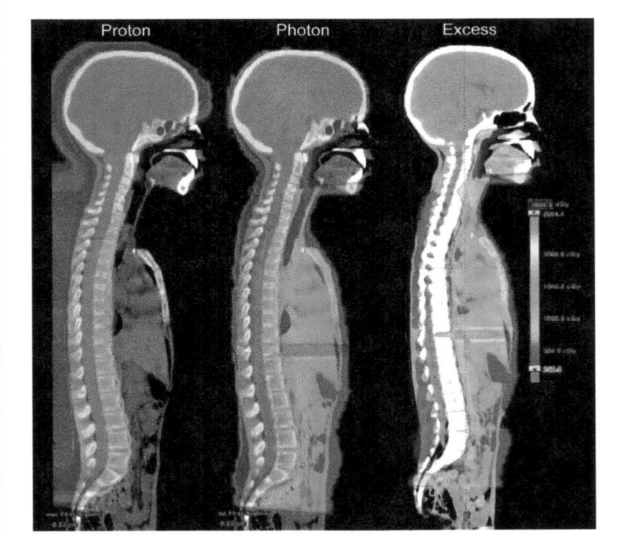

4DCT phases), where the HU of the IGTV volume are replaced (overridden) by a higher-density value close to tumor, is used for treatment planning. The override provides a conservative estimate of the densities within the IGTV volume for treatment planning purposes and ensures that the designed beamline covers the target distally in all 4DCT phases. Standard treatment planning techniques to design beamlines, apertures, and compensators as discussed in this chapter are then used to develop the treatment plan.

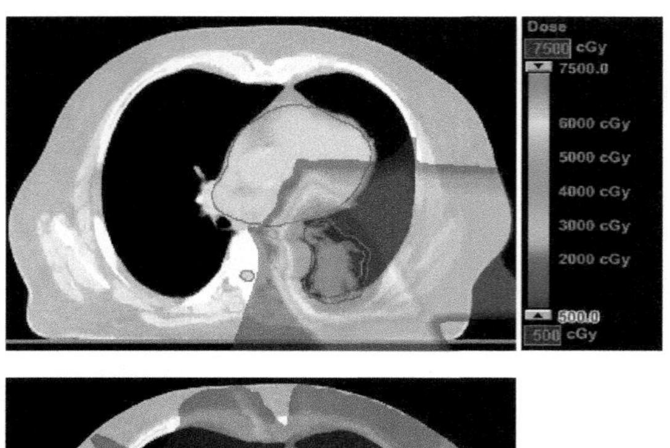

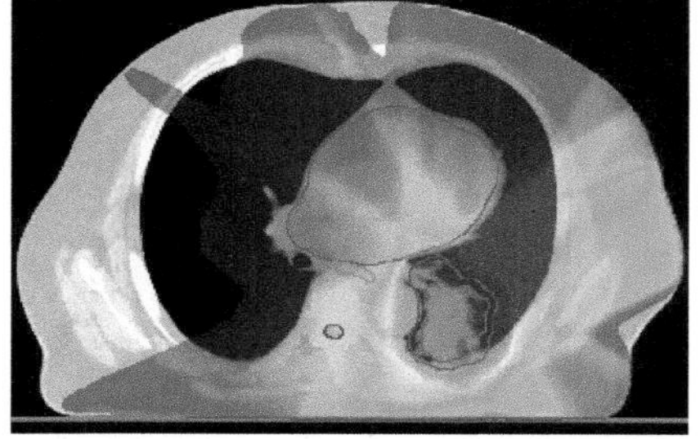

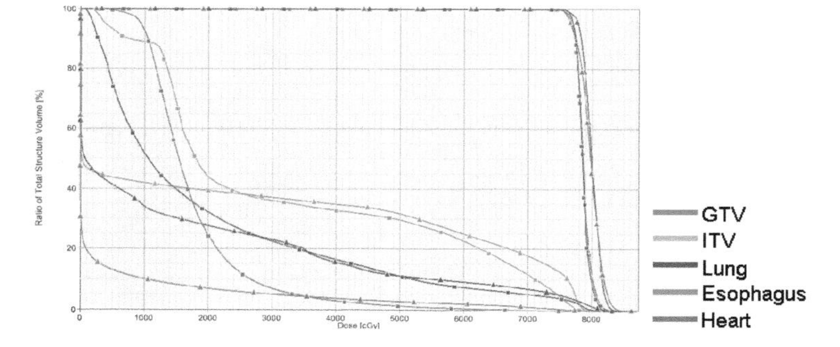

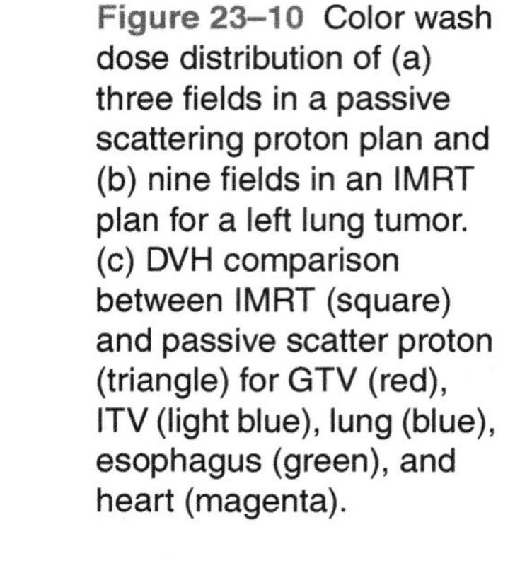

Figure 23–10 Color wash dose distribution of (a) three fields in a passive scattering proton plan and (b) nine fields in an IMRT plan for a left lung tumor. (c) DVH comparison between IMRT (square) and passive scatter proton (triangle) for GTV (red), ITV (light blue), lung (blue), esophagus (green), and heart (magenta).

Figure 23–10 shows an example of a three-field passive scattering proton plan and a nine-field IMRT plan for a lung tumor. Doses to the lung, heart, and esophagus are reduced using the passive scattering proton plan, although IMRT offers better conformity.

23.7.4 Simultaneous Integrated Boost with Passive Scattering Beams

The concept of simultaneous integrated boost (SIB) was originally proposed for IMRT using photons [37] and has been widely used since. For a patient with multiple target volumes to be treated at different prescription dose levels, one could develop multiple treatment plans using uniform fields to treat the patient sequentially. Alternatively, all fields could be treated within the same treatment session every day to

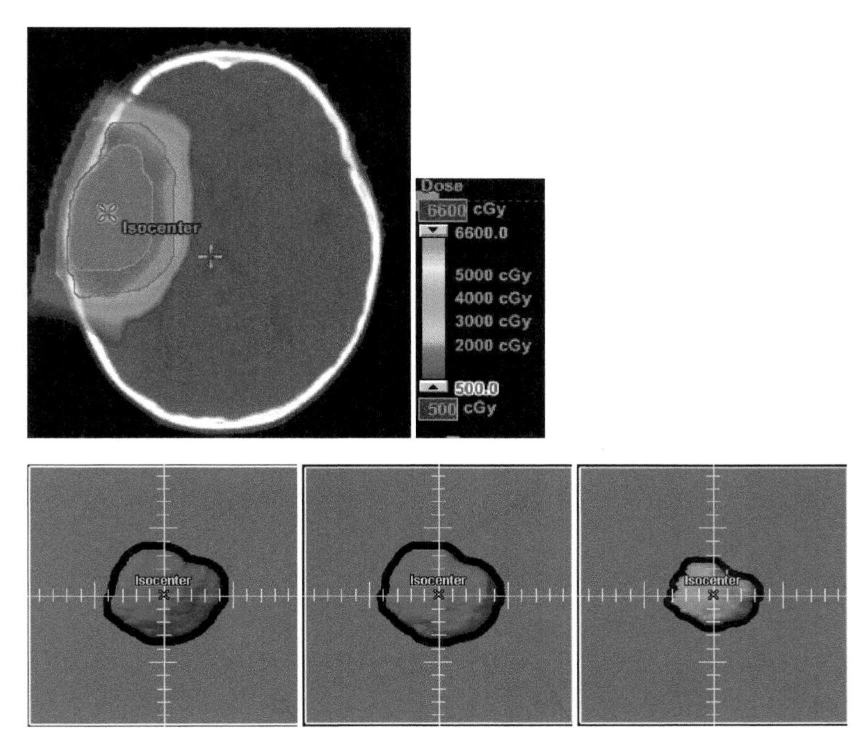

Figure 23–11 A passive scattering proton plan using the simultaneous integrated boost (SIB) technique to treat a right temporal lobe tumor. (a) Color wash dose distribution of a three-field SIB passive scattering proton plan. The prescription is 60 Gy to PTV60 (red contour) and 50 Gy to PTV50 (blue contour). BEV aperture designs for primary with (b) 270° beam angle and (c) 305° beam angle and boost with (d) 305° beam angle are shown.

achieve a higher biological effective dose, owing to the higher fractional dose to the GTV and better spare the normal tissue.

Figure 23–11 is an example of SIB using three passive scattering beams. The prescription for this patient is PTV60 (geometrical expansion of GTV with SM of 3 mm) to receive 60 Gy and PTV50 (geometrical expansion of CTV with SM of 3 mm) to receive 50 Gy. As shown in the BEV, the first two apertures were designed to cover the PTV50 laterally with the dosimetric margin, and the boost beam covers only the PTV60. The beamline and compensators were also designed using different volumes for the primary and boost fields, i.e., CTV and GTV, respectively.

SIB using passive beam usually works for only relatively simple geometry, and the conformity could be inferior to IMPT. However, since the aperture can be placed near the surface of the patient, the lateral spilled dose to the patient can be minimized and may benefit selected patients.

23.8 Summary

Treatment planning with passive scattering and uniform scanning beams in proton therapy has been well established. Passive scattering and uniform scanning proton plans can reduce integral dose to the body and dose to surrounding normal tissues compared with photon radiotherapy. However, it is critical for radiation oncologists, medical physicists, and treatment planners to understand the uncertainties involved in proton therapy and adequately take these uncertainties into account during the treatment planning process. While IMPT might improve the conformity and quality of treatment plans, passive scattering and uniform scanning proton therapy are likely to be applied to many patients because of the superior robustness of the treatment plans.

References

1. Koehler AM, Schneider RJ, Sisterson JM. Range modulators for protons and heavy ions. *Nucl Instrum Methods* 1975;131:437–440.
2. Suit HD, Goitein M, Tepper JE, Verhey L, Koehler AM, Schneider R, Gragoudas E. Clinical experience and expectation with protons and heavy ions. *Int J Radiat Oncol Biol Phys* 1977;3:115–125.
3. Koehler AM, Schneider RJ, Sisterson JM. Flattening of proton dose distributions for large-field radiotherapy. *Med Phys* 1977;4:297–301.
4. Farr J, Mascia A, Hsi W-C, Allgower C, Jesseph F, Schreuder A, Wolanski M, Nichiporov D, Anferov V. Clinical characterization of a proton beam continuous uniform scanning system with dose layer stacking. *Med Phys* 2008;35:4945–4954.
5. Paganetti H. Proton therapy physics. Proton therapy physics series. In: Medical physics and biomedical engineering, Paganetti H, Ed. CRC Press, 2011;1.
6. ICRU. Report 50: Prescribing, recording, and reporting photon beam therapy. The International Commission on Radiation Units and Measurements, 1993.
7. ICRU. Report 62: Prescribing, recording and reporting photon beam therapy. The International Commission on Radiation Units and Measurements, 1999.
8. ICRU. Report 78: Prescribing, recording, and reporting proton-beam therapy. The International Commission on Radiation Units and Measurements, 2007.
9. Thomas SJ. Margins for treatment planning of proton therapy. *Phys Med Biol* 2006;51:1491–1501.
10. Thomas SJ. Margins between clinical target volume and planning target volume for electron beam therapy. *Br J Radiol* 2006;79:244–247.
11. Olch AJ, Gerig L, Li H, Mihaylov I, Morgan A. Dosimetric effects caused by couch tops and immobilization devices: Report of AAPM task group 176. *Med Phys* 2014;41:061501.
12. Moyers MF, Miller DW. Range, range modulation, and field radius requirements for proton therapy of prostate cancer. *Tech Canc Res Treat* 2003;2:445–447.
13. Moyers MF, Miller DW, Bush DA, Slater JD. Methodologies and tools for proton beam design for lung tumors. *Int J Radiat Oncol Biol Phys* 2001;49:1429–1438.
14. Yu Z, Bluett J, Zhang Y, Zhu X, Lii M, Mohan R, Dong L. Impact of daily patient setup variation on proton beams passing through the couch edge. *Med Phys* 2010;37:3294–3294.
15. van Herk M. Errors and margins in radiotherapy. *Semin Radiat Oncol* 2004;14:52–64.
16. Park PC, Zhu XR, Lee AK, Sahoo N, Melancon AD, Zhang L, Dong L. A beam-specific planning target volume (PTV) design for proton therapy to account for setup and range uncertainties. *Int J Radiat Oncol Biol Phys* 2012;82:e329–336.
17. Urie M, Goitein M, Wagner M. Compensating for heterogeneities in proton radiation therapy. *Phys Med Biol* 1984;29:553–566.

18. Goitein M. Compensation for inhomogeneities in charged particle radiotherapy using computed tomography. *Int J Radiat Oncol Biol Phys* 1978;4:499–508.
19. Ju SG, Kim MK, Hong CS, Kim JS, Han Y, Choi DH, Shin D, Lee SB. New technique for developing a proton range compensator with use of a 3-dimensional printer. *Int J Radiat Oncol Biol Phys* 2014;88:453–458.
20. Rietzel E, Bert C. Respiratory motion management in particle therapy. *Med Phys* 2010;37:449–460.
21. Taylor M, Liao Z, Bluett J, Kerr M, Li H, Sahoo N, Gillin M, Zhu X, Zhang X. Improving passive scattering proton therapy plan quality by optimizing compensator parameters. *Med Phys* 2013;40:416.
22. Li Y, Zhang X, Dong L, Mohan R. A novel patch-field design using an optimized grid filter for passively scattered proton beams. *Phys Med Biol* 2007;52:N265.
23. Mahajan A. Proton craniospinal radiation therapy: Rationale and clinical evidence. *Int J Particle Ther* 2014;1:399–407.
24. Silber JH, Littman PS, Meadows AT. Stature loss following skeletal irradiation for childhood cancer. *J Clin Oncol* 1990;8:304–312.
25. Chin D, Sklar C, Donahue B, Uli N, Geneiser N, Allen J, Nirenberg A, David R, Kohn B, Oberfield SE. Thyroid dysfunction as a late effect in survivors of pediatric medulloblastoma/primitive neuroectodermal tumors. *Cancer* 1997;80:798–804.
26. Constine LS, Woolf PD, Cann D, Mick G, McCormick K, Raubertas RF, Rubin P. Hypothalamic-pituitary dysfunction after radiation for brain tumors. *N Engl J Med* 993;328:87–94.
27. Oeffinger KC, Mertens AC, Sklar CA, Kawashima T, Hudson MM, Meadows AT, Friedman DL, Marina N, Hobbie W, Kadan-Lottick NS. Chronic health conditions in adult survivors of childhood cancer. *N Engl J Med* 2006;355:1572–1582.
28. Grau C, Overgaard J. Postirradiation sensorineural hearing loss: A common but ignored late radiation complication. *Int J Radiat Oncol Biol Phys* 1996;36:515–517.
29. Yuh GE, Loredo LN, Yonemoto LT, Bush DA, Shahnazi K, Preston W, Slater JM, Slater JD. Reducing toxicity from craniospinal irradiation: Using proton beams to treat medulloblastoma in young children. *Cancer J* 2004;10:386–390.
30. Jimenez RB, Sethi R, Depauw N, Pulsifer MB, Adams J, McBride SM, Ebb D, Fullerton BC, Tarbell NJ, Yock TI. Proton radiation therapy for pediatric medulloblastoma and supratentorial primitive neuroectodermal tumors: Outcomes for very young children treated with upfront chemotherapy. *Int J Radiat Oncol Biol Phys* 2013;87:120–126.
31. Lee CT, Bilton SD, Famiglietti RM, Riley BA, Mahajan A, Chang EL, Maor MH, Woo SY, Cox JD, Smith AR. Treatment planning with protons for pediatric retinoblastoma, medulloblastoma, and pelvic sarcoma: How do protons compare with other conformal techniques? *Int J Radiat Oncol Biol Phys* 2005;63:362–372.
32. Giebeler A, Newhauser WD, Amos RA, Mahajan A, Homann K, Howell RM. Standardized treatment planning methodology for passively scattered proton craniospinal irradiation. *Radiat Oncol* 2013;8:32.
33. Cheng C-W, Das IJ, Srivastava SP, Zhao L, Wolanski M, Simmons J, Johnstone PA, Buchsbaum JC. Dosimetric comparison between proton and photon beams in the moving gap region in cranio-spinal irradiation (CSI). *Acta Oncol* 2013;52:553–560.
34. Cheng C, Das I, Chen D. Field matching in craniospinal irradiation. *Br J Radiol* 1995;68:670–671.
35. Dinh JQ, Mahajan A, Palmer MB, Grosshans DR. Particle therapy for central nervous system tumors in pediatric and adult patients. *Trans Cancer Res* 2012;1:137–149.
36. Kang Y, Zhang X, Chang JY, Wang H, Wei X, Liao Z, Komaki R, Cox JD, Balter PA, Liu H, Zhu XR, Mohan R, Dong L. 4D proton treatment planning strategy for mobile lung tumors. *Int J Radiat Oncol Biol Phys* 2007;67:906–914.
37. Wu Q, Manning M, Schmidt-Ullrich R, Mohan R. The potential for sparing of parotids and escalation of biologically effective dose with intensity-modulated radiation treatments of head and neck cancers: A treatment design study. *Int J Radiat Oncol Biol Phys* 2000;46:195–205.

Chapter 24

Treatment Planning for Pencil Beam Scanning

Tony Lomax, Ph.D., Alessandra Bolsi, M.Sc., Francesca Albertini, Ph.D., and Damien Weber M.D.

Centre for Proton Therapy,
Paul Scherrer Institut,
Switzerland

24.1 Introduction

Pencil beam scanning (PBS) will soon be the most prevalent form of proton therapy world-wide [1]. As described in Chapter 9 (Field Shaping: Scanned Beams), PBS consists of the magnetic deflection of many, narrow proton pencil beams such as to deliver an iso-energy layer of Bragg peaks across the target volume. In combination with a step-wise change of energy between layers, Bragg peaks can then be delivered throughout the target volume in three dimensions. Given that each individually delivered Bragg peak can be modulated in intensity (or, more strictly put, fluence), PBS is an extremely flexible modality, allowing for many modes of delivery. Treatment planning for PBS then must be able to define, calculate, and exploit these capabilities.

As for any form of radiotherapy, treatment planning for PBS has three main roles: target and critical organ identification and delineation, treatment design, and the definition of machine control data for controlling the delivery. Of these three roles, only the second will be considered in this chapter.

In common with treatment planning for any form of external beam radiotherapy, one of the main tasks in the planning of PBS proton therapy is the definition of the number and direction of fields with which to irradiate the target volume. There the similarity ends, however, as the factors affecting the selection of field orientations will, in many cases, be different for proton therapy than for conventional therapy, as will the way in which the geometry and characteristics of the individual fields can be and are defined. Indeed, PBS proton therapy, by its very nature, provides enormous flexibility in the geometry and distributions delivered by individual fields, and much more than in photon therapy.

As an example, consider this simple case. Imagine that we are to treat a cubic target in a cubic water phantom using PBS. If the cubic target has the dimensions 10 x 10 x 10 cm, and individual Bragg peaks will be delivered to this target on a rectangular grid with spacings of 5 x 5 x 5 mm, in order to completely cover the target in three dimensions, something like 8000 individual Bragg peaks need to be delivered (400 for each of the 20 different energy levels required to cover the target completely in depth).

In addition, in order to achieve a homogenous dose across such a target, each BP must be individually weighted (i.e., its fluence must be varied) at least in a similar way as to that described for the production of spread-out Bragg peaks in Chapter 8 (Field Shaping: Scattered Beam). Thus, in this simple example, there are about 8000 degrees of freedom which can be varied and must, therefore, be defined for any given plan, without allowing for the additional possibility of varying the scanning pattern (i.e., altering the Bragg peak spacing). And this for a single field direction only. The possibilities multiply immensely when multiple fields are then considered.

As there are many detailed reference works on the algorithmic and methodological aspects of treatment planning for PBS proton therapy [2,3,4], not to mention detailed descriptions of many of these aspects in other chapters of this book, we will keep such descriptions brief and try to concentrate more on the practical and clinical aspects of treatment planning for PBS proton therapy. As such, a number of clinical case studies, all from our institute, will be presented in order that the reader can get an impression of the issues and thought processes behind practical (one could say "real-world") treatment planning as carried out at the Paul Scherrer Institut.

This is not to say that any of the planning concepts presented here are necessarily the only approach to such treatments, but it is hoped that the presentation of cases and planning concepts from our more than 18 year experience with PBS proton therapy is helpful and interesting to both novice and experienced proton practitioners alike.

24.2 First Things First—What You Should Know about PBS Proton Therapy

In the spirit of the educational basis of this chapter, we will start off by making a few general statements about issues and potential problems that should be considered when planning PBS proton treatments. These are the basis for many decisions made during the planning process and, as such, are important to keep in mind. While not intending to be complete, this is hopefully a relatively comprehensive list of the most relevant issues, and the main ones which drive many of the planning solutions that we adopt in our clinic.

24.2.1 Lateral Falloff

It is a common misnomer that proton therapy, and PBS in particular, can produce extremely sharp dose gradients in comparison with passive scattered proton or photon-based therapies. This is simply not the case. Although PBS treatments can achieve somewhat better lateral dose gradients than collimated and passively scattered proton fields at depths in water of greater than about 14–15 cm, un-collimated PBS fields have rapidly *increasing* lateral penumbras with *decreasing* range in water, resulting in penumbras that are considerably larger than that achieved with collimated proton or photon techniques. This has been elegantly shown in the work of Safai et al. [5], the results of which are shown in Figure 24–1. This shows calculated lateral penumbras (80–20%) for a simple target in water for both collimated broad (passively scattered)

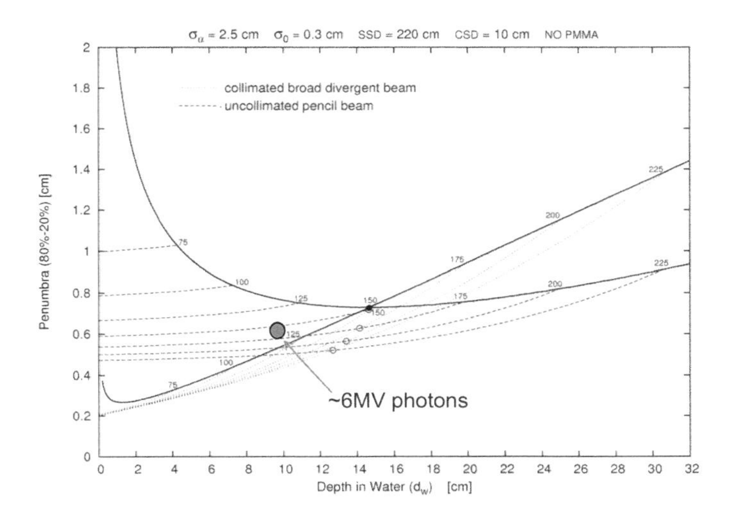

Figure 24–1 Plots of lateral penumbra (80–20%) as a function of energy and penetration depth in water for collimated and uncollimated (PBS) proton beams. The approximate penumbra for 6 MV photons at 10 cm depth is also indicated. (Based on figure 5 in Safai et al. 2008.)

and un-collimated PBS beams. The solid lines show the trend for the penumbra at the Bragg peak for both types of fields, whilst the dotted and dashed lines show how the lateral penumbra broadens as a function of depth on the way to the Bragg peak for collimated broad and PBS beams respectively.

Above about 14 cm water equivalent depth, PBS beams can indeed be sharper than passively scattered fields (due to the edge enhancing effect of pencil beam optimization, described in Chapter 21 (Plan Optimization). Below 14 cm, however, the penumbra of a PBS field rapidly increases to values of 1 cm or more at the Bragg peak for water equivalent ranges below about 5 cm. This can be compared to lateral penumbras of about 0.5–0.6 cm for passively scattered and photon fields at the same depth.

This is predominantly a problem of multiple coulomb scattering in nozzle elements intersecting with the beam and is a case of physics simply working against us. In addition, in Figure 24–1, a beam width in air of 0.3 cm (sigma) has been assumed. The lateral penumbra will be even worse if the initial beam width is larger, indicating the importance of specifying as small pencil beams as possible when purchasing a PBS proton therapy machine. It also indicates that there is a very good argument to request some form of field-specific collimation capability with PBS delivery machines, a feature sadly lacking on most commercial solutions at the moment.

In summary then, it is not to be expected that a PBS system will lead to very sharp gradients at the lateral edge of the field, or around critical structures when delivering Intensity Modulated Proton Treatments (see below) to superficial targets. Indeed, in many cases, these gradients will be *worse* than those expected for modern photon therapy. It is well worth keeping this in mind when planning and discussing PBS treatment plans in the clinic and with potential referring centers.

24.2.2 Superficial Spots and the Importance of the Air Gap

Superficial target volumes pose quite a problem for PBS proton therapy. To deliver Bragg peaks on or close to the surface of a patient, energies close to zero are required. For proton beam lines and gantries, it is more or less impossible to transport such low energies, with the minimum transportable energy for most gantries being around 70 MeV, equivalent to a water equivalent range in the patient of about 4 cm. To deliver Bragg peaks more superficially than this, the only solution is to introduce material in the nozzle, or between the nozzle and the patient, in order to degrade to the required lower energies.

So what's the problem? Well, it lies once again in physics, and is shown in Figure 24–2. Any material introduced before the patient will scatter the protons, making the beam somewhat divergent. Thus, if there is any gap between the degrader and the patient, the beam will widen across this gap for purely geometrical reasons (the air will also scatter the beam somewhat, but for gaps of a few tens of cm, this can be neglected). Consequently, unless the gap can be minimized or avoided, the effective pencil beam width will inevitably broaden across this gap, as shown in Figure 24–3. This figure shows the beam width as a function of the thickness of degrading material and the air gap (the gap between the degrader and the patient surface) and clearly shows the importance of minimizing the amount of material or keeping the air gap as small as possible if the pencil beam width is to be kept small. Thus, the best solution is to place such degrading material directly on the patient surface (when it is typically called a bolus), thus reducing the air gap to zero. Indeed, this was an approach we took for some of our early patient treatments at PSI.

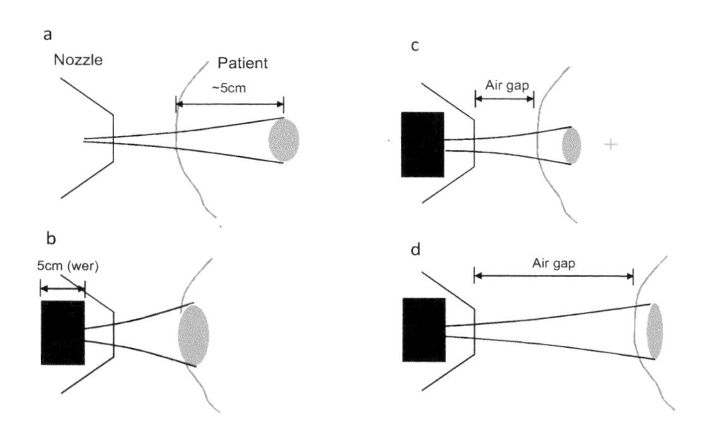

Figure 24–2 A schematic representation of the broadening of proton pencil beams and the effect of the air gap. (a) PBS broadening without pre absorber. (b) Broadening due to scattering in a nozzle mounted preabsorber. (c) Broadening with a "small" air gap. d) Broadening with a "large" air gap.

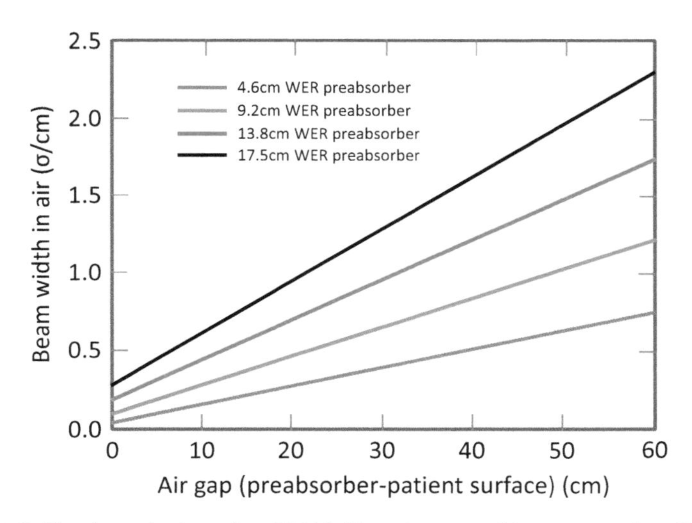

Figure 24–3 The broadening of a 177 MeV proton pencil beam as a function of pre-absorber thickness and distance from the pre-absorber in air.

In practice however, it is rather challenging to work with boluses. One of the main problems is that, ideally, any bolus material should be included at the time of the planning CT in order to best model its form on the patient surface, and therefore its effect on proton range. In practice this can be difficult, as it presupposes that the beam angles to be used are known up-front at the time of the fixation and planning CT. In addition, there can be major problems of reliably positioning such a bolus on a day-to-day basis, the worst case being that a planned bolus is forgotten on the treatment day. None of these are "no-gos," but they do make the regular and safe use of bolus quite challenging.

The more common solution is the use of a fixed pre-absorber as an add-on element in the treatment nozzle. This is certainly an easy and more reliable approach, but if pencil beam widths are to be kept small, the air-gap between the pre-absorber to the patient also needs to be kept small (see e.g., Figure 24–3). Although for some anatomical sites and beam angles this is possible (e.g., for apical/posterior beams in the head and posterior beams in the lower pelvic area) the shape of the patient, size of the nozzle and the treatment table, and patient fixation devices can often preclude minimizing the air gap below 10 cm or so. Although little can be done to change the patient or treatment nozzle, it is therefore worth considering carefully the shape and width of the treatment table and the fixation devices used. For instance, it makes a lot of sense to taper the width of the treatment table in the area where it will be supporting the patient's head such that its width reduces to that of a typical patient's head. This will potentially allow for smaller air gaps on lateral or near-lateral fields by being able to bring the treatment table closer to the treatment nozzle or vice versa. Alternatively, a solution somewhere between the bolus and nozzle-mounted pre-absorber has recently been reported by Both et al. [6], who have designed a U-shaped pre-absorber as an optional addition to the treatment couch. Such an approach avoids many of the disadvantages of a bolus, while bringing the pre-absorbing material closer to the patient. Despite all of these approaches, however, the optimal treatment of superficial tumors with current PBS machines remains an unsolved problem.

24.2.3 Calculating Proton Range *In Vivo*

It is well documented in other parts of this book (see e.g., chapters 3 (Particle Beam Interactions: Basic) and 4 (Particle Beam Interactions: Clinical), and in many publications and textbooks [7,8] that density heterogeneities in the path of the beam can have a major effect on the delivered dose distribution. In the simplest case, these will affect the overall range of protons in the patient due to the varying energy loss when traversing through materials of different densities. However, as long as good-quality CT data is available for treatment planning, this effect should be trivial to deal with. Nevertheless, it is worth remembering that the calculation of proton range from CT data is an indirect calculation. That is, the Hounsfield Units (HU) of the CT have no direct relationship to relative proton stopping power, and have to be first converted using an analytically derived calibration curve.

Surprisingly, despite the critical importance of this step, there is actually no universally accepted method of calibrating CT to proton stopping power (although an ini-

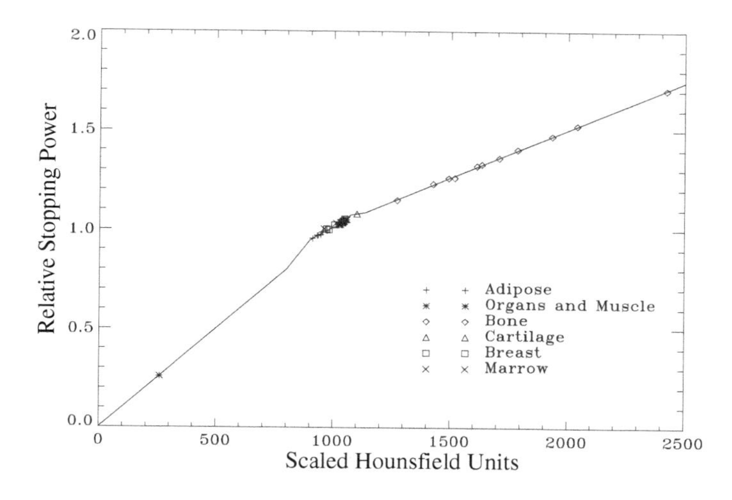

Figure 24–4 The stoichiometric CT Hounsfield unit to proton stopping power calibration curve used at PSI.

tiative in this direction would be very welcome). The main problem is that there is no unique conversion of CT HU values to proton stopping power. That is, different tissues or materials with the same HU value may well have different stopping powers, and vice versa. In addition, the calibration curve will be dependent on the acquisition parameters used for the CT, as HU values can vary as a function of tube voltage, field of view, and reconstruction algorithm. Consequently, if one doesn't want to define many different calibrations curves, the acquisition parameters used for CT acquisitions must be well defined and checked. Thus, it is relatively easy to find different solutions to this problem, and it is difficult, if not impossible, to say which one is correct. Indeed, given the nonuniqueness of the problem, there is simply *no* one correct answer.

In our clinic, we developed and still use the so-called stoichiometric approach as described by Schneider et al. [9] and as shown in Figure 24–4. Based on direct experimental validation of this approach, we could verify that this is accurate to about 1% in soft tissues and about 2% in bone [10]. Subsequently, this approach has become the most widely applied calibration technique. However, it should be stressed that we have optimized our calibration for biological tissues only, and the resulting curve is not necessarily valid for nonbiological substances and materials, which can have some consequences during the planning process.

So for simple geometries, and based on good-quality CT data sets which are artifact free and contain no "nonbiological" materials, ranges can be calculated quite accurately *in vivo*. Nevertheless, most proton therapy practitioners assume an overall accuracy of range calculations due to the calibration alone of about 3%. This may be somewhat conservative, but it is a sensible value to assume if one wants to be on the safe side with target coverage.

24.2.4 Density Heterogeneities

The calculation of overall proton range in the patient, however, is not enough to fully deal with the potential effects of density heterogeneities on proton pencil beams. An additional effect is the potential distortion of the Bragg peak due to beams passing through complex geometries of varying density.

As described in chapters 3 and 4 (Particle Beam Interactions: Basic and Particle Beam Interactions: Clinical), protons also scatter and can thus take slightly different paths through the patient to the end of their range. If these paths go through materials of different densities (say, for proton pencil beams for which the central axis is parallel to an air–bone interface in the nasal cavity), then their residual ranges will be very different, leading to a distortion or "spreading" of the Bragg peak. In the simplest case of a pencil beam passing along a long interface between materials of different density (e.g., see Figure 24–5), a "double" Bragg peak can result—a less deep one at the residual range for the protons predominantly passing through the higher density, and a somewhat deeper peak for those passing predominantly through the lower-density region. A schematic representation of this effect can be found in Figure 4 of Lomax [8], and a more clinical and experimental demonstration of the effect can be found in the seminal paper by Urie et al. [7].

Interestingly, such distorted Bragg peaks are probably much more the norm in clinical practice than the pristine shape shown in any introductory lecture on proton therapy. Indeed, there are very few density-homogenous regions of the human anatomy (the high cranium and eye being the exceptions), and so in almost all proton treatments, it could be said that there is no real Bragg peak. It will also be hopefully clear that, in order to make meaningful clinical decisions based on dose distributions, the dose calculation needs to be able to accurately model such distortions.

Although Monte Carlo (MC) approaches (see Chapter 19, Dose Calculation: Monte Carlo) are certainly the best and most accurate approach (although even their

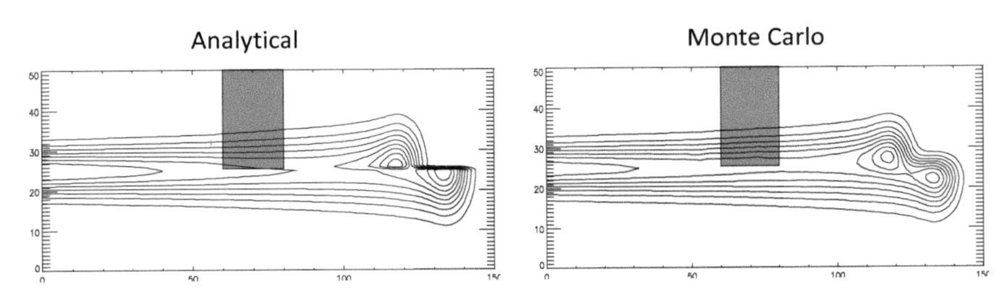

Figure 24-5 The difference between an analytical (ray casting) and Monte Carlo calculation of a single pencil beam passing long a parallel high-low density interface. (Adapted from Schaffner et al., 1999.)

accuracy is in the end limited by the resolution and quality of the CT data in which they are calculated), these are not currently fast enough for everyday clinical use. Although big advances are being made to increase the computational efficiency of MC calculations [11,12] therefore, most planning systems for PBS proton therapy still use analytical approaches (see Chapter 18, Dose Calculation: Analytical). In much the same way that there can be no perfectly correct calibration curve for determining proton stopping powers, no analytical calculation can be perfectly correct, either. Put another way, every analytical dose calculation must be "wrong" to a certain extent. As an example, the reader is referred to the paper by Schaffner et al. [13], which compared three different analytical algorithms for PBS proton therapy and clearly demonstrates the quite significant dose differences one can expect between the different algorithms (Figure 24–5).

Once again, there is no one "correct" approach to analytical dose calculations, and all have advantages and disadvantages. Much more important is to understand the algorithm you are using, where it works well and, perhaps most importantly, what its limitations are. At PSI, we still use the "ray-casting" algorithm as described by Schaffner et al. [13]. Although very simple, it is extremely fast (thus speeding up the optimization and planning process) and has the tendency to overestimate the effects of density heterogeneities in comparison to other algorithms (see Figure 24–5). Although the result is clearly different to the Monte Carlo calculation shown in the same figure, these differences are well understood and the effects of the density heterogeneities, if anything, are exaggerated (which, we believe, is better than it being underestimated). Nevertheless, when comparing the ray-casting approach to a Monte Carlo calculation for a number of clinical cases, the agreement was found to be surprisingly good, and better than would be expected given the simplicity of the approach (see Tourovsky et al. [14]), as was the case when experimentally validating the algorithm in a complex anthropomorphic phantom (Albertini et al. [15]). Such results show that, in the clinical setting, the algorithm is not too wrong. Nevertheless, its shortcomings still have to be recognized and understood.

24.2.5 Range Uncertainty

In Section 20.3.1.2 above, it was stated that, based on good-quality CT and a well-defined calibration curve, the uncertainty on the range calculation within the patient is of the order of about 3%. Unfortunately, however, there are a number of other sources of uncertainty that can also affect the accuracy with which proton range can be accurately determined in the patient, many of which are listed, and their magnitudes estimated, in the work of Paganetti [16]. Here we will just concentrate on CT calibration, the effects of metal implants, and anatomical changes of the patient.

It is important to understand that all range uncertainties resulting from the planning CT are systematic in nature. By their very nature, systematic errors will be present through the whole treatment course, and they are therefore somewhat of a worry. Thus, although uncertainties due to CT calibration are relatively small (3%), they are systematic in nature and, without improved imaging or direct calculation of proton SP from CT data, are impossible to avoid. The same applies for the effects of metal implants and, to a lesser extent, to anatomical changes. For both of these, it is extremely difficult to define magnitudes of the effects, as they will be very patient- and plan-specific. Nevertheless, range uncertainties of more than a centimeter are certainly possible as a result of CT artifacts (see below) or anatomical changes (e.g., weight changes, nasal cavity filling, or organ motion [17–19]). However, anatomical changes fall somewhere between systematic and random. Gradual weight changes could lead to systematically increasing range changes through the treatment course (see e.g., Albertini et al. [17]), whereas changes in bowel/intestine or nasal cavity fillings can change on a daily basis and so are neither random nor systematic. Unfortunately, such changes can also cause large range changes locally, and this possibility should certainly be taken into account during planning.

24.2.6 The Problem of the Table

Treating through the table can be challenging for photon treatments [20], but it is even more of a challenge for protons. First, if irradiating through the table, it will have a substantial influence on the residual range of the protons which, in order to accurately model, requires an accurate representation of the table in the treatment planning system. To address this problem at PSI, we perform our planning CTs using a CT-on-rails solution, with a customized table mounting system which allows us to perform the planning CT on exactly the same table as will be used for the treatments. Such a solution is not, however, available for most clinics, with the planning CT typically being performed on a different, CT compatible table. If one wants to then treat through the treatment table, clearly the CT table has to be removed and replaced with an accurate representation of the treatment table. In this case, it is important that the treatment planning system can support this in an efficient and accurate way.

The second problem is to know and be able to assign the correct stopping powers to the various parts of the table. If the table is homogenous, then the stopping power can be directly measured as part of the commissioning process. If it is heterogeneous, however, then this is more difficult, and it may be necessary to obtain the exact characteristics of the table from the manufacturer in order to be able to calculate the expected relative stopping power of the different materials. These may then need to be related back to the planning data by overwriting the proton stopping powers of the different parts of the table. However, even when correctly represented, heterogeneities in the table can still be problematic. For instance, if the patient's position needs to be corrected with respect to the beam by moving the table, then structures in the table will be in a different position with respect to the beam and patient as was assumed during the planning process.

As can be seen, accurately treating through the table is not necessarily trivial. On the other hand, there can also be advantages, at least if its proton-specific characteristics can be accurately represented. If it is a few centimeters thick and homogenous, then the table acts as a natural pre-absorber, which is naturally close to the patient surface (see Section 24.2.2) thus potentially minimizing pencil beam size when treating superficial target volumes.

24.2.7 Motion

Organ motion is potentially one of the most challenging problems for PBS proton therapy for two main reasons (see Chapter 24, Motion Management). The first is common to all proton therapy and is the substantial range changes that can occur due to motions of anatomical structures during treatment. The most extreme example is in the treatment of lung tumors, where motion of the "water-like" tumor within the "air-like" lung can lead to range differences at the edge of the field on the order of centimeters [18,19,21]. However, in addition, with PBS, there is the problem of the interplay effect. This is essentially an interference effect between the dynamics of the delivery (point-by-point delivery of individual pencil beams) and organ motion, which can lead to substantial deterioration of the delivered dose homogeneity across the target volume [22–25]. As discussed in other parts of this book (Chapter 24, Motion Management), there are many ways of mitigating these effects to a greater or lesser extent. Nevertheless, motion can still be a limiting factor for PBS proton therapy and should at least be considered when planning such treatments, for instance through the use of multiple fields (to smooth out the effects) or the selection of beam angles that are predominantly along the major axis of motion.

24.2.8 RBE

Finally, as described elsewhere in this book (e.g., Chapter 5, Relative Biological Effectiveness, RBE), relative biological effectiveness (RBE) is an important concept for particle therapy. Briefly stated, the RBE concept attempts to quantify differences in biological responses of different radiation qualities when the same physical dose is applied.

For many years, there has been controversy over the actual value of RBE for protons (see Chapter 5, Relative Biological Effectiveness and [26,27]). Despite this, it is still generally accepted that a global value of 1.1 is sufficient to deal with the issue clinically. Indeed, if clinical results from different proton clinics are to be meaningfully compared, it is of utmost important that a standard approach is taken. Nevertheless, there is plenty of *in vitro* evidence that the RBE can vary substantially from this value, and in particular could be significantly higher for very low-energy protons and, therefore, could be raised in the Bragg peak and distal falloff region. On the other hand, there is very little, if any, strong evidence that this is the case *in vivo* (see Chapter 5, Relative Biological Effectiveness and Paganetti [26]). Indeed, given the complexity of biological response at the micro and macro level in organs and tissues, it is this author's opinion that it will be extremely difficult to ever determine localized RBE effects in the patient as a result of outcomes analysis alone. Nevertheless, the possibility of higher RBE values at the distal end of a field should not be ignored, if only because of the potential range extension due to RBE [28]. For this reason, it is advisable, wherever possible, to avoid single fields "ranging-out" against critical structures that may be sensitive to localized regions of high dose (e.g., the spinal cord).

24.2.9 It Isn't All as Bad as it Sounds…

In the previous sections, we have outlined a number of important issues that the practitioner of PBS proton therapy should be aware off when planning and delivering treatments. At first glance, this may seem rather off-putting and give the impression that PBS proton therapy is, at best, difficult and challenging, and at worst, downright dangerous. Neither is correct. In fact, one of the big advantages of PBS is that, in comparison to passive scattered proton therapy, treatment planning is more automated and easier. Indeed, when using intensity-modulated proton therapy (see below), treatment planning becomes rather similar to treatment panning for IMRT and VMAT type techniques with photons. As such, the issues outlined above should be taken into consideration, but should not make the user paranoid.

As an example, it is oft stated (and perhaps reinforced by the discussions above) that the Bragg peak is never used to spare normal tissues due to uncertainties in the RBE and range. While this may be correct for *certain* critical structures (e.g., the spinal cord), it is certainly not always the case. As long as the Bragg peak is deposited somewhere in the patient, it is *always* being used to spare normal tissues, the only question being the significance of the normal tissue being spared. As will be seen in the case studies described below, many of our field arrangements are nonsymmetrical and certainly far from the rotational geometries currently in trend in conventional therapy. By taking these approaches, we are deliberately using the stopping potential of protons to spare large volumes of normal tissues and many critical structures on the contra-lateral side of the patient to the treatment volume. This could be a huge advantage, even if not all the normal tissues spared are "critical." In addition, when using protons to almost completely spare the contra-lateral hemisphere of a pediatric brain, the issue of range uncertainties of even a few millimeters to a centimeter may not be a clinically relevant issue, as long as such range uncertainties don't also affect the doses to more sensitive (and serially structured) organs, such as the optic nerves, chiasm, or brain stem. So in summary, try to keep things in perspective. There is no one perfect plan, and compromises will always have to be made. For instance, there could be compromises on the robustness of the plan in order to meet clinical constraints. If discussed properly, and with full awareness of the issues involved, such a compromise can be absolutely OK, and may well be the best for the patient. A-priori we can't know, and we can only use our best judgment to balance the two.

In addition, be prepared to make simple, "back-of-the-envelope" estimates of the potential effects of uncertainties. For instance, if the exact stopping power or structure of an object (either in the patient or outside) is not known, try to make a worst-case estimate to see what the possible effect will be. Imagine there is a nonbiological object (for example an implant of some form) in the path of the beam, and the stopping power can only be estimated within 50% (the stopping power is somewhere between 1 and 1.5, for instance). Does this mean we shouldn't treat through it? Not necessarily. Even if the object is a centimeter thick, the maximum uncertainty in the range will be about 5 mm, and less if the object is thinner. This may be perfectly acceptable depending on what is at the distal end of the target, and even more so if only one field of a three-field plan is affected.

In summary, don't be afraid of the Bragg peak. Use it with respect certainly, but also keep things in perspective.

24.3 Planning PBS Proton Therapy

After the cautionary nature of the previous section, we now move on to discuss the more mechanistic and methodological aspects of planning PBS proton therapy, taking into account target volumes, planning and optimization strategies, and finally some tips and tricks to help the planning process that we have learned in the past years.

24.3.1 What's the Target?

Before talking about defining and calculating treatment plans for PBS proton therapy, it is worth saying a few words about what the target is or should be. To begin with, it is important to state that based on ICRU reports 50 and 62 [29], there are *no* differences in the definition of the GTV and CTV between proton therapy and other radiation-based therapies. The GTV and CTV are purely patient- and tumor-specific, and they are completely independent of the treatment modality. So, the only differences for PBS proton therapy against other techniques will be in the definition of the PTV.

In order to deal with (mainly) positional uncertainties of the patient during the course of treatment, the use of a planning target volume (PTV) is standard practice in conventional therapy. Interestingly, however, there is a good reason why the PTV concept, at least as defined in the ICRU report, is not necessarily valid for proton therapy. Simply put, and in contrast to conventional therapy, there are *two* main sources of uncertainty that should be incorporated into a PTV—the positional uncertainties of the patient and delivery (as in a conventional PTV) and also uncertainty in proton range. There is little or no reason to believe that these are the same or even correlated, and thus, the margins calculated from these will generally be different. So, when defining an iso-tropic PTV margin, which of these uncertainties should be used? The problem is further complicated by the fact that positional and delivery errors are likely to be random in nature (i.e., they will change day-to-day over fractionated therapy) whereas, as discussed above, range uncertainty will almost certainly be systematic in nature. This has a significant effect on the calculated margins using standard margin recipes [30].

With passive scattering, this problem has been elegantly solved for many years by foregoing a PTV and building uncertainty mitigation directly into the design of the field through expansions of the aperture to deal with positional errors, while systematically reducing the thickness of the compensator to deliberately overshoot each field beyond the distal end of the GTV/CTV (e.g., see chapters 8, Field Shaping: Passive Scattering and 22, Treatment Planning). In addition, the compensator is "smoothed" to remove sharp range changes and add extra security against range changes due to positional errors. By these measures, positional and range errors are automatically dealt with, negating the necessity of a PTV. Due to the typical lack of apertures and compensators for PBS, however, this approach is not currently an option, and other solutions have to be used.

One approach could be to use field-specific PTVs. Such a volume must then be constructed essentially in the "beam's eye view," such that the margins lateral to the field are calculated using conventional margin recipes [30], but with a different distal margin calculated allowing for the range uncertainties and their systematic nature [31]. This concept has been expanded more recently into range-adapted PTVs as an additional tool to help in the calculation of plans under conditions of motion and to allow for the substantial range changes that can be present when treating lung tumors [32,33]. Unfortunately, few treatment planning systems currently support this concept, and performing such expansions manually is difficult and time consuming. A more elegant approach is to incorporate positional and range uncertainties directly into the optimization—so-called robust optimization [34]. This approach is described in detail in Chapter 21, Plan Optimization, and won't be expanded on here.

In our clinical practice, we have decided to stay with the standard PTV concept. Typically, for cranial tumors, we use an iso-tropic margin of 5 mm. This has been calculated using the van Herk formula, based on an analysis of our positioning accuracy alone [35]. However, along the beam direction, a 5 mm margin is also sufficient to allow for 3% to 3.5% range errors up to 15 cm in depth, which is a typical-to-long range for fields delivered to cranial targets. For extra-cranial cases, a 7 mm iso-tropic margin is used, again based on our positioning accuracy for such cases and also to allow for the somewhat longer ranges that may be necessary for these cases. This approach may not be optimal, but it has the beauty of simplicity and is borne out, we believe, by our clinical results.

24.3.2 Field Shaping

As described in Chapter 9, Field Shaping: Scanned Beam, (and more briefly above), a PBS proton field consists of the application of many individually weighted proton pencil beams distributed on a three-dimensional grid. In water equivalent space, (i.e., if delivered to a homogenous water phantom), this grid will generally be regularly spaced, with a constant spacing of the pencil beams in both directions orthogonal to the field direction, and either a regular or lightly varying spacing in the depth direction. Orthogonal to the beam, the pencil beam spacing can either be a fixed value or can vary as a function of pencil beam width. At PSI, we use an orthogonal spot spacing of 4–5 mm regardless of the pencil beam width (see Figure 24–6a).

A non-regular spacing in depth is typically due to the use of constant energy separation between energy levels (typically 2–4 MeV). Physical spacing at the higher energies is then somewhat larger than those at the lower energies, compensating for the differing widths of higher- and lower-energy Bragg peaks (due to range straggling, high-energy Bragg peaks are always broader than low-energy Bragg peaks), as shown in Figure 24–7.

This shows a set of Bragg peaks, separated with a constant energy of 2.5 MeV. The changing width of the Bragg peak as a function of depth (see Chapter 3, Particle Beam Interactions: Basic) is clearly seen, as is the gradual change in spatial separation from about 6 mm at the highest energies to 3 mm at the lowest.

Such spacings in depth will not, however, be regular in the patient due to the varying densities. For instance, if treating in low-density tissue (e.g., lung), the Bragg peak spacing in the lung will be much more than in water, and vice versa if depositing Bragg peaks in bone. Determination of Bragg peak positions in the patient therefore requires a radiological path length calculation through the planning CT, together with the calibration curve for converting CT-HU values to relative proton stopping powers (see Section 24.2.3 above).

The final step of field shaping is to select the subset of Bragg peaks that will be deposited inside the designated target volume, or a small distance outside, with the

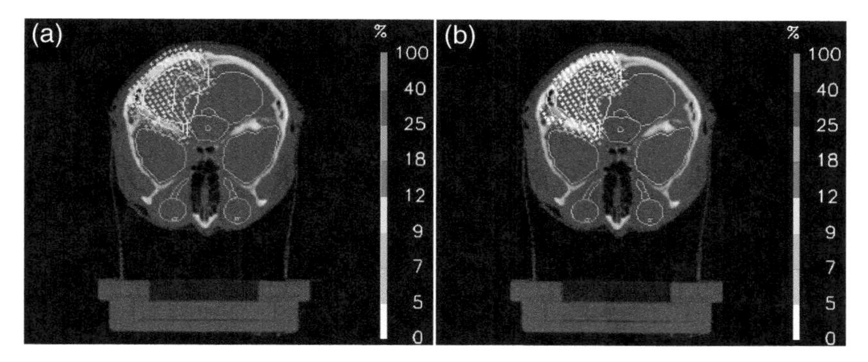

Figure 24–6 Example distributions of Bragg peak positions (crosses) and fluences for a single, SFUD field. (a) Unoptimised fluences and (b) fluences as a result of the optimization process. For both, the colors of the Bragg peaks represent their relative fluence as indicated by the color bar on the right hand side of each figure.

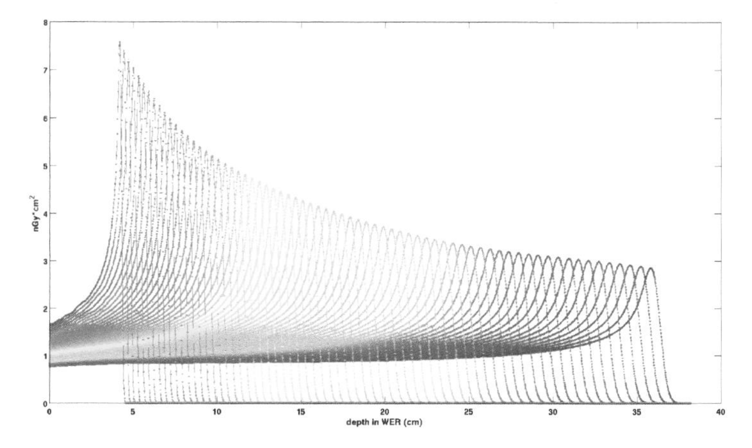

Figure 24–7 A set of Bragg peaks in water from 240 Mev down to 70 MeV in 2.5MeV steps.

extra "external" Bragg peaks being required in order to ensure high dose coverage at the edge of the volume. An example of such a set of Bragg peaks for a single field, planned to a skull base case, is shown in Figure 24–7b, with the colors showing the relative weights of the individual spots before and after optimization (see next section). For a more detailed description of this process, see Lomax [2].

24.3.3 Optimizing Pencil Beam Fluences

The field-shaping step described above ensures that the target volume is covered by Bragg peaks, and thus defines the spatial form of the delivered field. Once this geometrical distribution of Bragg peaks has been defined, fluences (weights) can then be calculated for all these individual peaks. Given the large number of individual Bragg peaks in a typical field (from a few hundred to many tens of thousands for larger volumes), the only tractable solution to defining Bragg peak specific fluences is by optimization. The theory and practice of optimization is covered in detail elsewhere in this book (see Chapter 21, Plan Optimization) and here we will just briefly describe some of the different modes by which fields and plans can be optimized.

Three main optimization modes can be identified for PBS proton therapy:

- SFUD (Single Field, Uniform Dose)
- SFO (Single Field Optimization)
- IMPT (Intensity-modulated Proton Therapy)

Briefly put, the first two are based on field-specific optimization, while the third utilizes simultaneous optimization of all fields in a plan. As such, the difference between SFUD and SFO is quite subtle, with SFUD being the mode of optimization in which the dose to the target volume from each field is optimized to be as homogenous as possible [3]. As such, no critical structure or normal tissue constraints are defined. Although the simplest approach to field and plan optimization, SFUD is nevertheless important, as it provides the only mode for PBS proton therapy that approximates to open field treatments with photons or SOBP/broad fields for passive scattered proton therapy. In contrast, with SFO treatments, the dose homogeneity requirement is relaxed somewhat with the addition of critical structure constraints to the optimization process for the individual field. For IMPT optimization on the other hand, no requirements are made for individual fields, with the optimization instead being performed for all Bragg peaks of all fields, with or without additional constraints on critical structures.

As such, IMPT is the direct equivalent for PBS proton therapy to IMRT in conventional therapy, and from this similarity, gets its name [4]. Details on the actual optimization algorithms used can be found in Chapter 21, Plan Optimization, and will not be covered further here.

24.3.4 Plan Design

24.3.4.1 Selecting Field Directions

The finite range of protons in tissue can have a major impact on the selection of field directions for treatments. Whereas for photon therapy, field selection is driven not only by the avoidance of critical organs, but also by the need to achieve homogenous doses across the target volume, the ability of proton therapy to deliver a homogenous dose to the target from a single field direction allows for more flexibility in defining field angles and plan geometry. For instance, one of the oldest and most successful forms of proton therapy, the treatment of uveal melanomas, is a single field technique which provides highly conformal and homogenous doses to the tumor, with correspondingly high rates of tumor control (e.g., 98% at our eye-line facility [36]). Thus, while the trend in photon therapy has been a move toward more, and evenly spaced fields (e.g., IMRT, Tomotherapy, VMAT, etc.), there is likely less necessity for such approaches for proton therapy, as we will see in the following discussion.

So what drives the selection of field directions for PBS proton therapy? The main points can be summarized as follows. It should be stressed, however, that none of these are, or can be, absolute rules and are, thus, provided only as guidelines.

- avoidance of critical structures,
- avoidance of "ranging-out" on critical structures,
- minimization of path length to target,
- avoidance of major density heterogeneities,
- avoidance of anatomical or other structures that may vary in filling or position, and
- minimization of motion effects.

Avoidance of critical structures is the most obvious and should be taken into account wherever possible. Indeed, if a homogenous dose to the target volume can be achieved using one or more fields which do not pass through particular critical organs, then why not?

One reason may be that, to do this, one or all fields may range-out on critical structures (i.e., critical structures lying directly at the distal end of the fields). For reasons described above, this is something that typically should be avoided due to potential range and RBE uncertainties (but see the case studies in sections 24.4.2 and 24.4.3 below). Second, the fields may have to take longer paths to the target than necessary. It should always be remembered that the main advantage of proton therapy is its ability to substantially reduce the doses to *all* normal tissues in relation to conventional therapy, through the reduction of the non-target integral dose. This reduction is typically maximized by selecting fields that have the shortest path length to the target volume. Thus, from the integral dose point of view, for a laterally positioned tumor, it will always be an advantage to treat from the ipsi-lateral side, rather than contra-laterally. It is for this reason that, in the author's opinion at least, full rotational delivery techniques will only ever have a limited role to play in proton therapy.

Third, major density heterogeneities should be, where possible, also avoided, for reasons outlined in Section 24.2.4 above. The same applies to anatomical or other structures which may change their density, filling, or move during the course of treatment and could thus cause major range changes. In some cases, these regions may be the same (e.g., nasal cavities). However, one should also be cautious treating through any gastro-intestinal organs such as the bowel/intestine, bladder, rectum, etc. which, although not necessarily major density heterogeneities, can have dramatic changes in filling and form on a day-to-day basis.

Finally, although maybe not a major issue for many indications, possible organ or patient motion may also play a role in selecting field directions. As described in Section 24.2.7 above, one way of mitigating the interplay effect of PBS proton therapy is to use field directions whose beam axes are close to the major axis of motion. For example, this approach is typically used in our clinic for the treatment of extra-cranial chordomas/chondrosarcomas in the thoracic, lumbar, and sacral regions (see the case study in Section 24.4.2) where breathing motion of the spine (as a result of the prone positioning of the patient) can be an issue. Similarly, en face fields for the treatment of breast cancer with PBS proton therapy have been proposed for reducing breast motions due to breathing. An interesting approach has also been taken by Chang et al. for the PBS treatment of lung cancers, where 4D-CT data is used to calculate the change in water-equivalent range (WER) as a function of field direction to help with field selection [37]. Based on this, field incidences with minimum WER changes at the distal margin of the PTV can be preferentially selected in order to minimize motion-induced range errors.

24.3.4.2 The Use of Multiple Fields

Many of the issues of field selection as described above are based on making PBS proton plans more robust against potential delivery uncertainties. However, one of the simplest and most effective methods of improving robustness is the simple and, of course, well-accepted approach of using multiple fields. This has been demonstrated clearly for motion mitigation, where the use of multiple SFUD fields can "smooth-out" dose heterogeneities resulting from inter-play effects, simply because the "hot" and "cold" regions resulting from the effect occur in different positions in the different fields [25]. However, the use of multiple fields is also often the main mitigation approach for many of the other issues outlined above. For instance, if range uncertainty of any form is a worry, then the use of multiple fields with a larger angular spacing can help. For instance, as long as the distal regions of dose uncertainty from each field don't substantially overlap, then the potential dose errors (in relation to the prescription dose) will be more or less reduced by a factor directly proportional to the number of fields (i.e., dose uncertainties for a three-field plan will be reduced by about a factor three). The same, of course, applies for reducing the effect of potential uncertainties in RBE and for the degrading effects of density heterogeneities. An example of this is shown in Figure 24–8.

Figures 24–8a and b show frontal and lateral SFUD fields calculated for a skull base chordoma using the PSI planning system. The degrading effects of the extreme density heterogeneities traversed by the frontal field can be clearly seen at the distal end of the field, where there is no sharp distal fall-off, but many narrow "tongues" of high dose penetrating through and beyond the brain stem (the critical structure directly distal to the target volume). This figure on its own is a nice demonstration of the problem of density heterogeneities and why the more extreme ones (as for this field direction) should be avoided (see section 24.2.4 above). A similar, but less extreme effect is also seen for the lateral field. So, neither of these fields on their own

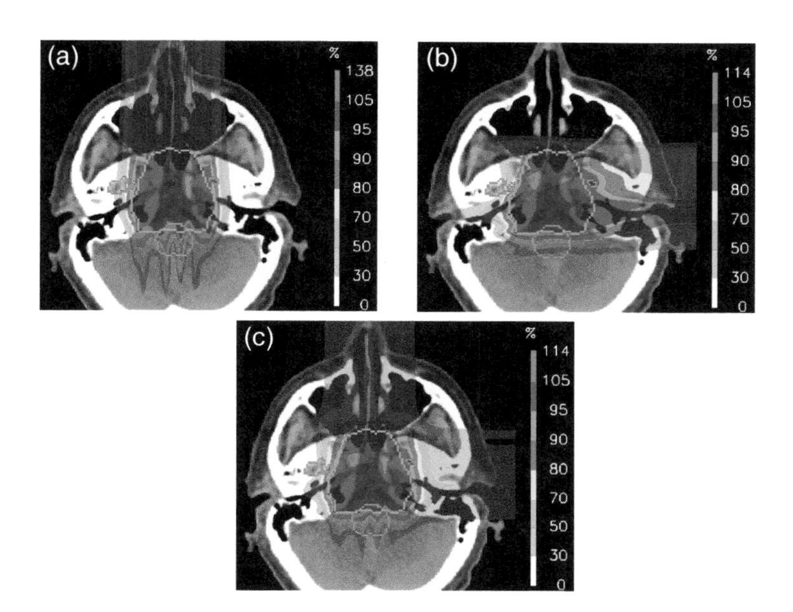

Figure 24–8 Two "unoptimal" fields and their combination to a skull base chordoma. (a) An anterior beam passing directly through the complex density heterogeneities of the nasal area. (b) A direct lateral field. (c) The dose distribution of both fields combined together with equal weighting. Such an approach is NOT used for treating such a tumor.

would be seen necessarily as "good" fields for treating such a target. However, when combined into a plan (with equal weighting to both fields), the resulting dose distribution suddenly appears to become much more acceptable (Figure 24–8c). This example nicely demonstrates the already forgiving nature of a simple two-field plan, even if we would *not* use such a plan clinically. The actual approach used for such a case will be described in detail in the case study in Section 24.4.1 below.

So, multiple fields can help treatments become more robust. Strictly speaking, however, the discussion above applies to SFUD and SFO plans only and not necessarily to IMPT plans. As the fields of an IMPT plan are not constrained to deliver homogenous doses to the target (see Section 24.3.2.2 above), they cannot automatically be relied on to mitigate uncertainties to the same level as SFUD plans. On the other hand, and as has been pointed out by Steven Dowdell [38], there is no absolute delineation between SFUD and IMPT, with the fields of an IMPT plan calculated with no constraints to any critical structures approaching those of SFUD plans from the point of view of dose homogeneity [3]. Nevertheless, if fields are deliberately planned in SFUD mode, single-field dose homogeneity is ensured, together with some of the mitigation effects described above. This is not necessarily the case for IMPT [3,39] and is the reason that our treatment planning system essentially defines SFUD and

IMPT as separate modes. Indeed, SFUD plans are used wherever possible at out institute, with IMPT being used only in cases where it is thought necessary, or in combination with SFUD as part of multiple series treatments (see Section 24.4 below).

As a final comment on the use of multiple fields, the potential *advantages* of IMPT (which, by definition, is a multiple field approach) should also be noted. Just like IMRT with photons, IMPT allows for more flexibility and automation in planning than SFUD/SFO and, due to the use of different incident fields, provides more potential for the sparing of critical structures overlapping the planning target.

In addition, and perhaps paradoxically, IMPT can be used to *improve* plan robustness. Indeed, the first IMPT treatment delivered at our institute in 2000 was planned so as to ensure a more robust plan than could be achieved with a single field SFUD approach [40]. Consequently, the flexibility of IMPT is now beginning to be exploited to make plans robust against either setup or range errors, either by incorporating robustness directly into the optimization [34] or through the manipulation of starting conditions for the IMPT optimization [41].

In summary then, and in addition to the more obvious reason such as reducing entrance and skin dose, there are many good reasons to use multiple fields for treatments, and very few to use a single field (cranial-spinal irradiations apart perhaps). On the other hand, it is our belief that there is less of a need to use as many fields as is currently the trend in photon therapy. Indeed, in our 18 years of clinical practice with PBS proton therapy, we have not felt it necessary to use more than four fields for any series of a treatment. On the other hand, this is also partly to do with the time to deliver a single field on our original PBS gantry, with which the majority of patients have been treated.

And this in itself indicates the other side of the coin for multiple-field treatments. The three-dimensional nature and discrete delivery of PBS proton therapy (i.e., that pencil beams are distributed and modulated in three dimensions for every field and are delivered one after the other) necessarily means that treatment times per field will be considerably higher than for modulated photon fields. Thus, and as always, a balance needs to be found between the mitigating and advantageous effects of using multiple fields, and the inevitable increase in treatment time resulting from their use.

24.3.5 Tips and Tricks

We will end this section with a few "tips and tricks" that we have found useful for helping PBS treatment planning. Although some of these may well be specific to our planning system, we hope that nevertheless they may provide some hints and ideas as to how to proceed if all seems lost in the planning of particularly complex cases.

24.3.5.1 Overwriting/Assigning Stopping Powers (Please Use This)

A first tip is to not be afraid of modifying the planning CT if this can make the plan more precise or robust—as long as it is done in controlled conditions and with caution. There are a number of reasons why you may wish to do this. For instance, one may wish to correct the HU or stopping power of nonbiological materials (e.g., bolus material) which do not lie on the standard HU-SP calibration curve (see section 24.2.3

above). In addition, to improve plan robustness, we have sometimes found it advantageous to override CT values in anatomical structures that we know will change during treatment, such as to create a "worst-case" condition on which to plan. One such example is if we have to treat through bowel gas on the planning CT. This is clearly an "artifact" in the planning CT which will change day-to-day in an unpredictable way. In these circumstances, we outline these regions and set the SP value to that of water. From the point of view of range and target coverage, this is the "worst case" condition as it assumes that the bowel loop will always be filled, and thus the residual range in the patient will be the shortest. If in this condition we have coverage of the distal end of the target, then the target will also be well covered even when the bowel loop is gas filled. A similar approach has been suggested by Soukup et al. [42] for prostate treatments. Note, however, that this is a worst case condition only for target coverage and not, of course, if the field is ranging-out on a critical structure that may be sensitive to small volumes of high dose (i.e., serially structured OARs).

Another application is for the assessment of the effects of nasal cavity filling on the delivered plan. For patients where we have fields passing through the nasal cavities, we perform regular but limited-range CT scans (i.e., slices just through these cavities) to assess this filling. If this is significantly different from that of the planning CT, then these slices are registered to the planning CT, the changes delineated, and then the corresponding voxels on the planning CT changed accordingly to either water if the cavity filling has increased, or air if it has decreased. The treatment plan is then simply re-calculated on the modified planning CT and the results compared to that of the nominal plan.

So changing the CT can help in the planning process and for assessing the potential consequences of anatomical changes. However, whenever the planning CT is modified, the planner must be aware of the consequences of these changes, and must also be aware of the non-uniqueness of the HU to proton stopping power conversion, as described in Section 24.2.3. Thus, if only CT HU values can be assigned to voxels, then it is imperative that the calibration curve is known, such that the HU corresponding to the desired stopping power can be assigned. In addition, the regions to be modified need to be as carefully defined as possible and generally in a conservative way. By this we mean that such regions should be defined on the small side, rather than too generously, particularly if the region to be modified is bordering with regions of very different stopping power. For instance, if outlining a bolus on the patient surface in order to correct its stopping power (typically bolus materials will not fit on a biologically defined calibration curve), then it is better to define the region of the bolus a little inside the visible border, rather than including air voxels into the region—the stopping power of the non-corrected bolus voxels will be closer to the correct value for the bolus than the stopping power of air voxels that may be included by mistake.

24.3.5.2 Dealing with Metal Implants and CT Artifacts

Metal implants of any type pose a considerable challenge to proton therapy. First, CT HUs typically saturate at some value above 3000 HU (the actual maximum HU value is CT dependent), and thus it is impossible to tell from these alone what the metal implant is or to automatically assign the correct stopping power without making some form of assumption about the material. Second, the implants themselves, due to their high densities, can cause considerable reconstruction artifacts in the CT, thus essentially corrupting large regions of the CT. Given the dependence of accurate proton dose calculations on good-quality CT data, the consequences of both problems are hopefully obvious.

There is currently no ideal way of dealing with this problem perfectly, and much work needs to be done to make proton treatments in the presence of metal a more exact science. Nevertheless, every proton therapy practitioner will be confronted with such implants, whether they are gold or amalgam tooth fillings in head and neck patients, stabilizing rods and screws in post-operative para-spinal irradiations, or hip prostheses in prostate patients. Thus a strategy of how to proceed needs to be defined. Although each center may come up with its own approach, it may nevertheless be helpful to describe how we mitigate this problem.

First, for the problem of assigning the correct stopping power to metal, as with many other things, we have taken a simple, and hopefully pragmatic, approach. In our clinical calibration curve, we by default have the stopping power of titanium assigned to the highest HU value provided by our planning CT (3095). Thus, any voxels with this HU value in the planning CT are automatically assigned the stopping power of titanium. As the majority of patients we treat with implants are post-operative para-spinal tumors, the implant is invariably titanium, and for such cases the stopping power of the implant itself is automatically defined without necessitating any manipulation of the planning CT data or having to delineate the metal structures. We have also found that this approach quite accurately models the shape and dimensions of the metal artifact itself.

As our current in-house developed planning system doesn't support the possibility of directly assigning stopping powers to pre-defined delineated regions of the planning CT, we simply cannot correctly support any other metal types. Thus, if we have patients with gold implants, we try to completely avoid treating through these or, if we must, they are outlined and assigned very low dose constraints in the optimization, such as to reduce as much as possible the weight of pencil beams passing through these (see also Section 24.3.4.4 below). For other implants, (e.g., amalgam or platinum coils), we estimate the thickness and stopping power from descriptions of the materials and perform estimates of the range error that we would expect. We make no claims that this is perfect or correct. On the other hand, the range errors from amalgam fillings (being typically only a few mm thick) are not large, whereas platinum coils, although having a stopping power more than three times that of titanium, tend to be sub-millimeter thick, and thus result in range errors of at most 1–2 mm.

In principle, directly assigning proton stopping powers to delineated regions of the planning CT (as provided by most commercial planning systems) should be more accurate, but this also necessitates the accurate definition of the implant. Whether this really helps in the end is an open question. For instance, a fine platinum coil is impossible to resolve completely in CT due to its small dimensions and high density (e.g., in our CT, the whole coil appears as a few voxels, all with HU values of 3095, whereas the actual coil wire is much smaller in dimension than a CT voxel). If all these voxels are assigned the stopping power of platinum, then very likely the stopping effect of the coil will be significantly over estimated. The bottom line then is that, if at all possible, fields passing through metal implants should be avoided.

The second and, in our opinion, more critical issue is the problem of reconstruction artifacts resulting from high-density metal implants. An example is shown in Figure 24–9a. It is clear from this figure that if treating on this uncorrected CT data set, the range errors will be considerable, even if the metal implant itself could be avoided, simply because large regions of the CT data have corrupted HU values. Our policy with these is to, as accurately as possible, delineate all artifacts that are deemed to be corrupting areas of soft tissue, and then assign these regions an average SP for soft tissue. This is either a standard value, or it can be estimated from similar anatomical areas on slices where there are no artifacts. Such a correction can be seen in Figure 24–9b.

In principle, bony regions affected by artifacts could be delineated as well, and then assigned an average bone value. We have decided not to do this, as accurately reconstructing bony structures under the artifacts is difficult. Also, in many anatomical areas, the amount of bone through which fields pass is considerably less than that of soft tissue. As such, we believe that the largest corrective effect is through the correction of the soft tissue regions, with the extra delineation of the much more complex

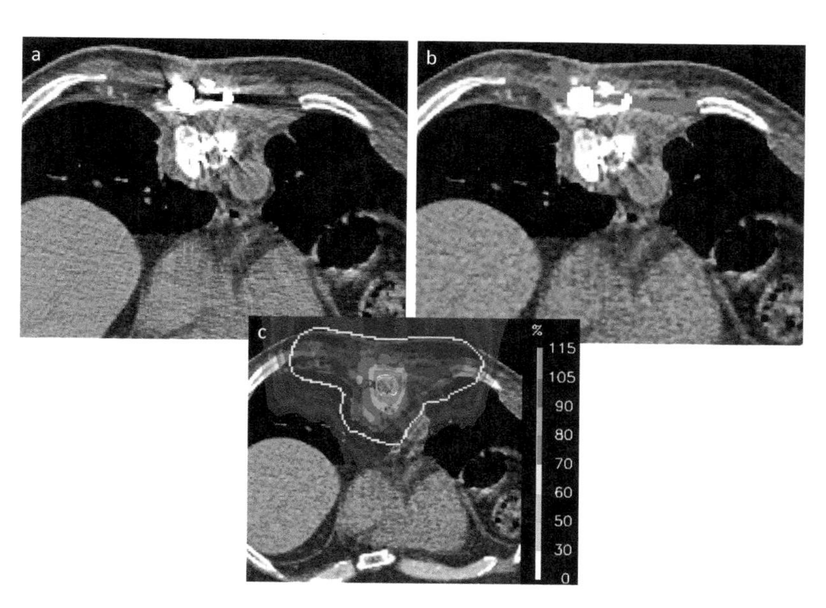

Figure 24–9 Metal artifact correction for a post-operative thoracic chordoma. (a) The nominal planning CT. Titanium implants are clearly visible, as are significant reconstruction artifacts resulting from these. (b) The CT on which the plan was calculated with the artifacts in the soft tissue regions corrected with an average HU value. (c) The two-field IMPT plan calculated and delivered to this patient.

bony structures not really being justified. An IMPT plan delivered to this patient, calculated with artifact corrections, is shown in Figure 24–9c.

So is this policy justified? In practice, it is very difficult to say. However, we have recently tried to validate the approach as accurately as possible using a customized anthropomorphic phantom with an implanted titanium rod [43]. Although this was a somewhat simplified representation of a real patient, we could experimentally show that the soft tissue correction method described above substantially improved the agreement between planned and delivered dose when measured in the phantom, and it even provided agreements close to those of a similar experiment performed in the same phantom in a region without metal implants or artifacts [15].

As a final comment, most modern CT scanners now provide reconstruction algorithms that reduce reconstruction artifacts. However, it is our experience that these result in altered HU values through the whole field-of-view of the CT, potentially rendering the CT-SP calibration invalid. It is currently not clear how dependent these changes are on the metal structures and artifacts in the image and, therefore, whether a specific HU-SP calibration could be defined for such CT sets. Whatever the case, CT data sets reconstructed using artifact reduction algorithms could help the delineation of the soft tissue regions using the artifact correction approach described above.

In summary, accurately treating patients with complex metal implants using proton therapy remains extremely challenging, and much work needs to be done to improve this. On the other hand, this is another case where common sense should prevail. If such patients have to be treated, then the best that can be done is to understand both the problem and the limitations of the solution, and then to assess the quality of the plan under these conditions. In addition, it may be worth considering extending the PTV margin to account for the extra uncertainties or using multiple fields to smooth out the potential effects of the inevitable residual uncertainties. Such approaches may degrade the overall quality of the plan (i.e., more normal tissue will be irradiated), but it will not necessarily mean that the advantages of proton therapy will disappear.

24.3.5.3 Tips for SFUD/SFO Planning

As discussed already, SFUD/SFO planning is in many ways more time consuming than IMPT, but for many indications it can still achieve excellent and robust treatments. As an example, the vast majority of all patients treated with PBS at our institute receive an SFUD plan as at least part of their multi-series treatment, and SFO—where critical organ constraints are also applied as part of the single-field optimization—certainly increases the applicability of this technique. Unfortunately, in our current TPS implementation, SFO optimization is not routinely supported. Instead, we work with so-called technical PTVs. These are copies of the actual clinical PTV which are locally modified in order to, for example, pull the dose out of neighboring or overlapping critical structures. An example for such a plan can be found in the case studies below, and although planned using the technical PTV approach, can be taken as an example of what can be achieved using SFO optimization.

Additional flexibility can also be achieved through varying the weight of individual SFUD/SFO fields when constructing plans. At first glance, varying the weights of

SFUD/SFO fields doesn't seem to make much sense. After all, each individual field delivers a homogenous dose across the target, so why then change the weights of the fields? There may be a number of reasons to do so. Perhaps one beam has been selected which passes directly through a critical structure, and this is just over the tolerance. A simple way of correcting this would be to simply reduce the weight of this field. Similarly, perhaps a contra-lateral beam has been defined for a rather lateral tumor. This helps target coverage or overall dose homogeneity, but, of course, adds to overall integral dose. In this case, reducing its weight may still improve target dose, while somewhat reducing integral dose.

There are other reasons to vary field weights. We have already referred to Figure 24–8a as a field that passes through extreme density heterogeneities, and we have pointed out the severely degraded distal falloff as a result of these. Another consequence is the rather poor overall dose homogeneity throughout the PTV. In other words, although SFUD optimization has been applied to this field, the degrading effects of the density heterogeneities on the individual pencil beams limits the dose homogeneity that is actually possible, indicating another reason why, in an ideal world, such a field should be avoided. However, it is simply not always possible to avoid all such heterogeneities. For instance, in the treatment of paranasal sinus tumors, it is extremely difficult to avoid such fields. In order to improve the overall dose homogeneity of the plan, therefore, it can be advantageous to reduce the weight of such fields in relation to other, more homogenous, fields.

In summary, field weighting of SFUD/SFO fields provides an additional degree of freedom in the planning process which can allow for the inclusion of individually "sub-optimal" beams into a plan while reducing their potential negative influence, giving somewhat increased flexibility to SFUD/SFO planning.

24.3.5.4 Tips for IMPT Planning

As has already been mentioned, IMPT planning is an inherently more automated process than SFUD/SFO planning. Nevertheless, although we talk about "optimization" in this process (as discussed in Chapter 21, Plan Optimization), there is, in fact, no such thing as an "optimal" plan. This is the reason that much work is now being done on multi-criteria optimization algorithms. In addition, given the complexity of the optimization problem for IMPT, it is not to be expected that current optimization algorithms will fully explore all possible solutions, and if the solution found by the optimizer is not quite what is required, it can be challenging to know what parameters need to be changed. So, despite the apparent fully automated nature of IMPT planning, it is also an iterative process, not just at the level of the optimization itself, but due to the fact that many different plans typically need to be calculated until a good clinical solution is found.

As such, we also include a number of tips and tricks for IMPT planning. It should be noted, however, that the success, or indeed necessity, of many of these may be dependent on the optimization algorithm used and, as such, may only be relevant for our institute. Nevertheless, we have decided to include these as possible pointers to try out if the IMPT optimization algorithm you are using is not giving quite the results you want. These sometimes work for us. Maybe they will work for you.

First, we have found that technical PTVs, as described for SFUD plans above, can also be useful for IMPT planning. For instance, for skull base cases (see the case study in Section 24.4.1) the chiasm and optic nerves often overlap with the cranial-most aspect of the PTV. These are small structures that are very close to each other, with separations smaller than typical pencil beam widths. Thus, trying to maximize PTV coverage in this region, while simultaneously sparing these structures, is extremely difficult. We may, therefore, modify the nominal PTV by simply removing the most cranial 1–2 contours of the PTV (i.e., those that overlap with the optic structures). Once again, this may not seem an optimal solution, but we have found that coverage of the nominal PTV is only marginally affected by this approach, whereas overall dose homogeneity of the plan can be improved. Technical PTVs can also help to spare OARs at the distal end of the nominal PTV. Again, we find that, at least with our optimizer, this can lead to somewhat more homogenous plans than planning directly on the nominal PTV and relying on the optimizer to reduce the dose on its own.

Another VOI-based technique is the use of what we call "virtual blocks," an approach similar to that already described in Section 24.3.4.2 for reducing the weight of pencil beams passing through gold teeth. A "virtual block" is a typically non-anatomic structure defined at the time of planning, to which dose constraints can be applied (sometimes even zero dose) in order to prevent, or reduce the weight, of pencil beams passing through. We find these particularly useful for head and neck cases where the treatment includes both lymph node chains, as shown in Figure 24–10. In these cases, laterally opposed beam arrangements are necessary for optimally treating both nodes (see Figure 24–10a), but they are *not* necessarily optimal for sparing the

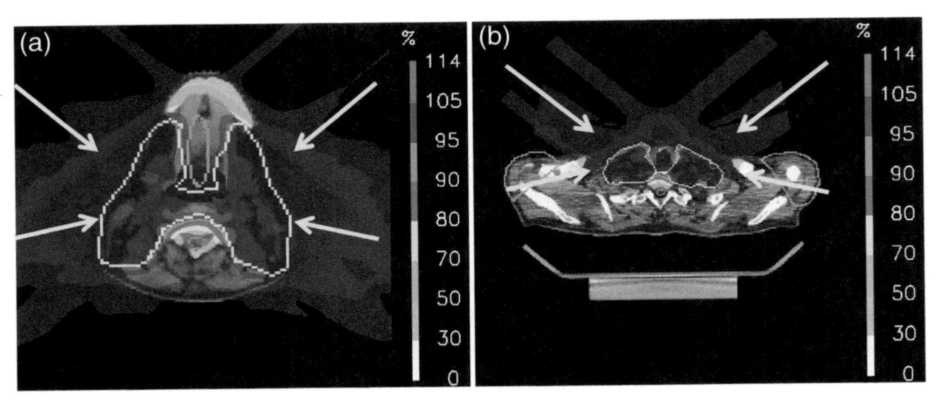

Figure 24–10 Example "virtual block" IMPT treatment for a head and neck case, with field directions as indicated by the yellow arrows. (a) A virtual block (green contour) in between the lymph nodes to reduce dose to the normal tissues. (b) "virtual blocks" (green contours) used to "block" out the lateral-posterior fields from passing through the shoulders.

normal tissues in between the nodes. For this reason, we defined a geometric virtual block (green contour) in the region between the nodes and give this a dose constraint of about 75% of the prescription dose. The result is that pencil beams from the left lateral field primarily treat the left nodes (the dose constraint on the "block" reduces the weight of pencil beams passing through it) and the right primarily to the right (see Figure 24–10a). The result is a reduced dose to the normal tissue between the nodes. A similar technique has been used by Widesott et al. [44] in order to spare large volumes of intestine in the treatment of advanced prostate cancer with just two lateral fields. In addition, as we use a "star" field arrangement for such cases (anterior- and posterior-lateral beams from both sides, see Figure 24–10) virtual blocks are also used in the treatment of the caudal aspects of the lymph nodes in order to "block out" the two posterior beams which would otherwise have to pass through the shoulders (see Figure 24–10b). In this case, constraints of zero dose were given to both blocks (green contours in Figure 24–9b) in order to completely block out these fields at this level.

As a final tip, we find that it helps to reach maximum dose constraints on critical organs if we optimize on an expanded version of the OAR, rather than on the OAR directly. Typically, we use a 3 mm iso-tropic expansion around particularly small OARs—such as the brain stem, spinal cord, and optical structures—and optimize to these structures. Occasionally we may also include somewhat lower constraints on the OAR itself, as we feel that this sometimes helps to provide a smoother dose gradient around the organ. This expansion should not be confused with the PRV concept recommended in ICRU 62, however. The expansion is not providing a safety margin around the OAR, but is simply a help for the optimization to reach the maximum dose constraints. Nevertheless, these expansions do have the tendency to pull the dose and dose gradient out from the OAR a little, thereby also providing a small safety margin. The use of expanded OARs for the optimization is a more or less standard approach for us now. In addition, we find additional dose–volume constraints applied to these expanded structures can also help to "shape" the dose distribution more precisely and lead to steeper dose gradients.

24.4 Case Studies

At the time of writing, we have treated more than 1000 patients with PBS proton therapy at PSI with encouraging clinical results. In this section, we have selected example plans from four of our most prevalent indications in order to provide a flavor of the planning strategies used. The description of each case starts with a brief description of the indication, followed by typical target prescriptions and OAR constraints, a description of the example plan, clinical outcomes for the indication (if available), and finally, a short summary.

24.4.1 Case Study 1—Skull Base Chordoma

24.4.1.1 Description of Indication

Chordomas are rare tumors of the bone that can occur anywhere along the neuro-axis. Due to their location close to many critical structures—such as the brain stem, optic

Table 24–1 Typical dose prescriptions and constraints used for skull base chordomas at Paul Scherrer Institut

Target/Structure	Prescription Dose (Gy (RBE))	Mean Dose (Gy (RBE))	D2 Dose (Gy (RBE))	Other Constraints (Gy (RBE))
PTV1 (CTV + 5 mm)	0–54	-	-	-
PTV2 (GTV +5 mm)	54–74	-	-	-
brain stem	-	-	64	-
brain stem center	-	-	53	-
Spinal cord	-	-	64	-
Spinal cord center	-	-	53	-
Chiasma	-	-	60	-
Optic nerves	-	-	60	-
Cochleas	-	-	50	-
Temporal lobes	-	-	≤105%	2 cc ≤74

structures, and cochleas—skull base chordomas are particularly challenging and have been a "standard" indication for proton therapy for many years.

24.4.1.2 Prescription

The typical dose prescriptions and constraints for skull base chordomas used at our institute are summarized in Table 24–1, and these are the ones adhered to for the case study described here.

24.4.1.3 Example Plan

Figure 24–11 shows the plans for a typical skull base chordoma treated at our institute. Three different series have been used—a three-field SFUD plan from 0–36 Gy (RBE) planned directly to PTV1, a four-field IMPT plan also to PTV1 from 36–54 Gy (RBE), and a final IMPT plan to boost PTV2 to 74 Gy (RBE).

For the 3-field SFUD plan (series 1), one quasi-lateral and two apical-lateral oblique fields have been used, with the quasi-lateral field being slightly angled frontally and also 10 degrees cranially. This small angulation and couch-kick are designed to avoid the field passing directly along the sella-plate of the skull base, which otherwise can present as a rather long interface between bone and soft tissue. But this approach also allows us to reduce the air gap a little by bringing the treatment nozzle into the angle between the shoulders and head of the patient. The two apical fields are used as they predominantly come through very homogenous parts of the patient, but also because they can help improve the dose gradient to the brain stem due to the small air gaps (we can get the nozzle very close to the patient from these directions). As this is an SFUD plan, no attempt has been made to spare the brain stem in this series, with full PTV1 coverage being the priority. This plan was delivered for almost half of the complete treatment.

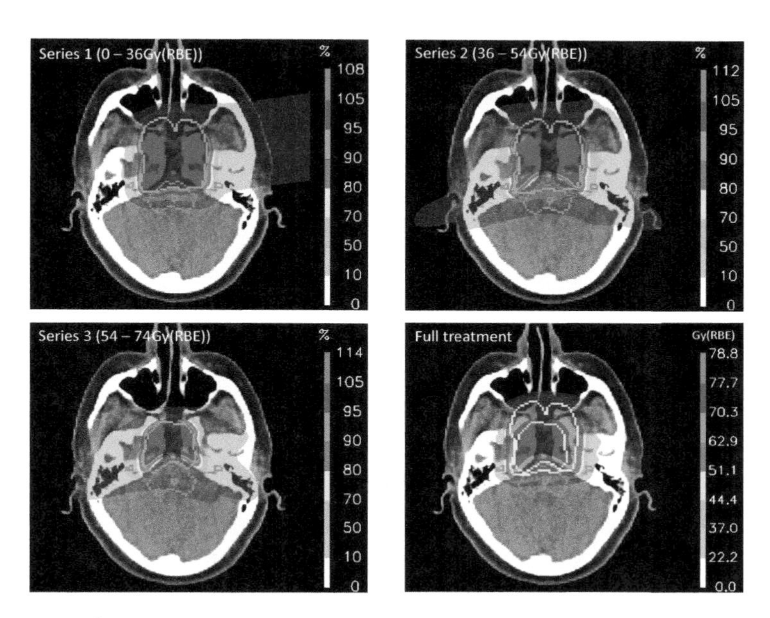

Figure 24–11 Case study 1—skull base chordoma.

The second series IMPT plan, also planned to PTV1, was introduced from 36 Gy (RBE) in order to start pulling the dose away from the main critical structures such as the brain stem (as seen in Figure 24–11b) and the optical structures (not seen at this slice level). For the IMPT plans, however, we change the field configuration to a shallow "star arrangement," using two lateral-oblique fields from both sides, angled 20 to 30 degrees away from the lateral direction. In addition, and again to help avoid coming directly parallel to the skull base bony structures, all four fields have couch kicks of 20 to 30 degrees. Although not seen at the slice level shown in the figure, this IMPT plan has also been helped by planning to a TechPTV, in which the most cranial two slices of the PTV (which overlap directly with the chiasm) have been deleted (as described above). Exactly the same field arrangement has been used for the final series, planned to the unmodified PTV2.

Figure 24–11d shows the combined dose distribution for the whole treatment, with PTV1 and PTV2 shown in yellow, and the color banding now displayed in absolute dose. The 51.1Gy (RBE) banding is 95% of 54 Gy, and the upper limit of the mauve color wash is also the tolerance level for the brain stem surface (63Gy (RBE)). One can see how this banding follows the form of the brain stem surface, while also covering the whole PTV1. The blue banding shows the 95% dose level to the PTV2 (77.7 Gy (RBE)) and also shows good coverage and conformity, except in the posterior portion, where the dose necessarily has been compromised in order to satisfy the D2 constraint to the brain stem surface. The brain stem center and both cochleas (both visible in the figures) are all spared and kept well below tolerance.

24.4.1.4 Clinical Outcomes

Ares et al. has reported on the clinical outcome of 196 patients treated for skull base chordomas (n=134) and chondrosarcomas (n=62) at PSI from 1998 to 2011 [45], the great majority of whom were treated using a strategy and planning approach similar to that described here. Actuarial five-year local control rate and overall survival was 80.6% and 85.5% respectively. Main toxicities (grade 3) were in the temporal lobes (n=10 (5%)), optic structures (n=5 (2.5%)), and unilateral hearing loss (n=2 (1%)). No incidences of brain stem or spinal cord toxicity have been observed. Note, the temporal lobe tolerances listed in the table above were only introduced in 2009 as a result of an analysis of the first 62 patients treated up to 2005 [46].

24.4.1.5 Summary

Skull base chordomas and chondroarcomas are the largest single indication that we have treated at PSI using PBS proton therapy. The planning approach described here has been the standard approach for many years, and it was selected in order to try to balance dose homogeneity to the PTV, plan robustness (hence the first series SFUD plan), and conformal avoidance of neighboring critical structures (hence the 2nd and 3rd series IMPT treatments).

24.4.2 Case Study 2—Sacral Chordoma

24.4.2.1 Description of Indication

Chordomas can also occur along the spinal axis, and they can result in rather large volumes surrounding the spinal cord, cauda equine, and nerve roots. As with skull base chordomas, such cases have been one of our major indications in the past years due to their size and challenging location.

24.4.2.2 Prescriptions

As with skull base chordomas, the prescription dose is typically 74 Gy, with the main critical structures being the cauda-equina, nerve roots, rectum, and intestine. A typical prescription is summarized in Table 24–2.

24.4.2.3 Example Plan

A typical sacral chordoma plan is shown in Figure 24–12. The treatment consisted of two series, the individual dose distributions of which are shown in figures 24–12a and b. As with most sacral chordomas, the initial PTV was large (1.4l) and extended 18 cm in the cranial–caudal direction. The slice shown is toward the cranial end of the volume.

Series 1 (0–36 Gy (RBE), Figure 24–12a) was a two-field SFUD plan, planned to a technical PTV derived from PTV1, slightly modified to pull the proximal portion of the PTV a little below the skin in order to reduce skin dose. A "narrow-angle" [47] approach has been used, using just two fields, angled 15 degrees away from the posterior aspect. Such a narrow approach has been adopted for two reasons. First, and as seen in Figure 24–12, this minimizes the path length to the target volume, thus minimizing integral dose to normal tissues and organs in the abdomen. However, a second

Table 24–2 Typical dose prescriptions and constraints used for sacral chordomas at Paul Scherrer Institut

Target/Structure	Prescription Dose (Gy (RBE))	Mean Dose (Gy (RBE))	D2 Dose (Gy (RBE))	Other Constraints (Gy (RBE))
PTV1 (CTV + 7 mm)	74	-	-	-
Cauda equine	-	-	<70	-
Nerve roots	-	-	<70	-
Small intestine	-	-	68	-
Posterior rectal wall	-	-	68	-
Anus	-	-	68	-
Skin	-	-	-	No hot spots

reason is to minimize the effects of patient motion. For technical reasons, all such patients at our institute have been treated in the prone position, which can lead to significant AP motion of the spinal column due to motion of the chest wall. The narrow angle approach used has, therefore, also been selected such that the residual AP motion is mainly *parallel* to the incident fields, thus minimizing potential inter-play effects during delivery. The dosimetric potential of protons and the very narrow approach angles for such cases are clearly evident, with excellent coverage and con-

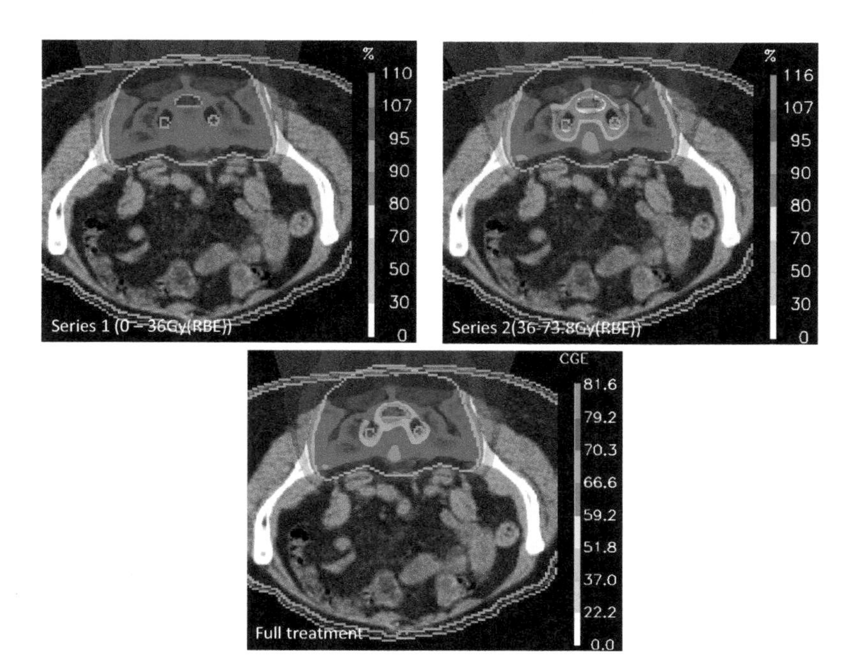

Figure 24–12 Case study 2—sacral chordoma.

formation of the dose and very little normal tissue involvement. Indeed, this is an excellent demonstration of the safe use of the stopping potential of protons.

For series 2, we then moved to an IMPT plan (36–74 Gy (RBE), Figure 24–12b), again using two posterior oblique fields, but this time with a wider separation (30 degrees away from the posterior direction). For this, constraints were applied primarily on the cauda-equina and nerve roots. The wider separation gives the IMPT optimizer more flexibility in constructing the dose "hole" around the cauda and nerve roots, and despite the use of only two fields with just a 60 degree separation, it is nevertheless possible to create a "donut" type distribution while completely sparing the anterior organs of the patient. This plan nicely demonstrates the power provided by the modulation of three dimensionally distributed Bragg peaks.

The full treatment for this patient is finally shown in Figure 24–12c. This shows an impressive coverage of the PTV with the 95% iso-dose line (70.3 Gy (RBE)) and excellent dose conformation to the PTV of not just the high dose, but at *all* dose levels. Also note the excellent sparing of the cauda-equina and nerve roots to doses of around 60–66 Gy (RBE), which more than satisfies the requirement that these should be kept below 70 Gy (RBE), despite these structures being in the middle of both PTVs.

24.4.2.4 Clinical Outcomes

In an initial evaluation of 40 extra-cranial chordomas treated at PSI [48], the five-year local control rate for the 19 patients *without* metal implants (11 of which were sacral chordomas) was 100%, with a median follow-up of 43 months. For this particular case study, the patient experienced grade 2 acute skin toxicity, but is locally controlled six years after therapy with no late side effects, including none in the intestine, small bowel, or rectum. Such low levels of late side effects in the GU structures are a standard for these patients using this treatment approach, as has been confirmed in the more recent follow-up of 31 patients (81% with extra cranial chordoma) published by Schneider et al. [49]. After a mean follow-up of 3.8 years, no late sequelae were observed in the small bowel for any of this cohort of patients.

24.4.2.5 Summary

This case study shows how the flexibility of PBS proton therapy in either SFUD or IMPT mode can be successfully utilized for the treatment of large target volumes in challenging anatomical regions. It also demonstrates the substantial advantage that the stopping characteristics of protons *can* bring safely and robustly. Indeed, it is likely because of this characteristic that we have been able to successfully treat a 1.4 liter target volume to 74 Gy with little or no late side effects, despite its proximity to many abdominal organs.

24.4.3 Case Study 3—Ependymomas

24.4.3.1 Description of Indication

Ependymomas are tumors of the central nervous system and occur mainly intra-cranially in pediatric patients. They make up about 10% of pediatric CNS tumors. They are the most common pediatric tumor that we have treated at our institute since we started our pediatric program in 2004.

Table 24–3 Typical dose prescriptions and constraints used for ependymomas at Paul Scherrer Institut

Target/Structure	Prescription Dose (Gy (RBE))	Mean Dose (Gy (RBE))	D2 Dose (Gy (RBE))	Other Constraints (Gy (RBE))
PTV1 (CTV + 5 mm)	54.0	-	-	-
PTV2 (GTV + 5 mm)	59.4	-	-	-
Brain stem	-	-	60	-
Center brain stem	-	-	53	-
Chiasm	-	-	60	-
Optic nerve	-	-	60	-
Cervical myelon	-	-	50	-
Inner ear	-	36	45	-

24.4.3.2 Prescriptions

A typical prescription is summarized in Table 24–3.

24.4.3.3 Example Plan

Figure 24–13 shows a typical ependymoma plan as treated at PSI. The full treatment consists of three series. The first series is a three-field SFUD plan calculated on a technical PTV (green) derived from PTV1 (outer yellow contour). The selected field

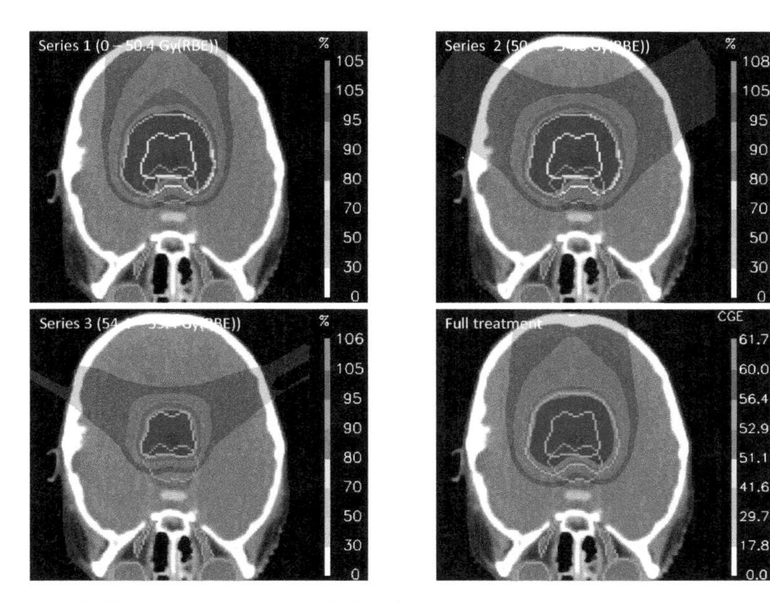

Figure 24–13 Case study 3—sacral chordoma.

directions were two posterior-lateral oblique fields (30 degrees from posterior) and a posterior field. All fields were weighted equally. The technical PTV has been used here to selectively reduce the dose to the center of the brain stem, which is the main dose-limiting structure at this level of the target volume. The second series plan is planned to a similar technical volume, but using just two lateral–posterior oblique fields (±60 degrees away from the posterior aspect), whereas the third series is planned on a technical volume derived from PTV2, but using the same field arrangement as series 2. For series 2 and 3, the technical volume has been modified at the level of the spinal cord to exclude this organ (not seen in the figure), in order to constrain the spinal cord to 50 Gy. The full treatment is shown in Figure 24–13d, with the 56.4 and 51.1 Gy (RBE) dose levels corresponding to 95% of the PTV2 and PTV1 prescription doses respectively. In addition, the 60 Gy (RBE) contour (constraint to brain stem surface) and 52.9 Gy (RBE) dose (constraint to brain stem center) contours are shown.

As is clear from the figure, with this planning approach (used to treat the vast majority of the more than 50 ependymoma patients treated at PSI), all fields actually range out inside the brain stem, which could be considered to go against received wisdom about safe planning with protons therapy (see above). However, in the first series, we use three fields, each which range out in somewhat different portions of the PTV, and then use different, more widely spaced angles for the last two series. In addition, a technical planning volume has been used already from the first series (which, remember, is our method of planning SFO treatments) to reduce the dose to the brain stem center from day one. This has the consequence that, on this slice at least, the total dose to the brain stem center is somewhat below the 53 Gy (RBE) constraint (see Figure 24–13), while the dose to the anterior aspect of the brain stem is only of the order 20 Gy (RBE). Finally, although at this level there is no apparent reason not to use more lateral fields, in more caudal aspects, their use would have the disadvantage of also increasing dose to the inner ear and also somewhat increasing the volume of normal brain irradiated.

24.4.3.4 Clinical Outcomes

The outcomes for 50 pediatric ependymoma patients treated at PSI using the above approach are currently being analyzed. Preliminary analysis shows comparable five-year survivals to published photon [50] and proton [51] series, even though 92% of the patients in the PSI series were high risk (grade 3) patients, compared to only about 50% in the other series. In addition, despite "ranging out" on the brain stem, incidences of grade 3 toxicity in this organ are also comparable to those published in comparable photon series (1.6% as reported in [50]).

24.4.3.5 Summary

The treatment of ependymomas is a good example of using the finite range of protons, together with a restricted spread of field incidences, to achieve excellent high dose conformity, complete sparing of the anterior optical structures, and reduction of the irradiated brain volume. However, using this approach, we are aware of the potential of range uncertainty and potentially increased RBE and have taken this into account by varying the incident angles in the different series, as well as compromising the PTV coverage in order to ensure that the brain stem and center stay below tolerance. Initial analysis of our clinical outcomes would seem to confirm the efficacy of this approach.

24.4.4 Case Study 4—Parameningeal Rhabdomyosarcoma

24.4.4.1 Description of Indication

Rhabdomysarcomas originate from skeletal muscle and are a major pediatric indication for radiotherapy. A subset of these are tumors originating in the para-meninges, which can result in superficial target volumes close to the optic structures eyes and lacrimal glands.

24.4.4.2 Prescriptions

A typical prescription is summarized in Table 24–4.

One of the over 40 parameningeal rhadomyaosarcomas that we have treated is shown in Figure 24–14. Once again, three series have been planned, but this time exactly following the prescription doses to the three PTVs defined for such cases. Series 1 is a three-field IMPT treatment to PTV1. It consists of a lateral and an anterior-lateral oblique field, plus a third lateral-posterior field with a 20 degree table kick in the cranial direction. IMPT has been used for this in order to already start sparing the ipsi-lateral orbit and lacrimal gland. The second series plan is also an IMPT plan which uses the same angles, but is planned directly on PTV2. The final series is a two-fraction SFUD boost to PTV3 using just two fields. The dose distribution for the full treatment is shown in Figure 24–14d.

This example shows once again how the use of fields all predominantly from the same quadrant can achieve good target coverage of all PTVs, while allowing for complete dose sparing of the contra-lateral anatomy (e.g., contra-lateral hemisphere, eye, lacrimal gland, etc.). In addition, there is significant sparing of the brain stem. However, dose sparing of the ipsi-lateral eye and lacrimal gland could only be achieved by compromising coverage of PTV1 (95% coverage of this PTV corresponds to the 42.7 Gy (RBE) dose level in the figure) due to the poor lateral falloff of the predominantly superficial Bragg peaks required to cover these targets.

It is evident in the figure that for treatment, the patient's head has been tilted roughly 20 degrees toward the contra-lateral side. This is predominantly to do with mechanical restrictions that we have with our Gantry 1, but it has also been done in order to allow for minimizing the air gap for the delivered fields. By tilting the head contra-laterally, an anatomically lateral field is achieved with only a 70 degree gantry rotation rather than 90 degrees. This means that the nozzle can get closer to the surface of the head as it will not be obstructed by the shoulders or structure of the patient couch. Given the relationship of beam width with air gaps for superficial Bragg peaks (see Section 24.2.2 above), such an arrangement helps improve lateral falloff.

Table 24–4 Typical dose prescriptions and constraints used for parameningeal rhabdomyosarcoma at Paul Scherrer Institut

Target/Structure	Prescription Dose (Gy (RBE))	Mean Dose (Gy (RBE))	D2 Dose (Gy (RBE))	Other Constraints (Gy (RBE))
PTV1 (CTV + 5 mm)	45.0	-	-	-
PTV2 (GTV 1 [initial tumor] + 5 mm)	50.4	-	-	-
PTV3 (GTV 2 [residual tumor] + 5 mm)	54.0	-	-	-
Brain stem	-	-	54	-
Center brain stem	-	-	50	-
Ipsilateral lacrimal gland	-	36	40	-
Ipsilateral optic nerve	-	-	54	-
Contra lateral lacrimal gland	-	20	30	-
Contra lateral optic nerve	-	-	50	-
Contra lateral lens	-	7	10	-

24.4.4.4 Clinical Outcomes

The outcome of 40 parameningeal rhabdomyosarcomas (39 Grade III/IV) treated with PBS at PSI are currently being analyzed. With a median follow-up of 4.5 years, pre-

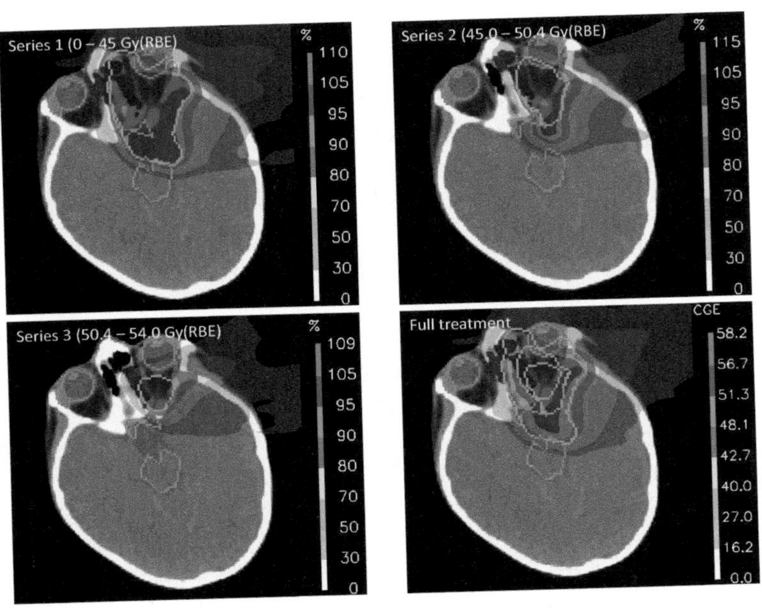

Figure 24–14 Case study 4—parameningeal rhabdomyosarcoma.

liminary analysis of five-year overall and progression free survival appear to be comparable to those reported for IMRT series with much shorter follow-up periods (e.g., two-year median FU in [52]). In addition, the incidence of grade 3 cataracts is also similar (roughly 10%) between our series and that reported by Wolden et al., which is to be expected given the proximity of the lens to the PTV in many of these cases [52].

24.4.4.5 Summary

Due to the often superficial position of these tumors, parameningeal rhabdomyosarcomas are a challenging indication for PBS proton therapy. On the other hand, the fact that they are superficial means that proton therapy can treat these tumors with very little dose to normal tissues and critical structures on the contra-lateral side. Given that many of the cases treated are young children (the median age in our series is 4.7 years old), we believe that this can have important consequences for reducing late sequelae as a result of treatment.

24.5 Future Directions

Despite its increasing popularity, PBS proton therapy and treatment planning is very much in its infancy. At the time of writing, only a few thousand patients have been treated world-wide. Thus, it is our belief that much still has to be learned, understood, and developed for the treatment planning of PBS. In the final section of this chapter, we will briefly outline some of the directions we believe such developments will take in the next years.

24.5.1 Improved Dose Calculations and Optimization

In the author's opinion, one of the great strengths of radiotherapy as a treatment modality is the fact that a patient's treatment can be quite accurately predicted *a-priori* in the treatment planning step. On the other hand, such a prediction can only be as good as the tools with which they are calculated, and for treatment planning, the limiting factor will always be the dose calculation engine. Although models required to represent the primary dose component of individual pencil beams (at least in water) are quite straightforward, accurate models for representing the effects of density heterogeneities and secondary contributions (secondary particles) are challenging. While the obvious approach to this is to move to Monte Carlo-based calculations, these will generally be inherently slower than analytical approaches, and there is considerable room for improvement in these latter types of algorithms. If these can be improved while conserving their considerable time advantage with respect to MC-based calculations, then this could have major benefits in plan optimization (the faster the dose calculation, the larger the "phase-space" of solutions that can be searched) and in moving to real-time planning for adaptive therapies (see below).

Optimization is another area where advances will surely be made in the next years. As has been pointed out at the beginning of this chapter, there are many degrees of freedom available to a PBS proton plan or field. Consequently, the optimization process can be highly degenerate (i.e., there are very many different solutions which, clinically, are more or less equivalent to each other) or can provide many different plans, each of which has advantages and disadvantages, but for which it is difficult to choose one single "optimum" plan. Thus, there is great potential to improve optimization procedures or to provide solutions by which the "phase-space" of possible solutions can be browsed or driven by the user. For the first approach, Multiple Criteria Optimization (MCO) tools—as described in Chapter 21, Plan Optimization (and references therein)—are being developed. These allow the planner to browse through an optimized subset of plans, from which the clinically most relevant can be found. On the other hand, it could be that the optimization can be "driven" to the desired solution through, for example, the use of *a-priori* defined "starting conditions" for the optimization, as described by Albertini et al. [41]. Whatever approach is taken, there is good reason to believe that the optimization of PBS proton therapy is itself ripe for optimization.

24.5.2 Robust Planning

A recurrent theme in proton planning, and one also discussed widely in this chapter, is "plan robustness." This is a healthy development, and it has been driven by the awareness that the extra "dimension" of proton range not only allows for substantial normal tissue dose reductions, but also brings additional uncertainties in the planning and delivery process. As such, robustness (i.e., the *in*sensitivity of a plan to potential delivery uncertainties) can be thought of as another parameter to be optimized, especially as a robust plan may not always be the most conformal, or even most clinically acceptable. This is the reason that robustness is now being included in the optimization process and also as part of MCO (multi-criteria optimization) regimes for proton therapy (see Chapter 21, Plan Optimization). This will certainly be a fruitful area of research in the future [34], either as an automated process, or also through other approaches, such as "smart" PTV margins that can deal with delivery uncertainties in a more flexible manner than conventional PTVs, or maybe through the manipulation of optimization starting conditions [41]. However, there is much work still to be done in more precisely defining metrics for plan robustness, the current techniques being, in our opinion, likely too conservative for clinical practice, especially for fractionated therapy. An exception may be LET-weighted optimization, as proposed by Giantsoudi et al. [53]. Although LET is only a surrogate for RBE, it can be much more precisely calculated and can be used to guide the optimization through the simple relationship that a raised LET can potentially lead to a raised (but not precisely known) RBE. By this approach, it could be possible to calculate biologically "safer" plans without necessarily knowing the myriad of parameters necessary to calculate local RBE values.

Further developments of robust planning are also required for dealing with anatomical changes of the patient, as discussed in Section 24.2.5 above. Such variations can cause range changes of such magnitude that we believe these cannot be sensibly included in an optimization process or by conventional PTV margins, at least not without treating unnecessarily large volumes of normal tissues. Therefore, alternative techniques may need to be used. We have already mentioned the concept of "virtual blocks" in Section 24.3.4.3 above, and such approaches could be developed further for potentially dealing with anatomical changes, whereby anatomical regions that could change during treatment are delineated and used to reduce the weight of pencil beams passing through them in order to limit the influence of potential range changes on the final dose distribution.

However, perhaps the ultimate robust "planning" approach will be the development of ultra-fast and automated planning procedures for daily adapted treatments, whereby the patient is imaged on a daily basis (immediately before therapy) and a plan of the day calculated, which will automatically take into account any anatomical (and positioning) changes for that day.

24.6 Summary

PBS proton therapy is a powerful and flexible radiotherapy modality that, to fulfill its potential, requires similarly powerful and flexible treatment planning tools. It also requires an understanding of the nuances of proton interactions and their interplay with the anatomical intricacies of the patient. In this chapter, we have therefore tried to convey the thought processes and rational for why we plan (and therefore treat) the way we do. What has been described, however, are not a set of rules, but more a few hints at what may (or may not be) good planning practice for PBS proton therapy. We believe that we have good reasons for planning the way we do, and we have tried to express these here. But as with the planning process itself, good practice may well be a highly degenerate problem to which there are many solutions.

24.7 References

1. Jermann M, PTCOG 53, Shanghai, 2014.
2. Lomax AJ. Physics of treatment planning using scanned beams. In: Proton Therapy Physics, Paganetti H, Ed. Boca Raton, FL:CRC Press, 2012;335–379.
3. Lomax AJ. Intensity modulated proton therapy. In: Proton and charged particle radiotherapy, Delaney T, Kooy H, Eds. Philadelphia:Lippincott, Williams and Wilkins, 2008.
4. Lomax AJ. Intensity modulated methods for proton therapy. *Phys Med Biol* 1999;44:185–205.
5. Safai S, Bortfeld T, Engelsman M. Comparison between the lateral penumbra of a collimated double-scattering beam and uncollimated scanning beam in proton radiotherapy *Phys Med Biol* 2008;21:1729–1750,
6. Both S, Shen J, Kirk M, Lin L, Tang S, Alonso-Basanta M, Lustig R, Lin H, Deville C, Hill-Kaser C, Tochner Z, McDonough J. Development and clinical implementation of a universal bolus to maintain spot size during delivery of base of skull pencil beam scanning proton therapy. *Int J Radiat Oncol Biol Phys* 2014;90:79–84.
7. Urie M, Goitein M, Holley WR, et al. Degradation of the Bragg peak due to inhomogeneities. *Phys Med Biol* 1986;31:1–15.
8. Lomax, A.J., Charged particle therapy: the physics of interaction. Cancer J. 2009;15:285–291.
9. Schneider U, Pedroni E, Lomax AJ. On the calibration of CT-Hounsfield units for radiotherapy treatment planning. *Phys Med Biol* 1996;41:111–124.
10. Schaffner B, Pedroni E. The precision of proton range calculations in proton radiotherapy treatment planning: Experimental verification of the relation between CT-HU and proton stopping power. *Phys Med Biol* 1998;43:1579–1592.
11. Ma J, Beltran C, Seum Wan Chan Tseung H, Herman MG. A GPU-accelerated and Monte Carlo-based intensity modulated proton therapy optimization system. *Med Phys* 2014;41:121707.
12. Jia X, Schuemann J, Paganetti H, Jiang SB, GPU-based fast Monte Carlo dose calculation for proton therapy. *Phy Med Biol* 2012;57:7783–97.

13. Schaffner B, Pedroni E, Lomax AJ. Dose calculation models for proton treatment planning using a dynamic beam delivery system: an attempt to include density heterogeneity effects in the analytical dose calculation. *Phys Med Biol* 1999;44:27–42.
14. Tourovsky A, Lomax AJ, Schneider U, Pedroni E. Monte Carlo dose calculations for spot scanned proton therapy. *Phys Med Biol* 2005;50:971–981.
15. Albertini F, Casiraghi M, Lorentini S, Rombi B, Lomax AJ. Experimental verification of IMPT treatment plans in an anthropomorphic phantom in the presence of delivery uncertainties. *Phys Med Biol* 2011;56:4415–4431.
16. Paganetti H. Range uncertainties in proton therapy and the role of Monte Carlo simulations. *Phys Med Biol* 2012;57:99–117.
17. Albertini F, Bolsi A, Lomax AJ, Rutz HP, Timmerman B, Goitein G. Sensitivity of intensity modulated proton therapy plans to changes in patient weight. *Radiother Oncol* 2008;86:187–194.
18. Mori S, Chen GT. Quantification and visualization of charged particle range variations *Int J Radiat Oncol Biol Phys* 2008;72:268–77.
19. Mori S, Dong L, Starkschall G, Mohan R, Chen GT. A serial 4DCT study to quantify range variations in charged particle radiotherapy of thoracic cancers. *Radiat Res* 2014;55:309–19.
20. Olch AJ, Gerig L, Li H, Mihaylov I, Morgan A. Dosimetric effects caused by couch tops and immobilization devices: report of AAPM Task Group 176. *Med Phys* 2014;41:061501.
21. Knopf A, Boye D, Lomax AJ, Mori S. Adequate margin definition for scanned particle therapy in the incidence of intra-fractional motion. *Phys Med Biol* 2013;58:6079–6094.
22. Phillips M, Pedroni E, Blattman H, Böhringer T, Coray A, Scheib S. Effects of respiratory motion on dose uniformity with a charged particle scanning method. *Phys Med Biol* 1992;37:223–34.
23. Grassberger C, Dowdell S, Lomax AJ, Sharp G, Shackleford J, Choi N, Willers H, Paganetti H. Motion interplay as a function of patient parameters and spot size in spot scanning proton therapy for lung cancer. *Int J Radiat Oncol Biol Phys* 2013;86:380–386.
24. Bert C, Durante M. Motion in radiotherapy: particle therapy. *Phys Med Biol* 2011;56:R113–R144.
25. Knopf A-C, Hong TS, Lomax AJ, Scanned proton radiotherapy for mobile targets—the effectiveness of re-scanning in the context of different treatment planning approaches and for different motion characteristics. *Phys Med Biol* 2011;56:7257–7271.
26. Paganetti H. Relative biological effectiveness (RBE) values for proton beam therapy. Variations as a function of biological endpoint, dose, and linear energy transfer. *Phys Med Biol* 2014;59:R419–72.
27. Paganetti H, van Luijk P. Biological considerations when comparing proton therapy with photon therapy. *Semin Radiat Oncol* 2013;23:77–87.
28. Paganetti H. Range uncertainties in proton therapy and the role of Monte Carlo simulations. *Phys Med Biol* 2012;57:99–117.
29. ICRU. Report 62: Prescribing, recording, and reporting photon beam therapy (Supplement to ICRU Report 50). Bethesda, MD:ICRU, 1999.
30. van Herk M, Remeijer P, Rasch C, Lebesque J V. The probability of correct target dosage: dose-population histograms for deriving treatment margins in radiotherapy. *Int J Radiat Oncol Biol Phys* 2000;47:1121–35.
31. Park PC, Zhu XR, Lee AK, Sahoo N, Melancon AD, Zhang L, Dong L. A beam-specific planning target volume (PTV) design for proton therapy to account for setup and range uncertainties. *Int J Radiat Oncol Biol Phys* 2012;82:329–36.
32. Graeff C, Durante M, Bert C. Motion mitigation in intensity modulated particle therapy by internal target volumes covering range changes. *Med Phys* 2012;39(10):6004–13.
33. Knopf A, Boye D, Lomax AJ, Mori S. Adequate margin definition for scanned particle therapy in the incidence of intra-fractional motion. *Phys Med Biol* 2013;58:6079–6094.
34. Chen W, Unkelbach J, Trofimov A, Madden T, Kooy H, Bortfeld T, Craft DL. Including robustness in multi-criteria optimization for intensity-modulated proton therapy. *Phys Med Biol* 2012;57:591–608.
35. Bolsi A, Lomax AJ, Pedroni E, Goitein G, Hug E. Experiences at the Paul Scherrer Institute with a remote patient positioning procedure for high-throughput proton radiation therapy. *Int J Radiat Oncol Biol Phys* 2008;71:1581–90.

36. Egger E, Schalenbourg A, Zografos L, Bercher L, Boehringer T, Chamot L, Goitein G, Maximising local tumour control and survival after proton beam radiotherapy of uveal melanoma. *Int J Radiat Oncol Biol Phys* 2001;51:138–47.
37. Chang JY, Li H, Zhu XR, Liao Z, Zhao L, Liu A, Li Y, Sahoo N, Poenisch F, Gomez DR, Wu R, Gillin M, Zhang X Clinical implementation of intensity modulated proton therapy for thoracic malignancies. *Int J Radiat Oncol Biol Phys* 2014;90:809–818.
38. Dowdell S, Grassberger C, Paganetti H. Four-dimensional Monte Carlo simulations demonstrating how the extent of intensity-modulation impacts motion effects in proton therapy lung treatments. *Med Phys* 2013;40:121713.
39. Albertini F, Lomax AJ, Hug EB. In regard to Trofimov et al: radiotherapy treatment of early-stage prostate cancer with IMRT and protons: a treatment comparison. *Int J Radiat Oncol Biol Phys* 2007;69:1333–1334.
40. Lomax AJ, Boehringer T, Coray A, Egger E, Goitein G, Grossmann M, Juelke P, Lin S, Pedroni E, Rohrer B, Roser W, Rossi B, Siegenthaler B, Stadelmann O, Stauble H, Vetter C, Wisser L. Intensity modulated proton therapy: A clinical example. *Med Phys* 2001;28:317–324.
41. Albertini F, Hug EB, Lomax AJ. The influence of the optimization starting conditions on the robustness of intensity-modulated proton therapy plans. *Phys Med Biol* 2010;55:2863–2878.
42. Soukup M, Söhn M, Yan D, Liang J, Alber M. Study of robustness of IMPT and IMRT for prostate cancer against organ movement. *Int J Radiat Oncol Biol Phys* 2009;75:941–49.
43. Dietlicher I, Casiraghi M, Ares C, Bolsi A, Weber DC, Lomax AJ, Albertini F. The effect of metal implants in proton therapy: Experimental validation using an anthropomorphic phantom. *Phys Med Biol* 2014;59:7181–7194.
44. Widesott L, Pirelli A, Fiorino C, Lomax AJ, Amichetti M, Cozzarini C, Soukup M, Di Muzio N, Calandrino R, Schwarz M. Helical Tomotherapy (HT) vs. intensity-modulated proton therapy (IMPT) for whole pelvis irradiation in high risk prostate cancer patients. *Int J Radiat Oncol Biol Phys* 2011;80:1589–160.
45. Ares C, Geismar JH, Albertini F, Koch KM, Schneider RA, Lomax AJ, Hug EB, Goitein G. Chordomas and chondrosarcomas of the skull base: clinical outcomes of the Paul Scherrer Institute Experience using spot scanning proton radiation therapy. ASTRO 2012.
46. Pehlivan B, Ares C, Lomax AJ, Stadelmann O, Goitein G, Timmermann B, Schneider RA, Hug EB. Temporal lobe toxicity analysis after proton radiation therapy for skull base tumors. *Int J Radiat Oncol Biol Phys* 2012;83:1432–144.
47. Rutz HP, Lomax AJ. Donut-shaped high dose configuration for proton beam radiation therapy. Strahlenther. Onkol. 2005;181:49-53.
48. Staab A, Rutz HP, Ares C, Timmermann B, Schneider R, Bolsi A, Lomax AJ, Goitein G, Hug EB. Spot-scanning-based proton therapy for extracranial chordoma Int J Radiat Oncol Biol Phys. 2011;81:489-496
49. Schneider, R., Vitolo, V., Albertini, F., Koch, T., Area, C., Lomax, A.J., Goitein, G., Hug, E.B., Small bowel toxicity after high dose spot scanning-based proton therapy for paraspinal/retroperitoneal neoplasms. *Strahlenther Onkol* 2013;189:1020–1025.
50. Merchant TE, Li C, Xiong X, Kun LE, Boop FA, Sanford RA. Conformal radiotherapy after surgery for paediatric ependymoma: a prospective study. *Lancet Oncol* 2009;10:258–266.
51. MacDonald SM, Sethi R, Lavally B, et al. Proton radiotherapy for pediatric central nervous system ependymoma: clinical outcomes for 70 patients. Neuro Oncol 2013;15:1552–9.
52. Wolden SL, Wexler LH, Kraus DH, et al. Intensity-modulated radiotherapy for head-and-neck rhabdomyosarcoma. *Int J Radiat Oncol Biol Phys* 2005;61:1432–1438.
53. Giantsoudi D, Grassberger C, Craft D, Niemierko A, Trofimov A, Paganetti H. Linear energy transfer-guided optimization in intensity modulated proton therapy: feasibility study and clinical potential. *Int J Radiat Oncol Biol Phys* 2013;87:216–222.

Chapter 25

Motion Management

Antje Knopf, Ph.D.[1] and Shinichiro Mori, Ph.D.[2]

[1]The Institute of Cancer Research
and The Royal Marsden NHS Foundation Trust,
London, UK

[2]National Institute of Radiological Sciences,
Research Center for Charged Particle Therapy,
Chiba, Japan

25.1 The Challenge of Motion for Proton Radiotherapy

The treatment of moving targets presents a special challenge in proton radiotherapy. Generally, all targets in the human body have to be considered as moving with variable amplitudes. Heartbeat, blood circulation, digestion, breathing, the perception of the environment, and the reaction to stimuli keeps the whole human body in permanent motion. Relevant to proton therapy treatments are motions that exceed approximately 0.5 cm and that happen within approximately 4 min (intra-fractional) or that change the patient geometry on a day-to-day basis (inter-fractional). Examples for intra-fractional motions are changes in the whole thorax that result from respiration or drifts of organs due to gravitational forces. Weight gain or target shrinkage as response to the treatment would for example classify as inter-fractional changes. In Section 25.2 of this chapter, motion characteristics of different target sites are described in detail.

External motion of the patient geometry can be accounted for by fixation devices. However, even if the external shape of a patient can be kept static, internal motion can still take place. That is especially the case in the thorax and in the abdomen, which do not represent rigid geometries and where no clear correlation between internal and external geometry exists. For proton treatments, motion in all three dimensions relative to the beam direction can have a significant impact. Due to the sensitivity of protons to the specific tissue densities they pass through, not only a rigid estimation, but the actual nonrigid deformation of the patient geometry, has to be considered. Motion can result in stopping positions of the proton beam different from what was planned. All possible effects of motion are discussed in Section 25.3 of this chapter.

Imaging plays a central role when treating moving targets. Prior to treatment, it is required to estimate the extent of motion and to build patient-specific motion models. During treatment delivery, it is important to monitor the motion and to correctly apply motion mitigation approaches. Implementations and challenges to realizing real-time 3D imaging prior to and during irradiation are discussed in Section 25.4 of this chapter. To establish a relation between images of the same patient taken at different time instances, image registration needs to be employed. Different implementations and resulting ambiguities are discussed in Section 15.5.

Patient-specific motion effects and motion mitigation strategies should be considered during treatment planning. To do this, a 4D treatment planning system is needed that is capable of modeling elapsing processes. Furthermore, for quality assurance, it is desirable to reconstruct the actual delivered dose based on the monitored motion during delivery and the treatment log files. Specifications of 4D treatment planning are discussed in Section 25.6.

In sections 25.7 through 25.9, different ideas on how to account for motion effects are introduced. First, concepts that can be applied to any kind of proton therapy—like the use of margins, the delivery during breath hold, or the employment of gating—are introduced. Then motion mitigation approaches specific to a scanned treatment delivery method and combined approaches are discussed.

Up to now, clinical proton practice has mainly considered inter-fractional changes in the patient geometry. A few proton centers equipped with passively scattered protons have started treating moving targets with gating. The application of different motion mitigation techniques for scanned proton therapy in clinical daily routine is still in its infancy. The current status of image-guided and adaptive protocols for moving targets in clinical daily routine for proton centers is reviewed in Section 25.10.

Lastly, Section 25.11 of this chapter summarizes the current challenge of motion for proton radiotherapy and gives an outlook on the most promising approaches to tackling this challenge in the future.

25.2 Organ Motion

An understanding of motion characteristics in radiotherapy planning is needed to determine internal margins and to optimize treatment parameters. An overview of organ motion and its management in radiotherapy is given by Langen et al., Korreman et al., and the AAPM Task Group 76 [1–3]. These are excellent sources that describe in detail different methods to investigate intra-fractional organ motion. In the following, we only briefly summarize intra-fractional organ and target motion for different treatment sites.

Most tumors move in a varying orbit during the respiratory cycle, showing hysteresis-like behavior. Others move in a closely regular orbit. For treatment sites in the lung, it is therefore important to evaluate tumor motion in 3D [4–6]. Seppenwoolde et al. investigated intra-fractional lung tumor motion using fluoroscopic imaging with an implanted marker [4]. Average lung tumor motion in the upper, middle, and lower lobes along the SI direction was found to be approximately 4 mm, 7 mm, and 9 mm by Plathow et al. [6]. Displacements in the lower lobe are significantly greater than in the other lobes. However, there exist exceptions, as shown in Figure 25–1, where a tumor in the lower lobe at the anterior side of the diaphragm does not move during respiration (less than 1mm). The treatment strategy for lung treatment sites should always be chosen after evaluating the patient-specific respiratory motion characteristics.

The liver is the largest organ displaying intra-fractional movement. It is located in the right upper abdominal region, and its position is significantly affected by respiration (Figure 25–1). Intra-fractional liver motion was reported to be approximately 10–20 mm [7–9]. Hallman et. al investigated intra-fractional liver motion with 4DCT [9]. They found a mean motion of 9.7 mm over all phases and 1 mm within a gating window of 30% duty cycle around exhalation.

The pancreas is located in the upper left abdomen and is typically 6–10 cm long, running in the left-to-right direction. The pancreatic head rests within the concavity of

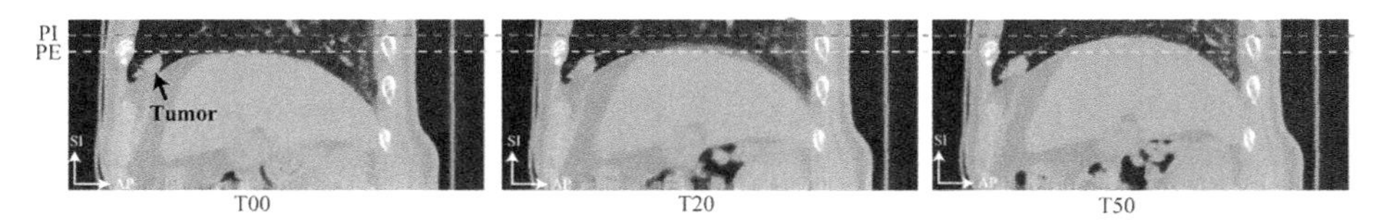

Figure 25–1 Sagittal 4DCT images for different respiratory phases (T00: peak inhalation, T20: mid-exhalation, T50: peak exhalation). Red and yellow dotted lines show the superior position of the diaphragm at peak inhalation and peak exhalation, respectively.

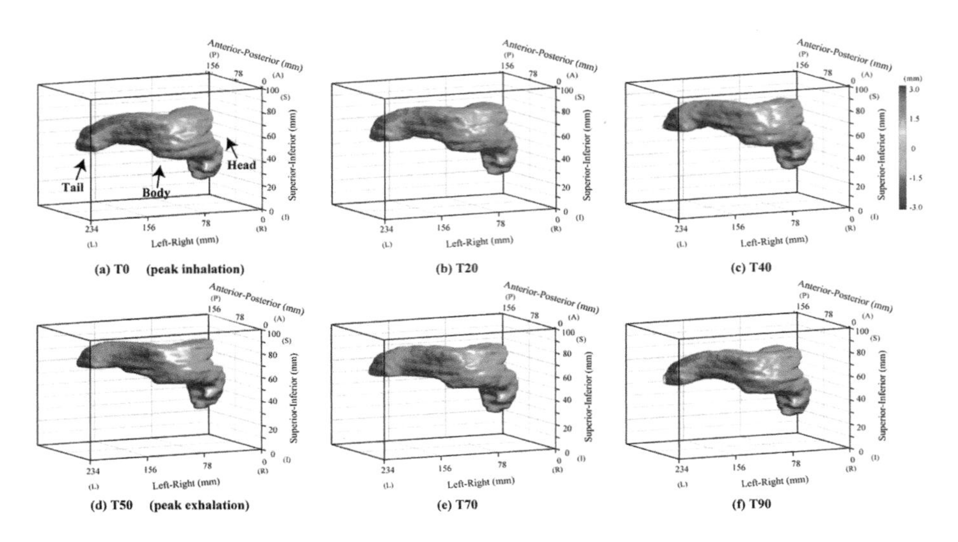

Figure 25–2 3D-visualized pancreas and the magnitude of geometrical variation from peak exhalation as a function of respiratory phase. (a) T0 (peak inhalation), (b) T20, (c) T40, (d) T50 (peak exhalation: reference), (e) T70, and (f) T90 [14].

the duodenum, the body is behind the base of the stomach, and the tail abuts the spleen. The geometrical variation is greater around the pancreas tail than the body and head regions (Figure 25–2). Quantifying pancreatic movement using fluoroscopic imaging is difficult, and imaging modalities with good image contrast—such as ultrasound (US), MRI, and CT—are generally used. Employing US, Suramo et al. [10,11] and Byran et al. [12] reported displacement in the SI direction of approximately 2 cm (range: 0–3 cm) under free breathing conditions. Horst et al. evaluated pancreas positional variation under breath-holding using multi-phase, contrast-enhanced spiral CT scans and reported that the most mobile part of the pancreas was the pancreatic tail [13]. In a 4DCT study, the average pancreas head, body, and tail displacement in the inferior direction was found to be 8.3 mm, 9.6 mm, and 13.4 mm, respectively (Figure 25–2).

A different type of intra-fractional motion occurs with the prostate. The prostate is located just below the urinary bladder and close to the rectum. The prostate position is, therefore, strongly affected by bladder volume and rectum filling. The prostate position is generally evaluated using CT, MRI, and US. The real-time monitoring of the prostate position with the help of an electromagnetic implanted marker was introduced by Langen et al. [15]. Several studies report that intra-fractional prostate movement is relatively small, approximately 2–4 mm in the AP direction and 1–2 mm in the SI direction [16,17].

Variations in the patient's body setup and internal anatomy are inevitable, even with some of the most thorough and technically advanced alignment and immobilization procedures [18]. Sources for inter-fractional changes—such as variations in tumor size, shape, and density—can include weight loss/gain, treatment response/disease progression, or setup variations. Positional changes of the tumor over a treatment course have also been reported.

Intra-fractional variations and inter-fractional changes can compromise the dose distribution and can affect the target dose conformity and the target dose coverage. Dosimetric effects of motion will be discussed in detail in the next section. Methods to monitor motion are highlighted in Section 25.4. Inter-fractional changes are generally easier to monitor and to account for than intra-fractional changes. For example, serial CT scans provide information about the patient geometry at each day of treatment, and adaptive protocols can be used to account for changes of the patient geometry. Strategies to account for intra-fractional changes are introduced in sections 25.7 to 25.9.

25.3 Effects of Motion

25.3.1 Target Miss

To date, there is no commercial treatment planning system available that allows for 4D dose calculations considering patient-specific as well as machine-specific elapsing processes during treatment delivery. Thus, assumptions are made about the actual treatment situation during treatment planning.

If motion is completely neglected for a moving target during treatment planning, the result will be target dose miss and over dosage of nearby tissue. This will be true for passively delivered treatments as well as for an actively scanned delivery. Assuming the target to be static, the dose distribution is calculated on a static representation of the patient geometry (reference phase), which differs from the moving patient geometry during irradiation. The degree of target miss or over-dosage of surrounding tissues due to this simplified assumption will depend on the degree of motion. Furthermore, dosimetric effects will depend on the timeline of delivery and target motion. For example, an almost instant delivery of a dose distribution during an exhalation phase, but which was planned at an inhale reference phase, can result in a complete target miss. However, if this plan is delivered over several breathing cycles, the negative motion effect is mitigated, and only parts of the target will be under dosed. This example shows how important it is to simulate the exact timeline of delivery in order to predict the effects when irradiating a moving target.

25.3.2 Undershoot and Overshoot

As dosimetry in proton beam therapy is highly sensitive to density changes, their penetration depth will change when densities in the beam path change. This is especially challenging in thoracic regions, where densities along the penetration path can vary from approximately 1.0 g/cm^3 to 0.3 g/cm^3 during tumor motion. Resulting effects on the dose distribution are illustrated in Figure 25–3. A prominent scenario resulting in density changes along the beam path is the motion of the ribcage along with breathing. During treatment delivery, ribs can enter the beam path of protons that would

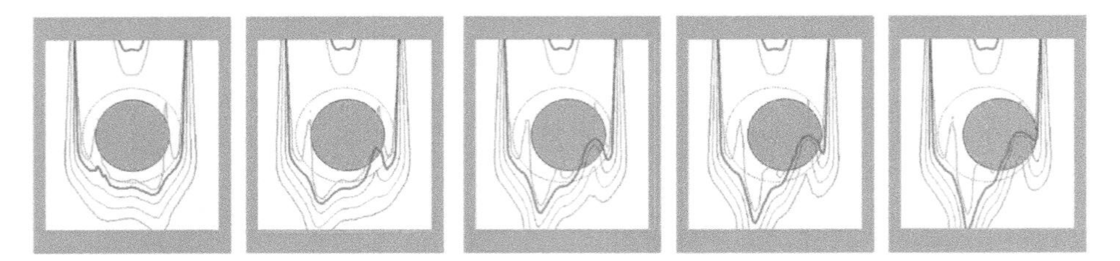

Figure 25–3 Dose distributions to a PTV (dashed circle), simulating a 10 mm amplitude motion of the target (grey circle). Dose distributions for five different motion phases with different target locations are shown. From outside inward the isodose lines are the 50, 80, 90, 95, and 100% isodose level, respectively. (Figure adapted from personal communication with M. Engelsman.)

have only traversed soft tissue in the static situation of the treatment plan. This results are a shortened penetration depth and potentially reduced dose in the target. Contrary to the above, proton beams that were planned to transverse ribs can overshoot into critical organs if they only face soft tissue in the actual delivery situation.

25.3.3 Blurring

Independent of the motion mitigation technique, motion will always result in blurring. In a static geometry, particle beams can achieve highly conformal dose distributions due to their sharp distal falloff and their distinct lateral penumbra. Depending on the relation between the beam direction and the direction of motion, mainly the lateral conformity will be compromised by motion. With regard to the scanned delivery, the effect of blurring is illustrated in Figure 25–4b. If the beam direction and the main motion direction are parallel, blurring will be minimal. Assuming a rigid motion, in this scenario only the air gap will change, which has a minimal influence on the proton dose distribution. However, if the beam direction and the motion direction are perpendicular to each other, the beam (or single proton pencil beams in the case of a scanned delivery) will be misplaced in the order of the motion magnitude. As a result, the lateral dose gradients will become shallower, and the conformity will be compromised. This example shows that for moving targets, the main motion direction should be determined and considered during beam angle selection in treatment planning [19].

25.3.4 Interplay Effect

An effect unique to actively scanned proton therapy treatments is the interplay between periodic anatomical changes and the timeline of treatment delivery. A scanning dose spot can be delivered to an unplanned location because the spot "sees" an anatomy that may be quite different from that in the planning reference CT images. As the patient geometry constantly changes, the difference between the planning situ-

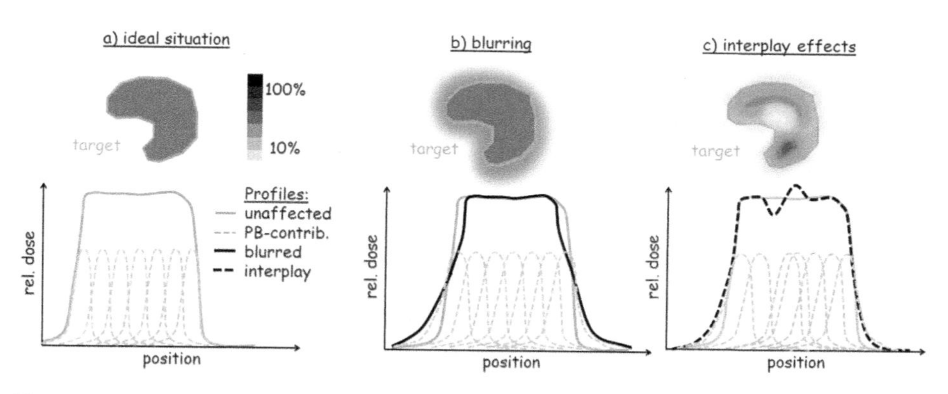

Figure 25–4 (a) Illustration of motion effects to conformal dose distribution. (b) For a passive as well as for a scanned delivery, motion leads to dose blurring. (c) The misplacement of single beam spots for a scanned delivery leads to interplay effects.

ation and the actual delivery situation is unique for each spot. Furthermore, it is unique to the starting time of the treatment and the relation between two timelines.

The result of misplaced beam spots is dose inhomogeneity in the target, as illustrated in Figure 25–4c. Rescanning (Section 25.8.1) is proposed to mitigate interplay effects by scanning a treatment field several times per fraction. If motion effects are random, i.e., if target motion and beam delivery are not correlated, rescanning should average out interplay effects over a number of scans. However, if rescanning is performed in a very regular fashion, the delivery timeline will be periodic. In this case, interference between the scanning and the motion can make rescanning ineffective. This can easily be conceived. Imagine that for 10 rescans, each single scan of the target volume takes exactly one patient motion period. Even if 10 rescans are applied in this case, nothing is gained compared to just one single scan because the scanning treatment beam will find 10 times the same geometry, and the same interplay effect will just accumulate 10 times.

Interplay effects depend on a lot of parameters. These include plan parameters (number of fields, beam directions, number of re-scans), motion parameters (amplitude, starting phase, regularity), and machine parameters (scanning speed, spot size). These parameters are difficult to predict for the actual treatment situation [20,21]. IMPT is especially susceptible for interplay effects. Dowdell et al. investigated how the extent of intensity modulation impacts the motion effects in proton therapy lung treatments [22]. For single-field uniform dose (SFUD) proton plans, Knopf et al. [19] proposed the following strategies to avoid interplay effects:

1. Single-field plans should be applied with a sufficient number of re-scans, and the field directions should be chosen with care. If clinically feasible, they should be aligned as parallel as possible to the main motion direction.

2. Multi-field plans should be constructed, for which any additional field gives additional robustness against motion effects. Due to the inherent rescanning effect of multi-field plans, only a few additional re-scans may then be necessary.

25.4 Imaging of Motion

In the early days of radiation therapy, imaging was solely used for tumor contouring and treatment planning in both particle beam therapy and photon beam therapy. The development of more precise delivery techniques—such as scanned beam irradiation and intensity-modulated particle therapy (IMPT) [23]—minimized excessive dose to healthy tissues, leading to good tumor control rates with less toxicity and improved quality of life [24,25]. These improvements in highly conformal irradiation techniques require highly accurate anatomical information before and during the course of treatment to ensure the correct treatment beam placement. Imaging is one approach to achieving this. The advantages of diagnostic imaging technology have found their way into radiotherapy treatments, helping oncology practitioners delineate tumors more accurately, spare normal tissues, and even modify treatment parameters [26,27]. Although the proverb "a picture is worth a thousand words" might not be applicable to all radiotherapy, it does reflect the role of image-guided radiotherapy (IGRT). IGRT allows us to explore medical images, which contain rich information on intra- and inter-fractional changes. Here we introduce medical imaging modalities and techniques used in radiotherapy, particularly those focused on capturing intra- and inter-fractional changes. More general information on imaging can be found in Chapter 7.

25.4.1 From Static to Dynamic Imaging

X-ray imaging provides anatomical patient information in 2D. It is frequently used for diagnostic and therapeutic purposes because it reveals internal anatomic structures of the patient with good image quality. Incorporation of flat panel detectors, also known as electronic portal imaging devices (EPID), into imaging systems provides better image quality and better geometrical accuracy than the use of film and image intensifier (II) imaging. EPID images are used for patient setup along with laser localizers [28]. Moreover, recent x-ray imaging can additionally provide 2D + time-based information (fluoroscopic imaging). Time-resolved 2D anatomical information can be obtained through the use of dynamic flat panel detectors (DFPD) [29]. To bypass the limitation of information in only one imaging plane, bi-plane x-ray imaging from two or more different directions can be used. This provides 3D positional information. Over the last few years, image processing techniques have progressed. In particular, machine learning has led to areas beyond conventional image processing. As an example, dual-energy x-ray images can be obtained from a single x-ray image with machine learning (Figure 25–5) [30], avoiding acquiring two images at different tube voltages [31].

3D anatomical information can be obtained from CT as well as MRI images. Since the fundamental CT reconstruction algorithm is slice-based, CT provides 3D

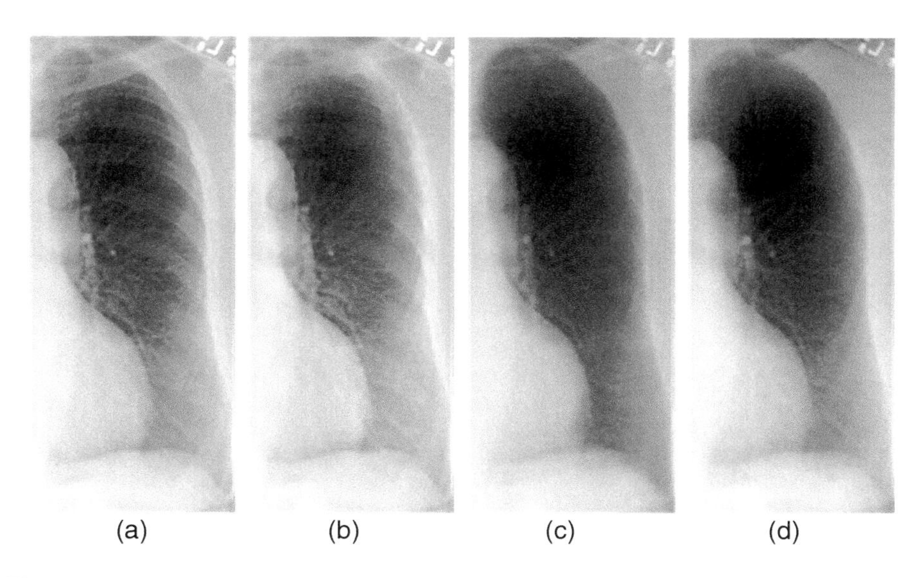

Figure 25–5 Result for a non-training normal chest radiograph. (a) Original normal chest radiograph, (b) A VDE soft tissue image obtained using our original means of massive training artificial neural network (MTANN) technique. (c) A virtual dual-energy soft-tissue image obtained using our new MTANN technique. (d) The corresponding "gold-standard" dual-energy soft-tissue image [32].

image data by stacking multiple CT images. This is useful to obtain the geometrical shape of organs as well as their position. CT images also provide electron density information, which can be converted to relative stopping power values, essential for proton treatment planning. However, since the CT reconstruction algorithm was developed for static objects, intra-fractional motion compromises the imaging quality. CT scanning under free breathing neither represents the time-averaged position nor the total moving distance of an imaged structure [33,34]; moreover, the artifacts induced by respiration can mimic disease symptoms [35]. In the diagnostic field, CT coronary angiography is already used to identify patients with significant coronary artery stenosis by ECG-correlated preprocessing [36]. Given that the heart beat (~1s) is faster than respiratory motion (~4s) and that coronary arteries (a few mm) are smaller than lung tumors (a few cm), the use of this diagnostic CT imaging technique should allow the visualization of intra-fractional motion. Accordingly, the demand for time-resolved 3DCT imaging has increased, and the 4DCT technique has been introduced [1,37]. If the acquisition of 4DCT images under free breathing is not done with care, geometrical shapes will not be correctly visualized (Figure 25–6a). However, when respiratory motion is accounted for, tumor and organ shapes can be correctly depicted (Figure 25–6b). It needs to be remembered that 4DCTs acquired with conventional multi-slice CT (MSCT) include geometric errors, for example, due to

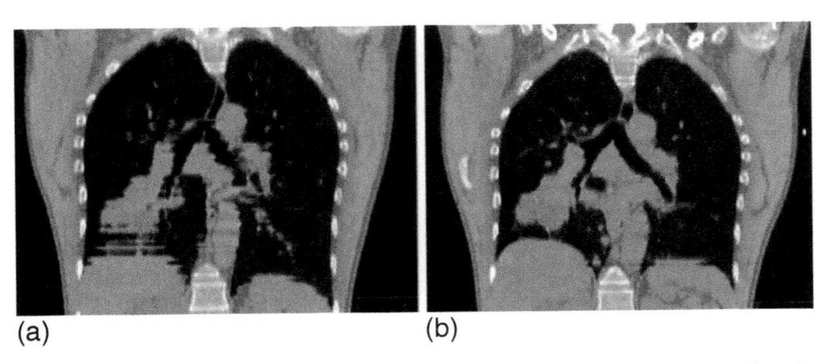

Figure 25–6 Coronal views of CT scans of the same patient taken (a) during free breathing (FB) and (b) with respiratory gated scanning at exhalation [1].

resorting [38]. To overcome these problems, state-of-the-art medical CT scanners acquire volumetric cine CT image data with over 256 detector rows. The volumetric cine CT data includes a coherent absolute time as well as a respiratory phase in all slices. Therefore, resorting of CT data at each slice position as a function of respiratory phase is not necessary [39].

The latest CT scanners incorporate several functions to reduce imaging dose while improving image quality. Nevertheless, increasing radiation dose remains problematic from a radiation protection point of view. An alternative approach to serial 4DCT imaging is 4DMRI. An advantage of MRI is that its images have higher contrast than CT and x-ray images. MRI is often used in the contouring process in treatment planning by being overlaid on a CT image. 4DMRI (cine MRI) obtains time-resolved 3D data by stacking retrospective 2D cine images without giving any imaging dose. For cardiac imaging, it is already in clinical use, but it remains at the research level for therapeutic purposes. Long acquisition times of about 1h enable the capture of many breathing cycles without any assumption of periodicity. 4DCTs can be simulated by combining 3DCT data and the motion information obtained from the 4DMRI. Deformation vector fields can be extracted from the 4DMRI by deformable image registration and then be wrapped to a conventional 3DCT [40]. 4DMRI data acquisition just prior to each treatment fraction enables investigating the influence of irregular breathing using 4D dose calculations. Furthermore, the availability of motion information over a long time interval also allows studying the effects of baseline drifts.

25.4.2 Real-time Range Monitoring

The advantage of charged particles is their finite beam range, which significantly reduces excessive dose past the end of range compared to photon beams. Beam range inaccuracies at specific beam positions can lead to dose inhomogeneities within the target volume. Thus, besides time-resolved geometrical information, the *in vivo* information about where the proton beam stops in the patient is of special interest. Ideally, this information could be obtained in real time and for each pencil beam position in the case of a scanned delivery. Prompt gamma or PET imaging might be approaches to this problem. More information about range verification methods is reviewed in Chapter 7.

25.5 Image Registration

The clinical merit of IGRT has led to substantial improvements in treatment accuracy. However, the use of 4D imaging and daily/weekly imaging increases the number of images, leading to major problems in clinical staff workload for managing medical images. One solution is to incorporate image registration techniques into the treatment workflow. Hill et al. have outlined the image registration approaches using medical images [41]. Here we introduce the major image registration approaches that can be used in radiotherapy.

25.5.1 Rigid

Recent improvements in conformal treatment requires a high degree of positional accuracy, but patient positioning takes several minutes to be completed manually. In

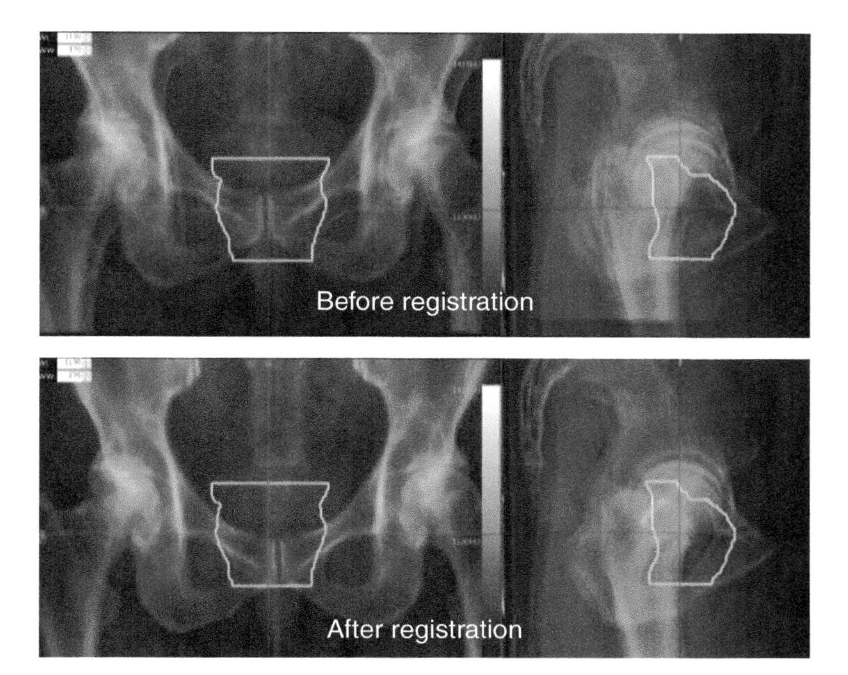

Figure 25–7 Flat panel detector images (blue layer) overlaid on digitally reconstructed radiography images (orange layer). Orthogonal images before registration and after registration for pelvic case.

particular, adjustment of the rotational component of the coordinate transformation is more difficult than that for coordinate transformation. Many organs can deform when the patient pose varies. However, in many cases we can assume that the body remains rigid during the positioning process, thanks to improvements in patient positional reproducibility using immobilization devices (see Chapter 18). Patient positioning can, therefore, be performed with a rigid registration approach (translation and rotation). 2D-3D image registration techniques have been introduced to improve positional accuracy. Comparison of 2D x-ray images with volumetric CT data used for treatment planning allows the positioning error between the treatment planning setup and treatment delivery setup to be calculated (Figure 25–7). Commercial systems such as Cyberknife and ExacTrac already provide a 2D-3D auto-registration function for patient setup in photon therapy [42]. Moreover, cone-beam CT imaging has recently come into routine use. The introduction of GPU-based computing to solve

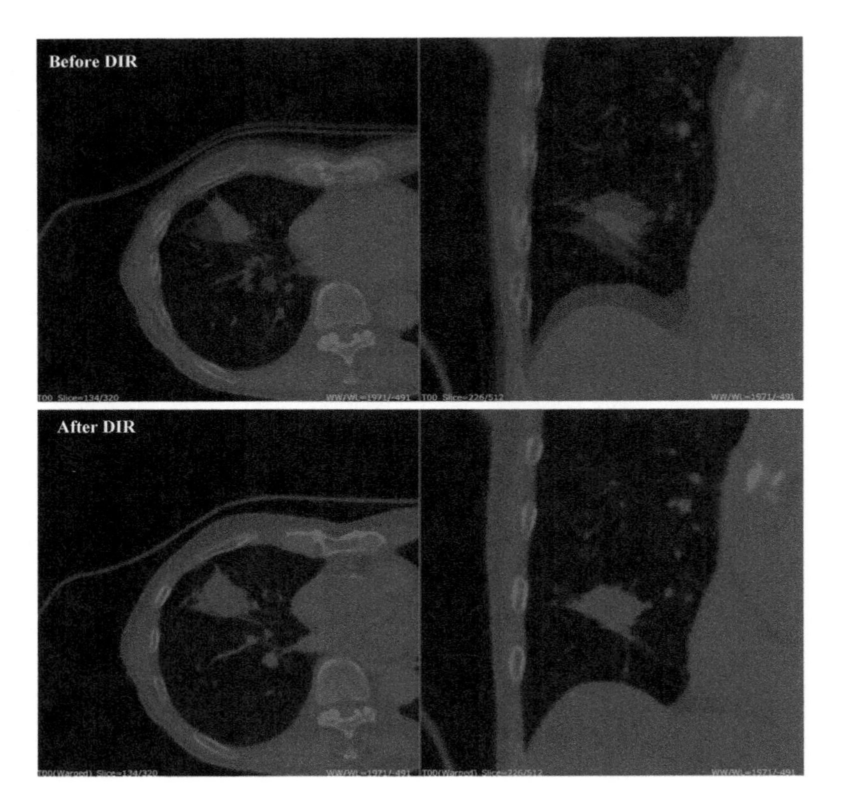

Figure 25–8 Upper panel: overlaid original CT images at T50: peak exhalation phase (orange) on T00: peak inhalation phase (blue). Lower panel: overlaid original CT image at T00 (blue) on a warped image from T50 to T00 (orange).

2D-3D registration has significantly shortened calculation times, and these are now acceptable from a clinical workflow perspective [43–45]. In the future, these image registration processes might be done automatically by computers, followed by final judgment by the medical staff.

25.5.2 Deformations

More realistic than rigid registration are deformable image registration (DIR) approaches. DIR is commercially available in some treatment planning software to propagate contours from a reference phase to other respiratory phases [46]. After an oncologist manually contours on one reference CT phase (generally peak exhalation), DIR calculates transformation maps based on the 4DCT data. These are then applied to the manual contours to transform them from the reference phase to the other respiratory phases. All counters at other respiratory phases are then automatically calculated. This process allows designing a target volume considering motion information, without a huge burden on the oncologist. An example of a registered lung CT using DIR is shown in Figure 25–8.

DIR is also needed for dose accumulation in the process of 4D treatment planning (also see Section 25.6). The most common approach in 4D treatment planning is to calculate the dose in all respective motion phases of a 4DCT and then use DIR to warp the resulting dose distributions back to a reference phase. DIR is a key factor in treatment planning with motion, and its accuracy can affect the resulting dose assessment. Several DIR algorithms have been developed, and the dependency of 4D dose calculations on these algorithms has to be discussed [46,47]. Zhang et al. reported that registration ambiguity-induced uncertainties in 4D dose distributions can be significant [48].

25.6 4D Dose Calculation

Dose calculation incorporating the time domain is typically referred to as 4D dose calculation or 4D treatment planning [49]. A basic approach is to convolve a static dose distribution by a probability density function, describing the probability that a volume element is found at a particular location [50,51]. This approach is also used to simulate the effects of daily setup errors. Such methods assume rigid motion, and besides blurring, the dose distribution will not be affected by the changes in internal anatomy that occur as a result of organ motion. More advanced 4D treatment planning has been investigated in parallel to the development of 4DCT. Several reviews have been published on this subject [52–56]. The combination of scanned particle beams and target motion presents a double-dynamic system and requires a dedicated solution for 4D treatment planning.

25.6.1 Phase-specific Calculations

Time-resolved volumetric imaging can be used to extend treatment planning capabilities for tumor sites influenced by respiratory motion [49,57]. A basic idea is to perform dose calculations per motion phase of a 4DCT dataset [56,58]. Deformation maps obtained by non-rigidly registering motion phases can then be used to transform the resulting sub-dose distributions to a reference motion phase for effective dose calculation by time-weighted summation. The simplest approach is to assume a one-to-one correspondence between un-deformed voxels at different phases and to use deformation vectors to determine the new reference voxel indices to which the dose is translated. However, the effect of applying these deformation vectors is to deform the voxels of the source image and, in the case of large deformations, the remapped, un-deformed reference voxels do not necessarily exactly overlap the voxels of the dose grid at the phase from which the dose is to be remapped. A more accurate estimate of the remapped dose can be obtained by trilinear interpolation of dose from voxels in the vicinity of the transformed center of mass of the reference voxel.

Any form of dose interpolation provides only an approximation of the remapped dose, and the accuracy of such methods in regions of large dose gradients remains uncertain. Furthermore, these methods ignore the changes in voxel density that must occur if mass is conserved. As different dose grids are used at each phase, the preservation of tissue type on a voxel-by-voxel basis is also neglected. Especially for proton dose calculations, the accuracy of these techniques in regions of large deformations may be limited.

25.6.2 Calculations on a Deforming Dose Grid

An alternative 4D dose calculation method executes on a deforming reference dose grid, where displacements of each voxel according to the anatomical deformations due to breathing are continuously taken into account [40,59,60]. In order to deal with time-varying changes in densities and, therefore, proton ranges, density variation maps derived for each motion phase have to be employed.

The motion and density extraction for the dose wrapping approach, as well as for calculations on a deforming dose grid, are challenged by the varying performance of different deformable registration algorithms. Registration ambiguity between different deformable registration approaches can be significant, especially within homogeneous regions, where there is little characteristic information about motion [46,48].

25.6.3 Dose Reconstruction

Reconstruction of the actual dose delivered to the patient based on delivery log file information and information from motion monitoring during treatment delivery can be valuable for quality assurance and for the optimization of motion mitigation techniques. Furthermore, knowledge of the already delivered dose within a treatment course will set the stage for adaptive concepts aimed at compensation of potential error in dose over the treatment course [61]. Dose reconstruction seems very suitable for tumor tracking treatments [62] since there the required knowledge of intra-treatment tumor motion is available, by definition, as it is a requirement for tracking. Dose reconstruction may, therefore, even be integrated as part of the routine workflow in tracking treatments, which could lead to better understanding of dose–response relationships by treatment follow-up. Many centers have implemented dose reconstruction as part of their quality assurance (QA) procedures [63] or within their adaptive strategies [61,64].

25.7 Approaches to Mitigating the Dosimetric Effects of Motion

Highly conformal irradiation treatment techniques mandate a need for precise target volume design. Target volume definitions for the GTV, CTV, ITV, and PTV can be found in ICRU Reports 50 and 62 [65,66] and are discussed in detail in chapter 23.

25.7.1 Margins

A moving target requires the definition of the expected magnitude of motion during treatment (internal margin), generally referred to as the internal target volume (ITV). In photon radiotherapy, the ITV is defined as "geometrically oriented" rather than "radiological path length oriented," and it does not take into account intra-fractional density changes.

Accordingly, the use of ITVs in particle beam therapy is limited, as discussed in [67]. Different approaches to incorporate the dependence of the particle dose distribution on density variation have been discussed [57,68–70]. One possible method is to adjust the Hounsfield units within the ITV to the average tumor density [68]. Another approach is to combine a geometrical target volume with an assumption of the occurring beam range variations. This assumption can be obtained for all phases of a 4DCT using maximum intensity projection (MIP) or average intensity projections (AIP) [71,72]. MIP and AIP can be performed in commercially available treatment planning systems and are already clinically available. Even though the MIP and AIP approaches were used clinically to deliver treatment to moving targets, they are limited. They combine 4DCT images into a single CT data set along the time axis. As a result, the anatomical density might be changed, especially in the thoracic region, and might cause underdosing within the target or overdosing in normal tissues. An alternative approach to account for varying beam ranges is to calculate the maximum and minimum beam range along each ray using 4DCT and adapt margins accordingly. This approach is useful to minimizing cold spots within the target, but it has not been used in commercial treatment planning systems [73].

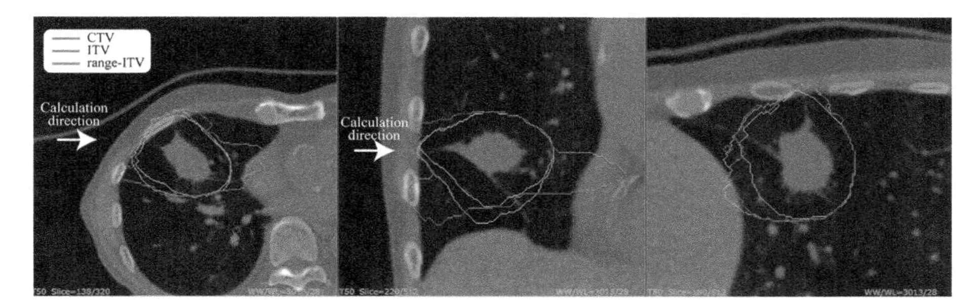

Figure 25–9 CT images at the reference phase and contours: blue, green and red lines show CTV, ITV and range-ITV, respectively.

Margin definition is especially critical for a scanned proton treatment delivery, as each beam position can be affected differently by motion. If the timeline of motion and the timeline of delivery is known, range-adapted ITVs (called range-ITV in Figure 25–9) can be constructed. The procedure is described in detail by Knopf et al. [74]. The use of range-ITVs in combination with rescanning has been shown to provide conformal dose distributions.

25.7.2 Breath Hold

The breath-hold approach with spirometers and valves, providing active respiratory control or deep inhalation breath holding, was introduced by Hanley et al. and Wong at al. [75,76]. Respiration control with valves ensures the reproducibility of lung volume and improves the reproducibility of motion. The use of breath hold has been shown to minimize dose to the heart and lung in breast cancer patients [77]. When treatments are delivered during breath hold, the intra-fractional tumor motion is very small, and internal margins can be minimized. When used in combination with gating, the remaining motion within the gating window is very small. The breath hold technique can also be useful during imaging to avoid artifacts. When acquiring images during inhalation and exhalation breath hold, an estimate for the motion amplitude can be obtained. However, valves and spirometers control and monitor only the amount of air inhaled and not the actual lung volume. Also, this technique is hard to tolerate for older patients or those with low lung function.

25.7.3 Gating

Respiratory gating was introduced into proton therapy by Ohara et al. [78], and it is now widely used in particle beam and photon beam therapy. A gated treatment involves the synchronization of irradiation with a gating signal, which is generally obtained by observing the patient respiration. Methods commonly used to measure patient respiration include monitoring patient surface motion via an external respiratory signal tracking system or direct monitoring of internal anatomy (internal target tracking). The external respiratory signal systems have tagging points with artificial markers that reflect or emit light, termed “passive” and “active.” The former type is routinely used and is already commercially available. Examples include RPM (Varian). Other external respiratory monitors obtain patient surface motion information (Vision RT, Catalyst) or use belt systems (Anzai). The gating window is usually focused on a set percentage level of the averaged respiratory amplitude (phase-based). However, all of these systems assume that the observed external respiratory signal is well correlated with the actual internal tumor motion, which is not always the case [79–81].

With regard to internal target tracking, systems using fluoroscopy (Cyberknife, Syntrax) and an electromagnetic implanted marker [82] are already commercially available. These systems have solved the tumor positional correlation problem mentioned above. Moreover, the gating window with these systems can be defined based on the tumor position rather than on the respiratory phase.

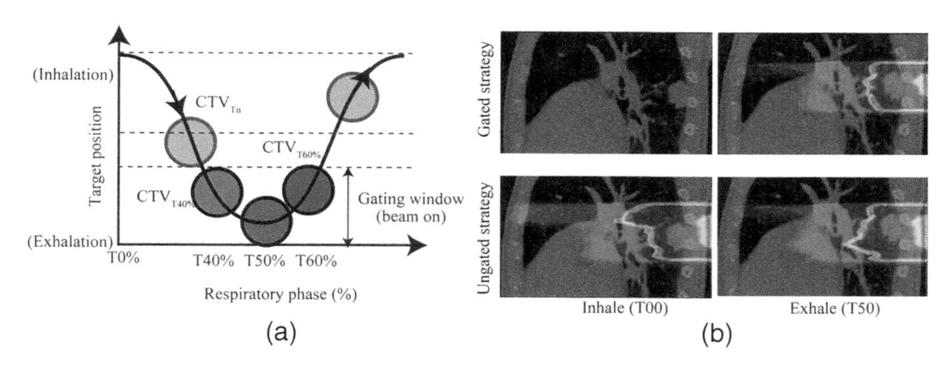

Figure 25–10 (a) Schematic drawing of the PTV as a function of respiratory phase. In gated treatment, the beam is on when the target is at T40%–T60% (= gating window). (b) Dose distribution at inhalation (T00) and exhalation (T50) for the ungated and gated strategies in sagittal section.

The gating strategy reduces the effect of intra-fractional motion and, as a result, helps to reduce the ITV margin (Figure 25–10a). For passive beam irradiation, the effect of residual target motion can be mitigated by the use of margins (Figure 25.10b). For scanning beam irradiation, the interplay effect of residual target motion can still lead to significant underdosing and, thus, requires further mitigation techniques, such as rescanning. In any case, gating minimizes excessive dose to healthy tissues for moving target geometries.

Several challenges in the gating strategy still remain, including the delay time between actual motion, the creation of the gating window, and the resulting reaction. Time latency is crucial to the success of gating, and it should be as low as possible. Even if the gating signal is sent to the irradiation system with short time latency, gating irradiation requires pulse coincidence between the accelerator extracted pulse and the gating window. To overcome this, the extended flattop operation was introduced in synchrotron-based carbon-ion beam irradiation. This way it is possible to hold an output current or voltage from the power supplies of the accelerator [83].

Generally, the gating window is selected to be around the exhalation phase due to its good reproducibility. On average, it is defined by a 30% peri-exhalation duty cycle (delivery efficiency, or the ratio of beam-on time to delivery time). Depending on the duty cycle, gating significantly increases the treatment time. The duty cycle with gating is approximately 70% than without gating. Intra-fractional residual organ motion during beam irradiation can be minimized with gating, but long treatment times may be uncomfortable for patients and might degrade the respiratory pattern reproducibility correlation between the skin surface and internal tumor motion across exhalations.

25.8 Motion Mitigation Techniques Specific to Scanned Delivery

Pencil beam scanning is the most advanced delivery technique for proton therapy. However, it is currently restricted at most centers to treatment of static tumors. Mobile

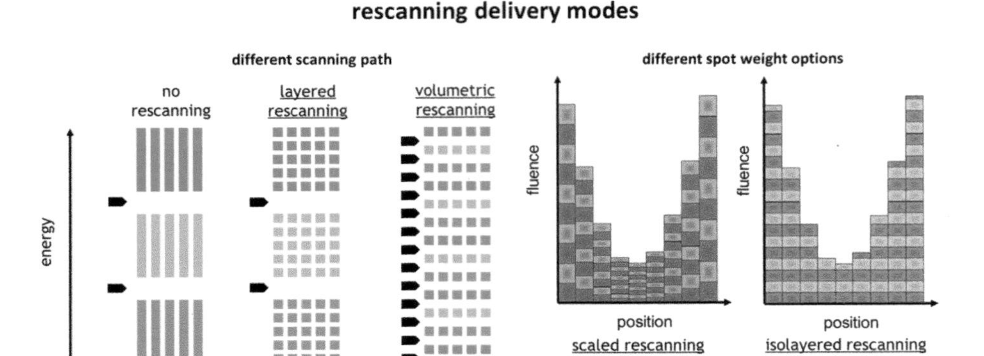

Figure 25–11 Different rescanning options. On the left, illustration of different scanning paths subject to different energy change sequences. On the right, illustration of different spot weight breakdowns over the different rescans. (Figure from personal communication with A. Schaetti.)

tumors pose a particular problem for scanned treatments because of large dose uncertainty due to interplay effects (interference between tumor motion and motion of the proton beam). Therefore, additional options are explored for motion mitigation in scanned delivery. Some motion mitigation approaches specific for a scanned delivery are introduced in the following.

25.8.1 Rescanning

The simplest delivery-based motion mitigation approach is rescanning [84], where each pencil beam spot position is visited several times. This repeated application of each delivered pencil beam intends to statistically smooth out interplay effects. However, rescanning will inevitably be at the cost of therapeutic dose in the internal margins that form the PTV due to blurring.

Multiple delivery modes for performing rescanning exist, with the most prominent distinction being whether the scans are first repeated within a 2D energy layer (layered rescanning) or the whole 3D volume is rescanned (volumetric rescanning) [70]. In both modes, different repainting strategies can be employed as, for example, scaled rescanning, for which the beam weight of each scan position is divided by the number of rescans or iso-layered rescanning, for which fixed beam weights are applied per rescan [85]. These basic rescanning delivery modes are illustrated in Figure 25–11.

When implementing rescanning, particular awareness has to be given to machine specifications—such as the scanning speed, the scanning method, the spot size, and the dose deliverable resolution—which can place limitations on the rescanning

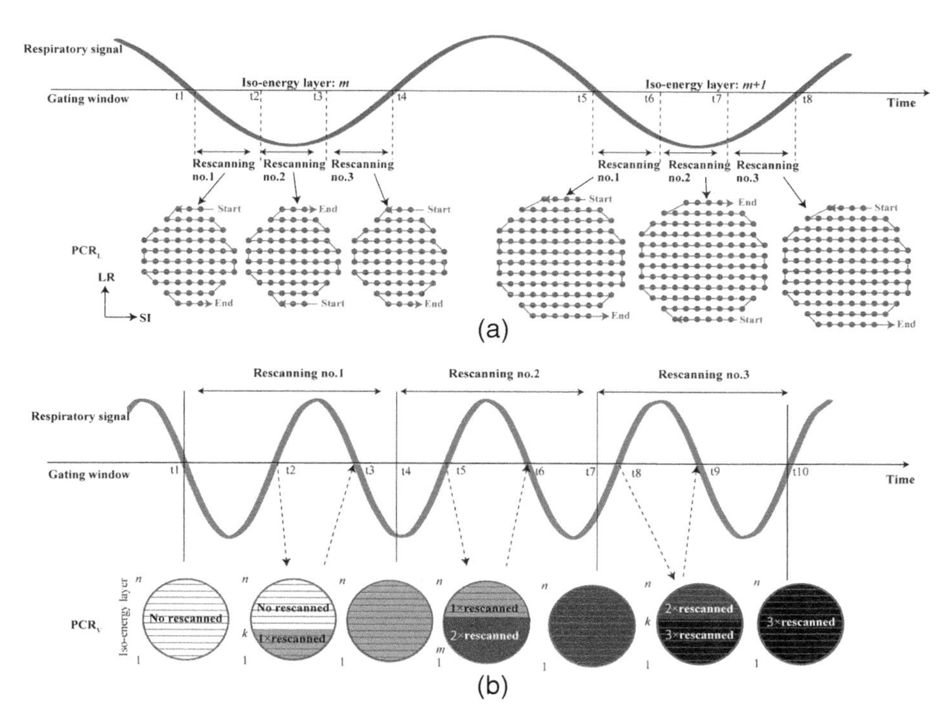

Figure 25–12 Time chart for (a) 3x layered PCR and (b) 3x volumetric PCR. (Upper panel) Curved line shows the respiratory signal. Gray points are scan spots. (Lower panel) White are = not yet rescanned, light gray area = 1x renned, gray area = 2x rescanned, black area = 3x rescanned [86].

approach. For the determination of the appropriate rescanning strategy, the fractionation scheme, number of fields, as well as the number and temporal sequence of the energy layers contributing to a beam position should be considered. All of these parameters have been shown to contribute an intrinsic rescanning effect, which should be considered in the decision of how many rescans and what type of rescanning should be employed [19–21].

Rescanning does not usually consider the respiratory cycle, and the scanning is done as rapidly as possible. If, however, the motion timeline is known, dose rate manipulation can be used to optimize the time needed to scan one layer or the whole target volume. This approach is known as phase-controlled rescanning (PCR) and was introduced by Mori et al. [86]. An example for layered and volumetric phase-controlled rescanning is shown in Figure 25–12.

Extensive simulations and experimental work is required to comprehensively evaluate rescanning effects for different patient and plan parameters. Synchronization effects due to the periodicity of target motion and individual rescans, especially for repetitive scaled rescanning, are a potential source of large underdosage in the CTV [87].

25.8.2 Tracking

Another motion mitigation technique unique to a scanned treatment delivery is tracking. Beam tracking for particle therapy refers to the lateral adaptation of pencil beam position, combined with appropriate energy changes to modify the position of the Bragg peak [88] in order to compensate for target motion in all directions. Like gating, tracking relies on motion monitoring; ideally, real-time 3D imaging should be employed or alternatively pre-treatment acquired 4DCT in combination with a motion surrogate. In current experiments, the adaptation parameters are determined prior to treatment delivery based on 4DCT data, which leads to concerns about variations in patient anatomy and respiratory motion between the 4DCT scan and the time of treatment. For tracking, it is particularly important to minimize the time lag between the detection of a geometry change and the corresponding adaptation of the beam. By minimizing the time between pre-treatment imaging, potential online plan adaptation, and treatment delivery, these uncertainties can be reduced, but there is need to investigate the robustness of beam tracking, particularly over the time scales of fractionated delivery. Despite an experimental implementation of tracking for carbon beam radiotherapy, beam tracking has not yet been used clinically [89]. The concern has been raised that tracking poses the danger of maximizing potential interplay effects and should, therefore, not be employed clinically.

25.9 4D Optimization/Combined Motion Mitigation Approaches

The objective of treatment planning is to find an "optimal" dose distribution that applies a high dose to the target and at the same time spares the close-by healthy tissue. In the first instance, treatment plan optimization refers to optimizing the target dose conformity. 4D optimization techniques have been proposed as a means of ensuring a conformal dose delivery to targets affected by respiratory motion by incorporating information about patient tissue motion into the plan optimization. In 4D plan optimization, instead of optimizing a static dose distribution, the expectation value of the dose received over all geometrical instances of the patient anatomy is optimized. Besides conformity, the robustness of the dose distribution against geometrical changes is an objective. Several studies have been performed comparing conventionally optimized treatment plans with sophisticated optimization approaches, testing their robustness against motion [90,91]. Graeff et al. investigated the synchronized delivery of pre-calculated fraction treatment plans based on specific motion phases [92]. Even though those complex motion mitigation approaches theoretically promise to be successful, they are limited by unpredictable variations of patient respiratory motion over the course of treatment. Heath et al. investigated probabilistic optimization, minimizing the dose variance in the target volume, and worst-case optimization, minimizing a weighted combination of the nominal and worst-case dose distributions, which occur in the presence of respiratory motion variation. The two 4D optimization approaches were compared with a margin-based mid-ventilation planning approach in five lung patients. Both robust planning methods were found suitable for automatic determination of treatment plans that ensure target dose conformity under respiratory motion variations, while minimizing the dose burden of healthy lung tissue [93].

Scanned particle therapy offers many parameters to be optimized in order to obtain a robust dose distribution. Conventional optimization parameters include the number of fields per plan and the field directions [19]. But in order to obtain dose distributions that are robust against motion, the spot size and the spot weight distribution can also be manipulated.

For static targets, the objective for the spot size and the spot weight optimization is target dose conformity. Spot weight optimization will result in highly unequal weight distributions. Spot weight analysis has shown that over all fields, only about 10% of delivered spots have a weight of more than 10% of the maximum in any given field. For static cases, this can be used to optimize the number of spots delivered per field.

For moving targets, different spot weight distributions can be beneficial, depending on which motion mitigation technique is applied. For spot weight distributions optimized for static situations with a few high-weighted spots and numerous low-weighted spots, geometrical variations result in severe dose errors in areas with high-weighted spots. For more smooth spot weight distributions, more positions will experience dose degradation in the case of motion, but the dosimetric effects will, in general, be less significant. When scaled rescanning is applied, low-weighted spots will become undeliverable, as the fraction of dose that should be delivered during each rescan will be very small. If iso-layered rescanning is applied to spot weight distributions with a few high-weighted spots and numerous low-weighted spots, the aim to break down regularity and to smooth out interplay effects will not be fulfilled. Rescanning will only take place at high-weighted spot positions, while low-weighted spot positions will not be revisited after the first scan. Thus, when employing rescanning, smoother spot weight distributions will be beneficial. Different spot weight distributions will have a significant effect on the robustness of scanned proton plans delivered to moving targets. Spot weight homogeneity is a possible alternative objective to target dose conformity for 4D treatment plan optimization in scanned particle therapy.

To obtain highly conformal dose distributions in a static geometry, small spot sizes are preferred. That will result in a high number of spots per field, with each spot position being subject to misplacements in the presence of motion. As motion will compromise the conformity due to blurring, alternative bigger spot sizes with increased spot overlap could be used instead. Larger beam spot sizes at identical raster grid spacing will result in more robust dose distributions for geometries affected by motion [20,21,94].

An optimal outcome for moving target geometries can also be obtained by combining different motion mitigation approaches. Slow tracking, combining breath hold and tracking, was proposed as a simpler version of real-time 3D tracking. Instead of tracking each beam position, the whole treatment field position would be adjusted for each breath hold if geometrical deviations are monitored. Mori et al. obtained good dose homogeneity and conformity by combining gating and rescanning [86]. Graeff et al. suggested the combination of gating and tracking [95].

As there exist many different motion scenarios, there is no unique, optimal way to account for motion effects for each scenario. In any case, adequately adapted margins have to be used to eliminate the possibility of target dose misses. Perhaps the ultimate solution for the treatment of moving targets with scanned particle beams would be the combination of real-time tracking and rescanning, so-called re-tracking. However, to implement this clinically, online on-board 3D imaging has to be improved. A lot of developments, simulations, and verifications have to be performed before such a technique could be used clinically.

25.10 Clinical Practice and Outcomes

Some motion management techniques have been routinely integrated into clinical protocols. Most proton treatments are still delivered using passive beam irradiation rather than scanning, and the most prominent motion mitigation technique used clinically is gating.

In general terms, the treatment workflow with motion management is not particularly different from the workflow without it. The main steps involve immobilization, CT acquisition, treatment planning, simulation, quality assurance, and treatment. CT image acquisition is used to generate respiratory-gated CT or 4DCT. If these respiratory-correlated CT images cannot be generated, other motion imaging approaches, such as fluoroscopic imaging, have to be used. Treatment planning with motion management has been introduced in Section 25.6. As only limited 4D tools are available in current commercially available treatment planning systems, planning for motion cases has to be performed with extra care. For example, it is important to select beam angles that avoid beam overshoot to OARs due to intra-fractional motion. Treatments are then delivered considering respiratory motion by applying gating or breath-hold techniques. Some treatment centers do not perform gated irradiation when an internal margin is added to compensate for intra-fractional motion, and they accept additional dose to tissue surrounding the tumor. Motion management typically increases the treatment workflow by several steps, especially treatment delivery. Workflow optimization as discussed in Chapter 17 is, therefore, especially desired.

Widesott et al. summarized clinical results (overall survival, local control, etc.) for lung treatments [96]. Most centers account for respiratory motion in lung patients by adding a constant margin or measured value, and then applying respiratory gating. However, these margins do not consider intra-fractional range changes as described in Section 25.7.1. The Loma Linda group reported that the three-year overall survival rate was dose-dependent, improving from 27% at 51 Gy(RBE) to 55% at 60 Gy(RBE).

Several treatment centers equipped with scanning beams have recently, or will soon, start irradiating tumors in the thoracic and abdominal region. For example, the Heidelberg Ion Therapy Center, which started carbon-ion treatments for liver patients in 2012, apply an abdominal compression technique to significantly minimize respiratory motion [97]. The National Institute of Radiological Sciences in Japan started

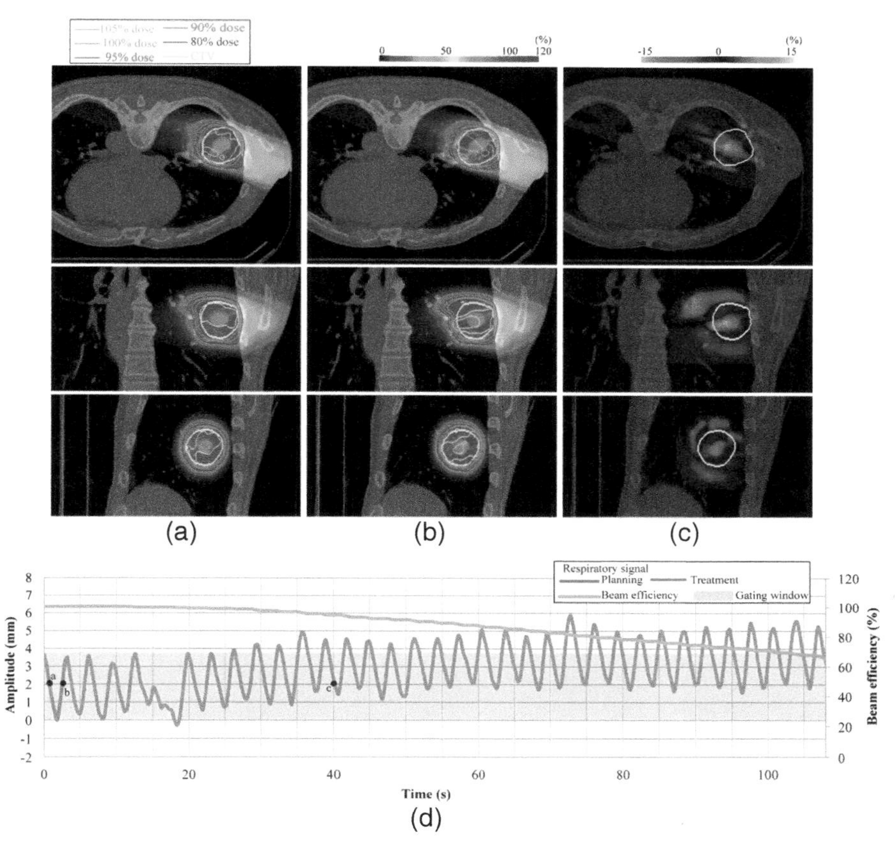

Figure 25–13 Lung dose distributions in axial, coronal, and sagittal view for a 110° field. (a) The planed dose. (b) The delivered dose. (c) The dose differences. In (d), the respiratory wave data are displayed. For planning, a respiratory cycle of 3.6 s was assumed, while the respiratory cycle during treatment was 3.1 ±0.6 s [98].

treatment under free breathing conditions using fluoroscopic image gating. They reported that the amplitude-based gating strategy with multiple rescans preserved dose distribution to moving targets, even though respiratory pattern were irregular (Figure 25–13) [98].

IGRT is a key factor in realizing highly conformal particle therapy, because quantitative information on intra-fractional/inter-fractional changes in the patient is not available if IGRT is not applied. Daily x-ray imaging has been performed in proton therapy since the 1970s. Several IGRT applications are now commercially available.

Online volumetric imaging with a high time resolution that provides good-contrast images is highly desirable for patient cases exhibiting tumor motion. In photon treatment facilities, cone-beam CT (CBCT) imaging is frequently used and directly integrated at most linear accelerators used in external beam therapy [99]. In particle beam therapy, however, the integration of online imaging tools in the treatment rooms is much more challenging due to the much more complex layout of the therapy delivering setup itself. An example of an innovative design is the beams eye view (BEV) imager at Gantry 2 at PSI (Villigen, Switzerland), which allows acquiring x-ray images in fluoroscopy mode during treatment delivery [100,101]. The Heidelberg ion therapy center (HIT) uses a robotic C-arm imaging system for CBCT acquisition [102]. One or two EPIDs have also been integrated on the rotating gantry to acquire CBCT imaging. An alternative technology for volumetric imaging involves in-room CTs. Diagnostic CT scanners are used in-room and provide better image quality and temporal resolution than CBCT. CBCT imaging can be acquired in the patient treatment position. To maintain the same position between imaging and treatment delivery, a robotic couch is needed to transfer the patient from the in-room CT to the treatment iso-center.

With regard to real-time imaging, several investigators have adopted tumor tracking using a dynamic FPD (DFPD) [103–105]. DFPDs have been shown to be useful for determining the respiratory gated beam-on time, with or without implanted fiducial markers. This system monitors the real-time tumor position and obtains a gating signal for beam on/off control during treatment. Moreover, it provides amplitude-based respiratory gating irradiation within the gating window defined by treatment planning, rather than phase-based gating irradiation. Since the implantation of fiducial markers into the patient is invasive, real-time tumor tracking without an implanted marker has been developed [106]. This is not yet commercially available, but will be used in clinical applications in the near future.

25.11 Summary and Outlook

The proton facilities delivering passively scattered proton therapy have nowadays often integrated motion management techniques into their clinical protocols in order to deal with moving target geometries. For scanned particle beam application, however, intra-fractional motion remains challenging, and it needs additional research and development. Currently, at centers with a strong research background, methods and techniques are being developed that allow for safe and more robust treatment deliveries.

Current research perspectives on 4D treatments include all critical aspects of time-resolved delivery, such as in-room imaging, motion detection, beam application, and quality assurance techniques. The adaptation of image guidance approaches from conventional radiotherapy into particle therapy needs to be pursued, and commercial treatment planning software has to be equipped with 4D capabilities.

In photon radiotherapy, the next generation of IGRT, replacing conventional x-ray guidance with treatment-integrated MRI-guided therapy is underway. This approach could also be a promising way forward for proton radiotherapy, taking into to account inter- as well as intra-fractional changes. The combination of a proton radiotherapy machine with magnetic resonance imaging (MRI) has been proposed by Raaijmakers at al. [107,108]. A hybrid proton-MR system could combine the advantages of a dose-free image, high 3D (or even 4D) soft tissue resolution, and a precisely steerable treatment beam.

References

1. Keall PJ, Mageras GS, Balter JM, Emery RS, Forster KM, Jiang SB, Kapatoes JM, Low DA, Murphy MJ, Murray BR, Ramsey CR, Van Herk MB, Vedam SS, Wong JW, Yorke E. The management of respiratory motion in radiation oncology report of AAPM task group 76. *Med Phys* 2006;33:3874–3900.
2. Langen KM, Jones DT. Organ motion and its management. *Int J Radiat Oncol Biol Phys* 2001;50:265–278.
3. Korreman SS. Motion in radiotherapy: Photon therapy. *Phys Med Biol* 2012;57:R161–191.
4. Seppenwoolde Y, Shirato H, Kitamura K, Shimizu S, van Herk M, Lebesque JV, Miyasaka K. Precise and real-time measurement of 3D tumor motion in lung due to breathing and heartbeat, measured during radiotherapy. *Int J Radiat Oncol Biol Phys* 2002;53:822–834.
5. Mori S, Endo M, Komatsu S, Yashiro T, Kandatsu S, Baba M. Four-dimensional measurement of lung tumor displacement using 256-multi-slice CT-scanner. *Lung Cancer* (Amsterdam, Netherlands) 2007;56:59–67.
6. Plathow C, Ley S, Fink C, Puderbach M, Hosch W, Schmahl A, Debus J, Kauczor H-U. Analysis of intrathoracic tumor mobility during whole breathing cycle by dynamic MRI. *Int J Radiat Oncol Biol Phys* 2004;59:952–959.
7. Weiss PH, Baker JM, Potchen EJ. Assessment of hepatic respiratory excursion. *J Nucl Med* 1972;13:758–759.
8. Shimizu S, Shirato H, Xo B, Kagei K, Nishioka T, Hashimoto S, Tsuchiya K, Aoyama H, Miyasaka K. Three-dimensional movement of a liver tumor detected by high-speed magnetic resonance imaging. *Radiother Oncol* 1999;50:367–370.
9. Hallman JL, Mori S, Sharp GC, Lu HM, Hong TS, Chen GT. A four-dimensional computed tomography analysis of multiorgan abdominal motion. *Int J Radiat Oncol Biol Phys* 2012;83:435–441.
10. Suramo I, Paivansalo M, Myllyla V. Cranio-caudal movements of the liver, pancreas and kidneys in respiration. *Acta Radiol Diagn (Stockh)* 1984;25:129–131.
11. Suramo I, Paivansalo M, Myllyla V. Cranio-caudal movements of the liver, pancreas and kidneys in respiration. *Acta Radiol Diagn (Stockh)* 1984;25:129–131.
12. Bryan P, Custar S, Haaga J, Balsara V. Respiratory movement of the pancreas: An ultrasonic study. *J Ultrasound Med* 1984;3:317–320.
13. Horst E, Micke O, Moustakis C, Schuck A, Schafer U, Willich NA. Conformal therapy for pancreatic cancer: Variation of organ position due to gastrointestinal distention—implications for treatment planning. *Radiology* 2002;222:681–686.
14. Mori S, Hara R, Yanagi T, Sharp GC, Kumagai M, Asakura H, Kishimoto R, Yamada S, Kandatsu S, Kamada T. Four-dimensional measurement of intrafractional respiratory motion of pancreatic tumors using a 256 multi-slice CT scanner. *Radiother Oncol* 2009;92:231–237.
15. Langen KM, Willoughby TR, Meeks SL, Santhanam A, Cunningham A, Levine L, Kupelian PA. Observations on real-time prostate gland motion using electromagnetic tracking. *Int J Radiat Oncol Biol Phys* 2008;71:1084–1090.
16. Padhani AR, Khoo VS, Suckling J, Husband JE, Leach MO, Dearnaley DP. Evaluating the effect of rectal distension and rectal movement on prostate gland position using cine MRI. *Int J Radiat Oncol Biol Phys* 1999;44:525–533.
17. Antolak JA, Rosen II, Childress CH, Zagars GK, Pollack A. Prostate target volume variations during a course of radiotherapy. *Int J Radiat Oncol Biol Phys* 1998;42:661–672.
18. Trofimov A, Nguyen PL, Efstathiou JA, Wang Y, Lu H-M, Engelsman M, Merrick S, Cheng C-W, Wong JR, Zietman AL. Interfractional variations in the setup of pelvic bony anatomy and soft tissue, and their implications on the delivery of proton therapy for localized prostate cancer. *Int J Radiat Oncol Biol Phys* 2011;80:928–937.
19. Knopf AC, Hong TS, Lomax A. Scanned proton radiotherapy for mobile targets—the effectiveness of re-scanning in the context of different treatment planning approaches and for different motion characteristics. *Phys Med Biol* 2011;56:7257–7271.

20. Dowdell S, Grassberger C, Sharp GC, Paganetti H. Interplay effects in proton scanning for lung: A 4D Monte Carlo study assessing the impact of tumor and beam delivery parameters. *Phys Med Biol* 2013;58:4137–4156.
21. Grassberger C, Dowdell S, Lomax A, Sharp G, Shackleford J, Choi N, Willers H, Paganetti H. Motion interplay as a function of patient parameters and spot size in spot scanning proton therapy for lung cancer. *Int J Radiat Oncol Biol Phys* 2013;86:380–386.
22. Dowdell S, Grassberger C, Paganetti H. Four-dimensional Monte Carlo simulations demonstrating how the extent of intensity-modulation impacts motion effects in proton therapy lung treatments. *Med Phys* 2013;40:121713.
23. Lomax A. Intensity modulation methods for proton radiotherapy. *Phys Med Biol* 1999;44:185–205.
24. Mohammed N, Kestin L, Ghilezan M, Krauss D, Vicini F, Brabbins D, Gustafson G, Ye H, Martinez A. Comparison of acute and late toxicities for three modern high-dose radiation treatment techniques for localized prostate cancer. *Int J Radiat Oncol Biol Phys* 2012;82:204–212.
25. Singh J, Greer PB, White MA, Parker J, Patterson J, Tang CI, Capp A, Wratten C, Denham JW. Treatment-related morbidity in prostate cancer: A comparison of 3-dimensional conformal radiation therapy with and without image guidance using implanted fiducial markers. *Int J Radiat Oncol Biol Phys* 2013;85(4):1018–23.
26. Verellen D, De Ridder M, Linthout N, Tournel K, Soete G, Storme G. Innovations in image-guided radiotherapy. *Nat Rev Cancer* 2007;7:949–960.
27. Chen GT, Sharp GC, Mori S. A review of image-guided radiotherapy. *Radiol Phys Technol* 2009;2:1–12.
28. Ma J, Chang Z, Wang Z, Jackie Wu Q, Kirkpatrick JP, Yin FF. Exactrac x-ray 6 degree-of-freedom image-guidance for intracranial non-invasive stereotactic radiotherapy: Comparison with kilo-voltage cone-beam CT. *Radiother Oncol* 2009;93:602–608.
29. Tanaka R, Sanada S, Fujimura M, Yasui M, Nakayama K, Matsui T, Hayashi N, Matsui O. Development of functional chest imaging with a dynamic flat-panel detector (FPD). *Radiol Phys Technol* 2008;1:137–143.
30. Suzuki K, Armato SG, Li F, Sone S, Doi K. Massive training artificial neural network (MTANN) for reduction of false positives in computerized detection of lung nodules in low-dose computed tomography. *Med Phys* 2003;30:1602.
31. Barnes GT, Sones RA, Tesic MM, Morgan DR, Sanders JN. Detector for dual-energy digital radiography. *Radiology* 1985;156:537–540.
32. Sheng C, Suzuki K. Separation of bones from chest radiographs by means of anatomically specific multiple massive-training ANNs combined with total variation minimization smoothing. *IEEE Trans Med Imaging* 2014;33:246–257.
33. Caldwell CB, Mah K, Skinner M, Danjoux CE. Can pet provide the 3D extent of tumor motion for individualized internal target volumes? A phantom study of the limitations of CT and the promise of PET. *Int J Radiat Oncol Biol Phys* 2003;55:1381–1393.
34. Gagné IM, Robinson DM. The impact of tumor motion upon ct image integrity and target delineation. *Med Phys* 2004;31:3378.
35. Tarver RD, Conces DJ, Godwin JD. Motion artifacts on CT simulate bronchiectasis. *Am J Roentogenol* 1988;151:1117–1119.
36. Flohr T, Ohnesorge B. Heart-rate adaptive optimization of spatial and temporal resolution for ECG-gated multi-slice spiral CT of the heart. *J Compt Assist Tomogr* 2001;25:907–923.
37. Mageras GS, Pevsner A, Yorke ED, Rosenzweig KE, Ford EC, Hertanto A, Larson SM, Lovelock DM, Erdi YE, Nehmeh SA, Humm JL, Ling CC. Measurement of lung tumor motion using respiration-correlated CT. *Int J Radiat Oncol Biol Phys* 2004;60:933–941.
38. Yamamoto T, Langner U, Loo BW, Jr., Shen J, Keall PJ. Retrospective analysis of artifacts in four-dimensional CT images of 50 abdominal and thoracic radiotherapy patients. *Int J Radiat Oncol Biol Phys* 2008;72:1250–1258.
39. Coolens C, Bracken J, Driscoll B, Hope A, Jaffray D. Dynamic volume vs respiratory correlated 4DCT for motion assessment in radiation therapy simulation. *Med Phys* 2012;39:2669–2681.
40. Boye D, Lomax T, Knopf A. Mapping motion from 4D-MRI to 3D-CT for use in 4D dose calculations: A technical feasibility study. *Med Phys* 2013;40:061702.
41. Hill DL, Batchelor PG, Holden M, Hawkes DJ. Medical image registration. *Phys Med Biol* 2001;46:R1–45.

42. Chang Z, Wang Z, Ma J, O'Daniel JC, Kirkpatrick J, Yin FF. 6D image guidance for spinal non-invasive stereotactic body radiation therapy: Comparison between Exactrac x-ray 6D with kilo-voltage cone-beam CT. *Radiother Oncol* 2010;95:116–121.
43. Kachelriess M, Knaup M, Bockenbach O. Hyperfast parallel-beam and cone-beam back-projection using the cell general purpose hardware. *Med Phys* 2007;34:1474–1486.
44. Neophytou N, Xu F, Mueller K. Hardware acceleration vs. algorithmic acceleration: Can GPU-based processing beat complexity optimization for CT? *Proc SPIE* 2007;6510.
45. Sharp GC, Kandasamy N, Singh H, Folkert M. GPU-based streaming architectures for fast cone-beam CT image reconstruction and demons deformable registration. *Phys Med Biol* 2007;52:5771–5783.
46. Brock KK. Results of a multi-institution deformable registration accuracy study (MID-RAS). *Int J Radiat Oncol Biol Phys* 2010;76:583–596.
47. Castadot P, Lee JA, Parraga A, Geets X, Macq B, Gregoire V. Comparison of 12 deformable registration strategies in adaptive radiation therapy for the treatment of head and neck tumors. *Radiother Oncol* 2008;89:1–12.
48. Zhang Y, Boye D, Tanner C, Lomax AJ, Knopf A. Respiratory liver motion estimation and its effect on scanned proton beam therapy. *Phys Med Biol* 2012;57:1779–1795.
49. Keall P. 4-dimensional computed tomography imaging and treatment planning. *Semin Radiat Oncol* 2004;14:81–90.
50. Lujan AE, Larsen EW, Balter JM, Ten Haken RK. A method for incorporating organ motion due to breathing into 3D dose calculations. *Med Phys* 1999;26:715–720.
51. Rosu M, Dawson LA, Balter JM, McShan DL, Lawrence TS, Ten Haken RK. Alterations in normal liver doses due to organ motion. *Int J Radiat Oncol Biol Phys* 2003;57:1472–1479.
52. Knopf A, Bert C, Heath E, Nill S, Kraus K, Richter D, Hug E, Pedroni E, Safai S, Albertini F, Zenklusen S, Boye D, Söhn M, Soukup M, Sobotta B, Lomax A. Special report: Workshop on 4D-treatment planning in actively scanned particle therapy—recommendations, technical challenges, and future research directions. *Med Phys* 2010;37:4608.
53. Bert C, Graeff C, Riboldi M, Nill S, Baroni G, Knopf AC. Advances in 4d treatment planning for scanned particle beam therapy - report of dedicated workshops. *Technol Cancer Res Treat* 2014;13:485-495.
54. Knopf A, Nill S, Yohannes I, Graeff C, Dowdell S, Kurz C, Sonke JJ, Biegun AK, Lang S, McClelland J, Champion B, Fast M, Wolfelschneider J, Gianoli C, Rucinski A, Baroni G, Richter C, van de Water S, Grassberger C, Weber D, Poulsen P, Shimizu S, Bert C. Challenges of radiotherapy: Report on the 4D treatment planning workshop 2013. *Phys Med* 2014;30:809–815.
55. Chen GT, Kung JH, Rietzel E. Four-dimensional imaging and treatment planning of moving targets. *Front Radiat Ther Oncol* 2007;40:59–71.
56. Keall PJ, Joshi S, Vedam SS, Siebers JV, Kini VR, Mohan R. Four-dimensional radiotherapy planning for DMLC-based respiratory motion tracking. *Med Phys* 2005;32:942.
57. Engelsman M, Rietzel E, Kooy HM. Four-dimensional proton treatment planning for lung tumors. *Int J Radiat Oncol Biol Phys* 2006;64:1589–1595.
58. Rietzel E, Chen GT, Choi NC, Willet CG. Four-dimensional image-based treatment planning: Target volume segmentation and dose calculation in the presence of respiratory motion. *Int J Radiat Oncol Biol Phys* 2005;61:1535–1550.
59. Trofimov A, Rietzel E, Lu HM, Martin B, Jiang S, Chen GT, Bortfeld T. Temporo-spatial imrt optimization: Concepts, implementation and initial results. *Phys Med Biol* 2005;50:2779–2798.
60. Heath E, Seuntjens J. A direct voxel tracking method for four-dimensional Monte Carlo dose calculations in deforming anatomy. *Med Phys* 2006;33:434–445.
61. Richter D, Saito N, Chaudhri N, Hartig M, Ellerbrock M, Jakel O, Combs SE, Habermehl D, Herfarth K, Durante M, Bert C. Four-dimensional patient dose reconstruction for scanned ion beam therapy of moving liver tumors. *Int J Radiat Oncol Biol Phys* 2014;89:175–181.
62. Riboldi M, Orecchia R, Baroni G. Real-time tumour tracking in particle therapy: Technological developments and future perspectives. *Lancet Oncol* 2012;13:e383–391.
63. Poulsen PR, Schmidt ML, Keall P, Worm ES, Fledelius W, Hoffmann L. A method of dose reconstruction for moving targets compatible with dynamic treatments. *Med Phys* 2012;39:6237–6246.

64. Ravkilde T, Keall PJ, Grau C, Hoyer M, Poulsen PR. Time-resolved dose reconstruction by motion encoding of volumetric modulated arc therapy fields delivered with and without dynamic multi-leaf collimator tracking. *Acta Oncol* 2013;52:1497–1503.
65. ICRU. Report 62. Prescribing, recording and reporting photon beam therapy (supplement to ICRU report 50). Bethesda, MD: International commission on radiation units and measurements. 1999.
66. ICRU. Report 50. Prescribing, recording and reporting photon beam therapy. Bethesda, MD: International commission on radiation units and measurements, 1993.
67. ICRU. Report 72. Prescribing, recording and reporting photon beam therapy (supplement to ICRU report 78): International commission on radiation units and measurements. DeLuca, PM, 2007.
68. Koto M, Miyamoto T, Yamamoto N, Nishimura H, Yamada S, Tsujii H. Local control and recurrence of stage I non-small cell lung cancer after carbon ion radiotherapy. *Radiother Oncol* 2004;71:147–156.
69. Bert C, Rietzel E. 4D treatment planning for scanned ion beams. *Radiat Oncol* 2007;2:24.
70. Rietzel E, Bert C. Respiratory motion management in particle therapy. *Med Phys* 2010;37:449–460.
71. Kang Y, Zhang X, Chang JY, Wang H, Wei X, Liao Z, Komaki R, Cox JD, Balter PA, Liu H, Zhu XR, Mohan R, Dong L. 4D proton treatment planning strategy for mobile lung tumors. *Int J Radiat Oncol Biol Phys* 2007;67:906–914.
72. Rietzel E, Liu AK, Doppke KP, Wolfgang JA, Chen AB, Chen GT, Choi NC. Design of 4D treatment planning target volumes. *Int J Radiat Oncol Biol Phys* 2006;66:287–295.
73. Mori S, Yanagi T, Hara R, Sharp GC, Asakura H, Kumagai M, Kishimoto R, Yamada S, Kato H, Kandatsu S, Kamada T. Comparison of respiratory-gated and respiratory-ungated planning in scattered carbon ion beam treatment of the pancreas using four-dimensional computed tomography. *Int J Radiat Oncol Biol Phys* 2010;76:303–312.
74. Knopf A-C, Boye D, Lomax A, Mori S. Adequate margin definition for scanned particle therapy in the incidence of intrafractional motion. *Phys Med Biol* 2013;58:6079–6094.
75. Hanley J, Debois MM, Mah D, Mageras GS, Raben A, Rosenzweig K, Mychalczak B, Schwartz LH, Gloeggler PJ, Lutz W, Ling CC, Leibel SA, Fuks Z, Kutcher GJ. Deep inspiration breath-hold technique for lung tumors: The potential value of target immobilization and reduced lung density in dose escalation. *Int J Radiat Oncol Biol Phys* 1999;45:603–611.
76. Wong JW, Sharpe MB, Jaffray DA, Kini VR, Robertson JM, Stromberg JS, Martinez AA. The use of active breathing control (ABC) to reduce margin for breathing motion. *Int J Radiat Oncol Biol Phys* 1999;44:911–919.
77. Nissen HD, Appelt AL. Improved heart, lung and target dose with deep inspiration breath hold in a large clinical series of breast cancer patients. *Radiother Oncol* 2013;106:28–32.
78. Ohara K, Okumura T, Akisada M, Inada T, Mori T, Yokota H, Calaguas MJ. Irradiation synchronized with respiration gate. *Int J Radiat Oncol Biol Phys* 1989;17:853–857.
79. Beddar AS, Kainz K, Briere TM, Tsunashima Y, Pan TS, Prado K, Mohan R, Gillin M, Krishnan S. Correlation between internal fiducial tumor motion and external marker motion for liver tumors imaged with 4D-CT. *Int J Radiat Oncol* 2007;67:630–638.
80. Ozhasoglu C, Murphy MJ. Issues in respiratory motion compensation during external-beam radiotherapy. *Int J Radiat Oncol* 2002;52:1389–1399.
81. Gierga DP, Brewer J, Sharp GC, Betke M, Willett CG, Chen GTY. The correlation between internal and external markers for abdominal tumors: implications for respiratory gating. *Int J Radiat Oncol* 2005;61(5):1551–8.
82. Litzenberg DW, Willoughby TR, Balter JM, Sandler HM, Wei J, Kupelian PA, Cunning AA, Bock A, Aubin M, Roach M, 3rd, Shinohara K, Pouliot J. Positional stability of electromagnetic transponders used for prostate localization and continuous, real-time tracking. *Int J Radiat Oncol Biol Phys* 2007;68:1199–1206.
83. Iwata Y, Kadowaki T, Uchiyama H, Fujimoto T, Takada E, Shirai T, Furukawa T, Mizushima K, Takeshita E, Katagiri K, Sato S, Sano Y, Noda K. Multiple-energy operation with extended flattops at HIMAC. *Nucl Instrum Meth A* 2010;624:33–38.
84. Phillips MH, Pedroni E, Blattmann H, Boehringer T, Coray A, Scheib S. Effects of respiratory motion on dose uniformity with a charged particle scanning method. *Phys Med Biol* 1992;37:223–234.

85. Zenklusen SM, Pedroni E, Meer D. A study on repainting strategies for treating moderately moving targets with proton pencil beam scanning at the new gantry 2 at PSI. *Phys Med Biol* 2010;55:5103–5121.
86. Mori S, Furukawa T, Inaniwa T, Zenklusen S, Nakao M, Shirai T, Noda K. Systematic evaluation of four-dimensional hybrid depth scanning for carbon-ion lung therapy. *Med Phys* 2013;40:031720.
87. Seco J, Robertson D, Trofimov A, Paganetti H. Breathing interplay effects during proton beam scanning: Simulation and statistical analysis. *Phys Med Biol* 2009;54:N283–294.
88. Grozinger SO, Bert C, Haberer T, Kraft G, Rietzel E. Motion compensation with a scanned ion beam: A technical feasibility study. *Radiat Oncol* 2008;3:34.
89. Bert C, Saito N, Schmidt A, Chaudhri N, Schardt D, Rietzel E. Target motion tracking with a scanned particle beam. *Med Phys* 2007;34:4768–4771.
90. Liu W, Liao Z, Schild SE, Liu Z, Li H, Li Y, Park PC, Li X, Stoker J, Shen J, Keole S, Anand A, Fatyga M, Dong L, Sahoo N, Vora S, Wong W, Zhu XR, Bues M, Mohan R. Impact of respiratory motion on worst-case scenario optimized intensity modulated proton therapy for lung cancers. *Pract Radiat Oncol* 2015;5(2):e77–86.
91. Stuschke M, Kaiser A, Pottgen C, Lubcke W, Farr J. Potentials of robust intensity modulated scanning proton plans for locally advanced lung cancer in comparison to intensity modulated photon plans. *Radiother Oncol* 2012;104:45–51.
92. Graeff C, Luchtenborg R, Eley JG, Durante M, Bert C. A 4D-optimization concept for scanned ion beam therapy. *Radiother Oncol* 2013;109:419–424.
93. Heath E, Unkelbach J, Oelfke U. Incorporating uncertainties in respiratory motion into 4D treatment plan optimization. *Med Phys* 2009;36:3059.
94. Bert C, Gemmel A, Saito N, Rietzel E. Gated irradiation with scanned particle beams. *Int J Radiat Oncol Biol Phys* 2009;73:1270–1275.
95. Graeff C, Constantinescu A, Luchtenborg R, Durante M, Bert C. Multigating, a 4D optimized beam tracking in scanned ion beam therapy. *Technol Cancer Res T* 2014;13:497–504.
96. Widesott L, Amichetti M, Schwarz M. Proton therapy in lung cancer: Clinical outcomes and technical issues. A systematic review. *Radiother Oncol* 2008;86:154–164.
97. Habermehl D, Debus J, Ganten T, Ganten MK, Bauer J, Brecht IC, Brons S, Haberer T, Haertig M, Jakel O, Parodi K, Welzel T, Combs SE. Hypofractionated carbon ion therapy delivered with scanned ion beams for patients with hepatocellular carcinoma—feasibility and clinical response. *Radiat Oncol* 2013;8:59.
98. Mori S, Inaniwa T, Furukawa T, Takahashi W, Nakajima M, Shirai T, Noda K, Yasuda S, Yamamoto N. Amplitude-based gated phase-controlled rescanning in carbon-ion scanning beam treatment planning under irregular breathing conditions using lung and liver 4DCTS. *J Radiat Res* 2014;55:948–958.
99. Jaffray DA, Siewerdsen JH, Wong JW, Martinez AA. Flat-panel cone-beam computed tomography for image-guided radiation therapy. *Int J Radiat Oncol Biol Phys* 2002;53:1337–1349.
100. Zhang Y, Knopf A, Tanner C, Boye D, Lomax AJ. Deformable motion reconstruction for scanned proton beam therapy using on-line x-ray imaging. *Phys Med Biol* 2013;58:8621–8645.
101. Pedroni E, Bearpark R, Bohringer T, Coray A, Duppich J, Forss S, George D, Grossmann M, Goitein G, Hilbes C, Jermann M, Lin S, Lomax A, Negrazus M, Schippers M, Kotle G. The PSI gantry 2: A second generation proton scanning gantry. *Zeitschrift fur Medizinische Physik* 2004;14:25–34.
102. Heiland M, Schulze D, Blake F, Schmelzle R. Intraoperative imaging of zygomaticomaxillary complex fractures using a 3D c-arm system. *Int J Oral Maxillofac Surg* 2005;34:369–375.
103. Neicu T, Berbeco R, Wolfgang J, Jiang SB. Synchronized moving aperture radiation therapy (SMART): Improvement of breathing pattern reproducibility using respiratory coaching. *Phys Med Biol* 2006;51:617–636.
104. Sharp GC, Lu HM, Trofimov A, Tang X, Jiang SB, Turcotte J, Gierga DP, Chen GT, Hong TS. Assessing residual motion for gated proton-beam radiotherapy. *J Radiat Res (Tokyo)* 2007;48 Suppl A:A55–59.
105. Murphy MJ. Tracking moving organs in real time. *Semin Radiat Oncol* 2004;14:91–100.

106. Cui Y, Dy JG, Sharp GC, Alexander B, Jiang SB. Multiple template-based fluoroscopic tracking of lung tumor mass without implanted fiducial markers. *Phys Med Biol* 2007;52:6229–6242.
107. Raaymakers BW, Raaijmakers AJ, Lagendijk JJ. Feasibility of MRI guided proton therapy: Magnetic field dose effects. *Phys Med Biol* 2008;53:5615–5622.
108. Raaijmakers AJ, Raaymakers BW, van der Meer S, Lagendijk JJ. Integrating a MRI scanner with a 6 MV radiotherapy accelerator: Impact of the surface orientation on the entrance and exit dose due to the transverse magnetic field. *Phys Med Biol* 2007;52:929–939.

Chapter 26

Monitor Unit (MU) Calculation

Timothy C. Zhu, Ph.D.[1], Haibo Lin, Ph.D.[2], and Jiajian Shen[3]

[1] Professor, Department of Radiation Oncology,
University of Pennsylvania,
Philadelphia, PA
[2]Medical Physicist, Department of Radiation Oncology,
University of Pennsylvania,
Philadelphia, PA
[3]Assistant Professor, Department of Radiation Oncology,
Mayo Clinic,
Phoenix, AZ

26.1 Introduction

A monitor unit (MU) is a measure of machine output of a proton accelerator. The MUs are measured by monitor chambers, which are ionization chambers that measure the dose delivered by a beam in a reference condition and are built into the treatment head of the proton beam line. For the proton radiation therapy, some centers use a dose per proton, not dose per MU that will be discussed later.

The treatment of MU calculation for broad uniform proton beams is different from that for pencil beam scanning (PBS), primarily because of the differences in proton therapy utilization: the former is dominated by a single spread-out Bragg peak (SOBP) beam, and the latter is dominated by 3D intensity-modulated proton therapy (IMPT).

The MU calculation for a PBS plan, similar to IMRT, is a backward calculation. The treatment planning system (TPS) needs to calculate the beam line parameters (e.g., the energy layers and the spot dose deposition positions). The optimization process is to find out the weights for all spots (i.e., MU), and the summation of all spot doses are conformal to target and to spare organs at risk. Since there are usually thousands of spots in a field, the solution of the spot weights would be highly degenerated. The robust optimization is implemented to find out the spot weights that are robust to the setup and range uncertainties. However, the beam line calculation and the spot weight (robust) optimization are not the scope of the second MU check. The dose calculation for the second MU check is based directly on the optimized treatment plan in TPS.

The second MU check can be implemented by the analytical (or factor-based) method and MC simulation. In this chapter, we will focus primarily on the analytical method because it is more straightforward, easier to implement, and has the fast calculation speed. Another advantage of the analytical method is that it usually does not need additional data since the data for commissioning TPS is typically enough for second MU check commissioning. The Monte Carlo simulation would provide accurate dose calculation, but it requires accurate modeling of the treatment nozzle, and the slower calculation speed is another concern.

26.2 Difference between Protons and Photons for Dose Per MU Verification

The fundamental difference between proton and photon beams is that while proton beams provide depth modulation, the photon beams do not. Thus, it is critical to verify the exact range of the proton beam. Even though the range of the proton beam is determined by its energy, the current consensus is that a given proton beam still has a range uncertainty of more than 1 mm, which is discussed in Chapter 21.

For broad proton beams used in conventional proton therapy, only SOBP beams with nominally flat beam lateral profiles are used for routine treatment. For dose per MU verification of the SOBP beam for a particular range R (or energy), verification of D/MU at one point (e.g., the middle of the SOBP beam) is sufficient if one can establish the relationship between D/MU there and the range R. Otherwise, it is necessary to verify both D/MU and the range R. One point D/MU check for proton beam is similar to what is currently used for photon beams in conventional 3D plans, where D/MU at one dose prescription point is used.

For PBS used for IMPT, it is really necessary to verify the dose distribution in 3D, unlike that for photon beam IMRT, where the 2D dose profile at a fixed depth is sufficient to verify an IMRT plan. The reason for this additional complexity is caused by the fact that a proton IMPT plan is made of pencil beam pristine peaks modulated in the entire 3D space.

26.3 Categories of Dose-to-MU Formalisms for Proton Beams

To perform MU calculation, it is necessary to determine the relationship between the absorbed dose and the measured quantity using fundamental physics quantities. This section establishes some of the basic relationships needed for MU calculation.

26.3.1 Factor-based Dose-to-MU Formalism

Factor-based methods determine absorbed dose per monitor unit by using the product of standardized dose ratio measurements. Successive dose ratio factors are multiplied for a chain of geometries, and thus the dose ratio factors are varied one by one until the geometry of interest is linked back to the reference geometry [1]:

$$\frac{D}{MU}(caseA) = \frac{(D/MU)_{caseA}}{(D/MU)_{caseB}} \times \frac{(D/MU)_{caseB}}{\cdots} \times \frac{\cdots}{(D/MU)_{ref}} \times \frac{D}{MU}(ref). \tag{26.1}$$

This equation is an identity equation. The strength of the formalism lies in that the calculations are simple and based on measured data. Obviously, one strives to use as few and as general factors as possible, where some factors might be modeled instead of measured (e.g., the inverse square factor).

In practice, Equation 26.1 can be implemented as a dose ratio (Equation 26.2a) or a dose-to-fluence ratio (Equation 26.2b):

$$\begin{aligned}\frac{D(c,s;z,d)}{D(c_{ref},s_{ref};z_{ref},d_{ref})} &= \frac{D(c,s;z,d)}{D(c,s;z,d_{ref})} \times \frac{D(c,s;z,d_{ref})}{D(c_{ref},s_{ref};z,d_{ref})} \times \frac{D(c_{ref},s_{ref};z,d_{ref})}{D(c_{ref},s_{ref};z_{ref},d_{ref})} \\ &= TPR(c,s;d) \times OF_w(c,s;z,d_{ref}) \times ISF(z)\end{aligned} \tag{26.2a}$$

$$\begin{aligned}\frac{D(c,s;z,d)}{D(c_{ref},s_{ref};z_{ref},d_{ref})} &= \frac{D(c,s;z,d)}{X(c,s;z)} \times \frac{X(c,s;z)}{X(c_{ref},s_{ref};z)} \times \frac{X(c_{ref},s_{ref};z)}{X(c_{ref},s_{ref};z_{ref})} \times \frac{X(c_{ref},s_{ref};z)}{D(c_{ref},s_{ref};z_{ref},d_{ref})} \\ &= TOR(c,s;d) \times H_p(c,s;z) \times ISF_{air}(z) \times \frac{1}{TOR(c_{ref},s_{ref};d_{ref})} \\ &= \frac{TOR(c,s;d)}{TOR(c_{ref},s_{ref};d_{ref})} \times H_p(c,s;z) \times ISF_{air}(z) \\ &= TPR(c,s;d) \times PSF(s;d_{ref}) \times H_p(c,s;z) \times ISF_{air}(z)\end{aligned} \tag{26.2b}$$

For simplicity, only the ratios on the central axis are shown, where c is the nozzle setting or MLC field setting, s is the field size defined at depth in phantom, z is the source-to-detector distance, and d is the depth in patient (or phantom). The in-air quantity $X(c,s,z) = D(c,s;z,0)$ is the dose measured at the entrance of the phantom, which is assumed to be the same as measurement in-air without any buildup. (This is a reasonable assumption if the backscattering of proton beam is negligible.) *TPR*, tissue-phantom ratio, is the depth–dose ratio taking out inverse square law, and it describes the dose variation (range) along depth direction. For proton, *TPR* is approximately equal to *PDD* divided by $ISF = SAD^2/(SSD+d)^2$ because the phantom scatter is negligible. Since in the proton field *PDD* is commonly used, likewise we will use *PDD* for the depth–dose variation in the rest of this chapter. *PSF*, the phantom scatter factor, describes the field size dependence due to proton phantom scattering. Unlike photon beams, this quantity for proton beam is more like lateral electron disequilibrium factor for the electron beam in that it saturates at a finite lateral field radius r and drops for small field sizes. H_p, proton head scatter factor, describes the field size dependence due to head scatter inside the accelerator; $OF_w=PSF*H_p$, the output factor in water, describes field size dependence due to both proton phantom and head scattering; and ISF_{air} and *ISF* describe the inverse square dependence determined in air or a water phantom, respectively. The advantage of the fluence-based, factor-based formulism (Equation 26.2b) is that the contribution from phantom scattering ($TPR*PSF$) and primary fluence (H_p*ISF_{air}) can be separated out.

26.3.2 Pencil-beam Kernel and Model-based Dose-to-Fluence Formalism

The absorbed dose resulting from proton irradiation is directly proportional to the particle fluence incident onto the patient. This makes normalization of calculated dose per particle fluence appealing, partially because the fluence is proportional to the dose quantity measured in air. The dose in the medium can be calculated as a pencil-beam convolution of a pencil-beam kernel, which can be calculated from MC simulation, with the incident proton fluence:

$$D = k \iint_{x',y'} p(x-x', y-y', z)\Phi(x', y')dx'dy' \tag{26.3}$$

The constant k in Equation 26.3 can be eliminated by repeating Equation 26.3 for the reference condition, where the MU is defined to equal to the dose ($MU = D_{ref}$) and then taking a ratio of D and D_{ref}, i.e.,

$$\frac{D}{MU} = \frac{D}{D_{ref}} = \frac{\iint_{x',y'} p(x-x', y-y', z)\Phi(x', y')dx'dy'}{\iint_{ref} p(x-x', y-y', z)\Phi_{ref}(x', y')dx'dy'} \tag{26.4}$$

Details on the determination of a pencil-beam kernel can be found in Chapter 19 (Dose Calculation) and in Section 26.4.

26.3.3 Monte Carlo Simulation

Monte Carlo (MC) simulation has been used to calculate the absorbed dose per monitor unit [2]. MC simulation platforms have been implemented using GEANT 4 [3] or MCNPX [4]. MC simulation has been used in a few clinics for dose per MU verification and is considered the gold standard in heterogeneous medium. However, this method is not available in most clinics because of its high labor cost and physics-intensive commissioning process. MC simulation is most suitable for relating the absorbed dose to the incident proton fluence directly. We will not include discussion of D/MU from MC in this chapter.

26.4 Pencil Beam-based Dose Per MU Algorithms

26.4.1 Passively Scattering and Uniform Scanning Broad Beams

Broad beam in proton therapy can be generated by pencil-beam kernel convolution to calculate dose, $D(x,y,z)$:

$$D(x,y,z) = k\left(\frac{SSD_v + z_n}{SSD_v + z}\right)^2 \times PDD(z) \times \iint f(x-x',y-y',z) \times \Phi(x',y')dx'dy' \tag{26.5}$$

where $\Phi(x,y)$ is the lateral proton primary fluence, usually treated as a constant, k is a constant proportional to MU, $PDD(z)$ is the measured depth–dose curve for infinitely large field and infinite SSD, and the term z_n is the normalization depth for PDD. The position of the virtual proton source, SSD_v, can be measured from in-air profile measurement. The pencil-beam kernel, $f(x,y,z)$, is usually expressed as a sum of two Gaussian functions with a range straggling correction, λ:

$$f(x,y,z) = \frac{\sum_{i=1}^{2} B_i \times e^{-\frac{x^2+y^2}{\lambda \times b_i \times \overline{r^2}}}}{\pi \times \overline{r^2} \times \lambda \sum_{i=1}^{2} B_i \times b_i}, \tag{26.6}$$

where the b_i and B_i, (i = 1, 2) are pre-calculated weighting parameters based on MC calculation in water and $\overline{r^2}$ is the mean square of radial spread. They are functions of the equivalent proton energy at water surface, E_{eq}, and the fractional depth, z/R_0, R_0 is the mean range of proton with energy E_{eq}. If one ignores the large angular multiple scattering, the pencil-beam kernel can be expressed as a single Gaussian:

$$f(x,y,z) = \frac{1}{\pi \times \overline{r^2}} e^{-\frac{x^2+y^2}{\overline{r^2}}}. \tag{26.7}$$

One can calculate the radial spread in the medium under Fermi–Eyges small angle approximation:

$$\overline{r^2}(z) = \overline{r_0^2}(L_0 + z) + \int_0^z (z-u)^2 T(u)du \tag{26.8}$$

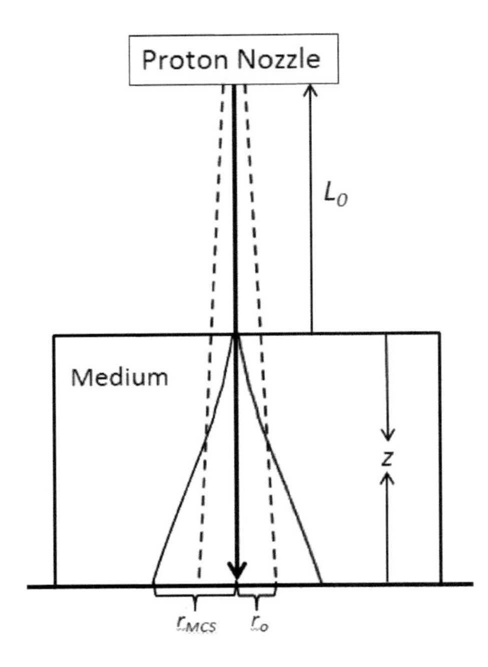

Figure 26–1 Schematics of the proton beam radial spread. $\overline{r_0}$ is the radial spread in air, and $\overline{r_{MCS}}$ is the radial spread due to the multiple Coulomb scattering medium.

where T is the scattering power, $r_0(L_0+z)$ is the in-air lateral spread at point of interest, L_0 is the distance between bottom of nozzle and patient surface, and z is the depth in the medium. The integral is also called the mean square multiple Coulomb scattering (MCS) radial spread $\overline{r_{MCS}^2}$ (as shown in Figure 26–1).

26.4.2 Pencil-beam Scanning

26.4.2.1 Dose per MU and MU Definition

For a scanning proton beam, the dose per MU relationship exists only for a single proton spot. There is no direct dose per MU relationship for a plan composed of thousands of spots unless the treatment volume and location is fully determined. Therefore, a 3D treatment target in the water phantom has to be defined first. Then the dose per MU relationship for scanning proton beam can be defined. In practice, a one liter cube volume is chosen by many centers to define dose per MU relationship. A homogenous dose to this cube is delivered by many evenly spaced spots with different proton energies. Thus, a "pseudo" broad beam is formed, and the dose per MU relationship can be established. The details of the dose per MU relationship is described in the IAEA TRS-398 protocol [5].

Following radiation therapy conventions, 1 cGy/MU is chosen by some centers for their proton PBS system [6]. However, depending on the MU definition, the dose per MU relationship could be some other artificial values (e.g., UPenn system, Scripps system). The essence of MU definition is to determine a conversion factor that defines a certain amount of charges generated in the monitor chamber as one MU. The charge generated in the monitor chamber by protons is proportional to the number of protons passing through the monitor chamber.

$$ion\ pair\ /\ proton = S_{air} \times L \times \rho_{air} \ / \ W_{air} \tag{26.9}$$

where S_{air} is the stopping power in air; L is the gap length of the dose monitor chamber ρ_{air} is the air density, and W_{air} is the mean energy to generate one ion pair in air. Since the PBS has an efficiency of 100%, the MU in PBS is directly related to the number of protons that enter into the patient. Using dose per MU is therefore equivalent to the dose per proton [7,8].

After MU is defined, the dose-to-MU relationship for a single proton spot at various energies can be established using integrated depth–dose (IDD). For PBS, IDD is used to characterize the vertical dose distribution, which is measured by a large parallel plate chamber (e.g., a Bragg Peak chamber). Since PBS spot is a narrow beam, the measured dose-to-MU relationship is inversely proportional to the active chamber area. In order to remove this chamber size effect, the dose-to-MU value is multiplied by the active chamber area. Thus, the dose-to-MU for PBS IDD has the unit of Gy*mm^2/MU. This leads to values of IDDs in Gy/MU, as if all of the doses were applied to a water column of 1 mm^2. The lateral dose distribution (i.e., spot profiles) is characterized by one or multiple Gaussians [9].

26.4.2.2 Dose Algorithm

Two analytical dose algorithms, ray casting and fluence convolution, have been used for PBS dose calculation [10]. The ray casting model of a single spot with nominal energy E and a single Gaussian lateral profile is shown below:

$$D_{spot}(x,y,z) = MU \times IDD(E,WET)\frac{1}{2\pi\sigma_x\sigma_y}\exp\left[-\frac{(x_0-x)^2}{2\sigma_x^{\ 2}}\right]\exp\left[-\frac{(y_0-y)^2}{2\sigma_y^{\ 2}}\right] \tag{26.10}$$

where *IDD* is the function of spot nominal energy E and water equivalent thickness (*WET*) at position (x, y, z). The rest of Equation 26.10 is the spot lateral dose distribution $LAT(x,y,z)$ in the medium (e.g., water), which is described by the single Gaussian distribution. Many studies reported that additional Gaussians are necessary to account for the low-dose halo existing over an extended distance of a few centimeter from the spot central peak [11–15]. It is reported that the lateral profile in medium would be best fit by two Gaussians with the addition of a modified Cauchy–Lorentz distribution [16]. When the single spot dose distribution is explicitly established, the total dose is a summation of the contribution from all spots.

$$D(x,y,z) = ISF \times \sum_{j=1}^{N}[MU_j \times IDD_j(E_j,WET) \times \iint LAT_j(x',y',z)dx'dy'] \tag{26.11}$$

where $ISF = \dfrac{SAD_{vx} * SAD_{vy}}{(SSD_{vx} + z) * (SSD_{vy} + z)}$ and LAT_j is the lateral dose distribution that is typically described by one or multiple Gaussians as referenced above. The inverse square factor (*ISF*) accounts for the diversions between spots. The spot divergence and the virtual source positions of the x, y magnets are shown in Figure 26–2. Without this factor, the ray casting model is valid only for a parallel beam with a small angular divergence. Since the source-to-axis distance for proton system is usually 2–3 meters, the angular divergence of scanning spots is small, and spots can be considered parallel [16].

The basic feature of fluence-based models is the convolution of fluence with an elemental pencil beam dose distribution or dose kernel. The fluence-based dose model, shown below, is used in Eclipse TPS [9,17]

$$D(x,y,z) = \sum\nolimits_{E_k} \sum\nolimits_{Beamlet\ j} \Phi_{E_k}(x_j, y_j, z) D_{E_k}^{Beamlet}(x - x_j, y - y_j, d(z)) \tag{26.12}$$

where $\Phi_{E_k}(x_j, y_j, z)$ is the proton fluence at the position of the beamlet for the kth energy layer (E_k), $D_{E_k}^{Beamlet}(x - x_j, y - y_j, d(z))$ is the dose distribution of beamlet (i.e., dose kernel), and *d(z)* is the WET of position z along the beamlet direction. The analytical dose kernel that is based on Monte Carlo simulation is used in Eclipse TPS [9].

The fluence model separates fluence and dose calculation. Its advantage is that the model is not vendor-specific since the beam-line-related issues (in-air fluence) are separated from general proton physics. However, the model requires an accurate dose

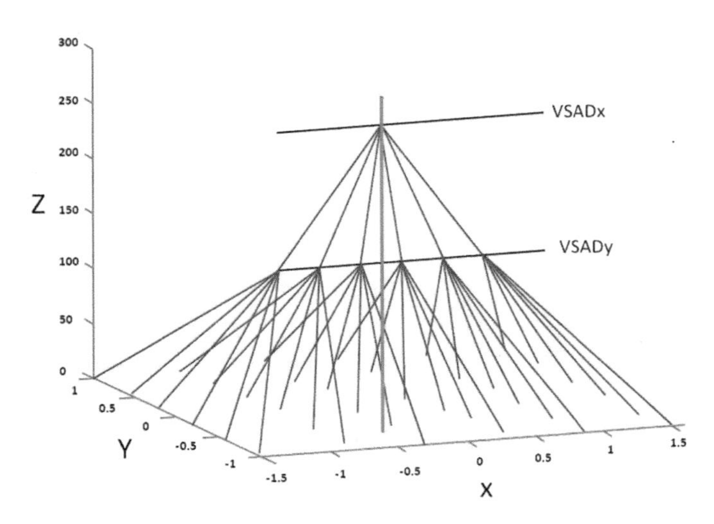

Figure 26–2 Spot divergence determined by virtual SAD of x, y magnets (SAD_{vx} and SAD_{vy}) for nonparallel spots.

kernel, which is not easy to achieve by analytical formulae. The ray casting model is based on the spot profiles in medium. Its advantage is that all contributions to the beam width—such as initial phase space, MCS, and nuclear interaction—are included.

26.5 Factor-based Dose Per MU Algorithms

26.5.1 Broad Beams

Factor-based monitor unit calculation algorithms are implemented by Massachusetts General Hospital (MGH) [18,19], MD Anderson Cancer Center [20], and others [21]. They can be separated to in-air fluence-based and in-water dose-based algorithms.

26.5.1.1 Fluence-based Formalism

Fluence-based formalism separates the measurement into in-air (fluence) and in water (dose) measurements [22]. The basic formalism can be expressed as [23]:

$$D = MU \cdot \frac{D_{ref}}{MU} \cdot PDD(r,z) \cdot PSF \cdot SF \cdot H_p \cdot ISF_{air} \cdot OAR_{air}(x,y) \quad (26.13)$$

where D_{ref}/MU is the dose per MU for each energy option measured under reference condition (e.g., for IBA machines, 10 x10 cm^2, SAD = 230 cm, d = 10 cm for SOBP or d_{max} for pristine peaks). *PSF* is the phantom scatter factor that account for lateral proton disequilibrium as a function of field size. *SF* is the snout factor that account for dose per MU variation for different snout sizes. H_p is the proton headscatter factor that accounts for the fluence variation for different blocked field for the same snout and *SSD*. OAR_{air} is the off-axis ratio measured in air, and ISF_{air} is the inverse-square factor measured in air. *PDD(r,z)* is the percentage depth–dose for a field with finite radius *r* and infinite SSD. In the analytical mode, calculation is made in the middle of SOBP PDD (PDD = 1). Alternatively, the energy modulation input file is used to calculate SOBP PDD using the expression:

$$PDD_{SOBP} = \sum_{I}^{N} w(R)_i PDD_{pris}(R)_i \quad (26.14)$$

where $w(R)_i$ is the weight for the pristine *PDD* with range R. *w(R)* can also be estimated using the maximum range *R* and modulation width *M* using a Cimmino algorithm [24].

The inverse-square factor, *ISF*, is calculated as:

$$ISF = \left(\frac{SAD - xv}{SSD + z - xv} \right)^2 \quad (26.15)$$

The ISF in Equation 26.15 is consistent with that in Equation 26.11 using SAD_v = *SAD*−*xv* and SSD_v = *SSD*−*xv*. The term *xv* is added to represent the virtual source shift (cm) due to the varied incident proton energy (*E*) and nozzle equivalent thickness

(*NET*) for pristine peaks. For SOBP, one usually sets $xv = 0$. The dependence of inverse square factor is instead lumped into the headscatter factor that follows the output factor formula [18]:

$$H_p = \frac{(CF/100)\times\psi_c \times D_c}{(1+a_1 \times r^{a_2})} \times (s_1 + s_2(R - R_L)) \times H_0(s) \tag{26.16}$$

where R is the range, M is modulation width, $m = 0.9$ is a conversion factor, CF is a constant to correct for the output change per option, and R_L is the minimal range of the option. In this expression for the headscatter factor, the effect of virtual source shift is included as the parameters s_1 and s_2 [18]. The expression $H_0(s)$ is the equivalent square dependence of the headscatter factor and can be determined experimentally.

26.5.1.2 Dose-based Formalism

The synchrotron accelerator of the Hitachi ProBeat machine (Hitachi, Ltd. Power Systems, Japan) at the University of Texas MD Anderson Proton Therapy Center in Houston (PTC-H) supplies passive scattering proton treatment beams. Each beam line was designed with three different snouts, namely, small, medium, and large to hold apertures and range compensators of sizes up to 10 cm x 10 cm, 18 cm x 18 cm, and 25 cm x 25 cm, respectively. Each beam line has options of eight discrete energies of 100, 120, 140, 160, 180, 200, 225, and 250 MeV. In order to spread the Bragg peak longitudinally for each beam, a specifically designed range-modulating wheel (RMW) is required. The combination of eight energies and three different snouts results in 24 RMWs per beam line.

The PTC-H has developed a conservative, measurement-based approach to calculating dose per monitor unit (D/MU) which closely follows the proven method that has been used for external beam photon therapy. The detailed information for this procedure can be found in studies by Sahoo [20]. This approach is similar to one that has been used at the proton therapy facility at the Loma Linda University Medical Center (LLMC) since 1991 [25]. According to these studies, the MU can be determined as

$$(\frac{D}{MU})_p = ROF \times SOBPF \times RSF \times SOBPOCF \times OCR \times RSF \times ISF \times CPSF \tag{26.17}$$

where $(D/MU)_p$ is *D/MU* at POI in the patient, and the other factors are explained below.

1. Relative Output Factor (*ROF*)

ROF represents the changes in D/MU relative to the reference calibration condition, such as different beam energies and modulations. Although the dose monitor uses a fixed amount of charge as a registration of 1 MU, the dose deposition pattern of beams with different ranges will change in water depending on the energy distribution of the range-modulated, passively scattered beams leading to different D/MU, as shown in Figure 26–3. *ROF* is used to represent this change. Measurements were

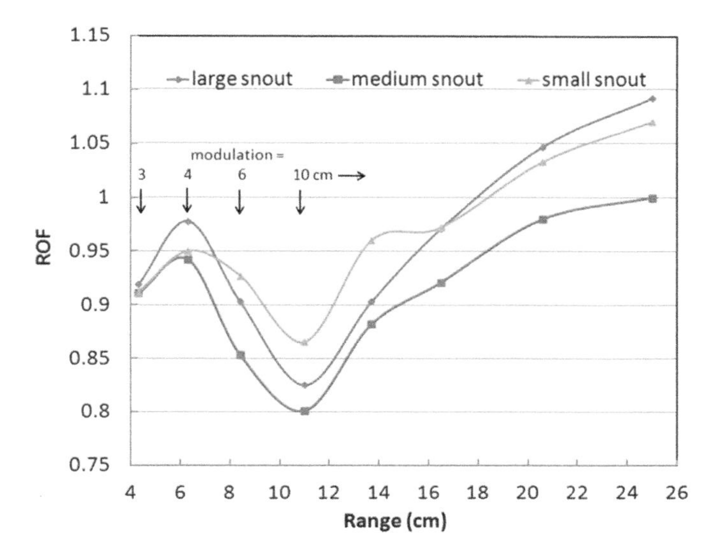

Figure 26–3 Example of ROF for passively scattered beams at PTCH. The modulations of the proton beams are 10 cm for those with their range larger than 10 cm and 6, 4, and 3 cm for the lowest three energies. Reprinted from [20] with permission.

taken at the center of a reference SOBP with various apertures and snout positions. *ROF* is one of the three major factors that can cause large changes in MU calculation. To reduce the uncertainty from this factor, the value of *ROF* should be measured for all energies.

2. SOBP Factor (*SOBPF*)

The *SOBPF* represents the change in D/MU with change in SOBP width relative the reference SOBP width used to determine the *ROF.* In order to spread the Bragg peak, the beam passes through a large number of modulating steps, and protons with lower energies are added to the beam to extend the plateau proximally. In addition to the perturbation from scattering by modulation materials, the relatively reduced weight of high-energy protons will lead to fewer protons reaching the point of measurement. As a result, it is expected that D/MU will decrease with an increase of the SOBP width. *SOBPF* is the second of the three major factors that can cause large changes in MU calculation. Measurements of *SOBPF* are time-intensive due to the large amount of data collection. The PTC-H group [20] has shown that *SOBPF* is a smooth function of SOBP width. Thus, measuring only a few SOBP widths and interpolating to get the other SOBP widths is feasible to save beam time and maintain accuracy. For example, the PTC-H group used a 2 cm interval for many of the high-energy beams, as shown in Figure 26–4. But when time permits, all available SOBP widths should be measured to confirm the accuracy of the interpolation.

3. Range Shifter Factor (*RSF*)

The *RSF* accounts for the output change due to the presence of the range shifters in the proton beam. The value of *RSF* is 1.0 without range shift in the beam path, and it

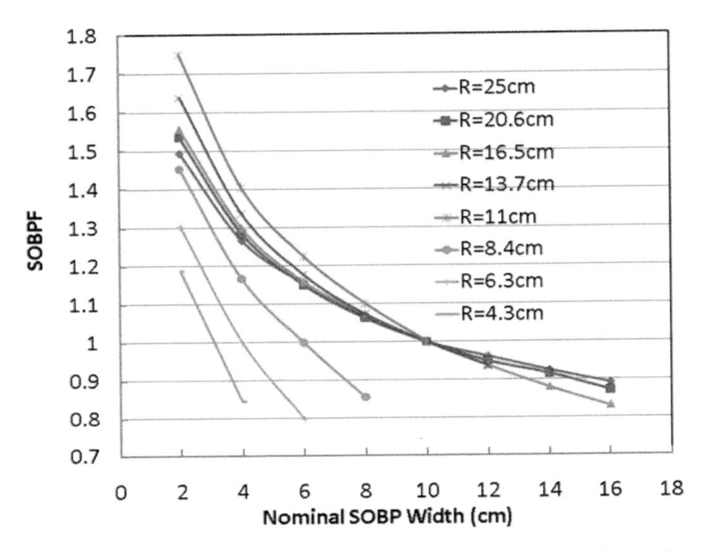

Figure 26–4 Example of *SOBPF* (PTC-H) for passively scattered beamlines with a large snout. Reprinted from [20] with permission.

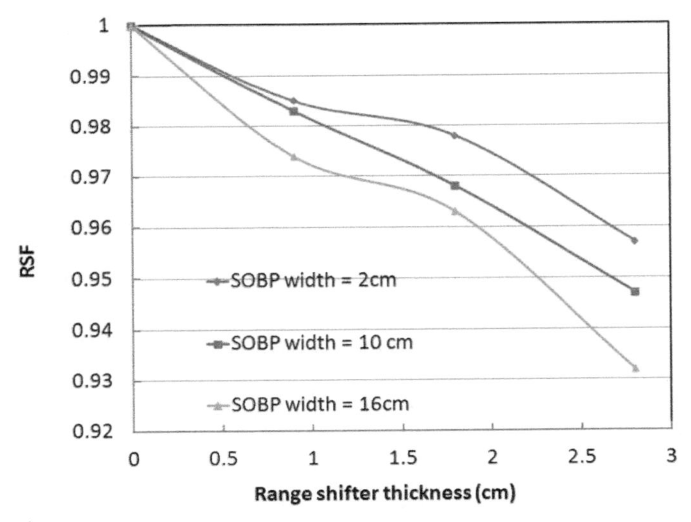

Figure 26–5 Example of *RSF* as a function of range shifter thickness and SOBP width from PTC-H. Reprinted from [20] with permission.

decreases with the increase of the thickness of the range shifter added to the beam. *RSF* is the third of the three major factors that can cause large changes in MU calculation. A complete *RSF* table requires a large number of measurements as well. As shown in Figure 26–5, the PTC-H group indicated the variation of *RSF* with range shifter thickness can be fitted to a polynomial curves for all of the beam options. The interpolation can reduce beam time for data collection, but in rare cases, interpolated *RSF* can cause more than 2% uncertainties. For these cases, either measured or calculated *RSF* factors from Monte Carlo simulation should be used to reduce the uncertainty.

4. SOBP Off-center Factor (*SOBPOCF*)

The *SOBPOCF* gives the relative change in D/MU when the location of the measurement point is away from the middle of the SOBP along the beam direction. Due to flat SOBP, the *PDD* is close to 100% along the plateau. However, since the measurements on *PDD* were made at a fixed source-to-surface distance (*SSD*), but the reference dose calibration was done with a fixed source-to-axial distance (*SAD*), the *SOBPOCF* is used to correct the differences caused by measurement geometry. The impact of the *SOBPOCF* is generally very small due to flat SOBP and large *SAD*.

5. Off-axis Ratio (*OAR*)

The *OAR* accounts for the change in D/MU when the location of the measurement point is located away laterally from the central axis of the field. It is also called the off-center ratio (*OCR*) [20]. This factor should be within the specifications of the flatness of the proton beam. *OAR* was typically within 1%, although the tolerance of flatness was 3% [20].

6. Field Size Factor (*FSF*)

The *FSF* gives the change in D/MU when the field size is different to the one used in *ROF* measurement. It is equal to the product of phantom scatter factor and the proton head scatter factor, $PSF*H_p$. For field sizes larger than 5 cm × 5 cm, the *FSF* is assumed to be 1.0. For smaller field sizes, the dose contribution from the aperture edge scatter may be larger, especially when the point of measurement is located in shallow depth. This aperture edge scatter also relies on the snout position [26], as indicated in Figure 26–6. It is recommended that *FSF* should be measured for field sizes smaller than 5 cm × 5 cm in order to reduce the uncertainty in D/MU calculation.

7. Inverse Square Factor (*ISF*)

The *ISF* converts the D/MU from the reference calibration distance at the *SAD* (270–325 cm) to any arbitrary *SAD* of interest (Equation 26.15, $SAD = SSD+z$).

8. Compensator and Patient Scatter Factor (*CPSF*)

The *CPSF* represents the change in D/MU at the depth of interest in a patient's treatment condition with a compensator relative to the same water-equivalent depth in uniform water phantom without a compensator. This factor considers the dose

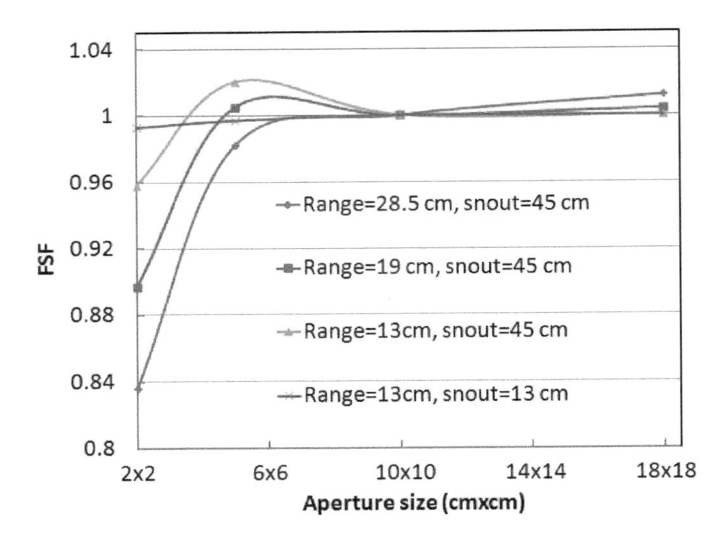

Figure 26–6 *FSF* of proton beams relative to the 10 cm × 10 cm field used in the medium snout for SOBP width of 10 cm (PTC-H). Reprinted from [20] with permission.

contributions of scatter from the compensator and the inhomogeneities in patient anatomy. The value of the *CPSF* is obtained from TPS by the ratio of dose at POI in a patient plan (D_{pp}) to the dose at the same WET in a verification plan (calculated in a uniform phantom) without compensator (D_{vpnc}) as:

$$CPSF = D_{pp} / \mathrm{D}_{\mathrm{vpnc}} \tag{26.18}$$

According to the definition of *CPSF*, Equation 26.18 can be converted to:

$$MU = D_{pp} / (D / MU)_p = D_{pp} / (CPSF \times (D / MU)_{wnc}) \tag{26.19}$$

where $(\mathrm{D/MU})_{\mathrm{wnc}}$ gives the D/MU in uniform water without a compensator.

Thus, by using Equation 26.19, the MU can be determined from

$$MU = D_{vpnc} (D / MU)_{wnc} \tag{26.20}$$

The *CPSF* factor is necessary only when the dose in a patient and the D/MU in a uniform water phantom is used to determine the MU. Alternatively, the same MU can be calculated by using the dose in the verification plan D/MU in a uniform water phantom. This allows us to determine the patient field MU without the factor of *CPSF*. In institutions where the patient MU is from measurement, MU can be measured in a uniform phantom without a compensator.

This MU calculation approach was developed at MD Anderson and has been used at PTC-H since 2006. However, this approach can be applied to any passive proton system (IBA, Varian, Hitachi, etc.). According to their analysis, the difference between the calculated and measured D/MU for 623 distinct treatment fields covered various treatment sites and concluded the averaged difference was within 1% (0.05% ±1.09%). All eight factors used in MU determination contribute to the differences between the measured and the calculated D/MUs except for *ISF* and *CPSF*, as *ISF* is the same for both measurement and calculation, and *CPSF* is not used as illustrated in Equation 26.16. This provides a tool to complete an independent MU check for each treatment field, and it helps us better understand the complexities associated with MU calculation in passive scattering proton therapy.

Uniform scanning delivers uniform-intensity proton beams layer by layer along the beam axis by adjusting the proton energy and sweeping the beam laterally using a scanning magnet system. Due to less beam degrading in the beam delivery system, a uniform scanning system generally has the ability to treat larger and deeper targets than a passive scattering system. In terms of MU calculation, the similar measurement-based approach discussed above has been applied to uniform scanting by different groups [27,28]. The output factor for a uniform scanning system was found as a function of parameters such as the range, modulation, scanning area, field size, and snout position. Considering the inverse square correction for the case where the center of SOBP is not at the isocenter, the author claimed excellent agreement between predicted and measured output factors for the prostate treatment field. This has led to the elimination of output measurement for prostate patients at their center.

The output of a uniform scanning beam increases with the energy of the proton beam for each modulation width and decreases with the modulation width, which are results similar to those reported by Hsi et al. [21] and Kim et al. [29]. The impact of

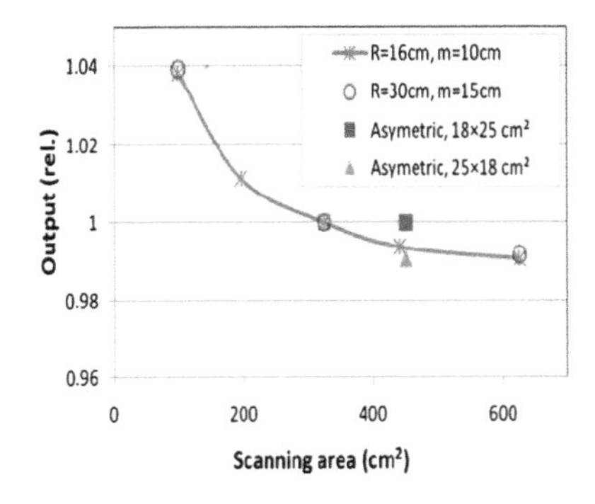

Figure 26–7 Output factor as a function of scanning area. Snout 25 was used with a 10 cm diameter aperture. From [28] with permission.

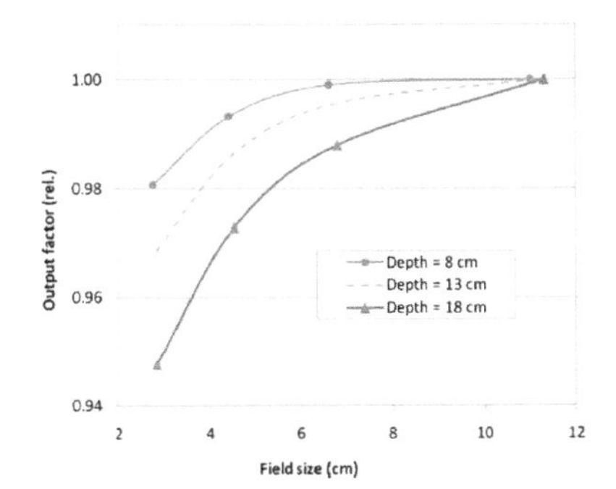

Figure 26–8 Output factor as a function of field size at different measurement depths. From [28] with permission.

scanning area on output is unique to a uniform scanning beam. The output varies significantly with scanning area, especially for small scanning areas, as shown in Figure 26–7. It was also found that the dependence on scanning area varies little with range and modulation of the proton beam, and the output behaves differently for an asymmetric scanning area.

The impact of field size on the output factor mainly depends on the depth of measurement, as shown in Figure 26–8 [27,30]. The results indicate that the output is more sensitive to small fields at large depths. In practice, accurate estimation of effective field size, especially for small fields with irregular field shapes, is critical for accurate output calculation. The output is also dependent on snout position and is complicated to model, similar to what was reported by another group [30]. Therefore, the impact of snout position is generally ignored. However, special attention should be paid when a small field was used for treatment.

26.5.2 Pencil-beam Scanning

Since a proton PBS plan is typically composed of a few thousand spots, small deviation to the singlet spot shape would cause significant dose deviations. Thus, the PBS dose model requires accurate modeling of each single proton spot. The proton spot profile usually has a laterally wide extended low-dose tail that is superimposed on the primary Gaussian. The low-dose tails are generated due to the large angle scattering in the beam line and in the medium. It has been reported that the field size effect was observed if the low-dose tails had not been modeled accurately. To directly measure the low-dose tails, it is reported by many centers that lateral extension to 10^{-4} of the central peak is required [11–15]. However, these measurements are very time-consuming and challenging. As an approximation, one could make a look-up table to correct the so-called field size effect.

26.6 Beam Data for Proton MU Calculations

26.6.1 Depth Dose

26.6.1.1 PDD for Broad Beam

For most broad beam applications, SOBP is used in practice. Typically, dose per MU is determined at the middle of SOBP where *PDD* = 1. However, for dose at other depths in an SOBP, *PDD* should be used. An example of SOBP *PDD* is shown in Figure 26–9.

26.6.1.2 IDD for Pencil Beam Scanning (PBS)

As shown in Section 26.4.2.2, both the ray casting model and the fluence model need *IDD* as input for dose calculation. Since the absolute dose per MU is included in *IDD*, it is required to measure *IDD* accurately. Any inaccuracy in *IDD* would be propagated to dose error in the dose calculation. An example is shown in Figure 26–10.

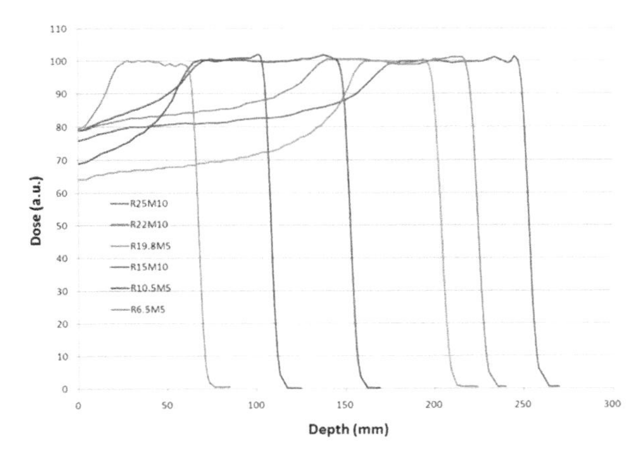

Figure 26–9 PDD for several SOBP of various ranges and modulation widths for a cyclotron-based IBA double scattering beam: R25M10, R22 M10, R19.8M5, R15M10, R10.5M5, and R6.5M5. The first number (e.g., R25) is the range in cm, the second number (e.g., M10) is the width of the SOBP in cm.

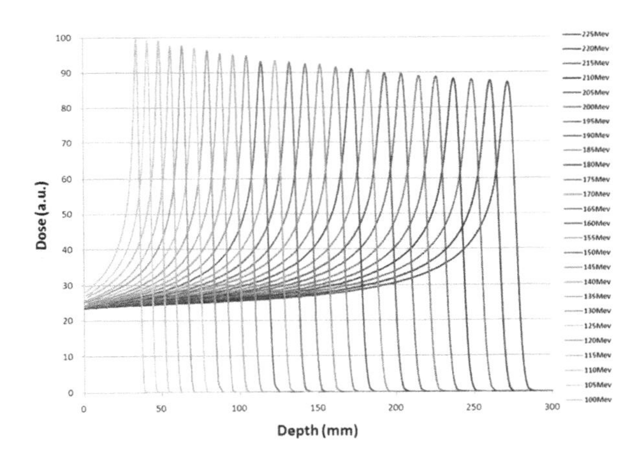

Figure 26–10 Dose per MU for integrated pristine peaks of PBS beams from a cyclotron-based IBA proton accelerator from 100 MeV to 225 MeV.

26.6.2 Output Factor

26.6.2.1 Output Factor for SOBP

The clinical utilization of proton therapy requires safe and efficient planning and delivery technologies. However, the calculation of output (D/MU) is not supported by proton therapy TPSs, such as Eclipse from Varian Medical Systems. This is due to the complexities and specifications of beam delivery systems. Historically, output was determined by measurement for each treatment field prior to treatment. This requires a significant amount of beam time and manpower.

An analytical expression for the depth–dose of an SOBP at infinite *SAD* was derived at MGH [31]. While applying that analytical expression to the surface of the SOBP, one can get a relationship between the surface dose and the SOBP dose.

$$D_{SOBP,0} \approx \frac{D_P}{1+0.44r^{0.6}} \tag{26.21}$$

where $r = (R-M)/M$, D_p is the dose in the plateau of the SOBP, and $D_{SOBP,0}$ is the dose at the surface of the SOBP. The proton MU is proportional to the proton fluence and the surface dose of the SOBP. Therefore, a relationship between MU and the SOBP plateau dose can be established by using the above analytical formula, and the proton dose-per-MU can be calculated analytically.

IBA double scattering has a total of eight treatment options, each of which is defined by a unique combination of hardware (e.g., scatterer and range modulation wheel), and covers a limited span of beam ranges. MGH applied the above analytical formula to each option of their IBA double scattering system, with the modifications to account for the effective source at finite distance and the effective source position change due to the change of fixed scattered materials [18,19], Equation 26.16. By measuring a few outputs with different range and modulation combinations, the free parameters (a_1, a_2, s) in Equation 26.16 could be fit. The fit has to be performed for each of the eight options. The selection of the output measurements for the model parameter fit should include the range and modulations that cover as large a range of r as possible. After the free parameters are fit, the output is just a function of the single variable r. The MGH group can predict outputs within 1.4% (one SD) of measurements.

26.6.2.2 Output Factor for PBS

For PBS, the output factor is not directly needed for the second MU check, which is different from photon and passive scattering proton. However, the output factor is directly related to the spot lateral dose distribution. Since the proton PBS plan is composed of thousands of spots, the small inaccurate modeling of the lateral dose distribution for individual spots will accumulate to a significant effect over thousands of spots. The output factor deviation would be up to 10% (as discussed later in Section 26.7.2). Therefore, PBS requires an accurate model of the lateral dose distribution. However, it is very challenging to measure the lateral dose accurately because of the extended low-dose tails. These low-dose tails are generated by two sources: the large angle scattering in the beam line when the protons pass through the devices in the nozzle, and the nuclear halo generated by non-elastic nuclear interaction when protons pass through the medium [32]. The large-size Bragg peak chamber is used to measure *IDD* so those laterally extended low-dose tails are not missed. However, it has been reported that even 8 cm diameter Bragg peak chambers are not big enough to collect 100% of spot doses due to these extended low-dose tails [14]. Monte Carlo simulations are used to correct the finite chamber size effect on the measured value of *IDD*.

The ray casting model needs both the in-air and in-water profiles as input. The profiles at a few depths from a subset of energies are required to fit the model parameters. The profiles at the other depths and energies would be predicted by the fit parameters. The fluence model needs an in-air profile as input and the dose kernel, but it does not need the in-medium profile. MCS would be used as a simple approximation to the dose kernel because proton beam propagation in a medium is mainly dominated by MCS. The MCS would be represented by analytical formulae, such as the generalized highland approximation [33] and the differential Molière method [34,35]. An accurate dose model should also include the nuclear scatter. An analytic approximation based on Monte Carlo simulations was proposed to account for the dose contributed from nuclear interaction [36].

26.6.3 Off-axis Ratio (*OAR*)

26.6.3.1 OAR for Broad Beam

The in-air profile of a broad proton beam is typically very flat, thus $OAR_{air} = 1$ is a good assumption for dose per MU calculation at off-axis points. An example profile of an SOBP profile in water is shown in Figure 26–11.

26.6.3.2 OAR for PBS

The in-air profile of the PBS spot is a Gaussian shape before it enters to the nozzle. After it enters into the nozzle, the protons are scattered by the devices and the air in the nozzle. Although the WET of the devices in the nozzle are manufactured to be within a minimum of a few tenths of cm^2/g, their effect to the spot shape should not be neglected due to their large distance from the isocenter. The devices in the nozzle produce a low-dose envelope that extends much farther than the single Gaussian pro-

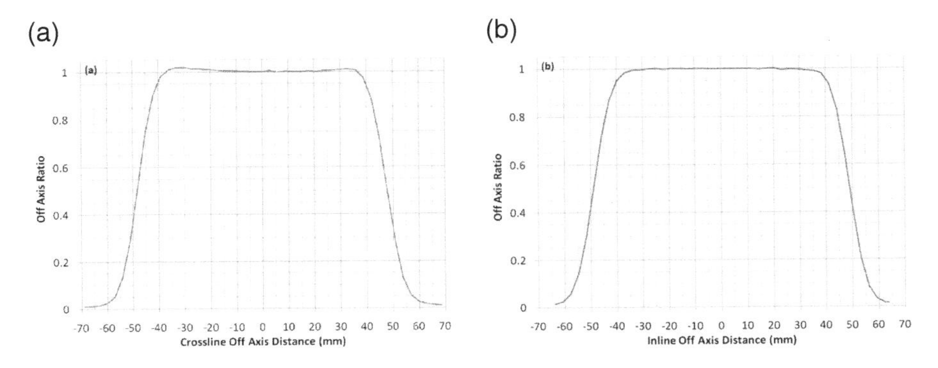

Figure 26–11 Cross line (a) and in-line (b) OAR for 10 cm × 10 cm proton beam of a SOBP beam with a range of 15 cm and a modulation of 10 cm.

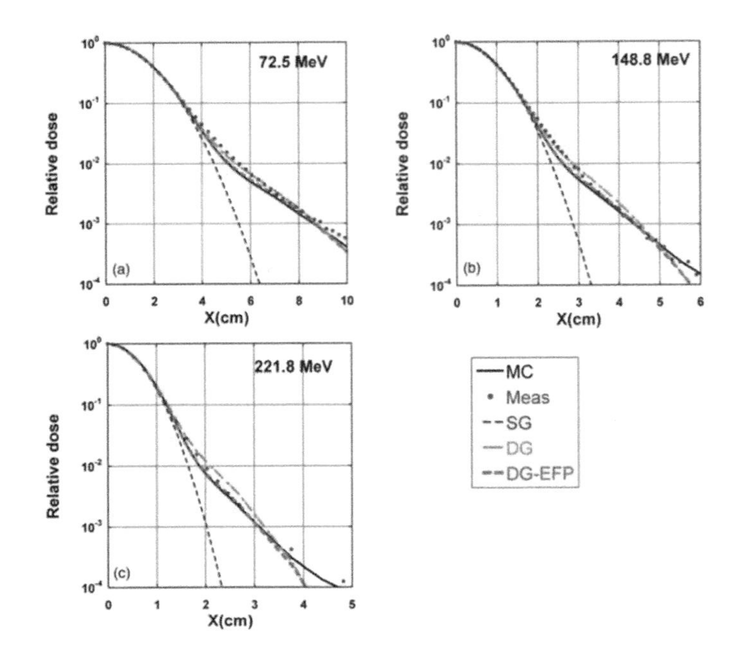

Figure 26–12 In-air lateral profiles for pencil beams with three different energies. Solid lines: Monte Carlo simulation; dots: measured data; dashed lines: calculated by single Gaussian fluence model; dashed-dotted lines: calculated by double Gaussian fluence model with empirical parameters; and dashed lines: calculated by double Gaussian Eclipse fit parameters. (a) 72.5, (b) 148.8, and (c) 221.8 MeV. Reprinted from [17] with permission.

file, which is called the halo effect, and which is more significant for the low-energy beam (as shown in Figure 26–12).

If the in-air halo effect is not modeled correctly in the TPS, it will cause dose deviation up to 10%. Therefore, the in-air profile should be described by an additional Gaussian to cover the halo effect (as shown in Figure 26–12). Among all the devices in the nozzle, the spot profile monitor that is positioned at the entrance of the nozzle (about 2–3 m away from the isocenter) is the major source of the in-air halo effect.

26.6.4 Inverse Square Law

26.6.4.1 Inverse Square Law for Broad Beam

The inverse square factor (*ISF*) for the broad beam is determined by the virtual *SAD* of each beam. Typically, the virtual shift from the nominal *SAD* of each beam is set to be zero.

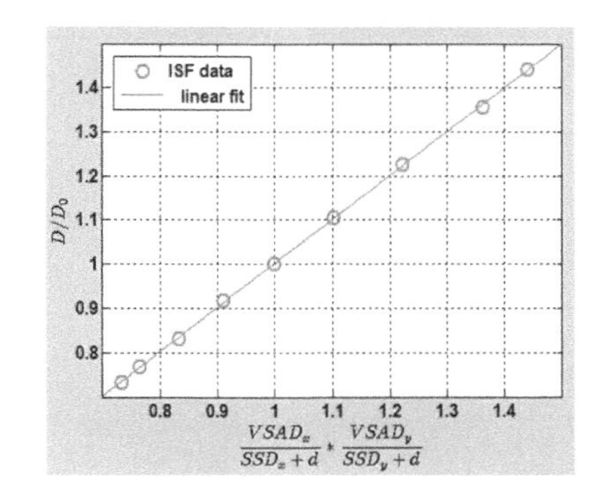

Figure 26–13 The inverse square factor defined in Equation 26.11 is shown for the IBA PBS system at the University of Pennsylvania. The solid line is the linear fit. The dose was extracted from an Eclipse TPS for various SSD, while keeping at the center of the SOBP for the same beam with range = 20 cm and modulation = 8 cm (i.e., at depth z = 16 cm).

26.6.4.2 Inverse Square Law for PBS

The *ISF* for the PBS system due to the nonparallel spots is determined by the virtual *SAD* of magnets x and y (as shown in figures 26–2 and 26–13). The virtual *SAD* could be derived by measuring the output variations with the different *SSD* at the broad beam geometry while keep the same beam line parameters. Figure 26–13 shows the *ISF* for the PBS at for the University of Pennsylvania's universal nozzle, which has the virtual *SAD* 230.6 and 192.5 for magnets x and y, respectively. The beam used for the test has a range of 20 cm, a modulation of 8 cm, and the dose was collected at SOBP (i.e., a depth of 16 cm). The derived virtual *SSD* is 210.2 cm.

26.6.5 Generic Beam Data

It is possible to develop generic beam data for PBS which, with appropriate tuning of the proton Gaussian energy spectrum, can be matched to a specific PT center [37].

26.7 Dose and MU Verification

26.7.1 Dose and MU Verification for SOBP Beams

Accuracy results (better than 3% agreement) were obtained by a number of cancer centers, as shown in Table 26–1 from MD Anderson [20,23].

Table 26–1 Results of an analysis of the difference between the calculated and measured D/MU for patient treatment fields [20]

Treatment Site	Number of Fields	Mean of Difference (%)	Standard Deviation (%)
Prostate	228	–0.35	0.83
Thorax	211	0.35	1.21
CNS	167	0.15	1.10
Total (including other treatment fields)	623	0.05	1.09

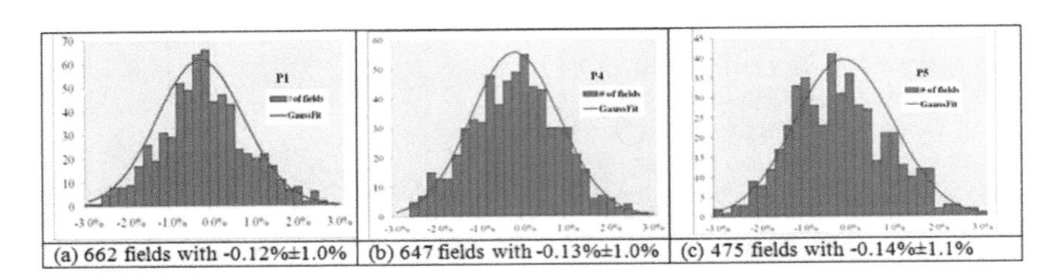

Figure 26–14 Percent difference between modeled and measured outputs for the three double scattering proton beamlines. Reprinted from [38] with permission.

The University of Pennsylvania uses an IBA proton therapy system that is similar to the one at MGH. By applying a linear-quadratic transformation to the nominal r, the MGH model was further improved at the University of Pennsylvania [38]. It has been reported that the model prediction is within 1% (one SD) after comparing 1784 patient-specific fields (see Figure 26-14).

26.7.2 Dose and MU Verification for PBS

The dose verification by measurements is applied to each clinic treatment plan [39,40]. A 2D array detector is usually used to collect the planar dose distribution. Since the treatment plan is composed of many independent spots, the same plan is delivered multiple times in order to verify the dose at multiple planes, which makes the PBS patient QA very time-consuming and tedious.

The analytical dose model proposed by Li et al. [16] was validated by measurements. The field size effects were accurately computed to within 2% of measurements at all depths, field sizes, and energies (see Figure 26–15). The model also achieved good dose accuracy for patient dose verification, with the gamma passing rates >99% for all tested clinic plans at 2% and 2 mm matching criteria.

Based on the success of the analytical model described above [16], MDACC developed an automated patient QA process called HPlusQA [39]. With HPlusQA, the 3D dose is calculated by its analytic dose engine, and dose profiles are extracted to compare with measurements by 2D ion arrays and calculations by TPS (refer to Figure 26–16). After comparisons of more than 100 planar dose distributions among measurements, second dose calculation, and the TPS, they found out that the second dose calculation is effective (79%) in predicting the dose deviation between measured dose planes and TPS-calculated dose planes. Therefore, second dose calculation would reduce the patient-specific QA measurements by 64%. This means that for more than half all patient plans, we only need to deliver it for one time to verify machine delivery, while the 3D dose comparisons would be alternatively generated between TPS and second dose calculation.

Figure 26–16 shows an example of the dose comparisons of the measurements (red dots), the TPS calculation (black line), and the HPlusQA dose calculation (blue line) for a thoracic tumor treatment with three beams. The dose comparisons in Figure 26–16 show that the HPlusQA dose calculations agree well with the measurements and the calculations by TPS. Therefore, the accurate analytic dose engine for PBS, such as HPlusQA, is an efficient way to reduce patient QA. Moreover, HPlusQA calculated 3D dose, while the measurements are usually planar 2D doses at multiple depths.

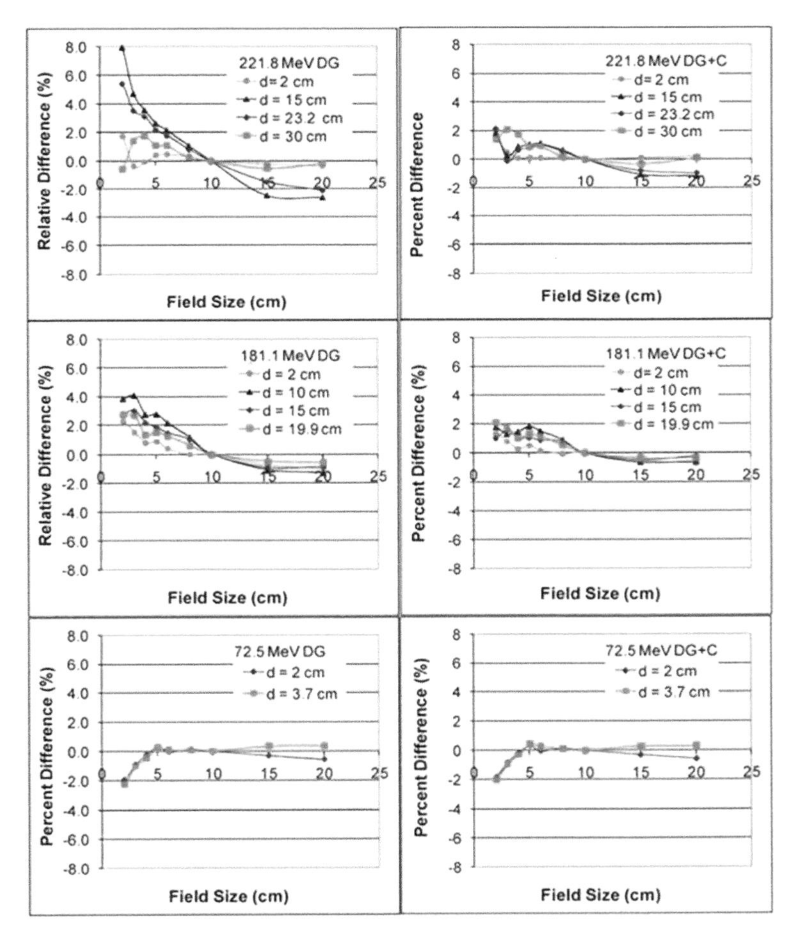

Figure 26–15 Percentage difference between the measured and calculated dose for three energies at the central axis under a series of square field sizes. The graphs in the left column show relative differences of calculations using two Gaussians (DG) only with measurements at various depths d. The graphs in the right column show relative differences of calculations using two Gaussian functions and the Cauchy–Lorentz component (DG+C) at various values of d. From [16] with permission.

26.7.3 Delivery Verification Using Log File

A treatment log file from a spot scanning beam (if available) is used for patient-specific quality assurance [41]. The treatment log files record the proton energy, MU, and lateral position of each delivered PBS spot. With an accurate second MU program, the dose delivered to the patient at each treatment fraction can be retrospectively recon-

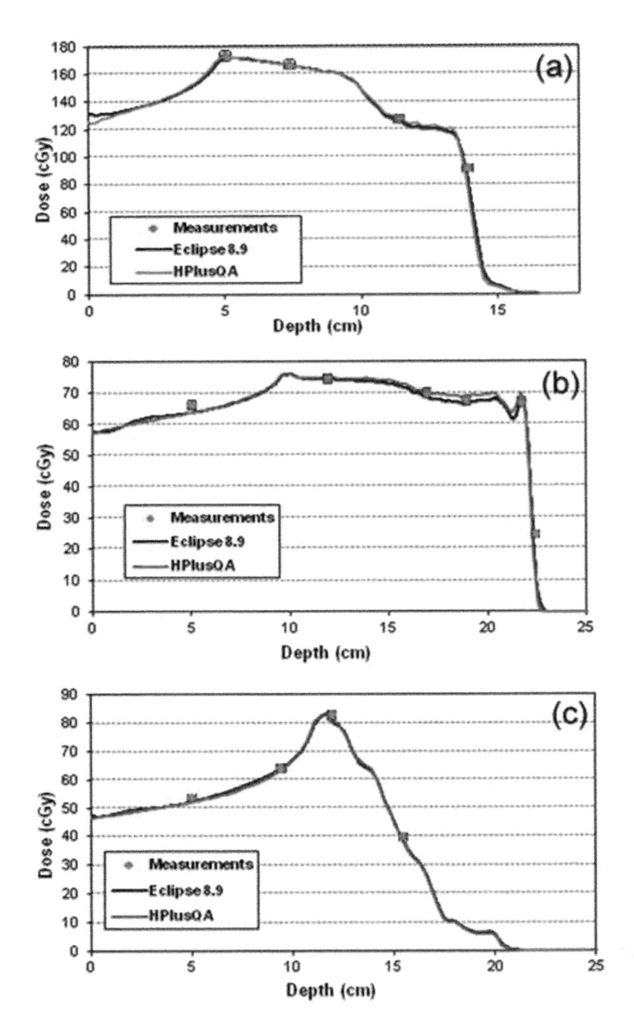

Figure 26–16 Depth–dose depth profiles measured using an ion chamber array and calculated using HPlusQA and Eclipse ver. 8.9. (a) Left anterior field. (b) Left posterior field. (c) Right posterior field used to treat a thoracic tumor. From [39] with permission.

structed, which is very helpful to track the delivered dose. With some efforts to automate the whole process, the dose delivered to all the patients can be tracked easily on a daily basis.

26.8 Conclusions

This chapter discussed formalisms used for the dose per MU calculation for broad (uniform, either passively scattering or uniform scanning) and pencil-beam scanning (PBS) proton beams. The algorithms are broadly separated into: factor-based and model- (or pencil-beam kernel) based methods, with the latter applicable to more general cases, e.g., intensity-modulated proton therapy (IMPT). For broad beams, agreements of 1% to 3% were generally observed between MU calculations and measurements. For PBS, acceptable agreements of IMPT 3D dose maps between MU calculation and measurements were observed in a uniform phantom, although further improvements can be made to improve the modeling of PBS lateral profiles at central and off-axis locations. Novel uses of the MU calculation program, e.g., validating IMPT delivery using log files, are under development to expand the area of its use.

References

1. Zhu TC, Ahnesjo A, Lam KL, Li XA, Ma C-MC, Palta JR, Sharpe MB, Thomadsen B, Tailor RC. Report of AAPM therapy physics committee task group 74: In-air output ratio, SC, for megavoltage photon beams. *Med Phys* 2009;36:5261–5291.
2. Koch N, Newhauser WD, Titt U, Gombos D, Coombes K, Starkschall G. Monte Carlo calculations and measurements of absorbed dose per monitor unit for the treatment of uveal melanoma with proton therapy. *Phys Med Biol* 2008;53:1581–1594.
3. Paganetti H, Jiang H, Parodi K, Slopsema R, Engelsman M. Clinical implementation of full monte carlo dose calculation in proton beam therapy. *Phys Med Biol* 2008;53:4825–4853.
4. Stankovskiy A, Kerhoas-Cavata S, Ferrand R, Nauraye C, Demarzi L. Monte Carlo modelling of the treatment line of the proton therapy center in Orsay. *Phys Med Biol* 2009;54:2377–2394.
5. Andreo P, Burns DT, Hohlfeld K, Huq MS, Kanai T, Laitano F, Smyth VG, Vynckier S. Absorbed dose determination in external beam radiotherapy: An international code of practice for dosimetry based on standards of absorbed dose to water. IAEA Technical Report Series No 398. Vienna: IAEA, 2000.
6. Gillin MT, Sahoo N, Bues M, Ciangaru G, Sawakuchi GO, Poenisch F, Arjomandy B, Martin C, Titt U, Suzuki K, Smith AR, Zhu XR. Commissioning of the discrete spot scanning proton beam delivery system at the University of Texas M.D. Anderson Cancer Center, proton therapy center, Houston. *Med Phys* 2010;37:154–163.
7. Pedroni E, Bacher R, Blattmann H, Bohringer T, Coray A, Lomax A, Lin S, Munkel G, Scheib S, Schneider U, et al. The 200-MeV proton therapy project at the Paul Scherrer Institution: Conceptual design and practical realization. *Med Phys* 1995;22:37–53.
8. Lorin S, Grusell E, Tilly N, Medin J, Kimstrand P, Glimelius B. Reference dosimetry in a scanned pulsed proton beam using ionisation chambers and a Faraday cup. *Phys Med Biol* 2008;53:3519–3529.
9. Ulmer W, Schaffner B. Foundation of an analytical proton beamlet model for inclusion in a general proton dose calculation system. *Rad Phys Chem* 2011;80:378–389.
10. Schaffner B, Pedroni E, Lomax A. Dose calculation models for proton treatment planning using a dynamic beam delivery system: An attempt to include density heterogeneity effects in the analytical dose calculation. *Phys Med Biol* 1999;44:27–41.
11. Lin L, Ainsley CG, McDonough JE. Experimental characterization of two-dimensional pencil beam scanning proton spot profiles. *Phys Med Biol* 2013;57:983–997.

12. Lin L, Ainsley CG, Solberg TD, McDonough JE. Experimental characterization of two-dimensional spot profiles for two proton pencil beam scanning nozzels. *Phys Med Biol* 2014;59:493–504.
13. Sawakuchi GO, Mirkovic D, Perles LA, Sahoo N, Zhu XR, Ciangaru G, Suzuki K, Gillin MT, Mohan R, Titt U. An MCNPX Monte Carlo model of a discrete spot scanning proton beam therapy nozzle. *Med Phys* 2010;37:4960–4970.
14. Sawakuchi GO, Zhu XR, Poenisch F, Suzuki K, Clangaru G, Titt U, Anand A, Mohan R, Gillin MT, Sahoo N. Experimental characterization of the low-dose envelope of spot scanning proton beams. *Phys Med Biol* 2010;55:3467–3478.
15. Schwaab J, Brons S, Fieres J, Parodi K. Experimental characterization of lateral profiles of scanned proton and carbon ion pencil beams for improved beam models in ion therapy treatment planning. *Phys Med Biol* 2011;56:7813–7827.
16. Li Y, Zhu XR, Sahoo N, Anand A, Zhang X. Beyond Gaussians: A study of single-spot modeling for scanning proton dose calculation. *Phys Med Biol* 2012;57:983–997.
17. Zhu XR, Poenisch F, Lii M, Sawakuchi GO, Titt U, Bues M, Song X, Zhang X, Li Y, Ciangaru G, Li H, Taylor MB, Suzuki K, Mohan R, Gillin MT, Sahoo N. Commissioning dose computation models for spot scanning proton beams in water for a commercially available treatment planning system. *Med Phys* 2013;40:041723.
18. Kooy HM, Rosenthal SJ, Engelsman M, Mazal A, Slopsema RL, Paganetti H, Flanz JB. The prediction of output factors for spread-out proton Bragg peak fields in clinical practice. *Phys Med Biol* 2005;50:5847–5856.
19. Kooy HM, Schaefer M, Rosenthal S, Bortfeld T. Monitor unit calculations for range-modulated spread-out bragg peak fields. *Phys Med Biol* 2003;48:2797-2808.
20. Sahoo N, Zhu XR, Arjomandy B, Ciangaru G, Lii M, Amos R, Wu R, Gillin MT. A procedure for calculation of monitor units for passively scattered proton radiotherapy beams. *Med Phys* 2008;35:5088–5097.
21. Hsi WC, Schreuder AN, Moyers MF, Allgower CE, Farr JB, Mascia AE. Range and modulation dependencies for proton beam dose per monitor unit calculations. *Med Phys* 2009;36:634–641.
22. Schaffner B. Proton dose calculation based on in-air fluence measurements. *Phys Med Biol* 2008;53:1545–1582.
23. Zhu TC, Li Z, Karunamuni R, Yeung D, Slopsema R. A fluence-based algorithm for MU calculation of proton beams. *Med Phys* 2007;34:2403–2403.
24. Zhu TC, Karunamuni R, Slopsema RL. Determination of SOBP PDD for therapeutic proton beams. *Med Phys* 2008;35:2766.
25. Moyers MF. Proton therapy. In: van Dyk J, Ed. Modern technology of radiation oncology. Madison, WI: Medical Physics Publishing, 1999, pp. 863–864.
26. Titt U, Zheng Y, Vassiliev ON, Newhauser WD. Monte Carlo investigation of collimator scatter of proton-therapy beams produced using the passive scattering method. *Phys Med Biol* 2008;53:487–504.
27. Zhao Q, Wu H, Wolanski M, Pack D, Johnstone PA, Das IJ. A sector-integration method for dose/MU calculation in a uniform scanning proton beam. *Phys Med Biol* 2010;55:N87–95.
28. Zheng Y, Ramirez E, Mascia AE, Ding X, Okoth B, Zeidan O, Hsi WC, Harris B, Schreuder AN, Keole S. Commissioning of output factors for uniform scanning proton beams. *Med Phys* 2011;38:2299–2306.
29. Kim DW, Lim YK, Ahn SH, Shin J, Shin D, Yoon M, Lee SB, Kim DY, Park SY. Prediction of output factor, range, and spread-out Bragg peak for proton therapy. *Med Dosim* 2011;36:145–152.
30. Daartz J, Engelsman M, Paganetti J, Bussiere MR. Field size dependence of the output factor in passively scattered proton therapy: Influence of range, modulation, air gap, and machine settings. *Med Phys* 2009;36:3205–3210.
31. Bortfeld T, Schlegel W. An analytical approximation of depth–dose distributions for therapeutic proton beams. *Phys Med Biol* 1996;41:1331–1339.
32. Sawakuchi GO, Titt U, Mirkovic D, Ciangaru G, Zhu XR, Sahoo N, Gillin MT, Mohan R. Monte Carlo investigation of the low-dose envelope from scanned proton pencil beams. *Phys Med Biol* 2010;55:711–721.
33. Gottschalk B, Koehler AM, Schneider RJ, Sisterson JM, Wagner MS. Multiple Coulomb scattering of 160 MevV protons. *Nucl Inst & Meth Sec B* 1993;74:467–490.

34. Shen J, Liu W, Anand A, Stoker JB, Ding X, Fatyga M, Herman MG, Bues M. Impact of range shifter material on proton pencil beam spot characteristics. *Med Phys* 2015;42:1335.
35. Gottschalk B. On the scattering power of radiotherapy protons. *Med Phys* 2010;37:352–367.
36. Soukup M, Fippel M, Alber M. A pencil-beam algorithm for intensity modulated proton therapy derived from Monte Carlo simulations. *Phys Med Biol* 2005;50:5089–5104.
37. Clasie B, Depauw N, Fransen M, Goma C, Panahandeh HR, Seco J, Flanz JB, Kooy HM. Golden beam data for proton pencil-beam scanning. *Phys Med Biol* 2012;57:1147–1158.
38. Lin L, Shen J, Ainsley CG, Solberg TD, McDonough JE. Implementation of an improved dose-per-MU model for double-scattered proton beams to address interbeamline modulation width variability. *J Appl Clin Med Phys* 2014;15(3):4748.
39. Mackin D, Li Y, Taylor MB, Kerr M, Holmes C, Sahoo N, Poenisch F, Li H, Lii J, Amos R, Wu R, Suzuki K, Gillin MT, Zhu XR, Zhang X. Improving spot-scanning proton therapy patient specific quality assurance with HPlusQA, a second-check dose calculation engine. *Med Phys* 2013;40:121708.
40. Zhu XR, Poenisch F, Song X, Johnson JL, Ciangaru G, Taylor MB, Lii M, Martin C, Arjomandy B, Lee AK, Choi S, Nguyen QN, Gillin MT, Sahoo N. Patient-specific quality assurance for prostate cancer patients receiving spot scanning proton therapy using single-field uniform dose. *Int J Radiat Oncol Biol Phys* 2011;81:552–559.
41. Li H, Sahoo N, Poenisch F, Suzuki K, Li Y, Li X, Zhang X, Lee AK, Gillin MT, Zhu XR. Use of treatment log files in spot scanning proton therapy as part of patient-specific quality assurance. *Med Phys* 2013;40:021703.

Chapter 27

Small Field Dosimetry: SRS and Eyes

Brian Winey, Ph.D. and Marc Bussiere, M.Sc.

Department of Radiation Oncology,
Massachusetts General Hospital and Harvard Medical School,
Boston, MA

27. 1 Clinical Background of Small Fields in SRS and Eyes

27.1.1 History

After Robert Wilsons's original suggestion of proton Bragg peaks [1] for radiation therapy in human populations, biological experiments with animals and human subjects were conducted at Berkeley in the 1950s [2–4]. Around the same time, the neurosurgeon Lars Leskell proposed and used proton therapy in the context of cranial stereotactic radiosurgery (SRS) [5–9] (his first patient was treated in 1957 at the Uppsala University cyclotron).

These earlier techniques used shoot through ("strålkniven", or Ray Knives) techniques which did not make use of the Bragg peak, but did result in very sharp penumbra (e.g., penumbra at a depth of 120 mm, 20% to 80%, was 1.0 mm for 340 MeV protons) [7,8]. In 1961 Massachusetts General Hospital in collaboration with the HCL was the first group to make use of the Bragg peak [10]. The earliest patients treated at the Harvard Cyclotron Laboratory (HCL) were SRS patients with pituitary diseases overseen by Raymond Kjellberg beginning in 1962 [10–14]. The primary advantages suggested by Robert Wilson, Leksell, Kjellberg, and others was the reduction of radiation dose to the brain and surrounding neural tissues [1–3,6–8,10]. In addition to the significant clinical sparing of neural tissues, cranial targets were optimal for original treatments due to the immobilization and ease of anatomical localization using surface surrogates (midline, ears, etc.) or orthogonal x-ray projections used at the HCL. Koehler performed some of the earliest measurements regarding the ranges of protons in tissue, specifically the human skull [15].

Since the earliest treatments of patients with protons and ions were performed within physics laboratories, the therapies were secondary to the primary aims of the laboratories to study particle interactions, including dosimetric measurement techniques. At the start of the HCL treatments, Koehler performed dosimetric measurements of the small treatment fields using a diode, based upon earlier beam measurement experiments in the context of the HCL.

Soon after cranial SRS treatments were documented, ocular treatments started from the 1970s onward at HCL in collaboration with Massachusetts General Hospital, Massachusetts Eye and Ear Infirmary in Boston, and many other locations throughout the world [18–29]. Ocular proton therapies are based upon similar dosimetric techniques of SRS treatments, utilizing small fields. Ocular proton therapy has been deployed widely due to high local control, being minimally invasive, preserving of vision, lower neural toxicity, and the requirement of lower proton energies without need of a gantry. Fixed beams and low-energy cyclotrons are cheaper and more widely available in physics research facilities and clinical settings.

With the growth of clinical proton therapy facilities of the past 20 years, a few centers have begun treating cranial lesions with SRS. More common is the rise of extracranial SBRT therapies with protons to sites in the lung, liver, and pancreas.

27.1.2 Patient Statistics

Even though cranial proton SRS is one of the oldest therapeutic uses of protons, ocular therapies are the most popular treatment sites, with more than 24,000 ocular patients treated to date [30]. According to the Proton Therapy Oncology Group (PTCOG) website, ocular patients have been distinguished from SRS patients despite their dosimetric principles being the same. The primary differences between the two procedures are the shallow depths for ocular treatments and immobilization techniques. Cranial SRS patients number much fewer than ocular proton therapy patients due to their being treated with more common photon radiation therapies such as linac, Gamma Knife, and CyberKnife, all widely available and accepted as clinically effective therapies. HCL and the Burr Proton Therapy Center have treated more than 5000 proton SRS patients since 1961.

Many proton centers have opened, but few routinely treat cranial SRS; thus, the rate of the number of patients treated is not increasing significantly. Proton SBRT patient numbers are likely in the hundreds due to the recent introduction of the techniques and limited adoption in clinical care.

Most ocular patients have been treated for uveal melanoma in 4–5 fractions to doses of 50–70 Gy. Various studies have reported the clinical outcomes, which compare similarly to other therapy options such as plaques, but few of these studies are randomized controlled trials.

SRS treatments in the cranial cavity began with pituitary diseases, such as adenomas and acromegaly disorders. This was done in a pre-CT imaging era, with the sella being simple to identify in planar imaging. Treatments could be simulated and delivered using anatomic landmarks for field definition and patient positioning. After the introduction of 2D angiogram techniques in neurosurgery, the angiogram frame was deployed for radiation therapy of arterio-venous malformations (AVM), including proton therapy.

Simultaneous to the rise of 3D imaging, first CT and then MRI, more treatment sites could be diagnosed, localized, simulated, and treated in the cranial cavity with proton SRS. With appropriate 3D imaging, proton therapy has been used to treat other neoplasms, including acoustic and other schwannomas, meningiomas, and sarcomas.

The strongest argument for proton therapy SRS in the brain is the reduced toxicity due to less integral dose to healthy brain tissues and optical structures. Hence, the traditional treatment targets listed above are benign neoplasms with high probability of long-term patient survival and increased risk of late effects of radiation therapy. Some institutions have treated metastatic lesions with similar clinical outcomes as photon therapy, but limited rationale for the proton therapy. When lesions are in close proximity to organs at risk, proton therapy may allow for higher therapeutic doses compared to photons, which may increase the local control for the same risk of acute neural toxicity.

There are currently many clinical protocols for proton extracranial SBRT sites in the lung, liver, pancreas, and prostate [31–42]. Most publications in proton therapy SBRT have involved lung lesions due to the volume toxicities associated with lung radiation therapy, such that minimization of integral dose significantly affects the clinical acute and chronic toxicities. Similarly, reduced liver toxicities have been the primary argument for protons due to the correlation between the volume of treated normal liver tissue and liver toxicities.

27.1.3 Biology

As with all proton therapies, biology is an open question, with limited data for dose levels and neoplasm cell lines treated with proton SRS, SBRT, and ocular therapies. Many institutions use a generic RBE factor of 1.1 to adjust for the biological effects of protons versus photons. However, the RBE factor is dependent upon dose levels, with a reduction of the enhancement when approaching ablative dose levels in single to few fractions. In essence, cells can only die once. Given enough radiation energy deposited in a small volume of cells, the cell-killing effect of the radiation decreases at higher doses as most of the cells have been ablated. While there is uncertainty regarding the biological dose inside the treatment volume, proton therapy reduces the physical and biological doses to normal tissues compared to photon therapy.

27.2 Physical Properties of Small Fields

While there is no well-defined limit that defines a small proton field from broad proton beams, small fields can be defined by the loss of electronic equilibrium along the central axis depth–dose. Similar to photon therapy, the loss of electronic equilibrium is due to the lateral penumbra approaching the central axis. Like in photons, the lateral penumbra location is primarily affected by the treatment field size as defined by the aperture or lateral spot positions and is affected by geometric divergence. A second effect to the electronic disequilibrium is the multiple Coulomb scattering (MCS) due to the beam delivery system and the patient or phantom. MCS can broaden the lateral penumbra, and the MCS effects are dependent upon field size and range. While MCS affects the lateral penumbra of all proton therapy fields, the effects complicate small proton therapy fields due the disequilibrium along the central axis. Additionally, small proton fields can be characterized by the detector specification for dosimetric measurements, as the size of the detector affects the robustness of the measurements.

27.2.1 Multiple Coulomb Scattering

The greatest difference between broad beam proton treatment fields and small proton treatment fields is the loss of electronic equilibrium due to multiple Coulomb scattering (MCS) in the treatment field when the field size become smaller than the scattering distance of the protons. Figure 27–1 details a planar dose distribution with the penumbra broadening visible as the depth increases. For proton beams, there is no definitive field size limit when equilibrium is lost, but small losses start to occur between ranges of 6–8 cm for large source size doubly scattered systems and between ranges of 4–6 cm for optimized single scattering systems [43]. The MCS is affected by the beam optics of the delivery system, including the effective source spot size and location, the location of the vacuum exit, and the materials used for modulating the

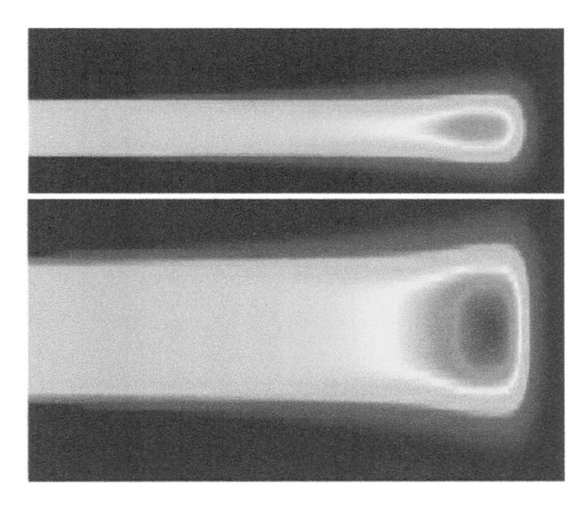

Figure 27–1 A planar dose distribution to detail the lateral dose scattering as it increases with range, as displayed in a small (top) and broad beam (bottom). The width of the dose distribution increases with depth. The scattering affects the penumbra and the central axis dose distribution. The top image of a small treatment beam displays the loss of dose in the SOBP region due to the effects of scatter along the central axis. The size of the penumbra and the increase of penumbra with depth have similar behavior between both the small and the larger field.

beam. Additionally, the air gap, range compensator thickness and material, and patient heterogeneities affect the MCS in the field.

In general, shallow fields have equilibrium even down to field sizes less than 1 cm diameter, especially when treating in mostly homogenous materials with ranges of a few centimeters. like the eye. When treating targets at depths greater than 10 cm, a field diameter of 1 cm will lose equilibrium along the central axis due to MCS and lateral spreading of the dose. Examples of this behavior can be found in multiple studies. including [43]. A particularly important note is that lateral integration measurements—such as plane parallel chambers, Bragg Peak chambers, and multi-layer ionization chambers (MLICs)—still display the typical Bragg peak dose distributions, even for small fields, due to the integration of dose along the entire planar surface of the measuring device. Greater details are provided below. Full appreciation of the central axis dose distributions and loss of equilibrium requires fine-resolution measuring devices, such as micro-ionization chambers, diodes, and film.

MCS dosimetric effects in smaller fields are larger in heterogeneous tissues due to the proximity of the heterogenetities to the central axis of the treatment beam. Studies have looked at the additional effects of heterogeneities in patients, particularly in the lung and cranial cavity [44]. Many treatment planning systems struggle to accurately model the effects of patient heterogeneities due to the limitations of pencil beam dose algorithms in capturing the effects of lateral perturbations of the dose calculations. Hence, many studies have relied upon Monte Carlo and physical measurements to determine the effects of heterogeneities on MCS and dosimetry [45].

In addition to the loss of dosimetric equilibrium due to patient heterogeneities, range compensators can increase the amount of MCS effects in the treatment field, especially when the range compensator is thick with steep ridges, which often occurs in tandem with greater patient heterogeneities. The thickness of the range compensa-

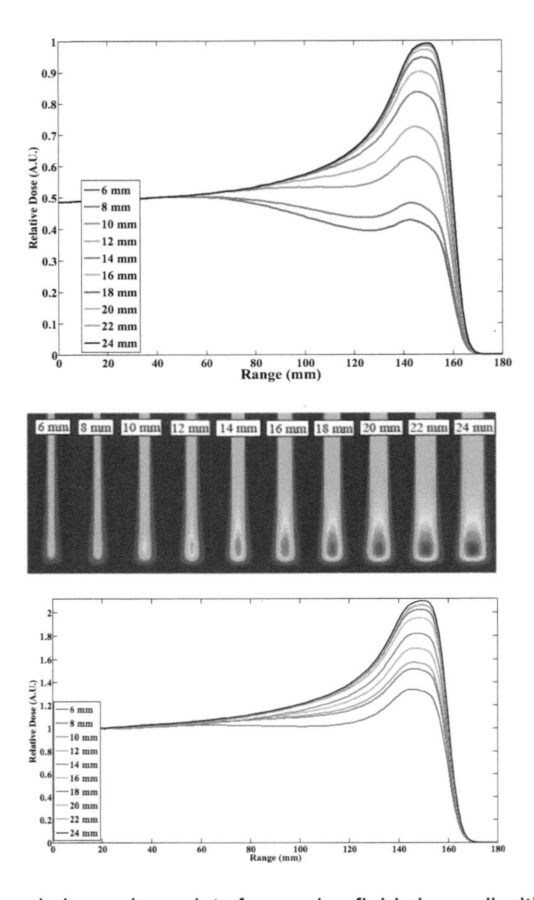

Figure 27–2 PDD and planar dose plots for varying field sizes, all with the same 90% range (160 mm) and d90-p90% modulation (10 mm). The field sizes are generated in the TPS and are close estimates as projected to the isocentric plane. The SOBP and Bragg peaks are degraded as the field size decreases due to the loss of central axis equilibrium. Additionally, the effective range of the SOBP also decreases. The bottom image displays the integrated depth–dose measurements of a large plane parallel chamber of the same dose distributions. Notice that the Bragg peak is degraded due to the geometric field size effect, but the IDD is much less pronounced compared to the PDD central axis profile since much of the dose scattered away from the CAX is recaptured.

tor may not be properly modeled in some treatment planning systems, resulting in this additional scattering contribution not being accounted for in the dose simulation.

When field sizes are small and the number of treatment fractions is limited, the dosimetric uncertainties due to MCS are amplified. Some aspects of the beam optics are fixed and cannot easily be adjusted, such as the location where the protons exit the vacuum and scattering system. Most approaches to minimizing the effects of MCS on the patient dosimetry involve treatment planning limits on field size and the number of fields, as well as the avoidance of beam paths through heterogeneous regions. These issues will be addressed below. Additionally, the limits of the treatment planning system must be understood using appropriate measurements of phantoms that closely model clinical scenarios and Monte Carlo simulations of patient scenarios using a Monte Carlo system that has been commissioned to accurately model MCS in patients [45].

27.2.2 Depth Doses

As alluded to in the previous sections, depth–dose distributions can be significantly different for smaller field sizes, especially at depths greater than 5–10 cm and field sizes with diameters less than 2 cm. Most important to depth–dose characteristics is the MCS. Figure 27–2 demonstrates the effects of MCS on the PDD as the field size decreases. MCS affects the central axis dose and degrades the Bragg peak along the central axis, and this scattering causes a loss of equilibrium. While the Bragg peak can be spread in the lateral direction, the entrance region can also be degraded for larger ranges and smaller field sizes. Similar effects can also be observed when there are large amounts of heterogeneities and small field sizes.

Most significant in clinical scenarios is an underestimation of the range required to treat a target if the treatment planning system does not accurately capture the MCS effects in the depth–dose resulting from air gap variations, patient heterogeneities, and range compensator scatter. The effects of air gap are displayed in Figure 27–3. In these cases, the MCS effects can reduce the range of the proton field, specifically the D98%, by several millimeters.

27.2.3 Lateral Doses

Lateral dose profiles in all proton fields are determined by beam optics, effective source spot size, effective SAD, aperture locations, and other beam delivery devices. When approaching small field sizes, the lateral dose distributions begin to affect the central axis dose distributions, and distal profiles as described in the section above. Additionally, when planning SRS and SBRT treatments, the lateral spreading of the penumbra that is common to all proton fields (Figure 27–1) can greatly affect the clinical outcomes of the proton therapy due to higher doses to normal tissues compared to other scenarios (shallower fields, better beam optics, etc.) such that the penumbra is sharper. One of the primary goals of hypofractionated radiotherapy is the geometrically precise delivery of high dose to avoid normal tissue toxicities. A wider penumbra can decrease the therapeutic ratio for the hypofractionated delivery.

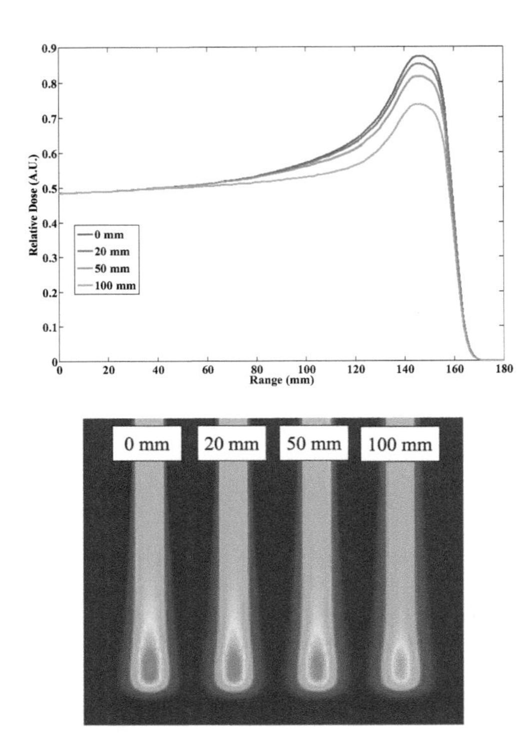

Figure 27–3 PDD and planar dose plots for varying air gaps of 0, 20, 50, and 100 mm with the same field size (20 mm) projection to isocenter, range (160 mm), and modulation (10 mm). The degradation of the SOBP can be seen as the air gap increases due to the additional air and distance between the aperture and the target. The additional geometric distance increases the penumbra and the effects of the scattering. The D98% can be seen to pull back in addition to the decrease of the SOBP intensity.

In patient treatments, the lateral distributions are affected by the scattering of protons in regions of heterogeneities. Specifically in the lung and base of skull regions, the heterogeneities can be large due to bone–tissue and lung–tissue interfaces that can affect the lateral dose profiles by broadening the dose distributions due to increased scatter and extended range of scattered protons and secondary electrons. Studies analyzing these effects have been conducted using Monte Carlo and anthropomorphic phantoms, where the full scatter effects are accounted for and measured [45].

27.2.4 Halo

Halo dose effects have been described in the literature with simulations and measurements. The halo dosimetric factors affect the central axis dose distributions and affect the central axis dose distribution at larger field sizes than the MCS effects due to the larger range of halo distributions of 8–10 cm. While MCS effects are very small for field sizes above 2–3 cm diameter, the halo can affect the central axis dose distribution for fields with diameters greater than 5 cm. Some studies have estimated the halo effects to be less than 10% [46], while others have estimated the effects to be somewhat larger [47]. There is additional debate as to the shape of the distribution required to model halo effects in the treatment planning dose model. While the halo may result in a 10% to 20% dose reduction along the central axis of smaller fields, the effect is minimally dependent upon energy and heterogeneities. Most centers use empirical monitor unit calculations, which should account for the halo effects. If a treatment planning system is used as the primary source of monitor unit calculations without scaling due to field-specific measurements, the halo must be accurately accounted for in the treatment planning dose algorithm. Measurement devices have been proposed, such as a larger Bragg peak chamber, with the intent to fully capture the halo dose distribution.

27.3 Measurements of Small Fields

Measurements of small fields are required for beam line commissioning, treatment planning system commissioning, regular quality assurance of the beam delivery system, patient field quality assurance, and Monte Carlo commissioning. Measurements include central axis profiles of pristine peaks and modulated fields, lateral profiles of pristine peaks and modulated fields, and point and profiles in both anthropomorphic phantoms and quality assurance phantoms.

27.3.1 Devices

For each of the measurements required for regular clinical operations of small fields therapy programs for eyes, SRS, and SBRT, there are devices that are optimally designed and fitted. Among the selections are Bragg peak chambers, cylindrical ionization chambers, micro ionization chambers, plane parallel plate ionization chambers, multi-layer ion chambers (MLICs), film, diodes, CCD cameras, and diamond detectors. Details about detectors can be found in Chapter 11.

27.3.2 Pristine Peaks

Pristine peak measurements can be conducted with a laterally integrating device such as a Bragg peak chamber or an MLIC. Some important notes are the ability of the device to capture the halo distribution, and some situations require the ability to generate an absolute or traceable relative dose distribution. The large plane parallel geometry integrates over the lateral dose distribution and does not capture the central axis dose distributions necessary for dose algorithm commissioning (Figure 27–2).

In addition to the integration chamber, an accurate measurement of the central axis dose requires other measuring devices. Due to their small sensitive area, diodes have been used, but studies have reported the LET dependence of the solid state materials can result in energy dependencies that skew the depth–dose profile [48]. The

LET variations in Bragg peaks occur at the end of range, and the diode should be characterized such that the LET dependence is understood and corrected. Diodes can also suffer from radiation damage, which can cause a drift in the diode response. Frequent cross calibrations to standard conditions can correct for any drift due to radiation damage.

Ionization chambers are also possible for central axis dose profile measurements, and they have little LET dependence. The size of the chamber can introduce volume averaging effects at small fields and at the end of range. Micro ionization chambers are often used for fields of small radii until the volume averaging interferes with the dose distribution, and plane parallel chambers are often used for PDD measurements when the volume averaging of cylindrical or micro chambers along the beam axis skews the measurement.

Film can be calibrated for absolute dosimetry, but there are LET concerns with film as well. Film has very fine resolution and can measure high dose levels, but the uncertainty of the dose measurements is difficult to reduce below 3% to 5% due to many well-documented effects of direction, scanner, molecular consistency, etc.

Pristine peak measurements are often measured in a relative dosimetry situation, such that the entrance dose, peak dose, and depth–dose shape are most important. Thus, LET dependence and volume effects can significantly affect the robustness of the measurements. It has been suggested to verify pristine measurements with Golden Beam [49] or Monte Carlo [45] that has been commissioned at larger field sizes against ionization chambers.

Detector selection requires sensitivity to the LET dependence of the detector. Greater detail of LET dependency of detectors is provided in Chapter 11. LET dependency can be characterized and corrected from measurements.

27.3.3 SOBP and Volume Doses

For absolute dosimetry and field calibration measurements, the most important dose measurements are spread-out Bragg peak (SOBP) and volume doses from a scanned beam delivery (e.g., 100 mm x 100 mm x 100 mm scanned volume). According to various absolute dosimetry protocols, measurements are performed in an SOBP or a scanned volume of a finite dimension (often 10 cm x 10 cm x 10 cm) at some representative range, i.e. 16 cm, with an ionization chamber (parallel plate or cylindrical) placed in the center of the irradiated volume. Such conditions will eliminate volume averaging effects and ensure equilibrium of MCS and halo effects.

When field sizes are reduced such that there is no longer equilibrium along the central axis and volume averaging affects the measurements, the measuring devices must be carefully selected. Similar to the pristine peak measurements, diodes and micro ionization chambers are typically used for smaller fields, with a cross calibration using diodes when the volume averaging affects the ionization chambers measurements. Common to all protocols is the use of a standard field for traceable absolute dosimetry and subsequent relative field-specific measurements using ionization chambers.

Also similar to pristine peak measurements is the use of plane parallel chambers and MLICs for depth–dose profiles, specifically the distal slope. Additionally, the same techniques for measuring the central axis depth–dose profile with diodes, ion chambers, and film can be used for SOBP and volumes [43,45,50].

Specific to scanned beam deliveries, volume profiles are difficult to measure in the proximal/distal direction due to the interplay of scanning the beam and scanning the detector. Many institutions use a single detector or an array of detectors placed at discrete depths and the same field delivered once per depth position, resulting in repeated deliveries of each scanned beam. MLICs can be used if the lateral integration of the dose is not consequential to the dose measurement.

27.3.4 Penumbra

The lateral penumbra of small fields used for SRS and SBRT are particularly important due to the large effects of MCS on the dose delivered to the target, with parts of the target likely being in or close to lateral regions of disequilibrium. Lateral dose profiles are typically measured using a diode scanned across the passively scattered field. Film has also been used for the small fields in a relative dose measurement technique. When using scanned proton beams, an integrating system—such as film, CCD camera device, or planar array of detectors—is required to capture the total delivered dose.

The penumbra of the fields should be sampled to several centimeters, particularly at large depths, due to the range of MCS and the halo that can extend to 10 cm beyond the field edge. Again, the LET dependencies of the measuring devices must be accounted for, but there is little LET variance over the lateral distribution [51] and lateral measurements are generally performed relative to the central axis.

The penumbra in regions of heterogeneities is more difficult to measure due to phantom design and device incorporation. Studies have used planar arrays of ion chambers and film [44] to assess the MCS effects on the lateral dose distributions in regions of heterogeneities.

27.3.5 Broad Beam Measurements

SRS and SBRT treatments can use fields with central axis dose equilibrium, but disequilibrium remains at the edges of the fields. Accurate dose measurements of these regions are still required, often using diodes scanned across the fields, planar arrays of ion chambers, CCD camera devices, or film. Planar ion chamber arrays often suffer from volume averaging from the chamber volumes and detector resolution. Planar arrays and film offer the advantage of integrating doses, while scanning a diode or ion chamber are limited when delivering a scanned beam pattern.

27.4 Treatment Planning Commissioning

27.4.1 Measurements

There are multiple treatment planning systems that have different requirements. Pristine peaks are the most fundamental measurement generally required for TPS commissioning. The peak measurements include both depth–dose and lateral profiles along the beam entrance. As a function of energy, a TPS generally fits the entrance dose/peak dose ratios, scatter angle cross section for penumbra, the effective SAD, and source sizes. Additionally for scanning systems, spot sizes should be measured for the range of clinical energies.

Specific to smaller treatment fields, the field size correction should be measured across representative clinical energies and field sizes. The TPS is not guaranteed to model the field size effects to the multiple sources of scatter, beam optics, and beam line geometries. An empirical model has been derived at Massachusetts General Hospital to supplement the broad beam field calibrations [43].

27.4.2 Check Measurements

Simple check measurements can quickly ascertain the robustness of a treatment planning system. Simple geometries in water phantoms can be calculated to simulate the commissioning measurements in both broad and small beams at various energies. The central axis and lateral profiles can be quickly compared to the commissioning measurements.

27.4.3 Check Plans

In order to validate a TPS model for small fields, SRS, SBRT, and ocular treatments, simple water phantom experiments utilizing the TPS workflow with clinically realistic beams should be tested first, measured with ionization chambers, and cross calibrated with diodes, detector arrays, or film. The geometries should be simple enough to perform a dose algorithm validation in water or solid water. For both passively scattered and scanned deliveries, check plans are typically measured with an ionization chamber or diode at various points in the target and planar dose measurements using film or detector arrays.

Second should be plans generated in an anthropomorphic phantom, again using beams designed to simulate actual clinical scenarios. Some phantoms include heterogeneities and the ability to incorporate planar dose measurements. Other phantoms allow for only a point verification. For cranial SRS, point measurements are generally sufficient, unless there are significant heterogeneities in the fields, such as AVM and base of skull targets.

When treating SBRT patients in the lung, more complicated phantoms should be used to verify the dose algorithms in regions of dose disequilibrium. Motion can also be assessed with one of the phantoms commercially available to evaluate dose delivery in a realistic lung target.

Monte Carlo is becoming a common technique to validate a treatment planning system using a commissioned beam line in the computer simulation. While Monte Carlo simulations can increase the flexibility of the TPS validation, the Monte Carlo also requires validation measurements, similar to a TPS commissioning, with more details provided in Chapter 20. Monte Carlo can allow for more accurate modeling of dose delivered in heterogeneous regions in patients and phantoms. Also, field size limitations of physical measurements using ionization chambers can be avoided with Monte Carlo.

27.4.4 Algorithms

The specific algorithm used in a treatment planning system must be fully documented and understood in regard to any treatment condition being commissioned. For SRS, SBRT, and ocular therapies using smaller fields with disequilibrium, a TPS algorithm might behave differently than a broad beam scenario. A pencil beam dose algorithm such as Hong et al. [52] can model smaller fields in homogeneous conditions. However, issues arise in heterogeneous regions.

Monte Carlo algorithms can be adapted to more diverse treatment scenarios, but they can be limited by the quality of the commissioning data and the measurements used to validate the beam models. Additionally, commercial Monte Carlo algorithms may employ some approximations or shortcuts to speed the dose calculation. Such approximations may impair the accuracy in the limits of the model often encountered in smaller field sizes, such as voxel size, the numbers of particles, scattering assumptions, and energy cutoffs. The use of Monte Carlo should be dependent upon a good understanding of the commissioning process and the approximations used in the calculations.

For scanning systems, the spot sizes in both lateral directions will vary with range, and many systems observe variability with treatment angle when using a gantry for delivery. The variables in the spot size might not be captured in all commercially available TPS, and the dose delivery will be affected by the uncertainty in the beam model spot sizes. Like many other treatment delivery variables, the beam spot size will have a larger effect on the dosimetry for smaller fields where there are fewer spots, and the target dose will be affected by the penumbra of each spot.

Overall, the TPS may require a different beam model specific to the field sizes and ranges used for ocular, SRS, or SBRT treatments. The limited number of parameters available in different TPS can reduce the optimization to a limited set of clinical scenarios, such that only small field sizes may be optimized or broad beams optimized, but not both.

27.4.5 Apertures/Range Compensators

When apertures and range compensators are included in the TPS, the machining accuracy of the hardware and the modeling of the dosimetric effects must be validated. The machining must be accomplished with sub-mm accuracy and compared to the treatment planning system request. Additionally, the TPS model should be able to accurately model the scatter and the energy straggling from the hardware. More details regarding quality assurance of hardware is presented in Chapter 16.

27.5 Clinical Workflow

Ocular simulations historically are performed after clips are attached to the surface of the eye in close proximity to the lesion. The simulation verifies the clip positions rela-

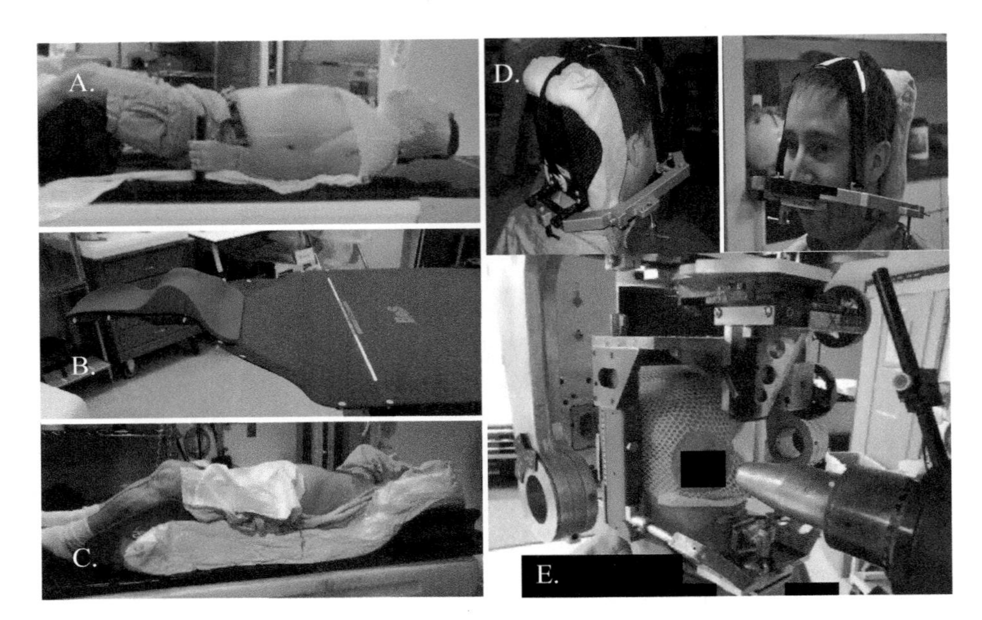

Figure 27–4 Immobilizations used for SRS, SBRT, and ocular treatments in proton beam therapy. The QFixBoS (A and B) and the mGTC [54] (D) are used for cranial and C-Spine targets. The vacuum bag (C) is used for more inferior spinal targets and torso lesions. The ocular system involves a mask (E) with an opening over the treated eye. The mask is secured directly to the beam line apparatus.

tive to the planned positions in the treatment planning system, verifies the gaze position, and ensures the patient can tolerate the procedure using eyelid retractors. Recently, some proton centers have proposed using a 3D CT simulation combined with a simulation in the treatment room using digitally reconstructed radiographs (DRRs) and a gaze fixation light or suction cup device. The advantages of using a 3D CT simulation is the common workflow with other treatment sites and the ability to more accurately model the dose delivered to the target with regard to local heterogeneities.

For SRS and SBRT, simulations are performed with a 3D CT simulation using the treatment immobilization devices in the CT. For cranial targets, the CT should always use a slice thickness of around 1 mm due to the accuracy of target definition, accuracy of patient positioning in the treatment room, and accuracy of the dose calculation due to heterogeneities that would suffer from slice averaging with a larger slice thickness. When treating extracranial targets, the slice thickness might be limited by the maximum number of slices accommodated in the clinical workflow. Additionally, there is a potential for reduced soft tissue contrast when using a reduced slice thickness requiring increased mAs. Thirdly, when acquiring a 4D CT, the reduced slice thickness can dramatically increase the size of the data sets.

When treating extracranial sites that undergo respiratory motion, multiple options are available for the CT simulation, but none to date are optimal for all considerations. Accurate tissue densities are required for all breathing phases for range and dose calculations. For scanning deliveries, the tissue density variations can interplay with the scanned beam [53] and, therefore, the delivery can be more sensitive to the breathing phase than passive scattering. Planning with every phase of the 4D CT is more complicated than most treatment planning systems can allow. It also requires the patient to breathe identically at each treatment fraction. For details on motion management, see Chapter 25.

An average, min, or max CT may model the tissue densities at different breathing phases, or the densities over multiple phases, which will reduce the effects of breathing on the beam range and compensator designs. However, the smearing will reduce the conformality of the treatment and require larger margins to account for density variations when delivering to the range of densities at different breathing phases. Using a single planning CT is the only option for most TPSs, and this represents a compromise between accuracy at each breathing phase and robustness of delivery when the treatment is delivered at a phase other than the planned phase. Extensive research is ongoing to determine the optimal combinations of CT, planning parameters, and delivery parameters to deliver the most accurate and robust treatment.

The goals of immobilizations in general are to secure the patient to a treatment table with limited intrafractional motion and to secure the patient in a highly reproducible manner [54]. For proton therapy specifically, the effects of the immobilization on scatter and beam attenuation add other considerations in the immobilization design. Most proton immobilizations are made of carbon fiber due to its low-Z nature and its thin, yet rigid and strong, mechanical properties. Some immobilizations still include metal, but the beam entrances should avoid the metal due to the high-Z scatter effects and the interfractional variability of the metal in the beam range variations. In addition to carbon fiber, many immobilizations utilize low-density vac-lock bags, custom cushions, or thin plastic masks. Figure 27–4 demonstrates various immobilizations employed for proton SRS, SBRT, and ocular treatments. Additional information can be seen in Chapter 18.

Ideally, all of the immobilization measures, including the treatment couch, will be included in the treatment planning CT to more accurately model the effects of the immobilizations on the treatment simulations. Specific to small fields, the bags and cushions, while low density, can significantly increase the MCS effects along the central axis dose calculation and at the distal edge. The immobilization devices can also increase the air gap between the aperture and the patient, which can increase the MCS effects in the treatment field.

With accurate pretreatment imaging, most immobilizations have been demonstrated to have 1–2 mm localization precision in the cranial cavity and 2–5 mm of precision in extracranial sites, depending upon the target localization method, ranging from fiducials, bony surrogates, or soft tissue alignment [55].

For limited fraction therapies, the intrafractional and interfractional patient positioning uncertainties must be fully quantified by the clinical team prior to treatment, such that the uncertainties are included in the margin equation to avoid near misses of targets. The immobilizations will affect the positioning uncertainty in addition to the localization method and imaging tools.

27.6 Treatment Planning

27.6.1 Eyes

While cranial targets are straightforward to localize due to the rigidity of the cranial bones and anatomical information, ocular lesions can be very difficult to localize. Due to the difficulties, larger margins are generally utilized, combined with surgical clips attached to the surface of the eye in close proximity to the lesion. Traditional ocular therapy institutions have relied upon ocular models instead of CT simulations. The eye is modeled as a sphere with specific anatomic landmarks, such as the optic disk and limbus. The target size, location relative to anatomic landmarks, and depth are also measured and transferred into the ocular model (Figure 27–5). The positions of the clips relative to the target are visualized using trans-illumination, measured at the time of attachment, and transferred into the ocular model. The model is completed with a gaze direction used to fix the position of the eye during treatments. The inclusion of daily varying portions of the eye lid in the beam path is an important source of uncertainty since treatment plans generally use a single enface treatment beam, therefore range and modulation margins of 3–4 mm are used. Lateral margins of 1–4 mm

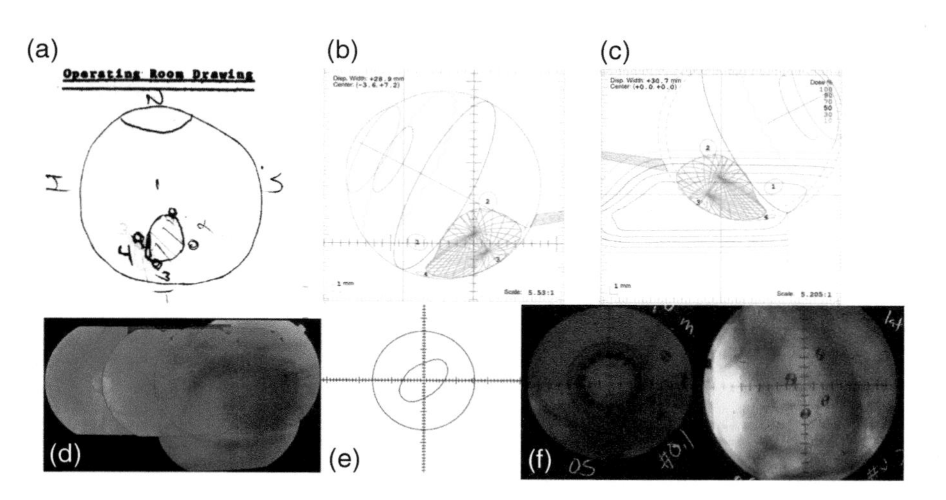

Figure 27–5 The eye treatment workflow involves an intraoperative exam (a and b) and report with the target manually transferred into the eye model (c). An aperture is created (d) and a range and modulation specified (e), all with appropriate margins. The surgical clips are identified in the model and visualized in the patient setup procedure (f).

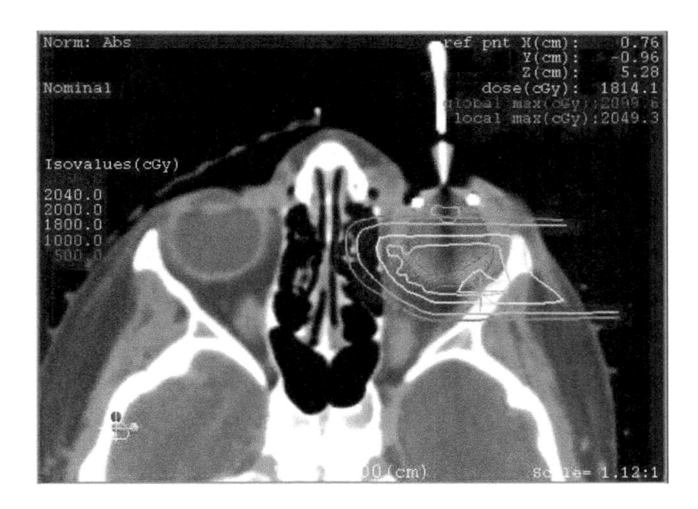

Figure 27–6 An example of a 3D plan for an eye target. The plan is performed with a single lateral beam with extra margins. Multiple beams could be used to increase the robustness of the delivered dose. In this example the eye vitreous density is corrected to minimize the impact of the altered Hounsfield numbers due to scatter caused by the stem that holds the eye straight.

are used to account for the multiple uncertainties in target definition and target localization (figures 27–4E and 27–5).

More recently, some institutions have begun to explore the use of 3D treatment planning and a gaze determination method using a light or eye suction cup to guide the position of the gaze in a reproducible manner. The workflow includes a thin-slice CT, gaze determination, and a treatment plan using a beam model in the TPS commissioned for field sizes and ranges used in ocular therapies. The field arrangements can vary from the traditional single enface beam with the use of lateral, superior, and superior oblique beams that match the geometry of the target and surrounding anatomy (Figure 27–6). The target can be localized in the CT using traditional methods or by using image fusion techniques to register the CT to other imaging modalities. Margins would still be required for such treatments, and multiple beams would be required to increase the robustness of the delivered dose.

27.6.2 SRS: Cranial

Stereotactic radiosurgery for cranial targets was one of the first uses of proton therapy. The current state of art consists of 3D CT simulations, fusions with MR for target localizations, and 3D dose simulations. Targets include benign neoplasms—such as acromegaly, meningiomas, AVMs, and schwannomas—as well as some metastatic lesions. Radiosurgery typically relies upon the geometric isolation of an ablative radi-

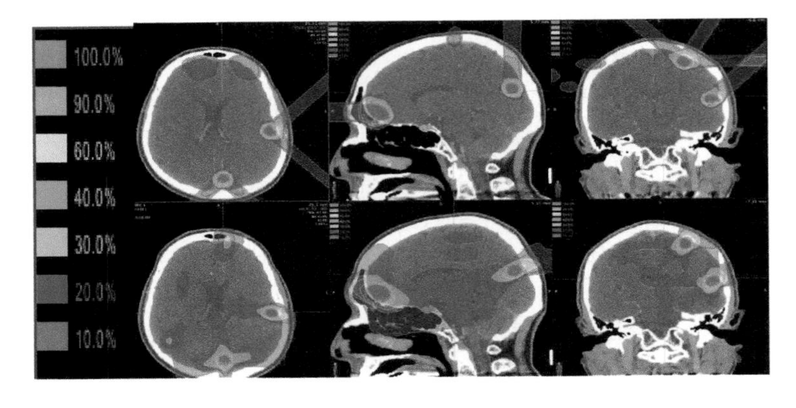

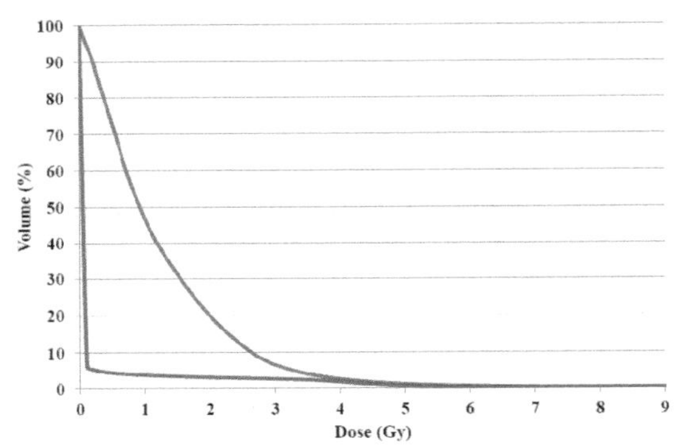

Figure 27–7 A dose comparison of a proton (top row) and a photon (bottom row) plan for multiple brain metastases. The DVH is for the normal brain and displays the difference between photons (blue) and protons (red) at the low doses. The convergence occurs around 700 cGy, or about 40% of the 1800 cGy prescription.

ation dose to a well-defined target such that the neural and optical toxicities are avoided by geometric sparing of the tissues. Due to the use of dosimetric sparing, protons can provide an excellent dose delivery option with low integral dose. With current technologies and associated proton uncertainties, the integral dose for proton cranial targets is typically less than photon deliveries at dose levels below the 40% isodose level relative to the prescription, i.e., the volume of healthy tissues receiving doses below 720 cGy for an 1800 cGy prescription will be less for proton versus photons (Figure 27–7). Additionally, the current state of the art proton range uncertainties [56] accommodated in planning protocols result in a less conformal volume at the prescription dose level.

Treatment planning is performed using a CT. Traditionally, an invasive stereotactic frame was used for patient and target localization in the CT, MR, and treatment room. More recently, noninvasive fixation using an upper jaw/occipital bone immobilization or a thermoplastic mask [54] is used for patient immobilization, combined with a treatment room imaging system that should be able to register the patient in six degrees of freedom to less than 1 mm accuracy.

To achieve the most robust and clinically effective plan for cranial SRS, there are several techniques that should be incorporated into the planning process, mostly associated with beam angle optimization typically in a manual mode.

27.6.2.1 Multiple Beams

For a majority of clinical cases, it is entirely feasible and recommended to use three treatment beams for a single or limited fraction SRS in the cranial cavity. The cranial cavity allows for beams entering in more planes than the axial plane, and target geometries often require using multiple planes of entrance to spare the most healthy tissues (figures 27–7 and 27–8). Using multiple beams generally increases the conformality of the proton treatment, most significantly for passive scattered deliveries, but also for scanning beam deliveries due to the reduced effects of range uncertainties along a single beam direction.

Multiple beams also increase the robustness of a treatment plan to both systematic and random uncertainties. Beams requiring various ranges and passing through different tissues will reduce the effects of range uncertainties and added margins to account for range uncertainties in single distal locations. Additionally, the multiple beams will reduce the overall patient motion uncertainties that are most pronounced in the cross-beam direction, since the lateral directions for multiple beams will be in different planes. MCS can also increase the dose uncertainty, most significantly in smaller fields and regions of heterogeneities. When these scenarios cannot be avoided, multiple beams can reduce the overall plan uncertainties resulting from the MCS effects in a single beam.

Scanning deliveries allow for full intensity modulation of the spots to create nonuniform dose distributions. In these cases, the robustness of the plans must be fully understood, since weakly robust plans suffer more in limited fraction treatments due to the greater importance of random uncertainties in the patient position and dose delivery. In order to create more robust plans, it is recommended to use uniform field delivery techniques.

27.6.2.2 Avoid Distal Sparing of Critical Structures

Due to range uncertainties and LET variations at the end of range, it is advisable to avoid using the distal edge of the treatment field to critical structures. In cases where the target is in close proximity to a critical structure, particularly along multiple sides, distal edge sparing can be unavoidable. This is another case where using multiple fields can increase the overall plan robustness when considering normal tissue complication probabilities in the case of range and patient positioning uncertainties. Figure 27–8 demonstrates some beam angle selections to increase plan robustness.

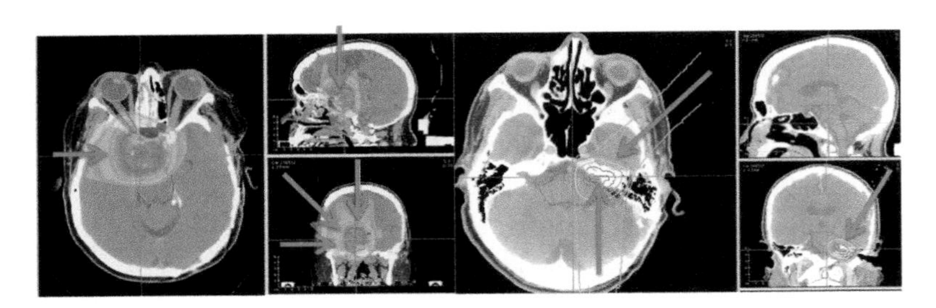

Figure 27–8 Beam angles for the cavernous sinus meningioma (left) and acoustic schwarnoma (right) were selected to avoid distal edge sparing and heterogeneous tissues.

27.6.2.3 Avoid Beams Entering Through or along Heterogeneous Boundaries

When possible, heterogeneous regions should be avoided due to the additional MCS and the subsequent effects on the dose delivery uncertainty. As stated above, small fields can be especially sensitive to MCS effects, which can broaden the lateral dose distribution, reduce the central axis dose, and reduce the range of the treatment field. The result can be increased normal tissue dose, reduced target dose, and a less robust treatment plan. In cases like targets in the base of skull, AVMs (Figure 27–9), or surgical resections, heterogeneities are difficult to avoid. In these cases, multiple treatment fields can reduce and spread the MCS effects from a single field on the total dose distributions. Additionally, margins can be added in both the distal/proximal and lateral directions in order to reduce the MCS effects in the treatment field.

27.6.2.4 Avoid Beam Field Sizes and Ranges that Extend beyond Commissioning Data

Adequate and accurate commissioning measurements are limited to specific field sizes and ranges due to time constraints, measurement device availability, etc. In these cases, it is best to have defined limits of distal ranges and target sizes that are accurately simulated in the commissioned beam model in the TPS. Also, the loss of the Bragg peak to lateral scattering in small fields and deep ranges can greatly reduce the advantage of proton beam therapy.

27.6.3 SBRT

Extracranial SRS and SBRT is a more recent treatment modality for proton therapy that required recent advancements in technology to facilitate the modality, including better 3D and 4D CT imaging, better in-room imaging, and clinical trials that have established SBRT as an effective treatment option for spine, lung, liver, and pancreas lesions. In general, protons are theoretically able to achieve higher therapeutic ratios for extracranial SBRT targets than typical photon treatments due to the sparing of nor-

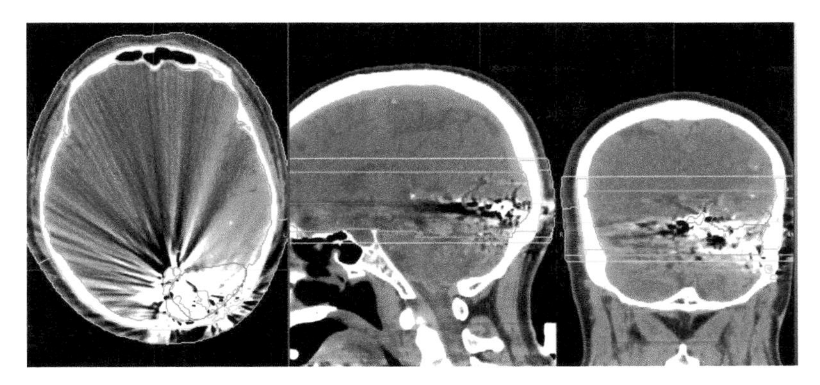

Figure 27–9 An example of heterogeneities and artifacts due to the embolization material in an AVM. Using multiple beams and avoiding beam angles parallel to the CT artifacts increases the plan robustness. In this example, the onyx, bone, brain, skin, and immobilization cushion are contoured, and the electron density is set to appropriate values in the treatment planning system to reduce the dosimetric errors due to the onyx induced CT artifacts.

mal tissues. Specifically, protons should be able to reduce lung V20 and V5 of both the involved and contralateral lungs. Protons should be able to reduce the normal liver volume receiving the radiation. Protons should be able to spare the cord due to limited treatment beam angles required for dose delivery.

Lung SBRT is well documented to be technically very difficult to treat with conformal proton therapy. The range uncertainties in lung greatly increase the normal lung dose and motion further increases the range uncertainties. There are the additional issues of CT simulations and dose calculations listed above. With these caveats, there remain promising studies that have presented clinical cases of proton therapy for lung SBRT. Most patients have motion that is less than 1.5 cm, especially in the upper lobes, and treatments with multiple beam angles in the axial plane can increase the dose conformality. The penumbra remains larger due to the low density of the lung, and lateral margins are recommended to be larger to compensate for the MCS effects and the motion effects. Figure 27–10 displays a lung SBRT treatment plan incorporating appropriate margins and planning methods to increase the robustness of the plan against motion and range uncertainties.

Setup uncertainties for lung treatments can be large when using bony surrogates. Setup uncertainties can be reduced when using CBCT for soft tissue alignment or fiducials.

For liver therapies, the treatments are also performed with 2–3 axial beams selected to increase the conformality to the target and avoid entering through radiation-sensitive organs, including the digestive tract and cardiac structures. Beam angles can also be selected to minimize the range, which can sharpen the lateral penumbra. The liver is also subject to respiratory motion, but the target has similar density to the

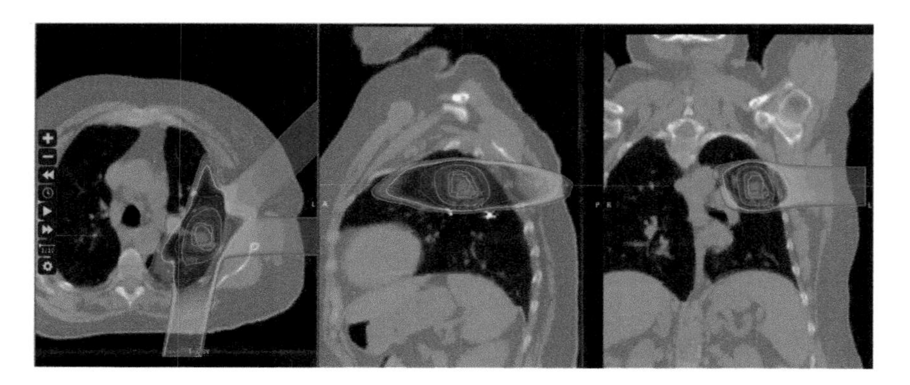

Figure 27–10 Axial, coronal, and sagittal views of a lung SBRT plan using multiple beams, attempting to minimize the effects of breathing motion and distal edge overshoot into the mediastinum. Margins were applied for motion, range uncertainty, and smearing.

liver tissue, resulting in little range variations due to motion. More importantly is the accurate internal target volume (ITV) assessment and caution when treating in the dome of the liver when the lung can receive dose during the inhale breathing phases.

Setup uncertainties for liver treatments can be large when using bony surrogates. Soft tissue is not a possible setup localizer due to the low contrast of tumor versus healthy liver tissue. Fiducials are commonly used for more precise target localization and a reduced setup margin.

Spine SRS and SBRT can be very difficult to treat with protons due to the typical concave geometries of vertebral lesions wrapping around the cord. Additional complications are the artifacts present in post-surgical patients, since surgery is a common first line of therapy for epidural and vertebral lesions.

Most important are the complex geometries often associated with spine lesions. Until the rise of scanning delivery systems, passive scattered deliveries cannot deliver a concave dose distribution without utilizing patching of treatment fields around the organ at risk. Such patching methods are not robust for single-fraction dose deliveries due to the risk of hot or cold spots in the target. With scanning delivery techniques, proton dose distributions can approach a concave geometry, particularly when using 2–3 posterior fields and intensity modulation. There is concern of plan robustness, which is a research topic. The use of posterior beams is most common due to the sharper penumbra of a more shallow range and the avoidance of most other critical organs, not inclusive of the cord.

Setup margins for spine targets are typically very small since the bony anatomy of the vertebral bodies is identically located to the target, or in very close proximity. The addition of surgical hardware can increase the numbers of surrogates highly correlated to the treatment target, but this may also increase range uncertainties.

27.6.4 Uncertainties

Treatments with small fields generally do not increase the typical range uncertainties. MCS uncertainties can increase the dose and range uncertainties. More important is the hypofractionation of ocular, SRS, and SBRT therapies. In these limited fraction cases, the random uncertainties—specifically patient position and patient anatomy variations—can have a larger effect on the dose delivery. When utilizing multiple beams with beam angle optimization to minimize range and MCS-induced uncertainties as detailed above, the effects of random variations can be reduced. Additionally, margins and increased image guidance precision can further reduce the effects of random uncertainties.

27.7 Quality Assurance

27.7.1 Daily

Daily QA is designed to test the beam delivery of essential devices, such as scatterers in passive scattered beam lines and magnets in scanning systems. Specific to ocular, SRS, and SBRT treatments, the beamlines likely have regular checks that verify the fidelity of the delivery system will not change in small fields or large dose fractions. If there is a beam delivery system, option, snout, etc. specific to ocular, SRS, or SBRT, an output and alignment check should be performed. If it is possible to achieve a more precise range and spot position measurement, the additional check would increase the reliability of the system.

Important to daily QA for SRS and SBRT is the patient alignment system. While image guidance systems must be validated daily for all treatments, SRS and SBRT deliveries require more accurate and precise patient positioning to reduce the margins and normal tissue toxicities. Similar to the Winston–Lutz test in photon SRS and SBRT [57,58], an isocentric check of the proton delivery system should be performed daily when SRS and SBRT treatments are delivered.

27.7.2 Monthly

Monthly tests will include spot checks of the commissioning scenarios of simple field arrangements in a simple water or solid water phantom. The measurements should include enface beams in basic geometries and representative, but limited, numbers of field sizes and ranges regularly used in clinical cases.

Additional monthly QA should include and meet the specifications as detailed in TG-142 for SRS and SBRT treatments, including dosimetric and image guidance tests (see Chapter 15).

27.7.3 Annual

Annual QA should include additional measurements to verify the beam model is correct for the specific clinical scenarios commissioned. For example, the smaller field sizes required for ocular treatments should be verified against an absolute dose standard and for specific clinical test cases, likely a subset of the clinical cases used for commissioning. Ideally, the measurements would include traceable absolute dose checks across the range of clinically relevant field sizes and ranges. Typically, TRS-398 is referenced for absolute dosimetry measurement protocols, as discussed earlier in this book.

Additional annual QA should include and meet the specifications as detailed in TG-142 for SRS and SBRT treatments, including dosimetric, image guidance, and planning system tests.

27.7.4 End-to-end Tests

Periodic, annual, or otherwise, end-to-end tests using anthropomorphic phantoms should be performed to validate the accuracy of the dose delivery in the entire clinical workflow. Using point measurements and film can repeat the commissioning measurements and continually validate the fidelity of the beam model, the image guidance system, and the dose delivery system.

27.8 Future Directions

27.8.1 Treatment Planning

27.8.1.1 Scanning Versus Scattering

Most of the deliveries for ocular, SRS, and SBRT treatments to date have been performed using passively scattered systems. Of the recently commissioned and developed therapy centers under construction, many are deciding to commission only scanning beam delivery systems. There are significant questions regarding the advantages and disadvantages of scanning versus scattering for ocular, SRS, and SBRT deliveries. For the smallest field sizes in ocular and SRS treatments, there is minimal volume for modulation of spots. If the scanning is performed without apertures, scattering systems with apertures generally have sharper penumbras. Additional research is needed to determine the treatment sites in SRS and SBRT that may benefit from scanning, such as difficult geometries.

27.8.1.2 Robustness

When moving toward more scanning beam deliveries for SRS and SBRT, the question of robustness of delivered dose becomes more significant. Much work regarding robust treatment planning has been performed for spine and lung treatment sites, and this work is more important for limited fraction deliveries. Most studies have analyzed the effects of patient setup uncertainties and range uncertainties on the robustness of the deliverable plan. Other studies have analyzed robustness in cases of respiratory motion. The questions regarding optimal and safe delivery techniques, particularly for scanning beam deliveries, is an active research area.

27.8.2 Image Guidance

Image guidance in proton therapy lags behind photon therapy when considering CT and CBCT systems. Many vendors are now developing and deploying CBCT systems to proton therapy centers, and the increased imaging ability in the treatment room should increase the accuracy of the dose delivery. Additional imaging modalities—including prompt gamma, PET, MRI, and diodes—are being developed and studied as potential solutions to range verification of proton therapy to reduce the range uncertainties.

27.8.3 Workflow Optimization

With less emphasis in the literature (but likely the greatest clinical impact), workflow optimization can greatly increase the numbers of patients treated with protons, increase the accuracy of delivered treatments, increase the safety of patient treatments, and speed the deployment of proton therapy to various clinical sites. There is ongoing research into more automated treatment planning, faster patient setup and immobilization, more unified software systems, and faster beam delivery techniques. For SRS and SBRT, the workflow can often require days of treatment planning due to the complexity of the targets and the complexity of the CT simulations and dose constraints. Additionally, SRS and SBRT patients require extensive time for immobilization and setup in the room, which may be reduced with more advanced positioning and imaging systems. Chapter 17 also discuss clinical workflow.

27.8.4 Beam Angle Optimization

As alluded to in the treatment planning section, significant time and expertise is required for optimizing the beam angles for SRS and SBRT treatments. The process is an open question for automation using various metrics of heterogeneity, field size limits, and other parameters that correlate with robust planning delivery. A full robust optimization algorithm will include beam angle optimization to account for the variables associated with the beam entrance angle.

27.9 Conclusions

Small proton fields have been utilized since the 1950s for proton SRS and ocular therapy. Small proton fields require careful selection of measurements, devices, and sensitivity to central axis electronic disequilibrium. When using small fields for proton therapy, the effects of MCS must be evaluated and incorporated into treatment planning and protocols. Proton therapy with small fields is useful for proton SRS and ocular proton therapy.

References

1. Wilson RR. Radiological use of fast protons. *Radiology* 1946;47:487–491.
2. Lawrence JH. Proton irradiation of the pituitary. *Cancer* 1957;10:795–798.

3. Tobias CA, Anger HO, Lawrence JH. Radiological use of high energy deuterons and alpha particles. *Am J Roentgenol Radium Ther Nucl Med* 1952;67:1–27.
4. Tobias CA, Van Dyke DC, Simpson ME, Anger HO, Huff RL, Koneff AA. Irradiation of the pituitary of the rat with high energy deuterons. *Am J Roentgenol Radium Ther Nucl Med* 1954;72:1–21.
5. Larsson B, Leksell L, Rexed B, Sourander P. Effect of high energy protons on the spinal cord. *Acta radiol* 1959;51:52–64.
6. Larsson B, Leksell L, Rexed B, Sourander P, Mair W, Andersson B. The high-energy proton beam as a neurosurgical tool. *Nature* 1958;182:1222–1223.
7. Leksell L. The stereotaxic method and radiosurgery of the brain. *Acta Chir Scand* 1951;102:316–319.
8. Leksell L, Larsson B, Andersson B, Rexed B, Sourander P, Mair W. Lesions in the depth of the brain produced by a beam of high energy protons. *Acta Radiol* 1960;54:251–264.
9. Leksell L, Larsson B, Andersson B, Rexed B, Sourander P, Mair W. Research on "Localized radio-lesions". VI. Restricted radio-lesions in the depth of the brain produced by a beam of high energy protons. *AFOSR TN United States Air Force Off Sci Res* 1960;60–1406:1–13.
10. Kjellberg RN, Sweet WH, Preston WM, Koehler AM. The Bragg peak of a proton beam in intracranial therapy of tumors. *Trans Am Neurol Assoc* 1962;87:216–218.
11. Kjellberg RN, Hanamura T, Davis KR, Lyons SL, Adams RD. Bragg-peak proton-beam therapy for arteriovenous malformations of the brain. *N Engl J Med* 1983;309:269–274.
12. Kjellberg RN, Kliman B. Bragg peak proton treatment for pituitary-related conditions. *Proc R Soc Med* 1974;67:32–33.
13. Kjellberg RN, Koehler AM, Preston WM, Sweet WH. Stereotaxic instrument for use with the bragg peak of a proton beam. *Confin Neurol* 1962;22:183–189.
14. Kjellberg RN, Shintani A, Frantz AG, Kliman B. Proton-beam therapy in acromegaly. *N Engl J Med* 1968;278:689–695.
15. Koehler AM, Dickinson JG, Preston WM. The range of protons in human skullbone. *Radiat Res* 1965;26:334–342.
16. Koehler AM, Preston WM. Protons in radiation therapy. Comparative dose distributions for protons, photons, and electrons. *Radiology* 1972;104:191–195.
17. Koehler AM, s Full N, Koehler AM. Dosimetry of proton beams using small silicon diodes. *Radiat Res Suppl 7:53–63, 1967 Radiat Res* 1967;7:53–63.
18. Aziz S, Taylor A, McConnachie A, Kacperek A, Kemp E. Proton beam radiotherapy in the management of uveal melanoma: Clinical experience in scotland. *Clin Ophthalmol* 2009;3:49–55.
19. Dendale R, Lumbroso-Le Rouic L, Noel G, Feuvret L, Levy C, Delacroix S, Meyer A, Nauraye C, Mazal A, Mammar H, Garcia P, D'Hermies F, Frau E, Plancher C, Asselain B, Schlienger P, Mazeron JJ, Desjardins L. Proton beam radiotherapy for uveal melanoma: Results of Curie Institut-Orsay proton therapy center (ICPO). *Int J Radiat Oncol Biol Phys* 2006;65:780–787.
20. Desjardins L, Lumbroso L, Levy C, Mazal A, Delacroix S, Rosenwald JC, Dendale R, Plancher C, Asselain B. Treatment of uveal melanoma with iodine 125 plaques or proton beam therapy: Indications and comparison of local recurrence rates. *J Fr Ophtalmol* 2003;26:269–276.
21. Gragoudas ES. The Bragg peak of proton beams for treatment of uveal melanoma. *Int Ophthalmol Clin* 1980;20:123–133.
22. Kincaid MC, Folberg R, Torczynski E, Zakov ZN, Shore JW, Liu SJ, Planchard TA, Weingeist TA. Complications after proton beam therapy for uveal malignant melanoma. A clinical and histopathologic study of five cases. *Ophthalmology* 1988;95:982-991.
23. Koch N, Newhauser WD, Titt U, Gombos D, Coombes K, Starkschall G. Monte carlo calculations and measurements of absorbed dose per monitor unit for the treatment of uveal melanoma with proton therapy. *Phys Med Biol* 2008;53:1581-1594.
24. Lawton AW. Proton beam therapy for uveal melanoma. *Ophthalmology* 1989;96:138–139.
25. Mosci C, Mosci S, Barla A, Squarcia S, Chauvel P, Iborra N. Proton beam radiotherapy of uveal melanoma: Italian patients treated in Nice, France. *Eur J Ophthalmol* 2009;19:654–660.
26. Naeser P, Blomquist E, Montelius A, Thoumas KA. Proton irradiation of malignant uveal melanoma. A five year follow-up of patients treated in Uppsala, Sweden. *Ups J Med Sci* 1998;103:203–211.

27. Petrovic A, Bergin C, Schalenbourg A, Goitein G, Zografos L. Proton therapy for uveal melanoma in 43 juvenile patients: Long-term results. *Ophthalmology*;121:898–904.
28. Spatola C, Privitera G, Raffaele L, Salamone V, Cuttone G, Cirrone P, Sabini MG, Lo Nigro S. Clinical application of proton beams in the treatment of uveal melanoma: The first therapies carried out in Italy and preliminary results (catana project). *Tumori* 2003;89:502–509.
29. Wollensak G, Zografos L, Perret C, Egger E, Fritz-Niggli H. Experimental study on the fractionation schedule for proton irradiation of uveal melanoma. *Graefes Arch Clin Exp Ophthalmol* 1990;228:562–568.
30. Particle Therapy Co-operative Group. PTCOG patient statistics. http://www.ptcog.ch/.
31. Kole TP, Nichols RC, Lei S, Wu B, Huh SN, Morris CG, Lee S, Tong M, Mendenhall NP, Dritschilo A, Collins SP. A dosimetric comparison of ultra-hypofractionated passively scattered proton radiotherapy and stereotactic body radiotherapy (SBRT) in the definitive treatment of localized prostate cancer. *Acta Oncol*:1–7.
32. Westover KD, Seco J, Adams JA, Lanuti M, Choi NC, Engelsman M, Willers H. Proton SBRT for medically inoperable stage I NSCLC. *J Thorac Oncol*;7:1021–1025.
33. Bush DA, Cheek G, Zaheer S, Wallen J, Mirshahidi H, Katerelos A, Grove R, Slater JD. High-dose hypofractionated proton beam radiation therapy is safe and effective for central and peripheral early-stage non-small cell lung cancer: Results of a 12-year experience at Loma Linda University Medical Center. *Int J Radiat Oncol Biol Phys*;86:964–968.
34. Bush DA, Slater JD, Shin BB, Cheek G, Miller DW, Slater JM. Hypofractionated proton beam radiotherapy for stage I lung cancer. *Chest* 2004;126:1198–1203.
35. Do SY, Bush DA, Slater JD. Comorbidity-adjusted survival in early stage lung cancer patients treated with hypofractionated proton therapy. *J Oncol*;2010:251208.
36. Fukumitsu N, Sugahara S, Nakayama H, Fukuda K, Mizumoto M, Abei M, Shoda J, Thono E, Tsuboi K, Tokuuye K. A prospective study of hypofractionated proton beam therapy for patients with hepatocellular carcinoma. *Int J Radiat Oncol Biol Phys* 2009;74:831–836.
37. Gomez DR, Gillin M, Liao Z, Wei C, Lin SH, Swanick C, Alvarado T, Komaki R, Cox JD, Chang JY. Phase 1 study of dose escalation in hypofractionated proton beam therapy for non-small cell lung cancer. *Int J Radiat Oncol Biol Phys*;86:665–670.
38. Hata M, Tokuuye K, Kagei K, Sugahara S, Nakayama H, Fukumitsu N, Hashimoto T, Mizumoto M, Ohara K, Akine Y. Hypofractionated high-dose proton beam therapy for stage I non-small-cell lung cancer: Preliminary results of a phase I/II clinical study. *Int J Radiat Oncol Biol Phys* 2007;68:786–793.
39. Kanemoto A, Mizumoto M, Okumura T, Takahashi H, Hashimoto T, Oshiro Y, Fukumitsu N, Moritake T, Tsuboi K, Sakae T, Sakurai H. Dose-volume histogram analysis for risk factors of radiation-induced rib fracture after hypofractionated proton beam therapy for hepatocellular carcinoma. *Acta Oncol*;52:538–544.
40. Kil WJ, Nichols RC, Jr., Hoppe BS, Morris CG, Marcus RB, Jr., Mendenhall W, Mendenhall NP, Li Z, Costa JA, Williams CR, Henderson RH. Hypofractionated passively scattered proton radiotherapy for low- and intermediate-risk prostate cancer is not associated with post-treatment testosterone suppression. *Acta Oncol*;52:492–497.
41. Kim YJ, Cho KH, Lim YK, Park J, Kim JY, Shin KH, Kim TH, Moon SH, Lee SH, Yoo H. The volumetric change and dose-response relationship following hypofractionated proton therapy for chordomas. *Acta Oncol*;53:563–568.
42. Kozak KR, Kachnic LA, Adams J, Crowley EM, Alexander BM, Mamon HJ, Fernandez-Del Castillo C, Ryan DP, DeLaney TF, Hong TS. Dosimetric feasibility of hypofractionated proton radiotherapy for neoadjuvant pancreatic cancer treatment. *Int J Radiat Oncol Biol Phys* 2007;68:1557–1566.
43. Daartz J, Engelsman M, Paganetti H, Bussiere MR. Field size dependence of the output factor in passively scattered proton therapy: Influence of range, modulation, air gap, and machine settings. *Med Phys* 2009;36:3205–3210.
44. Grassberger C, Daartz J, Dowdell S, Ruggieri T, Sharp G, Paganetti H. Quantification of proton dose calculation accuracy in the lung. *Int J Radiat Oncol Biol Phys*;89:424–430.
45. Bednarz B, Daartz J, Paganetti H. Dosimetric accuracy of planning and delivering small proton therapy fields. *Phys Med Biol*;55:7425–7438.
46. Pedroni E, Scheib S, Bohringer T, Coray A, Grossmann M, Lin S, Lomax A. Experimental characterization and physical modelling of the dose distribution of scanned proton pencil beams. *Phys Med Biol* 2005;50:541–561.

47. Gottschalk B, Cascio EW, Daartz J, Wagner MS. Nuclear halo of a 177 MeV proton beam in water: Theory, measurement and parameterization. 2014; arXiv:1409.1938. Cornell University Library, http://arxiv.org/abs/1409.1938.
48. Fidanzio A, Azario L, De Angelis C, Pacilio M, Onori S, Kacperek A, Piermattei A. A correction method for diamond detector signal dependence with proton energy. *Med Phys* 2002;29:669–675.
49. Clasie B, Depauw N, Fransen M, Goma C, Panahandeh HR, Seco J, Flanz JB, Kooy HM. Golden beam data for proton pencil-beam scanning. *Phys Med Biol*;57:1147–1158.
50. Fontenot JD, Newhauser WD, Bloch C, White RA, Titt U, Starkschall G. Determination of output factors for small proton therapy fields. *Med Phys* 2007;34:489–498.
51. Grassberger C, Trofimov A, Lomax A, Paganetti H. Variations in linear energy transfer within clinical proton therapy fields and the potential for biological treatment planning. *Int J Radiat Oncol Biol Phys*;80:1559–1566.
52. Hong L, Goitein M, Bucciolini M, Comiskey R, Gottschalk B, Rosenthal S, Serago C, Urie M. A pencil beam algorithm for proton dose calculations. *Phys Med Biol* 1996;41:1305–1330.
53. Dowdell S, Grassberger C, Sharp GC, Paganetti H. Interplay effects in proton scanning for lung: A 4D Monte Carlo study assessing the impact of tumor and beam delivery parameters. *Phys Med Biol*;58:4137–4156.
54. Winey B, Daartz J, Dankers F, Bussiere M. Immobilization precision of a modified GTC frame. *J Appl Clin Med Phys*;13:3690.
55. Engelsman M, Rosenthal SJ, Michaud SL, Adams JA, Schneider RJ, Bradley SG, Flanz JB, Kooy HM. Intra- and interfractional patient motion for a variety of immobilization devices. *Med Phys* 2005;32:3468–3474.
56. Paganetti H. Range uncertainties in proton therapy and the role of Monte Carlo simulations. *Phys Med Biol*;57:R99–117.
57. Lutz W, Winston KR, Maleki N. A system for stereotactic radiosurgery with a linear accelerator. *Int J Radiat Oncol Biol Phys* 1988;14:373–381.
58. Winston KR, Lutz W. Linear accelerator as a neurosurgical tool for stereotactic radiosurgery. *Neurosurgery* 1988;22:454–464.

Chapter 28

In Vivo Dosimetry for Proton Therapy

Narayan Sahoo, Ph.D. [1], Archana Singh Gautam, M.Sc. [2], Falk Poenisch, Ph.D. [3], X. Ronald Zhu, Ph.D. [1], Heng Li, Ph.D. [3], Xiaodong Zhang, Ph.D. [4], Richard Wu, M.S. [2], Sam Beddar, Ph.D. [1], and Michael T. Gillin, Ph.D.[1]

[1]Professor, Department of Radiation Physics,
UT MD Anderson Cancer Center,
Houston, TX

[2]Senior Medical Physicist, Department of Radiation Physics,
UT MD Anderson Cancer Center,
Houston, TX

[3]Assistant Professor, Department of Radiation Physics,
UT MD Anderson Cancer Center,
Houston, TX

[4]Associate Professor Department of Radiation Physics,
UT MD Anderson Cancer Center,
Houston, TX

28.1 Introduction

In vivo dosimetry (IVD) involves measurement of the dose or other dosimetric quantities while the patient receives the radiation therapy. This can be accomplished by various means. One can place a suitable dosimeter on or in the patient to measure the delivered dose to the point of interest where the dosimeter is placed. The dose can also be determined by using external dosimeters and using their response to calculate the dose using a suitable formalism. In either case, the goal of IVD is to verify that delivered dose at the measurement location conforms to the planned dose. IVD remains an important component of the quality assurance program in radiotherapy to eliminate possible dose delivery errors that may affect the clinical outcome or safe delivery of the planned course of the radiation therapy within the accepted tolerance level after various uncertainties are taken into consideration.

Although the goal is well intended, the IVD is not widely used in radiation therapy clinics, and its usefulness in preventing errors is still being debated [1]. Even though the availability of the online record and verify system as an integral part of the delivery system has reduced the possibility of the use of wrong radiation fields and beam parameters during the course of radiation therapy to a very low level, these systems cannot always detect the delivery system malfunctioning or use of the wrong devices. Additionally, the IVD can verify the calculated dose distribution in the treatment plan in the presence of complex inhomogeneities, which may have large uncertainties due to the limitation of dose calculation algorithms in the treatment planning system. Thus, the importance of IVD for dosimetric quality assurance is strongly advocated [2].

In situations where a critical organ or a life-saving device like a pacemaker or implanted cardiovascular defibrillator (ICD) is in the close proximity of the radiation field, it is often necessary to both determine and document the actual dose delivered to these organs or devices by IVD. The IVD for external beam dosimetry using photons and electrons are much more widely used and well documented in the literature [2,3]. Even though almost all of the dosimeters used in photon and electron beam IVD can be used for the same purposes in proton therapy with proper characterization of their response to proton beam irradiation, the application of IVD in proton therapy remains limited in clinical practice. The impediment of using IVD for proton therapy may have been caused by the more important concern about the *in vivo* range verification during proton therapy instead of the dose verification like the photon and electron beam therapy. Since the *in vivo* range verification remains a formidable challenge at this time, the use of dosimeters to verify entrance dose, out-of-field dose, and dose in the accessible regions of the body may still be useful for quality assurance and patient safety.

The methodology for *in vivo* entrance dose measurement using solid state detectors for external beam radiation of photons and electrons is well established and is also applicable for proton therapy. The main goal of such measurements is to detect possible changes in beam output outside the tolerance level, incorrect use of treatment field devices, and patient setup distances. Additionally, IVD for proton therapy is desirable for the following reasons. The entrance dose for modulated Bragg peak or spread-out Bragg peak (SOBP) is much higher than with photons, and this often leads to higher skin dose, leading to undesirable skin reaction. The skin dose calculated in the treatment planning system may not be accurate due to limitations of the analytical dose calculation models. Additionally, any change in delivery system functionality and the possible use of wrong devices during the course of treatment can lead to differences in the given dose from the planned dose. Thus, the verification of the entrance dose may provide the required information about the delivered entrance dose during the entire course, which may be useful in interpreting any adverse skin reaction observed in the clinic.

The entrance dose measurement can also be used to check the constancy of the SOBP delivery, as it is affected by the malfunctioning of the responsible components. The dose distribution in intensity-modulated proton therapy is greatly affected by small changes in the delivered spot pattern. The measurement of constancy of the entrance dose may provide additional quality assurance checks for the delivered dose. The entrance dose is usually high for proton therapy of shallow targets close to the skin. In this situation, the monitoring of the entrance dose with IVD will add to confidence in the accuracy of the delivered dose. It is also possible to place a dosimeter in body cavities like the rectum and nasal cavity to monitor the dose there. Additionally, range of the proton beam can also be monitored by placing suitable dosimeters at accessible locations in the body or monitoring the secondary detectable radiation from nuclear reaction products in the body. Unlike with photon beam, the exit dose cannot be directly monitored in proton beams used to treat deep-seated targets because all the protons in the beam stop inside the body.

The goal of this chapter is to review the practical aspects of IVD for proton therapy using suitable dosimeters and devices both for dose and range verification. Various dosimeters used for proton beam dosimetry were reviewed by Karger et al. [4] and are also described in Chapter 11. The use of some of these detectors for IVD will be reviewed in this chapter.

The IVD procedures can be divided into two groups. One group is the placement of dosimeters on or in the patient, and the other group is the use external devices to monitor the dose or range of the proton beam during patient irradiation. The first method includes the placement of detectors like thermoluminescent dosimeters (TLDs), diodes, optically stimulated luminescent dosimeters (OSLDs), radiophoto luminescent dosimeters (RPLDs), metal oxide field effect transistor (MOSFET) detectors, plastic scintillation detectors (PSDs), and films on or in the patient to monitor dose or range. The second method includes the placement of suitable detectors outside the patient or the use of imaging with positron emission tomography (PET) scanner, magnetic resonance imaging (MRI), and prompt gamma cameras to verify the dose or range of protons. The electronic portal imaging detector (EPID) used for monitoring the exit dose for photon beams cannot be used for proton therapy for the obvious reason of the lack of exist dose from proton beams. Both the IVD methods will be discussed in the following sections.

28.2 Detectors Suitable for In-contact *In Vivo* Dosimetry in Proton Beams

Diodes, MOSFETs, and PSDs have the advantage of providing real-time dose values when irradiated, whereas TLDs, OSLDs, RPLDs, and film require off-line analysis before dose values can be determined. However, all of these dosimeters are not absolute dosimeters and need calibration of their response to a given known dose. Additionally, the responses of these detectors change with dose given outside the calibration conditions and require correction factors to account for changed conditions. The changed conditions from the calibration that affect their response depend on the energy, field size, dose rate, geometrical conditions of irradiation (like SSD and angle), and environmental conditions (like temperature and pressure). Once the correction factors are properly quantified and the level of accuracy that can be achieved is determined, it becomes easy to use them in clinic for IVD.

The procedures for using the diode, MOSFET, PSD, TLD, OSLD, and film for radiation dosimetry in general, and for IVD in particular, are well established and described in the literature [2–6]. Therefore, these details will not be repeated in this chapter. Only the summary of their general dosimetry properties and proton beam-specific features that require special consideration for using them for IVD will be described.

28.2.1 Diodes

The diode dosimetry system consists of a suitable diode and an electrometer to measure the generated current during the irradiation. No external bias is needed for the flow of the generated charge because it takes place due the presence of an electric field created by the depletion layer in the diode itself. The details of the dosimetric characteristics of diodes for IVD in photon beams are described in detail in the AAPM Task Group Report 62 [3] and in an ESTRO publication [5].

The general characteristics of diodes for clinical proton beam dosimetry were studied by Grusell and Medin [7]. From the two types of diodes used for radiation dosimetry, the p-type diodes are found to be more suitable for IVD compared to the n-type diodes. The p-type diodes suffer less sensitivity loss with accumulated dose, and they are less dose rate dependent compared to n-type diodes. Highly doped Hi-p-type diodes are found to be most suitable for proton beam dosimetry. Some practical considerations for using diodes for IVD are:

- The leakage and precision should be small, within 1%.
- The diode system response is known to be linear in the dose range of interest for IVD. If found to be not so for a specific diode, a correction has to be applied if used outside the linear range.
- Pre-irradiated diodes are preferred because the decrease of sensitivity with accumulated dose is less compared to un-irradiated diodes.
- Proper account has to be taken for dose-rate dependence. It should be calibrated in the dose-rate range in which the patient will be irradiated.
- Change in the sensitivity of the diode response to temperature change should be studied and periodically checked for each diode. A temperature correction factor should be applied when contact temperature is found to be substantially different from the calibration conditions.
- Diodes have strong energy and angular dependence, so these should be accounted for when using them for IVD.

- Additionally, the diode response is known [4,7] to have linear energy transfer (LET) dependence in proton beam. However, when used for entrance dose measurement, the LET dependence may not be important if calibrated against entrance dose measured with an ionization chamber. Since most of the diodes have inherent buildup depth, the calibration factors should be determined by comparing diode response with the ionization chamber dose at the buildup depth. If the true entrance dose is desired from IVD measurement with diodes with build-up cap, it can be determined using the PDD curve measured with an ionization chamber. For the diodes to be used to measure dose at a location other than the entrance, then the calibration of diode response should be made by comparing with the dose at equivalent depth measured by an ionization chamber to minimize the LET dependence.

Diodes are more susceptible to radiation damage in proton therapy beams because of the neutron damage and the possibility of nuclear interaction. Thus, the constancy of the diode response under calibration conditions needs to be checked frequently.

28.2.2 Metal Oxide Semiconductor Field Effect Transistor (MOSFET) Detectors

The metal oxide semiconductor field effect transistor (MOSFET) is one of the cornerstones of modern semiconductor technology. The general structure is shown in Figure 28–1 and consists of a lightly doped *p*-type body or substrate (B), into which two regions—the source (S) and the drain (D)—both made of heavily doped *n*-type semiconductors, are embedded. Ionizing radiation generates electron-hole pairs in the insulating layer between the gate and the substrate. The holes drift toward the substrate under an appropriate bias voltage and are semi-permanently trapped at the interface, resulting in a shift in the gate voltage required for source-drain conductivity that is proportional to the radiation dose. Following exposure, the gate threshold voltage is measured by applying a constant source-drain current, and the cumulative dose is obtained using suitable calibration factors. MOSFETs can operate in active or passive mode.

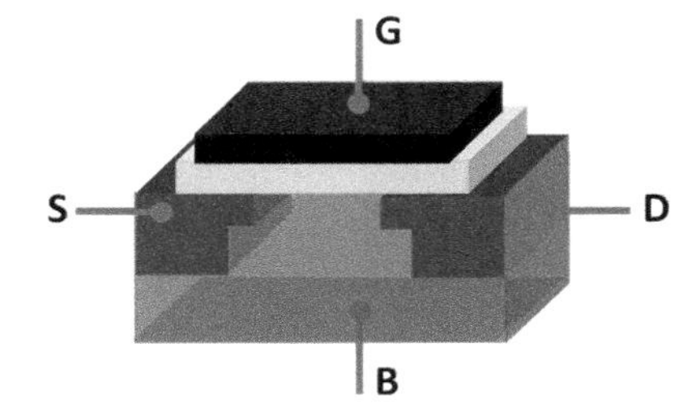

Figure 28–1 Basic MOSFET structure showing gate (G), body or substrate (B), source (S), and drain (D) terminals. The gate is separated from the body by an insulating layer (white). (From [8].)

The small size, instant readout, permanent storage of dose, and ease of use make the MOSFET a good choice as an *in vivo* dosimeter in certain radiotherapy treatments. It measures dose in real time. It can be used for online dosimetry by sampling in pre-defined time intervals [9]. There are different types of MOSFET detectors available:

- Single bias, single MOSFET.
- Dual-bias, dual-MOSFET. In this configuration, two MOSFETs on same silicon chip operate at two different gate biases. They have better sensitivity, reproducibility, and stability than a single MOSFET, and they have minimal temperature effects.
- Unbiased single MOSFET. These have temperature dependence and instability in response, and they have shorter linearity range than biased MOSFETs. They are frequently used as disposable detectors.

These are the MOSFET system available in the market:

- Mobile MOSFET system by BEST Medical, Ottawa, Canada (Thomson-Nielsen).
- One Dose and DVS (Dose Verification System) by Sicel Technologies, Inc. (Morrisville, NC, USA)

While other MOSFET detectors have been used clinically for IVD by placing them temporarily either on the surface or in a body cavity of the patient during treatment, the DVS detector is designed to be permanently implanted to measure dose over the entire course of treatment. The major advantages of this detector is its small physical size, excellent spatial resolution, and minimal beam perturbation. However, the drawbacks are finite lifetime (~100 Gy), energy dependence, and temperature dependence. The dosimeter is easy to use and wireless.

OneDose is a single-use MOSFET detector verification system. It is now used in radiotherapy treatment applications to measure the entrance and exit dose during the treatment. OneDose has a small detector size (300 μm x 50 μm x 0.4 μm, almost like a point detector). OneDose is factory calibrated, wireless, easy to use, and has relatively flexible design. The detector is attached with epoxy to a thin copper backing of dimensions 5.5 mm x 36 mm. The accuracy of the detectors, as specified by the manufacturer, is ±1 cGy for a dose of less than 20 cGy and ±5% for a dose of 20 to 500 cGy (Sicel Technologies, Inc). The OneDose MOSFET detector system is shipped in packages containing 32 detectors. Each detector is factory calibrated with a ^{60}Co beam with full build-up conditions. Instantaneous dose readout is obtained with the reader provided.

The dosimetric characterization of MOSFETs should be performed before they are used for IVD in patients. Correction factors need to be determined for the following conditions during IVD, which may be different from the calibration conditions:

- environment—temperature (no pressure correction)
- energy dependence
- accumulated dose

- dose rate
- field size
- SSD
- LET dependence for proton beams
- angular dependence

The calibration procedure depends on the intended use of the MOSFET. The calibration should be performed at body temperature (37 °C). The characteristics and commissioning of MOSFETs as *in vivo* dosimeters for high-energy photon external beam radiation therapy has been reported by Gopiraj et al. [10] and can be useful to develop the calibration procedure for proton beam.

Results of several studies [9] have reported the use of MOSFETs in x-ray and electron beams. It is possible to measure surface doses accurately using MOSFETs for radiotherapy with x-rays and electron beams [9].

Dosimetric characteristics of the OneDose system have been studied in proton beam by Cheng et al. [11] to determine reproducibility (consistency), the linearity with dose and dose rate, energy dependence, directional dependence, LET dependence, and fading (delay readout with time) of the dosimeter response. Their study shows that OneDose detectors exhibit pronounced energy dependence at depth and a large variation in dose response with LET. Results of the study by Cheng et al. [11] also show that detector response remains relatively constant (within 3%) at surface over a wide range of energies. There is also a slight angular dependence (about 2%) up to a 60° angle of incidence and small intra-batch variation (1%). Inter-batch variation is within 3%. OneDose have small dependency on energy, but most *in vivo* dosimetry involves entrance dose measurement. Thus, the small energy dependence of OneDose at the surface appears to be well suited for surface dose measurement.

Due to its copper backing, detector orientation should be such that the radiation field should not cover a large amount of the copper backing material. The angular dependence issue can be handled by establishing a correction factor for the variation of response by means of orientation of the detector. Since OneDose is a single-use detector, intra-batch consistency must be checked and verified before the remaining detectors from the same batch can be used for *in vivo* dosimetry. As it has small intra-batch variation, calibrating only a few detectors from each batch will be sufficient. To reduce the dosimetric uncertainty due to inter-batch variation, it is advisable that the detectors from the same batch should be used for the same application. Results of Cheng et al.'s study [11] showed that OneDose can provide an opportunity to measure *in vivo* dose in proton beam within acceptable clinical criterion of ±6.5%.

Kohno et al. [12,13] studied the use of the commercially available TN-502RD MOSFET detector with oxide thicknesses of 0.5 μm (Best Medical Canada, Ottawa, Canada) for proton dose measurement. They also studied a new MOSFET detector with a thickness of 0.25 μm (TN-252RD) with improved characterization of the MOSFET response in proton beams. They applied LET correction to the MOSFET response, comparing the MOSFET measured dose with that from Monte Carlo simulation, and they showed that MOSFET detectors can be used for accurate dose measurements in inhomogeneous media. Like diodes, the LET dependence for surface dose measurement by MOSFETs can be avoided by calibrating the detectors by comparing the surface dose measured with an ion chamber under similar conditions.

28.2.3 Plastic Scintillation Detectors (PSDs)

PSDs are relatively new dosimeters with great potential for IVD [2]. PSD consists of a plastic scintillating material of organic scintillating molecules in a polymerized solvent that emits light when exposed to ionizing radiation. The amount of emitted light is proportional to dose given to its sensitive volume. The PSDs can be used for real-time dose measurement as the light is emitted within nanoseconds of irradiation. The emitted light can be converted to an electric charge using a photon diode, which receives the emitted light through an optic fiber guide. The electric signal is then measured using an electrometer. The details of the construction of PSDs and their characterization for photon and proton beam dosimetry are described in papers by Archambault et al. [14] and Wang et al. [15] respectively. Some of the advantages of PSDs are 1) they can be reused, 2) they are water equivalent, 3) they have an energy-independent response to dose, 4) they have a linear response to dose, 5) they are resistant to radiation damage, and 6) they are small (1 mm diameter) leading to excellent spatial resolution.

Shortcomings of the PSDs that may require some special consideration are 1) the contributions of the radiation-induced light from Cerenkov emission and fluorescence arising in the optical fiber, 2) dependence of detector response on LET of the proton beam, and 3) the effect of ionization quenching on light emission. A recent study by Wootten et al. [16] has shown that these concerns can be addressed, and PSDs can be used for IVD of entrance dose for patients receiving proton therapy. It has been shown [16] that a PSD calibrated against the ionization chamber measured dose for a ^{60}Co beam remains valid for other photon beams, but when compared against entrance dose measured by the ionization chamber in proton beam, the under-response of PSDs can be between 6% to 10%, and higher for high-energy beams. The under-response does not depend on the SOBP width, and the contribution of the radiation-induced light is negligible. The relative lateral dose profiles agreed very well with that measured with EBT film. Thus, with the use of the proper quenching correction factor for proton beam to be used for patient entrance dose verification, PSDs can be used for accurate measurement of entrance dose in proton therapy.

An additional LET-dependent correction factor needs to be applied when PSDs will be used for IVD at some depth in the patient. A Monte Carlo calculation-based LET correction factor may have to be applied to accurately measure the proton beam dose in inhomogeneous media in the patient, and this remains to be investigated.

28.2.4 Thermo Luminescence Dosimeters (TLDs)

Thermo luminescence (TL) dosimetry is based on the physical process of formation of metastable electronic states in the TL material upon absorption of energy which, upon heating, is re-emitted as electromagnetic radiation, mainly in the visible wavelength. A photomultiplier tube detects the optical photons, and the light output is measured using a TL reader. The amount of emitted light depends on the amount of absorbed dose. The TL material can be restored with appropriate heating, which is called annealing, and can be reused.

The methodology for TL dosimetry and the use of TLDs for IVD in photon and electron beams are well established [2,5,17]. TLDs have also been found suitable for proton beam dosimetry [18,19], and they are routinely used by the Imaging and Radiation Oncology Core (IROC) Houston QA Center (formerly known a the Radiological Physics Center) for proton beam calibration check with similar level of accuracy as photon and electron beams. It has been also used for patient dose verification in proton therapy.

There are many TLD materials, and they are available in many forms, like chips, rods, ribbons, and powder. Rods and powder of TLD 100 made of LIF doped with traces of magnesium and titanium are most commonly used. Dose measurement within 2% to 3% of that measured by an ionization chamber is achievable with proper annealing, calibration, and sensitivity studies.

Advantages of TLD are 1) small size (1 mm x 1 mm x 1 mm), 2) wide linear range, 3) independence of response to clinically used dose rate, 4) no angular dependence, 5) reusability, 6) they require no power or cable during irradiation, and 7) response is not affected by electromagnetic interference and environmental factors like pressure and humidity.

Disadvantages of TLD are 1) instability in sensitivity, 2) susceptibility to surface contamination, and 3) susceptibility to structural damages. However, these issues can be overcome with careful handling and individual calibration.

TLDs exhibit supra-linearity in high-dose regions, and their response is energy dependent. TL efficiency also deceases with LET [18]. Thus, better precision can be achieved with calibration in the similar condition, same energy, dose, and LET condition in which the TLD will be irradiated. Additionally, because the density of the TLDs is higher than water, some form of correction is needed to convert the dose to water. The TLD 100 powder was used for patient plan dose verification in solid phantoms by Zullo et al. [19]. They showed that proton therapy dose measurements within 5% accuracy are possible with TLD 100.

28.2.5 Optically Stimulated Luminescent Dosimeters (OSLDs)

OSLDs have many features similar to TLDs, with light being used in place of heat for stimulation. The laser-stimulated emission is proportional to the absorbed dose. Two major advantages of OSLDs compared to TLDs are their greater sensitivity, faster post-irradiation readout (about 10 minutes), and the possibility of repeated readouts [20,21]. One major disadvantage of OSLDs is their sensitivity to light. However, this problem can be overcome by enclosing detectors in specially designed light-tight packages. Fading with time and depletion with repeated readings are found to be minimal, and this can be easily accounted for with appropriate characterization of this effect.

Commercially available carbon-doped aluminum oxide (Al_2O_3:C) OSLDs and Microstar OSL readers are used to characterize these dosimeters for therapeutic photon, electron, proton, and heavy charged particles [21]. For proton beam dosimetry, OSLD response was found to be energy independent, but it has LET dependence [22]. A reduction in the response of the OSLD at the high-LET region of proton beam dose distribution was observed. The LET dependence of the response was also found to depend on the emission band used to quantify the output signal in a study by Granville et al. [22]. It was shown that signals containing the blue band only are independent of LET, whereas UV band signals and mixed-band signals are LET dependent. One advantage of this finding is that one can use the ratio of UV/blue emission signal intensities to determine the LET of proton beams. Various protocols for readout have been developed, and the details of the principles and use of OSLDs for radiation therapy are described by Cygler and Yukihara [20].

Results of a study by Kerns et al. [23] show some amount of supralinearity of the response of OSLDs with proton dose, about 1% at 200 cGy and 5% at 1000 cGy. By using an average beam quality factor for proton beams with respect to its response to ^{60}Co beam and appropriate correction factors for fading, depletion, dose linearity, and individual detector sensitivity, it is possible [23] to measure dose in the center of SOBP within 1.2% of that measured with an ion chamber. It has been shown [23] that dose measurements are within 1% or better when appropriate methodology and readout equipment are used. The IROC, Houston QA center has been using these dosimeters for remote audit with phantom irradiation. Thus, OSLDs are suitable alternatives to TLDs as dosimeters for IVD for proton therapy.

28.2.6 Radiophotoluminescent Glass Dosimeters

GD-301 radiophotoluminescent glass dosimeters (AGC Techno Glass Corp, Shizuoka, Japan) have been evaluated for their suitability for proton beam dosimetry [24] and can be used for IVD in proton therapy. These silver-activated phosphate glass dosimeters are available as 1.5 mm diameter and 8.5 mm long cylinders, and they have effective atomic numbers of 12.039 and a density of 2.61 g/cm^3. The effective readout size is 1 mm in diameter and 0.6 mm in depth.

Stable radiophotoluminescence (RPL) centers are created in these dosimeters upon exposure to ionizing radiation. When the RPL centers are excited by a pulsed ultraviolet (UV) laser beam, orange luminescent photons are emitted, which can be read by an appropriate photodetection system. Commercial automatic readers (FGD-1000 from Asahi Techno Glass Corp., Japan) are available for use with these dosimeters. The amount of luminescent photons emitted by the RPL centers is proportional to the radiation dose given to the glass dosimeter. As the RPL centers are not removed after excitation by the UV laser, the readout can be repeated. RPL centers can only be removed by annealing them at high temperatures (400 °C for 1 h).

Characterization of GD-301 glass dosimeters for proton beam dosimetry was carried by Rah et al. [24]. Results of their feasibility study show that response of these dosimeters is linear with dose, and the reproducibility is within 1.5%. The dependence of glass dosimeter response on dose rate, energy, and fading was found to be about 1.5%, 3%, and 2% respectively. Angular dependence of glass dosimeters remains to be studied. The proton beam lateral profiles and depth–dose distribution measured by the glass dosimeters in modulated proton beam agreed well with the ionization chamber measurements, with differences of 4% in flatness, 0.3 mm in lateral penumbra width, 2.5% difference in peak to entrance dose ratio, 3% in PDD value at the end of plateau, and 1.9% difference in beam range. It was concluded [24] that the glass dosimeters have better dosimetric characteristics compared to TLD and MOSFET detectors and are suitable for use in IVD in clinical proton beams.

In a phantom study [24], the measured dose by the glass dosimeter—both at the surface and at a depth in a modulated beam—was found to be within 5% of the Eclipse (Varian Medical Systems, Palo Alto, CA) treatment planning system (TPS) when the dosimeter was calibrated at the center of the SOBP. The agreement between the measured and TPS-calculated dose was found to be better for glass dosimeters compared to TLDs. Overall uncertainty in proton beam dosimetry with glass dosimeters was estimated to be 4.3%.

28.2.7 Films

Films are used for proton beam dosimetry [4] and can, in principle, be used to measure 2D dose distribution in patients, both at the beam entrance and at a location in the body where they can be placed, as shown in some recent studies [25–27]. The self-developing EBT film is more suitable for IVD application because it does not need any light-tight cover, and because it can be cut into small pieces to be placed on or in patients.

The principles and procedures of film dosimetry are well established and are described in literature [28,29]. Because of strong dependence of film response to LET and beam quality of the proton beam, film is mostly limited to measuring relative dose distribution in the conditions of constant LET. It may be possible to achieve an accuracy of about 10% with film dosimetry for proton beams. IROC Houston QA Center uses the EBT film for dose verification in mail-in phantoms for site-specific credentialing of proton therapy centers. Thus, it is possible to use films for IVD application of 2D dose verification by using a suitable calibration and characterization procedure.

Like other dosimeters, the correction factors for various conditions that affect the film response to dose and LET need to be determined. The achievable accuracy of dose verification with film, as shown by various investigators [30,31], needs to be established before they are used for IVD in patients.

28.3 Dose Verification with Dosimeters Placed on Skin or Body Cavities

IVD in the current practice is predominantly performed by placing a suitable detector on the patient's skin at the entrance of the beam. This is useful if the entrance dose is monitored as part of the treatment program. Otherwise, its use can be considered as a quality assurance practice only. It provides a constancy check of the beam parameters of the treatment field. The entrance dose measurement provides very limited information about the target dose. It is possible to infer the target dose from the known ratio of the entrance and center of SOBP doses from a beam data table or treatment plan. However, no information about the possible changes in the dose value at the target due to anatomical changes or the presence of inhomogeneities can be obtained from the entrance dose measurements.

It is possible to implant dosimeters in accessible body cavities that are in the radiation field or close to the radiation field where dose measurement is desired. For example, TLDs, OSLDs, glass dosimeters, diodes, MOSFETs, and PSDs can be placed in the rectum or urethra to monitor dose in prostate cancer patients. Similarly, dosimeters can also be placed in the nasal cavity for head and cancer patients.

28.4 *In Vivo* Proton Beam Dosimetry by External Devices

28.4.1 Secondary Photon and Neutron Dose Monitoring

A novel method was recently explored to monitor the accuracy of the proton dose delivery by measuring the constancy of secondary radiation emitted during the beam delivery [32]. Secondary photon or neutron ambient dose equivalent per proton dose (H/D for photon or H*10/D for neutron) was found [32] to have a strong correlation with proton dose rate (dose/MU). It is therefore feasible to use a large ionization chamber at a fixed location in the room to measure these quantities and derive the proton beam dose rate of the delivered treatment to patient. Since dose/MU depend on the range, SOBP, and devices used in shaping the treatment field, it is possible to monitor their constancy during the entire course of treatment.

28.4.2 Dose Verification by Positron Emission Tomography (PET)

The PET method has shown big promise in dose verification for proton therapy [33]. When heavy charged particles with sufficient energy travel through tissue, inelastic interactions lead to activation of target nuclei. Positron emitters in those nuclei release positrons. Each of these positrons annihilate into a pair of gamma quanta separated by ~180 degrees. Those photons can be detected using diagnostic PET imaging devices. The resulting image is a 3D distribution of β^+ positron emitters. The measured activity distribution is correlated to the dose distribution, but not easily comparable with the planned dose distribution. Therefore, the measured activity is compared to the calculated activity, which is generally based on Monte Carlo simulations that model all known physical processes, such as fragmentation, propagation of positron emitter, annihilation, and photon interaction with tissue elements inside the patient.

The biggest challenges are the conversion of the measured activity into dose [33–35] and the lack of any activity when the proton energy falls below the threshold energy of 8 MeV for nuclear reaction toward the end of its range. The correlation between the activity and dose cannot be easily established due to a) the requirement of very high count rate and excellent signal-to-noise ratio may not be met, b) the differences in physical processes responsible for dose deposition and positron emitter production, and c) the reduction in measured activity in the patient tissue due to local perfusion.

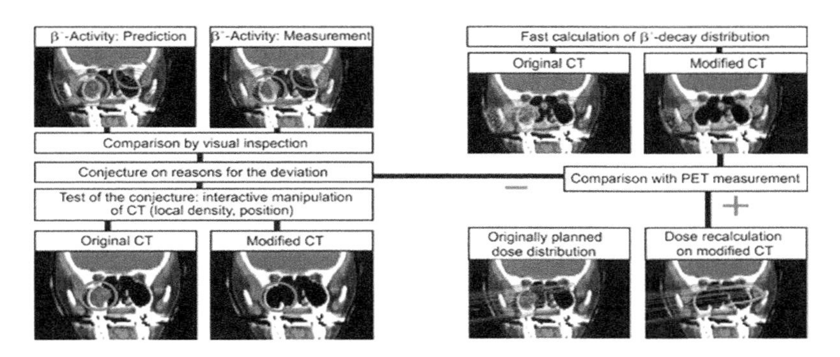

Figure 28–2 Steps for an interactive procedure used for a PET-guided quantification of local dose deviation from the planned dose. (Adapted from [34].)

As an alternative solution, an interactive procedure was developed [34] where first the measured and predicted β^+activity distributions are visually compared, as shown in Figure 28– 2 for a skull-based tumor irradiated from the right. Based on the deviation between the images, a hypothesis is made regarding the cause of the deviation—such as positioning errors or local density changes within the target volume—and the planning CT is artificially modified or translated. Consecutively, a revised β^+ activity is calculated using a fast calculation algorithm that allows interactive editing. Finally, a new dose distribution is then calculated and compared to the original calculated dose.

The technical realization of the PET method in radiotherapy can be divided into on-line, i.e., acquisition during therapeutic irradiation [36,37], or off-line after treatment, which can be further divided into in-room [35,38] or outside treatment room [39]. The latter method utilizes a commercial PET or PET/CT scanner. The advantage of the on-line solution is the higher count rate and, thus, shorter acquisition time, which reduces the effect of biological washout. Furthermore, since the measurements are done in treatment position, there is a very strong rigid physical correlation between patient position, delivered particles, and detected activity. This is similar to cone-beam CT for photon linacs. The disadvantage of any on-line solution is the limited time available in the treatment room and, thus, the potential to reduce patient treatment throughput.

28.5 *In Vivo* Range Verification in Proton Therapy

The main reason for choosing proton therapy over photon and electron beam therapy is the absence of any dose beyond the range of the proton beam. However, the range of the proton beam inside the patient cannot be determined accurately using the CT scans used for treatment planning. The estimated uncertainty [40] is about 3.5% of the range determined by the currently practiced CT number to relative stopping power conversion procedure. Because of this uncertainty, a generous margin is added to the target for planning to ensure the target coverage and pointing the beam toward an organ at risk is avoided in treatment planning. Thus, *in vivo* verification of the range of the proton beam is a very desirable goal in proton therapy.

But this goal still remains unrealized due to many challenges in measuring the range in patients. The current state of the *in vivo* range verification techniques is reviewed in a recent paper by Knopf and Lomax [41]. Brief descriptions of some of the techniques for *in vivo* range verification are given in this chapter. Like dose, range can be verified by implanted dosimeters in the patient or by using devices placed outside patient's body. The proton beam range information can be obtained by direct measurements or indirectly from the range-dependent changes that take place in the patient during or after the irradiation. It can also be measured online during the treatment or off-line after the treatment. These can be from single- or multiple-point measurements, 2D planar measurement, or 3D volumetric measurements. The following sections describe the current status of some of the techniques for *in vivo* range measurement.

28.5.1 Range Verification with Implanted Dosimeters

A novel device and procedure have been developed by Lu et al. [42] to verify the range with implanted dosimeters in the patient. They showed that by measuring the time-dependent dose rate that has a unique value at different depths with an implanted dosimeter, one can determine the residual range of the beam at the location of the dosimeter with submillimeter accuracy. Diodes were tested as possible dosimeters, and they can be attached to the rectal balloon for prostate cancer patients. A feasibility study performed by Bentefour et al. [43] in an inhomogeneous phantom provided promising results. An integrated system of hardware involving a matrix of 12 diodes, a special amplifier to capture low signals from small proton beam currents, and software has also been developed and commissioned [44] for *in vivo* dose and range verification based on this principle. However, it has not been used in patients.

Lu also proposed [45] an alternative simpler method for range verification with implanted dosimeters in which the conventional SOBP field is replaced by two complimentary fields with sloped depth–dose profile. The ratio of the dose from these two fields at any depth is very sensitive to the depth, unlike the flat SOBP field. Thus, the measurement of this ratio by an implanted dosimeter will provide the water equivalent depth or the beam range information at the location of the dosimeter. This technique can also be applied to proton fields of scanned pencil beam spots where the measured ratios of dose from individual pencil beam layers can be used. This method does not require measurement of any time-dependent parameter. Lu et al. [42] tested this method using a commercially available wireless dosimetry verification system (DVS) based on MOSFET dosimeters (Sicel Technologies, Morrisville, NC) to determine water equivalent thickness at four different depths in a water tank. They corrected the LET dependence of the MOSFET detectors using a columnar recombination model. It was shown that the WET can be determined with submillimeter accuracy. Although phantom studies have shown encouraging results, it has not been tested in actual clinical studies.

As pointed out by Knopf and Lomax [41], the effect of the distortion of the dose distribution due to the presence of inhomogeneity in patients compared to water phantom on the expected ratio of the dose from complimentary fields has not been studied. Furthermore, no information about the beam path can be obtained beyond the position of the implanted dosimeters, and it may not be possible to implant the dosimeter in many types of sites. Thus, this approach may have limited applications.

28.5.2 Range Verification with External Devices

28.5.2.1 Range Probe with Multilayer Ionization Chamber (MLIC)

The feasibility of using a multilayer ionization chamber (MLIC) as a range probe was investigated by Mumot et al. [46]. This approach is based on the principle that WET of the patient on the path of a mono-energetic proton pencil beam with sufficient range to pass through the patient can be determined from the measured range shift with an MLIC placed at the exit of beam. It was shown through Monte Carlo simulation that range resolution of 1 mm may be possible with a detector thickness of 4 mm for homogeneous regions of the patient anatomy. Since the energy of the range-probing pencil beam has to be such that the Bragg peak is formed outside the patient in the MLIC, the dose given to the patient from the plateau region is expected to be low. However, the measured WET with this method is for the entire path of the beam in the patient and does not provide any information about the range of the treatment field at any specific location of the patient, like the target or organs at risk (OAR). This method will only be useful for checking the accuracy of patient-specific CT numbers to relative stopping power conversion in the treatment planning system. Although the principle is simple and the measurements are feasible by using a commercially available MLIC, its usefulness for *in vivo* range verification remains to be tested.

28.5.2.2 Proton Radiography and Tomography

Proton radiography[47] uses the same principle as the range probe to determine the integral water equivalent path length (WEPL) on the path of the monoenergetic proton pencil beam passing through the patient. In proton radiography, a planar projection of the WEPL is created by allowing the beam to pass through different known locations in a 2D grid. With the knowledge of the entrance and exit coordinates of the pencil beam and the corresponding residual range, the proton radiography is created (see Chapter 7). Similarly, the measured integral WEPL in a 3D tomography mode can be used to reconstruct the 3D map of the relative stopping power (RSP) of the material in the path of the beam to create proton-computed tomographic (pCT) images [48]. Some details of the use of proton radiography and tomography for patient setup and range verification are described in Chapter 7. The integral range and RSP information from proton radiography and pCT respectively can potentially be used to verify the change in these values *in vivo* from their planned baseline values during the course of treatment. However, the proton radiography technology and pCT are still in research and developmental stage, and they are not available for routine clinical use.

28.5.2.3 Positron Emission Tomography (PET)

As mentioned earlier in Section 28.4.2 and in Chapter 7, PET imaging during or after the proton radiation provides the 3D distribution of positron-emitting isotope created during the radiation, which has correlation with the depth–dose distribution of the proton beam. Thus, PET imaging can, in principle, be used for *in vivo* proton beam range verification in 3D. The details of the use of PET for *in vivo* proton beam range verification are described in Chapter 7.

Published results [33] of a feasibility study for clinical use of PET for *in vivo* range verification are encouraging. Additionally, the accuracy of the range verification may be limited by the inherent large spatial resolution (4–5 mm) of the current PET imaging technology, as well as energy cutoff for the production on positron. Research and developmental work is ongoing to improve the accuracy of the range verification with PET so as to be useful for routine clinical use.

Verification of proton beam range from tissue β^+ emitter activation distribution on the beam's path alone is difficult for the following reasons.

1. Most patient tissue elements require relatively high proton beam energies to be activated (Litzenberg et al. [49]. As the proton beam loses its energy with depth, the amount activation of tissue elements gets gradually reduced at the regions of high energy loss near the distal end of the Bragg peak. Thus, the amount of β^+ emitters at the end of the proton range is comparatively small and nearly zero at 8 MeV.

2. Most of the β^+ emitters in the tissue decay relatively quickly, requiring an expensive in-beam PET scanner.

3. Biological washout of the activated tissue elements leads [33] to uncertainties in the determination of β^+ emitter distribution. As shown by Cho et al. [50], some of the above limitations can be overcome by using implantable markers that can be activated by the proton beam at a larger scale compared to the activation of the tissue elements. By measuring positron emission using PET from proton-activated material implanted in the regions of interest near the end of the range of the proton beam in the patient, proton range can be verified more accurately. It was shown by Cho et al. [50] that ^{18}O, ^{63}Cu, and ^{68}Zn are activated strongly at low proton energies, and their radioactive isotopes decay with relatively long half-lives suitable for off-line PET scanning. Bio-compatible implantable markers consisting of these isotopes were investigated [50] as potential candidates for range verification with PET scanners. These materials can also serve as radiographic fiducial markers. Results of Monte Carlo simulation and phantom measurements [50] showed that measured activities of these materials were 100 times greater than those of the β^+ emitters activated in tissue samples only in the distal dose falloff region of the proton Bragg peak. Thus, the accuracy of the *in vivo* proton range verification can be substantially improved [50] with implantable markers consisting of ^{18}O, ^{63}Cu, and ^{68}Zn.

28.5.2.4 Prompt Gamma Imaging

The prompt gamma emission due to the nuclear interaction of the protons in the tissue material has been found to have correlation with the depth–dose distribution of the proton beam [51]. Thus, quantitative monitoring of the gamma rays with suitable gamma and Compton camera systems can provide the range distribution of the proton beam. The potential of prompt gamma imaging for *in vivo* range verification in proton therapy is described in Chapter 7. Although a prototype with a slit camera has been tested successfully [52,53], prompt gamma imaging during proton therapy remains an area of active research and developmental work. It may become available in the future for routine clinical use.

28.5.2.5 Magnetic Resonance Imaging (MRI)

It has been observed that radiation-induced changes in the composition of human tissue can be visualized in MRI under certain conditions. For example, radiation-induced fatty conversion in the vertebral bone marrow has been observed [54] in MRI of the spine, although it is a late process that occurs three to four months after the completion of radiation therapy. Radiation-induced radiobiological changes have also been observed in MRI of liver tissue [55]. Therefore, it is possible in principle to use MRI for *in vivo* verification of proton dose distribution, especially the distal edge of the proton treatment field.

Gensheimer et al. [54] studied the MRI changes in 10 spine patients and established a planned dose versus signal intensity response curve for bone marrow in the lateral penumbral region in MRI scans of the sacrum. They then used this response curve to estimate the posterior–anterior proton beam range errors in the lumbar spine distal dose region, which was found to be on the average 1.9 mm more than the planned range. The inherent uncertainty in the range verification with their method is stated to be about the same as the average overshoot observed in the 10 patients. As the study was done at the end of treatment, the same methodology may not be useful for range verification MRI during the course of treatment because the radiation-induced changes may have time dependence, which remains unknown at this time. Other radiation-induced changes like marrow edema can be detected only in a few days after photon radiation therapy using short tau inversion recovery MRI. Therefore, by using the optimal imaging sequences that can detect possible early dose-dependent changes in the bone marrow, it may be possible to use MRI for *in vivo* dose and range verification.

A similar feasibility study was done by Yuan et al. [55] to determine the proton beam range for liver patients receiving proton therapy using MRI. Their study, using hepatocyte-specific MR functional imaging for five liver cancer patients, showed statistically significant correlation between the radiation doses in the superior/inferior penumbra region with MR signal intensity. The distal range of each proton beam in a treatment plan with two fields was determined by applying the dose and MR signal intensity response curve. The mean difference between the MRI estimated and planned beam range was found to be −2.18 ±4.89 mm for anterior–posterior beams and −3.90 ±5.87 mm for the lateral beams. This study was done for post-treatment range verification for a small number of patients, and the methodology may have inherent limitations leading to large uncertainties.

These two studies show the potential of MR imaging for *in vivo* dose and range verification. More research and developmental work is needed to make MRI a practical tool for IVD in proton therapy. Additionally, the accuracy of the range verification may be limited by the inherent large spatial resolution of the current MRI technology.

28.6 Clinical Experience with *In Vivo* Dosimetry in Proton Therapy

Hsi et al. [56] have published their clinical experience with IVD in proton therapy through the placement of TLDs on the surface of endorectal balloon under image guidance to monitor the rectal dose from proton therapy of prostate cancer patients. TLDs were placed at a fixed location (12 o'clock position) on the anterior balloon surface. The variation in the balloon orientation and the TLD displacement was monitored using radiopaque markers placed at the 3 o'clock and 9 o'clock positions on the balloon surface. Results of 81 *in vivo* TLD measurements in six patients show that 83% of all measured TLD doses were within −10% to 5% of the planned dose, with a mean of −2.1% and a standard deviation of 3.5%. This study also showed that it is possible to establish a practical image guidance method to confirm the correct placement of the endorectal balloon and to reduce the positional variation of TLDs during treatment delivery. This work showed that the use of TLD as an IVD technique to monitor anterior rectal dose with acceptable accuracy is feasible in clinical practice.

TLDs have also been used to monitor the out-of-field dose at pacemakers, the in-field dose of head and neck cancer patients, and the entrance dose of prostate cancer patients receiving proton therapy at the MD Anderson Cancer Center Proton Therapy Center in Houston. The calibration of the control TLDs were always carried out at the expected dose in the patient and by choosing the proton beam with the same energy that is expected to make the major contribution to the dose to the TLD placed on the patient. The TLD measured dose for about 20 patients was found to agree reasonably well (considering the ±5% uncertainties in TLD measured dose) with the expected dose calculated in the Eclipse treatment planning system (Varian Medical Systems, Palo Alto, CA).

In vivo surface dose measurements with glass dosimeters for six patients treated with proton therapy for various disease sites have been reported by Rah et al. [24]. The average difference between measured and Eclipse treatment planning system calculated dose was found to be 5.9% ±1.8%. The difference between the measured dose by glass dosimeter and TLDs was found to be within 2%.

Although characterization of various other detectors—like diode, OSLD, PSD and film—have been performed for IVD in proton therapy, there are no published data on their use for monitoring actual patient dose in the clinic. It is expected that some of these dosimeters will perform equally well as TLDs and glass dosimeters for surface dose measurements by using appropriate calibration factors and correction factors to account for their energy, LET, dose, and angular dependence. However, it appears from the paucity of published data that routine use of IVD in clinical practice is limited.

The clinical use of PET imaging for IVD in proton therapy has been reviewed by Zhu and Fakhari [33]. As described Section 28.4.2 of this chapter and in Chapter 7, the measured activity on the day of treatment is compared with a reference activity distribution for treatment verification. In a study at the National Cancer Center in Japan [57], in-beam PET activity images for 48 patients were taken every day of the treatment and were compared with the activity distribution of the first day of treatment (reference activity) to assess any change in the irradiated volume in the proton beam. It was possible to detect activity changes in the areas where partial tumor reduction was observed for head and neck cancer patients. The monitoring of activity during the course of treatment helped to identify three cases where replanning was necessary due tumor shrinkage or changes in the body shape.

Using the measured activity from the first day of treatment as the reference image will not provide any information about any deviation of the delivered dose from the planned dose. The confirmation of the delivery of the planned dose distribution and constancy of delivered dose distribution during the course of the treatment by PET imaging requires the availability of a reference activity distribution before the start of the treatment.

Monte Carlo-based and analytical calculation models are being used to calculate 3D activity distributions in patients [58,59]. However, such calculations have their limitations and lack the precision to be used for dose verification in actual patients by PET imaging. But the predicted and measured distal falloff in the activity distribution is considered to be precise enough to be used for proton range verification by PET imaging during the course of the treatment. Range verification by PET imaging was performed for 23 patients at Massachusetts General Hospital (MGH) in Boston using an off-line PET/CT scanner [60]. The feasibility of range verification was found to be highly dependent on the tumor location. It was found [60] that the method is more useful for intracranial and cervical spine patients, patients with arteriovenous malformation, and patients with metal implants compared to patients with abdominopelvic tumors. Proton beam range verification within 2 mm with PET imaging was found to be feasible for head and neck cancer patients [60]. Some of the factors contributing to the uncertainties in range verification PET imaging are [33] 1) reproducibility of the off-line PET, 2) the biological wash-out effect, 3) accuracy of the calculated activity distribution used as a reference, 4) organ motion, 5) image registration accuracy, and 6) beam arrangement.

It can be concluded from the available published data that the use of PET imaging for *in vivo* dose and range verification still requires major improvements in technology and methodology for performing and analyzing the scans [33,41].

There are two published papers [54,55] on the feasibility of the use of range verification using MR imaging, and this has been discussed earlier in this chapter. No other promising methods—like prompt gamma imaging or Compton camera or thermo-acoustic imaging (described in Chapter 7)—are yet ready for clinical use.

28.7 Concluding Remarks and Future Outlook

Although suitable dosimeters are available for IVD in proton therapy, their use in clinics is very limited. Entrance dose measurement may not be very attractive because it will not provide much needed information on the delivered dose to the target and OAR. Furthermore, their noninvasive use is limited to accessible regions in the patient, like skin and body cavities. Therefore, the use of external devices like PET is much more useful than implanted dosimeters for *in vivo* verification of proton beam dose and range.

One simple solution for quality assurance of daily treatment delivery is the measurement of induced secondary photon or neutron radiation in the room, as shown by Carnicer et al. [32]. Although PET imaging and prompt gamma imaging look very promising in principle, challenges remain to be overcome to make them feasible for routine clinical use. Research and developmental work on the use of PET is being pursued at many centers, and it is expected that on-line or off-line dose verification with PET will become feasible in the near future.

Similarly, research on the feasibility of range verification with PET and Prompt gamma cameras is being pursued and may become available for clinical application in the near future. As shown by Cho et al. [50], implantable markers containing ^{18}O, ^{63}Cu, and ^{68}Zn can substantially improve the accuracy of range verification with PET imaging. However, feasibility of their use in actual patient remains to be tested.

Development of proton radiography and proton CT is being pursued by some investigators [41]. It will become very useful in clinic for *in vivo* verification of proton RSP in 2D and 3D when it becomes available. Newer technologies like the use of proton-induced Bremsstrahlung radiation for range verification [61], proton-induced x-rays for range verification of proton beam used for ocular lesions [62], and thermoacoustic imaging for range verification of pulsed proton beams [63] are also promising and are in the research and development stage.

IVD of delivered dose to target and OAR is a desirable and worthy goal, especially in proton therapy where the dose distribution is more susceptible to perturbation due to the small changes in material in the beam path. However, this remains an elusive goal because of the lack of suitable devices for accurate, noninvasive, and robust proton beam dose and range measurements inside patients. With the growing interest in proton therapy and strong desire for *in vivo* dose and range verification to achieve the full potential of proton therapy to improve the therapeutic gain, the ongoing research and developmental effort will make it possible to bring some of the promising technology for IVD to clinics in the near future.

References

1. Essers, M, Mijnheer BJ. In vivo dosimetry during external beam radiotherapy. *Int J Radiat Oncol Biol Phys* 1999;43:245–249.
2. Mijnheer B, Beddar S, Izewska J, Reft C. In vivo dosimetry in external beam radiotherapy. *Med Phys* 2013;40:070903.
3. AAPM. Report of TG 62 of the Radiation Therapy Committee. Diode in vivo dosimetry for patients receiving external beam radiation therapy. American Association of Physicists in Medicine Report No. 87. Madison, WI: Medical Physics Publishing, 2005.

4. Karger CP, Jäkel O, Palmans H, Kanai T. Dosimetry for ion beam radiotherapy. *Phys Med Biol* 2010;55:R193–234.
5. European Society of Radiation Oncology. Methods for in vivo dosimetry in external radiotherapy. Van Dam J, Marinello G, Eds. ESTRO Booklet No. 1. Brussels, Belgium: European Society of Radiation Oncology, 2006.
6. International Atomic Energy Agency. Development of procedures for in vivo dosimetry in radiotherapy. IAEA Human Health Report No. 8. Vienna, Austria: International Atomic Energy Agency, 2011.
7. Grusell E, Medin J. General characteristics of the use of silicon diode detectors for clinical dosimetry in proton beams. *Phys Med Biol* 2000;45(9):2573–82.
8. Brews O. http://en.wikipedia.org/wiki/File:MOSFET_Structure.png.
9. Cygler JE, Scalchi P. MOSFET dosimetry in radiotherapy. In: Clinical dosimetry measurements in radiotherapy. Medical physics monograph No. 34. Rodgers DWO, Cygler JE, Eds. Madison, WI: Medical Physics Publishing, 2009.
10. Gopiraj A, Billimagga RS, Ramasubramaniam V. Performance characteristics and commissioning of MOSFET as an in-vivo dosimeter for high energy photon external beam radiation therapy. *Rep Pract Oncol Radiother* 2008;13:114–125.
11. Cheng C-W, Wolanski M, Zhao Q, Fanelli L, Gautam A, Pack D, Das IJ. Dosimetric characteristics of a single use MOSFET dosimeter for in vivo dosimetry in proton therapy. *Med Phys* 2010;37:4266–4273.
12. Kohno R, Nishio T, Miyagishi T, Hirano E, Hotta K, Kawashima M, Ogino T. Experimental evaluation of a MOSFET dosimeter for proton dose measurements. *Phys Med Biol* 2006;23:6077–86.
13. Kohno R, Hotta K, Matsubara K, Nishioka S, Matsuura T, Kawashima M. In vivo proton dosimetry using a MOSFET detector in an anthropomorphic phantom with tissue inhomogeneity. *J Appl Clin Med Phys* 2012;13:159–167.
14. Archambault L, Briere TM, Pönisch F, Beaulieu L, Kuban DA, Lee A, Beddar S. Toward a real-time in vivo dosimetry system using plastic scintillation detectors. *Int J Radiat Oncol Biol Phys* 2010;78:280–287.
15. Wang LL, Perles LA, Archambault L, Sahoo N, Mirkovic D, Beddar S. Determination of the quenching correction factors for plastic scintillation detectors in therapeutic high-energy proton beams. *Phys Med Biol* 2012;57:7767–81.
16. Wootton L, Holmes C, Sahoo N, Beddar S. Passively scattered proton beam entrance dosimetry with a plastic scintillation detector. *Phys Med Biol* 2015;60:1185–1198.
17. DeWerd LA, Bartol LJ, Davis SD. Thermoluminescent dosimetry. In: Clinical dosimetry measurements in Radiotherapy. Medical physics monograph No. 34. Rodgers DWO, Cygler JE. Madison, WI: Medical Physics Publishing, 2009.
18. Spurný F., Response of thermoluminescent detectors to charged particles and to neutrons. *Radiat Meas* 2004;38:407–412.
19. Zullo JR, Kudchadker RJ, Zhu XR, Sahoo N, Gillin MT. LiF TLD-100 as a dosimeter in high energy proton beam therapy—can it yield accurate results? *Med Dosim* 2010;35:63–66.
20. Cygler JE, Yukihara EG. Optically stimulated luminescence (OSL) dosimetry in radiotherapy. In: Clinical dosimetry measurements in radiotherapy. Medical physics monograph no. 34. Rodgers DWO, Cygler JE, Eds. Madison, WI: Medical Physics Publishing, 2009.
21. Reft CS. The energy dependence and dose response of a commercial optically stimulated luminescent detector for kilovoltage photon, megavoltage photon, and electron, proton, and carbon beams. *Med Phys* 2009;36:1690–1699.
22. Granville DA, Sahoo N, Sawakuchi GO. Linear Energy transfer dependence of Al_2O_3: Optically stimulated luminescence detectors. *Radiat Meas* 2014;71:69–73.
23. Kerns JR, Kry SF, Sahoo N. Characteristics of optically stimulated luminescence dosimeters in the spread-out Bragg peak region of clinical proton beams. *Med Phys* 2012;39:1854–1863.
24. Rah JE, Oh do H, Kim JW, Kim DH, Suh TS, Ji YH, Shin D, Lee SB, Kim DY, Park SY. Feasibility study of glass dosimeter for in vivo measurement: Dosimetric characterization and clinical application in proton beams. *Int J Radiat Oncol Biol Phys* 2012;84:e251–256.
25. Ganapathy K, Kurup PGG, Murali V, Muthukumaran M, Subramanian SB, Velmurugan J. A study on rectal dose measurement in phantom and *in vivo* using Gafchromic EBT3 film

in IMRT and CyberKnife treatments of carcinoma of prostate. *J Med Phys* 2013;38:132–138.

26. Rudat V, Nour A, Alaradi AA, Mohamed A, Altuwaijri S. *In vivo* surface dose measurement using GafChromic film dosimetry in breast cancer radiotherapy: comparison of 7-field IMRT, tangential IMRT and tangential 3D-CRT. *Radiat Oncol* 2014;9:156.
27. Liu H-W, Gräfe J, Khan R, Olivotto I, Barajas JEV. Role of *in vivo* dosimetry with radiochromic films for dose verification during cutaneous radiation therapy. *Radiat Oncol* 2015;10:12.
28. Niroomand-Rad A, Blackwell CR, Coursey BM, Gall KP, Galvin JM, McLaughlin WL, Meigooni AS, Nath R, Rodgers JE, Soares CG. Radiochromic film dosimetry: recommendations of AAPM radiation therapy committee task group 55. *Med Phys* 1998;25:2093–2115.
29. Pai S, Das IJ, Dempsey JF, Lam KL, Losasso TJ, Olch AJ, Palta JR, Reinstein LE, Ritt D, Wilcox EE. TG-69: radiographic film for megavoltage beam dosimetry. *Med Phys* 2007;34:2228–2258.
30. Zhao L, Das IJ. Gafchromic EBT film dosimetry in proton beams. *Phys Med Biol* 2010;55:N291–N301.
31. Arjomandy B, Tailor R, Anand A, Sahoo N, Gillin M, Prado K, Vicic M. Energy dependence and dose response of Gafchromic EBT2 film over a wide range of photon, electron, and proton beam energies. *Med Phys* 2010;37:1942–1947.
32. Carnicer A, Letellier V, G. Rucka G, Angellier G, Sauerwein W, Hérault J. Study of the secondary neutral radiation in proton therapy: Toward an indirect in vivo dosimetry. *Med Phys* 2012;39:7303–7315.
33. Xhu X, El Fakhri. Proton therapy verification with PET imaging. *Theranostics* 2013;3:731–740.
34. Enghardt W, Parodi K, Crespo P, Fiedler F, Pawelke J, Poenisch F. Dose quantification from in-beam positron emission tomography. *Radiother Oncol* 2004;73(Supplement 2):S96–S98.
35. Fourkal E, Fan J, Veltchev I. Absolute dose reconstruction in proton therapy using PET imaging modality: feasibility study. *Phys Med Biol* 2009;54:N217–N228.
36. Zhu X, España S, Daartz J, Liebsch N, Ouyang J, Paganetti H, Bortfeld TR, El Fakhri G. Monitoring proton radiation therapy with in-room PET imaging. *Phys Med Biol* 2011;56:4041–4057.
37. Pawelke J, Enghardt W, Haberer T, Hasch B G, Hinz R, Kramer M, Lauckner E, Sobiella M. In-beam PET imaging of the control of heavy-ion tumour therapy. *IEEE Trans Nucl Sci* 1997;44:1492–1498.
38. Shakirin G, Braess H, Fiedler F, Kunath D, Laube K, Parodi K, Priegnitz M, Enghardt W. Implementation and workflow for PET monitoring of therapeutic ion irradiation: A comparison of in-beam, in-room, and off-line techniques. *Phys Med Biol* 2011;56:1281–1298.
39. Parodi K, Paganetti H, Shih HA, Michaud S, Loeffler JS, DeLaney TF, Liebsch NJ, Munzenrider JE, Fischman AJ, Knopf A, Bortfeld T. Patient study of in vivo verification of beam delivery and range, using positron emission tomography and computed tomography imaging after proton therapy. *Int J Radiat Oncol Biol Phys* 2007;68:920–934.
40. Yang M, Zhu XR, Park PC, Titt U, Mohan R, Virshup G, Clayton JE Dong L. Comprehensive analysis of proton range uncertainties related to patient stopping-power-ratio estimation using the stoichiometric calibration. *Phys Med Biol* 2012;57:4095–4115.
41. Knopf AC, Lomax T. In vivo proton range verification: a review. *Phys Med Biol* 2013;58:R131–R160.
42. Lu H-M, Mann G, Cascio E. Investigation of an implantable dosimeter for single-point water equivalent path length verification in proton therapy. *Med Phys* 2010;37:5858–5866.
43. Bentefour EH, Shikui T, Prieels D, Lu H-M. Effect of tissue heterogeneity on an in vivo range verification technique for proton therapy. *Phys Med Biol* 2012;57:5473–5484.
44. Samuel D, Testa M, Schneider R, Park Y, Janssens G, Orban de Xivry J, Moteabbed M, Prieels D, Lu H, Bentefour E. Development and commissioning of a complete system for in vivo dosimetry and range verification in proton therapy. *Med Phys* (abstract)2014;41:326.
45. Lu HM. A point dose method for in vivo range verification in proton therapy. *Phys Med Biol* 2008;53:N415–N422.

46. Mumot M, Algranati C, Hartmann M, Schippers JM, Hug E, Lomax AJ. Proton range verification using a range probe: definition of concept and initial analysis. *Phys Med Biol* 2010;55:4771–4782.
47. Schneider U, Jürgen Besserer J, Pemler P, Dellert M, Moosburger M, Pedroni E , Kaser-Hotz B. First proton radiography of an animal patient. *Med Phys* 2004;31:1046–1051.
48. Penfold N, Rosenfeld AB, Schulte RW, Schubert KE. A more accurate reconstruction system matrix for quantitative proton computed tomography. *Med Phys* 2009;36:4511–4518.
49. Litzenberg DW, Roberts DA, Lee MY, Pham K, Vander Molen AM, Ronningen R, Becchetti FD. On-line monitoring of radiotherapy beams: experimental results with proton beams. *Med Phys* 1999;26:992–1006.
50. Cho J, Ibbott G, Gillin M, Gonzalez-Lepera C, Titt U, Paganetti H, Kerr M, Mawlawi O. Feasibility of proton-activated implantable markers for proton range verification using PET. *Phys Med Biol* 2013;58:7497–7512.
51. Polf J, Peterson S, Ciangaru G, Gillin M, Beddar S. Prompt gamma-ray emission from biological tissues during proton irradiation: a preliminary study. *Phys Med Biol* 2009;54:731–743.
52. Smeets J, Roellinghoff F, Prieels D, Stichelbaut F, Benilov A, Busca P, Fiorini C, Peloso R, Basilavecchia M, Frizzi T, Dehaes JC, Dubus A. Prompt gamma imaging with a slit camera for real-time range control in proton therapy. *Phys Med Biol* 2012;57:3371–3405.
53. Perali I, Celani A, Bombelli L, Fiorini C, Camera F, Clementel E, Henrotin S, Janssens G, Prieels D, Roellinghoff F, Smeets J, Stichelbaut F, Vander Stappen F. Prompt gamma imaging of proton pencil beams at clinical dose rate. *Phys Med Biol* 2014;59:5849–5871.
54. Gensheimer MF, Yock TI, Liebsch NJ, Sharp GC, Paganetti H, Madan N, Grant PE, Bortfeld T. In vivo proton beam range verification using spine MRI changes. *Int J Radiat Oncol Biol Phys* 2010;78:268–275.
55. Yuan Y, Andronesi OC, Bortfeld TR, Richter C, Wolf R, Guimaraes AR, Hong TS, Seco J. Feasibility study of in vivo MRI based dosimetric verification of proton end-of-range for liver cancer patients. *Radiother Oncol* 2013;106:378–382.
56. Hsi WC, Fagundes M, Zeidan O, Hug E, Schreuder N. Image-guided method for TLD-based in vivo rectal dose verification with endorectal balloon in proton therapy for prostate cancer. *Med Phys* 2013;40:051715.
57. Nishio T, Miyatake A, Ogino T, Nakagawa K, Saijo N, Esumi H. The development and clinical use of a beam ON-LINE PET system mounted on a rotating gantry port in proton therapy. *Int J Radiat Oncol Biol Phys* 2010;76:277–286.
58. Min CH, Zhu X, Winey BA, Grogg K, Testa M, El Fakhri G, Bortfeld TR, Paganetti H, Shih HA. Clinical application of in-room positron emission tomography for in vivo treatment monitoring in proton radiation therapy. *Int J Radiat Oncol Biol Phys* 2013;86:183–189.
59. Frey K, Bauer J, Unholtz D, Kurz C, Krämer M, Bortfeld T, Parodi K. TPS(PET)-A TPS-based approach for in vivo dose verification with PET in proton therapy. *Phys Med Biol* 2014; 59:1–21.
60. Knopf AC, Parodi K, Paganetti H, Bortfeld T, Daartz J, Engelsman M, Liebsch N, Shih H. Accuracy of proton beam range verification using post-treatment positron emission tomography/computed tomography as function of treatment site. *Int J Radiat Oncol Biol Phys* 2011;79:297–304.
61. Yamaguchi M, Torikai K, Kawachi N, Shimada H, Satoh T, Nagao Y, Fujimaki S, Kokubun M, Watanabe S, Takahashi T, Arakawa K, Kamiya T, Nakano T. Beam range estimation by measuring bremsstrahlung. *Phys Med Biol* 2012;57:2843–2856.
62. La Rosa V, Kacperek A, Royle G, Gibson A. Range verification for eye proton therapy based on proton-induced x-ray emissions from implanted metal markers. *Phys Med Biol* 2014;59:2623–2638.
63. Alsanea F, Moskvin V, Stantz KM. Feasibility of RACT for 3D dose measurement and range verification in a water phantom. *Med Phys* 2015;42:937–946.

APPENDIX: Nomenclature and Terminology in Proton Therapy

Radhe Mohan, Ph.D.[1] and Harald Paganetti, Ph.D.[2]

[1]MD Anderson Cancer Center,
Houston, TX

[2]Professor and Director of Physics Research, Department of Radiation Oncology,
Massachusetts General Hospital and Harvard Medical School
Boston, MA

29.1 Introduction

There is a lack of consistency in the terminology used in proton therapy. This sometimes negatively affects communication among physicians, physicists, biologists, treatment planners, dosimetrists, technicians, and vendors. The rationale for this appendix on nomenclature was not to redefine currently used terms. Rather, our goal was to remove ambiguities and point out redundancies.

We focus on those terms that are specific to proton radiation therapy, though most of them are applicable to particle therapy in general. We have used several sources in an attempt to ensure that the main terms are covered [M. F. Moyers, A. Mazal, and D. Nichiporov, private communication]. Our definitions may not all be fully in line with any of these prior efforts because we felt earlier definitions were inconsistent or that a different term would help avoid confusion.

We have tried not to redefine commonly used terms for consistency with the past literature. However, in several instances it was deemed appropriate to clarify and sometimes revise commonly used terms in an effort to provide terms that convey their meaning adequately. Terms are often used for historical reasons and are not necessarily logical, i.e., the *spread-out Bragg peak* is not really a peak and a Bragg peak often refers to a Bragg curve. Another example is the frequent confusion between pencil beam scanning delivery and a pencil-beam algorithm, the latter being a conceptual tool used in an algorithm to calculate dose distribution for a large field. Furthermore, we felt that definitions should be self-explanatory to the extent possible and should not be unnecessarily complicated.

The definitions given below are subjective. We do list alternative terms where we anticipate differences in opinion in our field. The list is not comprehensive as there are many more terms in use for various delivery or planning aspects. We did omit terms that are used at single institutions only or terms that the average proton practitioner rarely comes across. We acknowledge valuable feedback and suggestions from the authors of the chapters in this book as well as from members of the AAPM Working Group on Particle Beams.

29.2 Proton Beam Terminology

Recommended Term	Abbreviation/ Acronym	Definition	Alternate Term
Aperture		The non-attenuating opening in a collimator, shaped irregularly according to the shape of the target, to allow radiation to reach a target.	
Aperture Carriage		See Snout.	
Applicator Carriage		See Snout.	
Beam		A group of particles or rays traveling in the same direction in parallel or diverging from a point.	
Beam Current Modulation	BCM	Time-dependent modulation of the beam current.	
Beam Efficiency		Particle fluence at the treatment head exit as a percentage of particle fluence at the treatment head entrance.	
Beam Energy		Energy at the treatment head (nozzle) entrance. This is a generic term that may also specify the beam at positions other than the treatment head, in which case this should be stated.	
Beam Energy Spread		Energy spread (sigma) at the treatment head entrance. This is a generic term that may also specify the beam energy spread at positions other than the treatment head, in which case this should be stated.	
Beam Intensity		The number of particles entering the nozzle per unit time (e.g., giga protons per second). This is a generic term that may also specify the beam intensity at positions other than the treatment head, in which case this should be stated.	
Beam Line		A device or combination of devices used to deliver a beam from an accelerator to a specific location (e.g., a treatment room).	
Beam Tuning		Adjustments of parameter settings of an accelerator beamline magnetic devices for distribution of particles among different rooms to optimize the characteristics of the delivered beam.	
Beamlet		(For scanning beams and IMPT). A small cross-section, monoenergetic beam of particles entering the nozzle and reaching the patient or phantom without being scattered or modulated. (A range shifter may be inserted to reduce the range, to reduce the spot size or to broaden the Bragg peak.) Another term, i.e., pencil beam, has become common in the field of particle therapy. For instance, "pencil beam scanning," or PBS, is used to describe scanning beams. Therefore, either beamlet or pencil beam may be used. However, since the term pencil beam is also used as a conceptual tool in dose calculation algorithms, its use to describe the beam of particles entering the nozzle and used for scanned delivery should be clearly stated.	Pencil Beam
Beamlet Intensity		Beamlet or spot weight in scanning beam treatments specified in MUs per spot or number of giga protons per spot.	

Recommended Term	Abbreviation/ Acronym	Definition	Alternate Term
Beam-specific PTV		The planning target volume per beam in which the lateral margins are the same as those for traditional PTV, but the distal and proximal margins are computed based on range uncertainty. This term is used in passively scattered proton therapy (PSPT), single field uniform dose optimization, or in single field optimized intensity-modulated proton therapy (IMPT).	
Bolus		See Range Compensator.	
Bolus Expansion		See Smearing.	
Bragg Curve		The integral depth–dose curve for a beamlet or a pencil beam, or the depth–dose curve of a broad beam of a quasi-mono-energetic charged particles.	
Bragg Peak		The narrow, high-dose region adjacent to the maximum dose on a Bragg curve.	
Cobalt Gray Equivalent		See Gy(RBE).	
Continuous Beam		A beam with a continuous stream of particles as is the case for cyclotrons. In contrast, synchrotrons generate a pulsed beam.	
Control Point	CP	CP is a DICOM RT term. It describes a group of parameters of the treatment delivery device that change during delivery. For instance, in pencil beam scanning, CP specifies positions and weights for a collection of spots in a given energy layer.	
Cyclotron		A particle accelerator that accelerates charged particles to a fixed pre-determined energy by maintaining constant magnetic fields while increasing the particle energy by varying the accelerating potential. Cyclotrons produce a nominally continuous stream of particles.	
Distal Blocking		A process of reducing the dose to normal critical tissues distal to a target by modifying the PSPT range compensator. The same can be achieved by creating "pseudo-structures," i.e., modifying the target or normal structure contours appropriately.	
Distal Edge Tracking	DET	An IMPT optimization strategy in which the spots of the beamlets of each beam are placed only at the distal edge of the target volume.	
Distal Fall-off		The distal part (beyond the Bragg peak or SOBP) of the particle depth dose curve.	
Distal Penumbra		The distance from the depth of a specific percentage of the dose at the Bragg peak (or SOBP) and the depth of a lower percentage of the Bragg peak (or SOBP) dose. Typically, the distal penumbra is defined as the distance between the 80% and the 20% dose values.	
Double Scattering		In PSPT, double scattering is a method of spreading the beam laterally, in which a pair of specially designed scattering devices consisting of a flat scatterer (first scatterer) and a contoured scatterer (second scatterer) are placed on the beam central axis. This technique has a higher efficiency of beam usage compared to a technique that uses a single scatterer.	
Effective Source-to-Axis Distance	ESAD	The distance between a point upstream from the isocenter for a particle beam from which the inverse square law can be applied to calculate the change in dose rate as a function of distance.	

Recommended Term	Abbreviation/ Acronym	Definition	Alternate Term
Emittance		Beam emittance is the extent occupied by the particles of the beam in the physical and momentum phase space. Emittance is defined as the product of the beam radius and the beam angular divergence and is usually expressed in millimeters-milliradians. Since it is difficult to measure the full radius, or envelope, of the beam, a value of the radius that encompasses a specific percentage of the beam (for example, 95%) is specified. The emittance from these measurements is then referred to as the "95% emittance."	
Energy Absorber		A block of water-equivalent material of uniform thickness inserted across a scanning beam close to the patient to achieve one or more of the following goals: (1) reduce the range to treat shallow tumors, (2) reduce the spot size, and (3) widen the Bragg peak longitudinally to reduce the number of energy layers required. It may also be used for passive scattering to reduce the range to treat shallow tumors. (Note the distinction with "Range shifter.")	Pre-Absorber
Energy Degrader		The energy reducer in the beam line of cyclotron systems, typically at the cyclotron exit. The distinction between the energy absorber and the energy degrader is that the former is between the entrance to the nozzle and patient, and the latter is before, generally outside the treatment room.	Range Degrader or Range Shifter
Energy Layer		The region of irradiation with a scanned monoenergetic beam of particles.	
Energy Per Nucleon		The total kinetic energy of the ion divided by the number of nucleons in the nucleus.	
Energy Stacking		Step-by-step variation of the beam energy by the accelerator to deliver dose to different depth layers.	
Entrance-to-Peak Dose Ratio		The ratio of the absorbed dose on the radiation beam axis at the depth of 0.5 mm to the peak absorbed dose on the radiation beam axis, both measured in a phantom with its surface at a specified distance.	
Faraday Cup		A device designed to measure the intensity of a charged particle beam by absorbing the particles of the primary beam in the collector and measuring the accumulated charge.	
Field Patching		In PSPT, treating different parts of the target with different beams matched at the patch line. In other words, generating a conformal dose distribution with PSPT using a "through" field and a "patch" field, each treating parts of the target volume.	
Fragmentation		The disintegration of primary particles into lighter fragments as a result of nuclear collisions with a target nucleus. The remnants of the projectile nucleus emerge from the absorbing material with speeds similar to that of the original projectile nucleus. (The target nucleus may also break apart, but these fragments have relatively low energy and do not travel with the beam.)	
Fragmentation Tail		A tail in depth–dose distribution beyond the Bragg peak or SOBP formed by products of fragmentation.	
Full Width at Half Max	FWHM	The width of a beam or a beamlet/pencil beam profile (in air or in a uniform phantom) between points where the dose (or intensity) has dropped to one half of the maximum.	
Gy (RBE)	Gy (RBE)	Product of physical dose in Gy and RBE used as unit of dose.	Cobalt Gray Equivalent (CGE). Obsolete definition.

Recommended Term	Abbreviation/ Acronym	Definition	Alternate Term
Halo		The low-level secondary particle radiation field extending outside the lateral and distal penumbrae of a primary beam. (A low-dose halo may also occur when protons in a beamlet/pencil beam are scattered through multiple Coulomb scattering in the profile monitor.)	Nuclear Halo
Infinitesimal Pencil Beam		Monoenergetic beam of particles with an infinitesimal lateral dimension and angular emittance at the point of consideration (e.g., at the point of incidence on the patient).	Pencil Beam
Integral Depth Dose	IDD	Integral of dose on an infinite plane normal to the central axis of a beamlet or an infinitesimal pencil beam. It is represented as a function of depth.	
Intensity Modulated Particle Therapy	IMPT	One of multiple modes of particle therapy planning and delivery methods in which scanning beamlets of a sequence of energies and of optimized intensities are used to achieve an appropriate balance between the target dose and normal tissue doses. Examples (defined elsewhere in this document) include single-field optimized IMPT, multi-field optimized IMPT, distal-edge tracking, etc.	
Lateral Penumbra	LPxx-yy (e.g., LP80-20)	The distance from the point where the dose is a certain percentage of the central axis value and the point where the dose is at a lower percentage of the central axis value. Typically the penumbra is defined for the distance between the 80% and the 20% dose values.	
Light Ion Species		Family of ions with nuclei of atomic numbers less than or equal to that of neon ($Z \leq 10$) that is used to treat a patient and is specified by its number of protons, number of nucleons, and ionization state.	
Linear Energy Transfer	LET	Dose deposited per unit pathlength of a monoenergetic particle beam as it traverses a medium. It has multiple forms such as dose-averaged LET, track-averaged LET, fluence-averaged LET, etc.	
Match Line		See patch line.	
Modulation Width		Proximal distal dose fall-off position minus distal dose fall-off position in an SOBP in water. The position has to be explicitly stated, e.g., "98–90" reflecting 98% proximal distal dose fall-off position minus 90% distal dose fall-off position.	SOBP Width
Multi-Field Optimized IMPT	MFO or MFO-IMPT	The mode of IMPT in which the intensities of beamlets (spots) of all beams are optimized simultaneously to achieve the minimum of an objective function that appropriately balances the requirements of the targets and normal tissues.	
Multiple Coulomb Scattering	MCS	The primary scattering process, which is due to electrostatic interactions with nuclei. Multiple small-angle deflections of charged particles traversing a medium due to Coulomb scattering from nuclei results in the lateral spreading of proton beams.	
Multi-layer Faraday Cup	MLFC	A device similar to the Faraday cup, with the collector made of thin layers of conducting material separated with layers of dielectric. This design allows the rapid measurements of range to be made before the beam delivery to the patient.	
Multi-layer Ionization Chamber	MLIC	A device similar to a MLFC, but consisting of multiple layers of parallel plate ionization chambers. As with the MLFC, MLIC allows the rapid measurements of range to be made before the beam delivery to the patient. It is also used for off-line quality assurance purposes.	

Recommended Term	Abbreviation/ Acronym	Definition	Alternate Term
Nozzle		The part of a beam delivery system in the treatment room (attached to the gantry in rooms with gantries) into which the thin narrow beam of protons enter, and which houses the beam-shaping and dose-monitoring equipment, and from which the radiation emerges.	Treatment Head
Nuclear Buildup		A small increase in dose at the entrance of the Bragg curve caused by the accumulation of nuclear interaction products emitted predominantly in the forward direction.	
Nuclear Halo		See Halo.	
Passive Scattering		A method of producing a volume of uniform dose distributions of particles at a depth perpendicular to the beam central axis using scatterers and range modulators.	
Passively Scattered Proton Therapy	PSPT	A particle therapy technique in which one or more passively scattered (spread longitudinally and laterally) beams of particles are shaped using apertures and compensators to conform the dose distribution to the target shape.	
Patch Field		In PSPT, part of the field patching. A patch field treats part of the target volume that is not covered by the aperture outline of the through field. The range for the patch field is chosen so that the distal dose fall-off matches the lateral penumbra of the through field along the patch line. It is a field whose distal edge abuts the lateral edge of a shoot-through field to produce a nearly uniform dose profile across the junction.	
Patch Line		Line where the patch field and the through field meet in the target volume.	Match Line
Pencil Beam		A monoenergetic beam of particles with an infinitesimal lateral dimension and angular emittance at the point of consideration (e.g., at the point of incidence on the patient). The pencil beam is a conceptual tool used in algorithms to calculate dose distributions for large fields. (Note the distinction between a pencil beam and a beamlet. Many authors use the term pencil beam to mean a beamlet, i.e., a small cross-section beam. The pencil beam may be called the "Infinitesimal or Calculational Pencil Beam." To avoid confusion, the intended meaning of pencil beam should be clearly stated.)	
Pencil Beam Scanning	PBS	A technique for spreading the proton dose distribution laterally by magnetically scanning beamlets. It is used to deliver several different modes of scanned beam proton therapy, such as single-field optimized IMPT, multi-field optimized IMPT, wobbling, uniform scanning, etc.	
Plateau		The relatively uniform region of a depth–dose distribution between the surface and the SOBP of a range-modulated beam or the surface and the peak of a non-range-modulated (pristine monoenergetic) beam.	
Pre-absorber		See Energy Absorber.	
Pristine Beam		A monoenergetic, unscattered beam of particles. It may be a single beamlet or a beam composed of uniformly scanned beamlets of the same energy. The term may also be used for beams that have a narrow and symmetric energy spread.	
Propeller		See Range Modulator Wheel.	
Proton CT	pCT	A tomographic imaging modality that uses protons of sufficient energy to penetrate the patient in order to provide data of water equivalent pathlength integrated along the proton path from many directions. From a full scan and data acquisition, tomographic images of relative stopping power can be reconstructed.	

Recommended Term	Abbreviation/ Acronym	Definition	Alternate Term
Range	Rxx (e.g., R80 or R90)	The depth in a water phantom most distant from its surface at which the absorbed dose of a non-range-modulated beam is approximately 80% of the dose at the depth of peak dose. (For clinical purposes, range is usually defined as the depth of distal 90% of dose distribution in the patient.)	
Range Compensator		A range-modifying device for passively scattered proton therapy (PSPT) or for uniform scanning therapy that provides differential penetration laterally across a charged particle beam to shape the distal edge of the high dose region.	Bolus
Range Degrader		See Energy Degrader.	
Range Modulator		A device used to modulate the penetration of a beam into a patient to cover the intended target during the delivery of one portal. The device may consist of a propeller-shaped material that spins in the beam (see Range Modulator Wheel), a filter containing a repeating pattern of metallic or plastic ridges (see Ridge Filter), a cone or set of cones, or a set of uniform thickness blocks programmable in a binary fashion.	
Range Modulator Wheel	RMW	A range modulator wheel usually consists of a stack of triangular layers of plastic or low-Z material (e.g., aluminum or plastic) connected to a hub. The entire stack of material rotates, placing different thicknesses of the material in the path of a thin beam of monoenergetic protons, thereby modulating the penetrability of the beam and broadening the Bragg peak.	
Range Shifter		See Energy Degrader.	
Range Straggling		Spread in proton beam energies due to multiple Coulomb interactions in the medium.	
Range Uncertainty Margin		In PSPT, distal and proximal margins to account for uncertainty in the range of the particles.	
Raster Scanning		A beam delivery technique in which the scanning beam spot follows a path consisting of parallel rows. This technique results in a spatially and temporally constant scan pattern that is predefined for use with all patients. The beam flux may be constant or modulated while the spot moves along the scan pattern to produce uniform or modulated fluence distributions respectively.	
Relative Biological Effectiveness	RBE	The ratio of the photon dose, necessary to obtain a certain biological effect, to the particle physical dose needed to achieve the same effect. Typically, the RBE is defined for a certain dose level and certain biological endpoint.	
Relative Output Factor		The ratio of the dose at the center of a reference SOBP for different energies and field sizes to the dose for the standard proton beam that has been calibrated using a calibration protocol.	
Relative Stopping Power	RSP	The stopping power of tissues or material relative to that of a reference material, usually water, at the same proton energy. For practical purposes, the RSP is often assumed to represent an average value over a clinical proton energy range.	
Repainting		A scanning beam technique whereby the beam scans the same volume multiple times during a treatment with the intent of averaging the effects of patient motion or beam delivery errors. The term may also be used for scanning each energy layer multiple times before going to the next layer.	Re-scanning
Rescanning		See Repainting.	
Residual Range		The remaining range of particles at a point of interest, either in the phantom or in the patient. It is the difference in depths between the dosimetric range and a point of interest.	

Recommended Term	Abbreviation/ Acronym	Definition	Alternate Term
Ridge Filter		A range modulator that consists of several ridges and valleys that present different thicknesses of material to an incoming beam to vary the penetration into the patient. It can be designed so as to perform a similar function as the range modulator wheel (i.e., to construct an SOBP).	
Ripple Filter		A range modulator (typically a thin ridge filter) that produces just enough variation in light ion energies entering the patient so that a reduced number of accelerator energies may be used without producing ripple in the depth–dose distribution.	
Robust Optimization		An IMPT optimization method to account for uncertainties to make the resulting dose distribution resilient to uncertainties.	
Robustness		A measure of the resilience of a dose distribution to uncertainties	
Scan Pattern		A pattern of scanning, including spot positions, energies, and intensities	
Scanned Beam		A broad beam obtained by scanning the beamlets/pencil beams of multiple energies.	
Scatterer		In PSPT, a scatterer is a beam-spreading device with a flat or contoured profile that is placed on the beam axis in order to increase the beam's lateral size through the elastic interaction (scattering) of the beam particles with the device's material. Scatterers are usually fabricated of a material with a high atomic number.	
Single Scattering		Broadening of the beam in PSPT through the use of a single scatterer.	
Single-field Optimized IMPT	SFO or SFO-IMPT	A form of IMPT in which the intensities of the spots of each individual field are optimized with the intent to deliver a desired fraction of the prescribed uniform dose to one or more targets and a minimum dose to the surrounding normal anatomy. An extension of the SFO technique is the "single field integrated boost" (SFIB), which combines SFO with simultaneous integrated boost (SIB) to allow the delivery of different doses to different targets simultaneously.	SFUD
Single-field Uniform Dose IMPT	SFUD	See Single-field Optimized IMPT.	
Smearing		In PSPT, a technique that modifies a compensator, designed based on ray tracing, to account for multiple Coulomb scattering and anatomy misalignments due to setup uncertainties and inter- and intra-fractional anatomic variations.	Bolus Expansion
Smearing Radius		Radius of a circular region to smear the compensator.	
Snout		The distal-most part of the treatment head (nozzle) to which interchangeable beam applicators (e.g., apertures and compensators) are attached and which may extend toward and retract away from the isocenter. It moves parallel to the central axis of the beam.	Aperture Carriage
Spot		Terminal end of a beamlet around which most of its dose is deposited.	
Spot Intensity		See Beamlet Intensity.	Beamlet Intensity
Spot Scanning		A technique of creating a large field by scanning a beamlet spot across the target volume. The beamlet stops at each predetermined point and delivers a specified dose. Irradiation is usually switched off between the points of delivery.	Discrete Scanning

Recommended Term	Abbreviation/ Acronym	Definition	Alternate Term
Spot Size		One standard deviation of the beam profile of a spot, typically defined in air at the isocenter.	
Spot Spacing		In scanning beam treatments, lateral spacing between centers of uniformly spaced spots placed in the target volume.	
Spot Weight		See Beamlet Intensity.	Beamlet Intensity
Spread-out Bragg Peak	SOBP	Depth–dose distribution resulting from a combination of a set of quasi-monoenergetic beams. The intensities are chosen to produce a region of longitudinally and laterally flat dose distribution to cover a target of a finite size in depth. An SOBP may be produced by passive scattering with a scatterer and range modulator wheel, with a ridge filter or with a scanning beam.	
Spread-out Bragg Peak Width		See Modulation Width.	
Stopping Power Ratio	SPR	The stopping power of a medium for a particle of given energy and type is defined as the average energy loss of the particle per unit pathlength, typically in units of MeV/cm. The stopping power ratio is the stopping power of a medium relative to that of water.	Relative Stopping Power, RSP
Tail (or Fragmentation Tail)		The region of a particle depth–dose distribution beyond the nominal range of the primary particles consisting of dose only from secondary particle fragments.	
Through Field		In PSPT's field patching technique, the through field treats a part of the target volume from the proximal to distal edge.	Shoot Through Field
Treatment Head		See Nozzle.	
Uniform Scanning		A scanning mode in which beamlets/pencil beams of each of a sequence of energies is scanned laterally in a predefined pattern to produce a field large enough to treat a target such that the flux of the beam is not changed throughout the lateral scan. Depth modulation (energy stacking) is achieved using a range modulator to create an SOBP.	
Verification Dose Distribution		This term is used to describe two types of dose distributions: (1) Dose distribution computed in a phantom for the purpose of dosimetric verification using one beam at a time of the beam configuration identical to the beam used for treatment. (2) Dose distribution computed on a "verification image" acquired during the course of treatment to verify that the original treatment parameters are still appropriate.	
Verification Image		A repeat image (CT) acquired during the course of radiotherapy to assess whether the original treatment plan is still valid.	
Virtual Source-to-Axis Distance		The distance of a point on the central ray in space from which the radiation appears to originate (diverge). This point is used for fabricating patient-specific devices and for scaling field dimensions.	
Water Equivalent Depth	WED	See Water Equivalent Pathlength.	
Water Equivalent Pathlength	WEPL	The proton pathlength in water that leads to the same mean energy loss as observed for a proton traversing a slab of heterogeneous media.	Water Equivalent Depth
Water Equivalent Thickness	WET	The range of a proton beam in water that corresponds to the thickness of a specific material that leads to the same mean energy loss.	

Recommended Term	Abbreviation/ Acronym	Definition	Alternate Term
Wobbling		A uniform scanning method that uses a non- or slowly repeating pattern. One particular technique produces a circular pattern with a modulating radius. This pattern is generated by inputting current into two perpendicularly oriented scanning magnets with sine waves of identical frequencies but 90° out of phase. Another technique produces a Lissajous pattern by inputting sine waves of different frequencies (non-multiple) into the perpendicular magnets.	

29.2 Conclusion

We hope that this set of definitions will help remove some of the ambiguities that currently exist in proton therapy. The terminology given is a suggestion, and the terms are not binding. Authors in this book were free to use whatever terms they deemed appropriate. Those who chose to deviate from the suggested terms were asked to define their preferred terms in their text.

This list is certainly not complete, and the favorite definitions of many may have been left out. As technology advances in proton radiation therapy, there will be additional terms and abbreviations, and many of the infrequently used terms may become commonly accepted. We anticipate that this document will serve as a first step toward a "live document" that will evolve over the years based on consensus among the practitioners, researchers, and vendors.